Study Guide
to Accompany

Keller/Gettys/Skove

Physics SECOND EDITION

and

Physics: Classical and Modern SECOND EDITION

Study Guide
to Accompany

Keller/Gettys/Skove

Physics SECOND EDITION

and

Physics: Classical and Modern SECOND EDITION

Marllin Simon and G. Donald Thaxton
Auburn University

McGraw-Hill, Inc.
NEW YORK ST. LOUIS SAN FRANCISCO AUCKLAND BOGOTÁ CARACAS
LISBON LONDON MADRID MEXICO CITY MILAN MONTREAL
NEW DELHI SAN JUAN SINGAPORE SYDNEY TOKYO TORONTO

Study Guide to Accompany
Keller/Gettys/Skove: PHYSICS, Second Edition, and
PHYSICS: CLASSICAL AND MODERN, Second Edition

This book is printed on recycled paper containing a minimum of 50% total recycled fiber with 10% postconsumer de-inked fiber.

1 2 3 4 5 6 7 8 9 0 MAL MAL 9 0 9 8 7 6 5 4 3

ISBN 0-07-033909-0

The editors were Jack Shira and Maggie Lanzillo;
the production supervisor was Ann Marie D'Orazio.
Malloy Lithographing, Inc., was printer and binder.

CONTENTS

To

Kent D. Simon

and

Jonathan T. Simon

This study guide contains the following:

1. A comprehensive review of 194 of the principal concepts dealt with in the text **PHYSICS: CLASSICAL AND MODERN** by Keller, Gettys and Skove.
2. **1581** practice exercises with detailed solutions.
3. **213** example problems with an approach strategy and detailed solutions.
4. **934** practice test questions with answers.

ACKNOWLEDGMENTS

Fifteen months ago I enhanced a Macintosh 512 computer, purchased MacDraft and Microsoft WORD software, and started this study guide. The reader without desktop publishing experience cannot appreciate what we have accomplished and the reader with experience is probably laughing hilariously. I could not have produced this Study Guide in this time frame without the dedicated help of a number of people. It is appropriate that I acknowledge their contributions.

I am most grateful to the following:

Nancy H. Marsee, Technical Secretary at Auburn University for word processing chapters 1 through 20, 27, 30 through 38, and 41 through 43.

The Gnu's Room, For word processing chapters 25, 26, 28, 29, and 40.

Tim C. Richey, Physics Student at Auburn University for drawing most of the figures and for word processing chapters 21, 22, 23, 24, and 39.

James L. Graham, (Tank) Computer Science Student at Auburn University for editing, lay-out work, and production of the manuscript. When the rest of us thought we were through with a chapter, Tank went through it in a very painstaking and tireless manner and made it right!

George D. Thaxton, Associate Professor of Physics at Auburn University for writing chapters 20 through 30. When this project encountered time difficulties, Don Thaxton graciously agreed to bail me out by writing 11 chapters. I feel that Don's contribution is of such a magnitude that his name should also appear on the study guide.

Randy Rossignol, Senior Editor at McGraw-Hill for managing the project and providing encouraging words when the project encountered time difficulties.

Any errors which remain in the study guide are of course, fully my responsibility, and I welcome corrections and comments to improve future editions.

2/25/89 Marllin L. Simon

We are grateful for the opportunity to do a revision of this study guide. We hope we have eliminated most of the errors in the first 43 chapters and we have enjoyed adding two new chapters.

We gratefully acknowledge the dedicated work of Susan Tubb at McGraw-Hill for the production of this revised edition of The Study Guide.

12/31/92 Marllin L. Simon

HOW TO USE THIS STUDY GUIDE

RECALL FROM PREVIOUS CHAPTERS

Your success with each new chapter will depend partially on how well you have mastered previous chapters. This section of each chapter contains those concepts and equations from previous chapters most frequently used in the new chapter. Before starting your study of the new chapter, scan this list and briefly review anything you are not familiar with.

NEW IDEAS IN THIS CHAPTER

After reading a chapter in the text, check this section to make certain you are aware of the principal ideas introduced. At this point, it is not necessary that you feel comfortable with all these ideas, but you do need to be aware of them.

PRINCIPAL CONCEPTS AND EQUATIONS

This section lists the principal concepts and equations from the chapter. They are identified by the box around the number at the left-hand side of the page. After the concept is listed, some or all of the following are included to help you master the concept.

Review: Read this carefully to make sure you have picked up the main information in the chapter. While the review is fairly complete, it is not exhaustive. It is written with the understanding that you have read the text and will focus your attention on the important features before you work the problems. If you read the review sections at least twice a week, you will find that you are always on top of the physics knowledge required for your course, that you get more out of the lectures, that you can solve the problems faster and with less difficulty, and that you don't have to invest so much time preparing for exams.

Practice: Cover the right-hand side of this section with a sheet of paper, determine the desired quantity, and then compare your answer with the one in the study guide. If you understand a concept, the practice will prove that you do plus provide a little more practice. If you don't understand the concept, the practice section breaks each concept down into little segments and slowly builds your confidence and understanding to the point where you feel comfortable with that concept.

Examples: The examples should be approached as follows:

1. Read the problem.

2. Draw a diagram and then list all given quantities and those quantities to be determined. Compare with the study guide.

3. Outline a strategy for working the problem. When doing this, don't actually work the problem. Instead, consider what is given, what is to be determined,

the concepts involved and then decide how to manipulate the concepts to obtain the unknown from the known. Compare your strategy with that given in the study guide. Beginning physics students frequently experience a great deal of frustration and failure with exams. They find themselves reading a problem, muddling around with it for 10 minutes, never really obtaining an answer, and then going on to the next problem. Taking the time to acquire the skill of developing a problem strategy will help you solve problems in an efficient, confident manner. Compare your strategy with that in the study guide.

4. Work the problem. Compare your work with that in the study guide.

Notes: These notes should be given special attention. They point out common student errors and give study advice.

Related Text Exercises: Here are listed all text exercises related to the concept being studied. Work several of these in order to enhance your understanding of the concept and build your confidence. At the end of each chapter in the text you will find questions, exercises, and problems. You will benefit from working as many of these as possible, however only the exercises are referred to in the study guide.

PRACTICE TEST

Take and grade the practice tests. Doing so will allow you to determine any weak spots in your understanding of the concepts taught in this chapter. The following section prescribes what you should study further to strengthen your understanding. Answers to the practice test are found in Appendix I.

PRINCIPAL CONCEPTS AND EQUATIONS PRESCRIPTION

Your score on the practice test is an excellent measure of your understanding of the chapter. You should now use the chart in this section to write your own prescription for dealing with any weakness the practice test pointed out. Look down the leftmost column to the number of the question(s) you answered incorrectly, read across the row to find the concept and/or equation of concern, the section(s) of the study guide you should return to for further study, and then some suggested test exercises which you should work in order to gain additional experience.

Study Guide
to Accompany

Keller/Gettys/Skove

Physics SECOND EDITION

and

Physics: Classical and Modern SECOND EDITION

1 INTRODUCTION

NEW IDEAS IN THIS CHAPTER

Concepts and equations introduced	Text Section	Study Guide Page
Physical quantities, dimensions, units	1-1, 2	1-1
Unit conversion	1-2	1-2
Dimensional analysis	1-1	1-5
Significant figures	1-3	1-6

PRINCIPAL CONCEPTS AND EQUATIONS

1 Physical Quantities, Dimensions, Units (Sections 1-1 and 1-2)

Review: In this course we will study physical quantities and how they are related. We will divide physical quantities and how they are related into two groups, fundamental and derived. Fundamental physical quantities are defined in terms of comparisons with standards which have operational definitions and derived physical quantities are defined by an equation. During the mechanics portion of this course we will use the fundamental physical quantities mass, length, and time. In the SI system of units the standards for the fundamental physical quantities are kilogram (kg), meter (m), and second (s). The units of the derived physical quantities are combinations of the units for the fundamental physical quantities.

The dimension of a physical quantity denotes the nature of the physical quantity. Hence the dimensions of the fundamental physical quantities are the same as the quantities and the dimensions of the derived physical quantities are combinations of the dimensions of the fundamental physical quantities.

Table 1.1 summarizes information about the fundamental physical quantities.

Fundamental Physical Quantity	Dimension	Dimension Abbreviation	SI Unit Abbr.
Mass	Mass	[M]	kg
Length	Length	[L]	m
Time	Time	[T]	s

Table 1.1

Table 1.2 summarizes information about some derived physical quantities.

Derived Physical Quantity	Defining Equation	Dimension Abbreviation	SI Unit Abbr.
Area	$A = lw$	$[L][L] = [L^2]$	m^2
Volume	$V = lwh$	$[L][L][L] = [L^3]$	m^3
Mass Density	$\rho = M/V$	$[M][L^{-3}] = [ML^{-3}]$	$kg \cdot m^{-3}$
Mass Flow Rate	$R_M = M/t$	$[M][T^{-1}] = [MT^{-1}]$	$kg \cdot s^{-1}$
Volume Flow Rate	$R_V = V/t$	$[L^3][T^{-1}] = [L^3T^{-1}]$	$m^3 \cdot s^{-1}$
Speed	$v = \Delta x/\Delta t$	$[L][T^{-1}] = [LT^{-1}]$	$m \cdot s^{-1}$
Acceleration	$a = \Delta v/\Delta t$	$[LT^{-1}][T^{-1}] = [LT^{-2}]$	$m \cdot s^{-2}$

Table 1.2

Practice: Obtain the dimensions and SI units for each of the following derived physical quantities.

Physical Quantity	Defining Equation	Dimensions	Units
Area	$A = lw$	$[L][L] = [L^2]$	m^2
Volume	$V = lwh$	$[L][L][L] = [L^3]$	m^3
Mass Density	$\rho = M/V$	$[M][L^{-3}]=[ML^{-3}]$	$kg \cdot m^{-3}$
Mass Flow Rate	$R_M = M/t$	$[M][T^{-1}] = [MT^{-1}]$	$kg \cdot s^{-1}$
Volume Flow Rate	$R_V = V/t$	$[L^3][T^{-1}] = [L^3T^{-1}]$	$m^3 \cdot s^{-1}$
Speed	$v = \Delta x/\Delta t$	$[L][T^{-1}] = [LT^{-1}]$	$m \cdot s^{-1}$
Acceleration	$a = \Delta v/\Delta t$	$[LT^{-1}][T^{-1}] = [LT^{-2}]$	$m \cdot s^{-2}$
Force	$F = ma$	$[M][LT^{-2}] = [MLT^{-2}]$	$kg \cdot m \cdot s^{-2}$
Momentum	$p = mv$	$[M][LT^{-1}] = [MLT^{-1}]$	$kg \cdot m \cdot s^{-1}$
Force	$F = \Delta p/\Delta t$	$[MLT^{-1}][T^{-1}] = [MLT^{-2}]$	$kg \cdot m \cdot s^{-2}$
Impulse	$I = F\Delta t$	$[MLT^{-2}][T] = [MLT^{-1}]$	$kg \cdot m \cdot s^{-1}$
Work	$W = F\Delta x \cos\theta$	$[MLT^{-2}][L] = [ML^2T^{-2}]$	$kg \cdot m^2 \cdot s^{-2}$
Pressure	$P = F/A$	$[MLT^{-2}][L^{-2}] = [ML^{-1}T^{-2}]$	$kg \cdot m^{-1} \cdot s^{-2}$
Kinetic Energy	$K = mv^2/2$	$[M][LT^{-1}]^2 = [ML^2T^{-2}]$	$kg \cdot m^2 \cdot s^{-2}$

Related Text Exercises: Ex. 1-1 through 1-6.

2 Unit Conversion (Section 1-2)

Review: The two most common systems of units are the international system and the British system. When dealing with units, you need to be able to convert not only from one system to the other but also from one unit to another within a system. In order to convert a quantity from one unit to another, you need the relationship between the units (called the conversion factor). In SI, the conversion factor is easily obtained from the frequently used prefixes shown in Table 1.3.

Table 1.3

Prefix	Abbreviation	Power of ten
micro	μ	10^{-6}
milli	m	10^{-3}
centi	c	10^{2}
kilo	k	10^{3}
mega	M	10^{6}

Table 1.4 lists a number of typical conversions, the relationship between the units of interest and the conversion factor.

Desired Conversion	Relationship	Conversion Factor
Kilometers to meters	1 km = 10^3 m	10^3 m / 1 km = 1
Centimeters to meters	1 cm = 10^{-2} m	10^{-2} m / 1 cm = 1
Millimeters to meters	1 mm = 10^{-3} m	10^{-3} m / 1 mm = 1
Kilograms to grams	1 kg = 10^3 g	10^3 g / 1 kg = 1
Microseconds to seconds	1 μs = 10^{-6} s	10^{-6} s / 1 μs = 1

Table 1.4

Note: The principle involved in converting units is that you can multiply or divide any quantity by 1 without changing its value. In converting from kilometers to centimeters, for example, we simple multiply by 1 as many times as needed:

$$(50\ \text{km})\left(\frac{10^3\ \text{m}}{1\ \text{km}}\right)\left(\frac{1\ \text{cm}}{10^{-2}\ \text{m}}\right) = 50 \times 10^5\ \text{cm}$$

When dealing with units that are not based on powers of 10, you have to either memorize the conversion factors or obtain them from reference material. Table 1.5 gives some of the conversion factors you will use solving problems at the end of this chapter.

Desired Conversion	Relationship	Conversion Factor
Hours to seconds	1 h = 3600 s	3600 s / 1 h
Inches to feet	12 in = 1 ft	1 ft / 12 in
Feet to yards	3 ft = 1 yd	1 yd / 3 ft
Feet to miles	5280 ft = 1 mi	1 mi / 5280 ft

Table 1.5

In order to convert from the British system to SI and vice versa, you will need to obtain conversion factors from reference material. Those needed for the problems in Chapter 1 are given in Table 1.6.

Desired Conversion	Relationship	Conversion Factor
Meters to feet	1 m = 3.28 ft	3.28 ft / 1 m
Meters to yards	1 m = 1.09 yd	1.09 yd / 1 m
Meters to miles	1610 m = 1 mi	1 mi / 1610 m

Table 1.6

Conversion factors may be determined for derived physical quantities by combining the information found in tables 1.4, 1.5 and 1.6. For example we may convert from area in square meters to area in square inches as follows:

$$1\ m^2 = (1\ m^2)(3.28\ ft\ /\ 1\ m)^2\ (12\ in\ /\ 1\ ft)^2$$
$$= (1\ m^2)(10.76\ ft^2\ /\ 1\ m^2)(144\ in^2\ /\ 1\ ft^2) = 1550\ in^2$$

From this relationship, we obtain the conversion factor $1550\ in^2\ /\ 1\ m^2 = 1$. Knowing this, we may convert an area in square meters (say $5.00\ m^2$) to an area in square inches.

$$(5.00\ m^2)(1550\ in^2\ /\ 1\ m^2) = 7750\ in^2$$

Note: When converting units, be careful to insert the conversion factors so that all unwanted units cancel.

Practice: Using only the information in Tables 1.4, 1.5, and 1.6, determine the conversion factor for the following conversions:

1. Kilometers to inches	The relationship is $1\ km = (1\ km)(10^3\ m\ /\ 1\ km) \cdot (3.28\ ft\ /\ 1\ m)(12\ in\ /\ 1\ ft) = 39{,}400\ in$ The conversion factor is 39,400 in / 1 km.
2. Cubic inches to cubic meters	The relationship is $1\ in^3 = (1\ in^3)(1\ ft\ /\ 12\ in)^3 \cdot (1\ m\ /\ 3.28\ ft)^3 = 0.0000164\ m^3$ The conversion factor is $0.0000164\ m^3\ /\ 1\ in^3$.
3. Miles per hour to meters per second	The relationship is $1\ mi\ /\ 1\ hr = (1\ mi\ /\ 1\ hr) \cdot (1\ hr\ /\ 3600\ s)(1610\ m\ /\ 1\ mi) = 0.447\ m/s$ The conversion factor is (0.447 m/s) / (1 mi/hr).

Using the conversion factors just obtained, make the following conversions:

4. 10^6 in to kilometers	$(10^6 \text{ in})(1 \text{ km} / 39{,}400 \text{ in}) = 25.4 \text{ km}$
5. 500 in^3 to cubic meters	$(500 \text{ in}^3)(0.0000164 \text{ m}^3 / 1 \text{ in}^3)$ $= 0.00820 \text{ m}^3$
6. 55 mi/h to meters per second	(55 mi / h)(0.447 m / s) / (1 mi / h) = 24.6 m / s

If 1 zit = 2 zot; 3 zub = 4 zud, and 1 zud = (1/5) zot, determine the following:

7. 1 zot = ? zit	Given: 2 zot = 1 zit Then: 1 zot = (1/2) zit
8. 2 zud = ? zub	Given: 4 zud = 3 zub, or 1 zud = (3/4) zub Then: 2 zud = (2 zud)(3 zub / 4 zud) = (3/2) zub
9. 1 zit = ? zud	Given: 1 zit = 2 zot 1 zud = (1/5) zot, or 1 zot = 5 zud Then: 1 zit = (1 zit)(2 zot / 1 zit) • (5 zud / 1 zot) =10 zud
10. 1 zit = ? zub	Given: 1 zit = 2 zot, and 1 zud = (1/5) zot, or 1 zot = 5 zud 3 zub = 4 zud, or 1 zud = (3/4) zub Then: 1 zit = (1 zit)(2 zot / 1 zit) • (5 zud / 1 zot) • [(3/4) zub / 1 zud] = 7.5 zub

Related Text Exercises: Ex. 1-7 through 1-11.

3 Dimension Analysis (Section 1-1)

In the process of solving problems you will frequently end up with an algebraic expression for a Physical quantity. It is natural to want a little reassurance that your expression is correct. By analyzing the dimensions you can check on the validity of the expression. If the dimensions of each term in the expression are not consistent or if they are not the dimensions of the physical quantity under consideration, the expression is incorrect. The rules for dimensional analysis of an equation are:

(i.) assign a dimension to each symbol in the equation according to the physical property the symbol represents;
(ii.) multiply or divide dimensions using the rules of algebra;
(iii.) make sure the resulting dimension of each term in the equation is the same.

Practice: Perform a dimensional analysis for each of the following equations

1. $x = x_0 + v_0 t + 1/2 a t^2$	Term Dimensions x $[L]$ x_0 $[L]$ $v_0 t$ $[LT^{-1}][T] = [L]$ $1/2 a t^2$ $[LT^{-2}][T^2] = [L]$ Since each term in the equation has the same dimensions, the equation is dimensionally correct.
2. $v^2 = v_0^2 + 2a(x - x_0)$	Term Dimensions v^2 $[LT^{-1}]^2 = [L^2 T^{-2}]$ v_0^2 $[LT^{-1}]^2 = [L^2 T^{-2}]$ $2a(x - x_0)$ $[LT^{-2}][L] = [L^2 T^{-2}]$ Since each term in the equation has the same dimensions, the equation is dimensionally correct.
3. $v = 5at + x/t + at^2$	Term Dimensions v $[LT^{-1}]$ $5at$ $[LT^{-2}][T] = [LT]^{-1}$ x/t $[L][T^{-1}] = [LT^{-1}]$ at^2 $[LT^{-2}][T^2] = L$ Since the last term does not have the same dimensions as the other three, this expression is dimensionally inconsistent. Furthermore the last term (at^2) is the problem.

Related Text Exercises: Ex. 1-1 through 1-6.

4 Significant Figures (Section 1-3)

Review: In order to work physics problems efficiently, you need to know the following six things about significant figures.

1. How to determine which digits in a number are significant. This is done by scanning the number from left to right. The first nonzero digit and all following digits are significant figures. The only exception to this is a zero that just locates the decimal point.

Number	Significant figures
230.5	4
23.05	4
23.050	5
2.30	3
2300	4 if the zeroes are significant, that is if we know information to the units place and it is uncertain in the tenths place. We know the number is greater than 2299.5 but less than 2300.5.
2300	2 if the zeros just locate the decimal point, that is if we know information to the hundreds place but not to the tens place. We know the number is greater than 2250 but less than 2350.

Note: You will find it convenient to use scientific notation when writing numbers where the zeros are not significant. For example if we know that 2300 has two significant figures we can write 2.3×10^3. If 2300 has 3 significant figures we can write 2.30×10^3.

2. How to round off to the correct number of significant figures. If a series of values are used in a calculation, the value with the least number of significant figures determines the number of significant figures in the answer. To round off a number, look at the digit to the right of the last significant figure (i.e., the first unwanted digit). If it is 4 or less, retain the previous digit without change and drop all unwanted digits. If it is 5 or greater, increase the previous digit by one and drop all unwanted digits.

3. How to add and subtract using significant figures. In addition and subtraction, drop every digit in the result which falls in a column containing a nonsignificant figure. It is usually convenient to round off the number to the correct number of significant figures before adding or subtracting.

4. How to multiply and divide using significant figures. The result of multiplying or dividing should have no more significant figures than the least significant of the original numbers. When doing a series of calculations keep one extra digit so as not to accumulate errors in rounding off intermediate results. Then round off to the final result to the proper number of significant figures.

5. How to treat transcendental functions using significant figures. Transcendental functions such as the sine, arctangent, or exponential function are assumed to have the same number of significant figures as their argument.

6. How to treat a pure number. Pure numbers are considered to have an unlimited number of significant figures.

Practice:

1. Determine the number of significant figures in: (a) 205 (b) 452.0 (c) 116.72 (d) 0.03 (e) 0.043 (f) 1.030	(a) 3 (b) 4 (c) 5 (d) 1 (e) 2 (f) 4
2. Round off 147.6082 to: (a) 6 (b) 5 (c) 4 and (d) 3 significant figures.	(a) 147.608 (b) 147.61 (c) 147.6 (d) 148
3. Perform the indicated operation to the correct number of significant figures (a) 4.64 + 7.261 (b) 5.20 + 0.237 (c) 19.2 + 0.03 (d) 176.4 + 15.0 (e) 27.43 - 19.027 (f) 152.3 - 140. (g) 12.3 x 4.0 (h) 17.36 x 1.27 (i) 42.73 / 0.250 (j) sin 45.0° (k) ln(12.72) (l) $e^{2.0}$	(a) 4.64 + 7.26 = 11.90 (b) 5.20 + 0.24 = 5.44 (c) 19.2 + 0.0 = 19.2 (d) 176.4 + 15.0 = 191.4 (e) 27.43 - 19.03 = 8.40 (f) 152 - 140. = 12 (g) 12.3 x 4.0 = 49. (h) 17.36 x 1.27 = 22.0 (i) 42.74 / 0.250 = 171 (j) sin 45.0° = 0.707 (k) ln(12.72) = 2.543 (l) $e^{2.0}$ = 7.4

Related Text Exercises: Ex. 1-12 through 1-16 and 1-18 through 1-20.

PRACTICE TEST

Take and grade this practice test. Doing so will allow you to determine any weak spots in your understanding of the concepts taught in this chapter. The following section prescribes what you should study further to strengthen your understanding.

Table 1.7 gives information about several physical quantities.

Quantity	Distance	Mass	Time	Speed	Acceleration
Symbol	S	M	t	v	a
SI Abbreviation	m	kg	s	m/s	m/s^2

Table 1.7

Obtain the derived SI units for the quantities defined by the following expressions:

________	1.	Momentum:	$p = Mv$
________	2.	Kinetic energy:	$KE = Mv^2 / 2$
________	3.	Distance:	$S = at^2 / 2$
________	4.	Force:	$F = Ma$

Determine if the following equations are dimensionally correct:

________	5.	$F = Mv / t$	(see 4 above)
________	6.	$a = v^2 / 2s$	
________	7.	$v = at^2 - at$	
________	8.	$KE = Mas$	(see 2 above)

A car travels 900 km in 15.0 h and consumes 75.0 liter of gasoline that costs $24.00.

________ 9. How far can this car travel on 5.00 liters of gas?
________ 10. What would gasoline cost for a 2000 km trip?
________ 11. What was the distance rate of consumption of gasoline?
________ 12. What was the time rate of consumption of gasoline?
________ 13. What was the time rate of change of position?

A solid sphere has a mass of 2.107 kg and a radius of 0.500 m. Determine the following to the correct number of significant figures:

________	14.	Surface area:	$A = 4\pi r^2$
________	15.	Volume:	$V = 4\pi r^3 / 3$
________	16.	Mass density:	$\rho = m / V$

Make the following conversions:

________ 17. 40.0 mi/h to meters per second
________ 18. 200 in^2 to square centimeters
________ 19. 10.0 m/s to miles per hour
________ 20. 100 km to inches

Perform the indicated operations and express your answer in scientific notation to the correct number of significant figures:

________ 21. $1.20 \times 10^4 / 4.0 \times 10^{-3}$
________ 22. 110.02×0.025
________ 23. $102.56 + 0.0382$
________ 24. $27.43 - 19.027$

(See Appendix I for answers.)

PRINCIPAL CONCEPTS AND EQUATIONS PRESCRIPTION

Your score on the practice test is an excellent measure of your understanding of the chapter. You should now use the following chart to write your own prescription for dealing with any weaknesses the practice test points out. Look down the leftmost column to the number of question(s) you answered incorrectly, reading across that row you will find the concept and/or equation of concern, the section(s) of the study guide you should return to for further study, and some suggested text exercises and problems which you should work to gain additional experience.

Practice Test Questions	Concepts and Equations	Prescription	
		Principal Concepts	Text Exercises
1	Physical quantities and units	1	1-1, 2
2	Physical quantities and units	1	1-3, 4
4	Physical quantities and units	1	1-5, 6
5	Dimensional analysis	3	1-1, 2
6	Dimensional analysis	3	1-2, 3
7	Dimensional analysis	3	1-3, 4
8	Dimensional analysis	3	1-5, 6
9	Units and conversion factors	2	1-7, 8
10	Units and conversion factors	2	1-8, 9
11	Units and conversion factors	2	1-9, 10
12	Units and conversion factors	2	1-10, 11
13	Units and conversion factors	2	1-7, 8
14	Significant figures	4	1-12, 14
15	Significant figures	4	1-12, 16
16	Significant figures	4	1-14, 15
17	Unit conversion	2	1-7, 8
18	Unit conversion	2	1-8, 9
19	Unit conversion	2	1-9, 10
20	Unit conversion	2	1-10, 11
21	Scientific notation and significant figures	4	1-14, 15
22	Scientific notation and significant figures	4	1-12, 16
23	Scientific notation and significant figures	4	1-13, 14
24	Scientific notation and significant figures	4	1-15

2 VECTORS

NEW IDEAS IN THIS CHAPTER

Concepts and equations introduced	Text Section	Study Guide Page
Difference between vector and scalar quantities	2-1	2-1
Graphical treatment of vectors (representation, addition, subtraction, components)	2-2, 3	2-1
Analytical treatment of vectors (unit vectors, components, addition, subtraction)	2-4	2-8

PRINCIPAL CONCEPTS AND EQUATIONS

1 Difference Between Vector and Scalar Quantities (Section 2-1)

Review: Scalar quantities have magnitude only. For example, the mass of a bowling ball is 4.50 kg, a class period is 50.0 min, and a comfortable room temperature is 25.0° C. Vector quantities have direction as well as magnitude. For example, a quarterback's displacement is 5.00 yards down field, and the velocity of a car is 90.0 km/hr directed at an angle of 40.0° E of N.

2 Graphical Treatment of Vectors (Sections 2-2 and 2-3)

Review: In this section we will look at all aspects of the graphical treatment of vectors. While you will rarely solve a vector problem by graphical treatment, this method will help you learn to visualize vectors, see what is happening, and determine how to approach the problem analytically.

A. Graphical Representation of Vectors. A vector may be represented by an arrow. The orientation of the arrow shows the direction of the vector, and the length when drawn to the proper scale represents the magnitude of the vector. Figure 2.1 is the graphical representation of a displacement vector that has a magnitude of 10.0 m and is directed at an angle of 30.0° N of E.

Figure 2.1

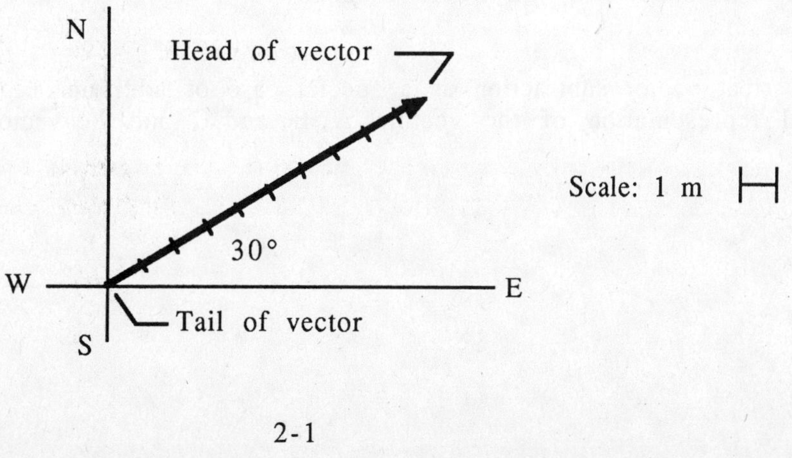

Note: You may slide a vector on the page as long as you don't compress or stretch it (i.e. change its magnitude) and as long as you don't re-orient it (i.e. change its direction).

B. Graphical Addition and Subtraction of Vectors by the Head-to-Tail Method. Figure 2.2 shows the graphical representation of three vectors **A**, **B**, and **C**.

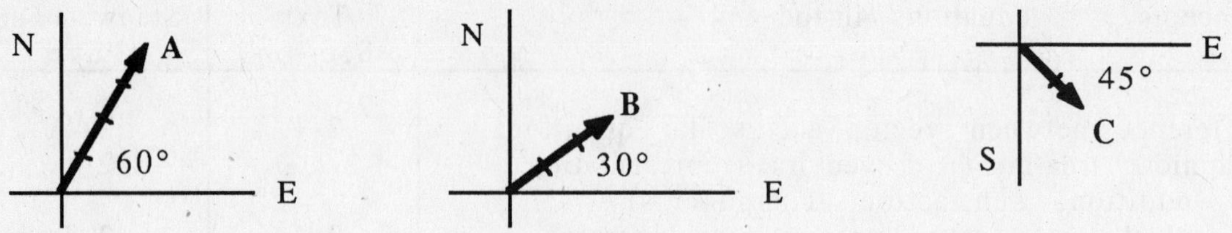

A = 4 m 60° N of E

B = 3 m 30° N of E

C = 2 m 45° S of E

Figure 2.2

Figure 2.3 shows that the sum of **A** and **B** is obtained by first drawing **A** and then "sliding" **B** around until its tail joins the head of **A**. It also shows that the result of adding vectors is independent of the order in which they are added.

Figure 2.3

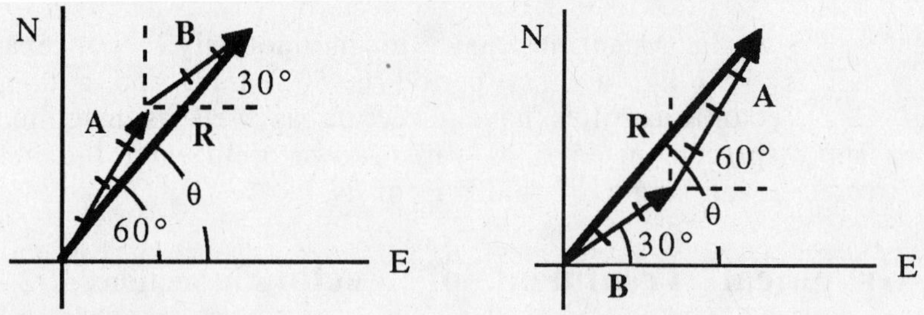

R = **A** + **B**
R = 6.76 m 47.2° N of E

R = **B** + **A**
R = 6.76 m 47.2° N of E

Since the result of adding **A** and **B** is **R**, we call **R** the resultant vector.

Note: Many students have trouble visualizing the result of adding **A** and **B**. Graphically if you want to find **A** + **B**, you just do what the expression (**A** + **B**) says to do. That is you take the vector **A** and you add to it the vector **B**. The resultant is the result of taking the vector **A** and adding to it the vector **B**. The resultant then must start where **A** starts and end where **B** ends.

We will treat vector subtraction as a special case of addition. Figure 2.4 shows the graphical representation of the vectors **A**, **B**, and **C** and the vectors -**A**, -**B** and -**C**.

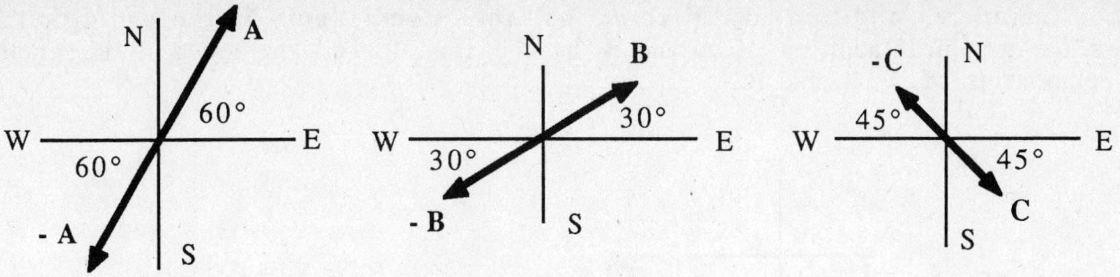

A = 4 m 60° N of E
- A = 4 m 60° S of W

B = 3 m 30° N of E
- B = 3 m 30° S of W

C = 2 m 45° S of E
-C = 2 m 45° N of W

Figure 2.4

The negative of a vector is a vector of the same magnitude but opposite direction. We can then treat **A** - **C** as the sum of the vectors **A** and **-C**. This is shown graphically in Figure 2.5.

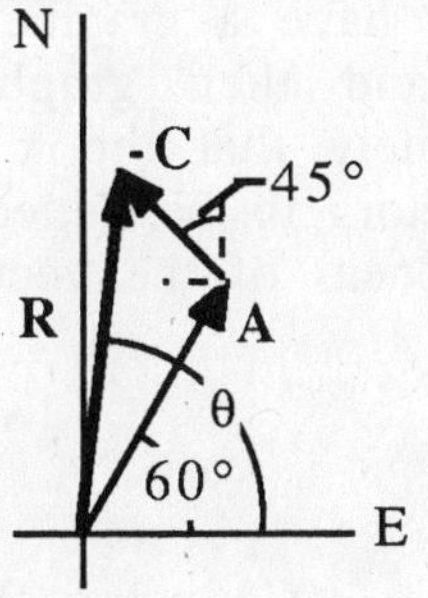

Figure 2.5

R = **A** - **C**
R = **A** + (**-C**)

C. Graphical Determination of the Components of a Vector. Figure 2.6 shows the graphical resolution of the vectors **A** and **-B** into components.

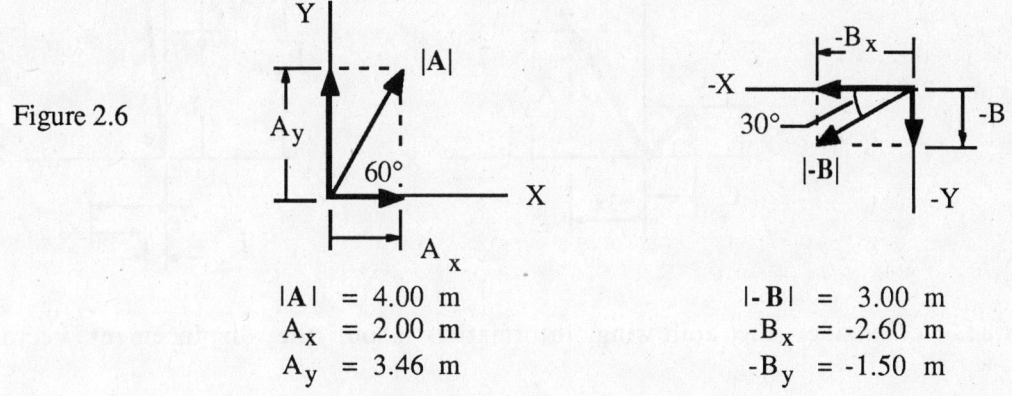

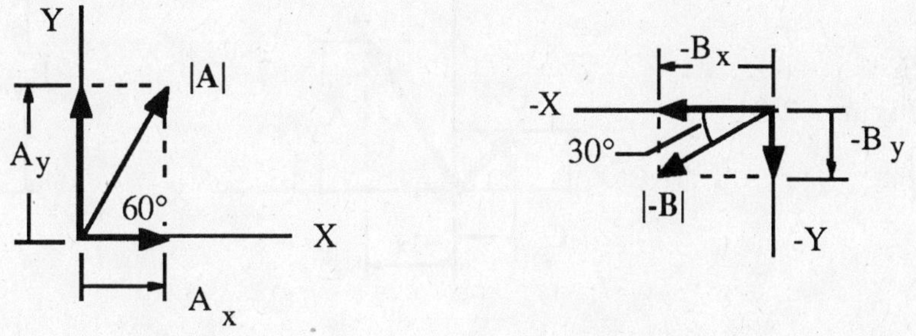

Figure 2.6

|**A**| = 4.00 m
A_x = 2.00 m
A_y = 3.46 m

|-**B**| = 3.00 m
$-B_x$ = -2.60 m
$-B_y$ = -1.50 m

As shown, the resolution is accomplished by projecting the vectors onto the axes. The projection is accomplished by drawing a line from the vector head perpendicular to the x axis for the x component and another perpendicular to the y axis for the y component.

NOTE: The components of a vector include the sign.

D. Graphical Addition of Vectors by the Component Method. Figure 2.7 shows the graphical addition of **A** and **B** by the Head-to-Tail method and the graphical components of **A**, **B**, and **R**.

Figure 2.7

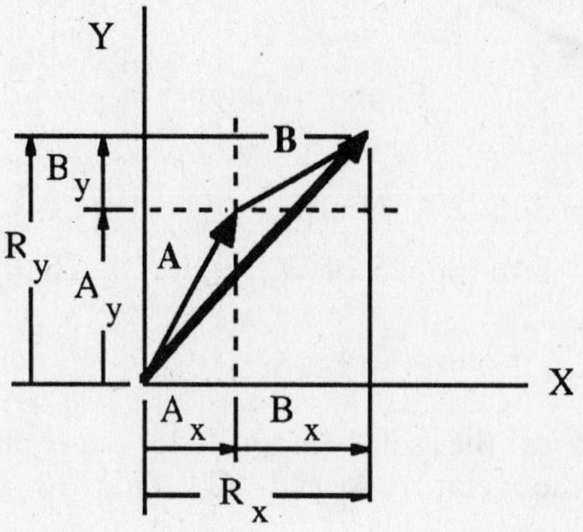

$R_x = A_x + B_x$
$R_y = A_y + B_y$

From Figure 2.7 we see that if we have a graphical representation of **A** and **B**, we can graphically find the components and then graphically determine the resultant by using what we just discovered, namely that the x component of the resultant is the sum of the x component of the vectors to be added and the y component of the resultant is the sum of the y component of the vectors to be added. This may be written as

$$R_x = A_x + B_x \qquad \text{and} \qquad R_y = A_y + B_y$$

Figure 2.8 shows the vector sum **A** + (-**C**) using this method.

Figure 2.8

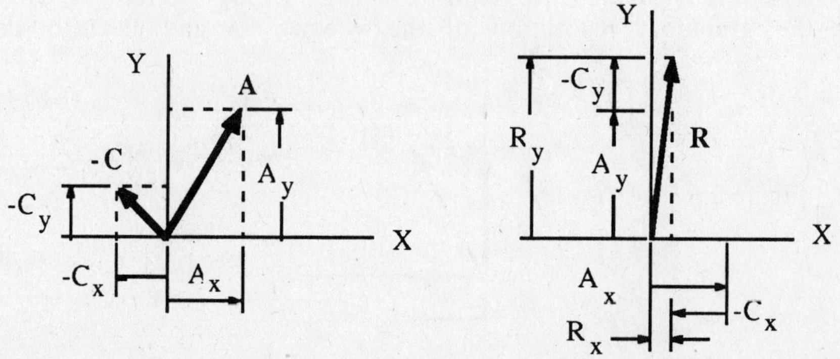

Practice: Consider the following information about the displacement vectors **A**, **B** and **C**.

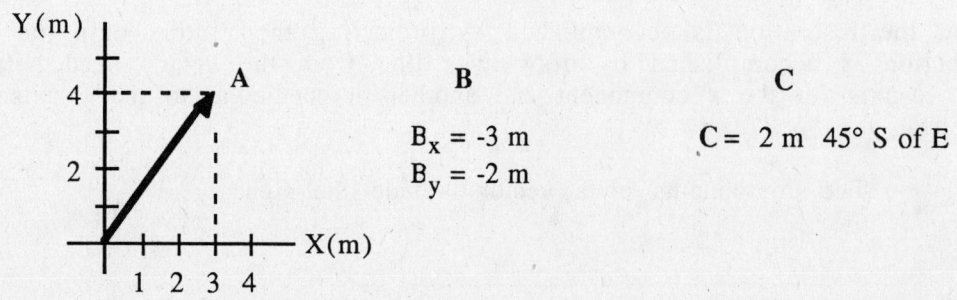

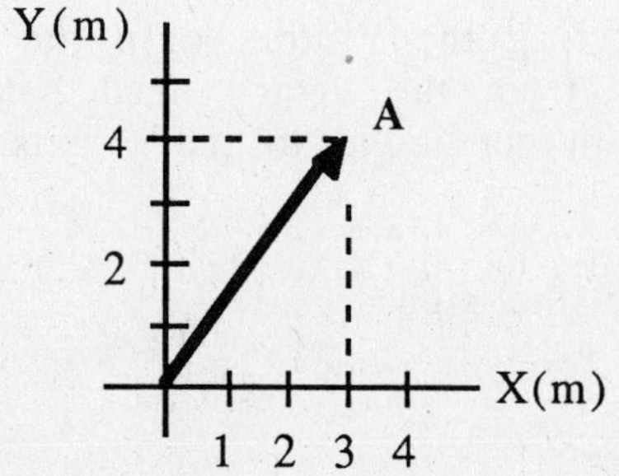

B

B_x = -3 m
B_y = -2 m

C

C = 2 m 45° S of E

Determine the following by graphical methods:

1. Representation of **B**	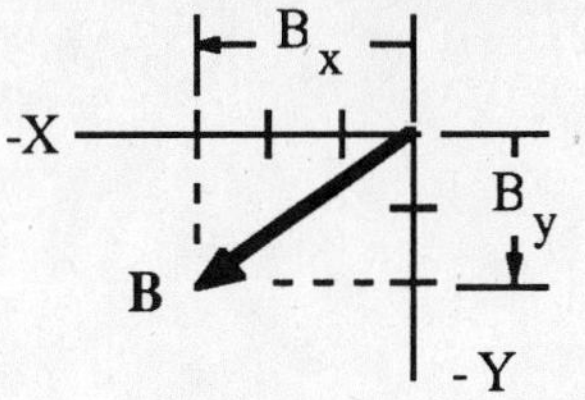 B_x = -3 m B_y = -2 m
2. Representation of **C**	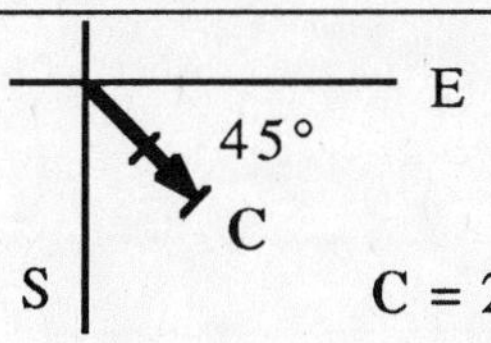 C = 2 m 45° S of E
3. The magnitude and direction of **A**	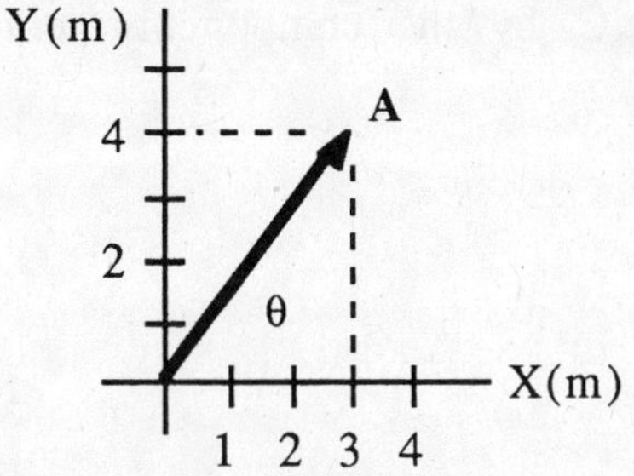 Using a linear scale, measure **A** A = 5.0 m Using a protractor, measure θ θ = 53.1° N of E or ccw form the +x axis
4. The magnitude and direction of **B**	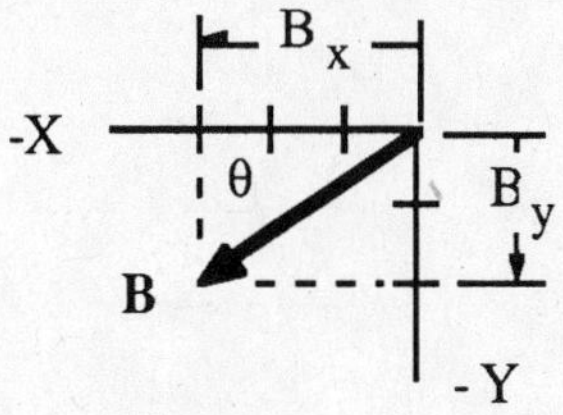 Using a linear scale, measure **B** B = 3.61 m Using a protractor, measure θ θ = 33.7° S of W or ccw from the -x axis.
5. Components of the vector **C**	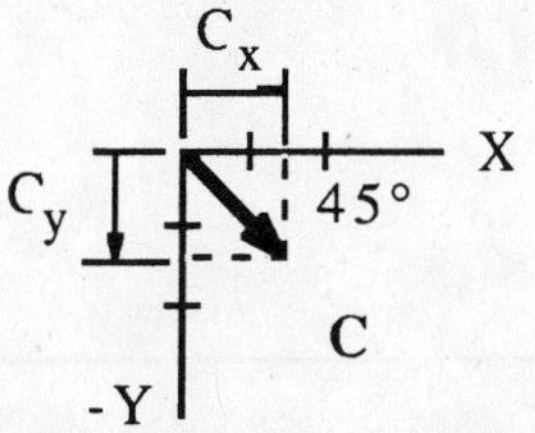 Project **C** onto the x and y axes and then measure the components. C_x = +1.41 m, and C_y = -1.41 m

6. **A** + **C** by the Head-to-tail method

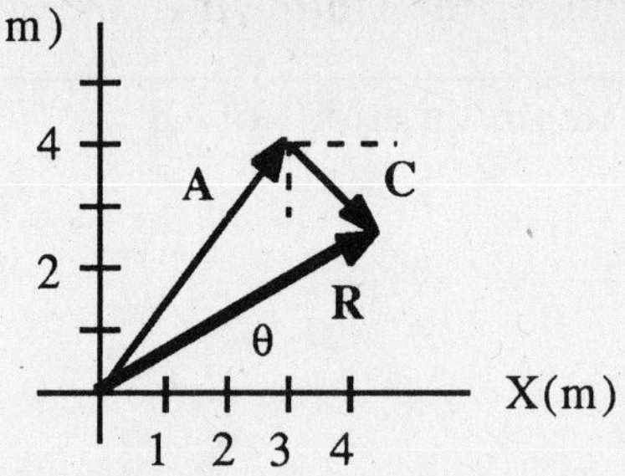

Using a linear scale, measure **R**
R = 5.11 m
Using a protractor, measure θ
θ = 30.4°
R = 5.11 m 30.4° N of E

7. **A** + **C** by the component method

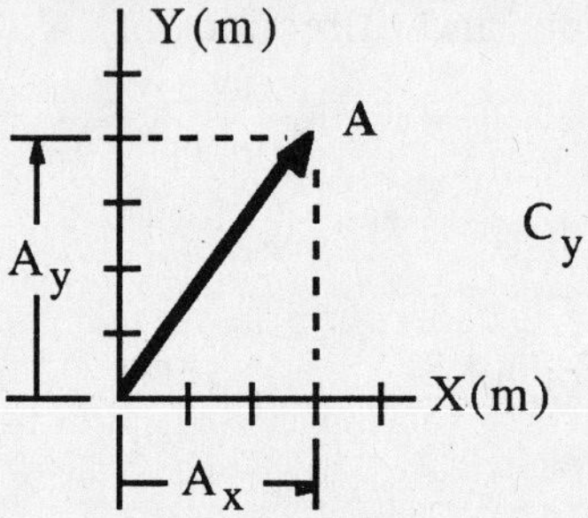

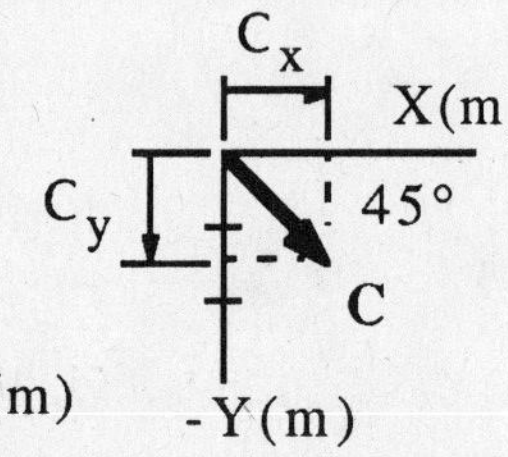

A_x = 3 m
A_y = 4 m

C_x = 1.41 m
C_y = -1.41 m

$R_x = A_x + C_x$ = 4.41 m
$R_y = A_y + C_y$ = 2.59 m

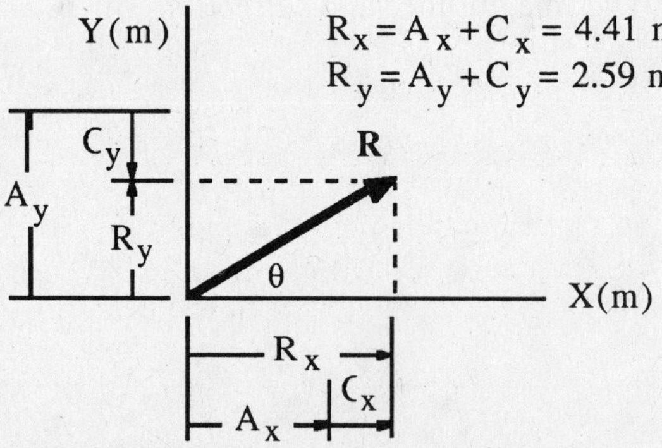

Using a linear scale, measure **R**
R = 5.11 m
Using a protractor, measure θ
θ = 30.4°
R = 5.11 m 30.4° N of E

8. **B** - **C** by the Head-to-tail method

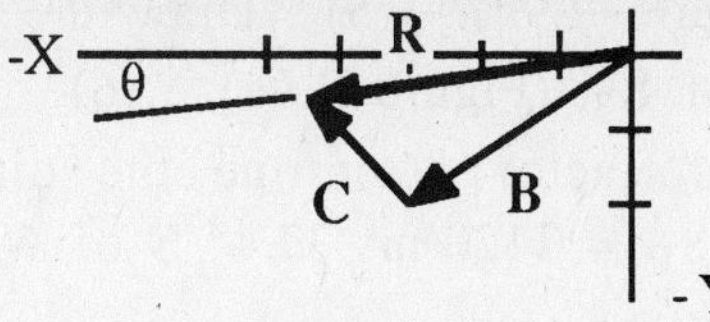

Using a linear scale, measure **R**.
Using a protractor, measure θ
R = 4.45 m   θ = 7.62°
R = 4.45 m 7.62° S of W

9. **B** - **C** by the component method

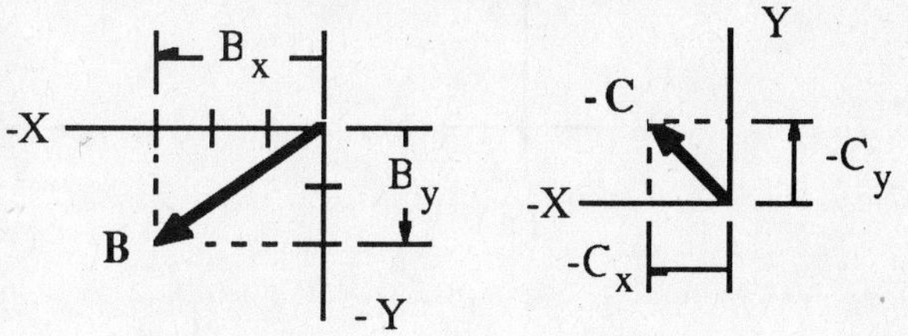

B_x = -3 m $-C_x$ = -1.41 m
B_y = -2 m $-C_y$ = 1.41 m

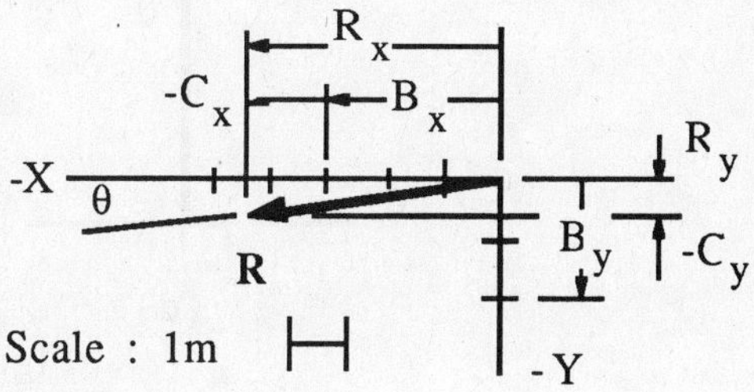

Using a linear scale, measure **R**.
Using a protractor, measure θ
R = 4.45 m   θ = 7.62°
R = 4.45 m 7.62° S of W

Example: A child walks to school by first walking 800 m N, next 600 m W, and then cuts across the vacant lot 400 m 30° S of W. Using graphical techniques, determine the child's resultant displacement vector.

Given: $\mathbf{S_1}$ = 800 m N, $\mathbf{S_2}$ = 600 m W, and $\mathbf{S_3}$ = 400 m 30° S of W.

Determine: $\mathbf{S_R}$, the child's resultant displacement vector.

Strategy: Locate the origin of the coordinate system at the child's home. Choose a convenient scale and draw the vector $\mathbf{S_1}$. Locate the tail of $\mathbf{S_2}$ at the head of $\mathbf{S_1}$ and then draw $\mathbf{S_2}$ (use the same scale.) Locate the tail of $\mathbf{S_3}$ at the head of $\mathbf{S_2}$ and then draw $\mathbf{S_3}$ (use the same scale). The resultant displacement $\mathbf{S_R}$ is from the tail of $\mathbf{S_1}$ to the head of $\mathbf{S_3}$.

Solution: (1) Establish the origin of the coordinate system and choose a convenient scale (Figure 2.9a). (2) Draw $\mathbf{S_1}$ (Figure 2.9b). (3) Draw $\mathbf{S_2}$ with its tail connected to head of $\mathbf{S_1}$ (Figure 2.9c). (4) Draw $\mathbf{S_3}$ with its tail connected to the head of S_2 (Figure 2.9d). (5) Draw the resultant S_R starting at the tail of S_1 and ending at the head of S_3 (Figure 2.9e). (6) Using a linear scale determine the magnitude of S_R, using a protractor determine the direction of S_R. The resultant vector may then be written as S_R = 1120 m 32.4° N of W.

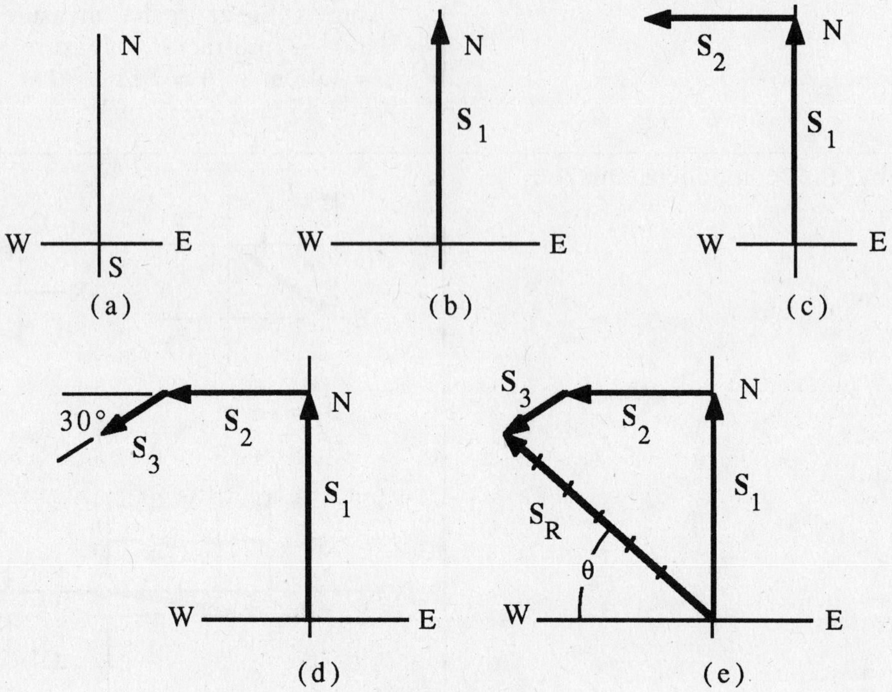

Figure 2.9

Related Text Exercises: Ex. 2-3 through 2-7.

3 Analytical Treatment of Vectors (Sections 2-3 and 2-4)

Review: In this section we will look at several aspects of the analytical treatment of vectors. Even though the treatment is analytical, it is essential that you have a clear mental picture of what is happening. To do this you will need to rely on your knowledge of graphical analysis.

A. Unit Vectors and Components. It is convenient to use **i**, **j**, and **k** to represent unit vectors along the x, y, and z axes respectively of a Cartesian coordinate system as shown in Figure 2.10.

Figure 2.10

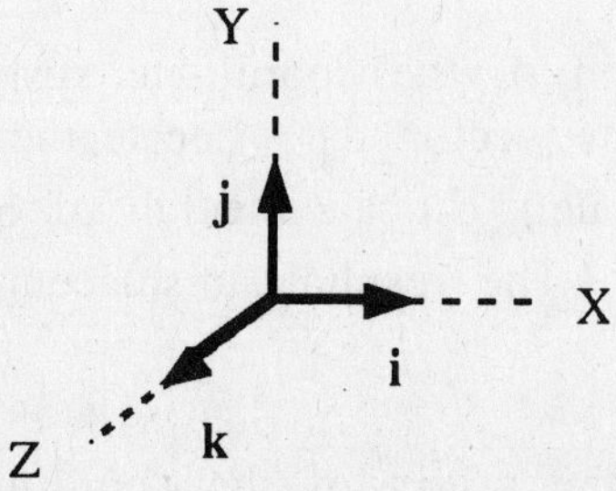

The vector **A** may be resolved into its components by projecting it onto the axes. As you will recall from your knowledge of graphical techniques this is accomplished as shown in Figure 2.11.

Figure 2.11

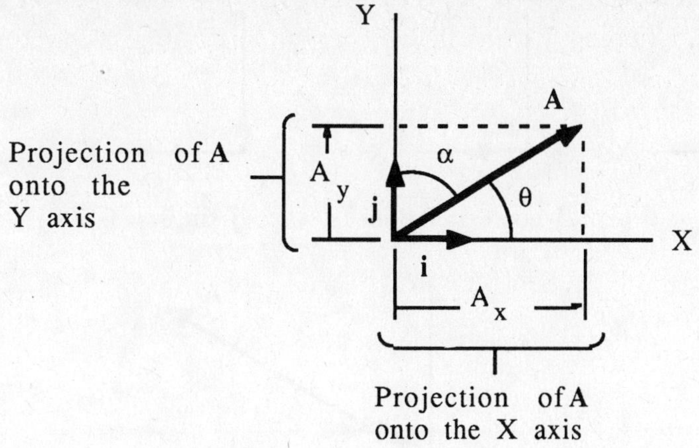

A is a vector of magnitude |**A**| or A and direction θ w.r.t. the +x axis and α w.r.t. the +y axis.

A_x is the x component of **A**, its value is $A_x = A \cos\theta = A \sin\alpha$

A_y is the y component of **A**, its value is $A_y = A \sin\theta = A \cos\alpha$

$A_x\mathbf{i}$ is a component vector of magnitude A_x in the direction of the unit vector **i**.

$A_y\mathbf{j}$ is a component vector of magnitude A_y in the direction of the unit vector **j**.

A is composed of the component vectors $A_x\mathbf{i}$ and $A_y\mathbf{j}$ as shown in Figure 2-12.

Figure 2.12

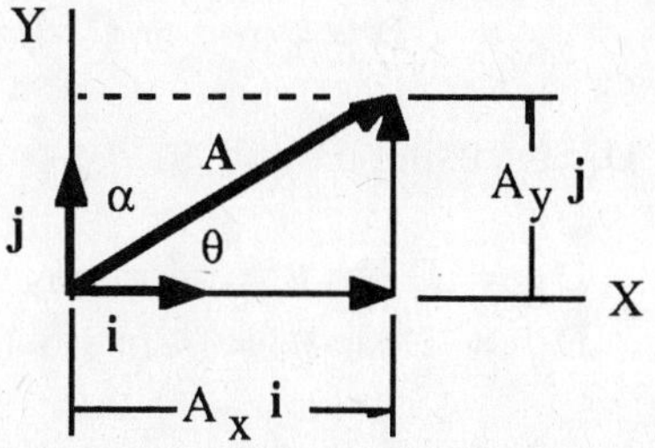

$\mathbf{A} = A_x\mathbf{i} + A_y\mathbf{j}$ is a mathematical statement of the fact that the vector **A** is composed of the component vectors $A_x\mathbf{i}$ and $A_y\mathbf{j}$.

Rather than giving information about **A** in the form A and θ, we can give it in the form of A_x and A_y. Then we obtain A and θ as follows

$$A = (A_x^2 + A_y^2)^{1/2} \quad \text{and} \quad \theta = \tan^{-1}(A_y / A_x)$$

We obtain a unit vector in the direction of **A** by dividing the vector **A** by its magnitude A.

$$\mathbf{n} = \mathbf{A} / A = (A_x\mathbf{i} + A_y\mathbf{j}) / A = (A_x / A)\mathbf{i} + (A_y / A)\mathbf{j} = \mathbf{i}\cos\theta + \mathbf{j}\sin\theta$$

We express **A** in terms of **n** as

$$\mathbf{A} = \mathbf{n} \cdot A \quad \text{where A is the magnitude of } \mathbf{A} \text{ and } \mathbf{n} = \mathbf{i}\cos\theta + \mathbf{j}\sin\theta$$

If A = 4 m and θ = 30° Figure 2.13 shows **n** and **A**

$i \cos \theta = 0.866i$

$j \sin \theta = 0.500j$

$n = i\cos \theta + j \sin \theta$
$n = 0.866i + 0.500j$

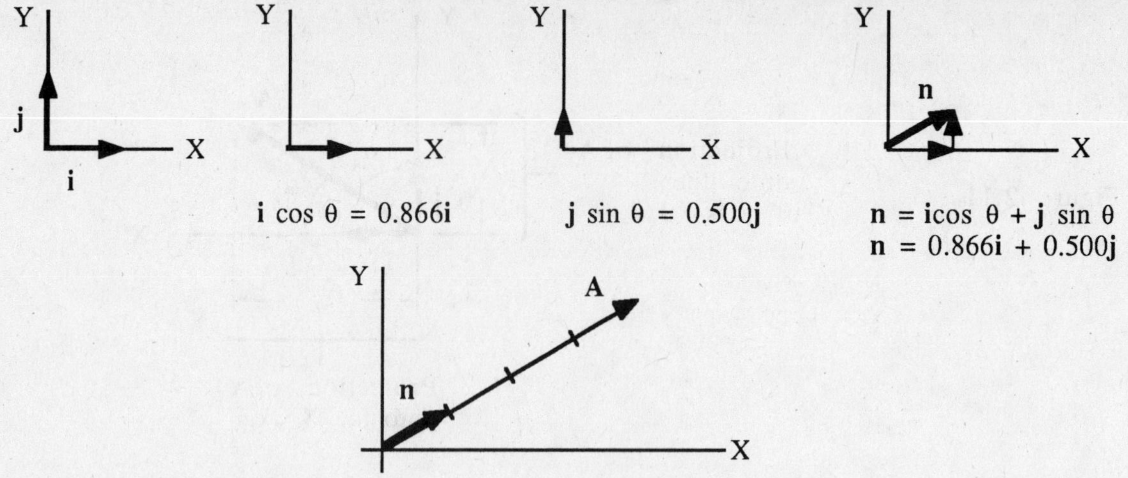

Figure 2.13

Many students have trouble with vectors that are not in the first quadrant. We can resolve this difficulty by graphically representing the vector. This allows us to visualize the vector and hence be sure of the sign of its components and its direction. To illustrate this, consider the vector **D**

$$\mathbf{D} = 4 \text{ m} \quad \text{and} \quad \theta = 150°$$

We may obtain D_x and D_y by using D and θ.

$$D_x = D \cos \theta = 4 \text{ m} \cos 150° = -3.46 \text{ m}$$
$$D_y = D \sin \theta = 4 \text{ m} \sin 150° = 2.00 \text{ m}$$

In the above we have just used $\theta = 150°$ and let the mathematics (performed by our calculators) tell us the sign of the components.

Now if we graphically represent the vector as shown in Figure 2.14 we see that D_x is negative and D_y is positive. We also see that the components may be obtained by using the other angles shown as in Figure 2.14.

Figure 2.14

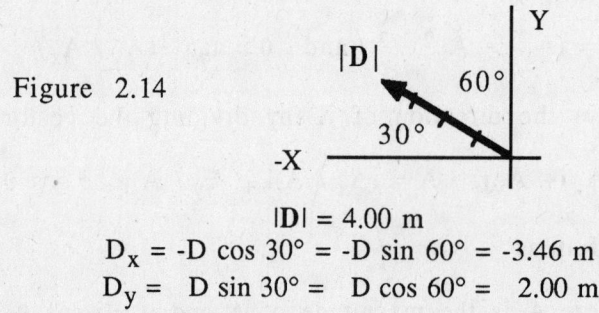

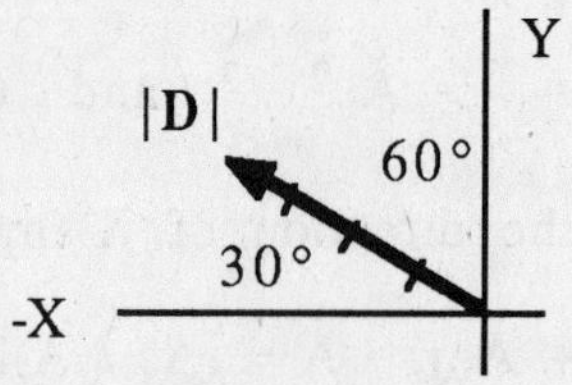

$|\mathbf{D}| = 4.00 \text{ m}$
$D_x = -D \cos 30° = -D \sin 60° = -3.46 \text{ m}$
$D_y = D \sin 30° = D \cos 60° = 2.00 \text{ m}$

If we are working from the components, the graphical representation has another advantage.

If $D_x = -3.46$ m and $D_y = 2.00$ m, then

$\theta = \tan^{-1}(\theta \pm 180°) = \tan \theta$, we have an ambiguity ($\theta$ may be -30°, -210° or 150°)

$\theta = -30°$ is in the fourth quadrant

θ = -210° and 150° are the same angle and are in the second quadrant

Only when we graphically represent **D** (on paper or mentally) do we realize that **D** is in the second quadrant. Hence the graphical representation allows us to resolve this ambiguity.

B. Analytical Addition and Subtraction of Vectors. Consider the displacement vectors **A, B, C** and **D**.

A = 4 m 60° N of E; **B** = 3 m 60° E of N; **C** = -1.5 m **i** + 1.5 m **j**; **D** = - 3 m **i** - 2 m **j**

We would like to use analytical methods to obtain the resultant vector **R**, where

$$\mathbf{R} = \mathbf{A} + \mathbf{B} - \mathbf{C} + \mathbf{D}$$

NOTE: When working vector problems you can just follow a mathematical prescription (i.e. just put numbers into an equation) and get an answer which might even be correct. However, the ideal situation involves you correctly seeing the vectors in your mind (or sketching them on paper) and then using this mental (or paper) image to show you what to do analytically. Using the latter approach, you are not reduced to merely plugging into equations but rather you understand the situation and because of this you know how to approach the problem analytically.

In order to facilitate your understanding, lets first make sure we agree on what these vectors look like. They are shown in Figure 2.15.

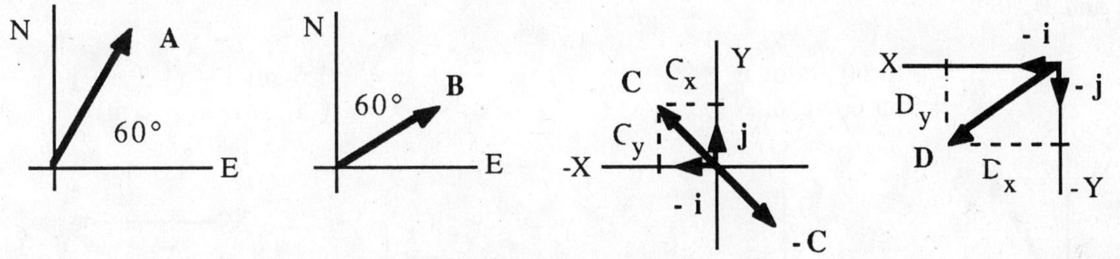

Figure 2.15

Next lets get some idea of what **A + B - C + D** looks like. We can easily do this by the graphical head-to-tail method shown in Figure 2.16.

Figure 2.16

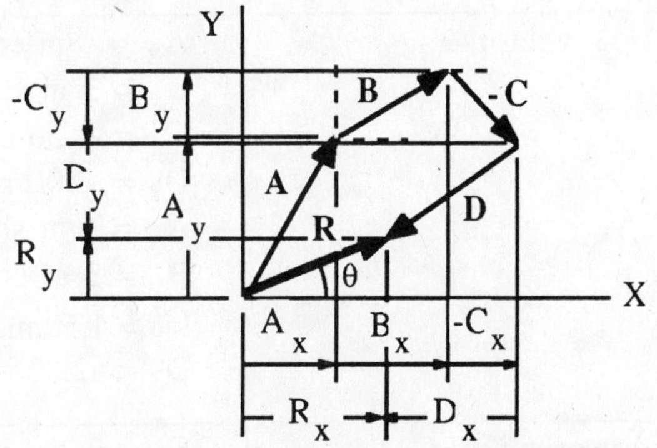

Using our knowledge of how to handle vectors graphically, we see what the result looks like and we also see how to proceed analytically. For instance we know the following:

A_x, B_x, and $-C_x$ are positive and D_x is negative
A_y and B_y are positive and $-C_y$ and D_y are negative
$R_x = A_x + B_x - C_x + D_x$ and R_x is positive
$R_y = A_y + B_y - C_y + D_y$ and R_y is positive
$R = (R_x^2 + R_y^2)^{1/2}$
$\theta = \tan^{-1}(R_y / R_x)$

Now that we have a good understanding of what these vectors and the resultant look like, we can confidently do the analytical work.

$A_x = A \cos 60° = 4 \text{ m} \cos 60° =$	2.00 m	$A_y = A \sin 60° = 4 \text{ m} \sin 60° =$	3.46 m
$B_x = B \sin 60° = 3 \text{ m} \sin 60° =$	2.60 m	$B_y = A \cos 60° = 3 \text{ m} \cos 60° =$	1.50 m
$-C_x =$	$= 1.50 \text{ m}$	$-C_y =$	$= -1.50 \text{ m}$
$D_x =$	$= -3.00 \text{ m}$	$D_y =$	$= -2.00 \text{ m}$
$R_x = A_x + B_x + (-C_x) + D_x =$	3.10 m	$R_y = A_y + B_y + (-C_y) + D_y =$	1.46 m
$R = (R_x^2 + R_x^2)^{1/2} =$	3.43 m	$\theta = \tan^{-1}(R_y / R_x)$	$= 25.2°$

We may express the resultant vector as

R = 3.43 m 25.2° N of E or **R** = (3.10 m) **i** + (1.46 m) **j**

Practice: Analytically determine the following for the displacement vectors **A**, **B**, **C** and **D**.

A = 4 m 60° N of E
B = 3 m 60° E of N

C = (-1.5 m) **i** + (1.5 m) **j**
D = (-3 m) **i** + (-2 m) **j**

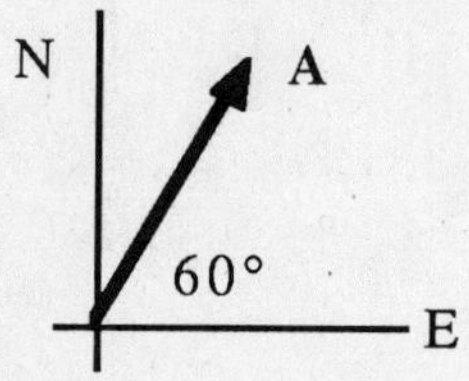

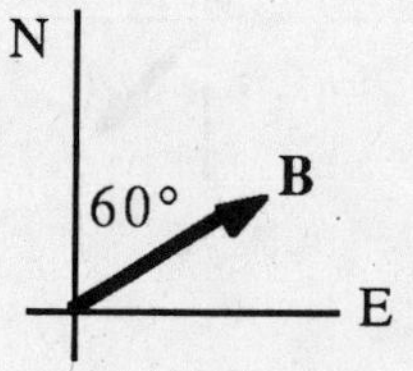

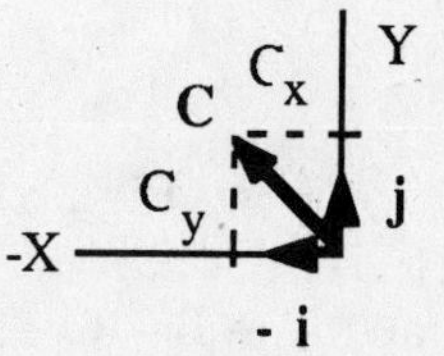

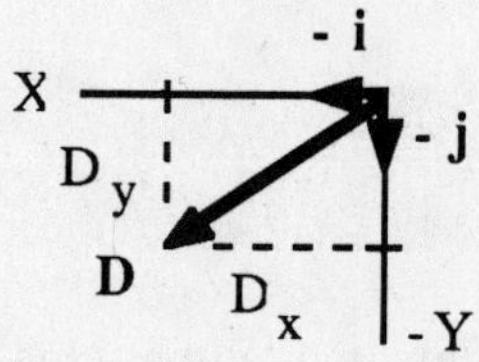

1. Components of all vectors	$A_x = 4 \text{ m} \cos 60° = 4 \text{ m} \sin 30° = 2.00 \text{ m}$ $B_x = 3 \text{ m} \sin 60° = 3 \text{ m} \cos 30° = 2.60 \text{ m}$ $C_x = -1.50 \text{ m}$ $D_x = -3.00 \text{ m}$ $A_y = 4 \text{ m} \sin 60° = 4 \text{ m} \cos 30° = 3.46 \text{ m}$ $B_y = 3 \text{ m} \cos 60° = 3 \text{ m} \sin 30° = 1.50 \text{ m}$ $C_y = 1.50 \text{ m}$ $Dy = -2.00 \text{ m}$
2. Magnitude and direction of **C**	$C = (C_x^2 + C_y^2)^{1/2} = 2.12 \text{ m}$ $\theta = \tan^{-1}(C_y / C_x) = \tan^{-1} 1 = 45°$ N of W

3. Resultant vector **A + B**	$R_x = A_x + B_x = 4.60$ m $R_y = A_y + B_y = 4.96$ m $R = (R_x^2 + R_y^2)^{1/2} = 6.76$ m Since R_x and R_y are positive, we know **R** is in the first quadrant $\theta = \tan^{-1}(R_y / R_x) = 47.2°$ N of E The resultant may be expressed as **R** = 6.76 m 47.2° N of E or **R** = (4.60 m) **i** + (4.96 m) **j**
4. Resultant vector **C + D**	$R_x = C_x + D_x = -4.50$ m $R_y = C_y + D_y = -0.50$ m $R = (R_x^2 + R_y^2)^{1/2} = 4.53$ m Since R_x and R_y are negative we know **R** is in the third quadrant $\theta = \tan^{-1}(R_y / R_x) = 6.34°$ S of W The resultant may be expressed as **R** = 4.53 m 6.34° S of W or **R** = (-4.50 m) **i** + (-0.50 m) **j**
5. Resultant vector **A - D**	$R_x = A_x - D_x = 5.00$ m $R_y = A_y - D_y = 5.46$ m $R = (R_x^2 + R_y^2)^{1/2} = 7.40$ m Since R_x and R_y are positive, we know **R** is in the first quadrant $\theta = \tan^{-1}(R_y / R_x) = 47.5°$ N of E The resultant may be expressed as **R** = 7.40 m 47.5° N of E or **R** = (5.00 m) **i** + (5.46 m) **j**
6. A unit vector in the direction of **C**	**C** = (-1.5 m) **i** + (1.5 m) **j** $C_x = -1.5$ m, $C_y = 1.5$ m $C = (C_x^2 + C_y^2)^{1/2} = 2.12$ m **n** = **C**/C = [(-1.5 m) **i** + (1.5 m) **j**] / 2.12 m **n** = (-0.708) **i** + (0.708) **j**

Example: An airplane is heading 30° N of E when it takes off at an angle of 20° above the horizontal. Its velocity is 100 m/s. (a) What is the vertical component of its velocity? (b) What is the horizontal component? (c) What is the component of the velocity toward the east? (d) What is the component of the velocity toward the north?

Given and diagram: $\alpha = 30°$ $\theta = 20°$, $|v| = 100$ m/s

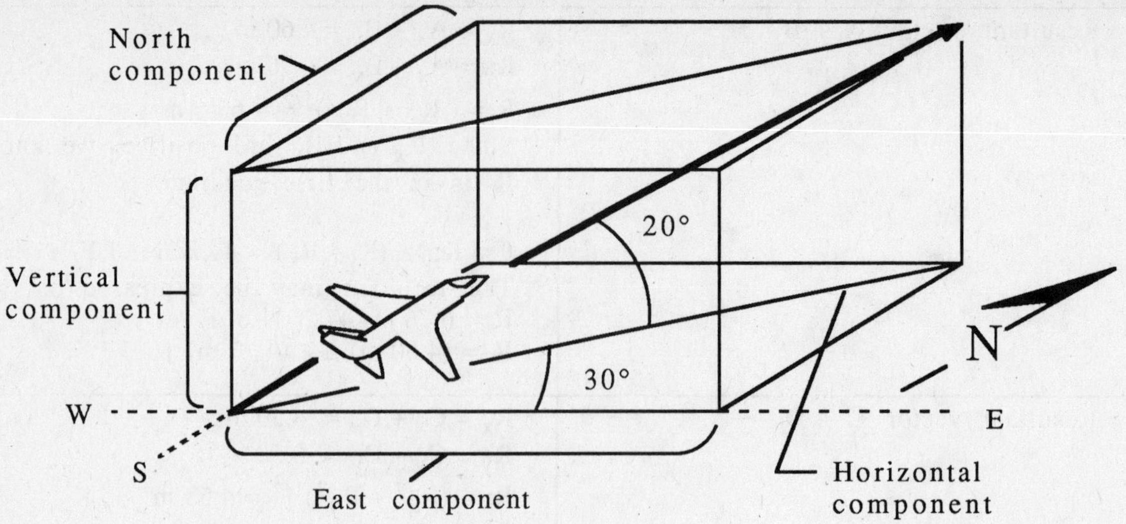

Determine: v_{vert}, v_{hor}, v_{east}, v_{north}

Strategy: We can determine v_{vert} and v_{hor} from v and θ. We can then use the values of v_{hor} and α to obtain v_{east} and v_{north}.

Solution:

$v_{vert} = v \sin\theta = (100 \text{ m/s}) \sin 20° = 34.2 \text{ m/s}$
$v_{hor} = v \cos\theta = (100 \text{ m/s}) \cos 20° = 94.0 \text{ m/s}$
$v_{east} = v_{hor} \cos\alpha = (94.0 \text{ m/s}) \cos 30° = 81.4 \text{ m/s}$
$v_{north} = v_{hor} \sin\alpha = (94.0 \text{ m/s}) \sin 30° = 47.0 \text{ m/s}$

Example: A reconnaissance plane leaves its home base and flies 100 mi N, 80 mi 60° W of N, and 150 mi 30° N of E. Determine how far the plane is from its home base and the direction it must fly in order to return straight to base.

Given: S_1 = 100 mi N, S_2 = 80 mi 60° W of N, S_3 = 150 mi 30° N of E

Determine: $|S_R|$, final distance of plane from base, and
θ, angle at which plane must fly to return to base

Strategy: Establish a coordinate system and resolve the displacement vectors into their components. From these components, determine the components and magnitude of the resultant displacement. This magnitude is the final distance of the plane from base. The direction in which the plane must fly on its return trip is opposite the direction of the resultant displacement.

Solution: First let's do a quick graphical representation to make sure we visualize the problem correctly. Then we can find the components of the various displacements and of the resultant displacement, the magnitude of the resultant displacement, and the direction of the return flight.

$S_{1x} = S_1 \sin 0° =$	0.0 mi	$S_{1y} = S_1 \cos 0° =$	100.0 mi
$S_{2x} = -S_2 \sin 60° =$	-69.3 mi	$S_{2y} = S_2 \cos 60° =$	40.0 mi
$S_{3x} = S_3 \cos 30° =$	130.0 mi	$S_{3y} = S_3 \sin 30° =$	75.0 mi
$S_{Rx} = S_{1x} + S_{2x} + S_{3x} =$	60.7 mi	$S_{Ry} = S_{1y} + S_{2y} + S_{3y} =$	215.0 mi

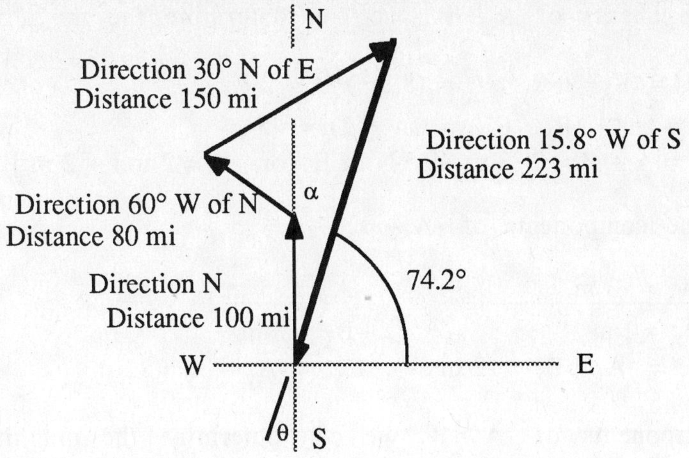

$$|S_R| = (S_{Rx}^2 + S_{Ry}^2)^{1/2} = [(60.7\text{ mi})^2 + (215\text{ mi})^2]^{1/2} = 223\text{ mi}$$
$$\alpha = \tan^{-1}(S_{Rx} / S_{Ry}) = 15.8° \text{ E of N}, \quad \text{hence} \quad \theta = 15.8° \text{ W of S}$$

At the end of the flight, the plane is 223 mi directed 15.8° E of N from the base. In order to return to the base it must fly 223 mi directed 15.8° W of S.

Example: Given the two dimensionless vectors $\mathbf{A} = 3\,\mathbf{i} + 5\,\mathbf{j}$ and $\mathbf{B} = -1\,\mathbf{i} - 3\,\mathbf{j}$, determine: (a) $\mathbf{A} + \mathbf{B}$, (b) $-\mathbf{A} - \mathbf{B}$, (c) a unit vector in the direction of $\mathbf{A} + \mathbf{B}$, and (d) a unit vector in the direction of $-\mathbf{A} - \mathbf{B}$.

Given: $\mathbf{A} = 3\,\mathbf{i} + 5\,\mathbf{j}$, $\mathbf{B} = -1\,\mathbf{i} - 3\,\mathbf{j}$

Determine: $\mathbf{A} + \mathbf{B}$, $-\mathbf{A} - \mathbf{B}$, unit vectors in the direction of $\mathbf{A} + \mathbf{B}$ and $-\mathbf{A} - \mathbf{B}$.

Strategy: Knowing the components of **A** and **B**, we can determine the components of $\mathbf{A} + \mathbf{B}$ and $-\mathbf{A} - \mathbf{B}$. Knowing these components we can express the resultant vectors in **i, j** and **k** notation. We can divide the resultant vectors by their magnitude to obtain the unit vectors. Finally, by noting that $-\mathbf{A} - \mathbf{B} = -(\mathbf{A} + \mathbf{B})$ we realize that we can check our work because $-\mathbf{A} - \mathbf{B}$ should have the same magnitude as $\mathbf{A} + \mathbf{B}$ but opposite direction.

Solution: First let's do a quick graphical representation of **A, B, A + B** and **-A - B**.

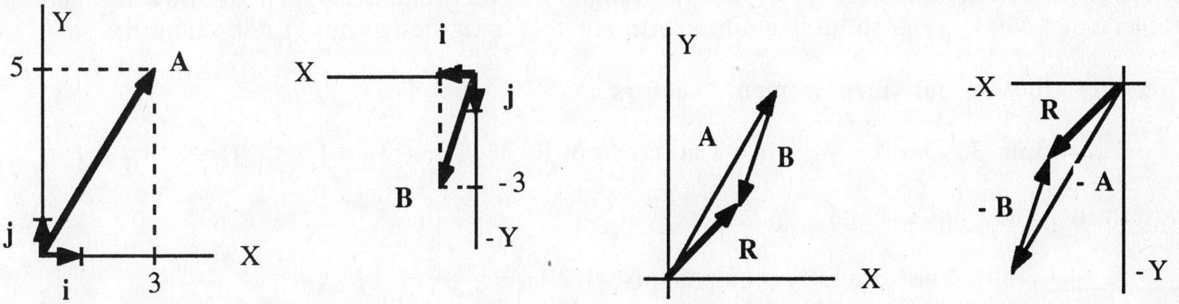

(a) Determine the components of **A, B** and **A + B**.

$A_x = 3$ m	$B_x = -1$ m	$R_x = A_x + B_x = 2$ m
$A_y = 5$ m	$B_y = -3$ m	$R_y = A_y + B_y = 2$ m

Knowing the components of **A** + **B**, we can determine the magnitude and direction.

$$|\mathbf{R}| = (R_x^2 + R_y^2)^{1/2} = (8\ m^2)^{1/2} = 2.83\ m$$
$$\theta = \tan^{-1}(R_y / R_x) = \tan^{-1}(1) = 45°$$
$$\mathbf{R} = \mathbf{A} + \mathbf{B} = 2.83\ m\ \ 45°\ N\ of\ E \quad or \quad \mathbf{R} = 2\ m\ \mathbf{i} + 2\ m\ \mathbf{j}$$

(b) Determine the components of -**A** - **B**.

$$-A_x = -3\ m \qquad -A_y = -5\ m$$
$$-B_x = 1\ m \qquad -B_y = 3\ m$$
$$R_x' = -A_y - B_x = -2\ m \qquad R_y' = -A_y - B_y = -2\ m$$

Knowing the components of -**A** - **B**, we can determine the magnitude and direction.

$$|\mathbf{R}'| = (R_x'^2 + R_y'^2)^{1/2} = (8\ m^2)^{1/2} = 2.83\ m$$
$$\theta' = \tan^{-1}(R_y' / R_x') = \tan^{-1}(1) = 45°$$
$$\mathbf{R}' = -\mathbf{A} - \mathbf{B} = 2.83\ m\ \ 45°\ S\ of\ W \quad or \quad \mathbf{R}' = -2\ m\ \mathbf{i} - 2\ m\ \mathbf{j}$$

Notice that **R** and **R'** have the same magnitude but opposite direction.

(c) The unit vectors are determined by dividing the vector by its magnitude

$$\mathbf{n} = \mathbf{R} / |\mathbf{R}| = (2\ m\ \mathbf{i} + 2\ m\ \mathbf{j}) / (2.83\ m) = 0.707\ \mathbf{i} + 0.707\ \mathbf{j}$$
$$\mathbf{n}' = \mathbf{R}' / |\mathbf{R}'| = (-2\ m\ \mathbf{i} - 2\ m\ \mathbf{j}) / (2.83\ m) = -0.707\ \mathbf{i} - 0.707\ \mathbf{j}$$

Notice that **n** and **n'** both have a magnitude of unity

$$|\mathbf{n}| = |\mathbf{n}'| = [(0.707)^2 + (0.707)^2]^{1/2} = 1.00$$

and they are in the opposite direction $\mathbf{n} = -\mathbf{n}'$

Related Text Exercises: Ex. 2-9 through 2-22 and 2-27 through 2-30.

PRACTICE TEST

Take and grade this practice test. Doing so will allow you to determine any weak spots in your understanding of the concepts taught in this chapter. The following section prescribes what you should study further to strengthen your understanding.

Consider these four displacement vectors

A = 5 m 30° N of E   **B** = 8 m 60° S of E   **C** = 2 **i** - 4 **j**   **D** = -3 **i** + 2 **j**

Determine the Following:

_________ 1. The x and y components (including sign) of **A**
_________ 2. The x and y components (including sign) of **B**
_________ 3. The x and y components (including sign) of **C**
_________ 4. The x and y components (including sign) of **D**
_________ 5. The x and y components (including sign) of **A** + **B**

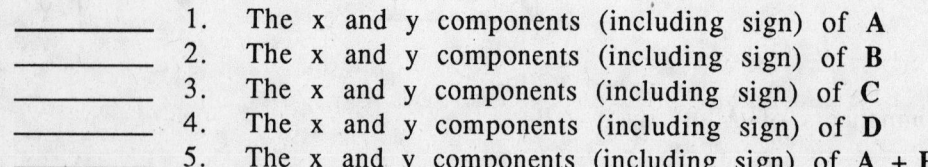

_________ 6. The magnitude of **A** + **B**
_________ 7. The direction of **A** + **B**
_________ 8. The angle between **A** and **B**
_________ 9. A unit vector in the direction of **A** + **B**
_________ 10. The magnitude of the vector of **C**
_________ 11. The direction of **C**
_________ 12. The angle between the direction of **C** and **D**
_________ 13. The x and y components of **C** - **D**
_________ 14. The magnitude of **C** - **D**
_________ 15. The direction of **C** - **D**
_________ 16. The x and y components of **A** + **B** + **C** - **D**
_________ 17. The magnitude of **A** + **B** + **C** - **D**
_________ 18. The direction of **A** + **B** + **C** - **D**

(See Appendix I for answers.)

PRINCIPAL CONCEPTS AND EQUATIONS PRESCRIPTION

Your score on the practice test is an excellent measure of your understanding of the chapter. You should now use the following chart to write your own prescription for dealing with any weaknesses the practice test points out. Look down the leftmost column to the number of the question(s) you answered incorrectly, reading across that row you will find the concept and/or equation of concern, the section(s) of the study guide you should return to for further study, and some suggested text exercises which you should work to gain additional experience.

Practice Test Questions	Concepts and Equations	Prescription	
		Principal Concepts	Text Exercises
1	Components of a vector - analytical	3A	2-11, 14
2	Components of a vector - analytical	3A	2-11, 14
3	Components of a vector - analytical	3A	2-15, 16
4	Components of a vector - analytical	3A	2-15, 16
5	Vector addition - components - analytical	3B	2-19, 20
6	Vector addition - magnitude - analytical	3B	2-19, 20
7	Vector addition - direction - analytical	3B	2-19, 20
8	Direction of a vector - analytical	3A	2-23
9	Unit Vector	3A	2-12, 13
10	Magnitude of a vector from comments	3A	2-18, 28
11	Direction of a vector from components	3A	2-16, 18
12	Direction of a vector - analytical	3A	2-23
13	Vector subtraction - components - analytical	3B	2-28, 29
14	Vector subtraction - magnitude - analytical	3B	2-28, 29
15	Vector subtraction - direction - analytical	3B	2-28, 29
16	Vector addition & subtraction - components	3B	2-22, 30
17	Vector addition & subtraction - magnitude	3B	2-22, 30
18	Vector addition & subtraction - direction	3B	2-22, 30

3 MOTION IN ONE DIMENSION

RECALL FROM PREVIOUS CHAPTERS

Previously learned concepts and equations frequently used in this chapter	Text Section	Study Guide Page
Analytical treatment of vectors (unit vectors, addition, subtraction)	2-3, 4	2-8

NEW IDEAS IN THIS CHAPTER

Concepts and equations introduced		Text Section	Study Guide Page
Position, displacement, distance		3-1	3-1
Average velocity	$\bar{\mathbf{v}} = \Delta\mathbf{r} / \Delta t$	3-2	3-5
Average speed	$\bar{v} = d / \Delta t$	3-2	3-5
Instantaneous velocity	$\mathbf{v} = \lim_{\Delta t \to 0}(\Delta\mathbf{r} / \Delta t) = d\mathbf{r} / dt$	3-2	3-5
Instantaneous speed	$v = \lvert\mathbf{v}\rvert = \lvert d\mathbf{r} / dt\rvert$	3-2	3-5
Average acceleration	$\bar{\mathbf{a}} = \Delta\mathbf{v} / \Delta t$	3-4	3-12
Instantaneous acceleration	$\mathbf{a} = d\mathbf{v} / dt = d^2\mathbf{r} / dt^2$	3-4	3-12
Equations for constant acceleration motion		3-5	3-13
$\bar{v}_x = \Delta x / \Delta t$	$v_x = v_{xo} + a_x t$		
$\bar{a}_x = \Delta v_x / \Delta t$	$\Delta x = v_{xo}t + \frac{1}{2}a_x t^2$		
$\bar{v}_x = \frac{1}{2}(v_x + v_{xo})$	$v_x^2 = v_{xo}^2 + 2a_x\Delta x$		

PRINCIPAL CONCEPTS AND EQUATIONS

1 Position, Displacement, Distance (Section 3-1)

Review: Figure 3.1 shows three positions (A, B, C) on a linear track. It also shows two different frames of reference with an appropriate scale.

Figure 3.1

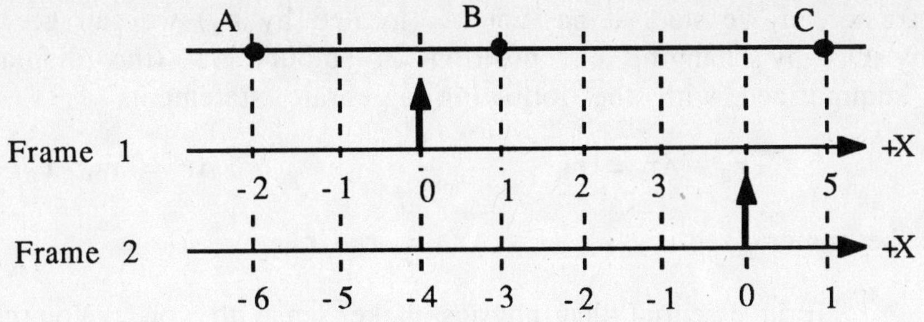

A position vector is a vector which locates a position with respect to a particular frame of reference. Shown below are the position vectors for positions A, B, and C for both frames.

Frame I	$\mathbf{r}_a = -2\mathbf{i}$	$\mathbf{r}_b = +1\mathbf{i}$	$\mathbf{r}_c = +5\mathbf{i}$
Frame II	$\mathbf{r}_a = -6\mathbf{i}$	$\mathbf{r}_b = -3\mathbf{i}$	$\mathbf{r}_c = +1\mathbf{i}$

A displacement vector describes a change in position. Suppose we are at position A and we want to go to position B. To see what must be done to get from position A to B, let's look at the initial position located by $\mathbf{r}_a$ and see how much we need to change our position (i.e. the displacement vector $\Delta\mathbf{r}$) to get to the position located by $\mathbf{r}_b$. Figure 3.2(a) shows the initial position A located by $\mathbf{r}_a$. Figure 3.2(b) shows the desired final position B located by $\mathbf{r}_b$. Figure 3.2(c) shows how much we must change our position (i.e. our displacement vector $\Delta\mathbf{r}$) to go from position A to B.

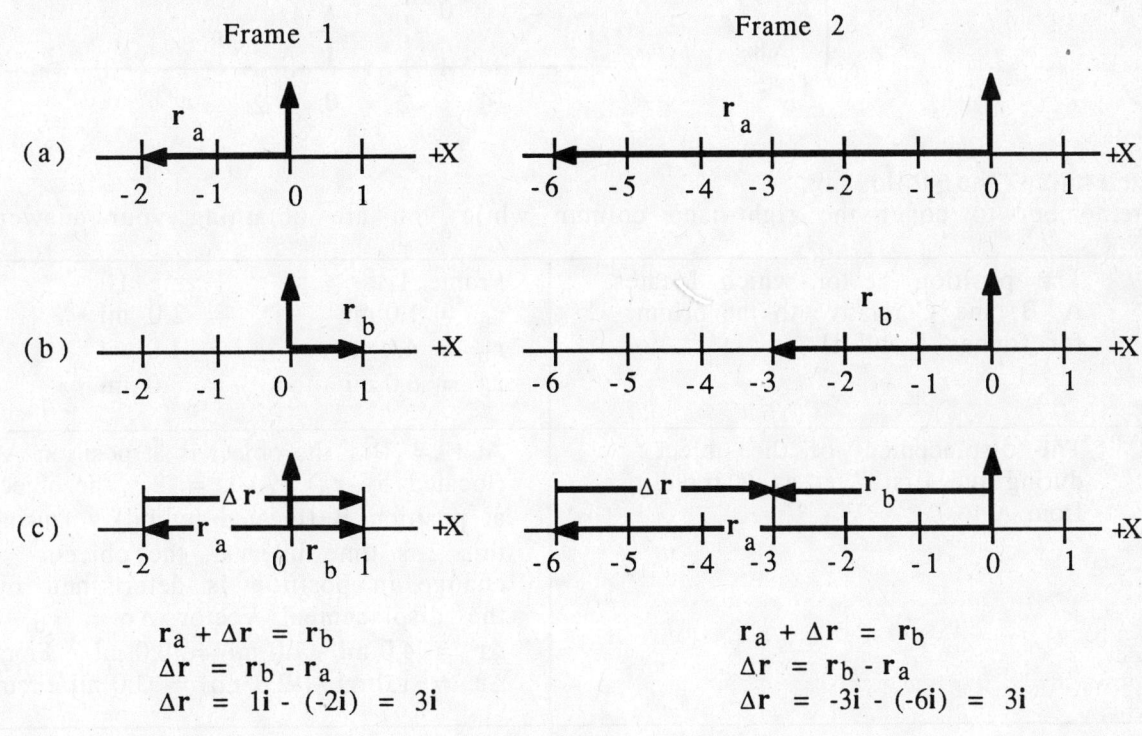

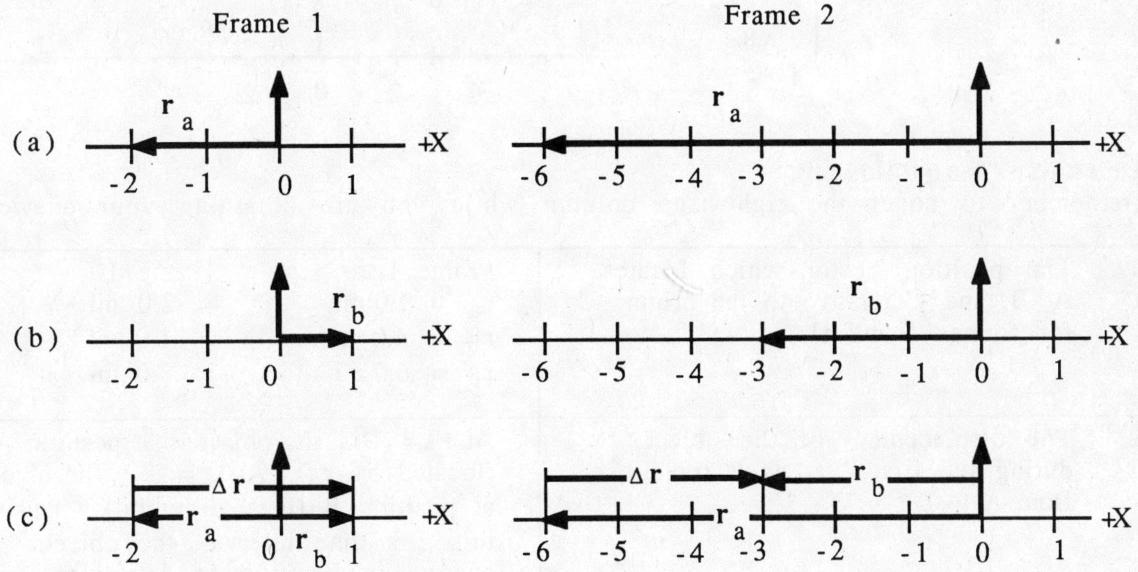

Frame 1:

$\mathbf{r}_a + \Delta\mathbf{r} = \mathbf{r}_b$

$\Delta\mathbf{r} = \mathbf{r}_b - \mathbf{r}_a$

$\Delta\mathbf{r} = 1\mathbf{i} - (-2\mathbf{i}) = 3\mathbf{i}$

Frame 2:

$\mathbf{r}_a + \Delta\mathbf{r} = \mathbf{r}_b$

$\Delta\mathbf{r} = \mathbf{r}_b - \mathbf{r}_a$

$\Delta\mathbf{r} = -3\mathbf{i} - (-6\mathbf{i}) = 3\mathbf{i}$

Figure 3.2

For both frames, if we start at position A (located by $\mathbf{r}_a$) we can get to position B (located by $\mathbf{r}_b$) by changing our position an amount $\Delta\mathbf{r}$ (the displacement vector). This is summarized with the following algebraic statements

$$\mathbf{r}_a + \Delta\mathbf{r} = \mathbf{r}_b \qquad \text{or} \qquad \Delta\mathbf{r} = \mathbf{r}_b - \mathbf{r}_a$$

This may be generalized by: $\Delta\mathbf{r} = \mathbf{r}_f - \mathbf{r}_i$

Note: It is essential that physics make sense to you. You can't expect to understand it if it doesn't make sense. Don't make the mistake of just memorizing the equation $\Delta\mathbf{r} = \mathbf{r}_f - \mathbf{r}_i$ but rather study Figure 3.2, see what is happening, and then realize that the equation $\Delta\mathbf{r} = \mathbf{r}_f - \mathbf{r}_i$ is just a mathematical statement of what the figure shows.

Practice: An object is traveling along the linear track shown in Figure 3.3. It travels from A to B to C, reverses direction, and travels back through B to A. An origin has been arbitrarily chosen and the x axis marked off in meters. In order to illustrate several ideas, two cases will be considered. The difference between the cases is the location of the origin. The time the object is at each position is as follows:

Pos	Time(s)
A	0
B	2
C	4
B	5
A	8

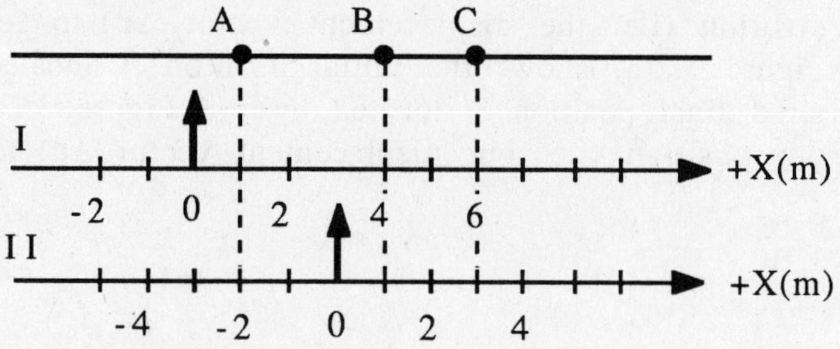

Figure 3.3

Determine the following:
(remember to cover the right-hand column while you are obtaining your answers).

1. The position vector which locates A, B, and C relative to the origin for frames I and II.	Frame I / Frame II $\mathbf{r}_a$ = 1.0 mi / = -2.0 mi $\mathbf{r}_b$ = 4.0 mi / = 1.0 mi $\mathbf{r}_c$ = 6.0 mi / = 3.0 mi
2. The displacement of the object during the first 2 s, as it travels from A to B	At t = 0 s, the object is at position A (located by $\mathbf{r}_a$). At t = 2 s, the object is at position B (located by $\mathbf{r}_b$). During this 2 s time interval, the objects change in position is determined by the displacement vector $\Delta\mathbf{r} = \mathbf{r}_b - \mathbf{r}_a$ $\Delta\mathbf{r}$ = 4.0 mi - 1.0 mi = 3.0 mi Frame I $\Delta\mathbf{r}$ = 1.0 mi -(-2.0 mi) = 3.0 mi Frame II

3. The distance traveled by the object during the first 2 s.	The distance traveled during the first 2 s is the magnitude of the displacement vector. $d = \|\Delta\mathbf{r}\| = \|3.0\text{ mi}\| = 3.0\text{ m}$ Frame I $d = \|\Delta\mathbf{r}\| = \|3.0\text{ mi}\| = 3.0\text{ m}$ Frame II
4. The position of the object after 5 s.	After 5 s, the object is a position B located by the position vector $\mathbf{r}_b$ $\mathbf{r}_b = 4.0\text{ mi}$ Frame I $\mathbf{r}_b = 1.0\text{ mi}$ Frame II
5. The displacement of the object during the first 5 s.	At t = 0 the objects position is located by $\mathbf{r}_a$. After 5 s it is located by $\mathbf{r}_b$. Hence the displacement vector is $\Delta\mathbf{r} = \mathbf{r}_b - \mathbf{r}_a$ $= 4.0\text{ mi} - 1.0\text{ mi} = 3.0\text{ mi}$ Frame I $= 1.0\text{ mi} - (-2.0\text{ mi}) = 3.0\text{ mi}$ Frame II
6. The distance traveled by the object during the first 5 s.	During the first 5 s the object traveled from A to B to C, reversed directions, and traveled back to B. The distance traveled is the sum of the magnitude of the displacement vectors $d = \|\Delta\mathbf{r}_{ac}\| + \|\Delta\mathbf{r}_{cb}\|$ $d = \|\mathbf{r}_c - \mathbf{r}_a\| + \|\mathbf{r}_b - \mathbf{r}_c\|$ For Frame I: $d = \|6\text{ mi} - 1\text{ mi}\| + \|4\text{ mi} - 6\text{ mi}\| = 7\text{ m}$ For Frame II: $d = \|3\text{ mi} - (-2\text{ mi})\| + \|1\text{ mi} -3\text{ mi}\| = 7\text{ m}$

Note: Numerous other questions regarding the position vector of the object at various times, the displacement vector for various changes in position, and the distance traveled for various changes in position may be asked for this situation. Using this practice as a guide, you should determine some of these questions and their answers.

Note: All position vectors include a direction (+ if the position of the object is to the right and - if to the left of the origin) and a magnitude (expressed in length units) that depends on the location of the origin. See steps 1. and 4. of the preceding practice.

Note: All displacement vectors include a direction (+ if the final position is to the right and - if to the left of the initial position) and a magnitude (expressed in length units) that does not depend on the location of the origin. See steps 2. and 5. of the preceding practice.

Note: The magnitude of the displacement vector is equal to the distance traveled if and only if the object does not change directions. All distances are positive and they do not depend on the location of the origin. See steps 2. and 3. of the preceding practice.

Note: The magnitude of the displacement vector is not equal to the distance traveled if the object changes directions. See steps 5. and 6. of the preceding practice.

Note: Don't try to memorize these notes. Understand the idea expressed in each note and then practice the physics until you are comfortable with the idea.

Related Text Exercises: Ex. 3-1 through 3-3.

2 Average Velocity, Average Speed, Instantaneous Velocity, Instantaneous Speed (Section 3-2)

Review: **Average velocity** - The average velocity of an object tells you the time rate of change of its position and the direction it is headed during some time interval. Since the displacement vector $\Delta\mathbf{r}$ describe the change in position, and Δt represents a change in time we may write the average velocity as

$$\bar{\mathbf{v}} = (\mathbf{r}_f - \mathbf{r}_i) / (t_f - t_i) = \Delta\mathbf{r} / \Delta t$$

In this expression, $\mathbf{r}_f$ and $\mathbf{r}_i$ are the position vectors that locate the object at times t_f and t_i respectively. The bar over the v is our way of representing an average quantity.

Note: Velocity is a vector quantity. When asked to determine a velocity you need to respond by giving both a magnitude and a direction.

Average Speed - The average speed of an object tells you the time rate of change of its distance traveled during some time interval. If we represent distance traveled during some time interval by d and the time interval by Δt we may write the average speed during that time interval by

$$\bar{v} = d / \Delta t$$

Note: Speed is not a vector, it is always a positive number and it gives you no information about a moving objects direction of motion.

Instantaneous Velocity - The instantaneous velocity of an object is its velocity at some instant of time. We already know how to obtain an average velocity over some time interval. If we determine the average velocity over a time interval so small that we are willing to call it an instant, then we have determined an instantaneous velocity. This may be written as:

$v = \bar{v}$	over a very short time interval
$= \bar{v}$ (as $\Delta t \to 0$)	just another way to say over a very short time interval
$= \lim\limits_{\Delta t \to 0} \bar{v}$	still another way to say over a very short time interval
$= \lim\limits_{\Delta t \to 0} (\Delta \mathbf{r} / \Delta t)$	since $\bar{v} = \Delta \mathbf{r} / \Delta t$
$= dr / dt$	recognizing a derivative

Instantaneous Speed - The instantaneous speed of an object is just the magnitude of the instantaneous velocity.

$$v = |\mathbf{v}| = |d\mathbf{r} / \Delta t|$$

Practice: Consider an object moving along a linear track. As the object moves, a measuring device collects position and time information. This information may be given to us by the three different methods shown in Figure 3.4

Method 1 (Data)

Time(s)	Position(m)
0	5
1	13
2	17
3	17
4	13
5	5

Method 3 (Equation)

$$\mathbf{x} = [5 \text{ m} + (10 \text{ m/s})t - (2 \text{ m/s}^2)t^2]\ \mathbf{i}$$

Method 2 (Graph)

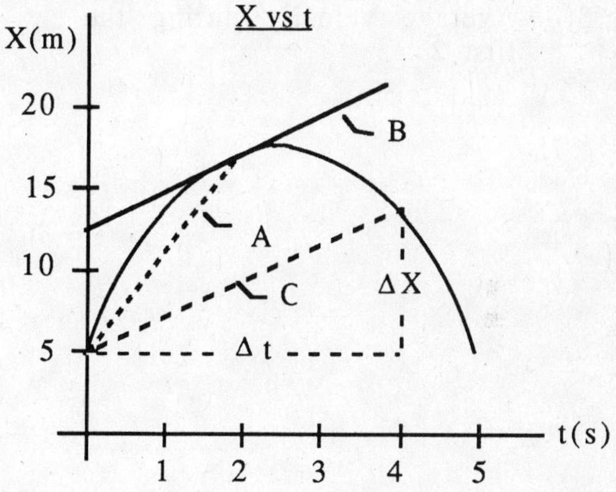

Figure 3.4

1. Show that the table of values and the graph give the same information	We can generate a table of values from the graph by reading the value of x for various values of t.

t(s)	0.0	0.5	1.0	1.5	2.0	2.5
x(m)	5.0	9.5	13.0	15.5	17.0	17.5

This agrees with the table of values.

Note: The table of values gives us the position at specific times. The plot of x vs t, even though generated from the table of values, gives not only continuous information but also a better visual representation of the objects actual motion. The plot method is limited by the accuracy with which we can get information off the plot.

2. Show that the table of values and the function give the same information.	Insert t = 0, 1, and 2 s into the function. $x = 5\text{ m} + (10\text{ m/s})t - (2\text{ m/s}^2)t^2$ t = 0.0 s x = 5 m t = 1.0 s x = 5 m + 10 m - 2 m = 13 m t = 2.0 s x = 5 m + 20 m - 8 m = 17 m These values agree with the table of values.

Note: The equation showing the position as a function of time is the most sophisticated, concise and convenient method for giving this information. From the equation, we can generate a table of values and hence a plot. However, the plot allows us to readily note that the object traveled out to 17.5 m in 2.5 s, stopped for an instant, and then returned to its starting position in another 2.5 s.

3. Average velocity during the first 2 s	Using the table of values, the plot, or the functional relationship, we may establish that at $t_i = 0$ s, $\mathbf{x}_i = 5$ mi and at $t_f = 2$ s, $\mathbf{x}_f = 17$ mi. The displacement and time interval are $\Delta\mathbf{x} = \mathbf{x}_f - \mathbf{x}_i = 17\text{ mi} - 5\text{ mi} = 12\text{ mi}$ and $\Delta t = t_f - t_i = 2\text{ s} - 0\text{ s} = 2\text{ s}$ The average velocity is $\bar{\mathbf{v}} = \Delta\mathbf{x} / \Delta t = (12\text{ m} / 2\text{ s})\mathbf{i} = (6\text{ m/s})\mathbf{i}$ We may also find the average velocity over this time interval by determining the slope of the line between the end points of the time intervals (dashed line A on the plot) $\bar{\mathbf{v}} = \text{slope} = \Delta\mathbf{x} / \Delta t = 12\text{ mi} / 2\text{ s}$ $= (6\text{ m/s})\mathbf{i}$
4. Average speed over the first 2 s.	Since the object does not change directions, the distance traveled is the magnitude of the displacement. Hence $\bar{v} = d / \Delta t = \lvert\Delta\mathbf{x}\rvert / \Delta t = \lvert 12\text{ mi}\rvert / 2\text{ s}$ $= (6\text{ m/s})$

5.	Instantaneous velocity at t = 2 s	This is easily obtained from the function $\mathbf{v} = dx / dt = [10\text{ m/s} - (4\text{ m/s}^2)t]\ \mathbf{i}$ At t = 2 s $\mathbf{v} = (10\text{ m/s} - 8\text{ m/s})\mathbf{i} = (2\text{ m/s})\mathbf{i}$ We may also find the instantaneous velocity at t = 2 s by determining the slope of the line tangent to the plot at t = 2 s (solid line B on the plot) $\mathbf{v} = \text{slope} = (8\text{ m} / 4\text{ s})\mathbf{i} = (2\text{ m/s})\ \mathbf{i}$
6.	Instantaneous speed at t = 2 s	The instantaneous speed is the magnitude of the instantaneous velocity $v = \lvert\mathbf{v}\rvert = \lvert(2\text{ m/s})\mathbf{i}\rvert = 2\text{ m/s}.$
7.	Average velocity during the first 4 s	Using the table of values, the plot, or the functional relationship, we may establish that at $t_i = 0$, $\mathbf{x}_i = 5$ mi and at $t_f = 4$ s, $\mathbf{x}_f = 13$ mi The displacement and time interval are $\Delta\mathbf{x} = \mathbf{x}_f - \mathbf{x}_i = 13\text{ mi} - 5\text{ mi} = 8\text{ mi}$, and $\Delta t = t_f - t_i = 4\text{ s} - 0\text{ s} = 4\text{ s}$ The average velocity is $\bar{\mathbf{v}} = \Delta\mathbf{x} / \Delta t = (8\text{ m} / 4\text{ s})\mathbf{i} = (2\text{ m/s})\mathbf{i}$ We may also find the average velocity over the first 4 s by taking the slope of the line between the end points of the time interval (dashed line C on the plot) $\bar{\mathbf{v}} = \text{slope} = \Delta\mathbf{x} / \Delta t = (8\text{ m} / 4\text{ s})\mathbf{i} = (2\text{ m/s})\mathbf{i}$
8.	Average speed during the first 4 s	From the graph it is obvious that the object travels a distance of 12.5 m in the +x direction and then 4.5 m in the -x direction for a total distance of 17 m. Hence the average speed is $\bar{v} = d / \Delta t = 17\text{ m} / 4\text{ s} = 4.25\text{ m/s}$

9. Average velocity for the entire 5 s	x_i = 5 mi, x_f = 5 mi, and $\Delta x = x_f - x_i = 0$ t_i = 0 s, t_f = 5 s, and $\Delta t = t_f - t_i$ = 5 s $\bar{v} = \Delta x / \Delta t$ = 0 m/s
10. Average speed for the entire 5 s	Using the plot, we see that the object traveled 12.5 m in the +x direction and then 12.5 m in the -x direction for a total distance of 25.0 m. Hence the average speed is $\bar{v} = d / \Delta t$ = 25 m / 5 s = 5 m/s

Example: A car travels around a circular track of circumference 6.00 x 10^3 m once every 90.0 s. What is (a) its average speed and (b) its average velocity for each lap?

Given:

d = 6.00 x 10^3 m	=	total distance traveled for each lap
Δ**r** = 0.0 m	=	displacement for each lap
Δt = 90.0 s	=	time of travel for each lap

Determine: Average velocity $\bar{\mathbf{v}}$ and average speed $\bar{v}$ for each lap.

Strategy: Knowing the displacement for each lap and the time for that displacement, we can determine the average velocity $\bar{\mathbf{v}}$. Knowing the total distance traveled for each lap and the time for that travel, we can determine the average speed $\bar{v}$.

Solution: (a) $\bar{\mathbf{v}} = \Delta \mathbf{r} / \Delta t$ = 0.0 m / 90.0 s = 0.0 m/s

(b) $\bar{v} = d / \Delta t$ = 6.00 x 10^3 m / 90.0 s = 66.7 m/s

Example: A car travels at a constant speed of 20.0 m/s for 2.00 x 10^3 s and then suddenly changes to a speed of 30.0 m/s for 1.00 x 10^3 s.

Determine: Average speed for the entire trip

Strategy: We can use v_1 and t_1 to determine d_1, the distance traveled in time t_1. In like manner, we can use v_2 and t_2 to determine d_2. We can then determine the total distance traveled, the total time of travel, and the average speed for the entire trip.

Solution:

$$d_1 = v_1t_1 = (20.0\ m/s)(2.00 \times 10^3\ s) = 4.00 \times 10^4\ m$$
$$d_2 = v_2t_2 = (30.0\ m/s)(1.00 \times 10^3\ s) = 3.00 \times 10^4\ m$$
$$d_{total} = d_1 + d_2 = 7.00 \times 10^4\ m$$
$$t_{total} = t_1 + t_2 = 3.00 \times 10^3\ s$$
$$\bar{v} = d_{total} / t_{total} = 7.00 \times 10^4\ m / 3.00 \times 10^3\ s = 23.3\ m/s$$

Note: Since the acceleration was not constant ($a_1 = 0$ for the time t_1, then some instantaneous a, and finally $a_2 = 0$ for the time t_2) we can't just average the speeds to get the average speed.

Example: An expression for the position vector locating an object traveling along the x-axis as a function of time is

$$\mathbf{x}(t) = [20\ m - (4\ m/s)t + (2\ m/s^2)t^2]\ \mathbf{i}$$

Determine the following:

(a) Average velocity of the object over the time interval $t = 1$ s to $t = 3$ s
(b) Average speed of the object over the time interval $t = 1$ s to $t = 3$ s
(c) Instantaneous velocity at $t = 1$, 2 and 3 s
(d) Instantaneous speed at $t = 1$, 2 and 3 s

Given: $\mathbf{x}(t) = [20\ m - (4\ m/s)t + (2\ m/s^2)t^2]\ \mathbf{i}$

Determine:

(a) Average velocity $\bar{\mathbf{v}}$ over the time interval $t = 1$ s to $t = 3$ s
(b) Average speed $\bar{v}$ over the time interval $t = 1$ s to $t = 3$ s
(c) Instantaneous velocity $\mathbf{v}$ at $t = 1$, 2 and 3 s
(d) Instantaneous speed v at $t = 1$, 2 and 3 s

Strategy:

(a) Knowing the function $\mathbf{x}(t)$ we can determine the position vector at $t = 1$ s and $t = 3$ s. Knowing the position vectors, we can determine the displacement and the average velocity.
(b) Knowing the position vector at $t = 1$ s and $t = 3$ s, we can determine the total distance traveled and then the average speed.
(c) We can differentiate the function $\mathbf{x}(t)$ to obtain the function $\mathbf{v}(t)$. We can then evaluate this function at $t = 1$, 2 and 3 s.
(d) Knowing $\mathbf{v}(t)$ at $t = 1$, 2 and 3 s, and knowing that $v = |\mathbf{v}|$, we can determine v at $t = 1$, 2 and 3 s.

Solution:

(a) $\mathbf{x}(t) = [20\ m - (4\ m/s)t + (2\ m/s^2)t^2]\ \mathbf{i}$

the position vector at $t = 1$ s and $t = 3$ s is

$\mathbf{x}(t = 1) = (20\ m - 4\ m + 2\ m)\mathbf{i} = 18\ m\mathbf{i}$

$\mathbf{x}(t = 3) = (20\ m - 12\ m + 18\ m)\mathbf{i} = 26\ m\mathbf{i}$

$\Delta\mathbf{x} = \mathbf{x}(t = 3) - \mathbf{x}(t = 1) = 26\ m\mathbf{i} - 18\ m\mathbf{i} = 8\ m\mathbf{i}$, and $\Delta t = 3$ s

$\bar{\mathbf{v}} = \Delta\mathbf{x} / \Delta t = (8\ m / 3\ s)\mathbf{i} = (2.67\ m/s)\mathbf{i}$

(b) Notice that at $t = 0$ s, $x = 20$ m; at $t = 1$ s, $x = 18$ m; and at $t = 3$ s, $x = 26$ m. Obviously the object is initially traveling in the -x direction and finally in the +x direction. The distance traveled is the same as the magnitude of the displacement vectors for left and then right travel. We can determine these displacement vectors if we know when and where the object changes directions. The object changes directions when $\mathbf{v} = 0$.

$\mathbf{x} = [20\text{ m} - (4\text{ m/s})t + (2\text{ m/s}^2)t^2]\,\mathbf{i}$ gives $\mathbf{v} = d\mathbf{x}/dt = [-4\text{ m/s} + (4\text{ m/s}^2)t]\,\mathbf{i}$

We can determine when the object changes direction by setting $\mathbf{v} = 0$ and solving for t.

$$0 = [-4\text{ m/s} + (4\text{ m/s}^2)t]\,\mathbf{i} \quad \text{or} \quad t = 1\text{ s}$$

We can determine where the object changes directions by inserting $t = 1$ s into $\mathbf{x}(t)$.

$$\mathbf{x}(t = 1\text{ s}) = (20\text{ m} - 4\text{ m} + 2\text{ m})\mathbf{i} = 18\text{ mi at } t = 1\text{ s}$$

Summarizing

$$\begin{aligned} \mathbf{x}(t = 0\text{ s}) &= 20\text{ mi} & \mathbf{v}(t = 0\text{ s}) &= (-4\text{ m/s})\mathbf{i} \\ \mathbf{x}(t = 1\text{ s}) &= 18\text{ mi} & \mathbf{v}(t = 1\text{ s}) &= 0 \\ \mathbf{x}(t = 3\text{ s}) &= 26\text{ mi} & \mathbf{v}(t = 3\text{ s}) &= (8\text{ m/s})\mathbf{i} \end{aligned}$$

At $t = 0$ s the object is at $\mathbf{x}(t = 0) = 20$ mi traveling to the left with a velocity of $\mathbf{v}(t = 0) = (-4\text{ m/s})\mathbf{i}$. After traveling for 1 s the object is at $\mathbf{x}(t = 1) = (18\text{ m/s})\mathbf{i}$ and it is momentarily at rest. Finally at $t = 3$ s the object is at $\mathbf{x}(t = 3) = 26$ mi and it is traveling to the right with a velocity of $\mathbf{v}(t = 3) = (8\text{ m/s})\mathbf{i}$.

The distance traveled is the sum of the magnitude of the displacement vectors.

$$\begin{aligned} d &= |\Delta\mathbf{x}_{left}| + |\Delta\mathbf{x}_{right}| \\ &= |\mathbf{x}(t = 1\text{ s}) - \mathbf{x}(t = 0\text{ s})| + |\mathbf{x}(t = 3\text{ s}) - \mathbf{x}(t = 1\text{ s})| \\ &= |18\text{ mi} - 20\text{ mi}| + |26\text{ mi} - 18\text{ mi}| = 10\text{ m} \end{aligned}$$

Finally we obtain the average speed

$$\bar{v} = d/\Delta t = 10\text{ m}/3\text{ s} = 3.33\text{ m/s}$$

(c) $\mathbf{v} = d\mathbf{x}/dt = [-4\text{ m/s} + (4\text{ m/s}^2)t]\,\mathbf{i}$ Inserting the values $t = 1$, 2 and 3 s obtain

$$\begin{aligned} \mathbf{v}(t = 1\text{ s}) &= 0.0\text{ m/s} \\ \mathbf{v}(t = 2\text{ s}) &= (-4\text{ m/s} + 8\text{ m/s})\mathbf{i} = (4\text{ m/s})\mathbf{i} \\ \mathbf{v}(t = 3\text{ s}) &= (-4\text{ m/s} + 12\text{ m/s})\mathbf{i} = (8\text{ m/s})\mathbf{i} \end{aligned}$$

(d) $v = |\mathbf{v}|$

$$\begin{aligned} v(t = 1\text{ s}) &= |\mathbf{v}(t = 1\text{ s})| = 0.0\text{ m/s} \\ v(t = 2\text{ s}) &= |\mathbf{v}(t = 2\text{ s})| = |(4\text{ m/s})\mathbf{i}| = 4\text{ m/s} \\ v(t = 3\text{ s}) &= |\mathbf{v}(t = 3\text{ s})| = |(8\text{ m/s})\mathbf{i}| = 8\text{ m/s} \end{aligned}$$

Related Text Exercises: Ex. 3-4 through 3-17.

3 Average and Instantaneous Acceleration (Section 3-4)

Review: If an object is changing its velocity, it is accelerating. Acceleration is a description of how an object is changing its velocity, both in magnitude and direction.

Average acceleration - The average acceleration tells you how much the object changes its velocity in a certain time interval. We express this as

$$\bar{\mathbf{a}} = \Delta\mathbf{v} / \Delta t = (\mathbf{v}_f - \mathbf{v}_i) / (t_f - t_i)$$

Instantaneous acceleration - The instantaneous acceleration of an object is its acceleration at some instant of time. If we determine the average acceleration over a time interval so small that we are willing to call it an instant, then we have determined an instantaneous acceleration. This may be written as

$$\mathbf{a} = \lim_{\Delta t \to 0}(\bar{\mathbf{a}}) = \lim_{\Delta t \to 0}(\Delta\mathbf{v} / \Delta t) = d\mathbf{v} / dt = d^2\mathbf{r} / dt^2$$

Practice: An expression for the position vector of an object traveling along the x-axis is

$$\mathbf{x}(t) = [2.0 \text{ m} + (10 \text{ m/s})t + (5 \text{ m/s}^2)t^2 + (3 \text{ m/s}^3)t^3]\,\mathbf{i}$$

1. Determine an expression for $\mathbf{v}(t)$	$\mathbf{v}(t) = dx / dt$ $= [10 \text{ m/s} + (10 \text{ m/s}^2)t + (9 \text{ m/s}^3)t^2]\,\mathbf{i}$
2. Determine the instantaneous velocity $\mathbf{v}(t)$ at t = 1 s and t = 3 s	$\mathbf{v}(t = 1 \text{ s}) = [(10 + 10 + 9) \text{ m/s}]\,\mathbf{i}$ $= (29 \text{ m/s})\mathbf{i}$ $\mathbf{v}(t = 3 \text{ s}) = [(10 + 30 + 81) \text{ m/s}]\,\mathbf{i}$ $= (121 \text{ m/s})\mathbf{i}$
3. Determine the average acceleration of the object during the time interval t = 1 s to t = 3 s	$\bar{\mathbf{a}} = \Delta\mathbf{v} / \Delta t$ $\Delta\mathbf{v} = \mathbf{v}(t = 3 \text{ s}) - \mathbf{v}(t = 1 \text{ s})$ $= [(121 - 29) \text{ m/s}]\,\mathbf{i} = (92 \text{ m/s})\mathbf{i}$ $\Delta t = 2 \text{ s}$ $\bar{\mathbf{a}} = \Delta\mathbf{v} / \Delta t = [(92 \text{ m/s}^2) / 2 \text{ s}]\,\mathbf{i}$ $= (46 \text{ m/s}^2)\mathbf{i}$
4. Determine an expression for $\mathbf{a}(t)$	$\mathbf{a}(t) = dv / dt = d^2\mathbf{x} / dt^2$ $\mathbf{a}(t) = [10 \text{ m/s}^2 + (18 \text{ m/s}^3)t)]\,\mathbf{i}$
5. Determine the instantaneous acceleration at t = 0, 1, and 3 s	$\mathbf{a}(t) = [10 \text{ m/s}^2 + (18 \text{ m/s}^3)t]\,\mathbf{i}$ $\mathbf{a}(t = 0 \text{ s}) = (10 \text{ m/s}^2)\mathbf{i}$ $\mathbf{a}(t = 1 \text{ s}) = [(10 + 18) \text{ m/s}^2]\,\mathbf{i}$ $= (28 \text{ m/s}^2)\mathbf{i}$ $\mathbf{a}(t = 3 \text{ s}) = [(10 + 54) \text{ m/s}^2]\,\mathbf{i}$ $= (64 \text{ m/s}^2)\mathbf{i}$

Related Text Exercises: Ex. 3-18 through 3-24.

4 Constant Acceleration Motion (Section 3-5)

Review: If an object is in motion with constant acceleration, the following equations may be used to analyze the motion

$$v_x = v_{xo} + a_x t$$
$$x = x_0 + v_{xo}t + a_x t^2 / 2$$
$$v_x^2 = v_{xo}^2 + 2a_x(x - x_0)$$

A special case of motion with constant acceleration is free fall. For the case of free fall, the constant acceleration is the acceleration due to gravity. The magnitude of the acceleration is represented by the symbol g and g = 9.8 m/s^2. Since most students prefer to represent vertical motion with the coordinate y and to have up as positive, the direction of this acceleration is -j. This allows us to write the acceleration due to gravity as

$$\mathbf{a} = -g\,\mathbf{j}$$

The above equations for one dimensional motion (in the y direction) under the influence of the constant acceleration due to gravity may be rewritten as

$$v_y = v_{yo} - gt$$
$$y = y_0 + v_{yo}t - gt^2 / 2$$
$$v_y^2 = v_{yo}^2 - 2g(y-y_0)$$

Note: In solving problems in physics, it is important to read the problem and decide the following:

(a) which quantities are known
(b) which quantities are unknown
(c) which equations to use to in order to express the unknown quantities in terms of the known quantities. After this step and perhaps a little algebra, you will be able to insert numbers and obtain a value for the desired unknown quantities.

Practice: Case I. An object moving in a straight line with an initial speed v_o undergoes a constant acceleration "a" for some time t. What is the object's final speed, change in position and average speed while accelerating?

Determine the following:

1. The known quantities	v_0, a, t
2. The unknown quantities	v, Δx, $\bar{v}$
3. The equation to determine final velocity?	$v = v_0 + at$
4. The equation to determine the change in position due to the initial speed	$\Delta x_{vo} = v_0 t$

5. The equation to determine the change in position due to the acceleration	$\Delta x_a = at^2 / 2$
6. The equation to determine the total change in position due to the initial speed and the acceleration	$\Delta x = \Delta x_{v0} + \Delta x_a$ $\Delta x = v_0 t + at^2 / 2$
7. The equation to determine average speed	$\bar{v} = \Delta x / \Delta t$
8. A method to check your work	Determine Δx with the equation $v^2 - v_0^2 = 2a\Delta x$ Determine $\bar{v}$ with the equation $\bar{v} = (v + v_0) / 2$

Case II. An object moving with an initial speed v_0 undergoes a constant acceleration "a" while changing its position an amount Δx. What is the speed after acceleration, the time elapsed during acceleration and the average speed during acceleration?

Determine the following:

1. The known quantities	v_0, a, Δx
2. The unknown quantities	v, t, $\bar{v}$
3. The equation to determine velocity after the acceleration	$v^2 - v_0^2 = 2a\Delta x$
4. The equation to determine time elapsed during acceleration	$\Delta x = v_0 t + at^2 / 2$ (This method uses only given information but requires solving a quadratic equation.) $v = v_0 + at$ (This method uses a previously calculated quantity but avoids solving a quadratic equation.)
5. The equation to determine average speed	$\bar{v} = \Delta x / \Delta t$
6. A method to check your work	Use the equation in step 4 not previously used to determine t. Determine $\bar{v}$ with the equation $\bar{v} = (v + v_0) / 2$

Example: A train traveling at a speed of 100 m/s is braked uniformly to rest in 20.0 s. Determine the distance the train travels while coming to rest.

Given: v_0 = 100 m/s; v = 0.00 m/s (since the train stops); t = 20.0 s

Determine: d, the distance the train travels while coming to rest.

Strategy: Using v_0, v and t, we can determine the acceleration of the train. We should expect the acceleration to be negative since the train is decelerating (i.e. slowing down). Once the acceleration is known, we can determine the trains change in position while slowing down. Assuming the track is linear and knowing the train does not change direction while slowing down, we know the distance the train travels while coming to rest is just its change in position.

Solution: Using the given information, we can determine the acceleration of the train.

$$a = (v - v_0) / t = [(0.00 - 100) \text{ m/s}] / 20.0 \text{ s} = -5.00 \text{ m/s}^2$$

Now that the acceleration is known, we can determine the change in position of the train while it is slowing down.

$$\Delta x = v_0 t + at^2 / 2 = (100 \text{ m/s})(20.0 \text{ s}) + (-5.00 \text{ m/s}^2)(20.0 \text{ s})^2 / 2 = 1.00 \text{ x.}10^3 \text{ m}$$

or

$$\Delta x = (v^2 - v_0^2) / 2a = -v_0^2 / 2a = -(100 \text{ m/s})^2 / [2(-5.00 \text{ m/s}^2)] = 1.00 \times 10^3 \text{ m}$$

Since the track is linear and the train does not change directions, we can write

$$d = \Delta x = 1.00 \times 10^3 \text{ m}$$

Example: A student holds a water balloon outside a fifth-story window (15.0 m off the ground), tosses it vertically upward and then quickly retracts his hand. The balloon hits the ground in 3.00 s. Determine the initial velocity of the balloon and its maximum displacement with respect to the ground.

Given and Diagram:

y_0 = +15.0 m (initial displacement)
y = 0.0 m (final displacement)
Δy = y - y_0 = -15.0 m
(total change in displacement)
Δy_{up} = (upward change in displacement)
a_y = -g = -9.80 m/s^2
(acceleration due to gravity)
t = 3.00 s (time of flight)

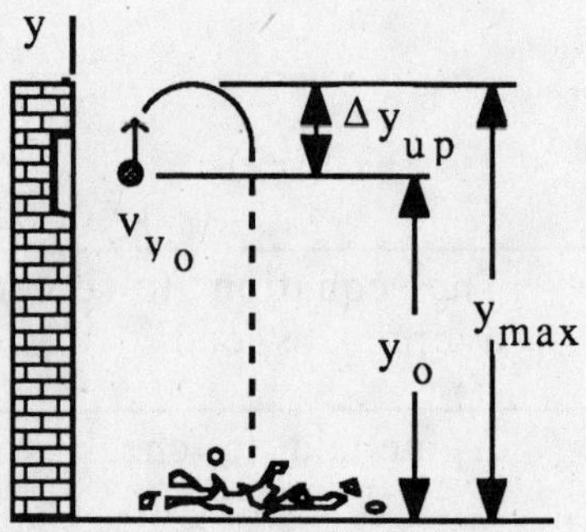

Determine: y_{max}, maximum displacement of the balloon with respect to the ground; and v_{yo}, the initial velocity of the balloon.

Strategy: Knowing Δy, t and g, we can determine v_{yo}. Knowing v_{yo} and that the balloon will travel upward until $v_y = 0$ m/s (at this point it momentarily stops and then starts downward), we can determine Δy_{up}. Adding Δy_{up} and the initial displacement y_o, we can obtain the maximum displacement of the balloon with respect to the ground.

Solution: Since this is one dimensional motion (in the y direction) with constant acceleration $a_y = -g$, we may write

$$\Delta y = y - y_o = v_{yo}t - gt^2 / 2 \quad \text{or}$$

$$v_{yo} = (\Delta y + gt^2 / 2) / t = [-15.0 \text{ m} + (9.80 \text{ m/s}^2)(3.00 \text{ s})^2 / 2] / 3.00 \text{ s} = 9.70 \text{ m/s}$$

Knowing that the balloon decelerates to $v = 0.00$ m/s while traveling upward we can determine the upward displacement.

$$v_y^2 = v_{yo}^2 - 2g\Delta y$$

knowing that $\Delta y = \Delta y_{up}$ when $v_y = 0$ we obtain $0 = v_{yo}^2 - 2g\Delta y_{up}$ or

$$\Delta y_{up} = v_{yo}^2 / 2g = (9.70 \text{ m/s})^2 / 2(9.80 \text{ m/s}^2) = 4.80 \text{ m}$$

Finally, the maximum displacement of the balloon with respect to the ground is

$$y_{max} = y_o + \Delta y_{up} = 15.0 \text{ m} + 4.80 \text{ m} = 19.8 \text{ m}$$

Example: A motorist traveling at a constant velocity of +5.00 m/s runs a red light. A police officer sitting at the interaction observes this and starts off after the motorist. If the patrol care can accelerate 1.00 m/s² until the motorist is caught, determine the time it takes to catch the motorist, the velocity of the patrol car when the motorist is caught, and the displacement of the two cars with respect to the intersection.

Given and Diagram:

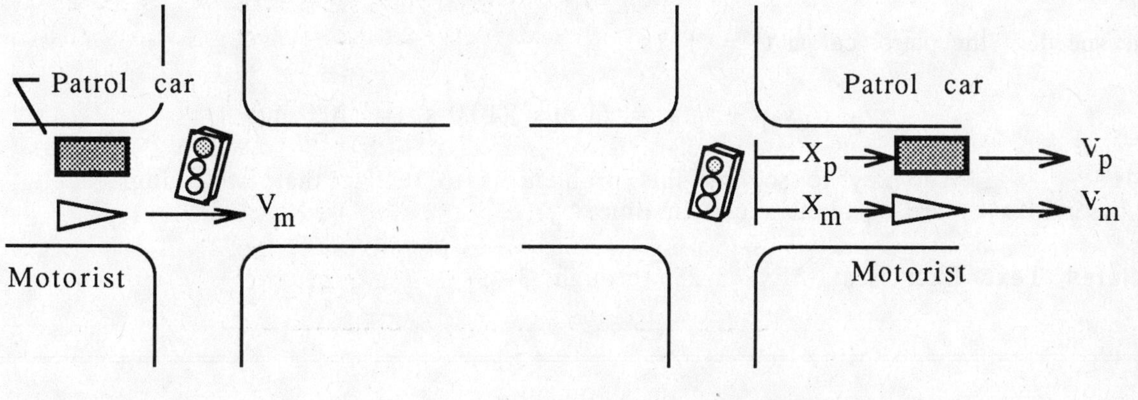

(a) Motorist runs red light

(b) Patrol car catches motorist ($x_p = x_m$)

$v_m = +5.00$ m/s (constant velocity of motorist)
$v_{po} = 0.0$ m/s (initial velocity of patrol car)
$a_p = +1.00$ m/s^2 (acceleration of the patrol car)

Determine: t_p time for patrol care to catch motorist; v_p, velocity of patrol car when motorist is caught; and x_p and x_m, displacement of both cars when patrol car catches motorists.

Strategy: We can write expressions for the change in displacement of the motorists and patrol car at any time. Since these two changes in displacement are equal when $t = t_p$ we will equate them and solve for t_p. The distance from the intersection may be determined by inserting t_p into the change in displacement expression of either the motorist or the patrol car. Knowing the acceleration of the patrol car and t_p we can determine the velocity of the patrol car when the motorist is caught.

Solution: The changes in displacement of motorist and patrol car at any time are:

$$\Delta x_m = v_m t \quad \text{and} \quad \Delta x_p = a_p t^2 / 2$$

Since $\Delta x_m = \Delta x_p$ when $t = t_p$ we may write

$$v_m t_p = a_p t_p^2 / 2 \quad \text{or} \quad t_p = 2v_m / a_p = 10.0 \text{ s}$$

The patrol car will catch the motorist in 10.0 s.

Now insert t_p into the equation for either Δx_m or Δx_p to obtain the change in displacement of either car.

Using the expression for the motorist's change in displacement at $t = t_p$ obtain

$$\Delta x_m = v_m t_p = (5.00 \text{ m/s})(10.0 \text{ s}) = 50.0 \text{ m}$$

Using the expression for the patrol car's change in displacement at $t = t_p$ obtain

$$\Delta x_p = a_p t_p^2 / 2 = (1.00 \text{ m/s}^2)(10.0 \text{ s})^2 / 2 = 50.0 \text{ m}$$

The speed of the patrol car at $t = t_p$ is

$$v_p = v_{op} + a_p t_p = (1 \text{ m/s}^2)(10.0 \text{ s}) = 10.0 \text{ m/s}$$

Note: The key to solving this problem is to realize that two things (Δx_p and Δx_m) are equal at a certain time (t_p).

Related Text Exercises: Ex. 3-25 through 3-49.

PRACTICE TEST

Take and grade this practice test. Doing so will allow you to determine any weak spots in your understanding of the concepts taught in this chapter. The following section prescribes what you should study further to strengthen your understanding.

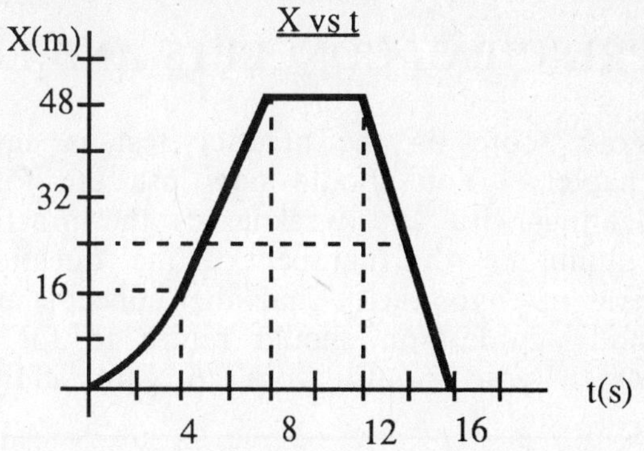

An object starts from rest and moves along the x axis as shown in this plot of displacement vs time.

Determine the following:

______ 1. Position at t = 14.0 s
______ 2. Distance traveled during the first 14.0
______ 3. Displacement during the time interval t = 4.00 s to t = 16.0
______ 4. Distance traveled during the time interval t = 4.00 s to t = 16.0 s
______ 5. Average velocity during the last 12.0 s
______ 6. Average velocity for the entire trip
______ 7. Average speed during the last 12.0 s
______ 8. Average speed for the entire trip
______ 9. Instantaneous velocity at t = 14.0 s
______ 10. Instantaneous speed at t = 14.0 s
______ 11. Acceleration at t = 6.00 s
______ 12. Acceleration during the first 4.00 s

An expression for the coordinate of an object is $\mathbf{x}(t) = (10 \text{ m} + (5 \text{ m/s})t + (4 \text{ m/s})t^2)\mathbf{i}$
Determine:

______ 13. The position vector which locates the object at t = 2.00
______ 14. The displacement of the object during the time interval t = 2.00 s to t = 4.00 s
______ 15. The distance traveled by the object during the first 3.00 s
______ 16. Instantaneous velocity of the object at t = 2.00 s.
______ 17. Instantaneous speed of the object at t = 2.00 s
______ 18. Average velocity of the object during the time interval t = 2.00 s to t = 4.00 s
______ 19. Average speed of the object during the first 3.00 s
______ 20. Acceleration of the object at all times

A hot-air balloon released from the ground at t = 0 s accelerates upward at the rate of 0.100 m/s^2. When it is at an altitude of 100 m, a camera is accidentally dropped overboard. Determine the following:

______ 21. The velocity of the balloon when the camera is dropped
______ 22. The upward change in displacement of the camera after it is dropped
______ 23. The time the camera travels upward after it is dropped
______ 24. The acceleration of the camera at the top of its trajectory

_________ 25. The time it takes the camera to hit the ground after it is dropped
_________ 26. The impact velocity of the camera

(See Appendix I for answers.)

PRINCIPAL CONCEPTS AND EQUATIONS PRESCRIPTION

Your score on the practice test is an excellent measure of your understanding of the chapter. You should now use the following chart to write your own prescription for dealing with any weaknesses the practice test points out. Look down the leftmost column to the number of the question(s) you answered incorrectly, reading across that row you will find the concept and/or equation of concern, the section(s) of the study guide you should return to for further study, and some suggested text exercises which you should work to gain additional experience.

Practice Test Questions	Concepts and Equations	Prescription	
		Principal Concepts	Text Exercises
1	Position	1	3-1, 2
2	Distance	1	3-2, 3
3	Displacement	1	3-1, 3
4	Distance	1	3-2, 3
5	Average velocity $\bar{v} = \Delta \mathbf{x} / \Delta t$	2	3-4, 12
6	Average velocity $\bar{v} = \Delta \mathbf{x} / \Delta t$	2	3-4, 13
7	Average speed $\bar{v} = d / \Delta t$	2	3-12, 13
8	Average speed $\bar{v} = d / \Delta t$	2	3-12, 13
9	Instantaneous velocity	2	3-12, 13
10	Instantaneous speed $v = \lvert v \rvert$	2	3-12, 13
11	Acceleration	3	3-19, 21
12	Acceleration	3	3-28, 29
13	Position	1	3-1, 3
14	Displacement	1	3-1, 3
15	Distance	1	3-1, 3
16	Instantaneous velocity $\mathbf{v} = d\mathbf{x} / dt$	2	3-11, 14
17	Instantaneous speed $v = \lvert v \rvert$	2	3-11, 16
18	Average velocity $\bar{v} = \Delta \mathbf{x} / \Delta t$	2	3-6, 11
19	Average speed $\bar{v} = d / \Delta t$	2	3-8, 9
20	Acceleration $\mathbf{a} = d\mathbf{v} / dt = d^2\mathbf{x} / dt^2$	3	3-23, 24
21	$v_y^2 - v_{yo}^2 = 2a_y\Delta y$	3	3-46, 47
22	$v_y^2 - v_{yo}^2 = 2g\Delta y$	4	3-47, 48
23	$v_y = v_{yo} - gt$	4	3-46, 48
24	$a_y = -g$	4	3-48, 49
25	$\Delta_y = v_{yo}t - gt^2 / 2$	4	3-35, 37
26	$v_y = v_{yo} - gt$ or $v_y^2 = v_{yo} - 2g\Delta y$	4	3-46, 47

4 MOTION IN TWO DIMENSIONS

RECALL FROM PREVIOUS CHAPTERS

Previously learned concepts and equations frequently used in this chapter	Text Section	Study Guide Page
Analytical treatment of vectors (unit vectors, components, addition, subtraction)	2-4	2-8
Position, displacement, distance	3-1	3-1
Instantaneous and average velocity	3-2	3-5
Instantaneous and average speed	3-2	3-5
Instantaneous and average acceleration	3-4	3-12
Equations for constant acceleration motion	3-5	3-13
$\bar{v}_x = \dfrac{\Delta x}{\Delta t}$ $\qquad v_x = v_{xo} + a_x t$		
$\bar{a}_x = \dfrac{\Delta v}{\Delta t}$ $\qquad \Delta x = v_{xo}t + \dfrac{1}{2}a_x t^2$		
$\bar{v}_x = \dfrac{v_x + v_{xo}}{2}$ $\qquad v_x^2 = v_{xo}^2 + 2\,a_x\Delta x$		

NEW IDEAS IN THIS CHAPTER

Concepts and equations introduced	Text Section	Study Guide Page
Two dimensional motion - position, displacement, velocity, speed and acceleration	4-1	4-1
Projectile motion	4-2	4-5
Uniform circular motion $a = \dfrac{v^2}{r}$	4-3	4-9
Relative velocity	4-4	4-11

PRINCIPAL CONCEPTS AND EQUATIONS

1 Two Dimensional Motion (Section 4-1)

Review: Figure 4.1 shows the two dimensional motion of an object.

Figure 4.1

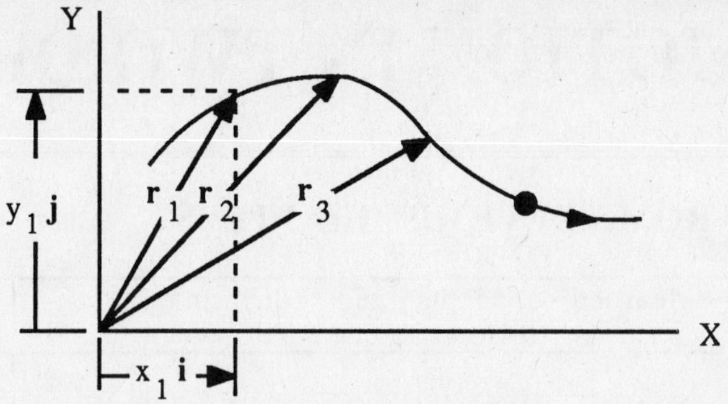

The position of the object is located by the position vectors r_1, r_2 and r_3 at the times t_1, t_2 and t_3 respectively. These times are separated by small but equal time intervals. That is

$$\Delta t = t_2 - t_1 = t_3 - t_2 = \text{a small time interval}$$

Each position vector has an x and y component (only those for $\mathbf{r}_1$ are shown), hence we may write in general that

$$\mathbf{r} = x\mathbf{i} + y\mathbf{j}$$

Figure 4.2 shows two successive displacement vectors $\Delta\mathbf{r}_1$ and $\Delta\mathbf{r}_2$. Each displacement vector has an x and y component (only those for $\Delta\mathbf{r}_1$, are shown), hence we may write in general that

$$\Delta\mathbf{r} = \Delta x\mathbf{i} + \Delta y\mathbf{j}$$

Figure 4.2

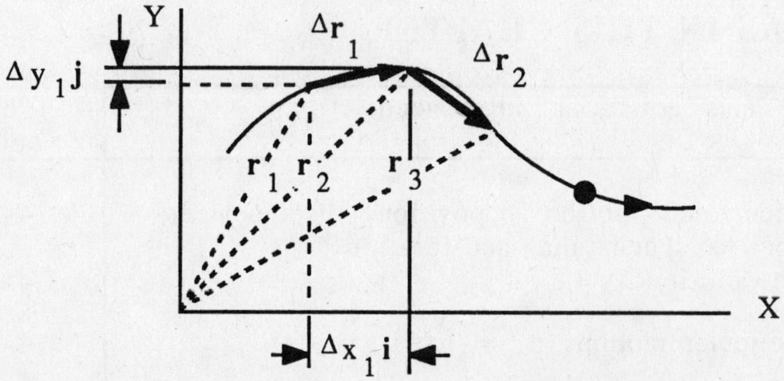

Since Δt is a small time interval, $\Delta\mathbf{r}_1$ and $\Delta\mathbf{r}_2$ are displacements over short time intervals and the instantaneous velocities at t_1 and t_2 are respectively given by

$$\mathbf{v}_1 = \lim_{\Delta t \to 0}\left(\frac{\Delta\mathbf{r}_1}{\Delta t}\right) \quad \text{and} \quad \mathbf{v}_2 = \lim_{\Delta t \to 0}\left(\frac{\Delta\mathbf{r}_2}{\Delta t}\right)$$

Since the $\Delta\mathbf{r}$'s are vectors and the Δt's scalars, the $\mathbf{v}$'s are vectors parallel to the $\Delta\mathbf{r}$'s. Recalling from chapter two that vectors may be slid around on the page, we may draw $\Delta\mathbf{r}_1$, and $\Delta\mathbf{r}_2$ as shown in Figure 4.3(a). Knowing that the $\mathbf{v}$'s are parallel to the $\Delta\mathbf{r}$'s we may draw Figure 4.3(b).

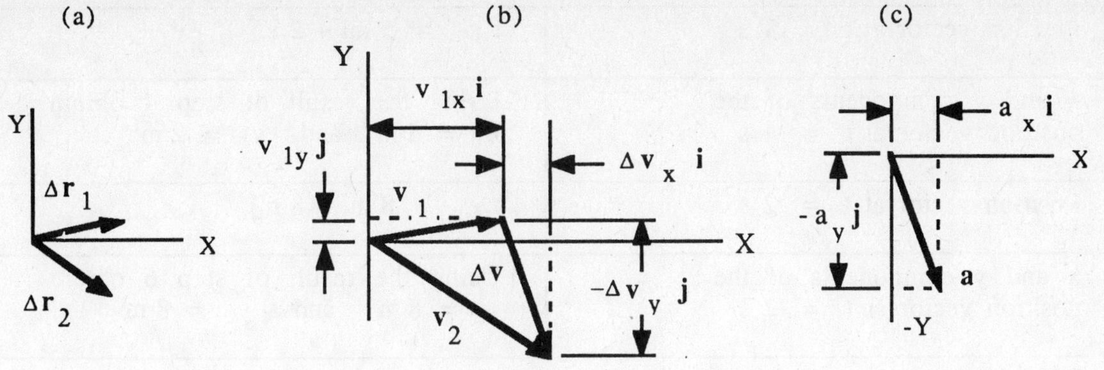

Figure 4.3

Each velocity vector has an x and y component (only those for $\mathbf{v}_1$ are shown), hence we may write in general that

$$\mathbf{v} = v_x\mathbf{i} + v_y\mathbf{j}$$

Also notice that $\Delta\mathbf{v}$ has an x and y component, hence we write in general that

$$\Delta\mathbf{v} = \Delta v_x\mathbf{i} + \Delta v_y\mathbf{j}$$

Since $\mathbf{v}_1$ and $\mathbf{v}_2$ are instantaneous velocities a small time interval apart, $\Delta\mathbf{v} = \mathbf{v}_2 - \mathbf{v}_1$ occurs over the very short time interval Δt and the instantaneous acceleration is given by

$$\mathbf{a} = \lim_{\Delta t \to 0}\left(\frac{\Delta\mathbf{v}}{\Delta t}\right)$$

Since $\Delta\mathbf{v}$ is a vector and Δt a scalar, $\mathbf{a}$ is a vector parallel to $\Delta\mathbf{v}$. Figure 4.3(c) shows the vector $\mathbf{a}$. Note that it is parallel to $\Delta\mathbf{v}$. The acceleration has an x and y component, hence we may write in general that

$$\mathbf{a} = a_x\mathbf{i} + a_y\mathbf{j}$$

Practice: Consider the following vector which gives the position of an object in 2-D space at any time t.

$$\mathbf{r} = [(3\text{ m/s})t + 2\text{ m}]\,\mathbf{i} + \left[(2\text{ m/s}^2)t^2\right]\mathbf{j}$$

Determine the following:

1. x and y components of the position vector at any time	$x = (3\text{ m / s})t + 2\text{ m}$ and $y = (2\text{ m / s}^2)t^2$
2. Position vector at t = 0 s	$\mathbf{r} = 2\text{ m}\mathbf{i}$
3. x and y components of the position vector at t = 0 s	Using the results of step 1 or step 2 obtain $x_0 = 2\text{ m}$ and $y_0 = 0\text{ m}$

4. Position vector at t = 1 s	$\mathbf{r}_1 = 5\,m\mathbf{i} + 2\,m\mathbf{j}$
5. x and y components of the position vector at t = 1 s	Using the result of step 4 obtain $x_1 = 5\,m$ and $y_1 = 2\,m$
6. Position vector at t = 2 s	$\mathbf{r}_2 = 8\,m\mathbf{i} + 8\,m\mathbf{j}$
7. x and y components of the position vector at t = 2 s	Using the result of step 6 obtain $x_2 = 8\,m$ and $y_2 = 8\,m$
8. Displacement vector for the displacement between t = 1 s and t = 2 s	$\Delta\mathbf{r} = \mathbf{r}_2 - \mathbf{r}_1$ $= 3\,m\mathbf{i} + 6\,m\mathbf{j}$
9. x and y components of the displacement vector between t = 1 s and t = 2 s	Using the result of step 8 obtain $\Delta x = 3\,m$ and $\Delta y = 6\,m$
10. Magnitude of the displacement vector between t = 1 s and t = 2 s	$\lvert\Delta\mathbf{r}\rvert = (\Delta x^2 + \Delta y^2)^{1/2}$ $= (9\,m^2 + 36\,m^2)^{1/2} = 6.71\,m$
11. Average velocity vector for the time interval t = 1 s to t = 2 s	$\bar{\mathbf{v}} = \Delta\mathbf{r} / \Delta t = (3\,m\mathbf{i} + 6\,m\mathbf{j}) / 1\,s$ $= (3\,m/s)\mathbf{i} + (6\,m/s)\mathbf{j}$
12. x and y components of the average velocity for the time interval t = 1 s to t = 2 s	$\bar{v}_x = 3\,m/s$ and $\bar{v}_y = 6\,m/s$
13. Expression for the velocity vector at any time	$\mathbf{v} = d\mathbf{r} / dt = (3\,m/s)\mathbf{i} + (4\,m/s^2)t\mathbf{j}$
14. x and y components of the velocity vector at any time	Using step 13 obtain $v_x = 3\,m/s$ and $v_y = (4\,m/s^2)t$ Using step 1 obtain $v_x = dx/dt = 3\,m/s$ and $v_y = dy/dt = (4\,m/s^2)t$
15. Instantaneous velocity and its x and y components at t = 0 s	$\mathbf{v}_o = (3\,m/s)\mathbf{i}$ $v_{xo} = 3\,m/s$ and $v_{yo} = 0\,m/s$
16. Instantaneous velocity and its x and y components at t = 1 s	$\mathbf{v}_1 = (3\,m/s)\mathbf{i} + (4\,m/s)\mathbf{j}$ $v_{x1} = 3\,m/s$ and $v_{y1} = 4\,m/s$
17. Instantaneous velocity and its x and y components at t = 2 s	$\mathbf{v}_2 = (3\,m/s)\mathbf{i} + (8\,m/s)\mathbf{j}$ $v_{x2} = 3\,m/s$ and $v_{y2} = 8\,m/s$

18. Expression for the acceleration vector at any time	**a** = d**v** / dt = (4 m / s^2)**j** Notice that the acceleration is independent of time t.
19. x and y components of the acceleration at all times	a_x = 0 m / s^2 and a_y = 4 m / s^2 These values are constant, independent of time.
20. Comment on v_x in light of the value of a_x	Since a_x = 0, we expect v_x to be a constant. Looking back to steps 15, 16 and 17 note that $v_{xo} = v_{x1} = v_{x2}$ = 3 m / s
21. Comment on v_y in light of the value of a_y	Since a_y = 4 m / s^2, we expect v_y to increase by 4 m / s every second. Looking at steps 15, 16 and 17 we note that v_{yo} = 0 m / s ; v_{y1} = 4 m / s ; and v_{y2} = 8 m / s . Indeed v_y is increasing by 4 m / s every second.

Related Text Exercises: Ex. 4-1 through 4-10.

2 Two-Dimensional Projectile Problems (Section 4-2)

Review: In Chapter 3, we studied one-dimensional (linear) motion and found that we could answer any question about constant-acceleration motion by using the equations shown below:

$$v = v_0 + at \qquad \bar{v} = \frac{v + v_0}{2} \qquad \bar{v} = \frac{\Delta x}{\Delta t} \qquad x = x_0 + v_0 t + \frac{1}{2} at^2 \qquad 2a\Delta x = v^2 - v_0^2$$

In this chapter, we have found that for projectile motion the x and y components are independent of one another. Consequently, we may write a set of equations (similar to those shown above) for the x and y directions. The most useful of these equations for solving projectile motion problems are:

$$v_x = v_{xo} + a_x t \qquad x = x_0 + v_{xo} t + \frac{1}{2} a_x t^2 \qquad 2a_x \Delta x = v_x^2 - v_{xo}^2$$

$$v_y = v_{yo} + a_y t \qquad y = y_0 + v_{yo} t + \frac{1}{2} a_y t^2 \qquad 2a_y \Delta y = v_y^2 - v_{yo}^2$$

After the two one-dimensional problems are solved, the results may be combined to answer questions about the two-dimensional motion. For example once we know v_x and v_y, we can combine them to determine **v**.

In solving two-dimensional motion problems, we will use the following prescription:

1. Make a sketch of the situation under consideration.
2. Pick a convenient coordinate system and use the standard notation that +y is up, -y is down, +x is to the right, and -x is to the left. This means that an acceleration, velocity, or displacement directed upward or to the right is positive and one directed downward or to the left is negative.
3. Write down the general x and y component equations.
4. Write down the initial conditions.
5. Insert the initial conditions into the general equations to obtain component equations that are unique for the situation being considered.
6. Read the question and establish the physics of the situation. For example, if asked to find the maximum height of a projectile, you must recognize that it obtains its maximum height when the y component of the velocity is zero. This may be stated algebraically as $y = y_{max}$ when $v_y = 0$.
7. Decide which of the unique component equations to use.
8. Manipulate the unique component equations to obtain expressions for the desired quantities.

Practice: A projectile is fired off a cliff at a height H. It is fired with a speed v_0 at an angle θ above the horizontal.

Known quantities:

v_0, H, θ, and g

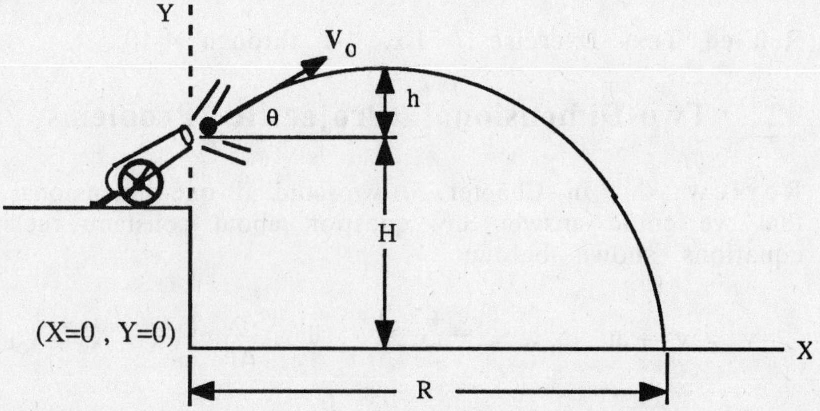

Determine the following:

1. The general equations for the x and y components of the motion	$v_x = v_{xo} + a_xt$; $v_y = v_{yo} + a_yt$ $\Delta x = v_{xo}t + a_xt^2 / 2$; $\Delta y = v_{yo}t + a_yt^2 / 2$ $2a_x\Delta x = v_x^2 - v_{xo}^2$; $2a_y\Delta y = v_y^2 - v_{yo}^2$
2. The initial conditions	$a_x = 0$; $v_{xo} = v_o \cos\theta$; $x_o = 0$ $a_y = -g$; $v_{yo} = v_0 \sin\theta$; $y_0 = H$
3. An expression for the x component of the velocity at any time	$v_x = v_{xo} + a_xt$; $v_{xo} = v_0 \cos\theta$; $a_x = 0$ $v_x = v_0 \cos\theta$
4. An expression for the y component of the velocity at any time	$v_y = v_{yo} + a_yt$ $v_{yo} = v_0 \sin\theta$ $a_y = -g$ $v_y = v_0 \sin\theta - gt$

5. An expression for the x component of the position vector at any time	$\Delta x = v_{xo}t + a_x t^2 / 2$ $\quad v_{xo} = v_0 \cos\theta$ $a_x = 0$ $\quad \Delta x = x - x_0$ $\quad x_0 = 0$ $x = v_0 t \cos\theta$
6. An expression for the y component of the position vector at any time	$\Delta y = v_{yo}t + a_y t^2 / 2$ $\quad v_{yo} = v_0 \sin\theta$ $a_y = -g$ $\quad \Delta y = y - y_0$ $\quad y_0 = H$ $y = H + v_0 t \sin\theta - gt^2 / 2$
7. Unique equations for the x and y components of the projectile's acceleration, velocity and position at any time	$a_x = 0$ $\quad v_x = v_0 \cos\theta$ $\quad x = v_0 t \cos\theta$ $a_y = -g$ $\quad v_y = v_0 \sin\theta - gt$ and $y = H + v_0 t \sin\theta - \frac{1}{2}gt^2$

Note: Step 7 is just a summary of steps 3 through 6.

8. The expression obtained when time is eliminated in expressions for v_y and Δy	$v_y = v_{yo} + a_y t$ $\quad \Delta y = v_{yo}t + a_y t^2 / 2$ Eliminate t to obtain $2a_y\Delta y = v_y^2 - v_{yo}^2$ Insert initial conditions to obtain $-2g\Delta y = v_y^2 - v_0^2 \sin^2\theta$

Note: You should be able to answer any question about the projectile's motion by manipulating the equations in 7 and 8.

9. An expression for maximum height (h) of projectile above cliff	$\Delta y = h$ when $v_y = 0$, and so use $-2g\Delta y = v_y^2 - v_0^2 \sin^2\theta$, which gives $-2gh = -v_0^2 \sin^2\theta$ or $h = \frac{1}{2g}(v_0^2 \sin^2\theta)$
10. An expression for the time (t_{up}) it takes projectile to reach maximum height	$v_y = 0$ when $t = t_{up}$, and so use $v_y = v_0 \sin\theta - gt$, which gives $0 = v_0 \sin\theta - gt_{up}$ or $t_{up} = (v_0 \sin\theta) / g$
11. An expression for the time (t') it takes projectile to get back down to height y = H	$\Delta y = 0$ when $t = t'$, and so use $\Delta y = v_0 t \sin\theta - gt^2 / 2$, which gives $0 = v_0 t' \sin\theta - \frac{g(t')^2}{2}$ or $t' = \frac{(2v_0 \sin\theta)}{g}$ Note that this is twice t_{up}.

12. An expression for the time of flight (t_f)	$\Delta y = -H$ or $y = 0$ when $t = t_f$, and so use $y = H + v_0 t \sin\theta - gt^2 / 2$, which gives $0 = H + v_0 t_f \sin\theta - g t_f^2 / 2$ or $t_f^2 - (2 v_0 t_f / g) \sin\theta - 2H / g = 0$ Solving this quadratic for t_f obtain $t_f = \frac{(v_0 \sin\theta)}{g} + \left(\left[\frac{v_0 \sin\theta}{g} \right]^2 + \frac{2H}{g} \right)^{1/2}$
13. An expression for the range R of the projectile	$x = R$ when $t = t_f$ and so use $x = v_0 t \cos\theta$, which gives $R = v_0 t_f \cos\theta$ t_f has already been determined.

Note: Do not memorize these equations! If any one detail is changed, these expressions will not work. They are correct only for this particular set of initial conditions and for this specific set of known information.

Example: A place kicker wishes to kick a field goal straight at the middle of the goal post from the 30-yard line. He kicks the ball up at an angle of 45° and with a speed of 21.2 yards / s. If the horizontal bar of the goal post is 3 yards off the ground, does he score a field goal? Note that the goal post is 10 yards behind the 0-yard line.

Given: $\Delta x = 40$ yards $\theta = 45°$ $v_0 = 21.2$ yards / s

$\Delta y_{min} = 3$ yards $g = 10.7$ yards / s^2

Determine: If a field goal is scored. That is, for a ball kicked at the given angle and speed, we must establish the value of Δy for $\Delta x = 40$ yards. If Δy is greater than 3 yards a field goal is scored.

Strategy: Obtain expressions for Δy and Δx of the ball at any time. Eliminate time in these two equations and solve for Δy as a function of Δx, Insert values for Δx, v_0, g, and θ to determine Δy. If Δy is greater than 3 yards, a field goal will be scored.

Solution: The general equations for the components of the motion are:

$$v_x = v_{xo} + a_x t \qquad \Delta x = v_{xo} t + \frac{1}{2} a_x t^2 \qquad v_y = v_{yo} + a_y t \qquad \Delta y = v_{yo} t + \frac{1}{2} a_y t^2$$

For this problem, the initial conditions are:

$$a_x = 0 \qquad v_{xo} = v_0 \cos\theta \qquad x_0 = 0 \qquad a_y = -g \qquad v_{yo} = v_0 \sin\theta \qquad y_0 = 0$$

When the initial conditions are inserted into the general equations, we obtain a set of equations for the components of the motion that are unique for this situation:

$$v_x = v_0 \cos\theta \qquad \Delta x = v_0 t \cos\theta \qquad v_y = v_0 \sin\theta - gt, \qquad \Delta y = v_0 t \sin\theta - \frac{1}{2} g t^2$$

Since we want to know the value of Δy for a particular Δx, we eliminate time from the expressions for Δy and Δx. This leaves us with an expression for Δy in terms of Δx:

$$\Delta x = v_0 t \cos\theta, \text{ which leads to } t = \Delta x / (v_0 \cos\theta)$$

$$\Delta y = v_0 t \sin\theta - \frac{1}{2} g t^2 = v_0\left(\frac{\Delta x}{v_0 \cos\theta}\right) \sin\theta - \frac{1}{2} g\left(\frac{\Delta x}{v_0 \cos\theta}\right)^2 = \Delta x \tan\theta - \frac{g \Delta x^2}{2 v_0^2 \cos^2\theta}$$

Now insert values for Δx, v_0, g, and θ to obtain $\Delta y = 1.91$ yards. Since $\Delta y = 1.91$ yards is less than 3 yards, a field goal is not scored.

Related Text Exercises: Ex. 4-11 through 4-21.

Note: After you have worked a number of projectile motion problems, you should notice that they fall into one of two classes:

I. Projected off a cliff or building at some angle
II. Launched from ground level at some angle

These problems may also be modified by varying the landing surface, but they are still essentially the same as the problems we have been working.

Note: Don't memorize the equations for a particular situation. Even if you are successful at thinking up all possible problems and their modifications, your instructor can create a new problem by changing the given information. Learn how to do the problems rather than memorizing a set of equations for a situation you probably won't be asked about.

3 Uniform Circular Motion (Section 4-3)

Review: If an object is traveling in a circle of radius r with a speed v, it must be experiencing a central or centripetal acceleration of magnitude

$$a = \frac{v^2}{r}$$

This acceleration is at right angles to the motion and is the result of the velocity vector continually changing directions. Figure 4.4 shows **r**, **v** and **a** at some t.

Figure 4.4

$$\mathbf{r} = r\mathbf{i}$$
$$\mathbf{v} = v\mathbf{j}$$
$$\mathbf{a} = -a\mathbf{i} = -(v^2/r)\mathbf{i}$$

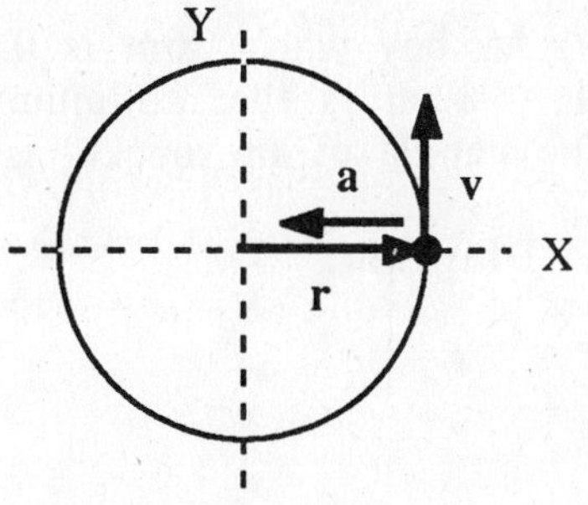

Practice: A conical pendulum is rotating in a horizontal plane as shown in the figure and the following quantities are known:

$L = 2.00$ m = length of the pendulum
$\theta = 30.0°$ = angle shown
$T = 5.00$ s = time for one revolution

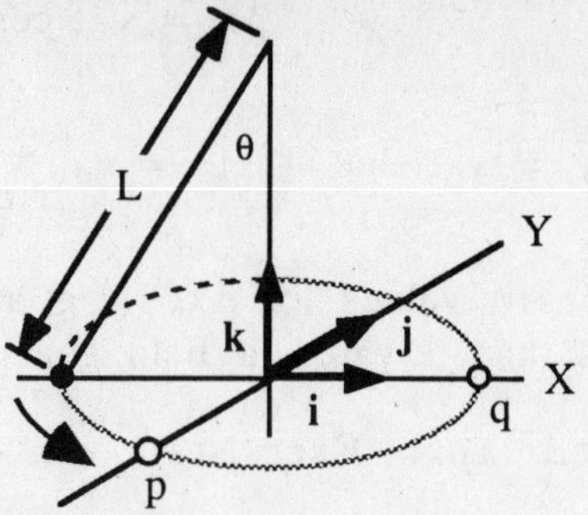

Determine the following:

1. Radius of the circular path of the pendulum bob	$r = L \sin\theta = 1.00$ m
2. Average speed of the pendulum bob	$\bar{v}$ = distance / time $\bar{v} = 2\pi r / T = 1.26$ m / s
3. Instantaneous speed of the pendulum bob	Since the motion is uniform the speed is constant and $v = \bar{v} = 1.26$ m / s
4. Magnitude of the acceleration at any time	$a = v^2 / r = 1.59\ \text{m / s}^2$
5. **r**, **v**, and **a** of the pendulum bob when it is at position p	**r** = -r**j** = (-1.00 m)**j** **v** = v**i** = (1.26 m / s)**i** **a** = a**j** = (1.59 m / s^2)**j**
6. **r**, **v**, and **a** of the pendulum bob when it is at position q	**r** = r**i** = (1.00 m)**i** **v** = v**j** = (1.26 m / s)**j** **a** = -a**i** = (-1.59 m / s^2)**i**

Example: A boy whose arm is 0.600 m long wishes to swing a bucket of water in a vertical circle. What is the minimum speed of the bucket in order for no water to spill out. The handle of the bucket is 0.150 m high.

Given and Diagram:

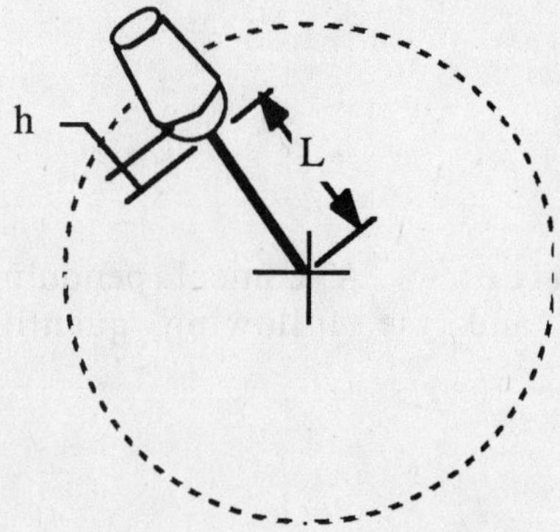

$L = 0.600$ m = length of boys arm

$h = 0.150$ m = height of bucket handle

Determine: v_{min} = the minimum speed of the bucket in order for no water to spill out.

Strategy: When the bucket is directly overhead, the water has a central acceleration of g. If the bucket does not have this same central acceleration as a result of its orbital speed, the water will spill out. Hence the minimum orbital speed of the bucket is the speed which will result in a central acceleration of g.

Solution: $r = h + L = 0.750$ m

$a = g = v^2 / r$ and hence

$v = (gr)^{1/2} = [(9.80 \text{ m} / \text{s}^2)(0.750 \text{ m})]^{1/2} = 2.71 \text{ m} / \text{s}$

Related Text Exercises: Ex. 4-22 through 4-32.

4 Relative Motion In Two Dimensions (Section 4-4)

Review: Consider the river and four swimmers (A, B, C, and D) shown in Figure 4.5.

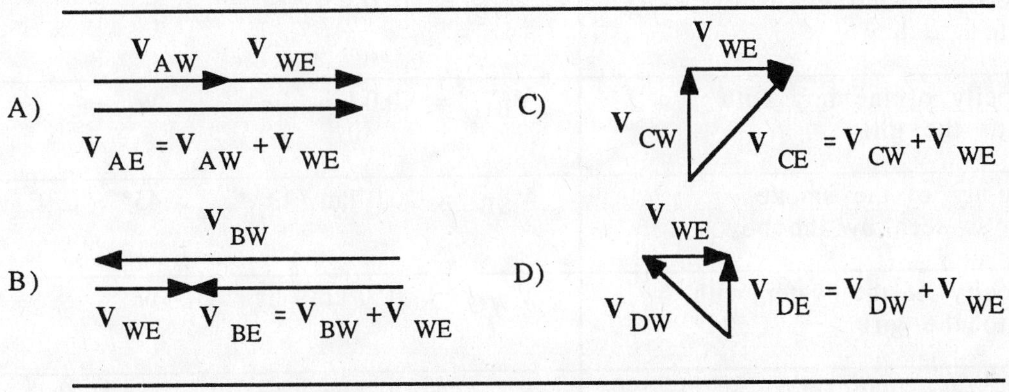

V_{WE} = velocity of the water with respect to (w.r.t.) the earth

V_{AE}, V_{BE}, V_{CE} and V_{DE} = velocity of the swimmers w.r.t. the earth

V_{AW}, V_{BW}, V_{CW} and V_{DW} = velocity of the swimmers w.r.t. the water

Figure 4.5

Practice: A steam ship is heading due east at 10.0 km / hr through a narrow channel and a steady wind is blowing from the southwest at 20.0 km / hr. A girl and a boy standing on the ship and the shore respectively watch the smoke particles. The water in the channel is at rest.

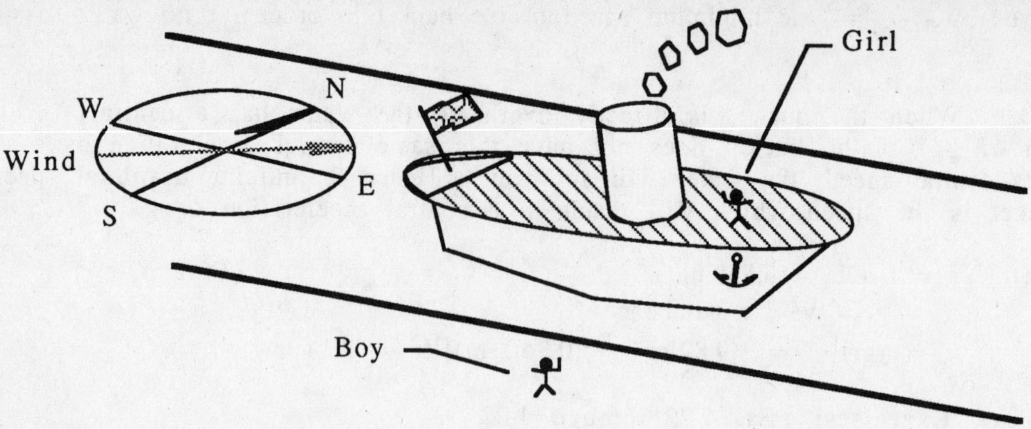

Hint: The smoke particles take up the motion of the air as soon as they leave the smoke stack.

Determine the following:

1. The velocity of the girl with respect to the boy	V_{GB} = 10.0 km / hr	E
2. The velocity of the boy with respect to the girl	V_{BG} = 10.0 km / hr	W
3. The velocity of the smoke particles as seen by the boy	V_{SB} = 20.0 km / hr	45° N of E
4. The velocity of the water with respect to the girl	V_{WG} = 10.0 km / hr	W
5. The velocity of the smoke particles with respect to the water	V_{SW} = 20.0.km / hr	45° N of E
6. The magnitude of the velocity of the smoke particles as seen by the girl	$V_{SG} = V_{SW} + V_{WG}$ 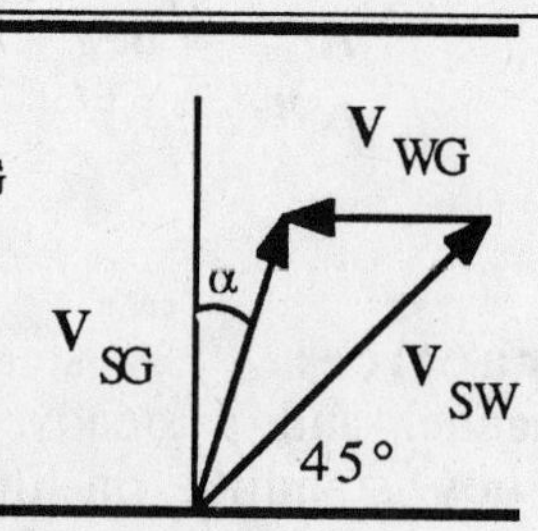 $V_{SGx} = V_{SW} \cos 45° - V_{WG}$ $= 4.14$ km / hr $V_{SGy} = V_{SW} \sin 45° = 14.1$ km / hr $V_{SG} = [(V_{SG})_x{}^2 + (V_{SG})_y{}^2]^{1/2}$ $= 14.7$ km / hr	
7. Determine the direction of the smoke particles as seen by the girl	$\alpha = \tan^{-1} (V_{SGx} / V_{SGy})$ $= \tan^{-1} (4.14 / 14.1) = 16.4°$ E of N	

Example: A child is riding a tricycle West along a straight stretch of sidewalk at 1.00 m / s. A small flag attached to the handlebars stands 10° off to the right of the center bar as seen by the child. A similar flag attached to a stationary wagon indicates that the true wind direction is 30° N of E. What is the magnitude of the winds actual velocity?

Given: V_{TG} = 1.00 m / s W - velocity of the tricycle w.r.t. the ground
α = 10° - angle between the center bar of the tricycle and the flag
θ = 30° N of E - direction of the wind

Determine: The magnitude of the winds actual velocity, that is the speed of the air w.r.t. the ground V_{AG}.

Strategy: Draw a vector diagram showing the velocity of the tricycle w.r.t. the ground V_{TG}, the velocity of the air w.r.t. the ground V_{AG} and the velocity of the air w.r.t. the tricycle V_{AT}. Then using our knowledge of the components of vectors solve for V_{AG}.

Solution: We can set this problem up by either of the following methods.

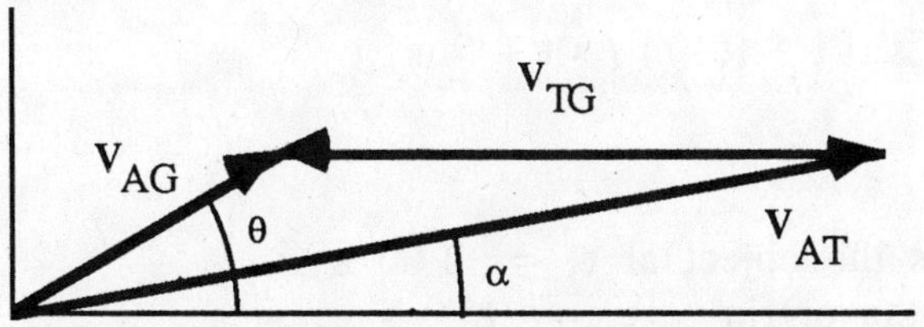

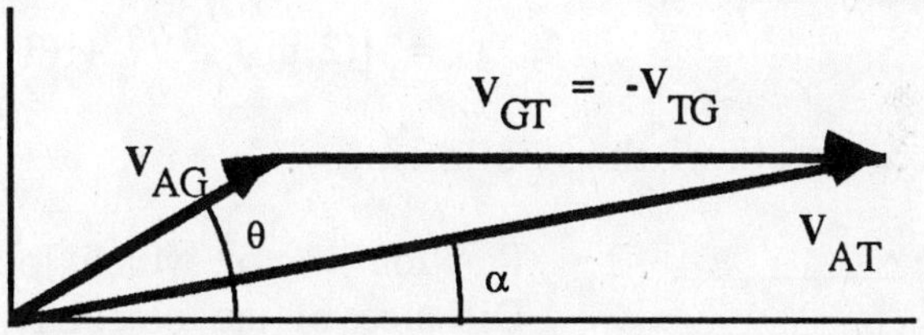

$V_{AG} = V_{AT} + V_{TG}$ $\qquad\qquad$ $V_{AT} = V_{AG} + V_{GT}$

Notice that the y-components of V_{AG} and V_{AT} are the same, hence we may write the following expression:

$$\text{(i)} \quad V_{AG} \sin 30° = V_{AT} \sin 10°$$

From the x-components we may write the following expression:

$$\text{(ii)} \quad V_{AG} \cos 30° + V_{GT} = V_{AT} \cos 10°$$

We now have two equations (i and ii) and two unknowns (V_{AG} and V_{AT}). Since we wish to find the V_{AG} we will eliminate V_{AT}.

Solve (i) for V_{AT}:

$$V_{AT} = V_{AG} \sin 30° / \sin 10°$$

Insert this expression for V_{AT} into eq. (ii) to obtain:

$$V_{AG} \cos 30° + V_{GT} = V_{AG} \sin 30° \cos 10° / \sin 10°$$

Solving for V_{AG} one may obtain:

$$V_{AG} = \frac{-V_{GT}}{\cos 30° - \sin 30° \cot 10°} = \frac{(-1\ m / s)}{(0.866) - (0.500)(5.67)} = 0.508\ m / s$$

Related Text Exercises: Ex. 4-33 through 4-37.

PRACTICE TEST

Take and grade this practice test. Doing so will allow you to determine any weak spots in your understanding of the concepts taught in this chapter. The following section prescribes what you should study further to strengthen your understanding.

Consider the following vector which gives the position of an object in 2-D space at any time t.

$$\mathbf{r} = [(2\ m / s^2)t^2 + (3\ m / s)t]\ \mathbf{i} + [(4\ m / s)t + 5\ m]\ \mathbf{j}$$

Determine the Following:

________ 1. Position vector which locates the object at t = 1.00 s
________ 2. Distance of the object from the origin at t = 1.00 s
________ 3. Angular location of the object at t = 1.00 s
________ 4. Position vector which locates the object at t = 3.00 s
________ 5. Displacement vector for the object during the time interval t = 1.00 s and t = 3.00 s
________ 6. Average velocity of the object during the time interval t = 1.00 s and t = 3.00 s
________ 7. Instantaneous velocity of the object at t = 1.00 s
________ 8. Instantaneous speed of the object at t = 1.00 s
________ 9. Instantaneous acceleration of the object at t = 1.00 s

A baseball is thrown off a dormitory roof. The building is 50.0 m high and the ball is thrown with a velocity v_o = 30.0 m / s at 60.0° above the horizontal.

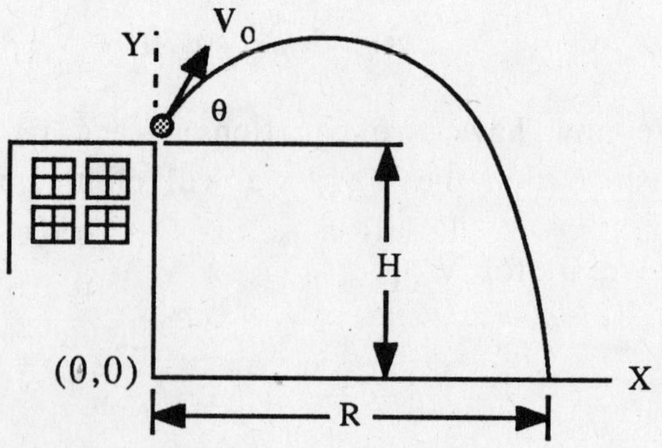

Determine the following:

_________ 10. The x and y components of the initial velocity
_________ 11. The x and y components of the ball's acceleration the instant it is released
_________ 12. The time it takes the ball to reach the top of its trajectory
_________ 13. The maximum height (with respect to the ground) of the ball
_________ 14. The ball's velocity at the top of its trajectory
_________ 15. The ball's acceleration at the top of its trajectory
_________ 16. The time it takes the ball to get back down to its initial height
_________ 17. The x and y components of the ball's velocity when it is back at the initial height
_________ 18. The time of flight
_________ 19. The magnitude of the impact velocity
_________ 20. The direction of the impact velocity
_________ 21. The range

Consider an object initially traveling in a circle of radius r_0 with a speed of v_0. Determine expressions for the following in terms of r_0 and v_0.

_________ 22. Central acceleration of the object when it is traveling with a speed v_0 in a circle of radius r_0
_________ 23. Central acceleration of the object if the radius decreases by a factor of two and the speed stays the same
_________ 24. Central acceleration of the object if the speed doubles and the radius stays the same
_________ 25. Central acceleration of the object if the speed decreases by a factory of four and the radius doubles

A girl can row a boat 2.20 m / s in still water. She wishes to cross a 100 m wide river at a point where the flow rate is 1.50 m / s. If she keeps the boat headed straight across, determine the following:

_________ 26. The time it takes her to get across the river
_________ 27. The distance down stream where she lands
_________ 28. Her velocity relative to the ground while she is crossing the river

If she wants to land directly across the river from her starting position, determine the following:

_________ 29. Heading of the boat as she crosses the river
_________ 30. Her velocity relative to the ground while she is crossing the river

(See Appendix I for answers.)

PRINCIPAL CONCEPTS AND EQUATIONS PRESCRIPTION

Your score on the practice test is an excellent measure of your understanding of the chapter. You should now use the following chart to write your own prescription for dealing with any weaknesses the practice test points out. Look down the leftmost column to the number of the question(s) you answered incorrectly, reading across that row you will find the concept and/or equation of concern, the section(s) of the study guide you should return to for further study, and some suggested text exercises which you should work to gain additional experience.

Practice Test Questions	Concepts and Equations	Prescription	
		Principal Concepts	Text Exercises
1	Position: $\mathbf{r}$	1	4-1, 7
2	Distance: $d = \lvert\mathbf{r}\rvert = (x^2 + y^2)^{1/2}$	1	4-1, 2
3	Position: $\theta = \tan^{-1}(y / x)$	1	4-1, 2
4	Position: $\mathbf{r}$	1	4-1, 7
5	Displacement: $\Delta\mathbf{r} = \mathbf{r}_f - \mathbf{r}_i$	1	4-1, 2
6	Average velocity: $\bar{\mathbf{v}} = \Delta\mathbf{r} / \Delta t$	1	4-3
7	Instantaneous velocity: $\mathbf{v} = d\mathbf{r} / dt$	1	4-3, 8
8	Instantaneous speed: $v = \lvert\mathbf{v}\rvert$	1	4-4, 8
9	Instantaneous acceleration: $\mathbf{a} = d\mathbf{v} / dt = d^2\mathbf{r} / dt^2$	1	4-8
10	Components of a vector - analytical	2	4-9, 10
11	Components of a vector - analytical	2	4-11
12	$v_y = v_{yo} + a_y t$	2	4-16, 19
13	$2a_y \Delta y = v_y^2 - v_{yo}^2$	2	4-16, 19
14	$v_x = v_{xo} + a_x t$	2	4-11, 13
15	Components of a vector - analytical	2	4-11
16	$\Delta y = v_{yo}t + a_y t^2 / 2$	2	4-11, 12
17	$v_x = v_{xo} + a_x t$ and $v_y = v_{yo} + a_y t$	2	4-12, 13
18	$\Delta y = v_{yo}t + a_y t^2 / 2$	2	4-14, 16
19	$v_x = v_{xo} + a_x t$ and $v_y = v_{yo} + a_y t$	2	4-11, 13
20	Direction of a vector - analytical	2	4-12
21	$\Delta x = v_{xo}t + a_x t^2 / 2$	2	4-17, 18
22	Central acceleration: $a = v^2 / r$	3	4-22, 23
23	Central acceleration: $a = v^2 / r$	3	4-27, 28
24	Central acceleration: $a = v^2 / r$	3	4-29, 30
25	Central acceleration: $a = v^2 / r$	3	4-31, 32
26	Average speed: $\bar{v}_{bw} = w / t_{across}$	4	3-6, 7
27	Average speed: $d = \bar{v}_{wg}\, t_{across}$	4	3-8, 9
28	Relative velocity: $\mathbf{v}_{bg} = \mathbf{v}_{bw} + \mathbf{v}_{wg}$	4	4-33, 35
29	Relative velocity: $\theta = \sin^{-1}(v_{wg} / v_{bs})$	4	4-36, 37
30	Relative velocity: $\mathbf{v}_{bg} = \mathbf{v}_{bw} + \mathbf{v}_{wg}$	4	4-36, 37

5 NEWTON'S LAWS OF MOTION

RECALL FROM PREVIOUS CHAPTERS

Previously learned concepts and equations frequently used in this chapter	Text Section	Study Guide Page
Resolving vectors into components analytically	2-3, 4	2-8
Adding vectors analytically	2-3, 4	2-8

NEW IDEAS IN THIS CHAPTER

Concepts and equations introduced	Text Section	Study Guide Page
Newton's first law of motion: If $\Sigma\mathbf{F} = 0$, then $\mathbf{a} = 0$	5-2	5-1
Newton's second law of motion: $\mathbf{F} = m\mathbf{a}$	5-4	5-5
Newton's third law of motion: $\mathbf{F}_{AB} = -\mathbf{F}_{BA}$	5-5	5-7

PRINCIPAL CONCEPTS AND EQUATIONS

1 Newton's First Law (Section 5-2)

Review: Newton's First Law of motion states that if the net force ($\Sigma\mathbf{F}$) on an object is zero, the object's acceleration ($\mathbf{a}$) is zero. The net force is obtained by vectorially adding all the forces acting on the object. That is:

$$\Sigma\mathbf{F} = \mathbf{F}_1 + \mathbf{F}_2 + \mathbf{F}_3 + \ldots$$

An object with zero acceleration can be either at rest or moving with a constant velocity.

NOTE: You should use the following procedure when working problems involving forces:

1. Construct a diagram showing all forces acting on an object.
2. Establish a convenient coordinate system. If motion is a consideration, you will find it convenient to choose a coordinate system parallel and perpendicular to the possible direction of motion.

3. Replace the object under consideration with a point and draw in all forces roughly to scale. This is called a free-body diagram, and it results in a simpler, less cluttered drawing.
4. Resolve all forces into components along the chosen coordinate system. When you replace one of the original vectors by its components, be sure to indicate this so that you don't use both a vector and its components in a subsequent calculation.

Practice: A 200-N crate is at rest on a horizontal frictionless surface. The forces $\mathbf{F}_1$ and $\mathbf{F}_2$ are applied as shown in Figure 5.1 below.

Figure 5.1

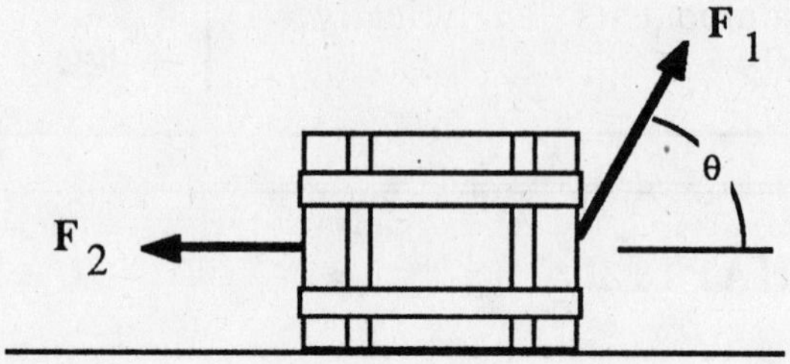

$F_1 = 50.0$ N
$F_2 = 25.0$ N
$\theta = 60.0°$

1. Draw a diagram showing all forces acting on the crate and an appropriate coordinate system.

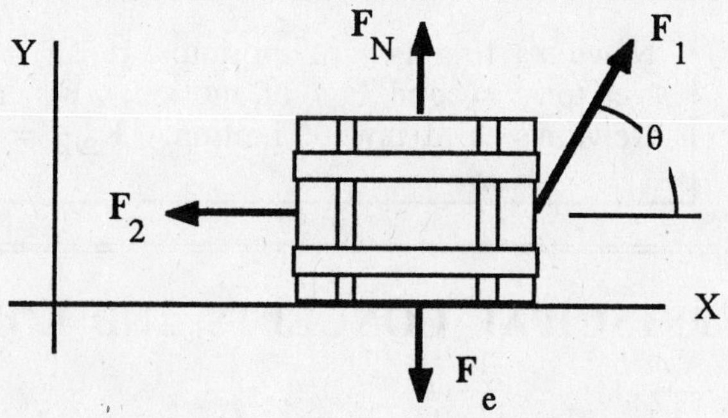

$\mathbf{F}_e$ = force of earth on crate (weight)
$\mathbf{F}_N$ = normal or perpendicular force the surface exerts on the crate
$\mathbf{F}_1$ and $\mathbf{F}_2$ = applied forces

2. Draw a free-body diagram with all forces resolved into components along the coordinate system.

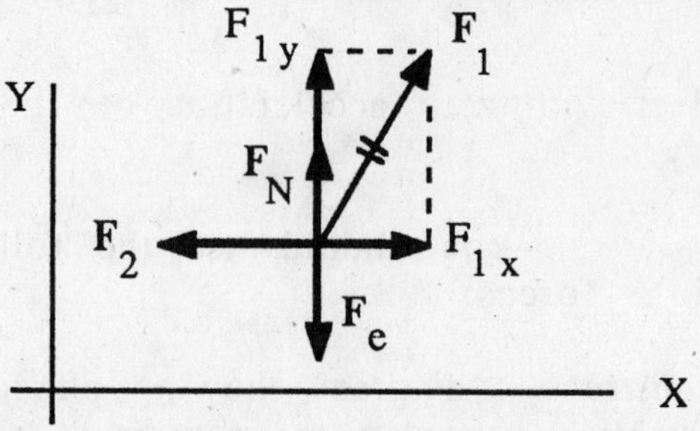

The two short lines through $\mathbf{F}_1$ indicate that this vector has been replaced by its components.

Determine the following:

3. The x component of $\mathbf{F}_1$	$F_{1x} = F_1 \cos\theta = (50.0\text{ N}) \cos 60.0° = 25.0\text{ N}$
4. The x component of $\mathbf{F}_2$	$F_{2x} = -F_2 = -25.0\text{ N}$
5. The net force ΣF_x in the x direction	$\Sigma F_x = F_{1x} + F_{2x} = 25.0\text{ N} - 25.0\text{ N} = 0$
6. Any change in motion in the x direction.	Since no net force exists in the x direction, the crate will not change its motion in this direction. It will remain at rest at the same location.
7. The y component of $\mathbf{F}_1$ and $\mathbf{F}_e$	$F_{1y} = F_1 \sin\theta = (50.0\text{ N}) \sin 60.0° = 43.3\text{ N}$ $F_{ey} = -F_e = -200\text{ N}$
8. The magnitude of $\mathbf{F}_N$	Since $F_{1y} < F_{ey}$, we expect no change in the crate's motion in the y direction. If there is no change in motion, we expect the net force in the y direction to be zero. That is: $F_N + F_{1y} + F_{ey} = \Sigma F_y$, or $F_N = \Sigma F_y - F_{ey} - F_{1y}$ $F_N = 0\text{ N} + 200\text{ N} - 43.3\text{ N} = 157\text{ N}$

Example: A crate is pulled up a frictionless ramp inclined at an angle of 30.0° (Figure 5.2). If the crate weighs 400 N, what is the magnitude of the force **F** needed to pull it up the ramp at a constant speed of 1.00 m/s? The angle between the force and the incline is 10.0°.

Figure 5.2

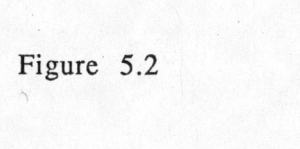

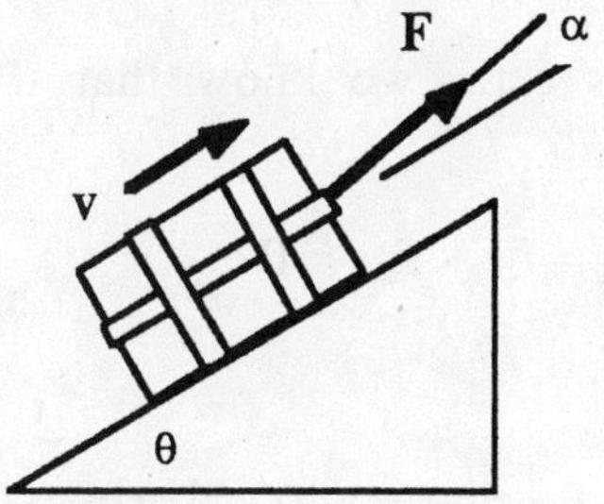

$F_e = 400\text{ N}$
$\theta = 30.0°$
$\alpha = 10.0°$
$v = 1.00\text{ m/s}$

Determine: F, the magnitude of the force that will pull the crate up the ramp at a constant speed of 1.00 m/s.

Strategy: Construct a diagram showing all forces acting on the crate and choose a convenient coordinate system. Construct a free-body diagram and resolve all vectors into components along the coordinate system. Since the speed is constant, we know that the acceleration parallel to the ramp and hence the net force parallel to the ramp are both zero. If the net force parallel to the ramp is zero, the component of **F** up the ramp must equal the component of $\mathbf{F_e}$ down the ramp. Knowing $\mathbf{F_e}$ and θ, we can determine the component of $\mathbf{F_e}$ down the ramp and hence the component of **F** up the ramp. Knowing the component of **F** and α, we can determine **F**.

Solution:

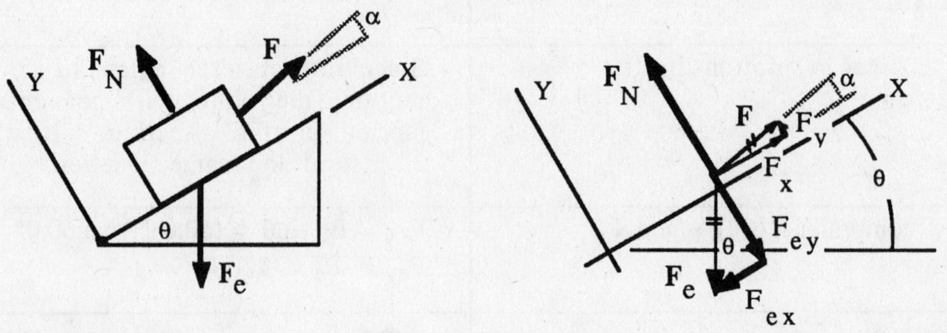

(a) All forces and a convenient coordinate system

(b) A free-body diagram with all forces resolved into components along the coordinate system

Figure 5.3

A convenient coordinate system is one that has its axes parallel and perpendicular to the motion. The decision to put the positive x axis up or down the ramp is one of personal preference. Most students like to orient the axes so that the distance of the moving object from the origin increases with time. For this reason, we put the positive x axis up the ramp.

The fact that v is constant allows us to write the following:

$$v_x = \text{constant} \Rightarrow \Delta v_x = 0 \Rightarrow a_x = 0 \Rightarrow \Sigma F_x = 0$$

Since the net force in the x direction is zero, we know that if we sum all the forces in the x direction, they must add to zero:

$$\Sigma F_x = F_x + F_{ex} = 0$$

From this, we may write:

$$F_x = -F_{ex} = -(-F_e \sin\theta) = F_e \sin\theta = (400 \text{ N}) \sin 30.0° = 200 \text{ N}$$

Finally, we may determine the magnitude of **F** as follows:

$$F_x = F \cos\alpha \quad \text{or} \quad F = F_x / \cos\alpha = (200 \text{ N}) / \cos 10.0° = 203 \text{ N}$$

Related Text Exercises: Ex. 5-5 through 5-10.

2 Newton's Second Law (Section 5-3)

Review: Newton's second law of motion may be written in symbols as

$$\Sigma \mathbf{F} = m\mathbf{a}$$

where m is the mass of the object, $\Sigma \mathbf{F}$ is the net force acting on the object ($\Sigma \mathbf{F} = \mathbf{F}_1 + \mathbf{F}_2 + \mathbf{F}_3 + \ldots$), and **a** is the acceleration of the object due to the net force.

Practice: A 100 kg box is being pushed across the floor by a force $\mathbf{F}_1$. This force has a magnitude of 100 N and is directed 60.0° below the horizontal. The force of friction exerted on the box by the floor is 25.0 N. Note that the frictional drag opposes the motion of the box.

1. Draw a figure that shows all forces and an appropriate coordinate system.	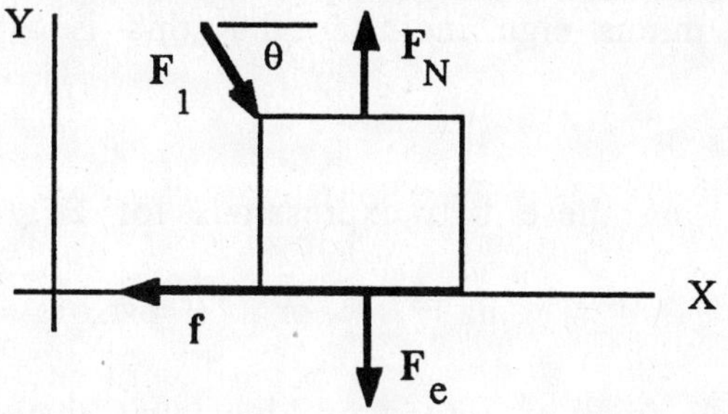
2. Draw a free-body diagram with all forces resolved into components	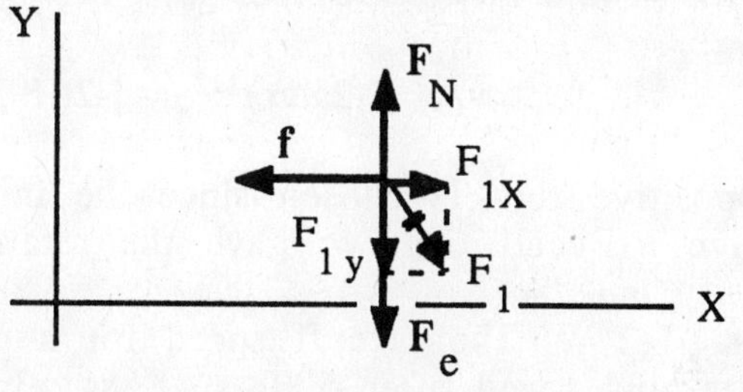

Now determine the following:

3. The component of $\mathbf{F}_1$ and f in the x direction	$F_{1x} = F_1 \cos\theta = (100\text{ N}) \cos 60.0° = 50.0\text{ N}$ $f_x = -f = -25.0\text{ N}$
4. The x component of the net force acting on the box	$\Sigma F_x = F_{1x} + f_x = 50.0\text{ N} - 25.0\text{ N} = 25.0\text{ N}$ This force is in the positive x direction.
5. The magnitude of the acceleration of the box	$\Sigma F_x = ma_x$ or $a_x = \Sigma F_x / m = 0.250\text{ m/s}^2$

Example: A 0.200 kg hockey puck slides 40.0 m down the rink before coming to rest. If the frictional drag is 0.100 N, what was the puck's initial speed?

Given: $v_f = 0$, $\Delta x = 40.0$ m, $f = 0.100$ N, $m = 0.200$ kg

Determine: v_o, the initial speed of the puck

Strategy: The net force on the puck is due to friction. Since the net force and m are known, we can use Newton's second law to determine the acceleration. Knowing v_f, a and Δx, we can use our knowledge of kinematics ($v^2 - v_o^2 = 2a\Delta x$) to determine v_o.

Solution: Since the net force on the puck is due to friction, we can write

$$\Sigma F = -f$$

The minus sign indicates that this is a retarding force. According to Newton's second law,

$$\Sigma F = ma$$

Equating these two expressions for ΣF we obtain

$$ma = -f \quad \text{or} \quad a = -f / m = -(0.100 \text{ N}) / 0.200 \text{ kg} = -0.500 \text{ m/s}^2$$

Recall from our study of kinematics that $v^2 - v_o^2 = 2a\Delta x$. Since $v_f = 0$, $a = -0.500$ m/s^2 and $\Delta x = 40.0$ m, this reduces to

$$v_o = (-2a\Delta x)^{1/2} = [-2(-0.500 \text{ m/s}^2)(40.0 \text{ m})]^{1/2} = 6.32 \text{ m/s}$$

The positive root is chosen since the initial direction of the motion was taken to be positive. Recall that we gave the retarding force a negative sign.

Example: The takeoff speed for a particular 50,000-kg jet is 200 km/h, and it can acquire this speed after 50.0 s. Assuming that the thrust force exerted by the jet is constant, calculate this thrust and the minimum length of runway needed.

Given: $v_o = 0$ km/h, $v_f = 200$ km/h, $m = 5.00 \times 10^4$ kg, $t = 50.0$ s

Determine: F, the thrust force of the jet engines
Δx, the minimum length of runway needed for takeoff

Strategy: Knowing v_o and v_f, determine Δv. Knowing Δv and t, determine a. Determine F from a and m. Knowing v_o, v and a, use your knowledge of kinematics ($v^2 - v_o^2 = 2a\Delta x$) to determine Δx.

Solution:

$$\Delta v = v - v_0 = (2.00 \times 10^2 \text{ km/h})(10^3 \text{ m} / 1.00 \text{ km})(1.00 \text{ h} / 3.60 \times 10^3 \text{ s}) = 55.6 \text{ m/s}$$

$$a = \Delta v / \Delta t = (55.6 \text{ m/s}) / 50.0 \text{ s} = 1.11 \text{ m/s}^2$$

$$F = ma = (5.00 \times 10^4 \text{ kg})(1.11 \text{ m/s}^2) = 5.55 \times 10^4 \text{ N}$$

$v^2 - v_0^2 = 2a\Delta x$, which may be solved for Δx to obtain

$$\Delta x = v^2 / 2a = (55.6 \text{ m/s})^2 / (2)(1.11 \text{ m/s}^2) = 1.39 \times 10^3 \text{ m}$$

Related Text Exercises: Ex. 5-11 through 5-18.

3 Newton's Third Law (Section 5-4)

Review: Newton's third law of motion states that whenever one object exerts a force on a second object, the second object exerts on the first a force equal in magnitude and opposite in direction.

Remember that the two forces to which the third law refers (often called an action-reaction pair) always act on different bodies.

If we let $\mathbf{F}_{AB}$ represent the force that body A exerts on body B and $\mathbf{F}_{BA}$ represent the force that body B exerts on body A, then we can write Newton's third law in symbols as

$$\mathbf{F}_{AB} = -\mathbf{F}_{BA}$$

Practice: A monkey hangs from a branch as shown in Figure 5.4. All forces on the monkey, tree and earth have been drawn in and labeled.

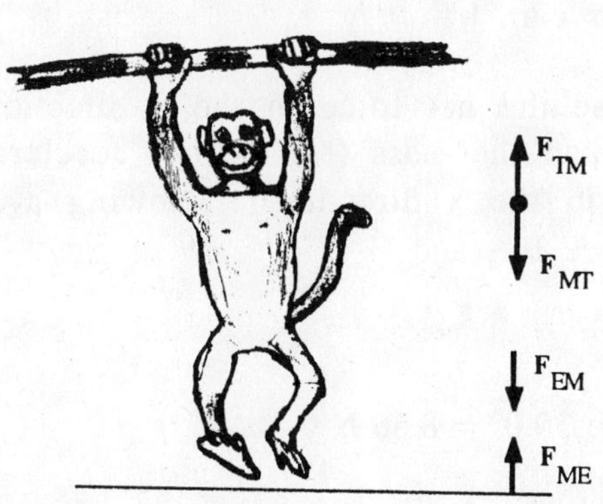

F_{EM} = force the earth exerts on the monkey (wt. of the monkey)

F_{ME} = force the monkey exerts on the earth

F_{TM} = force the tree exerts on the monkey

F_{MT} = force the monkey exerts on the tree

Figure 5.4

Determine the following:

1. Newton's third-law equation involving $\mathbf{F}_{MT}$	$F_{MT} = -F_{TM}$ One force is on the tree ($\mathbf{F}_{MT}$) and the other is on the monkey ($\mathbf{F}_{TM}$).
2. Newton's third-law equation involving $\mathbf{F}_{EM}$	$\mathbf{F}_{EM} = -\mathbf{F}_{ME}$ One force is on the monkey ($\mathbf{F}_{EM}$) and the other is on the earth ($\mathbf{F}_{ME}$).
3. Since the monkey's motion is not changing, we may write $\mathbf{F}_{EM} = -\mathbf{F}_{TM}$. Is this another Newton's third-law equation?	No, because both $\mathbf{F}_{EM}$ and $\mathbf{F}_{TM}$ are acting on the monkey. Forces of a third-law pair act on different objects. This particular force statement says that the unbalanced force on the monkey is zero; hence its motion will not change.

Example: Two carts, 1 and 2 are pulled across a horizontal frictionless surface by an applied force $\mathbf{F}_a$ as shown in Figure 5.5.

Figure 5.5

$M_1 = 4.00$ kg
$M_2 = 2.00$ kg
$F_a = 10.0$ N
$\theta = 30.0°$

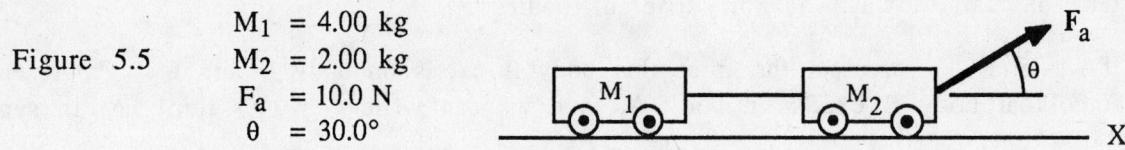

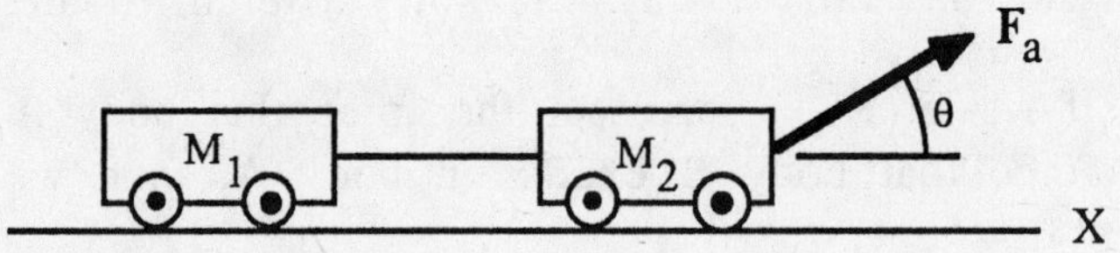

Determine the force that each cart exerts on the other.

Given: M_1, M_2, F_a and θ

Determine: F_{12}, the force that cart 1 exerts on cart 2
F_{21}, the force that cart 2 exerts on cart 1

Strategy: Knowing F_a and θ we can determine the net force in the x direction (ΣF_x). Knowing the net force in the x direction and the mass ($M_1 + M_2$) accelerated by this force, we can determine the acceleration in the x direction. Knowing a_x, M_1 and M_2, we can determine ΣF_{x1} and ΣF_{x2} and hence F_{12} and F_{21}.

Solution: The net force in the x direction is

$$\Sigma F_x = F_a \cos\theta = (10.0 \text{ N}) \cos 30.0° = 8.66 \text{ N}$$

The acceleration in the x direction is

$$a_x = \Sigma F_x / (M_1 + M_2) = 1.44 \text{ m/s}^2$$

Looking at the free-body diagram for M_2 we can determine the force M_1 exerts on M_2 (F_{12}).

Free-body diagram for M_2

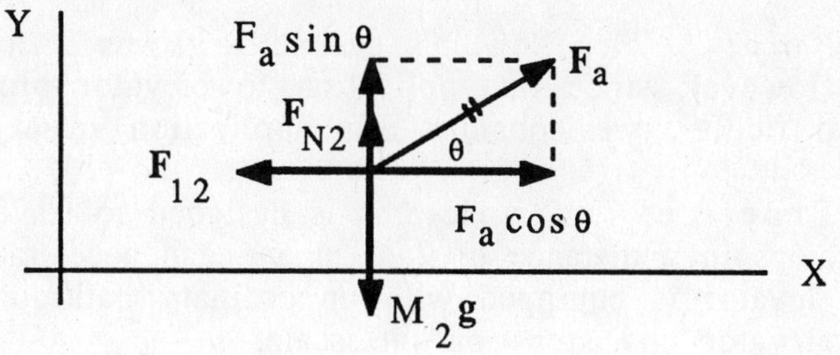

$$\Sigma F_{2x} = F_a \cos\theta + F_{12} = M_2 a_x$$

or

$$F_{12} = M_2 a_x - F_a \cos\theta = 2.88 \text{ N} - 8.66 \text{ N} = -5.78 \text{ N}$$

$$\mathbf{F}_{12} = -5.78 \text{ N } \mathbf{i}$$

According to Newton's third law the force that M_2 exerts on M_1 is equal in magnitude but opposite in direction to the force that M_1 exerts on M_2. Hence

$$\mathbf{F}_{21} = -\mathbf{F}_{12} = -(-5.78 \text{ N})\mathbf{i} = (5.78 \text{ N})\mathbf{i}$$

As a quick check on our work we can look at a free-body diagram for M_1 and determine $\mathbf{F}_{21}$.

Free-body diagram for M_1

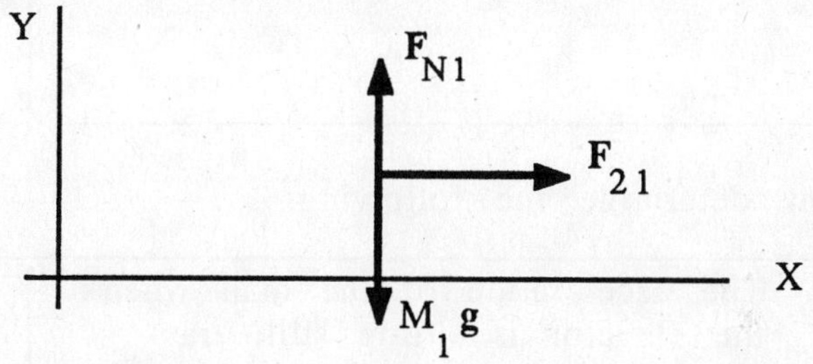

$$\Sigma F_{1x} = F_{21} = M_1 a_x$$

or

$$F_{21} = M_1 a_x = 5.76 \text{ N}$$

$$\mathbf{F}_{21} = (5.76 \text{ N})\mathbf{i}$$

The slight difference between the two values for F_{21} is due to rounding off.

Related Text Exercises: Ex. 5-19 through 5-23.

4 Newton's Second Law Applied to Elevator Problems
(Section 5-5)

Note: Newton's second law was introduced and reviewed in Section 5-3. However, since its application to elevator problems tends to give students some difficulty, we consider this application in a separate section.

Practice: An elevator is designed to travel at a constant speed of 4.00 m/s. It stops in a distance of 2.00 m and can accelerate to its cruising speed in 4.00 m. The elevator is equipped with an ordinary bathroom scale. A 60.0-kg man gets in the elevator and steps on the scale.

1. Sketch this situation, show a convenient coordinate system and all forces acting on the man.	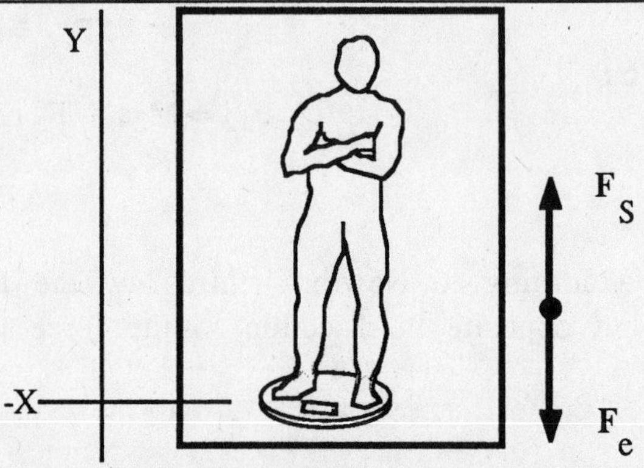 $\mathbf{F_s} = F_s\,\mathbf{j}$ = force the scale exerts on the man (scale reading) $\mathbf{F_e} = -F_e\,\mathbf{j} = -mg\,\mathbf{j}$ = force the earth exerts on the man (the man's wt.)

Now determine the following:

2. The acceleration of the man when the elevator is sitting still, traveling upward at a constant speed, or traveling downward at a constant speed	When the elevator is sitting still or traveling at a constant speed, its acceleration and hence that of the man are zero.
3. The acceleration of the man when the elevator accelerates upward to its cruising speed	$v_y^2 - v_{yo}^2 = 2a_y\Delta y$ or $a_y = v_y^2 / 2\Delta y$ $a_y = (4.00\text{ m/s})^2 / 2(4.00\text{ m}) = 2.00\text{ m/s}^2$ or using vector notation $\mathbf{a} = a_y\,\mathbf{j} = 2.00\text{ m/s}^2\,\mathbf{j}$
4. The acceleration of the man when the elevator accelerates downward to its cruising speed	This acceleration will have the same magnitude as in step 3 but will be in the opposite direction, hence $\mathbf{a} = -2.00\text{ m/s}^2\,\mathbf{j}$.

5. The acceleration of the man as the elevator slows to a stop while traveling upward	$v_y^2 - v_{yo}^2 = 2a_y\Delta y$ or $a_y = -v_{yo}^2 / 2\Delta y$ $a_y = -(4.00\ m/s)^2 / 2(2.00\ m) = -4.00\ m/s^2$ or using vector notation $\mathbf{a} = a_y\,\mathbf{j} = -4.00\ m/s^2\,\mathbf{j}$
6. The acceleration of the man as the elevator slows to a stop while traveling downward	This acceleration will have the same magnitude as in step 5 but will be in the opposite direction, hence $\mathbf{a} = 4.00\ m/s^2\,\mathbf{j}$.
7. An expression for the force the scale exerts on the man at any time	By Newton's second law, $\Sigma\mathbf{F} = m\mathbf{a}$. But by the drawing in step 1 we see that the net force on the man at any time is $\Sigma\mathbf{F} = \mathbf{F}_s + \mathbf{F}_e$. Equate these expressions for $\Sigma\mathbf{F}$ and solve for $\mathbf{F}_s$. $\mathbf{F}_s = m\mathbf{a} - \mathbf{F}_e$ Since $\mathbf{F}_e = -mg\,\mathbf{j}$, this may be written as $\mathbf{F}_s = m\mathbf{a} + mg\,\mathbf{j}$.
8. The scale reading when the elevator is at rest or traveling at a constant speed	$\mathbf{F}_s = m\mathbf{a} + mg\,\mathbf{j}$ (step 7) For this case, $\mathbf{a} = 0$ (step 2) $\mathbf{F}_s = 588\ N\,\mathbf{j}$
9. The scale reading when the elevator is accelerating upward to cruising speed	$\mathbf{F}_s = m\mathbf{a} + mg\,\mathbf{j}$ (step 7) For this case, $\mathbf{a} = 2.00\ m/s^2\,\mathbf{j}$ (step 3) $\mathbf{F}_s = 708\ N\,\mathbf{j}$
10. The scale reading when the elevator slows to a stop while traveling upward	$\mathbf{F}_s = m\mathbf{a} + mg\,\mathbf{j}$ (step 7) For this case, $\mathbf{a} = -4.00\ m/s^2\,\mathbf{j}$ (step 5) $\mathbf{F}_s = 348\ N\,\mathbf{j}$
11. The scale reading when the elevator is accelerating downward to cruising speed	$\mathbf{F}_s = m\mathbf{a} + mg\,\mathbf{j}$ (step 7) For this case, $\mathbf{a} = -2.00\ m/s^2\,\mathbf{j}$ (step 4) $\mathbf{F}_s = 468\ N\,\mathbf{j}$
12. The scale reading when the elevator slows to a stop while traveling downward	$\mathbf{F}_s = m\mathbf{a} + mg\,\mathbf{j}$ (step 7) For this case, $\mathbf{a} = 4.00\ m/s^2\,\mathbf{j}$ (step 6) $\mathbf{F}_s = 828\ N\,\mathbf{j}$

Related Text Exercises: Ex. 5-29 and 5-32.

5 Newton's Second Law Applied to a System of Masses
(Section 5-6)

Note: Newton's second law was introduced, reviewed and applied to individual masses in principal concept 2. However, since the application of the second law to a system of masses is somewhat more complicated, we will consider this application in a separate section.

Practice: Consider the situation shown in Figure 5.6.

Figure 5.6 $M_1 = 10.0$ kg
$M_2 = 20.0$ kg
$g = 9.80$ m/s^2

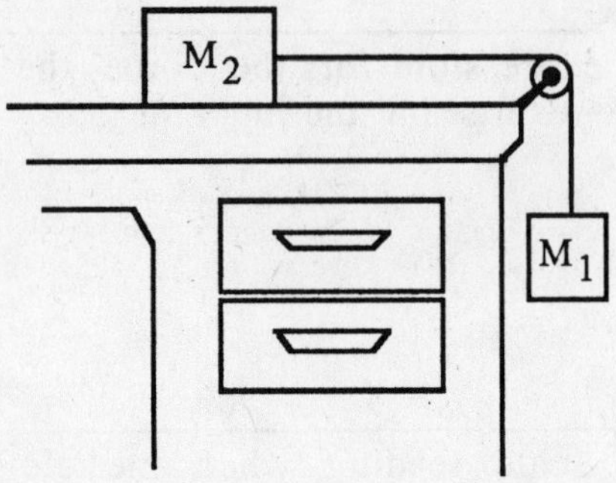

Determine the following:

<table>
<tr><td>1. Free body diagram for M_1 and M_2</td><td>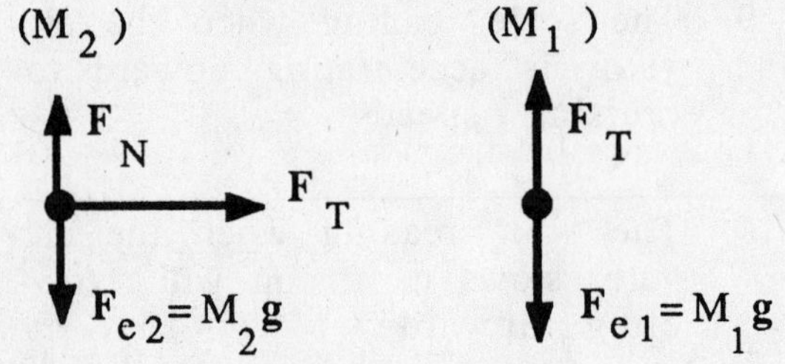
</td></tr>
<tr><td>2. Newton's second law equation for M_1 and M_2</td><td>(M_1) $M_1a = M_1g - F_T$
(M_2) $M_2a = F_T$</td></tr>
<tr><td>3. An expression for the acceleration of the system</td><td>At this point we have two equations (see step 2) and two unknowns (a and F_T). If we add these two equations, we can obtain an expression for a.
$M_1a = M_1g - F_T$ Eq. for M_1
$M_2a = F_T$ Eq. for M_2
$(M_1 + M_2)a = M_1g$ Eqs. added
or
$a = M_1g / (M_1 + M_2)$</td></tr>
</table>

4. An expression for the tension in the cord connecting M_1 and M_2	We can insert the expression for the acceleration back into either of the Newton's second law equations in step 2. Using the equation for M_1 obtain $F_T = M_1(g - a)$ $= M_1(g - \frac{M_1 g}{M_1 + M_2})$ $= M_1 M_2 g / (M_1 + M_2)$ Using the equation for M_2, obtain $F_T = M_2 a$ $= M_1 M_2 g / (M_1 + M_2)$
5. Numerical values for the acceleration of the system and the tension in the cord connecting M_1 and M_2	From step 3 $a = M_1 g / (M_1 + M_2) = 3.27\ m/s^2$ From step 4 $F_T = M_1 M_2 g / (M_1 + M_2) = 65.3\ N$
6. A quick check on your work	Consider the system $M_1 + M_2$ According to Newton $\Sigma F_{syst} = M_{syst} a$, where $M_{syst} = M_1 + M_2$ $\Sigma F_{syst} = M_1 g$ Combining and solving for a, obtain $a = M_1 g / (M_1 + M_2)$ This is the same result as step 3.

Example: Figure 5.7 shows an object of mass M_2 connected by a lightweight cord to mass M_1. The ramp is frictionless and is inclined at an angle θ w.r.t. the horizontal. The system is moving in such a manner that M_2 accelerates up the ramp. Find the tension in the cord and the acceleration of the system ($M_1 + M_2$).

Figure 5.7

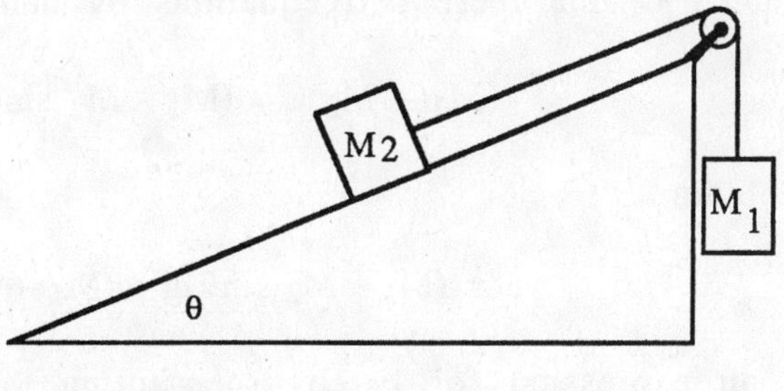

Given: M_1, M_2, θ and M_2 accelerates up the ramp

Determine: Expressions for the tension in the cord (F_T) between M_1 and M_2, and the acceleration (a) of the system.

Strategy: Draw a free-body diagram for each object, pick a convenient coordinate system (use + as the direction of motion), resolve all forces into components along the coordinate system, and write the Newton's second law equation for each object. These two equations will include the two unknowns (a and F_T), hence we can obtain expressions for a and F_T.

Solution:

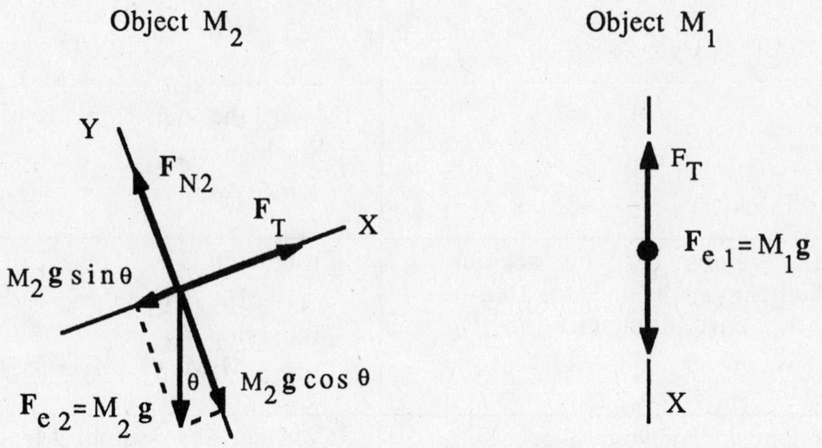

Newton said that $\Sigma F = ma$ for any object, hence for M_1 and M_2 we have

$$\Sigma F_2 = M_2 a \qquad\qquad \Sigma F_1 = M_1 a$$

Newton left it to us to look at our specific problem and determine ΣF. For M_1 and M_2 we have

$$\Sigma F_2 = F_T - M_2 g \sin\theta \qquad\qquad \Sigma F_1 = M_g - F_T$$

Combining these two expressions for ΣF_1 and ΣF_2, obtain

$$\text{(i)}\ M_2 a = F_T - M_2 g \sin\theta \qquad\qquad \text{(ii)}\ M_1 a = M_1 g - F_T$$

We may eliminate F_T from these two equations by adding them.

$$(M_1 + M_2)a = (M_1 - M_2 \sin\theta)g$$

solving for a, obtain

$$a = (M_1 - M_2 \sin\theta)g \,/\, (M_1 + M_2)$$

We may obtain an expression for F_T by substituting the expression for a into (i) or (ii). Let's use (ii).

$$F_T = M_1(g - a) = M_1\left[g - \frac{(M_1 - M_2 \sin\theta)\,g}{(M_1 + M_2)}\right] = M_1 M_2 g \frac{(1 + \sin\theta)}{(M_1 + M_2)}$$

Related Text Exercises: Ex. 5-38.

PRACTICE TEST

Take and grade this practice test. Doing so will allow you to determine any weak spots in your understanding of the concepts taught in this chapter. The following section prescribes what you should study further to strengthen your understanding.

A 500-N crate is pulled up a frictionless ramp inclined at an angle at an angle of 30.0° by a 300-N force that is parallel to the ramp.

Determine the following:

_________ 1. Component of the crate's weight perpendicular to the ramp
_________ 2. Normal or perpendicular force the surface of the ramp exerts on the crate
_________ 3. Component of the crate's weight acting down the ramp
_________ 4. The net force on the crate up the ramp
_________ 5. Acceleration of the crate up the ramp
_________ 6. Speed of the crate after it has been pulled 2.00 m up the ramp
_________ 7. Time it takes to pull the crate 2.00 m up the ramp

Three crates (M_1, M_2 and M_3) are pulled across a horizontal frictionless surface as shown below.

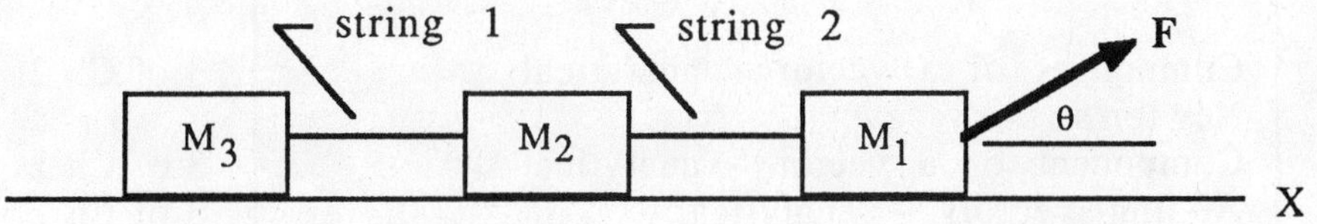

$M_1 = M_2 = M_3 = 100$ kg, $\theta = 30.0°$, $|F| = 100$ N

Determine the following:

_________ 8. The net force on the system (M_1, M_2 and M_3) in the +x direction
_________ 9. The acceleration of the system
_________ 10. Tension in string 1
_________ 11. Tension in string 2

A 70.0-kg man is standing on an ordinary bathroom scales in an elevator. The elevator is designed to cruise at 4.00 m/s, stop in 2.00 m, and accelerate to its cruising speed in 3.00 m.

Determine the following:

_________ 12. The scale reading when the elevator is cruising up or down at 4.00 m/s
_________ 13. The scale reading as the elevator accelerates upward to the cruising speed

________ 14. The scale reading as the elevator accelerates downward to the cruising speed

________ 15. The scale reading as the elevator decelerates to a stop while traveling upward

________ 16. The scale reading as the elevator decelerates to a stop while traveling downward

(See Appendix I for answers.)

PRINCIPAL CONCEPTS AND EQUATIONS PRESCRIPTION

Your score on the practice test is an excellent measure of your understanding of the chapter. You should now use the following chart to write your own prescription for dealing with any weaknesses the practice test points out. Look down the leftmost column to the number of the question(s) you answered incorrectly, reading across that row you will find the concept and/or equation of concern, the section(s) of the study guide you should return to for further study, and some suggested text exercises which you should work to gain additional experience.

Practice Test Questions	Concepts and Equations	Prescription	
		Principal Concepts	Text Exercises
1	Component of a vector - analytical	3 of Ch 2	2-11, 14
2	Newton's third law	3	5-22, 23
3	Component of a vector - analytical	3 of Ch 2	2-15, 16
4	Adding vectors - analytical	3 of Ch 2	2-19, 20
5	Newton's second law **F** = ma	2	5-7, 10
6	$v^2 - v_0^2 = 2 a\Delta x$	4 of Ch 3	3-46, 47
7	$v = v_0 + at$ or $\Delta x = v_0 t + at^2 / 2$	4 of Ch 3	3-35, 48
8	Net force	1	5-11, 12
9	Newton's second law **F** = ma	2	5-30, 31
10	Newton's second law	2	5-31, 38
11	Newton's second law	2	5-31, 32
12	Newton's first law	1	5-9, 10
13	Newton's second law	2	5-26, 29
14	Newton's second law	2	5-29, 32
15	Newton's second law	2	5-26, 32
16	Newton's second law	2	5-26, 29

6 APPLICATIONS OF NEWTON'S LAWS OF MOTION

Applications

RECALL FROM PREVIOUS CHAPTERS

Previously learned concepts and equations frequently used in this chapter	Text Section	Study Guide Page
Resolving vectors into components analytically	2-4	2-8
Adding vectors analytically	2-4	2-8
Central acceleration: $a_c = v^2 / R$	4-3	4-9
Newton's second law: $\Sigma F = Ma$	5-4	5-5

NEW IDEAS IN THIS CHAPTER

Concepts and equations introduced	Text Section	Study Guide Page
Normal force: $\mathbf{F_N}$	6-1	6-1
Static and kinetic friction: $f_s \leq \mu_s F_N$, $f_k = \mu_k F_N$	6-1	6-3
Contact force: $F_c = (F_N^2 + f^2)^{1/2}$	6-1	6-3
Central force: $F_c = Mv^2 / R$	6-2	6-7

PRINCIPAL CONCEPTS AND EQUATIONS

1 Normal Force for Different Situations (Section 6-1)

Review: The normal force $\mathbf{F_N}$ is the perpendicular force that a surface exerts on an object with which it is in contact.

Practice: Consider the following three situations.

The mass M is being pulled across the frictionless floor

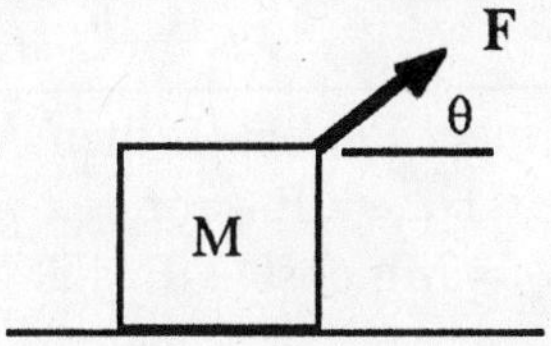

Known: M, F, θ, g

The mass M is being pushed across the frictionless floor

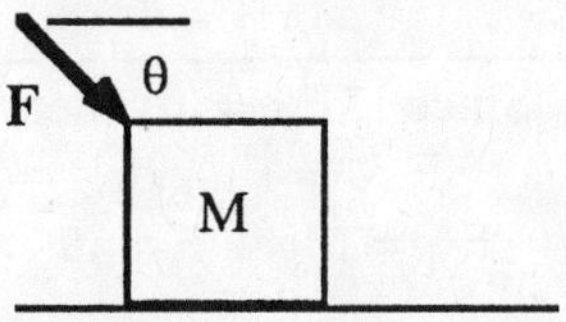

Known: M, F, θ, g

The mass M is being pushed up the frictionless ramp

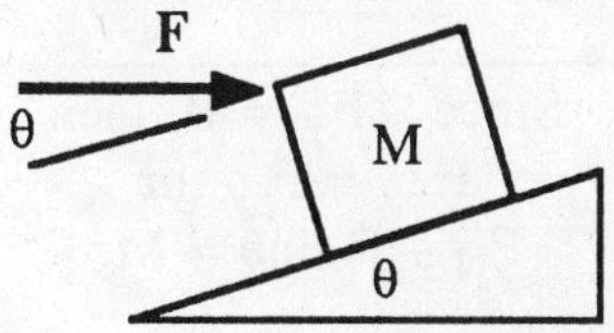

Known: M, F, θ, g

1. For each situation, draw a figure showing all forces acting on M and a convenient coordinate system.

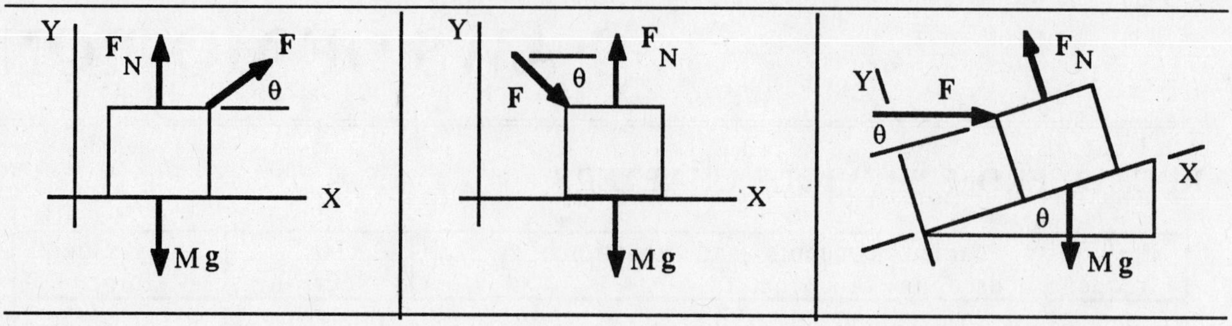

2. For each situation, draw a free-body diagram with all the forces resolved into components.

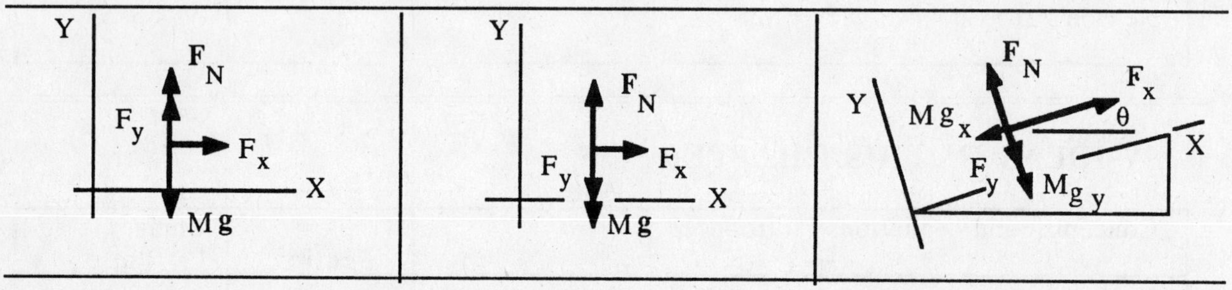

Now determine the following:

3. The total force in the -y direction for each situation

Mg	Mg + F sinθ	Mg cosθ + F sinθ

4. The total force in the +y direction for each situation

F_N + F sinθ	F_N	F_N

5. The net force in the y direction for each situation (assuming M does not leave the surface)

$\Sigma F_y = 0$	$\Sigma F_y = 0$	$\Sigma F_y = 0$

6. Any relationship that exists between the total force in the y direction and the total force in the -y direction

Since $\Sigma F_y = 0$, then $F_{+y} = F_{-y}$ or $F_N + F \sin\theta = Mg$	Since $\Sigma F_y = 0$, then $F_{+y} = F_{-y}$ or $F_N = Mg + F \sin\theta$	Since $\Sigma F_y = 0$, then $F_{+y} = F_{-y}$ or $F_N = Mg \cos\theta + F \sin\theta$

7. An expression for the normal force for each situation

$F_N = Mg - F \sin\theta$	$F_N = Mg + F \sin\theta$	$F_N = Mg \cos\theta + F \sin\theta$
The surface doesn't have to support the full weight of the object since **F** has an upward component.	The surface must support not only the weight of the object but also the downward component of **F**.	The surface must support a component of the weight and a component of **F**.

Related Text Exercises: Ex. 6-1 through 6-19.

2 Static and Kinetic Friction (Section 6-1)

Review: Friction always acts so as to oppose the start or the continuance of motion. Static friction opposes the start of motion and is given by

$$f_s \leq \mu_s F_N$$

Kinetic friction opposes the continuance of motion and is given by

$$f_k = \mu_k F_N$$

where μ_s and μ_k are the coefficients of static and kinetic friction.

Practice:

A 100-kg crate sits at rest on the floor and a horizontal force **F** is applied as shown in the figure to the right.
$\mu_s = 0.400$, $\mu_k = 0.200$, $M = 100$ kg

1. Draw a figure showing all forces acting on the crate and a convenient coordinate system.

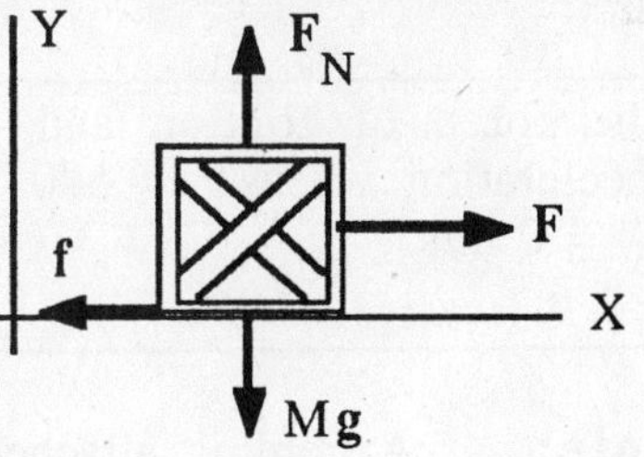

2. Draw a free-body diagram with all forces resolved into components.

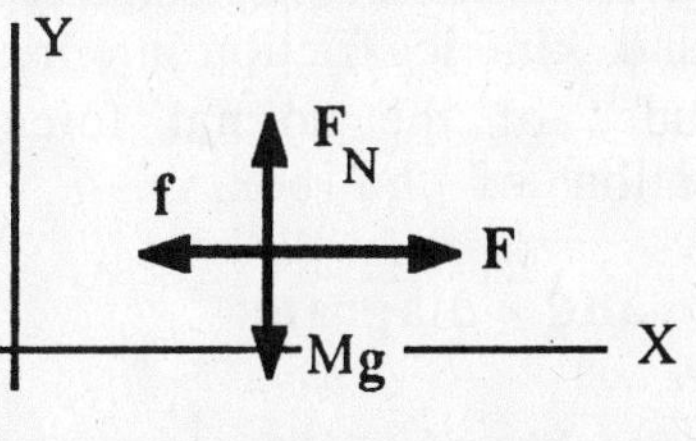

Now determine the following:

3. The normal force the floor exerts on the crate	Since the crate is not changing its motion in the y direction, we know that $\Sigma F_y = 0$, hence $F_N = Mg = (100\ kg)(9.80\ m/s^2) = 980\ N$
4. The maximum force of static friction	$f_{s\text{-}max} = \mu_s F_N = (0.400)(980\ N) = 392\ N$
5. The force of friction when F = 100 N	$f_s \leq \mu_s F_N = 392\ N$ That is, f_s can have any value between 0 and 392 N. In this case, its value needs to be only 100 N in order to keep the crate stationary.
6. The acceleration of the crate when F = 100 N	$\Sigma F_x = F_x - f_s = 100\ N - 100\ N = 0\ N$ $Ma_x = \Sigma F_x = 0$, hence $a_x = 0\ m/s^2$
7. The force of friction and the acceleration when F = 200 N	The force f_s can have any value between 0 and 392 N. Since F = 200 N, we need $f_s = 200$ N. Since $F = f_s$ and they are in opposite directions, $\Sigma F_x = 0$ and hence $a_x = 0$.
8. The force of friction and the acceleration when F = 392 N if $v_0 = 0$	Since F = 392 N, we need $f_s = 392$ N, which gives $\Sigma F_x = 0$ and $a_x = 0$.
9. The force of friction and the acceleration when F = 392 N if $v_0 = 0.100$ m/s	Since the object is moving, use μ_k. $f_k = \mu_k F_N = (0.200)(980\ N) = 196\ N$ $\Sigma F_x = F - f_k = 392\ N - 196\ N = 196\ N$ $a_x = \Sigma F_x / M = 196\ N / 100\ kg = 1.96\ m/s^2$
10. The force of friction and the acceleration when F = 500 N if $v_0 = 0$ m/s	$f_k = \mu_k F_N = (0.200)(980\ N) = 196\ N$ $\Sigma F_x = F - f_k = 500\ N - 196\ N = 304\ N$ $a_x = \Sigma F_x / M = 304\ N / 100\ kg = 3.04\ m/s^2$

Example: A cord is attached to a 20.0 kg box and pulled upward at an angle of 30.0° relative to the horizontal. The tension in the cord is 100 N. The coefficients of static and kinetic friction are respectively $\mu_s = 0.200$ and $\mu_k = 0.100$. Determine the magnitudes of the normal force, the frictional force, the contact force and the acceleration of the box.

Given and diagram:

$M = 20.0$ kg $\mu_s = 0.200$
$\theta = 30.0°$ $\mu_k = 0.100$
$F_T = 100$ N

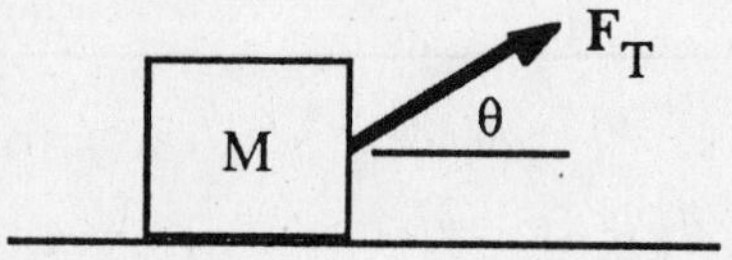

Determine: F_N = The normal force acting on the box
f = The frictional force acting on the box
F_c = The contact force acting on the box
a = The acceleration of the box

Strategy: Draw a free-body diagram showing all forces acting on the box. Pick a convenient coordinate system (+x in the direction of any anticipated motion) and resolve all forces into components along this coordinate system. Determine F_N from the vertical components of the forces. Determine if the box moves by comparing the horizontal component of F_T to the maximum static force of friction. Once we know whether or not the box moves, we can determine the static or kinetic frictional force. Knowing F_N and f, we can determine F_c. Finally, by summing the forces in the x direction we can determine the net force moving the box and hence its acceleration.

Solution:

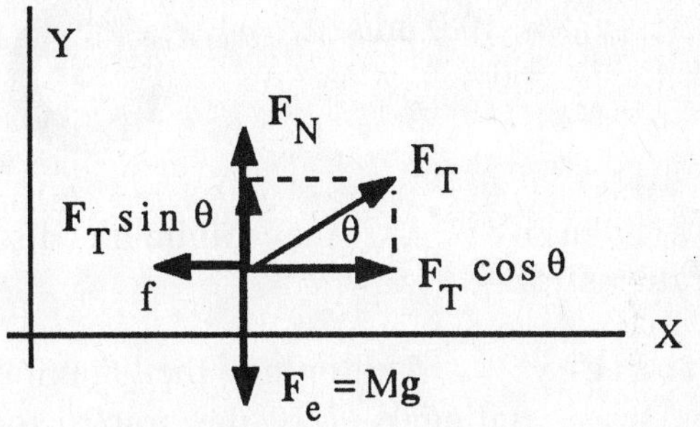

$$\Sigma F_y = F_N + F_T \sin\theta - Mg = 0$$

or

$$F_N = Mg - F_T \sin\theta = (20.0 \text{ kg})(9.80 \text{ m/s}^2) - 100 \text{ N} \sin 30.0° = 146 \text{ N}$$

The horizontal component of the tension is

$$F_T \cos\theta = 100 \text{ N} \cos 30.0° = 86.6 \text{ N}$$

The maximum static friction is

$$f_{s\text{-max}} = \mu_s F_N = (0.200)(146 \text{ N}) = 29.2 \text{ N}$$

Since the horizontal component of the tension is greater than the maximum static frictional force, the box moves and the kinetic frictional force is

$$f_k = \mu_k F_N = (0.100)(146 \text{ N}) = 14.6 \text{ N}$$

The magnitude of the contact force is

$$F_c = (f_k{}^2 + F_N{}^2)^{1/2} = [\,(14.6 \text{ N})^2 + (146 \text{ N})^2\,]^{1/2} = 147 \text{ N}$$

The net force acting to accelerate the box across the horizontal surface is

$$\Sigma F_x = F_T \cos\theta - f_k = 86.6 \text{ N} - 14.6 \text{ N} = 72.0 \text{ N}$$

The acceleration of the box is

$$a_x = F_x / M = 72.0 \text{ N} / 20.0 \text{ kg} = 3.60 \text{ m/s}^2$$

Example: A 100-kg crate sits on a flatbed truck as shown in the figure below. The truck is traveling 10.0 m/s and the coefficient of static and kinetic friction between the crate and the truckbed are $\mu_s = 0.300$ and $\mu_k = 0.200$. What is the minimum distance in which the truck can stop so that the crate will not slide?

Given:

$M = 100$ kg
$g = 9.80$ m/s^2
$\mu_s = 0.300$
$\mu_k = 0.200$
$v_o = 10.0$ m/s
$v_f = 0$

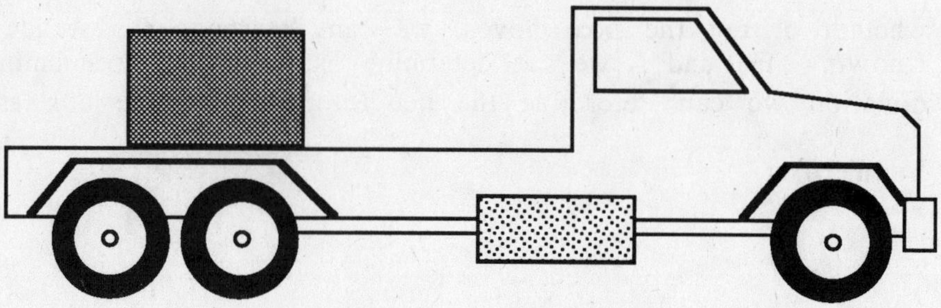

Determine: Δx, the minimum distance in which the truck can stop without the crate sliding.

Strategy: Determine the maximum force ($f_{s\text{-}max}$) static friction can exert to keep the crate stationary. If the truck stops so fast that a force bigger than $f_{s\text{-}max}$ is needed to keep the crate stationary, it will slide forward; consequently $f_{s\text{-}max}$ dictates the maximum deceleration (-a). Use $f_{s\text{-}max}$ in Newton's second law to determine this maximum deceleration. Knowing v_o, v and a_{max}, we can use our knowledge of kinematics ($v^2 - v_o{}^2 = 2a\Delta x$) to determine the minimum distance (Δx) the truck can stop in without the crate sliding forward.

Solution: The normal force the truckbed exerts on the crate is

$$F_N = Mg = (100 \text{ kg})(9.80 \text{ m/s}^2) = 980 \text{ N}$$

The maximum force static friction can exert to hold the crate stationary is

$$f_{s\text{-}max} = \mu_s F_N = (0.300)(980 \text{ N}) = 294 \text{ N}$$

The maximum deceleration this force can provide is obtained from Newton's second law:

$$a_f = -f_{s\text{-}max} / M = -294 \text{ N} / 100 \text{ kg} = -2.94 \text{ m/s}^2$$

If the truck decelerates more than this, the crate will slide. Since we don't want it to slide, we must require that

$$a_{Tmax} = a_f = -2.94 \text{ m/s}^2$$

The distance the truck can stop in with this deceleration is

$$v^2 - v_0^2 = 2a\Delta x, \quad \text{where} \quad v = 0, \quad v_0 = 10.0 \text{ m/s} \quad \text{and} \quad a = -2.94 \text{ m/s}^2; \quad \text{hence}$$

$$\Delta x = (v^2 - v_0^2) / 2a = -(10.0 \text{ m/s})^2 / (2)(-2.94 \text{ m/s}^2) = 17.0 \text{ m}$$

Related Text Exercises: Ex. 6-1 through 6-19.

3 Uniform Circular Motion (Section 6-2)

Review: If an object is traveling in a circle of radius R with a constant speed v (i.e. uniform circular motion), it must be experiencing a central or centripetal acceleration of magnitude

$$a_c = v^2 / R$$

The acceleration is the result of the velocity vector continually changing direction. If the object has a mass m, it must be experiencing a net central or centripetal force of

$$|\Sigma F| = F_c = mv^2 / R$$

If the speed is v and the radius R, a net central force of mv^2 / R will hold it in orbit; a net central force less than mv^2 / R will allow the object to spiral outward, and a net central force greater than mv^2 / R will cause the object to spiral inward.

Note: Up to this point we have represented a net force by ΣF. In this section we are introducing a net force that causes circular motion. For uniform circular motion this net force is directed toward the center of the circle. Subsequently we will call it a net central or centripetal force and represent it by F_c.

Note: When working circular-motion problems it will prove convenient to call any force directed toward the center positive and any force directed away from the center negative.

Note: Centripetal-force problems involve four physical quantities: F_c, M, v and r. You will usually be given three of these quantities and be asked to determine the fourth.

Practice:

For a car that is traveling around a flat circular curve the following quantities are known: g, μ, M and R.

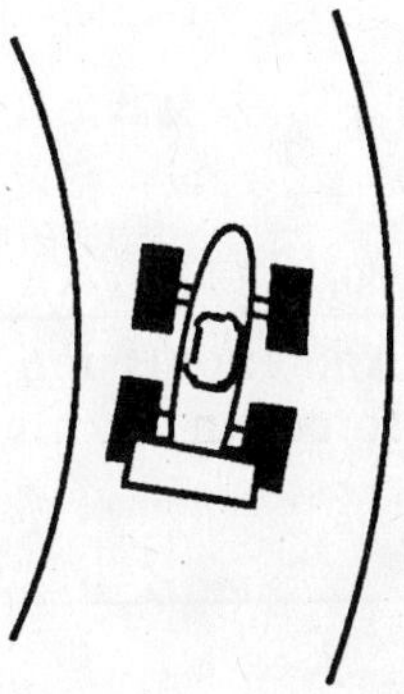

Determine the following:

1. Those forces that provide the centripetal force	Static Friction 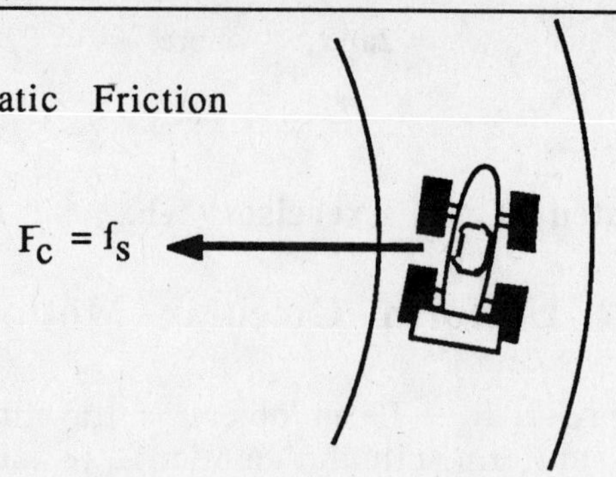
2. An expression for the maximum centripetal force in terms of known quantities	$F_c = f_s = \mu F_N = \mu Mg$
3. An expression for the speed limit of the curve in terms of known quantities	$F_c = \mu Mg$ and $F_c = Mv^2 / R$ Equate these expressions for F_c and solve for v: $v = (\mu gR)^{1/2}$

A conical pendulum is rotating in a horizontal plane, and the following quantities are known: g, L, θ and M.

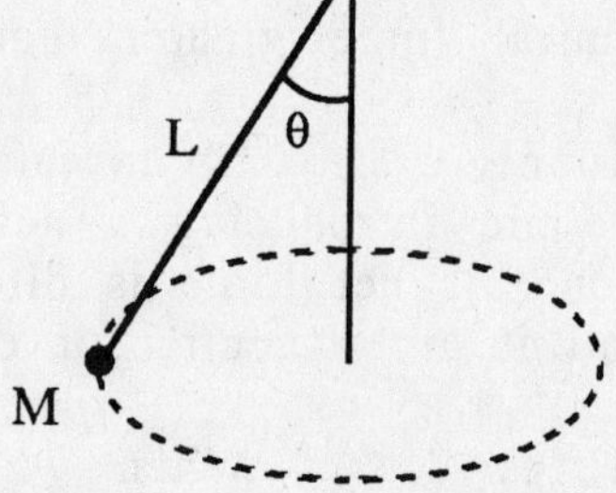

Determine the following:

1. Those forces that provide the centripetal force	The horizontal component of the tension F_T 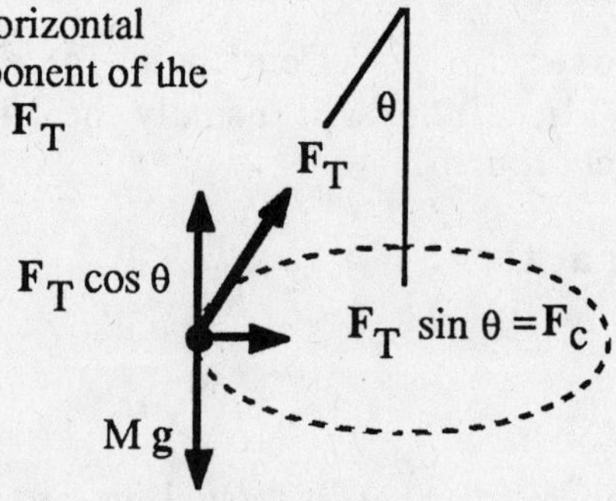
2. An expression for the centripetal force in terms of known quantities	$F_c = F_T \sin\theta$ and $Mg = F_T \cos\theta$ Eliminate F_T to obtain $F_c = Mg \sin\theta / \cos\theta = Mg \tan\theta$

3. An expression for the speed of the orbiting pendulum bob in terms of known quantities	$F_c = Mg \tan\theta$ $F_c = Mv^2 / R = Mv^2 / L \sin\theta$ Equate these expressions for F_c and solve for v: $v = \sin\theta(Lg / \cos\theta)^{1/2}$

For a car traveling around a banked frictionless curve the following quantities are known: g, θ, v and $\mu = 0$.

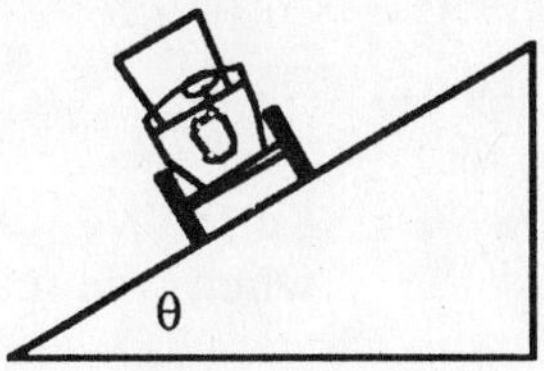

Determine the following:

<table>
<tr><td>1. Those forces that provide the centripetal force</td><td>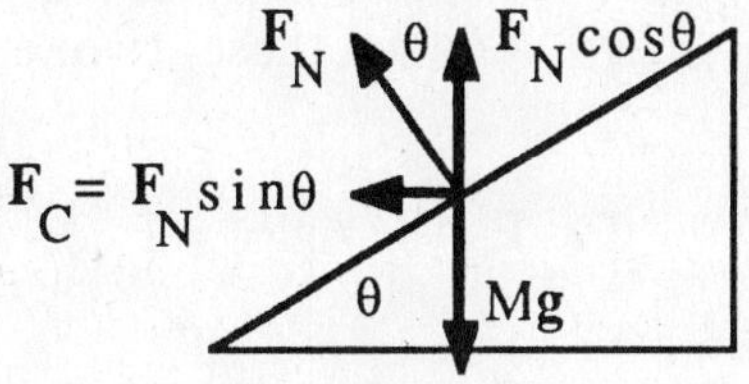

The horizontal component of the normal force provides the centripetal force.</td></tr>
<tr><td>2. An expression for the centripetal force in terms of known quantities</td><td>$F_c = F_N \sin\theta$ and $Mg = F_N \cos\theta$
Eliminating F_N, obtain
$F_c = (Mg / \cos\theta) \sin\theta = Mg \tan\theta$</td></tr>
<tr><td>3. An expression for the radius of curvature of the banked curve in terms of known quantities</td><td>$F_c = Mg \tan\theta$ and $F_c = Mv^2 / R$
Equate these expressions for F_c and solve for R: $R = v^2 / (g \tan\theta)$</td></tr>
</table>

Example: A 400-N girl sits on a bathroom scale while riding a roller coaster. When her car is in a valley that has a radius of curvature of 20.0 m, she notices that the scale reads 600 N. When her car tops a hill of 15.0-m radius, she notices that the scale reads 0 N. What is the speed of her car in the valley and at the top of the hill?

Given and Diagram:

$F_e = 400$ N
$R_v = 20.0$ m
$R_h = 15.0$ m
$F_{Nv} = 600$ N
$F_{Nh} = 0$ N
$g = 9.80$ m/s^2

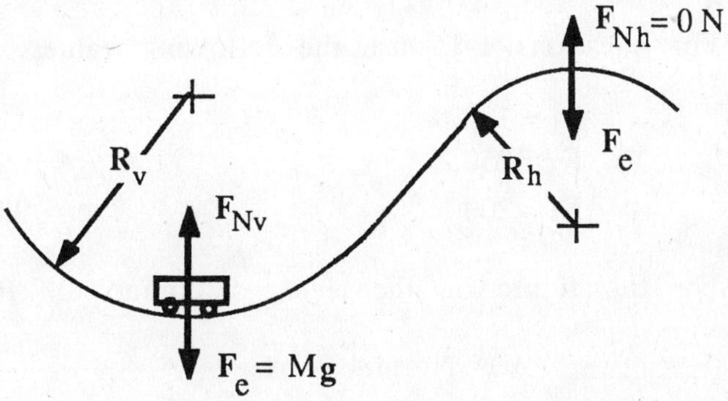

Determine: V_v and V_h

Strategy: Write an expression for the centripetal force in the valley in terms of F_{Nv} and F_e. Then equate this to MV_v^2 / R_v and solve for V_v. Write an expression for the centripetal force on the hill in terms of F_{Nh} and F_e. Then equate this to MV_h^2 / R_h and solve for V_h.

Solution:

Valley: $F_c = F_{Nv} - F_e$, where F_{Nv} = the scale reading = the normal force on the girl when the car is in the valley

$$F_c = MV_v^2 / R_v = F_e V_v^2 / gR_v$$

Equate these two expressions for F_c and solve for V_v:

$$F_e V_v^2 / gR_v = F_{Nv} - F_e \quad \text{or} \quad V_v = \left[gR_v[(\frac{F_{Nv}}{F_e}) - 1] \right]^{1/2} = 9.90 \text{ m/s}$$

Hill: $F_c = F_e - F_{Nh}$, where F_{Nh} = the scale reading = the normal force on the girl when the car is on the hill

$$F_c = MV_h^2 / R_h = F_e V_h^2 / gR_h$$

Equate these two expression for F_c and solve for V_h:

$$F_e V_h^2 / gR_h = F_e \text{ (since } F_{Nh} = 0) \quad \text{or} \quad V_h = (gR_h)^{1/2} = 12.1 \text{ m/s}$$

Related Text Exercises: Ex. 6-22, 6-23 and 6-25 through 6-33.

PRACTICE TEST

Take and grade this practice test. Doing so will allow you to determine any weak spots in your understanding of the concepts taught in this chapter. The following section prescribes what you should study further to strengthen your understanding.

For questions 1-15 use the following values:

$F_1 = 100$ N	$\mu_s = 0.200$	M = 100 kg
$F_2 = 250$ N	$\mu_k = 0.100$	$\theta = 30.0°$
$F_3 = 200$ N	$g = 9.80$ m/s^2	

For the figure on the right, determine the following:

_______ 1. Normal force

_______ 2. Force of friction

_______ 3. Net force in the x direction

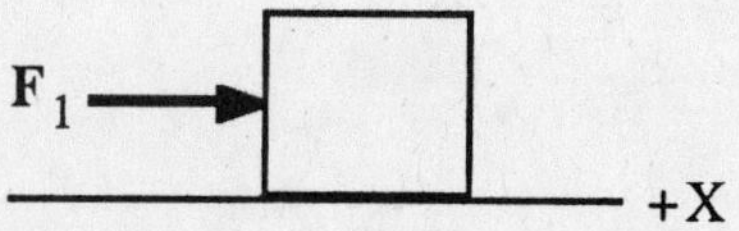

For the figure on the right, determine the following:

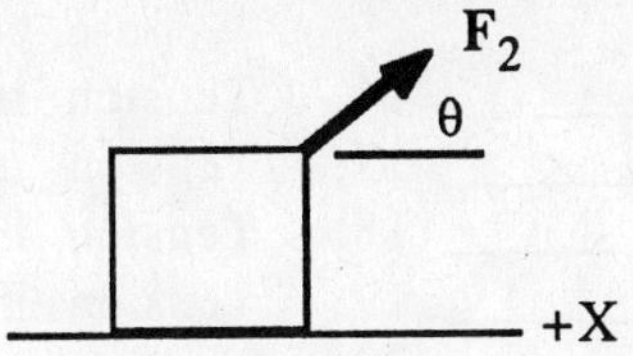

_________ 4. Normal force
_________ 5. Force of friction
_________ 6. Net force in the x direction

For the figure on the right, determine the following:

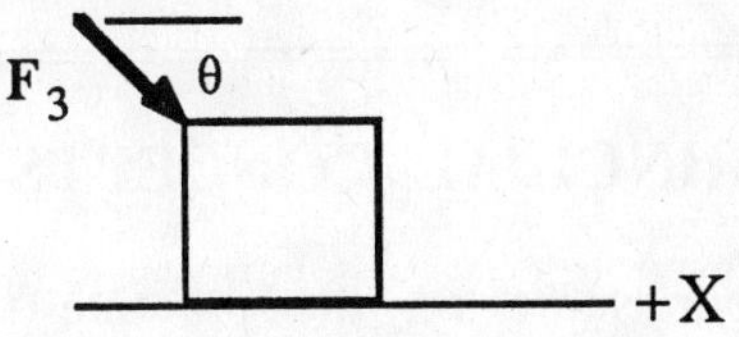

_________ 7. Normal force
_________ 8. Force of friction
_________ 9. Net force in the x direction

For the figure on the right, determine the following:

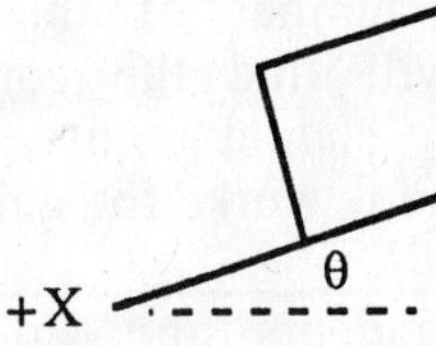

_________ 10. Normal force
_________ 11. Force of friction
_________ 12. Net force in the x direction

For the figure on the right, determine the following:

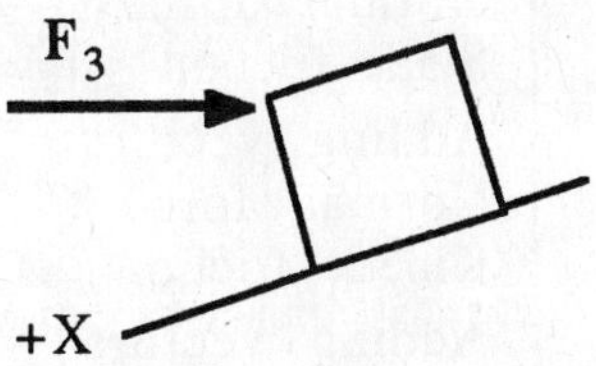

_________ 13. Normal force
_________ 14. Force of friction
_________ 15. Net force in the x direction

An object tied to a cord swings in a vertical circle. In general, the speed of the object will vary, and we know its value at the five labeled positions.

Given:

$r = 1.00$ m		$v_1 = 6.00$ m/s	
$M = 0.500$ kg		$v_2 = 5.50$ m/s	
$\theta = 30.0°$		$v_3 = 5.00$ m/s	
$\gamma = 60.0°$		$v_4 = 4.50$ m/s	
$g = 9.80$ m/s^2		$v_5 = 4.00$ m/s	

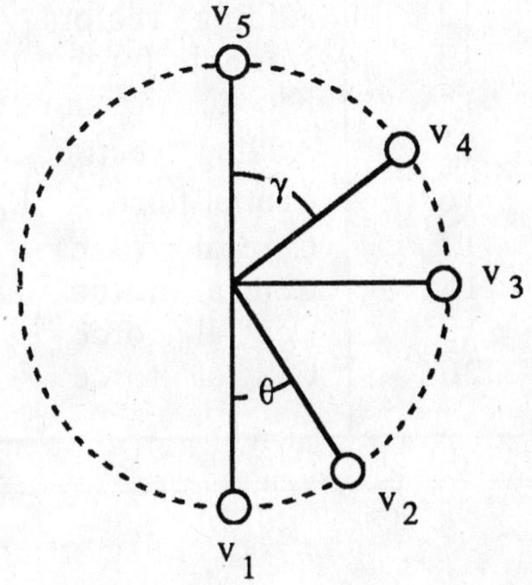

Based on this information determine the following:

_________ 16. Tension in the cord for position 1
_________ 17. Tension in the cord for position 2
_________ 18. Tension in the cord for position 3
_________ 19. Tension in the cord for position 4
_________ 20. Tension in the cord for position 5

(See Appendix I for answers.)

PRINCIPAL CONCEPTS AND EQUATIONS PRESCRIPTION

Your score on the practice test is an excellent measure of your understanding of the chapter. You should now use the following chart to write your own prescription for dealing with any weaknesses the practice test points out. Look down the leftmost column to the number of the question(s) you answered incorrectly, reading across that row you will find the concept and/or equation of concern, the section(s) of the study guide you should return to for further study, and some suggested text exercises which you should work to gain additional experience.

Practice Test Questions	Concepts and Equations	Prescription	
		Principal Concepts	Text Exercises
1	Normal force	1	6-1, 2
2	Static friction $f_s = 0 \rightarrow \mu_s F_N$	2	6-3, 4
3	Adding vectors - analytical	3 of Ch 2	2-15, 16
4	Normal force	1	6-4, 15
5	Kinetic friction $f_k = \mu_k F_N$	2	6-4, 15
6	Adding vectors - analytical	3 of Ch 2	2-18, 19
7	Normal force	1	6-16
8	Static friction $f_s = 0 \rightarrow \mu_s F_N$	2	6-16
9	Adding vectors - analytical	3 of Ch 2	2-20, 21
10	Normal force	1	6-14
11	Kinetic friction $f_k = \mu_k F_N$	2	6-14
12	Adding vectors - analytical	3 of Ch 2	2-15, 18
13	Normal force	1	6-15
14	Kinetic friction $f_k = \mu_k F_N$	2	6-15
15	Adding vectors - analytical	3 of Ch 2	2-16, 19
16	Central force $F_c = M_v^2 / R$	3	6-22, 23
17	Central force	3	6-25, 26
18	Central force	3	6-27, 28
19	Central force	3	6-29, 30
20	Central force	3	6-31, 32

7 NEWTON'S LAW OF UNIVERSAL GRAVITATION

RECALL FROM PREVIOUS CHAPTERS

Previously learned concepts and equations frequently used in this chapter	Text Section	Study Guide Page
Analytical treatment of vectors (unit vectors, components, addition)	2-4	2-8
Average speed: $\bar{v} = d/t$	3-2	3-5
Uniform circular motion: $a_c = v^2/r$	4-3	4-9
Newton's second law of motion: $\mathbf{F} = m\mathbf{a}$	5-4	5-5
Newton's third law of motion: $\mathbf{F}_{AB} = -\mathbf{F}_{BA}$	5-5	5-7

NEW IDEAS IN THIS CHAPTER

Concepts and equations introduced	Text Section	Study Guide Page
Newton's Law of Gravitation: $\mathbf{F}_{21} = Gm_1m_2\hat{\mathbf{r}}/r^2$	7-1	7-1
Acceleration due to gravity on the surface of planet x: $g_x = Gm_x/R_x^2$	7-1, 2, 4	7-5
Acceleration due to gravity at a distance r from the center of planet x: $g_x(r) = Gm_x/r^2 = g_xR_x^2/r^2$	7-1, 2, 4	7-5
Gravitational field: $\mathbf{g}_1(r) = \mathbf{F}_{12}/m_2 = -(Gm_1/r^2)\hat{\mathbf{r}}$	7-5	7-10

PRINCIPAL CONCEPTS AND EQUATIONS

1 Newton's Law of Universal Gravitation for Particles
(Section 7-1)

Review: Suppose we have two particles of mass m_1 and m_2 a distance r apart as shown in Figure 7.1.

Figure 7.1

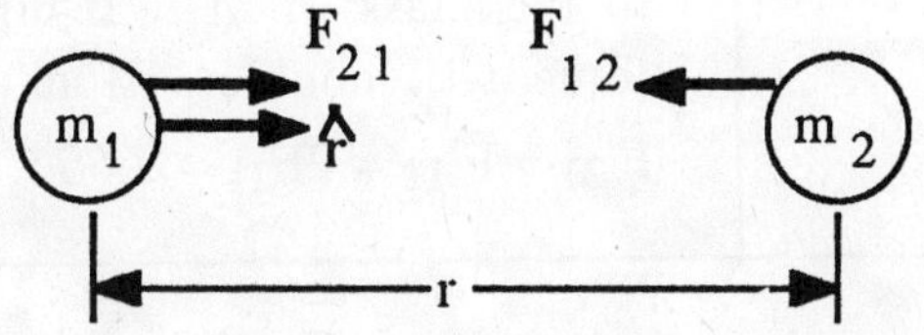

$\mathbf{F}_{21}$ = force by 2 on 1
$\mathbf{F}_{12}$ = force by 1 on 2

We are at liberty to place the origin of our coordinate system any place that we find convenient. When working problems involving the interaction of two particles, it is convenient to place the origin of the coordinate system on one of the particles. Let's arbitrarily place the origin on m_1, hence the unit vector $\hat{r}$ originates on m_1.

In Figure 7.1 particles 1 and 2 attract each other with a force equal in magnitude but opposite in direction.

$$\mathbf{F}_{21} = -\mathbf{F}_{12}$$

This equation expresses the fact that the force by 2 on 1 ($\mathbf{F}_{21}$) is equal in magnitude but opposite in direction (indicated by the minus sign) to the force by 1 on 2 ($\mathbf{F}_{12}$).

Newton determined that the gravitational force of attraction depends on the mass of each particle and their separation as follows:

$$\mathbf{F}_{21} = Gm_1m_2\hat{r}/r^2 \quad \text{and} \quad \mathbf{F}_{12} = -Gm_1m_2\hat{r}/r^2$$

We can use this expression for the gravitational force of attraction any time the objects are small compared to their separation.

Practice: Consider the situation shown in Figure 7.2.

Figure 7.2

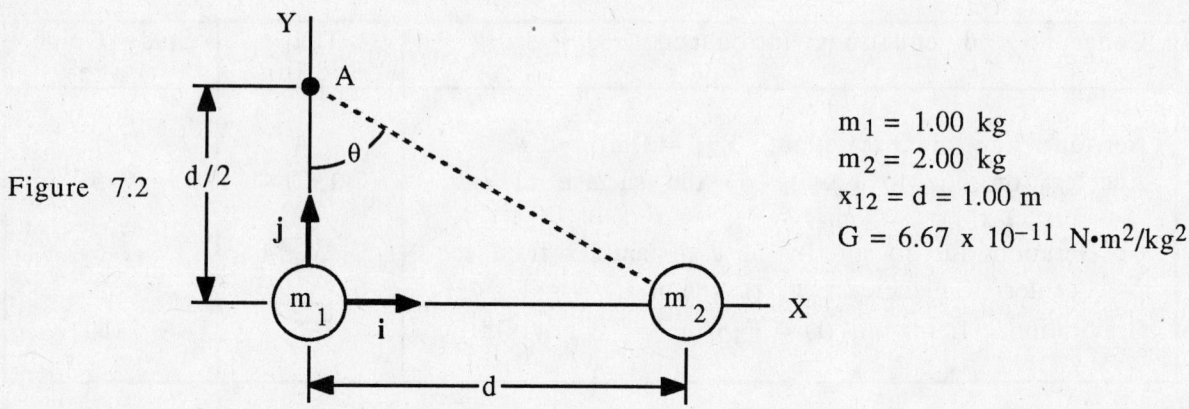

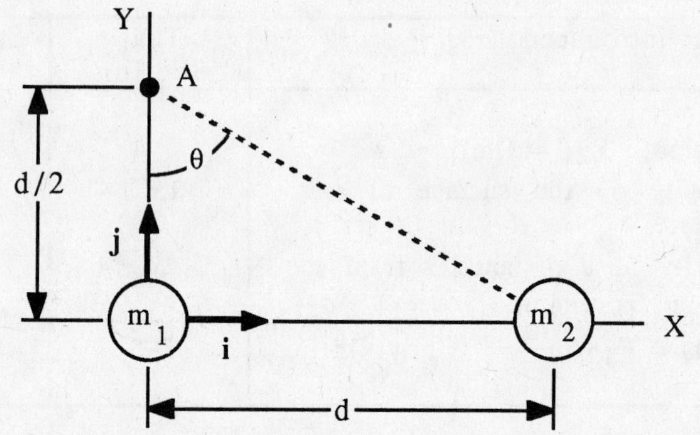

$m_1 = 1.00$ kg
$m_2 = 2.00$ kg
$x_{12} = d = 1.00$ m
$G = 6.67 \times 10^{-11}$ N•m^2/kg^2

Determine the following:

1. The force by 2 on 1 ($\mathbf{F}_{21}$)	$\mathbf{F}_{21} = (Gm_1m_2/d^2)\mathbf{i}$ $\mathbf{F}_{21} = (1.33 \times 10^{-10}\text{ N})\mathbf{i}$
2. The force by 1 on 2 ($\mathbf{F}_{12}$)	$\mathbf{F}_{12} = -\mathbf{F}_{21} = (-1.33 \times 10^{-10}\text{ N})\mathbf{i}$
3. The magnitude of the gravitational attractive force if m_1 doubles in mass ($m'_1 = 2m_1$)	The force is directly proportional to the mass of m_1. If m_1 doubles, the force will double. If $m'_1 = 2m_1$, then $F'_{21} = F'_{12} = 2F_{21} = 2.66 \times 10^{-10}$ N

4. The magnitude of the gravitational force if m_2 triples in mass ($m'_2 = 3m_2$)	The force is directly proportional to the mass of m_2. If m_2 triples, the force will triple. If $m'_2 = 3m_2$, then $F'_{21} = F'_{12} = 3F_{21} = 3.99 \times 10^{-10}$ N
5. The magnitude of the gravitational force if the value of d is quadrupled ($d' = 4d$)	The force is inversely proportional to the square of the separation (d). If $d' = 4d$, then $(d')^2 = 16d^2$ and $F'_{21} = F'_{12} = F_{21}/16 = 8.31 \times 10^{-12}$ N

Note: One of the skills you are to acquire from this course is the ability to look at an equation, translate it into English (i.e. tell what is says), and make predictions (i.e. answer the what if questions). For a straight forward expression (such as Newton's Law of Gravitation) this may be done in an intuitive manner by just looking at the expression (as we did in steps 3, 4 and 5 above). For more difficult equations, we may have to approach the problem analytically. In order to see how this is done, let's do steps 3, 4 and 5 analytically.

$m'_1 = 2m_1 \quad F'_{21} = Gm'_1m_2/d^2 = G(2m_1)m_2/d^2 = 2(Gm_1m_2/d^2] = 2F_{21}$

$m'_2 = 3m_2 \quad F'_{21} = Gm_1m'_2/d^2 = Gm_1(3m_2)/d^2 = 3(Gm_1m_2/d^2) = 3F_{21}$

$d' = 4d \quad F'_{21} = Gm_1m_2/(d')^2 = Gm_1m_2/(4d)^2 = (Gm_1m_2/d^2)/16 = F_{21}/16$

6. The magnitude of the gravitational force if $m'_1 = 2m_1$, $m'_2 = 3m_2$ and $d' = 4d$	First let's take an intuitive gravitational approach: $m'_1 = 2m_1$ causes the force to double $m'_2 = 3m_2$ causes the force to triple $d' = 4d$ causes the force to decrease by a factor of 16, hence $F'_{21} = (2 \cdot 3/16)F_{21} = 4.99 \times 10^{-11}$ N Second let's take an analytical approach: $F'_{21} = Gm'_1m'_2/(d')^2$ $= G(2m_1)(3m_2)/(4d)^2$ $= (2 \cdot 3/16)Gm_1m_2/d^2$ $= 0.375\ F_{21} = 4.99 \times 10^{-11}$ N
7. The position on the line between m_1 and m_2 where a particle of mass m_4 would feel no net force	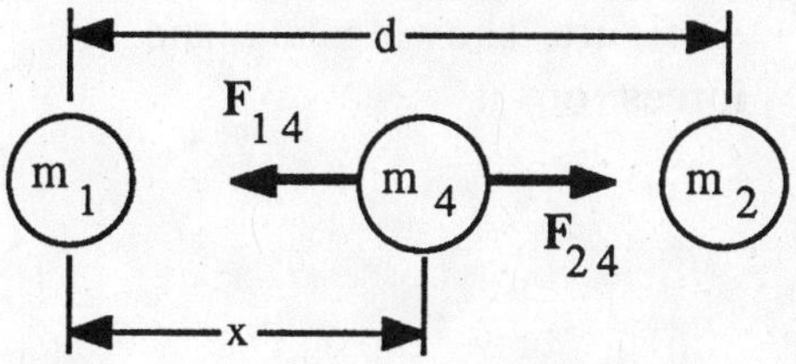

	The particle m_4 will feel no net force when: $$F_{14} = F_{24}$$ $$Gm_1m_4/x^2 = Gm_2m_4/(d - x)^2$$ $$(d - x)^2 = x^2(m_2/m_1)$$ $$d - x = \pm x(m_2/m_1)^{1/2}$$ $$x = d/[1 \pm (m_2/m_1)^{1/2}]$$ $$= 0.414 \text{ m, or } -2.41 \text{ m}$$ This says that m_4 will feel no net force when it is 0.414 m to the right of m_1 and when it is 2.41 m to the left of m_1. Since we want the position between m_1 and m_2, we choose $x = 0.414$ m.

Now introduce a third mass (m_3 = 3.00 kg), position it at site A, and then determine the following:

8. The force by 2 on 1 ($\mathbf{F}_{21}$) and the force by 1 on 2 ($\mathbf{F}_{12}$)	The introduction of this new mass does not effect the interaction of m_1 and m_2 with each other. It will create a new net force on m_1 and m_2 but it will not effect $\mathbf{F}_{21}$ and $\mathbf{F}_{12}$. Hence we still have $\mathbf{F}_{21} = -\mathbf{F}_{12} = (1.33 \times 10^{-10}\ \text{N})\mathbf{i}$
9. The force by 1 on 3 ($\mathbf{F}_{13}$)	$\mathbf{F}_{13} = -[Gm_1m_3/(d/2)^2]\ \mathbf{j}$ $= (-8.00 \times 10^{-10}\ \text{N})\mathbf{j}$
10. The distance between m_3 and m_2 (x_{23})	$x_{23} = (x_{12}^2 + x_{13}^2)^{1/2}$, $x_{12} = d$, $x_{13} = d/2$ $x_{23} = (5/4)^{1/2}\ d = 1.12$ m
11. The magnitude of the force by 2 on 3 (F_{23})	$F_{23} = Gm_2m_3/x_{32}^2 = 3.19 \times 10^{-10}$ N
12. The angle between x_{13} and x_{12} (θ)	$\theta = \tan^{-1}(x_{12}/x_{13}) = 63.4°$
13. A figure showing m_3 and all forces on it	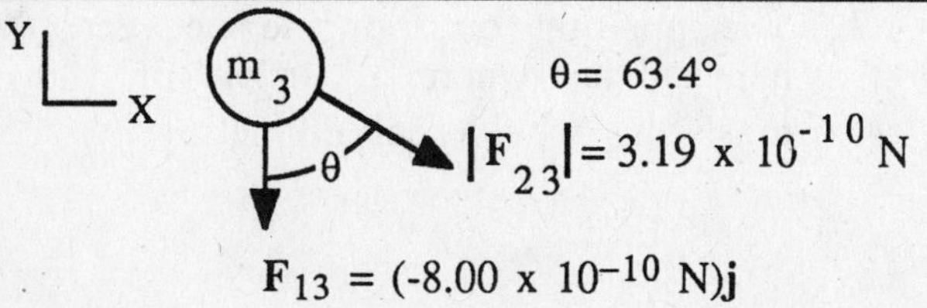
14. The net force on m_3 in the y direction	$\Sigma F_{3y} = -F_{13} - F_{23}\cos\theta$ $= (-8.00 - 3.19\cos 63.4°) \times 10^{-10}$ N $= -9.43 \times 10^{-10}$ N

15. The net force on m_3 in the x direction	$\Sigma F_{3x} = F_{23} \sin\theta = 3.19 \times 10^{-10}$ N $\sin 63.4°$ $= 2.85 \times 10^{-10}$ N
16. Magnitude of the net force on m_3	$F_3 = (F_{3x}^2 + F_{3y}^2)^{1/2} = 9.85 \times 10^{-10}$ N
17. Direction of the net force on m_3	Let's indicate the direction of $\mathbf{F}_3$ by giving the angle between $\mathbf{F}_3$ and x_{13} $\alpha = \tan^{-1}(F_{3x}/F_{3y}) = 16.8°$
18. A figure showing m_3, $\mathbf{F}_{13}$, $\mathbf{F}_{23}$ and $\mathbf{F}_3$	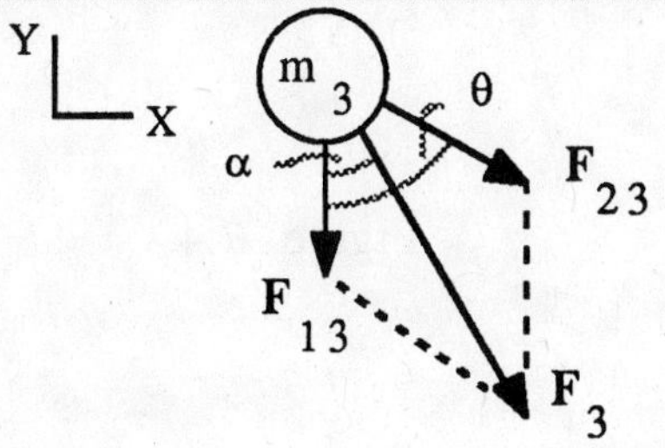 Notice that $\mathbf{F}_{13} + \mathbf{F}_{23} = \mathbf{F}_3$
19. Magnitude of the initial acceleration of m_3, if it is released	$a_3 = \Sigma F_3/m_3 = 3.28 \times 10^{-10}$ m/s^2

Related Text Exercises: Ex. 7-4 through 7-12, 7-16 through 7-21, 7-23 and 7-24.

2 Newton's Law of Universal Gravitation and a Planet Satellite System (Sections 7-1, 7-2 and 7-4)

Review: We have seen that Newton's Law of Universal Gravitation between two particles of mass m_1 and m_2 separated by a distance r (see Figure 7.1) is

$$\mathbf{F}_{21} = -\mathbf{F}_{12} = Gm_1m_2\hat{r}/r^2$$

We can also use this expression for extended objects (such as a planet) if they are spherical. In this case (as we shall see in a later chapter) the extended object acts like a particle of the same mass located at the center of the sphere.

In this section we will use a plane polar coordinate system. As shown in Figure 7.3, this system is convenient when considering the motion of one particle about another in a plane. In our case, the phenomenon of interest depends on the coordinate r and is independent of the coordinate θ.

Figure 7.3

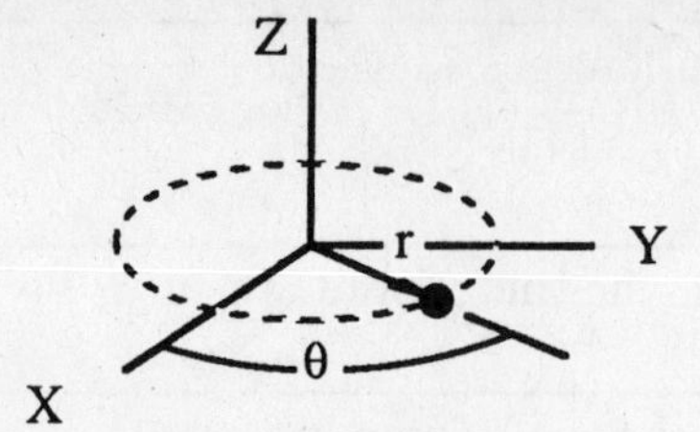

If an object of mass m_o is placed on the surface of the earth (mass m_e and radius R_e), then the earth and object attract each other as shown in Figure 7.4.

Figure 7.4

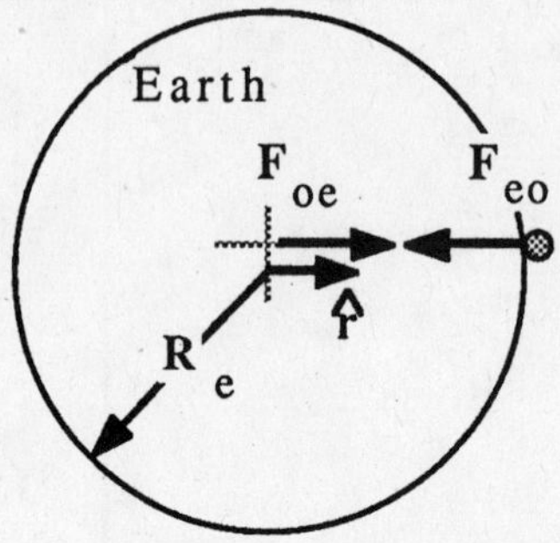

$$F_{oe} = Gm_em_o\hat{r}/R_e^2 \qquad \text{and} \qquad F_{eo} = -Gm_em_o\hat{r}/R_e^2$$

The gravitational attraction of the earth for an object is called the weight of the object and is expressed as

$$F_{eo} = -m_og_e\hat{r}$$

Comparing expressions for the force the earth exerts on the object (F_{eo}), we see that the acceleration due to gravity on the surface of the earth is

$$g_e = Gm_e/R_e^2$$

This says that the acceleration due to gravity on the surface of the earth depends on m_e and R_e but is independent of the mass of the object, hence objects of different mass experience the same acceleration due to gravity.

Note: It is essential that we appreciate the fact that g_e varies with distance from the center of the earth. On the surface of the earth the value for g_e is

$$g_e = Gm_e/R_e^2$$

At a distance r from the center of the earth ($r = R_e + h$, h = altitude above the earth's surface) the value for the acceleration due to gravity is $g_e(r)$, where

$$g_e(r) = Gm_e/r^2 = Gm_eR_e^2/r^2R_e^2 = g_eR_e^2/r^2 \qquad \text{where} \qquad r = R_e + h$$

If an object of mass m_o orbits the earth at a distance r from its center, we may say the following:

(a) The mass of the object on the surface of the earth is m_0
(b) The mass of the object in its orbit is m_0
(c) The weight of the object on the surface of the earth is

$$F_{eo} = m_0 g_e = m_0 G m_e / R_e^2$$

(d) The weight of the object in its orbit is

$$F_{eo}(r) = m_0 g_e(r) = m_0 G m_e / r^2 = m_0 g_e R_e^2 / r^2$$

(e) The central acceleration of the object in its orbit is

$$a_c = g_e(r) = G m_e / r^2 = g_e R_e^2 / r^2$$

(f) The central force on the object in its orbit is

$$F_c = F_{eo}(r) = m_0 a_c = m_0 g_e(r) = m_0 G m_e / r^2 = F_{eo} R_e^2 / r^2$$

(g) The speed of the object in its orbit about the earth is found as follows:

$$F_c = F_{eo}(r)$$

$$m_0 v_e^2 / r = m_0 G m_e / r^2$$

$$v_e = (G m_e / r)^{1/2}$$

(h) The period of the object in this orbit about the earth is found as follows:

$$v_e = 2\pi r / T_e \quad \text{or} \quad T_e = 2\pi r / v_e = 2\pi r^{3/2} / (G m_e)^{1/2}$$

Note: The above expressions may be adapted to any planet x (mass m_x and radius R_x) by substituting m_x and R_x for m_e and R_e respectively.

Practice: Consider three planets (earth, x and y) each with an artificial satellite of mass m_s orbiting at a radius r. The satellite orbits the earth with a period T_e. The information about these planets is as follows:

Planet	Mass	Radius	Surface Gravity
earth	m_e = known	R_e = known	g_e = known
x	$m_x = 2m_e$	$R_x = 2R_e$	g_x = unknown
y	$m_y = m_e/3$	R_y = unknown	$g_y = 2g_e$

Determine the following in terms of m_e, R_e, g_e, m_s, r and T_e.

1. Mass of the satellite on each planet and in orbit about each planet	$(m_s)_e = (m_s)_x = (m_s)_y = m_s$ The mass of the satellite is a constant independent of which planet it is on or its orbital radius.
2. Acceleration due to gravity on the surface of each planet	earth g_e = given as known x $g_x = G m_x / R_x^2 = G(2m_e)/4R_e^2 = G m_e / 2R_e^2 = g_e/2$ y $g_y = 2g_e$ also given as known
3. Weight of the satellite on the surface of each planet	$F_{es} = m_s g_e$ $F_{xs} = m_s g_x = m_s g_e / 2 = F_{es}/2$ $F_{ys} = m_s g_y = m_s(2g_e) = 2F_{es}$

4. Central acceleration of the satellite in its orbit about each planet (i.e. the value of g for the satellite's orbit)	$(a_c)_e = Gm_e/r^2 = g_eR_e^2/r^2$ $(a_c)_x = Gm_x/r^2 = 2Gm_e/r^2 = 2(a_c)_e$ $(a_c)_y = Gm_y/r^2 = Gm_e/3r^2 = (a_c)_e/3$
5. Acceleration due to gravity experienced by the satellite in orbit about each planet	The acceleration due to gravity experienced by the satellite in orbit is just the central acceleration of the satellite. Hence $g_e(r) = (a_c)_e = Gm_e/r^2 = g_eR_e^2/r^2$ $g_x(r) = (a_c)_x = Gm_x/r^2 = 2Gm_e/r^2$ $= 2g_e(r)$ $g_y(r) = (a_c)_y = Gm_y/r^2 = Gm_e/3r^2$ $= g_e(r)/3$
6. Force of attraction the satellite experiences (i.e. its' weight) as it orbits each planet	$F_{es}(r) = m_s(a_c)_e = m_sg_eR_e^2/r^2$ $F_{xs}(r) = m_s(a_c)_x = 2F_{es}(r)$ $F_{ys}(r) = m_s(a_c)_y = F_{es}(r)/3$ The weight of the satellite in its orbit about x and y is respectively twice and one third its weight when in orbit about the earth.
7. Radius of the planet y	Compare the acceleration due to gravity on the surface of the earth and y. $g_y = Gm_y/R_y^2 = Gm_e/3R_y^2$ $g_e = Gm_e/R_e^2$ Given $g_y = 2g_e$, hence $Gm_e/3R_y^2 = 2Gm_e/R_e^2$, or $R_y = R_e/(6)^{1/2}$
8. Orbital speed of the satellite about each planet	Since the central force is supplied by the gravitational attraction, $m_sv_e^2/r = Gm_em_s/r^2$ or $v_e = (Gm_e/r)^{1/2}$ $v_x = (Gm_x/r)^{1/2} = (2Gm_e/r)^{1/2}$ $= (2)^{1/2} v_e$ $v_y = (Gm_y/r)^{1/2} = (Gm_e/3r)^{1/2}$ $= v_e/(3)^{1/2}$
9. Period of the satellite about each planet	T_e, given or by using $v = 2\pi r/T$ obtain $T_e = 2\pi r/v_e = 2\pi(r^3/Gm_e)^{1/2}$ $T_x = 2\pi r/v_x = 2\pi(r^3/2Gm_e)^{1/2}$ $= T_e/(2)^{1/2}$ $T_y = 2\pi r/v_y = 2\pi(3r^3/Gm_e)^{1/2}$ $= T_e(3)^{1/2}$

10. Altitude of the satellite above the surface of each planet	Realizing that the central force is supplied by the gravitational attraction and that $v_e = 2\pi r/T_e$, obtain $r = h_e + R_e = (Gm_eT_e^2/4\pi^2)^{1/3}$, or $h_e = (Gm_eT_e^2/4\pi^2)^{1/3} - R_e$ In like manner $h_x = (Gm_xT_x^2/4\pi^2)^{1/3} - R_x$ $= (Gm_eT_e^2/4\pi^2)^{1/3} - 2R_e$, and $h_y = (Gm_eT_e^2/4\pi^2)^{1/3} - R_e/(6)^{1/2}$
11. The value for the orbital radius about each planet for the weight of the satellite to be one fourth its value in the orbit of radius r.	The weight of the satellite in its orbit is just the gravitational force of attraction and is inversely proportional to the radius of the orbit squared. This force may be decreased by a factor of four by doubling the radius of the orbit ($r_{new} = 2r_{old}$) This result is the same for each planet since the law of gravitation is the same for each planet.

Example: An artificial satellite with a mass of 20.0 kg is set in orbit around the planet Mars. The radius of the orbit is 10.2×10^6 m, the altitude of the orbit is 6.80×10^6 m and the period of revolution for the satellite is 3.12×10^4 s. Determine: (a) the radius of Mars, (b) the gravitational acceleration of the satellite in its orbit, (c) the gravitational acceleration on the surface of Mars, (d) the weight of the satellite on earth, on Mars, and in its orbit about Mars and (e) the mass of the planet Mars.

Given:

$m_s = 20.0$ kg	mass of the satellite
$r = 10.2 \times 10^6$ m	radius of the satellites orbit
$h = 6.80 \times 10^6$ m	altitude of the satellites orbit
$T = 3.12 \times 10^4$ s	period of the satellite

Determine:

R_m	radius of Mars
$g_m(h)$	gravitational acceleration a distance h above the surface of Mars
g_m	gravitational acceleration on the surface of Mars
	Weight of the satellite:
F_{es}	on the surface of the earth,
F_{ms}	on the surface of Mars,
$F_{ms}(r)$	and in orbit about Mars.
M_m	mass of Mars

Strategy:

(a) Knowing r and h, we can determine R_m.

(b) Knowing r and T, we can determine the speed v of the satellite in its orbit. Knowing v and r, we can determine the central acceleration of the satellite in its orbit. Then recognizing that $a_c = g_m(r)$, we have the gravitational acceleration of the satellite in its orbit.

(c) Knowing $g_m(r)$, r and R_m, we can determine g_m.

(d) Knowing g_e, g_m and $g_m(r)$, we can determine the weight of the satellite on earth (F_{es}), on Mars (F_{ms}) and in its orbit about Mars [$F_{ms}(r)$].

(e) Knowing g_m and R_m, we can determine M_m; alternately knowing $g_m(h)$ and r, we can determine M_m.

Solution:

(a) $r = R_m + h$, hence $R_m = r - h = 3.40 \times 10^6$ m

(b) $v = 2\pi r/T$, hence $g_m(r) = a_c = v^2/r = 4\pi^2 r/T^2 = 0.414$ m/s^2

(c) $g_m = GM_m/R_m{}^2$ and $g_m(r) = GM_m/r^2$,
Combining these, obtain $g_m = g_m(r)r^2/R_m{}^2 = 3.73$ m/s^2

(d) $F_{es} = m_s g_e = 196$ N, $F_{ms} = m_s g_m = 74.6$ N, $F_{ms}(r) = m_s g_m(r) = 8.28$ N

(e) $g_m = GM_m/R_m{}^2$, hence $M_m = g_m R_m{}^2/G = 6.46 \times 10^{23}$ kg or
$g_m(r) = GM_m/r^2$, hence $M_m = g_m(r)r^2/G = 6.46 \times 10^{23}$ kg

Related Text Exercises: Ex. 7-3, 7-9, 7-13, 7-22, 7-25 through 7-29, 7-32 through 7-35 and 7-38.

3 The Gravitational Field (Section 7-5)

Review: Shown in Figure 7.5 is a mass m_1 and a small test mass m_o.

Figure 7.5

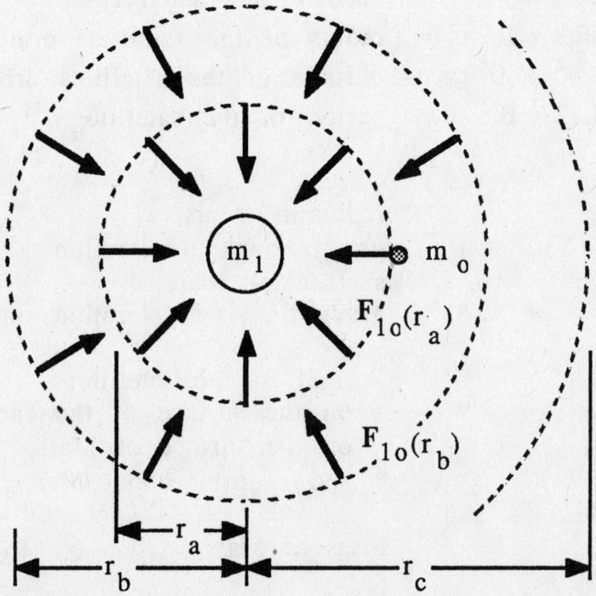

If the test mass is placed on a sphere of radius r_a or r_b it experiences a radially inward force respectively

$$\mathbf{F}_{10}(r_a) = -Gm_1m_0\hat{r}/r_a^2 \quad \text{and} \quad \mathbf{F}_{10}(r_b) = -Gm_1m_0\hat{r}/r_b^2$$

By measuring the gravitational force on m_0 at various locations, we map a gravitational force field about m_1. Notice that this force field has magnitude and direction at all points in space, hence we call it a vector field.

Now suppose we ask several physics students to grab their test mass and measure the force field at r_c. Unfortunately they might all get different answers unless they get together and agree on the size of test mass (m_0) to use in the measurement (as you can see, F depends on m_0). We could eliminate this potential problem by asking each student to determine not the force on m_0 but rather the force per unit mass $\mathbf{F}_{10}(r_c)/m_0$. This will give us a new vector field

$$\mathbf{g}_1(r_c) = \mathbf{F}_{10}(r_c)/m_0 = -(Gm_1m_0\hat{r}/r_c^2)/m_0 = -Gm_1\hat{r}/r_c^2$$

We call this new vector field a gravitational field. Note that the gravitational field depends only on the mass creating the field and how far we are from it.

Practice: Consider the two masses shown in Figure 7.6.

Figure 7.6

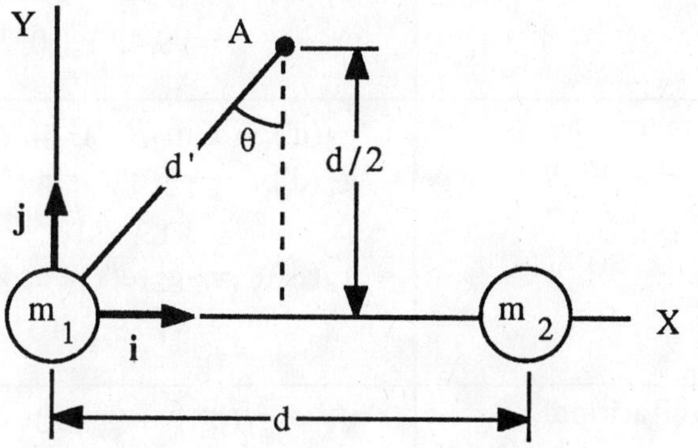

$m_1 = 1.00$ kg
$m_2 = 2.00$ kg
$x_{12} = d = 1.00$ m

Determine the Following:

1. The gravitational field at m_2 due to m_1	$\mathbf{g}_1(d) = -(Gm_1/d^2)\mathbf{i}$ $= -(6.67 \times 10^{-11}\ \text{m/s}^2)\mathbf{i}$
2. The gravitational field at m_1 due to m_2	$\mathbf{g}_2(d) = (Gm_2/d^2)\mathbf{i}$ $= (1.33 \times 10^{-10}\ \text{m/s}^2)\mathbf{i}$
3. Gravitational field at m_2 due to m_1 if m_1 is doubled ($m'_1 = 2m_1$)	If m_1 is doubled, the magnitude of the gravitational field due to m_1 doubles at all points in space $\mathbf{g}'_1(d) = 2\mathbf{g}_1(d) = (-1.33 \times 10^{-10}\ \text{m/s}^2)\mathbf{i}$

4. Gravitational field at m_2 due to m_1 if m_2 is doubled ($m'_2 = 2m_2$)	The gravitational field at m_2 due to m_1 is independent of m_2, hence $\mathbf{g}_1(d)$ remains the same as calculated in step 1. $\mathbf{g}_1(d) = -(6.67 \times 10^{-11}\ m/s^2)\mathbf{i}$
5. Gravitational field at m_2 due to m_1 if d is decreased by a factor of 2 ($d' = d/2$)	$\mathbf{g}_1(d/2) = -(Gm_1/d'^2)\mathbf{i}$ $= -[Gm_1/(d/2)^2]\ \mathbf{i} = \mathbf{g}_1(d)/4$ $= -(1.67 \times 10^{-11}\ m/s^2)\mathbf{i}$
6. The point between m_1 and m_2 where $\mathbf{g} = 0$	Since $\mathbf{g} = \mathbf{F}/m$, $\mathbf{g}$ will be zero when $\mathbf{F}$ is zero. In step 7 of the practice for principal concept 1. of this chapter we found this to be the point 0.414 to the right of m_1.
7. The x and y components of the gravitational field at A due to m_1	$d' = [2(d/2)^2]^{1/2} = d/(2)^{1/2} = 0.707$ m $g_1(d') = Gm_1/(d')^2 = 1.33 \times 10^{-10}\ m/s^2$ $\theta = \tan^{-1}[(d/2)/(d/2)] = \tan^{-1}(1) = 45.0°$ $\mathbf{g}_1(d')_x = -g_1(d')\ \sin 45.0°$ $= -9.40 \times 10^{-11}\ m/s^2$ $\mathbf{g}_1(d')_y = -g_1(d')\ \cos 45.0°$ $= -9.40 \times 10^{-11}\ m/s^2$
8. The x and y components of the gravitational field at A due to m_2	$g_2(d') = Gm_2/(d')^2 = 2.67 \times 10^{-10}\ m/s^2$ $g_2(d')x = g_2(d')\ \sin 45.0°$ $= 1.89 \times 10^{-10}\ m/s^2$ $g_2(d')_y = -g_2(d')\ \cos 45.0°$ $= -1.89 \times 10^{-10}\ m/s^2$
9. Magnitude of the gravitational field at A	$g_x = g_1(d')_x + g_2(d')_x = 9.50 \times 10^{-11}\ m/s^2$ $g_y = g_1(d')_y + g_2(d')_y = -2.83 \times 10^{-10}\ m/s^2$ $g = (g_x{}^2 + g_y{}^2)^{1/2} = 2.99 \times 10^{-10}\ m/s^2$
10. Magnitude of the force a mass m_3 = 300 kg would initially experience if released at A	$g = F/m_3$, hence $F = gm_3 = 8.97 \times 10^{-8}$ N

Related Text Exercises: Ex. 7-28 through 7-32.

PRACTICE TEST

Take and grade this practice test. Doing so will allow you to determine any weak spots in your understanding of the concepts taught in this chapter. The following section prescribes what you should study further to strengthen your understanding.

Consider two masses located on the x-axis as follows:

$m_1 = 2.00$ kg located at $x = 2.00$ m
$m_2 = 4.00$ kg located at $x = 4.00$ m
$m_3 = 3.00$ kg location to be determined

Determine the following:

_________ 1. Force m_1 exerts on m_2
_________ 2. Force m_2 exerts on m_1
_________ 3. If m_1 were to double in mass, the new force m_1 exerts on m_2
_________ 4. If m_2 were to triple in mass, the new force m_1 exerts on m_2
_________ 5. If m_2 is moved to $x = 6.00$ m, the new force m_1 exerts on m_2
_________ 6. If m_1 were to double in mass, m_2 were to triple in mass, and m_2 was moved to $x = 6.00$ m, the new force m_1 exerts on m_2
_________ 7. Location on the x-axis where we could place $m_3 = 3.00$ kg and have it feel no net force

Now let's locate m_3 at the site between m_1 and m_2 where the net gravitational force it will experience is zero.

Determine the following:

_________ 8. Gravitational field at m_3 due to m_1
_________ 9. Gravitational field at m_3 due to m_2
_________ 10. Gravitational field at m_3 due to m_3
_________ 11. Net gravitational field at m_3

A distant planet has a radius 1/4 that of the planet earth ($R_X = R_E/4$) and a mass 1/10 that of the earth ($M_X = M_E/10$).

Determine the following:

_________ 12. Acceleration due to gravity on planet X
_________ 13. Weight of a 70.0 kg astronaut on planet X
_________ 14. If we wish to put a space station into orbit with a period of 2.00×10^3 h around planet X, what is the required size of it's orbit?
_________ 15. How much would the astronaut in question 13 weigh if placed in the orbiting space station of question 14?

(See Appendix I for answers.)

PRINCIPAL CONCEPTS AND EQUATIONS PRESCRIPTION

Your score on the practice test is an excellent measure of your understanding of the chapter. You should now use the following chart to write your own prescription for dealing with any weaknesses the practice test points out. Look down the leftmost column to the number of the question(s) you answered incorrectly, reading across that row you will find the concept and/or equation of concern, the section(s) of the study guide you should return to for further study, and some suggested text exercises which you should work to gain additional experience.

Practice Test Questions	Concepts and Equations	Prescription	
		Principal Concepts	Text Exercises
1	Newton's law of gravitation	1	7-4, 5
2	Newton's law of gravitation	1	7-6, 7
3	Newton's law of gravitation	1	7-8, 9
4	Newton's law of gravitation	1	7-11, 16
5	Newton's law of gravitation	1	7-4, 5
6	Newton's law of gravitation	1	7-6, 7
7	Newton's law of gravitation	1	7-19, 20
8	Gravitational field	3	7-28, 29
9	Gravitational field	3	7-29, 30
10	Gravitational field	3	7-30, 31
11	Adding vectors	3 of Ch 2	2-19, 22
12	Acceleration due to gravity on planet X $g_X = GM_X/R_X^2$	2	7-28, 29
13	Weight on planet X $W_X = mg_X$	2	7-9
14	Relation between size and period of orbit $r = (GM_XT_X^2/4\pi^2)^{1/3}$	2	7-33, 38
15	Weight in orbit $W(r) = mg(r) = mg_XR_X^2/r^2$	2	7-9, 26

8 WORK AND ENERGY

RECALL FROM PREVIOUS CHAPTERS

Previously learned concepts and equations frequently used in this chapter	Text Section	Study Guide Page
Analytical treatment of vectors (unit vectors, addition, components, magnitude, direction)	2-4	2-8
Constant acceleration motion: $v_x^2 = v_{xo}^2 + 2a_x\Delta x$	3-5	3-13
Newton's Second Law: $\mathbf{F} = m\mathbf{a}$	5-4	5-5
Normal force: $\mathbf{F}_N$	6-1	6-1
Friction: $f_k = \mu_k F_N$	6-1	6-3

NEW IDEAS IN THIS CHAPTER

Concepts and equations introduced	Text Section	Study Guide Page
The dot product: $\mathbf{A} \cdot \mathbf{B} = AB\cos\theta$	8-2	8-1
Work by a constant force: $W = \mathbf{F} \cdot \mathbf{L}$	8-1	8-4
Work by a variable force: $W = \int_i^f \mathbf{F} \cdot d\mathbf{r}$ or $W = \int_i^f (F_x dx + F_y dy + F_z dz)$	8-3	8-7
Hook's Law $F = -kx$	8-3	8-7
Kinetic Energy: $K = mv^2 / 2$	8-4	8-13
Work-Energy Theorem: $W_{net} = \Delta K = K_f - K_i$	8-4	8-13
Average Power: $\bar{P} = \Delta W/\Delta t = \mathbf{F} \cdot \mathbf{v}$	8-5	8-15
Instantaneous Power: $P = dW/dt$	8-5	8-15

PRINCIPAL CONCEPTS AND EQUATIONS

1 The Dot Product (Section 8-2)

Review: Figure 8.1 shows two vectors **A** and **B** and the angle between their directions.

Figure 8.1

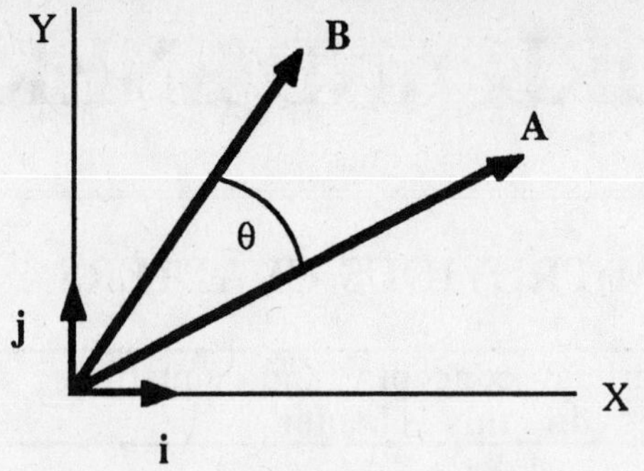

If we know the magnitude of **A** and **B** and the angle θ between their directions, we may obtain their dot or scalar product by

$$\mathbf{A} \bullet \mathbf{B} = AB \cos\theta$$

If we know the components of **A** and **B**, we may obtain their dot or scalar product by

$$\mathbf{A} \bullet \mathbf{B} = A_x B_x + A_y B_y$$

The dot product is commutative: $\mathbf{A} \bullet \mathbf{B} = \mathbf{B} \bullet \mathbf{A}$

The dot product is distributive: $\mathbf{A} \bullet (\mathbf{B} + \mathbf{C}) = \mathbf{A} \bullet \mathbf{B} + \mathbf{A} \bullet \mathbf{C}$

The dot product may be multiplied by a scalar (s)

$$s(\mathbf{A} \bullet \mathbf{B}) = (s\mathbf{A}) \bullet \mathbf{B} = \mathbf{A} \bullet (s\mathbf{B}) = (\mathbf{A} \bullet \mathbf{B})s$$

Also

$$\mathbf{i} \bullet \mathbf{i} = \mathbf{j} \bullet \mathbf{j} = \mathbf{k} \bullet \mathbf{k} = 1$$

and

$$\mathbf{i} \bullet \mathbf{j} = \mathbf{i} \bullet \mathbf{k} = \mathbf{j} \bullet \mathbf{k} = \mathbf{j} \bullet \mathbf{i} = \mathbf{k} \bullet \mathbf{i} = \mathbf{k} \bullet \mathbf{j} = 0$$

Practice: Consider the vectors **A** = 3i + 2j, **B** = 1i + 3j, **C** = -2i - 4j and then determine the following:

1. **A • B**	**A • B** = (3i + 2j) • (1i + 3j) = 3i • i + 9i • j + 2j • i + 6j • j = 3 + 6 = 9
2. **B • A**	**B • A** = (1i + 3j) • (3i + 2j) = 3i • i + 2i • j + 9j • i + 6j • j = 3 + 6 = 9

Note: Comparing the result of step 1 and 2, we see that **A • B = B • A**. We have demonstrated the commutative nature of the dot product.

3. **A • C**	**A • C** = (3i + 2j) • (-2i - 4j) = -14
4. **A • (B + C)**	**B + C** = -1i - 1j **A • (B + C)** = (3i + 2j) • (-1i - 1j) = -5

5. **A • B + A • C**	**A • B** = 9 (step 1) **A • C** = -14 (step 3) **A • B + A • C** = -5

Note: Comparing the results of steps 4 and 5, we see that
A • (B + C) = A • B + A • C.
We have demonstrated the distributive nature of the dot product.

6. 3(**A • B**)	**A • B** =9 (step 1) 3(**A • B**) = 27
7. (3**A**) • **B**	3**A** = 3(3**i** + 2**j**) = 9**i** + 6**j** (3**A**) • **B** = (9**i** + 6**j**) • (1**i** + 3**j**) = 27

Note: Comparing the results of steps 6 and 7, we see that
3(**A • B**) = (3**A**) • **B**.
We have demonstrated how to multiply a dot product by a scalar.

8. The angle between **A** and the x axis	**i • A** = **i** • (3**i** + 2**j**) = 3.00 $\|\mathbf{A}\| = A = (A_x{}^2 + A_y{}^2)^{1/2} = 3.61$ **i • A** = A cosα = 3.61 cosα comparing expressions for **i • A** obtain 3.61 cosα = 3.00 or $\alpha = \cos^{-1}(3.00 / 3.61) = 33.8°$
9. The angle between **B** and the x axis	**i • B** = **i** • (1**i** + 3**j**) = 1.00 $\|\mathbf{B}\| = B = (B_x{}^2 + B_y{}^2)^{1/2} = 3.16$ **i • B** = B cosβ = 3.16 cosβ comparing expressions for **i • B** obtain 3.16 cosβ = 1.00 or $\beta = \cos^{-1}(1.00 / 3.16) = 71.6°$
10. The angle between **A** and **B**	**A • B** = 9.00 (step 1) |**A**| = A = 3.61 (step 8) |**B**| = B = 3.16 (step 9) **A • B** = AB cosθ = 11.4 cosθ Comparing expressions for **A • B** 11.4 cosθ = 9.00 or $\theta = \cos^{-1}(9.00 / 11.4) = 37.9°$ We may also use α and β from steps 8 and 9 respectively to obtain θ θ = β – α = 71.6° - 33.8° = 37.8°

Related Text Exercises: Ex. 8-7 through 8-12.

2 Work Done By a Constant Force (Section 8-1)

Review: If a constant force **F** acts on an object (Figure 8.2) while it undergoes a displacement **L**, the work done by **F** is

$$W = \mathbf{F} \cdot \mathbf{L} = FL\cos\theta$$

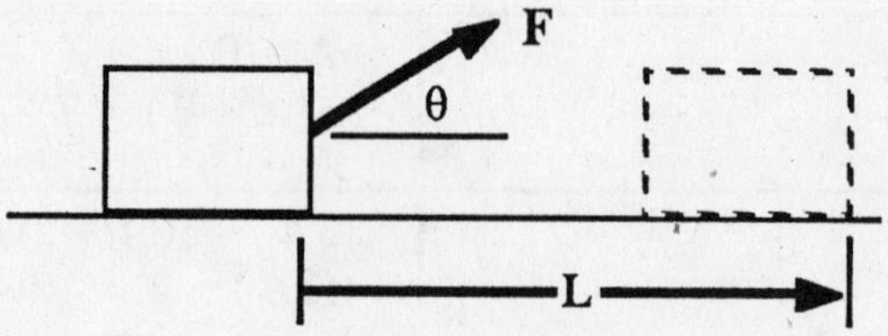

Figure 8.2

The net work done by several forces acting on an object may be found by

(a) obtaining the work done by each force and then adding, or
(b) obtaining the net force and then the work done by it.

Practice: Consider the three situations and the data shown below:

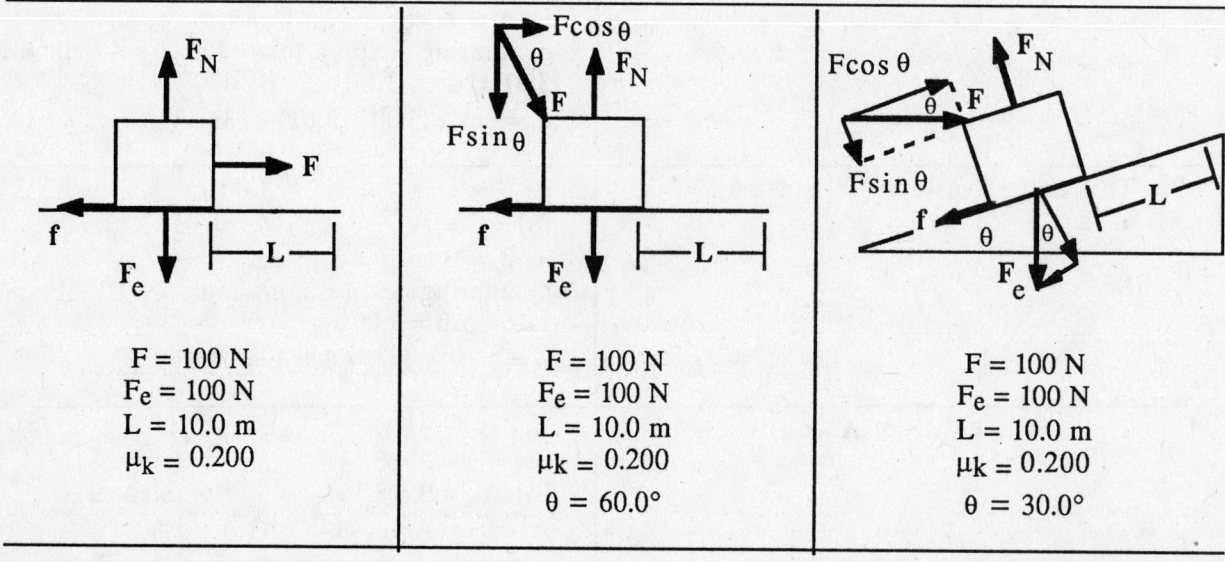

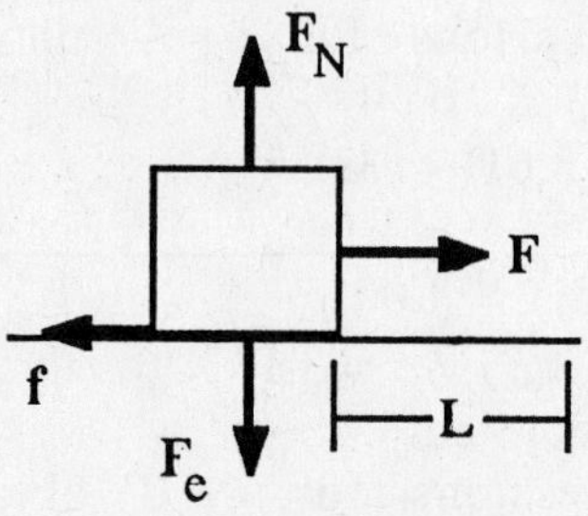

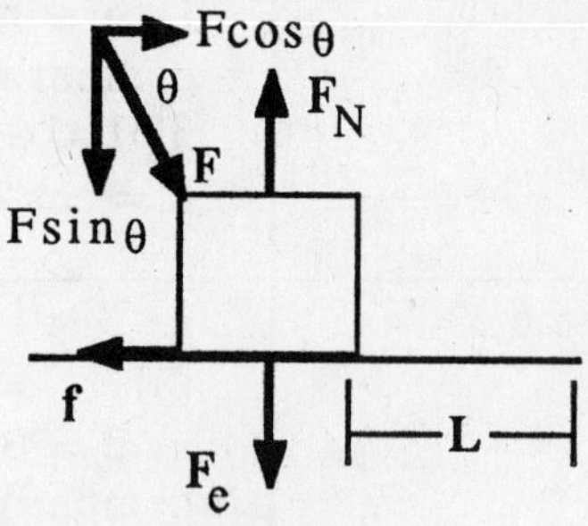

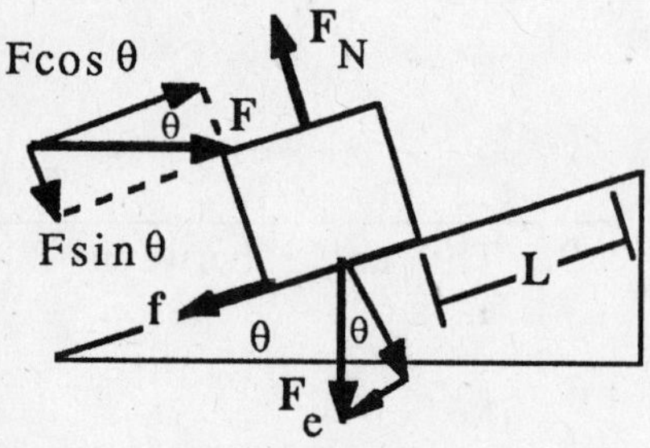

$F = 100$ N $F_e = 100$ N $L = 10.0$ m $\mu_k = 0.200$	$F = 100$ N $F_e = 100$ N $L = 10.0$ m $\mu_k = 0.200$ $\theta = 60.0°$	$F = 100$ N $F_e = 100$ N $L = 10.0$ m $\mu_k = 0.200$ $\theta = 30.0°$

Determine the following:

1. The normal force F_N acting on each object

$F_N = F_e = 100$ N	$F_N = F_e + F\sin\theta$ $= (100 + 100\sin 60.0°)$ N $= 187$ N	$F_N = F_e\cos\theta + F\sin\theta$ $= (100\cos 30.0° + 100\sin 30.0°)$ N $= 137$ N

2. The force of gravity F_e acting on each object

$F_e = 100$ N	$F_e = 100$ N	$F_e = 100$ N

3. The force of friction acting on each object

$f = \mu_k F_N = 20.0$ N	$f = \mu_k F_N = 37.4$ N	$f = \mu_k F_N = 27.4$ N

4. The work W_F done by the applied force F for each case

$W_F = \mathbf{F} \cdot \mathbf{L}$ $= FL \cos 0°$ $= 1000$ J	$W_F = \mathbf{F} \cdot \mathbf{L}$ $= FL \cos 60.0°$ $= 500$ J	$W_F = \mathbf{F} \cdot \mathbf{L}$ $= FL \cos 30.0°$ $= 866$ J

5. The work W_f done by the force of friction f for each case

$W_f = \mathbf{F} \cdot \mathbf{L}$ $= fL \cos 180°$ $= -200$ J	$W_f = \mathbf{F} \cdot \mathbf{L}$ $= fL \cos 180°$ $= -374$ J	$W_f = \mathbf{F} \cdot \mathbf{L}$ $= fL \cos 180°$ $= -274$ J

6. The work W_g done by the force of gravity F_e for each case

$W_g = \mathbf{F_e} \cdot \mathbf{L}$ $= F_e L \cos 90.0° = 0$ J	$W_g = \mathbf{F_e} \cdot \mathbf{L}$ $= F_e L \cos 90.0° = 0$ J	$W_g = \mathbf{F_e} \cdot \mathbf{L}$ $= F_e L \cos 120° = -500$ J

7. The work W_N done by the normal force F_N for each case

$W_N = \mathbf{F_N} \cdot \mathbf{L}$ $= F_N L \cos 90.0° = 0$ J	$W_N = \mathbf{F_N} \cdot \mathbf{L}$ $= F_N L \cos 90.0° = 0$ J	$W_N = \mathbf{F_N} \cdot \mathbf{L}$ $= F_N L \cos 90.0° = 0$ J

8. Net work done by all the forces acting on the object

$W_{net} = W_F + W_g$ $+ W_N + W_f$ $= 800$ J	$W_{net} = W_F + W_g + W_N + W_f$ $= 126$ J	$W_{net} = W_F + W_g + W_N + W_f$ $= 92.0$ J

9. Net horizontal force acting on the object for each case

$\Sigma F = F_{net}$ $= F - f$ $= 80.0$ N	$\Sigma F = F_{net}$ $= F \cos\theta - f$ $= 12.6$ N	$\Sigma F = F_{net}$ $= F \cos 30.0°$ $- F_e \sin 30.0° - f$ $= 9.20$ N

10. Net work W_{net} done by the net force for each case

$W_{net} = \mathbf{F_{net}} \cdot \mathbf{L}$ $= F_{net} L \cos 0°$ $= 800$ J	$W_{net} = \mathbf{F_{net}} \cdot \mathbf{L}$ $= F_{net} L \cos 0°$ $= 126$ J	$W_{net} = \mathbf{F_{net}} \cdot \mathbf{L}$ $= F_{net} L \cos 0°$ $= 92.0$ J

Example: A block of mass 20.0 kg is pulled up a rough incline at a constant speed by a rope that makes an angle of 30.0° with the incline. If the tension in the rope is 250 N and the coefficient of kinetic friction 0.200, determine the net work done on the block in the process of pulling it 10.0 m up the incline.

Given and Diagram:

Figure 8.3

$M = 20.0$ kg
$L = 10.0$ m
$\theta = 30.0°$
$T = 250$ N
$\mu_k = 0.200$
v = constant

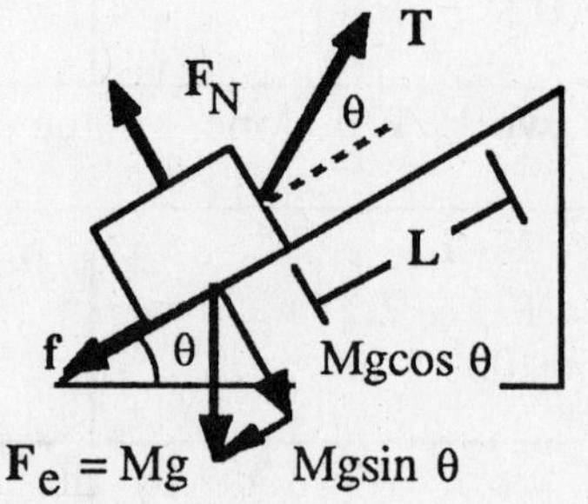

Determine: W_{net} - The net work done on the block in the process of pulling it 10.0 m up the incline.

Strategy: From the given information we can determine the magnitude of all the forces (T, F_e, F_N and f) and hence the net force (F_{net}) acting on the object. Knowing the magnitude of these forces, the magnitude of the displacement vector, and the angle between the forces and the displacement vector, we may proceed by either of the following methods.

Method A: Determine the work done by each force and then add these to get the net work.

Method B: Determine the net force and then the net work by this net force.

Solution:

Method A:

$W_T = \mathbf{T} \cdot \mathbf{L} = TL\cos 30.0° = 2165$ J

$W_f = \mathbf{f} \cdot \mathbf{L} = fL\cos 180° = -89.4$ J

$W_g = \mathbf{F_e} \cdot \mathbf{L} = M_gL\cos 120° = -980$ J

$W_{FN} = \mathbf{F_N} \cdot \mathbf{L} = F_NL\cos 90.0° = 0$ J

$W_{net} = W_T + W_f + W_g + W_{FN} = 1.10 \times 10^3$ J

Method B:

$F_{net} = \Sigma F = T\cos\theta - Mg\sin\theta - f$

$f = \mu F_N = \mu(Mg\cos\theta - T\sin\theta)$

$F_{net} = T\cos\theta - Mg\sin\theta - \mu Mg\cos\theta + \mu T\sin\theta = 110$ N

$W_{net} = \mathbf{F_{net}} \cdot \mathbf{L} = F_{net}L\cos 0° = 1.10 \times 10^3$ J

Related Text Exercises: Ex. 8-1 through 8-6, 8-8 and 8-9.

3 Work Done By A Variable Force (Section 8-3)

Review: If a constant force **F** which has only an x component (i. e. $\mathbf{F} = F_x\mathbf{i}$) acts on an object for a displacement $\mathbf{L} = (x_f - x_i)\mathbf{i}$, a plot of F_x vs x is shown in Figure 8.4.

Figure 8.4

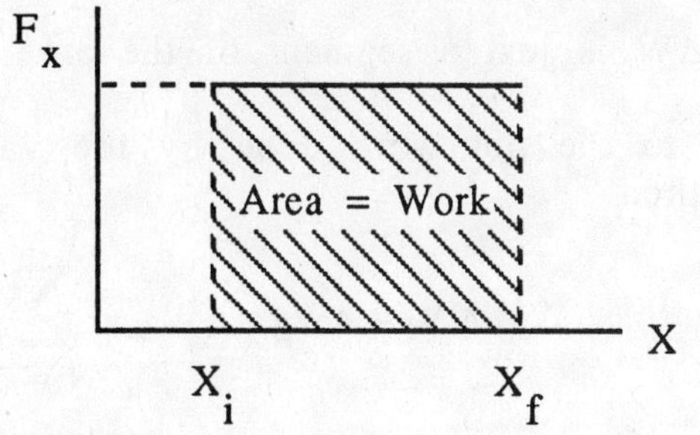

The work done by **F** during this displacement **L** is

$$W = \mathbf{F} \cdot \mathbf{L} = F_x\mathbf{i} \cdot (x_f - x_i)\mathbf{i} = F_x(x_f - x_i)\mathbf{i} \cdot \mathbf{i} = F_x(x_f - x_i)$$

Notice that this is just the area under the curve in the F_x vs x plot shown in Figure 8.4.

Consider a variable force **F** having only an x component and this x component depends only on the x coordinate (i. e. $\mathbf{F} = F_x(x)\mathbf{i}$). If **F** acts on an object for a displacement $\mathbf{L} = (x_f - x_i)\mathbf{i}$, a plot of F_x vs x is as shown in Figure 8.5(a).

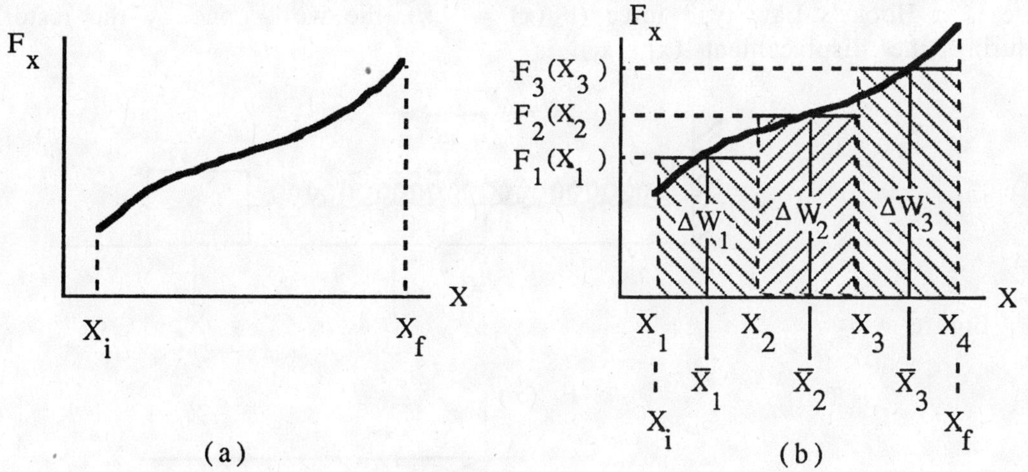

Figure 8.5

In order to determine the work done by the variable force **F** during the displacement **L**, let's make use of what we know about determining the work done by a constant force (namely that the work is the area under the curve of the F_x vs x plot). To do this let's divide the displacement $(x_f - x_i)\mathbf{i}$ into three smaller displacements $\Delta x_1\mathbf{i}$, $\Delta x_2\mathbf{i}$ and $\Delta x_3\mathbf{i}$. Then let's say that we have a constant force acting over each small displacement. Since the force is varying with the coordinate x, let's use the value at the midpoint $(\bar{x}_1, \bar{x}_2, \bar{x}_3)$ of each small displacement $\left(F_1(\bar{x}_1), F_2(\bar{x}_2), F_3(\bar{x}_3)\right)$. Then the work done during each small displacement is determined as shown below.

Displacement	Force	Work
$\Delta x_1 \mathbf{i}$	$F_{x1}(\bar{x}_1)\mathbf{i}$	$\Delta W_1 = F_{x1}(\bar{x}_1)\mathbf{i} \cdot \Delta x_1\mathbf{i} = F_{x1}(\bar{x}_1)\Delta x_1$
$\Delta x_2 \mathbf{i}$	$F_{x2}(\bar{x}_2)\mathbf{i}$	$\Delta W_2 = F_{x2}(\bar{x}_2)\mathbf{i} \cdot \Delta x_2\mathbf{i} = F_{x2}(\bar{x}_2)\Delta x_2$
$\Delta x_3 \mathbf{i}$	$F_{x3}(\bar{x}_3)\mathbf{i}$	$\Delta W_3 = F_{x3}(\bar{x}_3)\mathbf{i} \cdot \Delta x_3\mathbf{i} = F_{x3}(\bar{x}_3)\Delta x_3$

Notice that each ΔW is just a segment of the area under the curve (see Figure 8.4(b)).

An approximation to the net work done by the variable force during the displacement is then

$$W \approx \Delta W_1 + \Delta W_2 + \Delta W_3 = \sum_{N=1}^{3} F_{xN}(\bar{x}_N)\Delta x_N$$

As you can see, this approximation can be made better by dividing the displacement $(x_f - x_i)\mathbf{i}$ into more segments (i. e. larger N or smaller Δx). In fact in the limit, as $N \to \infty$ and $\Delta x \to 0$, the result becomes exact. This allows us to write the work done by the variable force **F** during the displacement $\mathbf{L} = (x_f - x_i)\mathbf{i}$ as

$$W = \lim_{\Delta x \to 0} \sum_{N} F_{xN}(\bar{x}_N)\Delta x_N = \int_{x_i}^{x_f} F_x(x)dx$$

An example of a variable force which is a function of displacement is the force of a spring. If $x = 0$ is the equilibrium position (Figure 8.6) of the spring and the restoring force is a Hooke's Law type force ($F_x(x) = -kx$), the work done by the restoring force during the displacement $(x_f - x_i)\mathbf{i}$ is

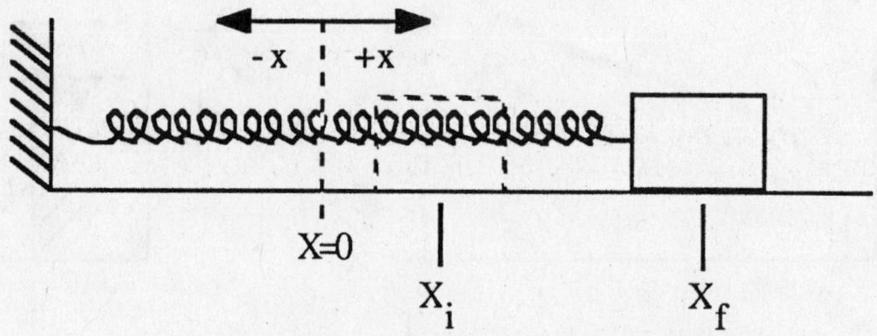

Figure 8.6

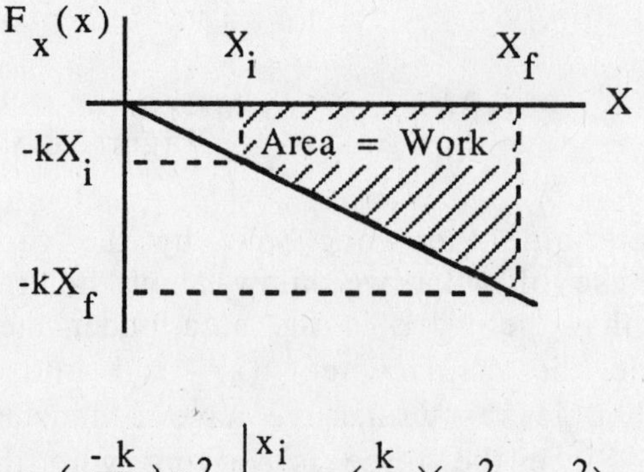

$$W = \int_{x_i}^{x_f} F_x(x)dx = \int_{x_i}^{x_f} -kxdx = \left(\frac{-k}{2}\right)x^2 \Big|_{x_f}^{x_i} = -\left(\frac{k}{2}\right)(x_f^2 - x_i^2)$$

Finally, the most general form of the expression for the work done by a force **F** (variable or constant) during any displacement is

$$W = \int_i^f \mathbf{F} \cdot d\mathbf{r}$$

or if we know the components of F

$$W = \int_i^f \left(F_x dx + F_y dy + F_z dz\right)$$

Practice: A spring and block are at the equilibrium position as shown in Figure 8.7(a).

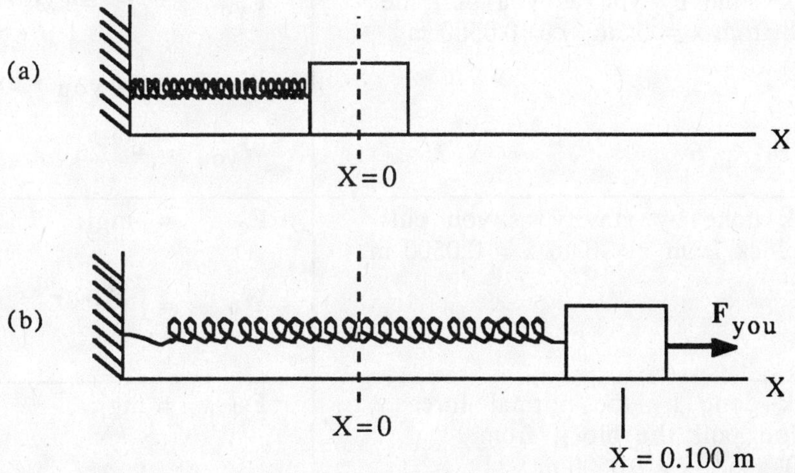

Figure 8.7

You grasp the block and pull it to the right at a constant speed. You find that you can hold the block fixed at x = 0.100 m (Figure 8.7(b)) by exerting a force F_{you} = 100 N. The surface is frictionless (μ = 0).

Determine the following:

1. The restoring force the spring exerts on the block as you hold it at x = 0.100 m	Since the block is in equilibrium, the restoring force has the same magnitude but opposite direction as the force you exert. $F_r = -F_{you} = -(100\ \mathrm{N})\mathbf{i}$
2. The spring constant	$F_r = -kx$ $F_r = -100$ N when $x = 0.100$ m $k = -F_r / x = 1000$ N / m

3. The restoring force the spring exerts on the block when $x = 0.0500$ m	$F_r = -kx = -(1000 \text{ N / m})(0.0500 \text{ m})$ $F_r = -50.0$ N
4. The force you would have to exert to hold the block fixed at $x = 0.0750$ m	$F_{you} = -F_r = -(-kx) = kx$ $= (1000 \text{ N/m})(0.0750 \text{ m}) = 75.0$ N
5. Work done by the restoring force on the block as the block moves from $x = 0$ to $x = 0.0500$ m	$\mathbf{F}_r = -kx\mathbf{i}$; $\mathbf{dr} = dx\mathbf{i}$ $W_r = \int \mathbf{F}_r \cdot \mathbf{dr} = -k \int_0^{0.05 \text{ m}} x dx(\mathbf{i} \cdot \mathbf{i})$ $W_r = \left(\frac{-k}{2}\right)x^2 \Big\|_0^{0.05 \text{ m}} = -1.25$ J
6. Work done by you as you pull the block from $x = 0$ to $x = 0.0500$ m	$\mathbf{F}_{you} = -\mathbf{F}_r = kx\mathbf{i}$; $\mathbf{dr} = dx\mathbf{i}$ $W_{you} = \int \mathbf{F}_{you} \cdot \mathbf{dr} = k \int_0^{0.05 \text{ m}} x dx(\mathbf{i} \cdot \mathbf{i})$ $W_{you} = +1.25$ J
7. Work done by gravity as you pull the block from $x = 0$ to $x = 0.0500$ m	$\mathbf{F}_e = -mg\mathbf{j}$; $\mathbf{dr} = dx\mathbf{i}$ $W_e = \int \mathbf{F}_e \cdot \mathbf{dr} = -mg \int_0^{0.05 \text{ m}} x dx(\mathbf{j} \cdot \mathbf{i}) = 0$
8. Work done by the normal force as you pull the block from $x = 0$ to $x = 0.0500$ m	$\mathbf{F}_N = mg\mathbf{j}$; $\mathbf{dr} = dx\mathbf{i}$ $W_N = \int \mathbf{F}_N \cdot \mathbf{dr} = mg \int_0^{0.05 \text{ m}} \mathbf{j} \cdot \mathbf{i} dr = 0$
9. Net work done on the block as you pull it from $x = 0$ to $x = 0.0500$ m	$W_{Net} = W_r + W_{you} + W_e + W_N$ $= -1.25 \text{ J} + 1.25 \text{ J} + 0 \text{ J} + 0 \text{ J} = 0$ J
10. Net force on block and work done by the net force as you pull the block to the right	Since you pull the block at a constant speed, then $F_{net} = \Sigma F = 0$, and $W_{net} = 0$

Example: A 2.00 kg ball is attached to a light string and whirled in a vertical circle of radius 1.50 m. (a) Determine the work done by the earth's gravitational force as the ball moves from the lowest point to the highest point in the circle. (b) Determine the work done on the ball by the tension in the string for one complete revolution.

Given and Diagram:

$m = 2.00$ kg
$R = 1.50$ m
$\theta_i = 0°$
$\theta_f = \pi$

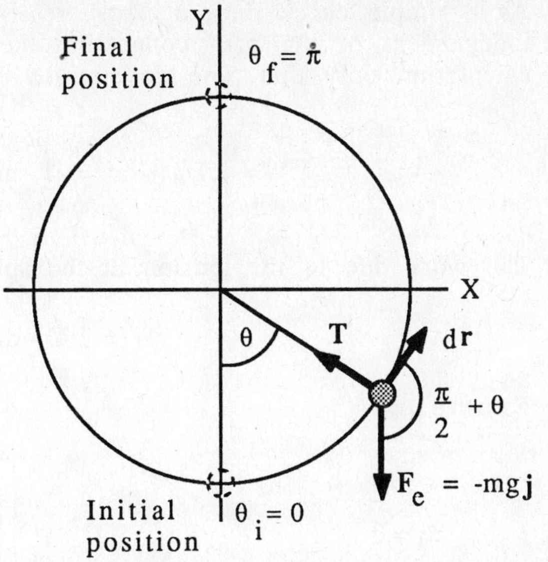

Determine: The work done by the force of gravity (F_e) as the ball travels from θ_i to θ_f and the work done by the tension in the string (T) as the ball travels one revolution.

Strategy: We can use the expression $W_e = \int \mathbf{F_e} \cdot \mathbf{dr}$ to calculate the work done by $\mathbf{F_e}$ if we can determine the magnitude of $\mathbf{F_e}$ and dr and the angle between the direction of $\mathbf{F_e}$ and dr. We can use this same expression to calculate the work done by the tension **T**. However, this calculation is trivial since **T** is at right angles to **dr**.

Solution: First let's find the magnitude of $\mathbf{F_e}$ and dr.

$$F_e = mg \quad \text{and} \quad dr = Rd\theta$$

The angle between **dr** and $\mathbf{F_e}$ is $[\theta + (\pi / 2)]$.
The work done by $\mathbf{F_e}$ as the ball goes from $\theta_i = 0$ to $\theta_f = \pi$ is

$$W = \int \mathbf{F_e} \cdot \mathbf{dr} = \int F_e dr \cos\left(\theta + \frac{\pi}{2}\right)$$

Recall that $\cos(\theta + \pi / 2) = \cos\theta \cos(\pi / 2) - \sin\theta \sin(\pi / 2) = -\sin\theta$.

$$W = \int_{\theta_i = 0°}^{\theta_f = \pi} F_e dr(-\sin\theta) = -mgR \int_{\theta_i = 0°}^{\theta_f = \pi} (\sin\theta \, d\theta)$$

$$W = mgR \cos\theta \Big|_{\theta_i = 0°}^{\theta_f = \pi} = mgR(\cos\pi - \cos 0°) = -2mgR$$

As a simple check on our work, recall that the work done by the gravitational force is independent of the path connecting the initial and final positions. Let's do the same calculation only this time go straight up the y axis.

$$W = \int \mathbf{F_e} \cdot d\mathbf{r} = \int_{-R}^{+R} -mg\mathbf{j} \cdot dy\mathbf{j} = -mgy \Big|_{-R}^{+R} = -2mgR$$

The work due to the tension in the springs is

$$W = \int \mathbf{T} \cdot d\mathbf{r} = \int T dr \cos 90.0° = 0$$

Example:

$x_i = 0$

$x_f = 0.300$ m

$y_i = 0$

$y_f = 0.200$ m

$k_1 = 200$ N / m

$k_2 = 400$ N / m

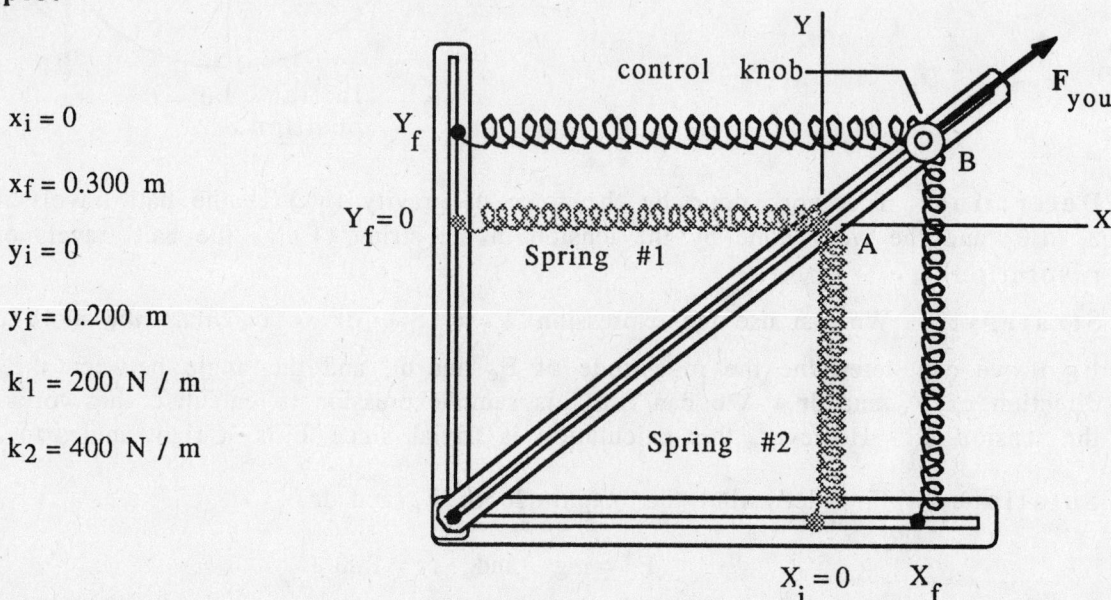

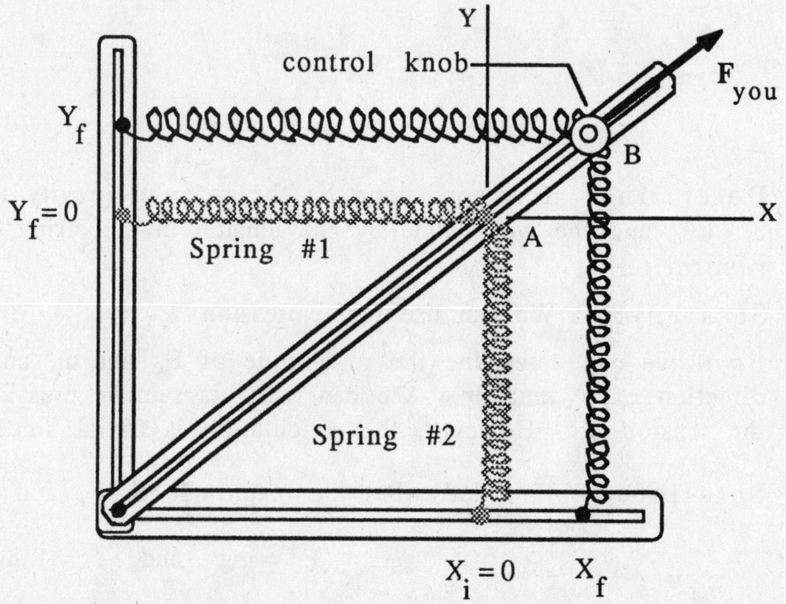

The figure above shows a slotted frame work, two springs and a control knob. When the control knob is at position A each spring is at its equilibrium length. Determine the amount of work you must do to pull the control knob out to position B. As the control knob is moved along its slotted frame work, the other end of each spring slides along the slotted frame work parallel to the x and y axis.

Given: $x_i = 0$, $x_f = 0.300$ m, $y_i = 0$, $y_f = 0.200$ m, $k_1 = 200$ /m, $k_2 = 400$ N/m

Determine: The amount of work you must do in order to pull the knob from position A to position B.

Strategy: Knowing that the springs exert a Hooke's Law type restoring force, we can find the x and y components of the variable restoring force and hence the variable force that you must exert. Knowing the components of the force you must exert and the limits of the displacement, we can determine the amount of work you must do to move the knob from position A to position B.

Solution: Springs 1 and 2 respectively supply the x and y components of the restoring force. The force you must exert is equal in magnitude but opposite in direction to the restoring force.

$$\mathbf{F}_{you} = -\mathbf{F}_r = -(F_{rx}\mathbf{i} + F_{ry}\mathbf{j}) = -(-k_1x\mathbf{i} - k_2y\mathbf{j}) = k_1x\mathbf{i} + k_2y\mathbf{j}$$

$$W = \int \mathbf{F}_{you} \cdot d\mathbf{r} = \int \left(F_{you-x}dx + F_{you-y}dy\right) = k_1 \int_{x_i}^{x_f} x\,dx + k_2 \int_{y_i}^{y_f} y\,dy$$

$$W = \left(\frac{k_1}{2}\right)x^2\Big|_{x_i}^{x_f} + \left(\frac{k_2}{2}\right)y^2\Big|_{y_i}^{y_f} = \left[\frac{k_1x_f^2 + k_2y_f^2}{2}\right]$$

$$W = [(200\ N / m)(0.300\ m)^2 + (400\ N / m)(0.200\ m)^2] / 2 = [18.0\ J + 16.0\ J] / 2 = 17.0\ J$$

Related Text Exercises: Ex. 8-13 through 8-21.

4 Work-Energy Theorem (Section 8-4)

Review: The work-energy theorem states that the net work done on an object is equal to the change in kinetic energy of that object.

$$W_{net} = \Delta K = K_f - K_i = mv_f^2 / 2 - mv_1^2 / 2$$

Practice: Consider the two situations and the data shown below.

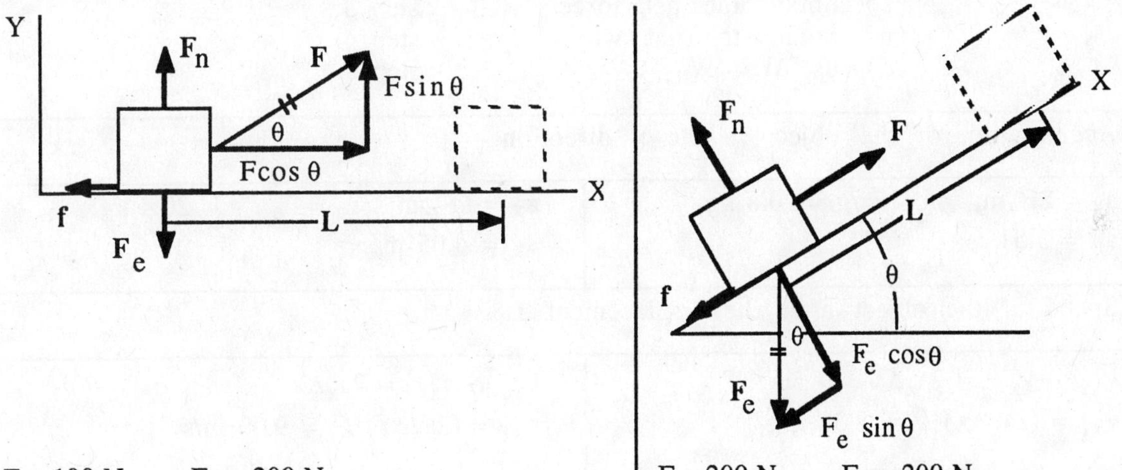

Case 1			Case 2		
F = 100 N,	F_e = 200 N		F = 200 N,	F_e = 200 N	
L = 10.0 m,	θ = 30.0°,	μ_k = 0.100	L = 10.0 m,	θ = 30.0°,	μ_k = 0.100

Determine the following for each case

1. The magnitude of the normal force F_n

$\Sigma F_y = F_n + F \sin\theta - F_e = 0$ $F_n = F_e - F \sin\theta = 150\ N$	$\Sigma F_y = F_n - F_e \cos\theta = 0$ $F_n = F_e \cos\theta = 173\ N$

2. The magnitude of the force of friction

$f = \mu_k F_n = 15.0\ N$	$f = \mu_k F_n = 17.3\ N$

3. The net force in the x direction

$F_{net} = \Sigma F_x = F \cos\theta - f = 71.6\ N$	$F_{net} = \Sigma F_x = F - F_e \sin\theta - f = 82.7\ N$

4. The net work done on the object during the displacement **L**

$W_{net} = \mathbf{F}_{net} \bullet \mathbf{L} = F_{net} L \cos 0° = 716\ J$	$W_{net} = \mathbf{F}_{net} \bullet \mathbf{L} = F_{net} L \cos 0° = 827\ J$

5. Kinetic energy of the object when the displacement is 10.0 m

$W_{net} = \Delta K = K_f - K_i$ $K_i = m v_i^2 / 2 = 0\ J$ $K_f = W_{net} = 716\ J$	$Wnet = \Delta K = K_f - K_i$ $K_i = m v_i^2 / 2 = 0\ J$ $K_f = W_{net} = 827\ J$

Note: We obtained the kinetic energy using the following simple scheme.

(a) obtain the net force - step 3.
(b) obtain the net work - step 4.
(c) use $\Delta K = W_{net}$ - step 5.

6. Acceleration of the object in the x direction

$a_x = \Sigma F / m$; $m = F_e / g = 20.4\ kg$ $a_x = 3.51\ /s^2$	$a_x = \Sigma F_x / m$; $m = F_e / g = 20.4\ kg$ $a_x = 4.05\ m/s^2$

7. Speed of the object after the displacement **L**

$v_{xf}^2 - v_{xi}^2 = 2 a_x \Delta x$ $v_{xf} = (2 a_x \Delta x)^{1/2} = 8.38\ m/s$	$v_{xf}^2 - v_{xi}^2 = 2 a_x \Delta x$ $v_{xf} = (2 a_x \Delta x)^{1/2} = 9.00\ m/s$

8. Kinetic energy of the object when the displacement is 10.0 m

$K_f = m v_{xf}^2 / 2 = 716\ J$	$K_f = m v_{xf}^2 / 2 = 827\ J$

Note: We obtained the kinetic energy using the following scheme.

(a) obtain the net force - step 3.
(b) obtain the acceleration - step 6.
(c) obtain the final speed - step 7.
(d) obtain the final kinetic energy - step 8.

Most students will consider obtaining the final kinetic energy by the $W_{net} = \Delta K$ method to be the most straight forward and the least amount of work.

Related Text Exercises: Ex. 8-22 through 8-33.

5 Power (Section 8-5)

Review: Power relates work to the time interval in which it is done. Power is the time rate at which work is performed. Average power $\bar{P}$ for a time interval Δt during which work ΔW is performed is defined as

$$\bar{P} = \Delta W / \Delta t \quad \text{also} \quad \bar{P} = (\mathbf{F} \cdot \Delta\mathbf{r}) / \Delta t = \mathbf{F} \cdot (\Delta\mathbf{r} / \Delta t) = \mathbf{F} \cdot \mathbf{v}$$

The instantaneous rate at which work is performed is the instantaneous power P and is defined as

$$P = \lim_{\Delta t \to 0} \left(\frac{\Delta W}{\Delta t}\right) = \frac{dW}{dt}$$

Practice: The crate shown in the figure below is pulled 10.0 m up a ramp at a constant speed by means of a cord that is attached to the shaft of a motor.

v = 0.100 m/s = speed of the crate up the ramp
μ_k = 0.200 = coefficient of kinetic friction
F_e = 100 N = weight of the crate
θ = 30° = angle of incline for the ramp
L = 10.0 m = distance the crate moves

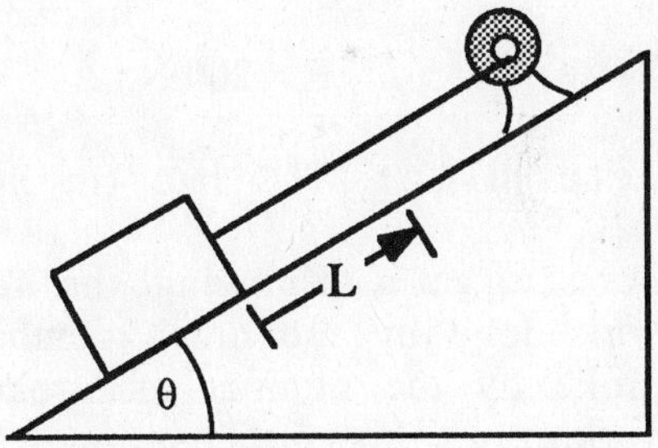

Determine the following:

1. Work done by gravity as the crate moves 10.0 m up the ramp	$W_g = \mathbf{F_e} \cdot \mathbf{L} = F_eL \cos 120° = -500$ J
2. Force of friction as the crate moves up the ramp	$F_n = F_e \cos 30° = 86.6$ N $f = \mu_k F_n = 17.3$ N
3. Work done by friction as the crate moves up the ramp	$W_f = \mathbf{f} \cdot \mathbf{L} = fL \cos 180° = -173$ J

4. Net force acting on the crate as it moves up the ramp	$F_{net} = \Sigma F = Ma$ since v = constant, then $a = 0$ and $F_{net} = 0$
5. Net work done on the object as it moves up the ramp	$W_{net} = F_{net} L \cos\theta = 0$ since $F_{net} = 0$
6. Work done on the crate by the motor	$W_{net} = W_{motor} + W_g + W_f$ $W_{motor} = W_{net} - W_g - W_f = 673$ J
7. Time it takes the motor to pull the crate 10 m up the ramp	$\Delta t = L / v = 100$ s
8. Average power supplied by the motor as it pulls the crate up the ramp	$\bar{P} = W_{net} / \Delta t = 6.73$ W
9. The force that the motor must supply in order to pull the crate up the ramp	$\Sigma F = F_{motor} - F_e \sin\theta - f = 0$ $F_{motor} = F_e \sin\theta + f = 67.3$ N
10. Average power supplied by the motor as it pulls the crate up the ramp (using a method other than that used in step 8.)	$\bar{P} = F_{motor} \bullet v = F_{motor} v \cos 0° = 6.73$ W

Example: A crew member of a racing shell can exert an average force of 20.0 N on his oar, working at the rate of 20.0 strokes / min. If his hands move back a distance of 0.700 m with each stroke, at what rate (in hp) is he working?

Given: $F = 200$ N, $n = 20.0$ strokes / min, $L = 0.700$ m

Determine: The rate (in hp) at which the crew member is working.

Strategy: Knowing the force and distance of each stroke, we can determine the work done by the crew member per stroke. We can multiply the work done per stroke by the number of strokes per minute to obtain the rate at which he is working. This rate can be converted from joules per minute to hp.

Solution:

$$W / \text{stroke} = F(L / \text{stroke}) = (200 \text{ N})(0.700 \text{ m} / \text{stroke}) = 140 \text{ J} / \text{stroke}$$

$$W / \text{min} = (140 \text{ J} / \text{stroke})(20.0 \text{ strokes} / \text{min}) = 2.80 \times 10^3 \text{ J} / \text{min}$$

$$P = \left(\frac{2.80 \times 10^3 \text{ J}}{\text{min}}\right)\left(\frac{1 \text{ min}}{60.0 \text{ s}}\right)\left(\frac{\text{W}}{\text{J} / \text{s}}\right)\left(\frac{1 \text{ hp}}{746 \text{ W}}\right) = 6.26 \times 10^{-2} \text{ hp}$$

Related Text Exercises: Ex. 8-34 through 8-38.

PRACTICE TEST

Take and grade this practice test. Doing so will allow you to determine any weak spots in your understanding of the concepts taught in this chapter. The following section prescribes what you should study further to strengthen your understanding.

A 100 N crate is pushed 10.0 m up a ramp by an applied 150 N force. The ramp is inclined at an angle of 30.0° and the force makes an angle of 30.0° with respect to the ramp. The coefficient of kinetic friction is 0.100.

$F_e = 100$ N
$F = 150$ N
$L = 10.0$ m
$\theta = 30.0°$
$\mu = 0.100$
$g = 9.80$ m/s^2

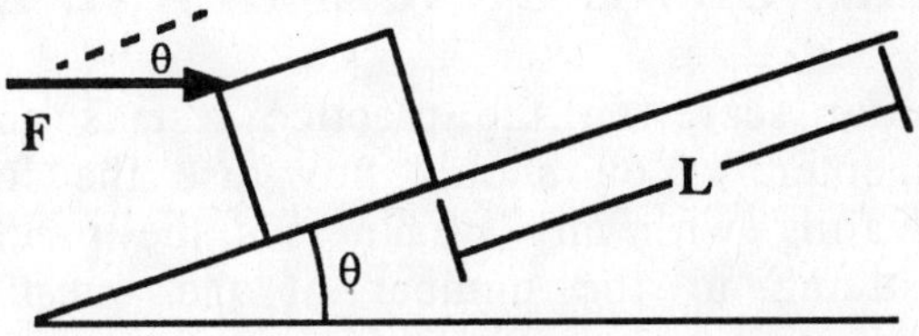

Determine the Following:

________ 1. Normal force acting on the crate
________ 2. Force of friction acting on the crate
________ 3. Work done by gravity
________ 4 Work done by friction
________ 5. Work done by the normal force
________ 6. Work done by the applied force
________ 7. Net work done on the crate, using your answers to questions 3, 4, 5 and 6
________ 8. Net force acting on the crate
________ 9. Net work done on the crate, using your answer to question 8
________ 10. Change in kinetic energy of the crate
________ 11. Speed of the crate after 10.0 m
________ 12. Acceleration of the crate up the ramp
________ 13. Time it takes the crate to travel 10.0 m up the ramp
________ 14. Average power delivered to the crate by the applied force

The displacement of an object is given by **r** = (2 m/s)t**i**. The object is acted on by the constant force **F** = 5 N**i** + 3 N**j**.

Determine the Following:

________ 15. Displacement of the object at t = 1.00 s
________ 16. Displacement of the object at t = 2.00 s
________ 17. Change in the displacement during the time interval t = 1.00 s to t = 2.00 s

________ 18. Average velocity of the object during the time interval t = 1.00 s to t = 2.00 s

________ 19. Work done on the object during the time interval t = 1.00 s to t = 2.00 s

________ 20. Average power delivered to the object during the time interval t = 1.00 s to t = 2.00 s

(See Appendix I for answers.)

PRINCIPAL CONCEPTS AND EQUATIONS PRESCRIPTION

Your score on the practice test is an excellent measure of your understanding of the chapter. You should now use the following chart to write your own prescription for dealing with any weaknesses the practice test points out. Look down the leftmost column to the number of the question(s) you answered incorrectly, reading across that row you will find the concept and/or equation of concern, the section(s) of the study guide you should return to for further study, and some suggested text exercises which you should work to gain additional experience.

Practice Test Questions	Concepts and Equations	Prescription	
		Principal Concepts	Text Exercises
1	Normal force: $\mathbf{F}_N$	1 of Ch 6	6-14, 15
2	Force of friction: $f = \mu_k F_N$	2 of Ch 6	6-18, 19
3	Work : $W_g = \mathbf{F}_e \cdot \mathbf{L}$	2	8-1, 2
4	Work : $W_f = \mathbf{f} \cdot \mathbf{L}$	2	8-2, 4
5	Work : $W_N = \mathbf{F}_N \cdot \mathbf{L}$	2	8-4, 5
6	Work : $W_F = \mathbf{F} \cdot \mathbf{L}$	2	8-5, 6
7	Net work : $W_{net} = W_g + W_f + W_N + W_F$	2	8-5, 42
8	Net force: $\mathbf{F}_{net} = \Sigma\mathbf{F}$	1 of Ch 5	5-8, 10
9	Net work : $W_{net} = \mathbf{F}_{net} \cdot \mathbf{L}$	2	8-42
10	Net work : $W_{net} = \Delta K$	4	8-24, 28
11	Kinetic Energy: $K = mv^2 / 2$	4	8-23, 32
12	Newton's Second Law: $\mathbf{F}_{net} = m\mathbf{a}$	2 of Ch 5	5-11, 14
13	Constant acceleration motion: $\Delta s = at^2/2$ or $v = at$	4 of Ch 3	3-28, 29
14	Average Power: $\overline{P} = \Delta W/\Delta t$, $\overline{P} = \mathbf{F} \cdot \mathbf{v}$	5	8-35, 36
15	Displacement vector	1 of Ch 3	3-1, 2
16	Displacement vector	1 of Ch 3	3-2, 3
17	Change in displacement: $\Delta\mathbf{r} = \mathbf{r}_2 - \mathbf{r}_1$	1 of Ch 3	3-1, 3
18	Average velocity: $\Delta\mathbf{v} = \Delta\mathbf{r}/\Delta t$	2 of Ch 3	3-4, 11
19	Work: $W = \mathbf{F} \cdot \Delta\mathbf{r}$	2	8-40, 41
20	Average Power: $\overline{P} = \Delta W/\Delta t$, $\overline{P} = \mathbf{F} \cdot \mathbf{v}$	5	8-37, 38

9 CONSERVATION OF ENERGY

RECALL FROM PREVIOUS CHAPTERS

Previously learned concepts and equations frequently used in this chapter	Text Section	Study Guide Page
Kinetic friction: $F_k = \mu_k F_N$	6-1	6-3
Work by a constant force: $W = \mathbf{F} \cdot \mathbf{L} = \lvert\mathbf{F}\rvert \cdot \lvert\mathbf{L}\rvert \cos\theta$	8-1	8-4
Work by a variable force: $W = \int_i^f \mathbf{F} \cdot d\mathbf{r}$ or $W = \int_i^f \left[F_x dx + F_y dy + F_z dz\right]$	8-3	8-7
Kinetic Energy: $K = mv^2 / 2$	8-4	8-13
Work - Energy Theorem: $W_{net} = \Delta K = K_f - K_i$	8-4	8-13

NEW IDEAS IN THIS CHAPTER

Concepts and equations introduced	Text Section	Study Guide Page
A conservative force is derivable from a potential energy function: $F_x = -dU(x) / dx$	9-1	9-1
A force is conservative if $W_{net} = 0$	9-1	9-1
A force is nonconservative if $W_{net} \neq 0$	9-1	9-1
For conservative forces: $W_{net} = -\Delta U$	9-1	9-1
Total energy: $E = K + U$	9-1	9-1
Conservation of energy: $E_f = E_i$	9-1, 4	9-8
Modified work-energy theorem: $E_f = E_i + W_{non}$	9-5	9-12
Graphical analysis of conservative systems	9-2	9-17

PRINCIPAL CONCEPTS AND EQUATIONS

1 Conservative and Nonconservative Forces (Section 9-1)

Review: Conservative Forces. A force is conservative if it meets the following conditions:

1. It does no net work on an object for any round trip.

2. It is derivable from a potential energy function U(x).

$$F(x) = -dU(x) / dx$$

Gravitational Force: Figure 9.1 shows an object moving from position 1 to position 2 and then back to position 1 while acted on by the gravitational force $\mathbf{F_e}$.

Figure 9.1

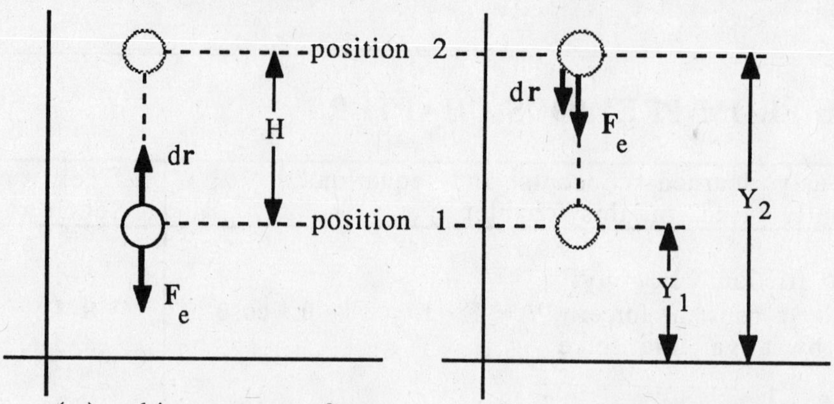

(a) object moves from y_1 to y_2 (θ = 180°)

(b) Object moves from y_2 to y_1 (θ = 0°)

Let's calculate the work done by the gravitational force $\mathbf{F_e}$ as the object moves from position 1 to position 2, from position 2 back to position 1, and the net work for the entire trip. Note that the work may be determined by

$$W = \int \mathbf{F} \cdot \mathbf{dr}$$

However since the force is constant this may be rewritten as

$$W = \mathbf{F} \cdot \Delta \mathbf{r} = |\mathbf{F}|\ |\Delta \mathbf{r}| \cos\theta$$

$$W_{1\to2} = \mathbf{F_e} \cdot \Delta \mathbf{y} = (mg)(H) \cos 180° = -mgH$$
$$W_{2\to1} = \mathbf{F_e} \cdot \Delta \mathbf{y} = (mg)(H) \cos 0° = mgH$$
$$W_{net} = W_{1\to2} + W_{2\to1} = -mgH + mgH = 0$$

The potential energy for the object shown in Figure 9.1 is

$$U(y) = mgy$$

The gravitation force then is

$$F(y) = -dU(y)\ /\ dy = -d(mgy)\ /\ dy = -mg$$

We have shown that when a gravitational force acts on an object the net work done for a round trip is zero and the force is derivable from a potential energy function. Since the gravitational force meets both of the above conditions, we conclude that it is a conservative force.

Another conservative force we will be using in this chapter is the force a spring exerts on an object.

Nonconservative Forces: A force is nonconservative if it meets the following conditions:

1. It does nonzero net work on an object for any round trip.

2. It is not derivable from a potential energy function.

Friction: Figure 9.2 shows an object moving from position 1 to position 2 and then back to position 1 while acted on by the force of kinetic friction.

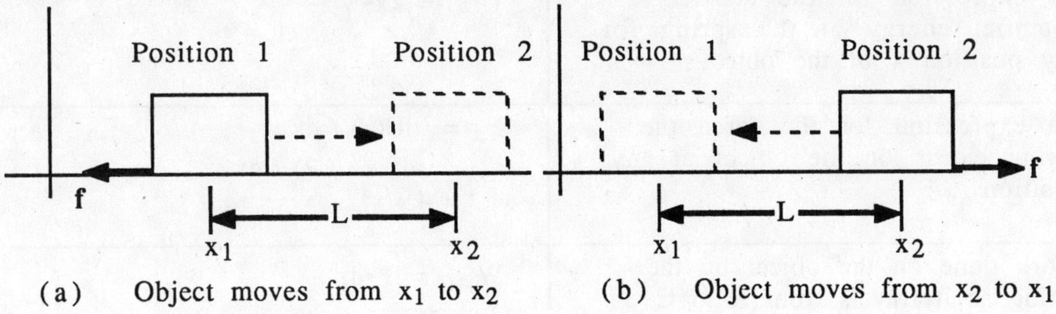

Figure 9.2

Let's calculate the work done by **f** as the object moves from position 1 to position 2, from position 2 to position 1, and the net work for the entire trip.

$$W_{1\to2} = \mathbf{f}\cdot\Delta\mathbf{x} = fL\cos180° = -\mu mgL$$
$$W_{2\to1} = \mathbf{f}\cdot\Delta\mathbf{x} = fL\cos180° = -\mu mgL$$
$$W_{net} = W_{1\to2} + W_{2\to1} = -\mu mgL + (-\mu mgL) = -2\mu mgL$$

We can find no potential energy function for this force such that $F(x) = -dU(x) / dx$.

We conclude that the force of kinetic friction is a nonconservative force since it does nonzero net work for a round trip and since it is not derivable from a potential energy function.

Another nonconservative force we will be using in this chapter is the applied force that you or some other external agent exerts on an object.

Practice: Figure 9.3(a) shows an object on the end of a spring at its equilibrium position. Figure 9.3(b) shows the object displaced to the right an amount x = 0.200 m where it is released from rest. The object then oscillates between positions A and D on the horizontal frictionless surface.

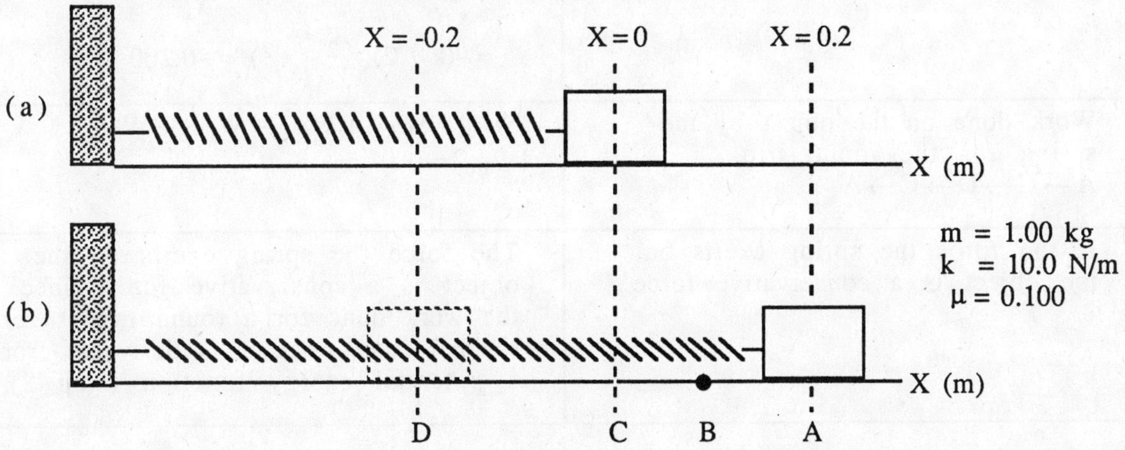

Figure 9.3

Determine the following:

1. An expression for the elastic potential energy of the spring for any position x of the object	$U(x) = kx^2 / 2$
2. An expression for the force the spring exerts on the object at any position x	$F_x = -dU(x) / dx$ $= -d(kx^2 / 2) / dx$ $= -kx$
3. Work done on the object by the spring as it travels from A to C	$W_{AC} = \int \mathbf{F} \cdot d\mathbf{r}$ $= \int_{x_i}^{x_f} [-\mathbf{i} k x] \cdot [\mathbf{i} d x] = -k \int_{x_i}^{x_f} x dx$ $= -(k / 2)(x_f^2 - x_i^2) = 0.200$ J
4. Work done on the object by the spring as it travels from C to D	$W_{CD} = \int \mathbf{F} \cdot d\mathbf{r}$ $= \int_{x_i}^{x_f} [-\mathbf{i} k x] \cdot [\mathbf{i} d x] = -k \int_{x_i}^{x_f} x dx$ $= -(k / 2)(x_f^2 - x_i^2) = -0.200$ J
5. Work done on the object by the spring as it travels from D to C	$W_{DC} = \int \mathbf{F} \cdot d\mathbf{r}$ $= \int_{x_i}^{x_f} [-\mathbf{i} k x] \cdot [\mathbf{i} d x] = -k \int_{x_i}^{x_f} x dx$ $= -(k / 2)(x_f^2 - x_i^2) = 0.200$ J
6. Work done on the object by the spring as it travels from C to A	$W_{CA} = \int \mathbf{F} \cdot d\mathbf{r}$ $= \int_{x_i}^{x_f} [-\mathbf{i} k x] \cdot [\mathbf{i} d x] = -k \int_{x_i}^{x_f} x dx$ $= -(k / 2)(x_f^2 - x_i^2) = -0.200$ J
7. Work done on the object by the spring for the round trip $A \rightarrow C \rightarrow D \rightarrow C \rightarrow A$	$W_{net} = W_{AC} + W_{CD} + W_{DC} + W_{CA}$ $W_{net} = 0$ J
8. If the force the spring exerts on the object is a conservative force	The force the spring exerts on the object is a conservative force since the work done for a round trip is zero (step 7) and since it is derivable from a potential energy function (step 2).

9. Work done on the object by the spring as the object moves from its equilibrium position to position B	$W = \int \mathbf{F} \cdot d\mathbf{r}$; $x_i = 0$, $x_f = 0.100$ m $= \int_{x_i}^{x_f} [-\mathbf{i} k x] \cdot [\mathbf{i} dx] = -k \int_{x_i}^{x_f} x dx$ $= -(k / 2)(x_f^2 - x_i^2) = -0.0500$ J
10. Potential energy of the spring at position B	The potential energy of the spring when the object is at position B is just the negative of the work done by the spring as the object moves from equilibrium position to position B $\Delta U_{CB} = -W_{CB}$ $\Delta U_{CB} = U_B - U_C$; $U_C = 0$ $U_B = -W_{CB} = 0.0500$ J
11. Total energy of the spring when the object is at position B	$E_B = E_A = k x_A^2 / 2 = 0.200$ J
12. Kinetic energy of the object when it is at position B	$E_B = U_B + K_B$ $K_B = E_B - U_B = 0.150$ J
13. Total energy, potential energy and kinetic energy of the system (spring and object) when the object is at the equilibrium position	$E_C = E_A = k x_A^2 / 2 = 0.200$ J $U_C = 0$ $K_C = E_C - U_C = 0.200$ J

Figure 9.4 shows an object being pushed across a rough horizontal surface by an applied force $\mathbf{F}_a$.

Figure 9.4

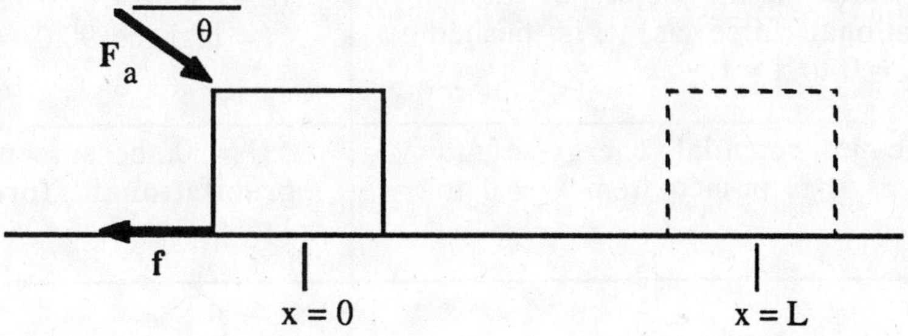

$L = 10.0$ m, $m = 100$ kg, $g = 9.80$ m/s^2, $\mu = 0.100$, $F_a = 200$ N, $\theta = 30.0°$

Determine the following:

1. The component of $\mathbf{F}_a$ which causes the object to move and the component which contributes to the normal force	The horizontal component of $\mathbf{F}_a$ causes the object to move. $F_a \cos\theta = (200 \text{ N}) \cos 30.0° = 173$ N The vertical component of $\mathbf{F}_a$ contributes to the normal force. $F_a \sin\theta = (200 \text{ N}) \sin 30.0° = 100$ N

2. The magnitude of the normal force which acts on the object	$N = mg + F_a \sin\theta = 1080$ N
3. Force of kinetic friction which opposes the motion of the object	$\mathbf{f} = -\mu N\mathbf{i} = (-108 \text{ N})\mathbf{i}$
4. Work done on the object by $\mathbf{F}_a$ as it is pushed from x = 0 to x = L, from x = L to x = 0, and for the round trip	$W_{0\to L} = \mathbf{F}_a \cdot \Delta\mathbf{x}$; $\lvert\Delta\mathbf{x}\rvert = \lvert\mathbf{x}_f - \mathbf{x}_i\rvert = L$ $= F_a L \cos 0° = 1730$ J $W_{L\to 0} = \mathbf{F}_a \cdot \Delta\mathbf{x} = F_a L \cos 0° = 1730$ J $W_{\text{round trip}} = W_{0\to L} + W_{L\to 0} = 3460$ J
5. Whether the applied force $\mathbf{F}_a$ is a conservative or nonconservative force	$\mathbf{F}_a$ is a nonconservative force because the net work done by this force for a round trip is nonzero.
6. Work done on the object by **f** as it is pushed from x = 0 to x = L	$W = \mathbf{f} \cdot \Delta\mathbf{x} = \lvert\mathbf{f}\rvert\, \lvert\Delta\mathbf{x}\rvert \cos 180°$ $\lvert\mathbf{f}\rvert = 108$ N; $\lvert\Delta\mathbf{x}\rvert = L = 10.0$ m $W = (108 \text{ N})(10.0 \text{ m})(-1) = -1080$ J
7. Net work done on the object as it is pushed from x = 0 to x = L	$W_{net} = W_{F_a} + W_f = 650$ J
8. Change in kinetic energy of the object as it is pushed from x = 0 to x = L	$\Delta KE = W_{net} = 650$ J
9. Speed of the object at x = L (assume it is at rest at x = 0)	$\Delta KE = KE_{final} - KE_{initial} = 650$ J $KE = mv^2 / 2$ or $v = \pm(2KE / m)^{1/2} = +3.61$ m/s
10. Work done on the object by the gravitational force as it is pushed from x = 0 to x = L	$W = \mathbf{F}_e \cdot \Delta\mathbf{x}$ $= mgL \cos 90.0° = 0$ J
11. Change in potential energy of the object as it is pushed from x = 0 to x = L	$\Delta U = 0$, because no work is done by the gravitational force.

Example: A 1.00 kg object is acted on by the force

$$\mathbf{F} = [(-10.0 \text{ N/m})x + (3.00 \text{ N/m}^2)x^2]\mathbf{i}$$

where x is in meters. Determine: (a) If this force is conservative or nonconservative. (b) The potential energy function if the force is conservative. (c) If the object has a speed of 12.0 m/s in the negative x direction when it is at x = 6.00 m, what is its speed at x = 2.00 m and as it passes the origin.

Given: $\mathbf{F} = [(-10.0 \text{ N/m})\, x + (3.00 \text{ N/m}^2)x^2]\mathbf{i}$
$m = 1.00$ kg
$\mathbf{v}(x = 6.00 \text{ m}) = -(12.00 \text{ m/s})\mathbf{i}$

Determine:

(a) If the force is conservative or nonconservative
(b) The potential energy function if the force is conservative
(c) The speed of the object at x = 2 m and x = 0 m

Strategy:

(a) We can calculate the net work done on the object by this force for a round trip ($x_i \rightarrow x_f \rightarrow x_i$). If the net work for the round trip is zero the force is conservative.

(b) Knowing that a conservative force is derivable from a potential energy function ($F_x = -dU(x)/dx$), we can obtain the potential energy function from the force.

(c) Knowing the speed of the object at 6.00 m, and the potential energy function, we can determine the kinetic energy, potential energy and total energy at x = 6.00 m. If the force is conservative, energy is conserved. Hence we know the total energy at any other position and we can calculate the potential energy and then the kinetic energy. Knowing the kinetic energy we can determine the speed.

Solution:

(a) Determine the work to go from x_i to x_f, from x_f back to x_i, and then the net work for the round trip.

$$W_{i\rightarrow f} = \int \mathbf{F}\bullet d\mathbf{r} = \int_{x_i}^{x_f} F_x dx = \int_{x_i}^{x_f} \left[-10x + 3x^2\right]dx$$

$$= \left(-5x^2 + x^3\right)\Big|_{x_i}^{x_f} = 5(x_i^2 - x_f^2) - (x_i^3 - x_f^3)$$

$$W_{f\rightarrow i} = \int \mathbf{F}\bullet d\mathbf{r} = \int_{x_f}^{x_i} F_x dx = \int_{x_f}^{x_i} \left[-10x + 3x^2\right]dx$$

$$= \left(-5x^2 + x^3\right)\Big|_{x_f}^{x_i} = 5(x_f^2 - x_i^2) - (x_f^3 - x_i^3)$$

$$W_{net} = W_{i\rightarrow f} + W_{f\rightarrow i} = 0$$

Since $W_{net} = 0$, the force is conservative.

(b) Knowing that a conservative force is derivable from a potential energy function ($F_x = -dU(x)/dx$), we may obtain the potential energy function by

$$U(x) = -\int F_x dx = -\int \left[(-10\ \text{N/m})x + \left(3\ \text{N/m}^2\right)x^2\right] dx$$

$$= (5\ \text{N/m})x^2 - \left(1\ \text{N/m}^2\right)x^3$$

(c) If the speed of the object at 6.00 m is 12.0 m/s in the negative x direction, then

$K(x = 6) = mv^2 / 2 = 72.0$ J
$U(x = 6) = (5.00 \text{ N/m})(6.00 \text{ m})^2 - (1.00 \text{ N/m}^2)(6.00 \text{ m})^3 = -36.0$ J
$E(x = 6) = K + U = 36.0$ J

Since energy is conserved,

$E(x = 2) = E(x = 6) = 36.0$ J
$U(x = 2) = (5 \text{ N/m})(2 \text{ m})^2 - (1 \text{ N/m}^2)(2 \text{ m})^3 = 12.0$ J
$K(x = 2) = E(x = 2) - U(x = 2) = 24.0$ J
$V(x = 2) = \pm(2K / m)^{1/2} = \pm 6.93$ m/s
$V(x = 2) = -6.93$ m/s as the object is traveling in the negative x direction

$E(x = 0) = E(x = 6) = 36.0$ J
$U(x = 0) = (5.00 \text{ N/m})(0 \text{ m})^2 - (1.00 \text{ N/m}^2)(0 \text{ m})^3 = 0$ J
$K(x = 0) = E(x = 0) - U(x = 0) = 36.0$ J
$V(x = 0) = \pm(2K / m)^{1/2} = \pm 8.49$ m/s
$V(x = 0) = -8.49$ m/s as the object is traveling in the negative x direction

Related Text Exercises: Ex. 9-6, 9-8, 9-16, 9-17 and 9-27.

2 Energy Calculations and Conservative Forces (Section 9-1, 4, 6)

Review: From the work-energy theorem, the change in the kinetic energy of an object is the net work done on the object.

$$\Delta K = K_f - K_i = W_{net}$$

If only conservative forces act on an object, the work done is independent of the path and is equal to the negative of the change in potential energy.

$$W_{net} = -\Delta U = -(U_f - U_i) = U_i - U_f$$

Combining the above two expressions for the net work, we obtain

$$W_{net} = K_f - K_i = U_i - U_f \quad \text{or}$$

$$K_f + U_f = K_i + U_i$$

Since the total energy is the sum of the kinetic and potential energy, we have an expression for the conservation of mechanical energy (if only conservative forces act)

$$E_f = E_i$$

Practice: An object with a mass of 1.00 kg is projected vertically upward with a speed of 30.0 m/s. Let's agree to set U = 0 at the level from which the object is projected.

Determine the following:

1. The initial kinetic energy, potential energy and total energy	$K_o = mv_o^2 / 2 = 450$ J, $U_o = 0$ J $E_o = K_o + U_o = 450$ J
2. Maximum height of the projected object	We can determine this two ways: Method I (using kinematics, Ch 4) $2a_y \Delta y = v_y^2 - v_{oy}^2$ $a_y = -g$ and $\Delta y = \Delta h_{max}$ when $v_y = 0$ Inserting values obtain: $-2g \Delta h_{max} = -V_{oy}^2$ $\Delta h_{max} = v_{oy}^2 / 2g = 45.9$ m Method II (using conservation of energy) At the top of the trajectory $E_T = E_o = 450$ J $K_T = 0$, so $\Delta K = -450$ J $\Delta U = -\Delta K = 450$ J $\Delta U = mg\ \Delta h$ $\Delta h = \Delta U / mg = 45.9$ m
3. Work done by gravity as the object travels half-way to the top	$y_i = 0$ m; $y_f = \Delta h / 2 = 23.0$ m $\|\Delta y\| = y_f - y_i = 23.0$ m $\|\mathbf{F}_e\| = mg = 9.80$ N $W_g = \mathbf{F}_e \bullet \Delta \mathbf{y}$ $= \|\mathbf{F}_e\|\ \|\Delta y\| \cos 180°$ $= -mg\ \Delta h / 2 = -225$ J
4. Change in gravitational potential energy of the object as it travels from the projected position to half-way to the top	$\Delta U = -W_g = 225$ J
5. Total energy (E'), potential energy (U'), and kinetic energy (K') of the object when it is half-way to the top	$E' = E_o = 450$ J $\Delta U = U' - U_o = U' = 225$ J $K' = E' - U' = 225$ J
6. Speed (v') of the object when it is half-way to the top	We can determine this by two methods: Method I (using kinematics, Ch 4) $2a_y\Delta y = v_y^2 - v_{oy}^2$; $a_y = -g$ $v_{oy} = 30.0$ m/s; $\Delta y = \Delta h / 2 = 23.0$ m $2(-g)\Delta h / 2 = v_y^2 - v_{oy}^2$ $v_y = (v_{oy}^2 - g\Delta h)^{1/2} = 21.2$ m/s Method II (using energy) $K' = M(v')^2 / 2$ $v' = \pm(2K' / m)^{1/2} = +21.2$ m/s

Practice: A 10.0 kg object is moving across a rough horizontal surface as shown in Figure 9.5. At position 1 it has a speed of 4.00 m/s, it slows down due to friction (μ = 0.100), and it comes to rest at position 2. Position 2 is 8.16 m from position 1.

Figure 9.5

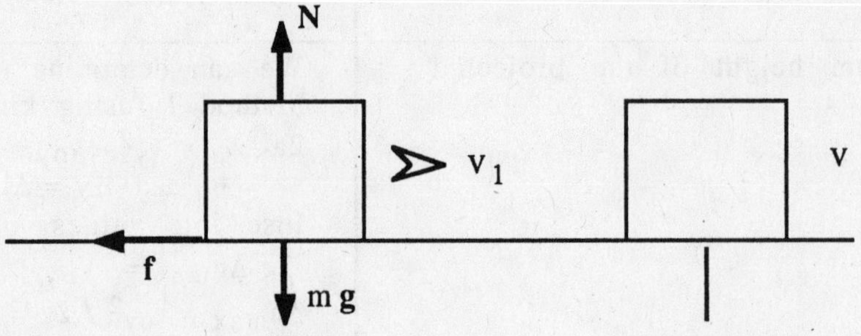

Determine the following:

1. Kinetic energy (K_1), potential energy (U_1), and total energy E_1 of the object at position 1	$K_1 = mv_1^2 / 2 = 80.0$ J $U_1 = 0$ (our choice) $E_1 = K_1 + U_1 = 80.0$ J
2. Magnitude of the forces acting on the object as it moves from position 1 to position 2	Gravity $F_e = mg = 98.0$ N Normal $N = F_e = 98.0$ N Friction $f = \mu N = 9.80$ N
3. Work done by each of these forces as the object travels from position 1 to position 2	$W_g = \mathbf{F_e} \cdot \Delta\mathbf{x} = (mg)(s)\cos 90.0° = 0$ J $W_N = \mathbf{N} \cdot \Delta\mathbf{x} = (mg)(s)\cos 90.0° = 0$ J $W_f = \mathbf{f} \cdot \Delta\mathbf{x} = \mu Ns \cos 180° = -80.0$ J
4. Total work done by conservative and nonconservative forces	$W_{con} = W_g = 0$ J $W_{non} = W_N + W_f = -80.0$ J
5. Final energy of the system	$E_f = E_i + W_{non}$ $E_i = 80.0$ J (Step 1) $W_{non} = -80.0$ J (Step 4) $E_f = 80.0$ J + (-80.0 J) = 0 J

Example: A 10.0 kg mass is at rest at the top of a long frictionless ramp inclined at an angle of 30.0°. At the base of the ramp is a spring of spring constant 2000 N/m at its equilibrium position. The mass is initially 5.00 m up the ramp from the upper end of the spring's equilibrium position. Determine (a) the speed of the mass as it collides with the spring and (b) the maximum compression of the spring.

Given and Diagram:

$m = 10.0$ kg
$k = 2000$ N/m
$\mu = 0$
$L = 5.00$ m
$\theta = 30.0°$
$g = 9.80$ m/s^2

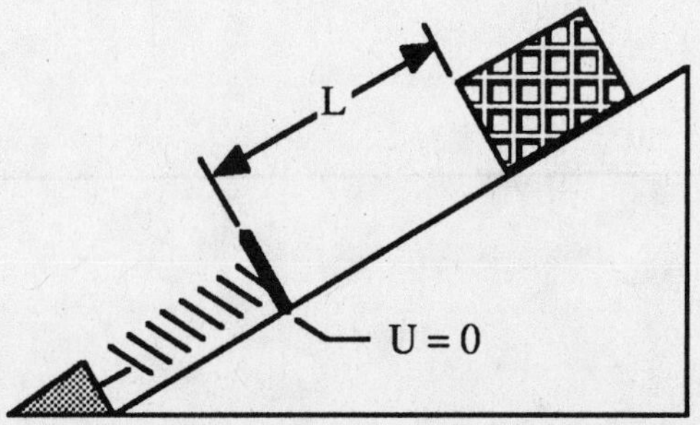

Determine: (a) The speed of the mass as it collides with the spring and (b) the maximum compression of the spring.

Strategy: All forces doing work on the mass m are conservative, hence we may use conservation of energy. For convenience lets choose zero gravitation potential energy to be at the higher end of the spring when the spring is at is equilibrium length. Since energy is conserved, as the mass slides down the ramp its potential energy becomes kinetic energy. At the top of the ramp the energy is all potential, at the spring's equilibrium position the energy is all kinetic. The kinetic energy of the mass at the spring is equal to its potential energy at the top. Knowing the kinetic energy of the mass as it collides with the spring, we can determine the speed of the mass. When the spring is at its maximum compression the energy is elastic and gravitational potential energy. All of the kinetic energy of the mass has been converted momentarily into potential energy. Knowing the elastic potential energy of the spring, we can calculate the maximum compression of the spring.

Solution: The kinetic, potential and total energy of the mass at the top of the ramp is

$U_{top} = mg\,h = mg\,L\,\sin 30.0° = 245\ J$
$K_{top} = 0\ J$
$E_{top} = K_{top} + U_{top} = 245\ J$

Since energy is conserved the energy of the mass at the bottom of the ramp is

$$E_{bot} = E_{top} = 245\ J$$

Since we chose the bottom of the 5.00 m length of ramp to be zero gravitation potential energy, the kinetic energy of the mass at this position on the ramp is

$$K_{bot} = E_{bot} - U_{bot} = 245\ J - 0\ J = 245\ J$$

The speed at the bottom of this section of the ramp is determined by

$$K_{bot} = mv_{bot}^2 / 2 \text{ or } v_{bot} = \pm(2K_{bot} / m)^{1/2} = +7.00\ m/s$$

Now as the spring compresses the mass gives up this kinetic energy and some gravitational potential energy for elastic potential energy of the spring. This may be written as

$$K_{bot} = kx^2 / 2 - mgx \sin\theta$$

This may be rewritten as a quadratic in x as

$$1000x^2 - 49.0x - 245 = 0$$

Which has the solution

$$x = \frac{-(-49.0) \pm \left[(-49.0)^2 - 4(1000)(-245)\right]^{1/2}}{2000} = [\ +0.520\ m, -0.471\ m]$$

The maximum compression is x = -0.471 m. The negative root is used because zero is the equilibrium position of the spring, plus is up the ramp and minus is down the ramp.

Related Text Exercises: Ex. 9-1, 9-2, 9-3, 9-5, 9-7, 9-9, 9-10, 9-15 and 9-18 through 9-26.

3 Energy Calculations and Nonconservative Forces (Section 9-5)

Review: The work-energy theorem states that the net work done on an object is equal to its change in kinetic energy

$$W_{net} = \Delta K = K_f - K_i$$

This is valid for both conservative and nonconservative forces.

If the net work is accomplished by both conservative and nonconservative forces, we may write

$$W_{net} = W_{con} + W_{non}$$

The work done by conservative forces is just the negative of the change in potential energy.

$$W_{con} = -\Delta U = -(U_f - U_i) = U_i - U_f$$

Combining the above, we have

$$W_{net} = W_{con} + W_{non}$$
$$\Delta K = -\Delta U + W_{non}$$
$$K_f - K_i = U_i - U_f + W_{non}$$
$$K_f + U_f = K_i + U_i + W_{non}$$

This is called the modified work energy theorem.

Practice: An object which has a mass of 10.0 kg is sliding across a rough (μ = 0.100) horizontal surface as shown in Figure 9.6. At the initial position it has a speed of 2.00 m/s and after sliding a distance of 2.04 m it comes to rest.

Figure 9.6

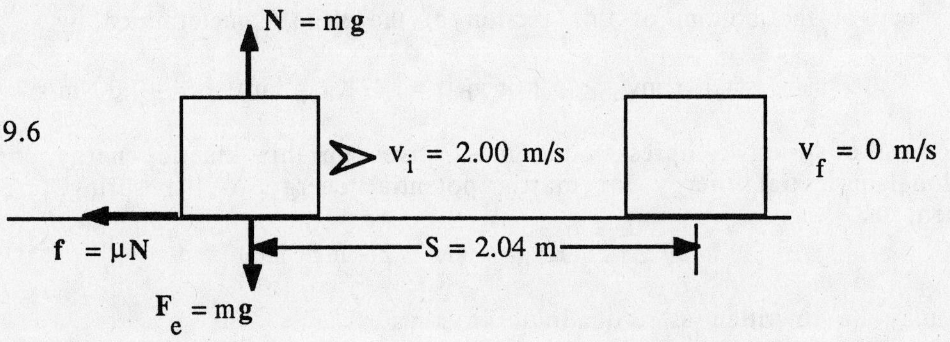

Determine the following:

1. Initial kinetic, potential and total energy	$K_i = mv_i^2 / 2$ $= (10.0\text{ kg})(2.00\text{ m/s})^2 / 2 = 20.0\text{ J}$ $U_i = 0$ J $E_i = K_i + U_i = 20.0$ J
2. Work done on the object by F_e, N and f	$W_{F_e} = F_e \cdot \Delta x = mgS \cos 90.0° = 0$ $W_N = N \cdot \Delta x = NS \cos 90.0° = 0$ $W_f = f \cdot \Delta x = fs \cos 180° = -20.0$ J

3. Final kinetic, potential and total energy	$K_f = mv_f^2 / 2 = 0$ J $U_f = 0$ J $E_f = K_f + U_f = 0$ J
4. Change in kinetic energy of the object	$\Delta K = K_f - K_i = -20.0$ J
5. Net work done on the object	$W_{net} = W_{F_e} + W_N + W_f = -20.0$ J Also $W_{net} = \Delta K = -20.0$ J
6. Conservative and nonconservative forces	F_e is a conservative force. N and f are nonconservative forces.
7. Work done by conservative and nonconservative forces	$W_{con} = W_{F_e} = 0$ (step 2) $W_{non} = W_N + W_f = -20.0$ J (step 2)
8. The final energy of the object, using the modified work energy theorem	$E_f = E_i + W_{non}$ $E_i = 20.0$ J (step 1) $W_{non} = -20.0$ J (step 2) $E_f = 20.0$ J - 20.0 J = 0

Practice: You push a crate up an incline as shown in Figure 9.7.

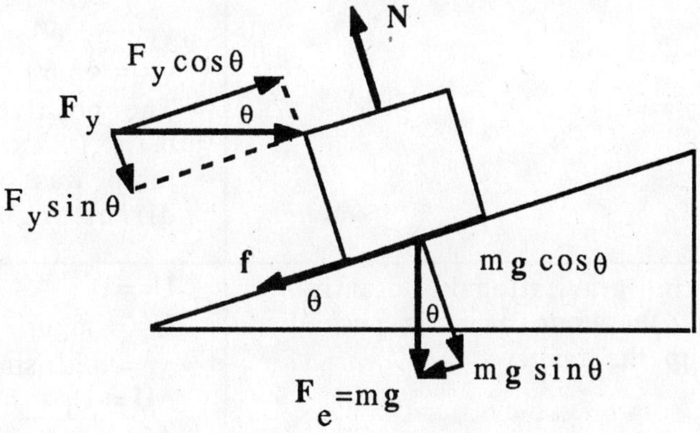

Figure 9.7

L	= 10.0 m	distance the crate is pushed up the incline
m	= 10.2 kg	mass of the crate
F_e	= mg = 100 N	force of gravity
μ	= 0.100	coefficient of kinetic friction
F_y	= 80.0 N	force you exert on the crate
θ	= 30.0°	angle of incline and between incline and force

Determine the following:

1. Normal force on the crate	$N = F_y \sin\theta + mg \cos\theta$ $= (80.0\text{ N}) \sin 30.0° + (100\text{ N}) \cos 30.0°$ $= 127$ N

2. Force of kinetic friction acting on the crate	$f = \mu N = 12.7$ N
3. Force of gravity acting on the crate	$F_e = mg = 100$ N
4. Work done on the crate by you, the normal force, the force of gravity, and kinetic friction	$W_y = \mathbf{F_y} \cdot \Delta\mathbf{x}$ $= F_y L \cos 30.0° = 693$ J $W_N = \mathbf{N} \cdot \Delta\mathbf{x}$ $= NL \cos 90.0° = 0$ J $W_g = \mathbf{F_e} \cdot \Delta\mathbf{x}$ $= F_e L \cos 120° = -500$ J $W_f = \mathbf{f} \cdot \Delta\mathbf{x}$ $= fL \cos 180° = -127$ J
5. Net work done on the crate as it moves up the ramp	$W_{net} = W_y + W_N + W_g + W_f = 66.0$ J
6. Change in kinetic energy of the crate as it moves up the ramp	$\Delta K = W_{net} = 66.0$ J
7. Speed of the crate after it has been pushed 10.0 m up the ramp	$\Delta K = K_f - K_i,\ K_i = 0$ $K_f = \Delta K = 66.0$ J $K_f = mv_f^2 / 2$ $v_f = \pm(2K_f / m)^{1/2}$ $v_f = \pm 3.60$ m/s $v_f = +3.60$ m/s The positive root is chosen because the crate is moving up the ramp and up the ramp was chosen to be the positive direction.
8. Change in gravitational potential energy of the crate as it is pushed 10.0 m up the ramp	$U_i = 0$ $U_f = mgh_f$ $= mgL \sin\theta = 500$ J $\Delta U = U_f - U_i = 500$ J Also note that $\Delta U = -W_g = -(-500 \text{ J}) = 500$ J (See step 4)
9. Total work done by nonconservative forces	$W_{non} = W_y + W_N + W_f = 566$ J
10. Initial kinetic, potential and total energy of the crate	$K_i = 0, \quad U_i = 0$ $E_i = K_i + U_i = 0$

11. Final energy of the crate after it has been pushed 10.0 m up the ramp	$K_f = 66.0$ J (step 6) $U_f = 500$ J (step 8) $E_f = K_f + U_f = 566$ J Also by the modified work energy theorem: $E_i = 0$ (step 10) $W_{non} = 566$ J (step 9) $E_f = E_i + W_{non} = 566$ J

Example: A 2.00 kg object is attached to a spring of constant k = 200 N/m. The spring is stretched 0.150 m and then released. If the coefficient of kinetic friction between the object and the horizontal surface it is sliding on is 0.100 determine: (a) the speed of the object the first time it goes through the equilibrium position and (b) the compression of the spring the first time it is compressed.

Given and Diagram:

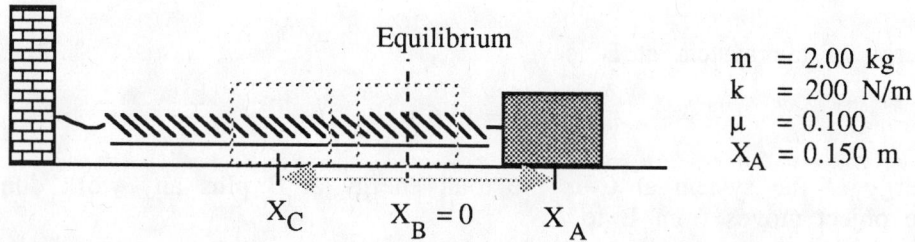

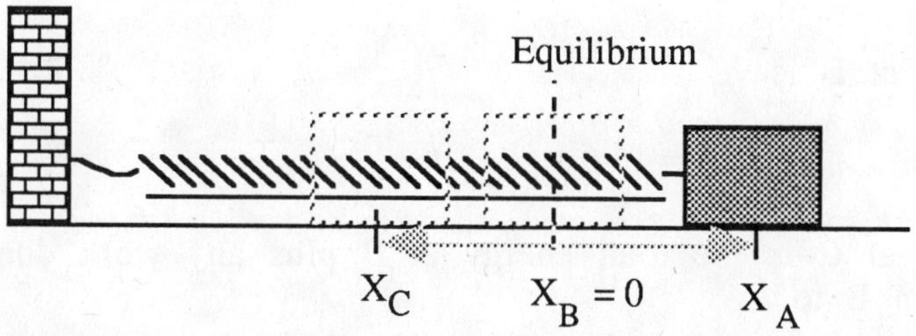

$m = 2.00$ kg
$k = 200$ N/m
$\mu = 0.100$
$X_A = 0.150$ m

Determine:

(a) The speed (v_B) of the object the first time it goes through equilibrium
(b) The compression (x_C) of the spring on its first oscillation

Strategy: (a) Knowing x_A and k, we can determine the total energy of the system at A. Knowing x_A, m, μ and g we can determine the work done on the system by friction as it moves from A to B. Knowing the total energy of the system at A and the work done on the system by friction as the object moves from A to B, we can determine the kinetic energy and hence the speed of the object as it passes through equilibrium (position B). (b) Knowing the total energy of the system at A, that all the energy is potential energy at C, and that the total energy at C is the total energy at B plus the work done on the system by friction as the object moves from B to C, we can determine the compression of the spring (x_C).

Solution: (a) The total energy of the system when the object is at A is

$$E_A = U_A = kx_A^2 / 2 = 2.25 \text{ J}$$

The work done on the system by friction as the object moves from A to B is

$$W = \int \mathbf{F} \cdot d\mathbf{r}, \quad \mathbf{F} = \mathbf{i}\mu mg, \quad d\mathbf{r} = \mathbf{i}dx, \quad x_i = x_A, \quad x_f = x_B = 0$$

$$W = \int_{x_A}^{x_B} (\mathbf{i}\mu m g) \cdot (\mathbf{i}dx) = \mu m g \int_{x_A}^{x_B} dx = \mu mg(x_B - x_A) = -0.294 \text{ J}$$

The kinetic energy of the system as the object goes through the equilibrium position is

$$E_B = E_A + W_f = 2.25 \text{ J} - 0.294 \text{ J} = 1.96 \text{ J}$$

Since the potential energy (U_B) is zero at equilibrium, we have

$$K_B = E_B - U_B = E_B = 1.96 \text{ J}$$

The speed of the object as it passes through equilibrium is

$$K_B = mv_B^2 / 2 \quad \text{or} \quad v_B = (2K_B / m)^{1/2} = 1.40 \text{ m/s}$$

(b) The work done on the system by friction as the object travels from B to C is

$$W = \int_{x_B}^{x_C} (\mathbf{i}\mu m g) \cdot (\mathbf{i}\, dx) = \mu m g \int_{x_B}^{x_C} dx = \mu m g(x_C - x_B) = (1.96 \text{ N})x_C$$

Note the x_C is a negative quantity.

The total energy of the system at B is

$$E_B = 1.96 \text{ J}$$

The total energy of the system at C is the total energy at B plus any work done on the system as the object moves from B to C.

$$E_C = E_B + W = 1.96 \text{ J} + (1.96 \text{ N})x_C$$

The total energy of the system at C is all potential energy.

$$E_C = U_C = kx_C^2 / 2 = (100 \text{ N/m})x_C^2$$

Equating the last two expressions for E_C

$$(100 \text{ N/m})x_C^2 = 1.96 \text{ J} + (1.96 \text{ N})x_C$$

Rearranging we obtain a quadratic in x_C

$$(100 \text{ N/m})x_C^2 - (1.96 \text{ N})x_C - 1.96 \text{ J} = 0$$

We may solve this equation to obtain the two roots

$$x_C = -0.131 \text{ m} \quad \text{and} \quad x_C = +0.150 \text{ m}$$

Since we know then x_C must have a negative value less than x_A, the correct physical solution is

$$x_C = -0.131 \text{ m}$$

Related Text Exercises: Ex. 9-27 through 9-29.

4 Graphical Analysis for a One-Dimensional Conservative System (Section 9-2)

Review: For a one-dimensional conservative system, the division of conserved mechanical energy into changing amounts of kinetic and potential energy can be displayed graphically. Such a graphical display also allows us to determine turning points, points of stable and unstable equilibrium, and the potential and kinetic energy of the system at any point.

In order to proceed with a graphical analysis we need a graphical representation of the potential energy U(x) as a function of x. This plot may be obtained as follows:

1. If we are given a potential energy function such as the one below, [e.g. $U(x) = (5.00 \times 10^4 \text{ J/m}^2)x^2$], we can insert values of x, calculate values of U(x), and then plot U(x) vs x.

2. If we are given a force function [e.g. $F(x) = -(1.00 \times 10^5 \text{ N/m})x$], we can determine the potential energy function by

$$U(x) = -\int F(x)dx$$

and then proceed as in step 1. to get a plot of U(x) vs x.

Next let's review the kind of information we can obtain about the system under consideration from the graphical representation of the potential energy (i.e. a plot of U(x) vs x). Suppose that the graphical representation of the potential energy appears as shown in Figure 9.8.

Figure 9.8

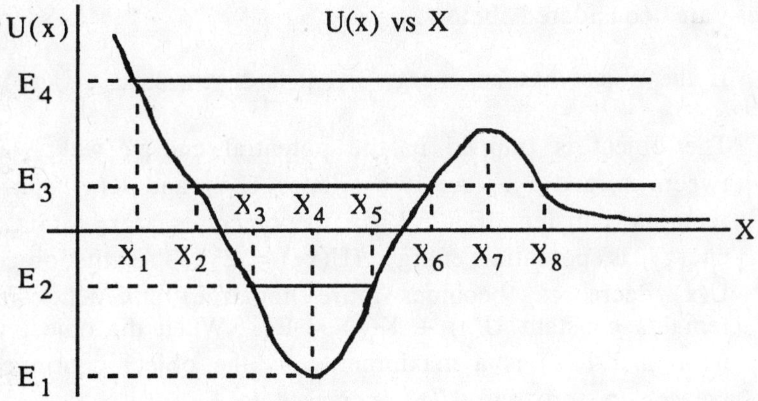

1. An object whose motion is determined by the potential energy function U(x) will be in equilibrium when the net force on the object is zero. This will occur when

$$F(x) = -dU(x) / dx = 0$$

Note that this occurs when the slope of the tangent to U(x) = 0. For this case, that is at positions x_4 and x_7.

2. A point of equilibrium is stable if, for any small displacement of the object from that point the force tends to return the object to the equilibrium point.

For this case, note that if the object is at x_4 and it is displaced slightly towards greater values of x, the slope (dU(x) / dx) of the tangent to U(x) is positive; hence the force is negative (recall $F(x) = -dU(x) / dx$). This implies a restoring force back towards x_4.

In like manner, if the object is displaced slightly towards smaller values of x, the slope of the tangent to U(x) is negative; hence the force is positive which implies a restoring force towards x_4.

Since small displacements (positive or negative) of the object from x_4 result in a restoring force towards x_4, we may conclude that x_4 is a point of stable equilibrium.

3. A point of equilibrium is unstable if, for any small displacement of the object from that point, the force tends to move the object from the equilibrium point. For this case, note that if the object is at x_7 and it is displaced slightly towards greater values of x, the slope of the tangent line to U(x) is negative; hence the force is positive. This implies a force away from x_7 and towards larger values of x.

   In like manner, if the object is displaced slightly towards smaller values of x, the slope of the tangent to U(x) is positive; hence the force is negative which implies a force away from x_7 and towards smaller values of x.

   Since small displacements (positive or negative) of the object from x_7 result in a force that tends to move the object away from x_7, we may conclude that x_7 is a point of unstable equilibrium.

4. Knowing the total energy of the object, we can describe its motion. Several cases are considered below.

   E_1: If the object has an energy E_1, it is at rest at x_4. $U(x_4) = E_1$ and $K(x_4) = 0$.

   E_2: The object is trapped in the potential energy well. It will oscillate between x_3 and x_5. As it oscillates U(x) and K(x) change but their sum remains constant E_2. When the object is at x_3, it is at rest hence all of its energy is potential energy ($U(x_3) = E_2$). As the object moves toward x_5, U(x) decreases (becomes more negative) and K(x) increases but the sum remains constant $U(x) + K(x) = E_2$. When the object is at x_4, U(x) is a minimum and K(x) is a maximum. As the object continues towards x_5, K(x) decreases to zero and U(x) increases to E_2.

   E_3: The object may either be in the well oscillating between x_2 and x_6 or to the right of the potential energy hill (i.e. beyond x_8). The motion in the well would generally be like that described for the preceding case. If the object is to the right of x_8 and approaching x_8, as U(x) increases KE(x) will decrease; the object will stop momentarily at x_8, turn around and then move continuously in the +x direction.

   E_4: If the particle is released at rest from x_1, it will move continuously in the +x direction. It will speed up, have its maximum KE at x_4, slow down as it travels over the potential energy hill, and then increases its KE as it continues to travel in the +x direction. If the particle is initially traveling to the left, it will slow down over the hill, speed up over the valley, stop momentarily and turn around at x_1, and then travel continuously in the +x direction.

5. Turning points occur when the total mechanical energy is equal to the potential energy. The following chart gives several different Energies and their respective turning points.

Energy	Turning Points
E_2	x_3 and x_5
E_3	x_2 and x_6 or x_8
E_4	x_1

Practice: Consider the potential energy function

$$U(x) = (1\ \text{J/m}^4)x^4 - (8\ \text{J/m}^2)x^2 + 5\ \text{J}$$

and then determine the following:

1. All points where $U(x) = 0$	$U(x) = x^4 - 8x^2 + 5 = 0$ This may be rewritten as a quadratic in x^2 as $(x^2)^2 - 8x^2 + 5 = 0$ $x^2 = \dfrac{-(-8) \pm \left[(-8)^2 - 4(1)(5)\right]^{1/2}}{2(1)}$ or $x^2 = \{ 7.32,\ 0.685 \}$ or $x = \pm(7.32)^{1/2} = \pm 2.71$ $x = \pm(0.685)^{1/2} = \pm 0.828$
2. All maxima and minima for this function	Max. and min. occur when $dU / dx = 0$, $dU / dx = 4x^3 - 16x = 0$ or $x(x^2 - 4) = 0$ with solutions at $x = 0$ and $x = \pm 2$ To determine which are max. or min. take a second derivative and evaluate it at $x = 0$ and $x = \pm 2$. $d^2U / dx^2\big\|_{x=0} = 12x^2 - 16\big\|_{x=0} = -16$ The negative implies a max. occurs at $x = 0$ $d^2U / dx^2\big\|_{x=\pm 2} = 12x^2 - 16\big\|_{x=\pm 2} = +32$ The positive implies a min. occurs at $x = \pm 2$

3. A rough sketch of the potential energy function	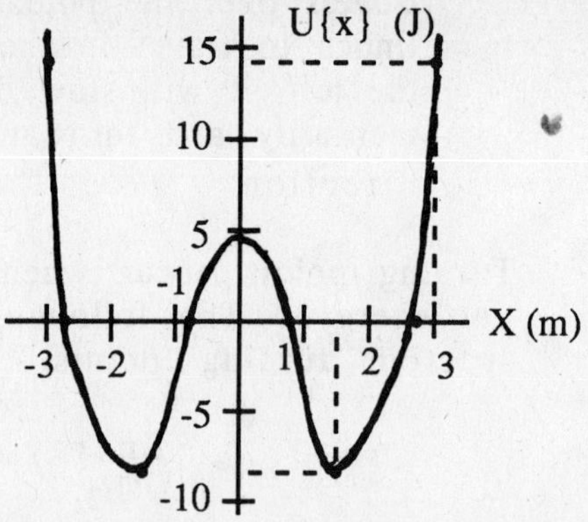
4. The turning points if E_{total} = 14.0 J	The turning points occur when $U(x) = E_{total} = 14.0$ J $14 = x^4 - 8x^2 + 5$, or $(x^2)^2 - 8x^2 - 9 = 0$, with solutions $x^2 = (+9.00\ m^2, -1.00\ m^2)$, or $x = \pm 3.00$ m (the only real solutions)
5. When the force on an object experiencing this potential energy function is zero	$F(x) = -dU(x) / dx = 0$ when the slope of the potential energy function is zero. This occurs when x = 0, and x = ±2.00 m.
6. Points of stable equilibrium	x = ±2.00 m
7. Points of unstable equilibrium	x = 0 m
8. Points where an object experiencing this U(x) would be traveling the fastest	The object will travel the fastest when the KE is a max. Since E = K + U = constant, this will occur when U is a min. This occurs for x = ±2.00 m.
9. Kinetic and potential energy of the object when it is at x = ±1.50 m	We can either read the value for U(±1.50 m) from the plot shown in step 3 or calculate it from the potential energy function. Either way U(±1.50 m) = -7.94 J. Hence K = E - U = 14.0 J - (-7.94 J) = 21.9 J
10. Kinetic and potential energy of the object when it is at x = ±2.80 m	Using either the potential energy function U(x) or the graphical representation of this function (step 3) obtain U(±2.80 m) = +3.75 J K = E - U = +10.25 J
11. Force on the object at x = +1 m	$F(x) = (-dU / dx)\Big\vert_{x=1}$ $= -\left(4x^3 - 16x\right)\Big\vert_{x=1}$ = +12 N

12. Where a particle experiencing this U(x) will have maximum kinetic energy	Since E = K + U = constant, maximum K will occur for minimum U, minimum U occurs at x = ±2 m.

Example: A particle moves along the x axis under the influence of a conservative force given by F = -(2 N/m)x. If the particle has a total energy of E = 25.0 J determine: (a) the turning points, (b) where the particle has its maximum speed, and (c) the potential, kinetic and total energy at x = 2.00 m.

Given: F = -(2 N/m)x and E = 25.0 J

Determine: (a) The turning points
(b) The value of x where the particle has its maximum speed
(c) The potential, kinetic and total energy at x = 2.00 m

Strategy: (a) Knowing the force, we can determine the potential energy function. The turning points will occur when U = E. (b) Once the potential energy function is determined, we can determine the value of x when U is a minimum and hence K and v have maximum values. (c) Knowing the force is conservative, we can determine the total energy at any point. Knowing the potential energy function we can determine the potential energy at any point. Knowing the total energy and the potential energy at some point we can determine the kinetic energy at that point.

Solution:

(a)

$$F_x = -dU(x) / dx$$

or

$$U(x) = -\int F_x dx = \int (2\ \text{N/m})x dx = \frac{(2\ \text{N/m})x^2}{2} + C$$

Where C is a constant of integration. Note that when we insert x = 0 we get C = U(0). This says that the potential energy at x = 0 is an arbitrary constant; for convenience let's choose it to be zero. Hence the potential energy function is

$$U(x) = (1\ \text{N/m})x^2$$

The turning points occur where U(x) = E = 25.0 J

$$(1\ \text{N/m})x^2 = 25.0\ \text{J} \quad \text{or} \quad x = \pm 5.00\ \text{m}$$

(b) The particle will have its maximum speed where the kinetic energy is a maximum. This will occur where the potential energy is a minimum. The minimum potential energy occurs where dU / dx = 0.

$$U(x) = (1\ \text{N/m})x^2$$
$$dU(x) / dx = (2\ \text{N/m})x = 0$$

Note that dU(x)dx = 0 at x = 0, hence the potential energy is a minimum and the kinetic energy a maximum at x = 0. The particle will have its greatest speed where the kinetic energy is a maximum. Hence the particle will have its greatest speed at x = 0.

(c)

$$E(2) = 25.0 \text{ J}$$
$$U(2) = (1 \text{ N/m})(2 \text{ m})^2 = 4.00 \text{ J}$$
$$K(2) = E(2) - U(2) = 21.0 \text{ J}$$

Related Text Exercises: Ex. 9-11 through 9-14, 9-46 and 9-47.

PRACTICE TEST

Take and grade this practice test. Doing so will allow you to determine any weak spots in your understanding of the concepts taught in this chapter. The following section prescribes what you should study further to strengthen your understanding.

A 1.00×10^3 N crate is sitting at position A, as shown in the figure below. You push the crate from A to B with a constant 400 N force. At B you release the crate, and it slides down the incline and across the horizontal section until it stops. Let's agree to choose zero gravitational potential energy at the level of the horizontal section. The coefficient of kinetic friction for all surfaces is $\mu_k = 0.100$.

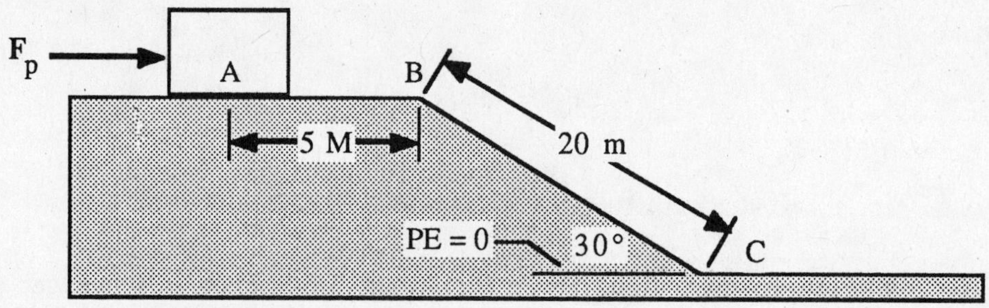

Determine the Following:

________ 1. The gravitational potential energy at position A
________ 2. The work done on the crate by you as you push it from A to B
________ 3. The work done on the crate by gravity as you push it from A to B
________ 4. The change in gravitational potential energy as you push the crate from A to B
________ 5. The work done on the crate by friction as you push it from A to B
________ 6. The net work done on the crate as you push it from A to B
________ 7. The change in kinetic energy of the crate as you push it from A to B
________ 8. The speed of the crate as it passes B
________ 9. The total energy of the crate at B
________ 10. The work done on the crate by gravity as it slides down the incline
________ 11. The change in gravitational potential energy of the crate as it slides down the incline
________ 12. The work done on the crate by friction as it slides down the incline
________ 13. The net work done on the crate as it slides down the incline
________ 14. The change in kinetic energy of the crate as it slides down the incline

________ 15. The kinetic energy of the crate at C

________ 16. The decelerating force acting on the crate as it slides across the final horizontal section

________ 17. The deceleration of the crate as it slides across the final horizontal section

________ 18. The distance the crate slides along the final horizontal section before coming to rest

(See Appendix I for answers.)

PRINCIPAL CONCEPTS AND EQUATIONS PRESCRIPTION

Your score on the practice test is an excellent measure of your understanding of the chapter. You should now use the following chart to write your own prescription for dealing with any weaknesses the practice test points out. Look down the leftmost column to the number of the question(s) you answered incorrectly, reading across that row you will find the concept and/or equation of concern, the section(s) of the study guide you should return to for further study, and some suggested text exercises which you should work to gain additional experience.

Practice Test Questions	Concepts and Equations	Prescription	
		Principal Concepts	Text Exercises
1	Gravitational potential energy	4	9-1, 2
2	Work	1	9-7, 15
3	Work	1	9-7, 15
4	Gravitational potential energy	4	9-1, 2
5	Work	1	9-7, 15
6	Work	1	9-27
7	Work-energy theorem	3	9-28, 32
8	Kinetic energy	3	9-1, 2
9	Work-energy theorem	3	9-32, 33
10	Work	1	9-29
11	Gravitational potential energy	4	9-1, 2
12	Work	1	9-27
13	Work	1	9-24
14	Work-energy theorem	3	9-28, 32
15	Work-energy theorem	3	9-32, 33
16	Friction	2 of Ch 6	2-4, 15
17	Newton's second law	2 of Ch 5	5-30, 31
18	Work-energy theorem	3	9-28, 32

10 MOMENTUM AND THE MOTION OF SYSTEMS

RECALL FROM PREVIOUS CHAPTERS

Previously learned concepts and equations frequently used in this chapter	Text Section	Study Guide Page
Resolving vectors into components	2-4	2-8
Adding vectors - analytically	2-4	2-8
Velocity: $\mathbf{v} = d\mathbf{r}/dt$	3-2	3-5
Acceleration: $\mathbf{a} = d^2\mathbf{r}/dt^2 = d\mathbf{v}/dt$	3-4	3-12
Newton's Second Law of motion: $\mathbf{F} = m\mathbf{a}$	5-4	5-5
Kinetic Energy: $K = mv^2/2$	8-4	5-7

NEW IDEAS IN THIS CHAPTER

Concepts and equations introduced	Text Section	Study Guide Page
Center of mass of a system of particles: $\mathbf{r}_{cm} = \Sigma\, m_i\mathbf{r}_i/M$	10-1	10-2
Center of mass of an extended object: $\mathbf{r}_{cm} = \int \mathbf{r}\,dm/M$	10-1	10-2
Velocity of the CM of a system of particles: $\mathbf{v}_{cm} = d(\mathbf{r}_{cm})/dt = \sum_i m_i\mathbf{v}_i/M$	10-2	10-6
Acceleration of the CM of a system of particles: $\mathbf{a}_{cm} = \dfrac{d(\mathbf{v}_{cm})}{dt} = \sum_i m_i\mathbf{a}_i/M = \sum \mathbf{F}_{ext}/M$	10-2	10-6
Momentum: $\mathbf{p} = m\mathbf{v}$	10-3	10-10
Conservation of momentum: If $\mathbf{F}_{ext} = 0$, then $\Delta\mathbf{p} = 0$ and $\mathbf{p}_i = \mathbf{p}_f$	10-5	10-10
Impulse: $\Delta\mathbf{p} = \mathbf{p}_f - \mathbf{p}_i = \int \mathbf{F}(t)dt = \mathbf{J}$	10-4	10-14
In elastic collisions, momentum and energy are conserved	10-7	10-18
In inelastic collisions only momentum is conserved	10-7	10-18
Collisions in two dimensions	10-8	10-23

PRINCIPAL CONCEPTS AND EQUATIONS

1 Center of Mass (Section 10-1)

Review: The center of mass is a mass weighted position. The center of mass of a system of particles is given by

$$\mathbf{r}_{cm} = \Sigma\, m_i \mathbf{r}_i / M$$

In component form this becomes

$$x_{cm} = \Sigma\, m_i x_i / M \qquad y_{cm} = \Sigma\, m_i y_i / M \qquad z_{cm} = \Sigma\, m_i z_i / M$$

The center of mass of an extended object is given by

$$\mathbf{r}_{cm} = \int \mathbf{r}\, dm / M$$

In component form this becomes

$$x_{cm} = \int x \rho\, dV / M \qquad y_{cm} = \int y \rho\, dV / M \qquad z_{cm} = \int z \rho\, dV / M$$

Practice: Consider the objects shown in Figure 10.1.

10.1(a)

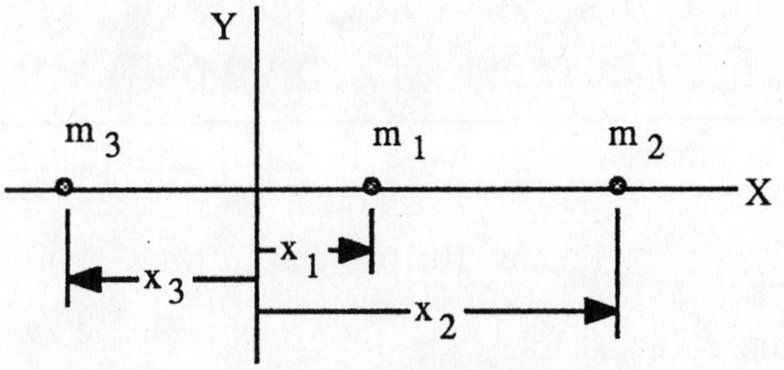

$m_1 = 1.00$ kg $\quad m_2 = 2.00$ kg $\quad m_3 = 0.500$ kg

$x_1 = 0.200$ m $\quad x_2 = 0.500$ m $\quad x_3 = -0.300$ m

10.1(b)

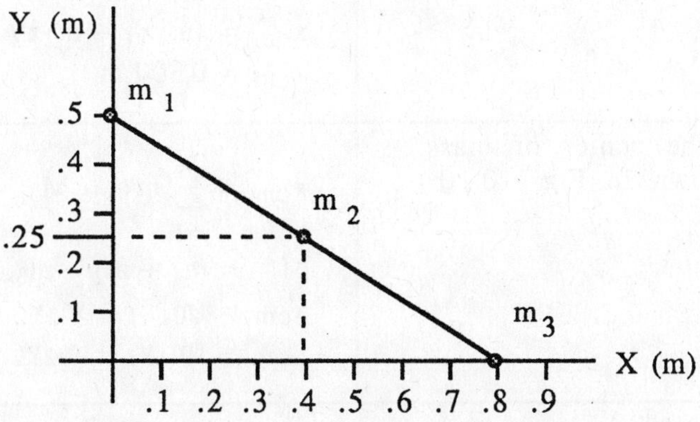

$m_1 = 1.00$ kg $\quad m_2 = 2.00$ kg $\quad m_3 = 0.500$ kg

Figure 10.1(a) and (b)

10.1(c)

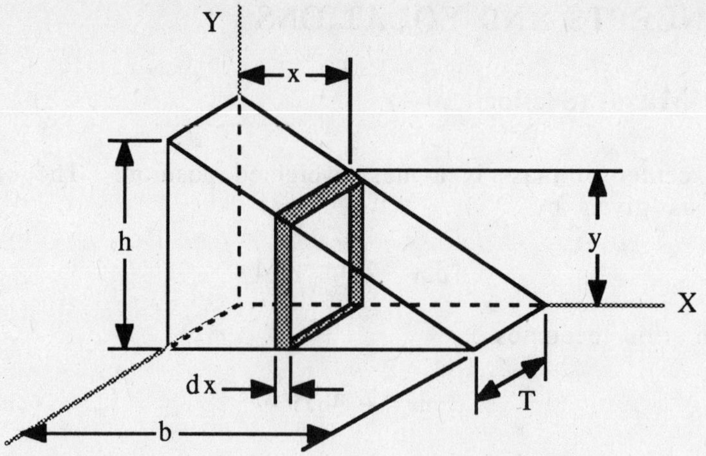

10.1(d)

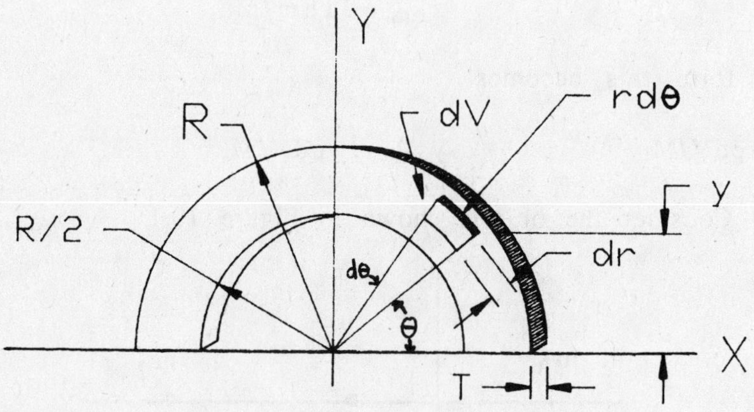

Figure 10.1(c) and (d)

Determine the following:

1. Center of mass of the object shown in Figure 10.1(a)	$x_{cm} = \sum_{i=1}^{3} m_i x_i / M$ $M = m_1 + m_2 + m_3 = 3.50$ kg $x_{cm} = (m_1x_1 + m_2x_2 + m_3x_3)/M$ $x_{cm} = 0.300$ m
2. Components of the center of mass of the object shown in Fig. 10.1(b)	$x_{cm} = \sum_{i=1}^{3} m_i x_i / M$ $M = m_1 + m_2 + m_3 = 3.50$ kg $x_{cm} = (m_1x_1 + m_2x_2 + m_3x_3)/M = 0.343$ m $y_{cm} = (m_1y_1 + m_2y_2 + m_3y_3)/M = 0.286$ m

3. Components of the center of mass of the object shown in Fig. 10.1(c)	$x_{cm} = \int_0^b x\rho dV/M$ $\rho = M/V = M/(hbT/2)\ , \quad dV = yTdx$ $x_{cm} = (\rho/M)\int_0^b xyTdx$ $y = (-h/b)x + h = h[1 - x/b]$ $x_{cm} = (2/hbT)\int_0^b xh[1 - x/b]\,Tdx$ $= (2/b)\int_0^b (x - x^2/b)dx$ $=(2/b)\left[x^2/2 - x^3/3b\right]\Big\|_0^b = b/3$ In like manner we get $y_{cm} = a/3$
4. Components of the center of mass of the object shown in Fig. 10.1(d)	$x_{cm} = 0$ by symmetry $y_{cm} = \int y\rho dV/M = (\rho/M)\int ydV$ $\rho/M = 1/V = 1/\pi T(R^2 - R^2/4) = 4/3\pi TR^2$ $y = r\sin\theta\ ,$ and $dV = Trd\theta dr$ $y_{cm} = \dfrac{4}{3\pi TR^2}\int_{R/2}^{R}\int_0^{\pi}(r\sin\theta)(Trd\theta dr)$ $= \dfrac{4}{3\pi R^2}\int_{R/2}^{R} r^2 \int_0^{\pi}(\sin\theta d\theta dr)$ $= \dfrac{4}{3\pi R^2}\int_{R/2}^{R} r^2\,(-\cos\theta)\Big\|_0^{\pi} dr$ $= \dfrac{8}{3\pi R^2}\int_{R/2}^{R} r^2 dr = (\dfrac{8}{3\pi R^2})(\dfrac{r^3}{3})\Big\|_{R/2}^{R}$ $= (\dfrac{8}{9\pi R^2})(R^3 - \dfrac{R^3}{8}) = 7R/9\pi$

Example: Determine the components of the center of mass of a right circular cone if the height of the cone is h and the radius of the base is R.

Given and Diagram: Right circular cone of height h and base of radius R.

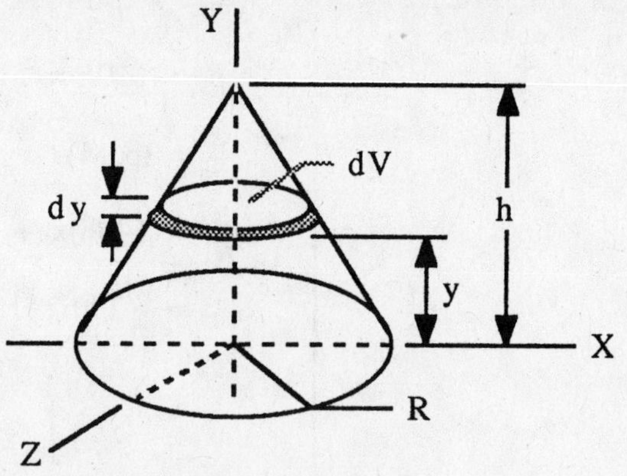

Determine: Components of the center of mass.

Strategy: We can easily determine x_{cm} and z_{cm} by symmetry. Since the density of the cone is constant, the equation for the y component of the center of mass is given by

$$y_{cm} = \int y\rho dV/M = (\rho/M) \int y dV = 1/V \int y dV$$

We can determine V the volume of the cone. If we use the infinitesimal volume element shown, dV will be a function of x (the radius of the element). As we add up all of the dV's we will be integrating over y. We can express x as a function of y by writing the equation for the straight line of the cone.

Solution: x_{cm} and z_{cm} are zero due to symmetry. The y component of the center of mass may be determined by

$$y_{cm} = \int y\rho dV/M = (\rho/M) \int y dV = 1/V \int y dV$$

The volume of the cone is: $V = \pi R^2 h/3$

The volume element is: $dV = \pi x^2 dy$

The equation for the straight line of the cone is

$$y = (-h/R)x + h \qquad \text{or} \qquad x = (h - y)R/h$$

Inserting the above into the expression for y_{cm}, we obtain

$$y_{cm} = \frac{3}{\pi R^2 h}\int_0^h y\pi x^2 dy = \frac{3}{R^2 h}\int_0^h y(h - y)^2\left(\frac{R^2}{h^2}\right)dy = \frac{3}{h^3}\int_0^h y(h^2 - 2hy + y^2)\,dy$$

$$= \left(\frac{3}{h^3}\right)\left(\frac{h^2 y^2}{2} - \frac{2hy^3}{3} + \frac{y^4}{4}\right)\Bigg|_0^h = \frac{h}{4}$$

The components of the center of mass are (0, h/4, 0).

Related Text Exercises: Ex. 10-1 through 10-6.

2 Motion of the Center of Mass (Section 10-2)

Review: The position of the center of mass of a system of particles is given by

$$\mathbf{r}_{cm} = \Sigma\ m_i\mathbf{r}_i/M$$

The velocity of the center of mass is given by

$$\mathbf{v}_{cm} = d(\mathbf{r}_{cm})/dt = \Sigma\ m_i\mathbf{v}_i/M$$

The acceleration of the center of mass is given by

$$\mathbf{a}_{cm} = d(\mathbf{v}_{cm})/dt = \Sigma\ m_i\mathbf{a}_i/M$$

The acceleration of the center of mass is determined by the external forces only.

$$\Sigma\ \mathbf{F}_{ext} = M\mathbf{a}_{cm}$$

Practice: Figure 10.2 shows two particles acted on by external forces at t = 0 s.

Figure 10.2

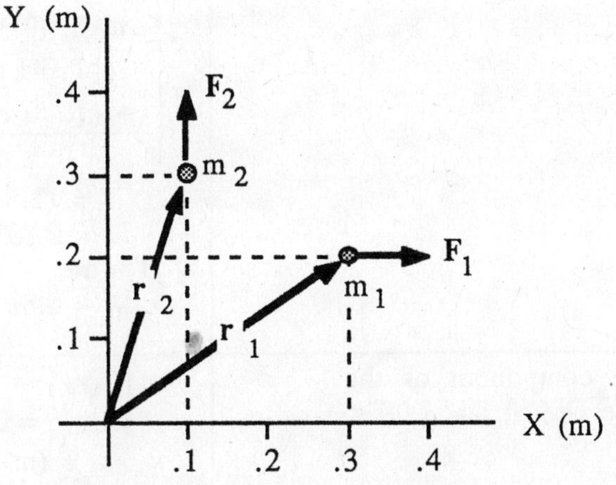

$m_1 = 1.00$ kg $\quad \mathbf{F}_1 = 4.00$ N $\mathbf{i}$ $\quad m_2 = 2.00$ kg $\quad \mathbf{F}_2 = 6.00$ N $\mathbf{j}$

$$\mathbf{r}_1 = [0.300\ \text{m} + (2.00\ \text{m/s}^2)t^2]\ \mathbf{i} + (0.200\ \text{m})\ \mathbf{j}$$

$$\mathbf{r}_2 = (0.100\ \text{m})\ \mathbf{i} + [0.300\ \text{m} + (0.500\ \text{m/s})\ t + (1.5\ \text{m/s}^2)t^2]\ \mathbf{j}$$

Determine the following:

1. Expressions for $\mathbf{v}_1$, $\mathbf{v}_2$, $\mathbf{a}_1$, and $\mathbf{a}_2$ at any time	$\mathbf{v}_1 = d\mathbf{r}_1/dt = (4.00\ \text{m/s}^2)t\ \mathbf{i}$ $\mathbf{v}_2 = \dfrac{d\mathbf{r}_2}{dt} = [(0.500\ \text{m/s}) + (3.00\ \text{m/s}^2)t]\ \mathbf{j}$ $\mathbf{a}_1 = d^2\mathbf{r}_1/dt^2 = (4.00\ \text{m/s}^2)\ \mathbf{i}$ $\mathbf{a}_1 = d^2\mathbf{r}_2/dt^2 = (3.00\ \text{m/s}^2)\ \mathbf{j}$
2. The x and y component of the position of each particle (m_1 and m_2) and the CM at t = 0	m_1: $x_1 = 0.300$ m $\quad y_1 = 0.200$ m m_2: $x_2 = 0.100$ m $\quad y_2 = 0.300$ m $x_{cm} = (m_1x_1 + m_2x_2)/(m_1 + m_2) = 0.167$ m $y_{cm} = (m_1y_1 + m_2y_2)/(m_1 + m_2) = 0.267$ m

3. The x and y component of the velocity of each particle and the CM at t = 0	m_1: $v_{x1} = 0$ $\quad v_{y1} = 0$ m_2: $v_{x2} = 0$ $\quad v_{y2} = 0.500$ m/s $v_{xcm} = (m_1v_{x1} + m_2v_{x2})/(m_1 + m_2) = 0$ $v_{ycm} = (m_1v_{y1} + m_2v_{y2})/(m_1 + m_2)$ $= 0.333$ m/s
4. The x and y component of the acceleration of each particle and the CM at t = 0	m_1: $a_{x1} = 4.00$ m/s^2 $\quad a_{y1} = 0$ m_2: $a_{x2} = 0$ $\quad a_{y2} = 3.00$ m/s^2 $a_{xcm} = (m_1a_{x1} + m_2a_{x2})/(m_1 + m_2)$ $= 1.33$ m/s^2 $a_{ycm} = (m_1a_{y1} + m_2a_{y2})/(m_1 + m_2)$ $= 2.00$ m/s^2
5. The x and y component of the position of each particle and the CM at t = 1.00 s	m_1: $x_1 = 2.30$ m $\quad y_1 = 0.200$ m m_2: $x_2 = 0.100$ m $\quad y_2 = 2.30$ m $x_{cm} = (m_1x_1 + m_2x_2)/(m_1 + m_2) = 0.833$ m $y_{cm} = (m_1y_1 + m_2y_2)/(m_1 + m_2) = 1.60$ m or $\mathbf{r}_{cm} = (m_1\mathbf{r}_1 + m_2\mathbf{r}_2)/M =$ $\left[\dfrac{([1.0\text{ kg}]\,[(2.3\text{ m})\mathbf{i} + (0.2\text{ m})\mathbf{j}]) + ([2.0\text{ kg}]\,[(0.1\text{ m})\mathbf{i} + (2.3\text{ m})\mathbf{j}])}{3\text{ kg}}\right]$ $= (2.5\text{ kg·m }\mathbf{i} + 4.8\text{ kg·m }\mathbf{j})/3\text{ kg}$ $= 0.833\text{ m }\mathbf{i} + 1.60\text{ m }\mathbf{j}$ Hence $x_{cm} = 0.833$ m and $y_{cm} = 1.60$ m
6. The x and y component of the velocity of the CM at t = 1.00 s	m_1: $v_{x1} = 4.00$ m/s $\quad v_{y1} = 0$ m_2: $v_{x2} = 0$ $\quad v_{y2} = 3.50$ m/s $v_{xcm} = (m_1v_{x1} + m_2v_{x2})/M = 1.33$ m/s $v_{ycm} = (m_1v_{y1} + m_2v_{y2})/M = 2.33$ m/s or $\mathbf{F}_{ext} = \mathbf{F}_1 + \mathbf{F}_2 = (4.00\text{ N})\mathbf{i} + (6.00\text{ N})\mathbf{j}$ $M = m_1 + m_2 = 3.00$ kg $\mathbf{a}_{cm} = \mathbf{F}_{ext}/M$ $= (1.33\text{ m/s}^2)\mathbf{i} + (2.00\text{ m/s}^2)\mathbf{j}$ $\mathbf{v}_{ocm} = (0.333\text{ m/s})\mathbf{j}$ (Step 3) $\mathbf{v}_{cm} = \mathbf{v}_{ocm} + \mathbf{a}_{cm}t$ $= (0.333\text{ m/s})\mathbf{j}$ $+ (1.33\text{ m/s})\mathbf{i} + (2.00\text{ m/s})\mathbf{j}$ $= (1.33\text{ m/s})\mathbf{i} + (2.33\text{ m/s})\mathbf{j}$ Hence $v_{xcm} = 1.33$ m/s and $v_{ycm} = 2.33$ m/s
7. External force acting on the CM at any time	$\mathbf{F}_{ext} = \mathbf{F}_1 + \mathbf{F}_2 = (4.00\text{ N})\mathbf{i} + (6.00\text{ N})\mathbf{j}$

8. Acceleration of the CM at any time	$a_{cm} = F_{ext}/M$ $= [(4.00\ N)i + (6.00\ N)j]/(3.00\ kg)$ $= (1.33\ m/s^2)i + (2.00\ m/s^2)j$
9. The acceleration of the CM at $t = 1.00$ s	$a_{cm} = (1.33\ m/s^2)i + (2.00\ m/s^2)j$ Since the expression for the acceleration is not a function of time, the acceleration is a constant.

Example: A 2.00 kg block slides down a 4.00 kg inclined plane which is 2.00 m long. The incline of the plane is 30.0° and the block slides without friction. Determine the location, velocity, and acceleration of the center of mass of the system (block and incline) 0.500 s after the block is released.

Given and Diagram:

m_b = 2.00 kg	mass of block	θ = 30.0°	angle of incline
m_p = 4.00 kg	mass of plane	μ = 0	coefficient of friction
L = 2.00 m	length of plane	t = 0.500 s	time of interest

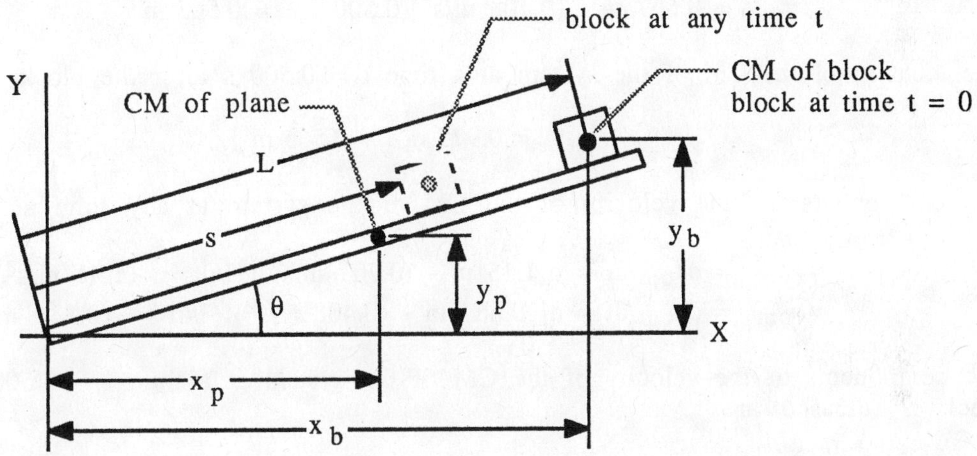

Figure 10.3

Determine: Location (r_{cm}), velocity (v_{cm}), and acceleration (a_{cm}) of the center of mass 0.500 s after the block is released from the top of the inclined plane.

Strategy: First let's write the components of the CM of the plane. Note that these will be constant in time since the plane is stationary. Next let's determine the components of the CM of the block as a function of time. Knowing the mass of the block and plane and the components of their CM at any time, we can determine expressions for the components of the CM of the system at any time. We can differentiate these expressions once to obtain expressions for the components of the velocity of the CM of the system at any time and twice to obtain expression for the components of the acceleration of the CM of the system at any time.

We can insert the value t = 0.500 s into these expression to obtain the components of the position, velocity, and acceleration of the CM of the system 0.500 s after it is released. Knowing the components at t = 0.500 s we can determine r_{cm}, v_{cm}, and a_{cm} at t = 0.500 s.

Solution: The components for the CM of the plane are

$$x_p = L\cos\theta/2 \qquad \text{and} \qquad y_p = L\sin\theta/2$$

The components for the CM of the block at any time are

$$x_b = (L - gt^2\sin\theta/2)\cos\theta \qquad \text{and} \qquad y_b = (L - gt^2\sin\theta/2)\sin\theta$$

The components for the CM of the system at any time are

$$\begin{aligned} x_{cm} &= (m_p x_p + m_b x_b)/(m_p + m_b) \\ &= [m_p L\cos\theta/2 + m_b(L - gt^2\sin\theta/2)\cos\theta]/(m_p + m_b) \\ &= 1.15\text{ m} - (0.707\text{ m/s}^2)t^2 \\ y_{cm} &= (m_p y_p + m_b y_b)/(m_p + m_b) \\ &= [m_p L\sin\theta/2 + m_b(L - gt^2\sin\theta/2)\sin\theta]/(m_p + m_b) \\ &= 0.667\text{ m} - (0.408\text{ m/s}^2)t^2 \end{aligned}$$

The components for the CM of the system at a time t = 0.500 s after the block is released are

$$\begin{aligned} x_{cm} &= 1.15\text{ m} - (0.707\text{ m/s}^2)(0.500\text{ s})^2 = 0.973\text{ m} \\ y_{cm} &= 0.667\text{ m} - (0.408\text{ m/s}^2)(0.500\text{ s})^2 = 0.565\text{ m} \end{aligned}$$

The location of the CM of the system at a time t = 0.500 s after the block is released is

$$\mathbf{r}_{cm} = 0.973\text{ m }\mathbf{i} + 0.565\text{ m }\mathbf{j}$$

The components for the velocity of the CM of the system at any time are

$$\begin{aligned} v_{xcm} &= dx_{cm}/dt = d[1.15\text{ m} - (0.707\text{ m/s}^2)t^2]/dt = -(1.41\text{ m/s}^2)t \\ v_{ycm} &= dy_{cm}/dt = d[\,0.667\text{ m} - (0.408\text{ m/s}^2)t^2]/dt = -(0.816\text{ m/s}^2)t \end{aligned}$$

The components of the velocity of the CM of the system at a time t = 0.500 s after the block is released are

$$\begin{aligned} v_{xcm} &= (-1.41\text{ m/s}^2)(0.500\text{ s}) = -0.705\text{ m/s} \\ v_{ycm} &= (-0.816\text{ m/s}^2)(0.500\text{ s}) = -0.408\text{ m/s} \end{aligned}$$

The velocity of the CM of the system at a time t = 0.500 s after the block is released is

$$\mathbf{v}_{cm} = (-0.705\text{ m/s})\,\mathbf{i} - (0.408\text{ m/s})\,\mathbf{j}$$

The components for the acceleration of the CM of the system at anytime are

$$\begin{aligned} a_{xcm} &= d\,v_{xcm}/dt = d[(-1.41\text{ m/s}^2)t]/dt = -1.41\text{ m/s}^2 \\ a_{ycm} &= d\,v_{ycm}/dt = d[(-0.816\text{ m/s}^2)t]/dt = -0.816\text{ m/s}^2 \end{aligned}$$

Note that the components of the acceleration of the CM are constant in time. This means that the acceleration of the block down the plane has a constant value which is independent of time. The acceleration of the CM of the system at any time (including the time t = 0.500 s) is

$$\mathbf{a}_{cm} = -1.41\text{ m/s}^2\,\mathbf{i} - 0.816\text{ m/s}^2\,\mathbf{j}$$

Related Text Exercises: Ex 10-7 through 10-12.

3 Conservation of Momentum (Section 10-3 and 10-5)

Review: The momentum **p** of a particle of mass m and velocity **v** is defined by

$$\mathbf{p} = m\mathbf{v}$$

Momentum is a vector which has the same direction as the velocity. If external forces act on a particle, its momentum is changed.

$$\Sigma \mathbf{F}_{ext} = d\mathbf{p}/dt$$

It should be recognized that this reduces to Newton's Second Law in the form we studied in chapter five.

$$\Sigma \mathbf{F}_{ext} = d\mathbf{p}/dt = d(m\mathbf{v})/dt = md\mathbf{v}/dt = m\mathbf{a}$$

Notice from the above that if $\Sigma \mathbf{F}_{ext} = 0$, then $d\mathbf{p}/dt = 0$ or $\Delta\mathbf{p} = \mathbf{p}_f - \mathbf{p}_i = 0$ or $\mathbf{p}_f = \mathbf{p}_i$.

That is the momentum of the particle is conserved if the external force on it is zero.

If we are dealing with a system of particles, the total momentum of the system is given by

$$\mathbf{P} = \sum_i \mathbf{P}_i$$

If a number of external forces act on the system, the total external force is given by

$$\mathbf{F}_{ext} = \sum_i (\mathbf{F}_{ext})_i$$

The external forces cause the momentum to change, and the change in momentum is given by

$$d\mathbf{P}/dt = d(\sum_i \mathbf{P}_i)/dt = d(\sum_i m_i\mathbf{v}_i)/dt$$

Recall that $\mathbf{v}_{cm} = \sum_i m_i\mathbf{v}_i/M$ or $\sum_i m_i\mathbf{v}_i = M\mathbf{v}_{cm}$

Then $d\mathbf{P}/dt = d(M\mathbf{v}_{cm})/dt = Md\mathbf{v}_{cm}/dt = M\mathbf{a}_{cm} = \mathbf{F}_{ext}$

Notice from the above that if $\mathbf{F}_{ext} = 0$, then $d\mathbf{P}/dt = 0$ or $\Delta\mathbf{P} = \mathbf{P}_f - \mathbf{P}_i = 0$ or $\mathbf{P}_f = \mathbf{P}_i$.

That is the momentum of the system of particles is conserved if the external force on the system is zero.

Practice: A 0.200-kg bullet penetrates a 2.00-kg block of wood initially at rest on a horizontal frictionless surface. The bullet is traveling 300 m/s before it hits the block and 200 m/s after it emerges from the block. Let's agree to call the direction of the bullet positive. Since this problem occurs in one dimension, it will be convenient to drop the vector notation and use "+" and "-" to indicate direction.

Determine the following:

1. Initial momentum of the bullet	$P_{bi} = m_b v_{bi} = 60.0$ kg•m/s
2. Final momentum of the bullet	$P_{bf} = m_b v_{bf} = 40.0$ kg•m/s
3. Resultant external force acting on the bullet and block as the bullet travels through the block	$F_{ext} = 0$. As the bullet penetrates the block, the bullet and block exert an internal force on each other, but no external forces exist.
4. Whether or not momentum is conserved	Yes, since $F_{ext} = 0$, then $\Delta P = 0$ $P_f = P_i$
5. Momentum of the block after it is hit by the bullet (Use the notation P_{bi} and P_{bf} for the initial and final momentum of the bullet, and use P_{wi} and P_{wf} for the initial and final momentum of the wood block.)	Given conservation of momentum, we write $P_i = P_f$ $P_{bi} = P_{bf} + P_{wf}$ (recall that $P_{wi} = 0$) $P_{wf} = P_{bi} - P_{bf} = 20.0$ kg•m/s
6. Speed of the block after it is hit by the bullet	$P_{wf} = m_w v_{wf}$ $v_{wf} = P_{wf}/m_w = 10.0$ m/s

Example: A child runs and jumps onto a stationary grocery cart. If the cart has a mass 2/3 that of the child, find the ratio of the final speed of the cart and child to that of the initial speed of the child.

Given: $m_g = (2/3)m_c$, where m_c and m_g represent the mass of the child and grocery cart respectively.

Determine: The ratio of the final speed of the grocery cart and child (v_{g+c}) to that of the initial speed of the child (v_c). That is, determine v_{g+c}/v_c.

Strategy: The time it takes the child to land on the grocery cart is so short that we can ignore external forces and consequently say that momentum is conserved. A statement of conservation of momentum contains both the initial speed of the child v_c and the final speed of the grocery cart and child v_{g+c}. Hence the ratio of v_{g+c}/v_c can be determined from a statement of conservation of momentum.

Solution: The initial momentum is just that of the child, $P_i = m_c v_c$. The final momentum is that of the grocery cart and child.

$$P_f = (m_g + m_c)v_{g+c} = [(2/3)m_c + m_c]v_{g+c} = (5/3)m_c v_{g+c}$$

Equating the initial and final momentum and solving for the ratio of the speeds, we obtain

$$v_{g+c}/v_c = 3/5$$

Example: A 60.0-kg astronaut has been spacewalking outside her spacecraft and realizes that she is stranded (stationary) 10.0 m from the spacecraft. In order to get back, she throws a 1.00-kg hammer directly away from the ship with a speed of 6.00 m/s. How long will it take her to reach the ship.

Given: $d_a = 10.0$ m $v_h = 6.00$ m/s $m_a = 60.0$ kg $m_h = 1.00$ kg

Determine: The time it takes the astronaut to return to the spacecraft.

Strategy: Take the positive direction to be radially outward from the ship and the negative direction to be radially inward. Since the initial radial momentum for the hammer and astronaut is zero and external forces are negligible, momentum in the radial direction is conserved. If the astronaut gives the hammer a radially outward momentum (positive), she will receive the same inward momentum (negative). From such a momentum statement, we can obtain the radially inward velocity (negative) of the astronaut. Knowing velocity of the astronaut and her displacement (negative), we can determine the time of travel.

Solution: $\mathbf{P_a} + \mathbf{P_h} = 0$ (conservative of momentum)

$\mathbf{P_a} = -\mathbf{P_h}$, $m_a v_a = -m_h v_h$, $v_a = -m_h v_h/m_a = -0.100$ m/s

$t_a = d_a/v_a = (-10.0 \text{ m})/(-0.100 \text{ m/s}) = 100$ s

Example: A cannon with a certain elevation has a range R when fired on level ground (Fig. 10.5(a)). If a shell explodes at the top of the trajectory in such a manner that 1/3 of the mass falls straight down and the other 2/3 continues onward (Fig. 10.5(b)), how far (in terms of R) will the 2/3 fragment land from the cannon?

Given and Diagram:

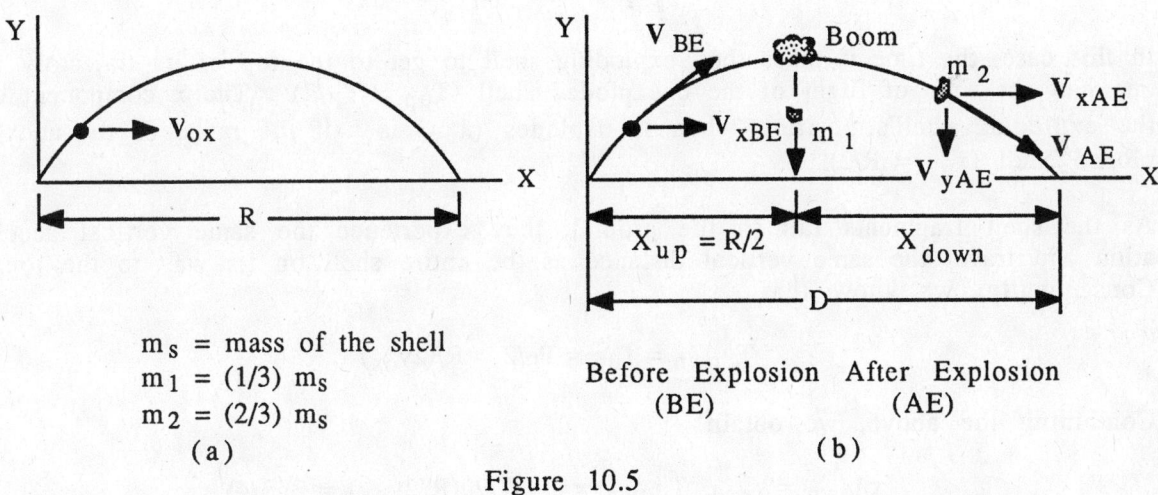

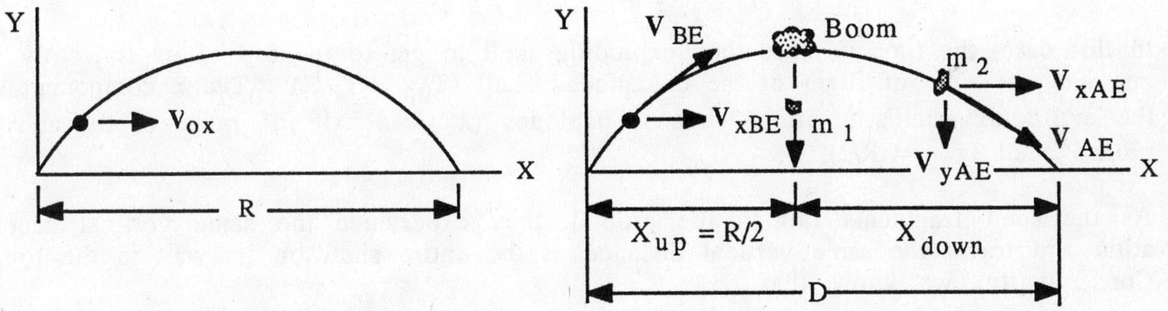

m_s = mass of the shell

$m_1 = (1/3)\ m_s$

$m_2 = (2/3)\ m_s$

(a)

Before Explosion (BE) After Explosion (AE)

(b)

Figure 10.5

Determine: How far (in terms of R) the mass m_2 lands from the cannon.

Strategy: Since no external forces act in the x direction (even during the explosion), momentum in the x direction is conserved. From a statement of conservation of momentum in the x direction, we can determine the x component of the velocity of

m_2 after the explosion in terms of the x component of the velocity of the entire shell before the explosion. Knowing the range of an unexploded shell, we can determine the time of flight of an unexploded shell in terms of the x component of its velocity. We can determine the time of flight for the exploded fragments from the time of flight of an unexploded shell. Knowing the x component of the fragment's velocity and the time of flight of the fragment, we can determine how far the fragment travels.

Solution: Since momentum is conserved in the x direction, we can equate the x component of the momentum before the explosion to the x component of the momentum after the explosion.

$$m_s v_{xBE} = m_2 v_{xAE_2} + m_1 v_{xAE_1}$$

Since $m_2 = (2/3)m_s$ and $v_{xAE_1} = 0$, this becomes

$$m_s v_{xBE} = (2/3)m_s v_{xAE_2} \quad \text{or} \quad v_{xAE_2} = (3/2)v_{xBE}$$

The x component of the velocity before the explosion (v_{xBE}) is equal to the x component of the velocity of an unexploding shell, v_{ox}. Hence the previous equation can be written as

$$v_{xAE_2} = (3/2)v_{ox}$$

The x component of the velocity, the range, and the time of flight of an unexploding shell are related by

$$R = v_{ox}T_F \quad \text{or} \quad T_F = R/v_{ox}$$

In this case, the time it takes the exploding shell to get to the top of its trajectory is one half the time of flight of the unexploded shell ($T_{up} = T_F/2$). The x component of the exploding shell's position when it explodes (x_{up}) is half the range of the unexploded shell ($x_{up} = R/2$).

As the shell fragments fall to the ground, they experience the same vertical acceleration and travel the same vertical distance as the entire shell on its way to the top. Consequently, we know that

$$T_{down} = T_{up} = T_F/2 = R/(2v_{ox})$$

Combining the above, we obtain

$$x_{down} = v_{xAE_2}T_{down} = (3v_{ox}/2)(R/2v_{ox}) = 3(R/4)$$

Finally

$$D = x_{up} + x_{down} = R/2 + 3(R/4) = 5(R/4)$$

Related Text Exercises: Ex. 10-13 through 10-15 and 10-21 through 10-26.

4 Impulse (Section 10-4)

Review: Forces which are exerted over a limited time are called impulsive forces. If a constant impulsive force **F** acts on an object for a time Δt, the change in momentum of the object and hence the impulse imparted to the object is given by:

$$\Delta \mathbf{p} = \mathbf{p}_f - \mathbf{p}_i = \mathbf{F}\Delta t = \mathbf{J}$$

If the impulsive force varies with time, the change in momentum of the object and hence the impulse imparted to the object is given by:

$$\Delta \mathbf{p} = \mathbf{p}_f - \mathbf{p}_i = \int_{t_i}^{t_f} \mathbf{F}(t)dt = \mathbf{J}$$

Note that the impulse is just the area under the force versus time plot. See Fig. 10.6.

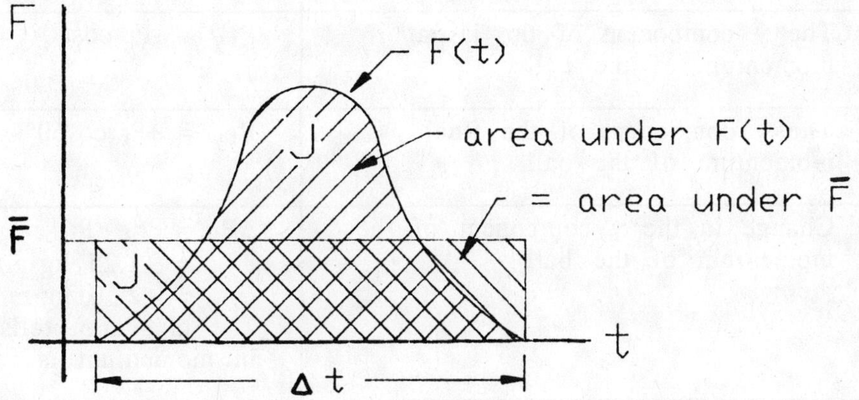

Figure 10.6

It is frequently difficult to know the force acting on the object as a function of time. When this is the case, we will use the constant average force $\bar{\mathbf{F}}$ which will deliver the same impulse. That is the area under the $\bar{\mathbf{F}}$ line in Figure 10.6 is the same as the area under the **F**(t) curve.

Practice: A 50.0 g golf ball bounces off a cement floor as shown in Figure 10.7. The ball hits the floor with a speed of 50.0 m/s and bounces off with no loss in speed. The ball and the floor are in contact for 0.0100 s.

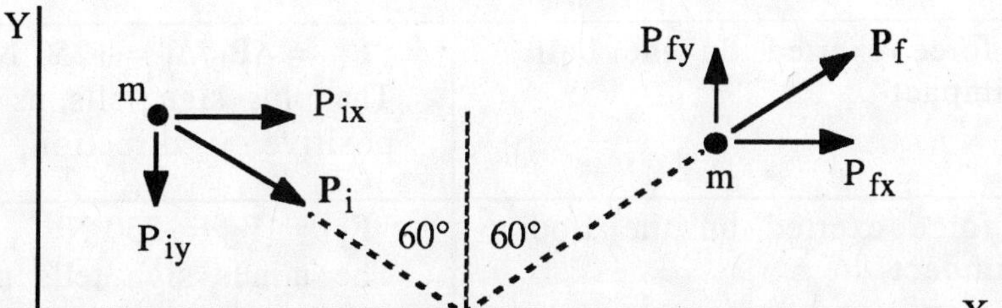

Figure 10.7

Determine the following:

1. Magnitude of the initial momentum of the ball	$P_i = mv_i = 2.50$ kg•m/s
2. Magnitude of the final momentum of the ball	$P_f = mv_f = 2.50$ kg•m/s
3. The x component of the initial momentum of the ball	$P_{ix} = P_i \sin 60° = 2.17$ kg•m/s
4. The x component of the final momentum of the ball	$P_{fx} = P_f \sin 60° = 2.17$ kg•m/s
5. Change in the x component of the momentum of the ball	$\Delta P_x = P_{fx} - P_{ix} = 0$
6. The y component of the initial momentum of the ball	$P_{iy} = -P_i \cos 60° = -1.25$ kg•m/s
7. The y component of the final momentum of the ball	$P_{fy} = +P_f \cos 60° = +1.25$ kg•m/s
8. Change in the y component of the momentum of the ball	$\Delta P_y = P_{fy} - P_{iy}$ $= +1.25$ kg•m/s $-$ (-1.25 kg•m/s) $= +2.50$ kg•m/s The plus sign tells us that the change in momentum is in the +y direction.
9. Vector diagram showing $\mathbf{P_i}$, $\mathbf{P_f}$ and $\Delta \mathbf{P}$	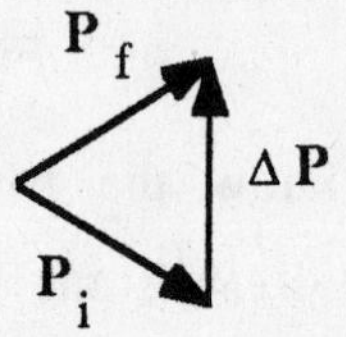 Notice that $\mathbf{P_f} = \mathbf{P_i} + \Delta \mathbf{P}$
10. Impulse imparted to the ball during the collision	$\text{Impulse}_b = \Delta \mathbf{P_b} = +2.50$ kg•m/s The impulse imparted to the ball is in the +y direction.
11. Impulse imparted to the floor during the collision	$\text{Impulse}_f = -\text{Impulse}_b = -2.50$ kg•m/s The impulse imparted to the floor is in the -y direction.
12. Average force exerted on the ball during impact	$\mathbf{F_b} = \Delta \mathbf{P_b}/\Delta t = +250$ N The plus sign tells us the $\mathbf{F_b}$ is in the positive y direction.
13. Average force exerted on the floor during impact	$\mathbf{F_f} = -\mathbf{F_b} = -250$ N The minus sign tells us the $\mathbf{F_f}$ is in the negative y direction.

Example: Two students are engaged in a snowball fight. One student throws a 0.250-kg snowball with a velocity of 10.0 m/s at the other student. The snowball sticks to the student and the impact takes place over a 0.0100 s time interval. Determine the collision impulse and the average impulse force experienced by the student.

Given: $m = 0.250$ kg $\quad v_i = 10.0$ m/s $\quad v_f = 0$ m/s $\quad \Delta t = 0.0100$ s

Determine: The collision impulse and the average impulse force experienced by the student.

Strategy: Call the initial direction of the snowball positive. Knowing the mass and initial speed of the snowball, determine its initial momentum. Since the final momentum of the snowball is zero, its change in momentum and hence its impulse are equal in magnitude but opposite in direction to its initial momentum. The impulse acting on the student is equal in magnitude to that acting on the snowball, but acts in the opposite direction. The average impulse force acting on the student can be determined from the impulse acting on the student and the time interval during which the impulse occurs. Since this problem occurs in one dimension, it will be convenient to drop the vector notation and use "+" and "-" to indicate the direction.

Solution:

Initial momentum of the snowball	$P_i = mv_i = +2.50$ kg•m/s
Final momentum of the snowball	$P_f = mv_f = 0$ kg•m/s
Change in momentum of the snowball	$\Delta P_{sb} = P_f - P_i = -2.50$ kg•m/s
Impulse delivered to the snowball	$\text{Impulse}_{sb} = \Delta P_{sb} = -2.50$ kg•m/s
Impulse delivered to the student	$\text{Impulse}_{st} = -\text{Impulse}_{sb} = +2.50$ kg•m/s
Impulse force acting on the student	$F_{st} = \Delta P_{st}/\Delta t = +250$ N

The positive sign tells us that the force acting on the student is in the same direction as the initial momentum of the snowball.

Example: Figure 10.8 shows the force acting on a 50.0 g ball thrown against a wall with a speed of 50.0 m/s. Find the total impulse experienced by the ball, the average force experienced by the ball, the final velocity of the ball, and the energy lost by the ball during the collision.

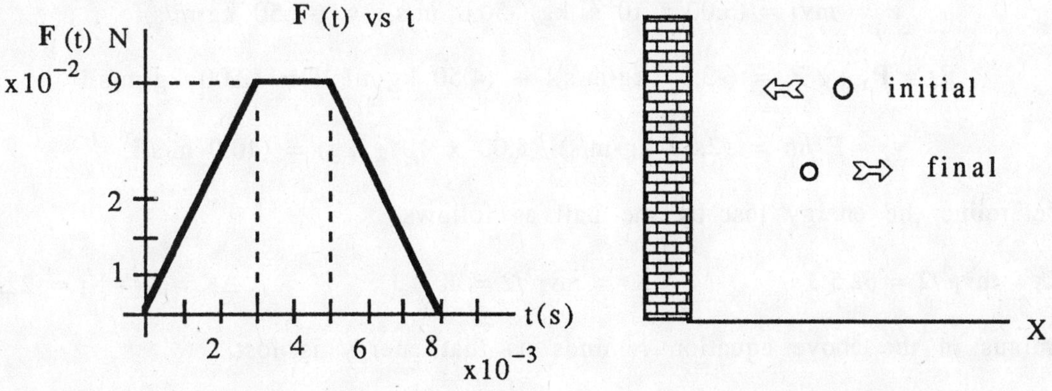

Figure 10.8

Given: $m = 5.00 \times 10^{-2}$ kg — mass of ball
$v_i = 50.0$ m/s — initial speed of ball
F(t) vs t plot — Fig. 10.8

Determine: J — total impulse experienced by the ball
$\bar{F}$ — average force experienced by the ball
v_f — final speed of the ball
ΔK — loss in energy during the collision

Strategy: Knowing that the area under the **F**(t) plot is the impulse, we may obtain the total impulse experienced by the ball and the average force experienced by the ball during the 8.00×10^{-3} s time interval. Knowing the impulse experienced by the ball, we know the change in momentum of the ball. Knowing the change in momentum, the mass and the initial speed, we can determine the final speed. Knowing the mass of the ball and its initial and final speed, we can determine the energy loss due to the ball hitting the wall.

Solution: We may determine the impulse experienced by the ball by calculating the area under the **F**(t) plot.

$$J = \left[\left[(9.00 \times 10^2 \text{ N})(3.00 \times 10^{-3} \text{ s})/2\right] + \left[(9.00 \times 10^2 \text{ N})(2.00 \times 10^{-3} \text{ s})\right] + \left[(9.00 \times 10^2 \text{ N})(3.00 \times 10^{-3} \text{ s})/2\right] \right]$$

$$J = 4.50 \text{ N•s or } \mathbf{J} = (4.50 \text{ N•s})\mathbf{i}$$

Now that we know the total impulse, we can determine the average force.

$$\bar{\mathbf{F}} = \mathbf{J}/\Delta t = (4.50 \text{ N•s})\mathbf{i}/8.00 \times 10^{-3} \text{ s} = 563 \text{ N}\mathbf{i}$$

Knowing the impulse, we know the change in momentum

$$\Delta\mathbf{P} = \mathbf{J} = (4.50 \text{ N.s})\mathbf{i}$$

We determine the final momentum and hence speed as follows

$$\mathbf{P}_i = m\mathbf{v}_i = (5.00 \times 10^{-2} \text{ kg})(-50.0 \text{ m/s})\mathbf{i} = (-2.50 \text{ kg•m/s})\mathbf{i}$$

$$\mathbf{P}_f = \mathbf{P}_i + \Delta\mathbf{P} = (-2.50 \text{ kg•m/s})\mathbf{i} + (4.50 \text{ kg•m/s})\mathbf{i} = (2.00 \text{ kg•m/s})\mathbf{i}$$

$$\mathbf{v}_f = \mathbf{P}_f/m = (2.00 \text{ kg•m/s})\mathbf{i}/(5.00 \times 10^{-2} \text{ kg}) = (40.0 \text{ m/s})\mathbf{i}$$

We determine the energy lost by the ball as follows

$$K_i = mv_i^2/2 = 62.5 \text{ J} \qquad K_f = mv_f^2/2 = 40.0 \text{ J} \qquad \Delta K = K_f - K_i = -22.5 \text{ J}$$

The minus in the above equation reminds us that energy is lost.

Related Exercises: Ex. 10-17 through 10-20.

5 Collisions In One Dimension (Section 10-6)

Review: Collisions may be elastic or inelastic.

Elastic Collisions: In an elastic collision, both momentum and kinetic energy are conserved. Hence we may write

(a) $m_1v_{1i} + m_2v_{2i} = m_1v_{1f} + m_2v_{2f}$ and

(b) $m_1v_{1i}^2/2 + m_2v_{2i}^2/2 = m_1v_{1f}^2/2 + m_2v_{2f}^2/2$

When these two equations are solved simultaneously for the two unknowns (v_{1f} and v_{2f}), we obtain

(c) $v_{1f} = (m_1 - m_2)\, v_{1i}/(m_1 + m_2) + 2m_2v_{2i}/(m_1 + m_2)$ and

(d) $v_{2f} = 2m_1v_{1i}/(m_1 + m_2) + (m_2 - m_1)\, v_{2i}/(m_1 + m_2)$

Inelastic Collisions: In an inelastic collision, momentum is conserved but energy is not. In this course the only type of inelastic collision we will consider is one in which the objects stick together after collision. For this case conservation of momentum gives

$m_1v_{1i} + m_2v_{2i} = (m_1 + m_2)\, v_f$ where $v_{1f} = v_{2f} = v_f$ or

(e) $v_f = (m_1v_{1i} + m_2v_{2i})/(m_1 + m_2)$

Practice: In Figure 10.9, object 1 is fired from the gun with a speed v towards object 2. The collision is inelastic and object 1 embeds itself in object 2 in a very short time interval. The surface is frictionless and the objects have identical masses.

Figure 10.9

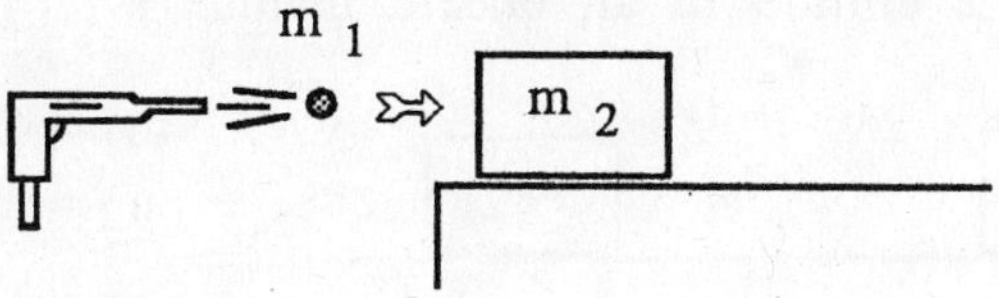

$m_1 = m_2$ $v_{1i} = v$

$\mu = 0$ $v_{2i} = 0$

Determine the Following:

1. If momentum is conserved	Since the surface is frictionless ($\mu = 0$), no external forces act on the system. Hence momentum is conserved.
2. If energy is conserved	Friction dissipates some of the energy, hence energy is not conserved.
3. Final velocity v_f of the system after collision	Since this is an inelastic collision with the objects sticking together, we may write $m_1v_{1i} + m_2v_{2i} = (m_1 + m_2)\, v_f$ But $v_{1i} = v$, $v_{2i} = 0$, and $m_1 = m_2$, hence $m_1v = 2m_1v_f$ or $v_f = v/2$.

In Figure 10.10, objects 1 and 2 collide in an elastic manner.

Figure 10.10

$m_1 = m_2$ $v_{1i} = v$

$\mu = 0$ $v_{2i} = 0$

4. Determine the velocity of objects 1 and 2 after the collision.	Since the collision is elastic, we may use equations (c) and (d) to determine v_{1f} and v_{2f}. When the given information is inserted into equations (c) and (d), we obtain $v_{1f} = v_{2i} = 0$ and $v_{2f} = v_{1i} = v$. Note that the two objects have traded velocities.

In Figure 10.11, objects 1 and 2 collide in an elastic manner.

Figure 10.11

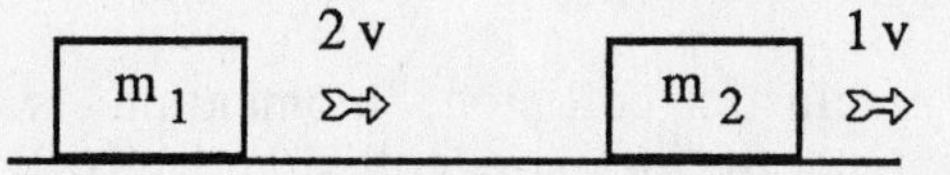

$m_1 = m_2$ $v_{1i} = 2v$

$\mu = 0$ $v_{2i} = v$

5. Determine the velocity of objects 1 and 2 after the collision.	Insert the given information into equations (c) and (d) to obtain $v_{1f} = v_{2i} = v$ and $v_{2f} = v_{1i} = 2v$. Note that the two objects have traded velocities.

Note: Any time two objects of equal mass undergo an elastic collision, they simply just trade velocities. (See steps 4 and 5 of this practice section.)

In Figure 10.12, objects 1 and 2 collide in an elastic manner

Figure 10.12

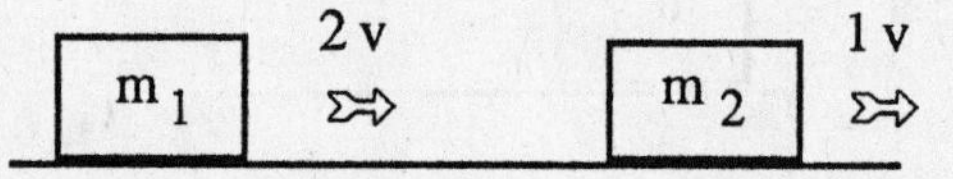

$m_1 = 2m$ $v_{1i} = 2v$

$m_2 = m$ $v_{2i} = v$

$\mu = 0$

6. Determine the velocity of objects 1 and 2 after the collision.	Insert the given information into equations (c) and (d) to obtain $v_{1f} = 4v/3$ and $v_{2f} = 7v/3$. Both objects continue moving to the right after the collision; however, object 1 is moving slower and object 2 is moving faster than before the collision.

In Figure 10.13 objects 1 and 2 collide in an elastic manner.

Figure 10.13

$m_1 = m$ $v_{1i} = v$

$m_2 = 4m$ $v_{2i} = v/3$

$\mu = 0$

7. Determine the velocity of objects 1 and 2 after the collision.	Insert the given information into equations (c) and (d) to obtain $v_{1f} = -v/15$ and $v_{2f} = 3v/5$ After the collision, m_1 moves slowly to the left and m_2 continues moving to the right but with an increased speed.

In Figure 10.14 objects 1 and 2 collide in an elastic manner.

Figure 10.14

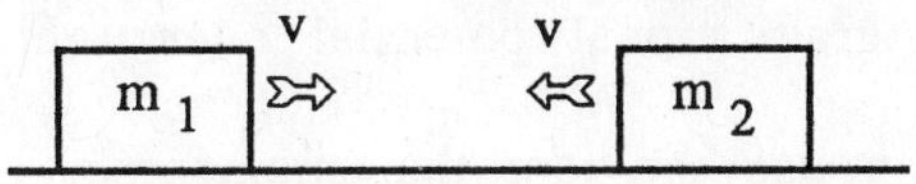

$m_1 = m$ $\quad$ $v_{1i} = v$

$m_2 = 2m$ $\quad$ $v_{2i} = -v$

$\mu = 0$

8. Determine the velocity of objects 1 and 2 after the collision.	Insert the given information into equations (c) and (d) to obtain $v_{1f} = -5v/3$ and $v_{2f} = v/3$ After the collision, object 1 moves to the left and object 2 moves to the right.

In Figure 10.15, objects 1 and 2 collide in an inelastic manner and stick together.

Figure 10.15

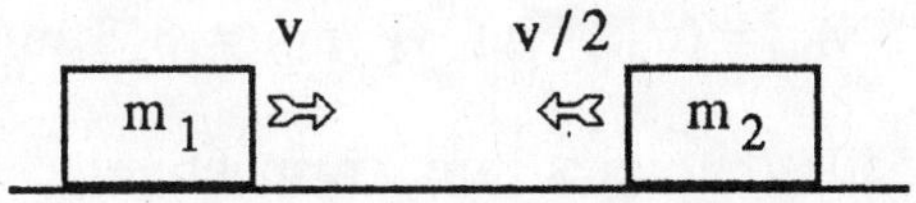

$m_1 = m$ $\quad$ $v_{1i} = v$

$m_2 = 2m$ $\quad$ $v_{2i} = -v/2$

$\mu = 0$

9. Determine the velocity of the combined object after the collision.	Insert the given information into equation (f) to obtain $v_f = 0$. After the collision, the objects are attached and at rest.

Example: A 0.0100-kg projectile is fired horizontally into and becomes embedded in a suspended block of wood of mass 0.990 kg, as shown in Figure 10.16. The block and the embedded projectile swing upward until the center of mass of the combination is raised by 0.500 m. (a) What is the velocity of the block and embedded projectile immediately after collision? (b) What is the initial velocity of the projectile? (c) What fraction of the initial kinetic energy is lost during the collision?

Diagram and Given:

Figure 10.16

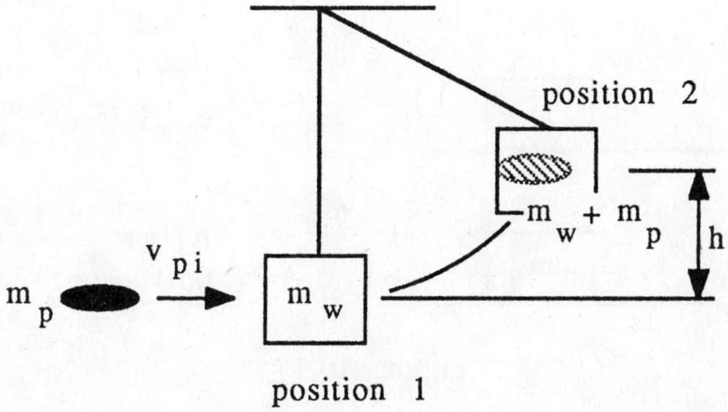

$m_p = 0.010$ kg

$m_b = 0.990$ kg

$h = 0.500$ m

Determine:

(a) The velocity of the block and embedded projectile (v_f) immediately after collision.
(b) The velocity of the projectile (v_{pi}) before collision.
(c) The fraction of the kinetic energy lost during the collision.

Strategy:

(a) Knowing that all of the kinetic energy of the block and embedded projectile at position 1 gets changed into gravitational potential energy at position 2, we can determine v_f.
(b) Now that v_f is known, we can use the fact that momentum is conserved during the collision to determine v_{pi}.
(c) Once all of the masses and velocities are known, we can determine the initial and final kinetic energy, the change in kinetic energy, and hence, the fraction of kinetic energy lost during the collision.

Solution:

(a) $\frac{1}{2}(m_p + m_b) v_f^2 = (m_p + m_b)gh$ or $v_f = (2gh)^{1/2} = 3.13$ m/s

(b) $m_p v_{pi} = (m_p + m_b) v_f$ or $v_{pi} = (m_p + m_b)v_f / m_p = 313$ m/s

(c) The initial and final kinetic energies are, respectively

$$KE_i = m_p v_{pi}^2/2 = 490 \text{ J} \quad \text{and} \quad KE_f = (m_p + m_b) v_f^2/2 = 5.00 \text{ J}$$

The change in the kinetic energy during the collision is obtained by

$$\Delta KE = KE_f - KE_i = -485 \text{ J}$$

The fraction of the energy lost is $f_{lost} = |\Delta KE/KE_i| = 0.990$

Example: A block of mass $m_1 = 500$ g slides with a velocity of 10.0 cm/s across a frictionless horizontal surface and makes an elastic head-on collision with a block of mass m_2, which is initially at rest as shown in Figure 10.17. After the collision, the 500-g block travels 5.00 cm/s in a direction opposite to its initial motion. Determine the mass of the second block and its velocity after the collision.

Diagram and Given:

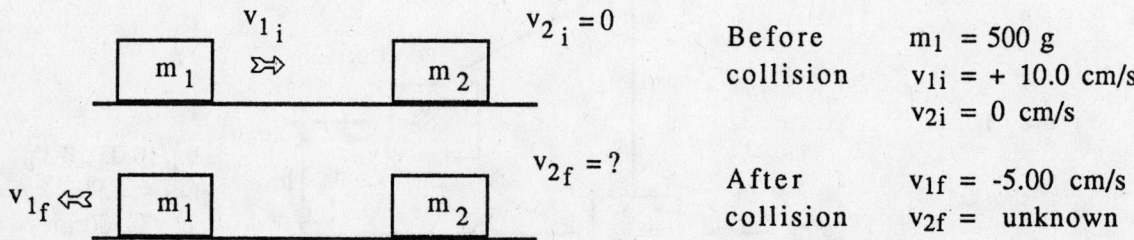

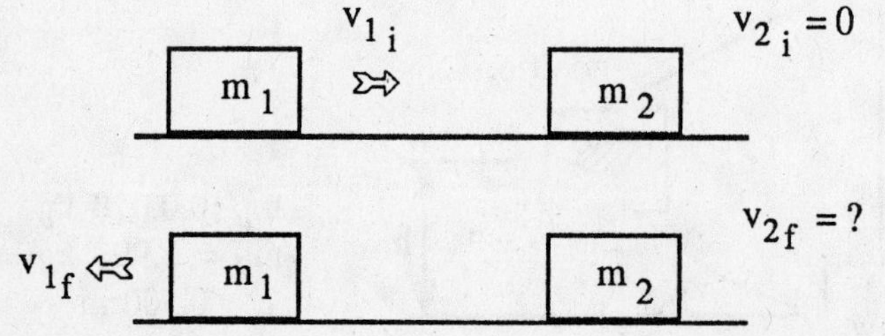

Before collision: $m_1 = 500$ g, $v_{1i} = +10.0$ cm/s, $v_{2i} = 0$ cm/s

After collision: $v_{1f} = -5.00$ cm/s, v_{2f} = unknown

Figure 10.17

Determine:

(a) The mass m_2 of the second block.

(b) The velocity v_{2f} of the second block after the collision.

Strategy: Since the collision is elastic, we know that momentum and kinetic energy are conserved. A statement of conservation of momentum will involve two unknowns, m_2 and v_{2f}. A statement of conservation of kinetic energy will involve these same two unknowns. Thus we can solve these two equations simultaneously for the two unknowns.

Solution: A statement of conservation of momentum for this situation is

$$m_1 v_{1i} + m_2 v_{2i} = m_1 v_{1f} + m_2 v_{2f}$$

Recalling that $v_{2i} = 0$ for this example, we can write

$$(\alpha) \qquad m_1(v_{1i} - v_{1f}) = m_2 v_{2f}$$

A statement of conservation of kinetic energy is

$$\frac{1}{2}(m_1 v_{1i}^2) + \frac{1}{2}(m_2 v_{2i}^2) = \frac{1}{2}(m_1 v_{1f}^2) + \frac{1}{2}(m_2 v_{2f}^2)$$

Since $v_{2i} = 0$, for this case we can write

$$m_1(v_{1i}^2 - v_{1f}^2) = m_2 v_{2f}^2$$

or

$$(\beta) \qquad m_1(v_{1i} + v_{1f})(v_{1i} - v_{1f}) = m_2 v_{2f}^2$$

Notice that we now have two equations, (α) and (β), and two unknowns, m_2 and v_{2f}. Eliminate m_2 by dividing (β) and (α):

$$\frac{(\beta)}{(\alpha)} = \frac{m_1(v_{1i} + v_{1f})(v_{1i} - v_{1f})}{m_1(v_{1i} - v_{1f})} = \frac{m_2 v_{2f}^2}{m_2 v_{2f}}$$

Solving for v_{2f}, we obtain $\quad v_{2f} = v_{1i} + v_{1f} = +5.00$ cm/s

The value for v_{2f} can be inserted into either equation (α) or (β) to obtain m_2. Since equation (α) looks easier, use it to obtain m_2.

$$m_2 = m_1(v_{1i} - v_{1f})/v_{2f} = 1.50 \text{ kg}$$

You can check this by inserting v_{2f} into equation (β) to obtain the same value for m_2.

Related Exercise: Ex. 10-27 through 10-40.

6 Collisions In Two Dimension (Section 10-8)

Review: If no external forces are involved, momentum will be conserved in two dimensional collisions. It will usually be convenient to conserve momentum in the x and y directions.

If the collision is elastic, we may also conserve energy.

In order to solve the problem we will need other information, usually θ_1 or θ_2 (i.e. the scattering angle for one of the two objects involved in the collision).

Practice: Two 0.200 kg hockey pucks are sliding across the ice as shown in Figure 10.18. One puck has been coated with instant glue so they will stick together after they collide. The initial velocities of the pucks are given by

$$v_{1i} = (20.0 \text{ m/s})\mathbf{i} \quad \text{and} \quad v_{2i} = (25.0 \text{ m/s})\mathbf{j}$$

Assume the surface is frictionless.

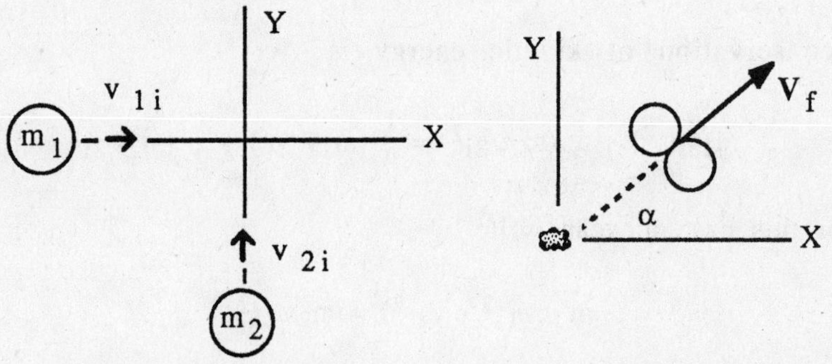

Figure 10.18

(a) Before Collision (b) After Collision

Determine the following:

1. Initial components of the momentum in the x and y directions	$p_{ix} = P_{1i} = m_1 v_{1i} = 4.00$ kg•m/s $p_{iy} = P_{2i} = m_1 v_{2i} = 5.00$ kg•m/s
2. Final components of the momentum in the x and y directions	Since the surface is frictionless, no external forces act on the system, hence momentum is conserved and $P_{fx} = P_{ix} = 4.00$ kg•m/s $P_{fy} = P_{iy} = 5.00$ kg•m/s
3. Magnitude of the final momentum	$P_f = [P_{fx}^2 + P_{fy}^2]^{1/2} = 6.40$ kg•m/s
4. Magnitude of the final velocity	$P_f = (m_1 + m_2)V_f$ $V_f = P_f/(m_1 + m_2) = 16.0$ m/s
5. The angle of scatter (α) of the two pucks after the collision	$\alpha = \tan^{-1} (P_{fy}/P_{fx}) = 51.3°$

Example: Two identical objects are involved in a glancing elastic collision. Object two is initially at rest and object one is incident at a speed of 10.0 m/s. After the collision object one is scattered 30.0° from the direction of its initial travel. Determine the speed of each object after the collision and the direction of scatter of object two.

Given and Diagram:

$m_1 = m_2 = m$
$v_{1i} = 10.0$ m/s
$v_{2i} = 0$
$\theta_1 = 30.0°$

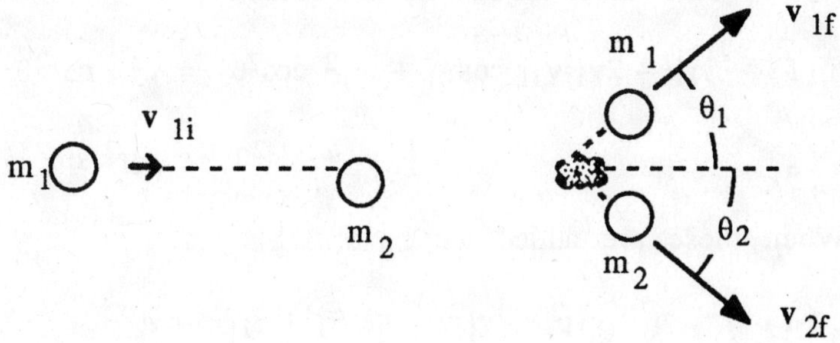

Determine: v_{1f} final speed of object one

v_{2f} final speed of object two

θ_2 scatter angle for object two

Strategy: Write the equations for conservation of momentum in the x and y directions and conservation of energy. These three equations will include the three unknowns v_{1f}, v_{2f}, and θ_2. Having three equations and three unknowns, we should be able to solve for the three unknowns.

Solution:

Conservation of momentum in the x direction:

$$m_1v_{1i} = m_1v_{1f}\cos\theta_1 + m_2v_{2f}\cos\theta_2$$

Conservation of momentum in the y direction:

$$0 = m_1v_{1f}\sin\theta_1 - m_2v_{2f}\sin\theta_2$$

Conservation of energy:

$$m_1v_{1i}^2/2 = m_1v_{1f}^2/2 + m_2v_{2f}^2/2$$

Since $m_1 = m_2$, these three equations may be rewritten as:

(a) $$v_{1i} = v_{1f}\cos\theta_1 + v_{2f}\cos\theta_2$$

(b) $$0 = v_{1f}\sin\theta_1 - v_{2f}\sin\theta_2$$

(c) $$v_{1i}^2 = v_{1f}^2 + v_{2f}^2$$

Notice that we now have three equations and three unknowns. Let's agree to first eliminate θ_2. This may be done by rewriting (a) and (b) as follows:

(d) $$v_{1i} - v_{1f}\cos\theta_1 = v_{2f}\cos\theta_2$$

(e) $$v_{1f}\sin\theta_1 = v_{2f}\sin\theta_2$$

Now square these equations to obtain:

(f) $$v_{1i}^2 - 2v_{1i}v_{1f}\cos\theta_1 + v_{1f}^2\cos^2\theta_1 = v_{2f}^2\cos^2\theta_2$$

(g) $$v_{1f}^2\sin^2\theta_1 = v_{2f}^2\sin^2\theta_2$$

When these are added we have:

(h) $$v_{1i}^2 - 2v_{1i}v_{1f}\cos\theta_1 + v_{1f}^2 = v_{2f}^2$$

Now eliminate v_{2f}^2 by solving equation (c) for v_{2f}^2 and inserting it into equation (h) to obtain:

(i) $$v_{1i}^2 - 2v_{1i}v_{1f}\cos\theta_1 + v_{1f}^2 = v_{1i}^2 - v_{1f}^2$$

Solving this for v_{1f} obtain:

(j) $$v_{1f} = v_{1i}\cos\theta_1 = (10.0\text{ m/s})\cos 30.0° = 8.66\text{ m/s}$$

Inserting (j) into (c) obtain:

(k) $$v_{2f}^2 = v_{1i}^2 - v_{1f}^2 = v_{1i}^2(1 - \cos^2\theta_1) = v_{1i}^2\sin^2\theta_1$$

or

(l) $$v_{2f} = v_{1i}\sin\theta_1 = (10.0\text{ m/s})\sin 30.0° = 5.00\text{ m/s}$$

Finally, we may obtain θ_2 by inserting (j) and (l) into (e):

(e) $$v_{1f}\sin\theta_1 = v_{2f}\sin\theta_2$$

$$v_{1i}\cos\theta_1\sin\theta_1 = v_{1i}\sin\theta_1\sin\theta_2$$

or

$$\sin\theta_2 = \cos\theta_1 = 0.866$$

or

$$\theta_2 = 60.0°$$

Notice that $\theta_1 + \theta_2$ is 90.0°. This is true for any elastic collision between objects of equal mass, if one of the objects is initially at rest.

Related Text Exercises: Ex. 10-41 through 10-46.

PRACTICE TEST

Take and grade this practice test. Doing so will allow you to determine any weak spots in your understanding of the concepts taught in this chapter. The following section prescribes what you should study further to strengthen your understanding.

Consider a system of three particles moving in the xy plane. Their positions are functions of time and are given by

$m_1 = 1.00$ kg $\quad\quad$ $m_2 = 2.00$ kg $\quad\quad$ $m_3 = 3.00$ kg

$\mathbf{r}_1 = [(5\ \text{m/s}^2)t^2 + 3\ \text{m}]\ \mathbf{i}$ $\quad$ $\mathbf{r}_2 = 3\ \text{m}\mathbf{i} + [4\ \text{m} + (2\ \text{m/s})t]\ \mathbf{j}$ $\quad$ $\mathbf{r}_3 = (3\ \text{m/s})t\mathbf{i} + (6\ \text{m/s}^2)t^2\mathbf{j}$

Determine the following:

_________ 1. Coordinates of the CM of the system after 2.00 s
_________ 2. Coordinates of the velocity of the CM of the system after 2.00 s
_________ 3. Coordinates of the acceleration of the CM of the system after 2.00 s
_________ 4. Coordinates of the net force on the system after 2.00 s

A 100-g ball moving at a speed of 40.0 m/s strikes a wall at an angle of 60.0° with respect to the normal. There is no change in the component of the velocity parallel to the wall and the ball leaves the wall with a speed of 38.0 m/s. The ball and wall are in contact for 0.100 s.

Determine the following:

_________ 5. Magnitude of the ball's initial momentum
_________ 6. Component of the ball's initial velocity perpendicular to the wall
_________ 7. Component of the ball's final velocity perpendicular to the wall
_________ 8. Angle with which the ball leaves the wall
_________ 9. Magnitude of the ball's final momentum
_________ 10. Magnitude of the impulse imparted to the ball
_________ 11. Magnitude of the average force exerted on the ball

m_1 is released from rest and collides elastically with m_2. The objects have the same mass, the horizontal surface is frictionless, the cord attached to m_1 is 1.00 m long, and θ is 30.0°.

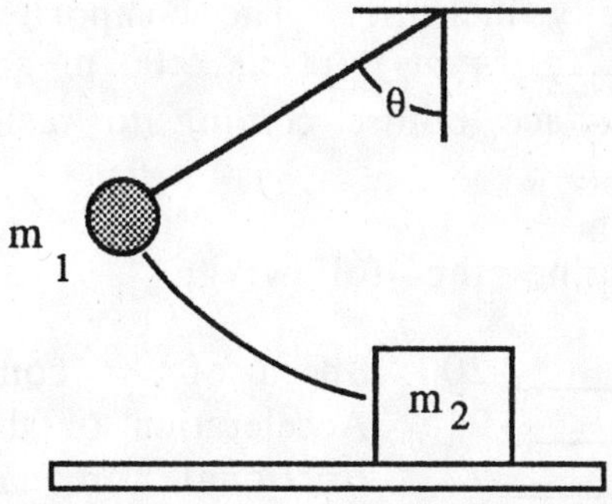

Determine the following:

_________ 12. Speed of m_1 the instant before the collision
_________ 13. Speed of m_1 after the collision
_________ 14. Speed of m_2 after the collision

m_1 and m_2 are traveling as shown in the figure below. m_1 slides down the ramp, overtaking and colliding elastically with m_2. The surface is frictionless.

$m_1 = 400$ g $h = 1.00$ m
$m_2 = 200$ g $\mu = 0$
$v_{1i} = v_{2i} = 1.00$ m/s

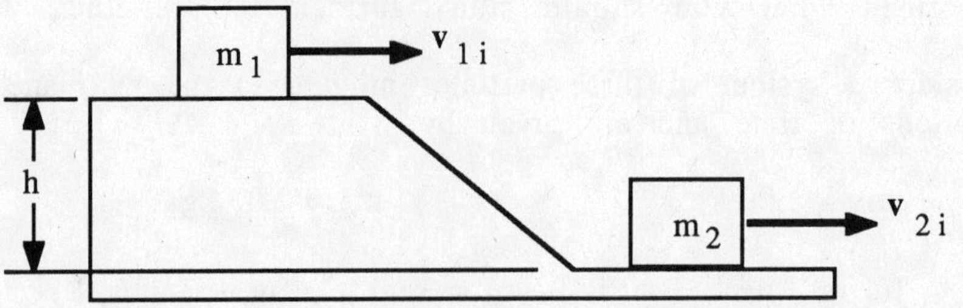

Determine the following:

________ 15. Speed of m_1 the instant before the collision
________ 16. Speed of m_1 after the collision
________ 17. Speed of m_2 after the collision

m_1 and m_2 travel toward each other before colliding inelastically, as shown in the figure at the right. The surface is frictionless.

$m_1 = 400$ g $v_{1i} = +1.00$ m/s
$m_2 = 200$ g $v_{2i} = -3.00$ m/s $\mu = 0$

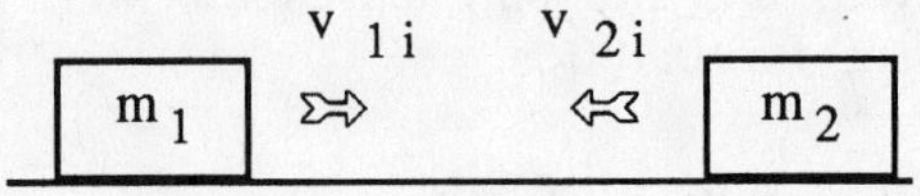

Determine the following:

________ 18. Velocity of m_1 after the collision
________ 19. Velocity of m_2 after the collision

m_1 is fired into m_2 which is initially at rest on a rough surface. The collision is totally inelastic. The composite object ($m_1 + m_2$) travels 5.00 m across the surface before coming to rest.

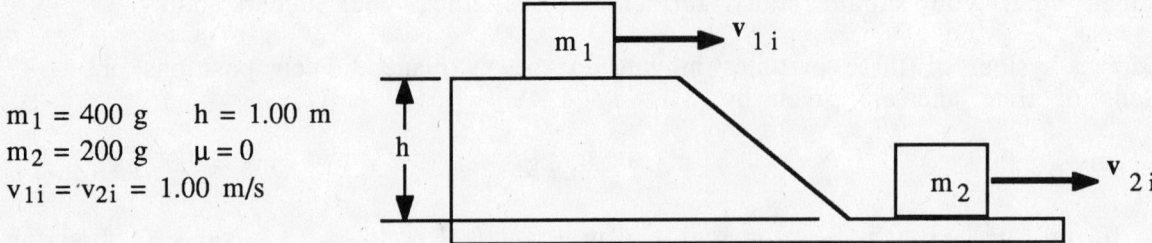

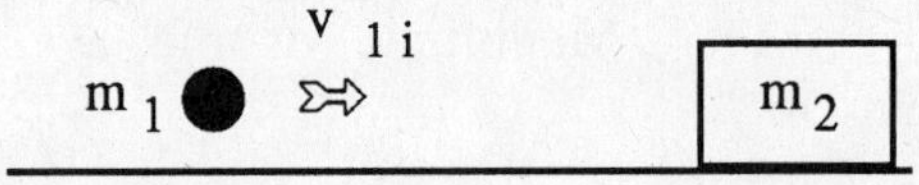

$m_1 = 20.0$ g $v_{1i} = 100$ m/s
$m_2 = 200$ g $\Delta s = 5.00$ m $v_{2i} = 0$ m/s

Determine the following:

________ 20. Speed of the composite object the instant after collision
________ 21. Acceleration of the composite object as it slides
________ 22. Frictional force decelerating the composite object
________ 23. Coefficient of friction for the composite object on the rough surface

(See Appendix I for answers.)

PRINCIPAL CONCEPTS AND EQUATIONS PRESCRIPTION

Your score on the practice test is an excellent measure of your understanding of the chapter. You should now use the following chart to write your own prescription for dealing with any weaknesses the practice test points out. Look down the leftmost column to the number of the question(s) you answered incorrectly, reading across that row you will find the concept and/or equation of concern, the section(s) of the study guide you should return to for further study, and some suggested text exercises which you should work to gain additional experience.

Practice Test Questions	Concepts and Equations	Prescription	
		Principal Concepts	Text Exercises
1	CM of a system: $\mathbf{r} = \Sigma\, m_i\mathbf{r}_i/M$	1	10-1, 3
2	Velocity of CM: $\mathbf{v} = \Sigma\, m_i\mathbf{v}_i/M$	2	10-7, 8
3	Acceleration of CM: $\mathbf{a} = \Sigma\, m_i\mathbf{a}_i/M$	2	10-9, 11
4	Net force on CM: $F_{cm} = Ma_{cm}$	2	10-11
5	Momentum: $\mathbf{p} = m\mathbf{v}$	3	10-14, 15
6	Component of a vector	3 of Ch 2	2-11, 14
7	Component of a vector	3 of Ch 2	2-14, 15
8	Component of a vector	3 of Ch 2	2-15, 16
9	Momentum: $\mathbf{p} = m\mathbf{v}$	3	10-14, 15
10	Impulse: $\mathbf{J} = \Delta\mathbf{p}$	4	10-17, 18
11	Impulse: $\mathbf{F} = \mathbf{J}/\Delta t$	4	10-17, 20
12	Conservation of mechanical energy	4 of Ch 8	8-4, 5
13	Elastic collision	5	10-34, 35
14	Elastic collision	5	10-34, 35
15	Conservation of mechanical energy	4 of Ch 8	8-4, 5
16	Elastic collision	5	10-34, 35
17	Elastic collision	5	10-34, 35
18	Inelastic collision	5	10-27, 30
19	Inelastic collision	5	10-32, 33
20	Inelastic collision	5	10-37, 41
21	$v^2 - v_0^2 = 2a\Delta s$	4 of Ch 3	4-46, 47
22	Newton's Second Law: $\mathbf{F} = m\mathbf{a}$	2 of Ch 5	5-30, 31
23	Coefficient of friction: $f_k = \mu_k F_N$	2 of Ch 6	6-14, 15

11 STATIC EQUILIBRIUM OF A RIGID BODY

RECALL FROM PREVIOUS CHAPTERS

Previously learned concepts and equations frequently used in this chapter	Text Section	Study Guide Page
Vectors - finding components	2-4	2-8
Vectors - addition	2-4	2-8

NEW IDEAS IN THIS CHAPTER

Concepts and equations introduced	Text Section	Study Guide Page
First condition of equilibrium: $\Sigma \mathbf{F}_{ext} = 0$	11-1	11-1
Magnitude of torque:	11-2	11-5
$\tau = rF \sin\theta$		
$\tau = F(\text{moment arm}) = rF \sin\theta$		
$\tau = rF_{\perp} = rF \sin\theta$		
Direction of torque:	11-2	11-5
$+\tau$ is counterclockwise		
$-\tau$ is clockwise		
Second condition of equilibrium: $\Sigma \tau_{ext} = 0$	11-3	11-9
Center of gravity: $x_{cg} = \Sigma (x_i)(m_i g)/\Sigma m_i g$	11-5	---
Torque and cross product: $\boldsymbol{\tau} = \mathbf{r} \times \mathbf{F}$	11-6	11-16

PRINCIPAL CONCEPTS AND EQUATIONS

1 Static Translational Equilibrium (Section 11-1)

Review: If the vector sum of all the forces acting on an object is zero, then the translational acceleration of the object is zero, and its translational velocity is constant. If the translational velocity is zero, the object is in static translational equilibrium. If the translational velocity is a nonzero constant, the object is in dynamic translational equilibrium. This may be summarized as follows:

If $\Sigma \mathbf{F}_{ext} = 0$, then $\mathbf{a} = 0$ and $\mathbf{v}$ = constant
If $\mathbf{v} = 0$, then static translational equilibrium exists.
If $\mathbf{v} \neq 0$ but is constant, then dynamic translational equilibrium exists.

When solving problems of this type, you will often find it convenient to write the condition for static translational equilibrium in component form:

$$\Sigma F_{x,ext} = 0 \qquad \Sigma F_{y,ext} = 0 \qquad \Sigma F_{z,ext} = 0$$

Practice: Consider the object shown in Figure 11.1 and the forces acting on it.

$|F_1| = 10.0$ N
$|F_2| = 15.0$ N
$|F_3| = 20.0$ N
$|F_4| = 10.0$ N
$|F_5| = 5.0$ N

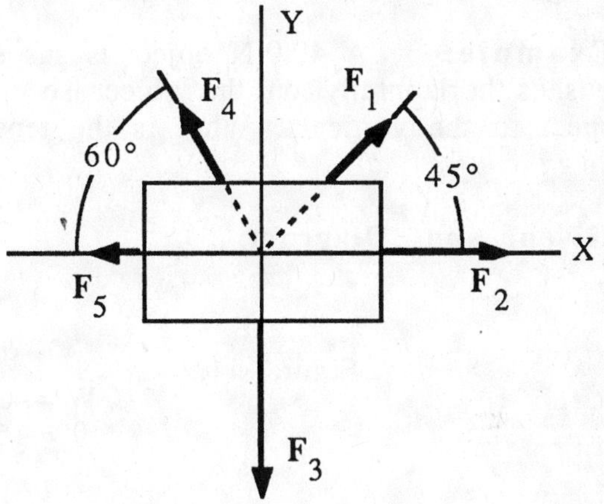

Figure 11.1

Determine the following:

1. The sum of all forces in the +y direction	$\Sigma F_{+y} = F_1 \sin 45° + F_4 \sin 60° = +15.7$ N
2. The sum of all forces in the -y direction.	$\Sigma F_{-y} = -F_3 = -20.0$ N
3. If the object is in equilibrium in the y direction	Since $\Sigma F_y = \Sigma F_{+y} + \Sigma F_{-y} = -4.30 \text{ N} \neq 0$ We conclude that the object is not in equilibrium in the y direction.
4. The sum of all forces in the +x direction	$\Sigma F_{+x} = F_2 + F_1 \cos 45° = +22.1$ N
5. The sum of all forces in the -x direction.	$\Sigma F_{-x} = -F_5 - F_4 \cos 60° = -10.0$ N
6. If the object is in equilibrium in the x direction.	Since $\Sigma F_x = \Sigma F_{+x} + \Sigma F_{-x} = +12.1 \text{ N} \neq 0$ We conclude that the object is not in equilibrium in the x direction.
7. The magnitude of the force that would put the object in translational equilibrium	$F = [(\Sigma F_x)^2 + (\Sigma F_y)^2]^{1/2} = 12.8$ N

8. The direction of the force that would put the object in translational equilibrium	Since the resultant force has components $\Sigma F_x = +12.1$ N and $\Sigma F_y = -4.30$ N, the force that would put the object in translational equilibrium must have the components $\Sigma F_x = -12.1$ N and $\Sigma F_y = +4.30$ N. The direction of this force is given by: $\theta = \tan^{-1} (\Sigma F_y/\Sigma F_x)$. $\theta = 19.6°$ clockwise from the direction of $\mathbf{F_5}$

Example: A 40.0-N object is suspended by a rope, as shown in Figure 11.2. A girl pushes horizontally on the object so that the rope makes an angle of 15.0° with respect to the vertical. What is the tension in the rope, and how hard is the girl pushing?

Given and Diagram:

Figure 11.2

$W = 40.0$ N
$\theta = 15.0°$

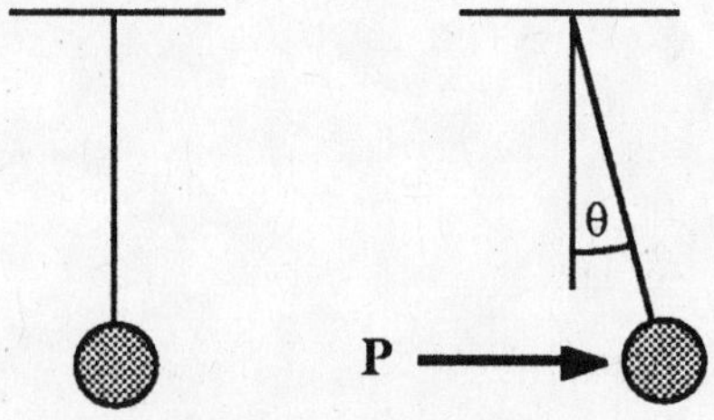

Determine: T, the tension in the rope while the girl is pushing
P, the magnitude of the girl's push

Strategy: Draw a free-body diagram showing all forces acting on the object. Establish a coordinate system, and resolve all forces into components. Use the fact that the object is in vertical equilibrium to determine T_y and, hence, T. Knowing T and θ, we can determine T_x. Since the object is in horizontal equilibrium, we can determine P from T_x.

Solution: Since the object is in vertical equilibrium, the sum of the forces in the y direction must be zero.

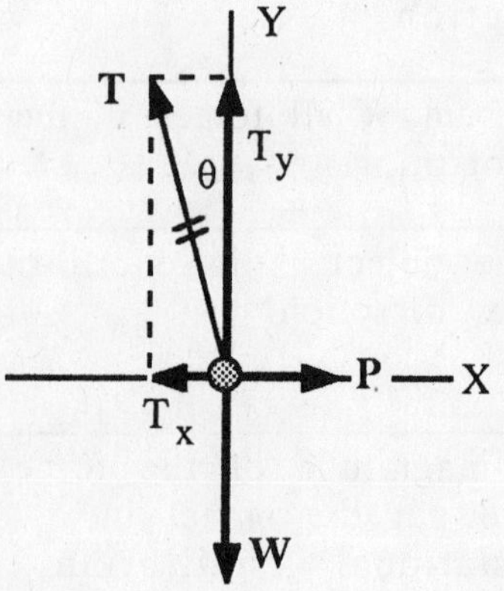

$\Sigma F_y = T_y - W = (T \cos\theta) - W = 0$ or

$T = W/\cos\theta = 40.0 \text{ N}/\cos 15° = 41.4$ N

Since the object is in horizontal equilibrium, the sum of the forces in the x direction must be zero.

$$\Sigma F_x = P - T_x = P - (T \sin\theta) = 0$$
or
$$P = T \sin\theta = (41.4 \text{ N}) \sin 15° = 10.7 \text{ N}$$

Example: A crate weighing 100 N sits on an incline that makes an angle of 10.0° with respect to the horizontal (Figure 11.3). The coefficient of kinetic friction between the crate and the incline is 0.200. What is the magnitude of the push parallel to the incline that will move the crate down the incline at a constant speed?

Given and Diagram:

$W = 100$ N
$\theta = 10.0°$
$\mu = 0.200$

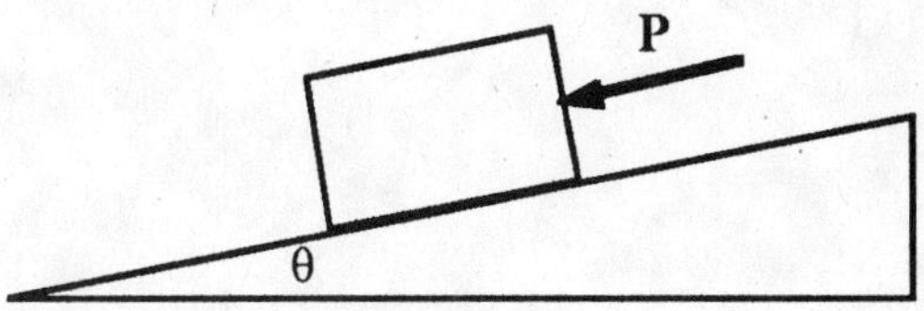

Figure 11.3

Determine: P, the magnitude of the push that will move the crate down the incline at a constant speed.

Strategy: Draw a free body diagram showing all forces acting on the crate. Establish a coordinate system and resolve all forces into components. Since the crate is to move down the ramp at a constant speed it will be in dynamic equilibrium. Consequently, the sum of the forces in the x direction must be zero. If we write a summation-of-forces statement for the x direction, we can solve for P if we know f. We can obtain f if we know F_N. We can obtain F_N by recognizing that the sum of the forces in the y direction must be zero.

Solution: Since the crate is in dynamic equilibrium in the x direction, the sum of forces in the x direction is zero.

$$\Sigma F_x = W_x + P - f = 0$$

or

$$P = f - W_x = f - W \sin\theta$$

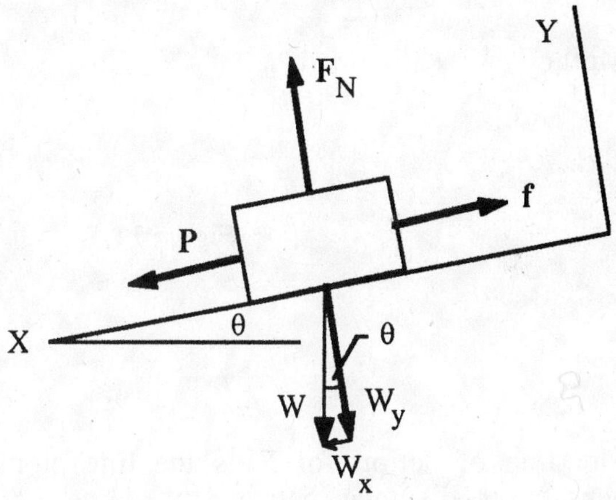

Since the crate is in static equilibrium in the y-direction, the sum of forces in the y direction is zero.

$$\Sigma F_y = +F_N - W_y = 0$$

or

$$F_N = W_y = W \cos\theta$$

Now that we know F_N, we can determine the friction by $f = \mu F_N = \mu W \cos\theta$. Then, substituting back into the expression for P, we obtain:

$$P = f - (W \sin\theta) = \mu(W \cos\theta) - (W \sin\theta) = W(\mu \cos\theta - \sin\theta) = 2.33 \text{ N}$$

Related Text Exercises: The concept of static translational equilibrium is used in Ex 11-3, 11-4 through 11-6, and 11-8 through 11-18. However since other concepts are involved, you cannot work these problems without further study.

2 Torque (Section 11-2)

Review: Torque is a vector quantity that has both magnitude and direction. The magnitude of the torque can be determined by three different methods. Figures 11.4, 11.5, and 11.6 illustrate these methods.

Method I

Figure 11.4

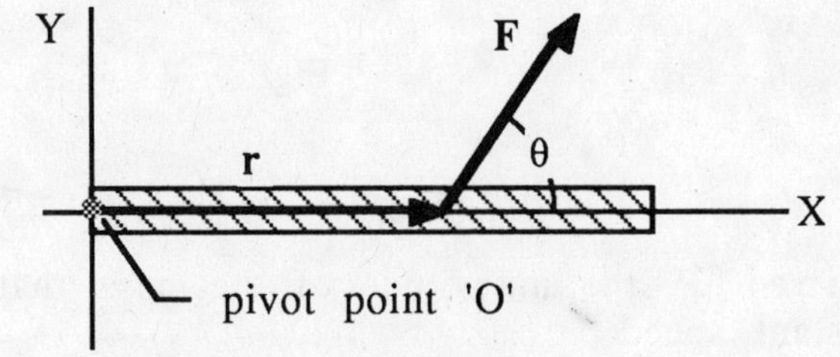

O = point we wish to determine the torque about
r = |**r**| = magnitude of the position vector that locates the point of application of the force
F = |**F**| = magnitude of the force that produces the torque
θ = angle between the direction of **r** and the direction of **F**

The magnitude of the torque is the product of the magnitude of **r**, the magnitude of **F**, and the sine of the angle between their directions.

$$\tau = rF\sin\theta$$

Method II

Figure 11.5

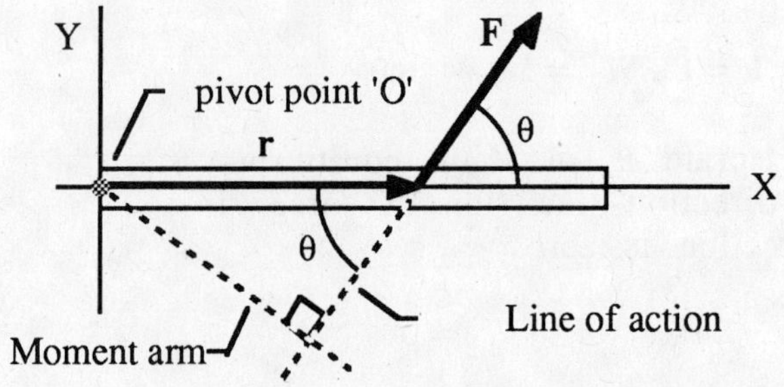

The line of action for **F** is the line along which **F** acts. The moment arm is the perpendicular distance between the line of action for **F** and the point about which we wish to determine the torque (O). Notice that the length of the moment arm is r sinθ. The magnitude of the torque is the product of the magnitude of **F** and the moment arm.

$$\tau = F(\text{moment arm}) = F(r\sin\theta) = rF\sin\theta$$

Notice that this is the same result as Method I.

Method III

Figure 11.6

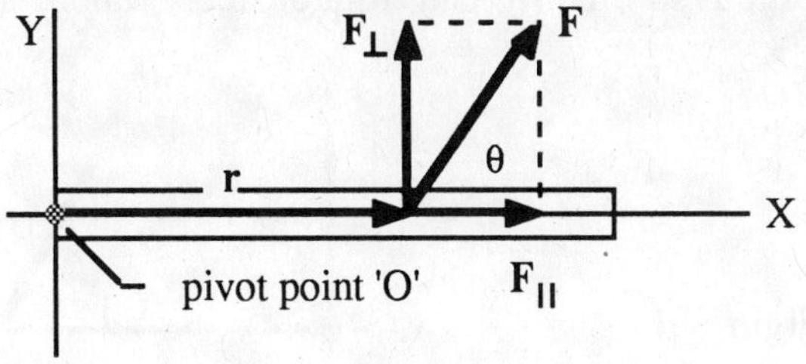

$F_{||} = F \cos\theta$, this force produces no torque about O
$F_\perp = F \sin\theta$, this force produces all the torque about O.

The magnitude of the torque is the product of the magnitude of **r** and $F_\perp$.

$$\tau = r F_\perp = rF \sin\theta$$

Notice that this is the same result as Methods I and II

If **r** and **F** are in the x-y plane, then the torque has only a z-component, hence

$$\tau_z = |\tau| = rF \sin\theta$$

The sign of the z-component of the torque (τ_z) is determined by the sense of rotation of the object when viewed from a point on the +z axis.

If counterclockwise then $\tau_z = +$, and if clockwise then $\tau_z = -$.

The sense of rotation is obtained by rotating the direction of **r** into the direction of **F** as shown in Figure 11.7.

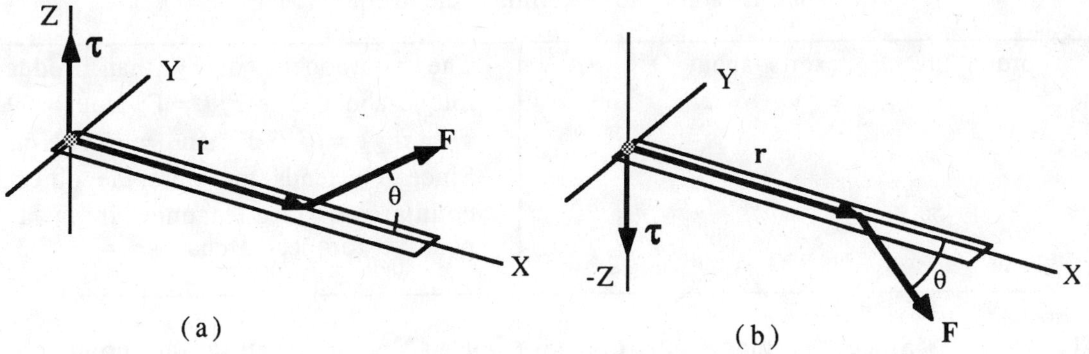

Figure 11.7

In Figure 11.7(a) as you view **r** and **F** from the +z axis, you see the rotation of the direction of **r** into the direction of **F** as a counterclockwise rotation. Hence τ is positive, that is it is directed in the +z direction. In Figure 11.7(b) the rotation of the direction of **r** into the direction of **F** is viewed as a counterclockwise rotation, hence τ is negative.

Practice: Consider the rectangle of width W and length L shown in Figure 11.8. It is mounted so that it can rotate freely about the axis O.

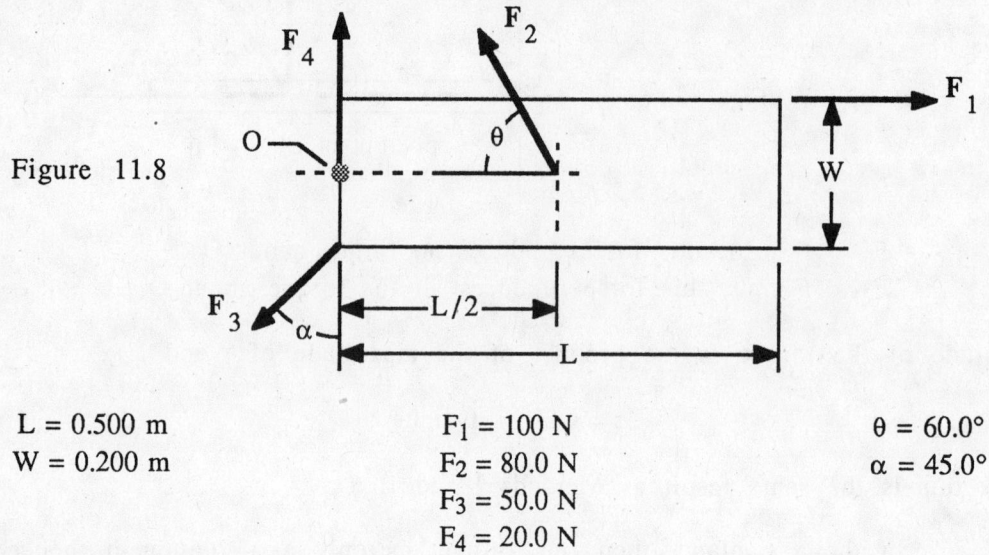

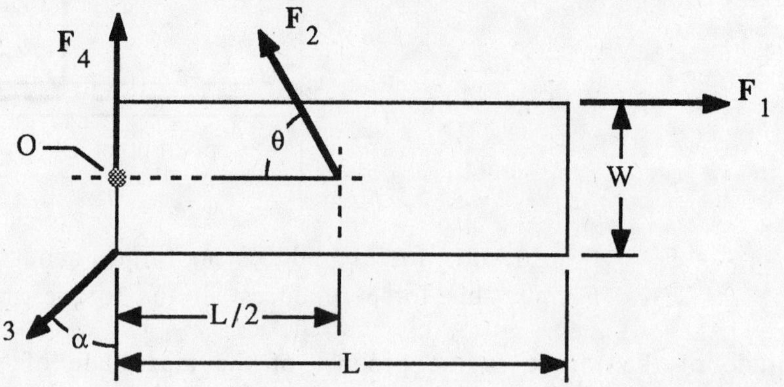

Figure 11.8

L = 0.500 m	F_1 = 100 N	θ = 60.0°
W = 0.200 m	F_2 = 80.0 N	α = 45.0°
	F_3 = 50.0 N	
	F_4 = 20.0 N	

Determine the following:

1. Torque that $\mathbf{F}_1$ exerts about O	The moment arm for $\mathbf{F}_1$ is W/2 = 0.100 m $\tau_1 = F_1(\text{moment arm})_1 = F_1W/2$ = 10.0 N•m Since $\mathbf{F}_1$ tends to rotate the object in a clockwise manner, it produces a negative torque. Hence τ_1 = -10.0 N•m.

Note: Since $\mathbf{F}_1$ and its moment arm were easily obtained, we used $\tau = F(\text{moment arm})$ to determine the torque (Method II).

2. Torque that $\mathbf{F}_2$ exerts about O	The component of F_2 that produces the torque about O is $F_{2\perp} = F_2 \sin\theta$ = 69.3 N $\tau_2 = rF_{2\perp} = (L/2)F_2 \sin\theta$ = 17.3 N•m Since $\mathbf{F}_2$ tends to rotate the object in a counterclockwise manner it is a positive torque. Hence τ_2 = +17.3 N•m.

Note: Since the magnitude of the position vector locating the point of application of the force and the component of the force perpendicular to the position vector were easily obtained, we use $\tau = rF_\perp$ to determine the torque (Method III).

3. Torque that $\mathbf{F}_3$ exerts about O.	$\tau_3 = r_3F_3 \sin\alpha$ = 3.54 N•m Since $\mathbf{F}_3$ tends to rotate the object in a clockwise manner, it is a negative torque. Hence τ_3 = -3.54 N•m.

Note: Since we know the magnitude of the position vector locating the point of application of the force, the magnitude of the force, and the angle between the direction of these two vectors, we use $\tau = rF \sin\theta$ to determine the torque (Method I).

4. Torque that $\mathbf{F}_4$ exerts about O.	Notice that the line of action for $\mathbf{F}_4$ passes through O, hence $\mathbf{F}_4$ can exert no torque about O. $\tau_4 = 0$.
5. What is the net or resultant torque about P?	$\tau_{net} = \tau_1 + \tau_2 + \tau_3 + \tau_4$ $= (-10.0 + 17.3 - 3.54)$ N•m = +3.76 N•m The positive value indicates that the object is experiencing a net counter-clockwise torque.

Note: We could have obtained τ_1, τ_2, τ_3 and τ_4 by using any one of the three methods. However, by virtue of the given information, Method II was easiest for $\mathbf{F}_1$, Method III for $\mathbf{F}_2$ and Method I for $\mathbf{F}_3$. You need to be familiar with all three methods in order to solve the problems in this and later chapters.

Example: A disk of 0.200-m radius is mounted on an axle, as shown in Figure 11.9. A rope is wrapped around the perimeter of the disk and then tied to a 0.500-kg mass. If the 0.500-kg mass is released from rest and travels 2.00 m in a time of 3.00 s, what is the torque tending to rotate the disk?

Given and Diagram:

R = 0.200 m = radius of the disk

M = 0.500 kg = mass tied to the rope

h = 2.00 m = distance M travels in 3.00 s

t = 3.00 s = time for M to travel 2.00 m

$v_0 = 0$ = speed of M at t = 0.00 s

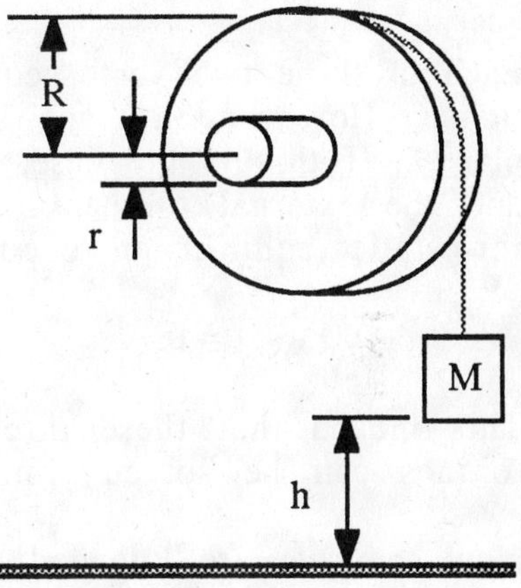

Figure 11.9

Determine: The torque being exerted on the disk.

Strategy: If we ignore friction, the only torque exerted on the disk is that due to the tension in the cord. If we know the tension in the cord, we can determine the torque by $\tau = RT$. We can write a Newton's Second Law equation for M and determine T if we know the acceleration of M. We can determine the acceleration of M, and hence T and then τ, from the fact that M starts from rest and travels the distance h in a time t.

Solution:

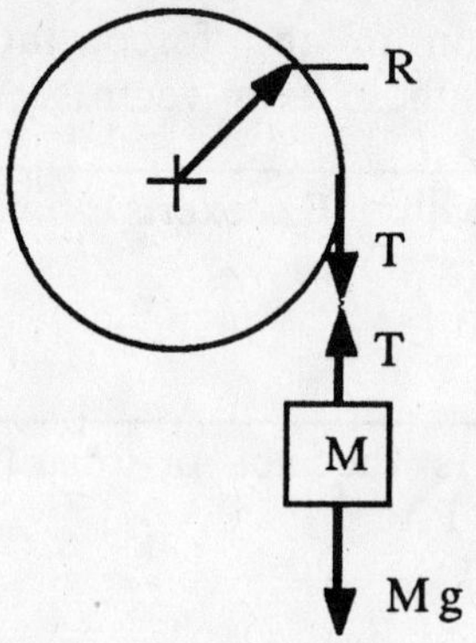

$\tau = -RT$

$Ma = Mg - T$ or $T = M(g - a)$

$\Delta y = v_0 t + \frac{1}{2} at^2$ or $a = \frac{2h}{t^2}$, since $\Delta y = h$ and $v_0 = 0$

$\tau = -RT = -RM(g - a) = -RM(g - 2h/t^2) = -0.936$ N•m

Related Text Exercises: Ex 11-1 through 11-4.

3 Conditions For Static Equilibrium (Section 11-3)

Review: For a rigid body to be in static equilibrium, the net external force must be zero and the net external torque must be zero. That is the following two conditions must be met:

$\Sigma F_{ext} = 0$, the first condition of equilibrium provides for translational static equilibrium.

$\Sigma \tau_{ext} = 0$, the second condition of equilibrium provides for rotational static equilibrium.

Each of these two vector equations has x, y, and z components for a total of six equations. However, in many situations, all external forces effectively lie in the x-y plane. If this is the case, then the external forces have only x and y components, and the external torques have only a z component. Subsequently, the conditions for static equilibrium become:

$\Sigma F_{x,ext} = 0 \qquad \Sigma F_{y,ext} = 0 \qquad \Sigma \tau_{z,ext} = 0$

This means that these three equations which contain forces, distances, angles, and torques can be solved simultaneously for at most three unknowns.

Now that we are knowledgeable about both conditions of equilibrium, we can work numerous problems that were previously not possible. For example Figure 11.10 shows a horizontal beam supported at one end by a wall brace and a pin and at the other end by a guy wire.

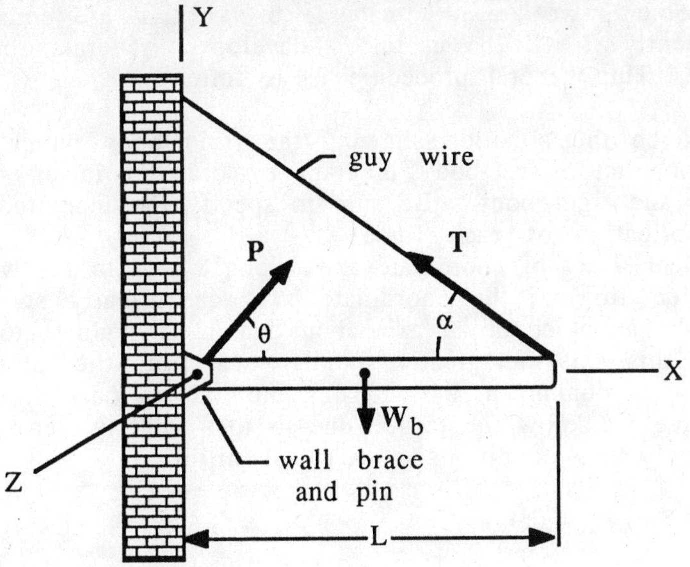

Figure 11.10

Suppose we are given the following values:

W_b = weight of the beam
α = angle between beam and guy wire
L = length of the beam

and we are asked to determine the following:

T = tension in the guy wire
P = push on the beam by the pin
θ = angle between P and the beam

Using only our knowledge of the first condition of equilibrium ($\Sigma F_{ext} = 0$, or since all external forces lie in the x-y plan $\Sigma F_{x,ext} = 0$ and $\Sigma F_{y,ext} = 0$) we may write:

(a) $\Sigma F_{x,ext} = P\cos\theta - T\cos\alpha = 0$
(b) $\Sigma F_{y,ext} = P\sin\theta + T\sin\alpha - W_b = 0$

At this point it should be noted that we have two equations [(a) and (b)] and three unknowns (T, P, and θ). If we had only the first condition of equilibrium at our disposal, we could not proceed further with this problem. However, by using the second condition of ($\Sigma \tau_{ext} = 0$, or since the torques have only a z component $\Sigma \tau_{z,ext} = 0$) we may write:

(c) $\Sigma \tau_{z,ext} = LT\sin\alpha - W_bL/2 = 0$

With the addition of (c) we now have three equations and three unknowns, hence we can solve the problems.

The problems we are now capable of working are considerably more complicated, subsequently it will be useful to develop a general procedure for obtaining the answer. This general procedure is as follows:

(i) Sketch the situation showing the rigid body which is in static equilibrium.
(ii) Construct a free-body diagram by drawing in all external force vectors acting on the rigid body. Be sure to specify the magnitude, direction, and point of application of each force.
(iii) Select a set of coordinate axes along which to resolve the forces. A judicious orientation of the coordinate axes can greatly simplify this resolution.
(iv) Make a choice of an axis about which to evaluate torques. A judicious choice of this axis can greatly simplify evaluating the torques. For example if a force is unknown, by choosing the axis to pass thru the line of action of this force we know the torque due to that force is zero.
(v) Apply the conditions of static equilibrium

$$\Sigma F_{x,ext} = 0; \qquad \Sigma F_{y,ext} = 0; \qquad \Sigma \tau_{z,ext} = 0$$

(vi) Solve these equations for up to 3 unknowns.

Practice: The plank shown in Figure 11.11 is in static equilibrium.

$W_1 = 100$ N

$W_2 = 75.0$ N

$W_P = 50.0$ N

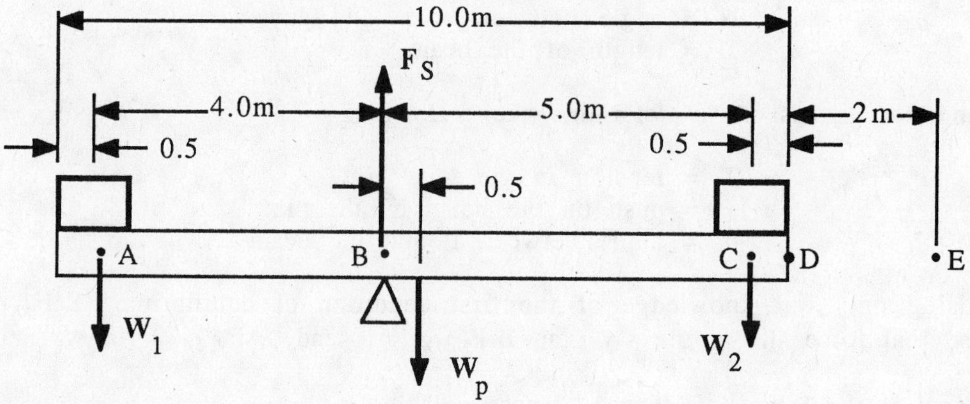

Figure 11.11

Determine the following:

1. Total cw torque about B	$\tau_{cwB} = -W_P(0.500 \text{ m}) - W_2(5.00 \text{ m})$ $= -400$ N•m
2. Total ccw torque about B	$\tau_{ccwB} = W_1(4.00 \text{ m}) = 400$ N•m
3. Total torque about B	$\tau_B = \tau_{cwB} + \tau_{ccwB} = 0$ N•m The plank is in rotational equilibrium about point B.
4. Total cw torque about A	$\tau_{cwA} = -W_P(4.5 \text{ m}) - W_2(9.00 \text{ m})$ $= -900$ N•m

5. Total ccw torque about A	$\tau_{ccwA} = F_s(4\text{ m})$ F_s = the force that the support must exert upward in order to maintain translational equilibrium $F_s = W_1 + W_2 + W_p = 225\text{ N}$ $\tau_{ccwA} = (225\text{ N})(4.00\text{ m}) = +900\text{ N}\bullet\text{m}$
6. Total torque about A	$\tau_A = \tau_{cwA} + \tau_{ccwA} = (-900 + 900)\text{ N}\bullet\text{m} = 0$ The plank is in rotational equilibrium about point A.
7. Total cw torque about D	$\tau_{cwD} = -F_s(5.50\text{ m}) = -1.24 \times 10^3\text{ N}\bullet\text{m}$
8. Total ccw torque about D	$\tau_{ccwD} = [\,W_1(9.50\text{ m}) + W_p(4.50\text{ m}) + W_2(0.500\text{ m})] = +1.24 \times 10^3\text{ N}\bullet\text{m}$
9. Total torque about D	$\tau_D = \tau_{cwD} + \tau_{ccwD} = 0\text{ N}\bullet\text{m}$ The plank is in rotational equilibrium about point D.

Note: From steps 3, 6 and 9 it should be evident that an object in rotational equilibrium about any one point is in rotational equilibrium about any other point. This point may be anywhere on or off the object. You should now convince yourself that the total torque about C and E is also zero.

Example: A painter's scaffolding plank is 3.00 m long and is resting on two saw-horses that are 0.500 m in from each end, as shown in Figure 11.12. The plank has a weight of 200 N, and the painter weighs 600 N. How far beyond the sawhorse can the painter stand before the plank starts to tip?

Given and Diagram:

$L = 3.00\text{ m}$ = length of plank
$W_{pk} = 200\text{ N}$ = weight of plank
$W_{pt} = 600\text{ N}$ = weight of painter

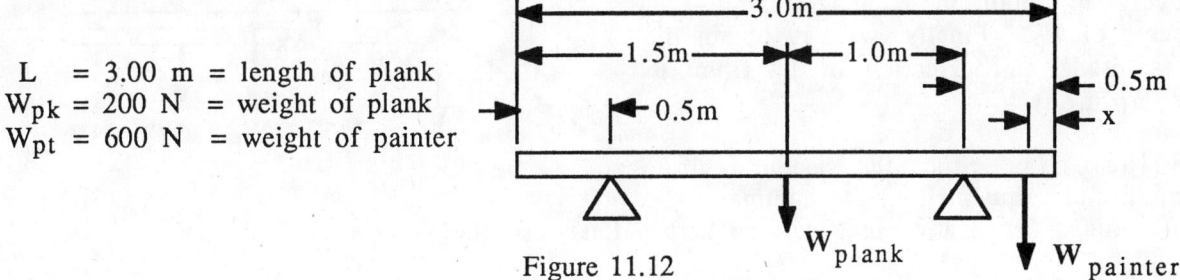

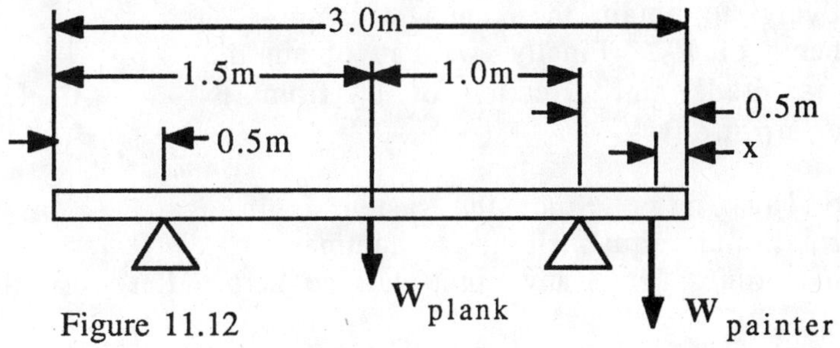

Figure 11.12

Determine: How far (distance x) beyond the sawhorse the painter can stand before the plank starts to tip.

Strategy: We want to know the largest value for distance x with the system maintaining static rotational equilibrium. When x has its largest value, the clock-wise torque due to the weight of the painter and the counterclockwise torque due to the weight of the plank will be equal, and the left sawhorse will be exerting no force on the plank. A summation of torques about the sawhorse on the right allows us to solve for the quantity x.

Solution: Since the system is in rotational equilibrium, we can write the following equation describing the torques about the sawhorse on the right.

$$\Sigma\, \tau_{z,ext} = \tau_{pk} + \tau_{pt} = W_{pk}(1.00\text{ m}) - W_{pt}(x) = 0$$

or

$$x = W_{pk}(1.00\text{ m})/W_{pt} = (200\text{ N})(1.00\text{ m})/(600\text{ N}) = 0.330\text{ m}$$

Example: For a sign suspended as shown in Figure 11.13, determine the tension in the cable and the force exerted on the beam by the pin.

Given and Diagram:

W_b = 100 N = weight of beam
W_s = 200 N = weight of sign
L = 2.00 m = length of beam

Figure 11.13

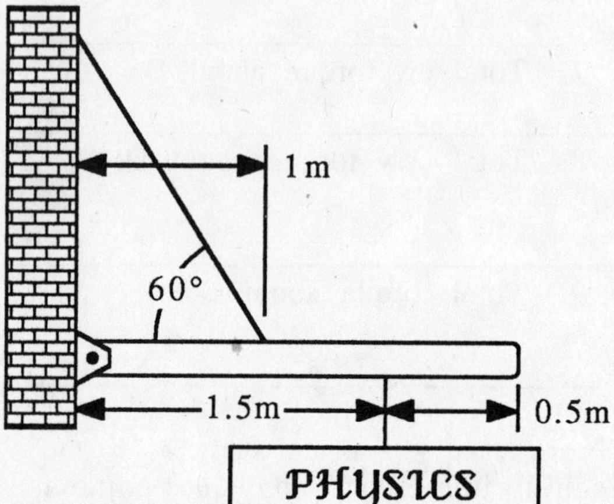

Determine: T, the tension in the cable
F_p, the force the pin exerts on the beam

Strategy: First draw a free-body diagram showing all forces acting on the beam. If we sum the torques about the pin P, we do not have to worry about the force F_p. The only unknown in a summation-of-torques statement about P is the tension in the cable. Once T is known, we can sum the x and y components of the forces respectively, to obtain the x and y components of F_p. Finally, we can obtain the magnitude and direction of F_p from its components.

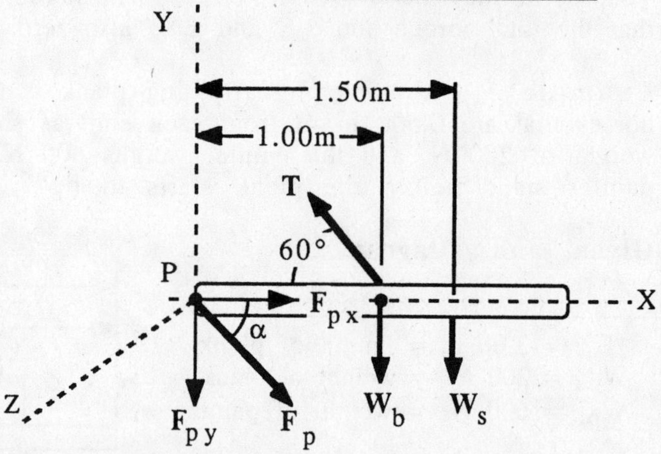

Solution: Since the system is in rotational equilibrium, a summation of torques about any point will be zero. Let's use the point P.

$$\Sigma\, \tau_{z,ext} = -W_s(1.50\text{ m}) - W_b(1.00\text{ m}) + T(1.00\text{ m})\sin 60° = 0$$

or

$$T = [(200\text{ N})(1.50\text{ m}) + (100\text{ N})(1.00\text{ m})]\,/(1.00\text{ m})\sin 60° = 462\text{ N}$$

Since the system is in translational equilibrium, a summation of forces in the x and y directions must be zero.

$$\Sigma F_{x,ext} = F_{px} - T\cos 60° = 0 \quad \text{or} \quad F_{px} = T\cos 60° = 231 \text{ N}$$

$$\Sigma F_{y,ext} = F_{py} + T\sin 60° - W_b - W_s = 0 \quad \text{or} \quad F_{py} = W_b + W_s - T\sin 60° = -100 \text{ N}$$

Note: When the above figure was drawn, the actual direction of $\mathbf{F_p}$ was unknown. Now that we know F_{px} and F_{py}, we can establish that $\mathbf{F_p}$ was drawn correctly. If $\mathbf{F_p}$ was not initially drawn correct, it could be redrawn at this time.

Now that we know the components of $\mathbf{F_p}$, we can determine the magnitude and direction of $\mathbf{F_p}$.

$$F_p = (F_{px}^2 + F_{py}^2)^{1/2} = 252 \text{ N and } \alpha = \tan^{-1}(F_{py}/F_{px}) = 23.4°$$

Example: A ladder 4.00 m long rests against a vertical, frictionless wall with its lower end 1.50 m from the wall, as shown in Figure 11.14. The ladder weighs 450 N, and its center of gravity is at its geometric center. The static coefficient of friction between the ladder and the ground is 0.300. How far up the ladder can a 650 N man climb before it starts to slip?

Given and Diagram:

Figure 11.14

$d = 1.50$ m
$L = 4.00$ m
$W_L = 450$ N
$W_M = 650$ N
$\mu = 0.300$

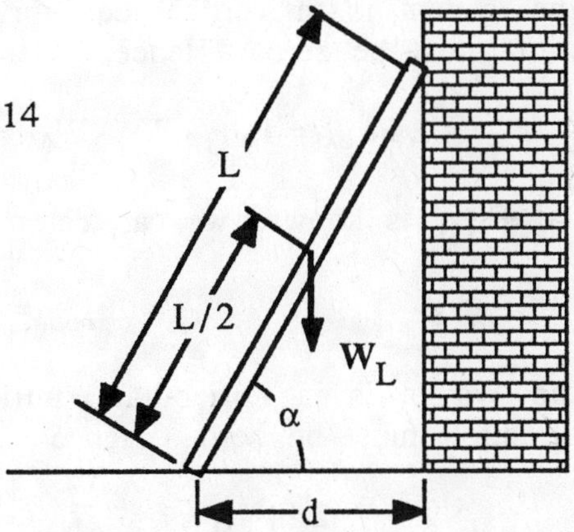

Determine: How far the 650-N man can climb up the ladder before it begins to slip.

Strategy: First draw a diagram showing all forces acting on the ladder. We wish to determine the maximum distance the man can climb up the ladder with the system maintaining static rotational equilibrium.

Since the system is in vertical equilibrium, we can determine the normal force that the ground exerts on the ladder (F_{NG}) from W_L and W_M. Knowing F_{NG} and μ, we can determine the frictional drag f_G that the ground exerts on the ladder. Since the system is in horizontal equilibrium, we can determine the normal force that the wall exerts on the ladder (F_{NW}) from f_G. If we use S to represent the distance that the man can go up the ladder without disrupting the rotational equilibrium and sum the torques about the base of the ladder, the only unknown quantity is S.

Solution:

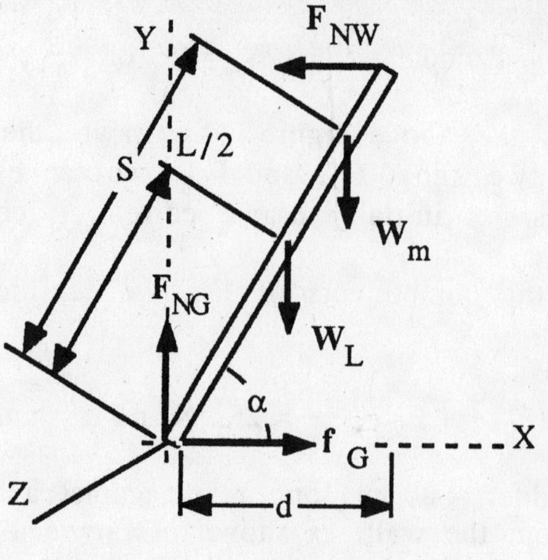

$$\alpha = \cos^{-1}\left(\frac{d}{L}\right) = \cos^{-1}\left(\frac{1.5}{4}\right) = 68.0°$$

Since the system is in vertical equilibrium, we know that the sum of the forces in the y direction must be zero. Hence,

$$\Sigma F_{y,ext} = F_{NG} - W_L - W_M = 0 \quad \text{or} \quad F_{NG} = W_L + W_M = 1100 \text{ N}$$

Now that F_{NG} is known, we can determine the frictional drag f_G.

$$f_G = \mu F_{NG} = (0.300)(1100 \text{ N}) = 330 \text{ N}$$

Since the system is in horizontal equilibrium, we know that the sum of the forces in the x direction must be zero. Hence

$$\Sigma F_{x,ext} = f_G - F_{NW} = 0 \quad \text{or} \quad F_{NW} = f_G = 330 \text{ N}$$

We wish to determine the largest possible value that S can have with the system maintaining rotational equilibrium. If the system is in equilibrium, the sum of the torques about the base of the ladder must be zero. Hence

$$\Sigma \tau_{z,ext} = -(W_L \frac{L}{2} \cos\alpha) - (W_M S \cos\alpha) + (F_{NW} L \sin\alpha) = 0$$

or

$$S = [(-W_L \frac{L}{2} \cos\alpha) + (F_{NW} L \sin\alpha)] / (W_M \cos\alpha) = 3.65 \text{ m}$$

Related Text Exercises: Ex. 11-5 through 11-18.

4 Torque And The Cross Product of Vectors (Section 11-6)

Review: Consider the two vectors **A** and **B** shown in Figure 11.15.

Figure 11.15

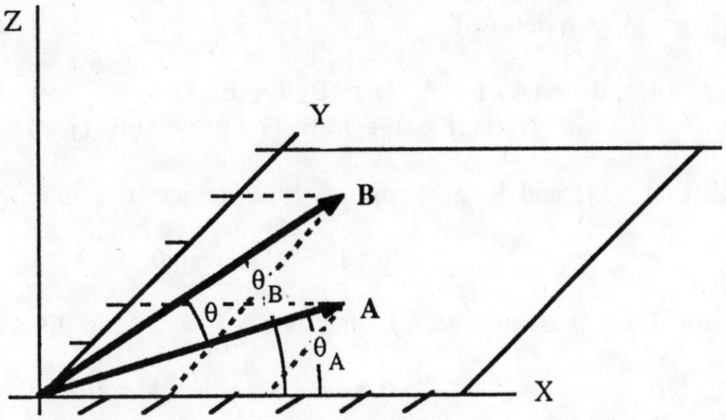

We may determine the magnitude and direction of the cross or vector product by two methods.

Method I.
Magnitude - The magnitude of **A** x **B** (|**A** x **B**|) is the product of the magnitude of **A** (|**A**| = A), the magnitude of **B** (|**B**| = B), and the sine of the angle between the direction of **A** and **B**. This may be written as follows:

$$|\mathbf{A} \times \mathbf{B}| = |\mathbf{A}|\,|\mathbf{B}| \sin\theta = AB \sin\theta$$

Direction - The direction of **A** x **B** is determined by the right hand rule which may be stated as follows:

(a) Extend the fingers of your right hand in the direction of **A**.
(b) Re-orient your right hand so that you may cross your fingers naturally from the direction of **A** to the direction of **B** (through the smallest angle).
(c) The thumb of your right hand points in the direction of **A** x **B** during the crossing process.

When the right rule is applied to the vectors **A** and **B** shown in Figure 11.16, we see that the direction of **A** x **B** is along the +z axis. Application of the right hand rule for **A** x **B** is shown in Figure 11.16.

Figure 11.16

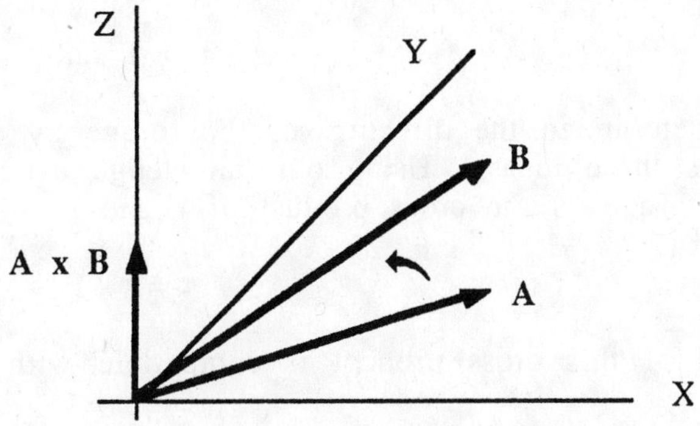

Method II.
Magnitude and Direction - The magnitude and direction of **A** x **B** are determined simultaneously when you perform the cross product in the component form. For the vectors **A** and **B** shown in Figure 11.16 we have

$$\mathbf{A} = A_x\mathbf{i} + A_y\mathbf{j} \qquad\qquad \mathbf{B} = B_x\mathbf{i} + B_y\mathbf{j}$$

$$\begin{aligned}\mathbf{A} \times \mathbf{B} &= (A_x\,\mathbf{i} + A_y\,\mathbf{j}) \times (B_x\,\mathbf{i} + B_y\,\mathbf{j}) \\ &= A_xB_x(\mathbf{i} \times \mathbf{i}) + A_xB_y(\mathbf{i} \times \mathbf{j}) + A_yB_x(\mathbf{j} \times \mathbf{i}) + A_yB_y(\mathbf{j} \times \mathbf{j})\end{aligned}$$

Now recall that **i**, **j**, and **k** are unit vectors, hence the magnitude of **i** x **i** is zero.

$$|\mathbf{i} \times \mathbf{i}| = (1)(1)\,\sin 0° = 0$$

So the vector **i** x **i** is a null vector, that is **i** x **i** = 0. In like manner we may write:

(1) $$\mathbf{i} \times \mathbf{i} = \mathbf{j} \times \mathbf{j} = \mathbf{k} \times \mathbf{k} = 0$$

The magnitude of **i** x **j** is unity.

$$|\mathbf{i} \times \mathbf{j}| = (1)(1)\,\sin 90° = 1$$

The direction of **i** x **j** is **k** (right hand rule). So we may write

$$\mathbf{i} \times \mathbf{j} = (1)\mathbf{k} = \mathbf{k}$$

In like manner we have

(2)	$\mathbf{i} \times \mathbf{j} = \mathbf{k}$	$\mathbf{j} \times \mathbf{k} = \mathbf{i}$	$\mathbf{k} \times \mathbf{i} = \mathbf{j}$
and	$\mathbf{j} \times \mathbf{i} = -\mathbf{k}$	$\mathbf{k} \times \mathbf{j} = -\mathbf{i}$	$\mathbf{i} \times \mathbf{k} = -\mathbf{j}$

When (1) and (2) of the above are inserted into the component form of **A** x **B**, we have

$$\mathbf{A} \times \mathbf{B} = A_xB_y(\mathbf{k}) + A_yB_x(-\mathbf{k}) = \mathbf{k}(A_xB_y - B_xA_y)$$

From the above, it is obvious that if **A** and **B** lie in the x-y plane (and in this case they do) then their cross product has a magnitude of $(A_xB_y - B_xA_y)$ and a direction of **k**.

In Concept 2 (Torque) we reviewed three methods for determining the magnitude of the torque. As you will recall, all three methods reduced to the same expression namely

$$\tau = rF\,\sin\theta$$

We also determined the direction of the torque by rotating **r** into **F**. Figure 11.7 summarizes these ideas. Using our knowledge of cross or vector products we may write the torque as the cross product of **r** and **F**.

$$\boldsymbol{\tau} = \mathbf{r} \times \mathbf{F}$$

The result of this cross product is compatible with all our previous expressions of torque.

Practice: Suspose the vectors **A** and **B** shown in Figure 11.16 are given by

$A = 4\ \mathrm{m}i + 2\ \mathrm{m}j$ and $B = 3\ \mathrm{m}i + 4\ \mathrm{m}j$

Determine the following:

1. The magnitude of **A** and **B**	$\lvert A\rvert = [(4\ m)^2 + (2\ m)^2]^{1/2} = 4.47$ m $\lvert B\rvert = [(3\ m)^2 + (4\ m)^2]^{1/2} = 5.00$ m
2. The direction of **A** (θ_A) and the direction of **B** (θ_B)	$\theta_A = \tan^{-1}(A_y/A_x) = \tan^{-1}(2/4) = 26.57°$ $\theta_B = \tan^{-1}(B_y/B_x) = \tan^{-1}(4/3) = 53.13°$
3. The angle (θ) between the directions of **A** and **B**	$\theta = \theta_B - \theta_A = 53.13° - 26.57° = 26.6°$
4. The magnitude of **A** x **B** by Method I	$\lvert A \times B\rvert = AB \sin\theta$ $= (4.47\ m)(5.00\ m) \sin 26.6°$ $= 10.0\ m^2$
5. The direction of **A** x **B** by Method I	Applying the right hand rule we see that when **A** is crossed into **B** our thumb points along the +z axis. The direction of **A** x **B** is +**k**.
6. The vector product of **A** x **B** by Method I	According to step 4 the magnitude is $\lvert A \times B\rvert = 10.0\ m^2$. According to step 5 the direction is **k**. The vector product of **A** x **B** then is **A** x **B** = $10.0\ m^2$ **k**.
7. The vector product of **A** x **B** by the component method (Method II)	$A \times B = (4\ m\ i + 2\ m\ j) \times (3\ m\ i + 4\ m\ j)$ $= \left[\begin{array}{l} 12\ m^2(i \times i) + 16\ m^2(i \times j) \\ + 6\ m^2(j \times i) + 8\ m^2(j \times j) \end{array}\right]$ Recall that $i \times i = j \times j = 0$, $i \times j = k$, and $j \times i = -k$ $A \times B = 16\ m^2(k) + 6\ m^2(-k)$ $= (16\ m^2 - 6\ m^2)k = 10.0\ m^2 k$
8. If Methods I and II give the same result for **A** x **B**.	Method I: **A** x **B** = $10.0\ m^2$ **k** (step 6) Method II: **A** x **B** = $10.0\ m^2$ **k** (step 7)
9. When it is best to use Method I.	If you are given the magnitude of **A**, the magnitude of **B**, and the angle between their directions θ, it is best to use Method I. When this is the case you just insert values into $AB \sin\theta$ and then apply the right hand rule.

10. When it is best to use Method II.	If you are given **A** and **B** in component form such as $\mathbf{A} = A_x\,\mathbf{i} + A_y\,\mathbf{j}$ and $\mathbf{B} = B_x\,\mathbf{i} + B_y\,\mathbf{j}$, it is best to use Method II. This should be apparent from the preceding practice. We were given **A** and **B** in component form, to determine **A** x **B** by Method I required steps 1 through 6, to determine **A** x **B** by Method II required only step 7.

Example: Consider the disk shown in Figure 11.17 which is acted on by the forces $\mathbf{F}_1$ and $\mathbf{F}_2$ applied at locations $\mathbf{r}_1$ and $\mathbf{r}_2$ respectively. Determine the net torque on the disk and the subsequent direction of rotation if the force and position vectors have the following values:

$\mathbf{F}_1 = 3\text{ Ni} + 5\text{ Nj}$ at $\mathbf{r}_1 = 4\text{ mi} + 3\text{ mj}$

$\mathbf{F}_2 = -4\text{ Ni} + 2\text{ Nj}$ at $\mathbf{r}_2 = -2\text{ mi} - 4\text{ mj}$

Figure 11.17

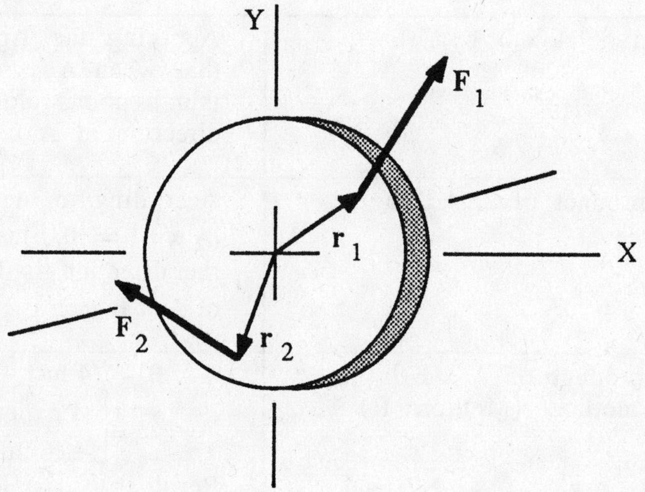

Given: $\mathbf{F}_1$, $\mathbf{F}_2$, $\mathbf{r}_1$ and $\mathbf{r}_2$

Determine: The net torque on the disk and the direction of rotation

Strategy: Knowing the components of the forces acting on the disk and the components of the vectors locating the points of application of each force, we can use the component cross product method to determine the torque τ_1 due to $\mathbf{F}_1$ acting at $\mathbf{r}_1$ and the torque τ_2 due to $\mathbf{F}_2$ acting at $\mathbf{r}_2$. We can determine the net torque and the subsequent sense of the rotation.

Solution: The torque τ_1 due to $\mathbf{F}_1$ acting at $\mathbf{r}_1$ is

$$\tau_1 = \mathbf{r}_1 \times \mathbf{F}_1 = (4\text{ mi} + 3\text{ mj}) \times (3\text{ Ni} + 5\text{ Nj})$$
$$= 12\text{ N}\cdot\text{m}\,(\mathbf{i} \times \mathbf{i}) + 20\text{ N}\cdot\text{m}\,(\mathbf{i} \times \mathbf{j}) + 9\text{ N}\cdot\text{m}\,(\mathbf{j} \times \mathbf{i}) + 15\text{ N}\cdot\text{m}\,(\mathbf{j} \times \mathbf{j})$$

Recall that $\mathbf{i} \times \mathbf{i} = \mathbf{j} \times \mathbf{j} = 0$, and $\mathbf{i} \times \mathbf{j} = -\mathbf{j} \times \mathbf{i} = \mathbf{k}$ to obtain

$$\tau_1 = (20\ \text{N}\bullet\text{m} - 9\ \text{N}\bullet\text{m})\mathbf{k} = 11\ \text{N}\bullet\text{m}\ \mathbf{k}$$

Since τ_1 is in the $+\mathbf{k}$ direction, $\mathbf{F}_1$ tends to rotate the disk in a counterclockwise direction.

The torque τ_2 due to $\mathbf{F}_2$ acting at $\mathbf{r}_2$ is

$$\begin{aligned}\tau_2 = \mathbf{r}_2 \times \mathbf{F}_2 &= (-2\ \text{m}\mathbf{i} - 4\ \text{m}\mathbf{j}) \times (-4\ \text{N}\mathbf{i} + 2\ \text{N}\mathbf{j}) \\ &= 8\ \text{N}\bullet\text{m}(\mathbf{i} \times \mathbf{i}) - 4\ \text{N}\bullet\text{m}(\mathbf{i} \times \mathbf{j}) + 16\ \text{N}\bullet\text{m}\ (\mathbf{j} \times \mathbf{i}) + -8\ \text{N}\bullet\text{m}\ (\mathbf{j} \times \mathbf{j}) \\ &= (-4\ \text{N}\bullet\text{m} - 16\ \text{N}\bullet\text{m})\mathbf{k} = -20\ \text{N}\bullet\text{m}\ \mathbf{k}\end{aligned}$$

Since τ_2 is in the $-\mathbf{k}$ direction, $\mathbf{F}_2$ tends to rotate the disk in a clockwise direction.

The net torque is $\tau_{net} = \tau_1 + \tau_2 = 11\ \text{N}\bullet\text{m}\ \mathbf{k} - 20\ \text{N}\bullet\text{m}\ \mathbf{k} = -9\ \text{N}\bullet\text{m}\mathbf{k}$

Since τ_{net} is in the $-\mathbf{k}$ direction, the net torque produces a clockwise rotation of the disk.

Related Text Exercises: Ex. 11-22 through 11-32.

PRACTICE TEST

Take and grade this practice test. Doing so will allow you to determine any weak spots in your understanding of the concepts taught in this chapter. The following section prescribes what you should study further to strengthen your understanding.

You push on a box resting on an incline, as shown in the following.

$W = 1.00 \times 10^3$ N = weight of the box
$P = 5.00 \times 10^2$ N = your push
$\mu_s = 0.300$ = static friction
$\mu_k = 0.100$ = kinetic friction

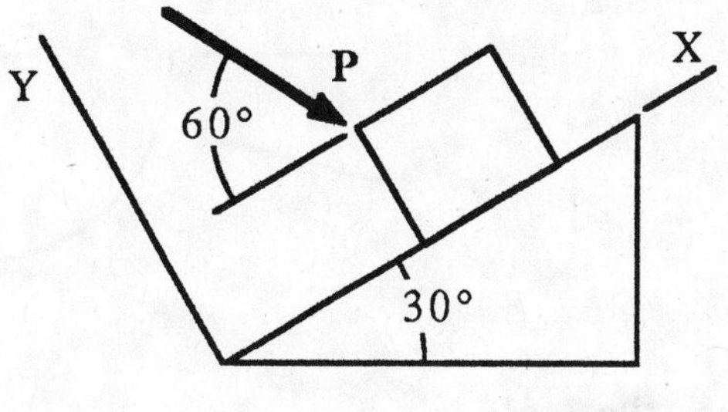

Determine the Following:

_______ 1. Component of the weight of the box perpendicular to the incline
_______ 2. Component of your push perpendicular to the incline
_______ 3. Normal force acting on the box
_______ 4. Component of the weight of the box parallel to the incline
_______ 5. Component of your push parallel to the incline
_______ 6. Force of friction acting on the box
_______ 7. Unbalanced force on the box parallel to the incline
_______ 8. Is the box in equilibrium?

A sign is suspended from a uniform pole as shown. The pole is attached to the wall by means of a bracket and pin.

W_s = 500 N = weight of the sign

W_p = 200 N = weight of the pole

L = 4.00 m = length of the pole

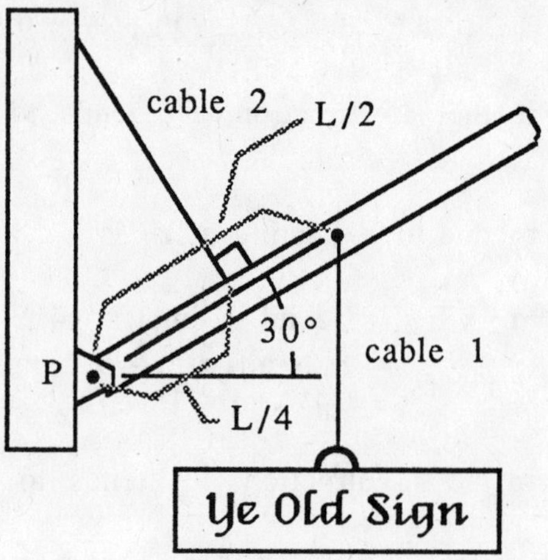

Determine the following:

________ 9. Torque about pin P due to the weight of the sign
________ 10. Torque about pin P due to the weight of the pole
________ 11. Torque about pin P due to cable 2
________ 12. The tension in cable 2
________ 13. Horizontal component of the force that pin P exerts on the pole
________ 14. Vertical component of the force that pin P exerts on the pole

A disk is mounted in a frictionless manner on an axle as shown.

F_1 = -10 Nj

r_1 = +4i - 3j

F_2 = -8 Ni

r_2 = -2i + 3j

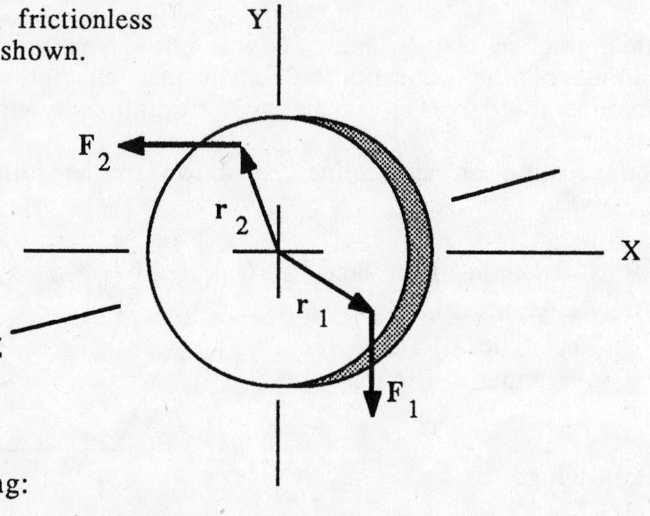

Determine the following:

________ 15. Magnitude of the torque due to F_1
________ 16. Direction F_1 tends to rotate the disk
________ 17. Magnitude of the torque due to F_2
________ 18. Direction F_2 tends to rotate the disk
________ 19. Net torque acting on the disk
________ 20. Direction of rotation of the disk

(See Appendix I for answers.)

PRINCIPAL CONCEPTS AND EQUATIONS PRESCRIPTION

Your score on the practice test is an excellent measure of your understanding of the chapter. You should now use the following chart to write your own prescription for dealing with any weaknesses the practice test points out. Look down the leftmost column to the number of the question(s) you answered incorrectly, reading across that row you will find the concept and/or equation of concern, the section(s) of the study guide you should return to for further study, and some suggested text exercises which you should work to gain additional experience.

Practice Test Questions	Concepts and Equations	Prescription	
		Principal Concepts	Text Exercises
1	Component of a vector	3 of Ch 2	2-11, 14
2	Component of a vector	3 of Ch 2	2-15, 16
3	First condition of equilibrium	1	11-5, 6
4	Component of a vector	3 of Ch 2	2-11, 14
5	Component of a vector	3 of Ch 2	2-15, 16
6	Force of friction	2 of Ch 6	6-15, 18
7	Addition of vectors	3 of Ch 2	2-19, 20
8	First condition of equilibrium	1	11-5, 6
9	Torque	2	11-9, 13
10	Torque	2	11-5, 7
11	Second condition of equilibrium	3	11-8, 9
12	Torque	2	11-33, 37
13	First condition of equilibrium	1	11-5, 6
14	First condition of equilibrium	1	11-5, 6
15	Torque: $\boldsymbol{\tau} = \mathbf{r} \times \mathbf{F}$ by component method	4	11-23, 27
16	Direction of torque: - is clockwise	2	11-1, 22
17	Torque: $\boldsymbol{\tau} = \mathbf{r} \times \mathbf{F}$ by component method	4	11-23, 27
18	Direction of torque: + is counterclockwise	2	11-1, 22
19	Adding vectors	3 of Ch 2	2-19, 20
20	Direction of torque: - is clockwise	2	11-1, 22

1 2 ROTATION I

RECALL FROM PREVIOUS CHAPTERS

Previously learned concepts and equations frequently used in this chapter	Text Section	Study Guide Page
Instantaneous velocity: $v = dx/\Delta t$	3-2	3-5
Instantaneous acceleration: $a = dv/dt = d^2x/dt^2$	3-4	3-12
Equations used to analyze translational motion:	3-5	3-13
$\bar{v}_x = \Delta x/\Delta t$ $v_x = v_{xo} + a_x t$		
$\bar{a}_x = \Delta v_x/\Delta t$ $\Delta x = v_{xo}t + a_x t^2/2$		
$\bar{v}_x = (v_x + v_{xo})/2$ $v_x^2 - v_{xo}^2 = 2a_x\Delta x$		
Translational kinetic energy: $K = mv^2/2$	8-4	8-13

NEW IDEAS IN THIS CHAPTER

Concepts and equations introduced	Text Section	Study Guide Page
Angular measure in radians, degrees and revolutions	12-2	12-2
Angular - coordinate: θ	12-3	12-5
speed: $\omega = d\theta/dt$		
acceleration: $\alpha = d\omega/dt = d^2\theta/dt^2$		
Equations used to analyze rotational motion:	12-4	12-8
$\bar{\omega} = \Delta\theta/\Delta t$ $\omega = \omega_0 + \alpha t$		
$\bar{\omega} = (\omega_0 + \omega)/2$ $\Delta\theta = \omega_0 t + \alpha t^2/2$		
$\alpha = \Delta\omega/\Delta t$ $2\alpha\Delta\theta = \omega^2 - \omega_0^2$		
Relation between linear and angular-	12-5	12-10
coordinates: $\Delta s = R\Delta\theta$		
speed: $v = R\omega$		
acceleration: $a_t = R\alpha$, $a_R = R\omega^2$		
Moment of Inertia	12-6, 7	12-13
Array of masses: $I = \Sigma\, m_i R_i^2$		
Continuous object: $I = \rho \int_V R^2 dv$		
Parallel-axis theorem: $I_p = I_{cm} + Md^2$		
Kinetic Energy of a rolling object	12-6, 8	12-18
$K = K_{rot} + K_{trans} = I_0\omega^2/2 + Mv^2/2$		

PRINCIPAL CONCEPTS AND EQUATIONS

1 Angular Measure in Degrees, Revolutions, or Radians
(Section 12-2)

Review: Your past experiences have left you familiar with angular measure in terms of degrees and revolutions. This knowledge is summarized in Figure 12.1.

Angle	Expressed in Degrees	Expressed in Revolutions
θ_1	90°	1/4 rev
θ_2	180°	1/2 rev
θ_3	270°	3/4 rev
θ_4	360°	1 rev

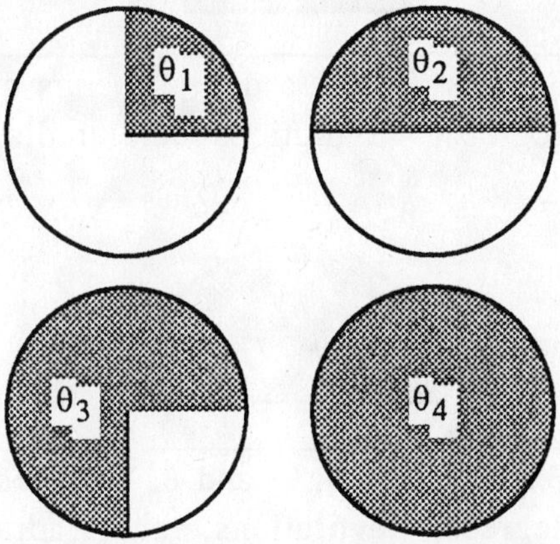

Figure 12.1

Once you start working problems, you will learn that a more convenient way to measure angles is in radians. One radian is the angle subtended by an arclength of one radius. As shown in Figure 12.2, exactly 6.28 (or 2π) radii fit on the circumference of a circle.

$$1 \text{ rad} = 57.3°$$

$$\pi \text{ rad} = 3.14 \text{ rad} = 180°$$

$$2\pi \text{ rad} = 6.28 \text{ rad} = 360°$$

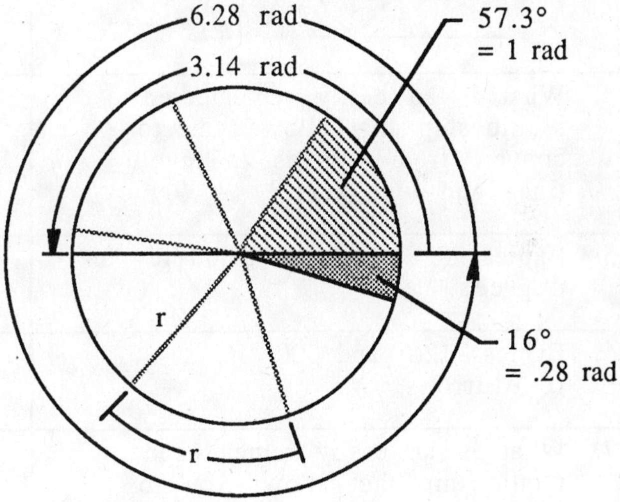

Figure 12.2

Since the number of radians in an angular measure is determined by dividing the arclength by the radius, we may write

$$\Delta\theta = \Delta s/r$$

Note: According to this definition, an angle expressed in radians has no units. However, we will carry the word "radian" (abbreviated rad) along in our calculations to confirm that we are using radian measure rather than some other angular measure.

Practice:

1. Express θ_1 and θ_2 in terms of degrees, revolutions, and radians.

Angle	Degrees	Rev	Rad
θ_1	90	1/4	$\pi/2$
θ_2	180	1/2	π

2. Express θ_1, θ_2, and θ_3 in terms of degrees, revolutions and radians.

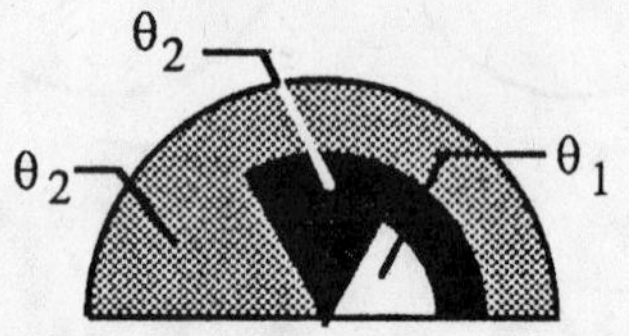

Angle	Degrees	Rev	Rad
θ_1	60	1/6	$\pi/3$
θ_2	120	1/3	$2\pi/3$
θ_3	180	1/2	π

3. Express θ_1, θ_2, θ_3 and θ_4 in terms of degrees, revolutions, and radians.

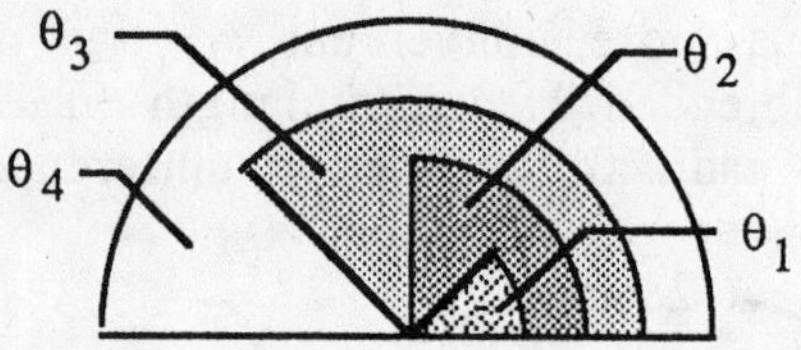

Angle	Degrees	Rev	Rad
θ_1	45	1/8	$\pi/4$
θ_2	90	1/4	$\pi/2$
θ_3	135	3/8	$3\pi/4$
θ_4	180	1/2	π

4. What is an easily remembered relationship that allows you to convert from degrees to revolutions and vice versa?

1 rev = 360°

5. Express (1/3) rev and 0.800 rev in degrees.

([1/3] rev)(360°/1 rev) = 120°
(0.800 rev)(360°/1 rev) = 288°

6. Express 120° and 300° in revolutions.

(120°)(1 rev/360°) = (1/3) rev
(300°)(1 rev/360°) = 0.833 rev

7. What is an easily remembered relationship that allows you to convert from degrees to radians and vice versa?

π rad = 180°

8. Convert 70.0° and 330° to radians.

70.0° (π rad/180°) = 1.22 rad
330° (π rad/180°) = 5.76 rad

9. Convert 0.700π rad and 5 rad to degrees.	$(0.700\pi \text{ rad})(180°/\pi \text{ rad}) = 126°$ $(5 \text{ rad})(180°/\pi \text{ rad} = 286°$
10. What is an easily remembered relationship that allows you to convert from revolutions to radians and vice versa?	$1 \text{ rev} = 2\pi \text{ rad}$
11. Convert (1/3) rev and 0.700 rev to radians.	$([1/3] \text{ rev})(2\pi \text{ rad}/1 \text{ rev}) = 2.09 \text{ rad}$ $(0.700 \text{ rev})(2\pi \text{ rad}/1 \text{ rev}) = 4.40 \text{ rad}$
12. Convert 6 rad and $(\pi/3)$ rad to revolution.	$(6.00 \text{ rad})(1 \text{ rev}/2\pi \text{ rad}) = 0.955 \text{ rev}$ $([\pi/3] \text{ rad})(1 \text{ rev}/2\pi \text{ rad}) = 0.167 \text{ rev}$
13. Determine how far a wheel with a radius of 0.500 m advances when it turns through the following angles: $\Delta\theta_1 = 0.800$ rev $\Delta\theta_2 = 0.300$ rad $\Delta\theta_3 = 200°$	Using $\Delta s = r\Delta\theta$ then obtain: $\Delta s_1 = (0.500 \text{ m})(0.800 \text{ rev})(\frac{2\pi \text{ rad}}{1 \text{ rev}})$ $= 2.51 \text{ m}$ $\Delta s_2 = (0.500 \text{ m})(0.300 \text{ rad}) = 0.150 \text{ m}$ $\Delta s_3 = (0.500 \text{ m})(200°)(\frac{\pi \text{ rad}}{180°}) = 1.74 \text{ m}$

Note: The calculation in step 13 should convince you that working with angular distance in radians is more convenient than working with revolutions or degrees.

Example: Two spools of rope are mounted on the same axle. One has a radius of 20.0 cm, and the other has a radius of 12.0 cm. As you pull on the rope wound on the larger spool, the smaller spool also unwinds. When you unwind 1.00 m of the rope from the large spool, through what angle will the smaller spool turn? Express your answer in radians, degrees, and revolutions.

Given and Diagram:

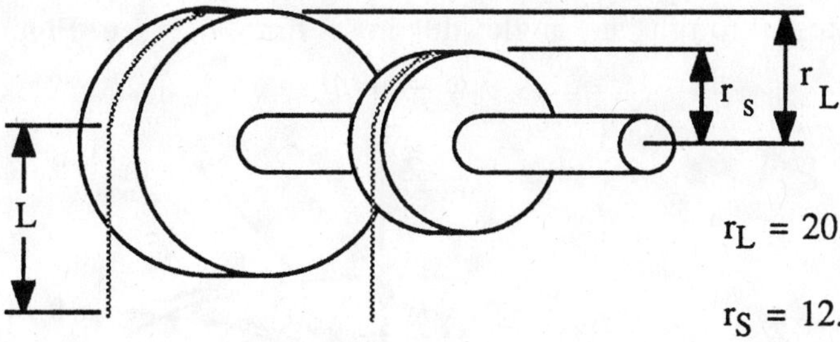

$L = 1.00 \text{ m}$

$r_L = 20.0 \text{ cm} = 0.200 \text{ m}$

$r_S = 12.0 \text{ cm} = 0.120 \text{ m}$

Figure 12.3

Determine: The angle through which the smaller spool rotates when you unwind 1 m of rope from the larger spool.

Strategy: Knowing the radius of the larger spool (r_L) and the length of rope unwound (L), we can determine the angle of rotation (in radians) of the larger spool. Since the spools are mounted on the same axle, both rotate through the same angle. When this angle is known in radians, we can convert it to degrees and revolutions.

Solution:

$$\Delta\theta_L = \frac{\Delta S}{r_L} = \frac{1.00\ \text{m}}{0.200\ \text{m}} = 5.00\ \text{rad}, \quad \text{then} \quad \Delta\theta_S = \Delta\theta_L = 5.00\ \text{rad}$$

$$\Delta\theta_S = 5.00\ \text{rad}\left(\frac{180°}{\pi\ \text{rad}}\right) = 286° \qquad \Delta\theta_S = 5.00\ \text{rad}\left(\frac{1\ \text{rev}}{2\pi\ \text{rad}}\right) = 0.796\ \text{rev}$$

Related Text Exercises: Ex. 12-1 through 12-5.

2 Angular Coordinate, Velocity and Acceleration (Section 12-3)

Review: Consider the rigid rotator shown in Figure 12.4. At time t = 0 the rotator is at $\theta = 0$ (along the x axis) and some time t later it is located by the angular coordinate θ.

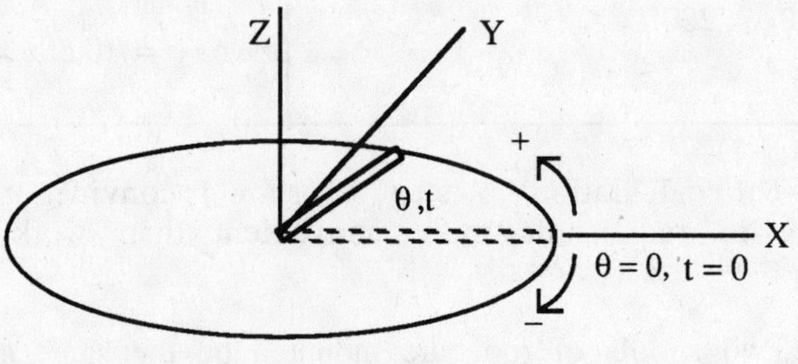

Figure 12.4

When viewed from the positive z direction, an angle measured counterclockwise is positive and an angle measured clockwise is negative. If you grasp the z axis with your right hand so that your thumb points in the +z direction, then your fingers curl in the positive θ sense.

If the rotator rotates through an angle $d\theta$ in a time dt (see Figure 12.5) the angular speed is given by

$$\omega = |d\theta/dt|$$

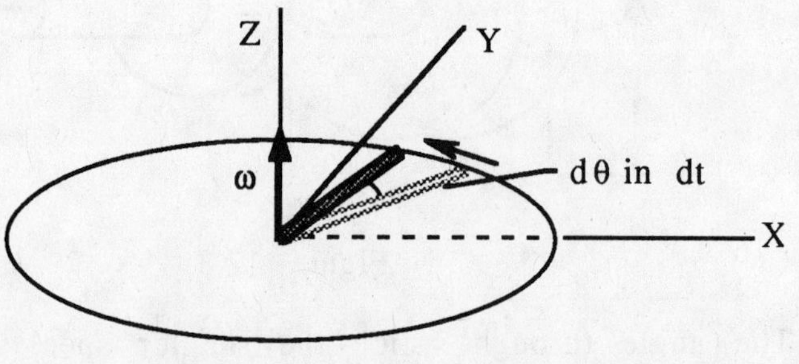

Figure 12.5

The angular velocity $\boldsymbol{\omega}$ is a vector along the axis of rotation. The magnitude of $\boldsymbol{\omega}$ is the angular speed and the direction of $\boldsymbol{\omega}$ is determined by the sense of rotation. If you imagine grasping the axis of rotation with your right hand such that your fingers curl in the sense of rotation, then your thumb points in the direction of ω. ω is toward +z if the rotation is positive (i.e. counterclockwise) and ω is toward -z if the rotation is negative (i.e. clockwise).

If as shown in Figure 12.6 the angular velocity is ω at some time t and $\omega + d\omega$ at the time t + dt, then the rotator has experienced an angular acceleration of magnitude

$$\alpha = |d\omega/dt| = |d^2\theta/dt^2|$$

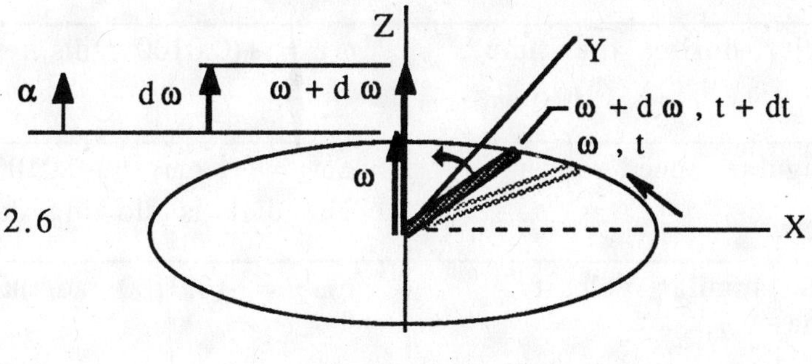

Figure 12.6

The angular acceleration is a vector along the axis of rotation. α is positive when ω is increasing and negative when ω is decreasing.

Note: You should be aware that θ, ω and α are analogous to x, v and a. This is summarized below:

	Linear	Angular
Coordinate	x	θ
Speed	$v = dx/dt$	$\omega = d\theta/dt$
Acceleration	$\alpha = dv/dt = d^2x/dt^2$	$\alpha = d\omega/dt = d^2\theta/dt^2$

Practice: Figure 12.7 shows a disk rotating on an axle.

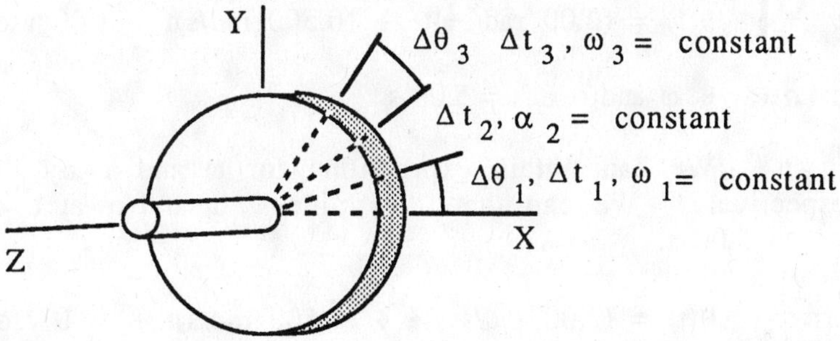

Figure 12.7

The disk rotates through the angle $\Delta\theta_1$ in a time Δt_1 at a constant angular speed ω_1. The disk then experiences constant angular acceleration α_2 for a time Δt_2. Finally the disk rotates through the angle $\Delta\theta_3$ in a time Δt_3 at a constant angular speed ω_3.

$\Delta\theta_1 = 0.0200$ rad $\qquad \Delta\theta_3 = 0.0100$ rad $\qquad \Delta t_1 = \Delta t_2 = \Delta t_3 = 1.00$ s

Determine the following:

1. Angular speed during the time Δt_1	$\omega_1 = \Delta\theta_1/\Delta t_1 = 0.0200$ rad/s
2. Angular velocity during the time Δt_1	$\boldsymbol{\omega}_1 = +(0.0200 \text{ rad/s})\mathbf{k}$
3. Angular speed during the time Δt_3	$\omega_3 = \Delta\theta_3/\Delta t = 0.0100$ rad/s
4. Angular velocity during the time Δt_3	$\boldsymbol{\omega}_3 = +(0.0100 \text{ rad/s})\mathbf{k}$
5. Change in angular speed during the time Δt_2	$\Delta\omega_2 = \omega_3 - \omega_1 = -0.0100$ rad/s The disk is slowing down.
6. Change in the angular velocity during the time Δt_2.	$\Delta\boldsymbol{\omega}_2 = -(0.0100 \text{ rad/s})\mathbf{k}$
7. Magnitude of the angular acceleration during the time Δt_2	$\alpha_2 = \Delta\omega_2/\Delta t_2 = -0.0100 \text{ rad/s}^2$
8. Angular acceleration during the time Δt_2	$\boldsymbol{\alpha}_2 = -(0.0100 \text{ rad/s}^2)\mathbf{k}$

Example: A particle is traveling in a circular orbit and its angular coordinate as a function of time is given by

$$\theta(t) = (2.00 \text{ rad/s}^2)t^2 + (0.500 \text{ rad/s})t + 1.00 \text{ rad}$$

Determine the angular coordinate, speed and acceleration at the time t = 5.00 s.

Given: $\theta(t) = (2.00 \text{ rad/s}^2)t^2 + (0.500 \text{ rad/s})t + 1.00 \text{ rad}$ and $t = 5.00$ s

Determine: θ, ω and α at $t = 5.00$ s

Strategy: We can obtain expressions for ω and α by differentiating θ once and twice respectively. We can then determine θ, ω and α at $t = 5.00$ s merely by inserting the value $t = 5.00$ s.

Solution:

$$\theta(t) = (2.00 \text{ rad/s}^2)t^2 + (0.500 \text{ rad/s})t + 1.00 \text{ rad}$$
$$\omega = d\theta/dt = (4.00 \text{ rad/s}^2)t + (0.500 \text{ rad/s})$$
$$\alpha = d^2\theta/dt^2 = 4.00 \text{ rad/s}^2$$

Inserting the value t = 5.00 s, obtain

$$\theta(t = 5.00 \text{ s}) = 50.0 \text{ rad} + 2.50 \text{ rad} + 1.00 \text{ rad} = 53.5 \text{ rad}$$
$$\omega(t = 5.00 \text{ s}) = 20.0 \text{ rad/s} + 0.500 \text{ rad/s} = 20.5 \text{ rad/s}$$
$$\alpha(t = 5.00 \text{ s}) = 4.00 \text{ rad/s}^2$$

Related Text Exercises: Ex. 12-6 through 12-10.

3 Kinematics of Rotation About A Fixed Axis (Section 12-4)

Review: The equations used to analyze translational motion and their rotational analogs are shown in the chart below:

Translational Motion	Rotational Motion
$\bar{v} = \Delta s/\Delta t$	$\bar{\omega} = \Delta\theta/\Delta t$
$\bar{v} = (v_0 + v)/2$ if $a = \text{const}$	$\bar{\omega} = (\omega_o + \omega)/2$ if $\alpha = \text{const}$
$a = \Delta v/\Delta t$	$\alpha = \Delta\omega/\Delta t$
$v = v_0 + at$	$\omega = \omega_0 + \alpha t$
$\Delta s = v_0 t + at^2/2$	$\Delta\theta = \omega_0 t + \alpha t^2/2$
$2a\Delta s = v^2 - v_0^2$	$2\alpha\Delta\theta = \omega^2 - \omega_0^2$

A little thought will reveal that, when analyzing an object's rotational motion, you are dealing with five physical quantities: ω_0, ω, α, t, and $\Delta\theta$. In general, you will be given three of these physical quantities and be asked to solve for the other two. If we are dealing with five quantities of which three are known, then ten possible cases exist. In the following practice, we will consider only three of these ten cases. You should consider the others on your own.

Physical Quantities		ω_0	ω	α	t	$\Delta\theta$
Given Quantities:	Case I	√	√		√	
	Case II	√	√			√
	Case III		√	√	√	

Practice: Case I. Given ω_0, ω, and t, determine α and $\Delta\theta$. A mounted wheel is initially rotating with an angular speed of ω_0. It accelerates for a time t and acquires a speed ω.

Determine expressions for the following:

1. The angular acceleration	$\alpha = \Delta\omega/\Delta t = (\omega - \omega_0)/t$
2. The angle it turns through during this acceleration	$\Delta\theta = \omega_0 t + \alpha t^2/2$ or $\Delta\theta = (\omega^2 - \omega_0^2)/2\alpha$

Case II. Given ω_0, ω, and $\Delta\theta$, determine α and t. A mounted wheel is initially rotating with an angular speed of ω_0. It decelerates with a constant deceleration and comes to rest after rotating through an angle $\Delta\theta$.

Determine expressions for the following:

1. The angular acceleration	$\alpha = (\omega^2 - \omega_0^2)/2\Delta\theta = -\omega_0^2/2\Delta\theta$
2. The time it takes to decelerate to a stop	$\omega = \omega_0 + \alpha t$ or $t = (\omega - \omega_0)/\alpha = -\omega_0/\alpha$

Case III. Given ω, α and t, determine ω_0 and $\Delta\theta$. A mounted wheel is rotating with some unknown angular speed. It decelerates with an angular acceleration α for a time t and has an angular speed ω.

Determine expressions for the following:

1. The initial angular speed	$\omega = \omega_0 + \alpha t$ or $\omega_0 = \omega - \alpha t$
2. The angle it turns through during this deceleration	$\Delta\theta = \omega_0 t + \alpha t^2/2$ or $\Delta\theta = (\omega^2 - \omega_0^2)/2\alpha$

Example: A rotary lawn mower blade requires 10.0 s to reach its maximum speed ω_m of 300 rpm. Calculate: (a) its angular acceleration in rad/s^2, (b) the angle through which it turns in the process of coming up to this speed, and (c) the time it takes to achieve a speed of 50.0 rad/s.

Given: $\omega_0 = 0$ $\quad \omega_m = \left(\frac{300 \text{ rev}}{\text{min}}\right)\left(\frac{2\pi \text{ rad}}{\text{rev}}\right)\left(\frac{\text{min}}{60 \text{ s}}\right) = 31.4 \text{ rad/s}$ $\quad t = 10.0$ s

Determine: α, $\Delta\theta$, and t to obtain ω = 50.0 rad/s

Strategy: We can determine α using ω_0, ω_m, and t in $\alpha = \Delta\omega/\Delta t$. We can then determine $\Delta\theta$ by either $\Delta\theta = \omega_0 t + \alpha t^2/2$ or $\Delta\theta = (\omega^2 - \omega_0^2)/2\alpha$. Finally, we can determine the time it takes to achieve a certain speed by $\omega = \omega_0 + \alpha t$.

Solution:

(a) $\alpha = \Delta\omega/\Delta t = (\omega_m - \omega_0)/(t - 0) = (31.4 \text{ rad/s})/10.0 \text{ s} = 3.14 \text{ rad/s}^2$

(b) $\Delta\theta = \omega_0 t + \alpha t^2/2 = (3.14 \text{ rad/s}^2)(10.0 \text{ s})^2/2 = 157 \text{ rad}$

or $\Delta\theta = (\omega^2_m - \omega_0^2)/2\alpha = \dfrac{(31.4 \text{ rad/s})^2}{2(3.14 \text{ rad/s}^2)} = 157 \text{ rad}$

(c) $\omega = \omega_0 + \alpha t$ or $t = (\omega - \omega_0)/\alpha = (50.0 \text{ rad/s})/(3.14 \text{ rad/s}^2) = 15.9 \text{ s}$

Related Text Exercises: Ex. 12-11 through 12-16.

4 | Relation Between Linear and Angular Coordinate, Linear and Angular Velocity, and Linear and Angular Acceleration
(Section 12-5)

Review: The relation between the linear (Δs) and angular ($\Delta \theta$) coordinates is shown in Figure 12.8(a). The relation between the tangential speed (v) and the angular speed (ω) is shown in Figure 12.8(b). The relation between the tangential acceleration (a_t), centripetal acceleration (a_R), total acceleration (a), and angular acceleration (α) is shown in Figure 12.8(c).

$$\Delta s = R \Delta \theta$$

(a)

$$v = R\omega$$

(b)

$$a_R = R\omega^2$$
$$a_t = R\alpha$$
$$a = R(\alpha^2 + \omega^4)^{1/2}$$

(c)

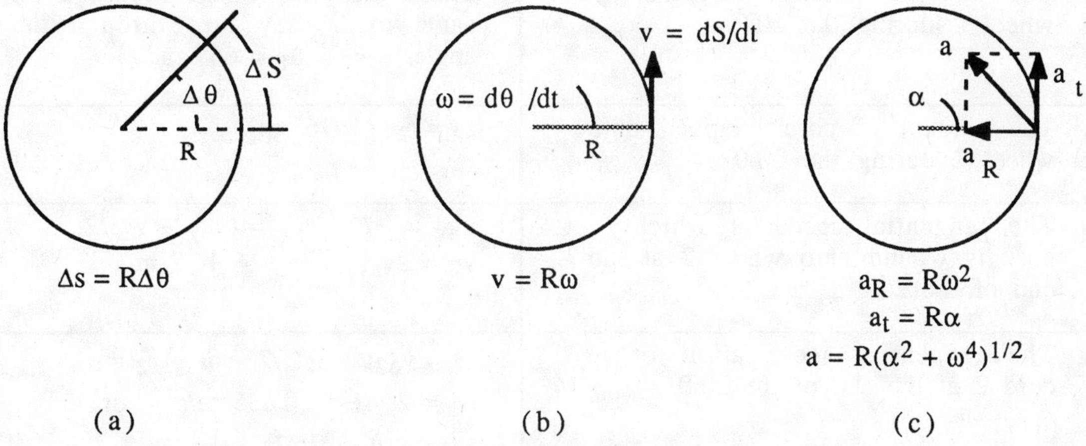

Figure 12.8

Practice: Consider the two wheels mounted on the axle shown in Figure 12.9.

$r_1 = 15$ cm = radius of wheel 1

$r_2 = 25$ cm = radius of wheel 2

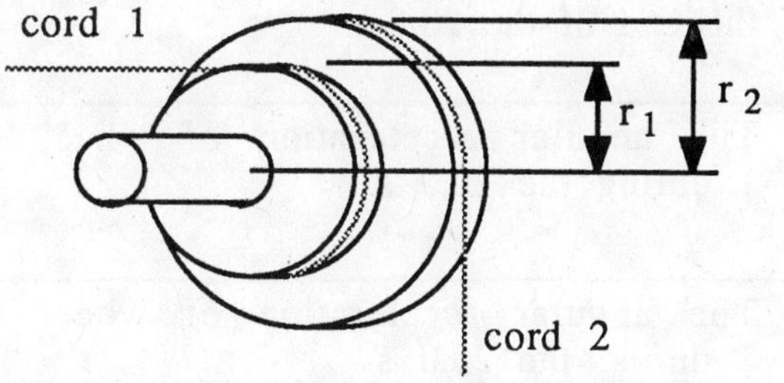

Figure 12.9

The system is initially at rest, and thin cords have been wrapped around both wheels. The cord wrapped around wheel 1 is pulled with a constant force for 20 s, and after that time the cord is unwound at the rate of 10 m/s. As the cord is unwound from wheel 1, another cord is wound onto wheel 2.

Determine the following:

1. The tangential acceleration of cord 1	$a_1 = \Delta v_1/t = (v_1 - v_{o1})/t = 0.500\ m/s^2$
2. The length of cord pulled off wheel 1 during the 20.0 s	$s_1 = a_1t^2/2 = 100\ m$
3. The angular distance traveled by wheel 1 during the 20.0 s	$\theta_1 = s_1/r_1 = 667\ rad$
4. The angular distance traveled by wheel 2 during the 20.0 s	Since the wheels are mounted on the same axle, they turn through the same angle. $\theta_2 = \theta_1 = 667\ rad$
5. The amount of cord wrapped onto wheel 2 during the 20.0 s	$s_2 = r_2\theta_2 = 167\ m$
6. The tangential speed at which cord is wound onto wheel 2 at the end of the 20.0 s	$v_{av} = s_2/t$; $v_{av} = (v_{o2} + v_2)/2$ $v_2 = 2v_{av} = 2s_2/t = 16.7\ m/s$
7. The tangential acceleration of cord 2 at the end of the 20.0 s	$s_2 = v_{o2}t + a_2t^2/2$ or $v_2 = v_{o2} + a_2t$ $a_2 = 2s_2/t^2 = 0.835\ m/s^2$ or $a_2 = v_2/t = 0.835\ m/s^2$
8. The angular speed of wheel 1 at the end of the 20.0 s	$\omega_1 = v_1/r_1 = 66.7\ rad/s$
9. The angular speed of wheel 2 at the end of the 20.0 s	$\omega_2 = \omega_1 = 66.7\ rad/s$ or $\omega_2 = v_2/r_2 = 66.7\ rad/s$
10. The angular acceleration of wheel 1 during the 20.0 s	$\alpha_1 = a_1/r_1 = 3.33\ rad/s^2$ or $\alpha_1 = \Delta\omega_1/t = 3.33\ rad/s^2$
11. The angular acceleration of wheel 2 during the 20.0 s	$\alpha_2 = \alpha_1 = 3.33\ rad/s^2$ We can also use either $\alpha_2 = a_2/r_2$ or $\alpha_2 = \Delta\omega_2/t$ to obtain the same answer.

Example: An automobile traveling 20.0 m/s accelerates to 30.0 m/s in a 15.0 s time interval. The outside diameter of the tires is 0.720 m. Determine the following physical quantities:

(a) The initial, final, and average angular speed of the wheels
(b) The tangential acceleration at the outside diameter of the tires and angular acceleration of the wheels
(c) The linear distance traveled and the number of revolutions turned by the wheels during the 15.0 s

Given: $v_0 = 20.0$ m/s $\quad v = 30.0$ m/s $\quad t = 15.0$ s $\quad d = 0.720$ m

Determine: ω_0, ω_f, ω_{av}, a, α, Δs, N

Strategy: Knowing v_0, v, and r, we can determine ω_0, and ω. We can then obtain ω_{av} from ω_0 and ω. Knowing v_0, v, and t, we can determine a and then α. Knowing v_0, a, and t, we can calculate Δs. Finally, we can use Δs to determine $\Delta\theta$ and then N.

Solution:

(a) $\omega_0 = v_0/r = (20.0 \text{ m/s})/0.720 \text{ m} = 27.8 \text{ rad/s}$
$\omega = v/r = (30.0 \text{ m/s})/0.720 \text{ m} = 41.7 \text{ rad/s}$
$\omega_{av} = (\omega_0 + \omega)/2 = [(27.8 + 41.7) \text{ rad/s}]/2 = 34.8 \text{ rad/s}$

(b) $a = \Delta v/\Delta t = (v - v_0)/t = [(30.0 - 20.0) \text{ m/s}]/15.0 \text{ s} = 0.667 \text{ m/s}^2$
$\alpha = a/r = (0.667 \text{ m/s}^2)/0.720 \text{ m} = 0.926 \text{ rad/s}^2$

or

$\alpha = \Delta\omega/\Delta t = (\omega - \omega_0)/t = [(41.7 - 27.8) \text{ rad/s}]/15.0 \text{ s} = 0.927 \text{ rad/s}^2$

(c) $\Delta s = v_0 t + at^2/2 = (20.0 \text{ m/s})(15.0 \text{ s}) + (0.667 \text{ m/s}^2)(15.0 \text{ s})^2/2 = 375 \text{ m}$

We can obtain $\Delta\theta$ by any one of the following

$$\Delta\theta = \Delta s/r \qquad \Delta\theta = \omega_0 t + \alpha t^2/2 \qquad \Delta\theta = (\omega^2 - \omega_0^2)/2\alpha$$

Using the first expression we obtain

$$\Delta\theta = \Delta s/r = (375 \text{ m})/0.720 \text{ m} = 521 \text{ rad}$$

The number of revolutions turned by the wheel is

$$N = \Delta\theta/(2\pi \text{ rad/rev}) = 521 \text{ rad}/(2\pi \text{ rad/sec}) = 82.9 \text{ rev}$$

Related Text Exercises: Ex. 12-17 through 12-20.

5 **Moment of Inertia** (Sections 12-6 and 12-7)

Review: The moment of inertia for an array of masses about an axis z can be determined by the expression

$$I = \Sigma\, m_i R_i^2$$

This quantity may be thought of as the body's resistance to angular acceleration. It is important to note that I depends not only on the mass but also its distribution about the axis of rotation.

For a continuous object, we can use the previous equation by considering the object to consist of many small pieces Δm_i. If the object has a uniform mass density ρ, each small piece has a mass

$$\Delta m_i = \rho \Delta V_i$$

where ΔV_i is the volume of the small piece Δm_i

$$I = \Sigma\, \Delta m_i R_i^2 = \Sigma\, \rho \Delta V_i R_i^2 = \rho\, \Sigma\, \Delta V_i R_i^2$$

As the volume ΔV_i approaches an infinitesimal, dV, the sum transforms into an integral:

$$I = \lim_{\Delta v_i \to 0} \rho\, \Sigma\, \Delta V_i R_i^2 = \rho \int_V R^2 dV$$

When this integral form of I is applied to various shapes about an axis through the center of mass, one obtains the results shown in Table 12-2 of the text.

If we know the moment of inertia about an axis through the center of mass, we can determine the moment of inertia about any axis parallel to the axis through the center of mass by the parallel-axis theorem. The parallel axis-theorem is illustrated in Figure 12.10 and may be stated as

$$I_p = I_{cm} + Md^2$$

Where
- M = mass of object
- I_{cm} = moment of inertia about an axis through the center of mass (cm)
- $axis_p$ = an axis parallel to the axis through the cm
- I_p = moment of inertia of the object about axis P
- d = distance between the two axes under consideration

Figure 12.10

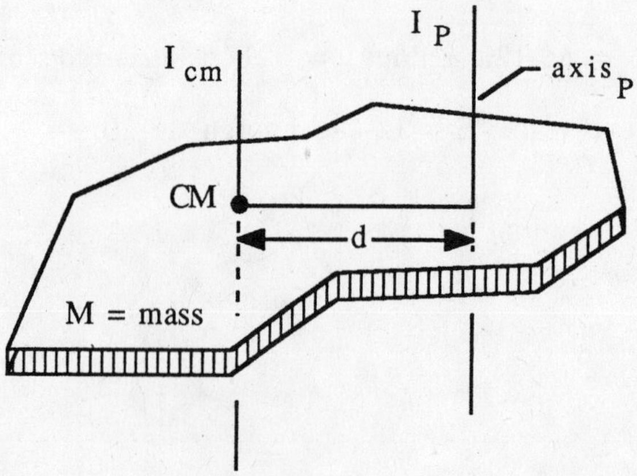

Practice: An array of three masses is located as shown in Figure 12.11.

m_1, m_2 and m_3 are tiny mass points

$m_1 = m_2 = m_3 = m$

Figure 12.11

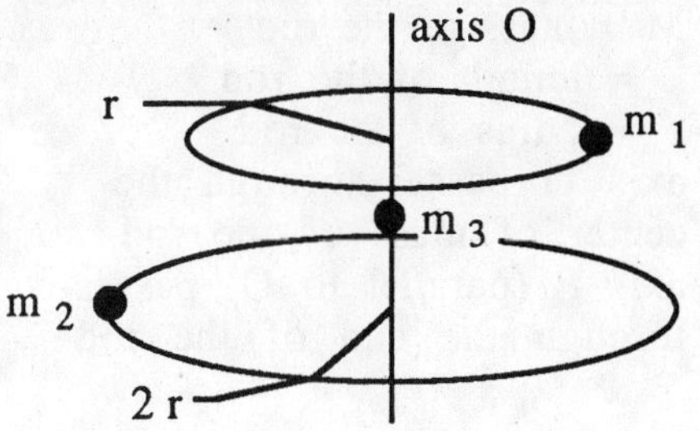

Determine the following:

1. Moment of inertia of m_1 about O	$I_1 = m_1r_1^2 = mr^2$
2. Moment of inertia of m_2 about O	$I_2 = m_2r_2^2 = m(2r)^2 = 4mr^2$
3. Moment of inertia of m_3 about O	$I_3 = m_3r_3^2 = m_3(0) = 0$
4. Total moment of inertia about O	$I_T = \sum_{i=1}^{3} m_ir_i^2 = m_1r_1^2 + m_2r_2^2 + m_3r_3^2$ $= I_1 + I_2 + I_3 = 5mr^2$

A child is riding on a merry-go-round (m-g-r) at a position that is one-half the distance to the rim.

Figure 12.12

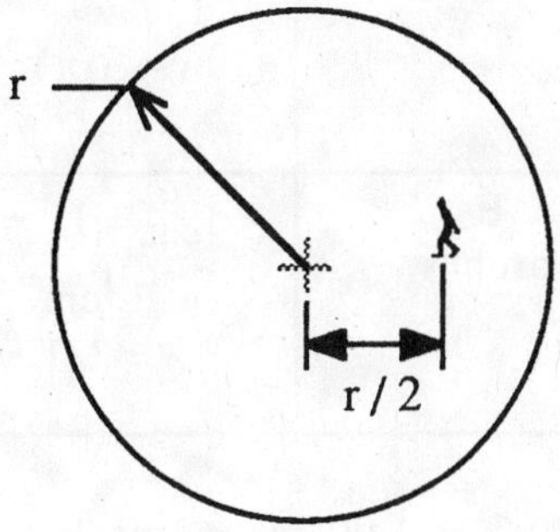

r = radius of the m-g-r

$r_c = r/2$ = position of child

m_m = mass of the m-g-r

m_c = mass of the child

Determine the following:

5. Moment of inertia of the m-g-r about O	$I_m = m_mr^2/2$ (Table 12-2 of Text)
6. Moment of inertia of the child about O	$I_c = m_cr_c^2 = m_c(r/2)^2 = m_cr^2/4$
7. Total moment of inertia about O	$I_T = I_m + I_c = (2m_m + m_c)r^2/4$

Consider the rod shown in Figure 12.13

M = mass of the rod
L = length of the rod
r = radius of the rod
axis O passes through the center of mass of the rod
axis P (parallel to O) passes through the end of the rod

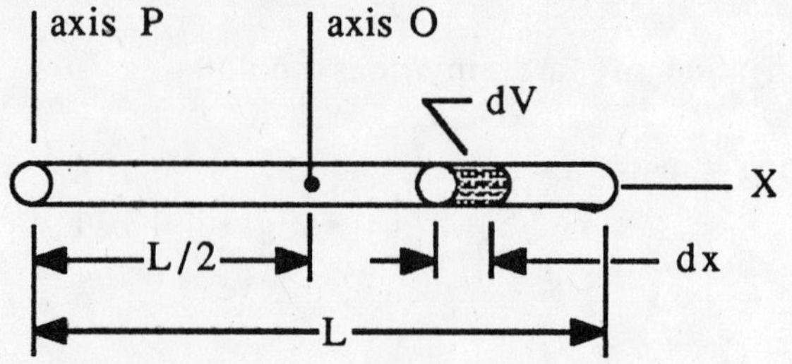

Figure 12.13

Determine the following:

1. Mass density (ρ) of the rod	$\rho = M/V = M/\pi r^2 L$
2. Volume of the small element shown	$dv = \pi r^2 dx$
3. Moment of inertia of the rod about axis P	$I_p = \rho \int x^2 dV = \rho\pi r^2 \int_0^L x^2 dx$ $I_p = \rho\pi r^2 (x^3/3) \Big\|_0^L = ML^2/3$
4. Moment of inertia of the rod about axis O	$I_0 = \rho \int x^2 dV = \rho\pi r^2 \int_{-L/2}^{+L/2} x^2 dx$ $I_0 = \rho\pi r^2 (x^3/3) \Big\|_{-L/2}^{+L/2} = ML^2/12$
5. Moment of inertia about axis P using the parallel-axis theorem	$I_{cm} = I_0 = ML^2/12$ $I_p = I_{cm} + M(L/2)^2$ $= ML^2/12 + ML^2/4 = ML^2/3$

Example: A cylinder of length L = 1.00 m, radius R = 1.00 x 10^{-2} m, and mass m_{cy} = 5.00 x 10^{-1} kg is mounted on an axle of radius r = 5.00 x 10^{-3} m and negligible mass, as shown in Figure 12.14. Two identical weights, each of mass m_w = 1.00 kg, are mounted on the cylinder 2.50 x 10^{-1} m from the axle. Where on the cylinder should the two weights be relocated if we wish to double the system's inertia about the axle.

Given and Diagram:

$L = 1.00$ m
$m_{cy} = 5.00 \times 10^{-1}$ kg
$m_w = 1.00$ kg
$r_{axle} = 5.00 \times 10^{-3}$ m
$r_{cy} = 1.00 \times 10^{-2}$ m
$I_{cy} = m_{cy}L^2/12$ (Table 12.2)
$d = 2.50 \times 10^{-1}$ m

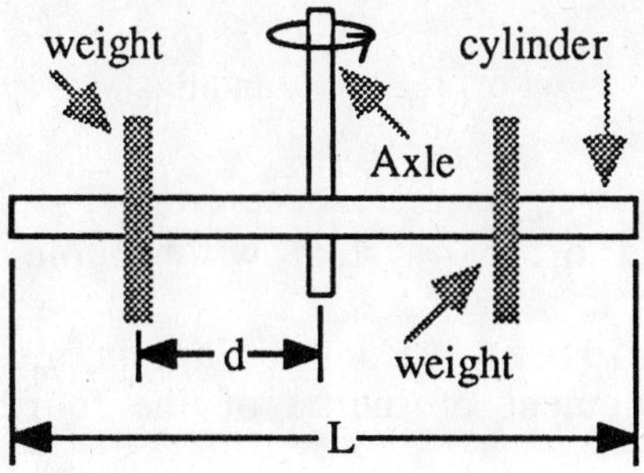

Figure 12.14

Determine: Where on the cylinder the two weights should be relocated in order to double the system's moment of inertia.

Strategy: With the given information, we can determine the moment of inertia of the cylinder about the axis, the moment of inertia of each weight (at its present position) about the axis, and hence the moment of inertia of the present system about the axis. We can then double this to determine the new moment of inertia of the system. Knowing the new moment of inertia of the system and the moment of inertia of the cylinder about the axis, we can determine the moment of inertia of the weights in their new location. Knowing the moment of inertia of the weights in their new location and the mass of the weights, we can determine the new location of the weights.

Solution: The moment of inertia of the cylinder about the axis is

$$I_{cy} = m_{cy}L^2/12 = (5.00 \times 10^{-1} \text{ kg})(1.00 \text{ m}^2)/12 = 4.17 \times 10^{-2} \text{ kg•m}^2$$

The moment of inertia of each weight in its present position about the axis is

$$I_w = m_w d^2 = (1.00 \text{ kg})(2.50 \times 10^{-1} \text{ m})^2 = 6.25 \times 10^{-2} \text{ kg•m}^2$$

The total moment of inertia of the system about the axis with the weights in their present positions is

$$I_{total} = I_{cy} + 2I_w = 16.7 \times 10^{-2} \text{ kg•m}^2$$

The new moment of inertia of the system is to be twice this value for I_{total}.

$$I_{total\ new} = 2I_{total} = 33.4 \times 10^{-2} \text{ kg•m}^2$$

The new total moment of inertia of the system about the axis has a contribution from the cylinder and from the weights in their new position.

$$I_{total\ new} = I_{cy} + 2I_{w\ new}$$

$$I_{w\ new} = (I_{total\ new} - I_{cy})/2 = 14.6 \times 10^{-2} \text{ kg•m}^2$$

Finally, we can determine the new location of the weights.

$$I_{w\ new} = m_w d_{new}^2 \qquad d_{new} = 3.82 \times 10^{-1}\ m$$

If the weights are moved from d = 2.50×10^{-1} m to d = 3.82×10^{-1} m the moment of inertia of the system will double.

Example: Consider a thin disk of thickness T, radius b, and mass M. Determine the moment of inertia of the fourth of the disk shown in Figure 12.15 about the z axis.

Given and diagram:

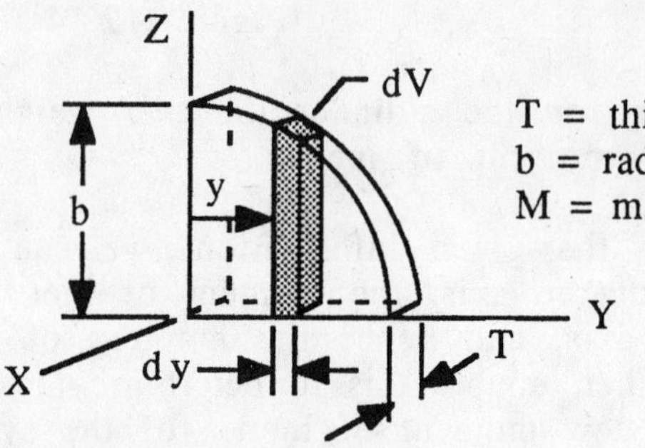

Figure 12.15

T = thickness of the disk
b = radius of the disk
M = mass of the disk

Determine: The moment of inertia of one fourth of the disk about the z axis.

Strategy: Determine the mass density of the disk. Determine the volume of the element dV and then use the expression for the moment of inertia of a continuous object to determine the moment of inertia of the object.

Solution:

Mass density of the object: $\rho = M/V = M/\pi b^2 T$

Volume of element: $dV = ZTdy$

The moment of inertia is $I = \rho \int_V y^2 dV = (M/\pi b^2 T)\int_0^b y^2 ZTdy$

The variables y and z are related by $y^2 + z^2 = b^2$, hence

$$z = (b^2 - y^2)^{1/2}$$

When this is inserted into the expression for I, obtain

$$I = (M/\pi b^2)\int_0^b y^2(b^2 - y^2)^{1/2} dy$$

When the integration is performed we get

$$I = (M/\pi b^2)\left\{-(y/4)(b^2 - y^2)^{3/2} + (b^2/8)[y(b^2 - y^2)^{1/2} + b^2 \sin^{-1}(y/b)]\right\}\Big|_0^b$$

$$I = (M/\pi b^2)\{(b^2/8)[b^2 \sin^{-1}(1)]\} = (M/\pi b^2)(b^4/8)(\pi/2) = Mb^2/16$$

Related Text Exercises: Ex. 12-25 through 12-32.

6 Translational and Rotational Motion Combined/Rolling Objects (Section 12-8)

Review: The motion of an object undergoing both translational and rotational motion can be considered from two points of view. The first considers the translational motion of the center of mass and the rotational motion about the center of mass. The second considers the instantaneous rotational motion about the instantaneously stationary point of contact between the rolling object and the surface on which it is rolling.

View 1
Translation of O and rotation about O

(a)

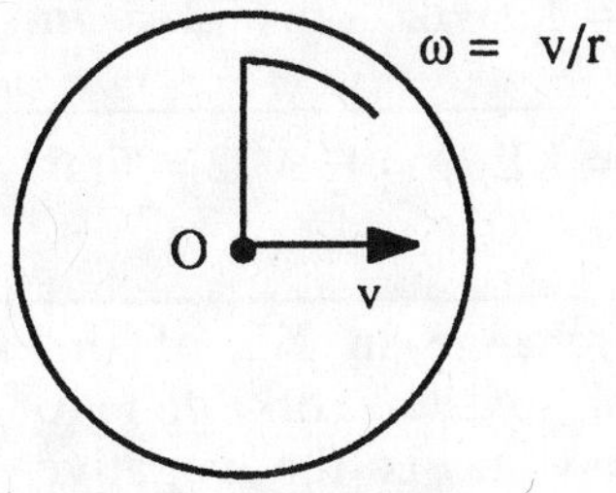

$$KE = KE_{R0} + KE_T$$
$$= I_0\omega^2/2 + mv^2/2$$
$$= I_0\omega^2/2 + m\omega^2 r^2/2$$
$$= (I_0 + mr^2)\omega^2/2$$

View 2
Rotation about P

(b)

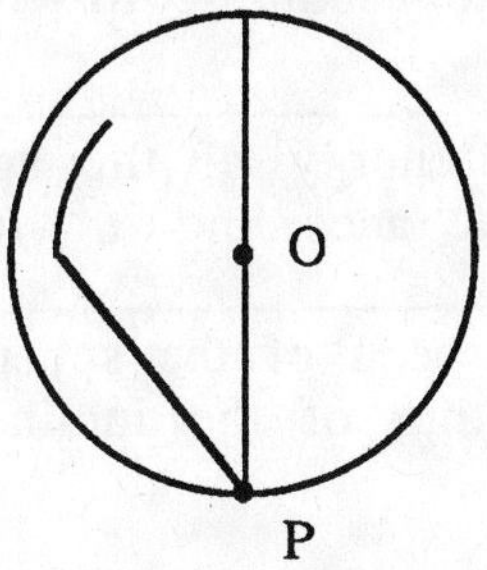

$$KE = KE_{RP}$$
$$= I_P\omega^2/2$$
$$= (I_0 + mr^2)\omega^2/2$$

Figure 12.16

Notice that both viewpoints lead to the same answer.

Practice: A sphere of mass m and radius r rolls without slipping down an incline of length L and inclination θ, as shown in Figure 12.17

m = mass of sphere

r = radius of sphere

L = Length of incline

θ = angle of inclination

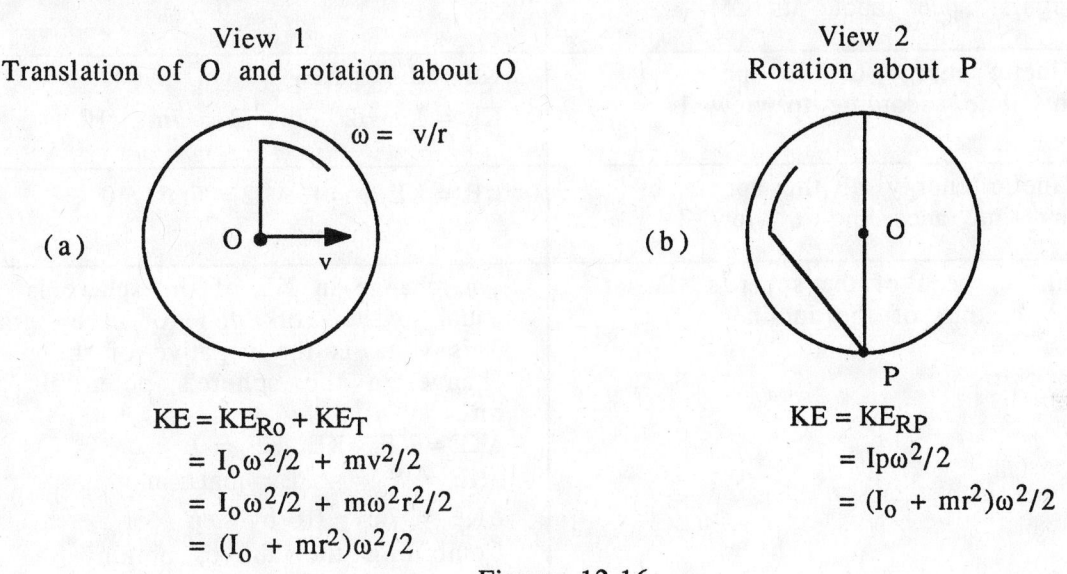

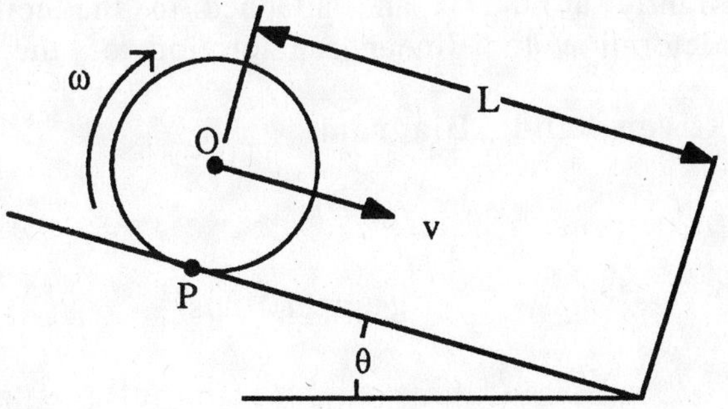

Figure 12.17

Determine an expression for the following:

1. Rotational moment of inertia of the sphere about O	$I_o = 2mr^2/5$
2. Rotational moment of inertia of the sphere about P	$I_P = I_o + mr^2 = 7mr^2/5$
3. Relationship between the linear speed of the sphere's CM and its angular speed about its CM	$v = \omega r$
4. Kinetic energy of the sphere at any time according to view 1	$KE = KE_{Ro} + KE_T$ $KE = I_o\omega^2/2 + mv^2/2 = 7mv^2/10$
5. Kinetic energy of the sphere at any time according to view 2	$KE = KE_{RP} = I_P\omega^2/2 = 7mv^2/10$
6. Linear speed of the sphere's CM at the bottom of the incline	The change in KE of the sphere is equal to the work done on it by gravity. This work is the negative of the change in the sphere's potential energy. $\Delta KE = KE_f - KE_i,\ KE_i = 0$ $\Delta KE = W_g = -\Delta PE = mgL \sin\theta$ $\Delta KE = 7mv_f^2/10$ by step 4 or 5 Combining the above, obtain $v_f = [10gL \sin\theta/7]^{1/2}$
7. Time it takes the sphere to reach the bottom of the incline	$v_{av} = (v_i + v_f)/2 = L/t$ $t = 2L/v_f = [14L/5g \sin\theta]^{1/2}$

Example: The solid cylinder in Figure 12.18 is supported by two cords wrapped tightly around it and attached to the ceiling. If the cylinder is released from rest, determine the linear acceleration of the CM.

Given and Diagram:

$v_o = 0$

$g = 9.80\ m/s^2$

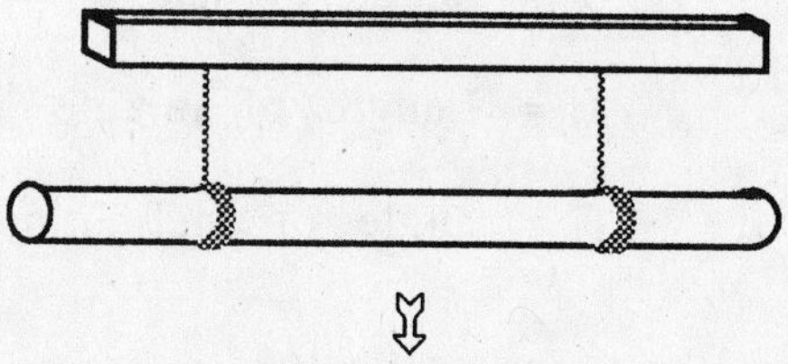

Figure 12.18

Determine: The linear acceleration of the cylinder's CM as it falls.

Strategy: Using view 1, we can state that the kinetic energy at any time is the translational kinetic energy of the CM plus the rotational kinetic energy about the CM. Using the work-energy theorem, we can also state that the change in kinetic energy of the cylinder is equal to the work done on it by gravity. These two statements can be combined to obtain the speed of the CM at any time. Finally we can use our knowledge of kinematics to determine the acceleration of the CM.

Solution: Using view 1, we write the kinetic energy at any time as

$$KE = mv^2/2 + I_0\omega^2/2 = mv^2/2 + (mr^2/2)(v/r)^2/2 = 3\ mv^2/4$$

The work done by gravity at any time is $W_g = F_gL = m_gL$, where L is the distance that the gravitational force F_g has acted at any time. According to the work-energy theorem, the net work done on the cylinder is equal to the change in its kinetic energy.

$$W_g = \Delta KE; \quad \Delta KE = KE_f - KE_i; \quad KE_f = 3mv_f^2/4; \quad KE_i = 0; \quad W_g = mgL$$

Substituting, we obtain

$$mgL = 3mv_f^2/4 \quad \text{or} \quad v_f^2 = 4gL/3$$

Finally, from our knowledge of kinematics, we write

$$v_f^2 - v_i^2 = 2a\Delta s; \quad v_i = 0; \quad v_f^2 = 4gL/3; \quad \Delta s = L$$

which when combined gives

$$4gL/3 = 2aL \quad \text{or} \quad a = 2g/3 = 6.53\ m/s^2$$

Related Text Exercises: Ex. 12-32 through 12-37.

PRACTICE TEST

Take and grade this practice test. Doing so will allow you to determine any weak spots in your understanding of the concepts taught in this chapter. The following section prescribes what you should study further to strengthen your understanding.

The disk shown below rotates through the angle $\Delta\theta_1$ in a time Δt_1 at a constant angular speed ω_1. The disk then experiences a constant angular acceleration α_2 for a time Δt_2. Finally the disk rotates through the angle $\Delta\theta_3$ in a time Δt_3 at a constant angular speed ω_3.

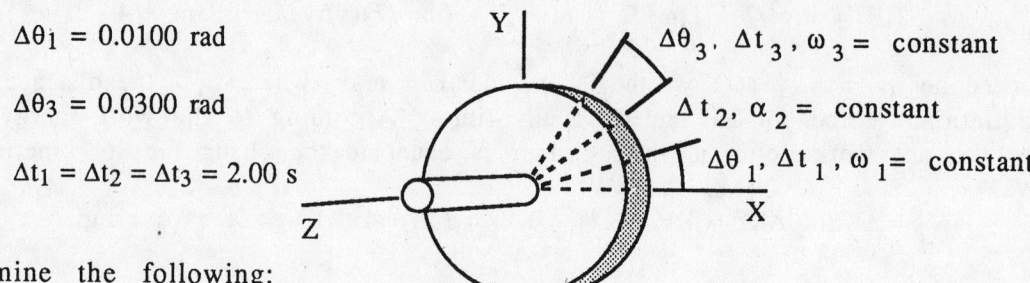

$\Delta\theta_1 = 0.0100$ rad

$\Delta\theta_3 = 0.0300$ rad

$\Delta t_1 = \Delta t_2 = \Delta t_3 = 2.00$ s

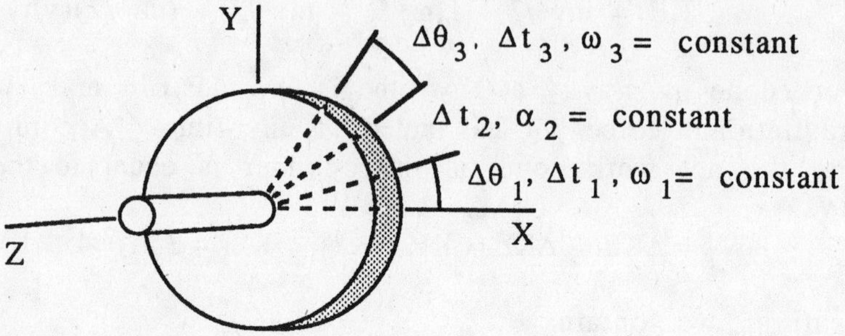

Determine the following:

________ 1. Angular velocity during the time Δt_1
________ 2. Angular velocity during the time Δt_3
________ 3. Angular acceleration during the time Δt_2
________ 4. Angular distance traveled by the disk during the time Δt_2

A particle is traveling in a circular orbit and its angular coordinate as a function of time is given by

$$\theta(t) = (3.00 \text{ rad/s}^2)t^2 + (2.00 \text{ rad/s})t + 1.00 \text{ rad}$$

Determine the following:

________ 5. The angular coordinate at $t = 2.00$ s
________ 6. The angular speed at $t = 2.00$ s
________ 7. The angular acceleration at $t = 2.00$ s

A wheel is mounted on an axle as shown:

$R = 0.200$ m
$r = 1.00$ cm
$h = 2.00$ m
$M_1 = 10.0$ kg
$M_2 = 0.100$ kg
$g = 9.80$ m/s^2
$t = 10.0$ s

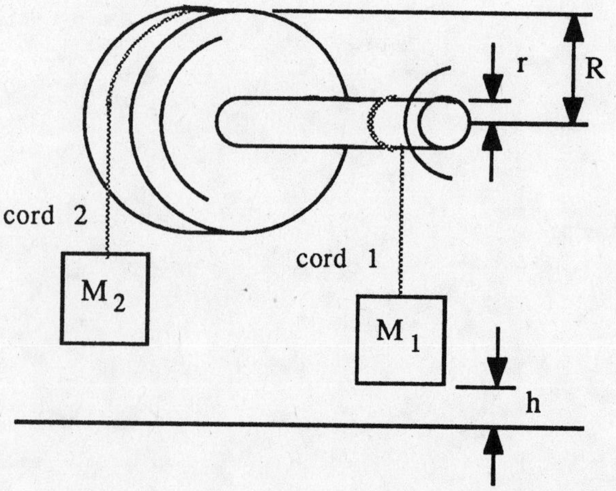

A cord is wrapped around the axle and attached to M_1. Another cord is wrapped around the wheel and attached to M_2. When the system is released, M_1 falls and unwraps cord 1; M_2 is lifted as cord 2 is wound on the wheel. M_1 falls the distance h in 10 s.

Determine the following:

_______ 8. The linear acceleration of M_1
_______ 9. The angular acceleration of the wheel and axle
_______ 10. The linear acceleration of M_2
_______ 11. The linear speed of M_1 after a time of 5.00 s
_______ 12. The angular speed of the wheel and axle after a time of 5.00 s
_______ 13. The linear speed of M_2 after a time of 5.00 s
_______ 14. The distance M_1 falls during the first 5.00 s of travel
_______ 15. The number of revolutions made by the wheel and axle during the first 5.00 s of travel
_______ 16. The distance M_2 is lifted during the first 5.00 s of travel

A rigid body of irregular shape is mounted on an axle of radius r and negligible mass, as shown in the figure below. A string is wrapped around the axle, and a block of mass m is attached to it. The system is released from rest, and the block descends a distance h in a time t.

m = mass of block
r = radius of axle
h = distance block descends in time t
t = time block takes to descend the distance h
g = acceleration due to gravity

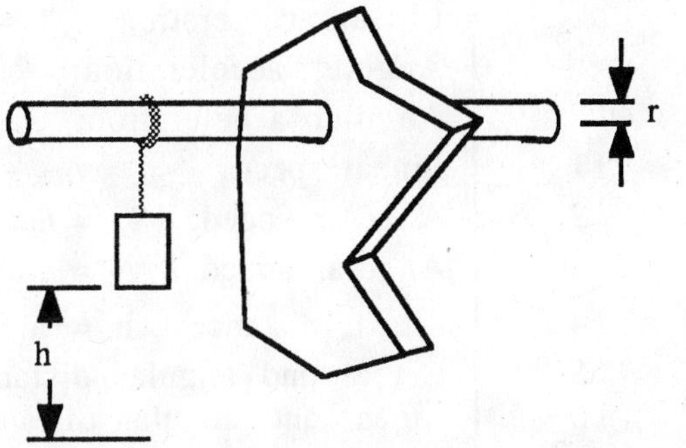

Determine expressions for the following physical quantities in terms of the known information only.

_______ 17. The final linear speed of the block (that is, its speed after it has fallen the distance h)
_______ 18. The final angular speed of the rigid body
_______ 19. The final translational kinetic energy of the block
_______ 20. The decrease in gravitational potential energy of the block as it descends the distance h
_______ 21. The final rotational kinetic energy of the rigid body
_______ 21. The moment of inertia of the rigid body obtained by using energy principles

(See Appendix I for answers.)

PRINCIPAL CONCEPTS AND EQUATIONS PRESCRIPTION

Your score on the practice test is an excellent measure of your understanding of the chapter. You should now use the following chart to write your own prescription for dealing with any weaknesses the practice test points out. Look down the leftmost column to the number of the question(s) you answered incorrectly, reading across that row you will find the concept and/or equation of concern, the section(s) of the study guide you should return to for further study, and some suggested text exercises which you should work to gain additional experience.

Practice Test Questions	Concepts and Equations	Prescription	
		Principal Concepts	Text Exercises
1	Angular velocity: $\omega = \Delta\theta/\Delta t$	2	12-7, 8
2	Angular velocity: $\omega = \Delta\theta/\Delta t$	2	12-7, 8
3	Angular acceleration: $\alpha = \Delta\omega/\Delta t$	2	12-14
4	Angular distance: $\Delta\theta = \omega_0 t + \alpha t^2/2$ $2\alpha\Delta\theta = \omega^2 - \omega_0^2$	3	12-12, 14
5	Angular coordinate	2	12-9, 11
6	Angular speed: $\omega = d\theta/dt$	2	12-12, 13
7	Angular acceleration: $\alpha = d^2\theta/dt^2$	2	12-9, 41
8	Linear acceleration: $h = a_1 t^2/2$	4 of Ch 3	3-35, 37
9	Angular acceleration: $a_1 = \alpha r$	4	12-19, 42
10	Angular acceleration: $a_2 = \alpha R$	4	12-19, 42
11	Linear speed: $v_1 = v_{01} + a_1 t$	4 of Ch 3	3-46, 48
12	Angular speed: $v_1 = \omega r$	4	12-17, 18
13	Angular speed: $v_2 = \omega r$	4	12-17, 18
14	Linear distance: $h = a_1 t^2/2$	4 of Ch 3	3-35, 37
15	Linear and angular distance: $h = r\theta$	4	12-3, 4
16	Linear and angular distance: $s = R\theta$	4	12-4, 5
17	Average speed: $\bar{v} = h/t = (v_o + v_f)/2$	2 of Ch 3	3-8, 9
18	Angular speed: $\omega = v/r$	4	12-17, 18
19	Translational kinetic energy: $K_T = mv^2/2$	4 of Ch 8	8-23, 32
20	Gravitational potential energy: $\Delta U = mgh$	1 of Ch 9	9-1, 2
21	Conservation of energy: $K + U =$ constant	2 of Ch 9	9-32, 33
22	Rotational kinetic energy: $K = I\omega^2/2$	6	12-34, 36

13 ROTATION II

RECALL FROM PREVIOUS CHAPTERS

Previously learned concepts and equations frequently used in this chapter	Text Section	Study Guide Page
Relation between force and acceleration: $\Sigma\mathbf{F} = M\mathbf{a}$	5-4	5-5
Work done by a force: $W = \int \mathbf{F}\cdot d\mathbf{s}$	8-3	8-7
Translational kinetic energy: $K_T = mv^2/2$	8-4	8-13
Net work due to forces: $W_{net} = \Delta K_T$	8-5	8-15
Average power delivered by a force: $\bar{P} = \Delta W/\Delta t = F\bar{v}$	8-5	8-15
Instantaneous power delivered by a force: $P = dW/dt = Fv$	8-5	8-15
Linear momentum: $\mathbf{p} = m\mathbf{v}$	10-3	10-10
Relation between force and momentum: $\mathbf{F} = d\mathbf{p}/dt$	10-3	10-10
Conservation of linear momentum: $\mathbf{p}_f = \mathbf{p}_i$ if $\mathbf{F}_{net} = 0$	10-5	10-10
Cross product: $\mathbf{A} \times \mathbf{B} = AB \sin\theta$	11-6	11-16
Relation between linear and angular distance, speed and acceleration: $s = r\theta$, $v = r\omega$, $a = r\alpha$	12-4	12-8

NEW IDEAS IN THIS CHAPTER

Concepts and equations introduced	Text Section	Study Guide Page
Angular momentum: $\mathbf{L} = \mathbf{r} \times \mathbf{p} = I\boldsymbol{\omega}$	13-1	13-2
Relation between torque and angular momentum: $\boldsymbol{\tau} = d\mathbf{L}/dt$	13-1	13-6
Relation between torque and angular acceleration: $\boldsymbol{\tau} = I\boldsymbol{\alpha}$	13-3	13-8
Work done by a torque: $W = \int \tau d\theta$	13-4	13-11
Rotational kinetic energy: $K_R = I\omega^2/2$	13-4	13-11
Net work due to torques: $W_{net} = \Delta K_R$	13-4	13-11
Average power delivered by a torque: $\bar{P} = \Delta W/\Delta t = \tau\bar{\omega}$	13-4	13-11
Instantaneous power delivered by a torque $P = dW/dt = \tau\omega$	13-4	13-11
Conservation of angular momentum $\mathbf{L}_f = \mathbf{L}_i$ if $\boldsymbol{\tau}_{net} = 0$	13-5	13-14

PRINCIPAL CONCEPTS AND EQUATIONS

1 Angular Momentum of a Particle (Section 13-1)

Review: Figure 13.1 shows an object of mass m moving in the x-y plane. At some instant it is located by the position vector **r** and has a momentum **p**. (Recall that **p** = m**v**.)

Figure 13.1

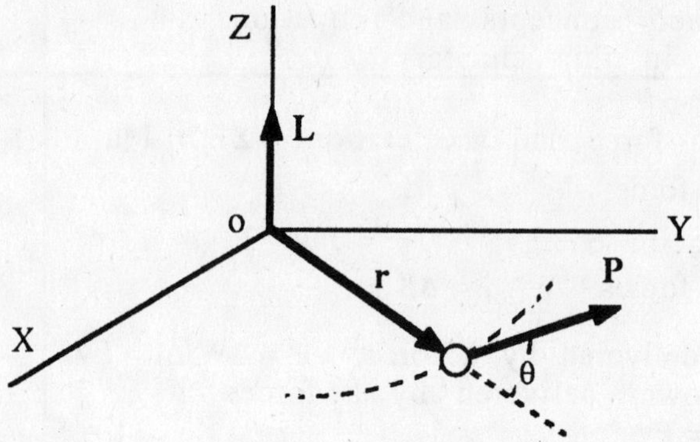

The magnitude of the angular momentum of the object about o may be determined by the following three methods.

Method I: $L = |\mathbf{L}| = |\mathbf{r} \times \mathbf{p}| = rp\sin\theta = rmv\sin\theta$

Method II:
$$L = rp_\perp = rp\sin\theta = rmv\sin\theta$$

Figure 13.2

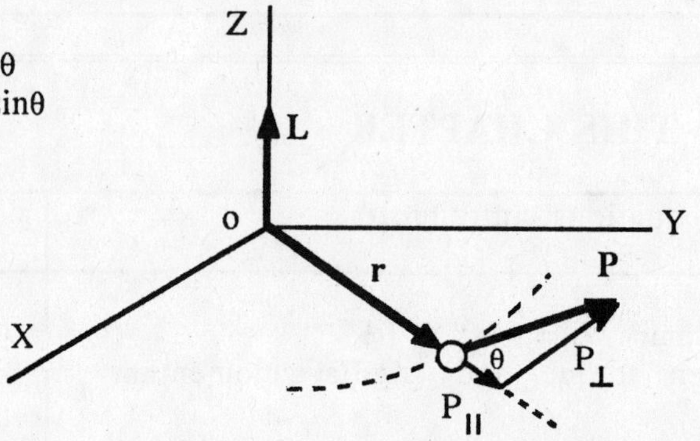

Method III:
$$L = r_\perp p = (r\sin\theta)p = rmv\sin\theta$$

Figure 13.3

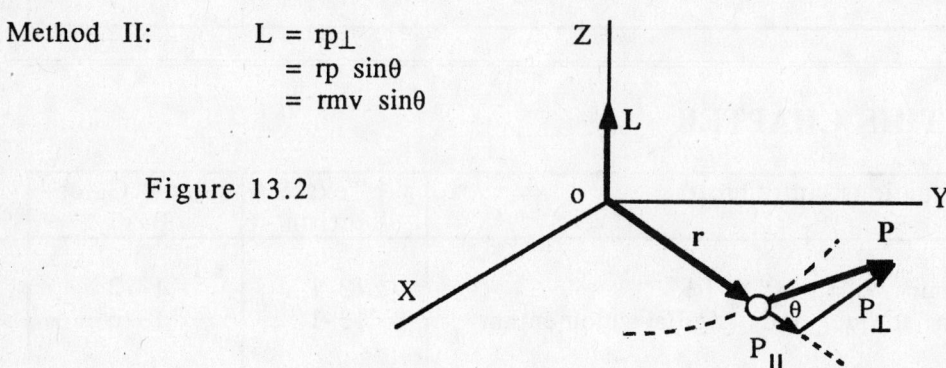

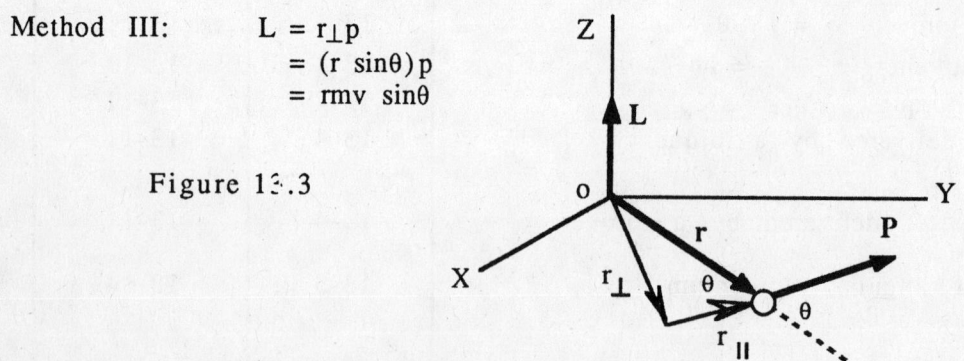

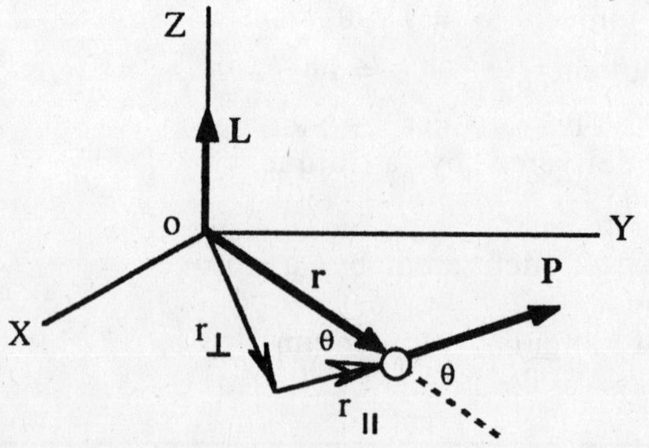

Notice that all three methods give the same end result.

The direction of the angular momentum is determined by the right hand rule for cross products. This process is summarized below:

1. Point the fingers of your right hand in the direction of **r**.
2. Orient your hand so that you can cross your fingers from the direction of **r** into the direction of **p**.
3. When you cross your fingers from the direction of **r**, through the angle θ, and into the direction of **p**, your thumb points in the direction of **L**. Note that θ is the angle between the direction of **r** and **p**.

If the object is acted on by an external torque τ, the angular momentum changes. In fact, the rate of change of the angular momentum is equal to the external torque, that is

$$\tau = dL/dt$$

Practice: Figure 13.4 shows three different cases. In all three cases the object has a mass of 2.00 kg and a constant speed of 5.00 m/s. In cases (b) and (c) the velocity is constant. At the instant shown, the magnitude of the displacement vector is 1.00 m.

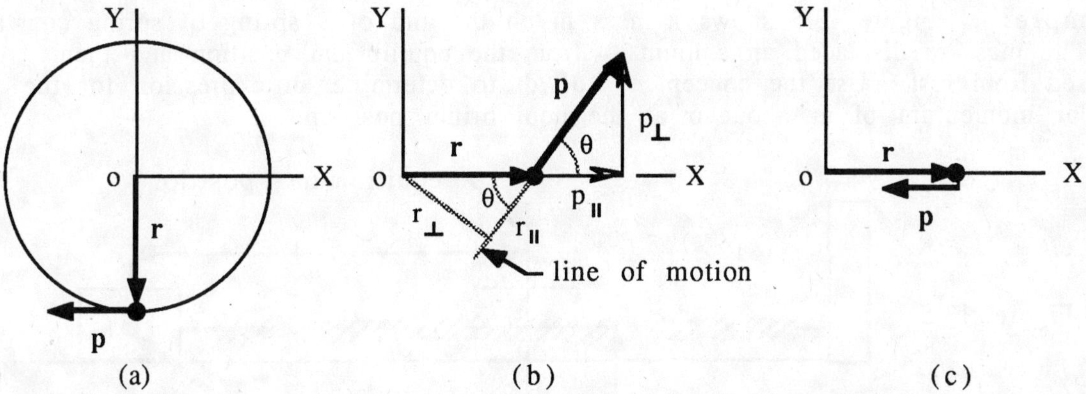

Figure 13.4

Determine the following:

1. Angular momentum about o for each case

For case (a)	For case (b)	For case (c)
Magnitude: $L = mvr\ \sin 90.0°$ $= 10.0\ kg{\cdot}m^2/s$ Direction: **-k** Vector: **L** = $(10.0\ kg{\cdot}m^2/s)$(-**k**)	$\theta = 60.0°$ $r_\perp = r \sin\theta = 0.866$ m $p = mv = 10.0\ kg{\cdot}m/s$ $p_\perp = p \sin\theta = 8.66\ kg{\cdot}m/s$ Magnitude: $L = r_\perp p = 8.66\ kg{\cdot}m^2/s$ $L = rp_\perp = 8.66\ kg{\cdot}m^2/s$ $L = rp \sin\theta = 8.66\ kg{\cdot}m^2/s$ Direction: **k** Vector: **L** = $(8.66\ kg{\cdot}m^2/s)$**k**	$r_\perp = 0,\ p_\perp = 0,\ \theta = 180°$ Magnitude: $L = r_\perp p = 0$ $L = r p_\perp = 0$ $L = rp \sin\theta = 0$ Vector: **L** = 0

2. Rate of change of the angular momentum at any time for each case.

For case (a)	For case (b)	For case (c)
Since m, v, and r are constant, the magnitude of **L** will not change. Since the sense of the angular motion does not change, the direction of **L** will not change. Hence dL/dt = 0.	Since **p** and $r_\perp$ are constant, the magnitude of **L** will not change. Since **v** is constant, the direction of **L** will not change. Hence dL/dt = 0.	Since $r_\perp$ is always zero, L is always zero. Since $p_\perp$ is always zero, L is always zero. Since θ is either 180° or 0°, L is always zero. If L is always zero, it is not changing. Hence dL/dt = 0.

3. Which cases have an external torque acting on the object?

(a) Since dL/dt = 0, τ = 0	(b) Since dL/dt = 0, τ = 0	(c) Since dL/dt = 0, τ = 0

Example: Figure 13.5 shows a mass m on the end of a spring of spring constant k. The mass is displaced an amount A from the equilibrium position and then released from rest. Use the concept τ = dL/dt to determine an expression for the angular momentum of m about o at the equilibrium position.

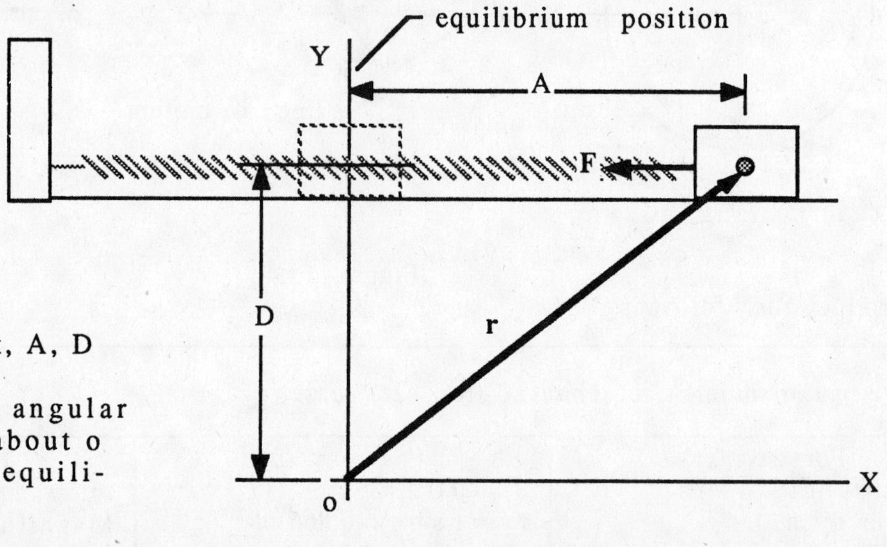

Figure 13.5

Given: m, k, A, D

Determine: The angular momentum of m about o when m is at the equilibrium position.

Strategy: Knowing that the spring exerts a Hooke's Law force on the mass, we can determine an expression for the torque acting on m. Knowing the torque, we can set up an integral to determine the angular momentum at any time. We can use conservation of energy to make a change of variable from t to x. Finally we can perform the integration to determine the angular momentum of the mass about o at equilibrium.

Solution: The force acting on m is given by Hooke's Law

$$F = -kx$$

The torque acting on m is $\tau = F(\text{moment arm}) = -kxD$

The angular momentum at the time $t = t_f$ may be determined by

$$L = \int_0^{t_f} \tau dt = -kD \int_0^{t_f} x\, dt$$

We can use conservation of energy to make a change of variable from t to x.

$$(\frac{1}{2})kA^2 = (\frac{1}{2})kx^2 + (\frac{1}{2})mv^2$$

or

$$v = dx/dt = (k/m)^{1/2} (A^2 - x^2)^{1/2}$$

hence

$$dt = \frac{dx}{[(k/m)^{1/2}] [(A^2 - x^2)^{1/2}]}$$

note also that

$$x = A \text{ at } t = 0 \quad \text{and} \quad x = 0 \text{ at } t = t_f$$

The angular momentum may now be written as

$$L = -kD(m/k)^{1/2} \int_A^0 (A^2 - x^2)^{-1/2} x dx$$

When this is integrated we obtain

$$L = -D(mk)^{1/2}\left[-(A^2 - x^2)^{1/2}\right]\Big|_A^0$$

Finally inserting the limits, we obtain

$$L = D(mk)^{1/2}A$$

As a quick check on our work, we can determine the speed of the mass at equilibrium by conservation of energy

$$(\frac{1}{2})kA^2 = (\frac{1}{2})mv^2$$

or

$$v = (k/m)^{1/2}A$$

The angular momentum at equilibrium then is

$$L = Dp \sin 90.0° = Dmv = Dm(k/m)^{1/2}A = D(mk)^{1/2}A$$

Note that this agrees with our previous expression for L.

Related Text Exercises: Ex. 13-1 through 13-6.

2 Angular Momentum of A System of Particles (Section 13-2)

Review: Suppose that we have a system of N particles (m_1, m_2, m_3, ... m_i, ... m_N) located by the position vectors ($\mathbf{r}_1$, $\mathbf{r}_2$, $\mathbf{r}_3$, ... $\mathbf{r}_i$, ... $\mathbf{r}_N$) and acted on by the forces ($\mathbf{F}_1$, $\mathbf{F}_2$, $\mathbf{F}_3$, ... $\mathbf{F}_i$, ... $\mathbf{F}_N$). We can determine torque acting on any particle by

$$\tau_i = \mathbf{r}_i \times \mathbf{F}_i$$

We can determine the momentum of any particle by

$$\mathbf{p}_i = m_i\mathbf{v}_i \quad \text{where} \quad \mathbf{v}_i = d\mathbf{r}_i/dt$$

We can determine the angular momentum of any particle by

$$\mathbf{L}_i = \mathbf{r}_i \times \mathbf{p}_i$$

We can determine the rate of change of the angular momentum of any particle by

$$d\mathbf{L}_i/dt$$

We can determine the total angular momentum by

$$\mathbf{L}_T = \mathbf{L}_1 + \mathbf{L}_2 + \mathbf{L}_3 + \ldots \mathbf{L}_i + \ldots \mathbf{L}_N = \Sigma\, \mathbf{L}_i$$

We can determine the rate of change of the total angular momentum by

$$d\mathbf{L}_T/dt = \sum_i d\mathbf{L}_i/dt$$

We can determine the total torque acting on the system by

$$\tau_T = \tau_1 + \tau_2 + \tau_3 + \ldots + \tau_i + \ldots \tau_N = \Sigma\, \tau_i$$

We would find that the rate of change of the total angular momentum and the total torque are related by

$$d\mathbf{L}_T/dt = \tau_T$$

Practice: Figure 13.6 shows two particles at t = 0 s, each moving under the influence of a constant force. The displacement vectors for the two particles are

$$\mathbf{r}_1 = (2\ \text{m/s}^2)t^2\mathbf{i} + 1\ \text{m}\mathbf{j} \quad \text{and} \quad \mathbf{r}_2 = (3\ \text{m/s}^2)t^2\mathbf{i} + [(-1\ \text{m}) + (3\ \text{m/s}^2)t^2]\,\mathbf{j}$$

The forces acting on the two particles are

$$\mathbf{F}_1 = (4\ \text{N})\mathbf{i} \qquad \text{and} \qquad \mathbf{F}_2 = (12\ \text{N})\mathbf{i} + (12\ \text{N})\mathbf{j}$$

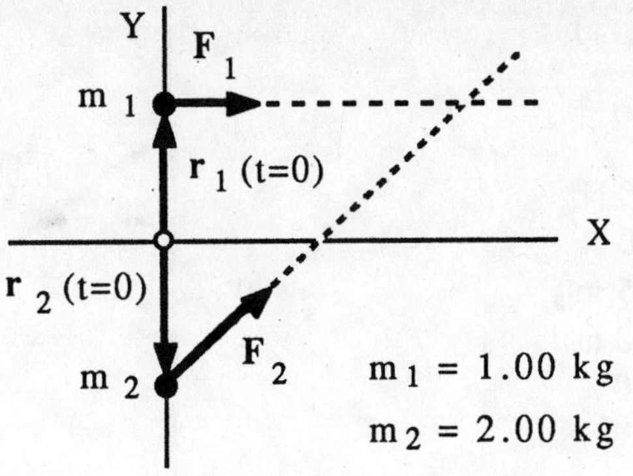

Figure 13.6

Determine the following:

1. Torque acting on each particle and the total torque acting on the system.	$\tau_1 = \mathbf{r}_1 \times \mathbf{F}_1$ = -(4 N•m)**k** $\tau_2 = \mathbf{r}_2 \times \mathbf{F}_2$ = (12 N•m)**k** $\tau_T = \tau_1 + \tau_2$ = (8 N•m)**k**
2. Angular momentum of each particle and of the system	$\mathbf{v}_1 = d\mathbf{r}_1/dt = (4\ \text{m•s}^{-2}\ t)\mathbf{i}$ $\mathbf{p}_1 = m_1\mathbf{v}_1 = (4\ \text{kg•m•s}^{-2}\ t)\mathbf{i}$ $\mathbf{L}_1 = \mathbf{r}_1 \times \mathbf{p}_1 = -(4\ \text{kg•m}^2\text{•s}^{-2}\ t)\mathbf{k}$ $\mathbf{v}_2 = d\mathbf{r}_2/dt = (6\ \text{m•s}^{-2}\ t)\mathbf{i} + (6\ \text{m•s}^{-2}\ t)\mathbf{j}$ $\mathbf{p}_2 = m_2\mathbf{v}_2$ $= (12\ \text{kg•m•s}^{-2}\ t)\mathbf{i} + (12\ \text{kg•m•s}^{-2}\ t)\mathbf{j}$ $\mathbf{L}_2 = \mathbf{r}_2 \times \mathbf{p}_2 = (12\ \text{kg•m}^2\text{•s}^{-2}\ t)\mathbf{k}$ $\mathbf{L}_T = \mathbf{L}_1 + \mathbf{L}_2 = (8\ \text{kg•m}^2\text{•s}^{-2}\ t)\mathbf{k}$
3. Rate of change of the angular momentum of each particle and of the system	$d\mathbf{L}_1/dt = -(4\ \text{kg•m}^2\text{•s}^{-2})\mathbf{k}$ $d\mathbf{L}_2/dt = (12\ \text{kg•m}^2\text{•s}^{-2})\mathbf{k}$ $d\mathbf{L}_T/dt = (8\ \text{kg•m}^2\text{•s}^{-2})\mathbf{k}$
4. If $\tau = dL/dt$ for each particle and the system	Comparing step 1 to step 3 we see that $\tau_1 = d\mathbf{L}_1/dt$ = (-4 N•m)**k** $\tau_2 = d\mathbf{L}_2/dt$ = (12 N•m)**k** $\tau_T = d\mathbf{L}_T/dt$ = (8 N•m)**k**

Example: Two particles, A and B, exert forces on one another of magnitude 20.0 N as shown in Figure 13.7. Determine the time rate of change of the angular momentum of the system about o.

Given and Diagram:

F_{AB} = 20.0 N
F_{BA} = 20.0 N
$\mathbf{r}_A$ = 1.15 m **j**
$\mathbf{r}_B$ = 2.00 m **i**
α = 30.0°
β = 60.0°

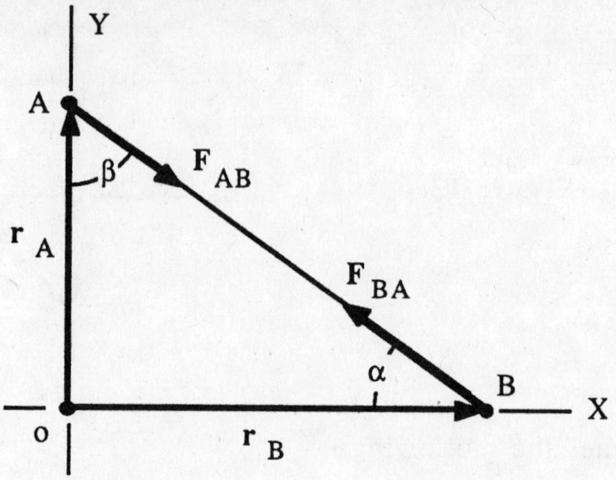

Figure 13.7

Determine: dL_T/dt

Strategy: Knowing the position vector for each particle and the force acting on each particle, we can determine the torque acting on each particle about o. Knowing the torque acting on each particle about o, we can determine the total torque about o and hence the rate of change of the angular momentum about o.

Solution: We can determine the torque acting on each particle about o.

$$\boldsymbol{\tau}_A = \mathbf{r}_A \times \mathbf{F}_{AB} = [r_A F_{AB} \sin(\pi - \beta)](-\mathbf{k}) = (r_A F_{AB} \sin\beta)(-\mathbf{k}) = -(20.0 \text{ N}\cdot\text{m})\mathbf{k}$$

$$\boldsymbol{\tau}_B = \mathbf{r}_B \times \mathbf{F}_{BA} = [r_B F_{BA} \sin(\pi - \alpha)](\mathbf{k}) = (r_B F_{BA} \sin\alpha)(\mathbf{k}) = (20.0 \text{ N}\cdot\text{m})\mathbf{k}$$

Knowing the torque acting on each particle about o, we can determine the total torque about o

$$\boldsymbol{\tau}_T = \boldsymbol{\tau}_A + \boldsymbol{\tau}_B = 0$$

Knowing the total torque about o, we can determine the rate of change of angular momentum of the system about o.

$$dL_T/dt = \boldsymbol{\tau}_T = 0$$

This answer should not surprise us since no external forces act on the system.

Related Text Exercises: Ex. 13-7 through 13-10.

3 Relationship Between Net External Torque Acting On A Body And Resulting Angular Acceleration (Section 13-3)

Review: If an object has a moment of inertia I about an axis and if it is spinning with an angular velocity ω about this axis, it has an angular momentum **L** about that axis given by

$$L = I\omega$$

Note: $L = I\omega$ is the rotational analog of $p = mv$.

If the object is subject to a torque τ, then the angular momentum changes and the object experiences an angular acceleration given by

$$\tau = dL/dt = d(I\omega)/dt = I d\omega/dt = I\alpha$$

Note: $\tau = I\alpha$ is the rotational analog of $F = ma$.

Practice: Shown in Figure 13.8 is a solid disk mounted on an axle. A constant tangential force is applied by means of a string that has been wrapped around the circumference of the disk. The disk is initially at rest, and it experiences a constant frictional torque while it is rotating.

m = 1.00 kg = mass of the disk
r = 2.00×10^{-1} m = radius of the disk
F = 2.00 N = constant force
τ_f = 1.00×10^{-1} N.m = frictional torque
ω_0 = 0 = initial angular speed
The axle is so small that we can ignore the hole made in the disk for the axle.

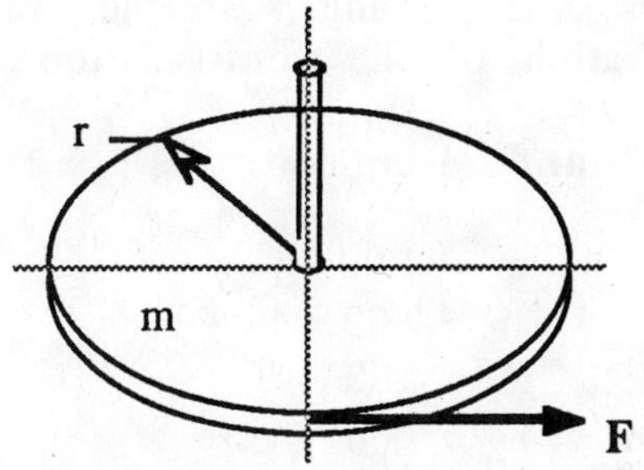

Figure 13.8

Determine the following:

1. Moment of inertia of the disk about the axle.	$I = mr^2/2 = 2.00 \times 10^{-2}$ kg•m²
2. Accelerating torque about the axle as a result of the constant force F	$\tau = rF \sin 90.0° = 4.00 \times 10^{-1}$ N•m
3. Net torque about the axle	$\tau_{net} = \tau - \tau_f = 3.00 \times 10^{-1}$ N•m
4. Angular acceleration of the disk about the axle	$\alpha = \tau_{net}/I = 15.0$ rad/s²
5. Angular speed of the disk after 5.00 s	$\omega = \omega_0 + \alpha t = 75.0$ rad/s
6. Angular momentum of the disk after 5.00 s	$L = I\omega = 1.50$ kg•m²/s
7. Rate of change of the angular momentum during the first 5.00 s.	$L_0 = 0$, $L_f = 1.50$ kg•m²/s $\Delta L = L_f - L_0 = 1.50$ kg•m²/s, $\Delta t = 5.00$ s $\Delta L/\Delta t = 3.00 \times 10^{-1}$ kg•m²/s²

8. A quick check on your work	$\tau_{net} = 3.00 \times 10^{-1}$ N•m (step 3) $\Delta L/\Delta t = 3.00 \times 10^{-1}$ kg•m²/s² (step 7) Since we determined τ_{net} and $\Delta L/\Delta t$ independently and obtained $\tau_{net} = \Delta L/\Delta t$, we can feel confident that our work is correct.

Example: Three 1.00-kg spheres are held in a horizontal equilateral triangle configuration by a Y-shaped frame, as shown in Figure 13.9. Each branch of the frame is 2.00×10^{-2} m long. This configuration is mounted on a vertical axis through the center of the frame, and forces of 1.00 N, 2.00 N, and 3.00 N are applied to the spheres at right angles to the branches of the Y. Determine the resulting angular acceleration of the configuration about the axis.

Given and Diagram:

$m = 1.00$ kg
$b = 2.00 \times 10^{-2}$ m
$F_1 = 1.00$ N
$F_2 = 2.00$ N
$F_3 = 3.00$ N

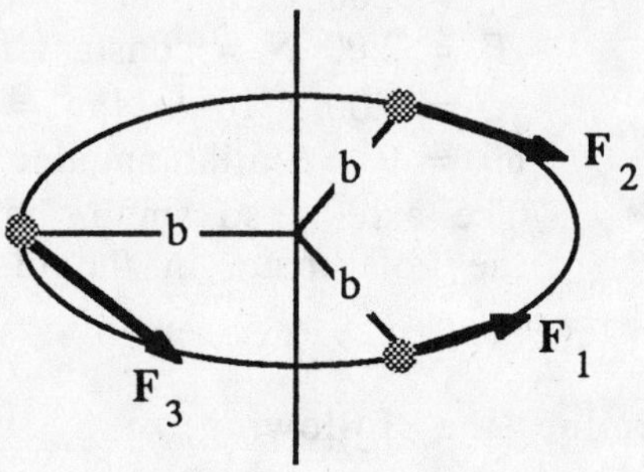

Figure 13.9

Determine: The resulting angular acceleration of the system about the axis.

Strategy: Knowing the mass of each sphere and its distance from the axis of rotation, we can determine the moment of inertia of the triangular configuration about the axis. Knowing the magnitude, direction, and moment arm for each force, we can determine the net torque acting on the system. Now that τ_{net} and I about the axis have been determined, we can obtain the angular acceleration α about that axis.

Solution: The moment of inertia of the system about the axis is

$$I = 3mb^2 = 12.0 \times 10^{-4} \text{ kg•m}^2$$

The net torque about the axis is

$$\tau_{net} = +\tau_1 + \tau_2 + \tau_3 = b(F_1 - F_2 + F_3) = 4.00 \times 10^{-2} \text{ N•m}$$

The resulting angular acceleration about the axis is

$$\alpha = \tau_{net}/I = 33.0 \text{ rad/s}^2$$

Related Text Exercises: Ex. 13-11 through 13-21.

4 Rotational Kinetic Energy, Work and Power (Section 13-4)

Review: If a constant torque τ is applied to a body, causing it to rotate about a fixed axis, the work done by the applied torque is

$$W = \int_{\theta_i}^{\theta_f} \tau d\theta = \tau\Delta\theta \text{ (if } \tau \text{ is constant)}$$

Note: $W = \int_{\theta_i}^{\theta_f} \tau d\theta$ is the rotational analog of $W = \int_{x_i}^{x_f} F_x dx$.

The rotational kinetic energy of a body about a fixed axis is

$$K_R = I\omega^2/2$$

Note: $K_R = I\omega^2/2$ is the rotational analog of $K_T = mv^2/2$.

The net work done on a body by the net torque is equal to the change in the rotational kinetic energy of the body. This may be stated algebraically as

$$W_{net} = \Delta K_R$$

Note: $W_{net} = \Delta K_R$ is the rotational analog of $W_{net} = \Delta K_T$.

When all the above are combined into one algebraic statement, we have

$$W_{net} = \tau_{net}\Delta\theta = \Delta K_R = K_{Rf} - K_{Ri} = I\omega_f^2/2 - I\omega_i^2/2$$

The average power delivered to a rotating rigid object by a torque is the rate at which work is done by the torque. This may be expressed algebraically as

$$\bar{P} = \Delta W/\Delta t = \tau\bar{\omega}$$

The power delivered to a rotating rigid object at any instant is

$$P = dW/dt = \tau\omega$$

Note: $\bar{P} = \Delta W/\Delta t = \tau\bar{\omega}$ is the rotational analog of $\bar{P} = \Delta W/\Delta t = F\bar{v}$ and
$P = dW/dt = \tau\omega$ is the rotational analog of $P = dW/dt = Fv$.

Practice: Shown in Figure 13.10 is a solid disk mounted on an axle. A constant tangential force is applied by means of a string that has been wrapped around the circumference of the disk. The disk is initially at rest, and it experiences a constant frictional torque while it is rotating. The force acts on the disk for a time t.

$m = 1.00$ kg = mass of disk
$r = 2.00 \times 10^{-1}$ m = radius of disk
$F = 2.00$ N = constant force
$\tau_f = 1.00 \times 10^{-1}$ N•m = frictional torque
$\omega_0 = 0$ = initial angular speed
$t = 5.00$ s = time the force acts
The axle is so small that we can ignore the hole made in the disk for the axle

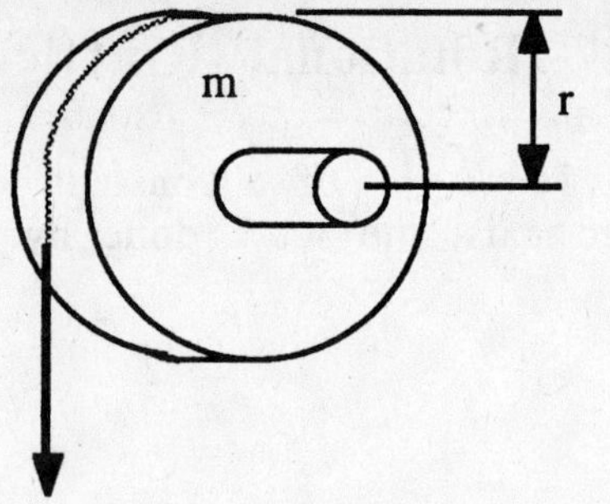

Figure 13.10

Determine the following:

1. Moment of inertia of the disk	$I = mr^2/2 = 2.00 \times 10^{-2}$ kg•m²
2. Accelerating torque about the axle as a result of the constant force F	$\tau = rF \sin 90.0° = 4.00 \times 10^{-1}$ N•m
3. Net torque about the axle	$\tau_{net} = \tau - \tau_f = 3.00 \times 10^{-1}$ N•m
4. Angular acceleration of the disk about the axle	$\alpha = \tau_{net}/I = 15.0$ rad/s²
5. Angular speed of the wheel at the end of the 5.00 s	$\omega = \omega_0 + \alpha t = 75.0$ rad/s
6. Angle through which the wheel rotates during this 5.00 s.	$\theta = \omega_0 t + \alpha t^2/2 = 188$ rad or $\theta = \omega^2/2\alpha = 188$ rad
7. Change in angular momentum of the wheel during this 5.00 s	$\Delta L = I\,\Delta\omega = 1.50$ kg•m²/s or $\Delta L = \tau\Delta t = 1.50$ kg•m²/s
8. Net work done on the wheel during this 5.00 s	$W_{net} = \tau_{net}\Delta\theta = 56.4$ J
9. The rotational kinetic energy of the wheel after the 5.00 s	$K = I\omega^2/2 = 56.3$ J or $W_{net} = \Delta K = K - K_0$, $K_0 = 0$ so $K = W_{net} = 56.4$ J
10. Average power delivered to the wheel by the Net torque during the 5.00 s	$\bar{P} = \Delta W_{net}/\Delta t = 11.3$ W or $\bar{\omega} = (\omega + \omega_o)/2 = 37.5$ rad/s $\bar{P} = \tau_{net}\bar{\omega} = 11.3$ W
11. Power delivered to the wheel at the instant t = 5.00 s	$P = \tau_{net}\omega_f = 22.5$ W

Example: A child pushes a disk-shaped merry-go-round (m-g-r) with a 100-N tangential force for 5.00 s. The m-g-r has a mass of 250 kg and a radius of 1.50 m. If it is initially at rest, determine its rotational kinetic energy at the end of the 5.00 s.

Given and Diagram:

F = 100 N = tangential force
t = 5.00 s = time force acts
m = 250 kg = mass of m-g-r
r = 1.50 m = radius of m-g-r
ω_0 = 0 = initial angular speed of m-g-r

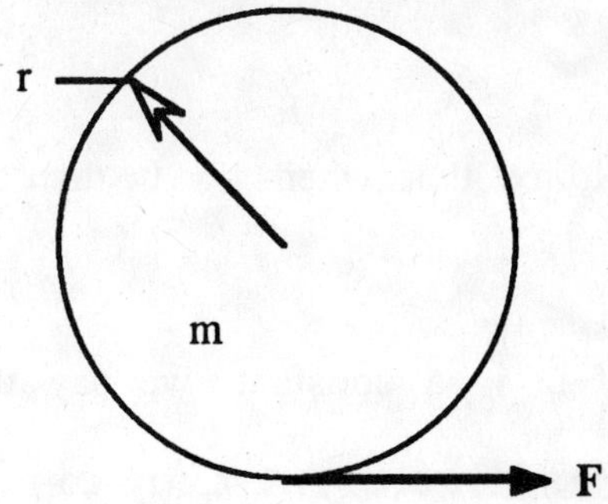

Figure 13.11

Determine: The rotational kinetic energy of the m-g-r after the child has been pushing for 5.00 s with a 100-N tangential force.

Strategy: Knowing m, r, and F, we can determine the torque τ acting on the m-g-r and its moment of inertia I. From I and τ, we can determine the angular acceleration α. Knowing the angular acceleration α and time t for that acceleration, we can determine the angular displacement $\Delta\theta$. From τ and $\Delta\theta$, we can determine the net work on the m-g-r and hence its rotational kinetic energy.

Solution: The net torque on the m-g-r about the cylindrical axis is

$$\tau = rF \sin 90.0° = (1.50 \text{ m})(1.00 \times 10^2 \text{ N}) = 1.50 \times 10^2 \text{ N•m}$$

The moment of inertia of the m-g-r about the cylindrical axis is

$$I = mr^2/2 = (2.50 \times 10^2 \text{ kg})(1.50 \text{ m})^2/2 = 281 \text{ kg•m}^2$$

The angular acceleration of the m-g-r is

$$\alpha = \tau/I = 1.50 \times 10^2 \text{ N•m}/281 \text{ kg•m}^2 = 5.34 \times 10^{-1} \text{ rad/s}^2$$

The angular displacement of the m-g-r during the push time is

$$\Delta\theta = \alpha t^2/2 = (5.34 \times 10^{-1} \text{ rad/s}^2)(5.00 \text{ s})^2/2 = 6.68 \text{ rad}$$

The net work done on the m-g-r is

$$W_{net} = \tau\Delta\theta = (1.50 \times 10^2 \text{ N•m})(6.68 \text{ rad}) = 1.00 \times 10^3 \text{ J}$$

The rotational kinetic energy at the end of the 5.00 s push is

$$K_{R_{final}} = \Delta K_R = W_{net} = 1.00 \times 10^3 \text{ J}$$

Related Text Exercises: Ex. 13-22 through 13-26.

5 Conservation of Angular Momentum (Section 13-5)

Review: We have previously learned that

$$\tau = dL/dt$$

Notice that when $\tau = 0$, then $dL/dt = 0$, and **L** is a constant. That is

$$L_f = L_i$$

If **L** is a constant, we say that angular momentum is conserved.

Practice: A merry-go-round, which can be considered to be a large flat disk of mass M and radius r, rotates about an axis through its center of mass with an angular speed ω_i. A child of mass m stands watching and then jumps radially onto the edge of the m-g-r, as shown in Figure 13.12.

$M = 150$ kg

$m = 25.0$ kg

$r = 1.50$ m

$\omega_i = 0.200$ rad/s

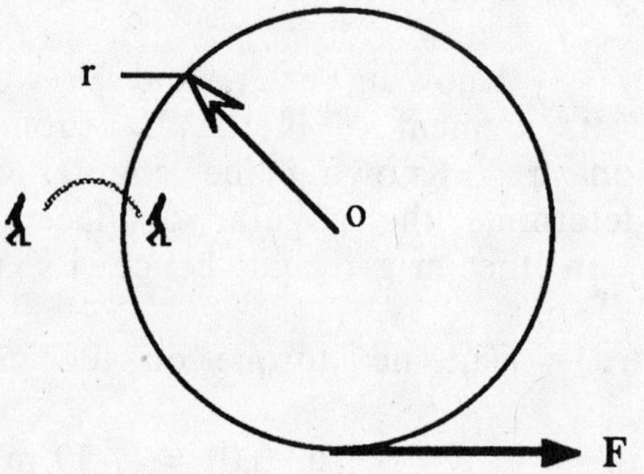

Figure 13.12

Determine the following:

1. Moment of inertia of the m-g-r about o	$I_M = Mr^2/2 = 169$ kg•m^2
2. Initial angular momentum of the m-g-r about o	$L_{Mi} = I_M\omega_i = 33.8$ kg•m^2/s
3. Initial angular momentum of the child about o	$L_{ci} = 0$ since $\omega_{ci} = 0$
4. Initial angular momentum of system (m-g-r + child) about o	$L_i = L_{Mi} + L_{ci} = 33.8$ kg•m^2
5. Net torque the child exerts on the m-g-r as she jumps onto it	$\tau_{net} = 0$
6. Change in angular momentum of the system (m-g-r + child)	Since $\tau_{net} = 0$, then $\Delta L = 0$ and $L_f = L_i = 33.8$ kg•m^2/s

7. Moment of inertia of the child after she jumps onto the m-g-r	$I_c = mr^2 = 56.3\ kg{\cdot}m^2$
8. Total moment of inertia of the final rotating system(m-g-r+child)	$I_f = I_M + I_c = 225\ kg{\cdot}m^2$
9. Final angular speed of the rotating system (m-g-r + child)	$\omega_f = L_f/I_f = 0.150$ rad/s

Example: The 0.100-kg puck in Figure 13.13 is sliding with a speed of 2.00 m/s in a circular orbit of radius 1.00 m on the horizontal frictionless surface. The centripetal force is provided by a student holding onto the cord. If the student pulls steadily downward on the string until the radius of the orbit is 0.500 m, determine the final angular speed of the puck and the work done by the student.

Given and Diagram: $m = 1.00$ kg, $r_1 = 1.00$ m, $r_2 = 0.500$ m, $v_1 = 2.00$ m/s

Figure 13.13

Determine: (a) The angular speed ω_2 of the puck after the student pulls on the cord until $r_2 = r_1/2$ and (b) the work done by the student.

Strategy: The force acts in such a manner that no external torque is applied to the system; consequently, angular momentum is conserved. Knowing that angular momentum is conserved, the definition of angular momentum, and the relationship between linear and angular speed, we can determine ω_2. Knowing the initial and final angular speed of the puck, we can determine its initial and final rotational kinetic energy, the change in its rotational kinetic energy, and hence the work done on it by the student.

Solution: Conserving angular momentum, we write

(a)	$L_2 = L_1$	conservation of angular momentum
	$I_2\omega_2 = I_1\omega_1$	definition of angular momentum
	$mr_2^2\omega_2 = mr_1^2\omega_1$	definition of moment of inertia
	$(r_1^2/4)\omega_2 = r_1^2\omega_1$	using $r_2 = r_1/2$
	$\omega_2 = 4\omega_1 = 4(v_1/r_1)$	relation between v and ω
	$\omega_2 = 4(2.00\ m/s)/1.00\ m = 8.00$ rad/s	

(b) The initial and final rotational kinetic energy are

$$K_1 = I_1\omega_1^2/2 = mr_1^2\omega_1^2/2$$
$$K_2 = I_2\omega_2^2/2 = mr_2^2\omega_2^2/2 = m(r_1^2/4)(4\omega_1)^2/2 = 2mr_1^2\omega_1^2$$
$$\Delta K = K_2 - K_1 = 3mr_1^2\omega_1^2/2 = 3mv_1^2/2 = 6.00 \text{ J}$$

The work done by the student is equal to the change in kinetic energy of the puck.

$$W = \Delta K = 6.00 \text{ J}$$

Related Text Exercises: Ex. 13-27 through 13-33.

6 Translational And Rotational Motion Combined (Sections 13-3, 4)

Review: The motion of an object undergoing both translational and rotational motion can be considered from two points of view. The first considers the translational motion of the center of mass and the rotational motion about the center of mass. The second considers the instantaneous rotational motion about the instantaneously stationary point of contact between the rolling object and the surface on which it is rolling.

View 1

Translation of O and rotation about O

View 2

Rotation about P

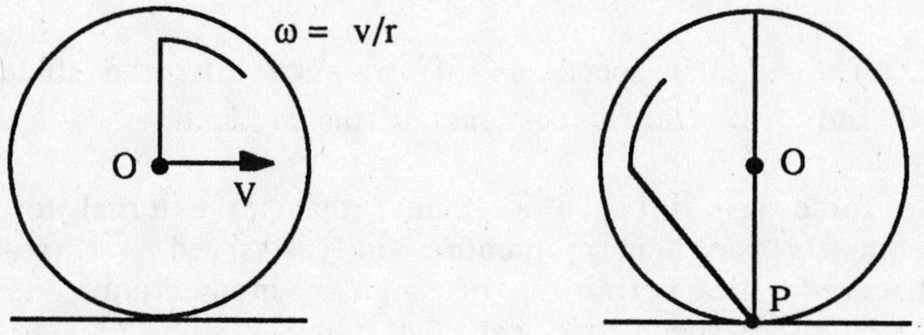

$$K = K_{Ro} + K_T$$
$$= I_o\omega^2/2 + mv^2/2$$
$$= I_o\omega^2/2 + m\omega^2r^2/2$$
$$= (I_o + mr^2)\omega^2/2$$

$$K = K_{RP}$$
$$= I_P\omega^2/2$$
$$= (I_o + mr^2)\omega^2/2$$

Figure 13.14

Notice that both viewpoints lead to the same answer.

Practice: A sphere of mass m and radius r rolls without slipping down an incline of length L and inclination θ, as shown in Figure 13.15.

m = mass of sphere

r = radius of sphere

L = length of incline

θ = angle of inclination

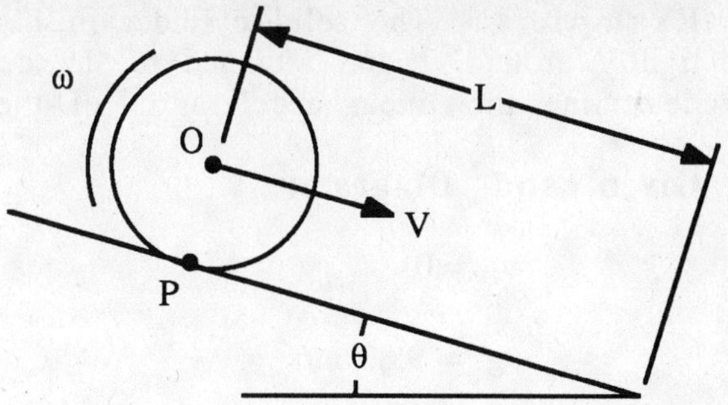

Figure 13.15

Determine an expression for the following:

1. Rotational moment of inertia of the sphere about O	$I_o = 2mr^2/5$ (Table 12-2 of Text)
2. Rotational moment of inertia of the sphere about P	$I_P = I_o + mr^2 = 7mr^2/5$
3. Relationship between the linear speed of the sphere's CM and its angular speed about its CM	$v = \omega r$
4. Kinetic energy of the sphere at any time according to view 1	$K = K_{Ro} + K_T$ $K = I_o\omega^2/2 + mv^2/2 = 7mv^2/10$
5. Kinetic energy of the sphere at any time according to view 2	$K = K_{RP} = I_P\omega^2/2 = 7mv^2/10$
6. Linear speed of the sphere's CM at the bottom of the incline	The change in K of the sphere is equal to the work done on it by gravity. This work is the negative of the change in the sphere's potential energy. $\Delta K = K_f - K_i$, $K_i = 0$ $\Delta K = W_g = -\Delta U = mgL \sin\theta$ $\Delta K = 7mv_f^2/10$ by step 4 or 5 Combining the above, obtain $v_f = [(10gL \sin\theta)/7]^{1/2}$
7. Time it takes the sphere to reach the bottom of the incline	$v_{av} = (v_i + v_f)/2 = L/t$ $t = 2L/v_f = (14L/5g \sin\theta)^{1/2}$

Example: The solid cylinder in Figure 13.16 is supported by two cords wrapped tightly around it and attached to the ceiling. If the cylinder is released from rest, determine the linear acceleration of the CM.

Given and Diagram:

$v_o = 0$

$g = 9.80\ m/s^2$

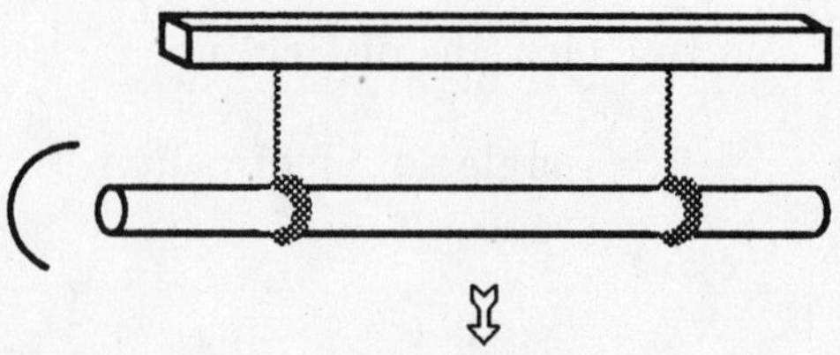

Figure 13.16

Determine: The linear acceleration of the cylinder's CM as it falls.

Strategy: Using view 1, we can state that the kinetic energy at any time is the translational kinetic energy of the CM plus the rotational kinetic energy about the CM. Using the work-energy theorem, we can also state that the change in kinetic energy of the cylinder is equal to the work done on it by gravity. These two statements can be combined to obtain the speed of the CM at any time. Finally we can use our knowledge of kinematics to determine the acceleration of the CM.

Solution: Using view 1, we write the kinetic energy at any time as

$$K = mv^2/2 + I_0\omega^2/2 = mv^2/2 + (mr^2/2)(v/r)^2/2 = 3mv^2/4$$

The work done by gravity at any time is $W_g = F_gL = mgL$, where L is the distance over which the gravitational force F_g has acted on the cylinder. According to the work-energy theorem, the net work done on the cylinder is equal to the change in its kinetic energy.

$$W_g = \Delta K \qquad \Delta K = K_f - K_i \qquad K_f = 3mv_f^2/4 \qquad K_i = 0 \qquad W_g = mgL$$

Substituting, we obtain

$$mgL = 3mv_f^2/4 \qquad \text{or} \qquad v_f^2 = 4gL/3$$

Finally, from our knowledge of kinematics, we write

$$v_f^2 - v_i^2 = 2a\Delta s \qquad v_i = 0 \qquad v_f^2 = 4gL/3 \qquad \Delta s = L$$

which when combined gives

$$4gL/3 = 2aL \qquad \text{or} \qquad a = 2g/3 = 6.53\ m/s^2$$

Related Text Exercises: Ex. 13-16 through 13-20, 13-25 and 13-26.

7 Both Energy and Torque Methods Can be Used to Solve Rotational Motion Problems (Sections 13-3, 4, 5)

Review: In the review section for principal concept 3 of this chapter, we saw that $\tau_{net} = I\alpha$. In the review section for principal concept 4, we saw that $W_{net} = \Delta K$. The concept introduced here is that both torque and energy methods can be used to solve rotational motion problems.

Practice: Two objects and a pulley, all of equal mass, are arranged as shown in Figure 13.17. The horizontal surface is frictionless.

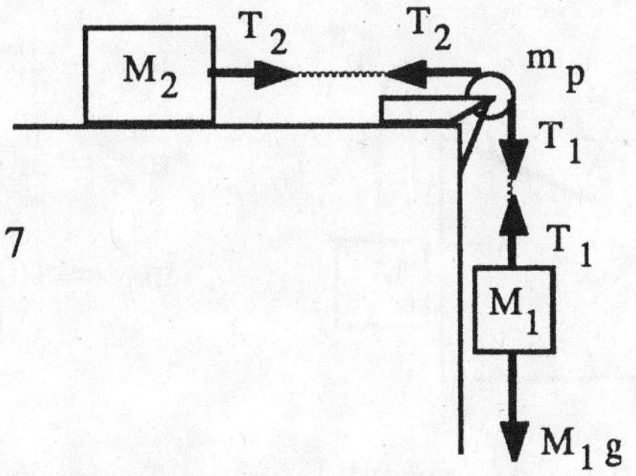

Figure 13.17

$m_1 = m_2 = m_p = m$

r = radius of pulley

Determine an expression for the following:

1. Kinetic energy of the system at any time	$K = m_1v^2/2 + m_2v^2/2 + I_p\omega^2/2$ $I_p = m_pr^2/2, \quad \omega = v/r$ $K = 5mv^2/4$
2. Work done by gravity when m_1 has been lowered by some distance h.	$W_g = m_1gh = mgh$
3. Translational speed of m_1 and m_2 after m_1 has been lowered by some distance h	$W_g = \Delta K, \quad W_g = mgh, \quad \Delta K = 5mv^2/4$ Substituting, we obtain $v = 2(gh/5)^{1/2}$
4. Translational acceleration of m_1 and m_2 at any time	$v_f^2 - v_i^2 = 2a\Delta s$ $v_i = 0, \quad v_f = 2(gh/5)^{1/2}, \quad \Delta s = h$ Substituting, we obtain $a = 2g/5$

Note: The preceding expression for acceleration was obtained using energy methods.

5. Tension T_1	By Newton's second law $m_1a = m_1g - T_1$ or $T_1 = m_1(g - a)$
6. Tension T_2	By Newton's second law, $m_2a = T_2$

7. Net torque on m_p	$\tau_{net} = r(T_1 - T_2) = rm(g - 2a)$
8. Translational acceleration of m_1 and m_2 at any time	$\tau_{net} = I\alpha = rm(g - 2a)$, $I = mr^2/2$ $\alpha = a/r$, Substituting and solving for a, we obtain $a = 2g/5$

Note: The preceding expression for acceleration was obtained using torque methods. It is exactly the same as the Step 4 value obtained using energy methods.

Example: Two objects and a pulley are arranged as shown in Figure 13.18.

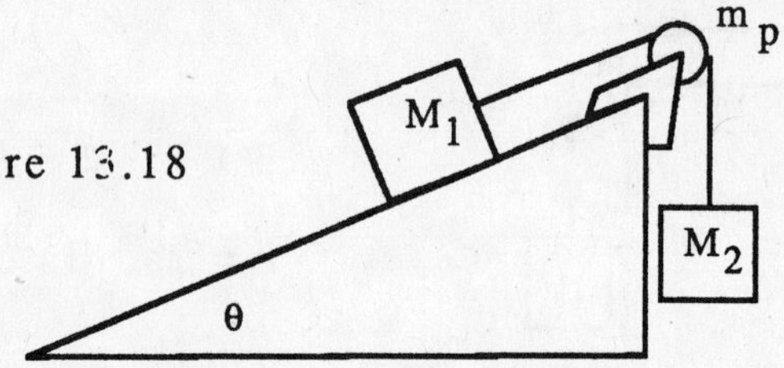

Figure 13.18

$m_1 = 2.00$ kg, $v_i = 0$

$m_2 = 5.00$ kg, $\mu = 0.100$

$m_p = 1.00$ kg, $g = 9.80\ m/s^2$

$\theta = 30.0°$

Calculate the translational acceleration of m_1 and m_2.

Determine: The translational acceleration of m_1 and m_2.

Strategy: Method I (energy method). First, obtain an expression for the kinetic energy of the system at any time. Second, obtain an expression for the net work done on the system when it moves some distance h. Finally, use the work-energy theorem to obtain an expression for v and use your knowledge of kinematics to obtain an expression for the acceleration.

Method II (torque method). First, use Newton's second law to obtain expressions for the tensions T_1 and T_2. Second, obtain an expression for the net torque acting on the system. Finally, use $\tau = I\alpha$ to obtain an expression for v and your knowledge of kinematics to obtain an expression for the acceleration.

Solution: Method I. Obtain an expression for the kinetic energy of the system.

$$\Delta K = K_f - K_i = K_f = m_1v^2/2 + m_2v^2/2 + I_p\omega^2/2 = (m_1 + m_2 + m_p/2)\,v^2/2$$

Obtain an expression for the net work done on the system when it moves some distance h.

$$W_{net} = W_g + W_f = m_2gh - (m_1gh\sin\theta) - (\mu m_1gh\cos\theta) = (m_2 - m_1\sin\theta - \mu m_1\cos\theta)gh$$

Now equate W_{net} and ΔK and solve for v^2. Then use $v_f^2 - v_i^2 = 2a\Delta s$ to obtain

$$a = (m_2 - m_1\sin\theta - \mu m_1\cos\theta)g/(m_1 + m_2 + m_p/2) = 0.522g = 5.11\ m/s^2$$

Method II. Use Newton's second law to obtain an expression for the tensions T_1 and T_2.

$$m_2a = m_2g - T_2 \quad \text{or} \quad T_2 = m_2(g - a)$$

$$m_1a = T_1 - m_1g \sin\theta - \mu m_1g \cos\theta \quad \text{or} \quad T_1 = m_1(a + g \sin\theta + \mu g \cos\theta)$$

The net torque acting on the pulley is

$$\tau_{net} = r(T_2 - T_1) = r[m_2(g - a) - m_1(a + g \sin\theta + \mu g \cos\theta)]$$

We can obtain another expression for τ_{net}:

$$\tau_{net} = I\alpha = (m_pr^2/2)a/r = m_par/2$$

Equating these expressions for τ_{net} and solving for the acceleration we obtain

$$a = (m_2 - m_1 \sin\theta - \mu m_1 \cos\theta)g/(m_1 + m_2 + m_p/2 = 5.11 \text{ m/s}^2$$

Related Text Exercises: Ex. 13-16 through 13-20.

PRACTICE TEST

Take and grade this practice test. Doing so will allow you to determine any weak spots in your understanding of the concepts taught in this chapter. The following section prescribes what you should study further to strengthen your understanding.

A disk of radius r and mass m (Figure 13.19) is rotating at an initial angular speed of ω_i. A sphere of putty of mass m is dropped from a height h onto the disk at a location of r/2. The putty sticks to the disk at the point of contact.

h = height from which putty is dropped

ω_i = initial angular speed

r = radius of disk

m = mass of putty = mass of disk

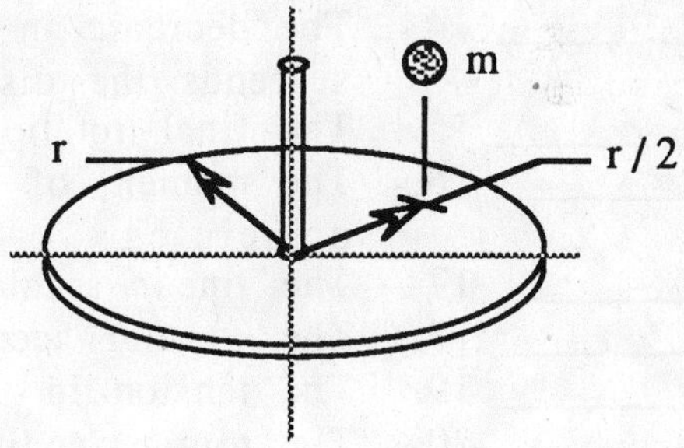

Figure 13.19

Develop expressions for the following physical quantities in terms of the known information only.

_________ 1. The initial moment of inertia of the rotating system
_________ 2. The initial angular momentum of the rotating system
_________ 3. The initial kinetic energy of the rotating system
_________ 4. The moment of inertia of the putty after it is dropped onto the disk
_________ 5. The final moment of inertia of the rotating system

________ 6. The final angular momentum of the rotating system
________ 7. The final angular speed of the rotating system
________ 8. The final kinetic energy of the rotating system
________ 9. The change in the kinetic energy of the rotating system as the putty falls onto the disk
________ 10. What happened to this decrease in kinetic energy

A rigid body of irregular shape is mounted on an axle of radius r and negligible mass, as shown in Figure 13.20. A string is wrapped around the axle, and a block of mass m is attached to it. The system is released from rest, and the block descends a distance h in a time t.

m = mass of block

r = radius of axle

t = time block takes to descend the distance h

h = distance block descends in time t

g = acceleration due to gravity

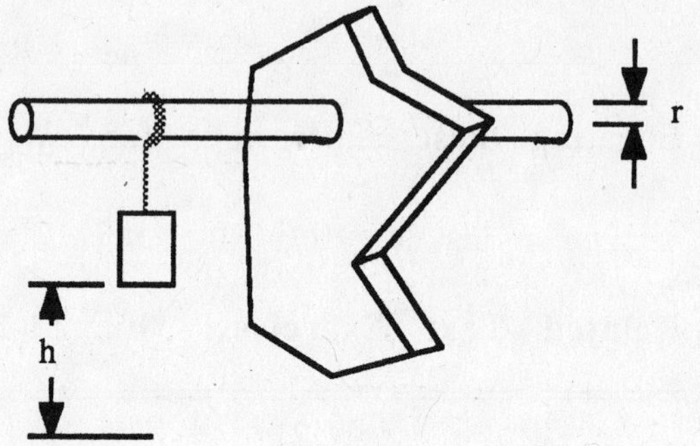

Figure 13.20

Determine expressions for the following physical quantities in terms of the known information only.

________ 11. The final linear speed of the block (that is, its speed after it has fallen the distance h)
________ 12. The final angular speed of the rigid body
________ 13. The final translational kinetic energy of the block
________ 14. The decrease in gravitational potential energy of the block as it descends the distance h
________ 15. The final rotational kinetic energy of the rigid body
________ 16. The moment of inertia of the rigid body obtained by using energy principles
________ 17. The linear acceleration of the block
________ 18. The angular acceleration of the rigid body
________ 19. The tension in the string that supports the block
________ 20. The torque tending to rotate the rigid body
________ 21. The moment of inertia of the rigid body obtained by using torque principles
________ 22. Average power delivered to the rotating system by the gravitational force

(See Appendix I for answers.)

PRINCIPAL CONCEPTS AND EQUATIONS PRESCRIPTION

Your score on the practice test is an excellent measure of your understanding of the chapter. You should now use the following chart to write your own prescription for dealing with any weaknesses the practice test points out. Look down the leftmost column to the number of the question(s) you answered incorrectly, reading across that row you will find the concept and/or equation of concern, the section(s) of the study guide you should return to for further study, and some suggested text exercises which you should work to gain additional experience.

Practice Test Questions	Concepts and Equations	Prescription	
		Principal Concepts	Text Exercises
1	Moment of inertia	5 of Ch 12	12-25, 26
2	Angular momentum	3	13-11
3	Rotational kinetic energy	4	13-16, 22
4	Moment of inertia	5 of Ch 12	12-24, 25
5	Moment of inertia	5 of Ch 12	---
6	Conservation of angular momentum	5	13-27, 28
7	Angular momentum	3	13-11
8	Rotational kinetic energy	4	13-16, 22
9	Rotational kinetic energy	4	13-16, 22
10	Inelastic collision	5 of Ch 10	10-27, 32
11	Average speed	2 of Ch 3	3-12, 13
12	Relation between linear and angular quantities	4 of Ch 12	12-17, 18
13	Translational kinetic energy	4 of Ch 8	8-23, 32
14	Gravitational potential energy	1 of Ch 9	9-1, 2
15	Rotational kinetic energy	4	13-16, 22
16	Work-energy theorem	4	13-22, 25
17	Analyzing translational motion	4 of Ch 3	3-46, 47
18	Relation between linear and angular quantities	4 of Ch 12	12-19, 20
19	Newton's second law	2 of Ch 5	5-31, 32
20	Torque	2 of Ch 11	11-3, 4
21	Relation between torque and angular acceleration	3	13-13, 14
22	Average power	4	13-22, 24

14 OSCILLATIONS

RECALL FROM PREVIOUS CHAPTERS

Previously learned concepts and equations frequently used in this chapter	Text Section	Study Guide Page
The relationship between linear and angular speed: $v = r\omega$	12-6	12-10
Centripetal acceleration: $a_c = v^2/r = \omega^2 r$	4-3	4-9
Kinetic energy: $K = mv^2/2$	8-4	8-13
The relationship between displacement, velocity, and acceleration: $v_x = dx/dt$, $a_x = d^2x/dt^2$	3-2, 4	3-5, 12
Spring constant: $k = -F_s/x$	9-1	9-1
Potential energy of a spring: $U = kx^2/2$	9-1	9-1

NEW IDEAS IN THIS CHAPTER

Concepts and equations introduced	Text Section	Study Guide Page
Equations to describe SHM: $x = A\cos(\omega t + \phi)$; $v_x = dx/dt = -\omega A \sin(\omega t + \phi)$; $a_x = d^2x/dt^2 = dv_x/dt = -\omega^2 A \cos(\omega t + \phi) = -\omega^2 x$	14-1	14-1
The relationship between period, frequency, and angular frequency: $\omega = 2\pi\nu = 2\pi/T$	14-1	14-1
Phase angle: $\phi = \tan^{-1}(-v_{x0}/\omega x_0)$	14-1	14-1
Condition for SHM: $a = -(k/m)x$	14-2	14-7
Relationship between angular frequency, spring constant and oscillator mass: $\omega^2 = k/m$	14-2	14-7
Total mechanical energy of a spring: $E = K + U = K_{max} = U_{max} = kA^2/2$	14-3	14-10
Reference circle	14-5	14-16

PRINCIPAL CONCEPTS AND EQUATIONS

1 Describing Simple Harmonic Motion (Section 14-1)

Review: Figure 14.1 shows a mass M on the end of a spring of constant k. The coordinate system is set up such that $x = 0$ is the equilibrium position for the spring.

The spring is pulled to the right to $x = A$ and then released.

(a) At $t = 0$

$x_0 = +A$

$v_{xo} = 0$

For this case $\phi = 0$

(b) Position at time t

$x = A\cos(\omega t + \phi)$

$= A\cos\omega t$ as $\phi = 0$

(c) Velocity and time t

$v_x = -\omega A\sin(\omega t + \phi)$

$= -\omega A\sin\omega t$ as $\phi = 0$

(d) Acceleration and time t

$a_x = -\omega^2 A\cos(\omega t + \phi)$

$= -\omega^2 A\cos\omega t$ as $\phi = 0$

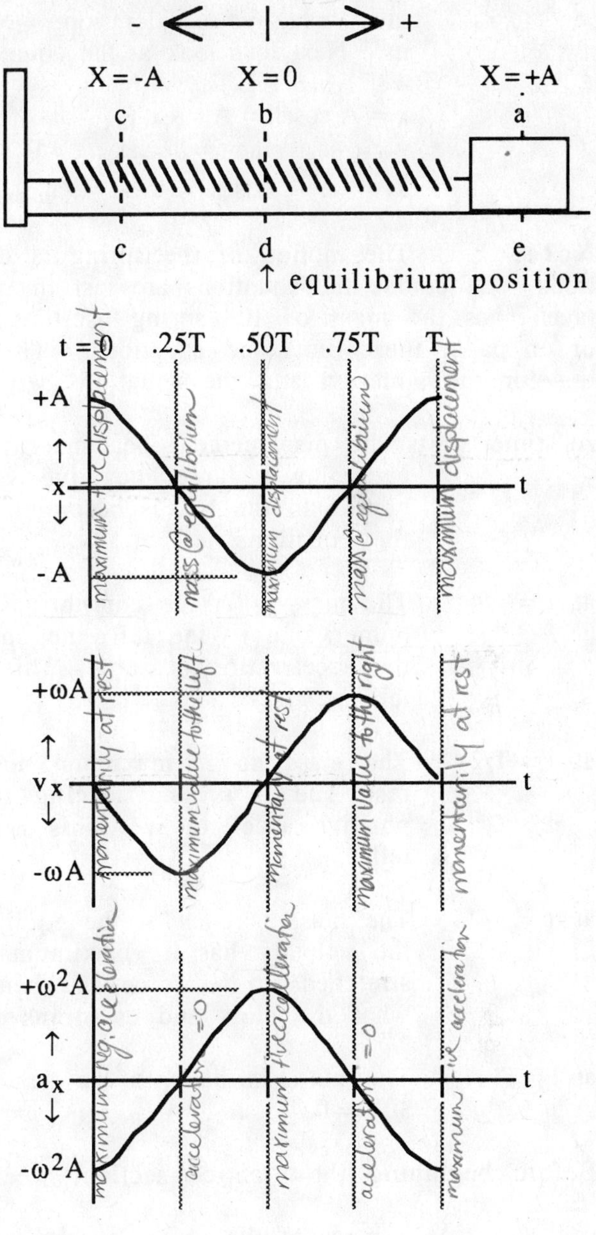

Figure 14.1

Note: It is critical that you learn to simultaneously relate the following three things:

(i) the mass on the spring oscillating back and forth

(ii) the simultaneous location on the x vs t, v_x vs t, and a_x vs t plots

(iii) the equations which describe x, v_x and a_x as a function of time

At t = 0   The mass has a <u>maximum positive displacement</u>, it is momentarily at <u>rest</u>, (and) it is experiencing the maximum negative force hence maximum negative acceleration. Note that this is exactly what the plots tell us. Next let's look at the equations at t = 0

$$x = A\cos\omega t = A = +x_{max} \quad \text{at} \quad t = 0$$
$$v_x = -\omega A \sin\omega t = 0 \quad \text{at} \quad t = 0$$
$$a_x = -\omega^2 A \cos\omega t = -\omega^2 A = -a_{max} \quad \text{at} \quad t = 0$$

Note: The motion of the spring is described by both the plots and the equations. Of course the equations are just the equations of the plots. It is to your advantage if as the mass on the spring oscillates back and forth you can visualize mentally or on paper these plots and equations. The plots will give you a good qualitative feeling for the situation and the equations will allow you to be quantitative.

As time starts the <u>displacement</u> becomes a <u>smaller positive value</u>, the velocity is becoming a larger negative value, and the force and hence the acceleration to the left is becoming less. This is exactly what the plots and the equations tell us.

at t = T/4   The <u>mass is</u> at the <u>equilibrium position</u> (x = 0), the <u>velocity</u> has its maximum value to the left, the spring is not stretched so the force and hence the acceleration is zero. This is exactly what the plots and the equations tell us.

=0.25T

at t = T/2   The <u>mass</u> has a <u>maximum negative displacement</u>, it is momentarily at rest, and it is experiencing the <u>maximum positive force</u> hence maximum positive acceleration. This is exactly what the plots and the equations tell us.

= 0.5T

at t = 3T/4   The mass is back at the equilibrium position (x = 0), however this time the velocity has its maximum value to the right, again the spring is not stretched so the force and hence the acceleration is zero. This is exactly what the plots and equations tell us.

= 0.75T

at t = T   The mass is back to the same displacement, velocity, and acceleration as at t = 0.

Before beginning the practice section, we need to agree on <u>nomenclature</u>

M = mass of the object on the end of the spring
k = spring constant for the spring (F = -kx)
A = amplitude of oscillation for the mass on the spring
x = displacement of the mass from equilibrium at any time
v = velocity of the mass at any time
a = acceleration of the mass at any time
T = period of the oscillation, time for one complete trip by the mass
ν = frequency of oscillations, number of oscillations per unit of time
     (ν = 1/T) *Period, time for one complete trip by the mass (frequency $\frac{\# of oscillations}{period}$*
ω = angular frequency of the mass (ω = 2πν = 2π/T)
φ = phase constant, depends on where we chose t = 0
(ωt + φ) = phase

It is also necessary to be aware of the following

$$x_{max} = \pm A$$
$$v_x = dx/dt$$

phase constant $\phi = \tan^{-1}(-v_{xo}/\omega x_0)$

$$A = \left[(x_0)^2 + \left[v_{xo}\right]^2/\omega^2\right]^{1/2}$$

$$v_{x\,max} = \pm\omega A$$
$$a_{x\,max} = \pm\omega^2 A$$
$$a_x = d^2x/dt^2 = dv_x/dt = -\omega^2 x$$

Practice: Consider the situation shown in Figure 14.1 and the following data:

A = 0.100 m   *amplitude*

ν = 0.500/s   *frequency of oscillations*

M = 1.00 kg   *Mass at end of spring*

Determine the following:

make sure your calculator is in Radians

1. The time for one complete oscillation $T = \frac{1}{\nu} = \frac{1}{0.500} = 2s$	The time for one oscillation is the period: T = 1/ν = 2.00 s
2. The angular frequency $\omega = 2\pi\nu = 2\pi(0.500s) = \pi \frac{rad}{s}$	$\omega = 2\pi\nu = 2\pi/T = \pi$ rad/s
3. The phase constant $x = A\cos(\omega t + \phi)$ @ $t=0$, $X_0 = A$ $A = A\cos(0 + \phi) \rightarrow A = A\cos\phi$ $\frac{A}{A} = \cos\phi$ $1 = \cos\phi$ *when t=0 & $U_0 = 0$* $V_x = -\omega A\sin(\omega t + \phi)$ $0 = -\omega A\sin(\phi)$ $0 = \sin\phi$ $\therefore \cos\phi = 1$ & $\sin\phi = 0$ when $\phi = 0$	$x = A\cos(\omega t + \phi)$ at t = 0, $x_0 = A$ $A = A\cos\phi$ or $\cos\phi = 1$ $v_x = -\omega A\sin(\omega t + \phi)$ at t = 0, $v_{xo} = 0$ $0 = -\omega A\sin\phi$ or $\sin\phi = 0$ $\cos\phi = 1$ and $\sin\phi = 0$ if $\phi = 0$
4. An expression for the displacement at any time $x = A\cos(\omega t + \phi) = (0.1m)\cos[\pi \frac{rad}{s} t]$	$x = A\cos(\omega t + \phi)$ $= (0.100\ m)\cos[(\pi\ rad/s)t]$
5. Displacement at the time 0.125 s $x = (0.1m)\cos[\pi \frac{rad}{s}][0.125s]$ $= 9.23 \times 10^{-2} m$	$x = (0.100\ m)\cos[(\pi\ rad/s)(0.125\ s)]$ $= 9.24 \times 10^{-2}$ m
6. Displacement at the time t = 0.500 s $x = (0.1m)\cos[\pi \frac{rad}{s}][0.5s]$ $= 0\ m$	$x = (0.100\ m)\cos[(\pi\ rad/s)(0.500\ s)] = 0\ m$ If the period is 2.00 s and the initial position is +A, then t = 0.500 s is T/4 = 0.25T hence x = 0
7. An expression for the velocity at any time $x = (0.1m)\cos[\pi \frac{rad}{s}][t]$ $V_x = -(\frac{\pi}{s})(0.1m)\sin([\pi \frac{rad}{s}][t])$ $= -(0.314 \frac{m}{s})\sin([\pi \frac{rad}{s}][t])$	$x = (0.100\ m)\cos[(\pi\ rad/s)t]$ $v_x = dx/dt$ $= -(0.100\ m)(\pi/s)\sin[(\pi\ rad/s)t]$ $= -(0.314\ m/s)\sin[(\pi\ rad/s)t]$
8. Velocity at t = 0.500 s $V_x = -(0.314 \frac{m}{s})\sin(\pi \frac{rad}{s} \cdot 0.5s)$ $= -0.314 \frac{m}{s} \rightarrow -3.14 \times 10^{-1} \frac{m}{s}$ *this max -ve velocity*	$v_x = -(0.100\ m)(\frac{\pi}{s})\sin[(\frac{\pi\ rad}{s})(0.500\ s)]$ $= -3.14 \times 10^{-1}$ m/s also if the period is 2.00 s then t = 0.200 s is T/4. At this time the velocity should be negative max $v_x = -\omega A = -3.14 \times 10^{-1}$ m/s

9. An expression for the acceleration at any time $a_x = -(\frac{\pi}{s})^2(0.1\,m)\cos[(\pi \frac{rad}{s})t]$ $= -(0.987\frac{m}{s^2})\cos[(\pi \frac{rad}{s})t]$	$a_x = d^2x/dt^2 = dv_x/dt$ $= -(0.100\ m)(\pi/s)^2 \cos[(\pi\ rad/s)t]$ $= -(0.987\ m/s^2) \cos[(\pi\ rad/s)t]$
10. Acceleration at t = 1.00 s $a_x = -(0.987\frac{m}{s^2})\cos[(\pi \frac{rad}{s})(1.00 s)]$ $= 0.987\frac{m}{s^2}$ this is at maximum positive acceleration.	$a_x = -(0.987\ m/s^2) \cos[(\pi\ rad/s)(1.00\ s)]$ $= 0.987\ m/s^2$ also if the period is 2.00 s then t = 1.00 s is T/2. (0.5T) At this time the acceleration should be positive max $a_x = +\omega^2 A = 0.987\ m/s^2$
11. The time until x = +0.0500 m and v_x is positive $x = (0.1m)\cos((\pi \frac{rad}{s})t)$ $\frac{0.05m}{0.1m} = \cos[(\pi \frac{rad}{s})t]$ $0.5 = \cos[(\pi \frac{rad}{s})t]$ $0.5 = \cos\theta$ $0.333s = \frac{1.04}{\pi}$ $= \theta = \frac{1.04}{2\pi} = 1.64$	$x = (0.100\ m) \cos[(\pi\ rad/s)t]$ $0.0500\ m = 0.100\ m \cos[(\pi\ rad/s)t]$ $0.500 = \cos[(\pi\ rad/s)t]$ this is true when $(\pi\ rad/s)t = \pi\ rad/3, \quad 5\pi\ rad/3$ or t = 0.333 s, 1.67 s at t = 0.333 s v_x is negative at t = 1.67 s v_x is positive
12. The phase constant if we choose t = 0 the first time the mass goes through the equilibrium position 3 parts	at this time $x_0 = 0,\ v_{xo} = -v_{max},$ and $a_{xo} = 0$ ← what we know / must know $x = A\cos(\omega t + \phi)$ $x_0 = 0$ at $t = 0$ $0 = A\cos\phi$ or $\cos\phi = 0$ $v_x = -\omega A \sin(\omega t + \phi)$ $v_{xo} = -v_{max} = -\omega A$ at $t = 0$ $-\omega A = -\omega A \sin\phi$ or $\sin\phi = 1$ $a_x = -\omega^2 A \cos(\omega t + \phi)$ $a_{xo} = 0$ at $t = 0$ $0 = -\omega^2 A \cos\phi$ or $\cos\phi = 0$ $\cos\phi = 0$ and $\sin\phi = 1$ at $\phi = \pi/2 = 90°$
13. The phase constant if we choose t = 0 when the mass has maximum negative displacement 3 parts again	At this time $x_0 = -A,\ v_{xo} = 0,\ a_{xo} = +a_{max}$ ← must know $x = A\cos(\omega t + \phi)$ $-A = A\cos\phi$ or $\cos\phi = -1$ $v_x = -\omega A \sin(\omega t + \phi)$ $0 = -\omega A \sin\phi$ or $\sin\phi = 0$ $a_x = -\omega^2 A \cos(\omega t + \phi)$ $+\omega^2 A = -\omega^2 A \cos\phi$ or $\cos\phi = -1$ $\cos\phi = -1$ and $\sin\phi = 0$ at $\phi = \pi = 180°$

Note: When determining the phase constant we get the same information out of the position and acceleration equations. As a result of this, we need to use only the position and velocity equations or the velocity and acceleration equations.

14. The phase constant if we choose $t = 0$ when the mass goes through the equilibrium position with a positive velocity.	At this time $x_0 = 0$ and $v_{x0} = +v_{max} = \omega A$ $x = A\cos(\omega t + \phi)$ $0 = A\cos\phi$ or $\cos\phi = 0$ $v_x = -\omega A\sin(\omega t + \phi)$ $\omega A = -\omega A\sin\phi$ or $\sin\phi = 3\pi/2 = 270°$ or $\phi = \tan^{-1}(-v_{x0}/\omega x_o)$ $= \tan^{-1}[-\omega A/\omega(0)]$ $= \tan^{-1}(-\infty) = 3\pi/2 = 270°$

Example: An object executes SHM with $A = 10.0$ cm, $\omega = 2.00$ rad/s, and $\phi = 0$. Write expressions for x, v_x and a_x. Determine the value of x, v_x and a_x at $t = 2.00$ s.

Given: $A = 10.0$ cm $\omega = 2.00$ rad/s $\phi = 0$ $t = 2.00$ s

Determine: Expressions x, v_x and a_x and the value of x, v_x and a_x at $t = 2.00$ s.

Strategy: Knowing A, ω, and ϕ, we can readily write down an expression for x. This expression may then be differentiated to obtain v_x and a_x. Once the expressions are known, we can easily insert the value $t = 2.00$ s to obtain x, v_x and a_x at $t = 2.00$ s.

Solution: The general form for x is

$$x = A\cos(\omega t + \phi)$$

Inserting values for A, ω and ϕ, obtain the expression for x

$$x = (10.0 \text{ cm})\cos[(2.00 \text{ rad/s})t]$$

Differentiating obtain expressions for v_x and a_x

$$v_x = dx/dt = -(10.0 \text{ cm})(2.00 \text{ rad/s})\sin[(2.00 \text{ rad/s})t] = -(20.0 \text{ cm/s})\sin[(2.00 \text{ rad/s})t]$$

$$a_x = d^2x/dt^2 = dv_x/dt = -(20.0 \text{ cm/s})(2.00 \text{ rad/s})\cos[(2.00 \text{ rad/s})t]$$
$$= -(40.0 \text{ cm/s}^2)\cos[2.00 \text{ rad/s})t]$$

Inserting the value $t = 2.00$ s, obtain

$$x = (10.0 \text{ cm})\cos[(2.00 \text{ rad/s})(2.00 \text{ s})] = (10.0 \text{ cm})\cos 4.00 \text{ rad} = -6.54 \text{ cm}$$
$$v_x = -(20.0 \text{ cm/s})\sin[(2.00 \text{ rad/s})(2.00 \text{ s})] = -(20.0 \text{ cm/s})\sin 4.00 \text{ rad} = 15.1 \text{ cm/s}$$
$$a_x = -(40.0 \text{ cm/s}^2)\cos[(2.00 \text{ rad/s})(2.00 \text{ s})] = -(40.0 \text{ cm/s}^2)\cos 4.00 \text{ rad} = 26.1 \text{ cm/s}^2$$

Related Text Exercises: Ex. 14-1 through 14-9.

2 Conditions For Simple Harmonic Motion (Section 14-2)

Review: Figure 14.2(a) shows a mass m resting on a horizontal frictionless surface and attached to an unstretched spring that has a spring constant k. In Figure 14.2(b), the mass is displaced by an amount ix. Since the spring is stretched an amount ix, it exerts a restoring force **F** = -(kx)**i**.

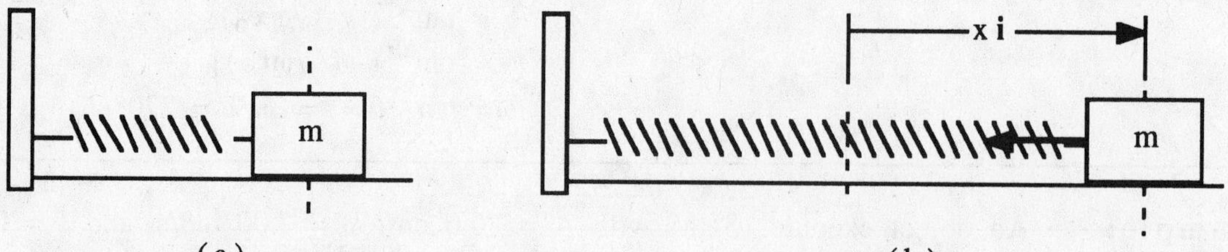

Figure 14.2

The negative sign is an explicit reminder that the restoring force and the displacement are in opposite directions. According to Newton's second law, the net force acting on the mass is equal to the product of the mass and its acceleration (i.e. Σ **F** = m**a**). In this case, the unbalanced force acting on the mass is the restoring force of the spring. Hence we may write

$$m a_x \mathbf{i} = -kx\mathbf{i} \quad \text{or} \quad a_x = -(k/m)x$$

A necessary condition for SHM then is that the acceleration is proportional to and in the opposite direction of the displacement.

In the previous section we saw that

$$x = A\cos(\omega t + \phi) \quad \text{and} \quad a_x = -\omega^2 A' \cos(\omega t + \phi) = -\omega^2 x$$

In order for our work with the restoring force to be compatible with our work on describing SHM, we must have

$$\omega = (k/m)^{1/2}$$

$m = 2.00$ kg
$L = 0.200$ m
$F_{pull} = 40.0$ N
$x = 0.100$ m

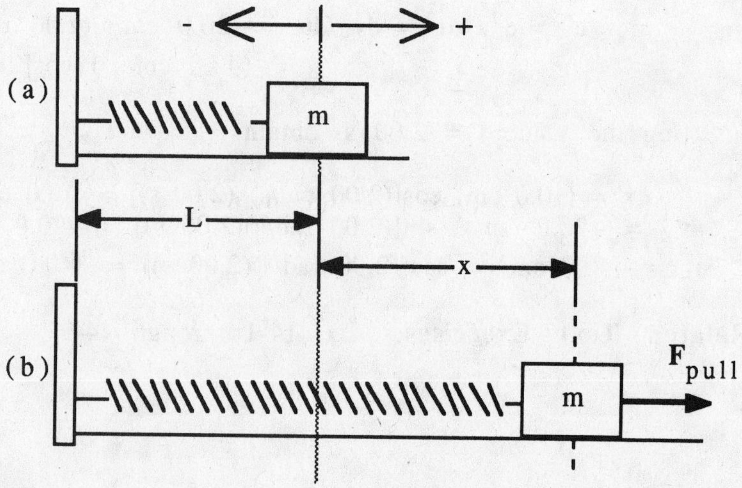

Figure 14.3

Practice: Figure 14.3(a) shows a 2.00-kg mass attached to a 0.200-m-long spring and resting on a horizontal frictionless surface.   Figure 14.3(b) shows that when you pull on the mass-spring system with a 40.0-N force, you can stretch the spring by 0.100 m.

$F = 40.0N$
$M = 2.00kg$
$x = 0.100m$

Determine the following:

1. The restoring force F_s exerted on the mass by the spring while you are exerting the 40.0-N force on the mass	Since the mass is in equilibrium (as long as you hold onto it), we can write $F_s = -F_{pull} = -40.0$ N. The minus sign tells us that the restoring force is to the left.
2. The spring constant K	$k = -F_s/x$ $= -(-40.0 \text{ N})/(0.100 \text{ m}) = 400 \text{ N/m}$
3. Acceleration of the mass the instant it is released a	$a = -kx/m = -20.0 \text{ m/s}^2$
4. Angular frequency of the oscillating system ω	$\omega = (k/m)^{1/2} = 14.1 \text{ rad/s}$
5. Linear frequency of the oscillating system $\checkmark$	$\nu = \omega/2\pi = 2.24 \text{ Hz}$
6. Period of oscillation T	$T = 1/\nu = 0.446 \text{ s}$
7. Time for 10 complete oscillations t	$t = 10 \text{ } T = 4.46 \text{ s}$
8. Force you would have to supply to compress the spring to a length of 0.050 m $F = -kx$	For this case, x = -0.150 m $F_{push} = -F_s = -(-kx) = -60.0$ N The minus sign confirms that you would have to push to the left in order to compress the spring.
9. Acceleration of the mass when released from the situation described in step 8	$a = -kx/m = +30.0 \text{ m/s}^2$ o r $a = F_s/m = +30.0 \text{ m/s}^2$ The positive sign tells us that the mass will accelerate to the right when released.
10. Period of oscillation of the system when released from the situation described in step 8. $T = \frac{1}{\nu} = \frac{1}{2.24 Hz} = 0.446 s$	$T = 0.446 \text{ s}$ Same as in step 6. Since m and k are not changed, the period is not changed $[T = 2\pi(m/k)^{1/2}]$.

Example: A 0.400-kg mass suspended by a spring stretches the spring 0.100 m. A 0.600-kg mass is attached to the 0.400-kg mass, the whole system is rotated onto its side, and the spring is set into oscillation on a horizontal frictionless surface by stretching it 0.150 m and then releasing it. Determine the maximum horizontal acceleration of the combined mass on the end of the spring and the period of oscillation.

Given: $m_1 = 0.400$ kg $\quad m_2 = 0.600$ kg $\quad x_1 = 0.100$ m $\quad x_2 = 0.150$ m

Determine: The maximum acceleration and period of oscillation of the 1.00-kg mass ($m_1 + m_2$) on the end of the spring after the spring is stretched horizontally 0.150 m and then released.

Strategy: Knowing m_1, we can determine the force needed to stretch the spring a distance x_1. When we know this force, we also know the restoring force exerted by the spring when it is stretched a distance x_1. Knowing the restoring force and the stretch, we can determine the spring constant k. Knowing that k does not change when the spring is placed in a horizontal position, the total mass attached to the spring, and the maximum horizontal stretch, we can determine the maximum acceleration of the combined mass and its period of oscillation.

Solution: Figure 14.4 shows the different situations involved in the problem.

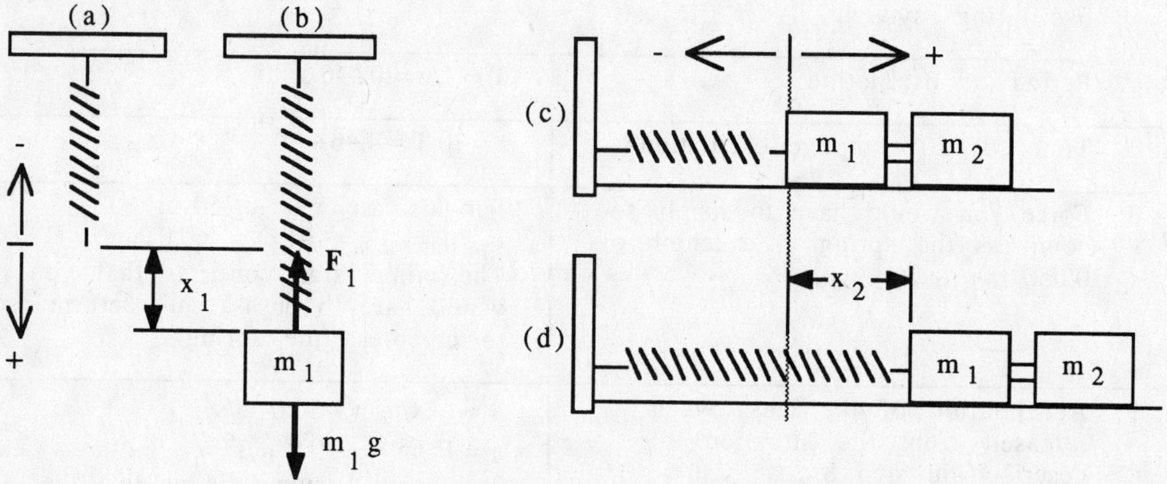

Figure 14.4

(a) Unstretched spring in vertical position
(b) Suspended mass m_1 stretches the spring a distance x_1
(c) Unstretched spring in the horizontal equilibrium position with mass $m_1 + m_2$ attached
(d) Maximum horizontal displacement from equilibrium with mass $m_1 + m_2$ attached

From Figure 14.4(b), we see that after m_1 is suspended from the spring, a new equilibrium is established. As a result, we can write

$$F_s = -m_1 g \quad \text{and} \quad F_s = -kx_1$$

Combining these expressions for the restoring force and solving for k, we obtain

$$k = m_1 g/x_1 = (0.400\ \text{kg})(9.80\ \text{m/s}^2)/(0.100\ \text{m}) = 39.2\ \text{N/m}$$

From Figure 14.4(d), we see that the maximum displacement from equilibrium is x_2. Hence the maximum acceleration of the combined mass is

$$a = -kx_2/(m_1 + m_1) = -(39.2\ \text{N/m})(0.150\ \text{m})/(1.00\ \text{kg}) = -5.88\ \text{m/s}^2$$

The period of oscillation is

$$T = 2\pi[m_1 + m_2)/k]^{1/2} = 2\pi(1.00\ \text{kg}/39.2\ \text{N/m})^{1/2} = 1.00\ \text{s}$$

Related Text Exercises: Ex. 14-10 through 14-16.

3 Energy Of A Simple Harmonic Oscillator (Section 14-3)

Review: Figure 14.5 shows a mass-spring system at rest on a horizontal frictionless surface at $t = 0$. When the mass is released, it undergoes SHM about the equilibrium position.

m = mass of oscillating object
k = spring constant
A = amplitude of oscillation
ϕ = phase angle

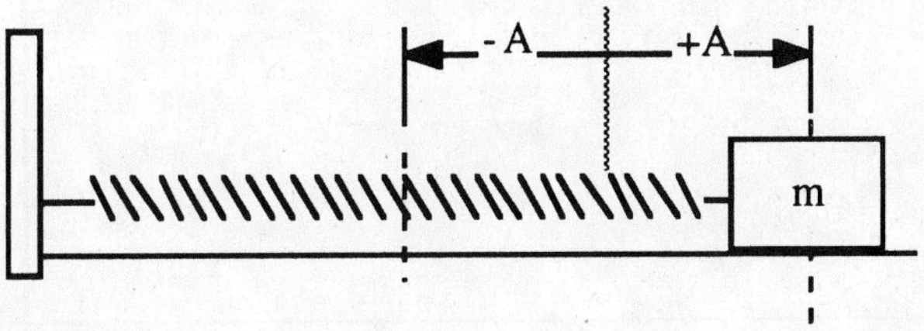

Figure 14.5

The position and speed of the mass at any time are $x = A\cos(\omega t + \phi)$ and $v = -\omega A \sin(\omega t + \phi)$, and its kinetic energy at any time is

$$K = (\tfrac{1}{2})mv^2 = (\tfrac{1}{2})m\omega^2 A^2 \sin^2(\omega t + \phi) = (\tfrac{1}{2})kA^2 \sin^2(\omega t + \phi) = K_{max} \sin^2(\omega t + \phi)$$

The potential energy of the spring at any time is

$$U = (\tfrac{1}{2})kx^2 = (\tfrac{1}{2})kA^2 \cos^2(\omega t + \phi) = U_{max} \cos^2(\omega t + \phi)$$

Notice that the kinetic energy and potential energy vary in time from 0 to $kA^2/2$. The total mechanical energy of the mass-spring system is the sum of the kinetic energy (K) of the mass and the potential energy (U) of the spring.

$$E = K + U = (\tfrac{1}{2})kA^2 \sin^2(\omega t + \phi) + (\tfrac{1}{2})kA^2 \cos^2(\omega t + \phi) = (\tfrac{1}{2})kA^2$$

Notice the that the total energy of the system is a constant in time.

Practice: A 2.00-kg object attached to the end of a spring with a 400 N/m spring constant is undergoing SHM on a horizontal frictionless surface. To start the system oscillating, the spring was compressed by 0.100 m and then released. The time record was started (t = 0) the first time the object went through the equilibrium position.

Determine the following:

1. Maximum kinetic energy	$K_{max} = kA^2/2 = 2.00$ J
2. Maximum potential energy	$U_{max} = kA^2/2 = 2.00$ J
3. Total mechanical energy at any time	$E = kA^2/2 = 2.00$ J
4. Potential energy when x = A/2	$U = kx^2/2 = kA^2/8 = 0.500$ J
5. Kinetic energy when x = A/2	$K = E - U = 1.50$ J
6. Phase angle	$x = A\cos(\omega t + \phi)$ at t = 0, $x_0 = 0$, hence we have $0 = A\cos\phi$ or $\cos\phi = 0$ $v_x = -\omega A\sin(\omega t + \phi)$ at t = 0, $v_{x0} = +max = \omega A$ hence we have $\omega A = -\omega A\sin\phi$ or $\sin\phi = -1$ $\cos\phi = 0$ and $\sin\phi = -1$ when $\phi = (3\pi/2)$ rad = 4.71 rad
7. Angular speed of a particle on the associated reference circle	$\omega = (k/m)^{1/2} = 14.1$ rad/s
8. Position of the oscillating object at t = 0.100 s	$\omega t + \phi = 1.41$ rad + 4.71 rad = 6.12 rad $x = A\cos(\omega t + \phi) = +0.0987$ m
9. Speed of the object at t = 0.100 s	$\omega t + \phi = 6.12$ rad $v = -\omega A\sin(\omega t + \phi) = +0.229$ m/s
10. Potential energy of the system at t = 0.100 s	$U = kx^2/2 = 1.95$ J
11. Kinetic energy of the system at t = 0.100 s	$K = mv^2/2 = 0.0500$ J
12. Total energy of the system at t = 0.100 s	$E = K + U = 2.00$ J, or $E = kA^2/2 = 2.00$ J
13. Position of the object when its K is equal to the U of the spring	$K + U = E$, $E = kA^2/2$, $U = K$ $2U = kA^2/2$, $U = kx^2/2$ $2(kx^2/2) = kA^2/2$, or $x = 0.0707$ m

Example: A 0.500-kg mass is attached to the end of a spring that has a spring constant of 50.0 N/m. This system is set into SHM on a horizontal frictionless surface by displacing it 0.100 m from equilibrium and then releasing it. The record of time (t = 0) is started when the mass has a maximum negative displacement. At t = 0.100 s, what is the system's (a) potential energy, (b) kinetic energy, (c) total energy?

Given: $m = 0.500$ kg, $k = 50.0$ N/m, $A = 0.100$ m, $t = 0.100$ s,
$t = 0$ when $x_0 = -A$ and $v_{x0} = 0$

Determine: The potential, kinetic, and total energy of the oscillating system 0.100 s after the clock is started.

Strategy: Knowing x_0 and v_{x0}, we can determine the phase angle ϕ. Knowing ϕ, we can obtain an expression for the position x and the velocity v at any time. We can use these expressions for x and v at any time to obtain expressions for the kinetic, potential and total energy at any time.

Solution: First let's determine the phase angle

$$x = A\cos(\omega t + \phi), \quad x_0 = -A \quad \text{hence} \quad -A = A\cos\phi \quad \text{or} \quad \cos\phi = -1$$
$$v = -\omega A\sin(\omega t + \phi), \quad v_{x0} = 0 \quad \text{hence} \quad 0 = -\omega A\sin\phi \quad \text{or} \quad \sin\phi = 0$$

$$\cos\phi = -1 \quad \text{and} \quad \sin\phi = 0 \quad \text{when} \quad \phi = \pi \text{ rad} = 180°$$

The position and velocity may be determined at $t = 0.100$ s as follows:

$$\omega = (k/m)^{1/2} = [(50.0 \text{ N/m})/0.500 \text{ kg}]^{1/2} = 10.0 \text{ rad/s}$$
$$x = A\cos(\omega t + \phi) = (0.100 \text{ m})\cos[(10.0 \text{ rad/s})(0.100 \text{ s}) + \pi \text{ rad}]$$
$$= (0.100 \text{ m})\cos(4.14 \text{ rad}) = -0.0542 \text{ m}$$
$$v = -\omega A\sin(\omega t + \phi) = -(10.0 \text{ rad/s})(0.100 \text{ m})\sin[(10.0 \text{ rad/s})(0.100 \text{ s}) + \pi \text{ rad}]$$
$$= -(1.00 \text{ m/s})\sin(4.14 \text{ rad}) = 0.841 \text{ m/s}$$

The kinetic, potential, and total energy at $t = 0.100$ s are

$$K = mv^2/2 = (0.500 \text{ kg})(0.841 \text{ m/s})^2/2 = 0.177 \text{ J}$$
$$U = kx^2/2 = (50.0 \text{ N/m})(-0.0542 \text{ m})^2/2 = 0.0734 \text{ J}$$
$$E = K + U = 0.250 \text{ J}$$

As a check on our work

$$E = kA^2/2 = (50.0 \text{ N/m})(0.100 \text{ m})^2/2 = 0.250 \text{ J}$$

Related Text Exercises: Ex. 14-17 through 14-21.

4 Examples of Simple Harmonic Motion (Section 14-4)

Review: We will review four different situations where an object undergoes SHM when displaced. While each situation is different, you should be intrigued by the fact that we will perform the same steps for each situation. These steps are as follows:

1. Recognize the displacement x or θ

2. Since the object undergoes SHM when released, we have

$$x = A\cos(\omega t + \phi) \quad \text{or} \quad \theta = \theta_{max}\cos(\omega t + \phi)$$
$$v_x = dx/dt = -\omega A\sin(\omega t + \phi) \quad \text{or} \quad \omega_z = -\omega\theta_{max}\sin(\omega t + \phi)$$
$$a_x = d^2x/dt^2 = -\omega^2 A\cos(\omega t + \phi) \quad \text{or} \quad \alpha_z = -\omega^2\theta_{max}\cos(\omega t + \phi)$$

3. As a result of the displacement there is a restoring force F_x due to x or a restoring torque τ_z due to θ.

4. The restoring force or torque causes an acceleration

$$a_x = F_x/m \quad \text{or} \quad \alpha_z = \tau_z/I$$

5. Since the object undergoes SHM the acceleration and displacement are related by

$$a_x = -\omega^2 x \quad \text{or} \quad \alpha_z = -\omega^2\theta$$

6. Subsequently we may determine ω, ν, and T

$$a_x = F_x/m = -\omega^2 x \quad \text{or} \quad \omega = (-F_x/mx)^{1/2}$$
$$\alpha_z = \tau_z/I = -\omega^2\theta \quad \text{or} \quad \omega = (-\tau_z/I\theta)^{1/2}$$
$$\nu = \omega/2\pi \quad \text{and} \quad T = 2\pi/\omega$$

7. The phase constant ϕ is determined by the initial conditions, that is we must know x_0 and v_{xo} or θ_0 and ω_{zo}.

Let's apply these seven steps to the following four situations.

Mass on the end of a spring

1. Displacement: x

2. $x = A\cos(\omega t + \phi)$
$v_x = dx/dt = -\omega A\sin(\omega t + \phi)$
$a_x = d^2x/dt^2 = -\omega^2 A\cos(\omega t + \phi)$

3. Restoring force: $F_x = -kx$

4. Acceleration due to F_x: $a_x = F_x/m = -kx/m$

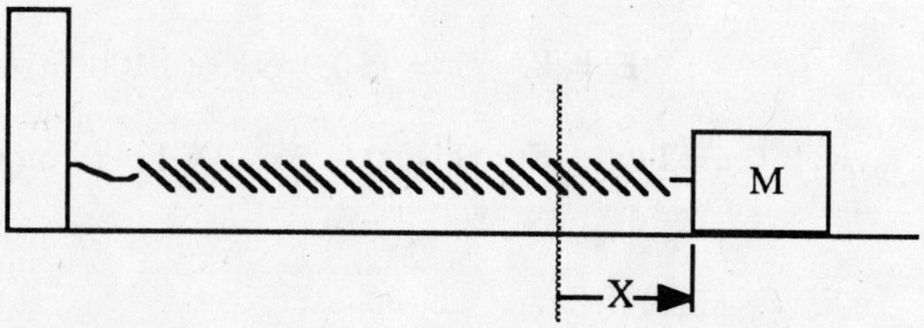

$x_0 = A$ and $v_{xo} = 0$

Figure 14.6

5. Relation between x and a_x: $a_x = -\omega^2 x$

6. Values for ω, ν, and T

$$\omega = (k/m)^{1/2} \qquad \nu = \omega/2\pi = (k/m)^{1/2}/2\pi \qquad T = 2\pi/\omega = 2\pi(m/k)^{1/2}$$

7. Phase constant: Since $x_0 = A$ and $v_{x0} = 0$, we get $\phi = 0$

Simple pendulum

1. Displacement: $x \approx S$

2. $x = A\cos(\omega t + \phi)$
   $v_x = dx/dt = -\omega A \sin(\omega t + \phi)$
   $a_x = d^2x/dt^2 = -\omega^2 A \cos(\omega t + \phi)$

3. Restoring force: $F_x \approx -Mg\sin\theta = -Mgx/L$

4. Acceleration due to F_x: $a_x = F_x/M \approx -gx/L$

5. Relation between x and a_x: $a_x = -\omega^2 x$

6. Values for ω, ν, and T

$$\omega = (g/L)^{1/2} \qquad \nu = \omega/2\pi = (g/L)^{1/2}/2\pi \qquad T = 2\pi/\omega = 2\pi(L/g)^{1/2}$$

7. Phase constant: Since $x_0 = A$ and $v_{x0} = 0$, we get $\phi = 0$

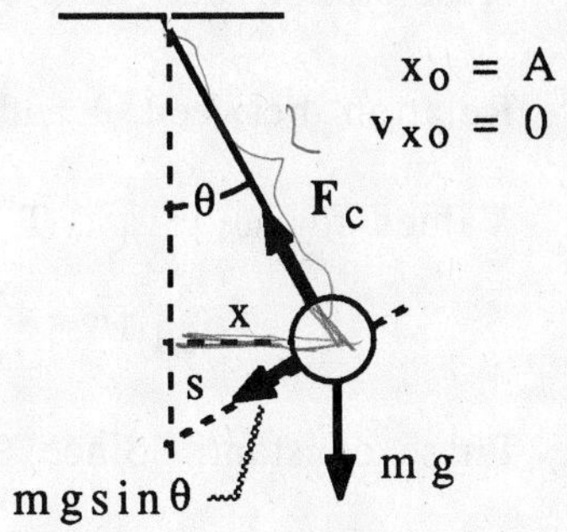

Figure 14.7

Physical pendulum

1. Displacement: θ

2. $\theta = \theta_{max}\cos(\omega t + \phi)$
   $\omega_z = d\theta/dt = -\omega\theta_{max}\sin(\omega t + \phi)$
   $\alpha_z = d^2\theta/dt^2 = -\omega^2\theta_{max}\cos(\omega t + \phi)$

3. Restoring torque: $\tau_z = -MgL\sin\theta \approx -MgL\theta$

4. Acceleration due to τ_z: $\alpha_z = \tau_z/I = -MgL\theta/I$

5. Relation between θ and α_z: $\alpha_z = -\omega^2\theta$

6. Values for ω, ν and T

$$\omega = (MgL/I)^{1/2} \qquad \nu = \omega/2\pi = (MgL/I)^{1/2}/2\pi \qquad T = 2\pi/\omega = 2\pi(I/MgL)^{1/2}$$

7. Phase constant: Since $\theta_0 = \theta_{max}$ and $\omega_{z0} = 0$, we get $\phi = 0$

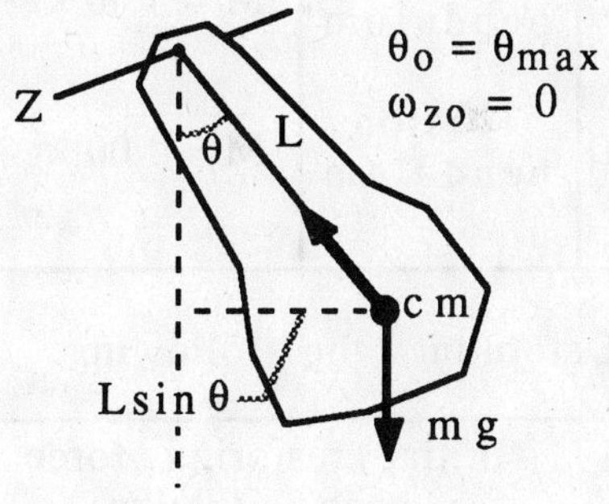

Figure 14.8

Torsional pendulum

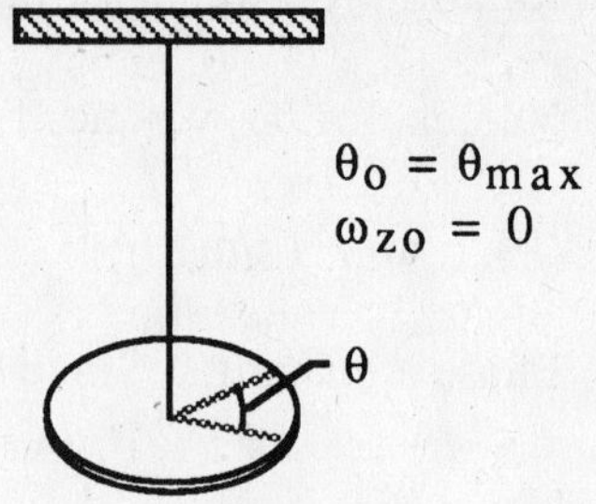

Figure 14.9

1. Displacement: θ

2. $\theta = \theta_{max} \cos(\omega t + \phi)$
   $\omega_z = d\theta/dt = -\omega\theta_{max} \sin(\omega t + \phi)$
   $\alpha_z = d^2\theta/dt^2 = -\omega^2\theta_{max} \cos(\omega t + \phi)$

3. Restoring torque: $\tau_z = -K\theta$

4. Acceleration due to τ_z: $\alpha_z = \tau_z/I = -K\theta/I$

5. Relation between θ and α_z: $\alpha_z = -\omega^2\theta$

6. Values for ω, ν, and T

   $\omega = (K/I)^{1/2}$ $\nu = \omega/2\pi = (K/I)^{1/2}/2\pi$ $T = 2\pi/\omega = 2\pi(I/K)^{1/2}$

7. Phase constant: Since $\theta_0 = \theta_{max}$ and $\omega_{zo} = 0$, we get $\phi = 0$

Practice: Consider the following data:

A	Mass on spring	$M = 1.00$ kg	$k = 200$ N/m	$A = 0.100$ m	$v_{xo} = 0$	
B	Simple pendulum	$M = 1.00$ kg	$L = 2.00$ m	$A = 0.050$ m	$v_{xo} = 0$	
C	Physical pendulum	$M = 1.00$ kg	$L = 0.50$ m	$\theta_{max} = \pi/20$ rad	$\omega_{zo} = 0$	$I_z = 10.0$ kgm^2
D	Torsion pendulum	$M = 1.00$ kg	$K = \dfrac{5.00 \text{ Nm}}{10^5 \text{ rad}}$	$\theta_{max} = \pi/30$ rad	$\omega_{zo} = 0$	$r = 0.100$ m

Determine the following:

1. Initial restoring force or torque for each situation	A) $F_{xo} = -kA = -20$ N B) $F_{xo} = -MgA/L = -0.245$ N C) $\tau_{zo} = -MgL\theta_{max} = -0.770$ Nm D) $\tau_{zo} = -K\theta_{max} = -5.24 \times 10^{-6}$ Nm
2. Angular frequency for each situation	A) $\omega = (k/M)^{1/2} = 14.1$ rad/s B) $\omega = (g/L)^{1/2} = 2.21$ rad/s C) $\omega = (MgL/I)^{1/2} = 0.700$ rad/s D) $\omega = (K/I)^{1/2} = [K/(Mr^2/2)]^{1/2}$ $= 0.100$ rad/s

3. Period for each situation	A) $T = 2\pi/\omega = 0.446$ s B) $T = 2.84$ s C) $T = 0.898$ s D) $T = 62.8$ s
4. Position after a time $t = T/8$ for each situation	$\omega t = (2\pi/T)(T/8) = \pi/4$ A) $x = A \cos\omega t = A \cos\pi/4 = 7.07 \times 10^{-2}$ m B) $x = A \cos\pi/4 = 3.54 \times 10^{-2}$ m C) $\theta = \theta_{max} \cos\pi/4 = 1.11 \times 10^{-1}$ rad D) $\theta = \theta_{max} \cos\pi/4 = 7.40 \times 10^{-2}$ rad
5. Velocity after a time $t = 3T/8$ for each situation	$\omega t = (2\pi/T)(3T/8) = 3\pi/4$ A) $v_x = -\omega A \sin 3\pi/4 = 9.97 \times 10^{-1}$ m/s B) $v_x = -\omega A \sin 3\pi/4 = -7.81 \times 10^{-2}$ m/s C) $\omega_z = -\omega\theta_{max} \sin 3\pi/4$ $= -7.78 \times 10^{-2}$ rad/s D) $\omega_z = -\omega\theta_{max} \sin 3\pi/4$ $= -7.40 \times 10^{-3}$ rad/s
6. Acceleration after a time $t = T/8$ for each situation	$\omega t = (2\pi/T)(T/8) = \pi/4$ A) $a_x = -\omega^2 A \sin\omega t = -14.1$ m/s^2 B) $a_x = -\omega^2 A \sin\omega t = -1.73 \times 10^{-1}$ m/s^2 C) $\alpha_z = -\omega^2 \theta_{max} \sin\omega t$ $= -5.44 \times 10^{-2}$ rad/s^2 D) $\alpha_z = -\omega^2\theta_{max} \sin\omega t$ $= -7.40 \times 10^{-4}$ rad/s^2

Related Text Exercises: Ex. 14-22 through 14-33.

5 Simple Harmonic Motion And The Reference Circle

(Section 14-5)

Review: The projection of the motion of a particle on a reference circle may be used as an aid in obtaining expressions for the position, velocity, and acceleration of an object undergoing SHM. It also makes determining the phase angle quite easy and straight forward.

Shown in Figure 14.10 is a mass on the end of a spring undergoing SHM and the associated reference circle. As a matter of convenience, instead of one reference circle a separate reference circle is shown for the position, velocity and acceleration of the mass.

We are viewing the mass a short time after its release.
At $t = 0$, $x = +A$ and $v = 0$
For these initial conditions, $\phi = 0$

The displacement of the mass in SHM is the same as the x component of the displacement of the particle traveling on the reference circle

$$x = A\cos\theta = A\cos\omega t$$

The velocity of the mass in SHM is the same as the x component of the velocity of the particle traveling on the reference circle

$$v = V_x = -V\sin\theta = -\omega A\sin\omega t$$

The acceleration of the mass in SHM is the same as the x component of the acceleration of the particle traveling on the reference circle

$$a = (a_c)_x = -a_c\cos\theta = -\omega^2 A\cos\omega t$$

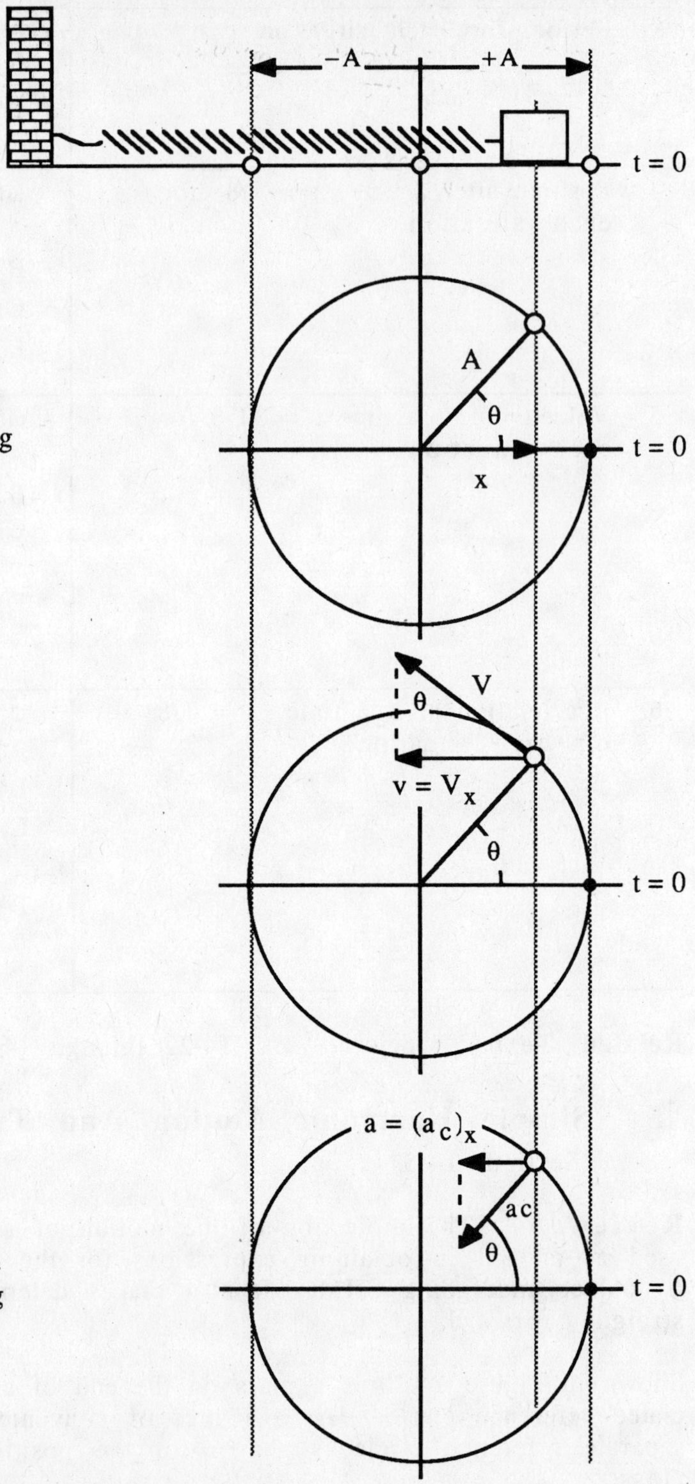

Figure 14.10

In order to show how convenient the reference circle is when determining x, v_x, a_x and ϕ, let's suppose the mass is pulled to the right, released, and the record of time is started (t = 0) the first time the object goes through equilibrium.

We may determine expressions for x, v_x and a_x simply by locating the particle on the reference circle and then writing the x component for its location, velocity and acceleration. Figure 14.11 shows the mass a short time after it is released.

At $t = 0$, $x = 0$ and $v_x = -\omega A$

$$V = \omega A$$
$$a_c = V^2/A = \omega^2 A$$
$$\theta = \omega t$$
$$x = -A_x = -A \sin\omega t$$
$$v_x = -V_x = -\omega A \cos\omega t$$
$$a_x = (a_c)_x = \omega^2 A \sin\omega t$$

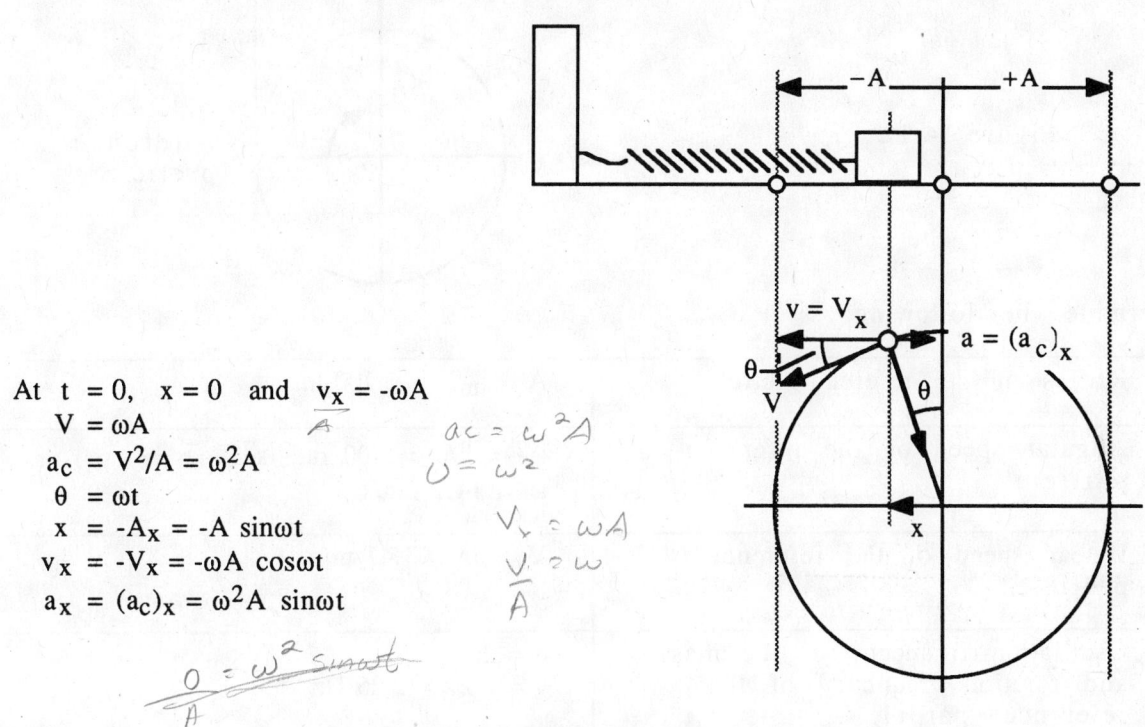

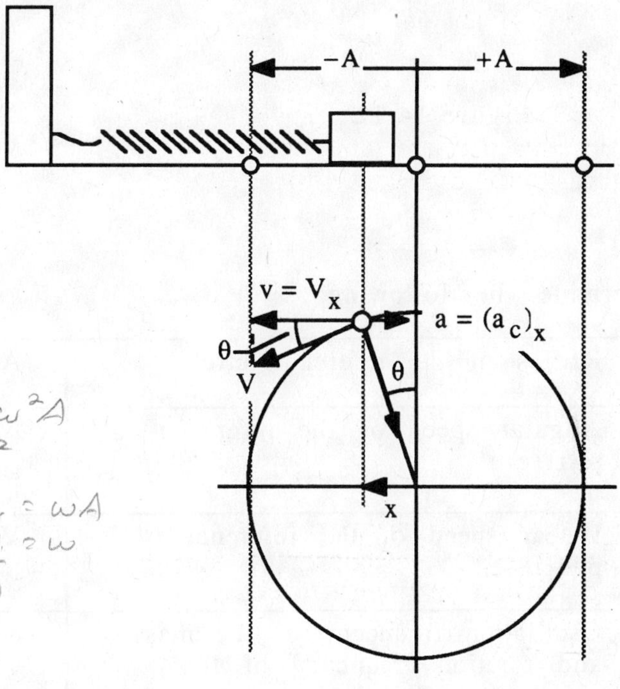

Figure 14.11

Or as an alternate approach we may proceed as follows. We know that if $t = 0$ when $x = +A$ and $v = 0$, then $\phi = 0$ and the equations are

$$x = A\cos\omega t \qquad v_x = -\omega A \sin\omega t \qquad a_x = -\omega^2 A \cos\omega t$$

But if we want $t = 0$ when $x = 0$ and $v_x = -V_{x\,max} = -\omega A$ then according to the reference circle we must choose $\phi = 90.0°$. The equations then become

$$x = A\cos(\omega t + \pi/2) = A[\cos\omega t \cos(\pi/2) - \sin\omega t \sin(\pi/2)] = -A \sin\omega t$$
$$v_x = -\omega A \sin(\omega t + \pi/2) = -\omega A[\sin\omega t \cos(\pi/2) + \sin(\pi/2) \cos\omega t] = -\omega A \cos\omega t$$
$$a_x = -\omega^2 A \cos(\omega t + \pi/2) = -\omega^2 A[\cos\omega t \cos(\pi/2) - \sin\omega t \sin(\pi/2)] = \omega^2 A \sin\omega t$$

Note that the above results agree exactly with the results obtained in Figure 14.11.

Practice: Figure 14.12(a) shows a 2.00-kg mass resting on a horizontal frictionless surface and attached to a spring of constant k = 400 N/m and length L = 0.200 m. Figure 14.12(b) shows the object displaced +0.100 m and released at t = 0.

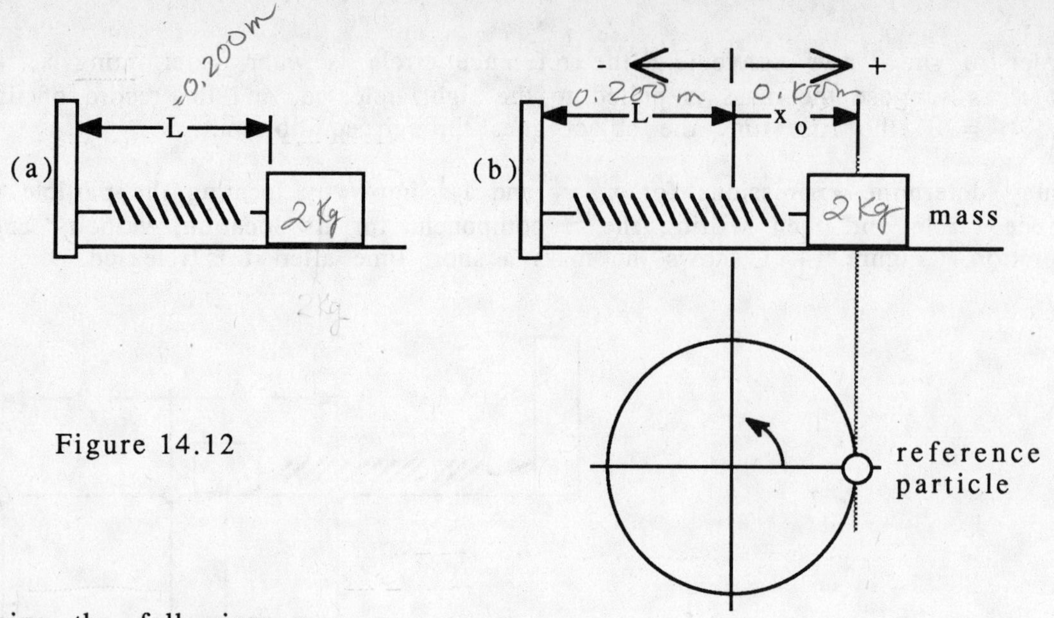

Figure 14.12

Determine the following:

1. Radius of the reference circle	$A = x_{max} = 0.100$ m
2. Angular speed of the reference particle	$\omega^2 = k/m = 200$ rad^2/s^2 $\omega = 14.1$ rad/s
3. Linear speed of the reference particle	$V = \omega A = 1.41$ m/s
4. Oscillation frequency of the mass and rotation frequency of the reference particle	$\omega = 2\pi\nu$ $\nu = \omega/2\pi = 2.25$ Hz
5. Period of oscillation of the mass and period of rotation of the reference particle	$T = 1/\nu = 0.444$ s
6. Magnitude of the reference particle's displacement, velocity and acceleration at any time	$r = A = 0.100$ m $V = \omega A = 1.41$ m/s $a_c = \omega^2 A = 20.0$ m/s
7. x component of the reference particle's displacement, velocity and acceleration at t = 0	$x = +0.100$ m $V_x = 0$ $(a_c)_x = -\omega^2 A$ $= -20.0$ m/s 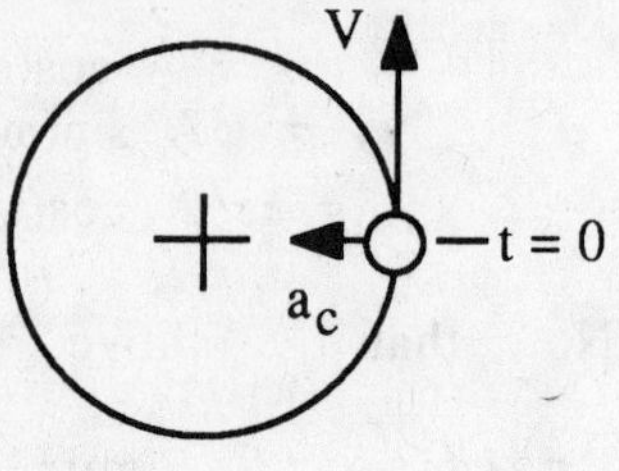
8. Displacement, velocity and acceleration of the mass at t = 0	$x = A \cos\omega t = +A = +0.100$ m $v_x = -\omega A \sin\omega t = 0$ $a_x = -\omega^2 A \cos\omega t = -\omega^2 A = -20.0$ m/s^2

9. x component of the reference particle's displacement, velocity and acceleration at t = T/4	$x = 0$ $V_x = -\omega A$ $= -1.41$ m/s $(a_c)_x = 0$ 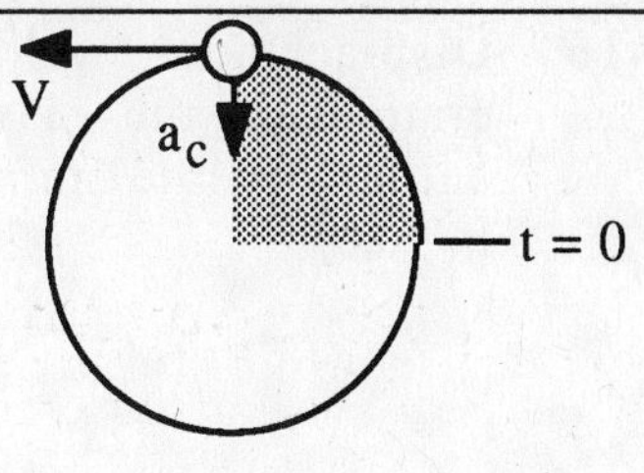
10. Displacement, velocity and acceleration of the mass at t = T/4	$x = A\cos\omega t = A\cos[(2\pi/T)(T/4)] = 0$ $v_x = -\omega A\sin\omega t = -\omega A\sin[(2\pi/T)(T/4)]$ $= -1.41$ m/s $a_x = -\omega^2 A\cos\omega t = -\omega^2 A\cos[(2\pi/T)(T/4)]$ $= 0$
11. x component of the reference particle's displacement, velocity and acceleration at t = T/2	$x = -A = -0.100$ m $V_x = 0$ $(a_c)_x = \omega^2 A$ $= 20.0$ m/s^2 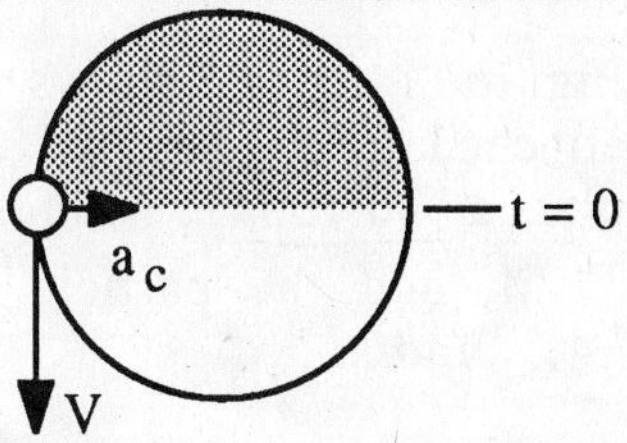
12. Displacement, velocity and acceleration of the mass at t = T/2	$x = A\cos\omega t = A\cos(2\pi/T)(T/2) = -0.100$ m $v_x = -\omega A\sin\omega t = -\omega A\sin(2\pi/T)(T/2) = 0$ $a_x = -\omega^2 A\cos\omega t = -\omega^2 A\cos(2\pi/T)(T/2)$ $= +20.0$ m/s^2
13. x component of the reference particle's displacement, velocity and acceleration at t = 3T/4	$x = 0$ $V_x = +\omega A$ $= +1.41$ m/s $(a_c)_x = 0$ 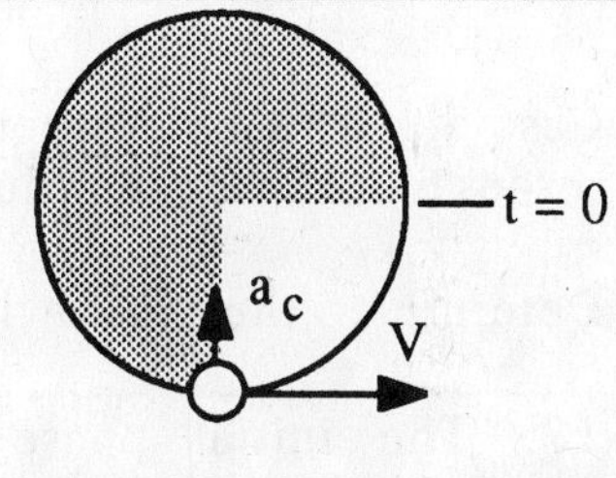
14. Displacement, velocity and acceleration of the mass at t = 3T/4 $\omega t = \omega\left(\frac{3T}{4}\right)$	If $t = 3T/4$, then $\omega t = (2\pi/t)(3T/4) = 3\pi/2$ $x = A\cos\omega t = A\cos 3\pi/2 = 0$ $v_x = -\omega A\sin\omega t = -\omega A\sin 3\pi/2 = 1.41$ m/s $a_x = -\omega^2 A\cos\omega t = -\omega^2 A\cos 3\pi/2 = 0$
15. Displacement, velocity and acceleration of the mass at t = 0.050 s	$x = A\cos\omega t$ $= (0.100\text{ m})\cos[(14.1\text{ rad/s})(0.050\text{ s})]$ $= 0.0762$ m $v_x = -\omega A\sin\omega t$ $= -(1.41\text{ m/s})\sin[(14.1\text{ rad/s})(0.050\text{ s})]$ $= -0.914$ m/s $a_x = -\omega^2 A\cos\omega t$ $= -(20.0\text{ m/s}^2)\cos[(14.1\text{ rad/s})(0.050\text{ s})]$ $= -15.2$ m/s^2

16. Displacement, velocity and acceleration of the mass when the angle on the reference circle is 120°	$\omega t = \theta = 120° = 2\pi$ rad/3 $x = A \cos\omega t = A \cos(2\pi \text{ rad}/3)$ $= -5.00 \times 10^{-2}$ m $v_x = -\omega A \sin\omega t = -\omega A \sin(2\pi \text{ rad}/3)$ $= -1.22$ m/s $a_x = -\omega^2 A \cos\omega t = -\omega^2 A \cos(2\pi \text{ rad}/3)$ $= +10.0 \text{ m/s}^2$

Note: The preceding practice section shows that the motion of an oscillating mass is the same as the x component of the motion of a particle on the reference circle. For example, compare your answer to step 7 with your answer to step 8, step 9 with step 10, 11 with 12, and 13 with 14.

Figure 14.13(a) shows a 2.00-kg object resting on a horizontal frictionless surface and attached to a spring of constant $k = 400$ N/m and length $L = 0.200$ m. The object is displaced +0.100 m (Figure 14.13(b)) and released. After its release, the object undergoes SHM, and the particle on the reference circle has an angular speed of $\omega = \sqrt{k/m} = 14.1$ rad/s.

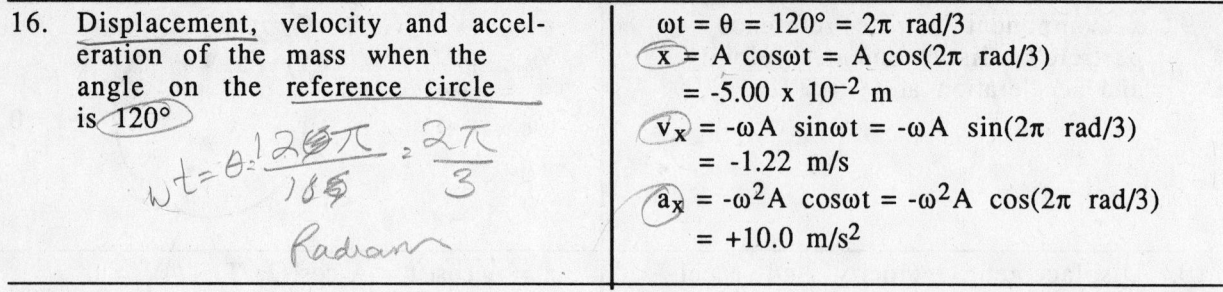

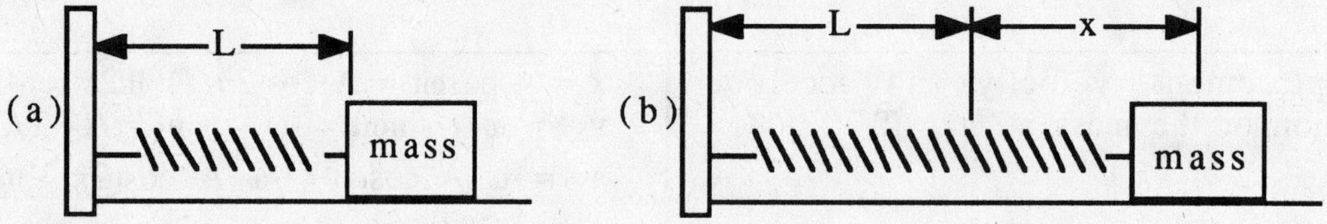

Figure 14.13

Case I. The time record is started (t = 0) the first time the oscillating object goes through the equilibrium position.

Determine the following:

17. The initial phase angle	$\theta_0 = 90° = \pi$ rad/2
18. The x component of the displacement, velocity and acceleration of a particle on the reference circle at t = 0	$x = 0$ $V_x = -\omega A = -(k/A)^{1/2}A = -1.41$ m/s $(a_c)_x = 0$
19. The displacement, velocity and acceleration of the oscillating object at t = 0	$x = A \cos(\omega t + \phi) = A \cos 90° = 0$ $v_x = -\omega A \sin(\omega t + \phi) = -\omega A \sin 90°$ $= -1.41$ m/s $a_x = -\omega^2 A \cos(\omega t + \phi) = -\omega^2 A \cos 90° = 0$
20. The x component of the displacement, velocity and acceleration of the reference particle at t = 3T/4	$x = +A = +0.100$ m $V_x = 0$ $(a_c)_x = -\omega^2 A = -20.0 \text{ m/s}^2$

21. The displacement, velocity and acceleration of the oscillating object at $t = 3T/4$	If $t = 3T/4$, then $\omega t = (2\pi/T)(3T/4) = 3\pi/2$ If $\phi = \pi/2$, then $\omega t + \phi = 2\pi$ $x = A\cos(\omega t + \phi) = A\cos 2\pi = +0.100$ m $v_x = -\omega A\sin(\omega t + \phi) = -\omega A\sin 2\pi = 0$ $a_x = -\omega^2 A\cos(\omega t + \phi) = -\omega^2 A\cos 2\pi$ $= -20.0\ m/s^2$

Case II. The time record is started ($t = 0$) when the oscillating object has its maximum negative displacement.

Determine the following:

22. The initial phase angle	$\phi = 180° = \pi$ rad
23. The x component of the displacement, velocity and acceleration of the reference particle at $t = 0$	$x = -A = -0.100$ m $V_x = 0$ $(a_c)_x = \omega^2 A = +20.0\ m/s^2$
24. The displacement, velocity and acceleration of the oscillating object at $t = 0$	$x = A\cos(\omega t + \phi) = A\cos\pi = -0.100$ m $v_x = -\omega A\sin(\omega t + \phi) = -\omega A\sin\pi = 0$ $a_x = -\omega^2 A\cos(\omega t + \phi) = -\omega^2 A\cos\pi$ $= +20.0\ m/s^2$
25. The x component of the displacement, velocity and acceleration of the reference particle at $t = T/4$	$x = 0$ $V_x = +\omega A = +1.41$ m/s $(a_c)_x = 0$
26. The displacement, velocity and acceleration of the oscillating object at $t = T/4$	If $t = T/4$, then $\omega t = (2\pi/T)(T/4) = \pi/2$ If $\phi = \pi$, then $\omega t + \phi = 3\pi/2$ $x = A\cos(\omega t + \phi) = A\cos(3\pi/2) = 0$ $v_x = -\omega A\sin(\omega t + \phi) = -\omega A\sin(3\pi/2)$ $= 1.41$ m/s $a_x = -\omega^2 A\cos(\omega t + \phi)$ $= -\omega^2 A\cos(3\pi/2) = 0$

Example: The SHM of a 0.200-kg mass attached to a spring and oscillating horizontally on a frictionless surface is described by $x = (0.200\ m)\cos(2\pi t + \pi/2)$. At time $t = 0.125$ s, what is the (a) displacement, (b) velocity and (c) acceleration of the mass?

Given: $x = (0.200\ m)\cos(2\pi t + \pi/2)$ $\quad \omega = 2\pi$ rad/s $\quad A = 0.200$ m
$m = 0.200$ k $\quad \phi = (\pi/2)$ rad $\quad t = 0.125$ s

Note: Values of A, ω and ϕ are determined by comparing the expression for x with $x = A\cos(\omega t + \phi)$.

Determine: The (a) displacement, (b) velocity and (c) acceleration of the mass at $t = 0.125$ s.

Strategy: First, by using $\theta = (\omega t + \phi)$, we can determine the location of the particle on the reference circle and get some idea of what to expect for values of x, v and a. Second, we can insert the value for $(2\pi t + \pi/2)$ and other given quantities into the appropriate expressions to determine x, v and a. Finally, we can verify our work by checking the signs on x, v and a to see if they are correct.

Solution: Let's first determine the location of the particle on the reference circle in order to get some idea of what to expect for values of x, v and a.

$\theta = (\omega t + \phi)$

$\theta = [(2\pi \text{ rad/s})(0.125 \text{ s}) + (\pi/2) \text{ rad}]$

$\theta = (3\pi/4) \text{ rad}$

Figure 14.14

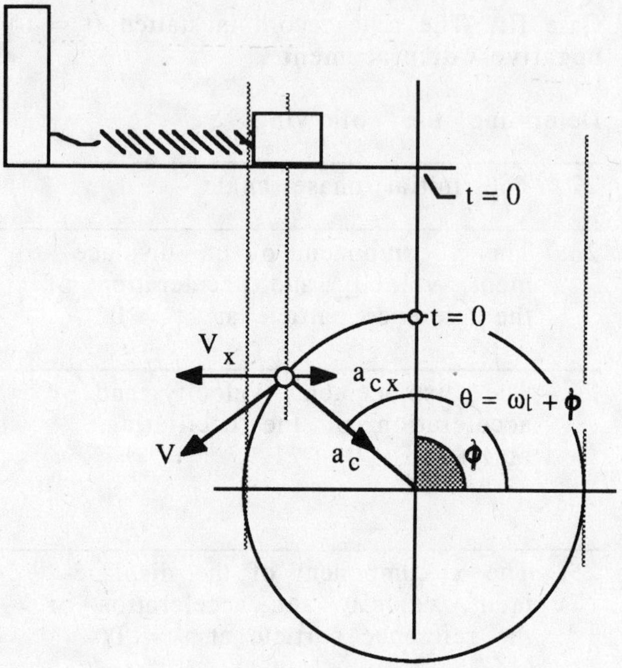

From Figure 14.14, we see that at t = 0.125 s, x and v are negative and a is positive.

Second, let's insert the value for $(2\pi t + \pi/2)$ and other given quantities into expressions for x, v and a.

(a) $x = (0.200 \text{ m}) \cos(2\pi t + \pi/2) = (0.200 \text{ m}) \cos(3\pi/4) = -0.141 \text{ m}$

(b) $v = -\omega A \sin(2\pi t + \pi/2) = -(2\pi)(0.200 \text{ m}) \sin(3\pi/4) = -0.889 \text{ m/s}$

(c) $a = -\omega^2 A \cos(2\pi t + \pi/2) = -(2\pi)^2(0.200 \text{ m}) \cos(3\pi/4) = +5.58 \text{ m/s}^2$

Notice that the signs on x, v and a agree with what we established using the reference circle. Finally as, a further check on our work, let's calculate the acceleration using another expression.

$$a = -\omega^2 x = -(2\pi)^2(-0.141 \text{ m}) = +5.58 \text{ m/s}^2$$

Example: A mass resting on a horizontal frictionless surface is connected to a fixed spring. The mass is displaced +0.200 m from its equilibrium position and released. The first time the oscillating mass goes through the equilibrium position with a positive speed, a record of time is started (i.e., t = 0). If at t = 0.500 s the displacement is x = +0.100 m, what is the period of the oscillating mass?

Given: $A = 0.200$ m $\quad x = +0.100$ m at $t = 0.500$ s $\quad t = 0$ when $x = 0$ m and $v = +\text{max}$

Determine: The period of the oscillating mass

Note: The value for t is derived from the fact that the time record starts the first time the oscillating mass goes through the equilibrium position with a positive speed.

Strategy: Use a reference circle to establish the initial phase angle ϕ. Use the expression for the position of the oscillating mass at any time to determine the value for ωt. Once the value for ωt is known, the value for ω and, subsequently, the value for T can be determined.

Solution: Establish a value for ϕ with the aid of a reference circle.

$\phi = (3\pi/2)$ rad

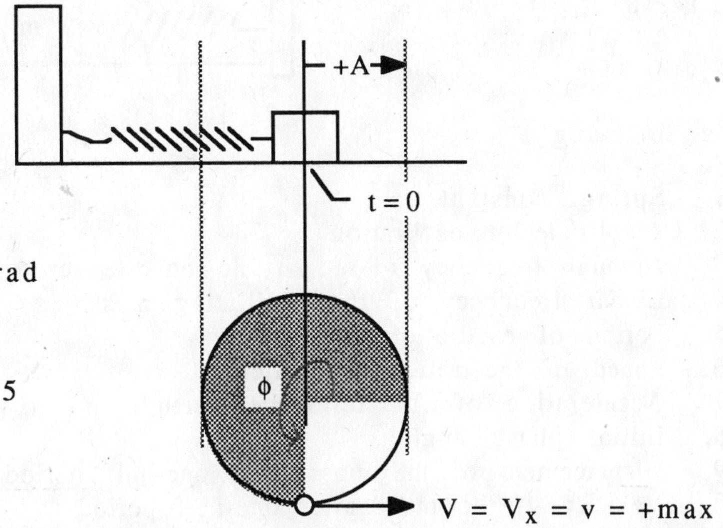

Figure 14.15

Use the expression $x = A\cos(\omega t + \phi)$ to determine a value for $\cos(\omega t + 3\pi/2)$.

$$0.100 \text{ m} = (0.200 \text{ m})\cos(\omega t + 3\pi/2) \qquad \text{or} \qquad \cos(\omega t + 3\pi/2) = +0.500$$

Obtain a value for $(\omega t + 3\pi/2)$ by taking the arccos of this last expression.

$$(\omega t + 3\pi/2) = \pi/3 \qquad \text{or} \qquad 5\pi/3$$

We choose the value $5\pi/3$ because the value $\pi/3$ results in a negative value for ω and hence T. Knowing a value for $(\omega t + \pi/2)$, we can determine a value for ωt and hence T.

$$\omega t + (3\pi/2) \text{ rad} = (5\pi/3) \text{ rad} \qquad \text{or} \qquad \omega t = (\pi/6) \text{ rad}$$

using $t = 0.500$ s, we obtain $\omega = (\pi/3)$ rad/s

using $\omega = 2\pi/T$, we obtain $T = 6.00$ s

$T = \frac{2\pi}{\omega}$

Related Text Exercises: Ex. 14-34 through 14-37.

PRACTICE TEST

Take and grade this practice test. Doing so will allow you to determine any weak spots in your understanding of the concepts taught in this chapter. The following section prescribes what you should study further to strengthen your understanding.

A 2.00-kg mass is attached to the end of a spring. The mass-spring system is placed on a horizontal frictionless surface, as shown in Figure 14.16. A force of 20.0 N is required to stretch the spring 0.100 m. You start the system oscillating by compressing the spring 0.200 m and then releasing it. You start your record of time (t = 0) the first time the oscillating mass goes through equilibrium.

Figure 14.16

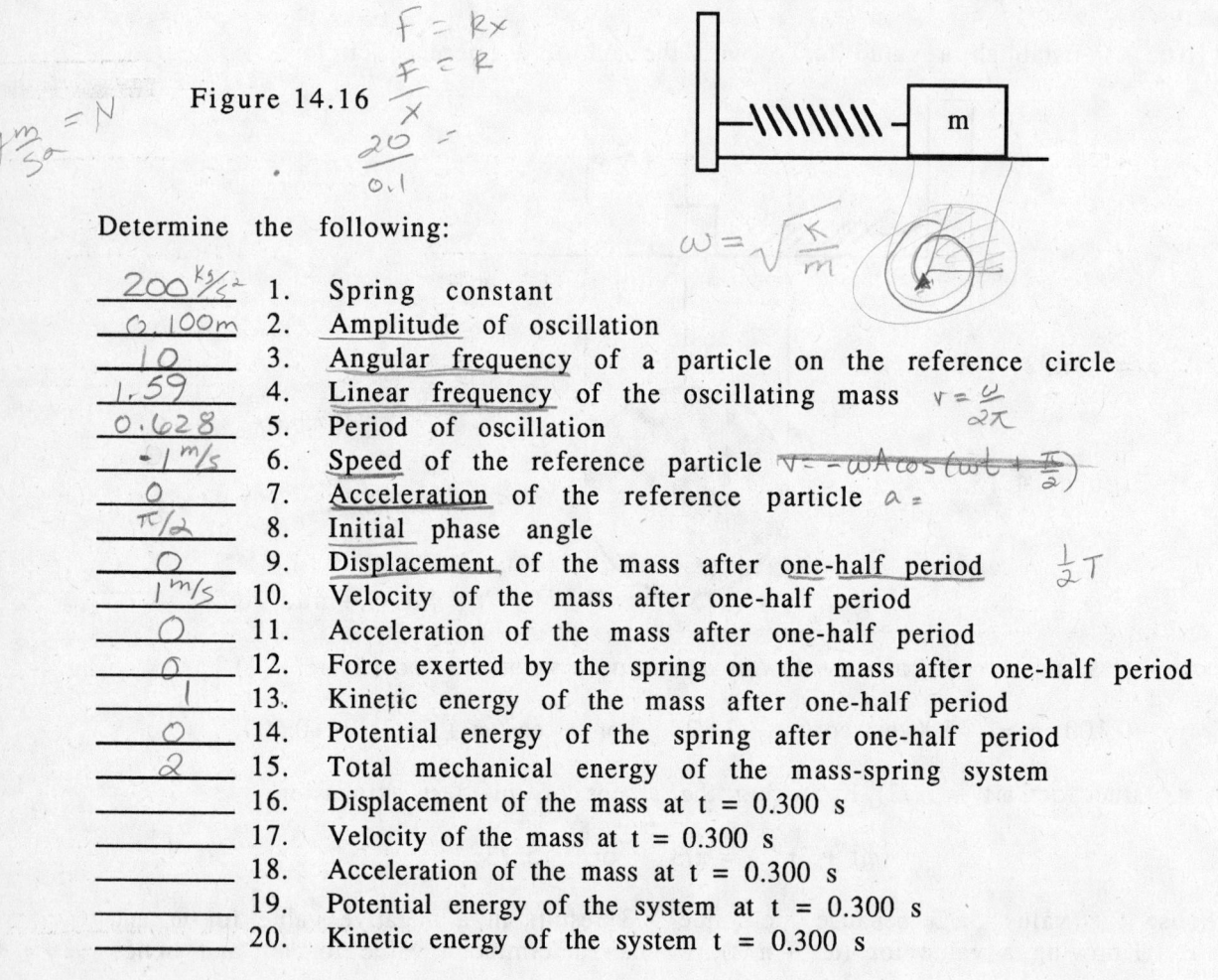

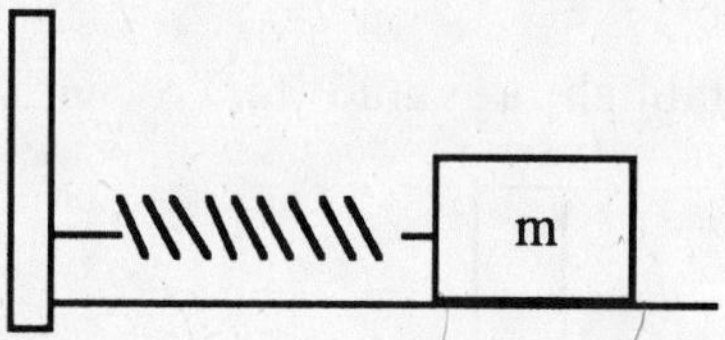

Determine the following:

______ 1. Spring constant
______ 2. Amplitude of oscillation
______ 3. Angular frequency of a particle on the reference circle
______ 4. Linear frequency of the oscillating mass
______ 5. Period of oscillation
______ 6. Speed of the reference particle
______ 7. Acceleration of the reference particle
______ 8. Initial phase angle
______ 9. Displacement of the mass after one-half period
______ 10. Velocity of the mass after one-half period
______ 11. Acceleration of the mass after one-half period
______ 12. Force exerted by the spring on the mass after one-half period
______ 13. Kinetic energy of the mass after one-half period
______ 14. Potential energy of the spring after one-half period
______ 15. Total mechanical energy of the mass-spring system
______ 16. Displacement of the mass at t = 0.300 s
______ 17. Velocity of the mass at t = 0.300 s
______ 18. Acceleration of the mass at t = 0.300 s
______ 19. Potential energy of the system at t = 0.300 s
______ 20. Kinetic energy of the system t = 0.300 s

(See Appendix I for answers.)

PRINCIPAL CONCEPTS AND EQUATIONS PRESCRIPTION

Your score on the practice test is an excellent measure of your understanding of the chapter. You should now use the following chart to write your own prescription for dealing with any weaknesses the practice test points out. Look down the leftmost column to the number of the question(s) you answered incorrectly, reading across that row you will find the concept and/or equation of concern, the section(s) of the study guide you should return to for further study, and some suggested text exercises which you should work to gain additional experience.

Practice Test Questions	Concepts and Equations	Prescription	
		Principal Concepts	Text Exercises
1	Spring constant: $F = -kx$	1	14-24
2	Amplitude of oscillation	1	14-4, 5
3	Angular frequency: $\omega = (k/m)^{1/2}$	1	14-10, 11
4	Linear frequency: $\nu = \omega/2\pi$	1	14-1, 11
5	Period of oscillation: $T = 1/\nu$	1	14-1, 11
6	Speed of reference particle: $v = \omega A$	5	14-34, 35
7	Acceleration of reference particle: $a = \omega^2 A$	5	14-34, 35
8	Phase angle	1	14-3, 8
9	Describing SHM (displacement): $x = A\cos(\omega t + \phi)$	1	14-3, 4
10	Describing SHM (velocity): $v = -\omega A \sin(\omega t + \phi)$	1	14-5, 7
11	Describing SHM (acceleration): $a = -\omega^2 x$	1	14-6, 8
12	Condition for SHM (force): $F = -kx$	2	14-24
13	Kinetic energy in SHM: $K = mv^2/2$	3	14-19, 21
14	Potential energy in SHM: $U = kx^2/2$	3	14-19, 21
15	Total Mechanical energy in SHM: $E = K + U$	3	14-17, 18
16	Describing SHM (displacement): $x = A\cos(\omega t + \phi)$	1	14-3, 4
17	Describing SHM (velocity): $v = -\omega A \sin(\omega t + \phi)$	1	14-5, 7
18	Describing SHM (acceleration): $a = -\omega^2 x$	1	14-6, 8
19	Potential energy in SHM: $U = kx^2/2$	3	14-19, 21
20	Kinetic energy in SHM: $K = mv^2/2$	3	14-19, 21

15 SOLIDS AND FLUIDS

RECALL FROM PREVIOUS CHAPTERS

Your success with this chapter will be affected by your understanding of previous chapters. However, you should be able to proceed with this chapter without reviewing any previous concepts and equations.

NEW IDEAS IN THIS CHAPTER

Concepts and equations introduced	Text Section	Study Guide Page
Tensile or compressive stress: $\sigma_t = F_n/A$	15-1	15-1
Shear stress: $\sigma_s = F_p/A$	15-1	15-1
Tensile or compressive strain: $\varepsilon_t = \Delta L/L$	15-1	15-1
Shear strain: $\varepsilon_s = \Delta x/H$	15-1	15-1
Pressure: $P = F_n/A$	15-1	15-1
Elastic Moduli:	15-1	15-1
Young's Modulus: $Y = \sigma_t/\varepsilon_t = (F_n/A)/(\Delta L/L)$		
Shear Modulus: $S = \sigma_s/\varepsilon_s = (F_p/A)/\Delta x/H)$		
Bulk Modulus: $B = \Delta P/(\Delta V/V)$		
Mass density: $\rho = m/V$	15-2	15-8
Pressure due to a fluid: $P = \rho g h$	15-3	15-10
Absolute pressure: $P_{abs} = P_o + \rho g h$	15-3	15-10
Archimedes principle: $F_B = W_{fd} = m_{fd}g = \rho_f V_{fd} g$	15-4	15-15
Continuity equation: $A_1 v_1 = A_2 v_2$	15-5	15-19
Bernoulli's equation: $P + \rho g h + \rho v^2/2 = \text{constant}$	15-5	15-19

PRINCIPAL CONCEPTS AND EQUATIONS

1 Stress, Strain and Elastic Moduli (Section 15-1)

Review: An object may be subjected to a tensile stress, compressive stress, shear stress, or uniform pressure as shown in Figures 15.1 through 15.4.

Tensile stress
$\sigma_t = F_n/A = F_n/HW$
F_n = force normal to surface
$A = HW$ = area to which F_n is applied

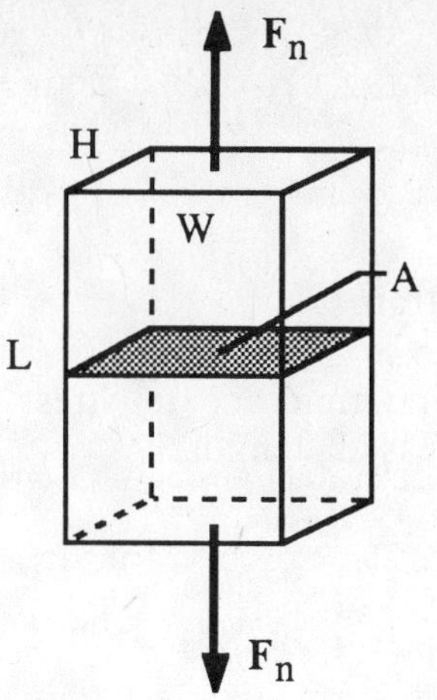

Figure 15.1

Compressive stress
$\sigma_t = F_n/A = F_n/HL$
F_n = force normal to surface
$A = HL$ = area to which F_n is applied

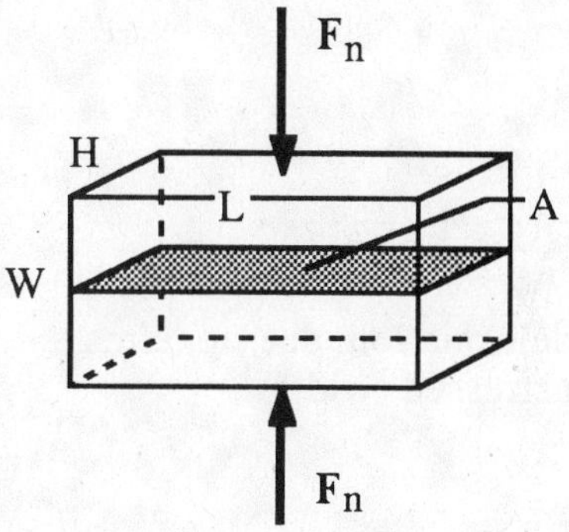

Figure 15.2

Shear stress
$\sigma_s = F_p/A = F_p/WL$
F_p = Force parallel to surface
$A = WL$ = area to which F_p is applied

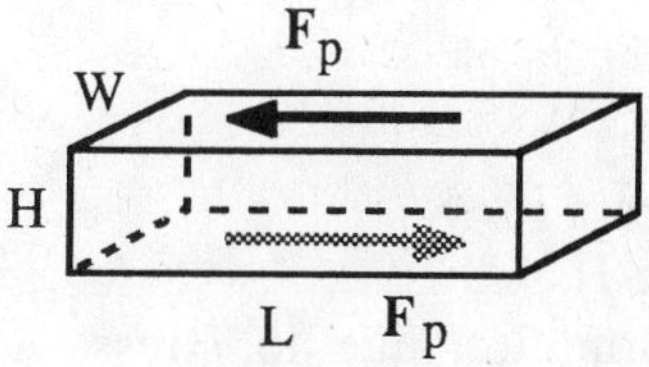

Figure 15.3

Uniform pressure
$P = F_n/A$
F_n = force normal to the surface
A = area to which F_n is applied

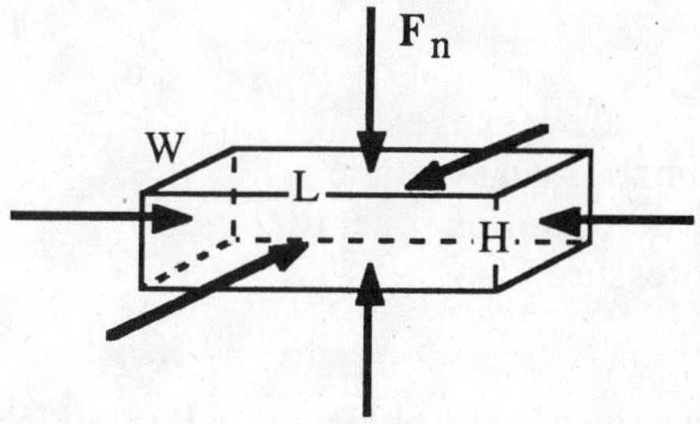

Figure 15.4

The strain of an object is a measure of its deformation under stress. An object subjected to a tensile, compressive, or shear stress experiences a tensile, compressive, or shear strain as shown in Figures 15.5 through 15.8.

Tensile strain
$\varepsilon_t = \Delta L/L$
ΔL = deformation due to stress
L = unstressed length

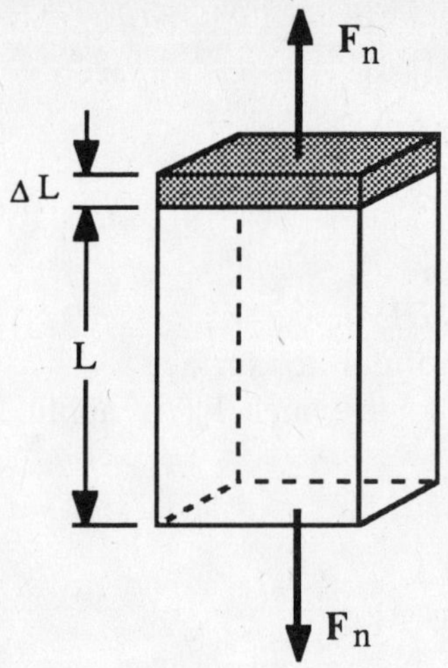

Figure 15.5

Compressive strain
$\varepsilon_c = \Delta W/W$
ΔW = deformation due to stress
W = unstressed width

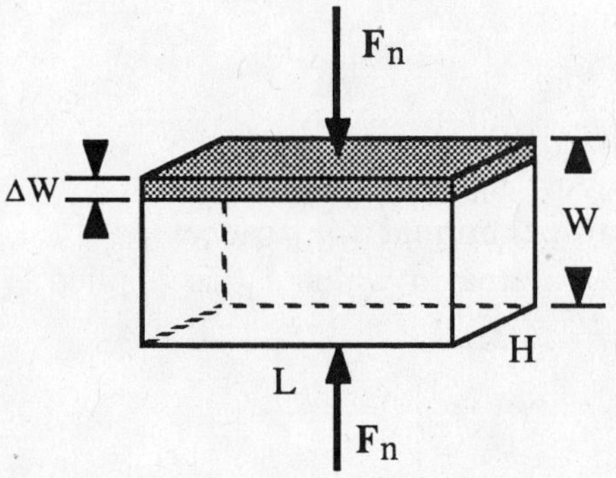

Figure 15.6

Shear strain
$\varepsilon_s = \phi \cong \Delta x/H$
Δx = deformation due to stress

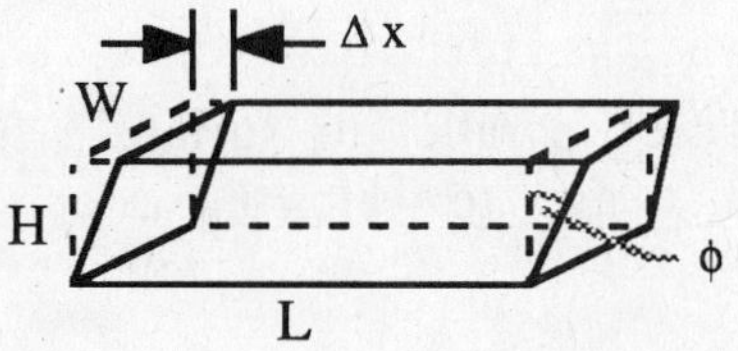

Figure 15.7

Uniform pressure
V = unstressed volume
ΔV = deformation due to stress
P = pressure

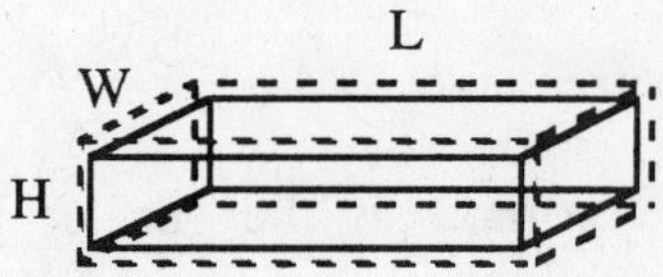

Figure 15.8

The strain of an object (i.e. its deformation) is directly proportional to the applied stress. For a tensile or compressive stress the proportionality constant is the reciprocal of Young's Modulus (Y), for a shear stress it is the reciprocal of the shear modulus (S), and for uniform pressure it is the reciprocal of the Bulk Modulus (B). This maybe written as follows:

$\varepsilon_t \propto \sigma_t$, $\quad \varepsilon_t = (1/Y)\sigma_t \quad$ or $\quad Y = \sigma_t/\varepsilon_t = (F_n/A)/(\Delta L/L)$

$\varepsilon_c \propto \sigma_c$, $\quad \varepsilon_c = (1/Y)\sigma_c \quad$ or $\quad Y = \sigma_c/\varepsilon_c = (F_n/A)/(\Delta L/L)$

$\varepsilon_s \propto \sigma_s$,	$\varepsilon_s = (1/S)\sigma_s$	or	$S = \sigma_s/\varepsilon_s = (F_p/A)/(\Delta x/L)$
$\Delta V/V \propto \Delta P$,	$\Delta V/V = (1/B)\Delta P$	or	$B = \Delta P/(\Delta V/V)$

Practice: The dimensions of the rectangular parallelapiped used in Figures 15.1 through 15.8 are

$$L = 0.200 \text{ m} \qquad W = 0.100 \text{ m} \qquad H = 0.0500 \text{ m}$$

The magnitude of the force used is $F_n = F_p = 400$ N.

Determine the following:

1. Tensile stress for the object in Figure 15.1	$\sigma_t = F_n/A = F_n/HW = 8.00 \times 10^4 \text{ N/m}^2$
2. Compressive stress for the object in Figure 15.2	$\sigma_c = F_n/A = F_n/HL = 4.00 \times 10^4 \text{ N/m}^2$
3. Shear stress for the object in Figure 15.3	$\sigma_s = F_p/A = F_p/WL = 2.00 \times 10^4 \text{ N/m}^2$
4. Pressure on the top surface of Figure 15.4	$P = F_n/A = F_n/WL = 2.00 \times 10^4 \text{ N/m}^2$

As a result of the stress, the following deformation occurs

$\Delta L = 8.00 \times 10^{-8}$ m for Figure 15.5 $\qquad \Delta x = 2.50 \times 10^{-8}$ m for Figure 15.7

$\Delta W = 3.00 \times 10^{-8}$ m for Figure 15.6 $\qquad \Delta V = 2.00\%V$ for Figure 15.8

Determine the following:

5. Tensile strain for the object in Figure 15.5	$\varepsilon_t = \Delta L/L = 4.00 \times 10^{-7}$
6. Compressive strain for the object in Figure 15.6	$\varepsilon_c = \Delta W/W = 3.00 \times 10^{-7}$
7. Shear strain for the object in Figure 15.7	$\varepsilon_s = \Delta x/H = 5.00 \times 10^{-7}$
8. Young's Modulus for the object in Figure 15.5	$Y = \sigma_t/\varepsilon_t = 2.00 \times 10^{11} \text{ N/m}^2$
9. Young's Modulus for the object in Figure 15.6	$Y = \sigma_c/\varepsilon_c = 1.33 \times 10^{11} \text{ N/m}^2$
10. Shear Modulus for the object in Figure 15.7	$S = \sigma_s/\varepsilon_s = 4.00 \times 10^{10} \text{ N/m}^2$

11. Bulk Modulus for the object in Figure 15.8 if the deformation is the result of an increase in pressure of $\Delta P = 1.60 \times 10^9$ N/m^2	$V = 1.00 \times 10^{-3}$ m^3 $\Delta V = 0.0200V = 2.00 \times 10^{-5}$ m^3 $B = \Delta P/(\Delta V/V) = 8.00 \times 10^{10}$ N/m^2

Example: The ultimate tensile strength of steel used to make elevator cable is 5.50×10^8 N/m^2. Find the maximum upward acceleration that can be given to a 2.50×10^3 kg elevator supported by a 1.00 cm radius cable while maintaining a safety factor of 6. In order to have a safety factor of six, the cable must be able to withstand a force six times the load.

Given:

$m = 2.50 \times 10^3$ kg = mass of elevator
$r = 1.00 \times 10^{-2}$ m = radius of cable
$(F/A)_{max} = 5.50 \times 10^8$ N/m^2 = ultimate tensile strength
$f = 6.00$ = safety factor

Determine: The maximum upward acceleration that can be given to the elevator while maintaining a safety factor of 6.

Strategy: Knowing the ultimate tensile strength and the safety factor, we can determine the maximum tensile stress allowed for the cable. Knowing this stress and the cross-sectional area, we can determine the maximum tension allowed. Finally, we can use this tension and the mass of the elevator to determine the maximum allowed acceleration.

Solution: Maximum tensile stress allowed in the cable is

$$\text{max stress allowed} = \frac{\text{ultimate stress}}{\text{safety factor}} = 9.17 \times 10^7 \text{ N/m}^2$$

$$\text{max tension allowed} = (\text{max stress allowed})(\pi r^2) = 2.88 \times 10^4 \text{ N}$$

Using the free-body force diagram shown in Figure 15.9, we can write an expression for F_{net} and then determine a.

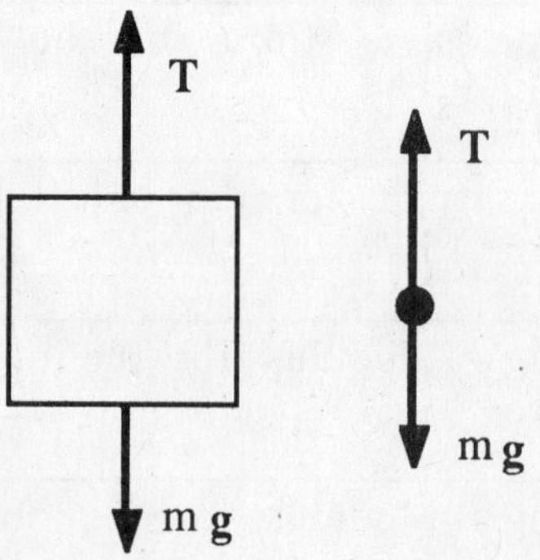

Figure 15.9

The net force on the elevator is

$$F_{net} = T - mg = ma \quad \text{or} \quad a = (T - mg)/m = 1.72 \text{ m/s}^2$$

Example: A contractor wishes to support a sagging floor with a steel pipe. The ultimate compressive strength of steel is 4.10×10^8 N/m^2. The pipe has an outside radius of 12.0 cm and inside radius of 10.0 cm. Determine the maximum load this pipe can support before buckling.

Given: $S_{max} = 4.10 \times 10^8$ N/m^2 = maximum compressive stress
$r_o = 12.0 \times 10^{-2}$ m = outside radius of pipe
$r_i = 10.0 \times 10^{-2}$ m = inside radius of pipe

Determine: The maximum load the pipe can support before buckling.

Strategy: Using r_o and r_i, we can determine the cross-sectional area of the pipe. Knowing this area and the maximum compressive stress of the pipe, we can determine the maximum load the pipe can support before buckling.

Solution: The cross-sectional area of the pipe is

$$A = \pi(r_o^2 - r_i^2) = 1.38 \times 10^{-2} \text{ m}^2$$

The maximum load the pipe can support is $F = S_{max}A = 5.66 \times 10^6$ N.

Example: A punching press that exerts a 2.50×10^4 N force is employed to punch circular holes 1.00 cm in diameter in a sheet of metal. If the metal can withstand a shear stress of 5.00×10^8 N/m^2, find the maximum thickness of a sheet of this metal that can be used.

Given and Diagram:

$F = 2.50 \times 10^4$ N
$d = 1.00 \times 10^{-2}$ m
shear stress = 5.00×10^8 N/m^2

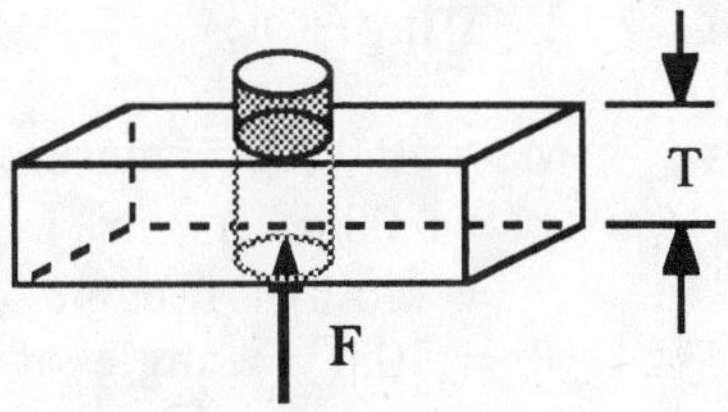

Figure 15.10

Determine: The maximum thickness T of a sheet of metal through which the press can punch a 1.00-cm-diameter hole.

Strategy: Using the shearing stress the metal can endure and the force exerted by the press, we can determine the maximum area the press can shear. Knowing this area and the diameter of the punch, we can determine the thickness T.

Solution: The area the press can shear is

$$A = F/\text{shear stress} = 5.00 \times 10^{-5} \text{ m}^2$$

The thickness of the metal may be determined by

$$A = \pi dT \quad \text{or} \quad T = A/\pi d = 1.59 \times 10^{-3} \text{ m}$$

Example: A steel wire 1.00 m long supports a load and is stretched 1.00×10^{-2} m. What is the strain in this wire, and how much would a 2.00-m steel wire stretch under the same load?

Given: $L = 1.00$ m $\quad \Delta L_T = 1.00 \times 10^{-2}$ m $\quad L' = 2.00$ m

Determine: The strain for the 1.00-m wire and the stretch $\Delta L'_T$ for the 2.00-m wire under the same load.

Strategy: Knowing the length L of the wire and the change in length ΔL_T, we can determine the strain. Knowing that the deformation is directly proportional to the original length (if the load is constant), we can determine $\Delta L'_T$.

Solution: The strain in the 1.00-m long wire is

$$\text{strain} = \Delta L_T/L = 1.00 \times 10^{-2}$$

For a constant load, the deformation is directly proportional to the original length. Consequently, if the length doubles, so does the stretch.

Since $L' = 2L$, then $\Delta L'_T = 2\Delta L_T = 2.00 \times 10^{-2}$ m.

Example: A 50.0-kg traffic light is suspended from a 5.00×10^{-3} m radius steal cable, as shown in Figure 15.11. The cable hangs 10.0° below the horizontal due to the weight of the light. Determine the fractional change in length of the cable.

Given and Diagram:

$M = 50.0$ kg = mass of light
$r = 5.00 \times 10^{-3}$ m
= radius of cable
$\theta = 10.0°$ = angle of cable
$Y_{steel} = 21.0 \times 10^{10}$ N/m^2
= Young's Modulus

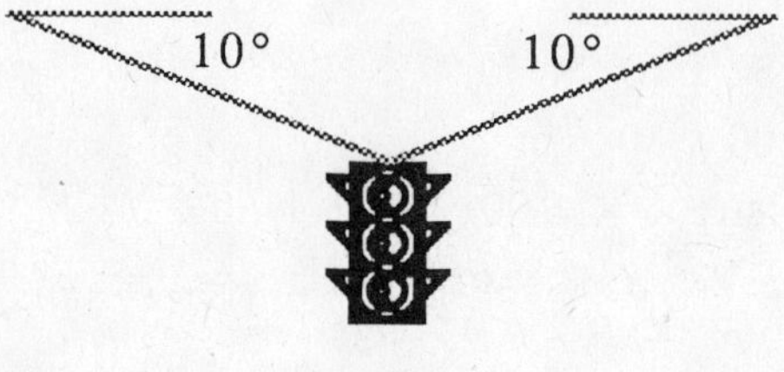

Figure 15.11

Determine: The fractional change in length of the cable.

Strategy: Construct a free-body diagram, choose a coordinate system and resolve the forces into components. Knowing M, g and θ, we can determine the tension in the cable. Knowing the tension in the cable and the radius of the cable, we can determine the tensile stress. Knowing the tensile stress and Young's modulus, we can determine the tensile strain. Finally we can recognize that the tensile strain is the fractional change in length of the cable.

Solution: First let's construct a free-body diagram.

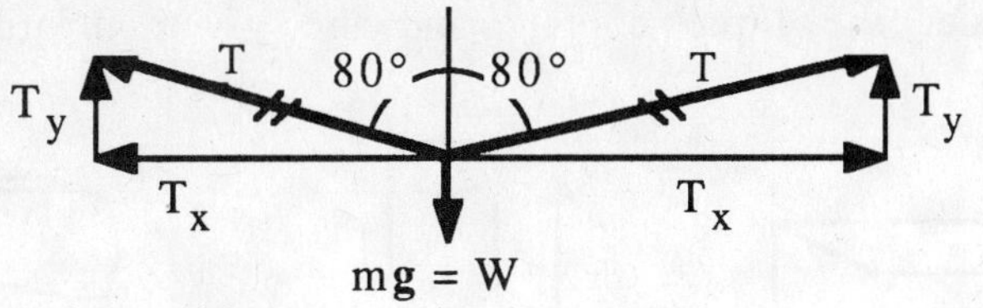

Determine the tension in the cable.

$$2T_y = 2T \sin 10.0° = W = Mg \quad \text{or} \quad T = (Mg/2) \sin 10.0° = 42.5 \text{ N}$$

The area of the cable is $A = \pi r^2 = 7.85 \times 10^{-5}$ m^2.

The tensile stress in the cable is $\sigma_t = T/A = 5.41 \times 10^5$ N/m^2.

The tensile strain in the cable is $\varepsilon_t = \sigma_t/Y = 2.58 \times 10^{-6}$.

Recall that strain is defined as $\varepsilon_t = \Delta L/L$, hence the fractional change in length of the cable is $\Delta L/L = 2.58 \times 10^{-6}$.

Example: Figure 15.12 shows a rectangular slab of the jello subjected to a shearing force at its upper surface. The force of static friction acting on the bottom surface is sufficient to keep the jello stationary. Calculate the shear stress, shear strain and shear modulus for jello.

$L = 1.00 \times 10^{-1}$ m $\Delta x = 1.00 \times 10^{-2}$ m
$W = 5.00 \times 10^{-2}$ m $F = 5.00 \times 10^{-1}$ N
$H = 3.00 \times 10^{-2}$ m

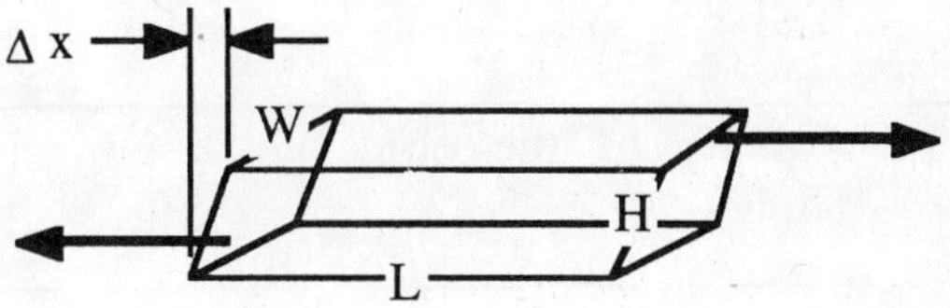

Figure 15.12

Given: L, W, H, F and Δx

Determine: Shear stress, shear strain and shear modulus.

Strategy: The shearing force F and the area LW can be used to obtain the shear stress. The deformation Δx due to the shearing force and the height of the slab can be used to obtain the shear strain. Finally, the shear modulus can be obtained from the stress and strain.

Solution: Shear stress = F/A = F/LW = 1.00×10^2 N/m^2
Shear strain = $\Delta x/H = 1.00 \times 10^{-1}$
Shear modulus = stress/strain = 1.00×10^3 N/m^2

Related Text Exercises: Ex. 15-1 through 15-8.

2 Mass Density (Section 15-2)

Review: The mass density ρ of a homogeneous substance is the mass of the material per unit volume.

$$\rho = m/V$$

Practice: Consider the three objects and the given information in Figure 15.13.

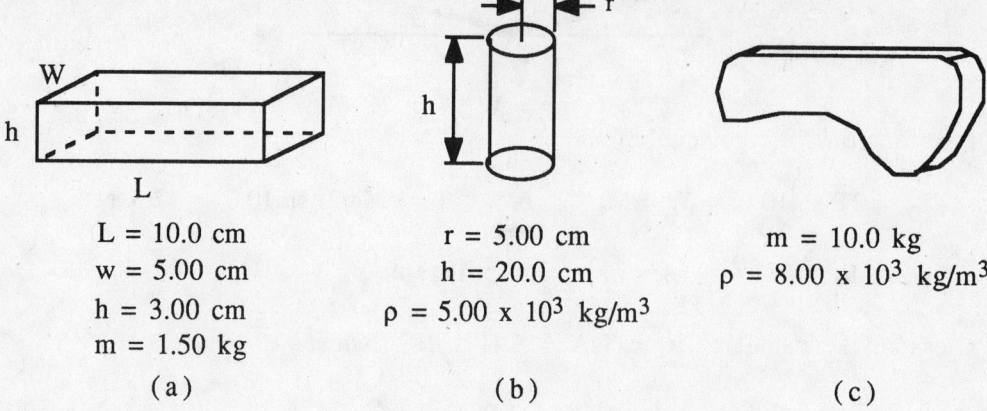

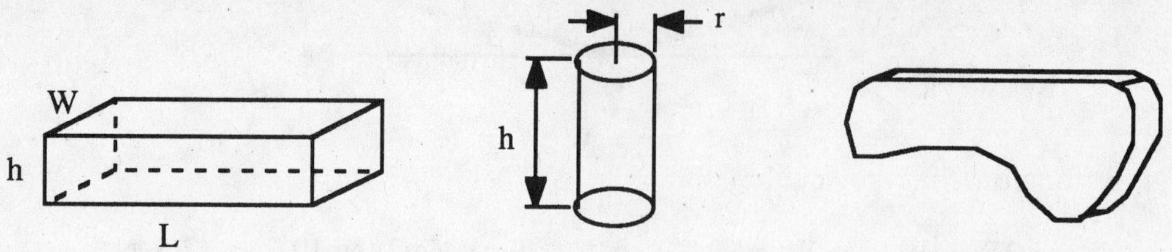

(a)	(b)	(c)
L = 10.0 cm w = 5.00 cm h = 3.00 cm m = 1.50 kg	r = 5.00 cm h = 20.0 cm $\rho = 5.00 \times 10^3$ kg/m^3	m = 10.0 kg $\rho = 8.00 \times 10^3$ kg/m^3

Figure 15.13

Determine the following:

1. Mass density of the object in Figure 15.13(a)	$V = Lwh = 1.50 \times 10^{-4}$ m^3 $\rho = m/V = 1.00 \times 10^4$ kg/m^3
2. Mass of the object in Figure 15.13(b)	$V = \pi r^2 h = 1.57 \times 10^{-3}$ m^3 $m = \rho V = 7.85$ kg
3. Volume of the object in Figure 15.13(c)	$V = m/\rho = 1.25 \times 10^{-3}$ m^3

Example: Which weighs more, 2.00 m^3 of brass or 1.50 m^3 of mercury?

Given: $V_b = 2.00$ m^3 = volume of brass
$V_{Hg} = 1.50$ m^3 = volume of mercury
We also know the density of brass and mercury (see Table in text)
$\rho_b = 8.67 \times 10^3$ kg/m^3
$\rho_{Hg} = 13.6 \times 10^3$ kg/m^3

Determine: Which of these two weighs more?

Strategy: Knowing the volume and density of each sample, we can determine the mass and hence the weight of each.

Solution: $w_b = m_b g = \rho_b V_b g = 1.70 \times 10^5$ N
$w_{Hg} = M_{Hg} g = \rho_{Hg} V_{Hg} g = 2.00 \times 10^5$ N

Related Text Exercises: Ex. 15-9 through 15-12.

3 Fluid Pressure (Section 15-3)

Review: Pressure is the magnitude of the force per unit area.

$$P = F/A$$

The pressure due to a height h of a fluid or at a depth h in a fluid of mass density ρ may be written as

$$P = \rho g h$$

The absolute pressure at a depth h in a liquid is equal to the sum of the pressure due to the liquid ($\rho g h$) and atmospheric pressure (P_A).

$$P_{abs} = P_A + \rho g h$$

Most pressure-measuring devices (called gauges) are calibrated to read zero when the pressure is equal to atmospheric pressure. That is, gauges indicate the pressure difference between inside and outside the container. Consequently, the absolute pressure inside the container is given by

$$P_{abs} = P_A + P_G$$

Pascal's principle states that when a change in pressure is applied to an enclosed fluid, the change is transmitted undiminished to every point in the fluid and to the walls of the container.

Practice: Figure 15.14(a) shows a cylindrical container equipped with a movable air-tight piston lid and a U-tube mercury manometer. Figure 15.14(b) shows that when the cylinder is half-filled with an unknown liquid, the difference between the level of mercury in the two sides of the U-tube is 1.47 cm. Figure 15.14(c) shows the same situation as Figure 15.14(b), except that the piston lid (m_1 = 1.00 kg) has been installed and a 9.00-kg mass (m_2) has been placed on it.

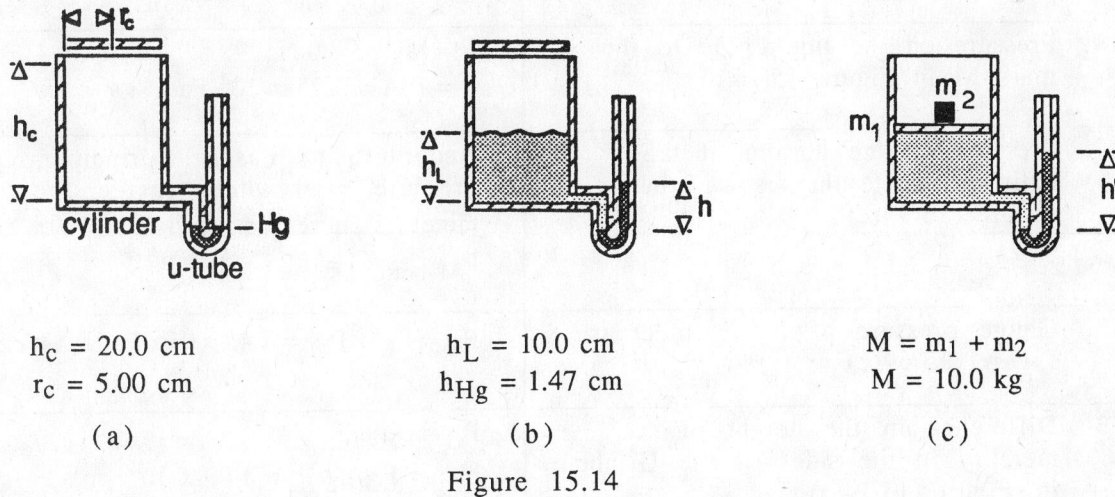

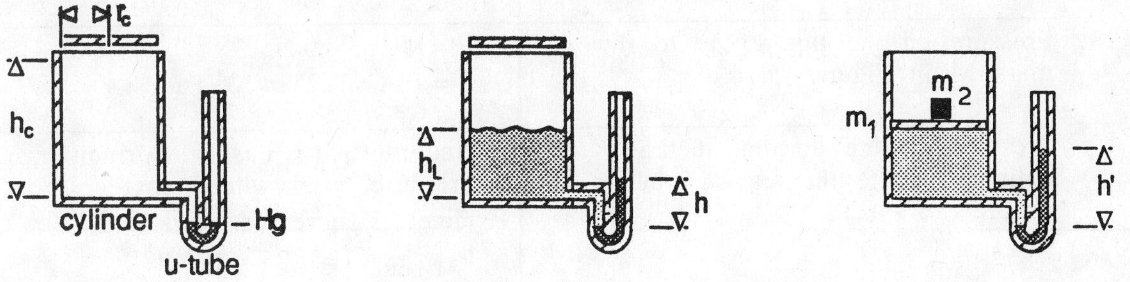

h_c = 20.0 cm
r_c = 5.00 cm

(a)

h_L = 10.0 cm
h_{Hg} = 1.47 cm

(b)

$M = m_1 + m_2$
M = 10.0 kg

(c)

Figure 15.14

Determine the following:

1. Gauge pressure of the cylinder in Figure 15.14(a)	$P_G = 0$
2. Absolute pressure of the cylinder in Figure 15.14(a)	$P_{abs} = P_A = 1.01 \times 10^5$ Pa
3. Force on the inside bottom of the cylinder in Figure 15.14(a)	$F = P_A A = P_A \pi r_c^2 = 793$ N
4. Gauge pressure of the cylinder in Figure 15.14(b)	$P_G = \rho_{Hg} g h_{Hg} = 1.96 \times 10^3$ Pa
5. Pressure at the bottom of the cylinder due to the liquid in Figure 15.14(b)	$P_{L\ bottom} = P_G = 1.96 \times 10^3$ Pa
6. Pressure due to the liquid at a depth of 5.00 cm in Figure 15.14(b)	Let h represent the depth $P_L = \rho_L g h$ since $h = h_L/2$ $P_h = P_{L\ bottom}/2 = 9.8 \times 10^2$ Pa
7. Density of the liquid in the cylinder.	$\rho_L = P_{L\ bottom}/g h_L$ $= 2.00 \times 10^3$ kg/m^3
8. Force on the bottom of the cylinder due to the liquid in Figure 15.14(b)	$F_L = P_{L\ bottom} A = 15.4$ N or $F_L = w_L = m_L g = \rho_L V_L g$ $= \rho_L \pi r^2 h_L g = 15.4$ N
9. Absolute pressure at the bottom of cylinder in Figure 15.14(b)	$P_{abs} = P_{L\ bottom} + P_A$ $= 1.96 \times 10^3$ Pa $+ 1.0 \times 10^5$ Pa $= 1.02 \times 10^5$ Pa
10. Pressure on the liquid due to the mass M in Figure 15.14(c)	$F = Mg = 98.0$ N $P = F/A = 1.25 \times 10^4$ Pa
11. Pressure on the bottom of the cylinder due to the mass M in Figure 15.14(c)	According to Pascal's principle, the pressure everywhere inside the container is increased by 1.25×10^4 Pa. $P_{M\ bottom} = 1.25 \times 10^4$ Pa
12. Gauge pressure of the cylinder in Figure 15.14(c)	$P_G = P_L + P_M = 1.45 \times 10^4$ Pa
13. Difference in the height of the mercury in the sides of the U-tube in Figure 15.14(c)	$P_G = \rho_{Hg} g h_{Hg}$ $h_{Hg} = P_G/\rho_{Hg} g = 1.09 \times 10^{-1}$ m

Example: The hydraulic system in Figure 15.15 is used to test the ultimate compressive stress of various samples. Stress is created on the sample by setting weights on piston 1.

Figure 15.5

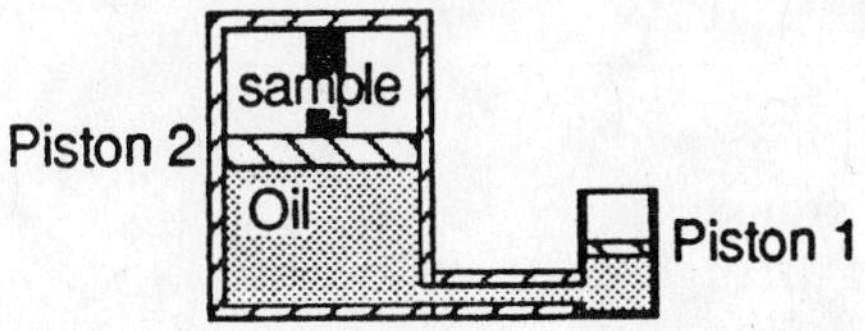

The samples are cylinders with a radius of 1.00 cm and a length of 10.0 cm. Determine the mass that must be placed on piston 1 in order to compress an aluminum sample by 1.00×10^{-5} m.

Given:

$A_1 = 1.00 \times 10^{-3}\ m^2$ = area of piston 1

$A_2 = 1.00 \times 10^{-1}\ m^2$ = area of piston 2

$Y_{Al} = 7.00 \times 10^{10}\ N/m^2$ = Young's Modulus for compression of Aluminum

$L = 1.00 \times 10^{-1}\ m$ = length of sample

$\Delta L = 1.00 \times 10^{-5}\ m$ = compression of the sample

$r = 1.00 \times 10^{-2}\ m$ = radius of sample

Determine: Mass to be placed on piston 1 to compress the sample by 1.00×10^{-5} m.

Strategy: Knowing Y, ΔL, L and r, we can determine the force (F_s) needed to compress the aluminum sample. Knowing F_s and the area of piston 2 (A_2), we can determine the necessary fluid pressure. From the fluid pressure and the area of piston 1 (A_1), we can determine the force and hence the mass needed at piston 1.

Solution: The expression for Young's modulus is $Y = (F/A)/(\Delta L/L)$. This may be solved for the force (F_s) needed to compress the sample.

$$F_s = YA(\Delta L/L) = Y(\pi r^2)\Delta L/L = 2.20 \times 10^3\ N$$

Since this force is to be supplied by piston 2, we obtain

$$F_2 = F_s = 2.20 \times 10^3\ N \quad \text{and} \quad P_2 = F_2/A_2 = 2.20 \times 10^4\ N/m^2$$

According to Pascal's law, $P_1 = P_2$. By the definition of pressure, $P_1 = F_1/A$. Combining these we obtain

$$F_1 = P_2A_1 = (2.20 \times 10^4\ N/m^2)(1.00 \times 10^{-3}\ m^2) = 22.0\ N$$

The mass required to supply this force is $m_1 = F_1/g = 2.24$ kg.

Example: An equilateral triangular cross section trough (Figure 15.16) is filled with water. The length of the trough is 2.00 m and the sides of the triangle are 0.500 m. Determine the magnitude of the force on each of the triangular ends and on each of the rectangular sides.

Given and Diagram:

$L = 2.00$ m - length of trough
$a = 0.500$ m - length of triangle side
$\rho = 1.00 \times 10^3$ - kg/m^3 density of water
$P_0 = 1.01 \times 10^5$ N/m^2 - atmospheric pressure

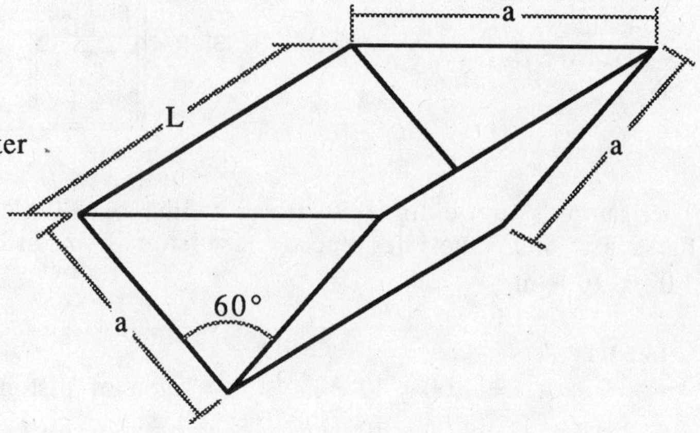

Figure 15.16

Determine: Magnitude of the force on each of the triangular ends and on each of the rectangular sides.

Strategy: Let's divide the problem into two parts (ends and sides). A convenient coordinate system can be established at one end of the trough's base. Now let's establish area elements da on the end and side of the trough. Next we can determine an expression for the pressure (due to atmosphere and water) on each dA. Finally, we can determine the force by summing all the dFs (dF = PdA) for each surface.

Solution: Figure 15.17 shows an end and side of the trough, a convenient coordinate system, and the area element dA for the end and side.

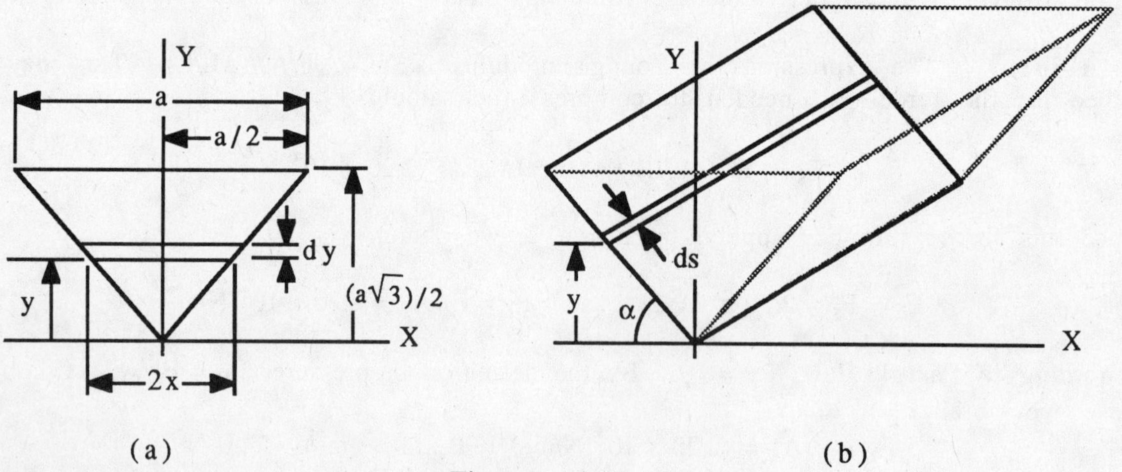

Figure 15.17

First let's work with the end.

$$dA = 2xdy \qquad \text{and} \qquad P = P_0 + \rho g[(\frac{\sqrt{3}}{2})a - y]$$

$$F = \int PdA = \int_0^{(a\sqrt{3})/2} \left[P_0 + \rho g[(\frac{a\sqrt{3}}{2}) - y] \right] 2x dy$$

The equation for the side of the trough is

$$y = (\sqrt{3})x \quad \text{or} \quad x = y/\sqrt{3} \quad \text{hence}$$

$$F = \frac{2}{\sqrt{3}} \left[\int_0^{(a\sqrt{3})/2} \left[P_0 + (\frac{\sqrt{3}}{2})\rho g a \right] y dy - \rho g \int_0^{(a\sqrt{3})/2} y^2 dy \right]$$

$$F = \frac{2}{\sqrt{3}} \left[\left(P_0 + (\frac{\sqrt{3}}{2})\rho g a \right) (\frac{y^2}{2}) \Big|_0^{(a\sqrt{3})/2} - \rho g (\frac{y^3}{3}) \Big|_0^{(a\sqrt{3})/2} \right]$$

Inserting the limits and simplifying we obtain

$$F = (\frac{\sqrt{3}}{4})P_0 a^2 + (\frac{1}{8})\rho g a^3 = 10.9 \times 10^3 \text{ N} + 0.153 \times 10^3 \text{ N} = 1.11 \times 10^4 \text{ N}$$

As a check, notice that the first term is just atmospheric pressure acting on the end of the trough.

$$F_{end} = P_0 A_{end} = P_0[(\frac{1}{2})a(\frac{a\sqrt{3}}{2})] = (\frac{\sqrt{3}}{4})P_0 a^2$$

Now working with the side

$$dA = Lds = Ldy/\sin\alpha = (2/\sqrt{3})Ldy \quad \text{and} \quad P = P_0 + \rho g(\frac{a\sqrt{3}}{2} - y)$$

$$F = \int PdA$$

$$F = \frac{2L}{\sqrt{3}} \int_0^{(a\sqrt{3})/2} \left[P_0 + \rho g[(\frac{a\sqrt{3}}{2}) - y] \right] dy$$

$$F = \frac{2L}{\sqrt{3}} \left[\int_0^{(a\sqrt{3})/2} \left[P_0 + (\frac{\rho g a\sqrt{3}}{2}) \right] dy - \rho g \int_0^{(a\sqrt{3})/2} y dy \right]$$

$$F = \frac{2L}{\sqrt{3}}\left[\left(P_0 + [\frac{\rho g a\sqrt{3}}{\sqrt{2}}] \right) y \Big|_0^{(a\sqrt{3})/2} - \left(\frac{\rho g y^2}{2}\right) \Big|_0^{(a\sqrt{3})/2} \right]$$

Inserting the limits and simplifying we obtain

$$F = P_0 La + (\frac{\sqrt{3}}{4})\rho g La^2 = 101 \times 10^3 \text{ N} + 2.12 \times 10^3 \text{ N} = 1.03 \times 10^5 \text{ N}$$

As a check, notice that the first term is just atmospheric pressure acting on the side of the trough.

$$F_{side} = P_0 A_{side} = P_0 La$$

Related Text Exercises: Ex. 15-13 through 15-17.

4 Archimedes' Principle (Section 15-4)

Review: Archimedes' principle states that a body immersed in a fluid is buoyed up by a force equal to the weight of the fluid displaced by the body.

$$F_B = w_{fd} = m_{fd}g = \rho_f V_{fd} g$$

Where F_B represents the buoyant force, w_{fd}, m_{fd} and V_{fd} are the weight, mass and volume of the fluid displaced; ρ_f is the density of the fluid and g is the acceleration due to gravity.

Practice: Figure 15.18(a) shows a block of wood, its length, cross-sectional area and mass. Figure 15.18(b) shows the block held under water by a weight. Figure 15.18(c) shows the block half submerged in an unknown liquid (called liquid 1). Figure 15.18(d) shows the result of pouring another liquid (liquid 2) on top of liquid 1). Note that liquids 1 and 2 do not mix.

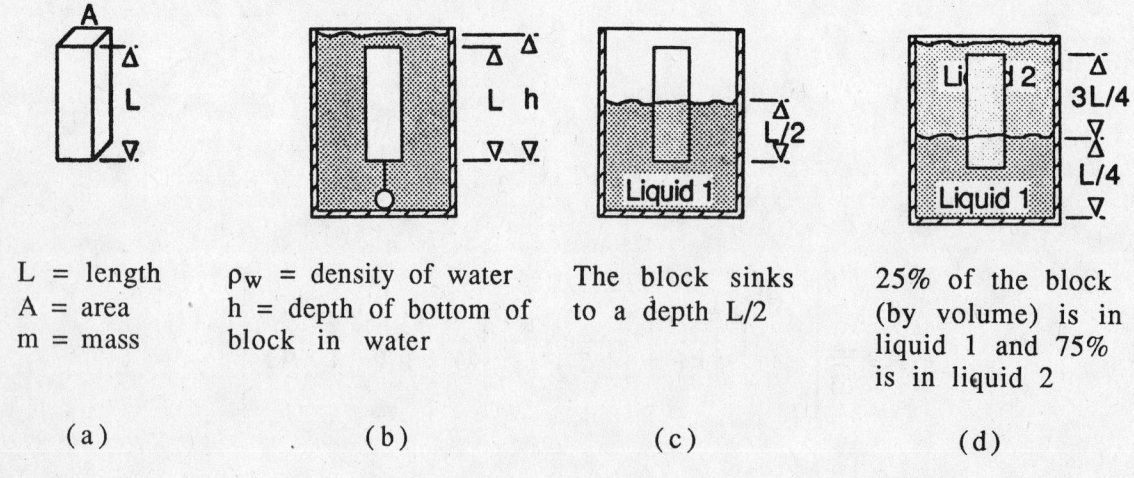

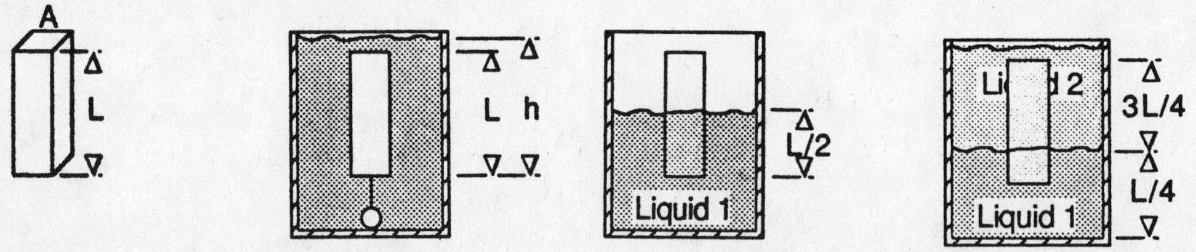

(a)	(b)	(c)	(d)
L = length A = area m = mass	ρ_w = density of water h = depth of bottom of block in water	The block sinks to a depth L/2	25% of the block (by volume) is in liquid 1 and 75% is in liquid 2

Figure 15.18

Determine an expression for the following in terms of given quantities:

1. Volume of the block	$V_b = AL$
2. Density of the block	$\rho_b = m/V_b = m/AL$
3. Pressure at the bottom of the block due to the water in Figure 15.18(b)	$P_{bottom} = \rho_w gh$
4. Pressure at the top of the block due to the water in Figure 15.18(b)	$P_{top} = \rho_w g(h - L)$
5. Pressure difference between the bottom and top of the block due to the water in Figure 15.18(b)	$\Delta P = P_{bottom} - P_{top}$ $= \rho_w gh - \rho_w g(h - L) = \rho_w gL$
6. Upward (buoyant) force on the block due to the water in Figure 15.18(b)	$F = \Delta PA = \rho_w gLA$
7. Weight of water displaced by the block in Figure 15.18(b)	$w_{wd} = m_{wd}g = \rho_w V_{wd}g = \rho_w LAg$

Note: The expression for the buoyant force in step 6 is the same as the expression for the weight of water displaced in step 7. You have just discovered the same thing that Archimedes discovered, namely that the bouyant force on an object is equal to the weight of the fluid displaced.

8. Mass density of liquid 1	Since the block is in equilibrium, the buoyant force due to liquid 1 must equal the weight of the block $F_{B1} = w_b$ $w_b = m_b g$ $F_{B1} = w_{1d} = m_{1d}g = \rho_1 V_{1d}g = \rho_1 ALg/2$ Substituting w_b and F_{B1} obtain $m_b g = \rho_1 ALg/2$ $\rho_1 = 2m_b/AL$
9. Mass density of liquid 2	$w_b = F_{B1} + F_{B2}$ $w_b = m_b g$ $F_{B1} = \rho_1 ALg/4 = m_b g/2$ (ρ_1 from step 8) $F_{B2} = 3\rho_2 ALg/4$ $m_b g = m_b g/2 + 3\rho_2 ALg/4$, or $\rho_2 = 2m_b/3AL$

Example: An object of volume 3.00×10^{-4} m^3 hangs by a cord from a spring balance, as shown in Figure 15.19. When the object hangs in air (Figure 15.19(a)), the balance reads 1.000 kg. When the object hangs in a liquid of unknown density (Figure 15.19(b)), the balance reads 0.550 kg. Determine the density of the liquid.

Balance reading
m = 1.000 kg
(a)

Balance reading
m' = 0.550 kg
(b)

Figure 15.19

Given: m = 1.000 kg = balance reading with the object suspended in air
m' = 0.550 kg = balance reading with the object suspended in liquid
$V = 3.00 \times 10^{-4}$ m^3 = volume of object

Determine: Density ρ_L of the liquid

Strategy: Since the object is in equilibrium, we can write a summation-of-forces statement that includes the tension in the cord, the buoyant force and the weight of the object. We can determine the weight of the object, the tension in the cord and hence the buoyant force from the given information. Once the buoyant force is known, we can use Archimedes' principle to determine the density of the liquid.

Solution: A free-body diagram for the object hanging in the liquid is shown in Figure 15.20. Since the object is in equilibrium, we can write the following summation-of-forces statement:

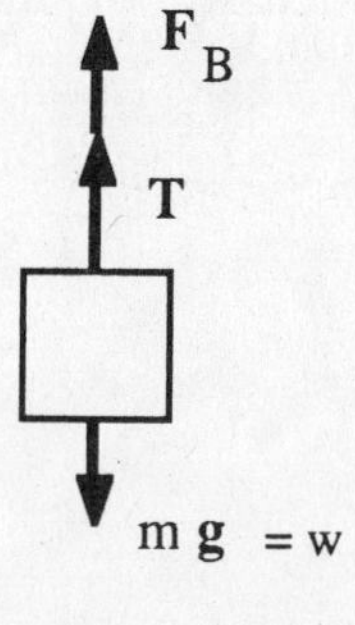

Figure 15.20

$$T + F_B = w$$

where T = m'g = cord tension when the object is in the liquid
w = mg = weight of the object
$F_B = w_{Ld} = \rho_L V_{Ld} g$ = buoyant force

When expressions for T, w and F_B are inserted into the summation-of-forces statement and it is solved for ρ_L, we obtain

$$\rho_L = (m - m')/V_{Ld}$$

Noting that $V_{Ld} = V$ and inserting values, we obtain

$$\rho_L = 1.50 \times 10^3 \text{ kg/m}^3$$

Example: A flat-bottomed barge has a length of 50.0 m, width of 10.0 m and a height of 2.00 m. The barge weighs 8.00×10^5 N, and we want it to float 0.500 m out of the water when loaded. What volume of coal ($\rho_c = 1.80 \times 10^3$ kg/m^3) can be loaded on the barge?

Given:

$L = 50.0$ m = length of barge
$W = 10.0$ m = width of barge
$h = 2.00$ m = height of barge
$d = 1.50$ m = depth of barge in water
$w_b = 8.00 \times 10^5$ N = weight of barge
$\rho_c = 1.80 \times 10^3$ kg/m^3 = density of coal
$\rho_w = 1.00 \times 10^3$ kg/m^3 = density of water

Determine: The volume of coal that can be loaded on to the barge with the barge floating 0.500 m out of the water.

Strategy: Knowing L, W and d, we can determine the volume of water displaced (V_{wd}) by the barge. Knowing V_{wd}, ρ_w and g, we can determine the weight of water displaced and hence the buoyant force F_B. Knowing that F_B must support the weight of the barge w_b and the weight of the coal w_c, we can determine w_c. Knowing w_c and ρ_c we can determine the volume of coal of V_c that can be loaded onto the barge.

Solution: The volume of water displaced is

$$V_{wd} = LWd$$

The weight of water displaced and hence the buoyant force are given by

$$F_B = w_{wd} = m_{wd}g = \rho_w V_{wd} g = \rho_w LWdg$$

This buoyant force must support the barge and coal, and so

$$F_B = w_b + w_c \quad \text{or} \quad w_c = F_B - w_b$$

Finally, the volume of the coal can be determined by

$$w_c = m_c g = \rho_c V_c g$$

$$V_c = \frac{w_c}{\rho_c g} = \frac{F_B - w_b}{\rho_c g} = \frac{\rho_w LWdg - w_b}{\rho_c g} = 371 \text{ m}^3$$

Related Text Exercises: Ex. 15-18 through 15-23.

5 Bernoulli's Equation (Section 15-5)

Review: Figure 15.21 shows a rigid pipe full of an incompressible fluid of density ρ.

Figure 15.21

The volume flow rate is given by $Q = Av$. Since the pipe is rigid and full of an incompressible fluid, the rate at which fluid flows past A_1 must equal the rate at which it flows past A_2. That is

$$Q_1 = Q_2 \quad \text{or} \quad A_1v_1 = A_2v_2$$

This expression is the continuity equation for an incompressible fluid. Bernoulli's equation states that if a rigid pipe is full of a flowing incompressible fluid, then

$$P + \rho gh + \rho v^2/2 = \text{constant}$$

Practice: Water enters a building in a pipe of inside diameter 5.00×10^{-2} m and is piped to various locations in pipes of inside diameter 2.00×10^{-2} m (Figure 15.22). At one of the faucets, it's discovered that a 2.00×10^{-2} m^3 container can be filled in 50.0 s.

$d_f = 2.00 \times 10^{-2}$ m
= inside diameter of the pipe at the faucet

$d_b = 5.00 \times 10^{-2}$ m
= inside diameter of the pipe entering the building

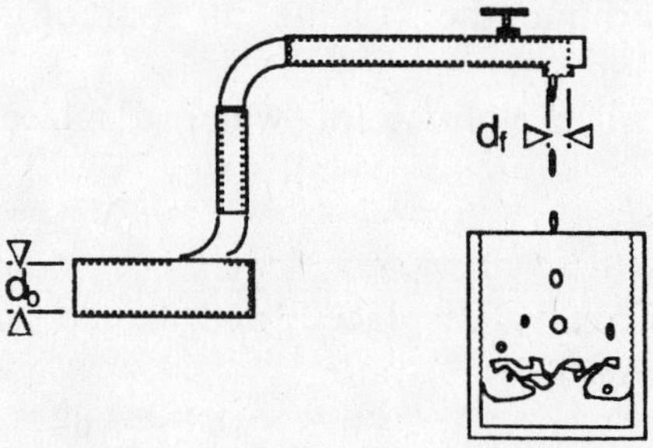

Figure 15.22

Determine the following:

1. Volume flow rate out of the faucet	Q_f = volume flow/time $Q_f = 2.00 \times 10^{-2}$ m^3/50.0 s $= 4.00 \times 10^{-4}$ m^3/s
2. Speed at which water leaves the faucet	$A_f = \pi d_f^2/4 = 3.14 \times 10^{-4}$ m^2 $v_f = Q_f/A_f = 1.27$ m/s
3. Volume flow rate at which water enters the building	$Q_b = Q_f = 4.00 \times 10^{-4}$ m^3/s
4. Speed at which water enters the building	$A_b = \pi d_b^2/4 = 19.6 \times 10^{-4}$ m^2 $v_b = Q_b/A_b = 0.204$ m/s

Practice: Consider the section of pipe and the given information in Figure 15.23. The pipe is full of water.

$r_1 = 1.00 \times 10^{-1}$ m
$r_2 = 3.00 \times 10^{-2}$ m
$v_2 = 2.00$ m/s
$P_2 = 1.00 \times 10^4$ Pa
$\rho_w = 1.00 \times 10^3$ kg/m^3
$h_1 = h_2$

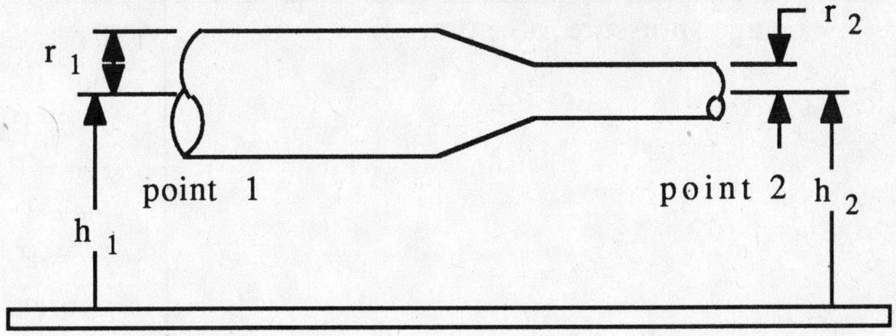

Figure 15.23

Determine the following:

1. Volume flow rate at point 2	$Q_2 = A_2v_2 = \pi r_2^2 v_2 = 5.65 \times 10^{-3}$ m^3/s
2. Volume flow rate at point 1	$Q_1 = Q_2 = 5.65 \times 10^{-3}$ m^3/s
3. Water speed at point 1	$v_1 = Q_1/A_1 = Q_1/\pi r_1^2 = 0.180$ m/s or $A_1v_1 = A_2v_2$ which gives $v_1 = (A_2/A_1)v_2 = (r_2/r_1)^2 v_2 = 0.180$ m/s
4. Water pressure at point 1	$P_1 + \rho g h_1 + \rho v_1^2/2 = P_2 + \rho g h_2 + \rho v_2^2/2$ For this case, $h_1 = h_2$; hence $P_1 = P_2 + \rho(v_2^2 - v_1^2)/2 = 1.20 \times 10^4$ Pa

Consider the section of pipe and the given information in Figure 15.24. The pipe is full of water with a mass flow rate of 4.00 kg/s.

$A_1 = A_2 = 3.00 \times 10^{-4}$ m^2
$h_1 = 0$
$h_2 = 4.00$ m
$m/t = 4.00$ kg/s
$\rho = 1.00 \times 10^3$ kg/m^3
$v_1 = v_2$

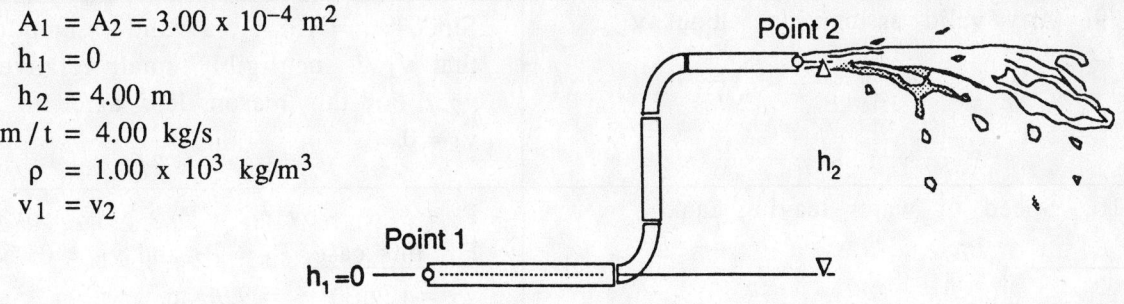

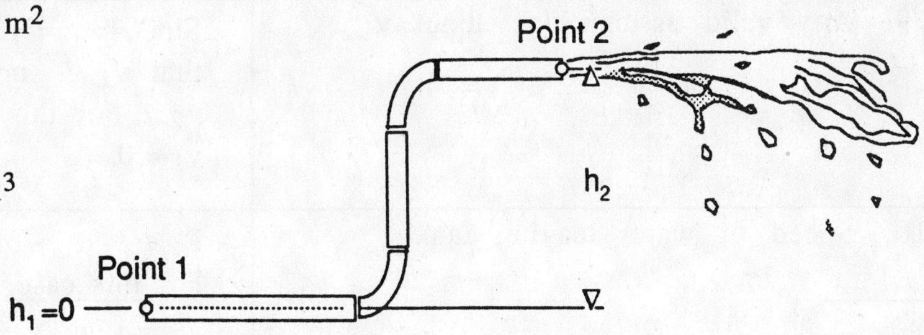

Figure 15.24

Determine the following:

5. Volume flow rate	$Q = V/t = (m/\rho)/t = (m/t)/\rho$ $= 4.00 \times 10^{-3}$ m^3/s

6. Water speed	$v = Q/A = 13.3$ m/s
7. Gauge pressure at point 1	$P_1 + \rho gh_1 + \rho v_1^2/2 = P_2 + \rho gh_2 + \rho v_2^2/2$ For this case, $v_1 = v_2$. Hence: $P_1 = P_2 + \rho g(h_2 - h_1)$. But $h_1 = O$ and $P_2 = P_A$ = atmospheric pressure. So $P_1 = P_A + \rho gh_2$. Also remember $P_1 = P_A + P_{1G}$. Comparing these expressions for P_1 obtain: $P_{1G} = \rho gh_2 = 3.92 \times 10^4$ Pa

Figure 15.25 shows a large cylindrical water tank open at the top and with a small hole in the bottom.

$r_1 = 2.00$ m
$r_2 = 2.00 \times 10^{-2}$ m
$h_1 = 5.00$ m
$P_1 = P_2 = P_A$

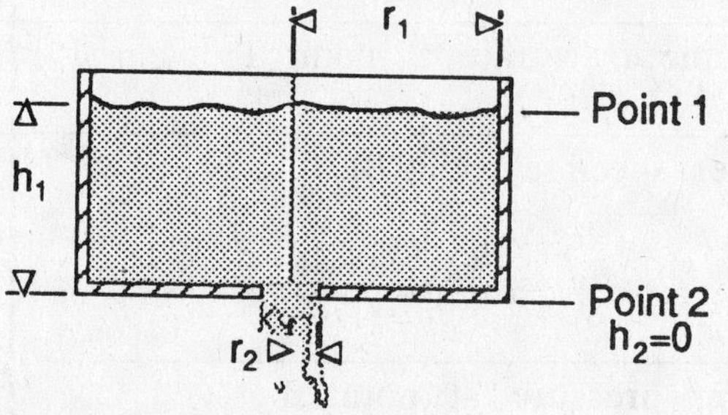

Figure 15.25

Determine the following:

8. The areas A_1 and A_2	$A_1 = \pi r_1^2 = 12.6$ m^2 $A_2 = \pi r_2^2 = 12.6 \times 10^{-4}$ m^2
9. Any valid assumptions about v_1	Since $A_1 = 10^4 A_2$, a valid assumption is that v_1 is negligibly small relative to v_2. For this reason, let's agree to set $v_1 = 0$.
10. Speed of water leaving tank	$P_1 + \rho gh_1 + \rho v_1^2/2 = P_2 + \rho gh_2 + \rho v_2^2/2$ For this case, $P_1 = P_2$ and $v_1 = 0$; hence $v_2 = (2gh_1)^{1/2} = 9.90$ m/s.
11. Volume of water leaving tank in 100 s	$Q_2 = A_2 v_2 = 1.25 \times 10^{-2}$ m^3/s $Q_2 = V_2/t$ $V_2 = Q_2 t = 1.25$ m^3

Example: The design of an airplane calls for a lift of 800 N per square meter of wing area. If the flow velocity is 100 m/s past the lower wing surface, what flow velocity past the upper wing surface will give the required lift? $\rho_a = 1.25\ kg/m^3$.

Given: Lift = $8.00 \times 10^2\ N/m^2$
= pressure difference between upper and lower surfaces
$v_L = 1.00 \times 10^2$ m/s = flow velocity past lower surface
$\rho_a = 1.25\ kg/m^3$ = density of air

Determine: Flow velocity past upper surface (v_u) to create the required lift.

Strategy: Since the wing thickness is small relative to the other distances involved, it is a good approximation to set $h_u = h_L$. The lift information gives us the required pressure difference between the lower and upper wing surfaces. We can use the pressure difference and v_L to determine v_u.

Solution: The lift information gives us the pressure difference between the two surfaces.

$$\text{lift} = P_L - P_u = 8 \times 10^2\ N/m^2$$

Inserting this information into Bernoulli's equation (and recalling that $h_u = h_L$)

$$P_L + \rho_a g h_L + \rho_a {v_L}^2/2 = P_u + \rho_a g h_u + \rho_a {v_u}^2/2$$

we obtain

$$v_u = [2(P_L - P_u)/\rho_a + {v_L}^2]^{1/2} = 106\ m/s$$

Related Text Exercises: Ex. 15-24 through 15-32.

PRACTICE TEST

Take and grade this practice test. Doing so will allow you to determine any weak spots in your understanding of the concepts taught in this chapter. The following section prescribes what you should study further to strengthen your understanding.

Lead blocks of length 2.00×10^{-1} m, width 1.00×10^{-1} m and height 2.00×10^{-2} m are stacked ten high. The mass of each block is 4.40 kg and Young's Modulus for lead is 1.60×10^{10} N/m^2.

Determine the following:

_________ 1. Stress on the bottom block
_________ 2. Strain on the bottom block
_________ 3. Compression of the bottom block

Two metal bars 0.500 cm thick are held together by rivets, as shown in the figure at the right. The rivets have a radius of 2.00×10^{-3} m, and the maximum shear stress they can withstand is 5.00×10^{8} N/m^2.

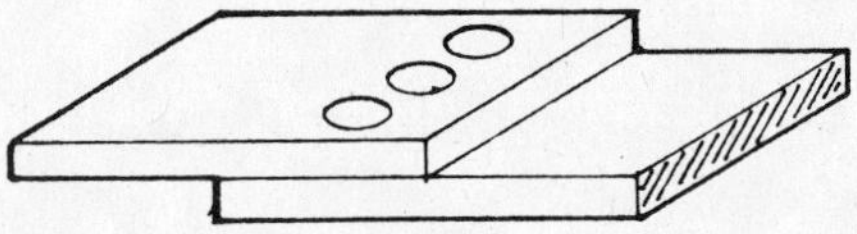

Determine the following:

_________ 4. The area that must be sheared in order to separate the bars by applying a force parallel to them
_________ 5. The force applied parallel to the bars that will shear the three rivets

The figure below shows a cylindrical water tank with an attached mercury U-tube manometer.

$\Delta h = 0.368$ m of mercury
$\rho_{Hg} = 13.6 \times 10^3$ kg/m^3
$\rho_{water} = 1.00 \times 10^3$ kg/m^3

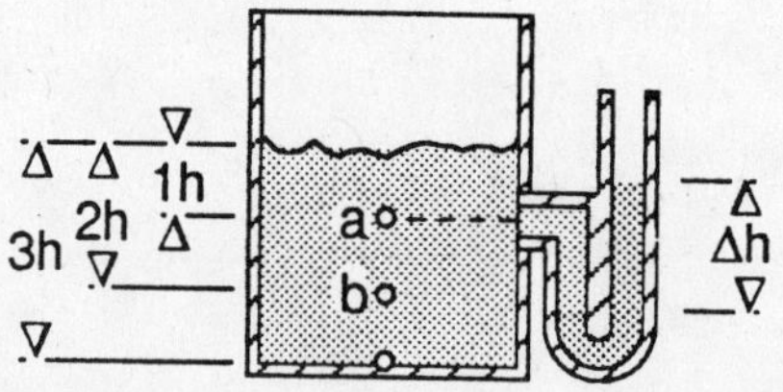

Determine the following:

_________ 6. Pressure due to water at level a (note the value of h is not given)
_________ 7. Pressure due to water at level b
_________ 8. Absolute pressure at bottom of tank
_________ 9. Distance h below surface where manometer is attached

The cylindrical object shown in the figure below floats in liquid 1 and hangs suspended by a cord in liquid 2. The tension in the cord is represented by T.

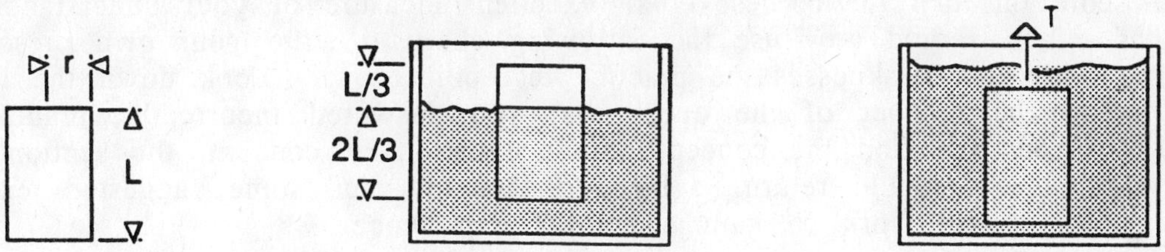

(a)	(b)	(c)
$L = 1.00 \times 10^{-1}$ m $r = 2.00 \times 10^{-2}$ m $m = 8.80 \times 10^{-2}$ kg	object floats in liquid 1	object suspended in liquid 2 $T = 2.45 \times 10^{-1}$ N

Determine the following:

_________ 10. Density of object
_________ 11. Weight of object
_________ 12. Buoyant force on object in liquid 1
_________ 13. Weight of liquid 1 displaced by object
_________ 14. Density of liquid 1
_________ 15. Buoyant force on object in liquid 2
_________ 16. Weight of liquid 2 displaced by object
_________ 17. Density of liquid 2

The figure below shows a rigid pipe full of a flowing incompressible fluid and an attached mercury manometer.

$d_1 = 1.00 \times 10^{-1}$ m = diameter of large pipe
$d_2 = 2.50 \times 10^{-2}$ m = diameter of small pipe
$\rho_f = 1.20 \times 10^3$ kg/m^3 = mass density of fluid
$m/t = 9.42$ kg/s = mass flow rate of fluid

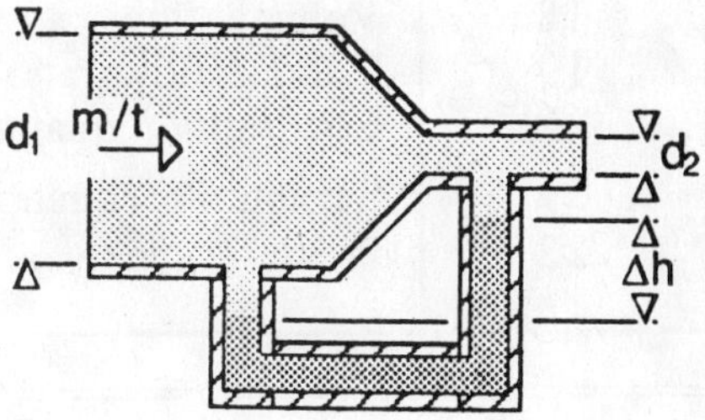

Determine the following:

_________ 18. Volume flow rate of fluid
_________ 19. Flow speed through large pipe
_________ 20. Flow speed through small pipe
_________ 21. Difference in pressure between the two pipes
_________ 22. Difference in mercury level in the sides of the U-tube

(See Appendix I for answers.)

PRINCIPAL CONCEPTS AND EQUATIONS PRESCRIPTION

Your score on the practice test is an excellent measure of your understanding of the chapter. You should now use the following chart to write your own prescription for dealing with any weaknesses the practice test points out. Look down the leftmost column to the number of the question(s) you answered incorrectly, reading across that row you will find the concept and/or equation of concern, the section(s) of the study guide you should return to for further study, and some suggested text exercises which you should work to gain additional experience.

Practice Test Questions	Concepts and Equations	Prescription	
		Principal Concepts	Text Exercises
1	Compressive stress: $\sigma_c = F_n/A$	1	15-1, 4
2	Young's modulus: $Y = \sigma_c/\varepsilon_c$	1	15-1, 4
3	Compressive strain: $\varepsilon_c = \Delta h/h$	1	15-5, 6
4	Area of shear stress	1	---
5	Shear stress: $\sigma_s = F_p/A$	1	---
6	Pressure due to a fluid: $P = \rho g h$	3	15-13, 15
7	Pressure due to a fluid: $P = \rho g h$	3	15-13, 15
8	Absolute pressure: $P = P_f + P_A$	3	---
9	Pressure due to a fluid: $P = \rho g h$	3	15-13, 15
10	Mass density: $\rho = m/v$	2	15-9, 10
11	Relation between weight and mass: $w = mg$	2 of Ch 7	7-9
12	First condition of equilibrium: $\Sigma F = 0$	1 of Ch 5	5-7, 8
13	Archimedes' principle: $F_B = W_{fd}$	4	15-18, 19
14	Archimedes' principle: $F_B = W_{fd}$	4	15-19, 20
15	First condition of equilibrium: $\Sigma F = 0$	1 of Ch 5	5-7, 8
16	Archimedes' principle: $F_B = W_{fd}$	4	15-39, 43
17	Mass density: $\rho = m/v$	2	15-9, 10
18	Volume flow rate: $Q = (m/t)/\rho$	5	15-24, 25
19	Volume flow rate: $Q = Av$	5	15-24, 25
20	Continuity equation: $A_1v_1 = A_2v_2$	5	15-25, 26
21	Bernoulli's equation: $P + \rho gh + \rho v^2/2 = \text{const.}$	5	15-30, 31
22	Pressure due to a fluid: $P = \rho g h$	3	15-13, 15

16 TEMPERATURE AND HEAT TRANSFER

RECALL FROM PREVIOUS CHAPTERS

Previously learned concepts and equations frequently used in this chapter	Text Section	Study Guide Page
Stress: $\sigma_t = F_n/A$	15-1	15-1
Strain: $\varepsilon_t = \Delta L/L$	15-1	15-1
Young's Modulus: $Y = \sigma_t/\varepsilon_t$	15-1	15-1

NEW IDEAS IN THIS CHAPTER

Concepts and equations introduced	Text Section	Study Guide Page
Relationship between temperature scales: $t_F = (9/5)t_C + 32$ $t_C = (5/9)(t_F - 32)$ $T = t_C + 273$ $T_R = (9/5)T$	16-3, 4	16-1
Thermal expansion Linear: $\Delta L = \alpha L_0 \Delta T$ Area: $\Delta A = 2\alpha A_0 \Delta T$ Volume: $\Delta V = 3\alpha V_0 \Delta T$	16-5	16-4
Heat Transfer: Conduction $H = \Delta Q/\Delta t = kA\Delta T/\Delta x$ Radiation $H = \Delta Q/\Delta t = \varepsilon\sigma A T^4$	16-6	16-7

PRINCIPAL CONCEPTS AND EQUATIONS

1 Temperature Scales (Section 16-3, 4)

Review: The temperature scales in common use, the degree abbreviation, the values for melting ice and boiling water, and the functional relationships are as follows:

Temperature Scale	Degree Abbr.	Value for Melting Ice	Value for Boiling Water	Functional Relationship
Fahrenheit	°F	32°F	212°F	$t_F = (9/5)t_C + 32$
Celsius	°C	0°C	100°C	$t_C = (5/9)(t_F - 32)$
Kelvin	K	273 K	373 K	$T = t_C + 273$
Rankine	°R	491°R	671°R	$T_R = (9/5)\ T$

Practice: Figure 16.1 shows five thermometers, each having different numerical values for the temperatures of melting ice and boiling water.

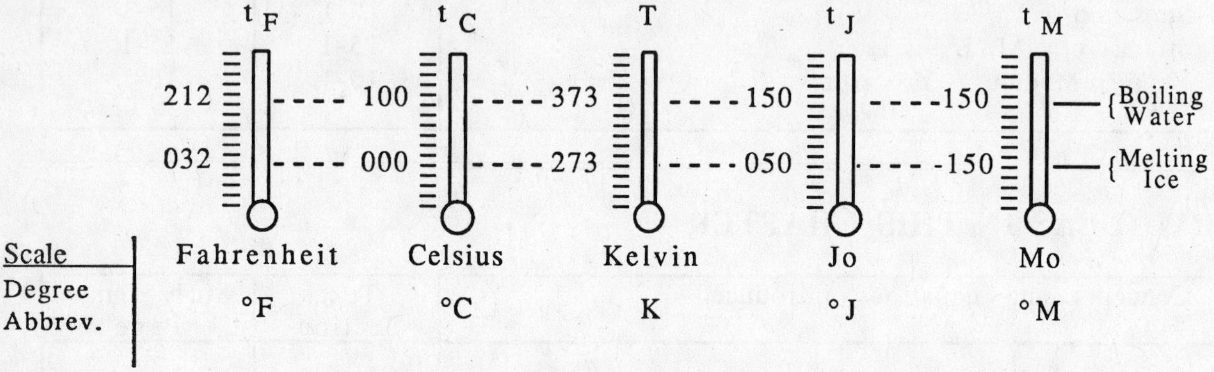

Figure 16.1

Note: Before starting the practice section, it is essential that we agree on notation.

$t_C = 10°C$ stands for ten degrees Celsius, a specific value on the Celsius scale.
$\Delta t_C = 10\ C°$ stands for a change in temperature of ten degrees on the Celsius scale (i.e. ten Celsius degrees).

When water is heated from its freezing point to its boiling point, it under goes a temperature change of 100 Celsius degrees (100 C°) on the Celsius scale, as compared with a change of 180 Fahrenheit degrees (180 F°) on the Fahrenheit scale. Express this as

$$100\ C° = 180\ F°$$

The specific value 0° on the Celsius scale is equal to the specific value 32° on the Fahrenheit scale. Express this as

$$0°C = 32°F$$

The specific value 100° on the Celsius scale is equal to the specific value 212° on the Fahrenheit scale. Express this as

$$100°C = 212°F$$

In summary use °C or °F for a specific temperature value (t) and use C° or F° for a change in temperature (Δt).

Determine the following:

1. The relationship between t_J and t_C	100 C° = 100 J°, or 1 C° = 1 J° Every time the Celsius scale advances 1 C°, Jo's scale advances 1 J°. Since Jo starts out 50° higher, we write $t_J = t_C + 50$. Notice that values of t_C equal to 0°C and 100°C, given values of t_J equal to 50°J and 150°J, respectively.
2. The relationship between t_M and t_C	100 C° = 300 M°, or 1 C° = 3 M° Every time the Celsius scale advances 1 C°, Mo's scale advances 3 M°. Since Mo starts out 150° lower, we write $t_M = 3t_C - 150$ Check: $t_C = 0°C \rightarrow t_M = -150°M$ $t_C = 100°C \rightarrow t_M = 150°M$
3. The relationship between T_J and t_M	We already know t_M as a function of t_C (step 2) and t_C as a function of t_J (step 1). We can combine these to obtain t_M as a function of t_J. $t_M = 3t_C - 150 = 3(t_J - 50) - 150 = 3t_J - 300$ (Perform the check.)
4. The relationship between t_F and t_C	100 C° = 180 F°, or 1 C° = (9/5) F° Every time the Celsius scale advances 1 C°, the Fahrenheit scale advances (9/5) F°. Since the Fahrenheit scale starts out 32° ahead, we write $t_F = (9/5)t_C + 32$ (Perform the check.)
5. The relationship between t_C and T	100 C° = 100 K° or 1 C° = 1 K° Every time the Celsius scale advances 1 C°, the Kelvin scale advances 1 K°. Since the Kelvin scale starts out 273° ahead, we write $T = t_C + 273$ (Perform the check.)
6. The relationship between t_F and T	$t_F = (9/5)t_C + 32$ (step 4) $t_C = T - 273$ (step 5) Combining these, we obtain $t_F = (9/5)(T-273) + 32$ $t_F = (9/5)T - 459.4$ (Perform the check.)
7. The temperature when the same number would be read from both the Celsius and Fahrenheit scales ($t_F = t_C$)	$t_F = (9/5)t_C + 32$, set $t_F = t_C$ $t_C = (9/5)t_C + 32$ or $t_C = -40°C = t_F$ Check: Insert $t_C = -40°C$ into the top expression to obtain $t_F = -40°F$

8. The temperature when the same number would be read from both the Jo and Mo scales ($t_J = t_M$)	First notice in Figure 16.1 that $t_J = t_M$ at 150°. However, let's prove this algebraically as follows: $t_M = 3t_J - 300$, set $t_M = t_J$ $t_J = 3t_J - 300$ or $t_J = 150°J$

Example: Using their thermometers, inhabitants of planet X report that ice melts at -100°X and water boils at 150°X. We wish to send them a container of liquid oxygen (which boils at -183°C) with instructions about its properties. What should we tell them the boiling point is in °X?

Given: Melting point of ice = -100°X
Boiling point of water = 150°X
Boiling point of oxygen = -183°C

Determine: The boiling point of oxygen in °X.

Strategy: Determine the number of X° equal to 1 C°, compare the melting point of ice in °X and °C, and then develop a functional relationship between t_X and t_C. Insert the boiling point of oxygen in °C to determine its value in °X.

Solution: 100 C° = 250 X° or 1 C° = 2.5 X° Every time the Celsius scale advances by 1 C°, the X-scale will advance 2.5 X°. Since the X scale starts out 100° behind, we write

$$t_X = 2.5t_C - 100$$

Check: $t_C = 0°C \rightarrow t_X = -100°X$, and $t_C = 100°C \rightarrow t_X = 150°X$

Since the check agrees with the calibration information (values for melting ice and boiling water), we are confident our functional relationship is correct. Finally, we insert $t_C = -183°C$ to obtain $t_X = -558°X$.

Related Text Exercises: Ex. 16-10 through 16-12.

2 THERMAL EXPANSION (Section 16-5)

Review: Most objects expand when heated. Figure 16.2 shows an object with dimensions L_0, W_0 and H_0 at temperature T and the change in these dimensions when the temperature is increased an amount ΔT.

Figure 16.2

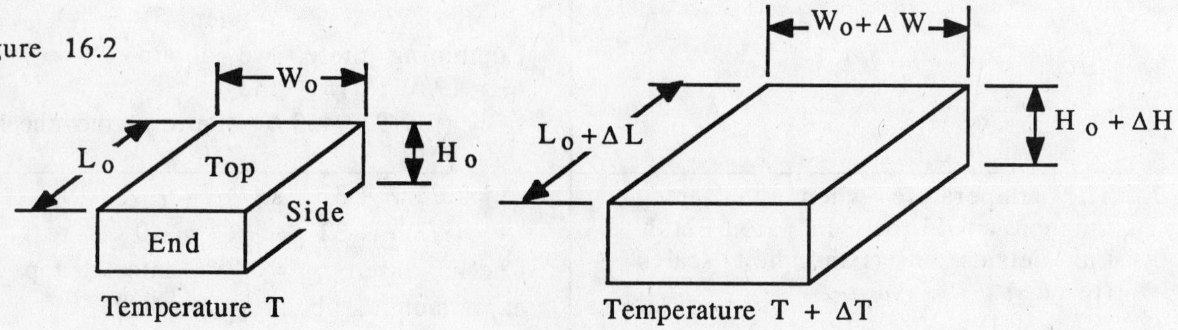

The linear dimensions are changed as follows:

L_0 is changed by an amount $\Delta L = \alpha L_0 \Delta T$
W_0 is changed by an amount $\Delta W = \alpha W_0 \Delta T$
H_0 is changed by an amount $\Delta H = \alpha H_0 \Delta T$

The quantity α (linear coefficient of thermal expansion) tells us the change in length (expansion or contraction) per unit length per degree change in temperature.

An area is changed by an amount $\Delta A = \gamma A_0 \Delta T$, where $\gamma = 2\alpha$.

The quantity γ is the area coefficient of thermal expansion.

The volume is changed by an amount $\Delta V = \beta V_0 \Delta T$, where $\beta = 3\alpha$

The quantity β is the volume coefficient of thermal expansion. When you work text problems 16-1 and 16-2, you will learn why $\gamma = 2\alpha$ and $\beta = 3\alpha$.

Practice: The following is known about the object shown in Figure 16.2:
$L_0 = 2.00 \times 10^{-1}$ m; $W_0 = 5.00 \times 10^{-2}$ m; $H_0 = 2.00 \times 10^{-2}$ m; $\alpha = 2.50 \times 10^{-5}\ K^{-1}$; $\Delta T = 100$ C°

Determine the following:

1. The change in the linear dimensions of the object	$\Delta L = \alpha L_0 \Delta T = 5.00 \times 10^{-4}$ m $\Delta W = \alpha W_0 \Delta T = 1.25 \times 10^{-4}$ m $\Delta H = \alpha H_0 \Delta T = 5.00 \times 10^{-5}$ m
2. The change in the area of the top of the object	$A_0 = L_0 W_0$, $\gamma = 2\alpha$ $\Delta A = \gamma A_0 \Delta T = 2\alpha L_0 W_0 \Delta T = 5.00 \times 10^{-5}\ m^2$
3. The change in volume of the object	$V_0 = L_0 W_0 H_0$, $\beta = 3\alpha$ $\Delta V = \beta V_0 \Delta T = 3\alpha L_0 W_0 H_0 \Delta T = 1.50 \times 10^{-6} m^3$

A walkway connecting two buildings is supported by two steel I-beams that are slipped into place with a perfect fit on a cold day (t_C = -10°C). Consider a day when the steel temperature reaches 90°C.

Figure 16.3

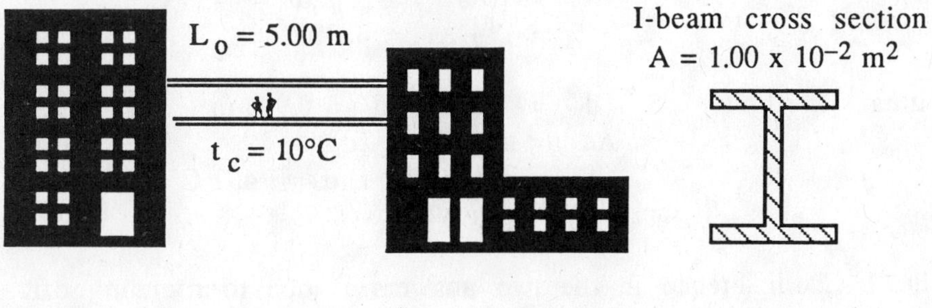

Determine the following:

4. The stress needed on each I-beam to prevent expansion.	Y = stress/strain $Y = 2.00 \times 10^{11}$ N/m^2, $\alpha = 1.10 \times 10^{-5}$/C° stress = (Y)(strain) = $Y\Delta L/L_o = Y\alpha\Delta T$ stress = 2.20×10^8 N/m^2
5. The force needed on each I-beam to prevent expansion	stress = F/A = 2.20×10^8 N/m^2 $A = 1.00 \times 10^{-2}$ m^2 F = (stress)(A) = 2.20×10^6 N
6. The total force the walls of the building must be able to withstand to prevent expansion of the beams	$F_T = 2F = 4.40 \times 10^6$ N. Since normal wall construction cannot stand up to forces of this magnitude, an expansion joint is highly recommended.

Example: The outside diameter of a wagon wheel is 1.00 m. A steel tire for this wheel has an inside diameter of 0.995 m at 25°C. To what temperature must the tire be heated in order for it to just slip onto the wheel?

Given: $d_w = 1.00$ m, $d_t = 0.995$ m at $t_C = 25$°C, $\alpha_{steel} = 1.10 \times 10^{-5}$/ C°

Determine: The temperature to which the tire must be raised so that its thermal expansion will allow it to slip onto the wheel.

Strategy: The solution to this problem depends on the knowledge that a cavity in a body expands or contracts with a change in temperature exactly as a solid object of the same size and composition would. We can treat this as either an area problem or a linear problem. In the area solution, we determine by how much we must change the area of the tire so that it is equal to the area of the wheel. Knowing ΔA, we can determine ΔT. In the linear solution, we determine how much we must change the diameter of the wheel. Knowing Δd_t, we can determine ΔT.

Solution:

Area Method

$$A_w = \pi r_w^2 \qquad A_t = \pi r_t^2$$
$$\Delta A_t = A_w - A_t = \pi(r_w^2 - r_t^2) = \gamma A_t \Delta T = 2\alpha\pi r_t^2 \Delta T$$
$$\Delta T = [(r_w/r_t)^2 - 1]/2\alpha = 458 \text{ C°}$$
$$T_f = T_i + \Delta T = 483\text{°C}$$

Linear Method

$$d_w = 1.00 \text{ m} \qquad d_t = 0.995 \text{ m}$$
$$\Delta d_t = d_w - d_t = \alpha d_t \Delta T$$
$$\Delta T = [(d_w/d_t) - 1]/\alpha = 457 \text{ C°}$$
$$T_f = T_i + \Delta T = 482\text{°C}$$

The 1 C° difference in the two answers is due to rounding off.

Related Text Exercises: Ex. 16-13 through 16-22.

3 Heat Transfer (Section 16-6)

Review: Heat is the energy transferred between systems or between a system and its surroundings due solely to a temperature difference.

Heat may be transferred between systems or between a system and its surroundings by conduction, convection or radiation. In conduction, which is heat transfer through a substance, energy is passed from one layer of molecules to another. The equation that describes the rate at which this process occurs is

$$H = \Delta Q/\Delta t = kA(T_2 - T_1)/L$$

Figure 16.4

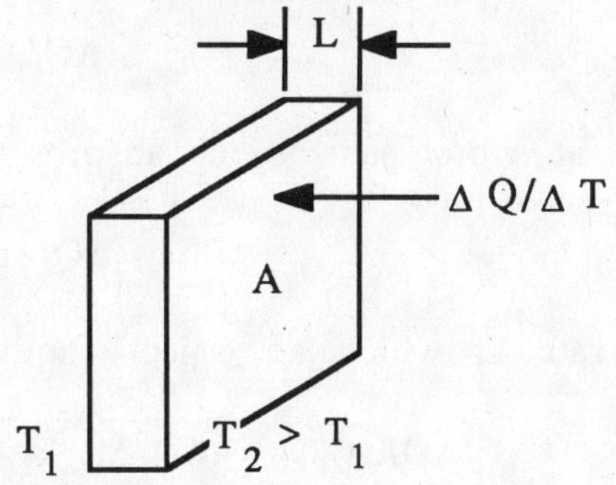

where T_1 and T_2 are temperatures, A is the cross-sectional area perpendicular to the heat transfer, L is the material thickness, k is the thermal conductivity and $\Delta Q/\Delta t$ is the rate of conduction of heat. If the heat is conducted through more than one type of material, this equation is modified to

$$H = \frac{\Delta Q}{\Delta T} = \frac{A(T_4 - T_1)}{\frac{L_1}{k_1} + \frac{L_2}{k_2} + \frac{L_3}{k_3}}$$

Figure 16.5

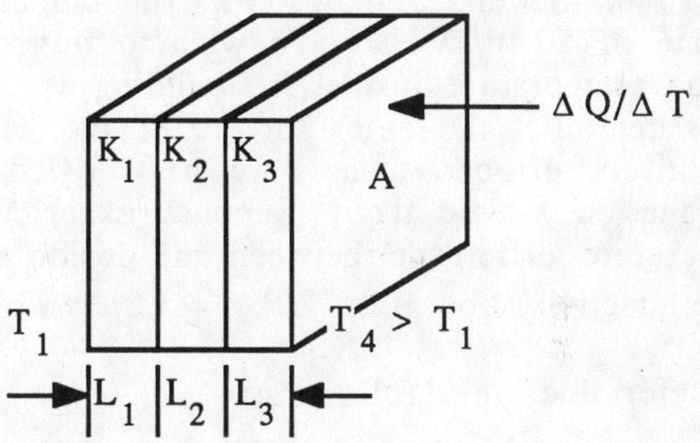

We can generalize the expression for heat conduction to

$$H = \Delta Q/\Delta t = kA\Delta T/\Delta x$$

We call the quantity $\Delta T/\Delta x$ the temperature gradient and the quantity $\Delta x/k$ the R-value. Hence the preceding equation may be written as

$$H = \Delta Q/\Delta t = A\Delta T/R$$

The R-value is useful in determining heat conduction through composite slab-type structures such as walls and floors. If the R-value and ΔT are known, we can easily determine the heat current per unit of area H/A (i.e. the heat conducted per unit of time per unit area $\Delta Q/\Delta tA$).

We may rewrite the expression for the rate of conduction through a composite material as

$$H = \Delta Q/\Delta t = A\Delta T/(R_1 + R_2 + R_3) = A\Delta T/R_{eff}$$

where ΔT is the temperature difference across the composite slab and the effective R-value for the composite slab is the sum of the individual R-values.

Radiation is heat transfer in the form of electromagnetic waves. All bodies are both emitting and absorbing radiant energy all the time. The rate at which an object emits radiant energy to its environment is

$$(\Delta Q/\Delta t)_{emit} = \varepsilon\sigma A T^4_{obj}$$

The rate at which an object absorbs radiant energy from the environment is

$$(\Delta Q/\Delta t)_{abs} = \varepsilon\sigma A T^4_{envir}$$

The net rate at which an object acquires energy is

$$(\Delta Q/\Delta t)_{net} = (\Delta Q/\Delta t)_{abs} - (\Delta Q/\Delta t)_{emit} = \varepsilon\sigma A(T^4_{envir} - T^4_{obj})$$

The net gain of energy will be positive if $T_{envir} > T_{obj}$ and negative if $T_{envir} < T_{obj}$.

Practice: (conduction) Single pane (SP) glass is 1/16 in. thick, and thermal pane (TP) glass is two single panes with an air gap of 1/4 in. A 1.00-m^2 window cost \$15 for SP and \$60 for TP. The building the window is to be placed in buys heat at the rate of \$0.10/kW•h. We wish to investigate the cost effectiveness of TP on a day when the temperature inside the building is 25°C and the temperature outside is -15°C. To assume that the outer surface of the window is -15°C and the inner surface is 25°C is entirely erroneous, as may be verified by touching the inner surface of a window pane on a cold day. Suppose experimentation determines that on such a day the temperature difference between the inside and outside of the window is 2.00°C. Thermal conductivity for glass is k_g = 1.00 W/(m•K) and for air is $k_a = 2.00 \times 10^{-2}$ W/(m•K).

Determine the following:

1. Rate at which heat is lost through SP	L_g = (1/16 in)(1 m/39.4 in) $= 1.59 \times 10^{-3}$ m $(\Delta Q/\Delta t)_{SP} = k_g A(T_1 - T_2)/L = 1.26 \times 10^3$ W
2. Amount of heat lost through SP in one day	$\Delta Q_{SP} = (\Delta Q/\Delta t)_{SP}t = (1.26 \times 10^3\ \text{W})(1\ \text{day})$ $= 1.09 \times 10^8$ J
3. Rate at which heat is lost through TP	$(\Delta Q/\Delta t)_{TP} = \dfrac{A(T_1 - T_2)}{(2L_g/k_g) + (L_a/k_a)}$ = 6.23 W

4. Amount of heat lost through TP in one day	$\Delta Q_{TP} = (\Delta Q/\Delta t)_{TP}t = (6.23\ W)(1\ day)$ $= 5.38 \times 10^5\ J$
5. Difference in the amount of heat lost in one day	$\Delta Q_{diff} = \Delta Q_{SP} - \Delta Q_{TP} = 1.08 \times 10^8\ J$
6. Cost of heat energy in dollar per joule	$Cost = (\frac{\$0.10}{kW \cdot h})(\frac{1\ h}{3600\ s})(\frac{1\ kW}{1000\ W})(\frac{1\ W}{1\ J/s})$ $= \$2.78 \times 10^{-8}/J$
7. Cost to supply difference in heat loss for one day	$Cost = (1.08 \times 10^8\ J)(\$2.78 \times 10^{-8}/J)$ $= \$3.00$
8. Cost difference between purchase of TP and SP	Cost difference = \$60.0 - \$15.0 = \$45.0
9. Number of days like the one being considered before TP would pay for itself	Cost/day = \$3.00/day Cost difference = \$45.00 N = \$45.00/(\$3.00/day) = 15.0 days

Example: (conduction) The outer walls of a house have a layer of wood 2.00 cm thick and a layer of styrofoam insulation 3.00 cm thick. The thermal conductivity of the wood and styrofoam, respectively, are 1.40×10^{-1} W/m•C° and 1.00×10^{-2} W/m•C°. The temperature of the interior wall is 20°C and that of the exterior wall -10°C. (a) How much heat flows through 1.00 m^2 of the wall in 1.00 h? (b) What is the temperature at the styrofoam-wood interface?

Given:

$L_w = 2.00 \times 10^{-2}\ m$ = thickness of wood
$L_s = 3.00 \times 10^{-2}\ m$ = thickness of styrofoam
$k_w = 1.40 \times 10^{-1}\ W/m \cdot C°$ = thermal conductivity of wood
$k_s = 1.00 \times 10^{-2}\ W/m \cdot C°$ = thermal conductivity of styrofoam
$T_{in} = 20°C$ = temperature of inside surface of styrofoam
$T_{out} = -10°C$ = temperature of outside surface of wood
$t = 1.00\ h$

Determine: (a) The amount of heat energy lost through 1.00 m^2 of the wall in 1.00 h. (b) The temperature of the styrofoam-wood interface.

Strategy: (a) From the given information we can determine the rate at which heat is conducted per square meter of the composite wall. Knowing the rate at which the heat is conducted and that the conduction time is 1.00 h, we can determine the amount of heat conducted through 1.00 square meter in 1.00 h. (b) Let's represent the temperature at the wood-styrofoam interface by T and then write an expression for the rate of transfer of heat per square meter for the wood and styrofoam separately. The rate of transfer of heat per square meter through either the styrofoam or the wood can be no different from that through the composite wall, and this latter value has been determined. Thus we can use either expression to obtain T.

Solution: (a) The rate at which heat is conducted per square meter of the composite wall is

$$\frac{(\Delta Q/\Delta t)}{A} = \frac{(T_1 - T_2)}{(L_w/k_w) + (L_s/k_s)} = 9.55 \text{ W/m}^2$$

(b) Represent the temperature at the interface by T, and write an expression for the rate of transfer of heat per square meter for each material.

$$\left(\frac{\Delta Q/\Delta t}{A}\right)_w = k_w[T - (-10°C)]/L_w \qquad \left(\frac{\Delta Q/\Delta t}{A}\right)_s = k_s(25°C - T)/L_s$$

The rate of transfer of heat per square meter through either the styrofoam or the wood is the same as that through the composite wall, which was determined in (a). We can use either of the preceding expressions to obtain T. Using the expression for wood, we obtain

$$T = \left[\left(\frac{\Delta Q/\Delta t}{A}\right)\left(\frac{L_w}{k_w}\right)\right] - 10°C = 3.64°C$$

You should use the expression for styrofoam to check our results.

Practice: (radiation) A student in a bathing suit lies in the sun, and the following information is known.

r_{sun}	$= 6.96 \times 10^8$ m	=	radius of sun
ε_{sun}	$= 1.00$	=	emissivity of sun
ε_{stu}	$= 0.980$	=	emissivity of student
A_{stu}	$= 1.50$ m^2	=	exposed area of student
d	$= 1.50 \times 10^{11}$ m	=	distance from sun to earth
T_{sun}	$= 5.80 \times 10^3$ K	=	surface temperature of sun
T_{stu}	$= 38°C = 311$ K	=	surface temperature of skin
σ	$= 5.67 \times 10^{-8}$ W/m^2•K^4	=	Stefan-Boltzmann constant
L_{per}	$= 2.43 \times 10^6$ J/kg	=	heat of vaporization for perspiration
$(\Delta Q/\Delta t)_{metab}$	$= 1.00 \times 10^3$ W	=	rate at which student's metabolism produces heat

Determine the following:

1. Rate at which the sun is emitting energy	$(\Delta Q/\Delta t)_{sun} = \sigma\varepsilon_{sun}A_{sun}T^4_{sun}$ $= 3.90 \times 10^{26}$ W
2. Area of the sun	$A_{sun} = 4\pi r^2_{sun} = 6.08 \times 10^{18}$ m^2
3. Rate at which energy is emitted per square meter of the sun's surface	$\frac{(\Delta Q/\Delta t)_{sun}}{A_{sun}} = 6.41 \times 10^7$ W/m^2

4. Rate at which energy from the sun arrives at a large sphere centered at the sun and with a radius equal to the distance from the sun to the earth	$(\frac{\Delta Q}{\Delta t})_{sphere} = (\frac{\Delta Q}{\Delta t})_{sun} = 3.90 \times 10^{26}$ W Energy arrives at this sphere at the same rate that it is emitted by the sun.
5. Rate per square meter at which energy arrives at the earth	The 3.90×10^{26} W is spread uniformly over the sphere of radius $d = 1.50 \times 10^{11}$ m. Consequently, the energy per square meter arriving at all points on this sphere is $(\Delta Q/\Delta t)_{sphere} = 3.90 \times 10^{26}$ W $A_{sphere} = 4\pi d^2 = 2.83 \times 10^{23}\ m^2$ $(\frac{\Delta Q/\Delta t}{A})_{sphere} = (\frac{\Delta Q/\Delta t_{sphere}}{A_{sphere}})$ $= 1.38 \times 10^3\ W/m^2$ Since the earth is on this large sphere of radius d, the rate per square meter at which energy arrives at the earth is also $(\frac{\Delta Q/\Delta t}{A})_{earth} = 1.38 \times 10^3\ W/m^2$.
6. Rate at which student absorbs radiant energy from the sun	$(\Delta Q/\Delta t)_{abs} = (\frac{\Delta Q/\Delta t}{A})_{earth} A_{stu}$ $= 2.07 \times 10^3$ W
7. Rate at which student radiates energy	$(\Delta Q/\Delta t)_{emit} = \sigma \varepsilon_{stu} A_{stu} T^4_{stu}$ $= 7.80 \times 10^2$W
8. Net rate at which heat is supplied to student via radiation	$(\Delta Q/\Delta t)_{rad} = (\Delta Q/\Delta t)_{abs} - (\Delta Q/\Delta t)_{emit}$ $= 1.29 \times 10^3$ W
9. Total rate at which heat is supplied to student via radiation and metabolism	$(\Delta Q/\Delta t)_{total} = (\Delta Q/\Delta t)_{rad} + (\Delta Q/\Delta t)_{metab}$ $= 2.29 \times 10^3$ W
10. The rate at which perspiration must evaporate in order for body heat to remain constant	$Q = mL_{per}$ $(\Delta Q/\Delta t) = (\Delta m/\Delta t)L_{per}$ $\Delta m/\Delta t = (\Delta Q/\Delta t)/L_{per} = 9.42 \times 10^{-4}$ kg/s

Example: (radiation) A glass of soda warms from 6.00°C to 10.0°C in 5.00 min. when the air temperature is 35.0°C. How long will it take to warm from 10.0°C to 15.0°C?

Given: $T_1 = 6.00°C = 279$ K $T_3 = 15.0°C = 288$ K $\Delta t = 5.00$ min
$T_2 = 10.0°C = 283$ K $T_a = 35.0°C = 308$ K

Determine: The length of time required for the soda to warm from T_2 to T_3.

Strategy: Write an expression for the net rate at which the soda absorbs energy by radiation while warming from T_1 to T_2 and again while warming from T_2 to T_3. These two expressions can be manipulated in such a manner as to eliminate all unknown quantities except the time to warm the soda from T_2 to T_3.

Solution: The rate at which the soda absorbs heat from the room while warming from T_1 to T_2 is

$$(\Delta Q/\Delta t)_{abs} = \sigma\varepsilon A T_a^4$$

The rate at which it loses heat to the room while warming from T_1 to T_2 is

$$(\Delta Q/\Delta t)_{emit} = \sigma\varepsilon A T_e^4$$

where $T_e = 8.00°C = 280$ K is halfway between T_1 and T_2.

The net rate at which the soda gains energy by radiation while warming from T_1 to T_2 is

$$(\Delta Q/\Delta t)_{net} = (\Delta Q/\Delta t)_{abs} - (\Delta Q/\Delta t)_{emit} = \sigma\varepsilon A(T_a^4 - T_e^4)$$

Since all of this heat goes into heating the soda, we can write

$$\Delta Q = mc\Delta T \quad \text{or} \quad \Delta Q/\Delta t = mc\Delta T/\Delta t, \quad \text{hence}$$

(α) $(\Delta Q/\Delta t) = \sigma\varepsilon A(T_a^4 - T_e^4) = mc\Delta T/\Delta t$, where $T_a = 308$ K and $T_e = 280$ K,
$\Delta T = T_2 - T_1 = 4.00$ K and $\Delta t = 500$ min

In like manner, as the soda warms from T_2 to T_3 we can write

(β) $(\Delta Q/\Delta t) = \sigma\varepsilon A(T_a^4 - T_e'^4) = mc\Delta T'/\Delta t'$, where $T_a = 308$ K and $T_e' = 285.5$ K,
$\Delta T' = T_3 - T_2 = 5.00$ K and $\Delta t'$ is to be determined

Divide (α) by (β) and solve for Δt'.

$$\Delta t' = \frac{(T_a^4 - T_e^4)}{(T_a^4 - T_e'^4)}\left(\frac{\Delta T'}{\Delta T}\right)\Delta t = \frac{(308^4 - 280^4)}{(308^4 - 285.5^4)}\left(\frac{5}{4}\right)(5.00 \text{ min}) = 7.57 \text{ min}$$

Related Text Exercises: Ex. 16-23 through 16-34.

PRACTICE TEST

Take and grade this practice test. Doing so will allow you to determine any weak spots in your understanding of the concepts taught in this chapter. The following section prescribes what you should study further to strengthen your understanding.

Sue decides to create her own temperature scale. She places an uncalibrated mercury thermometer into an ice-water bath and calls the mercury level 20.0°S. She then places the uncalibrated thermometer into boiling water and calls the mercury level 170.0°S. Finally, she divides the distance between these two levels into 150 equal lengths.

Determine the following:

________ 1. The reading on a Celsius thermometer when Sue's thermometer reads 95.0°S

________ 2. The reading on Sue's thermometer when a Fahrenheit thermometer reads 80.0°F

________ 3. The Kelvin temperature when Sue's thermometer reads 125°S

A steel cable 4.00 m long is stretched tightly across a driveway on a day when the temperature is 30.0°C. Determine the following on a day when the temperature is -10.0°C:

________ 4. The additional strain on the wire

________ 5. The additional stress on the wire

At room temperature (25.0°C), a copper sphere 3.0000 cm in diameter is 5. x 10^{-4} cm larger than the inside diameter of a steel ring.

________ 6. Determine the single temperature of both sphere and ring at which the sphere just slips through the ring.

A 150-cm^3 glass test tube (β_g = 2.00 x 10^{-5}°C^{-1}) is filled to the brim with glycerin at 0°C.

________ 7. Determine the amount of glycerin that will overflow if the test tube and contents are heated to 100°C.

Rods of aluminum, stainless steel, and copper are welded together to form a Y-shaped object as shown below. The free end of the copper is maintained at 100°C, and the free ends of the aluminum and stainless steel are maintained at 10.0°C. The rods are all 1.00 m long, have a cross-sectional area of 5.00 cm^2, and are insulated so that essentially no heat is lost from the surface.

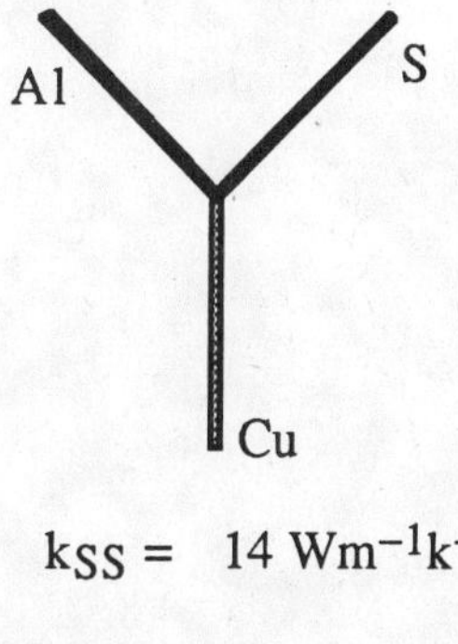

k_{Cu} = 401 $Wm^{-1}k^{-1}$ k_{Al} = 237 $Wm^{-1}k^{-1}$ k_{SS} = 14 $Wm^{-1}k^{-1}$

Determine the following:

_________ 8. Temperature at the junction
_________ 9. Rate (in cal/s) at which heat flows along copper rod
_________ 10. Temperature at midpoint of aluminum rod

Approximate the living space of a residence by a 50.0 ft by 50.0 ft floor and ceiling with 8.00 ft. walls. Suppose the interior is maintained at 70.0°F while the exterior surfaces of walls and ceiling are exposed to a steady 10.0°F and the exterior surface of the floor remains at 40.0°F. The wall structure has an effective R-value $R_w = 10$, the ceiling has $R_c = 15$, and the floor has $R_f = 8$, all in building industry units.

Determine the following:

_________ 11. Heat current in the ceiling
_________ 12. Heat current in the walls
_________ 13. Heat current in the floor
_________ 14. Heat loss to the outside in a 24.0 hr. period
_________ 15. Cost of heating for 24.0 hrs. if fuel is purchased at 5.00 cents per 10,000 BTU

A glass of soda warms from 8.00°C to 12.0°C in 5.00 min when the air temperature is 35.0°C. Assume the emissivity of the glass full of soda to be 0.800 and the area of the emitting surface to be 4.00×10^{-2} m^2.

Determine the following:

_________ 16. The rate at which the soda absorbs heat from the room while warming from 8.00°C to 12.0°C
_________ 17. The rate at which the soda loses heat to the room while warming from 8.00°C to 12.0°C
_________ 18. The net rate at which the soda gains energy by radiation while warming
_________ 19. Net heat gained by the soda is 10.0 min
_________ 20. Time for the soda to warm from 12.0°C to 18.0°C

(See Appendix I for answers.)

PRINCIPAL CONCEPTS AND EQUATIONS PRESCRIPTION

Your score on the practice test is an excellent measure of your understanding of the chapter. You should now use the following chart to write your own prescription for dealing with any weaknesses the practice test points out. Look down the leftmost column to the number of the question(s) you answered incorrectly, reading across that row you will find the concept and/or equation of concern, the section(s) of the study guide you should return to for further study, and some suggested text exercises which you should work to gain additional experience.

Practice Test Questions	Concepts and Equations	Prescription	
		Principal Concepts	Text Exercises
1	Temperature scales	1	16-10, 11
2	Temperature scales	1	16-11, 12
3	Temperature scales	1	16-10, 12
4	Linear thermal expansion: $\Delta L = \alpha L_0 \Delta T$	2	16-13, 15
5	Young's Modulus: $Y = \sigma_t/\varepsilon_t$	1 of Ch 15	15- 2, 5
6	Area thermal expansion: $\Delta A = 2\alpha A_0 \Delta T$	2	16-14, 15
7	Volume thermal expansion: $\Delta V = 3\alpha V_0 \Delta T$	2	16-18, 19
8	Conduction: $\Delta Q/\Delta t = kA\Delta T/\Delta x$	3	16-23, 25
9	Conduction	3	16-25, 26
10	Conduction	3	16-23, 25
11	Conduction: $\Delta Q/\Delta t = A\Delta T/R$	3	16-26, 27
12	Conduction	3	16-27, 28
13	Conduction	3	16-28, 29
14	Conduction	3	16-30, 31
15	Amount = (Rate)(time)	---	16-27, 31
16	Radiation: $\Delta Q/\Delta t = \varepsilon\sigma A T^4$	3	16-32, 33
17	Radiation	3	16-33, 34
18	Radiation	3	16-32, 34
19	Radiation	3	16-32, 33
20	Radiation	3	16-33, 34

17 THE FIRST LAW OF THERMODYNAMICS

RECALL FROM PREVIOUS CHAPTERS

Previously learned concepts and equations frequently used in this chapter	Text Section	Study Guide Page
Work by a constant force: $W = \mathbf{F} \cdot \mathbf{L}$	8-1	8-4
Pressure: $P = F/A$	15-1	15-1
Temperature scales: $T = t_c + 273$	16-3, 4	16-1

NEW IDEAS IN THIS CHAPTER

Concepts and equations introduced		Text Section	Study Guide Page
Equation of state:	$PV = nRT$	17-1	17-1
Specific heat:	$Q = mc\Delta T$	17-2	17-6
Latent heat:	$Q = mL$	17-2	17-6
Isochoric process: $\Delta Q = \Delta U = nc_v\Delta T$	$\Delta V = 0,\ W = 0,$	17-3	17-10
Isobaric process: $\Delta U = nc_v\Delta T,$	$\Delta P = 0,\ \Delta Q = nc_p\Delta T,$ $W = P\Delta V = nR\Delta T$	17-3	17-10
Isothermal process: $\Delta Q = W = nRT \ln(V_f/V_i)$	$\Delta T = 0,\ \Delta U = 0,$	17-3	17-10
Adiabatic process: $W = -\Delta U = [P_iV_i/\gamma - 1]\ [1 - (V_i/V_f)^{\gamma-1}]$	$\Delta Q = 0,\ \gamma = c_p/c_v,$	17-3	17-10
First law of thermodynamics:	$\Delta U = Q - W$	17-4	17-13

PRINCIPAL CONCEPTS AND EQUATIONS

1 Equation of State of an Ideal Gas (Section 17-1)

Review: When working with a gas the variables of importance are as follows:

P - pressure of the gas measured in Pascals, atmospheres or N/m^2
V - volume of the gas measured in m^3
T - temperature of the gas measured in Kelvin degrees
N - number of molecules of the gas
N_A - number of particals (i.e. atoms, molecules, ions) per mole of any given substance (Avogadro's number: 6.023×10^{23} mole^{-1})

M - molecular mass of the gas
M - mass of the gas
m - mass of a molecule of the gas - determined by $m = M/N$ or $m = \mathrm{M}/N_A$
n - number of moles of the gas - determined from the mass of the gas ($n = M/\mathrm{M}$) or the number of molecules of the gas ($n = N/N_A$)
ρ - density of the gas - determined by $\rho = M/V$

Experiments show that these variables are related as follows:

$P \propto 1/V$ or $PV = \text{const}$ if $T, n = \text{const}$

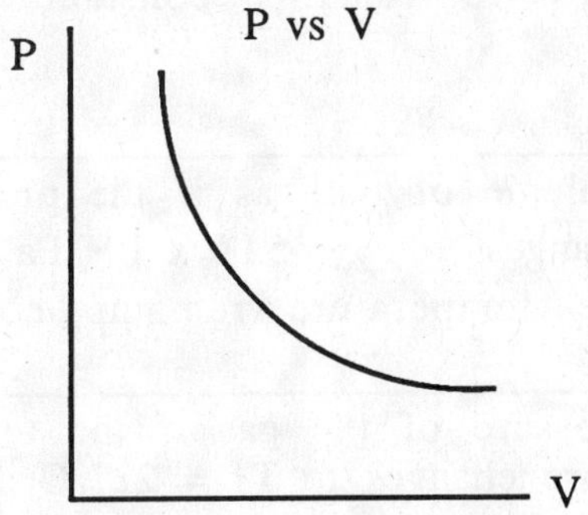

Figure 17.1

$P \propto T$ or $P/T = \text{const}$ if $V, n = \text{const}$

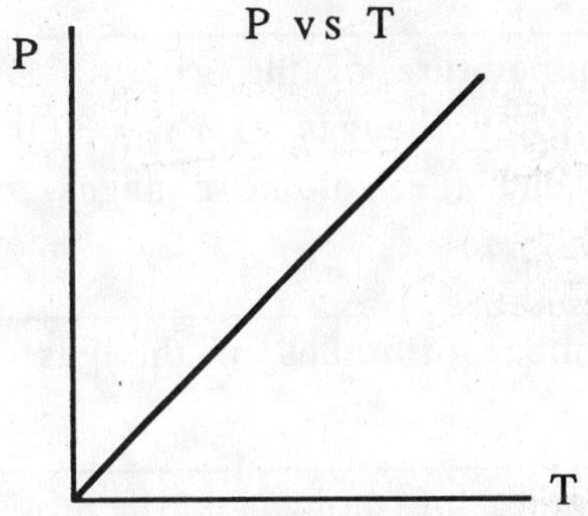

Figure 17.2

$P \propto n$ or $P/n = \text{const}$ if $T, V = \text{const}$

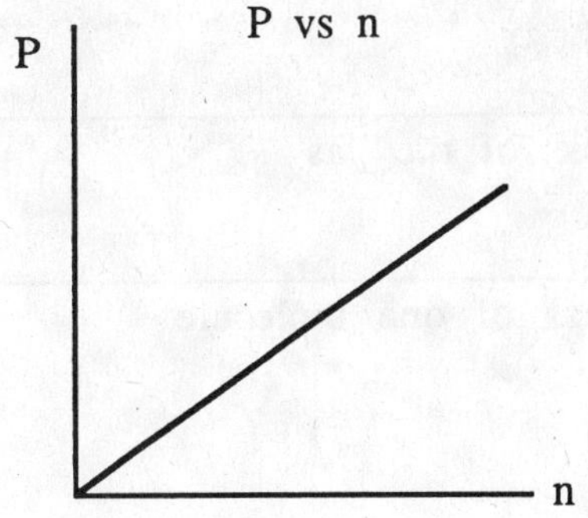

Figure 17.3

$PV \propto nT$ or $\dfrac{PV}{nT} = \text{const} = R$ a universal constant independent of the gas.

This last relationship may be written as

$$PV = nRT$$

and it is called the equation of state of an ideal gas.

Practice: The following is known about a quantity of an ideal gas.

$V_1 = 1.00 \times 10^{-2}\ m^3$ $T_1 = 27.0°C$ $P_1 = 2.00 \times 10^5\ Pa$ $\rho = 1.28\ kg/m^3$

Determine the following:

1. Volume of the gas if the temperature changes to $T_2 = 227°C$ and the pressure remains constant	$V_1/T_1 = V_2/T_2$ at constant P $V_2 = V_1T_2/T_1$ $V_2 = (1.00 \times 10^{-2}\ m^3)(500\ K/300\ K)$ $V_2 = 1.67 \times 10^{-2}\ m^3$
2. Volume of the gas if the pressure changes to $P_2 = 5.00 \times 10^5$ Pa and the temperature remains constant	$P_1V_1 = P_2V_2$ at constant T $V_2 = P_1V_1/P_2$ $V_2 = 4.00 \times 10^{-3}\ m^3$
3. Pressure of the gas if the temperature changes to $T_2 = 227°C$ and the volume remains constant	$P_1/T_1 = P_2/T_2$ at constant V $P_2 = P_1T_2/T_1 = 3.33 \times 10^5$ Pa
4. Temperature of the gas in °C if the pressure changes to $P_2 = 3.00 \times 10^5$ Pa and the volume changes to $1.50 \times 10^{-2}\ m^3$	$P_1V_1/T_1 = P_2V_2/T_2$ $T_2 = P_2V_2T_1/P_1V_1 = 675$ K $t_C = T - 273 = 402°C$
5. Number of moles of the gas	$PV = nRT$ $n = PV/RT = 0.802$ mole
6. Number of molecules of the gas	Every mole of the gas contains N_A molecules. $N = nN_A = 4.83 \times 10^{23}$ molecules
7. Mass of the gas	$\rho = M/V$ $M = \rho V = 1.28 \times 10^{-2}$ kg
8. Mass of one molecule	We just determined that N molecules have a mass M, so the mass per molecule is $m = M/N = 2.65 \times 10^{-26}$ kg/molecule
9. Molar mass of the gas	The molar mass is the mass of one mole or N_A molecules $M = mN_A = 1.60 \times 10^{-2}$ kg/mole.

Example: A scuba diver's 1.00×10^{-2} m^3 tank is filled with air at a gauge pressure of 1.50×10^7 Pa. If the diver uses 2.50×10^{-2} m^3 of air per minute at the same pressure as the water pressure at her depth below the surface, how long can she remain under water at a depth of 20.0 m in a fresh-water lake? The temperature of the water falls from 25.0°C at the top to 20.0°C as she dives.

Given:

$V_1 = 1.00 \times 10^{-2}$ m^3 = volume of the tank
$P_{1G} = 1.50 \times 10^7$ Pa = initial gauge pressure of the tank
$r = 2.50 \times 10^{-2}$ m^3/60 s = 4.17×10^{-4} m^3/s = rate of use of air
h = 20.0 m = depth of scuba diver
P_{2G} = pressure at which the diver uses oxygen
= same as the water pressure at her depth
T_1 = 25.0°C = water temperature at the surface
T_2 = 20.0°C = water temperature at 20.0 m
$\rho_w = 1.00 \times 10^3$ kg/m^3 = density of water

Determine: The length of time the diver can remain under water (at a depth of 20.0 m) with this tank of air.

Strategy: From the given information, we can determine the absolute pressure and temperature at the surface and at 20.0 m. Since the volume at the surface is given, we can determine the volume available to the diver at 20.0 m. Knowing the volume available and the rate of consumption, we can determine the time the tank of air will last.

Solution: First, let's determine the absolute pressure at the surface and at a depth of 20.0 m.

Surface $P_1 = P_{1G} + P_A = 150 \times 10^5 \text{ Pa} + 1.01 \times 10^5 \text{ Pa} = 151 \times 10^5 \text{ Pa}$
Depth of 20.0 m $P_2 = \rho gh + P_A = 1.96 \times 10^5 \text{ Pa} + 1.01 \times 10^5 \text{ Pa} = 2.97 \times 10^5 \text{ Pa}$

Next, let's determine the absolute temperature at the surface and at 20.0 m.

Surface $T_1 = t_{C1} + 273 = 298 \text{ K}$
Depth of 20.0 m $T_2 = t_{C2} + 273 = 293 \text{ K}$

Using the ideal gas law we can determine the amount of air available at 20.0 m.

$$P_1V_1/T_1 = P_2V_2/T_2 \quad \text{or} \quad V_2 = P_1V_1T_2/P_2T_1 = 0.500 \text{ m}^3$$

Because the tank is still full of air when it is all "used up" (no more air is available to the diver when the pressure inside the tank is the same as that outside the tank), the volume of air available is

$$V_{available} = V_2 - V_1 = 0.490 \text{ m}^3$$

The time this volume of air will last when being consumed at the rate r is

$$\text{Amount} = \text{rate x time} \qquad \text{or} \qquad t = V_{available}/r = 1.18 \times 10^3 \text{ s}$$

Example: The volume of an oxygen tank is 2.00×10^{-2} m^3. As oxygen is withdrawn from the tank, the reading on a pressure gauge drops from 10.0×10^5 Pa to 2.00×10^5 Pa and the temperature of the gas in the tank drops from 30.0°C to 15.0°C. (a) How many kilograms of oxygen were originally in the tank? (b) How many kilograms were withdrawn? (c) What volume would the withdrawn oxygen occupy at STP?

Given: $V_1 = 20.0 \times 10^{-3}$ m^3 $P_{1G} = 10.0 \times 10^5$ Pa $T_1 = 30.0°C$
$P_{2G} = 2.00 \times 10^5$ Pa $T_2 = 15.0°C$

Determine: (a) The mass of oxygen originally in the tank.
(b) The mass of oxygen withdrawn from the tank.
(c) The volume of the withdrawn oxygen at STP.

Strategy: Using the ideal gas law, we can determine the initial and final number of moles of oxygen in the tank. Knowing the number of moles and the molecular mass for oxygen, we can determine the initial and final mass of oxygen in the tank. Knowing the initial and final mass of oxygen in the tank, we can determine the mass withdrawn. Knowing the mass withdrawn, we can determine the number of moles withdrawn. Finally, we can use the ideal gas law to determine the volume of the withdrawn oxygen at STP.

Solution: We can find the initial and final number of moles in the tank by using the ideal gas law.

$$P_1 = P_{1G} + P_A = 11.0 \times 10^5 \text{ Pa} \qquad P_2 = P_{2G} + P_A = 3.01 \times 10^5 \text{ Pa}$$
$$T_1 = 30.0°C = 303 \text{ K} \qquad T_2 = 15°C = 288 \text{ K}$$
$$n_1 = P_1V_1/RT_1 = 8.74 \text{ moles} \qquad \text{and} \qquad n_2 = P_2V_2/RT_2 = 2.52 \text{ moles}$$

The molecular mass M (that is, the mass of one mole) of oxygen is 32.0×10^{-3} kg (recall that oxygen is diatomic O_2). The initial and final masses of oxygen gas in the tank are

$$M_1 = n_1M = 2.80 \times 10^{-1} \text{ kg} \qquad \text{and} \qquad M_2 = n_2M = 8.06 \times 10^{-2} \text{ kg}$$

The mass of gas withdrawn is $M_w = M_1 - M_2 = 1.99 \times 10^{-1}$ kg

The number of moles withdrawn is $n_w = n_1 - n_2 = 6.22$ moles

The volume the withdrawn oxygen occupies at STP is

$$V_w = n_wRT/P \qquad T = 0°C = 273 \text{ K} \qquad P = 1.01 \times 10^5 \text{ Pa} \qquad \text{hence} \qquad V_w = 1.40 \times 10^{-1} \text{ m}^3$$

Related Text Exercises: Ex. 17-1 through 17-7.

2 Specific Heat and Latent Heat (Section 17-2)

Review: The specific heat capacity at constant pressure c_p is defined as the amount of heat ΔQ needed to raise the temperature of a unit of mass by 1 K° at constant pressure.

$$c_p = \Delta Q/m\Delta T$$

The specific heat capacity at constant volume c_v is defined as the amount of heat ΔQ needed to raise the temperature of a unit of mass by 1 K° at constant volume.

$$c_v = \Delta Q/m\Delta T$$

For most solids and liquids $c_p \equiv c_v$ and we just represent it by c, hence

$$c = \Delta Q/m\Delta T$$

It will be more representative of what is taking place if we drop the Δ from ΔQ. The reason for this is that Q represents the heat in an energy transfer and ΔQ should be associated with a change in the amount of heat. This allows us to write the heat in an energy transfer due to a temperature change as

$$Q = mc\Delta T$$

Heat is measured in calories. One calory is the amount of heat needed to raise the temperature of 1 g of water 1 C°. This will be the case if the specific heat of water is

$$c_{water} = 1 \text{ cal g}^{-1} \text{ C}^{\circ -1}$$

Since heat is a form of energy it will be convenient to express calories in joules.

$$1 \text{ cal} = 4.186 \text{ J}$$

When a substance is changing phase, its temperature does not change even though we are supplying heat to it. The amount of heat per unit mass which must be added to or removed from a substance in order for it to change phase is called its latent heat.

$$Q = mL$$

If the phase change is between the solid and liquid phase $L = L_f$, the latent heat of fusion is used. If the phase change is between the liquid and vapor phase $L = L_v$, the latent heat of vaporization is used. The respective values are as follows:

$$L_f = 0.335 \text{ MJ/kg} = 80.0 \text{ cal/g}$$
$$L_v = 2.26 \text{ MJ/kg} = 540 \text{ cal/g}$$

Practice: A 40.0-kg steel engine expends energy at the rate of 200 W while idling. Twenty percent of this energy goes into heating the engine by internal friction. The design is such that the engine radiates 90.0% of all the heat produced. The specific heat of steel is 449 J/kg•K.

Determine the following:

1. Rate (in J/s) at which energy is being consumed by friction	$(E/t)_f = 0.200(E/t)$ $= 0.200(200\ W) = 40.0\ J/s$
2. Rate (in cal/s) at which heat is produced due to friction	$(Q/t)_f = (E/t)_f(1\ cal/4.184\ J)$ $= (40.0\ J/s)(1\ cal/4.184\ J)$ $= 9.56\ cal/s$
3. Total amount of heat (in cal) produced due to friction in 10.0 min	$Q_{total\ f} = (Q/t)_f t = (9.56\ cal/s)(10.0\ min)$ $= 5.74 \times 10^3\ cal$
4. Amount of heat (in cal) available to raise the temperature of the engine during the 10.0 min	Since the engine radiates 90% of the heat produced, 10% is available to heat the engine. $Q_{heat} = 0.100\ Q_{total\ f} = 5.74 \times 10^2\ cal$
5. Change in temperature of the engine during the 10.0 min	$m = 40.0\ kg \qquad c = 449\ J/kg \cdot K$ $Q_{heat} = (5.74 \times 10^2\ cal)(4.184\ J/cal)$ $= 2.40 \times 10^3\ J$ $\Delta T = Q_{heat}/mc = 0.134\ K = 0.134\ C°$
6. Amount of water that could undergo a temperature change of 100 C° during the 10.0 min if all the radiated heat was available for heating water	$Q_{rad} = 0.900\ Q_{total\ f} = 5.17 \times 10^3\ J$ $\Delta T = 100\ K,\ c = 4.18 \times 10^3\ J/kg \cdot K$ $m = Q_{rad}/c\Delta T = 1.24 \times 10^{-2}\ kg$

Figure 17.4 shows a plot of temperature vs. heat energy supplied to 1.00 kg of some substance. The substance is a solid at temperatures lower than -10.0°C and a gas at temperatures above 20.0°C.

Figure 17.4

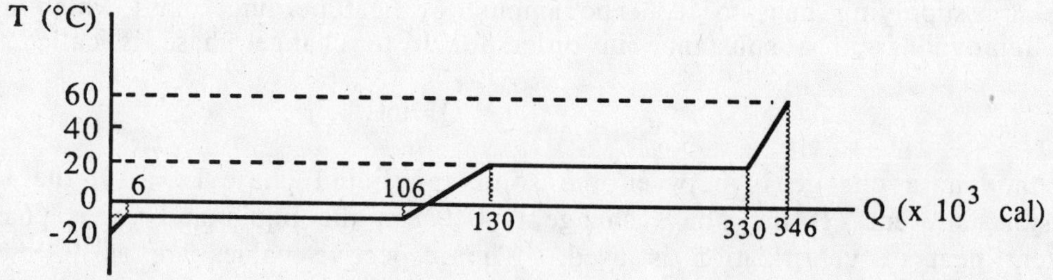

Determine the following:

1. Amount of heat energy needed to raise the temperature of the substance from -20.0°C to -10.0°C	$Q = Q_f - Q_i = 6.00 \times 10^3\ cal$ (Values are read from the graph.)

2. Specific heat of the substance as a solid	$Q = 6.00 \times 10^3$ cal $= 2.51 \times 10^4$ J $\Delta T = 10.0$ C° $= 10.0$ K°, $m = 1.00$ kg $c_{solid} = Q/m\Delta T = 2.51 \times 10^3$ J/kg•K
3. Melting point for the substance	$T_{melt} = -10.0°C$
4. Amount of heat required to melt the substance	$Q = Q_f - Q_i = 100 \times 10^3$ cal (Values are read from the graph.)
5. Latent heat of fusion for the substance	$Q = 1.00 \times 10^5$ cal $= 4.18 \times 10^5$ J $m = 1.00$ kg $L_f = Q/m = 4.18 \times 10^5$ J/kg
6. Specific heat of the substance as a liquid	$Q = 24.0 \times 10^3$ cal $= 1.00 \times 10^5$ J $\Delta T = 30.0$ C° $= 30.0$ K°, $m = 1.00$ kg $c_{liquid} = Q/m\Delta t = 3.33 \times 10^3$ J/kg•K
7. Latent heat of vaporization for the substance	$Q = 200 \times 10^3$ cal $= 8.37 \times 10^5$ J $m = 1.00$ kg $L_v = Q/m = 8.37 \times 10^5$ J/kg
8. Specific heat of the substance as a gas	$Q = 16.0 \times 10^3$ cal $= 6.69 \times 10^4$ J $\Delta T = 40.0$ C° $= 40$ K°, $m = 1.00$ kg $c_{gas} = Q/m\Delta T = 1.67 \times 10^3$ J/kg•K

Example: A lead sphere is dropped 10.0 m into a large tank of water. When the sphere enters the water, its kinetic energy is quickly dissipated through friction. If 80.0% of the work done by friction goes into heating the sphere, what is the change in temperature of the sphere?

Given: $\Delta h = 10.0$ m = distance sphere falls before entering water
0.800 = fraction of work done by friction that goes into heating sphere
$c = 128$ J/kg•K = specific heat of lead

Determine: The change in temperature of the sphere as its kinetic energy is dissipated through friction.

Strategy: Knowing the distance the sphere falls, we can write an expression for its decrease in gravitational potential energy. This decrease in gravitational potential energy is equal to the increase in kinetic energy, which in turn is equal to the amount of work done by friction on the sphere. Knowing that 80.0% of this work done by friction goes into heating the sphere, we can determine the heat supplied to the sphere and hence its temperature change.

Solution: Since the sphere falls a distance h before hitting the water, an expression for its change in gravitational potential energy is $\Delta PE = -mg\Delta h$. Since this decrease in gravitational potential energy is equal to the increase in kinetic energy, which is dissipated by doing work against friction, we can write

$$W_f = \Delta KE = -\Delta PE = mg\Delta h$$

Since 80.0% of this energy goes into heating the sphere, we can write

$$Q = 0.800W_f = 0.800mg\Delta h$$

We can then determine the temperature change as follows:

$$Q = mc\Delta T = 0.800mg\Delta h \quad \text{or} \quad \Delta T = 0.800g\Delta h/c = 0.613\ K° = 0.613\ C°$$

Example: A passive-solar house is to have storage facilities for 1.00×10^8 J of heat energy, to be stored in a salt with the following physical properties.

$T_{melt} = T_{freeze} = 35.0°C$ $\quad L_f = 3.00 \times 10^5$ J/kg $\quad \rho = 1.50 \times 10^3$ kg/m^3
$c_s = 2.00 \times 10^3$ J/kg•K = specific heat as a solid
$c_L = 3.00 \times 10^3$ J/kg•K = specific heat as a liquid

The minimum and maximum temperatures of the salt are, respectively, 25.0°C and 55.0°C. Determine the minimum storage space required.

Given: T_{melt}, L_f, ρ, c_s, c_L, Q_{total}, T_i and T_f

Determine: The minimum volume of storage space required for the salt.

Strategy: Under the most severe conditions, the salt will start the day as a solid at 25.0°C. During the day, it will absorb heat to (a) warm as a solid to its melting point, (b) melt (i.e., undergo a phase change from a solid to a liquid), and (c) warm as a liquid to 55.0°C. We can write an expression for the amount of heat required for each step and then equate the sum of these expressions to the total amount of heat to be stored. The only unknown in this equation is the mass of the salt. Once the mass of salt is known, we can use its density to determine the volume of salt and hence the minimum storage requirements.

Solution: Expressions for the heat required to warm the salt as a solid, melt it, and warm it as a liquid are

$$Q_s = mc_s\Delta T_s \qquad Q_{melt} = mL_f \qquad Q_L = mc_L\Delta T$$

The sum of these three expressions gives the total heat to be stored:

$$Q_{total} = mc_s\Delta T_s + mL_f + mc_L\Delta T_L$$

$$m = Q_{total}/(c_s\Delta T_s + L_f + c_L\Delta T_L) = 2.63 \times 10^2 \text{ kg}$$

The volume of salt and hence the minimum storage requirement is

$$V = m/\rho = (2.63 \times 10^2 \text{ kg})/(1.50 \times 10^3 \text{ kg/m}^3) = 0.175 \text{ m}^3$$

Related Text Exercises: Ex. 17-9 through 17-16.

3 Work (Section 17-3)

Review: Energy which is transferred between a system and its environment due to a temperature difference between them is heat. Energy which is transferred between a system and its environment by means independent of the temperature difference between them is work.

Figure 17.5

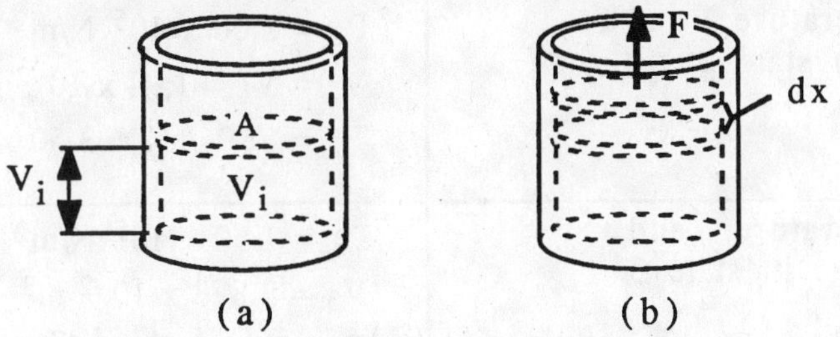

Figure 17.5(a) shows some amount of an ideal gas confined to a cylinder fitted with a leak-tight movable piston. If the gas exerts a force F on the piston and expands, (Figure 17.5(b)) moving the piston an infinitesimal amount dx, the work done by the gas is

$$dW = Fdx = PAdx = PdV$$

If $dV > 0$, then F is in the same direction as dx, and $dW > 0$
If $dV < 0$, then F is in the opposite direction of dx, and $dW < 0$.

The gas can do work for a number of different processes. This is summarized below.

Process			Work done by gas
Isochoric	or	$\Delta V = 0$	$W = 0$
Isoboric	or	$P = \text{constant}$	$W_{if} = P_i(V_f - V_i)$
Isothermal	or	$T = \text{constant}$	$W_{if} = nRT \ln(V_f/V_i)$
Adiabatic	or	$Q = 0$	$W_{if} = \dfrac{P_iV_i}{\gamma - 1}\left[1 - \left(\dfrac{V_i}{V_f}\right)^{\gamma-1}\right]$, $\gamma = c_p/c_v$

Practice: A cylinder contains 2.00 moles of helium gas. The P-V and P-T plots in Figure 17.6 show the thermodynamic processes and information for this gas as it goes through a complete cycle ($1 \rightarrow 2 \rightarrow 3 \rightarrow 1$).

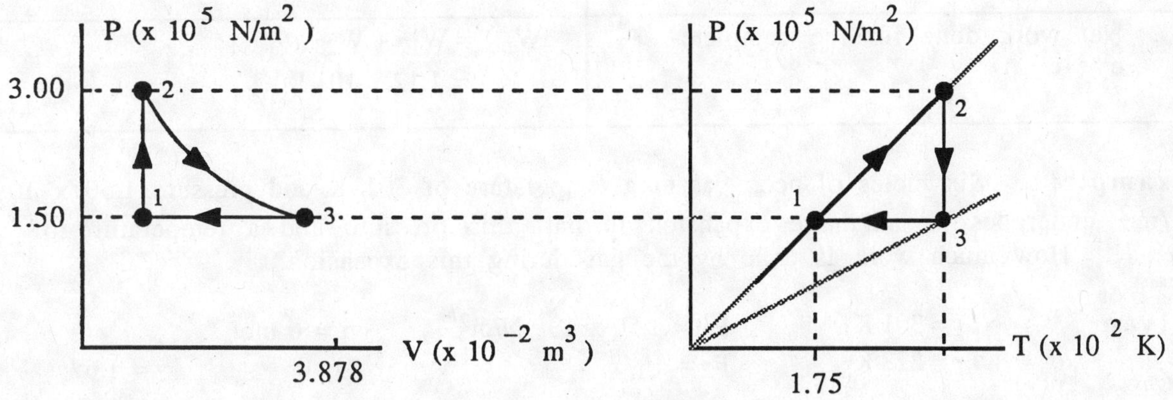

Figure 17.6

Determine the following:

1. Pressure, temperature and volume of the gas in state 1	$P_1 = 1.50 \times 10^5$ N/m^2 (P-V plot) $T_1 = 1.75 \times 10^2$ K (P-T plot) $V_1 = nRT_1/P_1 = 1.940 \times 10^{-2}$ m^3
2. Pressure, temperature, and volume of gas in state 2	$P_2 = 3.00 \times 10^5$ N/m^2 (P-V plot) $V_2 = V_1 = 1.94 \times 10^{-2}$ m^3 (P-V plot) $T_2 = V_2P_2/nR = 3.50 \times 10^2$ K
3. Pressure, temperature, and volume of the gas in state 3	$P_3 = 1.50 \times 10^5$ N/m^2 (P-V plot) $V_3 = 3.878 \times 10^{-2}$ m^3 (P-V plot) $T_3 = T_2 = 3.50 \times 10^2$ K (P-T plot)
4. Thermodynamic process the gas undergoes as it goes from state 1 to state 2	$\Delta V = 0 \Rightarrow$ Isochoric
5. Work done by the gas as it goes from state 1 to state 2	$W_{12} = P_1\Delta V = 0$
6. Thermodynamic process the gas undergoes as it goes from state 2 to state 3	$\Delta T = 0 \Rightarrow$ Isothermal
7. Work done by the gas as it goes from state 2 to state 3	$W_{23} = nRT_2 \ln(V_3/V_2)$ $= 4.03 \times 10^3$ J
8. Thermodynamic process the gas undergoes as it goes from state 3 to state 1	$\Delta P = 0 \Rightarrow$ Isobaric
9. Work done by the gas as it goes from state 3 to state 1	$W_{31} = P_3(V_1 - V_3) = -2.91 \times 10^3$ J
10. Net work done for the complete cycle	$W_{net} = W_{12} + W_{23} + W_{34}$ $= 1.12 \times 10^3$ J

Example: Six moles of neon gas at a temperature of 301 K and pressure 1.50×10^7 N/m^2 undergoes an adiabatic expansion to half this pressure and a temperature of 228 K. How much work is done by the gas during this expansion.

Given: $T_1 = 301$ K $P_1 = 1.50 \times 10^7$ N/m^2 $n = 6$ mol
$T_2 = 228$ K $P_2 = P_1/2$ $Q = 0$ $\gamma = 1.67$

Determine: The amount of work done by the gas during this expansion.

Strategy: Knowing T_1, P_1 and n we can use the equation of state to determine V_1. Knowing that the gas expands to half the initial pressure, we can determine the final pressure, P_2. Knowing P_2, T_2 and n we can use the equation of state to determine V_2. Knowing P_1, V_1, V_2 and γ, we can determine W_{12}.

Solution: Determine V_1 from the equation of state.

$$V_1 = nRT_1/P_1 = 1.00 \times 10^{-3}\ m^3$$

Knowing the gas expands to half the initial pressure, we can determine P_2

$$P_2 = P_1/2 = 7.50 \times 10^6\ N/m^2$$

Knowing n, P_2 and T_2, we can determine V_2 from the equation of state.

$$V_2 = nRT_2/P_2 = 1.51 \times 10^{-3}\ m^3$$

We may now determine W_{12}.

$$W_{12} = \frac{P_1V_1}{\gamma - 1}\left[1 - \left(\frac{V_1}{V_2}\right)^{\gamma - 1}\right] = 5.32 \times 10^3\ J$$

Related Text Exercises: Ex. 17-17 through 17-23.

4 The First Law of Thermodynamics (Section 17-4)

Review: A mathematical statement of the first law of thermodynamics is

$$\Delta U = Q - W$$

Q is the heat energy transferred between the system and its surroundings.

Q=+ if heat energy is added to the system
Q=- if heat energy is extracted from the system

The heat added to or extracted from a system in the process of going from an initial state to a final state depends on the process used to get from the initial to the final state.

W is the work done on or by the system.

W=+ if the system does work
W=- if work is done on the system

The work done on or by the system depends not only on the initial and final states of the system but also on how the process is performed.

U is the internal energy of the system. Notice that the internal energy of the system may be changed by adding or extracting heat and by work being done on or by the system.

If heat is added, $Q = +$, the internal energy of the system increases, and $\Delta U = +$.
If heat is extracted, $Q = -$, the internal energy of the system decreases, and $\Delta U = -$.
If the system does work, $W = +$, the internal energy of the system decreases, and $\Delta U = -$.
If work is done on the system, $W = -$, the internal energy of the system increases, and $\Delta U = +$.

ΔU depends only on the initial and final states and is independent of the process used to get form the initial to the final state. While Q and W depend on the process the quantity Q - W does not.

If a system starts out in some initial state, makes several successive changes, and returns to its original state, then $\Delta U = 0$.

Practice: Figure 17.7 shows two (path a and path b) of the numerous possible ways of taking an ideal gas from state 1 (P_1, V_1, T_1) to state 2 (P_2, V_2, T_2) and one (path c) of the numerous possible ways of taking a gas from state 2 to state 1.

Figure 17.7

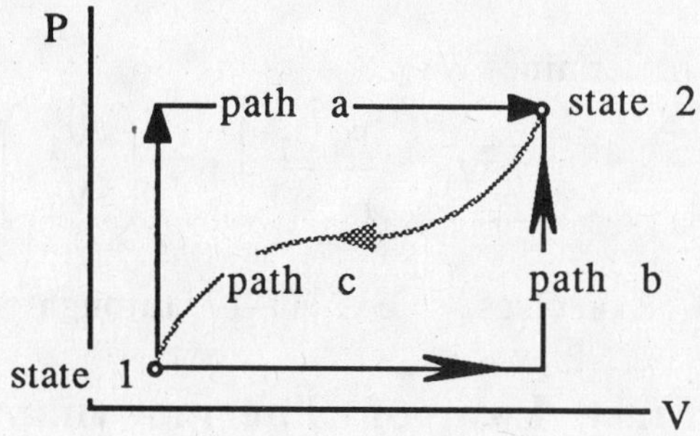

When the gas goes from state 1 to state 2 along path a, it absorbs 100 J of heat and does 50.0 J of work.

When the gas goes from state 1 to state 2 along path b, 80.0 J of heat flows into the system.

In order to return the gas to state 1 from state 2 along path c, 40.0 J of work must be done on the gas.

Determine the following:

1. Change in internal energy of the gas as it goes from state 1 to state 2 along path a	$\Delta U = Q - W$ $Q = +100$ J and $W = +50.0$ J (given) $\Delta U = +100\text{ J} - (+50.0\text{ J}) = +50.0\text{ J}$ The internal energy increases by 50.0 J
2. Change in internal energy of the gas as it goes from state 1 to state 2 along path b	ΔU = constant between any two states regardless of path; hence $\Delta U_b = \Delta U_a = +50.0\text{ J}$
3. Change in internal energy of the gas for a complete cycle (state 1 to state 2 back to state 1)	For a complete cycle, the final state of the gas is the same as the initial state. Hence: $\Delta U_{1\rightarrow 2\rightarrow 1} = 0$

4. Change in internal energy of the gas as it goes from state 2 to state 1 along path c	No matter how the gas goes from state 1 to state 2, $\Delta U_{1\to 2} = +50$ J For a complete cycle, $\Delta U_{1\to 2\to 1} = \Delta U_{1\to 2} + \Delta U_{2\to 1} = 0$ consequently, $\Delta U_{2\to 1} = -\Delta U_{1\to 2} = -50$ J No matter how the gas goes from state 2 to state 1, $\Delta U_{2\to 1} = -50$J hence $\Delta U_c = -50$J
5. Work done by the gas in going from state 1 to state 2 along path b	$\Delta U_b = +50.0$ J (step 2) $Q_b = +80.0$ J (given) $W_b = Q_b - \Delta U_b = +30.0$ J
6. Change in heat of the gas in going from state 2 to state 1 along path c	$\Delta U_c = -50.0$ J (step 4) $W_c = -40.0$ J (given) $Q_c = \Delta U_c + W_c = -90.0$ J
7. Net work for a complete cycle along paths a and c	$W_a = +50.0$ J and $W_c = -40.0$ J (given) $W_{net} = W_a + W_c = +10.0$ J
8. Net change in heat for a complete cycle along paths a and c	$Q_a = +100.0$ J (given) $Q_c = -90.0$ J (step 6) $Q_{net} = Q_a + Q_c = +10.0$ J Note $\Delta U_{net} = Q_{net} - W_{net}$ $\Delta U_{net} = 0$ (step 3) Hence $W_{net} = Q_{net}$ which agrees with step 7

Example: During a thermodynamic process, 300 cal of heat is added to a system and its internal energy decreases by 500 J. How much work is done on or by this system?

Given: $Q = +300$ cal $\Delta U = -500$ J

Determine: The amount of work done on or by this system.

Strategy: We can use the given information and the first law of thermodynamics to determine the work W. The sign of W will allow us to determine whether the system is doing work or if work is being done on it.

Solution: $\Delta U = -500$ J

$Q = (+300 \text{ cal})(4.184 \text{ J/cal}) = +1260$ J

$W = Q - \Delta U = (+1260 \text{ J}) - (-500 \text{ J}) = +1760$ J

The amount of work done by the system is +1760 J.

Related Text Exercises: Ex. 17-24 through 17-26.

5 Reversible Thermodynamic Processes (Section 17-5)

Review: A reversible process is one that takes place in such a way that the system is always in thermal equilibrium. The important reversible thermodynamic processes and their consequences can be treated graphically and analytically. The graphical treatment is made possible by the fact that reversible processes may be represented by curves on P-V and P-T graphs. The graphical treatment is especially useful because it helps us visualize what is taking place.

Isovolumetric process
(constant-volume)
$\Delta V = 0$
$W = P\Delta V = 0$
$Q = nc_V\Delta T$
$\Delta U = Q$
Isovolumetric is frequently shortened to isometric.

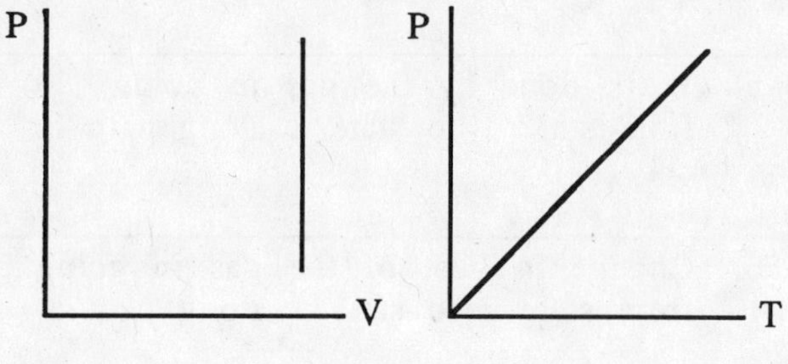

Figure 17.8

Isobaric process
(constant-pressure)
$\Delta P = 0$
$W = P\Delta V = nR\Delta T$
$Q = nc_p\Delta T$
$\Delta U = nc_V\Delta T$

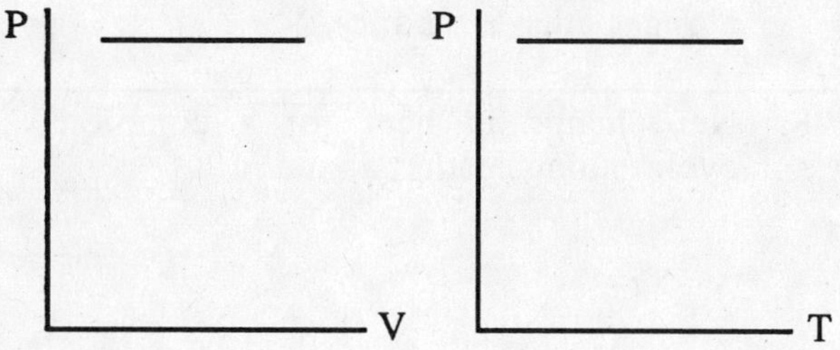

Figure 17.9

Isothermal process
(constant-temperature)
$\Delta T = 0$
$\Delta U = nc_V\Delta T = 0$
$Q = W = nRT \ln(V_2/V_1)$

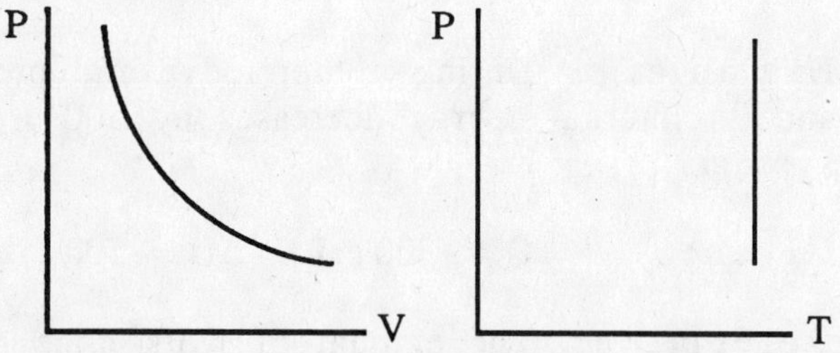

Figure 17.10

Adiabatic process
(no heat change)
$Q = 0$
$W = -\Delta U = -nc_V\Delta T$
$W = [P_iV_i/\gamma - 1]\,[1 - (V_i/V_f)^{\gamma-1}]$

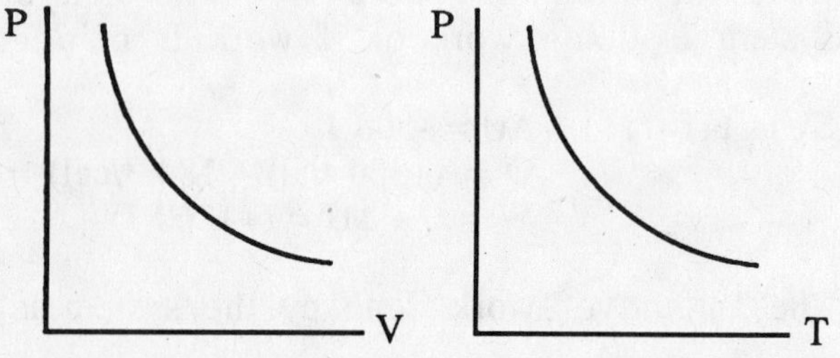

Figure 17.11

Practice: A cylinder contains 2.00 mole of helium gas (c_p = 20.77 J/mol•K; c_v = 12.46 J/mol•K). The P-V and P-T plots in Figure 17.12 show the thermodynamic processes and information for this gas as it goes through a complete cycle ($1 \rightarrow 2 \rightarrow 3 \rightarrow 4 \rightarrow 1$).

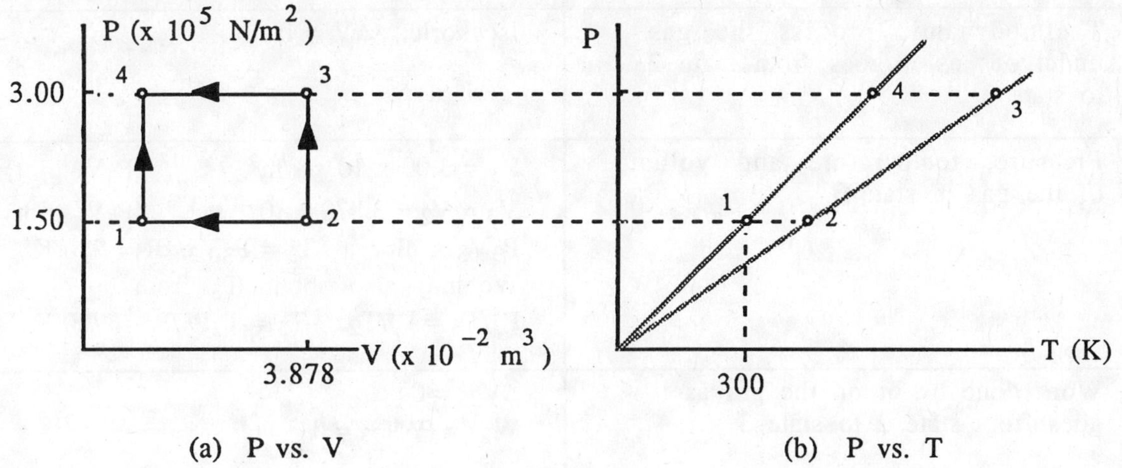

Figure 17.12

Determine the following:

1. Thermodynamic process the gas undergoes as it goes from state 1 to state 2	Isobaric: $\Delta P = 0$
2. Pressure, temperature and volume of the gas in state 1	$P_1 = 1.50 \times 10^5$ N/m^2 (P-V plot) $T_1 = 300$ K (P-T plot) $P_1V_1 = nRT_1$ $V_1 = nRT_1/P_1 = 3.324 \times 10^{-2}$ m^3
3. Pressure, temperature and volume of gas in state 2	$P_2 = P_1 = 1.50 \times 10^5$ N/m^2 (P-V plot) $V_2 = 3.878 \times 10^{-2}$ m^3 (P-V plot) $P_2V_2 = nRT_2$ $T_2 = P_2V_2/nR = 350$ K
4. Work done by the gas as it goes from state 1 to state 2	$\Delta T = T_2 - T_1 = 50.0$ K $P = 1.50 \times 10^5$ N/m^2 $\Delta V = V_2 - V_1 = 5.54 \times 10^{-3}$ m^3 $W_{12} = nR\Delta T = 8.31 \times 10^2$ J $W_{12} = P\Delta V = 8.31 \times 10^2$ J Note that W_{12} is the area under the curve on the P-V plot.
5. Heat supplied to the gas as it goes from state 1 to state 2	$\Delta T = 50.0$ K $Q_{12} = nc_p\Delta T_{12} = 2.08 \times 10^3$ J

6. Change in internal energy of the gas as it goes from state 1 to state 2	$\Delta T_{12} = 50.0$ K $\Delta U_{12} = nc_v\Delta T_{12} = 1.25 \times 10^3$ J $\Delta U_{12} = Q_{12} - W_{12} = 1.25 \times 10^3$ J
7. Thermodynamic process the gas undergoes as it goes from state 2 to state 3	Isochoric: $\Delta V = 0$
8. Pressure, temperature and volume of the gas in state 3	$P_3 = 3.00 \times 10^5$ N/m^2 (P-V plot) $V_3 = V_2 = 3.878 \times 10^{-2}$ m^3 (P-V plot) $P_3V_3 = nRT_3$; $T_3 = P_3V_3/nR = 700$ K We may also obtain T_3 from $P_3/T_3 = P_2/T_2$; $T_3 = T_2P_3/P_2 = 700$ K
9. Work done by or on the gas as it goes from state 2 to state 3	$\Delta V_{23} = 0$ $W_{23} = P\Delta V_{23} = 0$
10. Heat supplied to or taken from the gas as it goes from state 2 to state 3	$\Delta T_{23} = T_3 - T_2 = 350$ K $Q_{23} = nc_v\Delta T_{23} = 8.72 \times 10^3$ J This heat is supplied to the gas.
11. Change in internal energy of the gas as it goes from state 2 to state 3	$W_{23} = 0$; $Q_{23} = 8.72 \times 10^3$ J $\Delta U_{23} = Q_{23} - W_{23} = 8.72 \times 10^3$ J
12. Pressure, temperature and volume of the gas in state 4	$P_4 = P_3 = 3.00 \times 10^5$ N/m^2 (P-V plot) $V_4 = V_1 = 3.324 \times 10^{-2}$ m^3 (step 2) $P_4V_4 = nRT_4$; $T_4 = P_4V_4/nR = 600$ K We may also obtain T_4 from $V_3/T_3 = V_4/T_4$; $T_4 = T_3V_4/V_3 = 600$ K
13. Work done on the gas as it goes from state 3 to state 4	$\Delta V_{34} = V_4 - V_3 = -5.54 \times 10^{-3}$ m^3 $\Delta T_{34} = T_4 - T_3 = -100$ K $W_{34} = P_4\Delta V_{34} = -1.66 \times 10^3$ J $W_{34} = nR\Delta T = -1.66 \times 10^3$ J
14. Heat supplied to or taken from the gas as it goes from state 3 to state 4	$\Delta T_{34} = -100$ K $Q_{34} = nc_p\Delta T_{34} = -4.15 \times 10^3$ J This heat is taken from the gas.
15. Change in internal energy of the gas as it goes from state 3 to state 4	$\Delta T_{34} = -100$ K $\Delta U_{34} = nc_v\Delta T_{34} = -2.49 \times 10^3$ J $\Delta U_{34} = Q_{34} - W_{34} = -2.49 \times 10^3$ J

16. Work done on or by the gas as it goes from state 4 to state 1	$\Delta V_{41} = 0$ $W_{41} = P\Delta V_{41} = 0$
17. Heat supplied to or taken from the gas as it goes from state 4 to state 1	$\Delta T_{41} = -300$ K $Q_{41} = nc_v\Delta T_{41} = -7.48 \times 10^3$ J
18. Change in internal energy of the gas as it goes from state 4 to state 1	$\Delta T_{41} = -300$ K $\Delta U_{41} = nc_v\Delta T = -7.48 \times 10^3$ J $\Delta U_{41} = Q_{41} - W_{41} = -7.48 \times 10^3$ J
19. Net change in energy for the complete cycle	$\Delta U_{net} = 0$ $\Delta U_{net} = \Delta U_{12} + \Delta U_{23} + \Delta U_{34} + \Delta U_{41} = 0$
20. Net heat change for the complete cycle	$Q_{net} = Q_{12} + Q_{23} = Q_{34} + Q_{41}$ $Q_{net} = -8.30 \times 10^2$ J
21. Net work done for the complete cycle	$W_{net} = W_{12} + W_{23} + W_{34} + W_{41}$ $W_{net} = -8.30 \times 10^2$ J

A cylinder contains 1.00 mol of neon gas (c_p = 20.77 J/mol•K; c_v = 12.46 J/mol•K). The P-V and P-T plots in Figure 17.13 show the thermodynamic processes and information for this gas as it goes through a complete cycle ($1 \rightarrow 2 \rightarrow 3 \rightarrow 4 \rightarrow 1$).

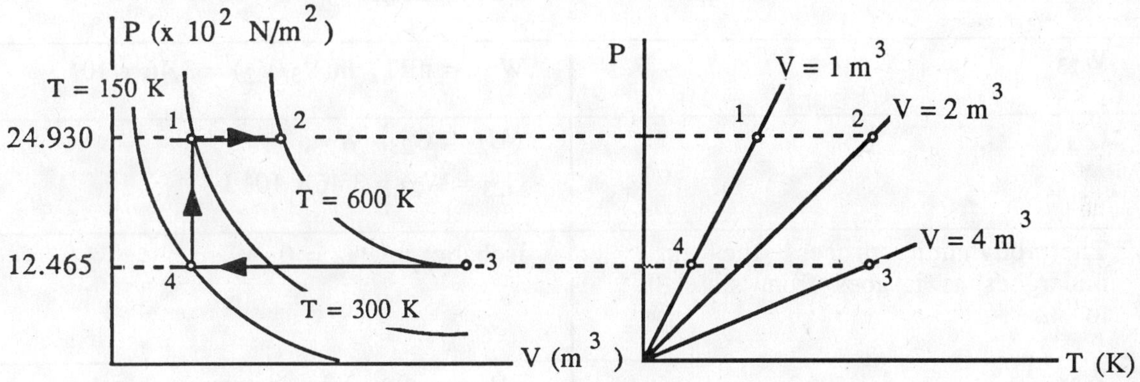

(a) P vs. V

(b) P vs. T

Figure 17.13

Determine the following:

1. P_1, V_1, T_1	$P_1 = 24.93 \times 10^2$ N/m²	(P-V or P-T plot)
	$V_1 = 1.00$ m³	(P-T plot)
	$T_1 = 300$ K	(P-V plot)

2. P_2, V_2, T_2	$P_2 = P_1 = 24.93 \times 10^2\ N/m^2$ (P-V plot) $V_2 = 2.00\ m^3$ (P-T plot) $T_2 = 600\ K$ (P-V plot)
3. Thermodynamic process the gas undergoes as it goes from state 1 to state 2	Isobaric: $\Delta P_{12} = 0$
4. W_{12}	$\Delta V_{12} = 1.00\ m^3$; $\Delta T_{12} = 300\ K$ $W_{12} = nR\Delta T_{12} = 2.49 \times 10^3\ J$ $W_{12} = P_1\Delta V_{12} = 2.49 \times 10^3\ J$
5. Q_{12}	$\Delta T_{12} = 300\ K$ $Q_{12} = nc_p\Delta T_{12} = 6.23 \times 10^3\ J$
6. ΔU_{12}	$\Delta T_{12} = 300\ K$ $\Delta U_{12} = nc_v\Delta T_{12} = 3.74 \times 10^3\ J$ $\Delta U_{12} = Q_{12} - W_{12} = 3.74 \times 10^3\ J$
7. Thermodynamic process the gas undergoes as it goes from state 2 to state 3	Isothermal: $\Delta T_{23} = 0$
8. ΔU_{23}	$\Delta U_{23} = nc_v\Delta T_{23} = 0$
9. W_{23}	$W_{23} = nRT_2 \ln(V_3/V_2) = 3.46 \times 10^3\ J$
10. Q_{23}	$\Delta U_{23} = Q_{23} - W_{23}$; $\Delta U_{23} = 0$ $Q_{23} = W_{23} = 3.46 \times 10^3\ J$
11. Thermodynamic process the gas undergoes as it goes from state 3 to state 4	Isobaric: $\Delta P_{34} = 0$
12. W_{34}	$\Delta V_{34} = -3.00\ m^3$; $\Delta T_{34} = -450\ K$ $W_{34} = P_3\Delta V_{34} = -3.74 \times 10^3\ J$ $W_{34} = nR\Delta T_{34} = -3.74 \times 10^3\ J$
13. Q_{34}	$\Delta T_{34} = -450\ K$ $Q_{34} = nc_p\Delta T_{34} = -9.35 \times 10^3\ J$
14. ΔU_{34}	$\Delta T_{34} = -450\ K$ $\Delta U_{34} = nc_v\Delta T_{34} = -5.61 \times 10^3\ J$ $\Delta U_{34} = Q_{34} - W_{34} = -5.61 \times 10^3\ J$

15. Thermodynamic process the gas undergoes as it goes from state 4 to state 1	Isovolumetric: $\Delta V_{41} = 0$
16. W_{41}	$W_{41} = 0$ since $\Delta V_{41} = 0$
17. Q_{41}	$\Delta T_{41} = 150$ K $Q_{41} = nc_v\Delta T_{41} = 1.87 \times 10^3$ J
18. ΔU_{41}	$\Delta U_{41} = Q_{41} - W_{41} = 1.87 \times 10^3$ J
19. ΔU_{net}	$\Delta U_{net} = \Delta U_{12} + \Delta U_{23} + \Delta U_{34} + \Delta U_{41} = 0$
20. Q_{net}	$Q_{net} = Q_{12} + Q_{23} + Q_{34} + Q_{41}$ $= 2.21 \times 10^3$ J
21. W_{net}	$W_{net} = W_{12} + W_{23} + W_{34} + W_{41}$ $= 2.21 \times 10^3$ J also $\Delta U_{net} = Q_{net} - W_{net} = 0$ $W_{net} = Q_{net}$ This agrees with the results of step 20

Related Text Exercises: Ex. 17-27 through 17-32.

PRACTICE TEST

Take and grade this practice test. Doing so will allow you to determine any weak spots in your understanding of the concepts taught in this chapter. The following section prescribes what you should study further to strengthen your understanding.

In the experiment shown in Figure 17.14 a 1.00-kg mass is allowed to fall 5.00 m at a constant speed while turning the paddles in water. Insulation prevents any heat from escaping from the system. The mass of water in the container is 0.500 kg.

Figure 17.14

$m = 1.00$ kg

$m_w = 0.500$ kg

$h = 5.00$ m

Determine the following:

________ 1. Energy (in J) delivered to water
________ 2. Energy (in cal) delivered to water
________ 3. Temperature change (in C°) of water

An ideal gas is pumped into a 1.00×10^{-2} m^3 container at 27.0°C until the gauge pressure reads 8.99×10^5 Pa. Determine the following:

_________ 4. Absolute pressure of the gas
_________ 5. Number of moles of the gas in the container
_________ 6. Pressure of the gas if the temperature is doubled
_________ 7. Pressure of the gas if the temperature is doubled and the volume of the container is reduced by a factor of three

Figure 17.15 shows the P-V and P-T diagrams for a system of Helium gas (c_p = 20.77 J/mol•K and c_v = 12.46 J/mol•K) as it goes the complete cycle 1 → 2 → 3 → 4 → 1.

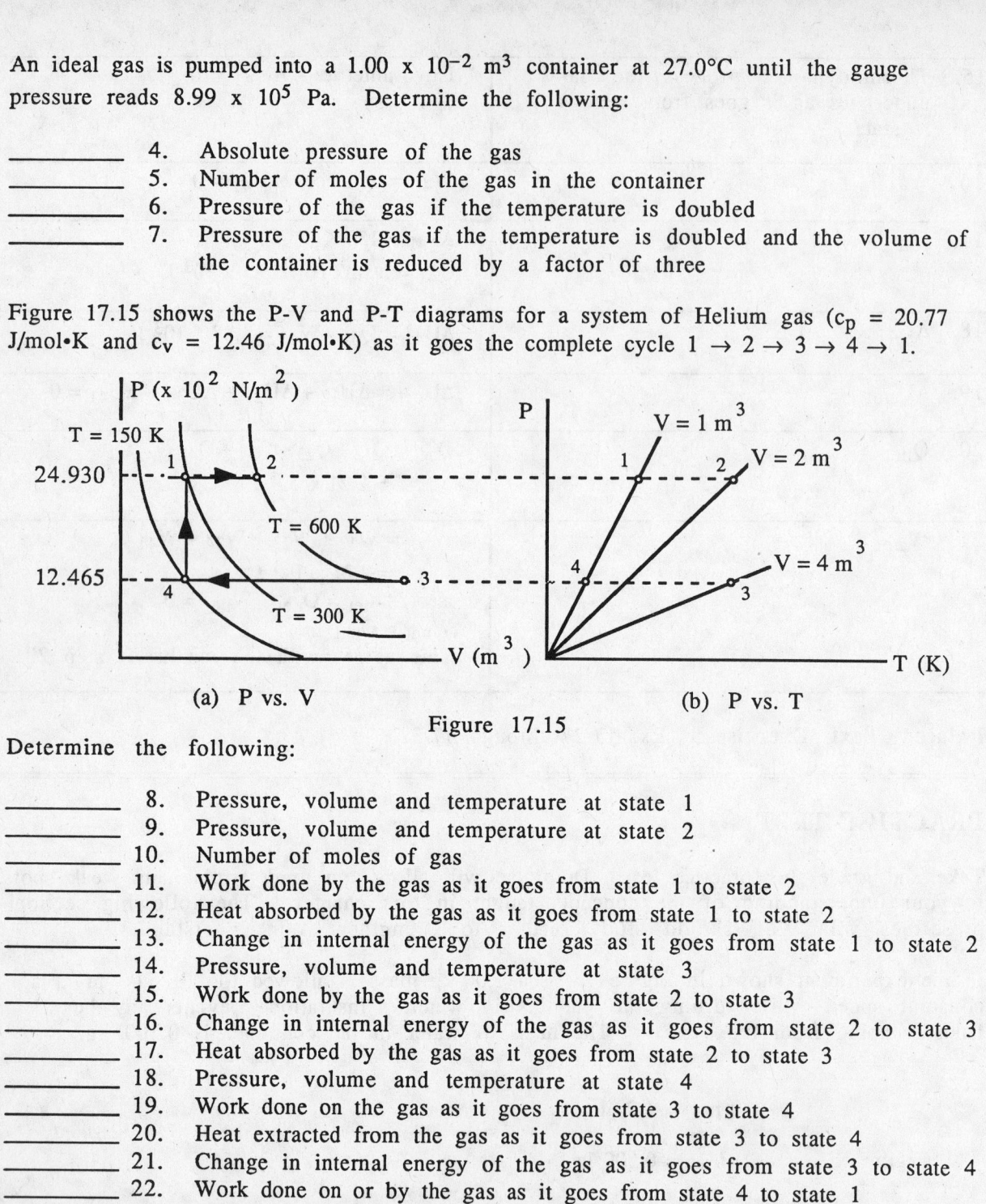

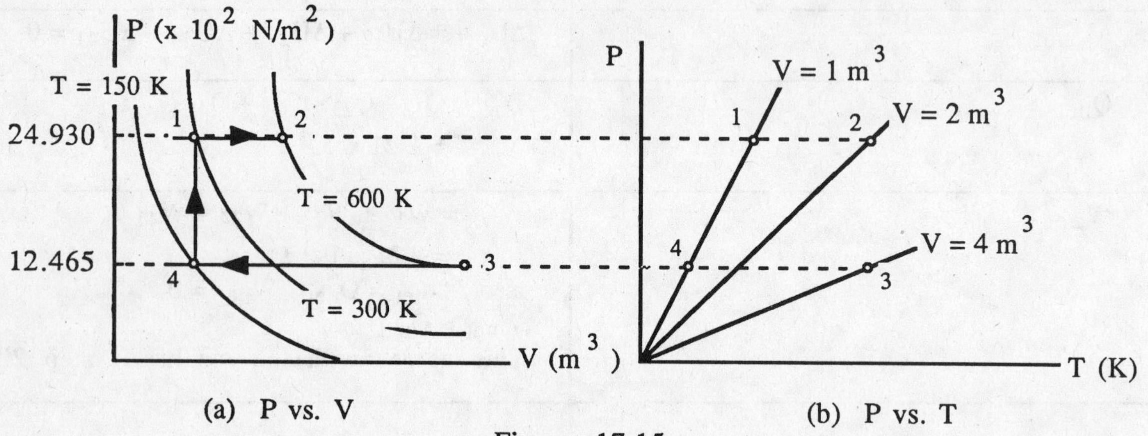

Figure 17.15

Determine the following:

_________ 8. Pressure, volume and temperature at state 1
_________ 9. Pressure, volume and temperature at state 2
_________ 10. Number of moles of gas
_________ 11. Work done by the gas as it goes from state 1 to state 2
_________ 12. Heat absorbed by the gas as it goes from state 1 to state 2
_________ 13. Change in internal energy of the gas as it goes from state 1 to state 2
_________ 14. Pressure, volume and temperature at state 3
_________ 15. Work done by the gas as it goes from state 2 to state 3
_________ 16. Change in internal energy of the gas as it goes from state 2 to state 3
_________ 17. Heat absorbed by the gas as it goes from state 2 to state 3
_________ 18. Pressure, volume and temperature at state 4
_________ 19. Work done on the gas as it goes from state 3 to state 4
_________ 20. Heat extracted from the gas as it goes from state 3 to state 4
_________ 21. Change in internal energy of the gas as it goes from state 3 to state 4
_________ 22. Work done on or by the gas as it goes from state 4 to state 1
_________ 23. Heat absorbed by the gas as it goes from state 4 to state 1
_________ 24. Change in internal energy of the gas as it goes from state 4 to state 1
_________ 25. Change in internal energy of the gas for the complete cycle
_________ 26. Net heat gained by the gas for the complete cycle
_________ 27. Net work done by the gas for the complete cycle

(See Appendix I for answers.)

PRINCIPAL CONCEPTS AND EQUATIONS PRESCRIPTION

Your score on the practice test is an excellent measure of your understanding of the chapter. You should now use the following chart to write your own prescription for dealing with any weaknesses the practice test points out. Look down the leftmost column to the number of the question(s) you answered incorrectly, reading across that row you will find the concept and/or equation of concern, the section(s) of the study guide you should return to for further study, and some suggested text exercises which you should work to gain additional experience.

Practice Test Questions	Concepts and Equations	Prescription	
		Principal Concepts	Text Exercises
1	Gravitational potential energy	4 of Ch 9	9-1, 2
2	Mechanical equivalent of heat	2	---
3	Specific heat: $Q = mc\Delta T$	2	17-11, 12
4	Absolute pressure: $P = P_G + P_A$	3 of Ch 15	15-13
5	Ideal gas law: $PV = nRT$	1	17-1, 2
6	Ideal gas law	1	17-3, 4
7	Ideal gas law	1	17-5, 6
8	Reading P-V and P-T plots	5	17-27, 30
9	Reading P-V and P-T plots	5	17-27, 30
10	Ideal gas law: $PV = nRT$	1	17-1, 2
11	Work for Isobaric Process: $W_{12} = nR\Delta T_{12}$ or $W_{12} = P_1\Delta V_{12}$	3	17-17, 18
12	Heat for Isobaric Process: $Q_{12} = nc_p\Delta T_{12}$	2	17-10
13	Change in internal energy for isobaric process: $\Delta U_{12} = nc_v\Delta T_{12}$ or $\Delta U_{12} = Q_{12} - W_{12}$	4	17-24, 27
14	Reading P-V and P-T plots	5	17-27, 30
15	Work for Isothermal Process: $W_{23} = nR\Delta T_2 \ln(V_3/V_2)$	3	17-19, 20
16	Change in internal energy for Isothermal process: $\Delta U_{23} = nc_v\Delta T_{23}$	5	17-25, 27
17	Heat for Isothermal Process: $\Delta U_{23} = Q_{23} - W_{23}$	4	17-24, 25
18	Reading P-V and P-T plots	5	17-27, 30
19	Work for Isobaric Process: $W_{34} = nR\Delta T_{34}$ or $W_{34} = P_3\Delta V_{34}$	3	17-17, 18
20	Heat for Isobaric Process: $Q_{34} = nc_p\Delta T_{34}$	2	17-10
21	Change in internal energy for isobaric process: $\Delta U_{34} = nc_v\Delta T_{34}$ or $\Delta U_{34} = Q_{34} - W_{34}$	5	17-10, 24
22	Work for Isovolumetric Process: $W_{41} = P\Delta V_{41}$	3	17-18
23	Heat for Isovolumetric Process: $Q_{41} = nc_v\Delta T_{41}$	2	17-10
24	Change in internal energy for Isovolumetric process: $\Delta U_{41} = Q_{41} - W_{41}$	4	17-24, 25
25	Change in internal energy for a cycle: $\Delta U = 0$	4	17-24, 25
26	Net heat gain for a cycle: $Q_{net} = \Sigma Q$	5	17-30
27	Net work for a cycle $W_{net} = \Sigma W = Q_{net}$	3	17-19, 30

18 KINETIC THEORY OF GASES

RECALL FROM PREVIOUS CHAPTERS

Previously learned concepts and equations frequently used in this chapter	Text Section	Study Guide Page
Equation of state: $PV = nRT$	17-1	17-1
Specific heat: $Q = mc\Delta T$	17-2	17-6
First law of thermodynamics: $\Delta U = Q - W$	17-4	17-13

NEW IDEAS IN THIS CHAPTER

Concepts and equations introduced	Text Section	Study Guide Page
Total mass of gas: $\mathcal{M} = nM = Nm$	18-1, 2	18-1
Mass of molecule: $m = M/N_A = \mathcal{M}/N$	18-1, 2	18-1
Number of molecules: $N = \mathcal{M}/m = nN_A$	18-1, 2	18-1
Average square of speeds of molecules: $\langle v^2\rangle = 3kT/m$	18-1, 2	18-1
Average KE per molecule: $\langle K\rangle = m\langle v^2\rangle/2 = 3kT/2$	18-1, 2	18-1
Average KE of gas: $\langle K_T\rangle = N\langle K\rangle = Nm\langle v^2\rangle/2 = \mathcal{M}\langle v^2\rangle/2 = 3NkT/2 = 3nN_AkT/2$	18-1, 2	18-1
Internal energy of the gas: $U = \langle K_T\rangle = 3nN_AkT/2 = 3nRT/2$	18-1, 2	18-1
Root mean square speed: $v_{rms} = (\langle v^2\rangle)^{1/2}$	18-1, 2	18-1
Relation between P, V and U: $PV = Nm\langle v^2\rangle/3 = 2N\langle K\rangle/3 = 2\langle K_T\rangle/3 = 2U/3$	18-1, 2	18-1
Average molecular mechanical energy: $E = \upsilon(kT/2)$	18-3	18-3
Molar heat capacities: $dQ = nC_vdT$, $dQ = nC_pdT$, $C_v = (1/n)dU/dT$, $C_p = C_v + R$	18-4	18-7
Adiabatic Process: $Q = 0$, $\Delta U = nC_v\Delta T$, $PV^\gamma = K$, $W = -\Delta U$, $W = [P_iV_i/(\gamma - 1)]\,[1 - V_i/V_f)^{\gamma-1}]$	18-5	18-10

PRINCIPAL CONCEPTS AND EQUATIONS

1 Molecular Model of An Ideal Gas (Sections 18-1, 2)

Review: If n moles of a monatomic gas of molecular mass M are placed in a container of volume V at a temperature T, then the following is true:

$\mathcal{M}$ = nM = Nm = total mass of gas

m = $M/N_A = \mathcal{M}/N$ = mass of a single molecule of gas

N = $\mathcal{M}/m = nN_A$ = number of molecules of gas

$<v^2> = 3kT/m$ = average of the square speeds of the molecules

$<K> = m<v^2>/2$ = average kinetic energy per molecule

$<K_T> = N<K> = Nm<v^2>/2 = \mathcal{M}<v^2>/2 = 3NkT/2 = 3nN_AkT/2$

= average kinetic energy of all the molecules of gas

U = $<K_T>$ = internal energy of the gas

$v_{rms} = (<v^2>)^{1/2}$ = root mean square speed of the molecules

$PV = Nm<v^2>/3 = 2N(m<v^2>/2)/3 = 2N<K>/3 = 2<K_T>/3 = 2U/3$

PV = nRT = equation of state for an ideal gas

U = 3nRT/2 = $3nN_AkT/2$ = internal energy of the gas

R = N_AK = relationship between constants

Practice: Six moles of helium gas are placed in a 2.00 x 10^{-3} m^3 container at a temperature of 27.0°C.

Determine the following:

1. Molecular mass of helium	$M = 4.00 \times 10^{-3}$ kg/mole
2. Mass of one helium molecule	$m = M/N_A = 6.64 \times 10^{-27}$ kg
3. Number of molecules of helium in the container	$N = nN_A = 3.61 \times 10^{24}$ molecules
4. Total mass of the gas	$\mathcal{M} = nM = 2.40 \times 10^{-2}$ kg, or $\mathcal{M} = Nm = 2.40 \times 10^{-2}$ kg
5. Average of the square of the speeds of the molecules	$<v^2> = 3kT/m = 1.87 \times 10^6\ m^2/s^2$
6. Root-mean-square speed of the molecules	$v_{rms} = (<v^2>)^{1/2} = 1.37 \times 10^3$ m/s
7. Average kinetic energy per molecule	$<K> = m<v^2>/2 = 6.21 \times 10^{-21}$ J, or $<K> = 3kT/2 = 6.21 \times 10^{-21}$ J
8. Average kinetic energy of the gas	$<K_T> = N<K> = 2.24 \times 10^4$ J, or $<K_T> = 3NkT/2 = 2.24 \times 10^4$ J
9. Internal energy of the gas	$U = <K_T> = 2.24 \times 10^4$ J, or $U = 3nRT/2 = 2.24 \times 10^4$ J
10. Pressure of the gas	$P = nRT/V = 7.47 \times 10^6$ Pa $P = Nm<v^2>/3V = 7.47 \times 10^6$ Pa $P = 2U/3V = 7.47 \times 10^6$ Pa

Example: At what temperature will the v_{rms} of nitrogen molecules be equal to that of helium molecules at 27.0°C?

Given: Helium gas at 27.0°C and nitrogen gas

Determine: The temperature at which nitrogen gas molecules will have the same v_{rms} as helium gas molecules at 27.0°C.

Strategy: We can write an expression for v_{rms} as a function of T for both gases. Since we want v_{rmsN_2} to equal v_{rmsHe}, we can equate these expressions and solve for T_{N_2}.

Solution:

$$v_{rmsN_2} = v_{rmsHe}$$

$$3kT_{N_2}/m_{N_2} = 3kT_{He}/m_{He} \quad \text{or} \quad T_{N_2} = T_{He}(m_{N_2}/m_{He})$$

Since $m_{N_2} = M_{N_2}/N_A$ and $m_{He} = M_{He}/N_A$, we can write

$$T_{N_2} = T_{He}(M_{N_2}/M_{He}) = (300\text{ K})(28.0/4.00) = 2.10 \times 10^3\text{ K}$$

Related Text Exercises: Ex. 18-1 through 18-15.

2 Equipartition of Energy (Section 18-3)

Review: For a system of molecules at temperature T, the average molecular mechanical energy $\langle E \rangle$ depends on the number of degrees of freedom and is given by the following expression

$$\langle E \rangle = \upsilon(kT/2)$$

Your understanding of υ, which is related to the number of degrees of freedom, will be developed in what follows.

The simplest way to establish the number of degrees of freedom for a particular molecule is to realize that it takes three coordinates to specify the location of each atom in the molecule. Therefore the number of degrees of freedom in a molecule with N' atoms is 3N'. In addition to determining the number of degrees of freedom it is also instructive to determine the energy contribution associated with the various degrees of freedom. Let's consider the molecules shown in Figure 18.1, determine the degrees of freedom and the energy contribution associated with each degree of freedom.

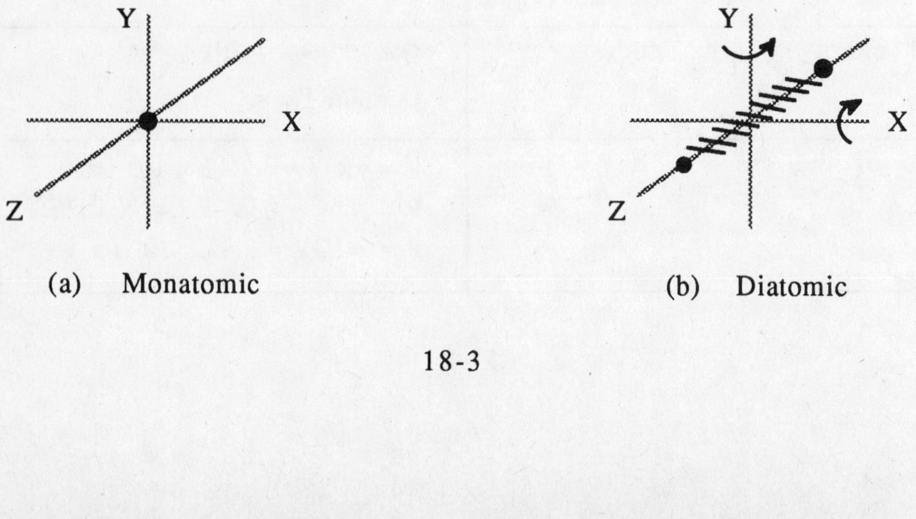

(a) Monatomic (b) Diatomic

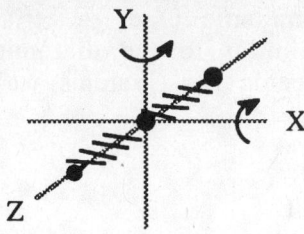

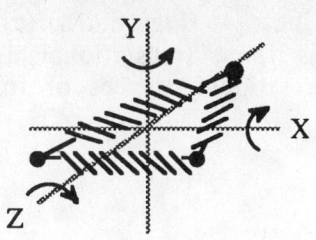

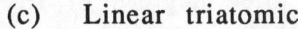

(c) Linear triatomic (d) Planar triatomic

Figure 18-1

Monatomic - (Figure 18.1(a)) - the molecule may move through space with x, y, and z components to its velocity. Subsequently it may have translational kinetic energy in the x, y, and z directions contributing to its total energy, hence it has three translational degrees of freedom. Note that N' = 1, and the total number of degrees of freedom is 3N' = 3. These three degrees of freedom are associated with the translational kinetic energy.

Diatomic - (Figure 18.1(b)) - similar to the monatomic case, the diatomic molecule has three translational degrees of freedom. If this molecule rotates about the x or y axes it has a significant amount of rotational kinetic energy (but not when it rotates about the z axis as the atoms are point masses), hence it has two rotational degrees of freedom. Using our rule that a diatomic (N' = 2) molecule has 3N' = 6 degrees of freedom, we see that we have one degree of freedom left for vibration.

$$3N' - t - r = v \qquad \text{inserting numbers obtain} \qquad 3(2) - 3 - 2 = 1$$

Where: N' = number of atoms in the molecule

t = 3 always — degrees of freedom associated with translational kinetic energy of the molecule as a whole

r = 0 for monatomic — degrees of freedom associated with rotational kinetic energy of the molecule as a whole

= 2 for linear

= 3 for nonlinear

v = degrees of freedom associated with vibration (potential and kinetic)

As the atoms vibrate, at times all the vibrational energy is potential, at times all the vibrational energy is kinetic (when the atoms are at their equilibrium position) and at times the vibrational energy is a combination of potential and kinetic. Averaged out over a period of time, the contribution to the vibrational mode of energy from potential and kinetic is equal. So for the diatomic molecule we have one degree of freedom (v = 1) associated with its vibration and we have a contribution of 2v associated with its energy. Subsequently we may rewrite υ as follows:

$$\upsilon = t + r + 2v \qquad \text{inserting values for the diatomic case} \qquad \upsilon = 3 + 2 + 2(1) = 7$$

While the above expression for υ was developed using a diatomic molecule, it is true for all molecules.

Triatomic Linear - (Figure 18.1(c)) - similar to the monatomic case, the triatomic linear molecule has three translational degrees of freedom. Similar to the diatomic molecule it has two rotational degrees of freedom. Since the molecule has 3 atoms we have

$$v = 3N' - t - r = 3(3) - 3 - 2 = 4$$

and

$$\upsilon = t + r + 2v = 3 + 2 + 2(4) = 13$$

Triatomic Planar - (Figure 18.1(d)) - this molecule has three degrees of freedom for translational kinetic energy, three for rotational kinetic energy, and three for vibrational energy ($v = 3N' - t - r$). Hence the value of υ is 12 ($\upsilon = t + r + 2v$).

The following table summarizes the above:

	Degrees of Freedom					
Molecule	Translation	Rotation	Vibration	Total (3N')	υ	Energy
Monatomic	3	0	0	3	3	3kT/2
Diatomic	3	2	1	6	7	7kT/2
Triatomic linear	3	2	4	9	13	13kT/2
Triatomic planar	3	3	3	9	12	6kT

Practice: A container holds 5.00 moles of a gas at 1200 K. The gas is composed of triatomic linear molecules of the type shown in Figure 18.1(c).

Determine the following:

1. Degrees of translational freedom	Since the molecule may move in the x, y, and z directions, it has 3 degrees of translational freedom
2. Degrees of rotational freedom	Since the molecule has a significant amount of rotational kinetic about the x and y axes, it has 2 rotational degrees of freedom.
3. Degrees of vibrational freedom	$v = 3N' - t - r = 3(3) - 3 - 2 = 4$
4. The value of υ for this molecule	$\upsilon = t + r + 2v = 3 + 2 + 2(4) = 13$
5. Energy per degree of translational and rotational freedom	$kT/2 = 8.28 \times 10^{-21}$ J
6. Average molecular mechanical energy	$\langle E \rangle = \upsilon(kT/2) = 13(kT/2) = 1.08 \times 10^{-19}$ J
7. Internal energy of the 5.00 moles of the gas	$U = N\langle E \rangle = nN_A\langle E \rangle = 3.25 \times 10^5$ J

Example: A nonlinear polyatomic gas at 800 K has 2.70 x 10^{-26} kg molecules each with 4 atoms. Determine the following:

(a) Number of vibrational degrees of freedom which are active.
(b) Average molecular mechanical energy
(c) Root mean square speed of the molecules

Given: T = 800 K N' = 4

Determine: (a) Number of vibrational degrees of freedom which are active
(b) Average molecular mechanical energy
(c) Root mean square speed of the molecules

Strategy: Knowing N' = 4, we can determine the total number of degrees of freedom. Knowing the number of degrees of freedom and t and r for a polyatomic molecule, we can determine the number of vibrational degrees of freedom (v). Knowing v, we can obtain υ and then the average molecular mechanical energy. Knowing that the root mean square speed of the molecule is associated with only the translational kinetic energy of the molecule, we can determine the root mean square speed of the molecules.

Solution: (a) The number of vibrational degrees of freedom may be obtained by

$$v = 3N' - t - r = 3(4) - 3 - 3 = 6$$

(b) Now that we know v, we may determine υ and then $\langle E \rangle$.

$$\upsilon = t + r + 2v = 3 + 3 + 2(6) = 18$$
$$\langle E \rangle = \upsilon(kT/2) = (18)(1.38 \times 10^{-23} \text{ J/k})(8.00 \times 10^{2} \text{ K})/2 = 9.94 \times 10^{-20} \text{ J}$$

(c) Since only the three translational degrees of freedom contribute to the average translational kinetic energy per molecule, we have

$$\langle K \rangle = 3(kT/2) = 1.66 \times 10^{-20} \text{ J}$$

But the average translational kinetic energy per molecule may also be written as:

$$\langle K \rangle = m\langle v^2 \rangle/2$$

This will allow us to determine the root mean square speed by

$$v_{rms} = (\langle v^2 \rangle)^{1/2} = (2\langle K \rangle/m)^{1/2} = 1.11 \times 10^{3} \text{ m/s}$$

Related Text Exercises: Ex. 18-16 and 18-17.

3 Heat Capacities of Ideal Gases (Section 18-4)

Review: When heat is added to a gas at constant volume, no work is done, the change in internal energy is equal to the heat added, and the heat added can be expressed in terms of the molar heat capacity at constant volume. This may be stated algebraically as follows:

$$\text{If } dV = 0 \text{ then } \begin{aligned} dW &= PdV = 0 \\ dU &= dQ + dW = dQ \\ dQ &= nC_V dT \end{aligned}$$

Combining these obtain an expression for the molar heat capacity at constant volume:

$$C_V = (1/n)dU/dT$$

When heat is added to a gas at constant pressure, the internal energy changes and work is performed. If this process has the same temperature change as a process at constant volume we may write the change in internal energy as follows:

$$dU_{\text{const } P} = dU_{\text{const } V} = dQ_{\text{const } V} = nC_V dT$$

The work done by the gas for a constant pressure process can be expressed in terms of the temperature change by differentiating the ideal gas equation of state.

$$PV = nRT$$
$$PdV + VdP = nRdT$$

Since $VdP = 0$ for constant pressure, we obtain the following for the work done at constant pressure.

$$dW = PdV = nRdT$$

Inserting these expressions for dU and dW at constant pressure into the differential form of the first law, we obtain the following:

$$dQ = dU + dW$$
$$dQ = nC_V dT + nRdT$$

but

$$dQ = nC_p dT$$

Combining the above we obtain an expression for the molar heat capacity at constant pressure.

$$nC_p dT = nC_V dT + nRdT$$
$$C_p = C_V + R$$

In general, the internal energy of a gas is given by:

$$U = N\langle E\rangle = nN_A \upsilon(kT/2) = n\upsilon(RT/2)$$

Using the above expression for C_V, C_p and U, we obtain the following chart.

System	υ	Internal Energy $U = \upsilon nRT/2$	dU/Dt	$C_v = (1/n)dU/dT$	$C_p=C_v+R$(gas) $C_p=C_v$(solid)
Monatomic gas	3	3nRT/2	3nR/2	3R/2	5R/2
Diatomic gas	7	7nRT/2	7nR/2	7R/2	9R/2
Polyatomic gas	$\upsilon = t + r + 2v$	$\upsilon nRT/2$	$\upsilon nR/2$	$\upsilon R/2$	$(\upsilon/2 + 1)R$
Solid	6	3nRT	3nR	3R	3R

In order to avoid confusion let's review terms again.

Heat capacity - the quantity of heat ΔQ supplied to a body for a corresponding temperature change ΔT.

$$C = \Delta Q/\Delta T$$

Specific heat - the heat capacity per unit mass - the quantity of heat that must be supplied per unit mass of the material to raise its temperature by one degree.

$$c = C/m = (\Delta Q/\Delta T)/m$$

Molar heat capacity - the quantity of heat that must be supplied per unit mole of the material to raise its temperature by one degree. It is essentially a specific heat where the mass in measured in moles.

$$C = (\Delta Q/\Delta T)/n$$

Two different molar heat capacities are of practical use for gases. These are the molar heat capacity at constant volume C_v and molar heat capacity at constant pressure C_p.

Practice: The temperature of 4.00 mol of a diatomic gas is increased by 100 K in a constant pressure process. The temperature range over which this process occurs is such that the vibrational degrees of freedom are not active.

Determine the following:

1. Degrees of freedom for the gas molecules	Since no vibration occurs, we will have three degrees of freedom associated with translation and two degrees of freedom associated with rotation. Hence we have $\upsilon = 5$.
2. Change in internal energy of the gas	$U = \upsilon nRT/2$ $\upsilon = 5$ and $n = 4.00$ mol, hence $U = 10RT$ $\Delta U = 10R\Delta T$, $\Delta T = 100$ K $\Delta U = 8.31 \times 10^3$ J
3. Molar heat capacity at constant volume	$\Delta U = nC_v \Delta T$ $C_v = \Delta U/n\Delta T = 20.8$ J/mol•K

4. Molar heat capacity at constant pressure	$C_p = C_v + R = 29.1$ J/mol•K
5. Heat added to the system	$Q = nC_p\Delta T = 1.16 \times 10^4$ J
6. Work done by the system	$\Delta W = nR\Delta T = 3.32 \times 10^3$ J
7. A check on the above work	$Q = 1.16 \times 10^4$ J (step 5) $\Delta U = 8.31 \times 10^3$ J (step 2) $\Delta W = 3.32 \times 10^3$ J (step 6) $Q = \Delta U + \Delta W = 1.16 \times 10^4$ J (1st law) which agrees with step 5

Example: The specific heat at constant volume of a monatomic gas is $c_v = 75.0$ cal/kg•K. Determine the mass of one molecule and the molar mass of this gas.

Given: Specific heat at constant volume $c_v = 75.0$ cal/kg•K
The gas is monatomic

Determine: (a) The mass of a molecule of the gas
(b) The molar mass of the gas

Strategy: Knowing that the gas is monatomic, we know its molar heat capacity at constant volume (i.e. the heat required to raise the temperature of one mole of the gas by one degree). Knowing the molar heat capacity and Avagadroe's number (number of molecules per mol) we can determine the heat required to raise a mass of one molecule by one degree. When this is set equal to the specific heat at constant volume, we can determine the mass of a molecule. Knowing the mass of a molecule and Avagadroe's number, we can determine the molar mass of the gas.

Solution: The molar heat capacity at constant volume for a monatomic gas is

$$C_v = 3R/2 = 12.5 \text{ J/mol•K}$$

The amount of heat needed to raise a mass equal to that of one molecule is just the molar heat capacity divided by N_A.

$$C_v/N_A = (12.5 \text{ J/mole•K})/(6.02 \times 10^{23} \text{ molecules/mol}) = 2.08 \times 10^{-23} \text{ J/molecule•K}$$

Now let's change the specific heat at constant volume to units of J and equate it to C_v/N_A

$$c_v = (75.0 \text{ cal/kg•K})(4.184 \text{ J/cal}) = 3.14 \times 10^2 \text{ J/kg•K}$$

$$c_v = C_v/N_A$$

$$3.14 \times 10^2 \text{ J/kg}\cdot\text{K} = 2.08 \times 10^{-23} \text{ J/molecule}\cdot\text{K}$$

or

$$\text{molecule} = 2.08 \times 10^{-23} \text{ kg}/3.14 \times 10^2 = 6.62 \times 10^{-26} \text{ kg}$$

The molar mass is the mass of one mole of the gas, but one mole has Avagadroe's number of molecules

$$M = mN_A = \left(\frac{6.62 \times 10^{-26} \text{ kg}}{\text{molecule}}\right)\left(\frac{6.02 \times 10^{23} \text{ molecules}}{\text{mol}}\right)\left(\frac{10^3 \text{ g}}{\text{kg}}\right) = 39.9 \text{ g/mole}$$

Note: The gas must be Argon.

Related Text Exercises: Ex 18-18 through 18-26.

4 Adiabatic Process (Section 18-5)

Review: An adiabatic process is one for which no heat is added to or lost from the system. The following is true for an adiabatic process:

$$Q = 0$$
$$\Delta U = nC_v\Delta T$$
$$W = [P_iV_i/(\gamma - 1)][1 - (V_i/V_f)^{\gamma-1}]$$
$$PV^\gamma = \text{constant}$$
$$W = -\Delta U$$

Practice: Consider the P-V diagram and given information shown in Figure 18.2.

$n = 1$ mol
$C_p = 20.77$ J/mol·K
$C_v = 12.46$ J/mol·K
$\gamma = C_p/C_v = 1.67$

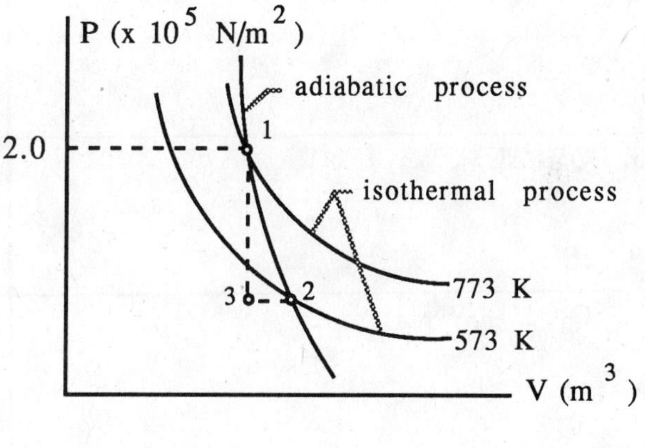

Figure 18.2

Determine the following:

1. P_1, V_1, T_1	$P_1 = 2.00 \times 10^5$ N/m^2 (P-V plot) $T_1 = 773$ K (isotherm on P-V plot) $V_1 = nRT_1/P_1 = 32.12 \times 10^{-3}$ m^3

2. P_2, V_2, T_2	T_2 = 573 K (isotherm on P-V plot) (a) $P_1V_1 = nRT_1$ (b) $P_2V_2 = n\ RT_2$ Divide (a) by (b) to obtain (c) $(P_1/P_2)(V_1/V_2) = (T_1/T_2)$ (d) $P_1V_1^{\gamma} = P_2V_2^{\gamma}$ or (e) $(V_2/V_1)^{\gamma} = P_1/P_2$ Insert (e) into (c) to obtain $(V_2/V_1)^{\gamma-1} = T_1/T_2$ or $V_2 = V_1(T_1/T_2)^{1/\gamma-1} = 50.2 \times 10^{-3}\ m^3$ $P_2 = nRT_2/V_2 = 0.9484 \times 10^5\ N/m^2$
3. P_3, V_3, T_3	$P_3 = P_2 = 0.9484 \times 10^5\ N/m^2$ $V_3 = V_1 = 32.12 \times 10^{-3}\ m^3$ $T_3 = P_3V_3/nR = 366$ K
4. $Q_{12}, \Delta U_{12}, W_{12}$	$Q_{12} = 0$ as it is an adiabatic process. $\Delta U_{12} = nC_v\Delta T = -2{,}492$ J $W_{12} = [P_1V_1/(\gamma -1)][1-(V_1/V_2)^{\gamma-1}]$ $= 2{,}492$ J Note that this agrees with the first law since $\Delta U_{12} = Q_{12} - W_{12}$.
5. $Q_{23}, \Delta U_{23}, W_{23}$	$Q_{23} = nC_p\Delta T_{23} = -4{,}299$ J $W_{23} = P\Delta V_{23} = -\ 1{,}708$ J $\Delta U_{23} = nC_v\Delta T_{23} = -\ 2{,}579$ J Note that this agrees with the first law since $\Delta U_{23} = Q_{23} - W_{23}$.
6. $Q_{31}, \Delta U_{31}, W_{31}$	$Q_{31} = nC_v\Delta T = 5{,}071$ J $W_{31} = 0$ $\Delta U_{31} = Q_{31} = 5{,}071$ J
7. $Q_T, \Delta U_T, W_T$	$\Delta U_T = \Delta U_{12} + \Delta U_{23} + \Delta U_{31}$ $= 0$ as it must for a complete cycle $Q_T = Q_{12} + Q_{23} + Q_{31} = 772$ J $W_T = W_{12} + W_{23} + W_{31} = 784$ J Note that this agrees with the first law since $\Delta U_T = Q_T - W_T$. The 12 J difference out of nearly 800 J is insignificant and is due to rounding off.

Example: Two moles of a monatomic ideal gas with $\gamma = 1.67$ undergo an adiabatic expansion from the initial state $P_i = 3.20 \times 10^5$ Pa, $V_i = 12.0$ L to a final volume $V_f = 18.0$ L. (a) Determine the final pressure of the gas. (b) Determine the initial and final temperature of the gas.

Given: $n = 2.00$ mol $P_i = 3.20 \times 10^5$ Pa $V_i = 12.0$.L $V_f = 18.0$ L $\gamma = 1.67$

Determine: (a) The final pressure of the gas
(b) The initial and final temperature of the gas

Strategy: Knowing P_i, V_i and V_f and that PV^γ is a constant for an adiabatic process, we can determine P_f. We may then use the equation of state for an ideal gas to determine T_i and T_f.

Solution: (a) For an adiabatic process we know that PV^γ is a constant, hence we may write

$$P_i V_i^\gamma = P_f V_f^\gamma$$

or

$$P_f = P_i(V_i/V_f)^\gamma = 3.20 \times 10^5 \text{ Pa})(12.0 \text{ L}/18.0 \text{ L})^{1.67} = 1.63 \times 10^5 \text{ Pa}$$

(b) We may then use the equation of state for an ideal gas to determine T_i and T_f.

$$P_i V_i = nRT_i \Rightarrow T_i = P_i V_i/nR = 231 \text{ K}$$

$$P_f V_f = nRT_f \Rightarrow T_f = P_f V_f/nR = 176 \text{ K}$$

We may also determine T_f by dividing the equation of state for the final state by the one for the initial state and the use PV^γ = constant. This method is shown below:

Dividing $P_f V_f = nRT_f$ by $P_i V_i = nRT_i$ we obtain

$$T_f = T_i(P_f/P_i)(V_f/V_i)$$

But for an adiabatic process we have

$$P_i V_i^\gamma = P_f V_f^\gamma \Rightarrow (P_f/P_i) = (V_i/V_f)^\gamma$$

When (P_f/P_i) is inserted into the expression for T_f we obtain

$$T_f = T_i(V_i/V_f)^\gamma(V_f/V_i) = T_i(V_i/V_f)^{\gamma-1} = 176 \text{ K}$$

Notice that this answer agrees with the one previously obtained.

Related Text Exercises: Ex 18-27 through 18-34.

PRACTICE TEST

Take and grade this practice test. Doing so will allow you to determine any weak spots in your understanding of the concepts taught in this chapter. The following section prescribes what you should study further to strengthen your understanding.

Figure 18.3 shows the P-V and P-T diagrams for a system of He gas as it undergoes the complete cycle 1 → 2 → 3 → 4 → 5 → 1. As the gas goes from state 2 to state 3, it absorbs 11.09×10^3 J of heat. As it goes from state 3 to state 4, it does 5.806×10^3 J of work.

$C_V = 12.46$ J/mol•K, $C_P = 20.77$ J/mol•K, $\gamma = 1.67$

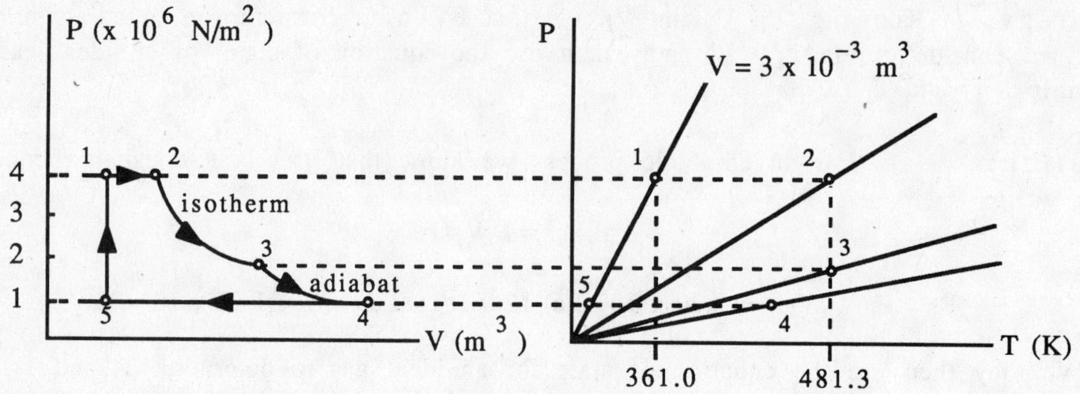

Figure 18.3

Determine the following:

_________ 1. Number of moles of gas in the system
_________ 2. Mass of each molecule of the gas
_________ 3. Number of molecules of the gas
_________ 4. Total mass of the gas
_________ 5. Average square of the speeds of the molecules in state 1
_________ 6. Average kinetic energy per molecule in state 1
_________ 7. Average kinetic energy of the gas in state 1
_________ 8. Internal energy of the gas in state 1
_________ 9. Volume of the gas in state 2
_________ 10. Work done by the gas as it goes from state 1 to state 2
_________ 11. Heat absorbed by the gas as it goes from state 1 to state 2
_________ 12. Change in internal energy of the gas as it goes from state 1 to state 2
_________ 13. Volume of the gas in state 3
_________ 14. Work done by the gas as it goes from state 2 to state 3
_________ 15. Volume of the gas in state 4
_________ 16. Temperature of the gas in state 4
_________ 17. Change in internal energy of the gas as it goes from state 3 to state 4
_________ 18. Work done on the gas as it goes from state 4 to state 5
_________ 19. Change in internal energy of the gas as it goes from state 4 to state 5
_________ 20. Heat absorbed by the gas as it goes from state 4 to state 5
_________ 21. Change in internal energy of the system for a complete cycle
_________ 22. Change in heat of the gas for a complete cycle
_________ 23. Net work done on or by the gas for a complete cycle

(See Appendix I for answers.)

PRINCIPAL CONCEPTS AND EQUATIONS PRESCRIPTION

Your score on the practice test is an excellent measure of your understanding of the chapter. You should now use the following chart to write your own prescription for dealing with any weaknesses the practice test points out. Look down the leftmost column to the number of the question(s) you answered incorrectly, reading across that row you will find the concept and/or equation of concern, the section(s) of the study guide you should return to for further study, and some suggested text exercises which you should work to gain additional experience.

Practice Test Questions	Concepts and Equations	Prescription	
		Principal Concepts	Text Exercises
1	Number of moles: $n = PV/RT$	1 of Ch 17	17-1, 2
2	Mass of molecule: $m = M/N_A$	1	---
3	Number of molecules: $N = nN_A$	1	---
4	Total mass of gas: $M = nM = Nm$	1	---
5	Average square of speeds: $<v^2> = 3kT/m$	1	18-8, 12
6	Average KE per molecule: $<K> = m<v^2>/2 = 3kT/2$	1	18-12, 13
7	Average KE of gas: $<K_T> = N<K>$	1	18-13, 15
8	Internal energy: $U = <K_T> = 3nRT/2$	1	18-12
9	Equation of state: $PV = nRT$	1 of Ch 17	17-1, 2
10	Work (isobaric process): $W_{12} = P_1\Delta V_{12}$; $W_{12} = nR\Delta T_{12}$	5 of Ch 17	17-17, 18
11	Molar heat capacity at const. V: $Q_{12} = nC_p\Delta T_{12}$	3	18-21, 25
12	Internal energy (isobaric process): $\Delta U_{12} = nC_v\Delta T_{12}$; $\Delta U_{12} = Q_{12} - W_{12}$	5 of Ch 17	17-24, 27
13	Heat absorbed (isothermal process): $Q_{23} = nRT_2 \ln(V_3/v_2)$	5 of Ch 17	17-24, 25
14	First law of thermodynamics: $\Delta U = Q - W$	4 of Ch 17	17-27, 28
15	Adiabatic process: PV^γ = constant	4	18-27, 30
16	Adiabatic process: $TV^{\gamma-1}$ = constant	4	18-29, 30
17	Energy change (adiabatic process): $\Delta U_{34} = nC_v\Delta T_{34}$	3	18-19, 25
18	Work (isobaric process): $W_{45} = P_5\Delta V_{45}$; $W_{45} = nR\Delta T_{45}$	5 of CH 17	17-17, 18
19	Energy change (isobaric process): $\Delta U_{45} = nC_v\Delta T_{45}$	5 of Ch 17	17-10, 24
20	Heat absorbed (isobaric process): $Q_{45} = nC_p\Delta T_{45}$	3	18-10
21	Energy change (total cycle): $\Delta U = \Sigma \Delta U = 0$	5 of Ch 17	17-24, 25
22	Heat change (total cycle): $Q_T = \Sigma Q$	5 of Ch 17	17-30
23	Net work (total cycle): $W_{net} = \Sigma W$; $W_{net} = Q_T - \Delta U_T$	5 of CH 17	17-19, 30

19 THE SECOND LAW OF THERMODYNAMICS

RECALL FROM PREVIOUS CHAPTERS

Previously learned concepts and equations frequently used in this chapter	Text Section	Study Guide Page
Equation of state: $PV = nRT$	17-1	17-1
First law of thermodynamics: $\Delta U = Q - W$	17-4	17-13
Molar heat capacities: $Q = nC_v\Delta T$ and $Q = nC_p\Delta T$	18-5	18-7
Q for the following processes:		
Isochoric: $Q = \Delta U = nC_v\Delta T$	17-3	17-10
Isobaric: $Q = nC_p\Delta T$	17-3	17-10
Isothermal: $Q = W = nRT \ln(V_f/V_i)$	17-3	17-10
Isothermal: $Q = W = nRT \ln(V_f/V_i)$	17-3	17-10
Adiabatic: $Q=0$	18-5	18-10
ΔU for the following processes:		
Isochoric: $\Delta U = Q = nC_v\Delta T$	17-3	17-10
Isobaric: $\Delta U = nC_v\Delta T$	17-3	17-10
Isothermal: $\Delta U = 0$	17-3	17-10
Adiabatic: $\Delta U = -W = -[P_iV_i/(\gamma - 1)][1 - (V_i/V_f)^{\gamma-1}]$	18-5	18-10
W for the following processes:		
Isochoric: $W=0$	17-3	17-10
Isobaric: $W = P\Delta V = nR\Delta T$	17-3	17-10
Isothermal: $W = Q = nRT \ln(V_f/V_i)$	17-3	17-10
Adiabatic: $W = -\Delta U = [(P_iV_i/(\gamma - 1)][1 - (V_i/V_f)^{\gamma-1}]$	18-5	18-10

NEW IDEAS IN THIS CHAPTER

Concepts and equations introduced	Text Section	Study Guide Page
Efficiency of a heat engine: $\eta = W/Q_H = 1 - \lvert Q_c\rvert/Q_H$	19-1, 2	19-1
Refrigerator coefficient of performance: $K = Q_c/(\lvert Q_H\rvert - Q_c)$	19-1, 2	19-1
Heat pump coefficient of performance: $K_{hp} = \lvert Q_H\rvert/W = \lvert Q_H\rvert/(\lvert Q_H\rvert - Q_c)$	19-1, 2	19-1
The Carnot cycle	19-3, 4	19-6
Entropy: $ds = dQ/T$ $\quad \Delta s = s_f - s_i = \int_i^f dQ/T$	19-5, 6	19-10

PRINCIPAL CONCEPTS AND EQUATIONS

1 Heat Engines and Refrigerators (Section 19-1, 2)

Review: A heat engine and a refrigerator are shown schematically in Figure 19.1.

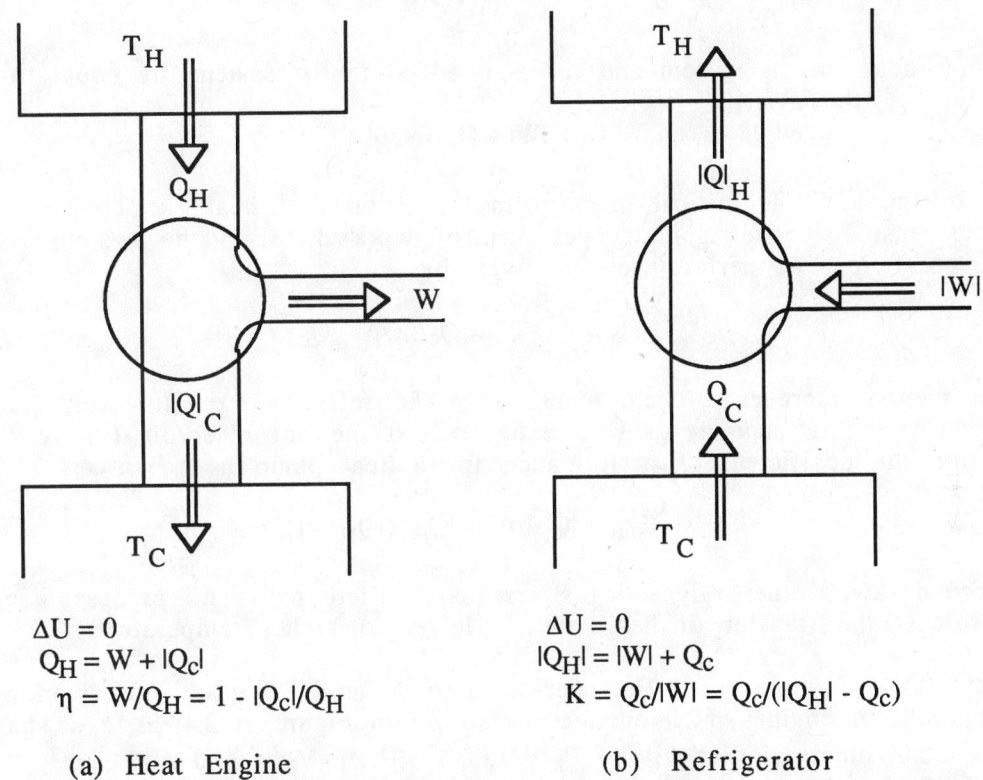

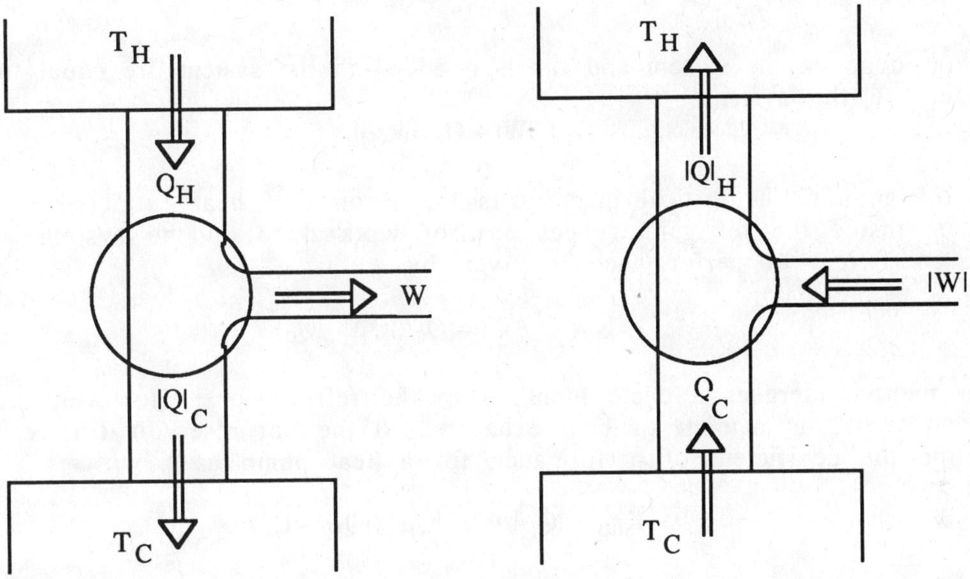

(a) Heat Engine

$\Delta U = 0$

$Q_H = W + |Q_c|$

$\eta = W/Q_H = 1 - |Q_c|/Q_H$

(b) Refrigerator

$\Delta U = 0$

$|Q_H| = |W| + Q_c$

$K = Q_c/|W| = Q_c/(|Q_H| - Q_c)$

Figure 19.1

Both devices operate in a complete cycle (hence $\Delta U = 0$) between a hot and a cold heat reservoir.

In the case of a heat engine, an amount of heat Q_H is added to the system (from the hot reservoir) at temperature T_H, the system does an amount of work W, and an amount of heat $|Q_c|$ is exhausted to the cold reservoir at temperature T_c. Since the system undergoes a complete cycle, we have

$$\Delta U = 0$$

The heat added to the system either does work or is exhausted to the cold reservoir.

$$Q_H = W + |Q_c|$$

The efficiency η is the amount of work we get out of the engine per unit of heat added.

$$\eta = W/Q_H = (Q_H - |Q_c|)/Q_H = 1 - |Q_c|/Q_H$$

The second law of thermodynamics states that: There exists no cycle which extracts heat from a reservoir at a single temperature and completely converts it into work.

In the case of a refrigerator, an amount of work W is done to extract an amount of heat Q_c from the cold reservoir at temperature T_c and an amount of heat $|Q_H|$ is exhausted to the hot reservoir at temperature T_H. Since the system undergoes a complete cycle, we have

$$\Delta U = 0$$

The work done on the system and the heat added to the system are equal to the heat exhausted by the system

$$|W| + Q_c = |Q_H|$$

For a refrigerator, the critical quantity is the amount of heat extracted from the cold reservoir (inside the refrigerator) per unit of work done on the system. Subsequently the coefficient of performance is given by

$$K = Q_c/|W| = Q_c/(|Q_H| - Q_c)$$

A heat pump undergoes a cycle identical to the refrigerator. However, since the critical quantity is the amount of heat exhausted to the hot reservoir (inside the house) we define the coefficient of performance for a heat pump as

$$K_{hp} = |Q_H|/W = |Q_H|/(|Q_H| - Q_c)$$

The second law of thermodynamics states that: There exists no process whose sole, net result is the transfer of heat from a lower to higher temperature

Practice: Two moles of He, considered to be an ideal gas, is the working substance in a heat engine which operates as shown in Figure 19.2. State a has a pressure and volume respectively of $P_a = 1.00 \times 10^5$ Pa, $V_a = 20.0 \times 10^{-3}$ m^3. $C_p = 20.77$ J/(mol•K) and $C_v = 12.46$ J/(mol•K)

Figure 19.2

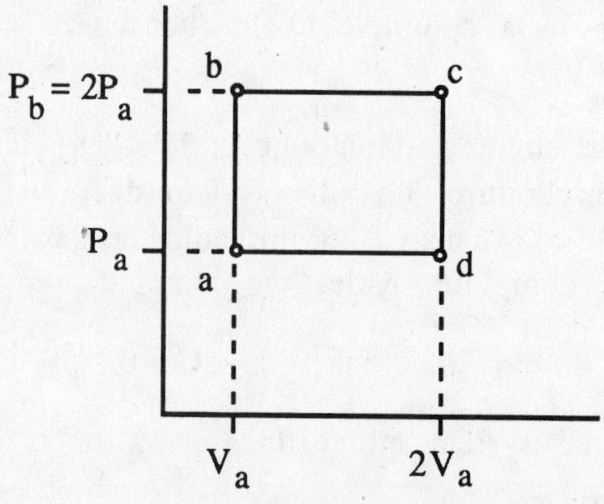

Determine the following:

1. Temperature of the gas in states a, b, c and d	$PV = nRT$ or $T_a = P_aV_a/nR = 120$ K In like manner $T_b = 240$ K, $T_c = 480$ K, $T_d = 240$ K

2. Heat loss or gain for each step of the process	$Q_{ab} = nC_v\Delta T_{ab}$, $\Delta T_{ab} = 120$ K° $= 2{,}990$ J $Q_{bc} = nC_p\Delta T_{bc}$, $\Delta T_{bc} = 240$ K° $= 9{,}970$ J $Q_{cd} = nC_v\Delta T_{cd}$, $\Delta T_{cd} = -240$ K° $= -5{,}980$ J $Q_{da} = nC_p\Delta T_{da}$, $\Delta T_{da} = -120$ K° $= -4{,}985$ J
3. Total heat gain, total heat loss and net heat for the entire cycle	$Q_{gain} = Q_{ab} + Q_{bc} = 12{,}960$ J $Q_{loss} = Q_{cd} + Q_{da} = -10{,}965$ J $Q_{net} = Q_{gain} + Q_{loss} = 1{,}995$ J
4. Work done on or by the system for each step of the process	$W_{ab} = 0$ $W_{bc} = P_b\Delta V_{bc}$, $\Delta V_{bc} = 20.0 \times 10^{-3}$ m³ $= 4.00 \times 10^3$ J $W_{cd} = 0$ $W_{da} = P_a\Delta V_{da}$, $\Delta V_{da} = -20.0 \times 10^{-3}$ m³ $= -2.00 \times 10^3$ J
5. Net work done by the heat engine in one cycle	$W_{net} = W_{ab} + W_{bc} + W_{cd} + W_{da} = 2.00 \times 10^3$ J
6. Change in internal energy of the working substance	Since the working substance goes through a complete cycle and ends up where it started, $\Delta U = 0$. We can also check this with the first law of thermodynamics. $\Delta U = Q - W$, $\Delta U = 0$ $Q_{net} = 1{,}995$ J (step 3) $W_{net} = 2.00 \times 10^3$ J (step 5) $Q = W$ (the difference is due to rounding off)
7. Efficiency of the heat engine	$\eta = W/\lvert Q_H\rvert$ $W = 2.00 \times 10^3$ J, $\lvert Q_H\rvert = 1.296 \times 10^4$ J $\eta = 0.154$

Note: Heat-engine problems involve the four quantities Q_H, Q_L, W, and η. You will usually be given two of them and asked to solve for the other two. Refrigerator problems involve the four quantities Q_L, W, Q_H, and K. You will usually be given two of them and asked to solve for the other two.

Example: A heat engine working with a thermodynamic efficiency of 0.400 exhausts 1.50×10^8 J of heat to a low-temperature reservoir. Calculate the work done by the engine.

Given: $\eta = 0.400$, $Q_L = 1.50 \times 10^8$ J

Determine: Work done by the engine

Strategy: Knowing the efficiency and the heat exhausted, we can determine the heat accepted from the high temperature reservoir and then the work done by the engine.

Solution: The heat accepted from the high-temperature reservoir is

$$\eta = 1 - (Q_L/Q_H); \quad Q_H = Q_L/(1 - \eta) = 2.50 \times 10^8 \text{ J}$$

The work done by the engine is

$$W = Q_H - Q_L = 2.50 \times 10^8 \text{ J} - 1.50 \times 10^8 \text{ J} = 1.00 \times 10^8 \text{ J}$$

Example: An ice-cube tray holding 400 ml of water at 20.0°C is placed in a freezer. If the unit is powered by a 0.333 hp motor and the coefficient of performance is 2.00, how long does it take the water to freeze? (Ignore the thermal effects of the tray).

Given: $V_w = 400$ ml $P = 0.333$ hp $T_{wi} = 20°C$ $K = 2.00$

Determine: Time required to freeze the water

Strategy: Knowing the volume and initial temperature of the water, we can determine the amount of heat that must be removed from the low-temperature reservoir in order to freeze the water. We can determine the rate at which work is done by the refrigerator from the known horsepower rating of the motor. We can then use this rate and the coefficient of performance to determine the rate at which heat is removed from the low-temperature reservoir. Knowing the amount of heat that must be removed and the rate at which it is being removed, we can determine the time needed to freeze the water.

Solution: First, let's determine the amount of heat that must be removed to freeze the water.

$$m_w = V_w \rho_w = (400 \text{ ml})(10^{-3} \text{ l/ml})(10^{-3} \text{ m}^3\text{/l})(10^3 \text{ kg/m}^3) = 4.00 \times 10^{-1} \text{ kg}$$
$$Q_L = Q \text{ (cool water to 0°C)} + Q \text{ (freeze water)}$$
$$Q_L = m_w c_w \Delta T_w + m_w L_f = m_w(c_w \Delta T_w + L_f)$$
$$Q_L = (4.00 \times 10^{-1} \text{ kg})[(4180 \text{ J/kg} \cdot \text{K})(20 \text{ K}) + 3350 \text{ J/kg}] = 3.48 \times 10^4 \text{ J}$$

Second, let's determine the rate at which heat is removed from the low-temperture reservoir.

$$P = W/t = (0.333 \text{ hp})(746 \text{ W/hp})[(\text{J/s})/\text{W}] = 249 \text{ J/s}$$

$$K = Q_L/W \Rightarrow Q_L = KW$$

which may be divided by t to obtain

$$Q_L/t = K(W/t) = 2.00(249 \text{ J/s}) = 498 \text{ J/s}$$

Finally let's determine the time required to freeze (that is, to remove the 3.48×10^4 J of heat from) the water.

$$t = Q_L/(Q_L/t) = (3.48 \times 10^4 \text{ J})/(4.98 \times 10^2 \text{ J/s}) = 6.99 \text{ s}$$

This number is unrealistically small because we have assumed that the motor is 100% efficient and that all of its work output goes into freezing the ice.

Related Text Exercises: Ex. 19-1 through 19-10.

2 The Carnot Cycle (Section 19-3, 4)

Review: The Carnot cycle is idealized in that it is a reversible cycle. The cycle can be reversed (from operating as an engine to operating as a refrigerator) by making only infinitesimal changes in external conditions. The cycle, operating as an engine, consists of the following four steps (see Figure 19.3):

Figure 19.3

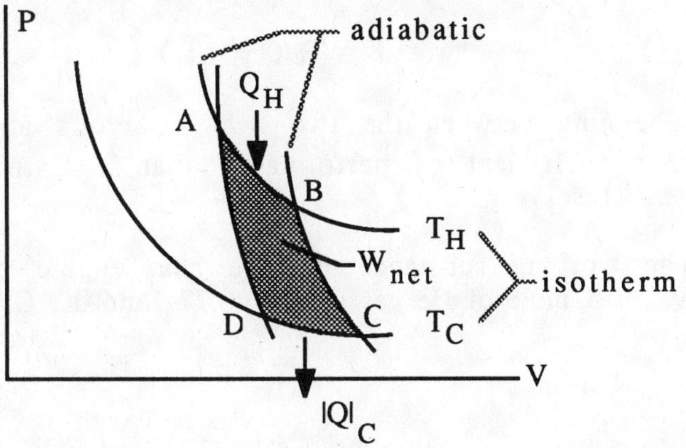

1. a reversible isothermal expansion at T_H, heat Q_H is added to the system (A → B in Figure 19.3)

2. a reversible adiabatic process, the temperature of the system drops from T_H to T_c (B → C in Figure 19.3)

3. a reversible isothermal compression at T_c, heat $|Q_c|$ is exhausted from the system (C → D in Figure 19.3).

4. a reversible adiabatic process to complete the cycle, the temperature of the system increases from T_c back to T_H (D → A in Figure 19.3).

Since the Carnot cycle consists of two isothermal and two adiabatic processes it will be beneficial to quickly review these two processes:

Isothermal Process:

$\Delta T_{if} = 0$	as the process is isothermal
$\Delta U_{if} = nC_V\Delta T_{if} = 0$	as ΔU depends on ΔT
$W_{if} = nRT \ln(V_f/V_i)$	$T = T_i = T_f$ as isothermal
$Q_{if} = \Delta U_{if} + W_{if} = W_{if}$	first law of thermodynamics

Adiabatic Process:

$Q_{if} = 0$	as the process is adiabatic
$\Delta U_{if} = nC_v\Delta T_{if}$	as ΔU depends on ΔT
$PV^\gamma =$ constant	as the process is adiabatic
$W_{if} = [P_iV_i/(\gamma - 1)]\ [1 - (V_i/V_f)^{\gamma-1}]$	

also

$W_{if} = Q_{if} - \Delta U_{if} = -\Delta U_{if}$	first law of thermodynamics

The efficiency of a heat engine operating in a Carnot cycle is

$$\eta = 1 - (T_c/T_H)$$

No heat engine operating between the two heat reservoirs at temperatures T_H and T_c can be more efficient than a heat engine operating in a Carnot cycle between these two reservoirs.

The coefficient of performance of a Carnot refrigerator is

$$K = T_c/(T_H - T_c)$$

No refrigerator operating between the two heat reservoirs at temperatures T_c and T_H can have a higher coefficient of performance than a Carnot refrigerator operating between these two reservoirs.

Practice: The working substance for the heat engine undergoing the Carnot cycle in Figure 19.4 is 1 mole of He gas (C_p = 20.77 J/mol•K; C_v = 12.46 J/mol•K).

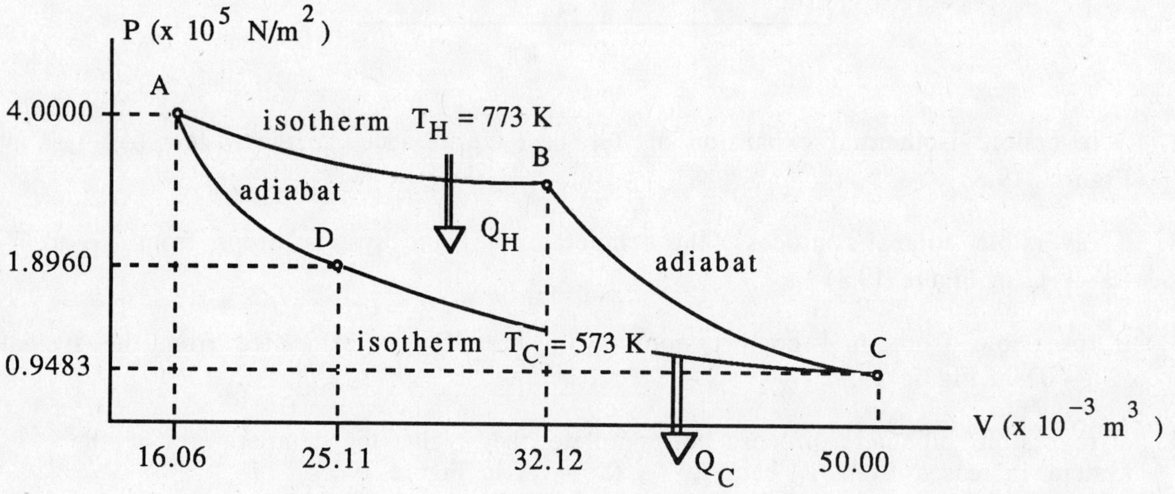

Figure 19.4

Determine the following:

1. Work done by the gas as it expands isothermally from A to B	n = 1 mol; $V_A = 16.06 \times 10^{-3}\ m^3$ R = 8.31 J/mol•K; $V_B = 32.12 \times 10^{-3}\ m^3$ T = 773 K $W_{AB} = nRT\ \ln(V_B/V_A) = 4.453 \times 10^3$ J

2. The change in internal energy of the gas as it goes from A to B	$\Delta T_{AB} = 0$ $\Delta U_{AB} = nC_V\Delta T_{AB} = 0$
3. Heat absorbed from the high-temperature reservoir as the gas goes from A to B	$Q_H = Q_{AB}$ $\Delta U_{AB} = Q_{AB} - W_{AB} = 0$ $Q_H = Q_{AB} = W_{AB} = 4.453 \times 10^3$ J
4. Q_{BC}	$Q_{BC} = 0$ since it is an adiabatic process
5. ΔU_{BC}	$\Delta T_{BC} = -200$ K $\Delta U_{BC} = nC_V\Delta T_{BC} = -2.492 \times 10^3$ J
6. W_{BC}	$\Delta U_{BC} = Q_{BC} - W_{BC}$; $Q_{BC} = 0$ $W_{BC} = -\Delta U_{BC} = 2.492 \times 10^3$ J
7. W_{CD}	$V_C = 50.22 \times 10^{-3}$ m^3 $V_D = 25.11 \times 10^{-3}$ m^3 $T_C = T_D = 573$ K $W_{CD} = nRT \ln(V_D/V_C) = -3.301 \times 10^3$ J
8. ΔU_{CD}	$\Delta U_{CD} = 0$ since $\Delta T_{CD} = 0$
9. Q_L	$\Delta U_{CD} = Q_{CD} - W_{CD} = 0$ $Q_L = Q_{CD} = W_{CD} = -3.301 \times 10^3$ J
10. Q_{DA}	$Q_{DA} = 0$ since it is an adiabatic process
11. ΔU_{DA}	$\Delta T_{DA} = +200$ K $\Delta U_{DA} = nC_V\Delta T_{DA} = +2.492 \times 10^3$ J
12. W_{DA}	$\Delta U_{DA} = Q_{DA} - W_{DA}$; $Q_{DA} = 0$ $W_{DA} = -\Delta U_{DA} = 2.494 \times 10^3$ J
13. W_{net}	$W_{net} = W_{AB} + W_{BC} + W_{CD} + W_{DA}$ $= 1.152 \times 10^3$ J $W_{net} = Q_H - Q_L = 1.152 \times 10^3$ J
14. Efficiency (η) for this Carnot cycle	$Q_H = 4.453 \times 10^3$ J (step 3) $W_{net} = 1.152 \times 10^3$ J (step 13) $T_H = 773$ K; $T_c = 573$ K $\eta = W_{net}/Q_H = 0.259$ $\eta = 1 - (T_c/T_H) = 0.259$

15. Coefficient of performance (K) of the associated Carnot refrigerator	$K = T_c/(T_H - T_c) = 2.87$

Example: A Carnot engine takes in heat from a 327°C reservoir and has an efficiency of 0.250. If the exhaust temperature remains the same and the efficiency increases to 0.400, what is the new temperature of the hot reservoir?

Given:
$T_{H1} = 327°C$ = temperature of the hot reservoir
$\eta_1 = 0.250$ = efficiency when $T_{H1} = 327°C$
$\eta_2 = 0.400$ = efficiency when reservoir is at T_{H1}

Determine: The temperature T_{H2} of the hot reservoir that will give this Carnot engine an efficiency of 0.400.

Strategy: Knowing T_{H1} and η_1, we can determine T_{c1}. Knowing η_2 and that $T_{c2} = T_{c1}$, we can determine T_{H2}.

Solution:
$\eta_1 = 1 - (T_{c1}/T_{H1})$; $T_{c1} = (1 - \eta_1)T_{H1} = 450$ K
$T_{c2} = T_{c1} = 450$ K
$\eta_2 = 1 - (T_{c2}/T_{H2})$; $T_{H2} = T_{c2}/(1 - \eta_2) = 750$ K

Example: A Carnot refrigerator operates between reservoirs at 23°C and -10°C. If it works long enough to convert a 200-g aluminum ice cube tray containing 400 g of water at 20°C to ice at -10°C, determine (a) the net amount of heat absorbed form the water and tray, (b) the coefficient of performance, (c) the work supplied to the refrigerator, and (d) the amount of heat exhausted to the room.

Given:
T_H = 23°C = temperature of hot reservoir
T_c = -10°C = temperature of cold reservoir
M_{Al} = 200 g = mass of aluminum tray
M_w = 400 g = mass of water
T_i = 20°C = initial temperature of water and tray

Determine:
Q_c = heat absorbed from cold reservoir (freezer)
K = coefficient of performance
W = work supplied to refrigerator
Q_H = amount of heat exhausted to hot reservoir (room)

Strategy: (a) Knowing the masses, temperatures, specific heats, and heat of fusion, we can determine the amount of heat absorbed from the water and tray. (b) Knowing the temperature of the hot and cold reservoirs, we can determine K. (c) Knowing K and Q_c, we can determine W. (d) Knowing W and Q_c, we can determine Q_H.

Solution: (a) The amount of heat absorbed from the cold reservoir (Q_c) is

$$Q_c = Q \text{ (cool and freeze water)} + Q \text{ (cool tray)}$$
$$Q_c = M_w c_w (20\text{ K}) + M_w L_f + M_{ice} c_{ice} (10\text{ K}) + M_{Al} c_{Al} (30\text{ K}) = 1.81 \times 10^5\text{ J}$$

(b) The coefficient of performance is

$$K = T_c/(T_H - T_c) = (263\text{ K})/(33\text{ K}) = 7.97$$

(c) The amount of work supplied to the refrigerator is

$$K = Q_c/W \Rightarrow W = Q_c/K = 2.27 \times 10^4\text{ J}$$

(d) The heat exhausted to the hot reservoir is

$$W = Q_H - Q_c \Rightarrow Q_H = W + Q_c = 20.4 \times 10^4\text{ J}$$

Related Text Exercises: Ex. 19-11 through 19-22.

3 Entropy (Section 19-5, 6)

Review: In previous sections we have seen that the variables of state P, V, T, and U have no net change for a complete cycle. That is

$$\Delta U = 0, \quad \Delta P = 0, \quad \Delta V = 0 \quad \text{and} \quad \Delta T = 0$$

In this section we will find a new variable of state, the entropy, which has $\Delta S = 0$ for a complete cycle. The entropy of a system changes an infinitesimal amount dS when an infinitesimal amount of heat dQ is added to the system at a temperature T. Expressed in differential form we have

$$dS = dQ/T$$

or expressed in integral form we have

$$\Delta S = S_f - S_i = \int_i^f dQ/T$$

and

$$\Delta S = 0 \qquad \text{for a cycle}$$

We may determine the entropy change associated with a phase change by

$$\Delta S = Q/T$$

Where Q is the amount of heat involved in the phase change and T is the temperature at which the phase change occurs.

We may determine the entropy change associated with a temperature change by

$$\Delta S = mC_p \ln(T_f/T_i)$$

We may determine the entropy change associated with a free expansion ($Q = 0$, $W = 0$, ΔU and hence $\Delta T = 0$) by

$$\Delta S = nR \ln(V_f/V_i)$$

We may determine the entropy change associated with the mixing of two substances and reaching an equilibrium temperature by

$$\Delta S = \Delta S_i + \Delta S_2$$

where

$$\Delta S_1 = m_1 C_p \ln(T_E/T_{1i})$$

and

$$\Delta S_2 = m_2 C_p \ln(T_E/T_{2i})$$

Practice: One mole of He gas goes through the cycle shown in Figure 19.5.

$P_A = 10.0 \times 10^5 \text{ N/m}^2$
$V_A = 5.00 \times 10^{-3} \text{ m}^3$
$n = 1$
$C_p = 20.77$ J/mol•K
$C_v = 12.46$ J/mol•K

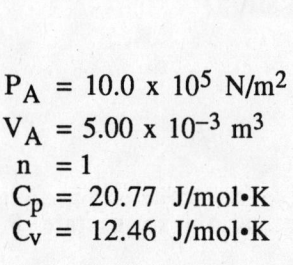

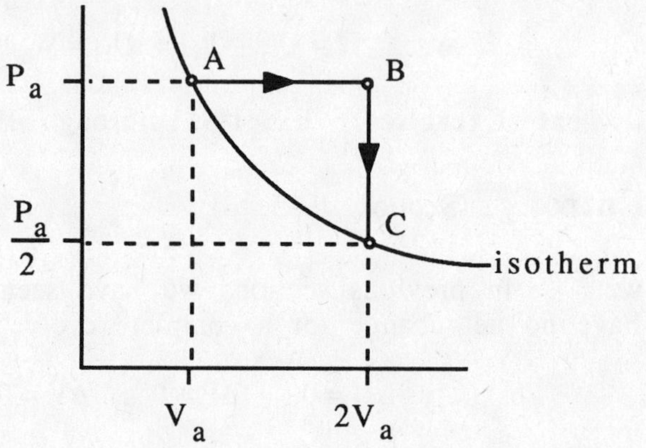

Figure 19.5

Determine the following:

1. Pressure, volume and temperature at state A, B and C	$P_A = P_B = 10.0 \times 10^5 \text{ N/m}^2$ $P_C = P_A/2 = 5.00 \times 10^5 \text{ N/m}^2$ $V_A = 5.00 \times 10^{-3} \text{ m}^3$ $V_B = V_C = 2V_A = 10.0 \times 10^{-3} \text{ m}^3$ Using the ideal gas law $T_A = P_A V_A/nR = 602$ K $T_C = T_A$ as on an isotherm $T_B = 2T_A = 1204$ K
2. ΔP, ΔV, ΔT, ΔU, Q, ΔW and ΔS for the gas as it goes from state A to state B	$\Delta P_{AB} = 0$ isobaric $\Delta V_{AB} = 5.00 \times 10^{-3} \text{ m}^3$ $\Delta T_{AB} = 6.00 \times 10^2$ K $\Delta U_{AB} = nC_v\Delta T_{AB} = 7{,}476$ J $Q_{AB} = nC_p\Delta T_{AB} = 12{,}462$ J $\Delta W_{AB} = P\Delta V_{AB} = 5{,}00$ J $\Delta S_{AB} = nC_p \ln(T_B/T_A) = 14.4$ J/K

3. ΔP, ΔV, ΔT, ΔU, Q, ΔW, and ΔS for the gas as it goes from state B to state C	$\Delta P_{BC} = 0.500 \times 10^5\ N/m^2$ $\Delta V_{BC} = 0$ $\Delta T_{BC} = -6.00 \times 10^2\ K$ $Q_{BC} = nC_V T_{BC} = -7{,}476\ J$ $W_{BC} = P\Delta V_{BC} = 0$ $\Delta U_{BC} = Q = -7{,}476\ J$ $\Delta S_{BC} = nC_v \ln(T_C/T_B) = -8.64\ J/K$
4. ΔP, ΔV, ΔT, ΔU, Q, ΔW and ΔS for the gas as it goes from state C to state A	$\Delta P_{CA} = 5.00 \times 10^5\ N/m^2$ $\Delta V_{CA} = -5.00 \times 10^{-3}\ m^3$ $\Delta T_{CA} = 0$ as on isotherm $\Delta U_{CA} = 0$ as $\Delta T_{CA} = 0$ $W_{CA} = Q_{CA} = nR \ln(V_A/V_C) = -3460\ J$ $\Delta S_{CA} = \frac{\Delta Q_{CA}}{T} = -5.76\ J/K$
5. ΔP, ΔV, ΔT, ΔU, Q, ΔW and ΔW and ΔS for the entire cycle	Adding the value from the previous states obtain $\Delta P_{AA} = 0$ $\Delta V_{AA} = 0$ $\Delta T_{AA} = 0$ $\Delta U_{AA} = 0$ $Q_{AA} = 1{,}526\ J$ $W_{AA} = 1{,}540\ J$ By the first law, we should have $Q_{AA} = W_{AA}$ and except for rounding errors we do. $\Delta S_{AA} = 0$

Example: A system of gas undergoes an isothermal process at 127°C. During this process, the gas does 3.00×10^4 J of work and the internal energy is lowered by 2.00×10^4 J. Determine the entropy change of the gas.

Given: $T = 127°C$ $W = 3.00 \times 10^4\ J$ $\Delta U = -2.00 \times 10^4\ J$

Determine: The entropy change (ΔS) of the gas

Strategy: Knowing ΔU and W, we can determine ΔQ. Once ΔQ and T are known, we can determine the entropy change.

Solution: Using the first law of thermodynamics, we can obtain ΔQ.

$$\Delta U = \Delta Q - W \Rightarrow \Delta Q = \Delta U + W = +1.00 \times 10^4\ J$$

The entropy change is

$$\Delta S = \Delta Q/T = (1.00 \times 10^4\ J)/400\ K) = 25.0\ J/K$$

Related Text Exercises: Ex. 19-23 through 19-32.

PRACTICE TEST

Take and grade this practice test. Doing so will allow you to determine any weak spots in your understanding of the concepts taught in this chapter. The following section prescribes what you should study further to strengthen your understanding.

A heat engine absorbs 2.00 x 10^3 J of heat from a high-temperature reservoir and exhausts 1.600 x 10^3 J to a low-temperature reservoir. When the same engine is run in reverse as a refrigerator between the same two reservoirs, 6.00 x 10^2 J of work is required to extract 1.40 x 10^3 J of heat from the low-temperature reservoir.

Determine the following:

________ 1. Amount of work done by the engine

________ 2. Efficiency of the engine

________ 3. Coefficient of performance for the refrigerator

________ 4. Amount of heat exhausted to the high-temperature reservoir by the refrigerator

On a winter day, the rate of heat loss from the inside of a house averages 8.00 KW. Suppose these losses are balanced by a heat pump with a 4.00 KW motor which runs 20.0 min each hour.

Determine the following:

________ 5. Coefficient of performance of the heat pump under these conditions

________ 6. Rate at which heat is extracted from the outside

________ 7. Cost for a full day of operation under these conditions assuming that electric energy may be purchased from the power company for \$0.10 per kW•h

Consider the Carnot cycle and information shown in Figure 19.6

$n = 1$ mol
$C_p = 20.77$ J/mol•K
$C_v = 12.46$ J/mol•K
$T_a = T_b = 773$ K
$T_c = T_d = 573$ K
$P_a = 4.00 \times 10^5$ N/m^2
$P_b = 2.00 \times 10^5$ N/m^2
$P_c = 0.9483 \times 10^5$ N/m^2
$P_d = 1.8960 \times 10^5$ N/m^2
$V_a = 16.06 \times 10^{-3}$ m^3
$V_b = 32.12 \times 10^{-3}$ m^3
$V_c = 50.22 \times 10^{-3}$ m^3
$V_d = 25.11 \times 10^{-3}$ m^3

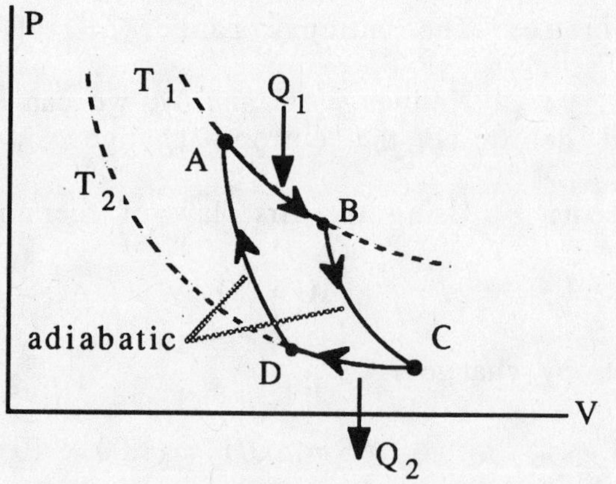

Figure 19.6

Determine the following:

________ 8. Heat absorbed by the gas from the high temperature reservoir
________ 9. Change in entropy of the gas at the high temperature reservoir
________ 10. Heat exhausted by the gas to the low temperature reservoir
________ 11. Change in entropy of the gas at the low temperature reservoir
________ 12. Change in entropy for the system during a cycle
________ 13. Efficiency of the Carnot engine
________ 14. Coefficient of performance of the associated Carnot refrigerator

In a calorimetry experiment, 200 g of water is placed in a 150 g copper calorimeter cup. The cup and water have an initial temperature of 20°C and are well insulated from the surroundings. After 200 g of lead shot is heated to 100°C it is dumped into the water and an equilibrium temperature obtained.

Determine the following:

________ 15. Final equilibrium temperature of the system
________ 16. Change in entropy of the water
________ 17. Change in entropy of the calorimeter cup
________ 18. Change in entropy of the led shot
________ 19. Change in entropy of the system
________ 20. Change in entropy of the surroundings
________ 21. Change in entropy of the universe

(See Appendix I for answers.)

PRINCIPAL CONCEPTS AND EQUATIONS PRESCRIPTION

Your score on the practice test is an excellent measure of your understanding of the chapter. You should now use the following chart to write your own prescription for dealing with any weaknesses the practice test points out. Look down the leftmost column to the number of the question(s) you answered incorrectly, reading across that row you will find the concept and/or equation of concern, the section(s) of the study guide you should return to for further study, and some suggested text exercises which you should work to gain additional experience.

Practice Test Questions	Concepts and Equations	Prescription: Principal Concepts	Prescription: Text Exercises
1	Work by engine: $Q_H = W + \lvert Q_c \rvert$	1	19-1, 2
2	Efficiency of engine: $\eta = W/Q_H = 1 - \lvert Q_c \rvert/Q_H$	1	19-2, 3
3	Coefficient of performance: $K = Q_c/\lvert W \rvert = Q_c/(\lvert Q_H \rvert - Q_c)$	1	19-7, 8
4	Heat exhaust by refrigerator: $\lvert W \rvert + Q_c = \lvert Q_H \rvert$	1	19-8, 9
5	Heat pump coefficient of performance: $K_{hp} = \lvert Q_H \rvert/W = \lvert Q_H \rvert/(\lvert Q_H \rvert - Q_c)$	1	19-9, 10
6	Heat pump rate of heat extraction: $\lvert W \rvert/t + Q_c/t = \lvert Q_H \rvert/t$	1	19-9, 10
7	Cost = (Rate)(Time)	1	19-9, 10
8	Isothermal process -- heat absorbed: $Q_{ab} = W_{ab} = nRT_a \ln(V_b/V_a)$	2	19-12, 13
9	Entropy: $\Delta S_{ab} = Q_{ab}/T_a$	3	19-23, 25
10	Isothermal process -- heat exhausted $Q_{cd} = W_{cd} = nRT_c \ln(V_d/V_c)$	2	19-12, 13
11	Entropy: $\Delta S_{cd} = Q_{cd}/T_c$	3	19-23, 25
12	Entropy change for cycle: $\Delta S = 0$	3	19-27
13	Efficiency of heat engine: $\eta = W/Q_H = 1 - (T_c/T_H)$	2	19-18
14	Heat pump coefficient of performance: $K_{hp} = \lvert Q_c \rvert/W = T_c/(T_H - T_c)$	2	19-20, 22
15	Calorimetry: $Q_{loss} = Q_{gain}$	2 of Ch 17	---
16 - 21	Entropy change: $\Delta S = mc_p \ln(T_E/T_i)$	3	19-27

20 COULOMB'S LAW AND THE ELECTRIC FIELD

RECALL FROM PREVIOUS CHAPTERS

Previously learned concepts and equations frequently used in this chapter	Text Section	Study Guide Page
Resolving vectors into components analytically	2-3, 4	2-8
Adding vectors analytically	2-3, 4	2-8

NEW IDEAS IN THIS CHAPTER

Concepts and equations introduced	Text Section	Study Guide Page
Electric charge	20-1	20-1
Insulators and conductors	20-2	20-2
Coulomb's Law: $\mathbf{F}_{ab} = k\left(\frac{q_a q_b}{r^2}\right)\hat{\mathbf{r}}$	20-3	20-2
Electric field for point charges	20-4, 5	20-5
Electric dipoles	20-5	20-7
Electric field for a continuous distribution	20-5	20-7
Electric field lines	20-6	20-11

PRINCIPAL CONCEPTS AND EQUATIONS

1 Charge (Section 20-1)

There are two different types of charge; these are designated as positive (+) charge and negative (-) charge. Macroscopic objects possess both types of charge and are said to be neutral if they possess equal amounts of positive and negative charge. If they have an excess of positive (negative) charge, then they are said to be positively (negatively) charged.

Experiments demonstrate:

i) objects that have charges of the same sign repel each other, and
ii) objects that have charges of different sign attract each other.

The unit of charge is the Coulomb (C), and it is defined in terms of magnetic forces.

We find from experiment that charge is quantized, i.e. there is a basic unit or size for charge. The basic unit is the amount of charge on a proton (+e) or an electron (-e), where $e = 1.6 \times 10^{-19}$ C.

2 Insulators and Conductors (Section 20-2)

A conductor is a material which readily allows charge to flow under the action of electric forces, and an insulator is a material for which it is difficult to cause a flow of charge.

3 Coulomb's Law (Section 20-3)

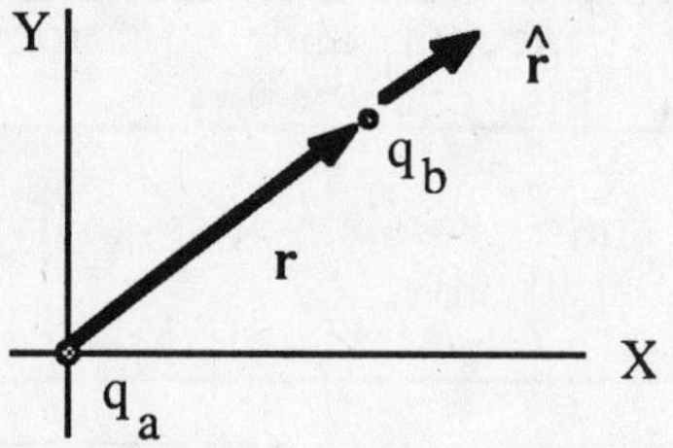

Figure 20.1

$$\mathbf{F}_{ab} = \left(\frac{k q_a q_b}{r^2}\right) \hat{r}, \text{ where } k = 9 \times 10^9 \ (\text{N}\cdot\text{m}^2)/\text{C}^2 \qquad (1)$$

The charges q_a and q_b are point charges and are specified in Coulombs; r is the distance between the charges and is specified in meters. As stated above $\mathbf{F}_{ab}$ is the force that point charge q_a exerts on point charge q_b.

Note: The statement of Coulomb's Law is in agreement with i) and ii) under Charge (section 1 - above).

If we have several point charges Q_i (i = 1, 2, ... , N) exerting an electric force on another point charge q_0, we simply use Coulomb's Law to get the separate electric forces on q_0 due to each of the other point charges Q_i (a total of N electric forces), and then add these forces vectorially to obtain the resultant electric force acting on the point charge q_0. Thus we write in compact notation:

2 different point charges.

$$\mathbf{F}_{q_0} = \sum_{i=1}^{N} \mathbf{F}_{iq_0} = \sum_{i=1}^{N} \left[\frac{kQ_i q_0}{r_2}\right] \hat{r}_i \qquad (2)$$

where $\mathbf{r}_i$ is the vector from the charge Q_i to the charge q_0.

Practice: A point charge q_0 = 2 μC is located at the origin and point charges Q_1 = -2 μC and Q_2 = 3 μC are located at the points (0,3) m and (2,0) m respectively.

Determine the following:

1. Draw a Cartesian coordinate system showing the locations of the charges 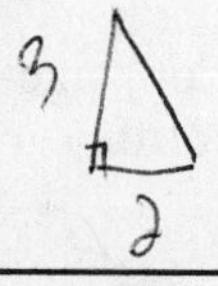	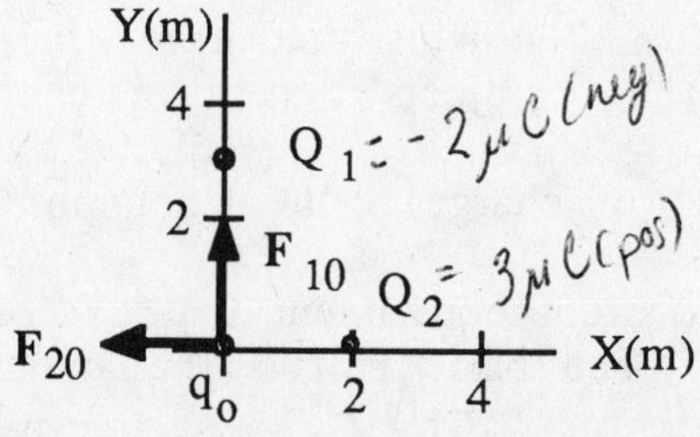

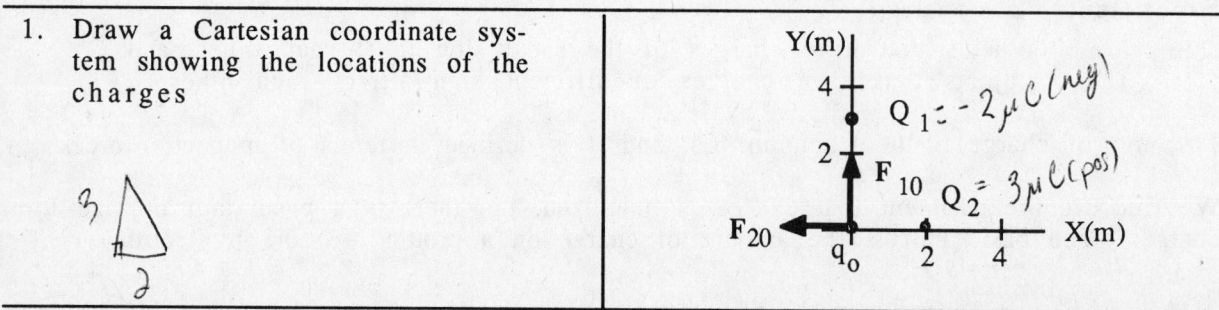

2. Calculate the magnitudes F_{1o} and F_{2o}	$F_{1o} = \left[\dfrac{k\lvert Q_1 q_0\rvert}{r_1^2}\right]$ in the y-direction $= (9 \times 10^9)(2 \times 10^{-6})(2 \times 10^{-6})/3^2$ $= 4.00 \times 10^{-3}$ N $F_{2o} = \left[\dfrac{k\lvert Q_2 q_0\rvert}{r_2^2}\right]$ in the x-direction $= (9 \times 10^9)(3 \times 10^{-6})(2 \times 10^{-6})/2^2$ $= 13.5 \times 10^{-3}$ N
3. Write the x and y components of each force	$F_{1ox} = 0$; $F_{1oy} = 4.00 \times 10^{-3}$ N $F_{2ox} = -13.5 \times 10^{-3}$ N; $F_{2oy} = 0$
4. Add components of F_{1o} and F_{2o} to obtain the components of $\mathbf{F}_{q_0}$	$F_{ox} = F_{1ox} + F_{2ox} = -13.5 \times 10^{-3}$ N $F_{oy} = F_{1oy} + F_{2oy} = 4.00 \times 10^{-3}$ N
5. Resultant electric force on q_0 due to Q_1 and Q_2	$\mathbf{F}_{q_0} = -13.5 \times 10^{-3}\ \text{N}\ \mathbf{i} + 4.00 \times 10^{-3}\ \text{N}\ \mathbf{j}$

Note: This completely specifies $\mathbf{F}_{q_0}$; we do not need to calculate the magnitude and an angle unless the statement of the problem specifically requires that we do.

Example: Given that point charge $Q_1 = 2\ \mu C$ is located at the origin, point charge $Q_2 = -3\ \mu C$ is located at the point (1,2) m, and that the point charge $q_0 = 4\ \mu C$ is located at the point (3,4) m, calculate the resultant electric force on q_0 due to Q_1 and Q_2.

Given: Point charge Q_1 is located at the origin, and $Q_1 = 2\ \mu C$
Point charge Q_2 is located at the point (1,2) m, and $Q_2 = -3\ \mu C$
Point charge q_0 is located at the point (3,4) m, and $q_0 = 4\ \mu C$

Determine: Resultant electric force on q_0 due to Q_1 and Q_2

Strategy:

1. Draw a set of coordinate axes and locate the charges.
2. Sketch the forces on q_0 due to each of Q_1 and Q_2; indicate the proper direction for each of these forces.
3. Calculate F_{1o} and F_{2o} and obtain the components of $\mathbf{F}_{1o}$ and $\mathbf{F}_{2o}$.
4. From the components of $\mathbf{F}_{1o}$ and $\mathbf{F}_{2o}$ obtain $\mathbf{F}_{q_0}$.

Solution:

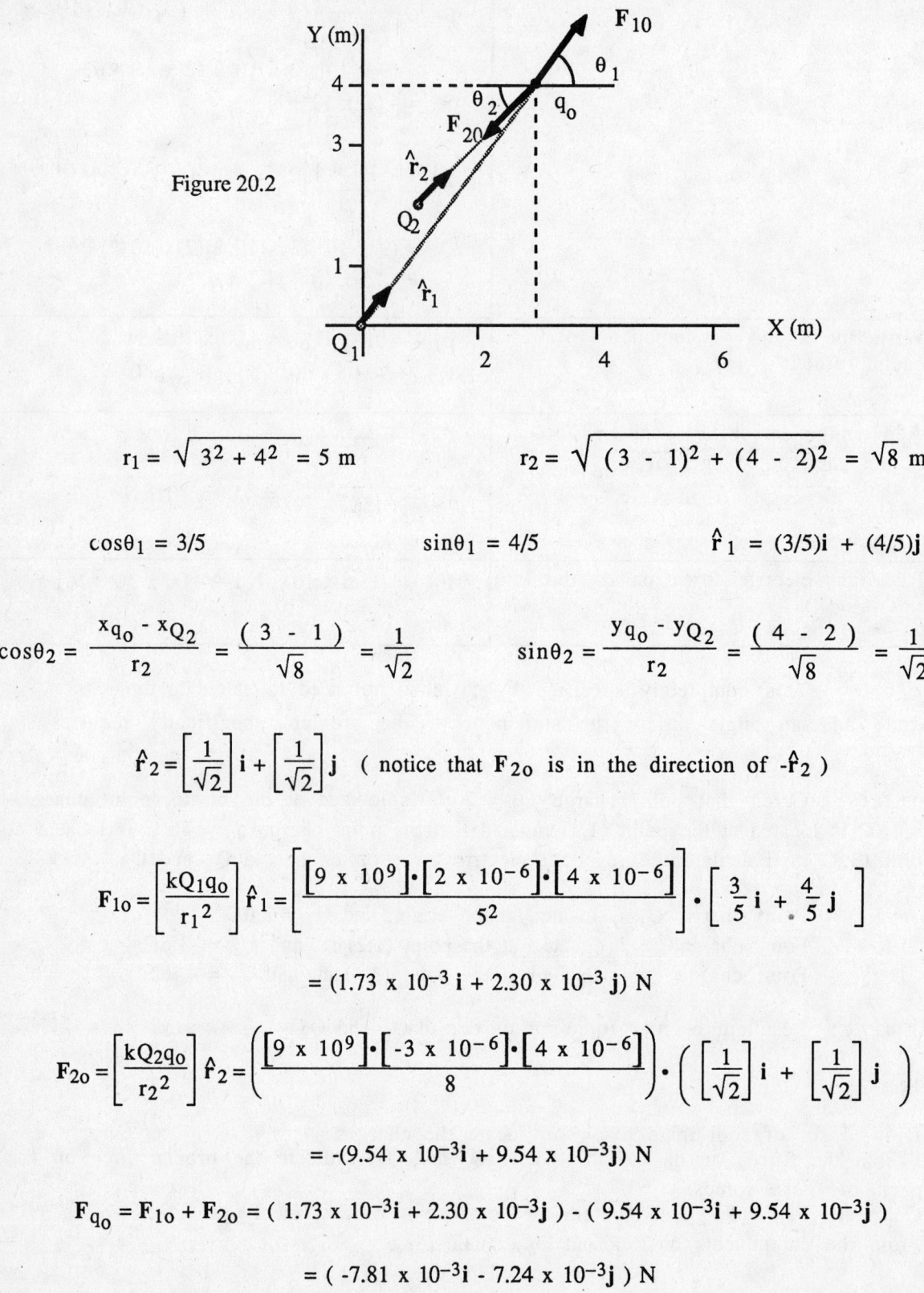

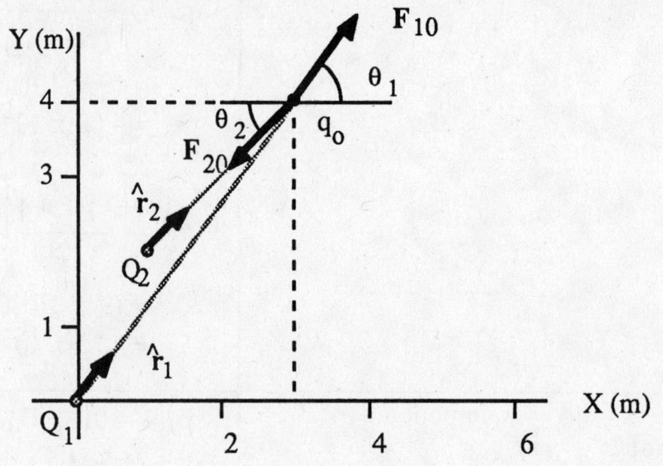

Figure 20.2

$$r_1 = \sqrt{3^2 + 4^2} = 5 \text{ m} \qquad r_2 = \sqrt{(3 - 1)^2 + (4 - 2)^2} = \sqrt{8} \text{ m}$$

$$\cos\theta_1 = 3/5 \qquad \sin\theta_1 = 4/5 \qquad \hat{r}_1 = (3/5)\mathbf{i} + (4/5)\mathbf{j}$$

$$\cos\theta_2 = \frac{x_{q_0} - x_{Q_2}}{r_2} = \frac{(3 - 1)}{\sqrt{8}} = \frac{1}{\sqrt{2}} \qquad \sin\theta_2 = \frac{y_{q_0} - y_{Q_2}}{r_2} = \frac{(4 - 2)}{\sqrt{8}} = \frac{1}{\sqrt{2}}$$

$$\hat{r}_2 = \left[\frac{1}{\sqrt{2}}\right]\mathbf{i} + \left[\frac{1}{\sqrt{2}}\right]\mathbf{j}$$ (notice that $\mathbf{F}_{2o}$ is in the direction of $-\hat{r}_2$)

$$\mathbf{F}_{1o} = \left[\frac{kQ_1q_0}{r_1^2}\right]\hat{r}_1 = \left[\frac{[9 \times 10^9]\cdot[2 \times 10^{-6}]\cdot[4 \times 10^{-6}]}{5^2}\right]\cdot\left[\frac{3}{5}\mathbf{i} + \frac{4}{5}\mathbf{j}\right]$$

$$= (1.73 \times 10^{-3}\,\mathbf{i} + 2.30 \times 10^{-3}\,\mathbf{j}) \text{ N}$$

$$\mathbf{F}_{2o} = \left[\frac{kQ_2q_0}{r_2^2}\right]\hat{r}_2 = \left(\frac{[9 \times 10^9]\cdot[-3 \times 10^{-6}]\cdot[4 \times 10^{-6}]}{8}\right)\cdot\left(\left[\frac{1}{\sqrt{2}}\right]\mathbf{i} + \left[\frac{1}{\sqrt{2}}\right]\mathbf{j}\right)$$

$$= -(9.54 \times 10^{-3}\mathbf{i} + 9.54 \times 10^{-3}\mathbf{j}) \text{ N}$$

$$\mathbf{F}_{q_0} = \mathbf{F}_{1o} + \mathbf{F}_{2o} = (1.73 \times 10^{-3}\mathbf{i} + 2.30 \times 10^{-3}\mathbf{j}) - (9.54 \times 10^{-3}\mathbf{i} + 9.54 \times 10^{-3}\mathbf{j})$$

$$= (-7.81 \times 10^{-3}\mathbf{i} - 7.24 \times 10^{-3}\mathbf{j}) \text{ N}$$

Related Text Exercises: Ex. 20-8 through 20-14.

4 Electric Field for Point Charges (Section 20-4)

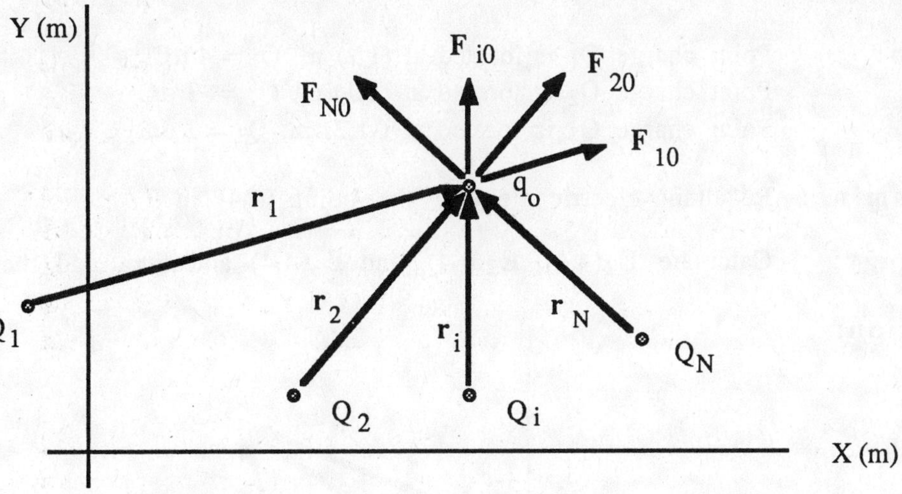

Figure 20.3

Recall equation (2): $\mathbf{F}_{q_0} = \sum_{i=1}^{N} \left[\frac{kQ_iq_0}{r_i^2}\right]\hat{r}_i = \left[\sum_{i=1}^{N} \left[\frac{kQ_i}{r_i^2}\right]\hat{r}_i\right] q_0$

Thus we see that $\mathbf{F}_{q_0}$ is the product of two terms. One is the force per coulomb due to the N point charges Q_1, Q_2, ... Q_N and the other term is the charge q_0 that the force is acting on. The first is called the electric field at the point P due to the N point charges. It depends on the spatial distribution of the charges relative to the point P. In essence it gives the force (magnitude and direction) on one coulomb of positive charge at the point P.

Thus $\mathbf{F}_{Pq_0} = q_0\mathbf{E}_P$, where $\mathbf{E}_P = \sum_{i=1}^{N} \left[\frac{kQ_i}{r_i^2}\right]\hat{r}_i$. From this we may write the electric field at a point P due to a spatial distribution of charges as $\mathbf{E}_P = \mathbf{F}_{Pq_0}/q_0$.

Practice: Consider Q_1, Q_2 and q_0 as described in the example of the previous section.

Determine the following:

1. The force on q_0 due to Q_1 and Q_2	In the example, this force was determined to be $\mathbf{F}_{q_0} = (-7.81 \times 10^{-3}\,\mathbf{i} - 7.24 \times 10^{-3}\,\mathbf{j})$ N
2. The electric field at the point x = 3.00 m, y = 4.00 m. i.e. $\mathbf{E}(3,4)$	We already know the force on q_0 at this point (step 1). Recall $q_0 = 4.00 \times 10^{-6}$ C. We may determine the electric field at this point by $\mathbf{E}(3,4) = \mathbf{F}_{q_0}(3,4)/q_0$ $= (-1.95 \times 10^3\,\mathbf{i} - 1.81 \times 10^3\,\mathbf{j})$ N/C

Example: Calculate the resultant electric field at the point (4,4) m due to point charges $Q_1 = 4\ \mu C$ located at (1,2) m, $Q_2 = -3\ \mu C$ located at (3,1) m and $Q_3 = 2\ \mu C$ located at (-1,2) m.

Given: Point charge Q_1 is located at (1,2) m, $Q_1 = 4\ \mu C$
Point charge Q_2 is located at (3,1) m, $Q_2 = -3\ \mu C$
Point charge Q_3 is located at (-1,2) m, $Q_3 = 2\ \mu C$

Determine: Resultant electric field at the point (4,4) m

Strategy: Calculate $\mathbf{E}_1(4,4)$, $\mathbf{E}_2(4,4)$, and $\mathbf{E}_3(4,4)$ and then add them vectorially.

Solution:

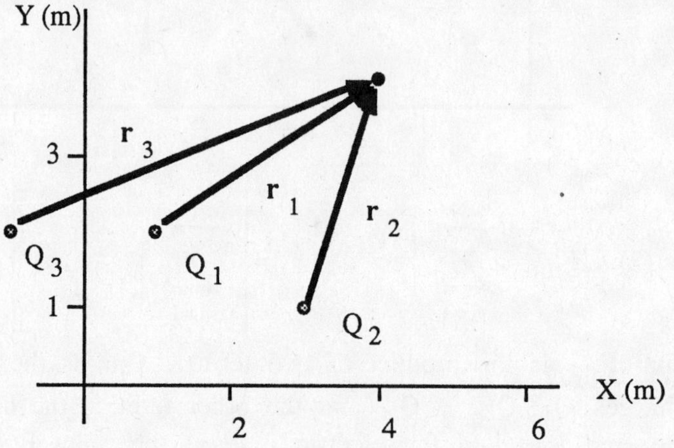

Figure 20.4

$$\mathbf{r}_1 = (4 - 1)\mathbf{i} + (4 - 2)\mathbf{j} = 3\mathbf{i} + 2\mathbf{j}$$

$$\mathbf{r}_2 = (4 - 3)\mathbf{i} + (4 - 1)\mathbf{j} = \mathbf{i} + 3\mathbf{j}$$

$$\mathbf{r}_3 = (4 - [-1])\mathbf{i} + (4 - 2)\mathbf{j} = 5\mathbf{i} + 2\mathbf{j}$$

$$\hat{\mathbf{r}}_1 = \frac{(3\mathbf{i} + 2\mathbf{j})}{\sqrt{3^2 + 2^2}} = \frac{(3\mathbf{i} + 2\mathbf{j})}{\sqrt{13}} \qquad \hat{\mathbf{r}}_2 = \frac{(\mathbf{i} + 3\mathbf{j})}{\sqrt{1^2 + 3^2}} = \frac{(\mathbf{i} + 3\mathbf{j})}{\sqrt{10}} \qquad \hat{\mathbf{r}}_3 = \frac{(5\mathbf{i} + 2\mathbf{j})}{\sqrt{5^2 + 2^2}} = \frac{(5\mathbf{i} + 2\mathbf{j})}{\sqrt{29}}$$

Therefore,

$$\mathbf{E}_1(4,4) = \left[\frac{kQ_1}{r_1^2}\right]\hat{\mathbf{r}}_1 = \left[\frac{(9 \times 10^9)\cdot(4 \times 10^{-6})}{13}\right]\cdot\left[\frac{(3\mathbf{i} + 2\mathbf{j})}{\sqrt{13}}\right]$$
$$= (2.30 \times 10^3\mathbf{i} + 1.54 \times 10^3\mathbf{j})\ \text{N/C}$$

$$\mathbf{E}_2(4,4) = \left[\frac{kQ_2}{r_2^2}\right]\hat{\mathbf{r}}_2 = \left[\frac{(9 \times 10^9)\cdot(-3 \times 10^{-6})}{10}\right]\cdot\left[\frac{(\mathbf{i} + 3\mathbf{j})}{\sqrt{10}}\right]$$
$$= (-0.850 \times 10^3\mathbf{i} - 2.56 \times 10^3\mathbf{j})\ \text{N/C}$$

$$\mathbf{E}_3(4,4) = \left[\frac{kQ_3}{r_3^2}\right]\hat{\mathbf{r}}_3 = \left[\frac{(9 \times 10^9)\cdot(2 \times 10^{-6})}{29}\right]\cdot\left[\frac{(5\mathbf{i} + 2\mathbf{j})}{\sqrt{29}}\right]$$
$$= (0.580 \times 10^3\mathbf{i} + 0.230 \times 10^3\mathbf{j})\ \text{N/C}$$

Then,

$$\mathbf{E}(4,4) = \mathbf{E}_1(4,4) + \mathbf{E}_2(4,4) + \mathbf{E}_3(4,4) = (2.03 \times 10^3\mathbf{i} - 0.790 \times 10^3\mathbf{j}) \text{ N/C}$$

Example: What force is exerted on a 2 μC charge placed at the point (4,4) m of the previous example.

Given: $\mathbf{E}(4,4) = (2.03 \times 10^3\mathbf{i} - 0.790 \times 10^3\mathbf{j})$ N/C
$q_o = 2.00 \times 10^{-6}$ C located at the point (4,4) m

Determine: The force exerted on q_o by **E**

Strategy: Knowing that q_o is at (4,4) m and knowing the electric field at this point, we can determine the force on q_o by **E**.

Solution:

$$\mathbf{F}_{q_o}(4,4) = q_o\mathbf{E}(4,4)$$
$$= (2.00 \times 10^{-6})(2.03 \times 10^3\,\mathbf{i} - 0.79 \times 10^3\,\mathbf{j}) \text{ N}$$
$$= (4.06 \times 10^{-3}\,\mathbf{i} - 1.58 \times 10^{-3}\,\mathbf{j}) \text{ N}$$

Related Text Exercises: Ex. 20-15 through 20-23 and 20-45.

5 Electric Dipoles (Section 20-4)

Figure 20.5

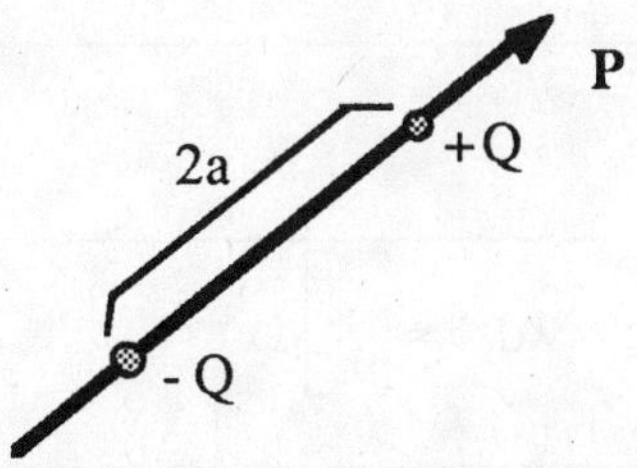

P is called the *electric dipole moment*

i.) The magnitude of **P** is 2aQ
ii.) The direction of **P** is along the line from -Q to +Q.

The field due to an electric dipole for special points which are far away from the dipole:

i.) The field point is in the xy plane a distance R (R » a) from the origin.
$\mathbf{E} \cong -(k\mathbf{P}/R^3)$ (see example 20-5 in the text)
ii.) The field point is on the +z axis a distance R (R » a) from the origin.
$\mathbf{E} \cong (2k\mathbf{P}/R^3)$ (see exercise 20-25 in the text)

Related Text Exercises: Ex. 20-24, 20-25, 20-31 and 20-46.

6 Electric Field for a Continuous Distribution of Charge (Section 20-4)

Review: If the charge distribution is continuous rather than a collection of point charges, then the sum of electric fields due to the point charges is replaced by an integral of the infinitesimal field contributions due to the infinitesimal elements of charge. If the charge is distributed along a line (not necessarily straight), we use a linear charge density λ in SI units of C/m. If the charge is distributed over a surface (not necessarily plane), we use a surface charge density σ in SI units of C/m^2. Finally if the charge is distributed throughout some volume, we use a volume charge density ρ in SI units of C/m^3. We should note that we may often avoid double (for surface distributions) and triple (for volume distributions) integrals if the distribution has sufficient symmetry so that it is a function of only one coordinate variable.

Practice: A charge Q is uniformly distributed along the x-axis from the origin to x = L. Find the electric field for points (x,0,0) on the x-axis such that x > L.

Determine the following:

1. A figure showing an infinitesimal element of charge and the field point.	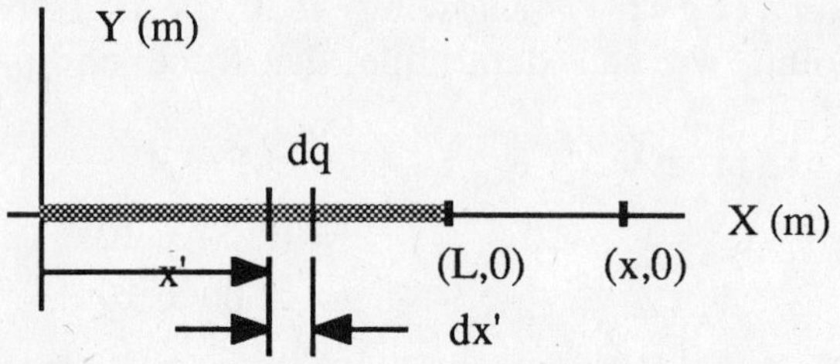 Observe that all infinitesimal elements of charge (dq) yield a contribution to only the x component of the electric field for points on the x-axis.
2. The linear charge density of the charge Q	$\lambda = Q/L$
3. The charge of the infinitesimal element of length dx'	$dq = \lambda dx' = \left[\frac{Q}{L}\right]dx'$
4. The electric field at x due to any infinitesimal element of charge	$dE_x = \frac{k(dq)}{r^2} = \left[\frac{k\left[\frac{Q}{L}\right]dx'}{(x - x')^2}\right]$
5. The x component of the electric field at x due to all of the infinitesimal elements of charge	$E_x = \int_0^L [dE_x] = k\left[\frac{Q}{L}\right] \cdot \int_0^L \frac{dx'}{(x - x')^2}$ $= \left[\frac{+k\left[\frac{Q}{L}\right]}{(x - x')}\right]\Bigg\|_0^L = +k\left[\frac{Q}{L}\right]\left[\frac{1}{(x - L)} - \frac{1}{(x)}\right]$ $= \left[\frac{+kQ}{L}\right]\left[\frac{L}{x(x - L)}\right] = \left[\frac{+kQ}{x(x - L)}\right]$

6. The electric field at (x,0,0)	Since the electric field has only an x component, we write: $$\mathbf{E} = E_x\mathbf{i} = \left[\frac{+kQ}{x(x - L)}\right]\mathbf{i}$$ for values of x > L

Example: Charge is distributed along the x-axis from x = -L to x = +L. The linear charge density is given by $\lambda(x) = bx$. Derive an expression for the electric field, due to this distribution of charge, for points on the +y-axis.

Given: Linear charge density $\lambda(x) = bx$ from x = -L to x = +L

Determine: An expression for the electric field due to this charge distribution, along the +y-axis.

Strategy: Since the charge distribution is symmetric about the origin, the x component of the electric field for points on the +y-axis will be zero and the y component is just double the contribution due to the charge along the positive segment of the x-axis.

Solution:

Figure 20.6

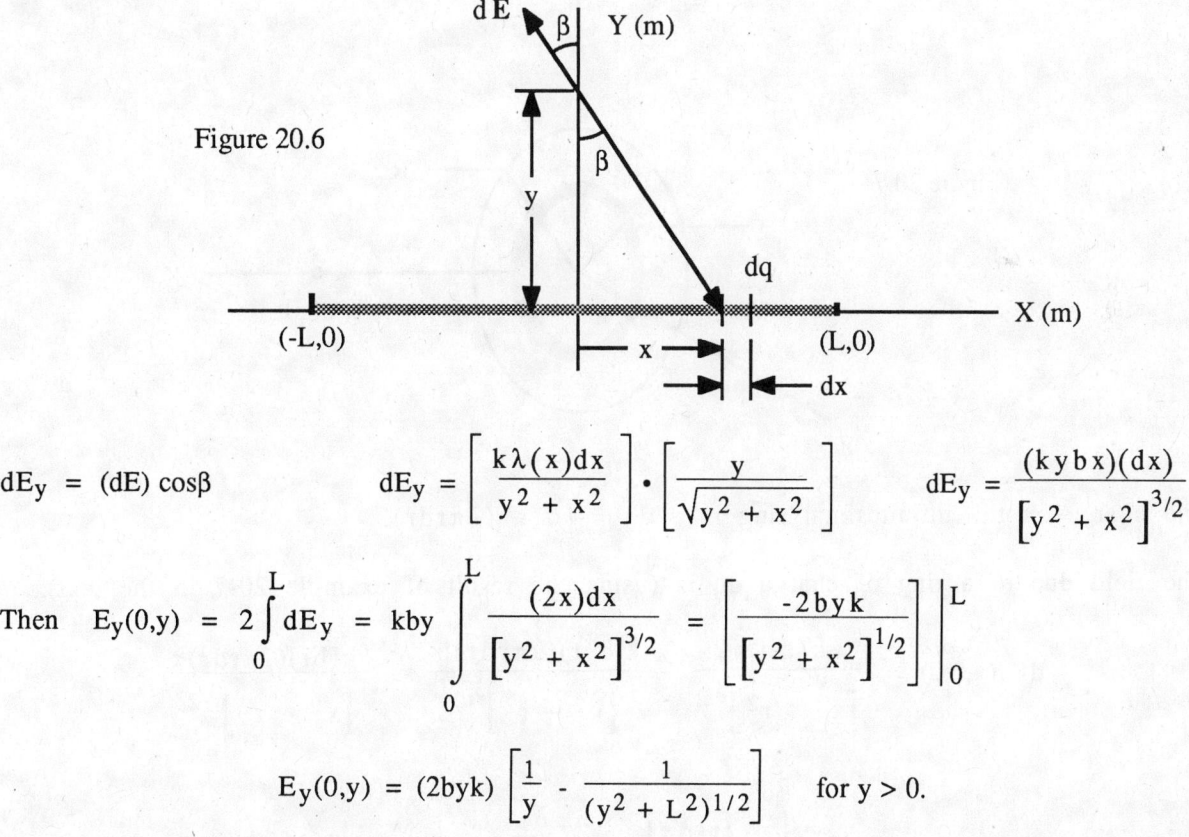

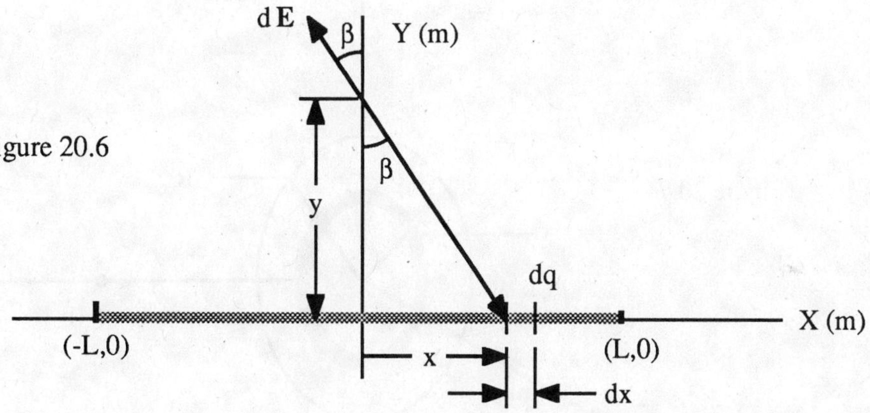

$$dE_y = (dE)\cos\beta \qquad dE_y = \left[\frac{k\lambda(x)dx}{y^2 + x^2}\right] \bullet \left[\frac{y}{\sqrt{y^2 + x^2}}\right] \qquad dE_y = \frac{(kybx)(dx)}{\left[y^2 + x^2\right]^{3/2}}$$

$$\text{Then } E_y(0,y) = 2\int_0^L dE_y = kby\int_0^L \frac{(2x)dx}{\left[y^2 + x^2\right]^{3/2}} = \left[\frac{-2byk}{\left[y^2 + x^2\right]^{1/2}}\right]_0^L$$

$$E_y(0,y) = (2byk)\left[\frac{1}{y} - \frac{1}{(y^2 + L^2)^{1/2}}\right] \quad \text{for } y > 0.$$

Example: The plane of a disc of radius R_0 is oriented perpendicular to the x-axis with its center at the origin (0,0,0). It carries a surface charge density $\sigma(r) = br$, where $r = \sqrt{y^2 + x^2}$, and b is a constant; notice that r measures the distance that a point on the disc lies from the center of the disc. Find the electric field at points on the x-axis.

Given: A charged disk in the y-z plane which is centered at the origin
R_0 = radius of the disk
$\sigma(r) = br$ = surface charge density on the disk

Determine: An expression for the electric field at points on the x-axis.

Strategy: Because of symmetry we need consider only points on the positive x-axis (E will have the same magnitude but opposite direction for positive and negative values of x). Also because of symmetry we can break the disc down into a sequence of infinitesimal rings; we can do this because the charge density $\sigma(r)$ doesn't vary over a ring (r has a constant value for a ring). Knowing the charge distribution, we can determine the charge on each infinitesimal ring. Knowing the charge on each ring, we can determine its contribution to the electric field at any point on the x-axis. Finally, we can determine the electric field at any point along the x-axis by summing the contribution from each ring.

Solution:

Figure 20.7

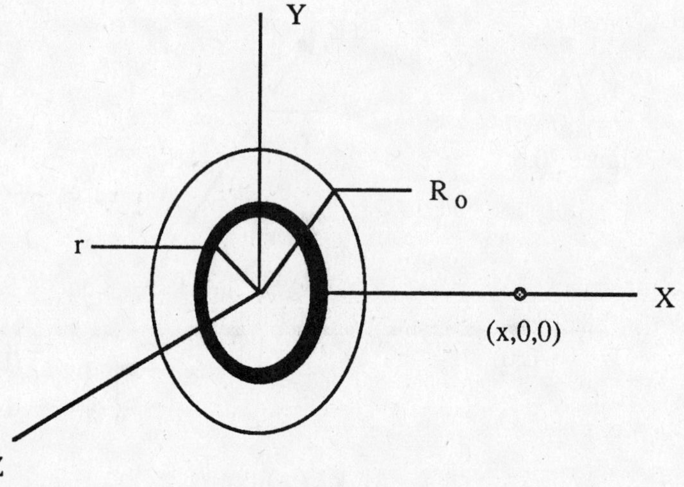

The charge on the infinitesimal ring is $dq = [\sigma(r)](2\pi r dr)$.

The field due to a ring of charge dq is (using the result of example 20-7 in the text):

$$dE_x(x,0,0) = \frac{k(dq)x}{\left[x^2 + r^2\right]^{3/2}} = \frac{k[\sigma(r)(2\pi r dr)]x}{\left[x^2 + r^2\right]^{3/2}} = \frac{k(br)(2\pi r dr)x}{\left[x^2 + r^2\right]^{3/2}}$$

Therefore $$E_x(x,0,0) = 2\pi bkx \int_0^{R_0} \frac{\left[r^2 dr\right]}{\left[x^2 + r^2\right]^{3/2}}$$

Notice that r is the variable of integration rather than x; x is constant during the integration.

The above integral may be evaluated using standard integral tables to obtain:

$$E_x = 2\pi bkx \left[\left(\frac{r}{2} \cdot \sqrt{x^2 + r^2}\right) - \left(\frac{x^2}{2} \cdot \log\left(r + \sqrt{r^2 + x^2}\right)\right)\right]\Bigg|_0^L$$

$$\text{or } E_x = 2\pi bkx \cdot \left(\left[\frac{L}{2} \cdot \sqrt{x^2 + L^2}\right] - \left[\frac{x^2}{2} \cdot \log\left(\frac{L + \sqrt{L^2 + x^2}}{x}\right)\right]\right) \quad \text{for } x > 0.$$

Related Text Exercises: Ex. 20-26 through 20-29, 20-32 through 20-35 and 20-47 through 20-49.

7 Electric Field Lines (Section 20-6)

Review: Lines are drawn which give a pictorial representation of the electric field. The salient features are:

i.) The direction of the electric field at a point is tangent to the field line through that point, and
ii.) The magnitude of the electric field at a point is proportional to the density (the number of lines per unit area perpendicular to the lines) of lines at the point.

The figure below shows the field line pattern for two point charges of equal positive charge.

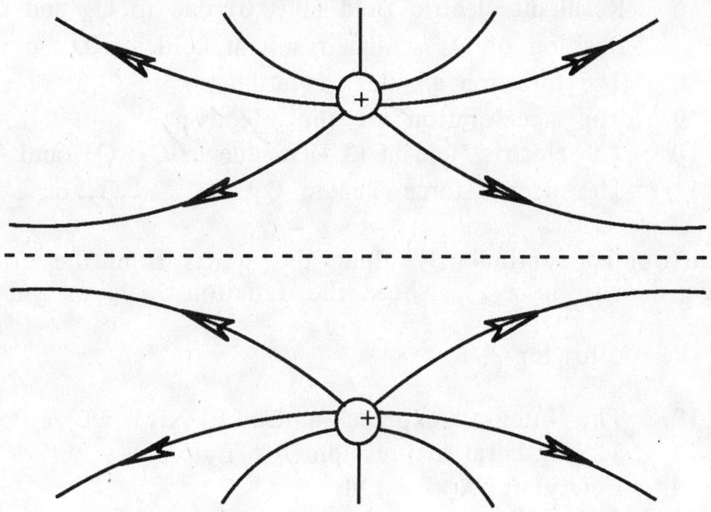

Figure 20.8

Related Text Exercises: Ex. 20-36 through 20-38.

PRACTICE TEST

Take and grade this practice test. Doing so will allow you to determine any weak spots in your understanding of the concepts taught in this chapter. The following section prescribes what you should study further to strengthen your understanding.

$Q_1 = 3\mu C$
$Q_2 = -2\mu C$
$Q_3 = 4\mu C$
$Q_4 = 2\mu C$

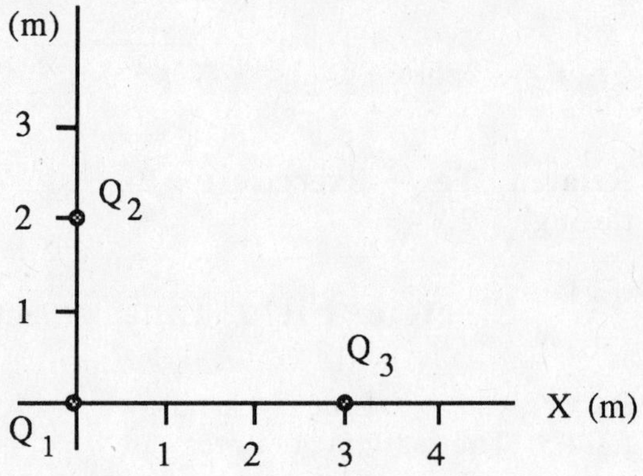

Determine the following:

_________ 1. Force on Q_3 due Q_1
_________ 2. Force on Q_1 due Q_3
_________ 3. Force on Q_2 due Q_1
_________ 4. Force on Q_1 due Q_2
_________ 5. Resultant force on Q_1 due to Q_2 and Q_3
_________ 6. Resultant electric field at (0,0) due to Q_2 and Q_3
_________ 7. Position of Q_4 so that resultant field at (0,0) due to Q_2, Q_3 and Q_4 is zero
_________ 8. The force on an electron at the point (0,0) due to Q_2 and Q_3
_________ 9. The acceleration of the electron
_________ 10. The electric field at (3,4) m due to Q_1, Q_2, and Q_3
_________ 11. The electric force, due to Q_1, Q_2, and Q_3, on a 2 μC at the point (3,4) m

Charge is distributed continuously along the x-axis from the origin to x = L. The linear charge density is $\lambda(x)$, where the function λ is as yet unspecified.

Determine the following:

_________ 12. The integral expression for $E_x(x,y)$ where (x,y) is an arbitrary point
_________ 13. The integral expression for $E_y(x,y)$
_________ 14. $E_x(x,y)$ for $\lambda(x) = Q/L$
_________ 15. $E_y(x,y)$ for $\lambda(x) = Q/L$

A disc of radius R_o has a surface charge density given by $\sigma(r) = br^2$. The disc is oriented so that its plane is perpendicular to the x-axis with its center at the origin.

Determine the following:

_________ 16. The electric field due to the disc at a point (x,0,0) on the x-axis.

(See Appendix I for answers.)

PRINCIPAL CONCEPTS AND EQUATIONS PRESCRIPTION

Your score on the practice test is an excellent measure of your understanding of the chapter. You should now use the following chart to write your own prescription for dealing with any weaknesses the practice test points out. Look down the leftmost column to the number of the question(s) you answered incorrectly, reading across that row you will find the concept and/or equation of concern, the section(s) of the study guide you should return to for further study, and some suggested text exercises which you should work to gain additional experience

Practice Test Questions	Concepts and Equations	Prescription	
		Principal Concepts	Text Exercises
1	Coulomb's Law	3	20-10, 11
2	Coulomb's Law	3	20-11, 12
3	Coulomb's Law	3	20-10, 12
4	Coulomb's Law	3	20-10, 11
5	Coulomb's Law	3	20-11, 12
6	Electric Field: $\mathbf{E} = \mathbf{F}/q_0$	4	20-15, 16
7	Electric Field	3, 4	20-14, 19
8	Force due to electric field	4	20-15
9	Electric force and acceleration	4	20-39
10	Electric Field	4	20-21, 22
11	Force due to electric field	4	20-15
12	Electric field for a continuous distribution	5	20-26, 27
13	Electric field for a continuous distribution	5	20-27, 28
14	Electric field for a continuous distribution	5	20-28, 29
15	Electric field for a continuous distribution	5	20-26, 27
16	Electric field for a continuous distribution	5	20-32

21 GAUSS'S LAW

RECALL FROM PREVIOUS CHAPTERS

Previously learned concepts and equations frequently used in this chapter	Text Section	Study Guide Page
Resolving vectors into components	2-3, 4	2-8
Vectors addition	2-3, 4	2-8
Coulomb's Law	20-3	20-2
Electric Fields	20-4, 5	20-5, 7
Electric Field Lines	20-6	20-11

NEW IDEAS IN THIS CHAPTER

Concepts and equations introduced	Text Section	Study Guide Page
Electric flux: $\Phi = \int_S \mathbf{E} \cdot d\mathbf{S}$	21-1, 2	21-1, 2
Gauss's Law: $\Phi = \oint_S \mathbf{E} \cdot d\mathbf{S} = \sum_{i=1}^{N} Q_i/\varepsilon_0$ for N charges inside S	21-2, 4	21-2, 8
Conductors	21-5	21-13

PRINCIPAL CONCEPTS AND EQUATIONS

1 Electric Flux (Section 21-1)

The net number of electric field lines that cut through a surface is called the electric flux through that surface. If the surface is closed, lines that cut outward are taken as *positive* and lines that cut inward are taken as *negative*. If the surface is not closed then an ambiguity exists and we must arbitrarily take those cutting through the surface in one direction as positive and those cutting in the opposite direction as negative.

The above is qualitative rather than quantitative until we specify how many lines originate (end) on one coulomb of positive (negative charge). If we draw $1/\varepsilon_0$ (approx. 1.113×10^{11}) lines per coulomb, then the number of lines per unit area perpendicular to **E** at a point turns out to be just the magnitude of the electric field at that point. This is easily seen to be true for the special case of a charge +Q at the center of a sphere of radius R.

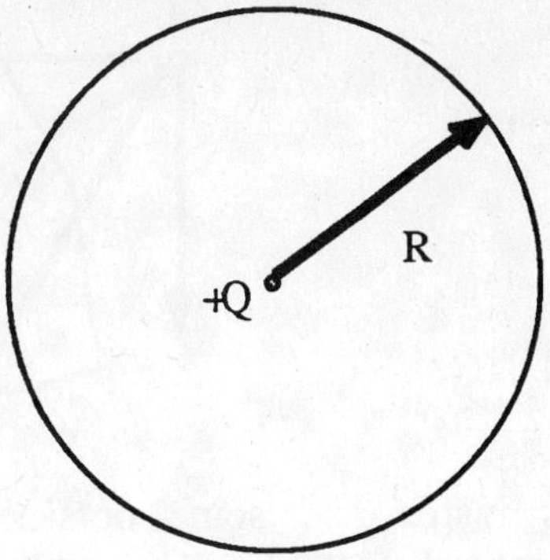

Figure 21.1

Then the number of lines cutting out through the sphere is Q/ε_0; note that all are perpendicular to the sphere.

Then the number per unit area is $\dfrac{Q/\varepsilon_0}{4\pi R^2} = \dfrac{kQ}{R^2}$ which is the magnitude of the field due to Q a distance R away.

2 Gauss's Law (Section 21-2)

Note the following observations:

(i) If a single charge Q <u>inside any closed surface</u> is positive, all Q/ε_0 lines leaving the charge pass outward through the surface. Hence the outward flux for any closed surface containing a positive charge Q is $\Phi = Q/\varepsilon_0$.

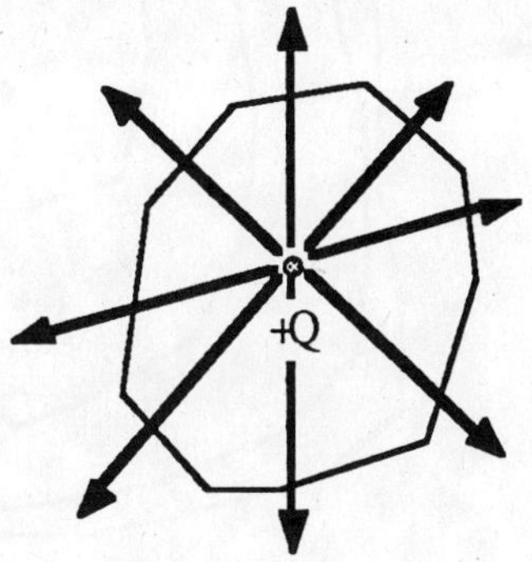

Figure 21.2

(ii) If a single charge <u>inside any closed surface</u> is negative, the flux lines all cut inward and the inward flux is Q/ε_0.

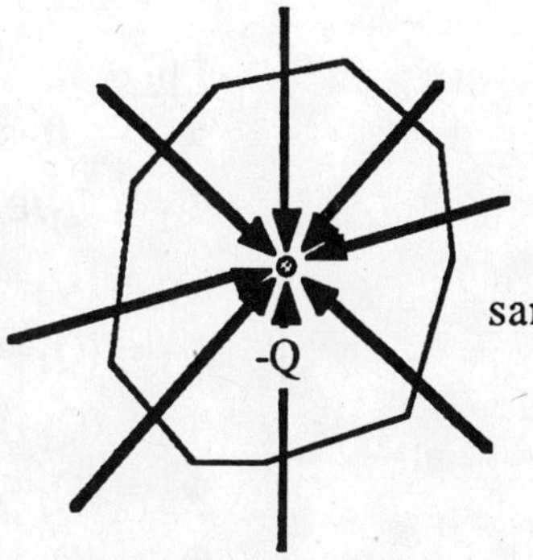

Figure 21.3

(iii) If any single charge (+ or -) is <u>outside any closed surface</u>, then the flux due to that charge through the surface is zero. This is because every line that cuts in or out through the surface does each the same number of times.

Figure 21.4

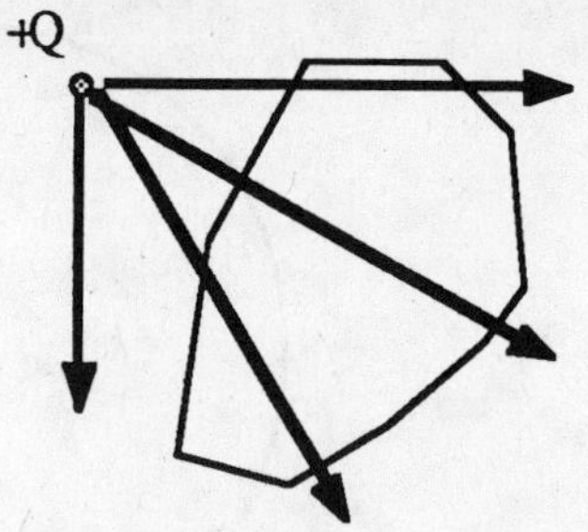

(iv) If there are some charges inside and some outside <u>any closed surface</u>, then the flux through the surface is determined solely by the charges inside the closed surface. Thus the flux is $\Phi = \frac{1}{\varepsilon_0} \sum_{i=1}^{N} Q_i$ where the sum is over the N charges inside.

Note the sign of the flux is + (-) if there is an excess of + (-) charge inside the closed surface. If the amount of + and - charge inside are the same, the flux is zero.

Practice: Shown in Figure 21.5 are several charges located inside different surfaces.

$Q_1 = 4\ \mu C$

$Q_2 = 7\ \mu C$

$Q_3 = -2\ \mu C$

$Q_4 = -6\ \mu C$

$Q_5 = -3\ \mu C$

$Q_6 = -1\ \mu C$

Figure 21.5

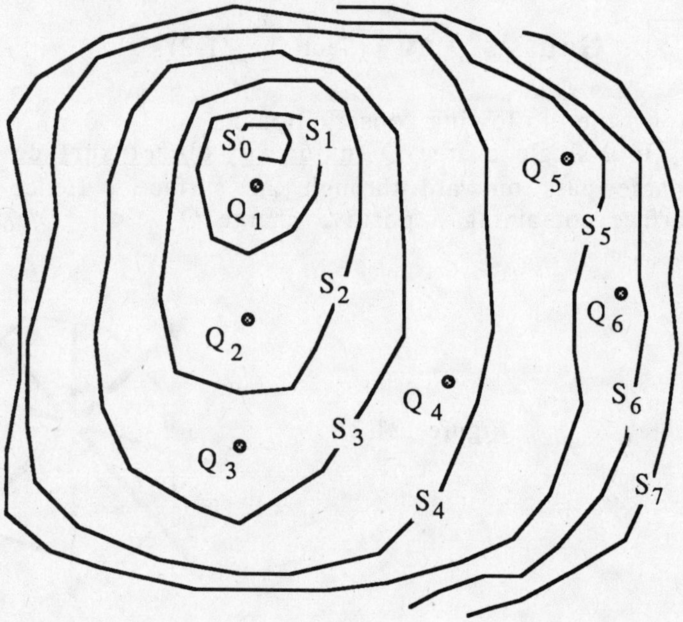

Determine the following:

1. Flux for each surface in Figure 21.5 by calculating the net charge inside the surface	
<u>Surface</u>:	<u>Flux</u>:
S_0	Φ_0 = 0 since there is no charge inside
S_1	$\Phi_1 = Q_1/\varepsilon_0 = 4.52 \times 10^5$ lines
S_2	$\Phi_2 = (Q_1+Q_2)/\varepsilon_0 = 1.24 \times 10^6$ lines
S_3	$\Phi_3 = (Q_1+Q_2+Q_3)/\varepsilon_0 = 1.02 \times 10^6$ lines
S_4	$\Phi_4 = (Q_1+Q_2+Q_3+Q_4)/\varepsilon_0 = 3.39 \times 10^5$ lines
S_5	$\Phi_5 = (Q_1+ ... Q_5)/\varepsilon_0 = 0$ lines
S_6	$\Phi_6 = (Q_1+ ... Q_6)/\varepsilon_0 = -1.13 \times 10^5$ lines
S_7	$\Phi_7 = \Phi_6 = -1.13 \times 10^5$ lines

We may write the above observations (all cases) in terms of an integral over the surface S of <u>any closed surface</u> . This is called Gauss's Law:

$$\oint_S \mathbf{E} \cdot d\mathbf{S} = \frac{Q_{\text{inside } S}}{\varepsilon_0} \qquad \text{for any } S$$

Note: $\oint_S$ means we are performing a surface integral over the closed surface S.

Figure 21.6

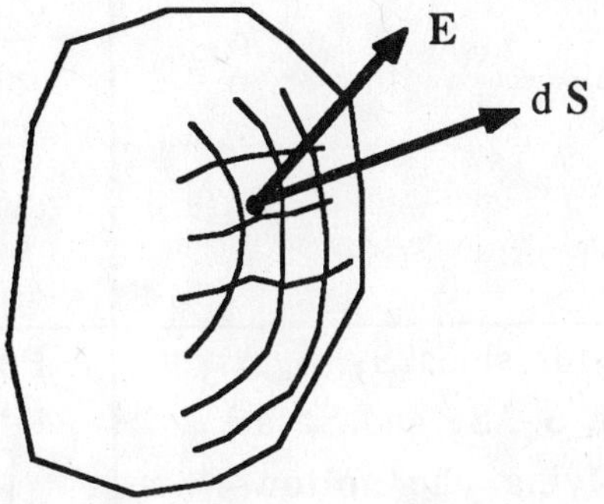

This means we must put a grid of lines over the surface which breaks it into small areas. In the limit of calculus each small area approaches zero and also becomes planer. The magnitude of dS is the area of one of these infinitesimal pieces and the direction of dS is perpendicular to the area and outward. Thus $\mathbf{E} \cdot d\mathbf{S}$ projects dS perpendicular to **E** so that $\mathbf{E} \cdot d\mathbf{S}$ is the electric flux through dS.

The integral $\oint_S \mathbf{E} \cdot d\mathbf{S}$ is the flux through the closed surface S, and we have just seen that this is determined by the net charge inside of S.

Note: This is a very powerful theorem. It states that even though the distribution of charge is changed (rearranged) and hence the electric field at every point on the surface could change, the integral will not change provided the net amount of charge inside doesn't change. The integral is zero if there is no net charge inside even though the field may be non-zero at every point on the surface. The full power of this theorem is most often used in calculating electric fields when there is enough symmetry so that we can easily choose a closed surface (which contains the field point) that allows us to (i) know the angle between **E** and dS at every point on the surface and (ii) make arguments about the magnitude of **E** over the surface so that even though **E** is unknown it can be factored through the integral (sum).

Practice: A uniform electric field is given by $\mathbf{E} = 2\mathbf{i}$ N/C. A cube 2.00 meters on a side is in the + octant with one corner at the origin, its faces are parallel to the coordinates planes.

Determine the following:

1. Draw a Cartesian coordinate system showing a sketch of the faces of the cube.	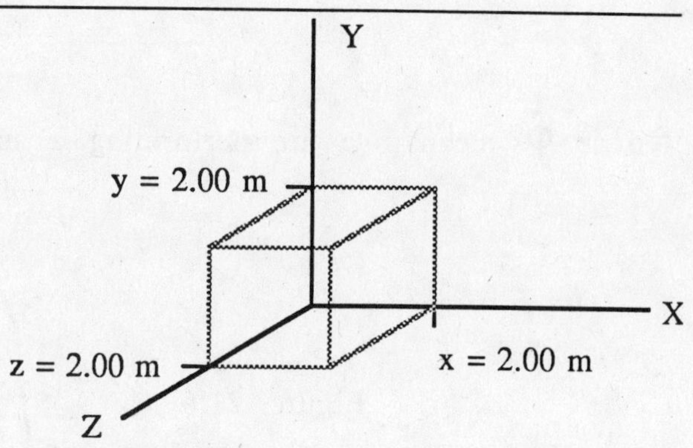
2. The electric flux for sides S_3, S_4, S_5 and S_6, where S_3, S_4, S_5 and S_6 are cube faces satisfying the following: S_3: $z = 0$ S_4: $y = 2m$ S_5: $z = 2m$ S_6: $y = 0$	For these surfaces $\mathbf{E}\cdot d\mathbf{S} = E(dS)\cos 90 = 0$, hence $\int_{S_3} \mathbf{E}\cdot d\mathbf{S}_3 = \int_{S_4} \mathbf{E}\cdot d\mathbf{S}_4 = \int_{S_5} \mathbf{E}\cdot d\mathbf{S}_5 = \int_{S_6} \mathbf{E}\cdot d\mathbf{S}_6 = 0$
3. The electric flux for sides S_1 and S_2 where: S_1: $x = 0$ S_2: $x = 2m$	$\mathbf{E}$ is constant on S_1 and S_2 $d\mathbf{S}_1$ is in the $-\mathbf{i}$ direction $d\mathbf{S}_2$ is in the $+\mathbf{i}$ direction $\int_{S_1} \mathbf{E}\cdot d\mathbf{S}_1 = \int_{S_1} (2.00\text{ N/C})\mathbf{i}\cdot(-\mathbf{i}dS)$ $= -2.00\text{ N/C} \int_{S_1} dS_1 = -2.00\text{ N/C } S_1$ $= (-2.00\text{ N/C})(4.00\text{ m}^2) = -8.00\text{ N}\cdot\text{m}^2/\text{C}$ In like manner $\int_{S_2} \mathbf{E}\cdot d\mathbf{S}_2 = \int_{S_2} (2.00\text{ N/C})\mathbf{i}\cdot(\mathbf{i}dS)$ $= 8.00\text{ N}\cdot\text{m}^2/\text{C}$
4. $\oint_{cube} \mathbf{E}\cdot d\mathbf{S}$	$\Phi_{cube} = \Phi_1 + \Phi_2 + \Phi_3 + \Phi_4 + \Phi_5 + \Phi_6$ $= (-8 + 8 + 0 + 0 + 0 + 0)\text{ N}\cdot\text{m}^2/\text{C} = 0$ This answer is expected since no charge exists inside the surface.

Example: A nonuniform electric field is given by **E** = (2.00x N/C•m)**i**. A cube 2.00 meters on a side is oriented as in the previous practice. Determine the electric flux for each of the six faces and for the entire surface of the cube.

Given: A nonuniform electric field **E** = (2.00x N/C•m)**i**. A cube 2.00 m on a side located in the + octant

Determine: The electric flux for each face and the entire surface of the cube.

Strategy: As in the practice (see Figure 20.7), **E** is perpendicular to S_3, S_4, S_5 and S_6 so there is no flux through any of these faces. Thus we must calculate the flux only for S_1 and S_2. Calculate the flux for each face. Finally, we can obtain the flux for the entire surface by summing the fluxes for the six faces.

Solution: $\Phi_3 = \Phi_4 = \Phi_5 = \Phi_6 = 0$ because **E** is perpendicular to each of these surfaces. The flux through S_1 and S_2 is determined by the definition of flux.

$$\Phi_1 = \int_{S_1} \mathbf{E}\cdot d\mathbf{S} = 0 \text{ since } \mathbf{E} = 0 \quad (x = 0 \text{ on all of } S_1)$$

$$\Phi_2 = \int_{S_2} \mathbf{E}\cdot d\mathbf{S} = \int_{S_2} (2.00x \text{ N/C}\cdot\text{m})\mathbf{i}\cdot dS\mathbf{i} = 4.00 \text{ N/C} \int_{S_2} dS \quad (x = 2 \text{ for all of } S_2)$$

$$= (4.00 \text{ N/C } S_2) = (4.00 \text{ N/C})(4.00 \text{ m}^2) = 16.0 \text{ N}\cdot\text{m}^2/\text{C}$$

Then

$$\Phi_{cube} = \Phi_1 + \Phi_2 + \Phi_3 + \Phi_4 + \Phi_5 + \Phi_6 = (0 + 16 \text{ N}\cdot\text{m}^2/\text{C} + 0 + 0 + 0 + 0) = 16 \text{ N}\cdot\text{m}^2/\text{C}$$

Example: Calculate the net charge inside the cube of Figure 20.7 for the electric fields given in the previous practice and the preceeding example.

Given: case (a) **E** = (2.00 N/C)**i** and $\Phi_{cube} = 0$ (practice)
case (b) **E** = (2.00x N/C•m)**i** and $\Phi_{cube} = 16$ N•m²/C (example)

Determine: The charge inside the cube for these two cases.

Strategy: Knowing that $\frac{Q_{inside\ S}}{\varepsilon_0} = \oint_S \mathbf{E}\cdot d\mathbf{S} = \Phi_{cube}$, we can determine the charge inside the cube.

Solution: case (a) $Q_{inside}/\varepsilon_0 = \Phi_{cube}$, we found $\Phi_{cube} = 0$; therefore $Q_{inside} = 0$. case (b) $Q_{inside}/\varepsilon_0 = \Phi_{cube}$, we found $\Phi_{cube} = 16.0$ N•m²/C; therefore $Q_{inside} = \Phi_{cube}\,\varepsilon_0$ = $(16.0 \text{ N}\cdot\text{m}^2/\text{C})(8.90 \times 10^{-12} \text{ C}^2/\text{N}\cdot\text{m}^2) = 1.42 \times 10^{-10}$ C.

Related Text Exercises: Ex. 21-9 through 21-12.

3 Using Gauss's Law to Calculate Electric Fields (Section 21-4)

If there is enough symmetry in the charge distribution, we can pick a suitable surface containing the point at which we want to determine the electric field. What characterizes a suitable surface? It must be one that takes advantage of the symmetry of the charge so that only the unknown magnitude of **E** at the field point is important to $\oint_S \mathbf{E}\cdot d\mathbf{S}$.

If we are able to find a "suitable surface", then the integral will become this unknown magnitude E times the area of some part or all of the chosen "suitable surface". If we can then determine the charge inside this "suitable surface" we will have succeeded in obtaining **E** at the field point. This will be illustrated in the practice and examples that follow. We will first consider uniform distributions of charge and then non-uniform distributions of charge. The only difference in the two cases is in the calculation of the charge inside a surface S. If the distribution is uniform we will take the appropriate fraction (ratio of volume inside S to the total volume occupied by the charge) of the total charge and if it is non-uniform we will need to perform an integration to determine the charge inside S.

Practice: An infinitely long line of charge carries a uniform linear density λ. Calculate the electric field due to this charge.

Determine the following:

1. A picture showing the wire and a "tin can" surface centered on the wire. The "tin can" exploits the cylindrical symmetry of the charge distribution.	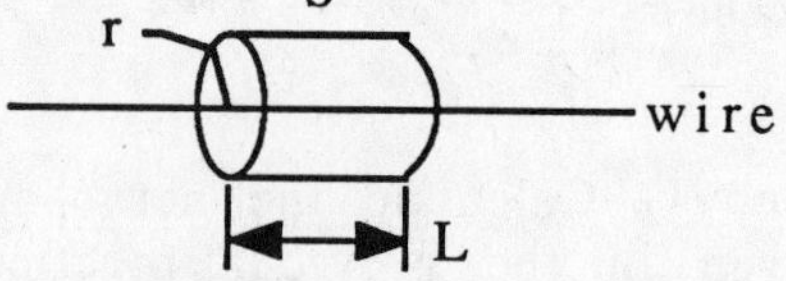 Figure 21.7
2. As much as possible about **E** on the surface of the "tin can", using symmetry	i) No part of **E** is perpendicular to the ends of the "tin can" ii) **E** is perpendicular to the curved surface iii) **E** has a constant magnitude over the curved surface
3. The left hand side of Gauss's Law	$\oint_S \mathbf{E}\cdot d\mathbf{S} = \int_{\text{ends}} \mathbf{E}\cdot d\mathbf{S} + \int_{\text{curved surface}} \mathbf{E}\cdot d\mathbf{S}$ $= 0 + E\int_{\text{curved surface}} dS$ $= 0 + E(2\pi rL)$
4. The charge inside the "tin can"	$Q_{inside} = \lambda L$

5. The electric field a distance r from the line of charge	$\oint_{\text{tin can}} \mathbf{E}\cdot d\mathbf{S} = \dfrac{Q_{\text{inside tin can}}}{\varepsilon_0}$ becomes: $E(2\pi rL) = \lambda L/\varepsilon_0$ and hence $E = \lambda/2\pi r\varepsilon_0$

Example: A long (infinite) cylinder of radius R_0 contains a uniform charge density ρ_0. Derive an expression for the electric field as a function of r, the perpendicular distance from the axis of the cylinder.

Strategy: We must consider points such that $r < R_0$ and $r > R_0$ separately. Again use a "tin can" as the closed surface because of the cylindrical symmetry.

Solution:

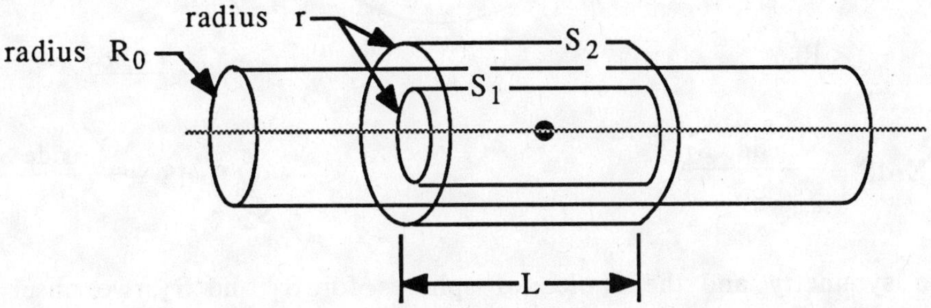

Figure 21.8

$\underline{r < R_0}$

$$\oint_{S_1} \mathbf{E}\cdot d\mathbf{S} = \frac{Q_{\text{inside } S_1}}{\varepsilon_0}$$

$$E(2\pi rL) = \frac{\rho_0(\pi r^2 L)}{\varepsilon_0}$$

$$E = \frac{\rho_0 r}{2\varepsilon_0}$$

$\underline{r > R_0}$

$$\oint_{S_2} \mathbf{E}\cdot d\mathbf{S} = \frac{Q_{\text{inside } S_2}}{\varepsilon_0}$$

$$E(2\pi rL) = \frac{\rho_0(\pi R_0^2 L)}{\varepsilon_0}$$

$$E = \frac{\rho_0(R_0^2)}{\varepsilon_0(2r)}$$

Notice that these two expressions give the same value at $r = R_0$. E plotted as a function of r appears as shown in Figure 21.9.

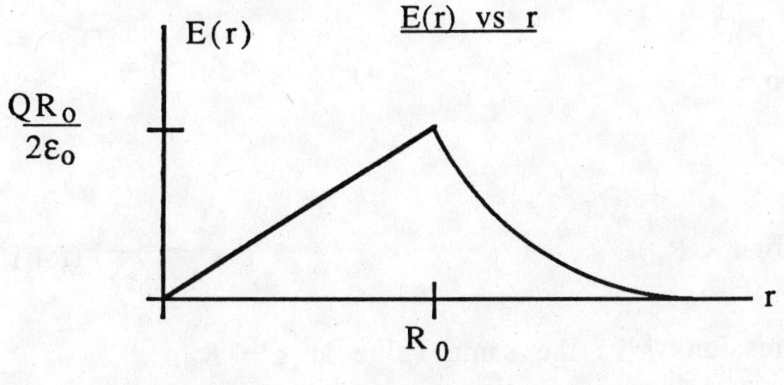

Figure 21.9

Example: A sphere of radius R_0 contains a charge Q_0 uniformly distributed throughout its volume. Determine the electric field as a function of r, the distance from the center of the sphere.

Strategy: Since we have spherical symmetry in the charge distribution we may use spheres as our closed surfaces and apply Gauss's Law to them. We must break space into two separate regions ($r < R_0$ and $r > R_0$) and treat them independently.

Solution:

Figure 21.10

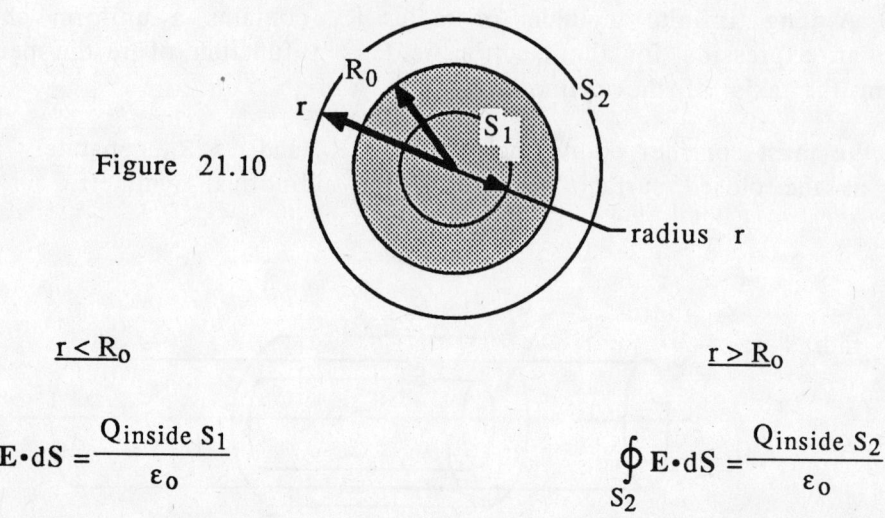

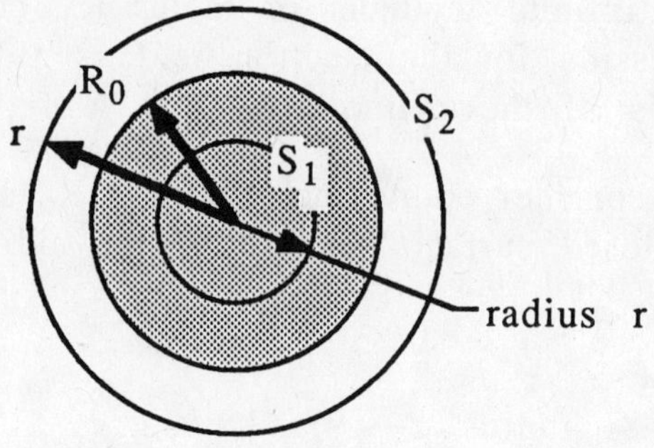

$r < R_0$

$$\oint_{S_1} \mathbf{E}\cdot d\mathbf{S} = \frac{Q_{\text{inside } S_1}}{\varepsilon_0}$$

$r > R_0$

$$\oint_{S_2} \mathbf{E}\cdot d\mathbf{S} = \frac{Q_{\text{inside } S_2}}{\varepsilon_0}$$

Because of the symmetry and the choice of spheres for S_1 and S_2, we observe that **E** is (i) always perpendicular to the spheres and (ii) constant in magnitude over the surface of each sphere. Thus we have,

$$E(4\pi r^2) = \frac{Q_{\text{inside } S_1}}{\varepsilon_0} \qquad\qquad E(4\pi r^2) = \frac{Q_{\text{inside } S_2}}{\varepsilon_0}$$

$$Q_{\text{inside } S_1} = \left(\frac{\text{charge}}{\text{volume}}\right)(\text{Volume inside } S_1) \qquad\qquad Q_{\text{inside } S_2} = Q_0$$

$$= \left(\frac{Q_0}{(\frac{4}{3})\pi R_0^3}\right)\left[(\frac{4}{3})\pi r^3\right] = \frac{Q_0 r^3}{R_0^3}$$

We may then write

$$E(4\pi r^2) = \frac{Q_0 r^3/R_0^3}{\varepsilon_0} \qquad\qquad E(4\pi r^2) = \frac{Q_0}{\varepsilon_0}$$

Solving for E, obtain

$$E = \frac{Q_0 r}{4\pi\varepsilon_0 R_0^3} \quad (\text{for } r < R_0) \qquad\qquad E = \frac{Q_0}{4\pi\varepsilon_0 r^2} \quad (\text{for } r > R_0)$$

Notice that these two expressions give the same value at $r = R_0$.

E plotted as a function of r appears as shown in Figure 21.11.

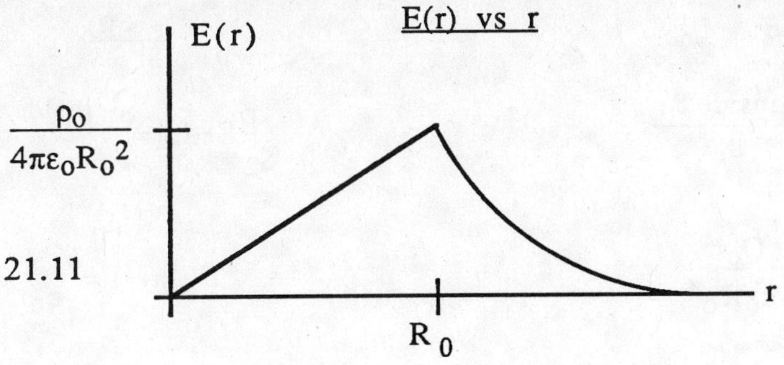

Figure 21.11

Example: The sphere of the previous example contains a charge Q_0 as before but this time the distribution is non-uniform with $\rho(r) = Q_0 r/\pi R_0^4$. Determine E(r).

Given: R_0 = radius of sphere
Q_0 = charge distributed throughout the sphere
$\rho(r)$ = a description of how the charge is distributed throughout the sphere

Determine: E(r) the electric field as a function of r for all values of r.

Strategy: Again use surfaces S_1 and S_2 as in Figure 21.10. Determine the charge inside S_1 and S_2. We must integrate to determine the charge inside S_1. Since $\rho(r)$ is a function only of r we may use a single integral (rather than a triple integral) to determine the charge inside S_1 if we use spherical shells to build up the volume enclosed by S_1. Figure 21.12 shows one of these spherical shells. Once we know the charge inside S_1 and S_2 we can use Gauss's Law to determine E(r).

The spherical shell has a radius r' (to distinguish it from r the radius of S_1) and thickness dr'. Then the infinitesimal volume element is

$$dV' = 4\pi r'^2 dr'$$

The charge in the shell is

$$dq' = \rho(r')dV' = \frac{Q_0 r'}{\pi R_0^4}(4\pi r'^2 dr')$$

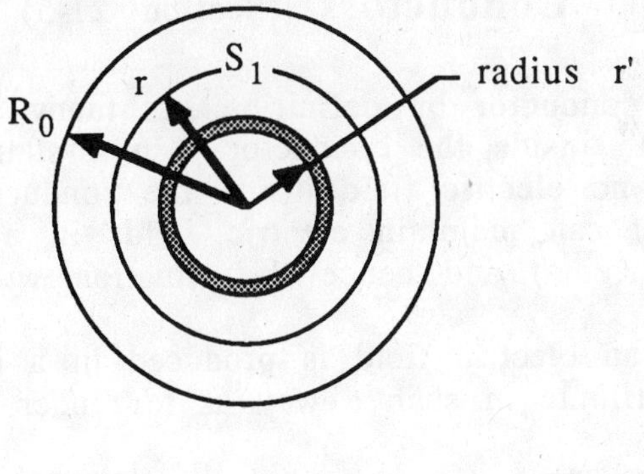

Figure 21.12

Solution:

$$Q_{inside\ S_1} = \int_0^r dq' = \int_0^r \left[\frac{Q_0 r'}{\pi R_0^4}\right](4\pi r'^2 dr') = \frac{4Q_0}{R_0^4}\int_0^r r'^3 dr' = \frac{Q_0 r^4}{R_0^4}$$

As a simple check on our work, note that when $r = R_0$ we have $Q_{inside} = Q_0$ (which was a given).

Proceeding with Gauss's Law in the two regions ($r < R_0$ and $r > R_0$) we have:

$\underline{r < R_o}$

$$\int_{S_1} \mathbf{E} \cdot d\mathbf{S} = \frac{Q_{\text{inside } S_1}}{\varepsilon_o}$$

$$E(4\pi r^2) = \frac{Q_o r^4}{\varepsilon_o R_o^4}$$

$$E = \frac{Q_o r^2}{4\pi\varepsilon_o R_o^4}$$

$\underline{r > R_o}$

$$\int_{S_2} \mathbf{E} \cdot d\mathbf{S} = \frac{Q_{\text{inside } S_2}}{\varepsilon_o}$$

$$E(4\pi r^2) = \frac{Q_0}{\varepsilon_0}$$

$$E = \frac{Q_o}{4\pi\varepsilon_o r^2}$$

E is plotted as a function of r in Figure 21.13.

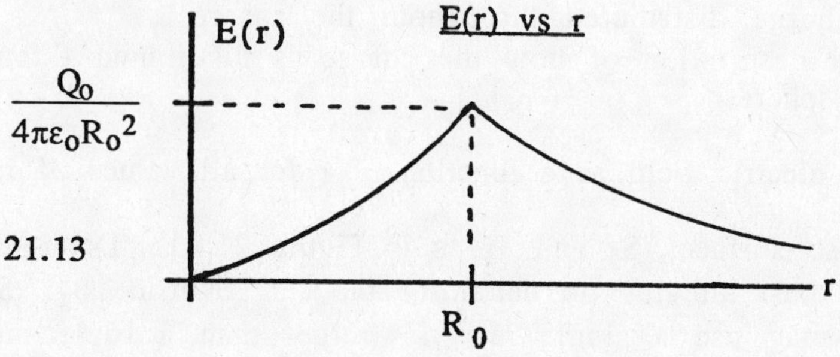

Figure 21.13

Related Text Exercises: Ex. 21-12 through 21-29.

4 Conductors (Section 21-5)

A conductor by definition has many charges that move easily if there is an electric field inside the conductor. In a static (no movement of charges) situation, there can be no electric field inside the conductor. When we consider devices such as batteries that can maintain electric fields in a conductor we will have moving charges (currents) and hence the situation will be one of dynamics rather than statics.

If an electric field is produced in a conductor the charges of the conductor will distribute in such a way as to render the electric field zero for statics. This means:

(i) There is no net charge density anywhere <u>inside</u> the conductor for a static situation. All charges will reside on the surfaces of the conductor.
(ii) The charge distribution will have to be such that the electric field is zero inside the conductor.
(iii) The field near the surface of a conductor is σ/ε_0, where σ is the density at the surface nearest the field point.

We can use Gauss's Law to prove (i) and (iii).

Proof of (i): Apply Gauss's Law to the surface which lies <u>entirely inside</u> the conductor as shown in Figure 21.14.

Figure 21.14

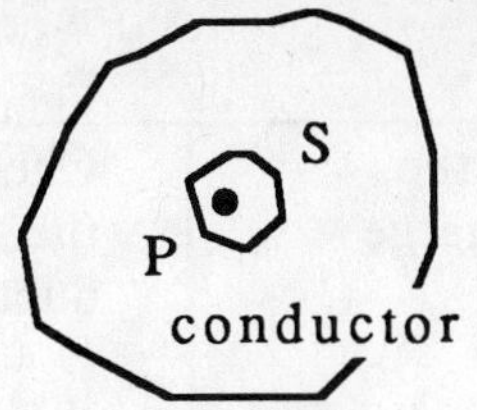

$$\int_S \mathbf{E}\cdot d\mathbf{S} = (Q_{\text{inside } S})/\varepsilon_0$$

Since E = 0 inside a conductor in a static situation, then $\int_S \mathbf{E}\cdot d\mathbf{S} = 0$ and $Q_{\text{inside } S} = 0$.

We may shrink S down around the point P shown inside S without changing the argument. Thus even in the limit of smaller volumes, $Q_{\text{inside } S} = 0$ and hence $\rho_p = 0$. Thus all charge must be on the surface or surfaces of the conductor.

Proof of (iii): Create a Gaussian surface which includes the point P near the surface of the conductors shown in Figure 21.15.

Figure 21.15

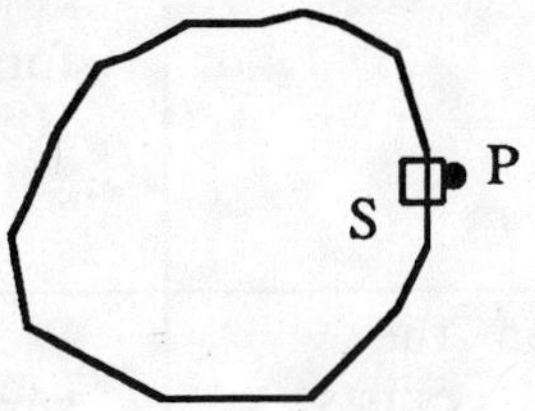

Since E = 0 inside the conductor there is no contribution to $\oint_S \mathbf{E}\cdot d\mathbf{S}$ for the part of S inside the conductor.

E is perpendicular to the conductor for a static situation, otherwise the component of E parallel to the surface of the conductor would cause a flow of charge contrary to the assumption of a static situation. Thus the only contribution to $\oint_S \mathbf{E}\cdot d\mathbf{S}$ is for the part of S outside the conductor and parallel to the surface of the conductor. Thus

$\oint_S \mathbf{E}\cdot d\mathbf{S} = EA$ where A is the area of this piece of S parallel to the conductor. The charge <u>inside</u> S is σA. Therefore

$$\oint_S \mathbf{E}\cdot d\mathbf{S} = \frac{Q_{\text{inside } S}}{\varepsilon_0} \quad \text{becomes} \quad EA = \frac{\sigma A}{\varepsilon_0} \quad \text{or} \quad E = \frac{\sigma}{\varepsilon_0}$$

Practice: A point charge Q_0 is at the center of a spherical conducting shell of inner radius 'a' and outer radius 'b' as shown in Figure 21.16.

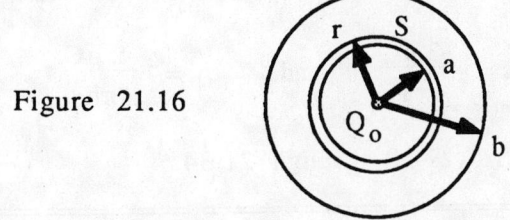

Figure 21.16

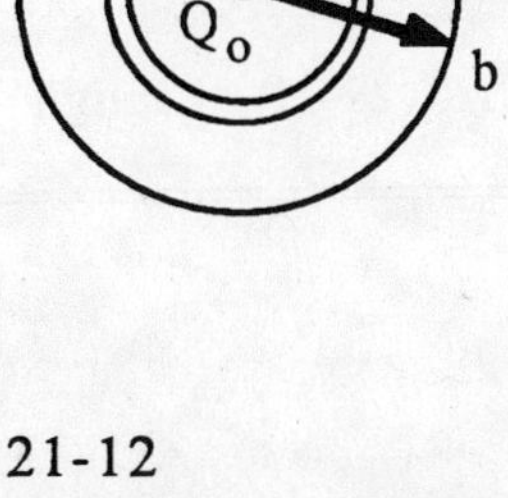

Determine the following:

1. The charge distribution on the sphere if there is no net charge on the sphere	Choose S as a sphere concentric with the conducting shell inside the conductor. Since S is inside the conductor, **E** = 0 over all of S. By Gauss's Law we then have $\oint_S \mathbf{E}\cdot d\mathbf{S} = \frac{Q_{inside\ S}}{\varepsilon_0} = 0$ or $Q_{inside\ S} = 0$. Since we have a charge Q_0 at the center of the conducting shell, we can have $Q_{inside\ S} = 0$ only if a charge $-Q_0$ is distributed uniformly over the inner surface at r = a. But the net charge on the shell must remain 0, hence a charge $+Q_0$ must be distributed uniformly over the outer surface at r = b. Hence $\sigma_a = -Q_0/4\pi a^2$ and $\sigma_b = (Q_0)/4\pi b^2$
2. The charge distribution on the sphere if a net charge Q_{net} exists on the surface	We must still have $-Q_0$ on the inner surface, but now there is a charge $+Q_0 + Q_{net}$ on the outer surface. Hence $\sigma_a = -\frac{-Q_0}{4\pi a^2}$ and $\sigma_b = -\frac{Q_0 + Q_{net}}{4\pi b^2}$

Example: A spherical conducting shell of inner radius "a" and outer radius "b" carries a net charge of -5 μC. A charge of 3 μC is uniformly distributed throughout the spherical region for r < a. Describe the charge distribution on the conducting sphere.

Given:
a = inner radius of conducting shell
b = outer radius of conducting shell
-5 μC = net charge on the shell
3 μC distributed uniformly throughout the region inside the shell (r < a)

Determine: The charge distribution on the conducting sphere

Strategy: Use Gauss's Law for a sphere inside the conductor

Solution: Construct a Gaussian surface S inside and concentric with the shell. Since E = 0 inside the conductor, Gauss' Law says $Q_{inside\ S}$ must be zero. This is possible if a charge -3 μC is on the inner surface (r = a). Since the net charge on the shell is -5 μC, this leaves -2 μC on the outer surface (r = b). Hence

$$\sigma_a = -\frac{-3\ \mu C}{4\pi a^2} \quad \text{and} \quad \sigma_b = -\frac{-2\ \mu C}{4\pi b^2}$$

Related Text Exercises: Ex. 21-30 through 21-34.

PRACTICE TEST

Take and grade this practice test. Doing so will allow you to determine any weak spots in your understanding of the concepts taught in this chapter. The following section prescribes what you should study further to strengthen your understanding.

Consider the point charges and the surfaces shown below

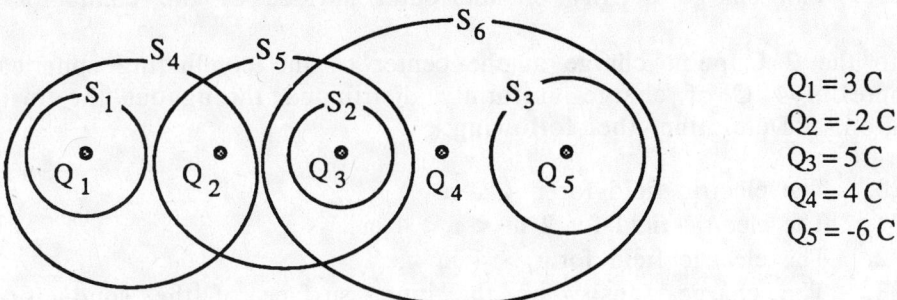

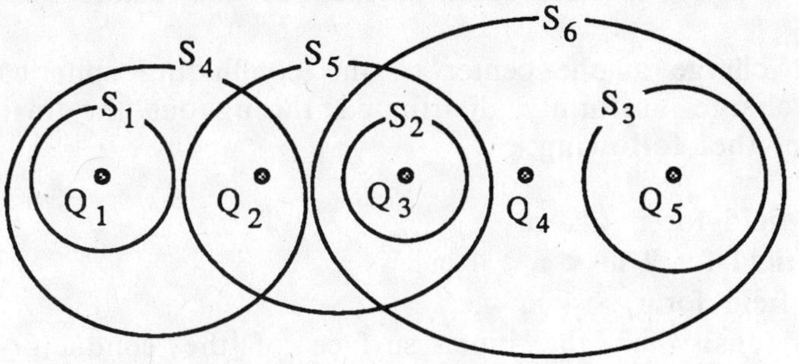

Determine the following:

________ 1. The flux Φ_1
________ 2. The flux Φ_2
________ 3. The flux Φ_3
________ 4. The flux Φ_4
________ 5. The flux Φ_5
________ 6. The flux Φ_6

A cube which has 2 meter edges is oriented with faces parallel to the xy, xz, and yz coordinate planes. It is entirely in the positive octant of space with one corner at the origin. There is an electric field given by $\mathbf{E}(x, y, z) = (x-1)\mathbf{i}$ in SI units. Determine the following:

________ 7. The flux for face in xy plane
________ 8. The flux for face in xz plane
________ 9. The flux for face in yz plane
________ 10. The flux for face parallel to one in xy plane
________ 11. The flux for face parallel to one in xz plane
________ 12. The flux for face parallel to one in yz plane
________ 13. The flux out through the cube
________ 14. The charge inside the cube

A spherical conducting shell with a 2 m inner and a 4 m outer radius carries a net charge of 5 C. A 2 C point charge is at the center. Determine the following:

_________ 15. The electric field for $r < 2$ m
_________ 16. The electric field for $2 \text{ m} < r < 4$ m
_________ 17. The electric field for $r > 4$ m
_________ 18. The charge density on the inner surface of the conductor
_________ 19. The charge density on the outer surface of the conductor

In addition to the 2 C point charge at the center of the conducting spherical shell just described, there is 4 C of charge uniformly distributed throughout the inside spherical region. Determine the following:

_________ 20. The electric field for $r < 2$ m
_________ 21. The electric field for $2 \text{ m} < r < 4$ m
_________ 22. The electric field for $r > 4$ m
_________ 23. The charge density on the inner surface of the conductor
_________ 24. The charge density on the outer surface

A charge Q_0 is distributed non-uniformly throughout a sphere of radius R_0 with a charge density given by $\rho(r) = \dfrac{5Q_0 r^2}{4\pi R_0^5}$. Determine expressions for the following:

_________ 25. The electric field for $r < R_0$
_________ 26 The electric field for $r > R_0$

(See Appendix I for answers.)

PRINCIPAL CONCEPTS AND EQUATIONS PRESCRIPTION

Your score on the practice test is an excellent measure of your understanding of the chapter. You should now use the following chart to write your own prescription for dealing with any weaknesses the practice test points out. Look down the leftmost column to the number of the question(s) you answered incorrectly, reading across that row you will find the concept and/or equation of concern, the section(s) of the study guide you should return to for further study, and some suggested text exercises which you should work to gain additional experience.

Practice Test Questions	Concepts and Equations	Prescription	
		Principal Concepts	Text Exercises
1	Electric flux from Q: $\Phi = \sum Q_i/\varepsilon_0$	1, 2	21-9, 10
2	Electric flux from Q	1, 2	20-10, 11
3	Electric flux from Q	1, 2	20-11, 12
4	Electric flux from Q	1, 2	20-9, 10
5	Electric flux from Q	1, 2	20-10, 11
6	Electric flux from Q	1, 2	20-11, 12
7	Electric flux from E: $\Phi = \oint_s \mathbf{E}\cdot d\mathbf{S} = Q/\varepsilon_0$	1	21-19, 20
8	Electric flux from E	1	21-1, 2
9	Electric flux from E	1	21-2, 3
10	Electric flux from E	1	21-3, 4
11	Electric flux from E	1	21-4, 5
12	Electric flux from E	1	21-5, 6
13	Electric flux from E	1, 2	21-6, 7
14	Electric flux and Gauss's Law: $\Phi = \oint_s \mathbf{E}\cdot d\mathbf{S} = Q/\varepsilon_0$	1, 2	21-7, 8
15	Gauss's Law to determine E	3	21-21, 22
16	Gauss's Law to determine E	3	21-22, 23
17	Gauss's Law to determine E	3	21-23, 24
18	Conductors	4	21-30
19	Conductors	4	21-30
20	Gauss's Law to determine E	3	21-24, 25
21	Gauss's Law to determine E	3	21-25, 26
22	Gauss's Law to determine E	3	21-26, 27
23	Conductors	4	21-30
24	Conductors	4	21-30
25	Gauss' Law to determine E	3	21-27, 28
26	Gauss' Law to determine E	3	21-28, 29

22 ELECTRIC POTENTIAL

RECALL FROM PREVIOUS CHAPTERS

Previously learned concepts and equations frequently used in this chapter	Text Section	Study Guide Page
Conservative forces and potential energy	9-1	9-1
Obtaining forces from potential energy	9-1	9-1

NEW IDEAS IN THIS CHAPTER

Concepts and equations introduced	Text Section	Study Guide Page
Electric potential energy	22-1	22-3
Electric potential	22-2	22-3
Electric potential difference	22-3	22-3
Relation between **E** and V	22-4	22-3
Equipotential surfaces (3-D) and lines (2-D)	22-5	---
Electrostatic potential of conductors	22-6	---

PRINCIPAL CONCEPTS AND EQUATIONS

1 Conservative Forces and Potential Energy (Sections 9-1 and 22-1)

Review: Recall from Ch. 9 that associated with a conservative force (vector) there is a potential energy function (scalar). The change in the potential energy function $U_B - U_A$ associated with a conservative force that acts on an object, is defined as the work done (by an external agent) against the conservative force as the object is moved from point A to B. Thus

$$U_B - U_A = -\int_A^B \mathbf{F} \cdot d\mathbf{l}$$

where **F** is the conservative force and the integral is evaluated along any path between points A and B. Recall that for a conservative force **F**, the work done is independent of the path between A and B. $U_B - U_A$ is called the change in the potential energy or the "potential energy difference" between points A and B.

We may define "the potential energy function" by choosing a particular "reference point A" and assigning zero as the value of the potential energy U at point A. Then the potential energy function is given by

$$U(\mathbf{r}) = -\int_A^{\mathbf{r}} \mathbf{F} \cdot d\mathbf{l}$$

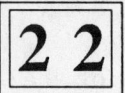

The potential energy $U(\mathbf{r})$ at $\mathbf{r}$ is the work done by an external agent against the conservative force $\mathbf{F}$ in moving the object from the reference point A to the arbitrary point $\mathbf{r}$. Any path connecting A and $\mathbf{r}$ may be used in determining $U(\mathbf{r})$.

If we have a number of objects and a number of conservative forces acting on them, then the potential energy for the entire system of objects is obtained by calculating the total work done against all conservative forces as we go from some reference configuration, where the potential energy is arbitrarily assigned the zero value, to some final configuration.

Practice: Consider an object with mass M_0 close to the surface of the earth as shown in Figure 22.1. Take the reference point to be at ground level.

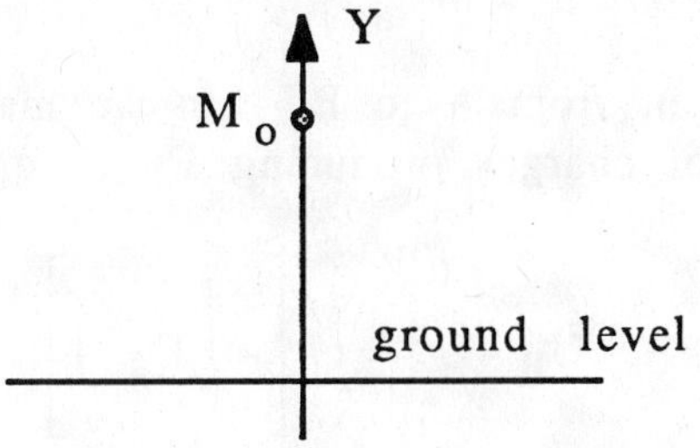

Figure 22.1

Determine the following:

1. The force exerted by the earth on M_0	$\mathbf{F} = -M_0 g\mathbf{j}$
2. Work done against gravity moving M_0 from the ground ($y = 0$) to y	$\mathbf{F} = -M_0 g\mathbf{j}$ $\qquad d\mathbf{l} = \mathbf{j}dy$ $W = -\int \mathbf{F}\cdot d\mathbf{l} = -\int_0^y -M_0 g\mathbf{j}\cdot\mathbf{j}dy = M_0 gy$
3. The gravitational potential energy $U(y)$. Take $U(0)$ as zero	$U(y) = M_0 gy$

As a continuation of our review recall that a conservative force is derivable from a potential function.

$$F_x = -(\frac{\partial U}{\partial x}) \qquad F_y = -(\frac{\partial U}{\partial y}) \qquad F_z = -(\frac{\partial U}{\partial z})$$

Using the gravitational potential energy just obtained for the mass M_0, determine the following:

4. $(\frac{\partial U}{\partial x})$, $(\frac{\partial U}{\partial y})$ and $(\frac{\partial U}{\partial z})$	$(\frac{\partial U}{\partial x}) = 0,\ (\frac{\partial U}{\partial y}) = M_0 g,\ (\frac{\partial U}{\partial z}) = 0$
5. $\mathbf{F} = -(\frac{\partial U}{\partial x})\mathbf{i} - (\frac{\partial U}{\partial y})\mathbf{j} - (\frac{\partial U}{\partial z})\mathbf{k}$	$\mathbf{F} = -M_0 g\mathbf{j}$

2 Electric Potential Energy, Electric Potential and Electric Potential Difference (Sections 22-1, 22-2, 22-3 and 22-4)

Review: Since the electric force is conservative, there is an electric potential energy function associated with the electric force. Thus

$$U_{E_B} - U_{E_A} = -\int_A^B \mathbf{F}_E \cdot d\mathbf{l}$$

is the work done by an external agent against the conservative electric force (or forces for a system of objects), $\mathbf{F}_E$.

If we move a single charge q_0 from A to B, then we may talk of the potential energy of q_0 due to the collection of charges producing $\mathbf{F}_E$ on q_0.
Then,

$$U_B - U_A = q_0\left[-\int_A^B \mathbf{E} \cdot d\mathbf{l}\right]$$

where **E** is the field due to the charges that produce the electric force on q_0.

We define the "change in electric potential" or "electric potential difference" as

$$\frac{U_B - U_A}{q_0} = \left(\frac{U_B}{q_0}\right) - \left(\frac{U_A}{q_0}\right) = V_B - V_A$$

Then the electric potential difference between points A and B is

$$V_B - V_A = -\int_A^B \mathbf{E} \cdot d\mathbf{l}$$

This is just the work done by an external agent per coulomb of positive charge moved from A to B. By choosing a reference point where V is defined as zero we get

$$V(\mathbf{r}) = -\int_A^{\mathbf{r}} \mathbf{E} \cdot d\mathbf{l}$$

$V(\mathbf{r})$ is called "the electric potential"; the reference point must be specified. We also have the relations:

$$E_x = -\left(\frac{\partial V}{\partial x}\right) \qquad E_y = -\left(\frac{\partial V}{\partial y}\right) \qquad E_z = -\left(\frac{\partial V}{\partial z}\right)$$

These relations connect **E** and V is the same way that **F** and U are connected.

Practice: Given $\mathbf{E}(x,y,z) = x\mathbf{i} + y\mathbf{j} + z\mathbf{k}$ with $V(0,0,0) \equiv 0$ and the curve $C_1 + C_2 + C_3$ between (0,0,0) and (x,y,z) as shown in Figure 22.2.

Figure 22.2

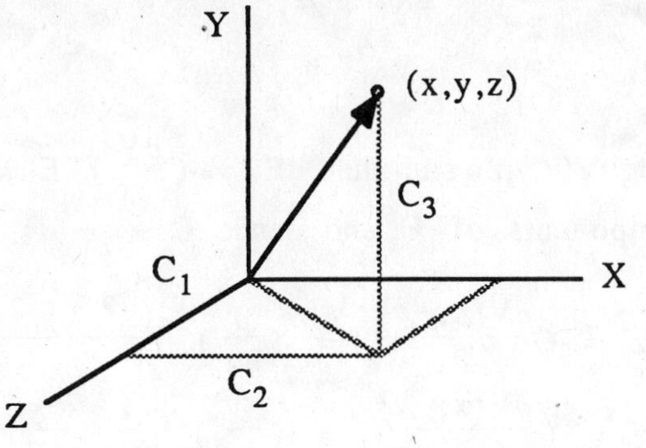

Determine the following:

1. The reference point	The origin (0,0,0)
2. The expression for V(x,y,z) using the curve $C_1 + C_2 + C_3$	Note: $d\mathbf{l} = d\bar{z}\mathbf{k}$ along C_1 $d\mathbf{l} = d\bar{x}\mathbf{i}$ along C_2 $d\mathbf{l} = d\bar{y}\mathbf{j}$ along C_3 Note: $\bar{x}, \bar{y}, \bar{z}$ are used for variables of integration to avoid confusion with x, y, z of the coordinate system $V(x,y,z) = -\int_{(0,0,0)}^{(x,y,z)} \mathbf{E}\cdot d\mathbf{l} \left\{ \begin{matrix} \text{along} \\ C_1 + C_2 + C_3 \end{matrix} \right\}$ $= -\int_{(0,0,0)}^{(0,0,z)} \mathbf{E}\cdot d\mathbf{l} - \int_{(0,0,z)}^{(x,0,z)} \mathbf{E}\cdot d\mathbf{l} - \int_{(x,0,z)}^{(x,y,z)} \mathbf{E}\cdot d\mathbf{l}$ along C_1 along C_2 along C_3
3. The value of the three integrals along $C_1 + C_2 + C_3$	$\int_{(0,0,0)}^{(0,0,z)} \left(\bar{x}\mathbf{i} + \bar{y}\mathbf{j} + \bar{z}\mathbf{k}\right)\cdot d\bar{z}\mathbf{k} = \int_0^z \bar{z}\,d\bar{z} = \frac{z^2}{2}$ along C_1 In like manner, when we integrate along C_2 and C_3 respectively we obtain $\left(\frac{x^2}{2}\right)$ and $\left(\frac{y^2}{2}\right)$
4. V(x,y,z)	$V(x,y,z) = -\left(\frac{1}{2}\right)(x^2 + y^2 + z^2)$
5. A quick check of our work	$E_x = -\left(\frac{\partial V}{\partial x}\right) = x$ Note that this agrees with the given.

Example: Starting with $V(x,y,z) = -(\frac{1}{2})(x^2 + y^2 + z^2)$ obtain $\mathbf{E}(x,y,z)$.

Given: $V(x,y,z) = -(\frac{1}{2})(x^2 + y^2 + z^2)$

Determine: $\mathbf{E}(x,y,z)$

Strategy: Knowing $V(x,y,z)$ and that $E_x = -(\frac{\partial V}{\partial x})$, $E_y = -(\frac{\partial V}{\partial y})$ and $E_z = -(\frac{\partial V}{\partial z})$, we can determine the components of **E** and hence **E**.

Solution:

$$E_x(x,y,z) = -(\frac{\partial V(x,y,z)}{\partial x}) = -(\frac{\partial}{\partial x})(\frac{-(x^2 + y^2 + z^2)}{2}) = x$$

$$E_y(x,y,z) = -(\frac{\partial V(x,y,z)}{\partial y}) = -(\frac{\partial}{\partial y})(\frac{-(x^2 + y^2 + z^2)}{2}) = y$$

$$E_z(x,y,z) = -(\frac{\partial V(x,y,z)}{\partial z}) = -(\frac{\partial}{\partial z})(\frac{-(x^2 + y^2 + z^2)}{2}) = z$$

Therefore $\mathbf{E}(x,y,z) = x\mathbf{i} + y\mathbf{j} + z\mathbf{k}$

Example: Given $\mathbf{E}(x,y,z) = x\mathbf{i} + y\mathbf{j} + z\mathbf{k}$ of the previous practice and example, obtain $V(x,y,z)$ with $V(0,0,0) \equiv 0$ using the straight line path C connecting (0,0,0) with (x,y,z).

Given: $\mathbf{E}(x,y,z) = x\mathbf{i} + y\mathbf{j} + z\mathbf{k}$ and $V(0,0,0) = 0$

Determine: $V(x,y,z)$ using the curve C as specified

Strategy: Knowing **E**, the path C and that $V = \int \mathbf{E}\cdot d\mathbf{l}$, we can determine $V(x,y,z)$.

Solution: The curve C can be parameterized using $\mathbf{l}(t) = xt\mathbf{i} + yt\mathbf{j} + zt\mathbf{k}$ for $0 \le t \le 1$. Then $d\mathbf{l}(t) = (x\mathbf{i} + y\mathbf{j} + z\mathbf{k})dt$ for $0 \le t \le 1$. $\mathbf{E}(\mathbf{r}(t)) = xt\mathbf{i} + yt\mathbf{j} + zt\mathbf{k}$ so that:

$$V(x,y,z) = -\int_{(0,0,0)}^{(x,y,z)} \mathbf{E}\cdot d\mathbf{l} = -\int_0^1 \mathbf{E}[\mathbf{r}(t)]\cdot[d\mathbf{l}(t)] = -\int_0^1 [xt\mathbf{i} + yt\mathbf{j} + zt\mathbf{k}]\cdot[(x\mathbf{i} + y\mathbf{j} + z\mathbf{k})dt]$$

$$= -(x^2 + y^2 + z^2)\int_0^1 t\,dt = -(x^2 + y^2 + z^2)(\frac{t^2}{2})\Big|_0^1 = -(\frac{1}{2})(x^2 + y^2 + z^2)$$

This is in agreement with the result using $C_1 + C_2 + C_3$ of our last example.

Example: Given $V(x,y,z) = -(\frac{1}{2})(x^2 + y^2 + z^2)$ of the previous examples, calculate the work done (by an external agent) in moving a 3 C charge from (1,2,3) to (2,3,4); take all units as metric.

Given: $V(x,y,z) = -(\frac{1}{2})(x^2 + y^2 + z^2)$, initial position = (1,2,3) m and final position = (2,3,4) m.

Determine: Work done against **E** in moving a 3 C charge from (1,2,3) m to (2,3,4) m.

Strategy: Knowing the form of V(x,y,z), we can determine V(1,2,3) and V(2,3,5). Knowing the relationship between electric potential difference and the work to move a charged particle between two points, we can determine the work.

Solution: $W_{(1,2,3)\to(2,3,4)} = Q[V(2,3,4) - V(1,2,3)]$

$$= 3\left[\left(\frac{-(2^2 + 3^2 + 4^2)}{2}\right) - \left(\frac{-(1^2 + 2^2 + 3^2)}{2}\right)\right] = -22.5 \text{ J}$$

Related Text Exercises: Ex. 22-1 through 22-32.

3 Electric Potential for Point Charges (Section 22-2)

Review: Consider the point charge Q located at the origin in Figure 22.3.

Figure 22.3

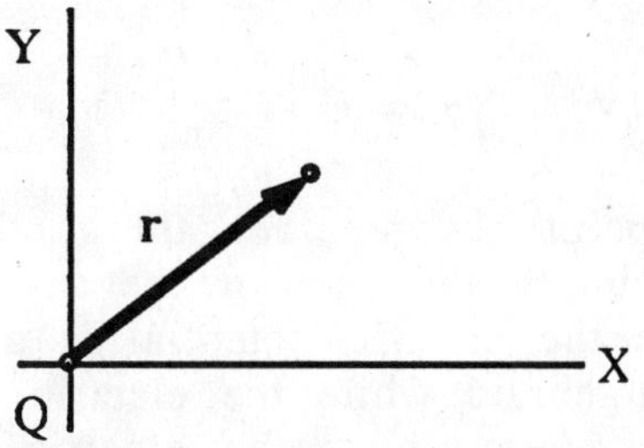

If we take our reference point $\mathbf{r}_0 = \infty$ and choose $V_\infty \equiv 0$, then the electric potential at **r** due to the point charge Q is

$$V(\mathbf{r}) = -\left(\frac{kQ}{r}\right)$$

Thus the potential is constant on a sphere of radius r and depends only on the distance from Q. Remember that V(**r**) is the work done (by an external agent) per coulomb of positive charge moved from far away (infinity) to the position denoted by **r**.

Practice: Consider a 2 C charge located at the origin.

Determine the following:

1. The distance r between the origin and the point (1,2,3) m	$r = (x^2 + y^2 + z^2)^{1/2} = (1^2 + 2^2 + 3^2)^{1/2}$ $= (14)^{1/2}$ m
2. The electric potential V(1,2,3) at the point (1,2,3) m	$V(1,2,3) = \frac{kQ}{r} = \frac{(9 \times 10^9)(2) \text{ J}}{(14)^{1/2} \text{ C}}$ $= 4.81 \times 10^9$ V We have used the volt (V), which is defined by V = J/C, as our unit for electric potential.

Example: A 2 μC charge and a 3 μC charge are separated by 5 meters. How much work is done against the electric repulsion in moving them closer together so that their separation is reduced to 2 meters.

Given: $Q_1 = 2\ \mu C$ $Q_2 = 3\ \mu C$ $r_I = 5\ m$ $r_F = 2\ m$

Determine: The amount of work done in moving the charges closer together

Strategy: We can work this problem either from the viewpoint of moving Q_1 in the **E** field due to Q_2 or from the viewpoint of moving Q_2 in the **E** field due to Q_1. Let's choose the former. Knowing Q_2, r_I and r_F, we can establish the initial and final electric potential due to Q_2. Knowing the relationship between the work to go from r_I to r_F and the electric potential difference between r_I and r_F, we can determine the work.

Solution: $W = \Delta U = U_F - U_I = Q_1 \Delta V_2$, where V_2 is the electric potential due to Q_2.
Note: We can also use $\Delta U = Q_2 \Delta V_1$ where V_1 is the electric potential due to Q_1.

$$\text{Then } W = Q_1 \Delta V_2 = Q_1(V_{2_F} - V_{2_I}) = Q_1\left[\left(\frac{kQ_2}{r_F}\right) - \left(\frac{kQ_2}{r_I}\right)\right] = kQ_1Q_2\left[\frac{1}{r_F} - \frac{1}{r_I}\right] = 0.0162\ J$$

If we have several point charges then the total electric potential at some point due to those charges is simply the sum of the individual potentials due to the individual point charges. Since the electric potential is a scalar whereas the electric field is a vector, this sum is algebraic while the electric field sum is vectorial. This can often make the problem of determining the electric potential easier then the problem of determining the electric field. Remember that if we can obtain either one we can get the other from it; integration of **E** yields V and differentiation of V yields **E**.

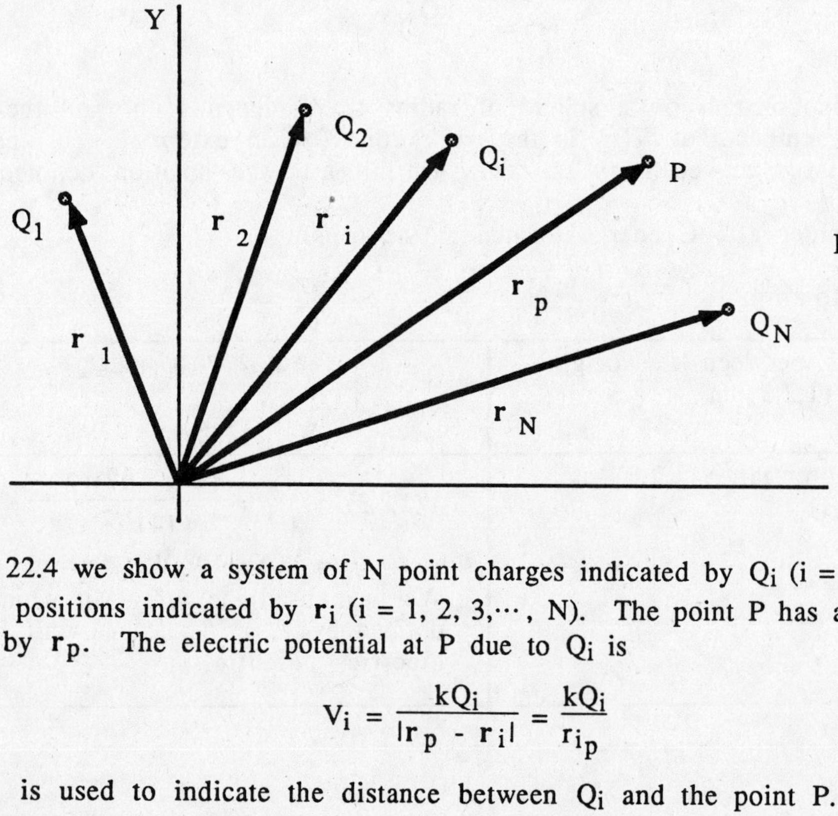

Figure 22.4

In Figure 22.4 we show a system of N point charges indicated by Q_i (i = 1, 2, 3,···, N) located at positions indicated by $\mathbf{r}_i$ (i = 1, 2, 3,···, N). The point P has a position indicated by $\mathbf{r}_p$. The electric potential at P due to Q_i is

$$V_i = \frac{kQ_i}{|\mathbf{r}_p - \mathbf{r}_i|} = \frac{kQ_i}{r_{ip}}$$

where r_{ip} is used to indicate the distance between Q_i and the point P.

Note: This notation differs slightly from that in the text. In the text r_i is the distance between Q_i and the point P, whereas here r_i is the distance of Q_i from the origin. The r_i of the text is replaced by r_{ip}.

The total electric potential at P is the sum of the individual potentials. Thus

$$V_p = \sum_{i=1}^{N} \frac{kQ_i}{r_{ip}}$$

Practice: A point charge $Q_1 = 2\ \mu C$ is located at (0,2) m, a point charge $Q_2 = -3\ \mu C$ is located at (0,0) m and a point charge $Q_3 = 4\ \mu C$ is located at (-1,0) m as shown in Figure 22.5.

Figure 22.5

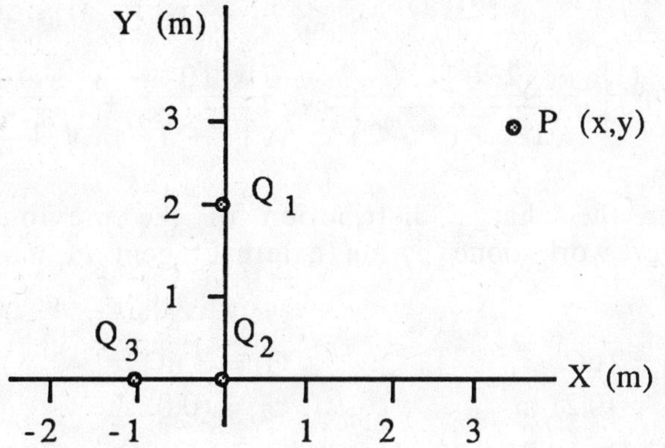

Determine the following:

1. The distance between Q_1 and P(i.e. r_{1p}), between Q_2 and P(i.e. r_{2p}), between Q_3 and P(i.e. r_{3p})	$r_{1p} = [(x - 0)^2 + (y - 2)^2]^{1/2}$ $= [x^2 + (y - 2)^2]^{1/2}$ $r_{2p} = [(x - 0)^2 + (y - 0)^2]^{1/2} = (x^2 + y^2)^{1/2}$ $r_{3p} = [(x - [-1])^2 + (y - 0)^2]^{1/2}$ $= [(x + 1)^2 + y^2]^{1/2}$
2. The electric potential at point P using: $V(x,y) = V_p(x,y) = \sum_{i=1}^{3} \frac{kQ_i}{r_{ip}}$	$V(x,y) = \sum_{i=1}^{3} \frac{kQ_i}{r_{ip}} = k\left[\frac{Q_1}{r_{1p}} + \frac{Q_2}{r_{2p}} + \frac{Q_3}{r_{3p}}\right]$ $=(9 \times 10^9)\left[\frac{2 \times 10^{-6}}{\sqrt{x^2+(y-2)^2}} - \frac{3 \times 10^{-6}}{\sqrt{x^2+y^2}} + \frac{4 \times 10^{-6}}{\sqrt{(x+1)^2+y^2}}\right]$

Example: For the charge distribution described in the previous practice problem determine the electric potential at the point (1,1) m.

Given: $Q_1 = 2\,\mu C$ $\quad r_1 = (0,2)$ m $\quad Q_2 = -3\,\mu C$ $\quad r_2 = (0,0)$ m $\quad Q_3 = 4\,\mu C$ $\quad r_3 = (-1,0)$ m

$$V(x,y) = (9 \times 10^9)\left[\frac{2 \times 10^{-6}}{\sqrt{x^2 + (y-2)^2}} - \frac{3 \times 10^{-6}}{\sqrt{x^2 + y^2}} + \frac{4 \times 10^{-6}}{\sqrt{(x+1)^2 + y^2}}\right]$$

Determine: The electric potential at the point (1,1) m

Strategy: Knowing the expression for V(x,y) and the components of $\mathbf{r}_1$, $\mathbf{r}_2$, $\mathbf{r}_3$, we can determine V(1,1)

Solution: We may use the result of the previous practice problem and substitute x = 1 m and y = 1 m. Thus

$$V(1,1) = (9 \times 10^9)\left[\frac{2 \times 10^{-6}}{\sqrt{1^2 + (1-2)^2}} - \frac{3 \times 10^{-6}}{\sqrt{1^2 + 1^2}} + \frac{4 \times 10^{-6}}{\sqrt{(1+1)^2 + 1^2}}\right] = 9.74 \times 10^3 \text{ V}$$

Example: For the charge distribution of the previous example problem determine the amount of work done by an external agent in moving a 2 μC charge from (1,1) m to (2,2) m.

Given: $Q_1 = 2\,\mu C$ $\quad r_1 = (0,2)$ m $\quad Q_2 = -3\,\mu C$ $\quad r_2 = (0,0)$ m $\quad Q_3 = 4\,\mu C$ $\quad r_3 = (-1,0)$ m

$V(1,1) = 9.74 \times 10^3$ V

$$V(x,y) = (9 \times 10^9)\left[\frac{2 \times 10^{-6}}{\sqrt{x^2 + (y-2)^2}} - \frac{3 \times 10^{-6}}{\sqrt{x^2 + y^2}} + \frac{4 \times 10^{-6}}{\sqrt{(x+1)^2 + y^2}}\right]$$

Determine: The amount of work required to move a 2 μC charge from (1,1) m to (2,2) m.

Strategy: Knowing the expression for V(x,y), we can determine V(1,1) and V(2,2). Knowing the relationship between work and electric potential difference, we can determine the work.

Solution: $V(1,1) = 9.74 \times 10^3$ V (from previous example)

$$V(2,2) = (9 \times 10^9)\left[\frac{2 \times 10^{-6}}{\sqrt{2^2 + (2-2)^2}} - \frac{3 \times 10^{-6}}{\sqrt{2^2 + 2^2}} + \frac{4 \times 10^{-6}}{\sqrt{(2+1)^2 + 2^2}}\right] = 9.45 \times 10^3 \text{ V}$$

$$W = Q[V(2,2) - V(1,1)] = (2 \times 10^{-6})(9.45 \times 10^3 - 9.74 \times 10^3) = -5.80 \times 10^{-4} \text{ J}$$

Related Text Exercises: Ex. 22-7 through 22-12.

4 Electric Potential for a Continuous Distribution of Charge
(Section 22-2)

Review: If the charge distribution is continuous rather than a collection of point charges, we must use calculus to determine the electric potential. We partition the continuous distribution into small pieces (small lengths for one dimensional distributions and small areas for two dimensional distributions and small volumes for three dimensional distributions) which are treated as point charges. The sum of point charges becomes an integral over the charge distribution in the limit of calculus (size of pieces → zero, number of pieces → ∞).

Case I (1-D)

$$V(\mathbf{r}) = \int_C \frac{k\lambda(\mathbf{r})\,dl}{|\mathbf{r} - \bar{\mathbf{r}}|}$$

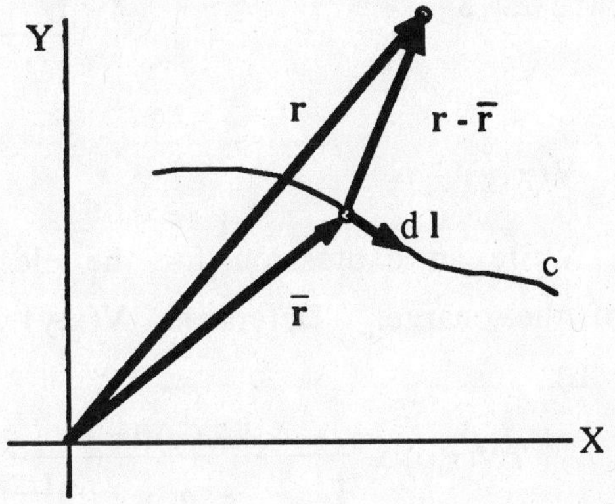

Figure 22.6

V(**r**) is the potential at **r** due to the charge distributed along C. The integration is over this charge.

Practice: A charge Q is distributed uniformly along the x-axis from x = 0 to x = L.

Determine the following:

1. A figure showing the charge distribution and the point (x,0) for x > L	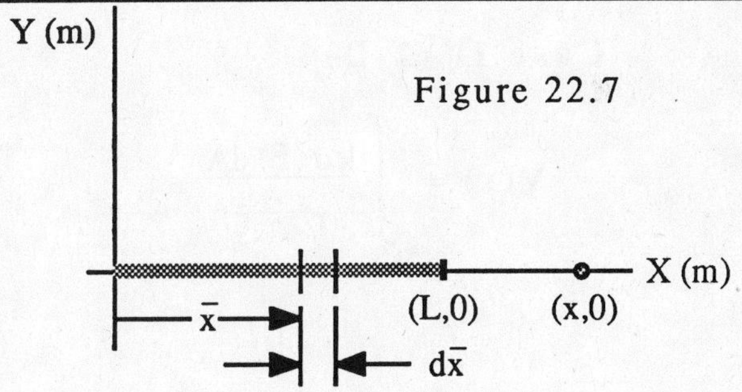 Figure 22.7
2. An expression for the contribution of the piece of charge at $\bar{x}$ to the electric potential	Write the linear density as $\lambda(\bar{x}) = Q/L$ $dV = \frac{k\lambda(\bar{x})d\bar{x}}{x - \bar{x}} = \frac{k(Q/L)d\bar{x}}{x - \bar{x}}$
3. V(x,0) for x > L	$V(x,0) = \int_0^L \frac{k(Q/L)d\bar{x}}{x - \bar{x}} = \left(\frac{kQ}{L}\right)\int_0^L \frac{d\bar{x}}{x - \bar{x}}$ $= -\left(\frac{kQ}{L}\right)\ln(x - \bar{x})\Big\vert_0^L = \left(\frac{kQ}{L}\right)\ln\left[\frac{x}{x - L}\right]$

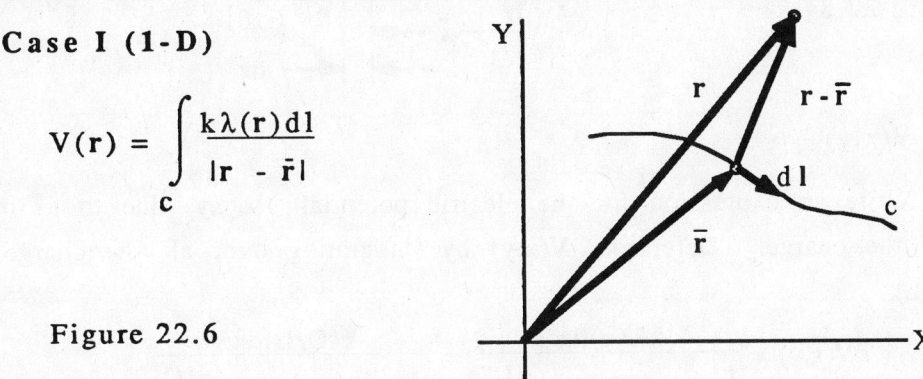

Example: Set up an integral expression (do not evaluate) for the electric potential at the point (x,y) for the charge distribution described in the previous practice problem.

Given and Figure:

$\lambda(x) = Q/L$ for
$0 \le x \le L$ and $y = 0$

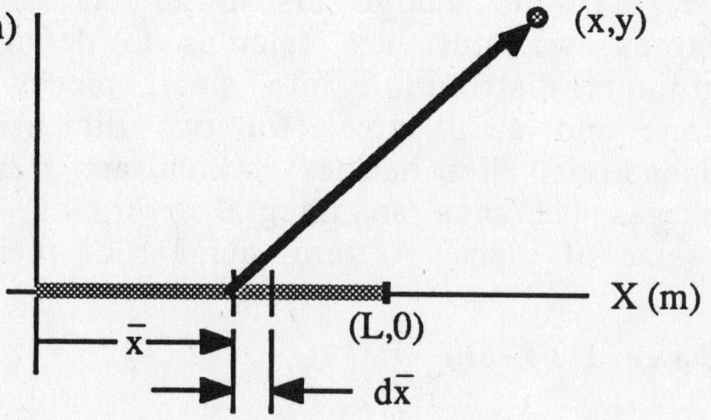

Figure 22.8

Determine: V(x,y)

Strategy: Write an expression for the electric potential dV(x,y) due to a small segment $d\bar{x}$ of the charge. Determine V(x,y) by integrating over all the charge (i.e. for $\bar{x} = 0$ to $\bar{x} = L$).

Solution:

$$dV(x,y) = \frac{k\lambda(\bar{x})d\bar{x}}{\left[(x - \bar{x})^2 + y^2\right]^{1/2}} = \frac{k(Q/L)d\bar{x}}{\left[(x - \bar{x})^2 + y^2\right]^{1/2}}$$

$$V(x,y) = \left(\frac{kQ}{L}\right)\int_0^L \frac{d\bar{x}}{[(x - \bar{x})^2 + y^2]^{1/2}}$$

Case II (2-D)

$$V(r) = \int_A \frac{k\sigma(\bar{r})\,d\bar{A}}{|r - \bar{r}|}$$

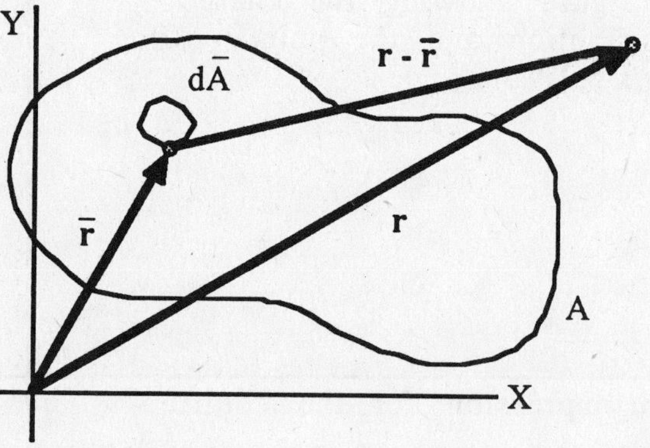

Figure 22.9

V(r) is the potential at **r** due to the charge distributed on the area A. The integration is over this charge.

Practice: A charge Q is uniformly distributed over a disc of radius R_0. The disk is in the yz plane and centered at the origin. Find the electric field for point P a distance x from the center of the disc measured out along the perpendicular to the disc axis.

Determine the following:

1. A figure showing the charge distribution and point P	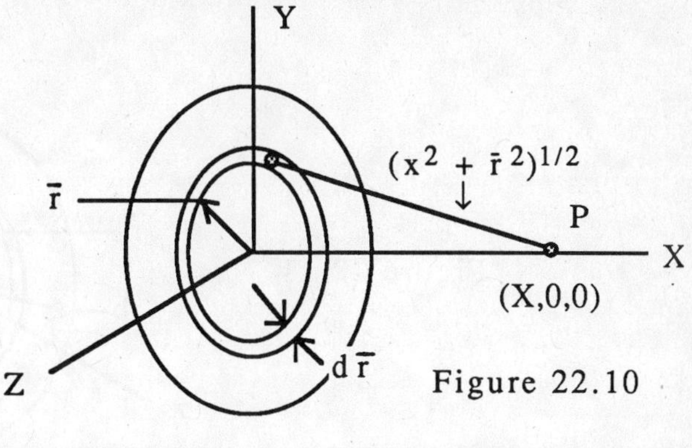 Figure 22.10
2. The charge dQ on the ring of radius $\bar{r}$	$dQ = \sigma(\bar{r})(2\pi\bar{r}d\bar{r}) = (\frac{Q}{\pi R_0^2})(2\pi\bar{r}d\bar{r})$
3. The potential at P due to the ring of radius $\bar{r}$	$dV_p = \frac{kdQ}{(x^2 + \bar{r}^2)^{1/2}} = \frac{(kQ/\pi R_0^2)[2\pi\bar{r}d\bar{r}]}{(x^2 + \bar{r}^2)^{1/2}}$
4. The potential $V_p(x,0,0)$ at any point on the x axis due to all the rings	$V_p(x,0,0) = (\frac{kQ}{R_0^2})\int_0^{R_0} \frac{2\bar{r}d\bar{r}}{(x^2 + \bar{r}^2)^{1/2}}$ $= (\frac{2kQ}{R_0^2})(x^2 + \bar{r}^2)^{1/2}\Big\|_0^{R_0}$ $= [\frac{2kQ}{R_0^2}][(x^2 + R_0^2)^{1/2} - (x^2)^{1/2}]$
5. The electric field $\mathbf{E}_p(x,0,0)$ for point P a distance x from the center of the disc	$E_x = -(\frac{\partial V}{\partial x}) = -\left[\frac{2kQ}{R_0^2}\right]\left[\frac{x}{(x^2 + R_0^2)^{1/2}} - 1\right]$ $E_y = 0$ $E_z = 0$ $\mathbf{E} = \left[\frac{2kQ}{R_0^2}\right]\left[1 - \frac{x}{(x^2 + R_0^2)^{1/2}}\right]\mathbf{i}$

Example: A charge Q is uniformly distributed over an annulus having inner radius a and outer radius b. Determine V_p for a point P a distance x out on the the axis which is perpendicular to the plane of the annulus and through the center of the annulus.

Given and Diagram: Q uniformly distributed over an annulus as shown in Figure 22.11

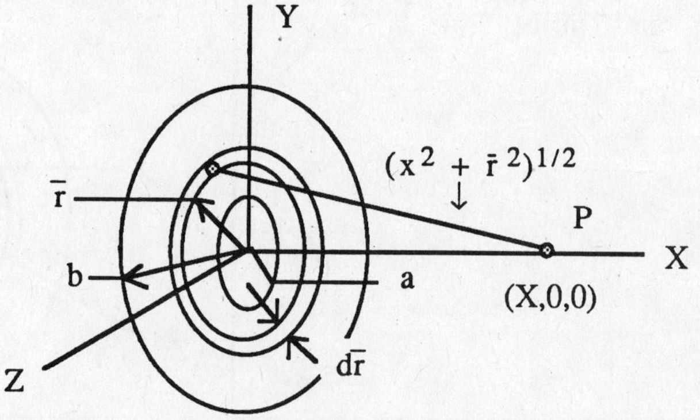

Figure 22.11

Determine: $V_p(x,0,0)$

Strategy: Knowing the charge on the annulus and the dimensions of the annulus, we can determine the area charge density (σ) on the annulus. Knowing σ, we can write dV_{ring} for a typical ring on the surface. We may then sum over all rings to get V_p due to the entire disc.

Solution:

$$\sigma \equiv \frac{Q}{\pi(b^2 - a^2)}$$

$$dV_{ring} = \frac{k\sigma(\bar{r})[2\pi\bar{r}d\bar{r}]}{(x^2 + \bar{r}^2)^{1/2}} = \frac{(kQ) \qquad (2\pi\bar{r}d\bar{r})}{\pi(b^2 - a^2)(x^2 + \bar{r}^2)^{1/2}}$$

$$V_p(x,0,0) = \frac{kQ}{(b^2 - a^2)} \int_a^b \frac{\bar{r}d\bar{r}}{(x^2 + \bar{r}^2)^{1/2}} = \left[\frac{2kQ}{(b^2 - a^2)}\right] [(x^2 + b^2)^{1/2} - (x^2 + a^2)^{1/2}]$$

Case III (3-D)

$$V(\mathbf{r}) = \int_V \frac{k\rho(\bar{r})\,d\bar{V}}{|\mathbf{r} - \bar{r}|}$$

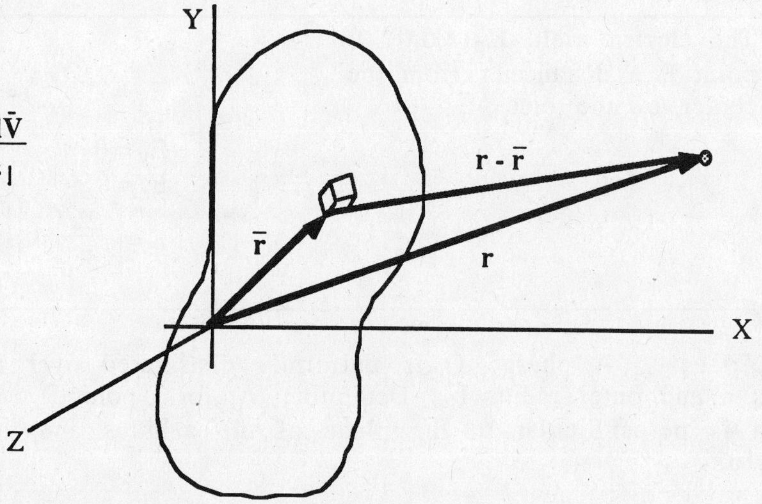

Figure 22.12

V(**r**) is the potential at **r** due to the charge distributed throughout the volume V. The integration is over this charge.

Practice: A charge Q is uniformly distributed over the surface of a sphere of radius R centered at the origin.

Determine the following:

1. A figure showing the charge distribution and a point P located at r such that r > R	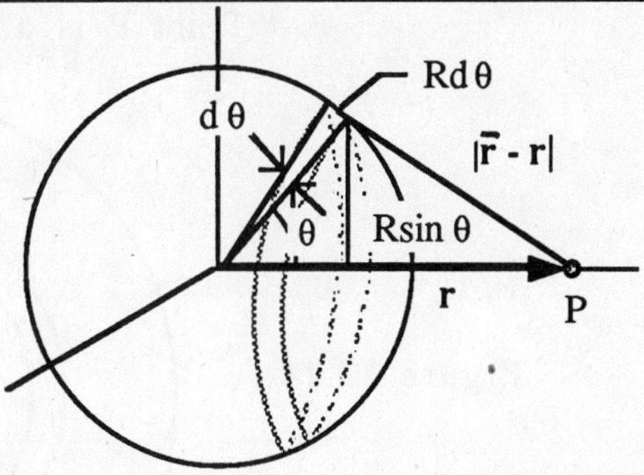 Figure 22.13
2. The charge on the ring of radius R sinθ	$dQ = \|\sigma\|(2\pi R \sin\theta)(R d\theta) \qquad \sigma = Q/4\pi R^2$ $= \dfrac{Q \sin\theta \, d\theta}{2}$
3. The distance of every point on this ring from the point P	All the points on the ring are the same distance from P. $\|\mathbf{r} - \bar{\mathbf{r}}\| = [\, R^2 \sin^2\theta + (r - R\cos\theta)^2 \,]^{1/2}$ $= (R^2 + r^2 - 2rR\cos\theta)^{1/2}$
4. The electric potential at P due to this ring of charge	$dV_{ring} = \dfrac{k\sigma(2\pi R^2 \sin\theta \, d\theta)}{(R^2 + r^2 - 2rR\cos\theta)^{1/2}}$ $= \dfrac{kQ \sin\theta \, d\theta}{2(R^2 + r^2 - 2rR\cos\theta)^{1/2}}$
5. The electric potential V_p at P due to all rings (i.e. the sphere)	$V_p = \left(\dfrac{kQ}{2}\right) \displaystyle\int_0^\pi \dfrac{\sin\theta \, d\theta}{(R^2 + r^2 - 2rR\cos\theta)^{1/2}}$ $= \left(\dfrac{kQ}{2rR}\right)(R^2 + r^2 - 2rR\cos\theta)^{1/2}\Big\|_0^\pi$ $= \left[\dfrac{kQ}{2rR}\right][(R + r) - (r - R)] = \dfrac{kQ}{r}$

Note: This is what we expect from Gauss's Law. The spherical distribution acts like a point charge Q at the center of the sphere for points such that r > R. Our result for V_p is that for a point charge Q at the center of the sphere.

Example: A charge is uniformly distributed throughout a sphere of radius R. Calculate the electric potential at point P a distance $r > R$ from the center of the sphere.

Given and Diagram: Q is uniformly distributed throughout a sphere of radius R. Point P is a distance $r > R$ from the center of the sphere.

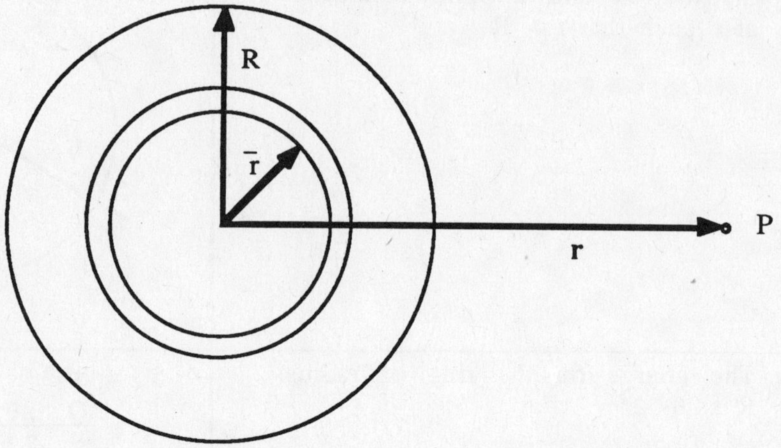

Figure 22.14

Determine: V_p for $r > R$

Strategy: Knowing Q and R, we can determine the charge density ρ. Knowing ρ we can determine the charge for a spherical shell and then the electric potential due to the shell. Finally we can sum over all shells to get the electric potential for the sphere.

Solution: Use the result of the last practice problem to get $dV_{p\ shell}$. First we need the charge contained in the shell. We get this from

$$dQ = (\rho)(4\pi\bar{r}^2 d\bar{r}) \quad \text{where } \rho = \frac{Q}{\left(\frac{4}{3}\right)\pi R^3}$$

Then

$$dV_{p\ shell} = \frac{k\rho[4\pi\bar{r}^2 d\bar{r}]}{r} = \frac{3kQ\bar{r}^2 d\bar{r}}{rR^3}$$

Therefore,

$$V_p = \int_0^R \frac{3kQ\bar{r}^2 d\bar{r}}{rR^3} = \left(\frac{3kQ}{rR^3}\right)\left(\frac{\bar{r}^3}{3}\right)\Bigg|_0^R = \frac{kQ}{r}$$

Again this is the expected result for $r > R$ since the charge acts like a point charge Q at the center of the sphere for such points.

Related Text Exercises: Ex. 22-14, 22-21, 22-22 and 22-24 through 22-29.

PRACTICE TEST

Take and grade this practice test. Doing so will allow you to determine any weak spots in your understanding of the concepts taught in this chapter. The following section prescribes what you should study further to strengthen your understanding.

Given the electric field $\mathbf{E}(x,y,z) = -y^2z\mathbf{i} - 2xyz\mathbf{j} - xy^2\mathbf{k}$, (in metric units) with $V(0,0,0) \equiv 0$, determine the following:

_________ 1. The electric potential $V(x,y,z)$

_________ 2. The work done by an external agent in moving a 2 C charge from (1,1,1) m to (2,2,2) m

Given the electric potential $V(x,y,z) = xye^{-z}$ in metric units, determine the following:

_________ 3. The electric field at (1,2,3) m

_________ 4. The force on a 2 C charge at (1,2,3) m

_________ 5. The work done by an external agent in moving a 2 C charge from (1,1,1) m to (1,2,3) m

Consider the situation shown in Figure 22.15.

$Q_1 = -200\ \mu C$ $m_1 = 2$ kg

$Q_2 = 300\ \mu C$ $m_2 = 2$ kg

$Q_3 = 400\ \mu C$ $m_3 = 2$ kg

$Q_4 = 300\ \mu C$ $m_4 = 2$ kg

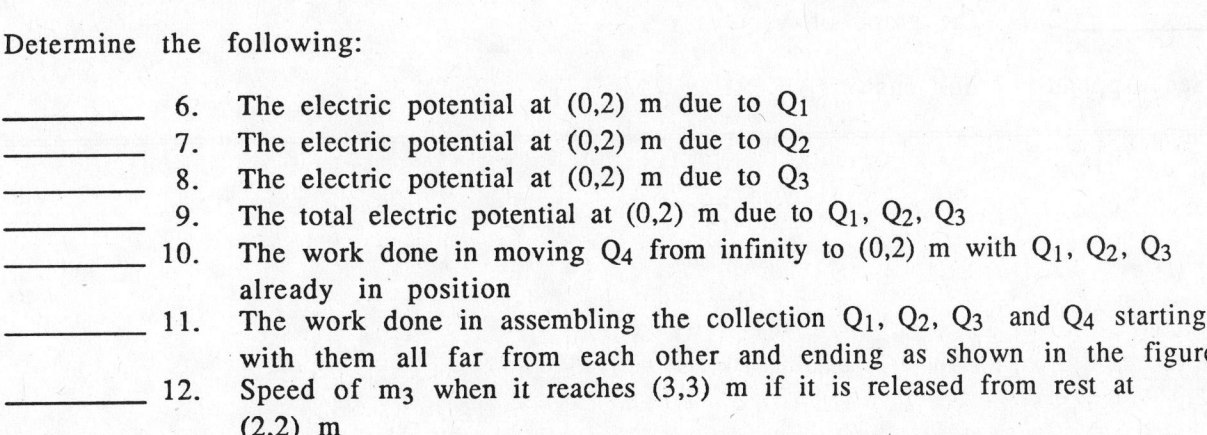

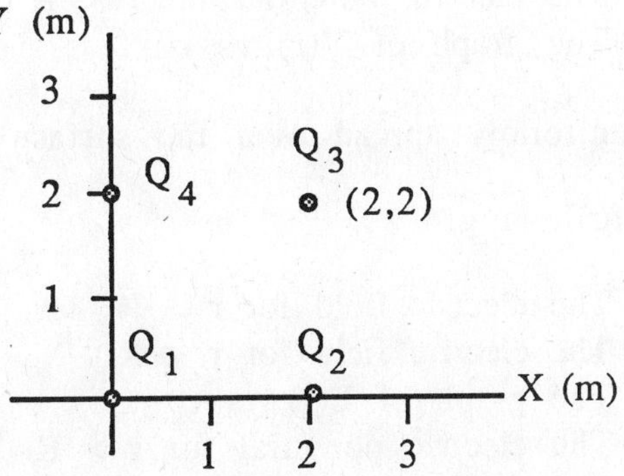

Figure 22.15

Determine the following:

_________ 6. The electric potential at (0,2) m due to Q_1

_________ 7. The electric potential at (0,2) m due to Q_2

_________ 8. The electric potential at (0,2) m due to Q_3

_________ 9. The total electric potential at (0,2) m due to Q_1, Q_2, Q_3

_________ 10. The work done in moving Q_4 from infinity to (0,2) m with Q_1, Q_2, Q_3 already in position

_________ 11. The work done in assembling the collection Q_1, Q_2, Q_3 and Q_4 starting with them all far from each other and ending as shown in the figure

_________ 12. Speed of m_3 when it reaches (3,3) m if it is released from rest at (2,2) m

_________ 13. The total electric potential at (x,y) m due to all four charges

A charge Q is uniformly distributed along the x-axis from $x = -L$ to $x = +L$, determine:

__________ 14. The electric potential at point P located at (0,y)

A non-uniform charge density is given by: $\lambda(x) = |x|$ for $-L \leq x \leq +L$, determine:

__________ 15. The electric potential at point P located at (0,y)

Charge is distributed over the surface of a disc of radius R according to $\sigma(r) = br$ for $0 \leq r \leq R$, determine:

__________ 16. The electric potential at point P located a distance x out along the axis of the disc.

A charge Q is uniformly distributed throughout a sphere of radius R

Determine the following:

__________ 17. The electric field for points such that $r > R$
__________ 18. The electric field for points such that $r < R$
__________ 19. The graph of E(r) vs r
__________ 20. The electric potential for $r > R$
__________ 21. The electric potential for $r < R$
__________ 22. The graph of V(r) vs r

A charge Q is uniformly spread over the surface of a conducting sphere of radius R

Determine the following:

__________ 23. The electric field for $r > R$
__________ 24. The electric field for $r < R$
__________ 25. The graph of E(r) vs r
__________ 26. The electric potential for $r > R$
__________ 27. The electric potential for $r < R$
__________ 28. The graph of V(r) vs r

(See Appendix I for answers.)

PRINCIPAL CONCEPTS AND EQUATIONS PRESCRIPTION

Your score on the practice test is an excellent measure of your understanding of the chapter. You should now use the following chart to write your own prescription for dealing with any weaknesses the practice test points out. Look down the leftmost column to the number of the question(s) you answered incorrectly, reading across that row you will find the concept and/or equation of concern, the section(s) of the study guide you should return to for further study, and some suggested text exercises which you should work to gain additional experience.

Practice Test Questions	Concepts and Equations	Prescription	
		Principal Concepts	Text Exercises
1	Electric potential: $V(r) = -\int \mathbf{E}\cdot d\mathbf{l}$	2	22-1, 2
2	Work: $W = qV$	2	22-9
3	Electric field: $E_x = -\partial V/\partial x$	2	22-25, 26
4	Electric field: $E = F/q$	4 of Ch 20	20-15, 16
5	Work: $W = qV$	2	22-9
6	Electric potential: $V(r) = kQ/r$	3	22-7, 8
7	Electric potential	3	22-8, 10
8	Electric potential	3	22-10, 12
9	Electric potential	3	22-7, 8
10	Work: $W = qV$	2	22-9
11	Work: $W = \Sigma\, kQ_iQ_j/r_{ij}$	3	22-9
12	Conservation of energy: $K + U = \text{const}$	3	22-3, 20
13	Electric potential: $V(r) = kQ/r$	3	22-7, 8
14	Electric potential: $V(\mathbf{r}) = \int (k\lambda(\mathbf{r})dl)/(\lvert\mathbf{r} - \bar{\mathbf{r}}\rvert)$	4	22-14, 24
15	Electric potential	4	22-14,
16	Electric potential: $V(\mathbf{r}) = \int (k\sigma(\bar{\mathbf{r}})\, d\bar{A})/(\lvert\mathbf{r} - \bar{\mathbf{r}}\rvert)$	4	22-29
17	Gauss's Law: $\int \mathbf{E}\cdot d\mathbf{s} = Q_i/\varepsilon_0$	2 of Ch 21	21-27, 28
18	Gauss's Law	2 of Ch 21	21-28, 29
19	Plotting	---	---
20	Electric potential: $V(\mathbf{r}) = \int (k\rho(\bar{\mathbf{r}})\, d\bar{V})/(\lvert\mathbf{r} - \bar{\mathbf{r}}\rvert)$	4	22-25, 26
21	Electric potential	4	22-27, 28
22	Plotting	---	---
23	Gauss's Law: $\int \mathbf{E}\cdot d\mathbf{s} = Q_i/\varepsilon_0$	2 of Ch 21	21-21, 22
24	Gauss's Law	2 of Ch 21	21-22, 23
25	Plotting	---	---
26	Electric potential: $V(r) = -\int \mathbf{E}\cdot d\mathbf{l}$	4	22-29
27	Electric potential	4	22-29
28	Plotting	---	---

23 CAPACITANCE, ELECTRIC ENERGY, AND INSULATION

RECALL FROM PREVIOUS CHAPTERS

Previously learned concepts and equations frequently used in this chapter	Text Section	Study Guide Page
Electric Fields: **E**	20-5	20-7
Electric Potential: V	22-2	22-3
Electric Potential Difference: $V_B - V_A = \Delta V$	22-3	22-3
Electric Potential Energy: $U = QV$	22-1	22-3
Electric Potential Energy Difference: $U_B - U_A = \Delta U = Q\Delta V$	22-1	22-3

NEW IDEAS IN THIS CHAPTER

Concepts and equations introduced	Text Section	Study Guide Page
Capacitors	23-1	23-2
Capacitance: $C = Q/V$	23-1	23-2
Equivalent Capacitance (a) Series: $1/C_S = 1/C_1 + 1/C_2$ (b) Parallel: $C_P = C_1 + C_2$	23-2	23-4
Energy stored in a capacitor: $U = \frac{1}{2}QV = \frac{1}{2}CV^2 = \frac{1}{2} \cdot \frac{Q^2}{C}$	23-3	23-7
Electrical Energy Density: $\mu = \frac{1}{2}\varepsilon_0 E^2$	23-3	23-7
Dielectrics: (a) Dielectric Constant: $K = C/C_o$ (b) Dielectric Strength	23-4, 5	23-9

PRINCIPAL CONCEPTS AND EQUATIONS

1 Capacitors and Capacitance (Sections 23-1, 2)

A capacitor is two conductors with +Q on one and -Q on the other.

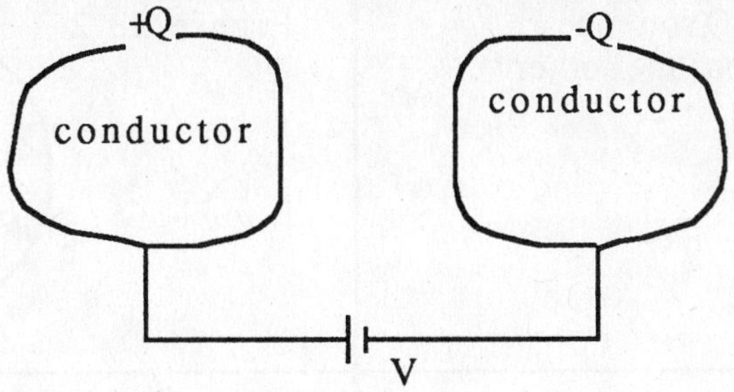

Figure 23.1

Often the two conductors are parallel plates; but we always refer to the conductors as the positive and negative plates regardless of their shapes.

Figure 23.1 shows two conductors connected by a battery producing a voltage difference between the plates. V is $(V_+) - (V_-)$, the difference in potential between the positive plate and the negative plate. Note: V is positive.

The battery produces this potential difference by causing +Q on the high voltage plate and -Q on the low voltage plate.

Once equilibrium is reached the distribution of the +Q and -Q charges is such as to make the electric field zero inside both conductors. Between the conductors there is an electric field and $V = (V_+) - (V_-) = -\int_{-\text{plate}}^{+\text{plate}} \mathbf{E} \cdot d\mathbf{r}$ for any path between the negative and positive plates.

It is obvious that if we double the charge distribution on each conductor then

(i) the field remains zero inside the conductors and
(ii) the field is doubled everywhere between them.

Thus V is doubled. Triple the charge distribution and you triple V. Therefore Q/V is constant; it depends only on the geometry of the two conductors. This ratio is called the *capacitance* (the capacity to store charge) and is denoted by:

$$C \equiv \frac{Q}{V}$$

The unit is coulomb/volt and is called the farad (F). As C increases the capacitor is able to hold more charge per volt.

Calculations of C are made by putting +Q and -Q on the conductors and calculating V. The notation Q/V gives C.

Practice: A capacitor is made of two concentric spherical shells with a vacuum between them. The outer radius of the inner shell is 'a' and the inner radius of the outer shell is 'b'.

Determine the following:

1. A figure showing +Q on one conductor and -Q on the other	Figure 23.2 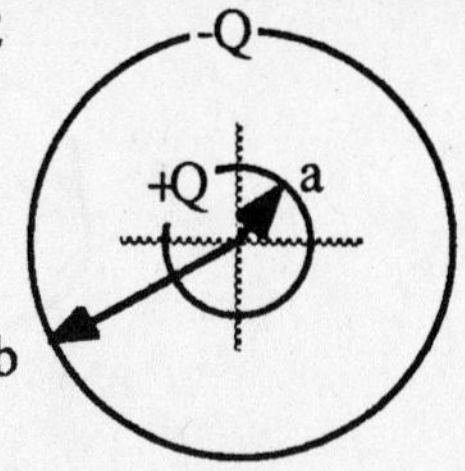
2. The electric field for the region between the shells	From Gauss' Law $\mathbf{E} = \dfrac{Q\hat{r}}{4\pi\varepsilon_0 r^2}$ for $a < r < b$,
3. The electric potential difference, $V = (V_+) - (V_-) = V_a - V_b$	$V = -\int_b^a \mathbf{E}\bullet d\mathbf{r} = -\int_b^a \left[\dfrac{Q\hat{r}}{4\pi\varepsilon_0 r^2}\right]\bullet dr\hat{r}$ $V = \dfrac{Q}{4\pi\varepsilon_0 r}\Bigg\vert_b^a = \dfrac{Q}{4\pi\varepsilon_0}\left[\dfrac{b\ -\ a}{a\,b}\right]$
4. The capacitance C from Q and V	$C = Q/V = \dfrac{4\pi\varepsilon_0 ab}{b\ -\ a}$

Note: C depends only on the geometry and ε_0 (this latter because the plates are separated by vacuum)

Example: Determine the capacitance of two parallel conducting plates each of area A and separated by a distance d. Assume d is very small compared to the linear dimension of the plates.

Given: Each plates has area A and the separation is d

Determine: The capacitance C

Strategy: Calculate V for +Q on one plate and -Q on the other; assume **E** is uniform since d is very small compared to the linear dimension of the plate. Knowing Q and V, we can obtain the capacitance.

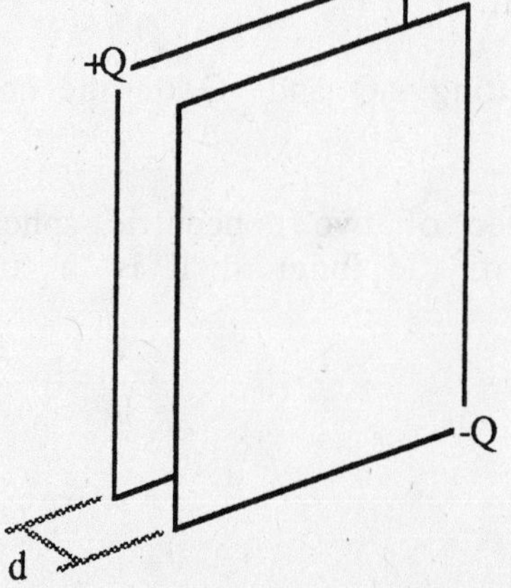

Figure 23.3

Solution: In between the plate $E = \sigma/\varepsilon_0$ for infinite plates. Since d is small we may use this value for E where σ is Q/A.

$$E = \frac{Q}{A\varepsilon_0}$$

Then $V = (V_+) - (V_-) = Ed = \frac{Qd}{A\varepsilon_0}$ and $C = Q/V = \frac{A\varepsilon_0}{d}$

Related Text Exercises: Ex. 23-1 through 23-9.

2 Equivalent Capacitance (Section 23-2)

Figure 23.4

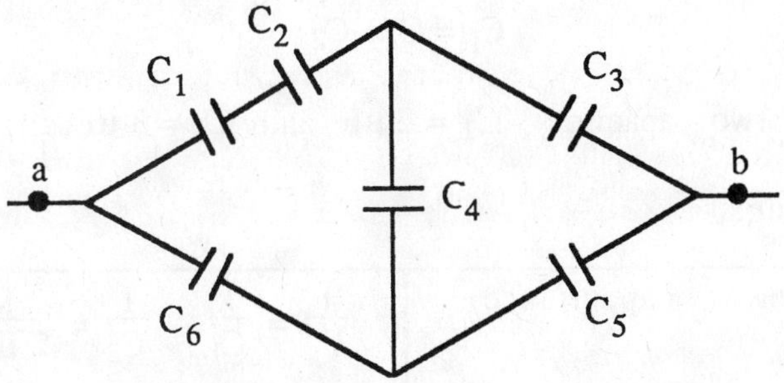

If the system of capacitors between points a and b can be replaced by a single capacitor of capacitance C_E so that no voltage, current or charge on any part of any circuit external to the system of capacitors is changed, then C_E is called the *equivalent capacitance* between points a and b. For a general case like that shown in Figure 23.4 this would be difficult, but there are two simple cases which are very useful.

(i) Capacitors in Series

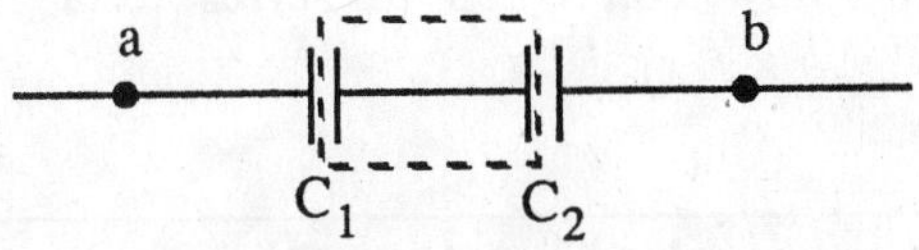

Figure 23.5

Because the region inside the dotted rectangle is isolated, the net charge on the two plates (one from each capacitor) is zero. Hence if one is +Q the other is -Q. Thus each capacitor has Q as its charge; see Figure 23.6.

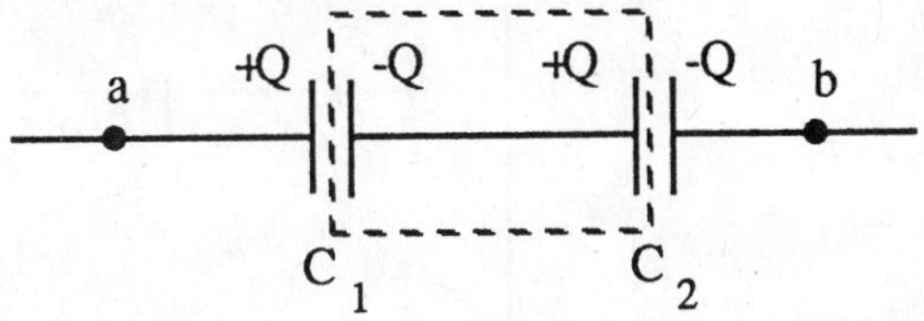

Figure 23.6

The charges will be the same on each of $C_1 + C_2$ and also on the the equivalent series capacitance C_s. The voltage across C_s is the sum of the voltages on C_1 and C_2. It can be shown (see Text 23-3) that

$$\frac{1}{C_s} = \frac{1}{C_1} + \frac{1}{C_2}$$

(ii) Capacitors in Parallel

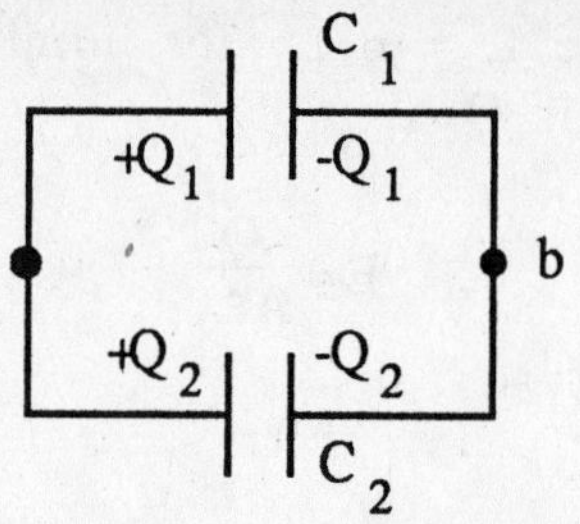

Figure 23.7

In this case the voltage between a and b and hence on the equivalent parallel capacitance C_P is the same as that on each of C_1 and C_2, but the charge Q_P on C_P is the sum of the charges on $C_1 + C_2$. Thus $Q_P = Q_1 + Q_2$. It can be shown (see Text 23-3) that

$$C_P = C_1 + C_2$$

Practice: Consider two capacitors $C_1 = 2\ \mu F$ and $C_2 = 3\ \mu F$.

Determine the following:

1. The equivalent series capacitance	$\frac{1}{C_s} = \frac{1}{C_1} + \frac{1}{C_2} = \frac{1}{2\ \mu F} + \frac{1}{3\ \mu F} = \frac{5}{6\ \mu F}$ Invert to get $C_s = \frac{6\ \mu F}{5} = 1.2\ \mu F$
2. The equivalent parallel capacitance	$C_P = C_1 + C_2 = 2\ \mu F + 3\ \mu F = 5\ \mu F$

The capacitors C_1 and C_2 are now connected in series to a 10 volt battery.

Determine the following:

3. A figure showing the original circuit and the equivalent circuit	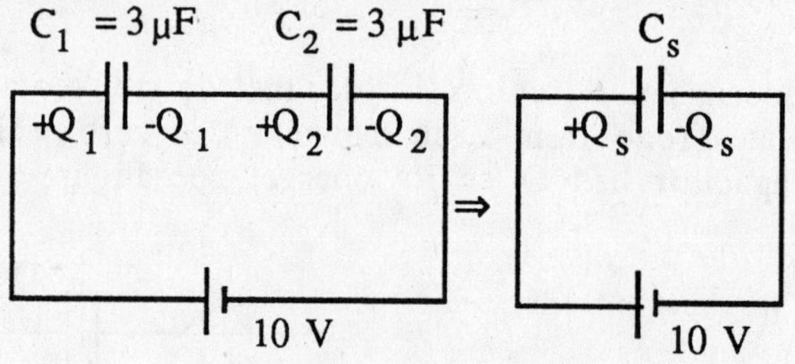 Figure 23.8
4. The equivalent series capacitance	$\frac{1}{C_s} = \frac{1}{2\ \mu F} + \frac{1}{3\ \mu F} = \frac{5}{6\ \mu F} \Rightarrow C_s = 1.2\ \mu F$
5. The charge on the plates of the equivalent capacitor	$Q_s = C_sV = (1.2 \times 10^{-6})(10) = 12\ \mu C$
6. The charge on the plates of C_1 and C_2	$Q_1 = Q_2 = Q_s = 12\ \mu C$

7. The electric potential difference between the plates of C_1 and C_2	$V_1 = \frac{Q_1}{C_1} = \frac{1.2 \times 10^{-6}}{2 \times 10^{-6}} = 6.0\text{ V}$ $V_2 = \frac{Q_2}{C_2} = \frac{1.2 \times 10^{-6}}{3} = 4.0\text{ V}$

Example: $C_1 = 2\ \mu F$ and $C_2 = 3\ \mu F$ are connected in parallel to a 10 volt battery. Find the charge and voltage on each capacitor and the total charge on the parallel combination.

Given: $C_1 = 2\ \mu F$ and $C_2 = 3\ \mu F$ and $V_P = V_1 = V_2 = 10$ volts

Determine: Q_1, Q_2, V_1, V_2, and Q_{total}

Strategy: Knowing V_1 and C_1, we may determine Q_1. Knowing V_2 and C_2, we may determine Q_2. Knowing Q_1 and Q_2, we may determine Q_P. We can check our work by using C_1 and C_2 to determine C_P. Then knowing C_P and V_P, we may determine Q_{total}.

Solution:

$$V_1 = V_2 = V = 10 \text{ volts}$$
$$Q_1 = C_1V_1 = (2 \times 10^{-6})(10) = 20\ \mu C$$
$$Q_2 = C_2V_2 = (3 \times 10^{-6})(10) = 30\ \mu C$$
$$Q_{total} = Q_1 + Q_2 = 20\ \mu C + 30\ \mu C = 50\ \mu C$$

As a check on our work we can determine Q_{total} by an alternate method

$$Q_{total} = C_PV = (5 \times 10^{-6})(10) = 50\ \mu C$$

Related Test Exercises: Ex. 23-10 through 23-20.

3 Energy Stored in a Capacitor and Energy Density (Section 23-3)

Review: Figure 23.9 shows a capacitor of capacitance C carrying a charge q. The voltage across the capacitor when it carries a charge q, denoted by V(q), is given by $V(q) = q/C$. The work done in increasing the charge by dq is:

$$dW = V(q)dq$$

Figure 23.9

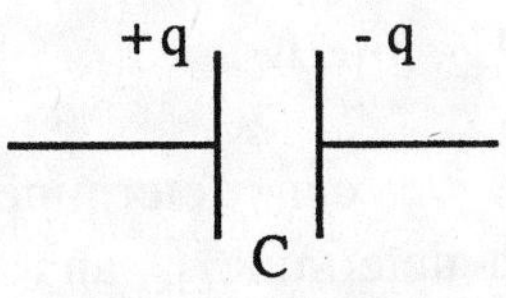

Then the energy stored in the capacitor when charged to a charge Q is

$$U = \int_0^Q dW = \int_0^Q V(q)dq = \int_0^Q \frac{q}{C}dq = \frac{1}{2} \cdot \frac{Q^2}{C}$$

Using C = Q/V, we may write

$$U = \frac{1}{2} \cdot \frac{Q^2}{C} = \frac{1}{2} VQ = \frac{1}{2} CV^2$$

If we view this energy as stored in the electric field which occupies the space between the conductors, then we can get the energy density (energy per unit volume between the conductors) for the special case of a parallel plate capacitor having area A and separation d. u is the symbol for energy density

$$u = \frac{U}{Vol} = \frac{\frac{1}{2} CV^2}{Ad} = \frac{(\frac{1}{2})(\frac{A\varepsilon_0}{d})(Ed)^2}{Ad} = \frac{1}{2}\varepsilon_0 E^2$$

Thus u is determined by the square of the electric field E. The relation $u = \frac{1}{2}\varepsilon_0 E^2$ turns out to be generally true for all electric fields in empty space even though the derivation was for a special case.

Practice: A 2 μF parallel plate capacitor with a plate separation of 10^{-5} m is connected to a 10 V battery.

Determine the following:

1. The energy stored in the capacitor	$U = \frac{1}{2} CV^2 = \frac{1}{2}(2 \times 10^{-6})(10^2) = 10^{-4}$ J
2. The electric field between the plates of the capacitor	$E = V/d = \frac{10}{10^{-5}} = 10^6$ V/m
3. The energy density	$u = \frac{1}{2}\varepsilon_0 E^2 = \frac{1}{2}(8.85 \times 10^{-12})(10^6)^2$ $= 4.43$ J/m^3

Example: A 2 μF capacitor and a 3 μF capacitor are charged in series from a 10 volt battery. Find the charge, voltage, energy and energy density for each capacitor if each is a parallel plate capacitor with a plate separation of 10^{-5} m.

Given: $C_1 = 2$ μF $C_2 = 3$ μF $V = V_1 + V_2 = 10$ volts $d_1 = d_2 = 10^{-5}$ m

Determine: Q_1, Q_2, V_1, V_2, U_1, U_2, u_1, u_2

Strategy: Knowing C_1 and C_2, we can determine the equivalent series capacitance C_s. Knowing C_s and V we can determine Q and hence Q_1 and Q_2 since the capacitors are in series. Knowing the value of Q and C for each capacitor, we can determine V for each capacitor. Knowing Q and V for each capacitor we can determine U for each capacitor. Knowing V and d for each capacitor, we can determine E and hence u for each capacitor.

Solution: $\frac{1}{C_s} = \frac{1}{C_1} + \frac{1}{C_2} = \frac{1}{2 \times 10^{-6}} + \frac{1}{3 \times 10^{-6}} = \frac{5}{6 \times 10^{-6}} \Rightarrow C_s = 1.20\ \mu F$

$Q_1 = Q_2 = Q_s = C_sV(1.2 \times 10^{-6})(10) = 12.0\ \mu C$

$V_1 = \frac{Q_1}{C_1} = \frac{12 \times 10^{-6}}{2 \times 10^{-6}} = 6.00\ V$ and $V_2 = \frac{Q_2}{C_2} = \frac{12 \times 10^{-6}}{3 \times 10^{-6}} = 4.00\ V$

Note: $V_1 + V_2 = 6 + 4 = 10\ V$ as it should.

$$U_1 = \frac{1}{2}Q_1V_1 = \frac{1}{2}(12 \times 10^{-6})(6) = 3.60 \times 10^{-5}\ J$$

$$U_2 = \frac{1}{2}Q_2V_2 = \frac{1}{2}(12 \times 10^{-6})(4) = 2.40 \times 10^{-5}\ J$$

$$E_1 = V_1/d = \frac{6}{10^{-5}} = 6.00 \times 10^5\ V/m$$

$$E_2 = V_2/d = \frac{4}{10^{-5}} = 4.00 \times 10^5\ V/m$$

Therefore

$$u_1 = \frac{1}{2}\varepsilon_0 E_1^2 = \frac{1}{2}(8.85 \times 10^{-12})(6.00 \times 10^5)^2 = 1.60\ J/m^3$$

$$u_2 = \frac{1}{2}\varepsilon_0 E_2^2 = \frac{1}{2}(8.85 \times 10^{-12})(4.00 \times 10^5)^2 = 0.710\ J/m^3$$

Related Text Exercises: Ex. 23-21 through 23-33.

4 Dielectrics (Section 23-4)

A dielectric material when inserted between the two conductors forming a capacitor increases the capacitance over what it would be if the conductors were separated by empty space. Two numbers characterize the dielectric material; the dielectric constant K and the dielectric strength. The first is the factor by which the capacitance is increased from its no dielectric value and the second is the electric field it can withstand before conduction. Both numbers are important in the design of a useful capacitor.

$$K = \frac{C}{C_0}$$

where C is the capacitance with the dielectric in and C_0 is the capacitance with the dielectric out.

Practice: The plates of a parallel-plate capacitor each have an area A, and their initial separation is d_0. We have a battery with a terminal potential difference V_0 available for charging the capacitor. We also have sheets of dielectric with any desired thickness and a dielectric constant K.

Determine the following:

1. The initial capacitance C_0	$C_0 = \varepsilon_0 A/d_0$

Now let's connect the capacitor to the battery.

2. The charge q_0 on each plate	$q_0 = C_0 V_0$
3. The magnitude of the electric field intensity E_0	$E_0 = V_0/d_0$
4. The energy U_0 stored in the capacitor U_0	$U_0 = C_0 V_0^2/2 = q_0 V_0/2$
5. The energy per unit volume u_0	$u_0 = U_0/Ad_0 = (1/2)\varepsilon_0 E_0^2$

Now disconnect the battery and pull the plates apart until $d = 2d_0$.

6. The new capacitance C'	$C' = \varepsilon_0 A/d' = \varepsilon_0 A/2d_0 = C_0/2$
7. The charge q' on each plate	Since the plates have been disconnected, no charge can flow either on or off them. Therefore: $q' = q_0$
8. The new potential difference V' across the plates	$V' = q'/C' = q_0/(C_0/2) = 2V_0$
9. The new electric field E'	$E' = V'/d' = 2V_0/2d_0 = E_0$
10. The new energy U' stored in the capacitor	$U' = C'V'^2/2 = (C_0/2)(2V_0)^2/2 = 2U_0$ $U' = q'V'/2 = q_0(2V_0)/2 = 2U_0$
11. The source of this increase in energy	As you pull the plates apart, you add your work to the system, thus giving it energy

Let's now go back to our initial capacitor, charge it, disconnect the battery, and insert a sheet of the dielectric that has a thickness d_0

12. The new capacitance C_d	$C_d = KC_0$
13. The charge q_d on each plate	Since the plates have been disconnected, no charge can flow on or off them $q_d = q_0$
14. The new potential difference V_d across the plates	$V_d = q_d/C_d = q_0/KC_0 = V_0/K$
15. The new electric field E_d	$E_d = V_d/d_d = (V_0/K)d_0 = E_0/K$
16. The new energy U_d stored in the capacitor	$U_d = C_d V_d^2/2 = (KC_0)(V_0/K)^2/2 = U_0/K$ $U_d = q_d V_d/2 = q_0(V_0/K)/2 = U_0/K$

17. The cause of this decrease in energy	As the dielectric material approaches the capacitor, the dipoles are lined up by the electric field. This gives rise to a net induced charge on the sides of the dielectric material causing it to be pulled into the region between the plates. The energy to do this comes out of the energy stored in the capacitor.

Example: The capacitance of a capacitor is 50.0 μF when filled with a dielectric of K = 5.00. If this capacitor is connected to a 25.0 V battery, charged, and the dielectric removed, calculate:

(a) the charge on the plates while the dielectric is between the plates
(b) the energy stored on the plates with the dielectric between the plates
(c) the energy on the plates after the dielectric is removed
(d) the energy stored after the dielectric is removed

Given:

C = 50.0 μF
= the capacitance of the capacitor when the dielectric is in place
K = 5.00 = the dielectric coefficient
V = 25.0 V = terminal potential difference of the battery

Determine:

(a) Q = charge on the plates while the dielectric is in place
(b) U = energy stored with the dielectric in place
(c) Q_0 = charge on the plates with the dielectric removed
(d) U_0 = energy stored with the dielectric removed

Strategy: Knowing C and V, we can determine Q and U. Knowing C and K and recognizing that $V_0 = V$, we can determine Q_0 and U_0

Solution: We can determine Q and U by setting up the following equations

$$Q = CV = (50.0 \times 10^{-6}\ F)(25.0\ V) = 1.25 \times 10^{-3}\ C$$
$$U = CV^2/2 = (50.0 \times 10^{-6}\ F)(25.0\ V)^2/2 = 1.56 \times 10^{-2}\ J$$

Since the battery remains connected, $V_0 = V$

Since the dielectric is removed, $C_0 = C/K$

We can determine the new charge Q_0 and energy stored U_0 by setting up the following equations

$$Q_0 = C_0V_0 = (C/K)V = q/K = (1.25 \times 10^{-3}\ C)/5 = 2.5 \times 10^{-4}\ C$$

$$U_0 = C_0V_0^2/2 = (C/K)V^2/2 = U/K = (1.56 \times 10^{-2}\ J)/5 = 3.12 \times 10^{-3}\ J$$

Related Text Exercises: Ex. 23-34 through 23-42.

5 Connecting Previously Charged Capacitors (Section 23-3)

Now that we have considered some of the characteristics of capacitors let's consider a single problem that is often difficult for the beginning student.

Example: A 2 μF and a 3 μF capacitor are charged in series from a 10 volt battery. While fully charged they are disconnected from the battery. The positive plates of the two capacitors are then joined by a conductor and the negative plates are also joined by a conductor. Find the final charge and voltage on each capacitor after they are joined.

Given: $C_1 = 2\ \mu F$ $\quad C_2 = 3\ \mu F$ $\quad V_{1_i} + V_{2_i} = V = 10\ V$

Determine: Q_{1_f}, Q_{2_f}, V_{1_f}, V_{2_f}

Strategy: Find the charges and voltages Q_{1_i}, Q_{2_i}, V_{1_i}, V_{2_i} when they are first joined. This is done by considering the series combination being charged by the 10 volt battery. Next notice that these voltages are generally not equal but they must be when a static situation is reached. Charge is transferred from one capacitor to the other until the voltages are equal (and opposing each other). The total charge for this connection must be $Q_{1_i} + Q_{2_i}$.

Solution:

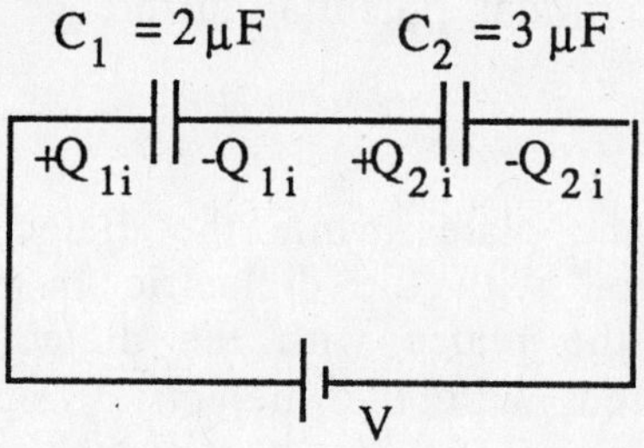

Figure 23.10

$$\frac{1}{C_s} = \frac{1}{C_1} + \frac{1}{C_2} = \frac{1}{C_s} = \frac{1}{2 \times 10^{-6}} + \frac{1}{3 \times 10^{-6}} = \frac{5}{6 \times 10^{-6}} \Rightarrow C_s = 1.2\ \mu F$$

$$Q_{1_i} = Q_{2_i} = Q_s = C_s V = (1.2 \times 10^{-6})(10) = 12\ \mu C$$

This means each capacitor has a charge of 12 μC when they are disconnected from the battery and joined to each other.

$$V_{1_i} = \frac{Q_{1_i}}{C_1} = \frac{12 \times 10^{-6}}{2 \times 10^{-6}} = 6\ V \qquad \text{and} \qquad V_{2_i} = \frac{Q_{2_i}}{C_2} = \frac{12 \times 10^{-6}}{3 \times 10^{-6}} = 4\ V$$

Figure 23.11 shows the arrangements when the capacitors are first connected to each other. This is the initial (i) arrangement.

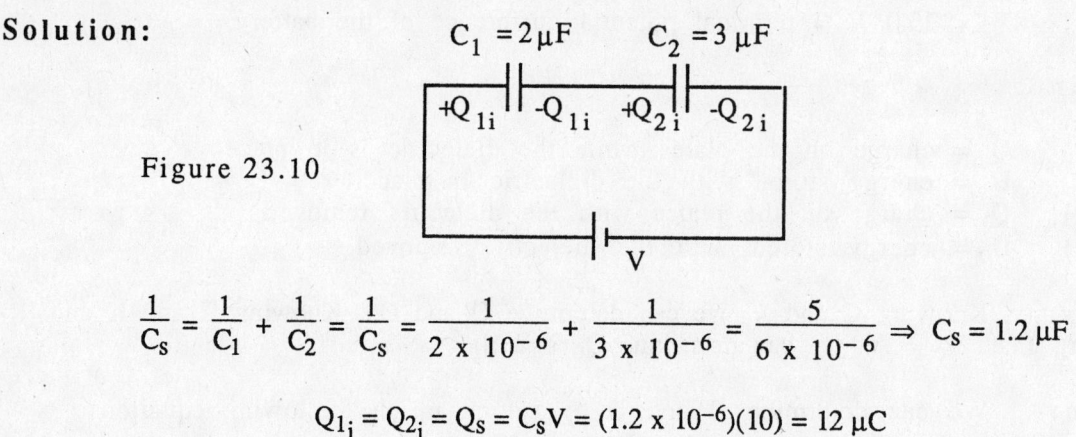

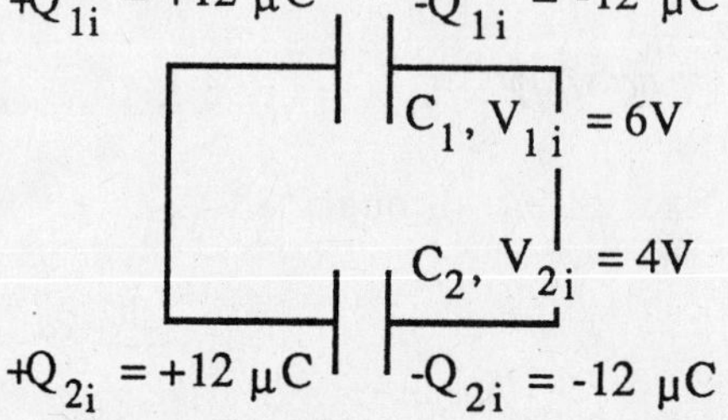

Figure 23.11

Note: (1) The two voltages are not equal
(2) The two voltages both increase to the left as shown
(3) The positive charge on the left plates is trapped there although it can redistribute between the two left plates and likewise the negative charges are trapped on the two right plates but they can (and do) redistribute

Figure 23.12 shows the final (f) arrangement.

Figure 23.12

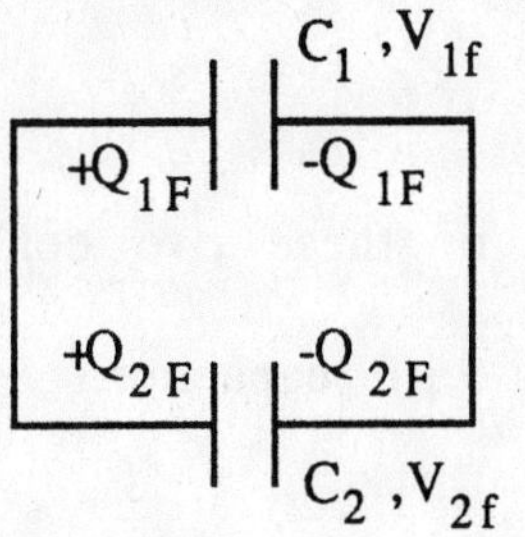

Since the charges redistribute we must have $Q_{1f} + Q_{2f} = Q_{1i} + Q_{2i} = 12\ \mu C + 12\ \mu C = 24\ \mu C$. This is one condition used in determining Q_{1f} and Q_{2f}. The other is that the two voltages V_{1f} and V_{2f} must oppose and be equal as shown.

$$V_{1f} = V_{2f}$$

$$\frac{Q_{1f}}{C_1} = \frac{Q_{2f}}{C_2} \Rightarrow \frac{Q_{1f}}{2\ \mu F} = \frac{Q_{2f}}{3\ \mu F} \Rightarrow Q_{2f} = \frac{3}{2} Q_{1f}$$

Therefore $Q_{1f} + Q_{2f} = 24\ \mu C$ becomes

$$Q_{1f} + \frac{3}{2} Q_{1f} = \frac{5}{2} Q_{1f} = 24\ \mu C$$

$$Q_{1f} = \frac{48}{5}\ \mu C \quad \text{and} \quad Q_{2f} = \frac{3}{2} Q_{1f} = \frac{72}{5}\ \mu C$$

Note: $Q_{1f} + Q_{2f} = \frac{48}{5}\ \mu C + \frac{72}{5}\ \mu C = 24\ \mu C$ as required

$$V_{1f} = Q_{1f}/C_1 = 4.8\ V \quad \text{and} \quad V_{2f} = Q_{2f}/C_2 = 4.8\ V$$

Note: $V_{1f} = V_{2f}$ as it should

We see that charge moved from C_1 to C_2 thus lowering V_1 and increasing V_2. This continued until the two voltages became equal (they already opposed each other). This was not the average of the two (5 volts) as some students are led to believe by faulty intuition. The problem must be dealt with in terms of total charge and equal voltages.

Related Text Exercises: Ex. 23-10 through 23-20.

PRACTICE TEST

Take and grade this practice test. Doing so will allow you to determine any weak spots in your understanding of the concepts taught in this chapter. The following section prescribes what you should study further to strengthen your understanding.

Two conducting spheres each have a radius R and have their centers separated by a distance D, where D >> R (assume that the charge on the conductors is uniformly distributed over the surface).

Determine the following:

________ 1. The capacitance of these two conducting spheres

You are given a 3 μF, a 4 μF and a 5 μF capacitor.

Determine the following:

________ 2. The equivalent capacitance when they are connected in series with a 10 V battery
________ 3. The charge on each capacitor when they are connected in series
________ 4. The voltage on each capacitor when they are connected in series
________ 5. The equivalent capacitance when they are connected in parallel with a 10 V battery
________ 6. The charge on each capacitor when they are connected in parallel
________ 7. The voltage on each capacitor when they are connected in parallel

Consider the combination of capacitors shown in Figure 23.13

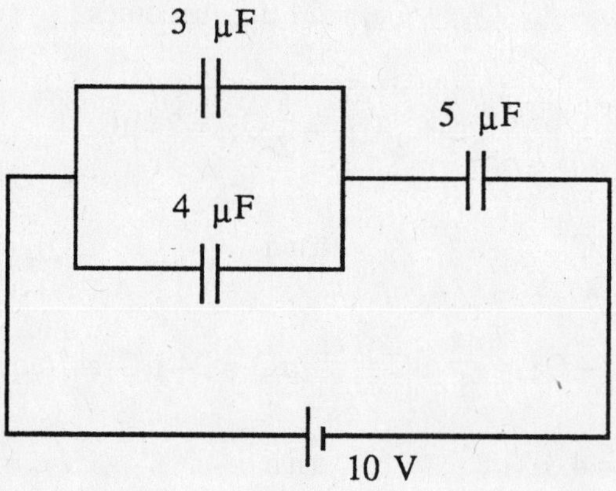

Figure 23.12

Determine the following:

________ 8. The equivalent capacitance across the 10 V battery
________ 9. The charge on each capacitor
________ 10. The voltage on each capacitor

Figure 23.14 summarizes a sequence of events involving a capacitor, battery, and a slab of dielectric material.

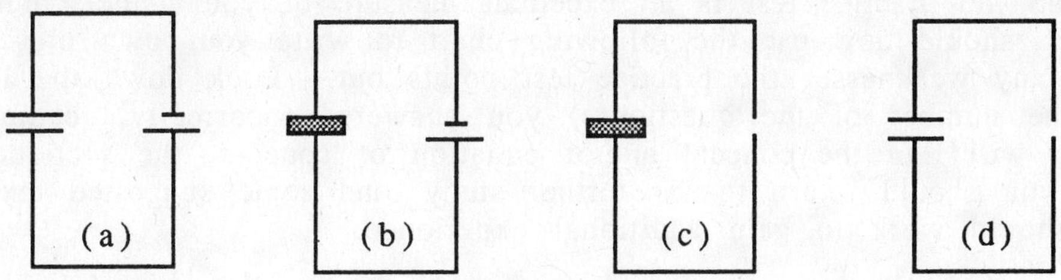

(a) The capacitor is connected to the battery and charged
(b) While the battery is connected, a dielectric slab is inserted
(c) The battery is disconnected
(d) The dielectric is withdrawn

Figure 23.14

The following information is known:

$A = 1.00 \times 10^{-2}$ m^2 = area of separation of the plates
$d = 2.00 \times 10^{-2}$ m = separation of the plates
$V_0 = 24$ V = voltage of the battery
$K = 5.00$ = dielectric constant

Determine the following:

________ 11. Capacitance of the capacitor in case (a)
________ 12. Charge on each plate of the capacitor in case (a)
________ 13. Capacitance of the capacitor in case (b)
________ 14. Charge on each plate of the capacitor in case (b)
________ 15. Charge on each plate in case (d)
________ 16. Potential difference across the plates in case (d)

A 3 μF and 4 μF capacitor.are connected in series and charged by a 100 volt battery. They are disconnected while fully charged and they are joined by conductors with their positive plates connected to each to each other and their negative plates connected to each other.

Determine the following:

________ 17. The charge on each capacitor when first disconnected from the battery
________ 18. The charge on each capacitor after being reconnected to each other without the battery
________ 19. The voltage on each capacitor after being reconnected
________ 20. The energy stored in each capacitor when first disconnected from the battery
________ 21. The energy stored in each capacitor after being reconnected to each other without the battery

(See Appendix I for answers)

PRINCIPAL CONCEPTS AND EQUATIONS PRESCRIPTION

Your score on the practice test is an excellent measure of your understanding of the chapter. You should now use the following chart to write your own prescription for dealing with any weaknesses the practice test points out. Look down the leftmost column to the number of the question(s) you answered incorrectly, reading across that row you will find the concept and/or equation of concern, the section(s) of the study guide you should return to for further study, and some suggested text exercises which you should work to gain additional experience.

Practice Test Questions	Concepts and Equations	Prescription	
		Principal Concepts	Text Exercises
1	Capacitance	1	23-6, 7, 23-8, 9
2-10	Equivalent Capacitance	2	23-3, 10, 23-11, 12, 23-18
11-12	Capacitance	1	23-2, 3
13-16	Dielectrics	4	23-34, 37, 23-38
17	Equivalent Capacitance	2	23-10, 11, 23-12, 18, 23-19
18-19	Connecting charged capacitors	5	23-14, 15, 23-20
20-21	Energy stored in capacitor	3	23-21, 22, 23-26, 27, 23-28

24 CURRENT AND RESISTANCE

RECALL FROM PREVIOUS CHAPTERS

Previously learned concepts and equations frequently used in this chapter	Text Section	Study Guide Page
Electric Field: $\mathbf{E} = \mathbf{F}/q = \sum_i (kQ_i/r_i^2)\,\hat{\mathbf{r}}_i$	20-4	20-5
Electric Potential: $V(\mathbf{r}) = -\int \mathbf{E}\cdot d\mathbf{l}$	22-3	22-3

NEW IDEAS IN THIS CHAPTER

Concepts and equations introduced	Text Section	Study Guide Page
Current: $I = dQ/dt = nqAv_d$	24-1	24-1
Current density: $j = I/A$; $\mathbf{j} = nq\mathbf{v}_d$	24-1	24-1
Resistance: $R = V/I$	24-2	24-5
Ohm's Law: R is constant and independent of V, and I	24-2	24-5
Resistivity: $R = \rho L/A$	24-2	24-5
Temperature Dependence of R: $R = R_0[1 + \alpha(T-T_0)]$	24-2	24-5
Conductivity: $\mathbf{j} = \sigma\mathbf{E}$; $\sigma = 1/\rho$	24-2	24-5
Semiconductors: Intrinsic and Doped	24-6	--
Charge carriers: electrons and holes	24-6	--
N and P type semiconductors	24-6	--
Equivalent Resistance:	24-3	24-10
Series $R_s = \sum_{i=1}^{N} R_i$ Parallel $1/R_p = \sum_{i=1}^{N} 1/R_i$		
Ammeters and Voltmeters	24-4	24-17

PRINCIPAL CONCEPTS AND EQUATIONS

1 Electric Current and Current Density (Section 24-1)

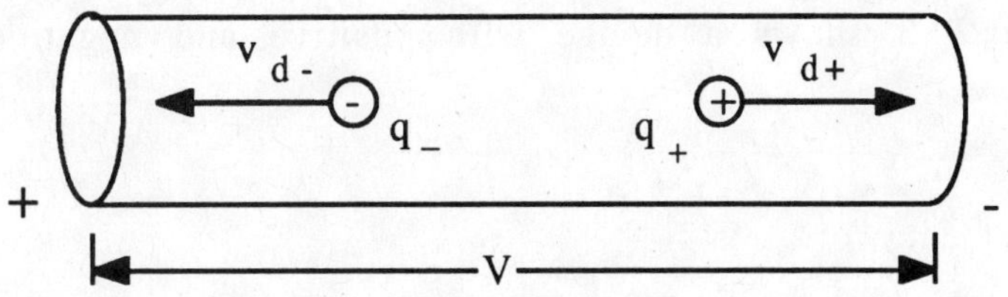

Figure 24.1

In Figure 24.1 we see a conductor with a potential difference V maintained across it; the left (right) end is positive (negative) relative to the right (left) end.

If the charges are free to move we get a current or flow of charge. The positive charges (q_+) are viewed as drifting to the right with an average drift speed v_{d+} and the negative charges to the left with an average drift speed v_{d-}. Either case leads to a conventional current I that is from left to right; i.e. conventional current is in the direction that positive charges move even if the actual current is due to negative charges moving in the opposite direction.

The current I, the time rate of flow of charge, is defined by: $I = dq/dt$
I is the number of coulombs passing through a cross-sectional area of the conductor in one second.

n_+ = number of positive charge carriers per unit volume
q_+ = charge on each carrier
A = cross-sectional area of the conductor
v_{d+} = drift speed of the positive charge carriers

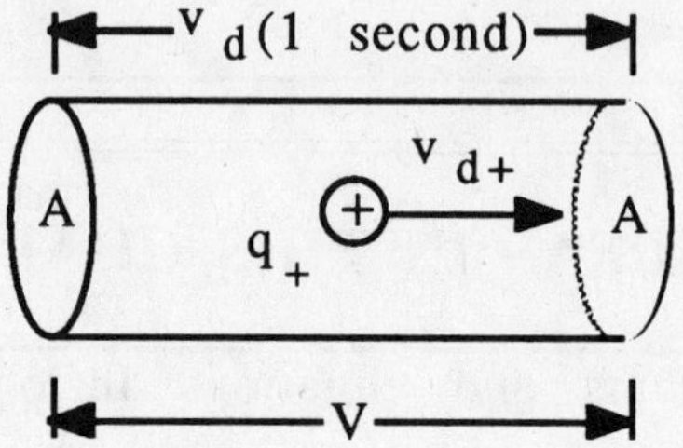

Figure 24.2

Examination of Figure 24.2 shows that the number of charges passing through the cross-sectional area A on the right side in one second is just the number of charges in the length (v_d)(1 s). Multiply this number of charges by q_+ and it yields the current. Therefore

$$I = n_+ q_+ A v_d$$

If we have N different types of charge carriers (which may be positive or negative) then,

$$I = \sum_{i=1}^{N} n_i |q_i| A v_{di}$$

The current density j (current per unit area) is just

$$j = \frac{I}{A} = \sum_{i=1}^{N} n_i |q_i| v_{di}$$

or making it into a vector

$$\mathbf{j} = \sum_{i=1}^{N} n_i q_i \mathbf{v}_{di}$$

Practice: Figure 24.3 shows a device with positive and negative charge carriers.

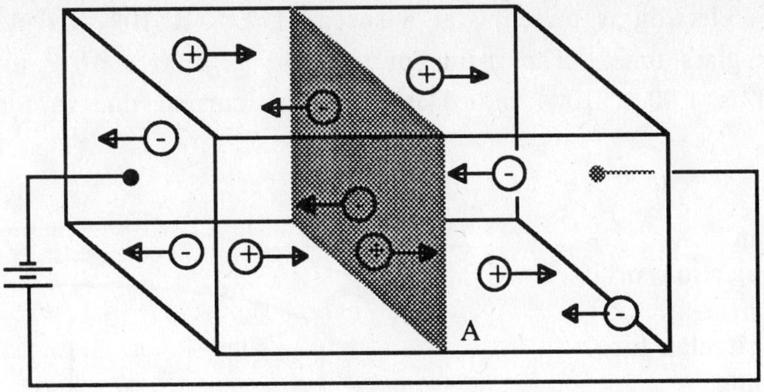

$q = 1.60 \times 10^{-19}$ C = magnitude of the charge on both positive and negative charge carriers
$A = 4.00 \times 10^{-4}\ m^2$ = cross-sectional area of the device
$n_+ = 6.00 \times 10^{24}\ /m^3$ = positive charge carriers per unit volume
$n_- = 4.00 \times 10^{26}\ /m^3$ = negative charge carriers per unit volume
$v_{d+} = 3.00 \times 10^{-5}$ m/s = drift speed of the positive charge carriers
$v_{d-} = 1.00 \times 10^{-4}$ m/s = drift speed of the negative charge carriers

Figure 24.3

Determine the following:

1. Number of positive and negative charge carriers crossing A per unit time	$N_+/t = n_+Av_{d+} = 7.20 \times 10^{16}$ /s $N_-/t = n_-Av_{d-} = 1.60 \times 10^{19}$ /s
2. Amount of positive and negative charge crossing A per unit time	$q_+/t = (N_+/t)q = 1.15 \times 10^{-2}$ C/s $q_-/t = (N_-/t)q = -2.56$ C/s
3. Net flow of charge across A per unit time	$Q_{net}/t = (q_+/t) + (\lvert q_-\rvert/t) = +2.57$ C/s
4. Current (magnitude and direction) across A	$I = Q_{net}/t = 2.57$ C/s left to right. From step 3 and Figure 24-3 we realize that a net charge of -2.57 C crosses A each second from right to left. Negative movement to the left is the same as positive movement to the right as far as conventional current is concerned since conventional current is in the direction of motion of the positive charge carriers. Therefore the current is 2.57 A left to right.
5. Time for a total charge of 10.0 C to drift across A	$I = \Delta q/\Delta t$ $\Delta t = \Delta q/I = 3.89$ s
6. The current density j	$j = I/A$ (left to right) $= \dfrac{2.57\ A}{4.00 \times 10^{-4}\ m^2}$ $= 6425\ A/m^2$ (left to right)

Example: An electron is traveling at a speed of 2.00×10^6 m/s in a small, evacuated, circular glass tube. The radius of the tube is 1.00×10^{-3} m and the radius of the electrons orbit is 1.00×10^{-2} m. Determine the current due to the electron's motion.

Given:

$r_o = 1.00 \times 10^{-2}$ m = radius of electron orbit

$r_t = 1.00 \times 10^{-3}$ m = radius of circular tube

$v = 2.00 \times 10^6$ m/s = speed of the electron

$q_e = -1.60 \times 10^{-19}$ C = charge on the electron

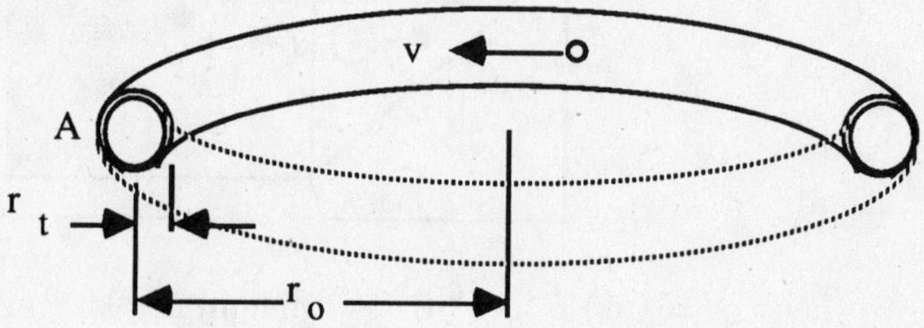

Figure 24.4

Determine: The current due to the electron's motion

Strategy: Knowing the electron's radius of orbit and speed, we can determine the time for one orbit. Knowing that an electron passes point A once every orbit and the time of the orbit, we can determine the current due to the electron's motion.

Solution: The length of the electron's orbit is

$$C = 2\pi r_o = 6.28 \times 10^{-2} \text{ m}$$

The time for the electron to make one orbit is

$$T = C/v = 3.14 \times 10^{-8} \text{ s}$$

Charge $|\Delta q| = 1.60 \times 10^{-19}$ C passes A every $\Delta t = 3.14 \times 10^{-8}$ s, hence

$$I = |\Delta q|/\Delta t = 5.10 \times 10^{-12} \text{ A}$$

Example: A wire carries a current of 1.00 A. How many electrons must pass through a cross-sectional area of the wire in 2.00 s to produce this current?

Given:

$I = 1.00$ A = current in the wire

$\Delta t = 2.00$ s = time of current flow

$q_e = -1.60 \times 10^{-19}$ C = charge on the electron

Determine: The number of electrons which must pass through a cross-sectional area of the wire every 2.00 s in order to have a current of 1.00 A.

Strategy: Knowing the current and time, we can determine the total charge that must pass A in order to sustain the current. Knowing the total charge and the charge on each charge carrier, we can determine the number of charge carriers needed.

Solution: The total charge that must pass A is

$$|\Delta q| = I\Delta t = 2.00 \text{ C}$$

The number of electrons needed to supply this charge is

$$N = |\Delta q|/|q_e| = 1.25 \times 10^{19}$$

Related Text Exercises: Exs. 24-1 through 24-6.

2 Resistance and Ohm's Law (Section 24-2)

Review:

Figure 24.5

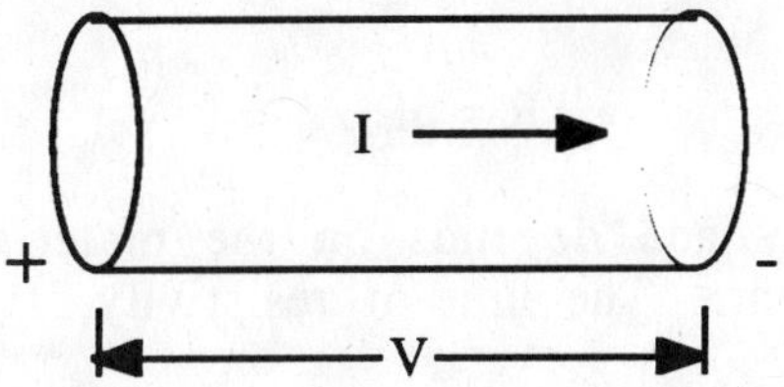

Figure 24.5 shows a device with a current I when the voltage across the device is V. The resistance R is defined by

$$R = V/I$$

The SI unit of resistance is the ohm (Ω). 1 ohm = 1 volt/amp

(i) Ohm's Law--Figure 24.6 shows a resistor connected to first one cell (a), then two cells (b), and finally three cells (c). The ammeter measures the current in the circuit. The voltmeter measures the potential difference across the resistor.

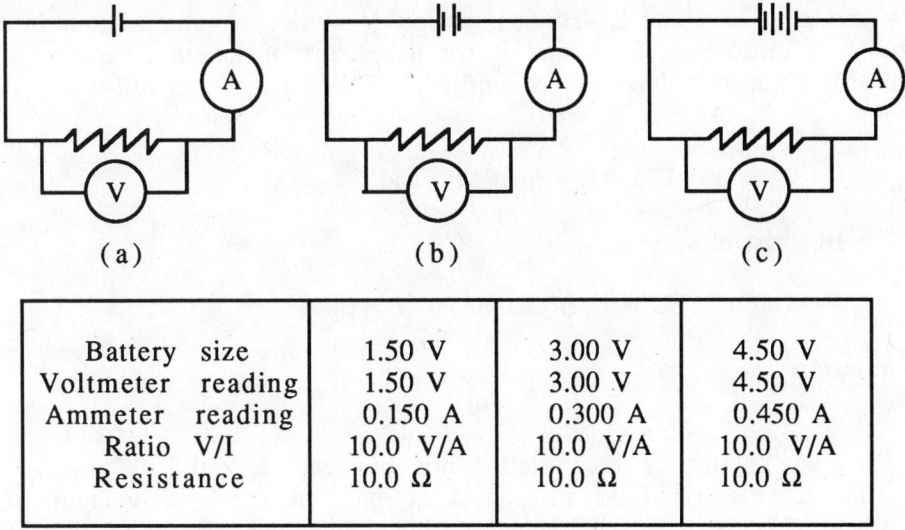

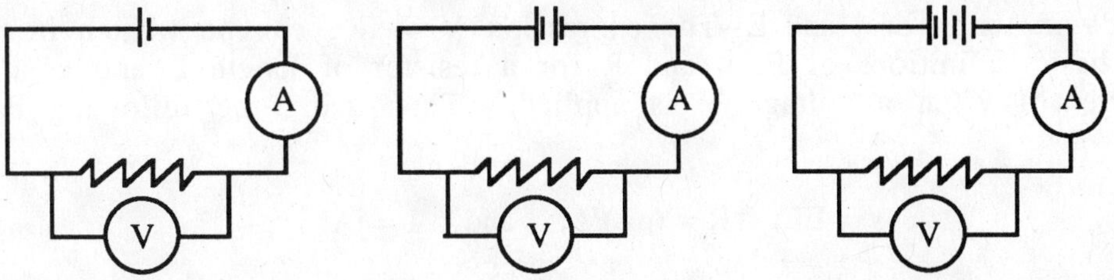

Battery size	1.50 V	3.00 V	4.50 V
Voltmeter reading	1.50 V	3.00 V	4.50 V
Ammeter reading	0.150 A	0.300 A	0.450 A
Ratio V/I	10.0 V/A	10.0 V/A	10.0 V/A
Resistance	10.0 Ω	10.0 Ω	10.0 Ω

Figure 24.6

Notice that as V increases, I also increases. That is I is proportional to V.

$$I \propto V$$

Notice also that the ratio of V/I, the resistance of the resistor, is a constant. Stated algebraically

$$R = V/I = \text{constant}$$

When R is constant then the resistor is said to be ohmic and to obey Ohm's Law. Unless told otherwise we will assume all resistances obey Ohm's Law.

Note that a plot of I versus V will give a straight line whose slope is 1/R. If I versus V is not a straight line then R is not constant and we say the resistance is non-ohmic.

Note: In all cases R= V/I because this is the definition of resistance.

(ii) Resistivity--If we have a resistor with a length L and a uniform cross-sectional area A then the resistance is

$$R = \rho L/A$$

where ρ is called the resistivity and depends on the material. If the material is ohmic then R and hence ρ is a constant. The unit of resistivity is $\Omega \cdot m$.

(iii) Temperature dependence of ρ and R--For most metals (good conductors) the resistivity and hence R vary nearly linearly with temperature over a wide temperature range (thus when the current causes heating, the resistance is not truly ohmic but only approximately ohmic). Thus we write

$$\rho = \rho_0[1+ \alpha(T - T_0)] \quad \text{or} \quad R = R_0[1+ \alpha(T - T_0)]$$

where ρ_0 and R_0 are the resistivity and resistance at temperature T_0 and ρ and R are the corresponding quantities at temperature T. α is called the temperature coefficient of resistivity and it depends on the material used for the resistor.

(iv) Ohm's Law in terms of **j** and **E**--The expression V = IR can be written in terms of E and j using basic definitions of E, j, and R for a resistor of length L and area A carrying a current I when a voltage V is applied. Thus using the following basic definitions

$$V = EL; \quad R = (\rho L)/A; \quad \text{and} \quad I = jA,$$

the expression V = IR becomes

$$EL = (jA)(\rho L/A) \Rightarrow E = \rho j.$$

Using vector notation

$$\mathbf{E} = \rho \mathbf{j}.$$

Note: In many materials the relationship between **E** and **j** is more complicated because an electric field in one direction can cause movement of charge in other directions, This is like squeezing a rubber ball and observing that it flattens and spreads out.

We may also write:

$$\mathbf{j} = \sigma \mathbf{E}$$

where $\sigma = 1/\rho$ is called the conductivity of the resistor material. If ρ and σ are constant then the resistor is ohmic and obeys Ohm's Law.

Practice: Figure 24.7 shows several simple circuits.

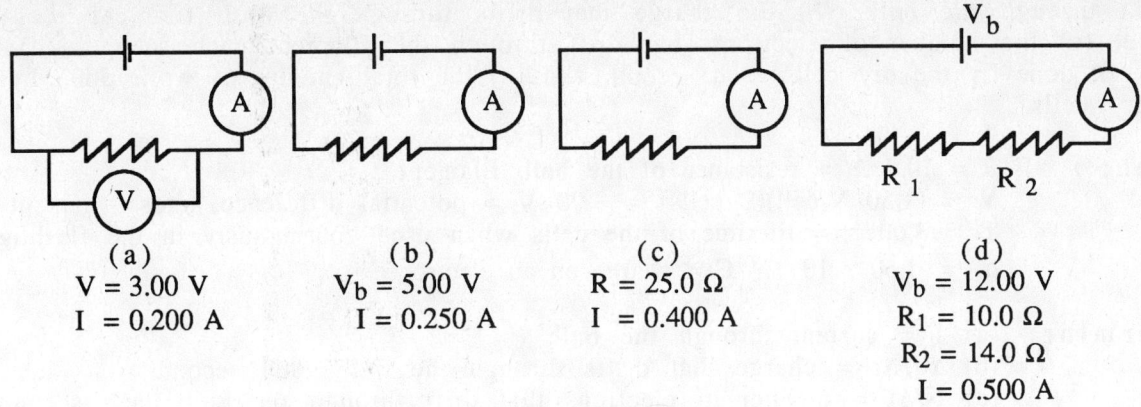

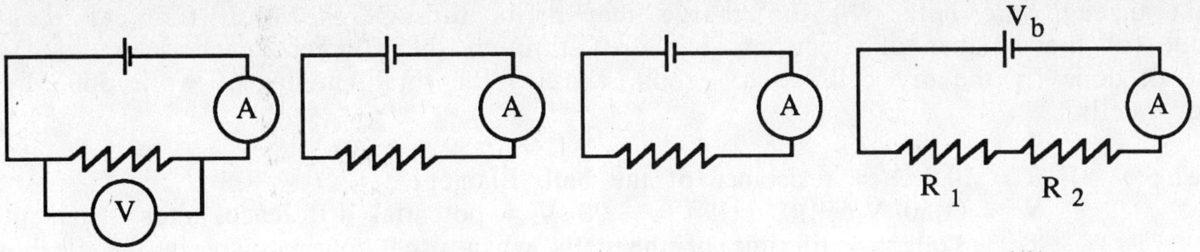

(a)	(b)	(c)	(d)
$V = 3.00$ V	$V_b = 5.00$ V	$R = 25.0\ \Omega$	$V_b = 12.00$ V
$I = 0.200$ A	$I = 0.250$ A	$I = 0.400$ A	$R_1 = 10.0\ \Omega$
			$R_2 = 14.0\ \Omega$
			$I = 0.500$ A

Figure 24.7

The notation used is as follows:

V = Voltmeter reading = potential difference across the resistor
I = ammeter reading = current in the circuit
R = size of resistor
V_b = battery size = terminal potential difference of the battery

Determine the following:

1. Terminal potential difference of the battery for case (a)	$V_b = V = 3.00$ V
2. Resistance of the resistor for case (a)	$R = V/I = 15.0\ \Omega$
3. Potential difference across the resistor for case (b)	$V = V_b = 5.00$ V
4. Resistance of the resistor for case (b)	$R = V/I = 20.0\ \Omega$
5. Potential difference across the resistor for case (c)	$V = IR = 10.0$ V
6. Terminal potential difference of the battery for case (c)	$V_b = V = 10.0$ V
7. Potential difference across R_1 for case (d)	$V_1 = IR_1 = 5.00$ V
8. Potential difference across R_2 for case (d)	$V_2 = IR_2 = 7.00$ V
9. The total potential difference across R_1 and R_2 for case (d)	$V_b = V_T = V_1 + V_2 = 12.0$ V

Example: A small flashlight uses a bulb that has a resistance of 10.0 Ω and two 1.50 V dry cells. The dry cells can last 3.00 h under continuous use. Determine (a) the current through the bulb, (b) the charge that drifts through the bulb filament each second, (c) the number of electrons that drift through the filament each second, (d) the work done by the dry cells each second, and (e) the total amount of work done by the dry cells.

Given: $R = 10.0\ \Omega$ = resistance of the bulb filament
$V = (1.50\ V/cell)(2\ cells) = 3.00\ V$ = potential difference across filament
$T = 3.00\ h$ = lifetime of the cells when used continuously in the flashlight
$q_e = -1.60 \times 10^{-19}\ C$ = charge on an electron

Determine: (a) I = current through the bulb
(b) $\Delta q/\Delta t$ = charge that drifts through the bulb each second
(c) $N/\Delta t$ = number of electrons that drift through the bulb each second
(d) $W/\Delta t$ = work done by the cells each second
(e) W_T = total amount of work done by the dry cells

Strategy: (a) Knowing V and R, we can determine I
(b) Knowing I, we also know $\Delta q/\Delta t$
(c) Knowing $\Delta q/\Delta t$ and q_e, we can determine $N/\Delta t$
(d) Knowing $\Delta q/\Delta t$ and V, we can determine $W/\Delta t$
(e) Knowing $W/\Delta t$ and T, we can determine W_T

Solution: (a) $I = V/R = 0.300\ A$
(b) $\Delta q/\Delta t = I = 0.300\ C/s$
(c) $\Delta q/\Delta t = N|q_e|/\Delta t$ or $N/\Delta t = (\Delta q/\Delta t)/|q_e| = 1.88 \times 10^{18}\ /s$
(d) $W/\Delta t = V(\Delta q/\Delta t) = 0.900\ J/s$
(e) $W_T = (W/\Delta t)T = 9.72 \times 10^3\ J$

Practice: A nichrome wire with the following characteristics is used as a heating element.

$\rho_{20} = 1.00 \times 10^{-6}\ \Omega \bullet m$	$L = 1.00\ m$	$T = 20°\ C$
$\alpha = 4.50 \times 10^{-4}\ /C°$	$A = 1.00 \times 10^{-8}\ m^2$	$T_0 = 0°\ C$

When the element is connected to a 120 V outlet, the current stabilizes at 0.900 A.

Determine the following:

1. Resistance of the heating element at T = 20° C	$R_{20} = \rho_{20}L/A = 100\ \Omega$
2. Resistance of the element at T = 0° C	$R_{20} = R_0[1 + \alpha(20°\ C)]$ $R_0 = R_{20}/[1 + \alpha(20°\ C)] = 99.11\ \Omega$
3. Resistance of the element at T = 50° C	$R_{50} = R_0[1 + \alpha(50°\ C)]$ Using R_0 from step 2, obtain $R_{50} = 101.3\ \Omega$
4. Resistance of the element at T = -20° C	$R_{-20} = R_0[1 + \alpha(-20°\ C)] = 98.22\ \Omega$

5. Final resistance of the wire (i.e., its resistance while hot)	$R_f = V/I = \dfrac{120}{0.900} = 133.3\ \Omega$
6. Final temperature of the wire	$R_f = R_0[1 + \alpha T]$ or $T = [(R_f / R_0) - 1]/\alpha = 767°\ C$

Example: A copper wire of length L = 2.00 m and cross-sectional area $A = 5.00 \times 10^{-8}\ m^2$ has a potential difference across it sufficient to produce a 5.00 A current at 30.0° C. If the temperature of the wire is increased to 300° C, by how much must the potential difference across the wire change in order to maintain the 5.00 A current?

Given:
$L = 2.00$ m = length of the wire
$A = 5.00 \times 10^{-8}\ m^2$ = cross-sectional area of the wire
$I = 5.00$ A = current in the wire
$T_i = 30.0°$ C = initial temperature of the wire
$T_f = 300°$ C = final temperature of the wire
$\rho = 1.67 \times 10^{-8}\ \Omega \bullet m$ = resistivity of copper (Table 24.1)
$\alpha = 3.90 \times 10^{-3}\ /C°$ = temperature coefficient of resistivity of copper (Table 24.2)

Determine: The amount by which the potential difference across the wire must change (ΔV) in order to maintain a 5.00 A current.

Strategy: Knowing the length, cross-sectional area, and type of material, we can calculate the resistance at the initial temperature. Knowing the resistance at the initial temperature, we can determine the resistance at the final temperature. We can determine the initial and final potential difference across the resistor from the current and the initial and final resistance. Finally, knowing the initial and final potential difference, we can determine the amount by which the potential difference must change in order to maintain a 5.00-A current.

Solution: The initial resistance of the wire is

$$R_{30} = \rho L/A = 0.668\ \Omega$$

The final resistance of the wire can be determined as follows:

$$R_{30} = R_0[1 + \alpha(30.0°\ C)] \quad \text{or} \quad R_0 = R_{30} /[1 + \alpha(30.0°\ C)]$$
$$R_{300} = R_0[1 + \alpha(300°\ C)] = R_{30}\ [1 + \alpha(300°\ C)] / [1 + \alpha(30.0°\ C)] = 1.29\ \Omega$$

The initial and final potential difference are

$$V_{30} = IR_{30} = 3.34\ V \quad \text{and} \quad V_{300} = IR_{300} = 6.46\ V$$

Finally, the amount by which the potential difference across the wire must be changed in order to maintain a 5.00 A current is

$$\Delta V = V_{300} - V_{30} = 3.12\ V$$

Related Text Exercises: Exs. 24-7 through 24-23.

3 The Drude Model of a Conductor (Section 24-5)

The Drude model gives the conductivity of a conductor as:

$$\sigma = ne^2\tau/m$$

where:
- n = carrier charge density
- e = electronic charge
- m = mass of charge carrier
- τ = mean time between collisions

This model assumes energy is gained by free electrons in the conductor from the applied electric field. This energy is gained during the time between the collisions that the electrons make with the ions at crystal sites. During these collisions the gained energy is given up to the lattice and equilibrium is reached where all electrons drift at speed V_d.

Practice: Take the resistivity of germanium at room temperature to be 5.00×10^{-1} $\Omega\cdot$m and assume that the mean free time τ for the carriers is 2.44×10^{-14} s.

Determine the following:

1. Formula for charge carrier density using the Drude Model	$n = m/\tau e^2\rho$ where $n = n_e + n_h$
2. Negative charge carrier density	Assume $n_e \cong n_h \cong n/2$, hence $n_e = m/(2\tau e^2\rho)$
3. Value for the negative charge carrier density	$n_e = (9.10 \times 10^{-31})/((2)(2.44 \times 10^{-14}) \cdot (-1.60 \times 10^{-19})^2(5.00 \times 10^{-1}))$ $= 1.46 \times 10^{21}$ carriers/m^3

Related Text Exercises: Exs. 24-34 through 24-37.

4 Resistors in Series and Parallel (Section 24-3)

Review: Figure 24.8 illustrates a series circuit.

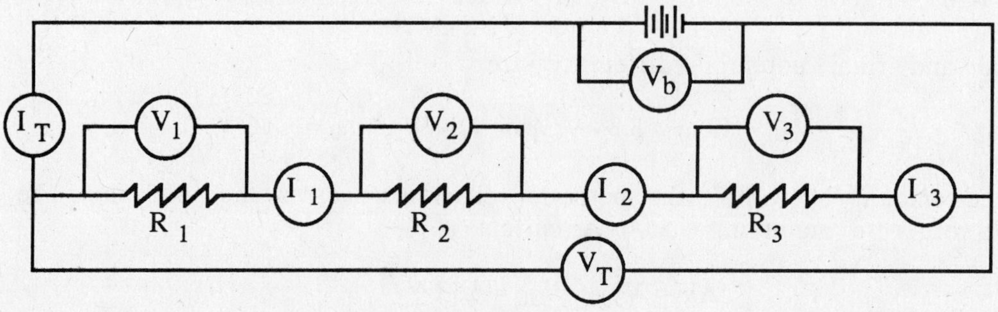

Figure 24.8

For this series circuit, the following is true:

$I_1 = I_2 = I_3 = I_T$	The current through each resistor is the same, and it is equal to the total current in the circuit.
$V_1 + V_2 + V_3 = V_T = V_b$	The sum of the losses in electric potential difference across the resistors is equal to the total loss in electric potential difference and to the gain in electric potential difference from the battery.
$R_1 + R_2 + R_3 = R_S$	Resistors add for a series circuit.
$V_1 = I_1R_1$; $V_2 = I_2R_2$ $V_3 = I_3R_3$; $V_T = I_TR_S$	Ohm's Law can be applied to each resistor individually, to any combination of resistors, and to the entire circuit.

Practice: The following data are given for the circuit in Figure 24.8.

$$I_T = 2.00\text{ A}; \quad V_1 = 2.00\text{ V}; \quad V_2 = 3.00\text{ V}; \quad V_3 = 4.00\text{ V}$$

Determine the following:

1. Current through each resistor	$I_1 = I_2 = I_3 = I_T = 2.00$ A
2. Resistance of each resistor	$R_1 = V_1/I_1 = 1.00\ \Omega$ $R_2 = V_2/I_2 = 1.50\ \Omega$ $R_3 = V_3/I_3 = 2.00\ \Omega$
3. Total potential difference across the three resistors	$V_T = V_1 + V_2 + V_3 = 9.00$ V
4. Potential difference for the battery	$V_b = V_T = 9.00$ V
5. Total resistance for the circuit	$R_S = R_1 + R_2 + R_3 = 4.50\ \Omega$ or $R_S = V_T/I_T = 4.50\ \Omega$

Review: Figure 24.9 illustrates a parallel circuit.

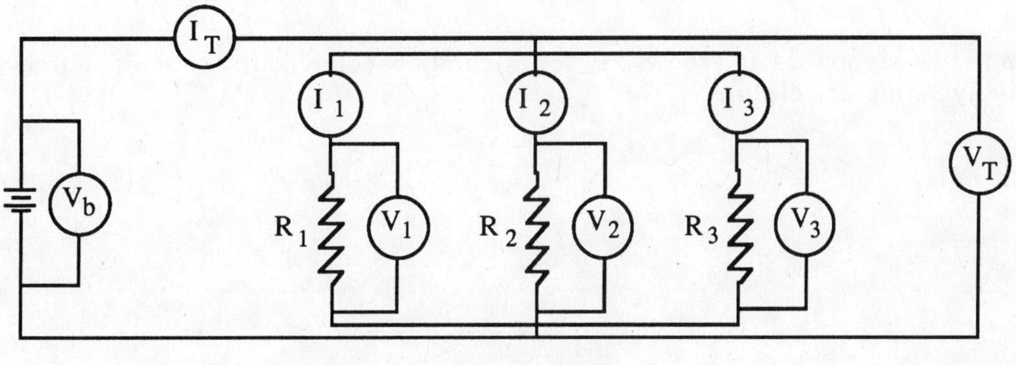

Figure 24.9

For this parallel circuit, the following is true:

$I_1 + I_2 + I_3 = I_T$	The sum of the currents through each branch is equal to the total current in the circuit
$V_1 = V_2 = V_3 = V_T = V_b$	The potential difference is the same across each resistor and is equal to the total potential difference across the combination of resistors and the potential difference of the battery
$\frac{1}{R_1} + \frac{1}{R_2} + \frac{1}{R_3} = \frac{1}{R_p}$	Resistors add in a reciprocal manner for parallel circuits
$V_1 = I_1R_1$; $V_2 = I_2R_2$ $V_3 = I_3R_3$; $V_T = I_TR_T$	Ohm's Law can be applied to each resistor individually, to any combination of resistors, and to the entire circuit.

Practice: The following data are given for the circuit in Figure 24.9.

$V_T = 2.00$ V; $I_T = 2.00$ A; $R_1 = 2.00\ \Omega$; $R_2 = 3.00\ \Omega$

Determine the following:

1. Total resistance of the circuit	$R_T = V_T/I_T = 1.00\ \Omega$
2. Resistance of R_3	$\frac{1}{R_p} = \frac{1}{R_1} + \frac{1}{R_2} + \frac{1}{R_3}$ $\frac{1}{R_3} = \frac{1}{R_p} - \frac{1}{R_1} - \frac{1}{R_2} = \frac{1}{6}$ or $R_3 = 6.00\ \Omega$
3. Potential difference across each resistor	$V_1 = V_2 = V_3 = V_T = 2.00$ V
4. Current through each resistor	$I_1 = V_1/R_1 = 1.00$ A $I_2 = V_2/R_2 = 0.667$ A $I_3 = V_3/R_3 = 0.333$ A Note: $I_1 + I_2 + I_3 = I_T$

Review: Figure 24.10 shows a series-parallel combination circuit redrawn into successively simpler circuits.

(a) Series-parallel combination

$I_2 = I_3$; $I_2 + I_4 = I_1$; $I_1 + I_5 = I_T$
$V_2 + V_3 = V_4$; $V_1 + V_4 = V_5 = V_T = V_b$

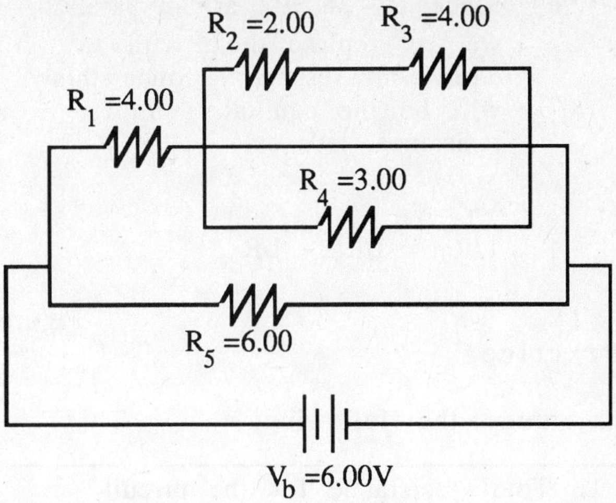

(b) Since R_2 and R_3 are in series, we can replace them with the equivalent resistor R_s

$I_s + I_4 = I_1$; $I_1 + I_5 = I_T$
$V_s = V_4$; $V_1 + V_4 = V_5 = V_T = V_b$
$R_s = R_2 + R_3$

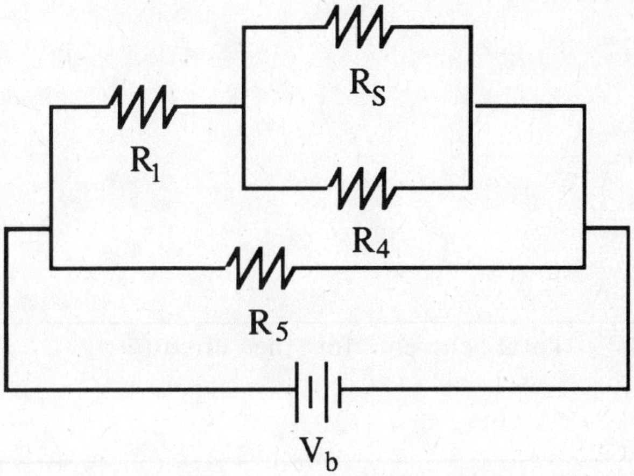

(c) Since R_s and R_4 are in parallel, we can replace them with the equivalent resistor R_p

$I_p = I_1$; $I_1 + I_5 = I_T$
$V_1 + V_p = V_5 = V_T = V_b$
$1/R_p = 1/R_s + 1/R_4$

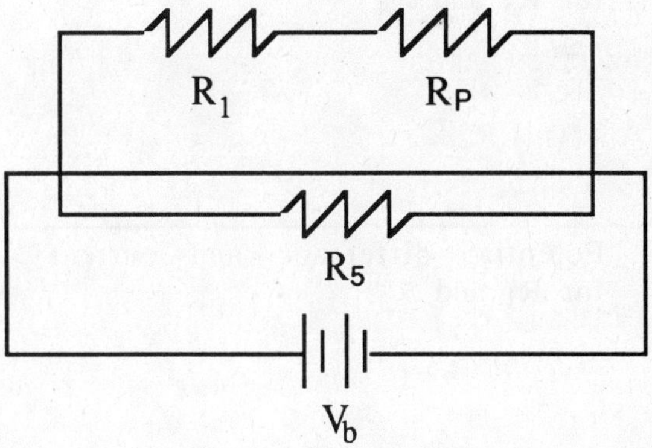

(d) Since R_1 and R_p are in series, we can replace them with the equivalent resistor R_s'

$I_s' + I_5 = I_T$
$V_s' = V_5 = V_T = V_b$
$R_s' = R_1 + R_p$

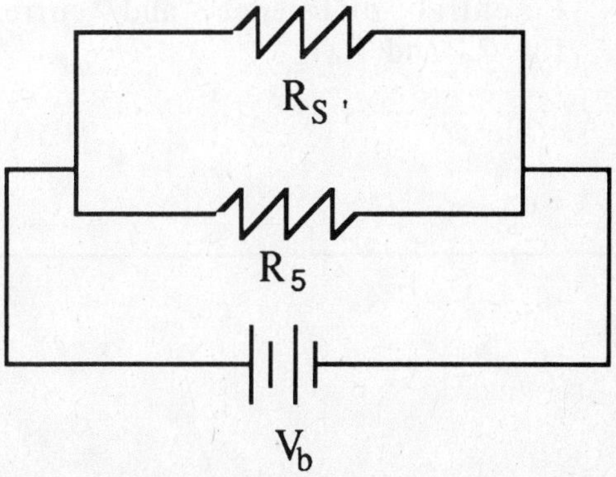

(e) Since R_s' and R_5 are in parallel, we can replace them with an equivalent resistor. Since this will be the equivalent total resistance, let's call it R_T

$V_T = V_b$
$1/R_T = 1/R_s' + 1/R_5$

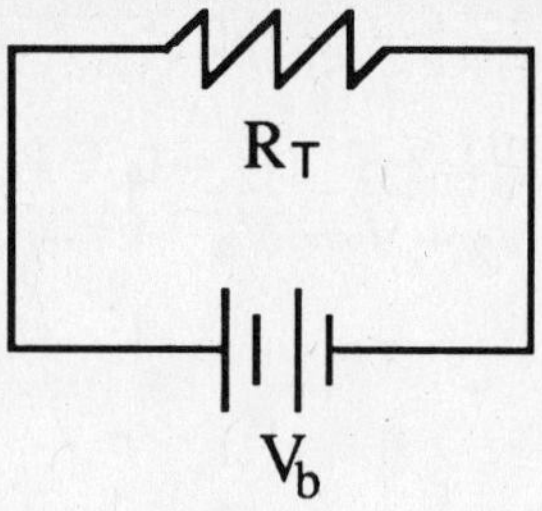

Figure 24.10

Practice:

Determine the following:

1. Total resistance for the circuit	$R_s = R_2 + R_3 = 6.00\ \Omega$ (Figs. 24.10a & b) $\frac{1}{R_p} = \frac{1}{R_s} + \frac{1}{R_4}$ or $R_p = 2.00\ \Omega$ (Figs. 24.10b & c) $R_s' = R_1 + R_p = 6.00\ \Omega$ (Figs. 24.10c & d) $\frac{1}{R_T} = \frac{1}{R_s'} + \frac{1}{R_5}$ or $R_T = 3.00\ \Omega$ (Figs. 24.10d & e)
2. Total current for the circuit	$V_T = V_b = 6.00$ V (Fig. 24.10e) $I_T = V_T/R_T = 2.00$ A
3. Potential difference and current for R_5 and R_s'	$V_s' = V_5 = V_T = V_b = 6.00$ V (Figs. 24.10e & d) $I_s' = V_s'/R_s' = 1.00$ A $I_5 = V_5/R_5 = 1.00$ A As a check, note that $I_s' + I_5 = I_T = 2.00$ A
4. Potential difference and current for R_1 and R_p	$I_1 = I_p = I_s' = 1.00$ A (Figs. 24.10c & d) $V_1 = I_1R_1 = 4.00$ V $V_p = I_pR_p = 2.00$ V As a check, note that $V_1 + V_p = V_5 = V_T = 6.00$ V
5. Potential difference and current for R_s and R_4	$V_s = V_4 = V_p = 2.00$ V (Figs. 24.10b & c) $I_s = V_s/R_s = 0.333$ A $I_4 = V_4/R_4 = 0.667$ A As a check, note that $I_s + I_4 = I_1 = 1.00$ A

6	Potential difference and current for R_2 and R_3	$I_2 = I_3 = I_s = 0.333$ A (Figs. 24.10a & b) $V_2 = I_2R_2 = 0.666$ V $V_3 = I_3R_3 = 1.33$ V As a check, note that $V_2 + V_3 = V_4 = 2.00$ V

The results of analyzing this circuit are summarized in the following.

Potential difference volts	Current amps	Resistance ohms	
$V_1 = 4.00$	$I_1 = 1.00$	$R_1 = 4.00$	Quantities you know or wish to know
$V_2 = 0.666$	$I_2 = 0.333$	$R_2 = 2.00$	
$V_3 = 1.33$	$I_3 = 0.333$	$R_3 = 4.00$	
$V_4 = 2.00$	$I_4 = 0.667$	$R_4 = 3.00$	
$V_5 = 6.00$	$I_5 = 1.00$	$R_5 = 6.00$	
$V_T = 6.00$	$I_T = 2.00$	$R_T = 3.00$	
$V_S = 2.00$	$I_S = 0.333$	$R_S = 6.00$	Intermediate quantities that must be determined
$V_P = 2.00$	$I_P = 1.00$	$R_P = 2.00$	
$V_s' = 6.00$	$I_S' = 1.00$	$R_S' = 6.00$	

Note: When working problems of this type, you will find it convenient to set up a grid like the one above. The advantages of such a grid are as follows:

1. It helps you keep track of what you know and what you need to determine.
2. Ohm's law can be applied to any row. Since $V = IR$, the I column times the R column should equal the V column.
3. You can check to see that resistors in series have the same current.
4. You can check to see that resistors in parallel have the same potential.

Example: Consider the circuit shown in Figure 24.11.

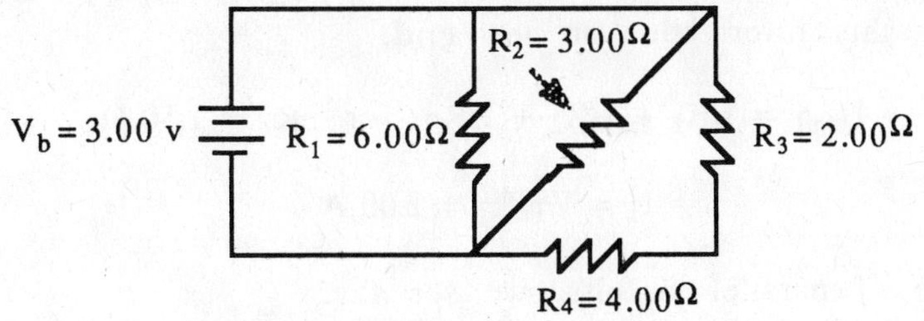

Figure 24.11

Given: $R_1 = 6.00\ \Omega$, $R_2 = 3.00\ \Omega$, $R_3 = 2.00\ \Omega$, $R_4 = 4.00\ \Omega$ and $V_b = 3.00$ V

Determine: The potential difference across each resistor and the current through each resistor.

Strategy: First, let's redraw the circuit so it is easier to see which resistors are in parallel and which resistors are in series. Next, we can determine R_T and I_T. Then, using our knowledge of series and parallel resistors, we can determine the desired quantities. We can also use our knowledge of series and parallel resistors to check our work. In order to keep track of what we know, let's construct a VIR grid.

Solution: The circuit in Figure 24.11 can be redrawn as shown in Figure 24.12.

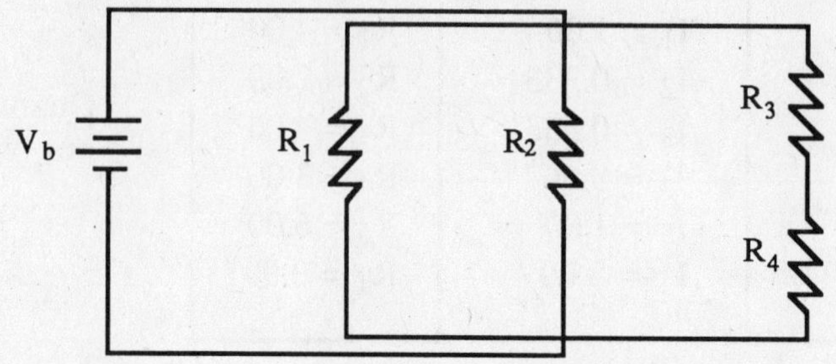

Figure 24.12

Since R_3 and R_4 are in series, we can replace them with the equivalent resistor $R_S = R_3 + R_4 = 6.00\ \Omega$. Enter this information in the grid. Next, we can redraw the circuit as shown in Figure 24.13.

Figure 24.13

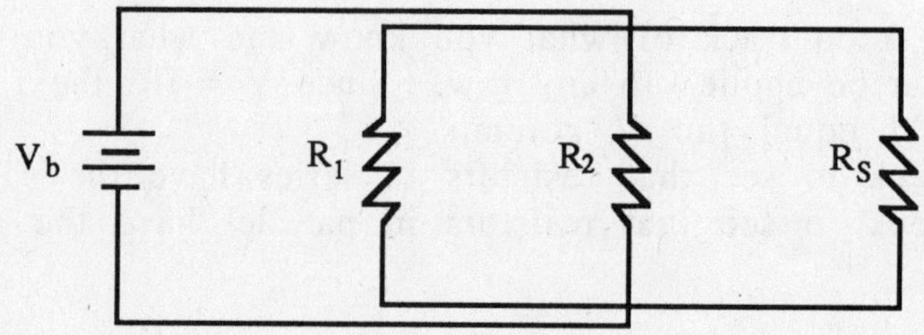

Since R_1, R_2 and R_S are in parallel, we can obtain the equivalent resistance and the total current. Enter this information in the grid.

$$1/R_T = 1/R_1 + 1/R_2 + 1/R_S \quad \text{or} \quad R_T = 1.50\ \Omega$$

$$I_T = V_T/R_T = 2.00 \text{ A}$$

Using our knowledge of parallel circuits, we see that

$V_1 = V_2 = V_S = V_b = 3.00$ V
$I_1 = V_1/R_1 = 0.500$ A, $I_2 = V_2/R_2 = 1.00$ A, $I_S = V_S/R_S = 0.500$ A
As a check, note that $I_1 + I_2 + I_S = I_T$.
Enter this information into the grid.

Using our knowledge of series circuits, we see that

$I_3 = I_4 = I_S = 0.500$ A
$V_3 = I_3R_3 = 1.00$ V, $V_4 = I_4R_4 = 2.00$ V
As a check note that $V_3 + V_4 = V_S = V_b$.
Enter this information in the grid.

At this point the grid should appear as follows:

V (volts)	I (amps)	R (ohms)
$V_1 = 3.00$	$I_1 = 0.500$	$R_1 = 6.00$
$V_2 = 3.00$	$I_2 = 1.00$	$R_2 = 3.00$
$V_3 = 1.00$	$I_3 = 0.500$	$R_3 = 2.00$
$V_4 = 2.00$	$I_4 = 0.500$	$R_4 = 4.00$
$V_b = 6.00$	$I_T = 2.00$	$R_T = 1.50$
$V_S = 2.00$	$I_S = 0.500$	$R_S = 6.00$

Just as a final check, note that

(a) for each row, V = IR,
(b) all resistors in series have the same current,
(c) all resistors in parallel have the same potential difference,
(d) currents for resistors in parallel add to give the total current, and
(e) potential difference for resistors in series add to give the total potential difference.

Related Text Exercises: Exs. 24-24 through 24-31.

5 Ammeters and Voltmeters (Section 24-4)

Review: A galvanometer has a fixed resistance R_m and is designed for a full-scale deflection with a current I_m, hence it can withstand a potential difference of $V_m = I_mR_m$. A galvanometer can be used to measure currents greater than I_m if a shunt resistor R_s is connected in parallel with it. Such a combination is called an ammeter (Figure 24.14).

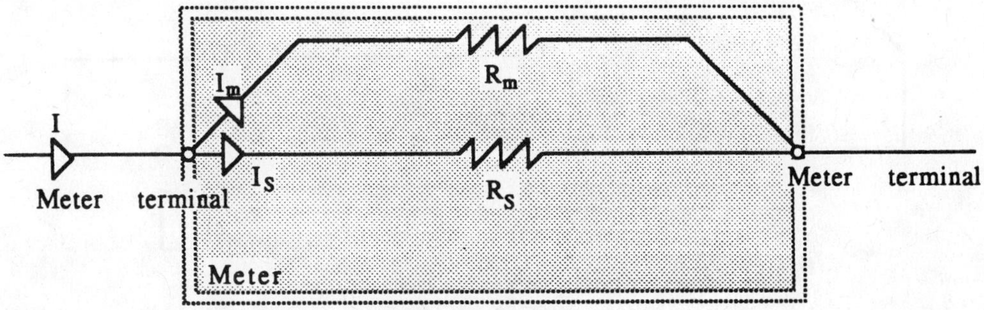

Figure 24.14

Since the resistors are in parallel, we can write

$$I_m + I_s = I$$
$$V_s = V_m \quad \text{but} \quad V_s = I_sR_s \quad \text{and} \quad V_m = I_mR_m \quad \text{(by Ohm's Law)}$$
$$I_sR_s = I_mR_m$$
$$(I - I_m)R_s = I_mR_m$$
$$R_s = I_mR_m/(I - I_m)$$

Generally I_m and R_m are known and we need to determine the size of the shunt resistor R_s needed to provide us with the capability of measuring a current I.

A galvanometer can be used to measure potential difference greater than $V_m = I_mR_m$ if a resistor is placed in series with it. Such a combination is called a voltmeter (Figure 24.15).

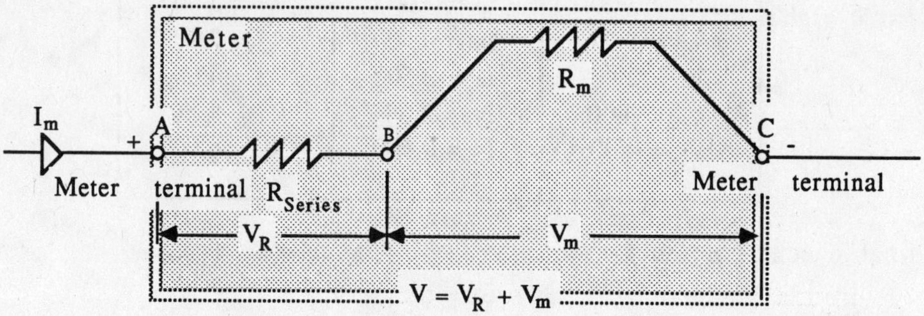

Figure 24.15

Since the resistors are in series, we can write

$$I_R = I_m$$
$$V_R + V_m = V$$
$$I_mR_{series} + I_mR_m = V$$
$$R_{series} = (V - I_mR_m)/I_m$$

Generally I_m and R_m are known, and we need to determine the size of the series resistor R_{series} needed to provide us with the capability of measuring a potential difference V.

A galvanometer can be used to measure resistance, as shown in Figure 24.16.

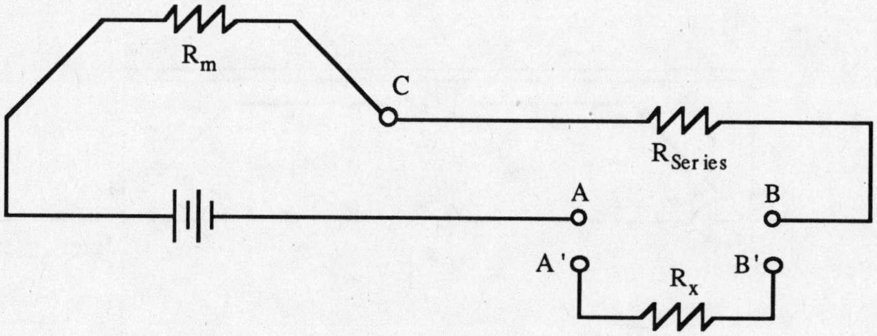

Figure 24.16

The resistor R_{series} is selected so that the galvanometer reads full scale (i.e., $I = I_m$) when A and B are connected, hence zero resistance ($R_x = 0$). When the circuit is open ($R_x = \infty$), there is no current, so the galvanometer reads zero. Thus, an ohmmeter reads backwards.

$R_x = 0$, gives a full-scale deflection

$R_x = \infty$, gives no deflection

For R_x in between these extremes, the meter gives a reading between 0 and full scale. Several known resistors can be used to calibrate the meter.

Practice: A galvanometer has a resistance $R_m = 30.0\ \Omega$ and is designed for a full scale deflection when $I_m = 1.00 \times 10^{-3}$ A.

Determine the following:

1. Shunt resistor needed to make a 0 to 1.00 A ammeter	$R_s = I_m R_m/(I - I_m)$ $R_s = 3.00 \times 10^{-2}\ \Omega$
2. Series resistor needed to make a 0 to 5.00 V voltmeter	$R_{series} = (V - I_m R_m)/I_m$ $R_{series} = 4.97 \times 10^3\ \Omega$
3. The current in the circuit if this galvanometer has a deflection one-half of full scale when $R_s = 6.00 \times 10^{-2}\ \Omega$	When $R_s = 6.00 \times 10^{-2}\ \Omega$, the maximum current the meter can handle is obtained by $R_s = I_m R_m/(I - I_m)$ or $I = I_m[1 + (R_m/R_s)]$ $I = 0.501$ A for a full scale, hence $I = 0.251$ A for a half scale
4. The potential difference being measured if this galvanometer is used with a $2.00 \times 10^3\ \Omega$ series resistor and has a one-fourth scale deflection	When $R_{series} = 2.00 \times 10^3\ \Omega$, the maximum potential difference the meter can handle is obtained by $R_{series} = (V - I_m R_m)/I_m$, or $V = I_m(R_{series} + R_m)$ or $V = 2.03$ V for a full scale, hence $V = 0.508$ V for a one fourth scale
5. The size of the resistor needed to make an ohmmeter if a 9.0 V cell is used with this galvanometer	$\mathcal{E} = I_m(R_m + R)$ or $R = (\mathcal{E}/I_m) - R_m = 8970\ \Omega$

Example: A galvanometer is converted into a 0-5.00 A ammeter with a 0.100-Ω shunt resistor and into a 0-5.00 V voltmeter with a 900 Ω series resistor. Determine the galvanometer characteristics (i.e. $R_m + I_m$)

Given: $R_s = 0.100\ \Omega$ = shunt resistor for a 0-5.00 A ammeter
$R_{series} = 900\ \Omega$ = series resistor for a 0-5.00 V voltmeter
$I = 5.00$ A = maximum current measured by the ammeter
$V = 5.00$ V = maximum potential difference measured by the voltmeter

Determine: I_m and R_m for the galvanometer

Strategy: Knowing R_s, R_{series}, V and I, we can write two equations (one for the voltmeter and one for the ammeter) containing the two unknowns I_m and R_m. These two equations can be solved simultaneously to obtain I_m and R_m.

Solution: For the ammeter and voltmeter, respectively, we can write

(a) $R_s = I_m R_m/(I - I_m)$ and (b) $R_{series} = (V - I_m R_m)/I_m$

Inserting values into (a) and (b) we obtain

(c) $0.100 = I_m R_m/(5.00 - I_m)$ and (d) $900 = (5.00 - I_m R_m)/I_m$

Manipulating the algebra in (c) and (d) we obtain

(e) $0.500 - 0.100I_m = I_m R_m$ and (f) $I_m R_m = 5.00 - 900I_m$

We can equate these expressions for $I_m R_m$, eliminate R_m and solve for I_m.

$0.500 - 0.100I_m = 5.00 - 900I_m$ or $I_m = 5.00 \times 10^{-3}$ A

We can insert I_m into (e) or (f) to obtain R_m

Using (e) we obtain: $0.500 - 0.100I_m = I_m R_m$ or $R_m = (0.500 - 1.00I_m)/I_m = 100\ \Omega$

Using (f) we obtain: $I_m R_m = 5.00 - 900I_m$ or $R_m = (5.00 - 900I_m)/I_m = 100\ \Omega$

Related Text Exercises: Ex. 24-32 and 24-33.

PRACTICE TEST

Take and grade this practice test. Doing so will allow you to determine any weak spots in your understanding of the concepts taught in this chapter. The following section prescribes what you should study further to strengthen your understanding.

Figure 24.17

$R_1 = 5.00\ \Omega$
$R_2 = 3.00\ \Omega$
$R_3 = 6.00\ \Omega$
$R_4 = 3.00\ \Omega$
$V_b = 2.50\ \Omega$

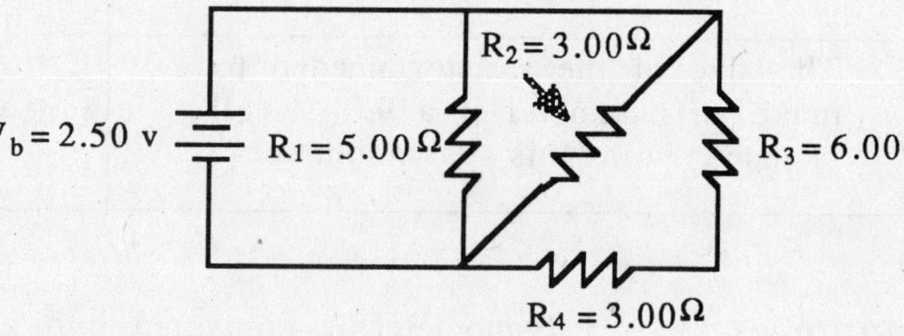

For the circuit shown in Figure 24.17, determine the following:

_______ 1. Total resistance for the circuit
_______ 2. Current through R_1
_______ 3. Current through R_4
_______ 4. Potential difference across R_2

A heating coil designed to be used with a standard 120 V wall outlet has the following characteristics:

$\rho = 1.00 \times 10^{-6}\ \Omega\cdot m$ at 25.0°C $\alpha = 5.00 \times 10^{-4}/°C$ $A = 2.00 \times 10^{-4}\ m^2$

If the current to the coil is monitored, one finds that when the coil is at room temperature (25.0°C), the current to the coil is 1.00 A and that after the coil has been on for a long time, the final current is 0.900 A.

Determine the following:

_________ 5. Resistance of the coil at 25.0°C
_________ 6. Length of the wire used in the coil
_________ 7. Resistance of the coil at 50.0°C
_________ 8. Resistivity of the coil material at -10.0°C
_________ 9. Resistance of the coil when it has been on for a long time
_________ 10. Final temperature of the coil

A galvanometer has a 30.0-Ω resistance and is designed for a full scale deflection at 1.00×10^{-3} A. Determine the following:

_________ 11. Shunt resistor needed to make 0-1 A ammeter
_________ 12. Series resistor needed to make a 0-5 V voltmeter
_________ 13. Current in a circuit if this galvanometer is used with a 6.00×10^{-2} Ω shunt resistor as an ammeter and it has a one half scale deflection.

A 10 gauge copper wire (radius 1.30 mm) carries a current of 20.0 A. Assume one free electron per copper atom, with mass density $\rho = 8.95 \times 10^3$ kg/m^3, molecular weight M = 63.5 g/mol, and a resistivity of 1.67×10^{-8} Ω•m. The mass of an electron is 9.11×10^{-31} kg. Determine the following:

_________ 14. The number density of free electrons
_________ 15. The drift speed of the free electrons
_________ 16. The mean free time

Consider a salt solution in a long insulating tube of inner radius 10.0 mm carrying a current of 2.00 A along the tube axis. The carriers in the solutions are single charged positive and negative ions with equal number densities: $n_+ = n_- = 6.00 \times 10^{25}$ ions/m^3. Assume the drift speed of the positive ions is twice that of the negative ions. Determine the following:

_________ 17. The current due to the positive ions
_________ 18. The current due to the negative ions
_________ 19. The drift speed of the positive ions
_________ 20. The drift speed of the negative ions
_________ 21. The contribution to uniform j of the positive ion
_________ 22. The contribution to uniform j of the negative ion

Assume the number density of the free electrons n_e and the number density of holes n_h in pure silicon is about 2.80×10^{18} carriers/m^3 at 20.0°C for each type of carrier. Assume the free electrons and holes have the same mass and mean free time. The resistivity of silicon is 4.30×10^3 Ω•m. Determine:

_________ 23. The mean free time for these carriers in silicon

(See Appendix I for answers.)

PRINCIPAL CONCEPTS AND EQUATIONS PRESCRIPTION

Your score on the practice test is an excellent measure of your understanding of the chapter. You should now use the following chart to write your own prescription for dealing with any weaknesses the practice test points out. Look down the leftmost column to the number of the question(s) you answered incorrectly, reading across that row you will find the concept and/or equation of concern, the section(s) of the study guide you should return to for further study, and some suggested text exercises which you should work to gain additional experience.

Practice Test Questions	Concepts and Equations	Prescription	
		Principal Concepts	Text Exercises
1	Equivalent resistance	4	24-24, 25
2	$V = IR$; $V_1 = V_2 = V_{34} = V_b$	2	24-24, 25
3	$V = IR$; $V_1 = V_2 = V_{34} = V_b$	2	24-24, 25
4	$V = IR$; $V_1 = V_2 = V_{34} = V_b$	2	24-24, 25
5	$R = V/I$	2	24-8
6	$R = \rho L/A$	2	24-11, 16
7	$R = R_0[1 + \alpha(T-T_0)]$	2	24-19, 20
8	$R = R_0[1 + \alpha(T-T_0)]$	2	24-19, 20
9	$R = V/I$	2	24-8
10	$R = R_0[1 + \alpha(T-T_0)]$	2	24-21
11	Ammeters	5	24-32, 33
12	Voltmeters	5	24-32, 33
13	Ammeters	5	24-32, 33
14	$n = N_A \rho_M/M$	1	24-5, 6
15	$I = n\lvert q\rvert Av_d$	1	24-4, 6
16	$\sigma = ne^2\tau/m$ and $\rho = 1/\sigma$	3, 2	24-34
17	$I = n_A\lvert q_A\rvert Av_{dA} + n_b\lvert q_b\rvert Av_{db}$	1	24-5
18	$I = n_A\lvert q_A\rvert Av_{dA} + n_b\lvert q_b\rvert Av_{db}$	1	24-5
19	$I = n_A\lvert q_A\rvert Av_{dA} + n_b\lvert q_b\rvert Av_{db}$	1	24-5
20	$I = n_A\lvert q_A\rvert Av_{dA} + n_b\lvert q_b\rvert Av_{db}$	1	24-5
21	$j = I/A$	1	24-5
22	$j = I/A$	1	24-5
23	$\sigma = ne^2\tau/m$; $\sigma = 1/\rho$	3	24-10

25 ENERGY AND CURRENT IN DC CIRCUITS

RECALL FROM PREVIOUS CHAPTERS

Previously learned concepts and equations frequently used in this chapter	Text Section	Study Guide Page
Potential difference: $V = U/q_0$	22-3	23-3
Current: $I = dQ/dt$	24-1	24-2
Resistance: $R = V/I$	24-2	24-5
Resistors in series and parallel: $R_s = \sum_{i=1}^{n} R_i$ $\quad 1/R_p = \sum_{i=1}^{n} (1/R_i)$	24-5	24-11

NEW IDEAS IN THIS CHAPTER

Concepts and equations introduced	Text Section	Study Guide Page
Emf and internal resistance of a battery	25-1	25-1
Electrical energy and power: battery $P_0 = \varepsilon I$; resistor $P_R = VI = I^2R = V^2/R$	25-2	25-4
Kirchoff's rules	25-3	25-7
R-C series circuits: charging capacitor $Q(t) = C\varepsilon(1 - e^{-t/\tau})$; discharging capacitor $Q(t) = Q_0 e^{-t/\tau}$; time constant $\tau = RC$	25-4	25-12

PRINCIPAL CONCEPTS AND EQUATIONS

1 Emf, Terminal Potential Difference and Internal Resistance of a Battery (Section 25-1)

Review: The emf of a battery (or other source of energy) is the chemical energy (or other form) transferred to each coulomb of charge passing through the battery from the negative to positive terminal (conventional current moves from the positive to negative terminal in the external circuit). The emf is specified in volts (joules/coulomb). Some of the energy is used in moving through the battery due to an internal resistance r and the remainder is used in the external circuit. The terminal voltage V_b of the battery is the emf ε less the voltage drop due to the internal resistance r. This may be stated algebraically as:

$$V_b = \varepsilon - ir.$$

The above equation describes the case where the battery is supplying energy. When the battery is being charged, the current moves through the battery from the "+" terminal to the "-" terminal. For this case, V_b, ε and r are related by:

$$V_b = \varepsilon + ir.$$

These ideas are illustrated in Figure 25.1 and the discussion that follows.

Notation:

ε = emf of the cell
R = external resistance
r = internal resistance
I = circuit current measured by A
V_R = potential difference across R
V_b = potential difference across the cell terminals

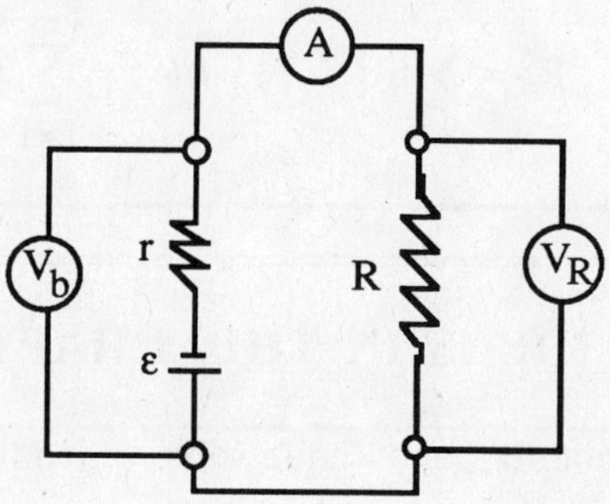

Figure 25.1

The cell does work on the charge carriers as they move through it from terminal to terminal. The energy given to the charge carriers by the cell is used up as they work their way through the external resistance R and the internal resistance r. If ε represents the emf of the cell or the energy given to each charge carrier by the cell, V_R represents the potential difference across the external resistance or the energy loss by each charge carrier as it passes through R, and V_r represents the energy loss by each charge carrier due to the internal resistance of the cell, then we can write

$$\varepsilon = V_R + V_r$$

This says that the energy supplied to the charge carriers by the cell is consumed as they go through circuit resistance (external and internal). Using Ohm's law, we can write V_R = IR and V_r = Ir.

The energy loss by the wires and ammeter in Figure 25.1 is negligible, hence

$$V_b = V_R$$

Combining the preceding equations we obtain an expression for the potential difference across the terminals of the cell when it is delivering a current I.

$$V_b = \varepsilon - Ir$$

In order to charge the cell, we must force a current (i.e., charge carriers) through the cell in the reverse direction. To do this, we must give the charge carriers enough energy to overcome the work done on them by the cell (ε) and the energy loss due to the internal resistance ($V_r = Ir$). Hence, the amount of energy we must supply each charge carrier, or the potential difference we must put across the terminals of the cell to be charged, is

$$V_b = \varepsilon + Ir$$

Practice: The same cell is connected across two different resistors, with voltmeter and ammeter readings as shown in Figure 25.2.

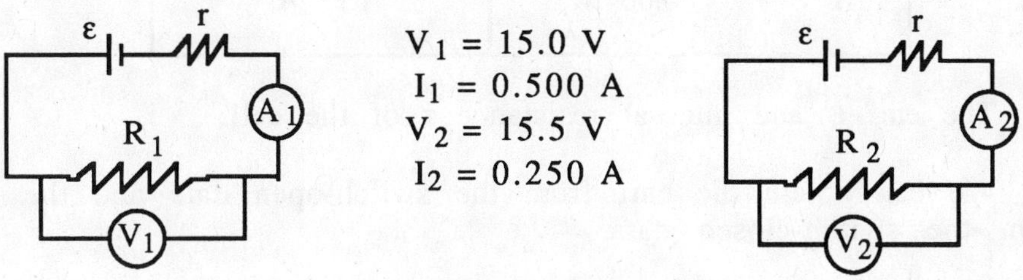

Figure 25.2

Determine the following:

1. Resistance of R_1 and R_2	$R_1 = V_1/I_1 = 30.0\ \Omega$ $R_2 = V_2/I_2 = 62.0\ \Omega$
2. Internal resistance of the cell	$\varepsilon = V_1 + I_1 r$ and $\varepsilon = V_2 + I_2 r$ Since it is the same cell, we can equate these expressions for ε. $V_1 + I_1 r = V_2 + I_2 r$ or $r = (V_2 - V_1)/(I_1 - I_2) = 2.00\ \Omega$
3. Emf of the cell	$\varepsilon = V_1 + I_1 r = 16.0$ V or $\varepsilon = V_2 + I_2 r = 16.0$ V
4. Potential difference needed to charge the cell at 1.00 A	$V_b = \varepsilon + Ir = 18.0$ V

Example: When switch S is open, the voltmeter V, connected across the terminals of the dry cell in Figure 25.3, reads 1.48 V. When the switch is closed, the voltmeter reading drops to 1.33 V, and the ammeter A reads 1.25 A. Find the emf and internal resistance of the cell. Assume the meters have no effect on the circuit.

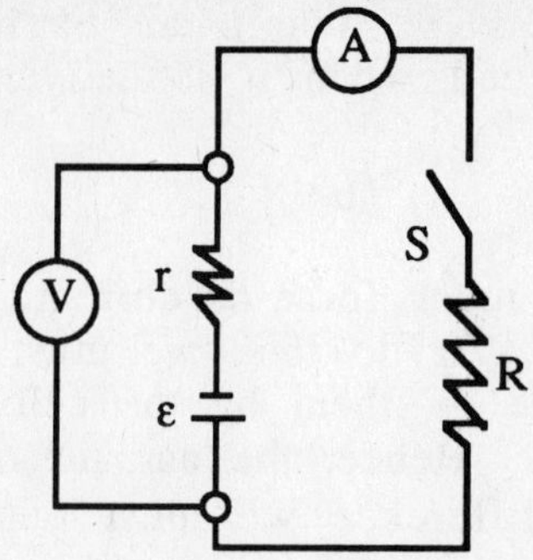

Figure 25.3

Given:

	Switch Open	Switch Closed
V	1.48 V	1.33 V
I	0.00 A	1.25 A

Determine: The emf ε and internal resistance r of the cell.

Strategy: We can obtain the emf from the switch-open data and the internal resistance from the switch-closed data.

Solution: When the switch is open, $I = 0.00$ A and $\varepsilon = V + Ir = V = 1.48$ V.

When the switch is closed, $I = 1.25$ A and $\varepsilon = V + Ir$ or $r = (\varepsilon - V)/I = 0.12\ \Omega$

Related Text Exercises: Ex. 25-1 through 25-5.

2 Energy and Power In Electric Circuits (Section 25-2)

Review: Figure 25.4 shows a simple circuit consisting of a cell, resistor, switch, voltmeter and ammeter.

Notation:

- ε = emf of the cell
- r = internal resistance of the cell
- I = current measured by A
- V = potential difference measured by V
- R = external resistance
- V_R = potential difference across R
- V_{AB} = terminal potential difference of cell
- S = switch

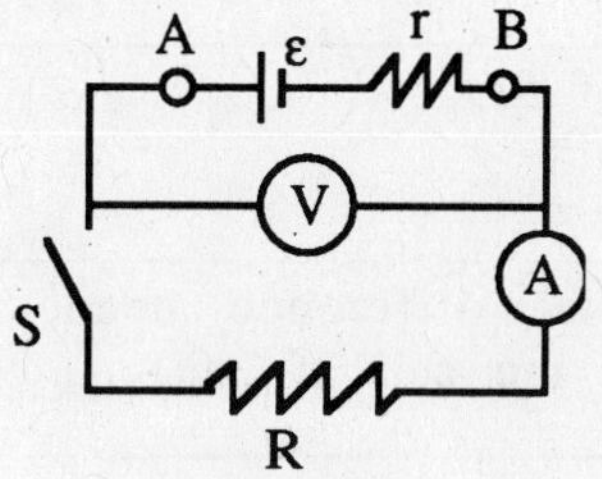

Figure 25.4

Using the given notation, we can write:

$P_\varepsilon = I\varepsilon$ = power delivered by the emf
$P_r = I^2r$ = power consumed by the internal resistance of the cell
$P_R = I^2R = IV_R = V_R{}^2/R$ = power consumed by the external resistor
$P_\varepsilon = P_r + P_R$ = the rate at which energy is supplied by the cell which is also equal to the rate at which energy is consumed by r and R.
$P_0 = P_\varepsilon - P_r = \varepsilon I - I^2r$ = the power output (to the external circuit) of the battery

Practice: The data below were collected with the circuit shown in Figure 25-4.

	Switch Open	Switch Closed
V	3.00 V	2.50 V
I	0.00 A	1.00 A

Determine the following:

1. Emf of the cell	$\varepsilon = V_{AB} + Ir$ When the switch is open, I = 0 and $\varepsilon = V_{AB} = V = 3.00$ V
2. Internal resistance of the cell	$\varepsilon = V_{AB} + Ir$ $r = (\varepsilon - V_{AB})/I = 0.500\ \Omega$
3. External resistance	$R = V_R/I = V/I = 2.50\ \Omega$
4. Energy supplied by cell in 100 s	$\Delta U_c = \varepsilon\Delta q = \varepsilon I\Delta t = 300$ J
5. Power supplied by the cell	$P_c = \Delta U_c/\Delta t = 3.00$ W or $P_c = \varepsilon I = 3.00$ W
6. Power consumed by R in 100 s	$\Delta U_R = V_R\Delta q = V_R I\Delta t = 250$ J or $\Delta U_R = I^2R\Delta t = 250$ J
7. Power consumed by R	$P_R = \Delta U_R/\Delta t = 2.50$ W or $P_R = I^2R = 2.50$ W $P_R = V_R I = 2.50$ W
8. Energy consumed by r in 100 s	$\Delta U_r = (\varepsilon - V)\Delta q = (\varepsilon - V)I\Delta t = 50.0$ J or $\Delta U_r = (Ir)I\Delta t = I^2\Delta t = 50.0$ J
9. Power consumed by r	$P_r = \Delta U_r/\Delta t = 0.500$ W or $P_r = (\varepsilon - V)I = 0.500$ W or $P_r = I^2r = 0.500$ W

10. Total power consumed by R and r	$P_R = 2.50$ W (Step 7) $P_r = 0.500$ W (Step 9) $P_T = P_R + P_r = 3.00$ W Note that P_T consumed is equal to the power supplied by the cell (Step 5).
11. Efficiency of the cell	ε = P(available for use)/P_T $\varepsilon = P_R/P_T = 0.833$

Example: Household circuits operate at 120 V and are commonly fused with 15-A fuses. A bathroom circuit consists of a single-bulb light fixture and a wall outlet. The wire for the circuit is 15.0 m of 18-gauge (1.02×10^{-3} m diameter) copper wire. (a) Determine the maximum wattage light bulb we can use and still run a 1650 W hair dryer without melting a fuse. (b) If we buy energy at a rate of 10¢ per kilowatt-hour (kW•h), how much does it cost us to run the hair dryer and a 60.0-W light for 0.250 h?

Given:

$V = 120$ Volts
$I = 15.0$ A
$L = 15.0$ m

$A = \pi r^2 = 8.17 \times 10^{-7}\ m^2$
$P_{dryer} = 1650$ W
$\rho = 1.67 \times 10^{-8}\ \Omega \cdot m$
$C = 10¢/kW \cdot h$

Determine: (a) the maximum wattage light bulb we can use and not burn a fuse. (b) The cost to run the hair dryer and a 60.0-W bulb for 0.250 h.

Strategy: (a) From the given information, we can determine the resistance of the circuit wire and hence the power consumed by the wire when the current is 15.0 A. Knowing I and V, we can determine the maximum power available. Knowing the maximum power available, the power consumed by the circuit wire, and the power consumed by the hair dryer, we can determine the maximum wattage light bulb we can use and not burn a fuse. (b) knowing expressions for the total power supplied and the power loss to the wiring and values for the power consumed by the dryer and the bulb, we can determine the current in the circuit when a 60-W bulb is used. Knowing the current and the line voltage we can determine the total power supplied to the circuit (wiring, dryer, and bulb). Knowing the total power supplied, the time it is supplied, and the cost of energy, we can determine the cost to run the dryer and 60-W bulb for 0.250 h.

Solution:

(a)

$R_{wire} = \rho L/A = 3.07 \times 10^{-1}\ \Omega$
$P_{wire} = I^2 R_{wire} = 69.1$ W
$P_{dryer} = 1650$ W
$P_{total} = IV = 1.8 \times 10^3$ W
$P_{light} = P_{total} - P_{dryer} - P_{wire} = 80.9$ W

(b) $P_{total} = IV \qquad P_{wire} = I^2R_{wire} \qquad P_{dryer} = 1650\ W \qquad P_{light} = 60\ W$

$$P_{total} = P_{wire} + P_{dryer} + P_{light}$$
$$IV = I^2R_{wire} + P_{dryer} + P_{light}$$
$$I(120\ V) = I^2(3.07 \times 10^{-1}) + 1650\ W + 60\ W$$

When this quadratic is solved for I we obtain

$$I = 14.8\ A$$
$$P_{total} = IV = (14.8\ A)(120\ V) = 1.78 \times 10^3\ W$$
$$E_{total} = P_{total}\Delta t = (1.78 \times 10^3\ W)(0.250\ h) = 445\ W{\bullet}h$$
$$\text{Cost} = CE_{total} = (10¢/kW{\bullet}h)(445\ W{\bullet}h)(kW/1000\ W) = 4.45¢$$

Related Text Exercises: Ex. 25-6 through 25-18.

3 Kirchoff's Rules (Section 25-3)

Review: Kirchoff's rules can be used to solve circuit problems that cannot be solved by the direct application of Ohm's law. Such a circuit is shown in Figure 25.5

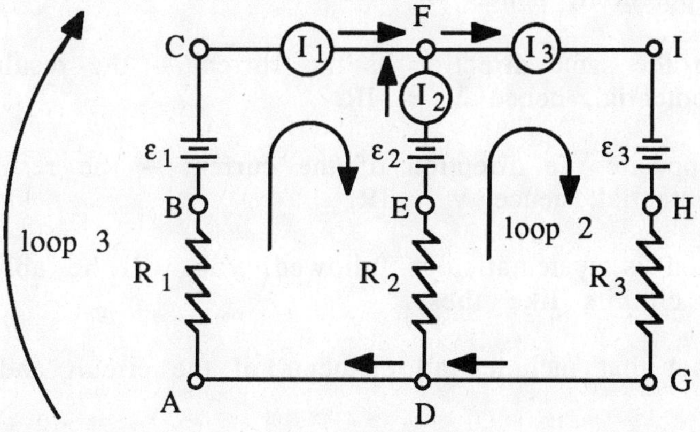

Figure 25.5

Kirchoff's first rule states that the algebraic sum of the currents into any branch point is equal to the sum of the currents away from the branch point. Mathematically, we can write

$$I_1 + I_2 = I_3 \qquad \text{for point F}$$

$$I_3 = I_1 + I_2 \qquad \text{for point D}$$

Notice that the previous two equations are identical. Even though we have two points, we can only have one point equation. In general, n points will result in n-1 different point equations.

Kirchoff's second rule states that the algebraic sum of the voltages in traversing a closed loop is zero. Mathematically, we write

$$\sum_i V_i = \varepsilon_1 - \varepsilon_2 + I_2R_2 - I_1R_1 = 0 \qquad \text{for loop \#1}$$

$$\sum_i V_i = \varepsilon_2 + \varepsilon_3 - I_3R_3 - I_2R_2 = 0 \qquad \text{for loop \#2}$$

$$\sum_i V_i = \varepsilon_1 + \varepsilon_3 - I_3R_3 - I_1R_1 = 0 \qquad \text{for loop \#3}$$

Notice that the previous three equations are not independent. The sum of the first two loop equations is equal to the third loop equation. Even though we have only two useful equations, the third one serves as a simple check on our work.

If the electric potential increases when traversing a circuit element, use a "+" sign. If it decreases use a "-" sign. This results in the following sign convention.

1. Traverse an emf from the negative to the positive terminal -- this results in an increase in electric potential, hence $+\varepsilon$.

2. Traverse an emf from the positive to the negative terminal -- this results in a decrease in electric potential, hence $-\varepsilon$.

3. Traverse a resistor in the same direction as the current -- the result is a decrease in electric potential, hence $V = -IR$.

4. Traverse a resistor opposite the direction of the current -- the result is an increase in electric potential, hence $V = IR$.

If the following prescription is systematically followed, you will be able to routinely solve problems involving circuits like these.

1. Draw a circuit diagram that includes all elements of the circuit and label all known quantities.

2. Assign a current direction to each branch. Choose a different symbol for each separate current.

3. Apply Kirchoff's rule to the branch points that yield independent equations.

4. Select enough closed loops from the circuit so that, including the equations from step 3, you have as many independent equations as currents. Be sure that every element in the circuit is included in at least one loop. By choosing *all* of the *small* loops you will automatically obtain independent equations.

5. Apply Kirchoff's loop rule to the closed loops selected in step 4, and obtain the rest of the needed independent equations.

6. Rewrite and organize all equations in a convenient form.

7. Solve the equations in step 6 as simultaneous equations.

Practice: Suppose that the following information is given for the circuit in Figure 25.5.

$\varepsilon_1 = 8.00\ V$ $\varepsilon_2 = 4.00\ V$ $\varepsilon_3 = 6.00\ V$ $R_1 = R_2 = R_3 = 2.00\ \Omega$

Determine the following:

1. The point equations for F and D	F: $I_1 + I_2 = I_3$ D: $I_3 = I_1 + I_2$ **Note:** The second point equation gives us no new information.
2. Loop equations for loops 1, 2 and 3	1: $\varepsilon_1 - \varepsilon_2 + I_2R_2 - I_1R_1 = 0$ 2: $\varepsilon_2 + \varepsilon_3 - I_3R_3 - I_2R_2 = 0$ 3: $\varepsilon_1 + \varepsilon_3 - I_3R_3 - I_1R_1 = 0$

Note: The third loop equation gives no new information. We can obtain the third loop equation by adding the first two.

3. Three equations which can be solved simultaneously to obtain I_1, I_2 and I_3	(a) $I_1 + I_2 = I_3$ (b) $\varepsilon_1 - \varepsilon_2 + I_2R_2 - I_1R_1 = 0$ (c) $\varepsilon_2 + \varepsilon_3 - I_3R_3 - I_2R_2 = 0$ **Note:** The point equation and any two of the three loop equations will work.
4. Obtain I_1, I_2 and I_3	(a) $I_1 + I_2 = I_3$ (b) $4.00\ V = 2I_1 - 2I_2$ or (b') $2\ V = I_1 - I_2$ (c) $10.0\ V = 2I_2 + 2I_3$ or (c') $5\ V = I_2 + I_3$ Insert I_3 from (a) into (c) to obtain (d) $5\ V = 2I_2 + I_1$ Subtract (b') from (d) to obtain $I_2 = 1.00\ A$ Insert I_2 into (c') to obtain $I_3 = 4.00\ A$ Insert I_2 and I_3 into (a) to obtain $I_1 = 3.00\ A$
5. Determine the electrical potential difference V_{CA}	Let's travel from A to C with a positive charge. We go through R_1 in the same direction as the current, hence B is at a lower potential than A. In fact, $V_{BA} = -I_1R_1 = -6.00\ V$. We go through ε_1 such that we gain potential. In fact, $V_{CB} = +8.00\ V$. Putting this together, we have $V_{CA} = V_{BA} + V_{CB} = -I_1R_1 + \varepsilon_1 = +2.00\ V$.

6. Determine the electrical potential difference V_{FD}	$V_{ED} = -I_2R_2 = -2.00\text{ V}$ $V_{FE} = +\mathcal{E}_2 = +4.00\text{ V}$ $V_{FD} = V_{ED} + V_{FE} = -I_2R_2 + \mathcal{E}_2 = +2.00\text{ V}$
7. Determine the electrical potential difference V_{IG}	$V_{HG} = +I_3R_3 = +8.00\text{ V}$ $V_{IH} = -\mathcal{E}_3 = -6.00\text{ V}$ $V_{IG} = V_{HG} + V_{GH} = I_3R_3 - \mathcal{E}_3 = +2.00\text{ V}$

Note: The three branches A to C, D to F, and G to I are in parallel; consequently, V_{CA}, V_{FD}, and V_{IG} should be equal. From steps 5, 6, and 7 of the preceeding, we see that this is indeed true. If any one of the currents were incorrect, we would not get $V_{CA} = V_{FD} = V_{IG}$. This simple check allows us to be certain we have correctly determined I_1, I_2, and I_3.

Example: For the circuit shown in Figure 25.6, find the current in each resistor.

$\mathcal{E}_1$ = 2.00 V R_1 = 2.00
Ω
$\mathcal{E}_2$ = 4.00 V R_2 = 4.00
Ω
$\mathcal{E}_3$ = 6.00 V R_3 = 6.00
Ω
$\mathcal{E}_4$ = 10.0 V

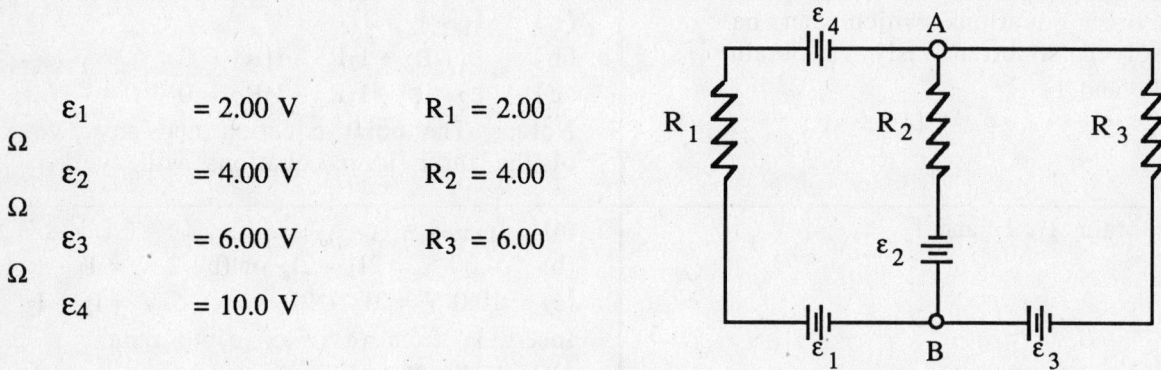

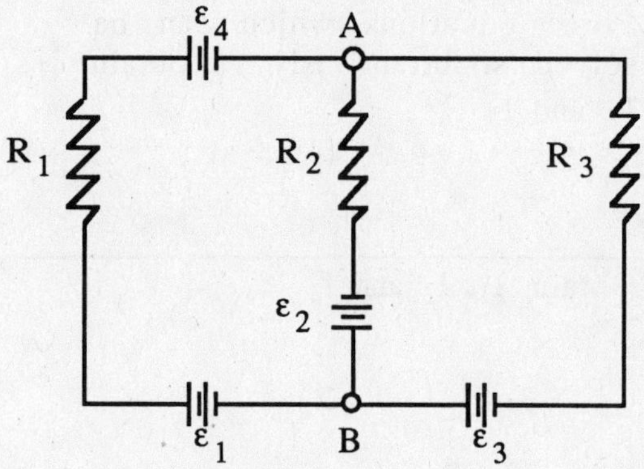

Figure 25.6

Given: $\mathcal{E}_1 = 2.00\text{ V}$ $\mathcal{E}_2 = 4.00\text{ V}$ $\mathcal{E}_3 = 6.00\text{ V}$ $\mathcal{E}_4 = 10.0\text{ V}$
$R_1 = 2.00\ \Omega$ $R_2 = 4.00\ \Omega$ $R_3 = 6.00\ \Omega$

Determine: The current in each resistor

Strategy: Assume a direction for the current in each branch and then write the point equations for A or B and two independent loop equations. These three equations contain the three unknown currents and hence can be solved simultaneously for the currents.

Solution: Assume the direction for the current in each branch as shown in Figure 25.7

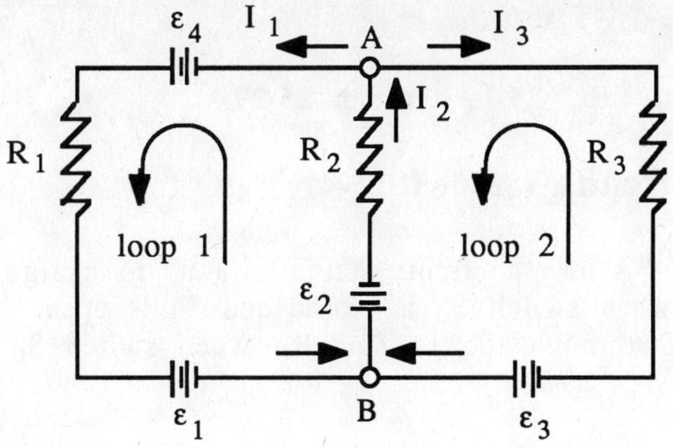

Figure 25.7

The point equation for B is

(a) $I_1 + I_3 = I_2$

The loop equations for loops 1 and 2, respectively are

(b) $\varepsilon_4 - I_1R_1 - \varepsilon_1 - \varepsilon_2 - I_2R_2 = 0$

(c) $\varepsilon_2 - \varepsilon_3 + I_3R_3 + I_2R_2 = 0$

Insert values into (b) and (c) to obtain

(b') $I_1 + 2I_2 = 2$

(c') $3I_3 + 2I_2 = 1$

Solve (a) for I_1 and substitute it into (b') to obtain

(b'') $-I_3 + 3I_2 = 2$

Multiply (b'') by 3 to obtain

(b''') $-3I_3 + 9I_2 = 6$

Add (b''') and (c') to eliminate I_3 and obtain I_2

$11I_2 = 7$ or $I_2 = 0.636$ A

Substitute I_2 into (b') and (c') to obtain, respectively, I_1 and I_3

From (b') $I_1 = 2 - 2I_2 = 0.728$ A
From (c') $I_3 = (1 - 2I_1)/3 = -0.0907$ A

As a check, substitute I_1 and I_3 into (a) to see if the same value for I_2 is obtained.

From (a) $I_2 = I_1 + I_3 = 0.637$ A

Related Text Exercises: Ex. 25-19 through 25-29.

4 RC - Series Circuits (Section 25-4)

Review: Figure 25.8 shows a circuit that allows us to charge a capacitor C through the resistor R_c when switch S_c is closed and S_d is open. We can then discharge the capacitor C through the resistor R_d when switch S_c is opened and S_d is closed.

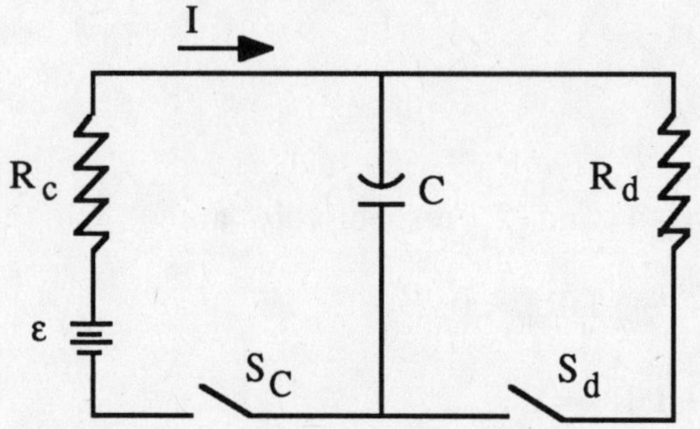

Figure 25.8

Charging the capacitor (S_c closed and S_d open): As the capacitor is charged, the charge on the capacitor (q), the current in the circuit (I), the potential difference across the resistor (V_{R_c}), and the potential difference across the capacitor (V_c) all vary in time as follows:

(a) $q = C\varepsilon(1 - e^{-t/R_cC}) = q_{max}(1 - e^{-t/R_cC})$, where $q_{max} = C\varepsilon$

Equation (a) and Figure 25.9 show how the charge on the capacitor varies in time.

$q = 0$ at $t = 0$
$q = q_{max}(1 - e^{-1}) = 0.63q_{max}$
at $t = R_cC = \tau_{c_c}$
$q \to q_{max}$ as $t \to \infty$

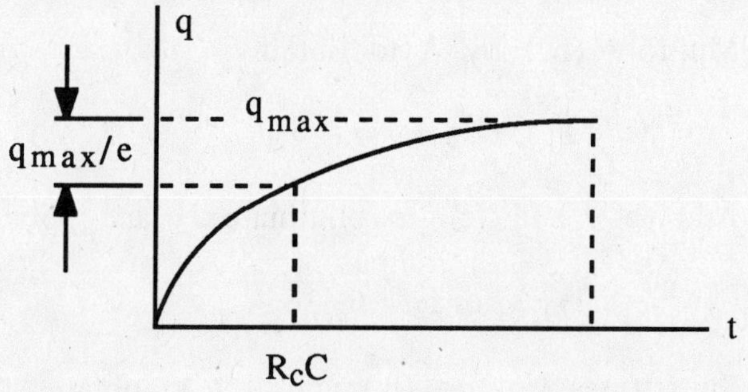

Figure 25.9

Using $I = dQ/dt$, we obtain

(b) $I = (C\varepsilon/R_cC)e^{-t/R_cC} = (\varepsilon/R_c)e^{-t/R_cC} = I_{max}e^{-t/R_cC}$

Equation (b) and Figure 25.10 show how the current in the circuit varies in time.

$I = I_{max} = \varepsilon/R_c$ at $t = 0$
$I = I_{max}/e = 0.37 I_{max}$
at $t = R_cC = \tau_{cc}$
$I \to 0$ as $t \to \infty$

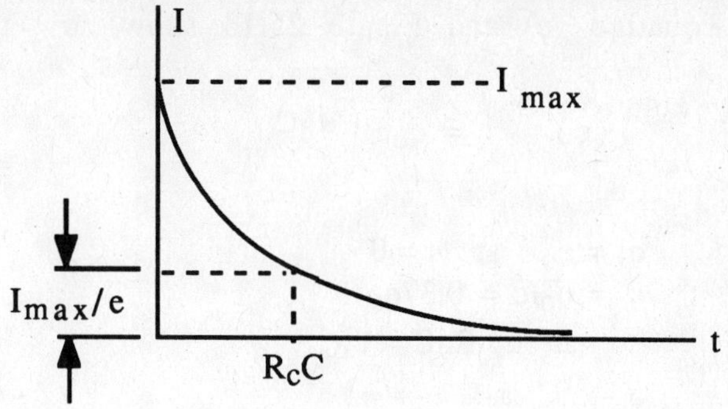

Figure 25.10

(c) $$V_{R_c} = IR_c = \frac{\varepsilon}{R_c} e^{-t/R_cC} R_c = \varepsilon e^{-t/R_cC}$$

Equation (c) and Figure 25.11 show how the potential difference across the resistor R_c varies in time.

$V_{R_c} = \varepsilon$ at $t = 0$
$V_{R_c} = \varepsilon/e = 0.37\varepsilon$
at $t = R_cC = \tau_{cc}$
$V_{R_c} \to 0$ as $t \to \infty$

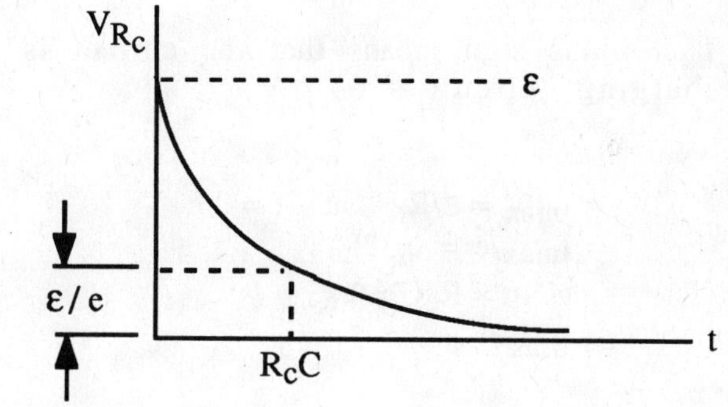

Figure 25.11

(d) $$V_c = q/C = C\varepsilon(1 - e^{-t/R_cC})/C = \varepsilon(1 - e^{-t/R_cC})$$

Equation (d) and Figure 25.12 show how the potential difference across the capacitor varies in time.

$V_C = 0$ at $t = 0$
$V_C = \varepsilon(1 - e^{-1}) = 0.63\varepsilon$
at $t = R_cC = \tau_{cc}$
$V_C \to \varepsilon$ as $t \to \infty$

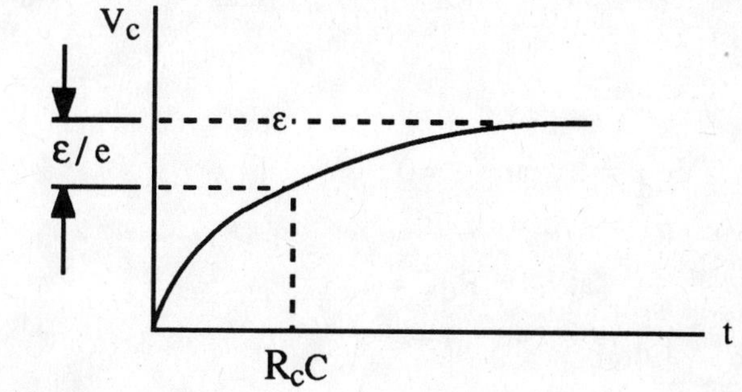

Figure 25.12

Discharging the capacitor (S_c open and S_d closed):

Equation (e) and Figure 25.13 show how the charge on the capacitor varies in time.

(e) $q = q_0 e^{-t/R_dC}$

$q = q_0$ at $t = 0$
$q = q_0/e = 0.37q_0$
at $t = R_dC = \tau_{cd}$
$q \to 0$ as $t \to \infty$

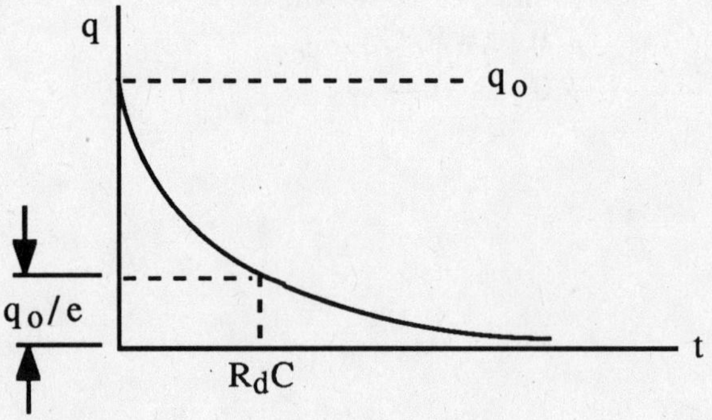

Figure 25.13

Equation (f) and Figure 25.14 show how the current in the circuit varies with time.

(f) $I = -\dfrac{q_0 e^{-t/R_dC}}{R_dC} = -\dfrac{C\varepsilon e^{-t/R_dC}}{R_dC} = -I_{max}\, e^{-t/R_dC}$

The minus sign means that the current is in the direction opposite to that of the charging circuit.

$I = I_{max} = \varepsilon/R_d$ at $t = 0$
$I = I_{max}/e = 0.37I_{max}$
at $t = R_cC = \tau_{cd}$
$I \to 0$ as $t \to \infty$

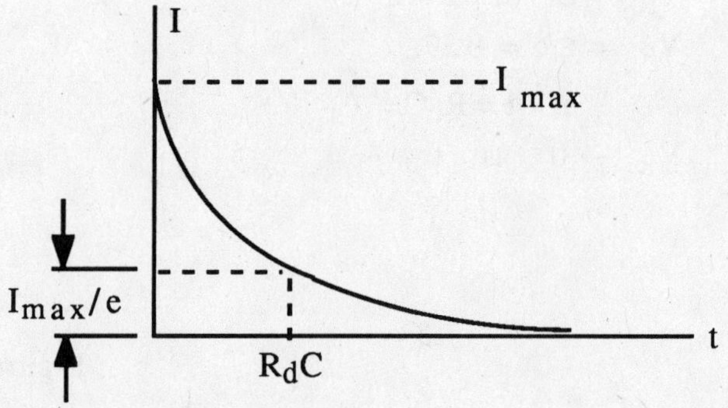

Figure 25.14

Equation (g) and Figure 25.15 show how the potential difference across the resistor varies with time.

(g) $V_{R_d} = IR_d = -I_{max}(e^{-t/R_dC})R_d = -\dfrac{\varepsilon}{R_d}(e^{-t/R_dC})R_d = -\varepsilon e^{-t/R_dC}$

$V_{R_d} = \varepsilon$ at $t = 0$
$V_{R_d} = \varepsilon/e = 0.37\varepsilon$
at $t = R_dC = \tau_{cd}$
$V_{R_d} \to 0$ as $t \to \infty$

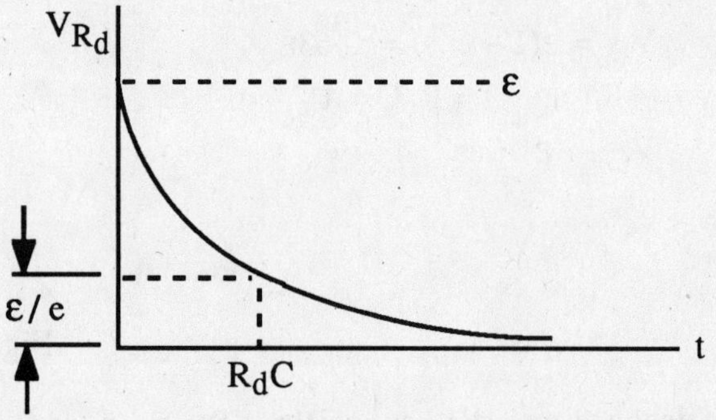

Figure 25.15

Equation (h) and Figure 25.16 show how the potential difference across the capacitor varies with time.

(h) $$V_c = \frac{q}{C} = \frac{q_0 e^{-t/R_dC}}{C} = \frac{C\varepsilon e^{-t/R_dC}}{C} = \varepsilon e^{-t/R_dC}$$

$V_C = \varepsilon$ at $t = 0$

$V_C = \varepsilon/e = 0.37\varepsilon$

at $t = R_dC = \tau_{c_d}$

$V_C \to 0$ as $t \to \infty$

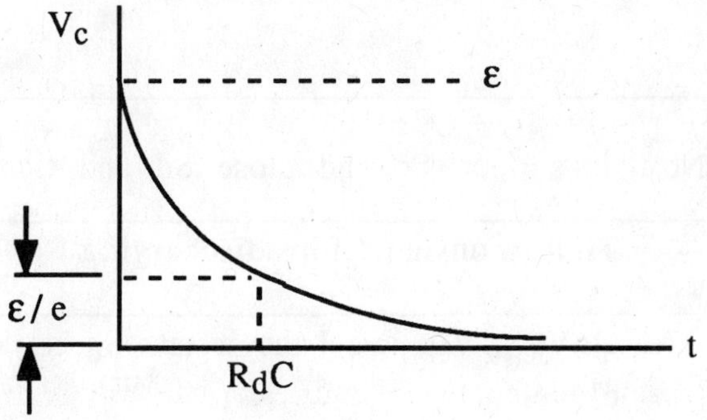

Figure 25.16

Practice: Consider Figure 25.8 with the following values.

$R_c = 100\ \Omega$ $R_d = 50.0\ \Omega$ $C = 5.00 \times 10^{-6}$ F $\varepsilon = 10.0$ V

First, let's close S_c and open S_d and then determine the following:

1. q, V_{R_c}, V_C and I at t = 0 (i.e., at the instant the switch is closed	$q = C\varepsilon(1 - e^{-t/R_cC}) = 0$ $I = I_{max}\, e^{-t/R_cC} = I_{max} = \varepsilon/R_C = 0.100$ A $V_{R_c} = \varepsilon e^{-t/R_cC} = V = 10.0$ V $V_C = \varepsilon(1 - e^{-t/R_cC}) = 0$
2. q, V_{R_c}, V_C and I at $t = R_cC$ (i.e., at the time equal to the time constant	$q = C\varepsilon(1 - e^{-1}) = 3.16 \times 10^{-5}$ C $I = I_{max}/e = (\varepsilon/R_c)/e = 3.68 \times 10^{-2}$ A $V_{R_c} = \varepsilon/e = 3.68$ V $V_C = \varepsilon(1 - e^{-1}) = 6.32$ V Note as a check that $V_{R_c} + V_C = \varepsilon$
3. q, V_{R_c}, V_C and I at a time equal to one tenth the time constant	$t = 0.100\ R_CC$ $q = C\varepsilon(1 - e^{-0.1}) = 4.76 \times 10^{-6}$ C $I = I_{max}\, e^{-0.1} = (\varepsilon/R_c)e^{-0.1} = 9.05 \times 10^{-2}$ A $V_{R_c} = \varepsilon e^{-0.1} = 9.05$ V $V_C = \varepsilon(1 - e^{-0.1}) = 0.950$ V Note as a check that $V_{R_c} + V_C = \varepsilon$

4. Time constant for charging	$\tau_c = R_cC = 5.00 \times 10^{-4}$ s
5. Time for the charge on the capacitor to reach one-half its maximum value	$q = q_{max}/2$ $q = q_{max}(1 - e^{-t/R_cC})$ $q_{max}/2 = q_{max}(1 - e^{-t/R_cC})$ $2 = e^{t/R_cC}$ $t = R_cC \ln 2 = 3.47 \times 10^{-4}$ s

Now let's open Sc and close Sd and then determine the following:

6. Time constant for discharging	$\tau_d = R_dC = 2.50 \times 10^{-4}$ S
7. q, V_{R_d}, V_C, and I the instant S_d is closed	$q = q_oe^{-t/R_dC} = q_o = C\varepsilon = 5.00 \times 10^{-5}$ C $V_{R_d} = -\varepsilon e^{-t/R_dC} = -\varepsilon = -10.0$ V $V_C = \varepsilon e^{-t/R_dC} = \varepsilon = 10.0$ V $I = -I_{max}e^{-t/R_dC} = -I_{max} = -\varepsilon/R_d$ $= -0.200$ A

Example: A capacitor is placed in series with a 25.0 V battery and a $6.00 \times 10^4\ \Omega$ resistor. It takes 10.0 s for the capacitor to charge from 0.0 V to 20.0 V. Calculate the value of the capacitor.

Given: $\varepsilon = 25.0$ V, $R = 6.00 \times 10^4\ \Omega$, $V_C = 20.0$ V at $t = 10.0$ s

Determine: C -- the value of the capacitor

Strategy: We can write an expression for the potential difference across the capacitor as a function of time. We can then solve this expression for C and substitute known values.

Solution: The potential difference across the capacitor as a function of time can be expressed as

$$V_C = \varepsilon\ (1 - e^{-t/RC})$$

This can be solved for C as follows

$$V_C/\varepsilon = (1 - e^{-t/RC})$$

$$e^{-t/RC} = 1 - V_C/\varepsilon$$

$$-t/RC = \ln\ [1 - V_C/\varepsilon]$$

$$C = -t/R\,[\,\ln(1 - V_C/\varepsilon)\,] = 1.04 \times 10^{-4}\ \text{F}$$

Related Text Exercises: Ex. 25-30 through 25-39.

PRACTICE TEST

Take and grade this practice test. Doing so will allow you to determine any weak spots in your understanding of the concepts taught in this chapter. The following section prescribes what you should study further to strengthen your understanding.

Figure 25.17 shows a circuit and data collected with that circuit.

	S Open	S Closed
V	3.00 V	2.00 V
I	0.00 A	1.00 A

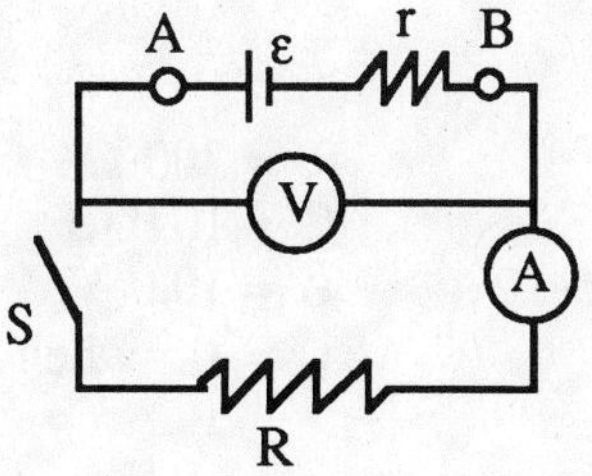

Figure 25.17

Determine the following:

_________ 1. Emf of the cell
_________ 2. Terminal potential difference of the cell with S closed
_________ 3. Potential difference across R with S closed
_________ 4. Resistance of R
_________ 5. Internal resistance of the cell
_________ 6. Charge drifting through R in 10.0 s
_________ 7. Power delivered by the cell
_________ 8. Power consumed by R
_________ 9. Power consumed by r
_________ 10. Efficiency of the cell
_________ 11. Work done by the cell in 10.0 s

Figure 25.18

$\varepsilon_1 = \varepsilon_2 = \varepsilon_3 = 6.00$ V
$R_1 = 4.00\ \Omega$
$R_2 = 2.00\ \Omega$
$R_3 = 8.00\ \Omega$

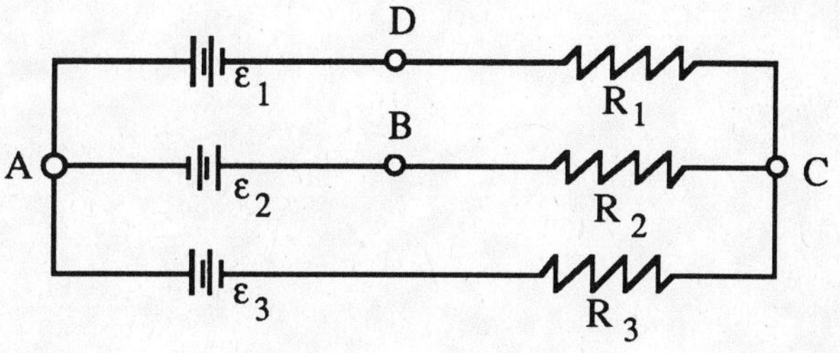

For the circuit shown in Figure 25.18, determine the following:

________ 12. Current through R_1
________ 13. Current through R_2
________ 14. Current through R_3
________ 15. Voltage $V_C - V_A$
________ 16. $P_{\varepsilon 1}$: the power supplied by ε_1
________ 17. P_{R_1}: the power dissipated by R_1
________ 18. $P_{\varepsilon 2}$: the power supplied by ε_2
________ 19. P_{R_2}: the power dissipated by R_2
________ 20. $P_{\varepsilon 3}$: the power supplied by ε_3
________ 21. P_{R_3}: the power dissipated by R_3

Figure 25.19

$R = 200\ \Omega$
$C = 10.0\ \mu F$
$\varepsilon = 10.0$ V
$t = 0$ when S is closed

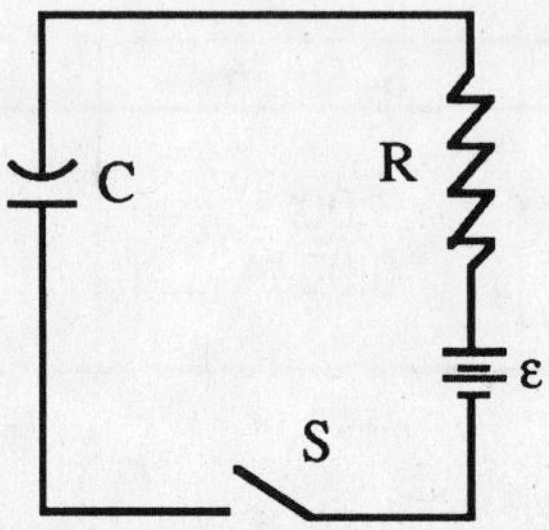

For the circuit shown in Figure 25.19, determine the following:

________ 22. Potential difference across R at t=0
________ 23. Time constant
________ 24. Current after one time constant
________ 25. Charge on the capacitor after two time constants

(See Appendix I for answers.)

PRINCIPAL CONCEPTS AND EQUATIONS PRESCRIPTION

Your score on the practice test is an excellent measure of your understanding of the chapter. You should now use the following chart to write your own prescription for dealing with any weaknesses the practice test points out. Look down the leftmost column to the number of the question(s) you answered incorrectly, reading across that row you will find the concept and/or equation of concern, the section(s) of the study guide you should return to for further study, and some suggested text exercises which you should work to gain additional experience.

Practice Test Questions	Concepts and Equations	Prescription	
		Principal Concepts	Text Exercises
1	Emf	1	25-1, 2
2	$V_T = \varepsilon - Ir$	1	25-1, 2
3	$V_R = V_T$	1	25-3, 4
4	Ohm's Law: $R = V/I$	2 of Ch 24	24-28, 29
5	Internal Resistance: $r = (\varepsilon - V)/I$	1	25-4, 5
6	Current: $\Delta Q = I\Delta t$	1 of Ch 24	24-1, 2
7	Power: $P_\varepsilon = \varepsilon I$	2	25-16, 17
8	Power: $P_R = I^2R$	2	25-6, 7
9	Power: $P_r = I^2r$	2	25-16, 7
10	Efficiency of cell: $\varepsilon = P_R/P_\varepsilon$	---	---
11	Power: $P = W/t$	5 of Ch 8	25-6, 15
12	Kirchoff's Rules	3	25-25, 26
13	Kirchoff's Rules	3	25-27, 28
14	Kirchoff's Rules	3	25-28, 29
15	Potential difference	3	25-24, 26
16	Power: $P_{\varepsilon_1} = \varepsilon_1 I_1$	2	25-16, 17
17	Power: $P_{R_1} = I_1^2R_1$	2	25-26
18	Power: $P_{\varepsilon_2} = \varepsilon_2 I_2$	2	25-16, 17
19	Power: $P_{R_2} = I_2^2R_2$	2	25-26
20	Power: $P_{\varepsilon_2} = \varepsilon_3 I_3$	2	25-16, 17
21	Power: $P_{R_3} = I_3^2R_3$	2	25-26
22	R-C series: potential difference $V_R(0) = \varepsilon/R$	4	25-32, 36
23	R-C series: time constant $\tau_C = RC$	4	25-31, 32
24	R-C series: current $I(t) = (\varepsilon/R)e^{t/RC}$	4	25-35, 37
25	R-C series: charge $Q(t) = C\varepsilon(1 - e^{-t/RC})$	4	25-33, 36

26 THE MAGNETIC FIELD

RECALL FROM PREVIOUS CHAPTERS

Previously learned concepts and equations frequently used in this chapter	Text Section	Study Guide Page
Central force: $F_c = mv^2/r$	6-2	6-7
Vector dot product: $\mathbf{A} \cdot \mathbf{B} = AB\cos\theta$	8-2	8-1
Potential energy	9-1	9-1
Vector cross product: $\lvert\mathbf{A} \times \mathbf{B}\rvert = AB\sin\theta$	11-6	11-16
Torque: $\boldsymbol{\tau} = \mathbf{r} \times \mathbf{F}$	11-6	11-16

NEW IDEAS IN THIS CHAPTER

Concepts and equations introduced	Text Section	Study Guide Page
Magnetic field **B**	26-1	26-1
Magnetic force on a moving charge: $\mathbf{F}_B = q\mathbf{v} \times \mathbf{B}$	26-1	26-1
Magnetic force on a current element: $\mathbf{F}_B = I\boldsymbol{l} \times \mathbf{B}$	26-2	26-4
Magnitude of magnetic moment: $m = NIA$	26-3	26-8
Right hand rule for direction of **m**	26-3	26-8
Torque on a plane current loop: $\boldsymbol{\tau} = \mathbf{m} \times \mathbf{B}$	26-3	26-8
Potential energy of a plane current loop: $U = -\mathbf{m} \cdot \mathbf{B}$	26-3	26-8
Charges moving in a magnetic field	26-4	26-11

PRINCIPAL CONCEPTS AND EQUATIONS

1 The Magnetic Field B and the Force on Moving Charges

(Section 26-1)

Review: Experiments show that moving charges exert magnetic forces on other moving charges just as fixed (or moving) charges exert electric forces on fixed (or moving) charges. We described the latter by attributing an electric field **E** to a collection of charges and then the force they exert on another charge q was found to be

$$\mathbf{F}_E = q\mathbf{E}$$

In the case of magnetic forces, we attribute a magnetic field **B** (to be discussed in the next chapter) to a collection of moving charges and find the force they exert on another moving charge q to be

$$F_B = q\mathbf{v} \times \mathbf{B}$$

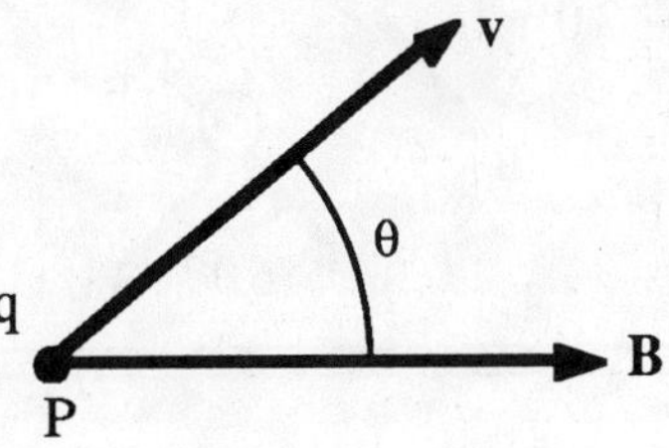

Figure 26.1

Figure 26.1 shows a charge q with a velocity **v** moving through point P where there is a magnetic field **B** as shown. **v x B** is into the page, hence if q is positive then $\mathbf{F}_B$ is into the page and if q is negative then $\mathbf{F}_B$ is out of the page. The magnitude of the force is

$$F_B = |\mathbf{F}_B| = |q|vB \sin\theta$$

Note: If **v** is along **B** ($\theta = 0°$ or $180°$) then $\mathbf{F}_B = 0$, hence the direction of **B** is along the line of no force.

The unit of B is the Tesla (T). If a 1 C charge is moving perpendicular to **B** at 1 m/s and experiences a magnetic force of 1 N, then B is 1 T.

Practice: An electron is moving with a velocity of $(3 \times 10^7\ \mathbf{i} + 4 \times 10^7\ \mathbf{j})$m/s through a point where there is a magnetic field given by $\mathbf{B} = (0.2\ \mathbf{j} - 0.3\ \mathbf{k})$ T.

Determine the following:

1. **v x B**	$\mathbf{v} \times \mathbf{B} = \begin{vmatrix} \mathbf{i} & \mathbf{j} & \mathbf{k} \\ 3 \times 10^7 & 4 \times 10^7 & 0 \\ 0 & 0.2 & -0.3 \end{vmatrix}$ $= (-1.2\mathbf{i} + 0.9\mathbf{j} + 0.6\mathbf{k}) \times 10^7$ m/s
2. Magnetic force on the electron	$\mathbf{F}_B = e(\mathbf{v} \times \mathbf{B})$, $e = 1.6 \times 10^{-19}$ C $\mathbf{F}_B = (1.9\ \mathbf{i} - 1.4\ \mathbf{j} - 1.0\ \mathbf{k}) \times 10^{-12}$ N
3. Acceleration of the electron	$\mathbf{a} = \mathbf{F}/m_e$, $m_e = 9.10 \times 10^{-31}$ kg $\mathbf{a} = (2.1\ \mathbf{i} - 1.5\ \mathbf{j} - 1.1\ \mathbf{k}) \times 10^{18}$ m/s^2

Figure 26.2 shows several charged particles moving in a magnetic field.

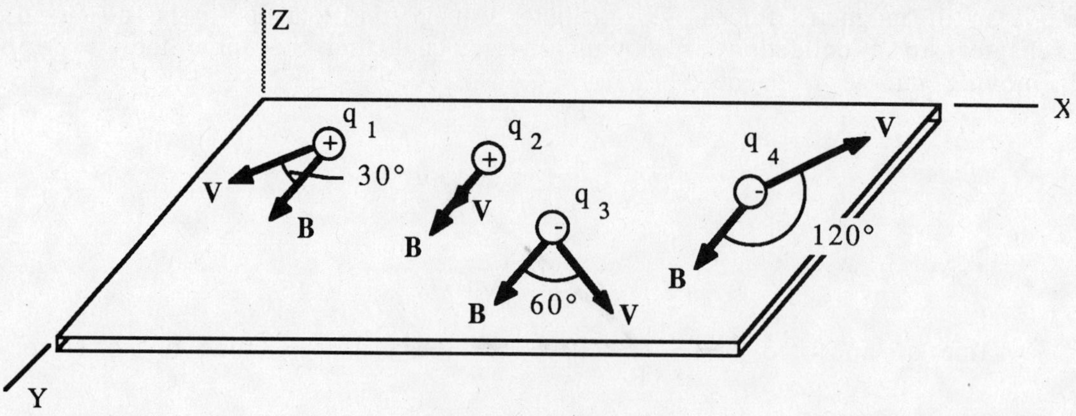

Figure 26.2

$q_1 = q_2 = +1.60 \times 10^{-19}$ C

$q_3 = q_4 = -1.60 \times 10^{-19}$ C

$v_1 = v_2 = v_3 = v_4 = 2.00 \times 10^3$ m/s

B = 1.00 T

Determine the following:

4. Magnitude and direction of the force on each charged particle	$F_1 = q_1v_1B \sin30° = 1.60 \times 10^{-16}$ N (+z) $F_2 = q_2v_2B \sin0° = 0$ $F_3 = q_3v_3B \sin60° = 2.77 \times 10^{-16}$ N (+z) $F_4 = q_4v_4B \sin120° = 2.77 \times 10^{-16}$ N (+z)

Example: An electron ($q_e = -1.60 \times 10^{-19}$ C) is traveling at 4.00×10^7 m/s due North in a horizontal plane through a point where the earth's magnetic field has a component to the North of 2.00×10^{-5} T and a downward component of 5.00×10^{-5} T. Calculate the magnetic force on the electron and its acceleration.

Given: $q_e = -1.60 \times 10^{-19}$ C -- charge on the electron

$\mathbf{v}_e = 4.00 \times 10^7$ m/s N -- velocity of the electron

$\mathbf{B}_E = (2.00 \times 10^{-5}$ North + 4.00×10^{-5} Down) T -- earth's magnetic field at site of interest

$m_e = 9.10 \times 10^{-31}$ kg -- mass of the electron

Determine: $\mathbf{F}_B$ -- the magnetic force on the electron

a -- the acceleration of the electron

Strategy: Knowing q_e, $\mathbf{v}_e$, and $\mathbf{B}_E$, we can determine the magnetic force, $\mathbf{F}_B$, on the electron. Knowing $\mathbf{F}_B$ and m_e, we can determine the acceleration **a** of the electron with Newton's Second Law.

Solution:

$$F_B = (-e)v_e \times B_E = (-1.6 \times 10^{-19}\ C)\begin{vmatrix} \text{North} & \text{East} & \text{Down} \\ 4 \times 10^{7}\ T & 0 & 0 \\ 2 \times 10^{-5}\ T & 0 & 5 \times 10^{-5}\ T \end{vmatrix}$$

$F_B = 3.20 \times 10^{-16}$ N East and $a = F_B/m_e = 3.50 \times 10^{14}$ m/s² East

Related Text Exercises: Ex. 26-1 through 26-7.

2 Force on a Current Element in a Magnetic Field (Section 26-2)

Review: Figure 26.3 shows a straight segment of conductor of length L carrying a current I in a uniform magnetic field **B**.

Figure 26.3

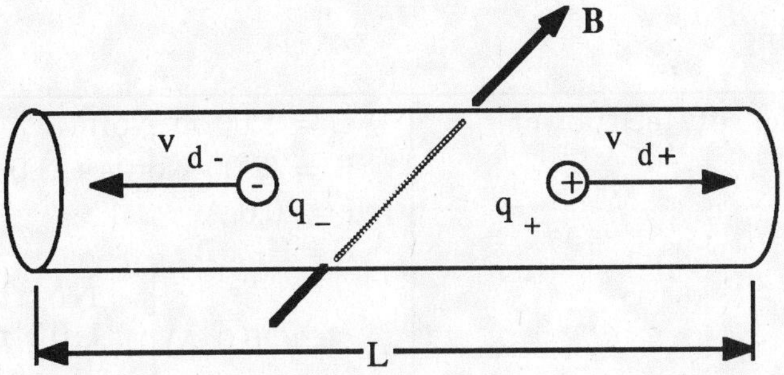

The current I is to the right, so we could have positive charges moving to the right or negative charges moving to the left. In either case, $\mathbf{F}_B = q\mathbf{v}_d \times \mathbf{B}$ will give a force that is out of the page. A current to the right (regardless of source) leads to a force out of the page. Using $\mathbf{F}_B = q\mathbf{v} \times \mathbf{B}$ for the force on the individual charges and the knowledge that $I = nqv_dA$, the text shows that

$$\mathbf{F} = I\mathbf{L} \times \mathbf{B}$$

where
- I = current in the conductor
- **L** = length in the direction of the current
- **B** = magnetic field at the site of the length L
- **F** = force on the current carrying wire due to the interaction of the magnetic field with the current

Note: The above expression ($\mathbf{F} = I\mathbf{L} \times \mathbf{B}$) is for a straight conductor in a uniform magnetic field **B**. If these conditions do not hold, we must integrate $d\mathbf{F} = Id\mathbf{L} \times \mathbf{B}$ over the length of the conductor as shown in Figure 26.4.

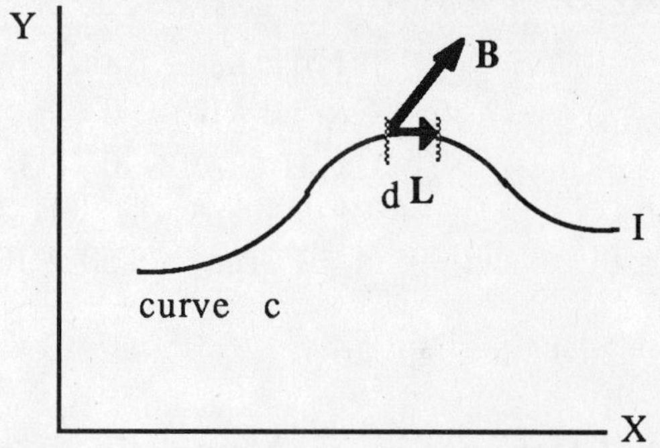

Figure 26.4

$$F_c = \int_{curve\ c} I\,dL \times B$$

Practice: A horizontal power line running North and South carries a current of 20.0 A North. The earth's magnetic field is uniform with a horizontal component of 2.00×10^{-5} T North and a vertical component of 5.00×10^{-5} T down.

Determine the following:

1. Force on one meter of the power line	**L** = 1.00 m North **B** = (2.00 North + 5.00 Down) x 10^{-5} T I = 20.0 A **F** = I**L** x **B** $= (20.0\ \text{A}) \begin{vmatrix} \text{North} & \text{East} & \text{Down} \\ 1.00\ \text{m} & 0 & 0 \\ 2.00 \times 10^{-5} & 0 & 5.00 \times 10^{-5} \end{vmatrix}$ = (20.0 A)(-5.00 x 10^{-5} m•T East) = 1.00 x 10^{-3} N West

Example: Figure 26.5 shows a conductor of length 2.00 m supported by two conducting wires. The wires are massless and the mass of the conductor is 2.00 kg. A magnetic field of 2.00 T is directed into the page as shown. What current (specify direction) will cause the tension in the wires to be zero.

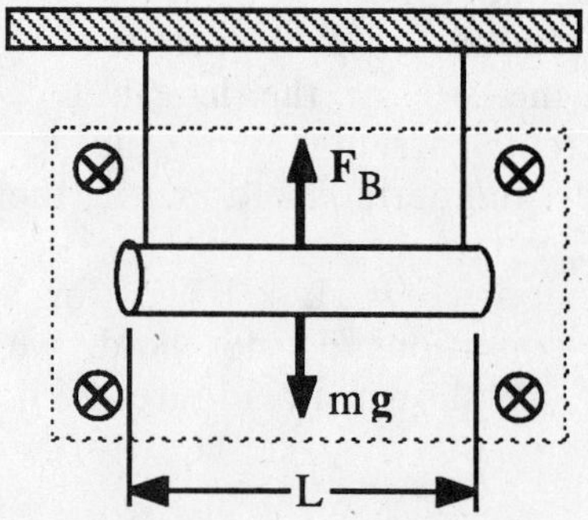

Figure 26.5

Given:

m = 2.00 kg = mass of the conductor of length L
L = 2.00 m = length of the conductor
B = 2.00 T (into page) = magnetic field
g = 9.80 m/s^2 (downward)
support wires are massless

Determine: The current (including direction) through the conductor which will cause zero tension in the support wires.

Strategy: The tension in the support wires will be zero when the magnetic force on the wire is equal in magnitude but opposite in direction to the gravitational force on the wire. Knowing m and g, we can determine the gravitational force and hence the magnetic force. Knowing the magnetic force, **B** and L, we can determine the current needed. Finally, we can use the right hand rule for the force on a current carrying wire in a magnetic field to determine the direction of the current in order for $\mathbf{F_B}$ to be directed vertically upward.

Solution:

$$F_B = F_g = mg$$

$$F_B = ILB \sin 90.0° = ILB$$

Equating and solving for I, we obtain

$$I = mg/LB = 4.90 \text{ A}$$

(left to right in Figure 26.5.)

Example: Figure 26.6 shows a conductor of arbitrary shape carrying a current I in a uniform magnetic field **B**. Show that **F** = IL' x **B**, where L' is the vector from one end of the conductor to the other.

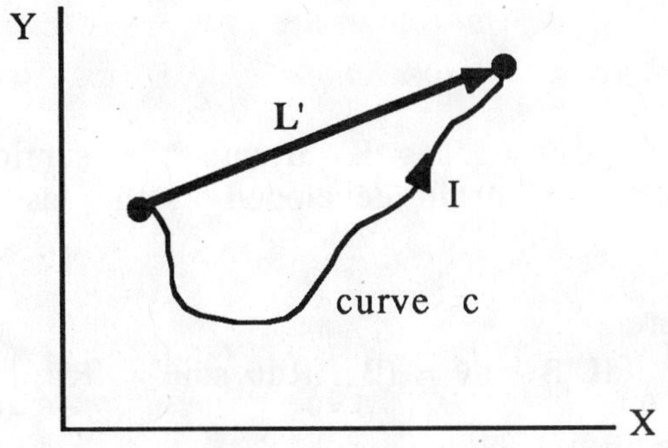

Figure 26.6

Given: A conductor C of arbitrary shape carrying a current I in a uniform magnetic field **B**.

Determine: That **F** = IL' x **B**

Strategy: Use dF = IdL x **B** and integrate over the length of the conductor C.

Solution: Since I and **B** are constant we may write

$$\mathbf{F} = \int_{\text{over } c} I d\mathbf{L} \times \mathbf{B} = I\left[\int_{\text{over } c} d\mathbf{L}\right] \times \mathbf{B} = I\mathbf{L}' \times \mathbf{B}$$

The last statement above may be made because the integration amounts to summing (tail to tip) all the little vectors making up the path c and this sum is just the resultant vector **L'**. This is a useful relationship as the next example shows.

Example: A semicircular conductor of 1.00 m radius carries a current of 3.00 A as shown in Figure 26.7. The conductor is in a uniform magnetic field **B** = (4.00 T) **k**. Calculate the magnetic force on the conductor.

Figure 26.7

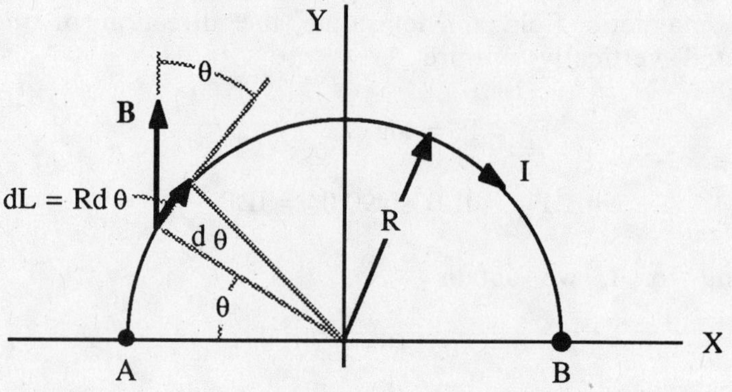

Given:	I = 3.00 A	=	current in the conductor
	B = (4.00 T) **j**	=	magnetic field
	r = 1.00 m	=	radius of the semicircular conductor

Determine: **F**, the magnetic force on the conductor

Strategy:

Method I - we can integrate d**F** = Id**L** x **B** around the semicircular conductor.
Method II - we can use the relationship developed in the last example (**F** = I**L'** x **B**).

Solution:

Method I.

$$F = |\mathbf{F}| = \int_{\theta=0}^{\pi} I|d\mathbf{L}|B \sin\theta = IB\int_{\theta=0}^{\pi} R d\theta \sin\theta = IRB\int_{\theta=0}^{\pi} \sin\theta\, d\theta$$

$$= IRB(-\cos\theta)\Big|_{\theta=0}^{\pi} = -IRB(\cos\pi - \cos 0) = 2IRB$$

Using the right hand rule, we see that the force is in the **k** direction. Hence

$$\mathbf{F} = 2IRB\,\mathbf{k} = (24.0\text{ N})\,\mathbf{k}$$

Method II. **F** = I**L'** x **B** **L'** = (2.00 m) **i** **B** = (4.00 T) **j**

$$\mathbf{F} = I\mathbf{L}' \times \mathbf{B} = (3.00\text{ A})(2.00\text{ m }\mathbf{i}) \times (4.00\text{ T }\mathbf{j}) = (24.0\text{ N})\,\mathbf{k}$$

Related Text Exercises: Ex. 26-8 through 26-13.

3 Magnetic Moment and Torque for a Current Loop (Section 26-3)

Review: The magnetic moment of a plane current loop is useful in describing the magnetic characteristics of the loop, i.e. the torque exerted on it in a uniform external magnetic field and the magnetic field it produces. The magnitude of the magnetic moment **m** denoted by m is the product of the current and area for a single turn loop.

$$m = IA$$

The magnetic moment increases by a factor of N for a coil with N turns

$$m = NIA$$

The direction of **m** is perpendicular to the plane of the coil and is determined by the following right hand rule.

> Right hand rule for determining the direction of **m** of a coil:
> Let the fingers of your right hand curl in the direction of the current in the coil. The thumb of your right hand points in the direction of **m**.

Consider the single turn loops shown in Figure 26.8.

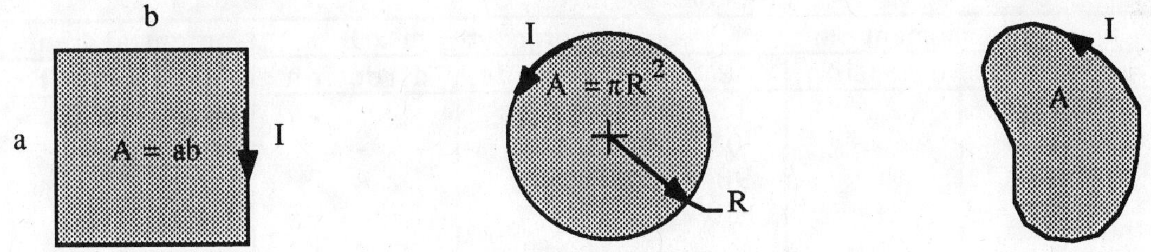

(a) (b) (c)

Figure 26.8

Using the above information regarding the magnitude and direction of the magnetic moment we obtain:

Case	Magnitude	Direction
a	$m = IA = Iab$	into page
b	$m = IA = I\pi R^2$	out of page
c	$m = IA$	out of page

If a current carrying loop is placed in a uniform magnetic field (**B**), there is no net force on the loop but it does experience a torque $\boldsymbol{\tau}$. This is conveniently expressed by:

$$\boldsymbol{\tau} = \mathbf{m} \times \mathbf{B}$$

There is a potential energy associated with the orientation of the loop in the field. This is conveniently expressed by:

$$U = -\mathbf{m} \cdot \mathbf{B}$$

Figure 26.9 shows a current carrying loop and its associated **m** oriented at various angles with respect to the external magnetic field.

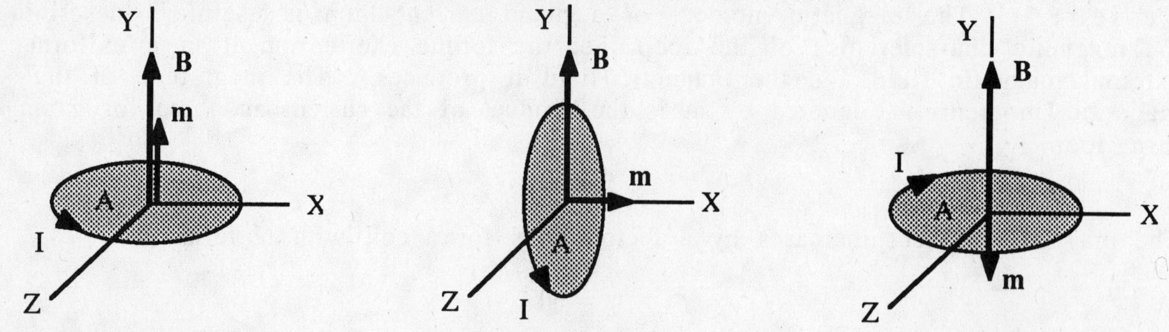

Loop perpendicular to **B**	Loop parallel to **B**	Loop perpendicular to **B**
(a)	(b)	(c)

Figure 26.9

The chart below summarizes such information about the loop's magnetic moment, orientation in the external field, torque, and potential energy.

	magnetic moment **m**			torque τ = **m** x **B**		potential energy
Case	magnitude	direction	θ	magnitude	direction	U = -**m** • **B**
a	IA	**j**	0°	0	-	-IAB
b	IA	**i**	90°	IAB	**k**	0
c	IA	-**j**	180°	0	-	IAB

Practice: A ten turn circular loop 2.00 m in radius carries a current of 3.00 A in a uniform magnetic field of 6.00 T.

Determine the following:

1. Magnitude of the magnetic moment	$m = NIA = NI\pi r^2 = 120\pi$ A•m^2
2. Maximum torque on the loop	$\boldsymbol{\tau}$ = **m** x **B** $\tau = mB \sin\theta$ Maximum torque exists when $\theta = 90.0°$ $\tau_{max} = mB = 720\pi$ N•m
3. Potential energy at the point of stable equilibrium	U = -**m** • **B** = -mB cosθ Stable equilibrium occurs at $\theta = 0°$ $U_{min} = -720\pi$ J

Example: The rectangular coil of wire shown in Figure 26.10 has a linear mass density of 2.00×10^{-2} kg/m and is pivoted about side a-d (a friction-less axis). The coil is 5.00 cm wide, 10.0 cm long, and has 20 turns. A uniform 0.200 T magnetic field is in the +y direction. Determine the angle between the plane of the coil and the **B** field when the current in the coil is 10.0 A.

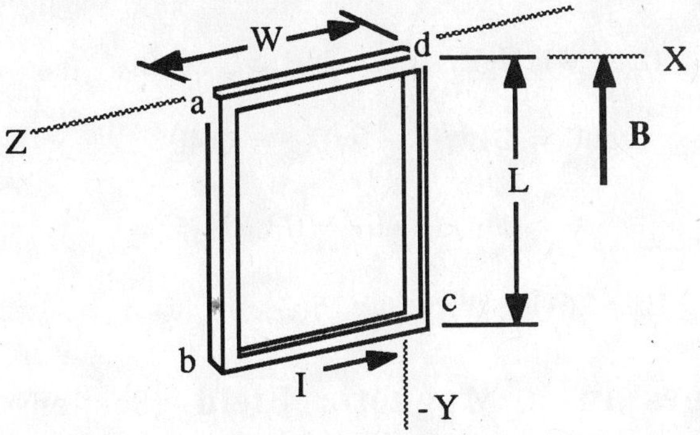

Figure 26.10

Given: $w = 5.00 \times 10^{-2}$ m, $L = 1.00 \times 10^{-1}$ m, $\lambda = M/L = 2.00 \times 10^{-2}$ kg/m, $B = 2.00 \times 10^{-1}$ T, $N = 20$, $I = 10.0$ A

Determine: The angle θ between the plane of the coil and the **B** field when the current is 10.0 A.

Strategy: The coil will rotate counterclockwise about the Z axis until the torque due to the **B** field is equal to the torque due to the weight of the wire. We can write and equate expressions for these torques and then solve for θ.

Solution: As shown in Figure 26.11, the coil rotates counterclockwise about the Z axis until the torque due to the **B** field is equal to the torque due to the weight of the coil.

$M = \lambda L N$
$M' = \lambda w N$

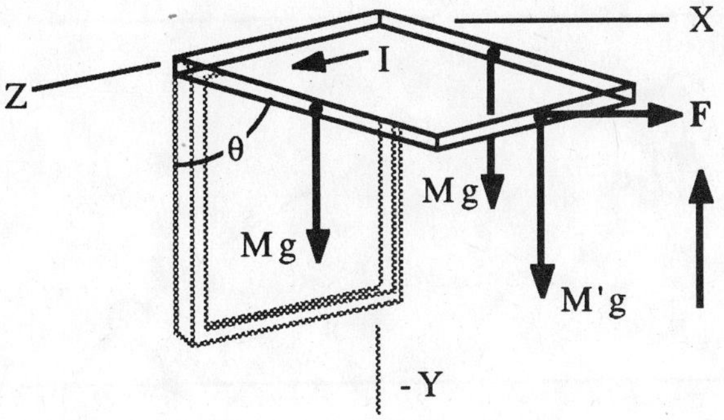

Figure 26.11

The clockwise torque due to the weight of the coil (two lengths each of mass M and one width of mass M') is

$$\tau_{cw} = 2(Mg)(\frac{L}{2} \sin\theta) + M'gL \sin\theta = (L + w)\lambda NgL \sin\theta$$

The counterclockwise torque due to the **B** field (acting on the width w) is

$$\tau_{ccw} = NFL\cos\theta = NBIwL\cos\theta$$

Since the coil is in rotational equilibrium, we can equate the clockwise and counter-clockwise torques and solve for θ

$$(L + w)\lambda NgL\sin\theta = NBIwL\cos\theta$$

$$\tan\theta = BIw/(L + w)\lambda g = 3.40$$

$$\theta = \tan^{-1}(3.40) = 73.6°$$

Related Text Exercises: Ex. 26-14 through 26-24.

4 Motion of Charges in a Magnetic Field (Section 26-4)

Review: If a charged particle moves along the direction of the **B** field then $\mathbf{F}_B = q\mathbf{v} \times \mathbf{B} = 0$, and there is no force on the particle. In a uniform field, the particle will continue to move along the field lines.

If a charged particle's motion is perpendicular to **B**, the magnetic force is at right angles to both **v** and **B**. You should recall from section 26.5 of the text that when this is the case and when **B** is uniform, the particle travels in a circle, with the magnetic force supplying the central force, If **v** has components parallel and perpendicular to **B**, the perpendicular component causes the particle to have circular motion in the field, the parallel component causes the particle to move along the field, and the result is a helical trajectory. Figure 26.12 shows a particle of charge +q entering a region where a magnetic field **B** is directed up out of the page. By applying the right-hand rule we see that the magnetic force supplies a central force causing the particle to move in a circle.

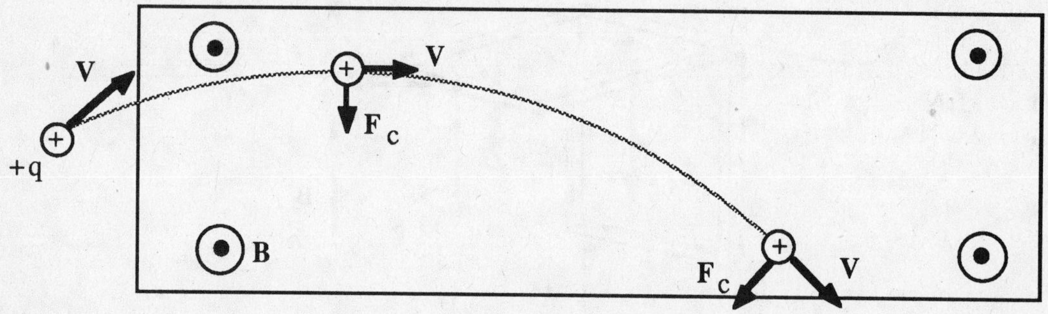

Figure 26.12

Since a particle of mass m and speed v is moving in a circle of radius r, it must be experiencing a central force of magnitude

$$F_c = mv^2/r$$

For this case, the central force is supplied by the magnetic force. Hence we can write

$$F_C = F_B = qvB \sin 90° = qvB$$

Combining these two expressions for F_C, we have

$$mv^2/r = qvB \qquad \text{or} \qquad mv = qBr$$

Note: This last expression involves five physical quantities (m, v, q, B and r). Consequently we can determine any one of these five quantities given the other four.

Practice: Figure 26.13 shows several charged particles traveling in a region where a magnetic field is directed vertically into the page.

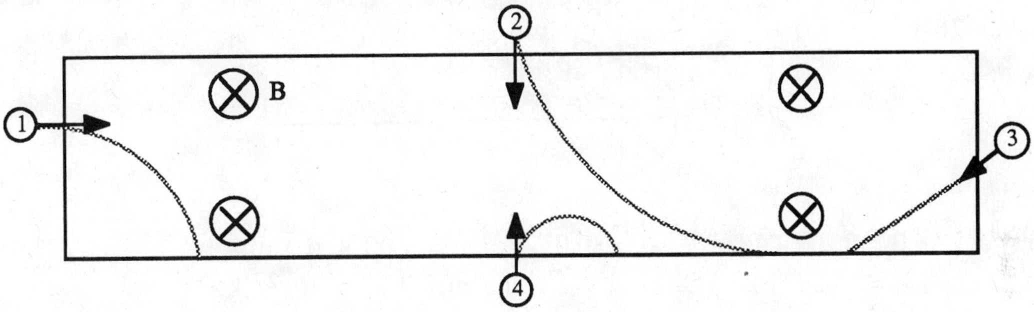

Figure 26.13

Determine the following:

1. The particle(s) with no charge	Particle 3 is not deflected, consequently it is experiencing no magnetic force. Since v, B, and θ are not zero, q must be zero
2. The particle(s) with a negative charge	Particles 1 and 4 (using the right-hand rule)
3. The particle(s) with a positive charge	Particle 2 (using the right-hand rule)
4. If particles 1 and 2 have the same mass and magnitude of charge, which one has the greater speed	$mv = qBr$ or $r = mv/qB$ Since q, B, and m are the same for each particle, a greater v results in a greater r. Subsequently, particle 2 must have the greater speed.
5. If particles 2 and 4 have the same speed and mass, which one has the greater charge	$mv = qBr$ or $r = mv/qB$ Since m, v, and B are the same for each particle, a greater q results in a smaller r. Subsequently, particle 4 must have the greater charge.

Example: Figure 26.14 shows a cylindrical tube aligned with a magnetic field of 0.200 T. A proton is injected into the cylinder with a velocity of 1.00×10^6 m/s directed 30.0° above the direction of the field. As shown, the particle will travel in a spiral trajectory. Determine the following:

a) central force experienced by the particle,
b) radius of the spiral trajectory of the particle,
c) time for the particle to make one complete revolution, and
d) length of the tube in order for the particle to make three complete revolutions

Figure 26.14

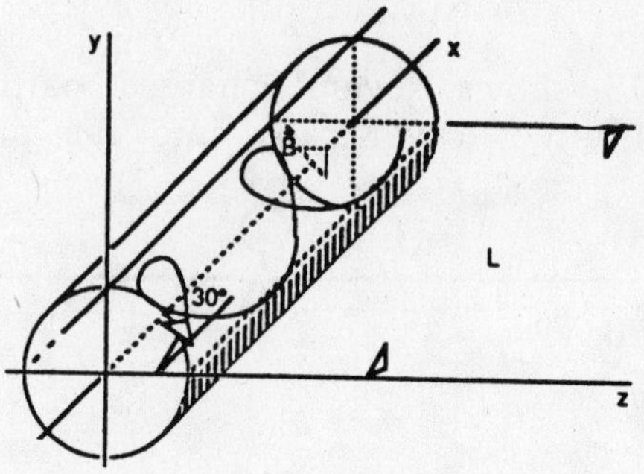

Given: $B = 0.200$ T $\quad \theta = 30.0°$ $\quad v = 1.00 \times 10^6$ m/s

$q = 1.60 \times 10^{-19}$ C $\quad m = 1.67 \times 10^{-27}$ kg

Determine:

(a) F_c -- central force experienced by the particle
(b) r -- radius of the spiral trajectory
(c) T -- time for the particle to make one complete revolution
(d) L -- length of the tube in order for the particle to make three revolutions

Strategy:

(a) Knowing q, v, θ, and B, we can determine the magnetic force and hence the central force F_c.
(b) Knowing F_c, m, v and θ, we can determine r.
(c) Knowing r, v, and θ, we can determine T.
(d) Knowing T, v, and θ, we can determine L.

Solution: First, let's determine the component of the velocity parallel and perpendicular to the magnetic field.

$v_{\parallel} = v \cos\theta = 8.66 \times 10^5$ m/s -- this component moves the particle down the tube

$v_{\perp} = v \sin\theta = 5.00 \times 10^5$ m/s -- this component causes the particle to experience a central force

(a) The central force is supplied by the magnetic force and can be determined by

$$F_C = F_B = qv_\perp B = 1.60 \times 10^{-14}\ \text{N}$$

(b) Now that the central force is known, the radius of the spiral trajectory can be determined by

$$F_C = mv_\perp^2/r \quad \text{or} \quad r = mv_\perp^2/F_C = 2.61 \times 10^{-2}\ \text{m}$$

(c) Using the radius of the curvature, the time for one revolution can be determined by

$$T = 2\pi r/v_\perp = 3.28 \times 10^{-7}\ \text{s}$$

(d) Finally, we can determine the length of the tube needed for the particle to undergo three revolutions.

Speed of the particle down the tube is $v_{||} = v \cos 30° = 8.66 \times 10^5$ m/s
Time for three revolutions is $t = 3T = 9.84 \times 10^{-7}$ s
Length of tube needed is $L = tv_{||} = 0.852$ m

Example: Singly charged positive ions traveling at various speeds are introduced into the experimental apparatus shown in Figure 26.15. The ions are first accelerated by the electric field established by placing a potential difference V_1 across the vertical parallel plates. Next, they pass through the velocity selector. Finally, they enter a region where a magnetic field of 0.200 T exists.

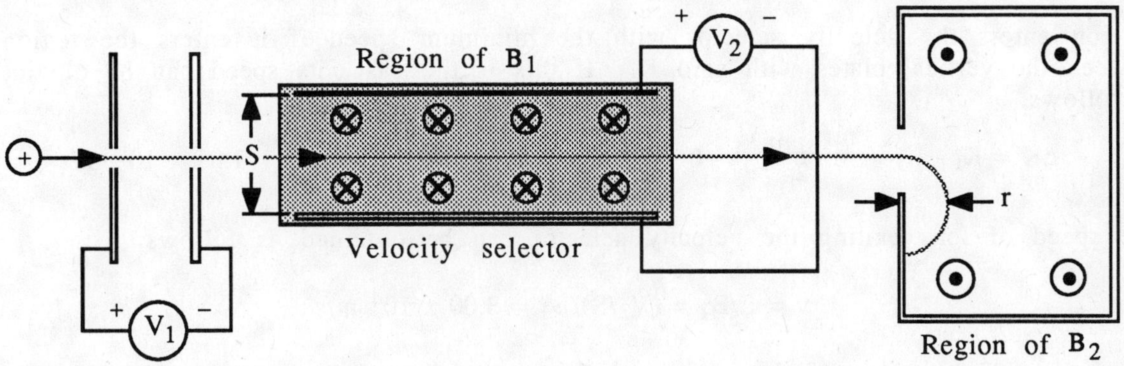

Figure 26.15

$m = 5.80 \times 10^{-26}$ kg -- mass of the ions
$q = 1.60 \times 10^{-19}$ C -- charge on the ions
$V_1 = 20.0$ V -- potential difference across the vertical parallel plates
$V_2 = 750$ V -- potential difference across the horizontal parallel plates in the velocity selector
$S = 5.00 \times 10^{-2}$ m -- separation of the plates in the velocity selector
$B_1 = 0.500$ T -- magnetic field in the velocity selector
$B_2 = 0.200$ T -- final magnetic field experienced by the ions

Calculate (a) the minimum speed of the ions as they enter the velocity selector, (b) the speed of ions as they exit the velocity selector, and (c) the radius of curvature of the ions in the final magnetic field $\mathbf{B}_2$.

Given: q, m, V_1, V_2, S, B_1, and B_2

Determine:

(a) v_{min} -- the minimum speed of ions entering the velocity selector
(b) v -- the speed of ions exiting the velocity selector
(c) r -- the radius of curvature of the ions in the magnetic field $\mathbf{B}_2$

Strategy: Knowing the charge on the ions and the potential difference V_1, we can determine the change in electric potential energy of the ions as they travel through the vertical parallel plates. Knowing this change in electric potential energy, we can determine the change in kinetic energy and hence the minimum speed of ions entering the velocity selector. Knowing V_2 and S, we can determine the **E**-field in the velocity selector. We can then use the magnitude of the **E**-field and **B**-field in the velocity selector to determine the exit speed of the ions. We can calculate the radius of curvature in the final **B**-field from B_2, v, q, and m.

Solution: Since no dissipative forces are involved, total energy is conserved ($\Delta K + \Delta U = 0$). Since energy is conserved, the increase in kinetic energy of the ions traveling through V_1 is equal to the negative of the decrease in electric potential energy.

$$\Delta K = -\Delta U = qV_1 = 3.20 \times 10^{-18} \text{ J}$$

An ion enters the velocity selector with the minimum speed if it enters the region between the vertical plates with zero K. If this is the case, its speed can be obtained as follows:

$$\Delta K = K_f - K_i = \frac{m(v_{min})^2}{2} - 0 \quad \text{or} \quad v_{min} = (2\Delta K/m)^{1/2} = 1.05 \times 10^4 \text{ m/s}$$

The speed of ions exiting the velocity selector can be obtained as follows:

$$v = E/B_1 = (V_2/S)/B_1 = 3.00 \times 10^4 \text{ m/s}$$

The radius of curvature of the ions in the magnetic field $\mathbf{B}_2$ can be obtained by combining $F_c = mv^2/r$ and $F_B = qvB_2$ to obtain

$$r = mv/qB_2 = 5.44 \times 10^{-2} \text{ m}$$

Related Text Exercises: Ex. 26-25 through 26-33.

PRACTICE TEST

Take and grade this practice test. Doing so will allow you to determine any weak spots in your understanding of the concepts taught in this chapter. The following section prescribes what you should study further to strengthen your understanding.

Figure 26.16 shows a region where a magnetic field is directed down into the page, several moving particles, and some information about the particles

Figure 26.16

$v_2 = v_3 = v_4 = 1.00 \times 10^6$ m/s
$m_2 = m_3 = 2.00 \times 10^{-26}$ kg
$|q_3| = |q_4| = 3.00 \times 10^{-16}$ C
$r_2 = r_4 = 10.0$ cm
$B = 0.500$ T

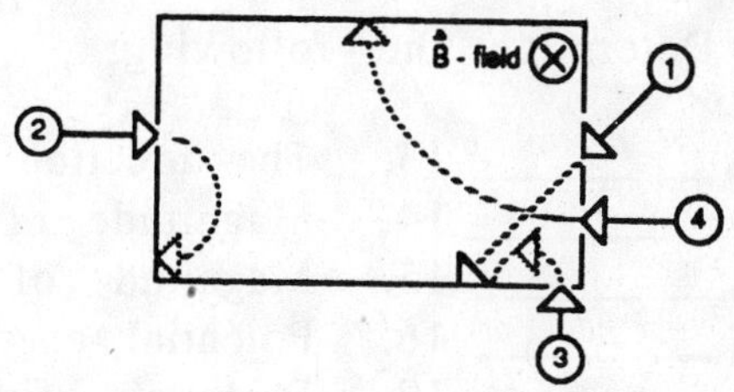

Determine the following:

________ 1. The particle(s) with no charge
________ 2. The particle(s) with positive charge
________ 3. The particle(s) with negative charge
________ 4. The central force on particle 2
________ 5. The charge of particle 2
________ 6. The radius of curvature for particle 3
________ 7. The momentum of particle 4
________ 8. The magnitude of the force on particle 3
________ 9. The magnitude of the force on particle 2
________ 10. The mass of particle 4

Figure 26.17 shows a 2.00 kg, 2.00 m long conductor in a 4.00 T magnetic field.

Figure 26.17

$m = 2.00$ kg
$L = 2.00$ m
$B = 4.00$ T
$\theta = 53.0°$

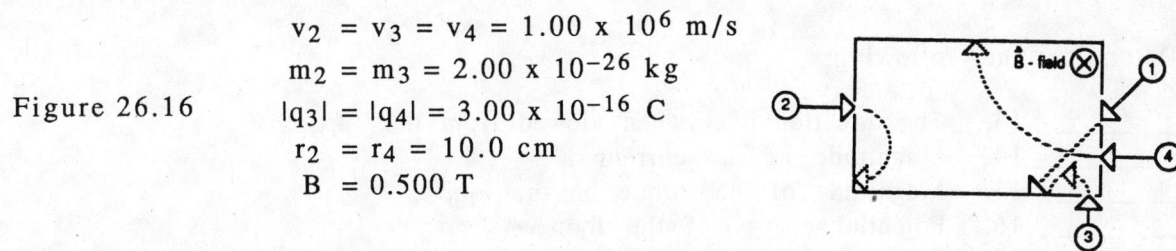

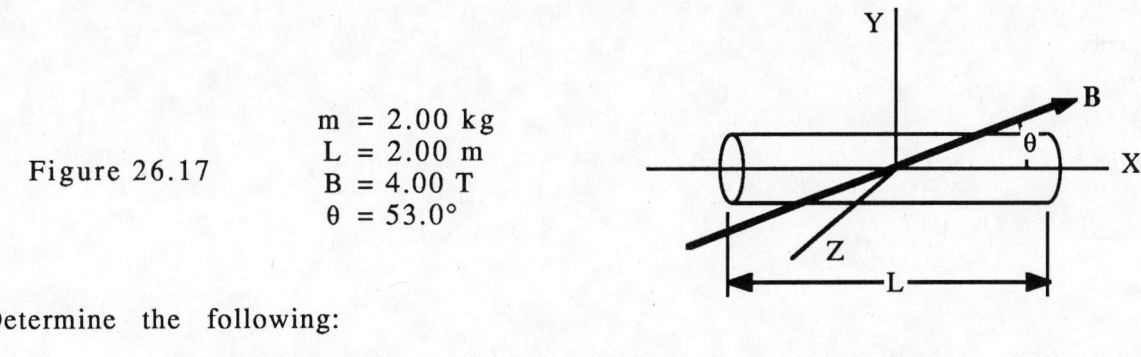

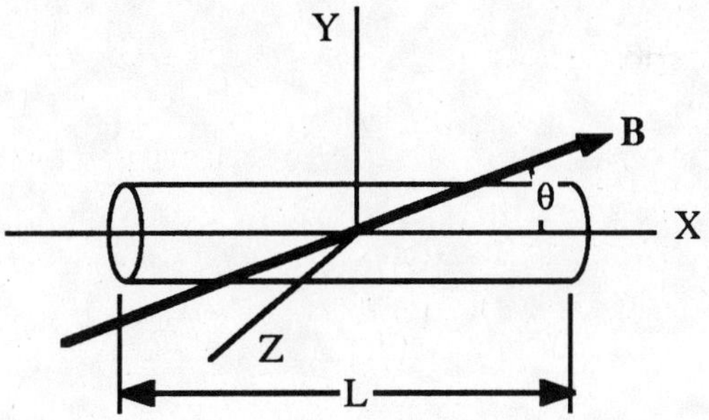

Determine the following:

________ 11. Magnitude of the current needed in the conductor in order for the magnetic force to support the weight
________ 12. Direction of the current

Consider the current carrying loop in a magnetic field as shown in Figure 26.18

Figure 26.18

$R = 2.00$ m
$m = 20.0\pi$ A•m^2
$B = 4.00$ T

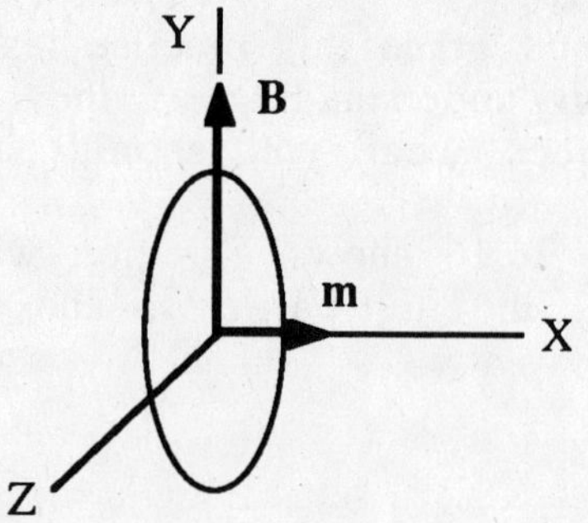

Determine the following:

_________ 13. The direction of I when viewed from the +x-axis
_________ 14. Magnitude of the current I
_________ 15. Magnitude of the torque on the loop
_________ 16. Potential energy of the loop
_________ 17. Direction of **m** when the loop is in the position of stable equilibrium
_________ 18. Direction of **m** when the loop is in the position of unstable equilibrium
_________ 19. Work done against the magnetic field in rotating the loop from the position of stable equilibrium to the position of unstable equilibrium

(See Appendix I for answers.)

PRINCIPAL CONCEPTS AND EQUATIONS PRESCRIPTION

Your score on the practice test is an excellent measure of your understanding of the chapter. You should now use the following chart to write your own prescription for dealing with any weaknesses the practice test points out. Look down the leftmost column to the number of the question(s) you answered incorrectly, reading across that row you will find the concept and/or equation of concern, the section(s) of the study guide you should return to for further study, and some suggested text exercises which you should work to gain additional experience.

Practice Test Questions	Concepts and Equations	Prescription	
		Principal Concepts	Text Exercises
1	Magnetic force: $\mathbf{F} = q\mathbf{v} \times \mathbf{B}$	1	26-2, 3
2	Magnetic force	1	26-2, 29
3	Magnetic force	1	26-3, 29
4	Central force: $F_c = mv^2/r$	3 of Ch 6	6-22, 23
5	Central force = magnetic force: $mv^2/r = \lvert q \rvert vB \Rightarrow \lvert q \rvert = mv/qr$	4	26-26
6	Central force = magnetic force: $mv^2/r = \lvert q \rvert vB \Rightarrow r = mv/\lvert q \rvert B$	4	26-26
7	Momentum: $p = mv = \lvert q \rvert rB$	4	26-25
8	Magnetic force: $\mathbf{F} = q\mathbf{v} \times \mathbf{B}$	1	26-2, 3
9	Magnetic force	1	26-2, 3
10	Central force = magnetic force: $mv^2/r = qvB \Rightarrow m = \lvert q \rvert rB/v$	4	26-25
11	Gravitational force = magnetic force $mg = BIL \Rightarrow I = mg/BL$	2	26-8, 9
12	Magnetic force: $\mathbf{F} = I\mathbf{L} \times \mathbf{B}$	2	26-9, 10
13	Right hand rule for **m**	3	26-14, 15
14	Magnetic moment: $m = IA$	3	26-19, 23
15	Torque of **m**: $\boldsymbol{\tau} = \mathbf{m} \times \mathbf{B}$	3	26-16, 19
16	Potential energy of **m**: $U = -\mathbf{m} \cdot \mathbf{B}$	3	26-23, 24
17	Potential energy of **m**	3	26-23, 24
18	Potential energy of **m**	3	26-23, 24
19	Work equals change in potential: $W = \Delta U$	3	26-23, 24

27 SOURCES OF THE MAGNETIC FIELD

RECALL FROM PREVIOUS CHAPTERS

Previously learned concepts and equations frequently used in this chapter	Text Section	Study Guide Page
Magnetic force on a moving charge: $\mathbf{F_B} = q\mathbf{v} \times \mathbf{B}$	26-1	26-1
Magnetic force on a current element: $\mathbf{F_B} = I\boldsymbol{l} \times \mathbf{B}$	26-1	26-4

NEW IDEAS IN THIS CHAPTER

Concepts and equations introduced	Text Section	Study Guide Page
Biot-Savart law: $d\mathbf{B} = (\mu_0 I/4\pi)d\boldsymbol{l} \times \hat{\mathbf{r}}/r^2$	27-1	27-1
Ampere's Law: $\oint_c \mathbf{B} \cdot d\mathbf{r} = \mu_0 \sum_i I_i$ linking c	27-2	27-7
Force per unit length between long wires: $F/l = \mu_0 I_1 I_2/2\pi R$	27-4	27-13
Magnetic flux: $\Phi_B = \int_s \mathbf{B} \cdot d\mathbf{S}$	27-5	27-20
Displacement current: $I_d = \varepsilon_0 d\Phi_E/dt$	27-6	---

PRINCIPAL CONCEPTS AND EQUATIONS

1 The Biot-Savart Law (Section 27-1)

Review: Charges are the sources of electric fields as discussed in chapter 20 of the text. Figure 27.1 shows a volume of charge and an infinitesimal part of this charge denoted by dq.

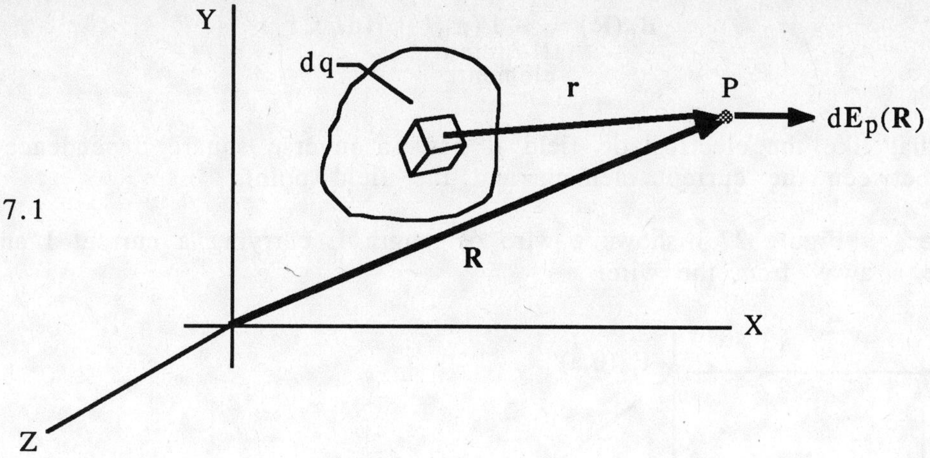

Figure 27.1

The charge dq makes a contribution to the electric field $\mathbf{E}_p(\mathbf{R})$ at point P given by:

$$dE_p(\mathbf{R}) = (1/4\pi\varepsilon_0)(dq\hat{\mathbf{r}}/r^2)$$

The total electric field at P is obtained by adding up the contributions due to all charges, this gives

$$\mathbf{E}_p(\mathbf{R}) = \int_{\substack{\text{all}\\\text{charge}}} (1/4\pi\varepsilon_0)dq\hat{\mathbf{r}}/r^2$$

The source of magnetic fields is found experimentally to be current elements. The Biot-Savart law gives the contribution to the magnetic field at point P due to a current carrying wire. Figure 27.2 shows the segment of a current carrying wire and the resulting contribution to the magnetic field.

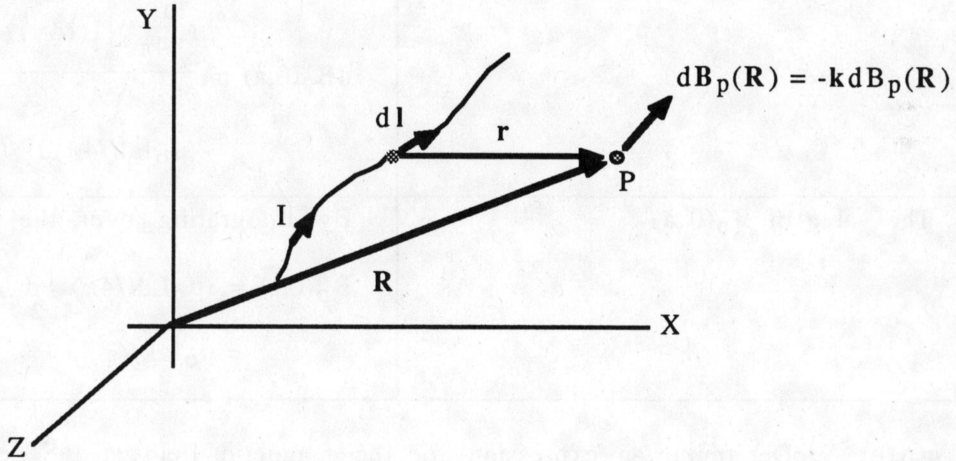

Figure 27.2

The current carrying segment $d\boldsymbol{l}$ makes a contribution to the magnetic field $B_p(\mathbf{R})$ at point P given by

$$d\mathbf{B}_p(\mathbf{R}) = (\mu_0/4\pi)(Id\boldsymbol{l} \times \hat{\mathbf{r}})/r^2$$

The total magnetic field at P is obtained by adding up the contributions due to all current carrying elements, this gives

$$B_p(\mathbf{R}) = \int_{\substack{\text{all current}\\ \text{elements}}} (\mu_0/4\pi) I d\mathit{l} \times \hat{r}/r^2$$

We see that like the electrostatic field it has an inverse square dependence on the distance between the current element and the field point.

Practice: Figure 27.3 shows a wire of length L carrying a current I and a point P a distance a away from the wire.

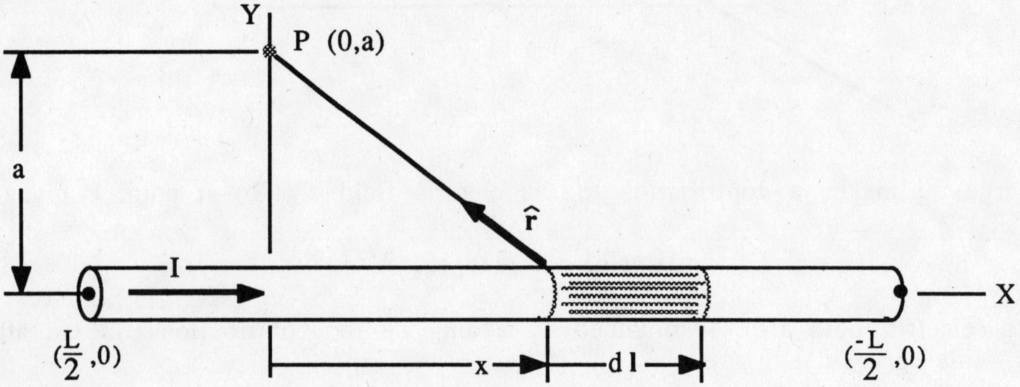

Figure 27.3

Determine the following:

1. The contribution to $\mathbf{B}_p(0,a)$ due to the segment $d\mathit{l}$ a distance x from the center of the wire	$d\mathbf{B}_p(\mathbf{R}) = [(\mu_0/4\pi)I(d\mathit{l}) \times (\hat{r})]/(r^2)$ $r^2 = x^2 + a^2$ and $d\mathit{l} = \mathbf{i}dx$ $\hat{r} = (-\mathbf{i}x + \mathbf{j}a)/(x^2 + a^2)^{1/2}$ $d\mathbf{B}_p(0,a) = \dfrac{(\frac{\mu_0}{4\pi})[I(\mathbf{i}dx)] \times \left[\frac{(-\mathbf{i}x + \mathbf{j}a)}{(x^2 + a^2)^{1/2}}\right]}{(x^2 + a^2)}$ $= (\mu_0 Ia\mathbf{k}/4\pi)dx/(x^2 + a^2)^{3/2}$
2. The value of $\mathbf{B}_p(0,a)$	By integrating over the wire, obtain $\mathbf{B}_p(0,a) = (\mu_0 Ia\mathbf{k}/4\pi)\int_{-L/2}^{L/2} dx/(x^2 + a^2)^{3/2}$ $= \mu_0 Ia\mathbf{k}/4\pi$

Example: Determine an expression for the magnetic field at the center of an arc of a circular loop of radius R carrying a current I. The arc subtends an angle θ at the center.

Given: I, R, and θ for the loop

Determine: An expression for B_c, the magnetic field at the center of the loop.

Strategy: Draw a figure and then use the Biot-Savart law to get $\mathbf{B}_c$.

Solution: Figure 27.4 shows the situation under consideration.

Figure 27.4

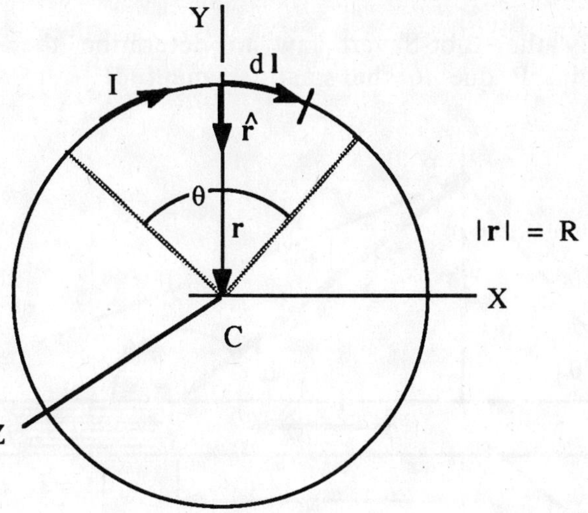

Applying the Biot-Savart Law we have

$$d\mathbf{B}_c = (\mu_0 I/4\pi)d\boldsymbol{l} \times \hat{\mathbf{r}}/R^2$$
$$= [(\mu_0 I/4\pi)dl\ \sin 90°/R^2](-\mathbf{k})$$

and

$$\mathbf{B}_c = \int d\mathbf{B}_c = (\mu_0 I/4\pi R^2)\int_0^{R\theta} dl$$
$$= (\mu_0 I\theta/4\pi R)(-\mathbf{k})$$

Note for a circular loop $\theta = 2\pi$, hence

$$\mathbf{B}_c = (\mu_0 I/2R)(-\mathbf{k})$$

Example: Consider the situation shown in Figure 27.5 and then determine the magnetic field at point P due to the line of current.

Figure 27.5

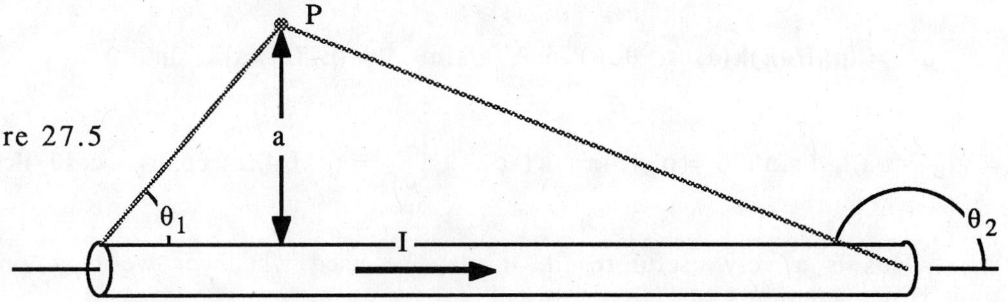

Given: The current shown in Fig. 27.5 and the point P located by a, θ_1, and θ_2.

Determine: The magnetic field at point p, $\mathbf{B}_p$

Strategy: Establish a coordinate system so that the point P is on the y axis and the line of current is on the x axis. Apply the Biot-Savart law to determine the contribution to the magnetic field at point P due to a small segment $d\mathbf{l}$. Integrate over the length of the line of current to determine the total magnetic field at P.

Solution: Apply the Biot-Savart law to determine the contribution to the magnetic field at point P due to the small segment $d\mathbf{l}$.

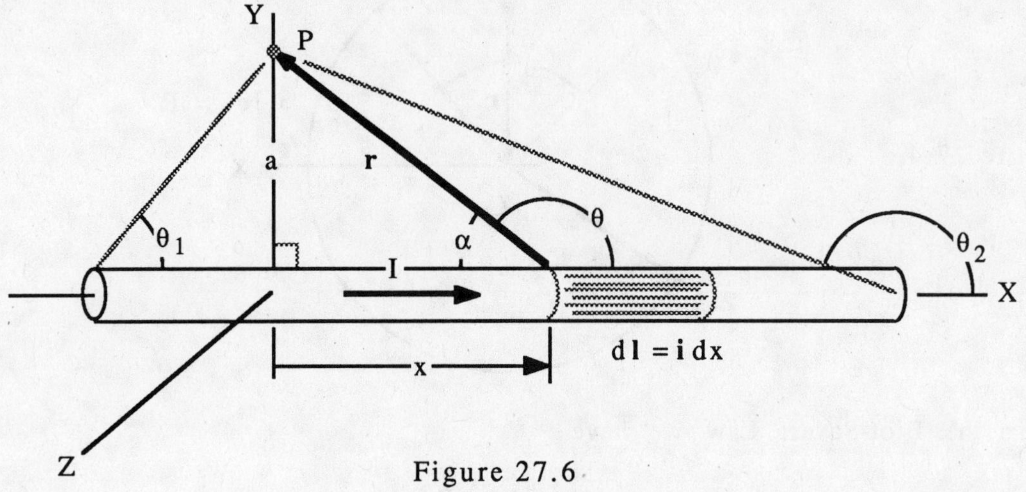

Figure 27.6

$$d\mathbf{B} = (\mu_0 I/4\pi)d\mathbf{l} \times \hat{\mathbf{r}}/r^2$$
$$= (\mu_0 I/4\pi)(\mathbf{k})dx \ \sin\theta/r^2$$

It is convenient to use θ rather than x as the variable of integration. We observe from Figure 27.6 that

$$a/x = \tan\alpha = -\tan\theta$$

or

$$x = -a \cot\theta$$

differentiating $dx = a \csc^2\theta d\theta.$

We also see that $\sin\alpha = \sin\theta = a/r$ or $r = a/\sin\theta$

$$d\mathbf{B} = (\mu_0 I/4\pi)\mathbf{k}(a \csc^2\theta d\theta) \sin\theta/(a/\sin\theta)^2 = (\mu_0 I/4\pi a)\mathbf{k} \sin\theta d\theta$$

$$\mathbf{B} = (\mu_0 I/4\pi a)\mathbf{k} \int_{\theta_1}^{\theta_2} \sin\theta d\theta = (\mu_0 I/4\pi a)\mathbf{k}(-\cos\theta)\Big|_{\theta_1}^{\theta_2} = (\mu_0 I/4\pi a)(\cos\theta_1 - \cos\theta_2)\mathbf{k}$$

Note 1: This is a very useful result. It may be used whenever we have one or more straight line segments.

Note 2: For a long wire and points P not near either end, $\theta_1 \to 0$ and $\theta_2 \to \pi$. Then

$$\mathbf{B} = (\mu_0 I/2\pi a)\mathbf{k}$$

Example: Consider the current carrying arrangement shown in Figure 27.7 and then determine an expression for the magnetic field at point C.

Figure 27.7

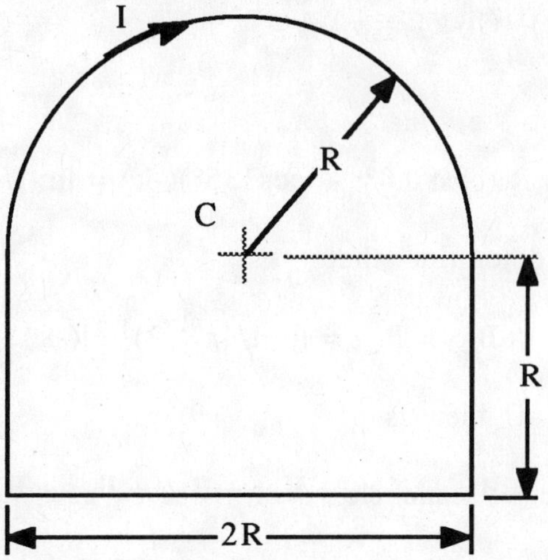

Given: The current I and the distance R

Determine: An expression for the magnetic field $\mathbf{B}_c$ at point C.

Strategy: Using the results of the last two examples we can obtain the contribution to $\mathbf{B}_c$ due to the semi-circle and due to each of four segments of length R which contribute equally to $\mathbf{B}_c$. We can then add the contributions to B_c due to these five segments and obtain the total magnetic field at point C.

Solution: The example for the arc of a circle gives

$$\mathbf{B}_{semicircle} = (\mu_0 I\theta/4\pi R)(-\mathbf{k}) \quad \text{but} \quad \theta = \pi$$
$$= (\mu_0 I/4R)(-\mathbf{k})$$

The contribution to $\mathbf{B}_c$ from the four segments of length R is obtained as follows:

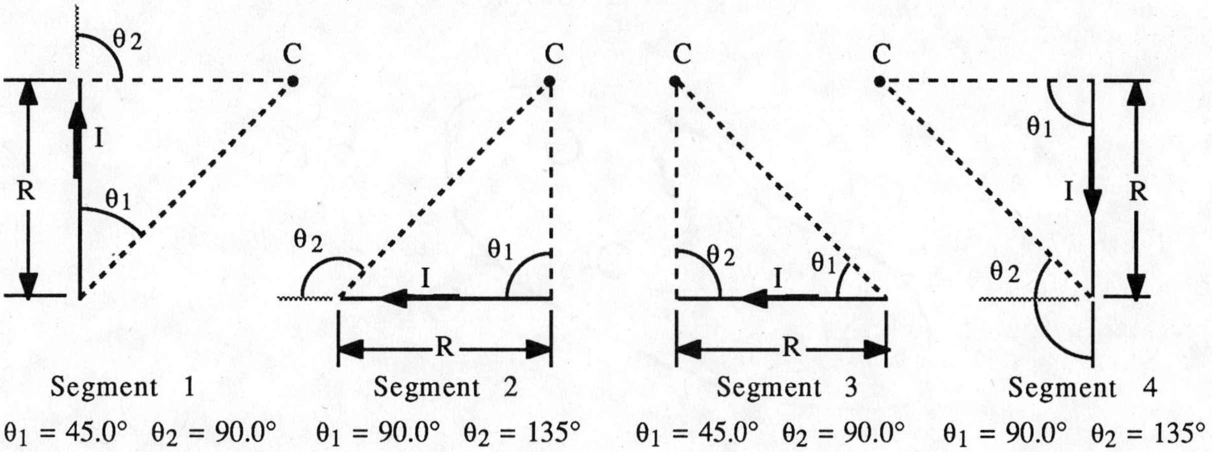

Figure 27.8

Working with segment 1 we have

$$\mathbf{B}_{c1} = (\mu_0 I/4\pi R)(\cos\theta_1 - \cos\theta_2)(-\mathbf{k}) = (\mu_0 I/4\pi R)(\cos 45.0° - \cos 90°)(-\mathbf{k})$$
$$= [\mu_0 I/4\pi R(2)^{1/2}](-\mathbf{k})$$

For segment 2 we have

$$\mathbf{B}_{c2} = (\mu_0 I/4\pi R)(\cos 90.0° - \cos 135°)(-\mathbf{k}) = [\mu_0 I/4\pi R(2)^{1/2}](-\mathbf{k})$$

In like manner we may determine that

$$\mathbf{B}_{c3} = \mathbf{B}_{c4} = [\mu_0 I/4\pi R(2)^{1/2}](-\mathbf{k})$$

The total magnetic field at C then is

$$\mathbf{B}_c = \mathbf{B}_{semicircle} + \mathbf{B}_{c1} + \mathbf{B}_{c2} + \mathbf{B}_{c3} + \mathbf{B}_{c4}$$

$$= (\mu_0 I/4R)(-\mathbf{k}) + 4[\mu_0 I/4\pi R(2)^{1/2}](-\mathbf{k})$$

$$= (\mu_0 I/4\pi R)[\pi + 2(2)^{1/2}](-\mathbf{k})$$

Related Text Exercises: Ex. 27-1 through 27-17.

2 Ampere's Law (Section 27-2)

Review: Recall that in the case for electric fields Gauss' Law was equivalent to Coulomb's Law. It was stated as

$$\oint_s \mathbf{E} \cdot d\mathbf{S} = Q_{inside\ s}/\varepsilon_0$$

A similar result holds for magnetic fields. Ampere's Law is equivalent to the Biot-Savart law and is stated as follows (see Figure 27.9)

$$\oint_c \mathbf{B} \cdot d\mathbf{r} = \mu_0 \sum_i I_i \text{ linking } c$$

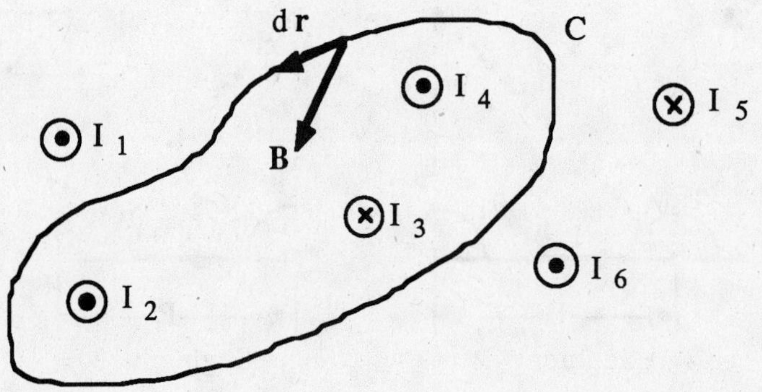

Figure 27.9

Ampere's Law states that although **B** at each point of C depends on all the currents, the integral of the part of **B** tangent to the curve C when integrated around the curve depends only on the currents passing though any area bounded by C. These currents are said to link the curve C. In Figure 27.9 the currents I_2, I_3 and I_4 link C. These currents are positive when coming toward you and negative when going away assuming that the direction of integration around C is viewed from a position so that it is counter-clockwise. In Figure 27.9 I_2 and I_4 are positive and I_3 is negative.

Practice: Consider Figure 27.10.

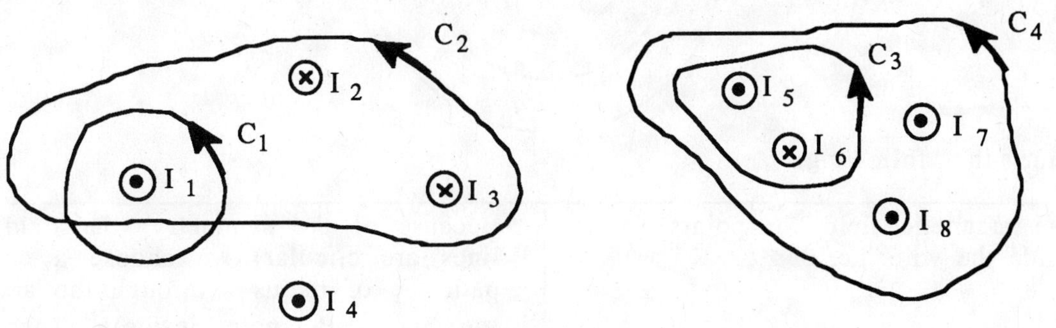

Figure 27.10

Determine the following:

1. Apply Ampere's Law to curve C_1	$\oint_{C1} \mathbf{B} \cdot d\mathbf{r} = \mu_0 I_1$
2. Apply Ampere's Law to curve C_2	$\oint_{C2} \mathbf{B} \cdot d\mathbf{r} = \mu_0(I_1 - I_2 - I_3)$
3. Apply Ampere's Law to curve C_3	$\oint_{C3} \mathbf{B} \cdot d\mathbf{r} = \mu_0(I_5 - I_6)$
4. Apply Ampere's Law to curve C_4	$\oint_{C4} \mathbf{B} \cdot d\mathbf{r} = \mu_0(I_5 - I_6 + I_7 + I_8)$

Related Text Exercises: Ex. 27-18 through 27-20.

3 Applications of Ampere's Law (Section 27-3)

Review: Just as Gauss' Law was useful for obtaining the electric field **E** for highly symmetric charge distributions, so also is Ampere's Law useful for obtaining the magnetic field **B** for current distributions with appropriate symmetry.

Practice: Figure 27.11 shows the cross section of a long current carrying wire of radius R. The current is uniform throughout the wire.

Figure 27.11

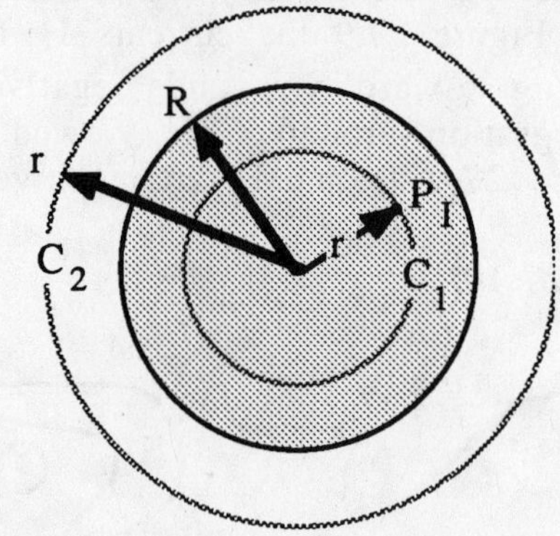

Determine the following:

1. The magnetic field for points inside the wire, i.e. $0 \le r \le R$	Because of the symmetry (the field lines are circular) we choose a circular path C_1 of radius r through the arbitrary point P_I (point inside). Then $\oint_{C1} \mathbf{B} \cdot d\mathbf{r} = \mu_0 I_{linking\ C1}$ Since **B** is tangent to C_1 and constant in magnitude along C_1, the left hand side of the above expression becomes $\oint_{C1} \mathbf{B} \cdot d\mathbf{r} = B(2\pi r)$. The right hand side of the above expression is μ_0 times the fraction of I through the area enclosed by C_1, hence $\mu_0 I_{linking\ C1} = \mu_0 I\pi r^2/\pi R^2$. Combining, we obtain $B(2\pi r) = \mu_0 I\pi r^2/\pi R^2$ or $B(0 \le r \le R) = \mu_0 Ir/2\pi R^2$.
2. The magnetic field for points outside the wire, i.e. $r \ge R$	Again we use a circle. This time it is C_2. The arguments about the left hand side of Amperes Law are unchanged. The current linking C_2 is now all of I. $\oint_{C2} \mathbf{B} \cdot d\mathbf{r} = \mu_0 I_{linking\ C2}$ $B(2\pi r) = \mu_0 I$ or $B(r \ge R) = \mu_0 I/2\pi r$ (It acts like a long wire for points outside the wire).

3. The magnetic field for points on the wire, i.e. $r = R$	Using the result from step 1 of this practice obtain $B(0 \le r \le R) = \mu_0 Ir/2\pi R^2$ or $B(r = R) = \mu_0 I/2\pi R$. Using the result from step 2 of this practice obtain $B(r \ge R) = \mu_0 I/2\pi r$ or $B(r = R) = \mu_0 I/2\pi R$. Notice that the solution inside the wire must equal the solution outside the wire at $r = R$.
4. A plot of B(r) vs r	B(r) vs r 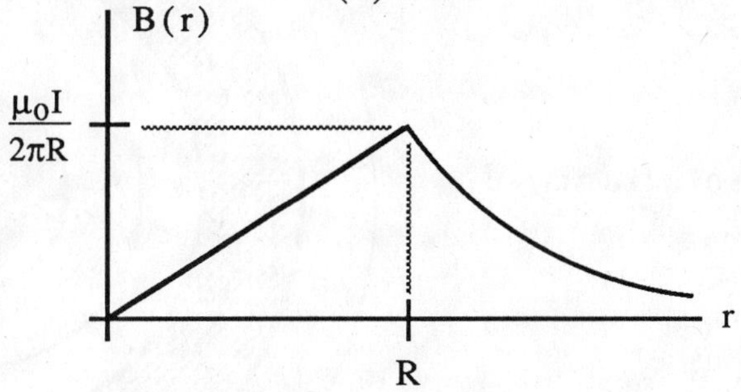 Figure 27.12

Example: Figure 27.13 shows a coaxial cable which consists of a solid conductor of radius "a" surrounded by an insulating cylinder of inner radius "a" and outer radius "b" and then surrounded by a conducting cylinder of inner radius "b" and outer radius "c". A current I goes down the wire (uniformly) in one direction and returns (uniformly) in the opposite direction in the outer conducting cylinder. Obtain an expression for B(r) for all regions, i.e. (a) $r \le a$, (b) $a \le r \le b$, (c) $b \le r \le c$, (d) $r \ge c$.

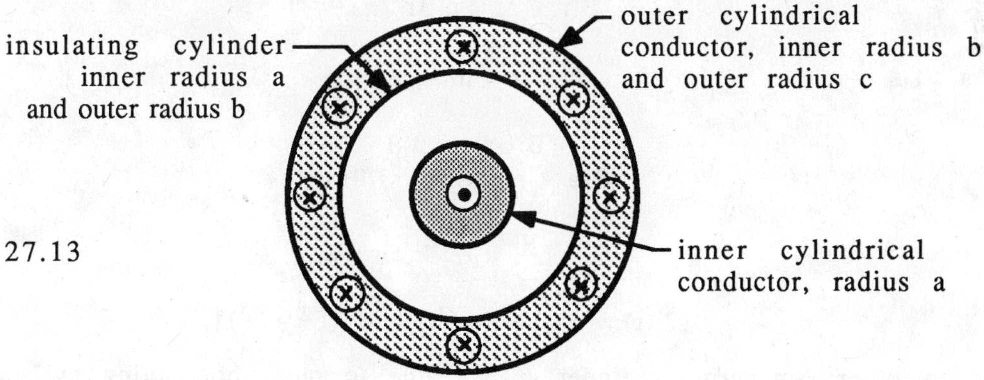

Figure 27.13

Given: The coaxial cable shown in Figure 27.13
I the current - out of the page for the inner conducting cylinder
- into the page for the outer conducting cylinder
a, b, c - various radii

Determine: An expression for B(r) for all regions in space, i.e.
(a) $r \le a$, (b) $a \le r \le b$, (c) $b \le r \le c$, (d) $r \ge c$.

Strategy: Since we are interested in four regions, we construct a circular path through an arbitrary point in each region and then apply Ampere's Law.

Solution:

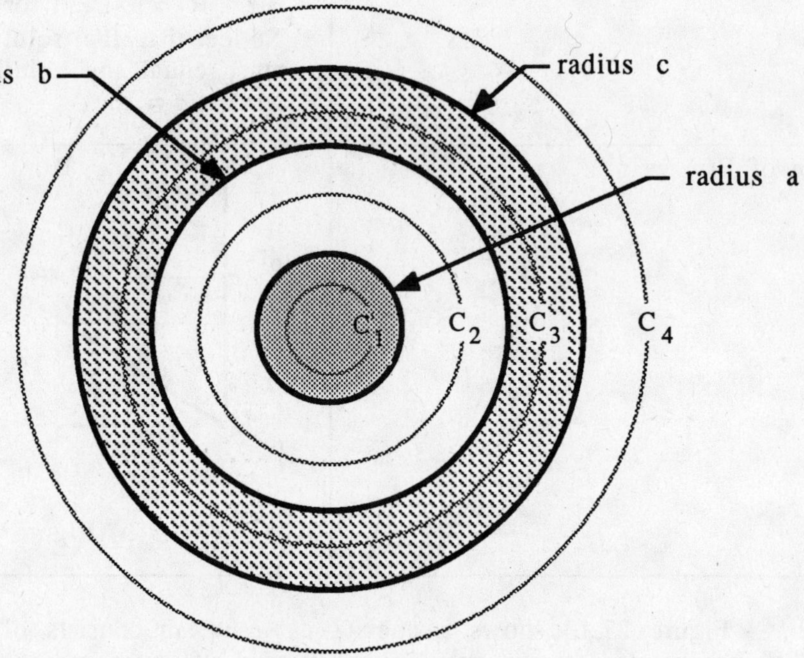

Figure 27.14

(a) $r \le a$

$$\oint_{C_1} \mathbf{B} \cdot d\mathbf{r} = \mu_0 I_{\text{linking } C_1}$$

$$B(2\pi r) = \mu_0 I \pi r^2/\pi a^2$$

$$B(r \le a) = \mu_0 I r/2\pi a^2$$

(b) $a \le r \le b$

$$\oint_{C_2} \mathbf{B} \cdot d\mathbf{r} = \mu_0 I_{\text{linking } C_2}$$

$$B(2\pi r) = \mu_0 I$$

$$B(a \le r \le b) = \mu_0 I/2\pi r$$

(c) $b \le r \le c$

$$\oint_{C_3} \mathbf{B} \cdot d\mathbf{r} = \mu_0 I_{\text{linking } C_3}$$

$$B(2\pi r) = \mu_0[I - I\pi(r^2 - b^2)/\pi(c^2 - b^2)]$$

due to inner conducting cylinder ↑ ↑ due to outer conducting cylinder

$$B(b \le r \le c) = (\mu_0 I/2\pi r)[(c^2 - r^2)/(c^2 - b^2)]$$

(d) $r \geq c$

$$\oint_{C4} \mathbf{B} \cdot d\mathbf{r} = \mu_0 I_{linking\ C4}$$

$$B(2\pi r) = \mu_0\ (I - I)$$

$$B(r \geq c) = 0$$

Example: A long wire of radius R carries a non uniform current with the current density having cylindrical symmetry. The current density is given by $j(r) = Kr$ for $0 \leq r \leq R$. Obtain an expression for B(r) for all regions, i.e. (a) $0 \leq r \leq R$ and (b) $r \geq R$.

Given: The current density $j(r) = Kr$ for $0 \leq r \leq R$ for a long wire of radius R.

Determine: An expression for B(r) for all regions, i.e. (a) $0 \leq r \leq R$ and (b) $r \geq R$.

Strategy: Use Ampere's Law in each of the two regions (a) $0 \leq r \leq R$ and (b) $r \geq R$. Since the current density is not uniform we will have to obtain the current linking our curves (circles) by integrating the current density over the area enclosed by the circles. See Figure 27.15.

Solution:

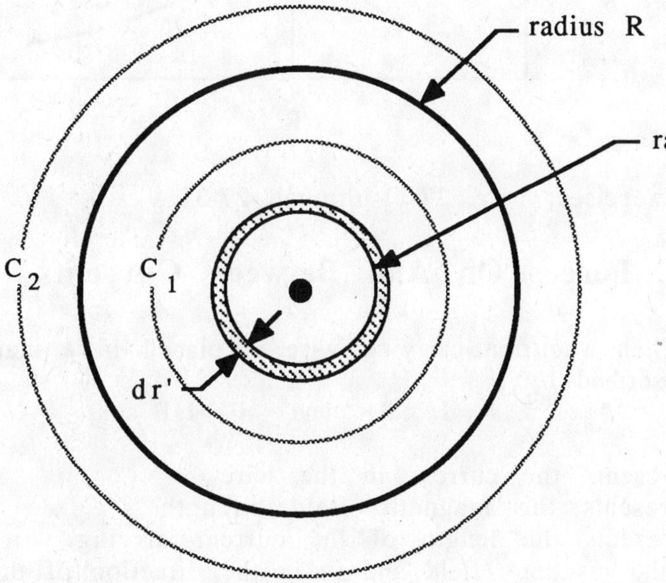

Figure 27.15

(a) $0 \leq r \leq R$

$$I_{linking\ C1} = \int_0^r j(r')2\pi r'dr' = \int_0^r (Kr')2\pi r'dr' = 2\pi K \int_0^r (r')^2 dr'$$

$$= 2\pi K (r')^3/3 \Big|_0^r = 2\pi K r^3/3$$

$$\oint_{C1} \mathbf{B} \cdot d\mathbf{r} = \mu_0 I_{linking\ C1}$$

$$B(2\pi r) = \mu_0(2\pi K r^3/3)$$

$$B(0 \leq r \leq R) = \mu_0 K r^2/3$$

(b) $r \geq R$ $\quad I_{\text{linking } C2} = \int_0^R j(r')2\pi r'dr' = \int_0^R (Kr')2\pi r'dr' = 2\pi K \int_0^R (r')^2 dr'$

$$= 2\pi K (r')^3/3 \Big|_0^R = 2\pi K R^3/3$$

$$\oint_{C2} \mathbf{B} \cdot d\mathbf{r} = \mu_0 I_{\text{linking } C2}$$

$$B(2\pi r) = \mu_0(2\pi K R^3/3)$$

$$B(r \geq R) = \mu_0 K R^3/3r$$

A plot of B(r) vs r is shown in Figure 27.16.

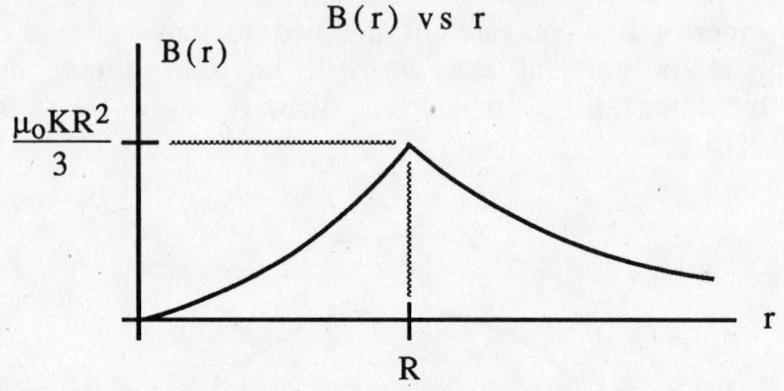

Figure 27.16

Related Text Exercises: Ex. 27-21 through 27-31.

4 Magnetic Forces On And Between Currents (Section 27-4)

Review: When a current-carrying wire is placed in a magnetic field, it experiences a force described by:

$$\mathbf{F} = I\mathbf{l} \times \mathbf{B} \quad \text{and} \quad F = IlB \sin\theta$$

where:

- I represents the current in the wire
- B represents the magnetic field strength
- l represents the length of the current-carrying wire that is actually in the magnetic field and is in the direction of the current
- θ represents the angle between the direction of the current and the direction of the **B**-field
- **F** represents the magnetic force on the current-carrying wire due to the **B**-field

The direction of the force is obtained by using the right-hand rule.

Figure 27-17 shows segments of two long parallel wires separated by distance "a" and carrying currents I_1 and I_2, respectively. Each wire creates a magnetic field that exerts a magnetic force on the current in the other wire.

For both cases:

$B_1 = \mu_0 I_1/2\pi a$
$B_2 = \mu_0 I_2/2\pi a$
$F_1 = B_2 I_1 l \sin 90.0°$
$= \mu_0 I_1 I_2 l/2\pi a$
$F_2 = B_1 I_2 l \sin 90.0°$
$= \mu_0 I_1 I_2 l/2\pi a$
$F/l = \mu_0 I_1 I_2/2\pi a$

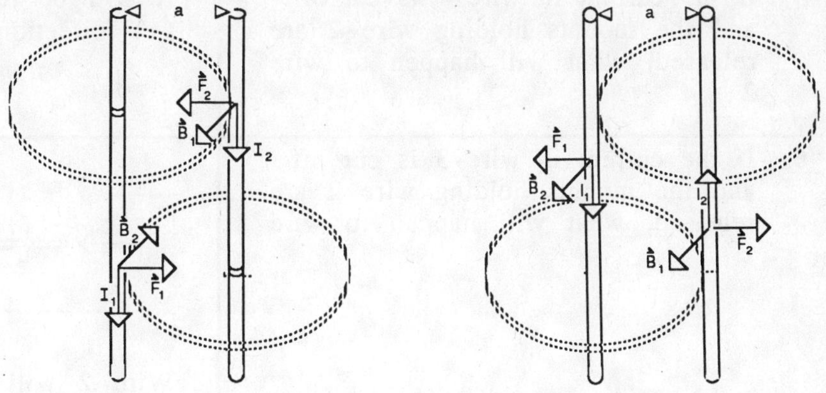

(a) Parallel currents (b) Anti-parallel currents

Figure 27.17

Practice: Figure 27.18 shows three current-carrying wires mounted so that they cannot move.

$I_1 = 1.00$ A
$I_2 = 2.00$ A
$I_3 = 3.00$ A
$d = 0.100$ m

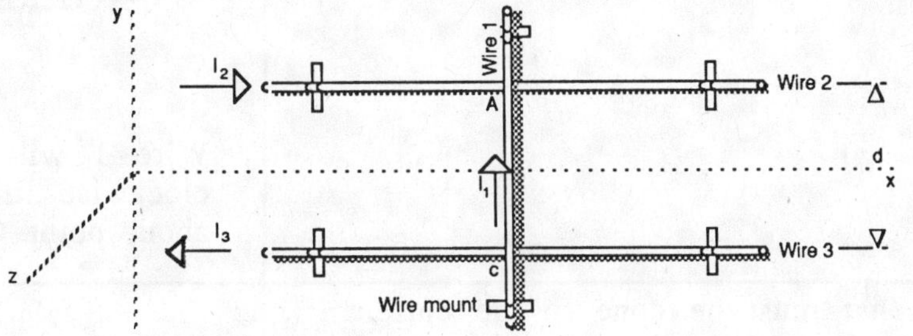

Figure 27.18

Determine the following:

1. Magnitude and direction of the **B**-field experienced by wire 2 due to the current in wire 3 (**B**$_3$)	$B_3 = \mu_0 I_3/2\pi d = 6.00 \times 10^{-6}$ T **B**$_3$ is in the -Z direction at wire 2.
2. Magnetic force per unit length felt by wire 2 due to wire 3 (**F**$_2/l$)	$F_2 = B_3 I_2 l \sin 90.0°$ $F_2/l = B_3 I_2 = 12.0 \times 10^{-6}$ N/m The force experienced by wire 2 (**F**$_2$) is in the +Y direction.
3. Magnitude and direction of the **B**-field experienced by wire 3 due to the current in wire 2 (**B**$_2$)	$B_2 = \mu_0 I_2/2\pi d = 4.00 \times 10^{-6}$ T **B**$_2$ is in the -Z direction at wire 3.
4. Magnetic force per unit length felt by wire 3 due to wire 2 (**F**$_3/l$)	$F_3 = B_2 I_3 l \sin 90.0°$ $F_3/l = B_2 I_3 = 12.0 \times 10^{-6}$ N/m The force experienced by wire 3 (**F**$_3$) is in the -Y direction.

5. If the current in wire 1 is cut off and the mounts holding wire 2 are released, what will happen to wire 2	It will be repelled upward (+Y direction) from wire 3
6. If the current in wire 3 is cut off and the mounts holding wire 2 are released, what will happen to wire 2	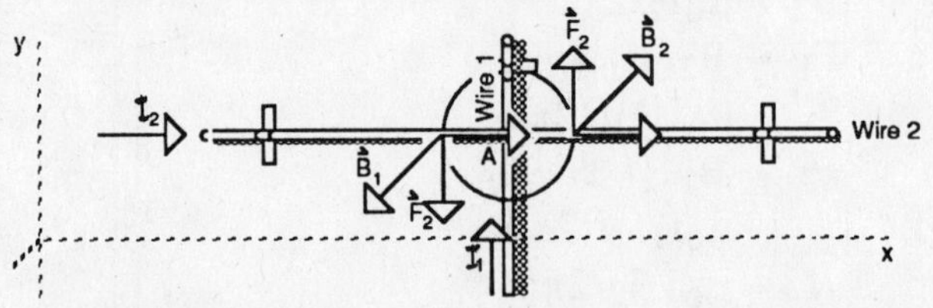 Wire 2 will initially rotate counter-clockwise in the plane of the paper about point A.
7. If the current in wire 2 is cut off and the mounts holding wire 1 are released, what will happen to wire 1	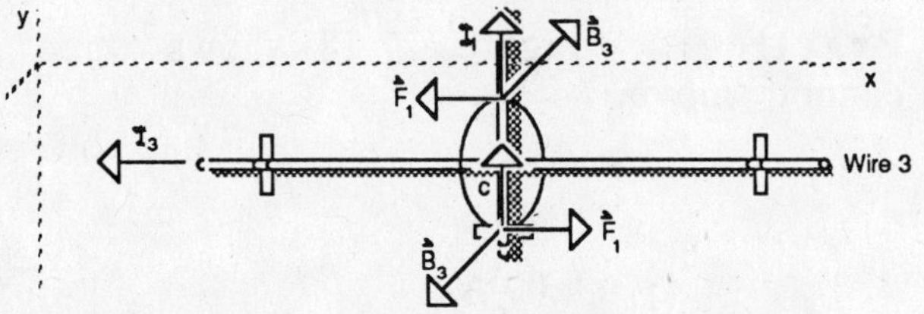 Wire 1 will initially rotate counter-clockwise in the plane of the paper about point C.
8. What must be done to the current in wire 2 in order for wires 2 and 3 to attract each other with a force per unit length of 2.40×10^{-5} N	$F/l = \mu_0 I_2 I_3 / 2\pi d$ $I_2 = (F/l)2\pi d/\mu_0 I_3 = 4.00$ A The current in wire 2 must double in magnitude and reverse in direction.

Example: A long straight wire has a current of 10.0 A. A square coil 20.0 cm on a side, having a resistance of 5.00 Ω and containing a 10.0-V battery, is positioned 10.0 cm from the wire, as shown in Figure 27.19. Calculate the net force experienced by the square coil.

Figure 27.19

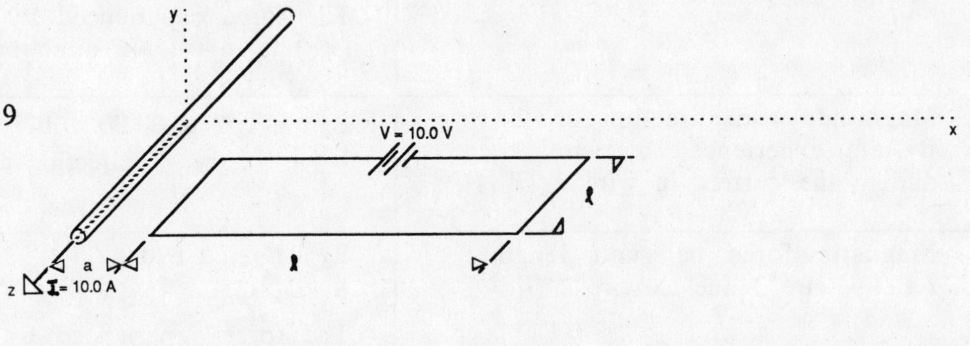

Given: $I = 10.0$ A = current in the long straight wire
$V = 10.0$ V = terminal potential difference of the battery
$R = 5.00\ \Omega$ = resistance of the coil
$l = 2.00 \times 10^{-1}$ m = length of the sides of the square coil
$a = 1.00 \times 10^{-1}$ m = distance of the coil from the wire

Determine: The net force on the square coil

Strategy: We can first establish the fact that we only need to be concerned with those segments of the square coil that are parallel to the long straight wire. Next, we can determine the magnitude and direction of the magnetic field experienced by the parallel segments of the coil. Knowing V and R, we can determine the current in the coil. Finally, we can combine the information about the magnetic field and the current to find the force on each segment and hence the net force on the coil.

Solution: First let's investigate the effect of the segments of the coil that are perpendicular to the long straight wire.

Figure 27.20

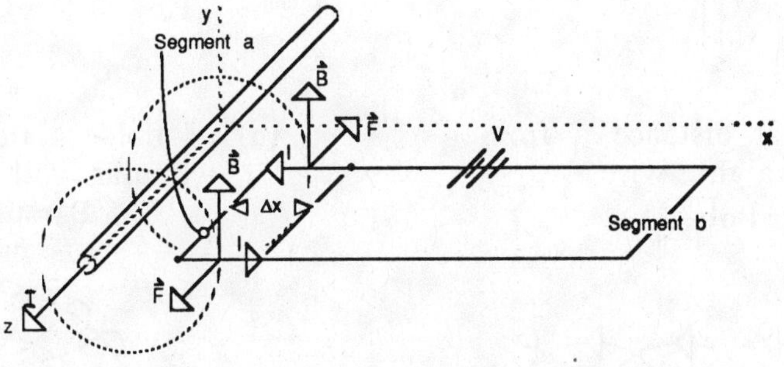

Let's look at the tiny segments of length Δx, shown in Figure 27.20. These two segments are an equal distance from the long straight wire, hence they experience the same **B**-field. Since these segments have the same current and experience the same **B**-field, they will feel magnetic forces equal in magnitude and opposite in direction. If we continue in this manner along the two perpendicular segments, we see that their net effect is zero (i.e., they cancel each other). For convenience, let's refer to the parallel segments as segment a and segment b. The magnetic field experienced by these segments is given by

$$\mathbf{B}_a = \mu_0 I/2\pi a = 2.00 \times 10^{-5}\ \text{T} \qquad +\text{Y direction}$$

$$\mathbf{B}_b = \mu_0 I/2\pi(a + b) = 6.67 \times 10^{-6}\ \text{T} \qquad +\text{Y direction}$$

The current in the coil is $I_c = V/R = 2.00$ A

The magnetic force experienced by each of these segments is

$$\mathbf{F}_a = -B_a I_c l \sin 90.0° = -8.00 \times 10^{-6}\ \text{N} \qquad (-\text{X direction})$$
$$\mathbf{F}_b = +B_b I_c l \sin 90.0° = +2.67 \times 10^{-6}\ \text{N} \qquad (+\text{X direction})$$
$$\mathbf{F}_{net} = \mathbf{F}_a + \mathbf{F}_b = -5.33 \times 10^{-6}\ \text{N} \qquad (-\text{X direction})$$

Related Text Exercises: Ex. 27-32 through 27-34.

5 Some Important Current Geometries (Sections 27-2, 3 and 4)

Note: We have discussed a number of special geometries so far but not all of the important ones; they are all discussed in the text however. Below is a review of some of these important special cases.

Review: A current-carrying wire creates a magnetic field. Figure 27-21 shows several different current-carrying wires, the resulting magnetic field, and expressions for the magnitude of the **B**-field.

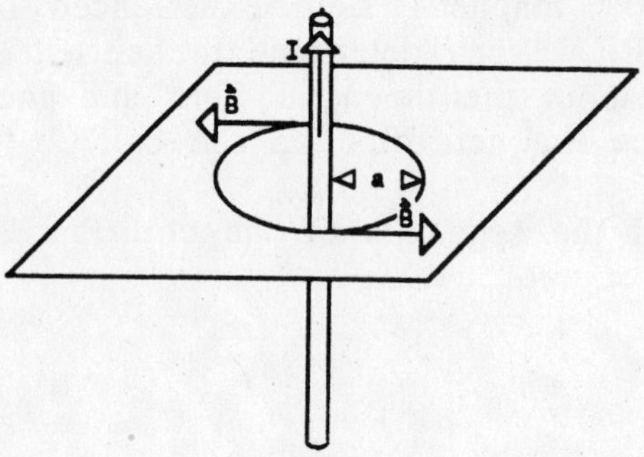

(a) **B**-field at distance a from long straight wire
$B = \mu_0 I/2\pi a$

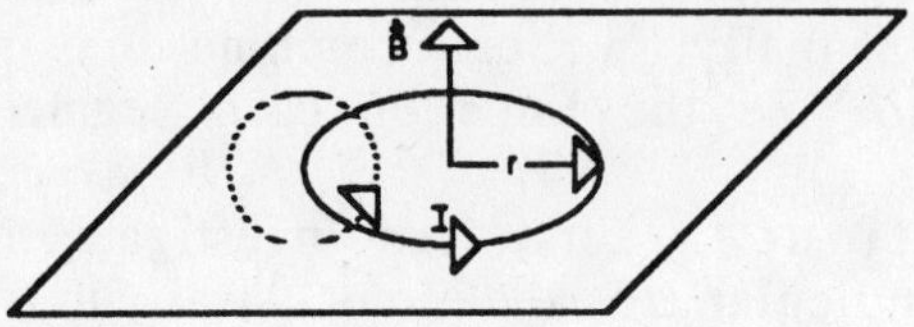

(b) **B**-field at the center of a flat circular coil of N turns
$B = \mu_0 NI/2r$

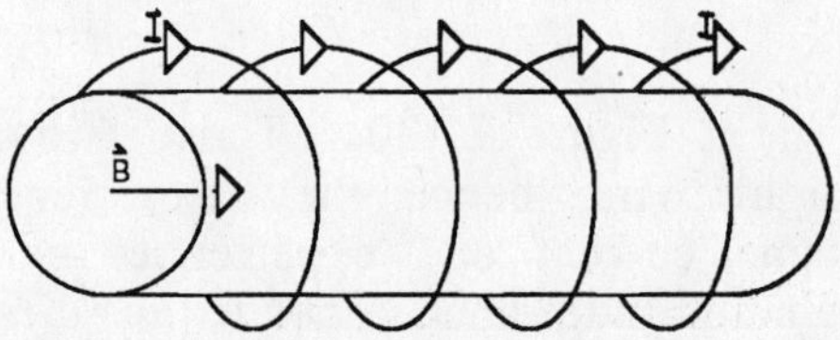

(c) **B**-field inside a long straight solenoid
$B = \mu_0 NI/L = \mu_0 nI$

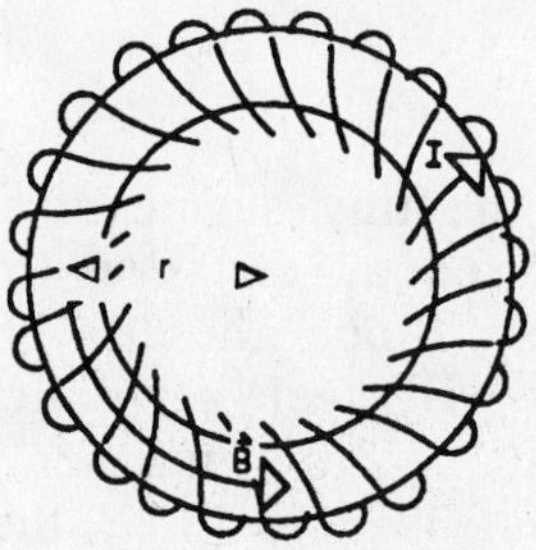

(d) **B**-field inside a toroid
$B = \mu_0 NI/2\pi r$

Figure 27-21

The direction of the magnetic field for each case shown in Figure 27.21 is determined as follows:

(a) Grasp the current-carrying wire with your right hand in such a manner that your thumb is in the direction of the current. The fingers of your right hand curl around the wire in the direction of the **B**-field.

(b) You can apply the preceding right-hand rule to the coil (see the small field circles) to determine that the **B**-field at the center is upward. You can also grasp the coil in your right hand in such a manner that the fingers of your right hand curl around the coil in the direction of the current. Your thumb then indicates the direction of the **B**-field.

(c) and (d) Apply either the right hand rule introduced in (a) or the right hand rule introduced in (b).

Practice: Consider the situation shown in Figure 27.22.

$v = 1.00 \times 10^5$ m/s
$q = 2.00 \times 10^{-10}$ C
$d = 2.00 \times 10^{-2}$ m
$I = 1.00$ A
+ = up out of page
- = down into page

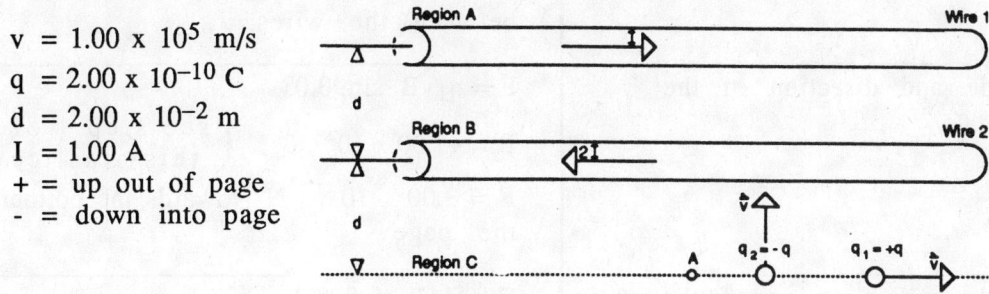

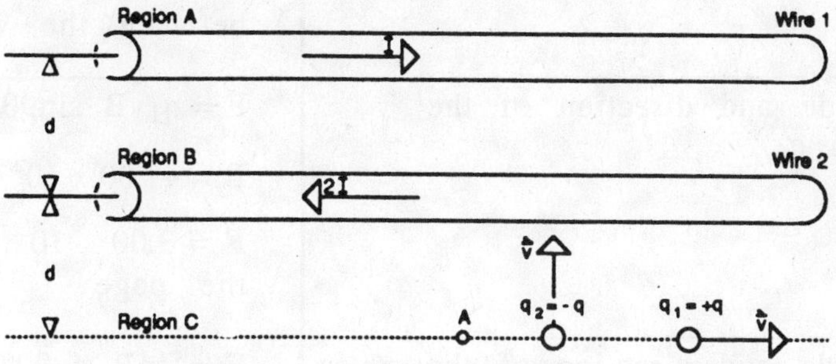

Figure 27.22

Determine the following:

1. Magnitude and direction of the magnetic field at A	The field at A due to wire 1 is $\mathbf{B}_1 = -\mu_0 I/2\pi(2d) = -5.00 \times 10^{-6}$ T. The field at A due to wire 2 is $\mathbf{B}_2 = +\mu_0(2I)/2\pi d = +2.00 \times 10^{-5}$ T. The net field at A is $\mathbf{B} = \mathbf{B}_1 + \mathbf{B}_2 = +1.50 \times 10^{-5}$ T. The + sign indicates the direction.
2. Magnitude and direction of the magnetic field midway between the wires	$\mathbf{B}_1 = -\mu_0 I/2\pi(d/2) = -2.00 \times 10^{-5}$ T $\mathbf{B}_2 = -\mu_0(2I)/2\pi(d/2) = -4.00 \times 10^{-5}$ T The net field is $\mathbf{B} = \mathbf{B}_1 + \mathbf{B}_2 = -6.00 \times 10^{-5}$ T.
3. Magnitude and direction of the magnetic field experienced by the current in wire 1	The magnetic field experienced by the current in wire 1 is due to the current in wire 2. $\mathbf{B}_2 = -\mu_0(2I)/2\pi d = -2.00 \times 10^{-5}$ T
4. The distance from wire 1 in region A where the magnetic field is zero	In region A the **B**-field due to wire 1 is up (+) and the **B**-field due to wire 2 is down (-). Let x equal the distance from wire 1, then $\mathbf{B}_1 + \mathbf{B}_2 = 0$. $+\mu_0 I/2\pi x - \mu_0(2I)/(2\pi)(d + x) = 0$ or $x = d = 2.00 \times 10^{-2}$ m
5. What should be done to the current in wire 1 to get zero magnetic field at A	$\mathbf{B}_1 = -\mu_0 I/2\pi(2d) = -\mu_0 I/4\pi d$ $\mathbf{B}_2 = +\mu_0(2I)/2\pi d = \mu_0 I/\pi d$ Note that B_2 is 4 times B_1. Consequently, we need to increase the current in wire 1 by a factor of 4 in order to get B = 0 at A.

6. What should be done to the current in wire 1 to get zero magnetic field midway between the two wires	$B_1 = -\mu_0 I/2\pi(d/2) = -\mu_0 I/\pi d$ $B_2 = -\mu_0(2I)/2\pi(d/2) = -2\mu_0 I/\pi d$ We need to reverse the direction and double the magnitude of the current in wire 1 in order to get B = 0 midway between the wires.
7. Magnitude and direction of the force on q_1	$F = q_1 vB \sin 90.0°$ $B = +1.50 \times 10^{-5}$ T (see step 1 of this practice) $F = 3.00 \times 10^{-10}$ N towards the bottom of the page
8. Magnitude and direction of the force on q_2	$F = qvB \sin 90.0°$ $F = 3.00 \times 10^{-10}$ N towards the left of the page

Example: Two flat circular coils are placed concentric to each other in the same plane as shown in Figure 27.23.

Figure 27.23

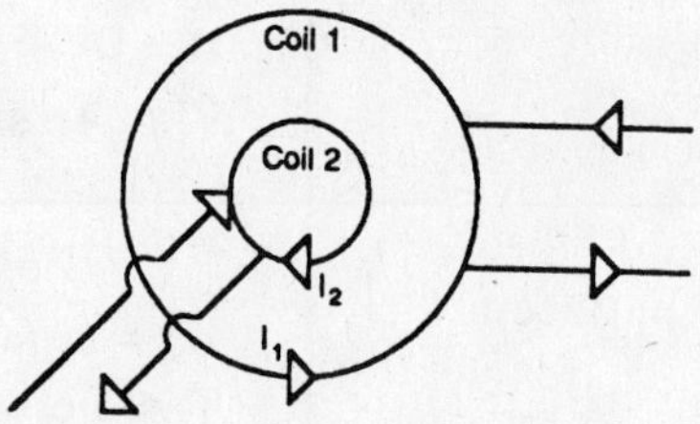

$r_1 = 2.00 \times 10^{-1}$ m $N_1 = 5$ turns $I_1 = 2.00$ A

$r_2 = 1.00 \times 10^{-1}$ m $N_2 = 10$ turns $I_2 = 4.00$ A

Determine the magnitude and direction of the magnetic field at the center of the coils.

Given: r_1 and r_2 -- radius of each flat circular coil
N_1 and N_2 -- number of turns in each coil
I_1 and I_2 -- current in each coil

Determine: Magnitude and direction of the net **B**-field at the center of the coil.

Strategy: From the given information, we can determine the magnitude and direction of the B-field at the center of the coils due to the current in each coil. We can add these results vectorially to obtain the net **B**-field.

Solution: The magnetic field at the center due to 1 turn of wire from each coil is determined as follows:

$$B_1 = +\mu_0 I_1/2r_1 = +6.28 \times 10^{-6}\ T$$
$$B_2 = -\mu_0 I_2/2r_2 = -25.1 \times 10^{-6}\ T$$

The magnetic field at the center due to the number of turns of wire (N) is

$$B_{1T} = +N_1B_1 = +31.4 \times 10^{-6}\ T \quad \text{and} \quad B_{2T} = N_2B_2 = -251 \times 10^{-6}\ T.$$

The net magnetic field at the center of the coil is

$$B_{net} = B_{1T} + B_{2T} = -2.20 \times 10^{-4}\ T.$$

The minus sign indicates that the net field is directed downward.

Related Text Exercises: Ex. 27-3 through 27-16 and 27-27 through 27-31.

6 Magnetic Flux (Section 27-5)

Review: In the discussion of Gauss' Law (Text Chapter 21) the concept of an electric flux through a closed surface played a central role. The electric flux for an arbitrary open surface is given by

$$\Phi_E = \int_S \mathbf{E} \cdot d\mathbf{S}.$$

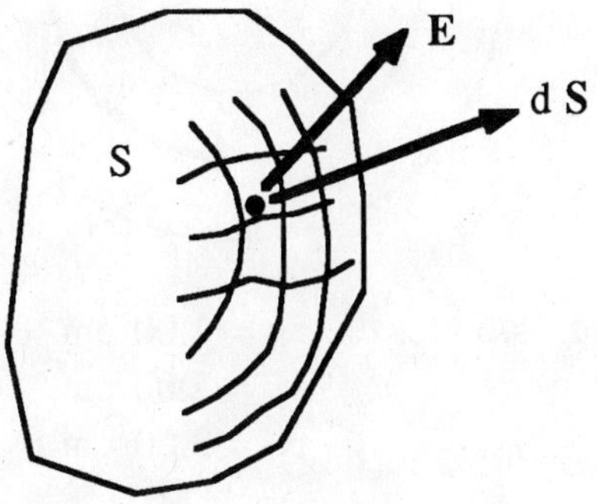

Figure 27.24

Note: The direction of dS is not completely determined for an open surface; it can be in the direction shown or opposite to that direction. If the surface is closed we remove the ambiguity concerning the direction by adapting the convention that the vector representing the area always points outward.

In later chapters we will need to calculate the magnetic flux through an open surface; the flux for a closed surface is always zero since the magnetic field lines are continuous rather than emanating from positive charges and ending on negative charges. The magnetic flux is defined by

$$\Phi_B = \int_S \mathbf{B} \cdot d\mathbf{S}.$$

Practice: Consider the magnetic field $\mathbf{B} = (3\mathbf{i} + 4\mathbf{j} - 2\mathbf{k})$T

Determine the following:

1. The flux for a square 2 meters on a side parallel to the xy-plane	$\Phi_B = \mathbf{B} \cdot \mathbf{S}$ (integral not needed) $\Phi_B = (3\mathbf{i} + 4\mathbf{j} - 2\mathbf{k}) \cdot (4\mathbf{k})$ $\Phi_B = -8 T \cdot m^2$
2. The flux for a square 2 meters on a side parallel to the xz-plane	$\Phi_B = \mathbf{B} \cdot \mathbf{S} = (3\mathbf{i} + 4\mathbf{j} - 2\mathbf{k}) \cdot (4\mathbf{j})$ $\Phi_B = 16\ T \cdot m^2$
3. The flux for a square 2 meters on a side parallel to the yz-plane	$\Phi_B = \mathbf{B} \cdot \mathbf{S} = (3\mathbf{i} + 4\mathbf{j} - 2\mathbf{k}) \cdot (4\mathbf{i})$ $\Phi_B = 12\ T \cdot m^2$

Example: A long solenoid having a radius of 2.00 cm and 2000 turns/meter is shown looking along the axis in Figure 27.25. A counterclockwise current of 4.00 A is maintained in the coils of the solenoid. Calculate the flux through the three circles shown.

Figure 27.25

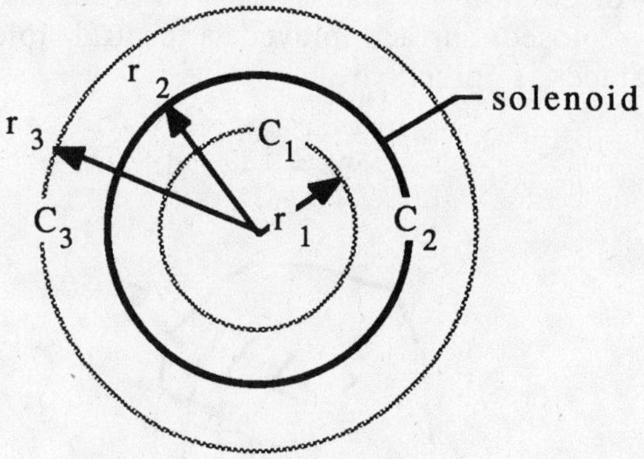

$r_1 = 1.00$ cm
$r_2 = 2.00$ cm
$r_3 = 3.00$ cm

Given: $n = 2000$ turns/meter, $R_s = 2.00$ cm, $I = 4.00$ A; $r_1 = 1.00$ cm, $r_2 = 2.00$ cm, $r_3 = 3.00$ cm

Determine: The flux through the circles C_1, C_2, and C_3.

Strategy: Use the magnetic field for the solenoid, $B_{inside} = \mu_0 n I_s$ and $B_{outside} = 0$, to obtain the flux.

Solution:

$$\Phi_1 = (\mu_0 n I_s)(\pi r_1^2) = (4\pi \times 10^{-7})(2000)(4)(\pi)(10^{-2})^2 = 31.6 \times 10^{-7}\ T \cdot m^2$$

$$\Phi_2 = (\mu_0 n I_s)(\pi r_2^2) = (4\pi \times 10^{-7})(2000)(4)(\pi)(2 \times 10^{-2})^2 = 126.4 \times 10^{-7}\ T \cdot m^2$$

$$\Phi_3 = \Phi_2 \quad \text{since} \quad B = 0 \quad \text{for} \quad r > R_s$$

Related Text Exercises: Ex. 27-35 through 27-37.

PRACTICE TEST

Take and grade this practice test. Doing so will allow you to determine any weak spots in your understanding of the concepts taught in this chapter. The following section prescribes what you should study further to strengthen your understanding.

Note: For this test take all magnetic fields out of page as + and all magnetic fields into the page as -.

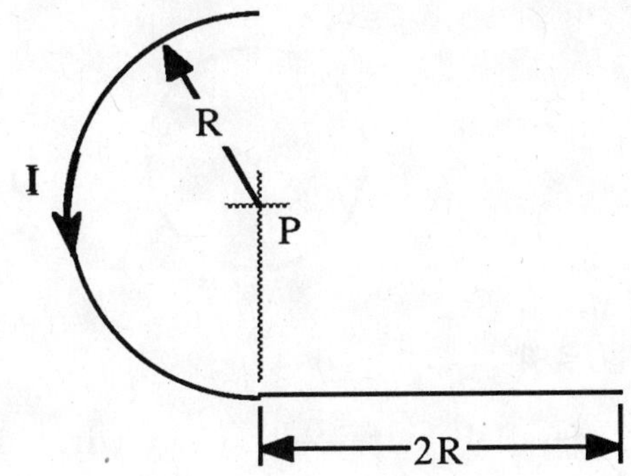

Figure 27-26

Determine the following:

_________ 1. Magnetic field at P due to semicircular arc
_________ 2. Magnetic field at P due to straight segment of length 2R
_________ 3. Total magnetic field at P

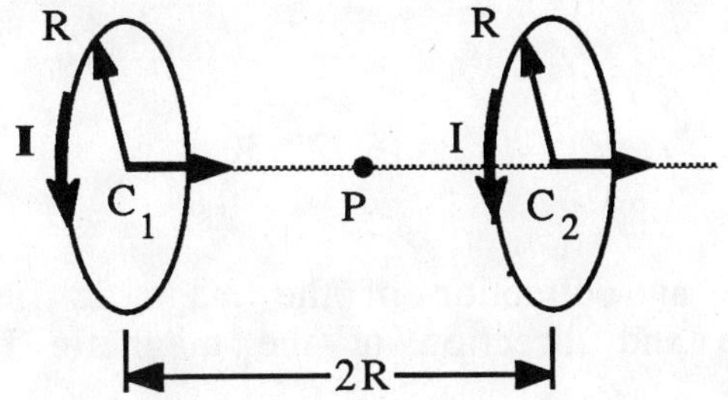

Figure 27.27

Determine the following:

_________ 4. Magnetic field at C_1
_________ 5. Magnetic field at C_2
_________ 6. Magnetic field at P

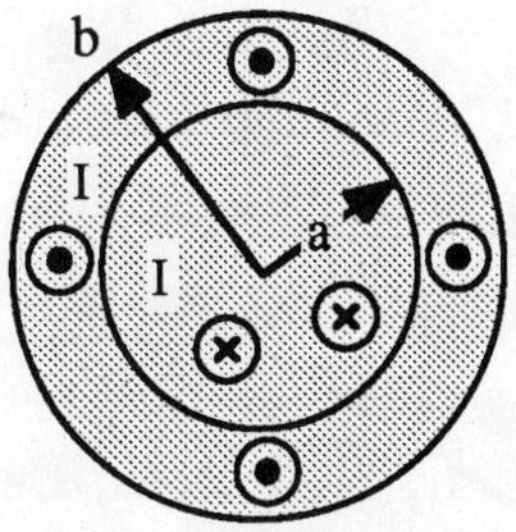

Figure 27-28

Figure 27-28 shows a long cylindrical conducting coaxial cable carrying a uniform current I into the page in the inner wire and returning it out of the page in the outer conductor.

Determine the following:

______ 7. B(r) for $r \le a$
______ 8. B(r) for $a \le r \le b$
______ 9. B(r) for $r \ge b$

Figure 27-29

$j(r) = Kr^2$ for $0 \le r \le a$

______ 10. B(r) for $r \le a$
______ 11. B(r) for $r \ge a$

Figure 27-30 shows two parallel current-carrying wires held rigidly in place by wire mounts.

$I_1 = 1.00$ A
$I_2 = 2.00$ A
$I_3 = 1.00$ A
$d = 2.00$ cm
$+$ = out of page
$-$ = into page

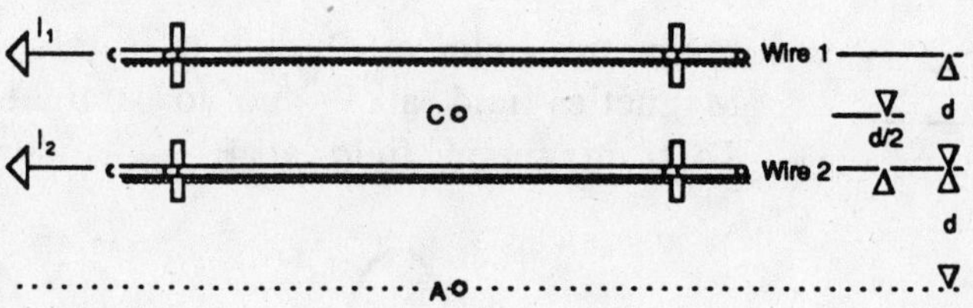

Figure 27-30

Determine the following:

______ 12. Magnitude and direction of the magnetic field at position A
______ 13. Magnitude and direction of the magnetic field midway between the wires
______ 14. Distance from wire 1 to where B = 0
______ 15. Magnitude and direction of the force per unit length on wire 1
______ 16. What would need to be done to the current in wire 1 in order to get B = 0 at position A

Figure 27.31

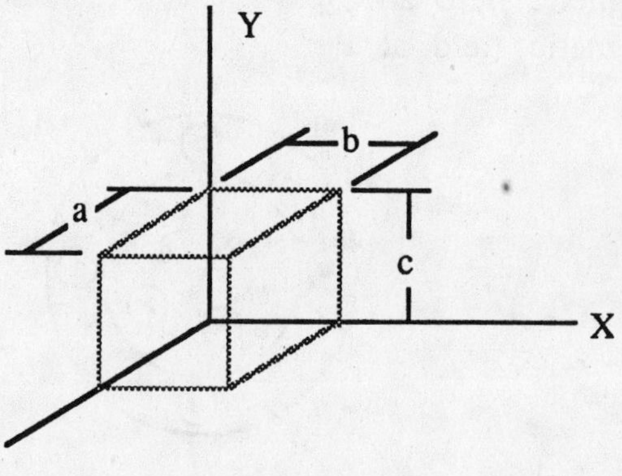

$\mathbf{B} = (2\mathbf{i} - \mathbf{j} - 3\mathbf{k})$ T

Consider Figure 27.31.

Determine the following:

________ 17. Magnetic flux for face in xy plane
________ 18. Magnetic flux for face in yz plane
________ 19. Magnetic flux for face in xz plane
________ 20. Magnetic flux for entire surface of rectangular box

Figure 27-32

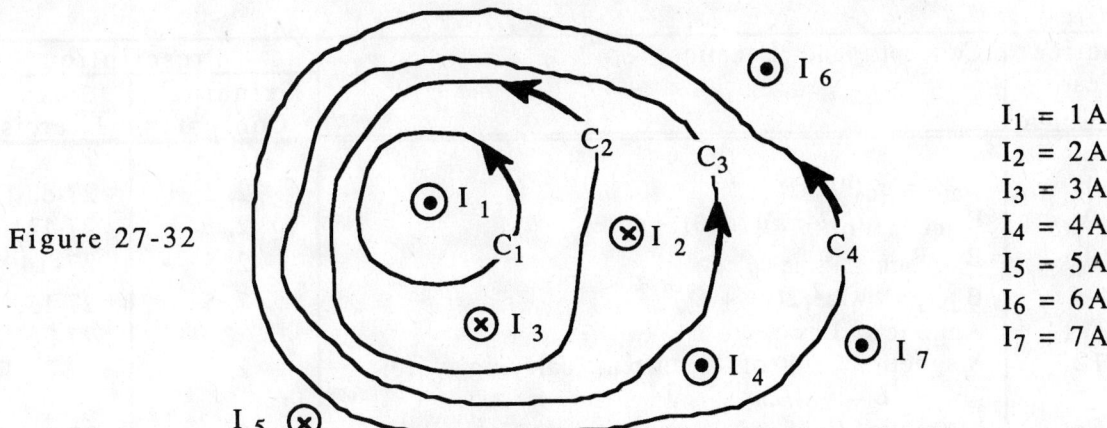

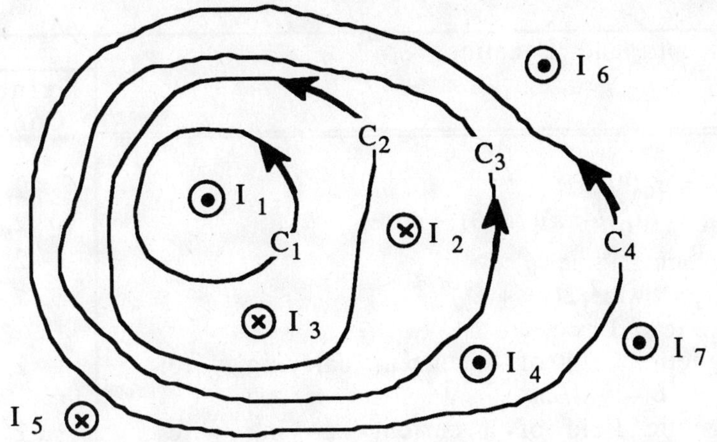

$I_1 = 1\text{A}$
$I_2 = 2\text{A}$
$I_3 = 3\text{A}$
$I_4 = 4\text{A}$
$I_5 = 5\text{A}$
$I_6 = 6\text{A}$
$I_7 = 7\text{A}$

For the paths and currents shown in Figure 27-32, determine the following:

________ 21. $\oint_{c_1} \mathbf{B} \cdot d\mathbf{r}$

________ 22. $\oint_{c_2} \mathbf{B} \cdot d\mathbf{r}$

________ 23. $\oint_{c_3} \mathbf{B} \cdot d\mathbf{r}$

________ 24. $\oint_{c_4} \mathbf{B} \cdot d\mathbf{r}$

(See Appendix I for answers.)

PRINCIPAL CONCEPTS AND EQUATIONS PRESCRIPTION

Your score on the practice test is an excellent measure of your understanding of the chapter. You should now use the following chart to write your own prescription for dealing with any weaknesses the practice test points out. Look down the leftmost column to the number of the question(s) you answered incorrectly, reading across that row you will find the concept and/or equation of concern, the section(s) of the study guide you should return to for further study, and some suggested text exercises which you should work to gain additional experience.

Practice Test Questions	Concepts and Equations	Prescription	
		Principal Concepts	Text Exercises
1	$B_{arc} = \mu_0 I\theta/2\pi R$	2, 5	27-6, 11
2	$B_{line} = (\mu_0 I/4\pi a)(\cos\theta_1 - \cos\theta_2$	2, 5	27-12, 13
3	$B = B_{arc} + B_{line}$	2, 5	27-14
4-6	$B_{axis} = \mu_0 Ia^2/2(x^2 + a^2)^{3/2}$	2, 5	27-15, 16
7-11	Ampere's Law	3	27-26, 39
12	Magnetic field of a current carrying wire: $B = \mu_0 I/2\pi r$	2, 5	27-3, 4
13	Magnetic field of a current carrying wire	2, 5	27-4, 5
14	Magnetic field of a current carrying wire	2, 5	27-7, 8
15	Force per unit length on a current carrying wire: $F/l = \mu_0 I_1 I_2/2\pi d$	4	27-32, 33
16	Magnetic field of a current carrying wire	2, 5	27-7, 8
17-20	Magnetic flux: $\Phi_B \int_S \mathbf{B} \cdot d\mathbf{S}$	6	27-35, 36
21-24	Amperes law: $\oint_C \mathbf{B} \cdot d\mathbf{r} = \mu_0 \sum_i I_i$	3	27-18, 19

28 FARADAY'S LAW

RECALL FROM PREVIOUS CHAPTERS

Previously learned concepts and equations frequently used in this chapter	Text Section	Study Guide Page
Magnetic force: $\mathbf{F} = q\mathbf{v} \times \mathbf{B}$	26-1	26-1
$\mathbf{B}_{long\ wire} = \mu_0 I/2\pi r$	27-1	27-1
$\mathbf{B}_{solenoid} = \mu_0 n I$	27-3	27-8
Magnetic flux: $\Phi_B = \int_s \mathbf{B} \cdot d\mathbf{s}$	27-5	27-20

NEW IDEAS IN THIS CHAPTER

Concepts and equations introduced	Text Section	Study Guide Page
Faraday's law: $\varepsilon_I = -d\Phi_B/dt$	28-1	28-1
Lenz's law: ε_I opposes the change producing it	28-1	28-1
Motional emf's: $\varepsilon_I = Blv$	28-2	28-4
Generators and alternators: $\varepsilon = NBS\omega \sin\omega t$	28-3	28-8

PRINCIPAL CONCEPTS AND EQUATIONS

1 Faraday's Law (Section 28-1)

Review: An experimental law discovered by Michael Faraday and also by Joseph Henry and known as Faraday's law relates the changing magnetic flux through any closed curve (conducting or otherwise) to an experimentally observed emf that is induced around the curve by the changing flux. See Figure 28.1 below. It is usually written as

$$\varepsilon_I = -d\Phi_B/dt$$

Figure 28.1

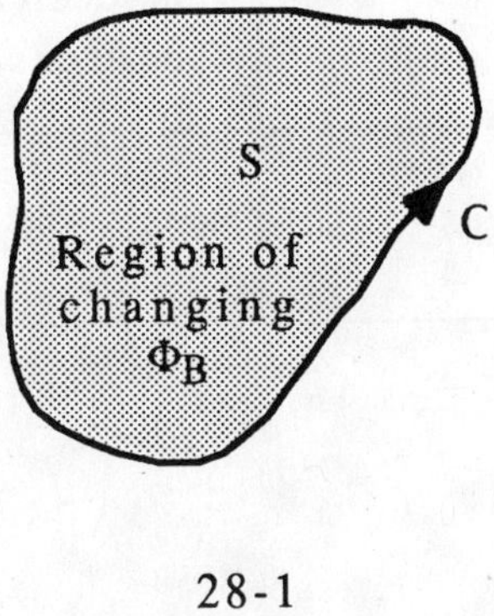

The "-" sign is really of no benefit, it is an attempt to remind us of the sense of the emf. The emf tends to oppose the change producing it; this statement is known as Lenz's law. It takes both Faraday's law and Lenz's law to determine both the magnitude and the sense of ε_I. In Figure 28.1, if **B** is into the page and increasing then ε_I tends to cause a current in the counterclockwise direction. If **B** is into the page and decreasing then ε_I tends to cause a current that is clockwise. We say "tends" because it can only oppose the change by creating a current. This can be done only for a conducting path; the induced emf ε_I is present for the path however whether or not it is conducting.

We summarize as follows:

1. Faraday's law: $|\varepsilon_I| = |d\Phi_B/dt|$ gives the magnitude of ε_I
2. Lenz's law: ε_I tends to oppose the change producing it

Practice: Consider the rectangular conducting loop and given information shown in Figure 28.2.

Figure 28.2

a = 1.00 m
b = 2.00 m
B = $3.00e^{-2t}$ T
R = 5.00 Ω

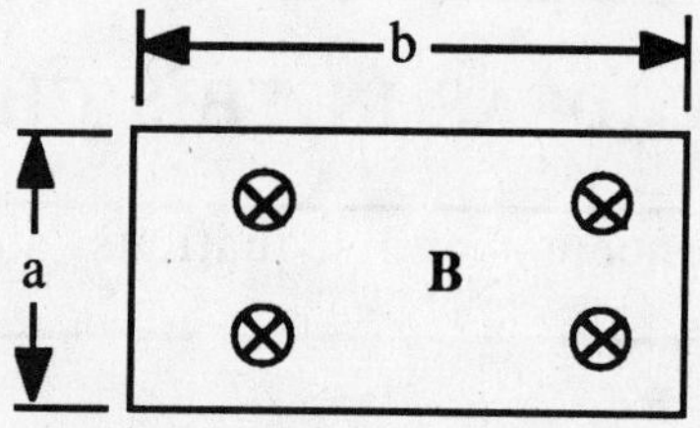

Determine the following:

1. The magnetic flux through the rectangular conducting loop at any time	$\Phi_B(t) = \mathbf{B}(t) \cdot \mathbf{S}$ $= (3.00e^{-2t})(2.00\ m^2)$ $= 6.00e^{-2t}\ T{\cdot}m^2$ This flux is "-" for **B** into the page as shown in the figure.
2. The magnitude of induced emf	$\lvert\varepsilon_I\rvert = \lvert d\Phi_B/dt\rvert = 12.0e^{-2t}$ V
3. The magnitude of the current due to ε_I	$I(t) = \lvert\varepsilon_I(t)\rvert/R = 2.40e^{-2t}$ A
4. The sense of the current	From Lenz's law the induced current must be clockwise to produce an induced magnetic field $\mathbf{B}_I$ into the page. We need $\mathbf{B}_I$ into the page because **B**(t) is decreasing and $\mathbf{B}_I$ must oppose this decrease.
5. The induced current in the rectangular conductor at t = 2.00 s	$I(t) = 2.40e^{-2t}$ A $= 2.40e^{-2(2.00)}$ A = 4.40×10^{-2} A

Example: The solenoid shown in Figure 28.3 is 20.0 cm long, has a 2.00 cm radius, and 1000 turns of wire. A flat, 10.0 turn circular coil with a 0.500 cm radius is mounted inside the solenoid in a concentric manner and attached to a 10.0 Ω resistor. If the current in the solenoid is increasing at the rate of 10.0 A/s, determine the magnitude and direction of the induced current through the resistor.

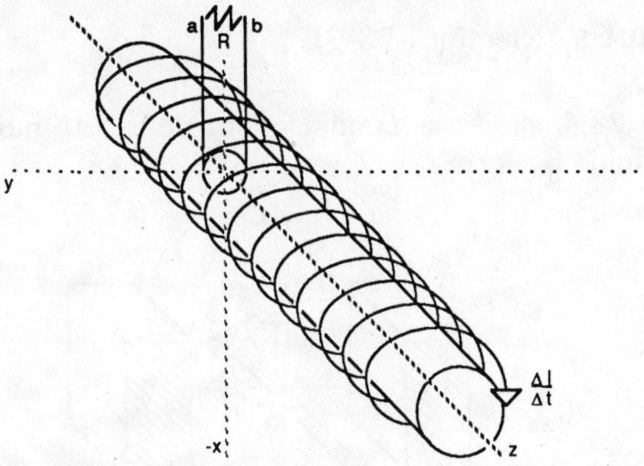

Figure 28.3

Given:

l = 20.0 cm -- length of the solenoid
r_s = 2.00 cm -- radius of the solenoid
N_s = 1000 -- turns of wire on the solenoid
N_c = 10 -- turns of wire on the coil
r_c = 0.500 cm -- radius of the coil
$\Delta I/\Delta t$ = 10.0 A/s -- rate of change in current in the solenoid
R = 10.0 Ω -- resistance of the resistor

Determine: Magnitude and direction of the induced current through the resistor.

Strategy: We can use our knowledge about the magnetic field of a solenoid and magnetic flux to develop an expression for the flux through the small coil as a function of the current in the solenoid. Using this expression and $\Delta I/\Delta t$, we can determine the induced emf in the coil. Knowing the induced emf and R, we can determine the induced current. We can use Lenz's law to determine the direction of the induced current.

Solution: The magnetic field inside the solenoid is given by

$$B = \mu_0 N_s I/l$$

The magnetic flux through the coil is

$$\Phi = BA_{coil} = (\mu_0 N_s I/l)\pi r_c^2$$

The time rate of change of the magnetic flux through the coil, and hence the emf induced in the coil, is

$$\varepsilon = N_c\Delta\Phi/\Delta t = (\mu_0 N_s N_c \pi r_c^2/l)(\Delta I/\Delta t) = 4.93 \times 10^{-5}\ \text{V}$$

The current induced in the coil is

$$I_c = \varepsilon/R = 4.93 \times 10^{-6}\ \text{A}$$

As the current in the solenoid increases the magnetic flux increases in the -z direction. The induced current can oppose this by rotating counterclockwise in the coil (b to a in the resistor).

Related Text Exercises: Ex. 28-1 through 28-8.

2 Motional Emf's (Section 28-2)

Review: Figure 28.4 shows a conductor of length l moving at a constant speed v through a magnetic field **B**.

Figure 28.4

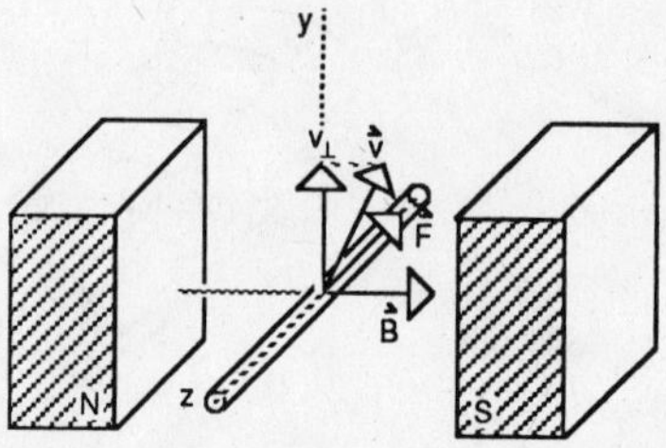

- v -- represents the velocity of the conductor
- l -- represents the length of the conductor in the **B**-field
- B -- represents the magnetic field
- θ -- represents the angle between the direction of **v** and **B**
- F -- represents the force on positive charges in the conductor
- $v_\perp$ -- represents the component of **v** perpendicular to **B** ($v_\perp = v \sin\theta$)

The induced motional emf is given by

$$\varepsilon = Blv_\perp$$

As shown in Figure 28.5, the magnetic flux over an area in space is the product of area times the component of the **B**-field perpendicular to that area.

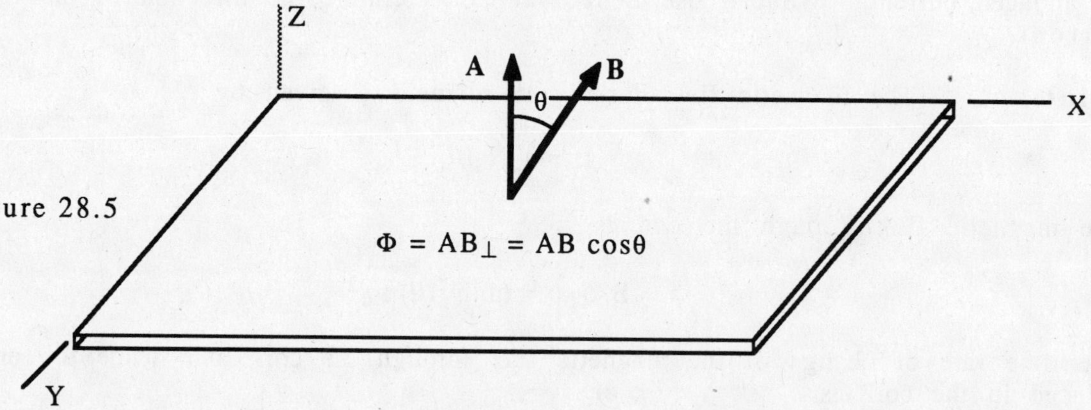

Figure 28.5

If a circuit experiences a time rate of change of magnetic flux, an emf will be induced. Faraday's law expresses this mathematically as

$$\varepsilon = -\Delta\Phi/\Delta t$$

The minus sign is due to Lenz and reminds us that the induced current in the circuit must be in such a direction as to oppose the change that caused it.

Practice: Suppose that the moving conductor slides along the U-shaped conductor in a uniform B-field, as shown in Figure 28.6

Figure 28.6

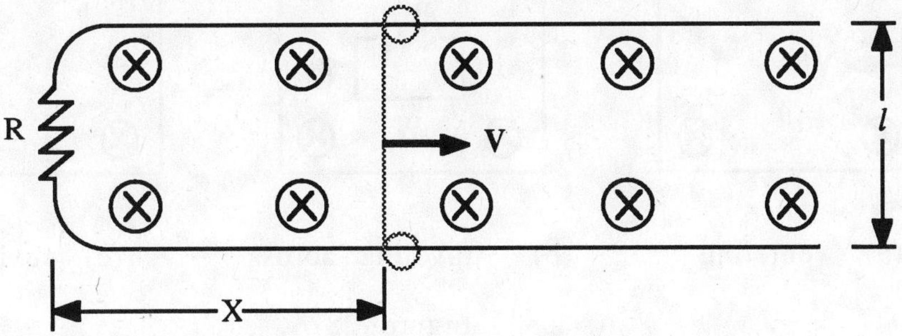

Determine expressions for the following:

1. Motional emf developed across the length of the sliding wire	$\varepsilon = Blv$
2. Magnetic flux through the circuit at any instant	$\Phi = BS = Blx$
3. Rate of change of magnetic flux in the circuit at any instant	$\frac{\Delta\Phi}{\Delta t} = B(\frac{\Delta S}{\Delta t}) = Bl(\frac{\Delta x}{\Delta t}) = Blv$

Note: We have just demonstrated Faraday's law, which states that the induced emf is equal to the time rate of change of magnetic flux.

4. The magnitude of the induced current	$I = \varepsilon/R = Blv/R$
5. The direction of the induced current	The induced current must oppose what caused it. It is caused by an increasing flux in the circuit. If the current is counterclockwise, it produces a magnetic flux that opposes this increase.
6. The magnitude of the force required to keep the sliding wire traveling at a constant speed	The magnetic force on the sliding wire due to the induced current interacting with the magnetic field is $F = IlB \sin 90.0° = (\frac{Blv}{R})lB = B^2l^2v/R$. A force of this magnitude parallel to **v** is required to keep the wire traveling at a constant speed.

Figure 28.7 shows a rectangular loop (L long and w wide) of wire with a resistance R traveling at a constant speed as it (a) enters, (b) travels across, and (c) exits the uniform **B**-field which is directed into the page.

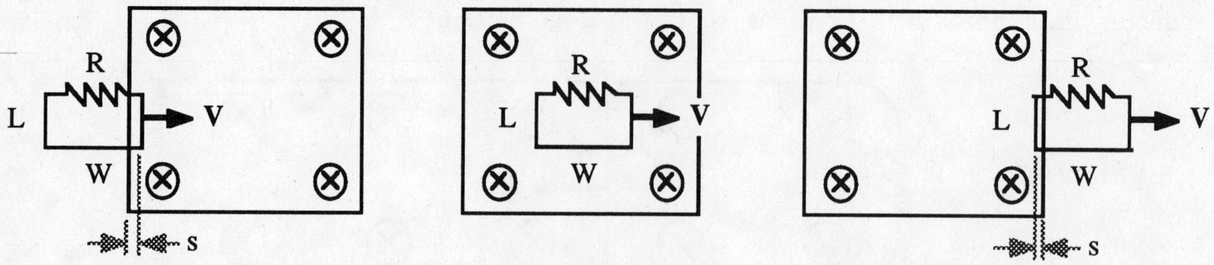

(a) entering (b) traveling across (c) exiting

Figure 28.7

Determine expressions for the following:

7. Magnetic flux through the coil at any instant while the coil is entering the **B**-field	$\Phi = BLs$ Note that the value of s is changing
8. Time rate of change of magnetic flux through the coil as it enters the **B**-field	$\frac{\Delta\Phi}{\Delta t} = BL(\frac{\Delta s}{\Delta t}) = BLv$
9. Emf induced as the coil enters the **B**-field	$\varepsilon = \frac{-\Delta\Phi}{\Delta t} = -BLv$

Note: The minus sign on the emf has nothing to do with the magnitude of the emf. The minus sign is just Lenz's way of reminding us that the induced emf is in such a direction as to oppose the change that caused it.

10. Direction of the induced current as the coil enters the **B**-field	Method A: The induced emf is the result of an increase in the downward magnetic flux in the coil. A counterclockwise current in the coil creates an upward magnetic flux, hence opposes the increase in downward magnetic flux. Method B: A counterclockwise current establishes a magnetic force ($F = BIL \sin 90.0°$) on the leading segment of the coil and this force opposes the motion of the coil.
11. Induced current as the coil enters the **B**-field	$I = \varepsilon/R = \frac{BLv}{R}$

12. Force required to move the coil into the B-field at a constant speed	The induced current causes a retarding force $F = BIL = B(\frac{\varepsilon}{R})L = B(\frac{BLv}{R})L = B^2L^2v/R$ A force of this magnitude is required to move the coil into the **B**-field at a constant speed.
13. The magnetic flux through the coil as it travels across the **B**-field	$\Phi = BLw$
14. The time rate of change of magnetic flux through the coil as it travels across the **B**-field	$\frac{\Delta\Phi}{\Delta t} = 0$ Consequently, the induced emf is zero.

Note: The magnetic flux through the coil is a constant as it travels across the magnetic field. Consequently, no emf is induced.

15. Induced emf as the coil leaves the **B**-field	$\Phi = Bls$ $\Delta\Phi/\Delta t = Bl\Delta s/\Delta t = Blv$ $\varepsilon = -\Delta\Phi/\Delta t = -Blv$
16. Direction of the induced current as the coil leaves the **B**-field	Method A: The induced emf is the result of a decrease in the downward magnetic flux in the coil. A clockwise current in the coil creates a downward magnetic flux, hence opposes the decrease in downward magnetic flux. Method B: A clockwise current establishes a magnetic force ($F = BIl \sin 90.0°$) on the trailing segment of the coil, and this force opposes the motion of the coil.

Example: Figure 28.8 shows a rod of mass m and resistance R that can slide down the frictionless, resistanceless vertical conducting track; the rod maintains contact as it slides. Determine the terminal velocity of the rod.

Figure 28.8

$m = 2.00$ kg
$L = 3.00$ m
$R = 3.00\ \Omega$
B $= 0.500$ T

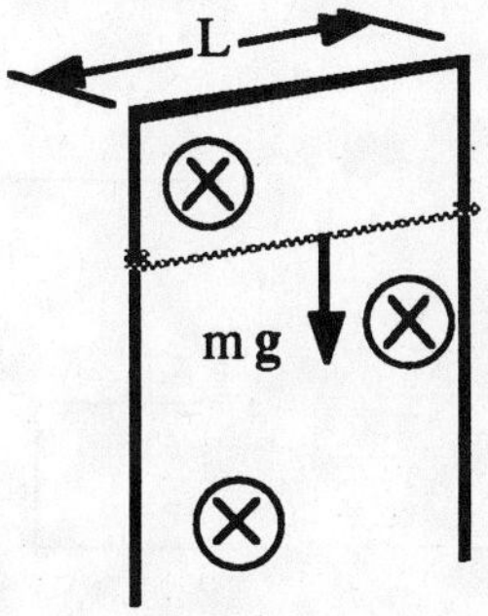

Given: $m = 2.00$ kg $\quad L = 3.00$ m $\quad R = 3.00\ \Omega \quad \mathbf{B} = 0.500$ T $\quad g = 9.80\ \text{m/s}^2$

Determine: v_T, the terminal velocity of the rod

Strategy: As the rod increases its speed, $|d\Phi_B/dt|$ increases and hence $|\varepsilon_I|$ increases. The increase in $|\varepsilon_I|$ leads to an increase in the induced current. This increase in induced current leads to an increase in the retarding magnetic force on the rod. In fact the speed and consequently the induced emf, current and force will increase until the induced magnetic force on the rod is equal to its weight. At this time the rod is in equilibrium and it has its terminal velocity. According to Lenz's law the induced force on the rod must be up -- this would oppose the force of gravity which is down.

Solution:

$$|\varepsilon_I| = BLv_T \Rightarrow |I| = |\varepsilon_I|/R = BLv_T/R$$

From Lenz's law the induced current must be from left to right through the rod. This leads to an upward magnetic force on the rod given by:

$$F_B = |I|LB = (\ BLv_T/R\ )LB = B^2L^2v_T/R$$

$$F_g = mg$$

The terminal velocity will be obtained when

$$F_g = F_B \Rightarrow Mg = B^2L^2v_T/R$$

or

$$v_T = MgR/B^2L^2) = 26.1 \text{ m/s downward}$$

Related Text Exercises: Ex. 28-9 through 28-18.

3 Electric Generators (Section 28-3)

Review: The N-Turn coil of wire in Figure 28.9 is mechanically turned at a constant angular speed ω in a uniform magnetic field **B**.

Figure 28.9

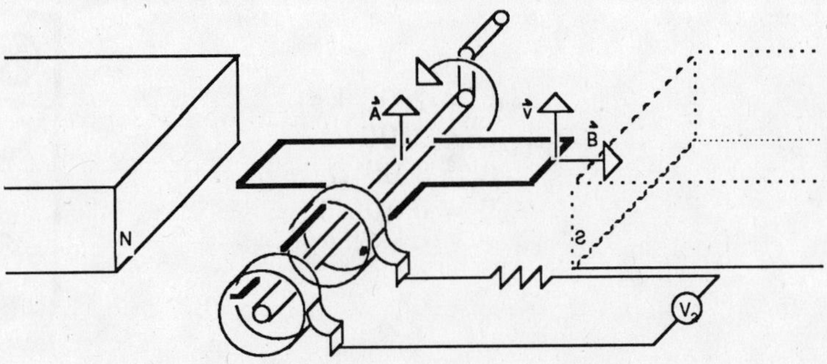

As the coil turns, the following quantities vary in time.

1. The magnetic flux through the coil (Φ)
2. The rate of change of magnetic flux through the coil ($\Delta\Phi/\Delta t$)
3. The induced emf (ε)
4. The induced current (I)
5. The angle between **v** and **B** (θ)
6. The angle between **S** and **B** (θ)
7. The angle between the plane of the coil and **B** (α)

If we agree to start our record of time when the angle between **v** and **B** is 0° (Figure 28.10), the magnetic flux, rate of change of magnetic flux, induced emf, and induced current are obtained as follows:

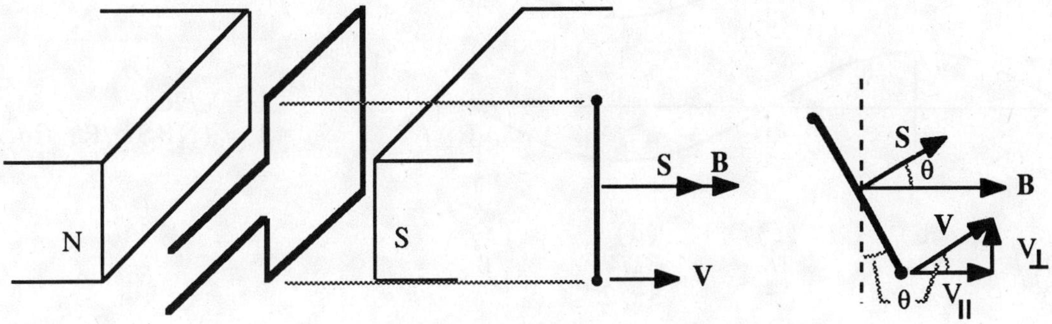

(a) t = 0 (b) t = 0 (c) t > 0

Figure 28.10

The magnetic flux at any time is $\Phi = BS\cos\theta = BS\cos\omega t$

The induced emf of one wire of side l of the coil is

$$\varepsilon_1 = Blv_\perp = Blv\sin\theta = Bl\omega(w/2)\sin\theta = (BS\omega/2)\sin\theta$$

The induced emf due to N turns of both sides is $\varepsilon = 2N\varepsilon_1 = NBS\omega\sin\theta$

The induced current is $I = \varepsilon/R = (NBS\omega/R)\sin\theta$

Figure 28.11 shows the coil in four different orientations and a graphical representation of Φ, $\Delta\Phi/\Delta t$, ε, and I.

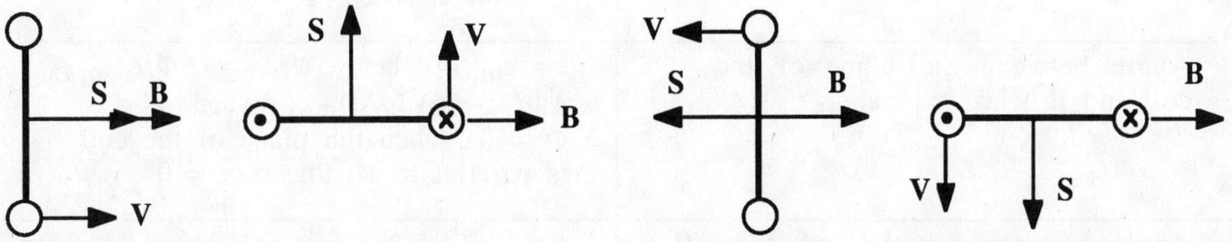

(a) $t = 0$, $\theta = \omega t = 0$
(b) $t = T/4$, $\theta = 90°$
(c) $t = T/2$, $\theta = 180°$
(d) $t = 3T/4$, $\theta = 270°$

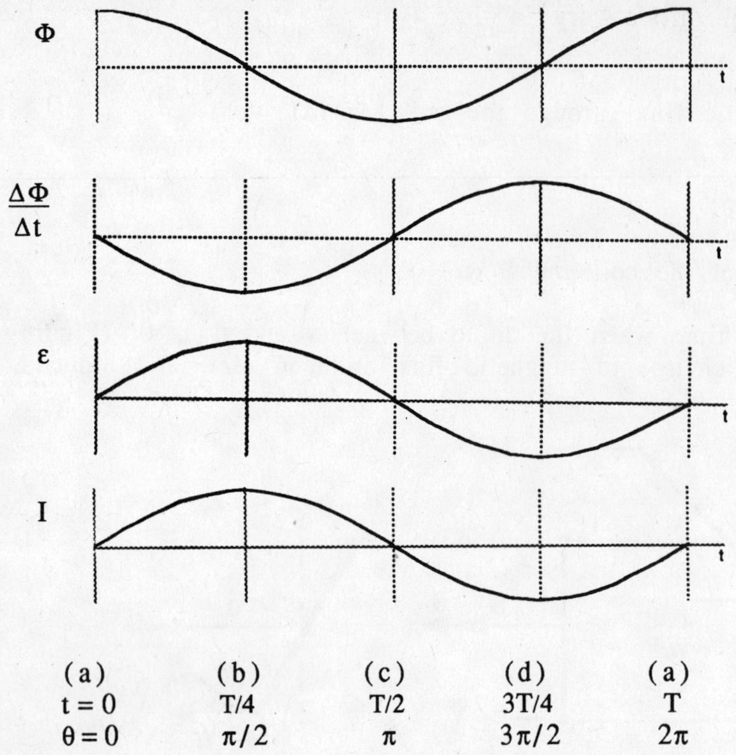

$\Phi = BS \cos\theta$
$\Phi = BS \cos\omega t$

$\dfrac{\Delta\Phi}{\Delta t} = -BS\omega \sin\omega t$

$\varepsilon = -N(\Delta\Phi/\Delta t)$
$\varepsilon = NBS\omega \sin\omega t$

$I = \varepsilon/R$
$I = (NBS\omega/R) \sin\omega t$

Figure 28.11

Practice: The following information is known for the coil in Figure 28.9

$S = 50.0\ cm^2$ $\quad B = 5.00\ T$ $\quad N = 100$ turns $\quad \omega = 10.0$ rad/s $\quad t = 0$ when $\theta = 0$

Determine the following:

1. Maximum emf generated by the coil	$\varepsilon = NBS\omega \sin\theta$ $\varepsilon_{max} = NBS\omega = 25.0$ V
2. Angular speed required to induce a maximum emf of 50.0 V	$\varepsilon_{max} = NBS\omega$ $\omega = \varepsilon_{max}/NBS = 20.0$ rad/s
3. Angle between **v** and **B** when $\varepsilon = \pm\varepsilon_{max}$	$\varepsilon = \varepsilon_{max} \sin\theta$ $\varepsilon = +\varepsilon_{max}$ when $\theta = \pi/2$ $\varepsilon = -\varepsilon max$ when $\theta = 3\pi/2$
4. Angle between the plane of the coil and **B** when $\varepsilon = \pm\varepsilon_{max}$	$\varepsilon = \varepsilon_{max}$ when $\Delta\Phi/\Delta t = (\Delta\Phi/\Delta t)_{max}$ $\Delta\Phi/\Delta t = (\Delta\Phi/\Delta t)_{max}$ when $\Phi = 0$ $\Phi = 0$ when the plane of the coil is parallel to **B**, that is $\alpha = 0°$
5. Angle between **v** and **B** when $\varepsilon = 0$	$\varepsilon = \varepsilon_{max} \sin\theta$, $\quad \varepsilon = 0$ when $\theta = 0$ or π The angle between **v** and **B** is also θ, hence the angle between **v** and **B** is $\theta = 0$ or π when $\varepsilon = 0$

6. Angle between the plane of the coil and **B** when $\varepsilon = 0$	$\varepsilon = 0$ when $\Delta\Phi/\Delta t = 0$ $\Delta\Phi/\Delta t = 0$ when $\Phi = \pm\Phi_{max}$ $\Phi = \pm\Phi_{max}$ when the plane of the coil is perpendicular to **B**, that is $\alpha = 90°$
7. Angle between **v** and **B** when $\varepsilon = 0.500\ \varepsilon_{max}$	$\varepsilon = \varepsilon_{max} \sin\theta$ $0.500\ \varepsilon_{max} = \varepsilon_{max} \sin\theta$ $\theta = \sin^{-1}(0.500) = 30°$ and $150°$
8. Magnetic flux through the coil when $\theta = 60°$	$\Phi = NBS \cos 60° = 1.25\ T{\bullet}m^2$
9. Emf induced in the coil when $\theta = 60°$	$\varepsilon = NBS\omega \sin\theta = 21.6\ V$
10. Rate of change of magnetic flux when $\theta = 60°$	$\varepsilon = -N(\Delta\Phi/\Delta t)$ $\Delta\Phi/\Delta t = -\varepsilon/N = -0.216\ V$

Example: In a model ac generator, a 100-turn rectangular coil of dimensions 10.0 cm by 20.0 cm rotates at 120 rev/min in a uniform magnetic field of 0.500 T. (a) What is the maximum emf induced in the coil? (b) What is the instantaneous value of the emf in the coil at $t = (1/16\pi)$ s? (c) If we choose $t = 0$ when the emf is zero, what is the smallest value of t for which the emf will have its maximum value?

Given: $N = 100$ turns $w = 10.0$ cm $l = 20.0$ cm $B = 0.500$ T $\omega = 120$ rev/min

Determine: (a) ε_{max}, (b) ε at $t = (1/16\pi)$ s, (c) time to go from $\varepsilon = 0$ to $\varepsilon = \varepsilon_{max}$

Strategy: Knowing the expression for the induced emf (ε) as a function of time, we can determine (a) ε_{max}, (b) ε at $t = (1/16\pi)$ s, and (c) the time to go from $\varepsilon = 0$ to $\varepsilon = \varepsilon_{max}$.

Solution: $\omega = (120\ \text{rev/min})(\text{min}/60\ \text{s})(2\pi\ \text{rad/rev}) = 4\pi\ \text{rad/s}$

(a) $\varepsilon = NSB\omega \sin\omega t = \varepsilon_{max} \sin\omega t$ where $\varepsilon_{max} = NBS\omega = 12.6\ V$

(b) $\varepsilon = \varepsilon_{max} \sin\omega t = \varepsilon_{max} \sin[(4\pi\ \text{rad/s})(\text{s}/16\pi)] = 3.12\ V$

(c) $\varepsilon = \varepsilon_{max}$ at $t = t'$

$\varepsilon_{max} = \varepsilon_{max} \sin\omega t'$

$\omega t' = \sin^{-1}(1) = 1.57$ rad or $t' = 1.57\ \text{rad}/\omega = 0.125$ s

Also, ε will go from 0 to ε_{max} in a time equal to one fourth period

$$\omega = 2\pi f = 2\pi/T \text{ or } T = 2\pi/\omega = 0.500 \text{ s and } t' = T/4 = 0.125 \text{ s}$$

Related Text Exercises: Ex. 28-19 through 28-25.

PRACTICE TEST

Take and grade this practice test. Doing so will allow you to determine any weak spots in your understanding of the concepts taught in this chapter. The following section prescribes what you should study further to strengthen your understanding.

In Figure 28.12, a changing magnetic field is into the page over the area of the circular loop.

Figure 28.12

$B = 5.00t$ T
$r = 2.00$ m
$R = 4.00\ \Omega$

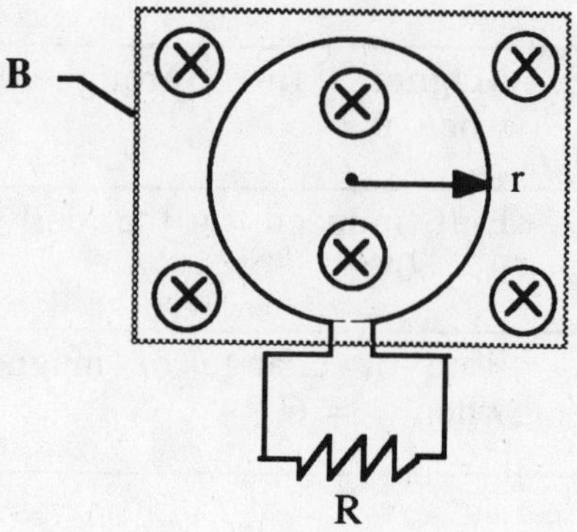

Determine the following:

_________ 1. The magnetic flux linking the circuit
_________ 2. The magnitude of the induced emf
_________ 3. The magnitude of the induced current through R
_________ 4. The direction of the induced current through R

Figure 28.13 shows a flat circular coil inside a solenoid and concentric to the axis of the solenoid.

Figure 28.13

$r_s = 4.00$ cm
$r_c = 1.00$ cm
$N_s = 1000$
$N_c = 10$
$L_s = 40.0$ cm
$R_s = 100\ \Omega$
$V_b = 25.0$ V

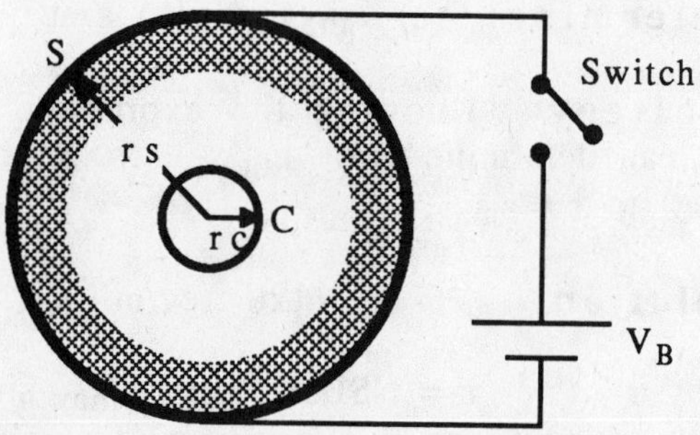

Determine the following:

_________ 5. Magnitude and direction of the final magnetic field established in the solenoid

After the switch is closed, the current through the solenoid goes from zero to its maximum value. At the instant when the current is changing at the rate of 1.00 A/s, determine the following:

_________ 6. The rate at which the magnetic field in the solenoid is changing
_________ 7. The rate at which the magnetic flux in each turn of the solenoid is changing
_________ 8. The rate at which the magnetic flux in each turn of the coil is changing
_________ 9. The emf induced in each turn of wire in the coil
_________ 10. The total emf induced in the coil

The rectangular loop shown in Figure 28.14 is moving to the right as shown.

L = 10.0 cm
W = 5.00 cm
v = 2.00 cm/s
B = 3.00 T
R = 2.00 Ω

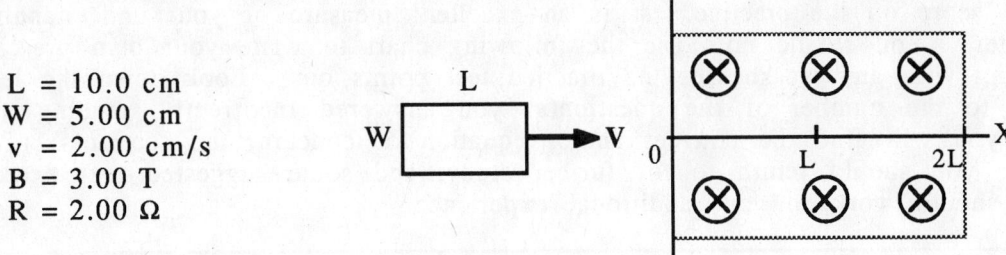

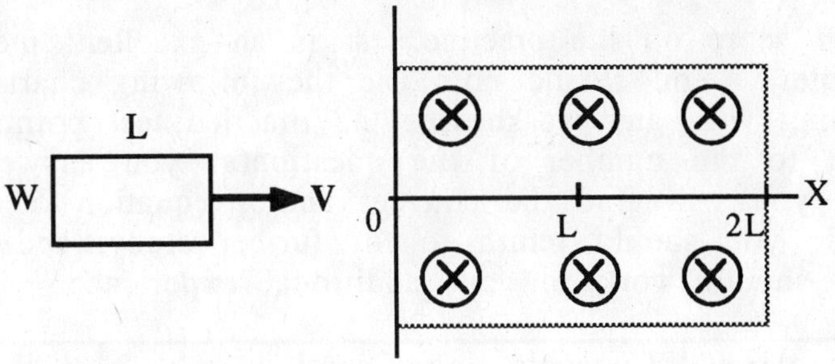

Figure 28.14

Determine the following:

________ 11. The emf (magnitude and direction) as the rectangular loop enters the magnetic field (i.e. when the leading edge of the loop is at x = 0

________ 12. The emf (magnitude and direction) when the leading edge of the loop is at $0 \le x \le L$

________ 13. The emf (magnitude and direction) when the leading edge of the loop is at $L \le x \le 2L$

________ 14. The emf (magnitude and direction) when the leading edge of the loop is at $2L \le x \le 3L$

________ 15. The emf (magnitude and direction) when the leading edge of the loop is at $x > 3L$

________ 16. Sketch the induced emf ε_I vs position x, take ε_I as "+" for the ccw sense

Determine the induced current in the loop or the force on the loop when the leading edge of the loop is at each specified position.

________ 17. Current in the loop (magnitude and direction) for $x < 0$
________ 18. Current in the loop (magnitude and direction) for $0 \le x \le L$
________ 19. Current in the loop (magnitude and direction) for $L \le x \le 2L$
________ 20. Current in the loop (magnitude and direction) for $2L \le x \le 3L$
________ 21. Current in the loop (magnitude and direction) for $x > 3L$
________ 22. Force on the loop (magnitude and direction) for $x < 0$
________ 23. Force on the loop (magnitude and direction) for $0 \le x \le L$
________ 24. Force on the loop (magnitude and direction) for $L \le x \le 2L$
________ 25. Force on the loop (magnitude and direction) for $2L \le x \le 3L$
________ 26. Force on the loop (magnitude and direction) for $x > 3L$

A generator consists of 20,000 turns of wire each having an area of 10.0 cm^2. The coil is rotated at 60.0 rev/s in a uniform magnetic field of 2.00 T.

Determine the following:

________ 27. The maximum voltage generated

(See Appendix I for answers.)

PRINCIPAL CONCEPTS AND EQUATIONS PRESCRIPTION

Your score on the practice test is an excellent measure of your understanding of the chapter. You should now use the following chart to write your own prescription for dealing with any weaknesses the practice test points out. Look down the leftmost column to the number of the question(s) you answered incorrectly, reading across that row you will find the concept and/or equation of concern, the section(s) of the study guide you should return to for further study, and some suggested text exercises which you should work to gain additional experience.

Practice Test Questions	Concepts and Equations	Prescription	
		Principal Concepts	Text Exercises
1	Magnetic flux: $\Phi_B = \int \mathbf{B} \cdot d\mathbf{s}$	6 of Ch 27	28-1
2	Induced emf: $\lvert\varepsilon_I\rvert = \lvert d\Phi_B/dt\rvert$	2	28-2, 4
3	Ohm's law: $I = \varepsilon/R$	3 of Ch 24	28-7
4	Lenz's law	2	28-1, 6
5	Magnetic field of solenoid: $B = \mu_0 nI$	4 of Ch 27	28-3, 5
6	Rate of change of magnetic field of solenoid: $dB/dt = \mu_0 n dI/dt$	2	28-3, 5
7	Magnetic flux: $\Phi_B = \int \mathbf{B} \cdot d\mathbf{s} = \mathbf{B} \cdot \mathbf{S}$	2	28-3, 5
8	Magnetic flux	2	28-3, 5
9	Induced emf: $\varepsilon_I = -d\Phi/dt$	2	28-3, 5
10	Induced emf: $\varepsilon_N = N\varepsilon_I$	2	28-3, 5
11-16	Induced emf: $\varepsilon = BLv$	3	28-15, 16
17-21	Ohm's law: $I = \varepsilon/R$	3 of Ch 24	28-13, 14
22-26	Magnetic force: $\mathbf{F} = I\mathbf{L} \times \mathbf{B}$	3 of Ch 26	28-13, 14
27	Generator: $\varepsilon_{max} = NSB\omega$	4	28-20, 22

29 INDUCTANCE

RECALL FROM PREVIOUS CHAPTERS

Previously learned concepts and equations frequently used in this chapter	Text Section	Study Guide Page
Magnetic field of a coil: $B = \mu_0 NI/2r$	27-1	27-1
Magnetic field of a solenoid: $B = \mu_0 NI/L$	27-3	27-8
Magnetic flux: $\Phi = BS \cos\theta$	27-5	27-20
Induced emf: $\varepsilon = -\Delta\Phi/\Delta t$	28-1	28-1

NEW IDEAS IN THIS CHAPTER

Concepts and equations introduced	Text Section	Study Guide Page
Self inductance: $L = -\varepsilon_1/(\Delta I_1/\Delta t) = \Phi_{11}/I_1$	29-1	29-1
Current in a RL circuit: $I = (V/R)(1 - e^{-[t/(L/R)]})$ and $I = I_0 e^{-[t/(L/R)]}$	29-2	29-6
Time constant for a RL circuit: $\tau = L/R$	29-2	29-6
Energy of an inductor: $U = LI^2/2$	29-3	29-9
Inductor energy density: $u = U/V = B^2/2\mu_0$	29-3	29-9
Mutual inductance: $M_{21} = -\varepsilon_2/(\Delta I_1/\Delta t) = \Phi_{21}/I_1$	29-4	29-1
Transformers: $V_s/V_p = N_s/N_p = I_p/I_s$	29-5	29-11

PRINCIPAL CONCEPTS AND EQUATIONS

1 Mutual and Self Inductance (Section 29-1 and 29-4)

Review: Figure 29.1 shows a coil of wire wrapped around a solenoid.

Figure 29.1

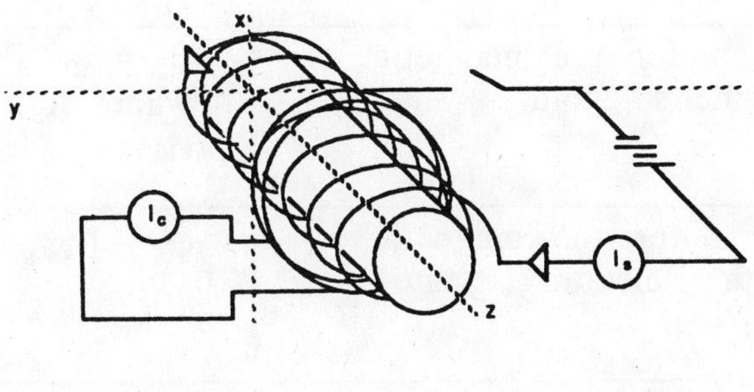

A changing current in the solenoid induces an emf in the coil and a changing current in the coil induces an emf in the solenoid. This property is called mutual inductance. The mutual inductance of the coil with respect to the solenoid is the emf ε_c induced in the coil per unit rate of change of current $\Delta I_s/\Delta t$ in the solenoid. This can be expressed mathematically as

$$M_{cs} = -\varepsilon_c/(\Delta I_s/\Delta t) = +\Phi_c/I_s$$

The mutual inductance of the solenoid with respect to the coil is the emf ε_s induced in the solenoid per unit rate of change of current $\Delta I_c/\Delta t$ in the coil. This can be expressed mathematically as

$$M_{sc} = -\varepsilon_s/(\Delta I_c/\Delta t) = +\Phi_s/I_c$$

The minus sign in the above expressions reminds us that the induced emf opposes the change that caused it.

The mutual inductance of the coil with respect to the solenoid is equal to the mutual inductance of the solenoid with respect to the coil. That is

$$M_{sc} = M_{cs}$$

When the current changes in the solenoid in Figure 29.1, the flux through the solenoid also changes, causing a self-induced emf in the solenoid. The self inductance of a coil in a circuit is equal to the self-induced emf ε_c in the coil per unit rate of change of current $\Delta I_c/\Delta t$ in the coil. This is expressed mathematically as

$$L = -\varepsilon_c/(\Delta I_c/\Delta t) = +\Phi_c/I_c$$

The minus sign in the above expression reminds us that the induced emf opposes the change that caused it.

Practice: Consider Figure 29.1 and the following data:

l_s = 20.0 cm -- length of the solenoid
N_s = 1000 -- turns of wire on the solenoid
r_s = 2.00 cm -- radius of the solenoid
$\Delta I_s/\Delta t$ = 10.0 A/s -- rate of change of current in the solenoid
N_c = 20 -- turns of wire on the coil
r_c = 4.00 cm -- radius of the coil
R_c = 20.0 Ω -- resistance of the coil

Determine the following:

1. An expression for the magnetic field inside the solenoid at any time	$B_s = \mu_0 N_s I_s/l_s$ where I_s represents the current at any time
2. An expression for the magnetic flux inside the solenoid at any time	$\Phi_s = B_s S_s = \mu_0 N_s I_s \pi r_s^2/l_s$

3. An expression for the magnetic flux through the coil at any time	$\Phi_c = \Phi_s = B_sS_s = \mu_o N_s I_s \pi r_s^2/l_s$
4. An expression for the time rate of change of flux through the solenoid and coil	$\Delta\Phi_c/\Delta t = \Delta\Phi_s/\Delta t = (\mu_0 N_s \pi r_s^2/l_s)(\Delta I_s/\Delta t)$
5. An expression for the emf induced in each turn of the wire of the coil and each turn of wire of the solenoid	$\varepsilon_c/\text{turn} = \varepsilon_s/\text{turn} = -\Delta\Phi_c/\Delta t$ $= -(\mu_0 N_s \pi r_s^2/l_s)(\Delta I_s/\Delta t)$
6. An expression for the total emf induced in the coil (i.e., in all N_c turns)	$\varepsilon_c = -N_c(\Delta\Phi_c/\Delta t) = N_c(\varepsilon_c/\text{turn})$ $\varepsilon_c = -N_c(\mu_0 N_s \pi r_s^2/l_s)(\Delta I_s/\Delta t)$
7. An expression for the total emf induced in the solenoid (i.e. in all N_s turns)	$\varepsilon_s = -N_s(\Delta\Phi_s/\Delta t) = N_s(\varepsilon_s/\text{turn})$ $\varepsilon_s = -N_s(\mu_0 N_s \pi r_s^2/l_s)(\Delta I_s/\Delta t)$
8. An expression for the mutual inductance of the coil with respect to the solenoid	$M_{cs} = -\varepsilon_c/(\Delta I_s/\Delta t) = N_c \mu_0 N_s \pi r_s^2/l_s$ **Note:** This expression shows that the mutual inductance can be expressed as a function of the parameters for the two circuits. This also shows $M_{cs} = \Phi_{cs}/I_s$
9. An expression for the self inductance of the solenoid	$L_s = -\varepsilon_s/(\Delta I_s/\Delta t) = N_s^2 \mu_0 \pi r_s^2/l_s$ Note: This expression shows that the self inductance can be expressed as a function of the parameters of the circuit. This also shows $L_s = \Phi_s/I_s$
10. Time rate of change of flux through the solenoid and coil	From step 4: $\Delta\Phi_c/\Delta t = \Delta\Phi_s/\Delta t = (\mu_0 N_s \pi r_s^2/l_s)(\Delta I_s/\Delta t)$ $= 7.90 \times 10^{-5}$ V
11. Emf induced in each turn of wire and in the solenoid and coil	$\varepsilon_s/\text{turn} = -\Delta\Phi_s/\Delta t = -7.90 \times 10^{-5}$ V $\varepsilon_c/\text{turn} = -\Delta\Phi_c/\Delta t = -7.90 \times 10^{-5}$ V
12. Emf induced in the entire coil (all N_c turns)	$\varepsilon_c = N_c(\varepsilon_c/\text{turn}) = -1.58 \times 10^{-3}$ V
13. The direction of the induced current in the coil	When the switch is closed, the flux through the solenoid and coil increases in the +z direction. An induced current flowing clockwise through the coil will oppose this change.

14. Current through the coil	$I_c = \varepsilon_c/R_c = 7.90 \times 10^{-5}$ A
15. Mutual inductance of the coil with respect to the solenoid	From step 8: $M_{cs} = \dfrac{-\varepsilon_c}{(\Delta I_s/\Delta t)} = 1.58 \times 10^{-4}$ V•s/A or $M_{cs} = N_c\mu_o N_s \pi r_s^2/l_s = 1.58 \times 10^{-4}$ V•s/A
16. Rate at which the current in the solenoid must change in order to induce a 1.00 x 10^{-3} V emf in the coil	$\Delta I_s/\Delta t = -\varepsilon_c/M_{cs} = 6.33$ A/s
17. Mutual inductance of the solenoid with respect to the coil	$M_{sc} = M_{cs} = 1.58 \times 10^{-4}$ V•s/A
18. Emf induced in the solenoid when the current in the coil is changing at the rate of 5.00 A/s	$M_{sc} = -\varepsilon_s/(\Delta I_c/\Delta t)$ or $\varepsilon_s = -M_{sc}\Delta I_c/\Delta t = -7.90 \times 10^{-4}$ V
19. Self induced emf in the solenoid	From steps 7 and 9: $L_s = -\varepsilon_s/(\Delta I_s/\Delta t) = N_s(\varepsilon_s/\text{turn})/(\Delta Is/\Delta t) = 7.90 \times 10^{-3}$ V•s/A or $L_s = N_s^2\mu_o\pi r_s^2/l_s = 7.90 \times 10^{-3}$ V•s/A
20. Rate at which the current in the solenoid must change in order to induce a 1.00 x 10^{-3} V emf in the coil	$\Delta I_s/\Delta t = -\varepsilon_s/L_s = 0.127$ A/s
21. Rate at which the current in the coil must change in order to induce a 1.00 x 10^{-3} V emf in the solenoid	$\Delta I_c/\Delta t = -\varepsilon_s/M_{sc} = 6.33$ A/s

Example: A 1.00 cm radius coil of 20 turns is placed inside a solenoid in a concentric manner. The solenoid has a 4.00 cm radius, 50.0 cm length, and 1000 turns of wire. Determine the emf induced in the coil when the current in the solenoid is changing at the rate of 10.0 A/s.

Figure 29.2

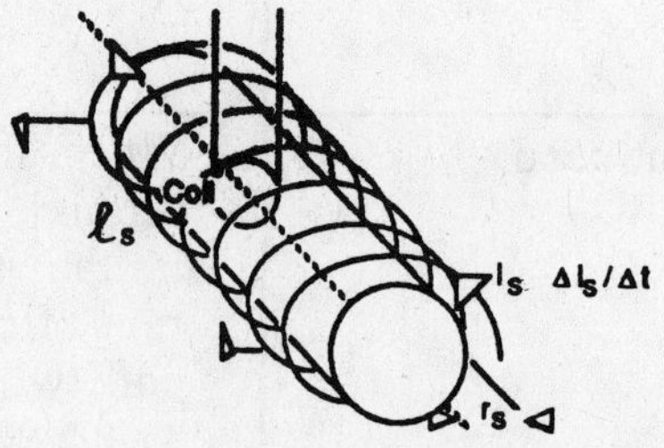

Given: r_c = 1.00 cm -- radius of the coil
r_s = 4.00 cm -- radius of the solenoid
l_s = 50.0 cm -- length of the solenoid
N_c = 20 -- turns of wire on the coil
N_s = 1000 -- turns of wire on the solenoid
$\Delta I_s/\Delta t$ = 10.0 A -- rate of change of current in the solenoid

Determine: The emf induced in the coil.

Strategy: We can write an expression for the magnetic field in the solenoid, and subsequently the flux through the coil, as a function of the current in the solenoid. From this expression, we can obtain an expression for the rate of change of flux in the coil as a function of the rate of change of current in the solenoid. We can now proceed by either of the following two methods.

(A) Knowing the rate of change of the flux through the coil and N_c, we can determine the emf induced in the coil.
(B) Knowing the rate of change of flux through the coil and definition of mutual inductance, we can determine the emf induced in the coil.

Solution: An expression for the magnetic field inside the solenoid is

$$B_s = \mu_0 N_s I_s / l_s$$

An expression for the magnetic flux through the coil is

$$\Phi_c = S_c B_c = S_c B_s = \pi r_c^2 \mu_0 N_s I_s / l_s$$

An expression for the time rate of change of magnetic flux through the coil is

$$\Delta\Phi_c/\Delta t = (\pi r_c^2 \mu_0 N_s / l_s)(\Delta I_s/\Delta t)$$

(A) The emf induced in the coil is

$$\varepsilon_c = -N_c(\Delta\Phi_c/\Delta t) = -(N_c \pi r_c^2 \mu_0 N_s / l_s)(\Delta I_s/\Delta t) = -1.58 \times 10^{-4} \text{ V}$$

(B) The mutual inductance of the coil with respect to the solenoid (see step 15 of the previous practice section).

$$M_{cs} = -\varepsilon_c/(\Delta I_s/\Delta t) = N_c \pi r_c^2 \mu_0 N_s / l_s \qquad \text{hence}$$

$$\varepsilon_c = -M_{cs}(\Delta I_s/\Delta t) = -(N_c \pi r_c^2 \mu_0 N_s / l_s)(\Delta I_s/\Delta t) = -1.58 \times 10^{-4} \text{ V}$$

Related Text Exercises: Ex. 29-1 through 29-9 and 29-16 through 29-30.

2 Inductors In Electric Circuits (Section 29-2)

Review: Figure 29.3 shows an inductor in a circuit with a resistor and a battery

$R = 20.0\ \Omega$
$L = 5.00$ H
$V = 10.0$ V

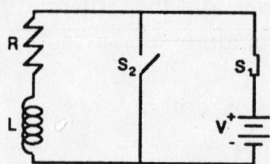

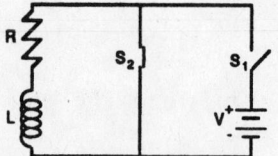

(a) S_1 closed and S_2 open (b) S_1 open and S_2 closed

Figure 29.3

If S_2 is left open and S_1 closed at t = 0 (Figure 29.3(a)), applying Kirchhoff's loop rule to the circuit we obtain

$$V - V_R - V_L = 0 \quad \text{where} \quad V_R = IR \quad \text{and} \quad V_L = L\,dI/dt$$

The current in the circuit at any time is

$$I = (V/R)(1 - e^{-[t/(L/R)]})$$

The time constant for the circuit is $\tau_L = L/R$

If after the current has been established S_1 is opened and S_2 closed (Figure 29.3(b)) applying Kirchhoff's loop rule we obtain

$$-V_R - V_L = 0$$

The current in the circuit at any time is

$$I = I_o e^{-[t/(L/R)]}$$

where I_0 is the current in the circuit the instant S_2 is closed.

Practice: For the situation shown in Figure 29.3, determine the following:

1. An expression for the potential drop across the resistor at any time for Figure 29.3(a)	$V_R = IR$ $V_R = V(1 - e^{-[t/(L/R)]})$
2. An expression for the potential drop across the inductor at any time for Figure 29.3(a)	$VL = V - V_R$ $V_L = Ve^{-[t/(L/R)]}$
3. Current at t = 0 for Figure 29.3(a)	$I = (V/R)(1 - e^{-[t/(L/R)]})$ $I = (V/R)(1 - e^0) = 0$

4. Current as $t \to \infty$ for Figure 29.3(a)	$I = V/R(1 - e^{-[t/(L/R)]})$ As $t \to \infty$, $e^{-t/(L/R)} \to 0$ and $I \to V/R = 0.500$ A
5. The current at t = L/R for Figure 29.3(a)	$I = (V/R)(1 - e^{-[t/(L/R)]})$ At t = L/R, we obtain $I = (V/R)(1 - e^{-1}) = 0.316$ A
6. Potential drop across the resistor and the inductor at $t = L/R = \tau_L$ for Figure 29.3(a)	$V_R = V(1 - e^{-1}) = 6.32$ V $V_L = Ve^{-1} = 3.68$ V, Note that $V_R + V_L = V$
7. Time constant for Figures 29.3(a) and 29.3(b)	$\tau_L = L/R = 0.250$ s
8. Current in the circuit after ten time constants for Figure 29.3(a)	$I = V/R(1 - e^{-10}) = 0.500$ A
9. Potential drop across the resistor and the inductor after ten time constants for Figure 29.3(a)	$V_R = V(1 - e^{-10}) = 10.0$ V $V_L = Ve^{-10} = 4.54 \times 10^{-4}$ V
10. Time t' for the current to reach one-half its maximum value for Figure 29.3(a)	$I_{max} = V/R$ (step 4) $I = I_{max}/2 = (V/R)/2$ when t = t' $I = V/R(1 - e^{-[t/(L/R)]})$ $(V/R)/2 = (V/R)(1 - e^{-[t'/(L/R)]})$ $e^{-[t'/(L/R)]} = 1/2$ $t' = (L/R)\ \ln(2) = 0.173$ s
11. Rate at which the current is changing at t = L/R for Figure 29.3(a)	After t = L/R, $V_L = Ve^{-1} = 3.68$ V $dI/dt = V_L/L = 0.736$ A/s
12. Current at t = 0 s for Figure 29.3(b)	$I = I_0 e^{-[t/(L/R)]}$ At t = 0, $I = I_0 = 0.500$ A
13. Current after 0.100 s for Figure 29.3(b)	$I_0 = 0.500$ A $I = I_0 e^{-[t/(L/R)]} = 0.335$ A
14. Potential drop across the resistor and the inductor at t = 0.100 s for Figure 29.3(b)	$V_R = IR = 6.70$ V $V_L = V - V_R = 0 - V_R = -6.70$ V
15. Rate at which the current is changing after t = 0.100 s for Figure 29.3(b)	$dI/dt = V_L/L = -1.34$ A/s The minus sign tells us that the current is decreasing.

Example: A 2.00 H inductor and a 4.00 Ω resistor are connected in series with a 1.50 V battery. At a time equal to one time constant, determine (a) the current in the circuit, (b) the rate of change of the current, (c) the rate at which energy is being stored in the magnetic field, (d) the rate which energy is being lost to joule heating, (e) the rate at which energy is being delivered by the battery.

Given: $L = 2.00$ H $R = 4.00\ \Omega$ $V = 1.50$ V $t = L/R$

Determine:

(a) I -- current in the circuit
(b) dI/dt -- rate of change of current
(c) dU_B/dt -- rate at which energy is being stored in the magnetic field
(d) dU_{joule}/dt -- rate at which energy is being lost due to joule heating
(e) $dU_{battery}/dt$ -- rate at which energy is being supplied by the battery

Strategy: Knowing V, R, L, and t, we can determine the current I. Once I is known, we can use Kirchhoff's loop rule to determine dI/dt. Knowing the current, we can also use our knowledge of power to determine the rate at which energy is stored in the magnetic field, the rate at which energy is lost to joule heating, and the rate at which energy is supplied by the battery. As a quick check on our work, we can use the fact that the rate at which energy is supplied by the battery equals the sum of the rate at which it is stored in the B field and lost to joule heating.

Solution: (a) The current in the circuit at a time equal to one time constant ($t = L/R$) is obtained by

$$I = (V/R)(1 - e^{-[t/(L/R)]}) = (V/R)(1 - e^{-1}) = 0.237 \text{ A}$$

(b) The rate of change of current can be obtained by applying Kirchhoff's loop rule to the circuit.

$$V - V_R - V_L = 0 \qquad \text{or} \qquad V - IR - L(\Delta I/\Delta t) = 0$$

$$dI/dt = (V - IR)/L = 0.276 \text{ A/s}$$

(c) The rate at which energy is being stored in the magnetic field is obtained as follows:

$$P = IV_L = IL(dI/dt) = 0.131 \text{ W}$$

(d) The rate at which energy is being lost to joule heating is

$$P = I^2R = 0.225 \text{ W}$$

(e) The rate at which energy is being supplied by the battery is

$$P = IV = 0.356 \text{ W}$$

As a check, note that the power supplied by the battery is the sum of the power stored in the magnetic field and the power lost due to joule heating.

Related Text Exercises: Ex. 29-10 through 29-19.

3 Energy Stored in an Inductor (Section 29-3)

Review: Figure 29.4 shows an inductor of inductance L in a simple dc circuit. When the switch is closed, the current in the inductor increases from 0 to its final value I, and the magnetic field increases from 0 to its final value B in a time Δt.

Figure 29.4

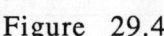

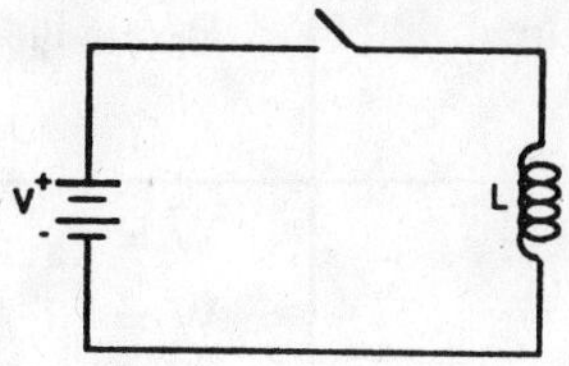

As the current is changing, an average induced emf ($\varepsilon = -L\Delta I/\Delta t$) opposite the direction of the applied voltage is produced across the inductor. The voltage supplied to the inductor by the power source is

$$V_L = -\varepsilon_L - L\Delta I/\Delta t$$

The average current and the change in current are

$$I_{av} = (0 + I)/2 = I/2 \quad \text{and} \quad \Delta I = (I - 0) = I$$

The average power delivered to the inductor is

$$P_{av} = I_{av} V_L = I_{av} L\Delta I/\Delta t = LI^2/2\Delta t$$

The energy stored in the inductor is the amount of work done on the inductor by the source to establish the B field.

$$U = W = P_{av}\Delta t = I_{av} L\Delta I = LI^2/2$$

The energy density (i.e., the energy per unit volume) in the inductor is

$$u = U/V = B^2/2\mu_o$$

Practice: The following is known for the situation shown in Figure 29.4.

The inductor is a solenoid
I = 3.00 A -- final current in the inductor
Δt = 1.00×10^{-5} s -- time to establish the final **B**-field
l = 0.100 m -- length of the solenoid
r = 0.500 cm -- radius of the solenoid
n = 2.00×10^3/m -- turns of wire on the solenoid per m of length

Determine the following:

1. Magnitude of the final magnetic field	$B = \mu_o nI = 7.54 \times 10^{-3}$ T

2. Inductance of the solenoid	$N = nl = 200$ $L = N\mu_0 nA = N\mu_0 n\pi r^2 = 3.95 \times 10^{-5}$ V•s/A
3. The potential difference across the solenoid	$V_L = \varepsilon_L = L(\Delta I/\Delta t) = 11.9$ V
4. Average power delivered to the solenoid	$P_{av} = I_{av}V_L = IV_L/2 = 17.8$ W
5. Work done by the power source to establish the final B field	$W = P_{av}\Delta t = 1.78 \times 10^{-4}$ J or $W = LI^2/2 = 1.78 \times 10^{-4}$ J
6. Energy stored in the magnetic field of the solenoid	$U = W = 1.78 \times 10^{-4}$ J
7. Energy density of the energy stored in the solenoid	$V = Sl = \pi r^2 l = 7.85 \times 10^{-6}$ m^3 $u = U/V = 22.7$ J/m^3 or $u = B^2/2\mu_0 = 22.6$ J/m^3

Example: We have a 20.0 cm long solenoid with 2.00 cm radius and 10^3 turns of wire. Determine the current in the solenoid to produce a magnetic energy density equal to that of a 10^4 V/m **E**-field.

Given: $l = 20.0$ cm $r = 2.00$ cm $N = 10^3$ turns $E = 10^4$ V/m $u_E = u_B$

Determine: The current in the solenoid so that $u_B = u_E$ if $E = 10^4$ V/m.

Strategy: We can equate expressions for the energy density of the **E** and **B** field and solve for the strength of the **B** field. Knowing B and the dimensions of the solenoid, we can determine the value for the current.

Solution: Expressions for the energy density of an **E** and **B** field are

$$u_E = \varepsilon_0 E^2/2 \quad \text{and} \quad u_B = B^2/2\mu_0$$

We want the value of B when $u_B = u_E$, hence

$$B^2/2\mu_0 = \varepsilon_0 E^2/2 \quad \text{or} \quad B = (\mu_0\varepsilon_0)^{1/2} E$$

The current required to give this value of B in the solenoid is obtained as follows:

$$B = \mu_0 NI/l \quad \text{or} \quad I = Bl/\mu_0 N = (\varepsilon_0/\mu_0)^{1/2} El/N = 5.30 \times 10^{-3} \text{ A}$$

Related Text Exercises: Ex. 29-20 through 29-25.

4 Transformers (Section 29-5)

Review: A schematic illustration of a transformer is shown in Figure 29.5.

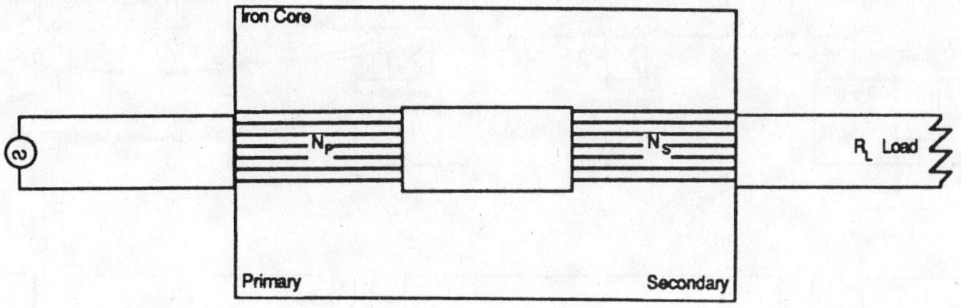

Figure 29.5

An alternating current source supplies a time-varying current to the primary coil. This current creates a time-varying magnetic flux in the iron core and hence through the secondary coil. The changing flux at the secondary induces a time-varying emf in the secondary, which is transmitted to the load. If we assume no flux losses and no energy losses, we obtain

$$V_s/V_p = \varepsilon_s/\varepsilon_p = N_s/N_p = I_p/I_s$$

Practice: The following data are given for the situation shown in Figure 29.5.

$$N_p = 100 \qquad N_s = 10{,}000 \qquad R_L = 1000\ \Omega$$

Determine the following:

1. Time rate of change of magnetic flux in the primary coil when $V_p = 10.0$ V	$(d\Phi/dt)_p = -V_p/N_p = 0.100\ \text{T}\cdot\text{m}^2/\text{s}$ $= 0.100$ V
2. Time rate of change of magnetic flux in the secondary coil when $V_p = 10.0$ V	$(d\Phi/dt)_s = (d\Phi/dt)_p = 0.100\ \text{T}\cdot\text{m}^2/\text{s}$ $= 0.100$ V
3. Emf induced in the secondary coil when $V_p = 10.0$ V	$V_s = -Ns(d\Phi/dt)_s = 1.00 \times 10^3$ V or $V_s = V_pN_s/N_p = 1.00 \times 10^3$ V
4. Current in the secondary coil when $V_p = 10.0$ V	When $V_p = 10.0$ V, $V_s = 1{,}000$ V $I_s = V_s/R_L = 1.00$ A
5. Current in the primary coil when $V_p = 10.0$ V	When $V_p = 10.0$ V,. $V_s = 1{,}000$ V and $I_s = 1.00$ A $I_p = V_sI_s/V_p = I_sN_s/N_p = 100$ A

Figure 29.6 shows a typical power distribution system as you would see it driving along the road (Figure 29.6(a)) and as the power plant engineer sees it on paper (Figure 29.5(b)). Also shown is a house fuse box (Figure 29.6(c)) and the circuit to the kitchen plug-ins (Figure 29.6(d)).

(a)

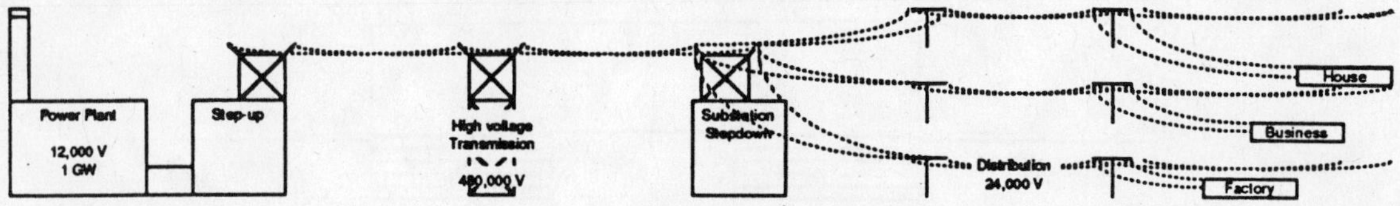

(b)

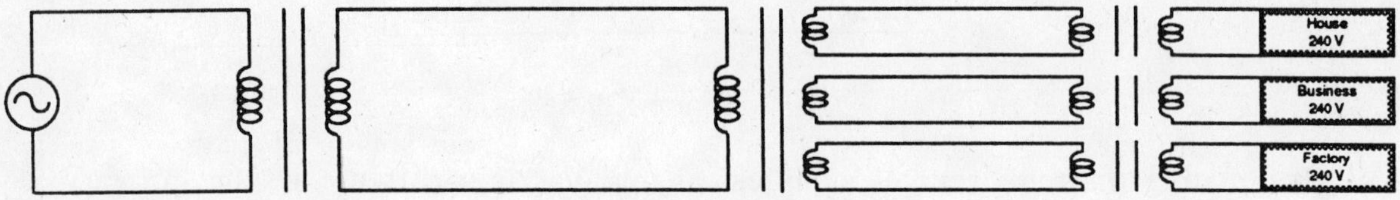

(c)

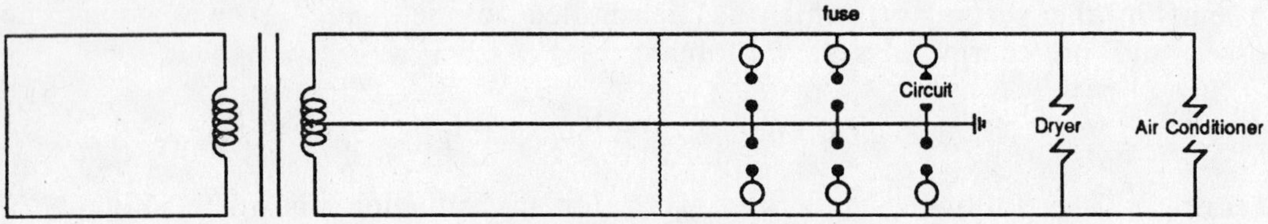

Fuse box at the house

(d)

R_1 -- Toaster (1200 W)
R_2 -- Radio (1000 W)
R_3 -- Mixer (1300 W)

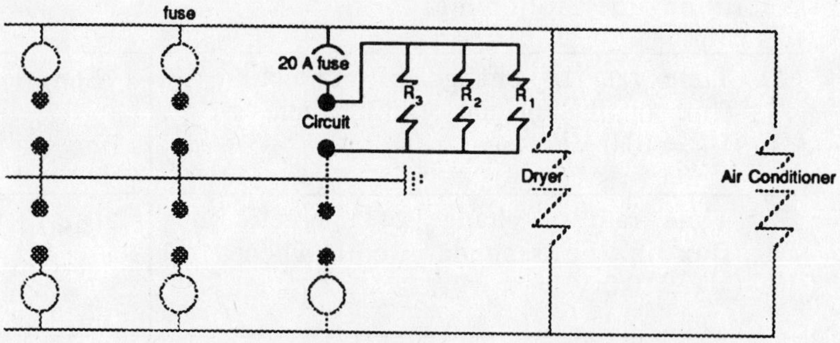

Circuit to kitchen plug-ins

Figure 29.6

As shown in Figure 29.6, the plant capacity is 1 GW (1 gigawatt = 1×10^9 W) at 12,000 V ac, stepped up to 480,000 V for transmission to the substation, distributed via utility poles at 24,000 V, and delivered to customers at 200 A and 240 V. It is 20 miles from the power plant to the substation, and the high-voltage transmission lines have a resistance of 0.150 Ω/mile. The power company sells energy at the rate of 10 ¢/kW•h.

Determine the following:

6. N_s/N_p for the step-up transformer at the plant	$N_s/N_p = V_s/V_p$ $= 4.80 \times 10^5$ V/1.20×10^4 V = 40/1
7. N_s/N_p for the step-down transformer at the substation	$N_s/N_p = V_s/V_p$ $= 2.40 \times 10^4$ V/4.80×10^5 V = 1/20
8. N_s/N_p for the transformer on the utility pole	$N_s/N_p = V_s/V_p$ $= 2.40 \times 10^2$ V/2.40×10^4 V = 1/100
9. Maximum power available to the homeowner	$P_{max} = IV$ $= (200\ A)(240\ V) = 4.80 \times 10^4$ W
10. Current in the service line to the house when the power usage is one half its maximum value	$P = P_{max}/2 = 2.40 \times 10^4$ W $I = P/V = 2.40 \times 10^4$ W/2.40×10^2 V = 100 A
11. Whether the fuse for the kitchen's plug-in circuit will blow	$P = P_1 + P_2 + P_3 = 3500$ W $I = P/V = 3500$ W/120 V = 29.2 A The 20.0 A fuse will blow
12. The current in high-voltage transmission lines at a time when the plant output is 500 megawatts	$I_p = P_p/V_p = 5.00 \times 10^8$ W/1.20×10^4 V $= 4.17 \times 10^4$ A $I_s = I_pN_p/N_s$ $= (4.17 \times 10^4\ A)(1/40) = 1.04 \times 10^3$ A or $P_s = P_p = 5.00 \times 10^8$ W and $I_s = P_s/V_s = 5.00 \times 10^8$ W/4.80×10^5 V $= 1.04 \times 10^3$ A
13. Rate of loss of energy due to joule heating in the high-voltage transmission lines. Assume the plant operates continually at half capacity	$R = \left(\frac{0.1500\ \Omega}{\text{mile}}\right)\left(\frac{20\ \text{miles}}{\text{wire}}\right)(2\ \text{wire})$ $= 6.00\ \Omega$ $P = I^2R = (1.04 \times 10^3\ A)^2(6.00\ \Omega)$ $= 6.49 \times 10^6$ W
14. Financial loss per year due to joule heating. Assume the plant operates continually at half its capacity	$E_{loss} = P_{loss}\Delta t$ $= (6.49 \times 10^6\ W)(8.76 \times 10^3\ h)$ $= 5.69 \times 10^{10}$ W•h Loss = $(E_{loss})(\text{Cost})$ $= \left[\begin{array}{l}(5.69 \times 10^{10}\ W\cdot h)(10\ ¢/kW\cdot h) \\ \times (kW/10^3\ W)(1\ \$/10^2\ ¢)\end{array}\right]$ $= \$5.69 \times 10^6$

Example: An electric generating plant produces electric energy at 12,000 V. We have a piece of electrical equipment that requires 75 A and 220 V. This equipment is housed a distance of 100 km from the generating plant. The transmission line used has a resistance of 1.50×10^{-4} Ω/m. Electrical energy costs 10.0 ¢/kW•h. We have two options for getting power to the equipment.

Option A — Transmit the power at 12,000 V and then use a step-down transformer at the equipment site.

Option B — Use a step-up transformer to transmit the power at 60,000 V and then use a step-down transformer at the equipment site.

Determine the annual financial loss to the power company if they use Option A over Option B.

Given:

$V_p = 1.2 \times 10^4$ V -- voltage output at the plant
$I_e = 75.0$ A -- current requirement of the equipment
$V_e = 220$ V -- voltage requirement of the equipment
$d = 1.00 \times 10^5$ m -- distance between plant and equipment
$R/l = 1.50 \times 10^{-4}$ Ω/m -- resistance per meter of transmission line
Cost = 10.0 ¢/kW•h -- cost of electrical energy
Option A -- transmit power at 1.20×10^4 V
Option B -- transmit power at 6.00×10^4 V

Determine: The annual financial loss to the power company if they use Option A over Option B.

Strategy: From the given information, we can determine the resistance of the transmission line and then the current in the transmission line for Option A and Option B. Knowing the resistance and the current for each option, we can establish the rate of loss of energy for each option. Knowing the rate of loss of energy, the time involved, and the price of energy to the customer, we can determine the financial loss to the power company if they choose Option A over Option B.

Solution: The resistance of the transmission line is

$$R = (2 \text{ wires})(R/l)(d) = (2)(1.50 \times 10^{-4}\ \Omega/\text{m})(1.00 \times 10^5\ \text{m}) = 30.0\ \Omega$$

Option A -- The power is transmitted at 1.20×10^4 V and stepped down to 2.20×10^2 V at the pole transformer. The current in the transmission line (I_A) is the same as the current to the primary (I_p) of the transformer at the equipment site. Since we have been given V_p, V_s, and I_s, we can determine I_p.

$$I_A = I_p = I_s(V_s/V_p) = (75.0\ \text{A})(2.20 \times 10^2\ \text{V}/1.20 \times 10^4\ \text{V}) = 1.38\ \text{A}$$

The rate of loss of energy due to joule heating is

$$P_{lossA} = I_A{}^2R = (1.38\ \text{A})^2(30.0\ \Omega) = 57.1\ \text{W}$$

Option B -- The power is transmitted at 6.00×10^4 V and stepped down to 2.20×10^2 V at the pole transformer. The current in the transmission line (I_B) is the same as the current to the primary (I_p) of the transformer at the equipment site. Since we have been given V_p, V_s, and I_s, we can determine I_p.

$$I_B = I_p = I_s(V_s/V_p) = (75.0\ \text{A})(2.20 \times 10^2\ \text{V}/6.00 \times 10^4\ \text{V}) = 0.275\ \text{A}$$

The rate of loss of energy due to joule heating is

$$P_{lossB} = I_B{}^2R = (0.275\ \text{A})^2(30.0\ \Omega) = 2.27\ \text{W}$$

The difference in the rate of loss of energy for these two options is

$$P_{loss} = P_{lossA} - P_{lossB} = 54.8\ \text{W}$$

The difference in energy loss for these two options over a year is

$$E_{loss} = P_{loss}\Delta t = (54.8\ \text{W})(1\ \text{yr})(365\ \text{d/y})(24\ \text{h/d}) = 4.80 \times 10^5\ \text{W}\cdot\text{h}$$

If this energy can be sold for 10 ¢/kW•h, the financial loss to the power company over a period of one year is

$$\$_{loss} = (E_{loss})(\text{Cost}) = (4.80 \times 10^5\ \text{W}\cdot\text{h})(10\ \text{¢/kW}\cdot\text{h})(\text{kW}/10^3\ \text{W})(\$1.00/100\ \text{¢}) = \$48.00$$

Related Text Exercises: Ex. 29-31 through 29-35.

PRACTICE TEST

Take and grade this practice test. Doing so will allow you to determine any weak spots in your understanding of the concepts taught in this chapter. The following section prescribes what you should study further to strengthen your understanding.

Figure 29.7 shows two different solenoids wrapped on the same coil form and some associated data.

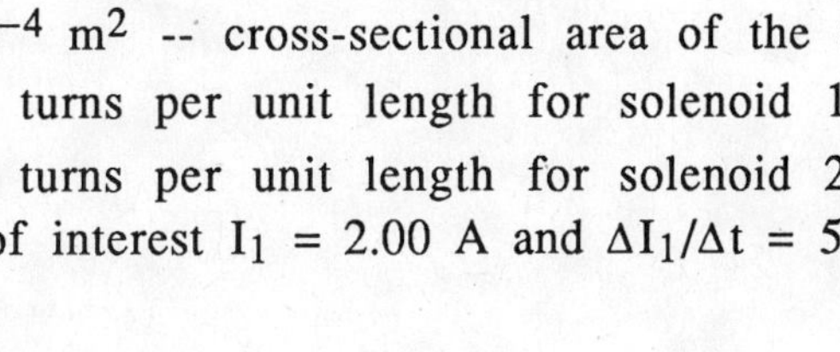

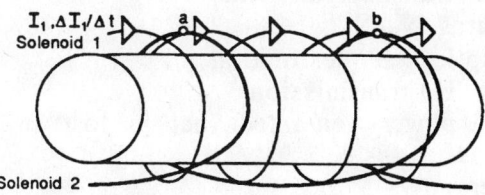

Figure 29.7

l = 10.0 cm -- length of the solenoids

A = 10.0×10^{-4} m^2 -- cross-sectional area of the solenoids

n_1 = 10^3/m -- turns per unit length for solenoid 1

n_2 = 10^2/m -- turns per unit length for solenoid 2

At the instant of interest I_1 = 2.00 A and $\Delta I_1/\Delta t$ = 5.00 A/s

Determine the following for the instant of interest:

_______ 1. Magnitude of the magnetic field due to solenoid 1
_______ 2. Magnetic flux inside solenoid 1 and solenoid 2
_______ 3. Time rate of change of magnetic flux through each solenoid
_______ 4. Emf induced in each turn of wire of solenoid 1 and 2
_______ 5. Total emf induced in solenoid 1
_______ 6. Total emf induced in solenoid 2
_______ 7. Mutual inductance of the solenoids
_______ 8. Self inductance of solenoid 1
_______ 9. Direction of the current induced in solenoid 2
_______ 10. Energy stored in the magnetic field of solenoid 1

Figure 29.8 shows an inductor in a circuit with a resistor and a battery.

Figure 29.8

$R = 30.0\ \Omega$
$L = 10.0$ H
$V = 5.00$ V

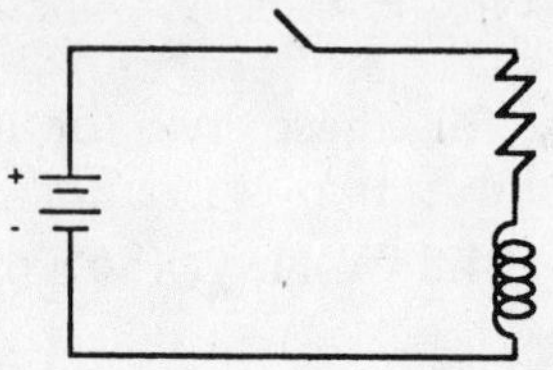

Determine the following:

_______ 11. Time constant for the circuit
_______ 12. Current in the circuit after two time constants
_______ 13. Potential difference across the resistor after two time constants
_______ 14. Potential difference across the inductor after two time constants
_______ 15. The rate of change of current in the inductor after two time constants

Power is transmitted 80.0 km at 22,000 V and is stepped down to 110 V at its destination. The transmission lines have a resistance of $1.50 \times 10^{-4}\ \Omega$/m for each of the two lines. The current required at the destination is 15.0 A.

Determine the following:

_______ 16. Current in transmission line
_______ 17. Ratio of turns N_p/N_s
_______ 18. Power supplied at destination
_______ 19. Power lost in transmission
_______ 20. Fraction of power generated that is lost in transmission

(See Appendix I for answers)

PRINCIPAL CONCEPTS AND EQUATIONS PRESCRIPTION

Your score on the practice test is an excellent measure of your understanding of the chapter. You should now use the following chart to write your own prescription for dealing with any weaknesses the practice test points out. Look down the leftmost column to the number of the question(s) you answered incorrectly, reading across that row you will find the concept and/or equation of concern, the section(s) of the study guide you should return to for further study, and some suggested text exercises which you should work to gain additional experience.

Practice Test Questions	Concepts and Equations	Prescription: Principal Concepts	Prescription: Text Exercises
1	Solenoid: $B_s = \mu_0 nI$	5 of Ch 27	27-27, 29
2	Magnetic flux: $\Phi_B = \int \mathbf{B} \cdot d\mathbf{s} = BS\cos\theta$	6 of Ch 27	29-8, 9
3	Rate of change of Flux: $\frac{d\Phi}{dt} = \mu_0 nS(\frac{dI}{dt})\cos\theta$	1	29-8
4	Induced emf: $\varepsilon = -d\Phi/dt$	1	29-8, 9
5	$\varepsilon = -Nd\Phi/dt$	1	29-8, 9
6	$\varepsilon = -Nd\Phi/dt$	1	29-8, 9
7	$M_{21} = \frac{-\varepsilon_2}{(dI_1/dt)}$	1	29-26
8	$L = \frac{-\varepsilon_1}{(dI_1/dt)}$	1	29-3, 4
9	Lenz's Law	1 of Ch 28	29-29
10	$U = (\frac{1}{2})LI^2$	3	29-22
11	$\tau_L = L/R$	2	29-11, 19
12	$I = (V/R)(1 - e^{-[t/(L/R)]})$	2	29-11, 14
13	$V_R = IR$	2 of Ch 24	24-28, 29
14	Kirchhoff's Rule: $V_L = V - V_R$	3 of Ch 25	29-17
15	$V_L = \varepsilon_L = -L(dI/dt)$	2	29-17
16	$V_pI_p = V_sI_s$	4	29-33
17	$N_p/N_s = V_p/V_s$	4	29-31, 34
18	$P = VI$	2 of Ch 25	29-31
19	$P_L = I_p^2R_{\text{transmission lines}}$	4	29-33
20	$F = \frac{P_L}{P_D + P_L} = \frac{P_L}{P_S + P_L} = \frac{P_L}{P_P + P_L}$	4	29-33

30 MAGNETIC FIELDS IN MATTER

RECALL FROM PREVIOUS CHAPTERS

Previously learned concepts and equations frequently used in this chapter	Text Section	Study Guide Page
Magnetic moment: $\mathbf{m} = NI\mathbf{S}$	26-3	26-7
Magnetic field of a solenoid: $\beta = \mu_0 nI$	27-3	27-17

NEW IDEAS IN THIS CHAPTER

Concepts and equations introduced		Text Section	Study Guide Page
Magnetization:	$\mathbf{M} = <\sum_i \mathbf{M}_i >/\Delta V$	30-1	30-1
Diamagnetism:	**M** and **B** are anti-parallel **M** is very small	30-2	30-2
Paramagnetism:	**M** and **B** are parallel **M** is small	30-3	30-3
Ferromagnetism:	**M** and **B** aren't linearly related **M** can be very large	30-4	30-3
Magnetic Intensity:	$\mathbf{H} = \mathbf{B}/\mu_o - \mathbf{M}$	30-5	30-4

PRINCIPAL CONCEPTS AND EQUATIONS

1 Magnetization (Section 30-1)

Review: Recall that for a plane current loop we define a magnetic moment vector **m**

$$\mathbf{m} = NI\mathbf{S}$$

where **S** is area vector of the loop
I is the current in the loop
N is the number of turns in the loop

The direction of **S** is obtained using the right hand rule.

In a planetary model of the atom, the electrons orbit the nucleus; the electrons also spin like a planet in this model. Using classical (non-relativistic and non-quantum mechanical) physics the orbital motion leads to a current. Thus there is a magnetic moment associated with the orbital motion of each electron. It can be shown that the magnetic moment associated with the orbital motion of the electron is related to the orbital angular momentum of the electron by

$$\mathbf{m}_L = -(e/2m_e)\mathbf{L}$$

where $\mathbf{m}_L$ is the magnetic moment associated with the angular motion
$-e$ is the charge on the electron
m_e is the mass of the electron
$\mathbf{L}$ is the orbital angular momentum of the electron

The spin angular momentum and the magnetic moment associated with the spin are related by

$$\mathbf{m}_s = -(e/m_e)\mathbf{S}$$

where $\mathbf{m}_s$ is the magnetic moment associated with the spin motion
$-e$ is the charge on the electron
m_e is the mass of the electron
$\mathbf{S}$ is the spin angular momentum of the electron

The total $\mathbf{m}$ for the atom is the sum of $\mathbf{m}_L$ and $\mathbf{m}_s$ over all the electrons in the atom plus any magnetic moment arising from the nucleus (which is smaller, hence usually neglected). Often $\mathbf{m} = 0$.

The above is an atomic or molecular description. For a macroscopic description we use a small volume ΔV (small on a macroscopic scale but large on a microscopic scale) and define the magnetization vector by

$$\mathbf{M} = \left\langle \sum_i \mathbf{m}_i \right\rangle / \Delta V$$

$\left\langle \sum_i \mathbf{m}_i \right\rangle$ is the average of the sum of the magnetic moments contained in ΔV.

There is a value for $\mathbf{M}$ at each point in space.

Related Text Exercises: Ex. 30-1 through 30-8.

2 Diamagnetism (Section 30-2)

Review: If we use the planetary model, then electrons moving in opposite directions in their orbits are equivalent to currents in opposite directions. The electrons tend to pair off so that their currents cancel and $\mathbf{M} = 0$. $\mathbf{M}$ is then due to any extra unpaired electrons.

When a magnetic field **B** is applied to the atom the induced (Faraday's Law and Lenz's Law) emf causes electrons orbiting the nucleus one way to speed up and those orbiting the nucleus in the opposite direction to slow down. A magnetic field is produced opposite to the applied field (Lenz's Law). Thus the atom makes a magnetic field opposite to the applied field. The atom has a magnetization now even if it was zero before. The effect is very small but leads to a magnetization opposite to **B**.

If either of the next two types of magnetization (Paramagnetism and Ferromagnetism) are present then the diamagnetism is usually neglected because it is small compared to these latter types. It is however present in all cases.

Related Text Exercises: Ex. 30-9 and 30-10.

3 Paramagnetism (Section 30-3)

Review: Paramagnetic materials have one or more unpaired electrons so that each atom has a non zero magnetic moment. However without an applied magnetic field **B** the magnetic moments of different atoms are randomly oriented and **M** = 0. An applied magnetic field causes some alignment of the atomic moments (the alignment increases with increasing **B**) hence we have **M** and **B** in the same direction. The magnetic field is increased because of the paramagnetic material. As in diamagnetic materials **M** and **B** are linearly related for isotropic paramagnetic materials. This is expressed in Curie's law:

$$\mathbf{M} = C\mathbf{B}/\mu_0 T$$

where C is Curie's constant - depends on the particular material
B is the applied magnetic field
μ_0 is the permeability of free space
T is the Kelvin temperature
M is the magnetization of a paramagnetic material due to an applied magnetic field **B** at temperature T

Related Text Exercises: Ex. 30-11 through 30-13.

4 Ferromagnetism (Section 30-4)

Review: Some materials (iron, cobalt and nickel) when aligned with an applied magnetic field will greatly enhance the field. Sometimes this alignment remains after the applied field is removed; **M** and hence **B** are not zero even in the absence of macroscopic currents. These ferromagnetic materials are said to be "hard". If the magnetization returns to zero when the applied field is removed the ferromagnetic materials are said to be "soft". All ferromagnetic materials become soft at high temperatures.

For ferromagnetic materials the relationship between **M** and **B** is not linear; in fact it depends on the magnetic history of the ferromagnetic material.

In ferromagnetic materials the atomic magnetic moments align over macroscopic regions called domains. Each domain has an associated magnetization **M**. The magnetization in different domains may not align. In "hard" materials, once the domains are aligned they stay aligned. This happens in permanent magnets. In "soft" materials the alignment quickly subsides once the applied **B** field is removed.

Related Text Exercises: Ex. 30-14 and 30-15.

5 Magnetic Intensity H (Section 30-5)

Review: In a vacuum, **B** depends only on macroscopic currents. If some material is present, **B** also depends on the magnetization **M**. In fact a permanent magnet is a case where there is a **B** field when there are no macroscopic currents. When material media are present it is sometimes difficult to control and measure **B**. The reason for this difficulty is that **M** both depends on and contributes to **B**.

We define another field **H**, the magnetic intensity, which in some cases is much easier to control and determine. **H** is defined by

$$\mathbf{H} = \mathbf{B}/\mu_0 - \mathbf{M}$$

we may also write

$$\mathbf{B} = \mu_0(\mathbf{H} + \mathbf{M})$$

For isotropic diamagnetic and paramagnetic materials **B**, **H** and **M** are linear and either parallel or anti-parallel. For ferromagnetic materials, no such linearity exists. When we have linearity we write

$$\mathbf{B} = \mu\mathbf{H}$$

where μ is the permeability of the material. For a vacuum the permeability is just μ_0. For diamagnetic materials $\mu < \mu_0$ and for ferromagnetic materials $\mu > \mu_0$, but the difference between μ and μ_0 for both of these cases is usually small. Except at low temperatures paramagnetic effects are small.

Boundary conditions (derived using Ampere's Law) that must be satisfied by **H** and **B** are very useful in solving real problems with specific geometries. These boundary conditions state that at any boundary:

i) the normal component of **B** must be continuous across the boundary, and
ii) the tangential component of **H** must be continuous across the boundary.

This last condition allows **H** to change directions at the edge of a permanent magnet while B cannot. See Figure 30.1.

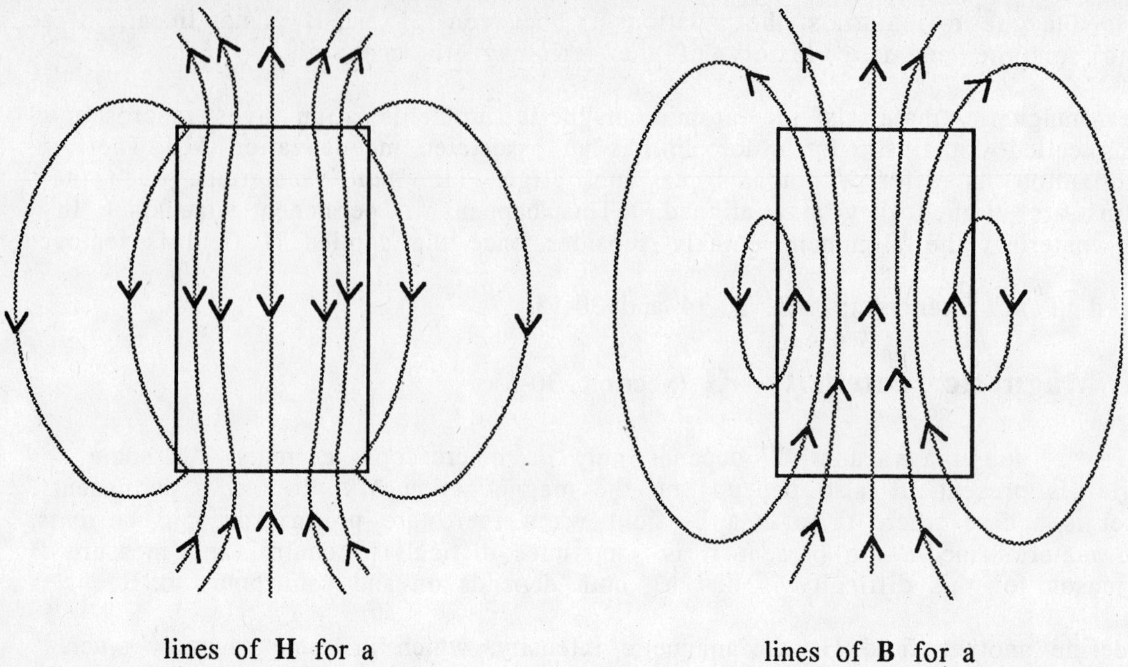

lines of **H** for a permanent magnet

(a)

lines of **B** for a permanent magnet

(b)

Figure 30.1

Figure 30.2 shows the relationship between the three vectors **H**, **B** and **M** for a point P_1 that is outside a permanent magnet and a point P_2 that is inside a permanent magnet.

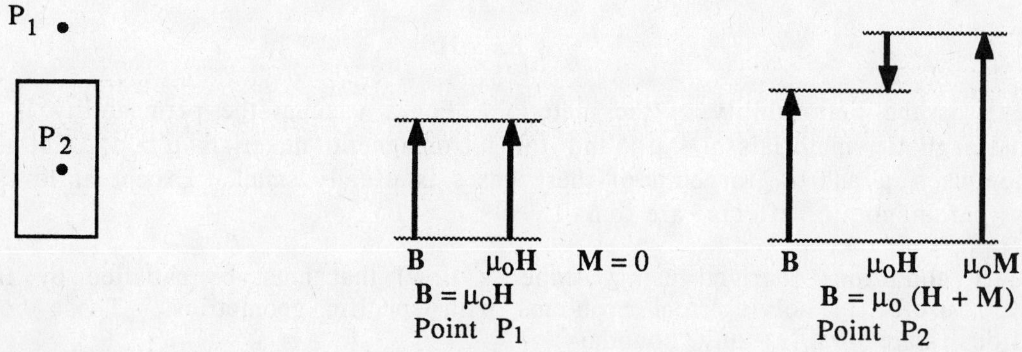

(a) (b) (c)

Figure 30.2

In ferromagnetic materials, the relationship between **B**, **H** and **M** depends on the prior history of the material. Figure 30.3 shows a typical curve of B verses H for a ferromagnet.

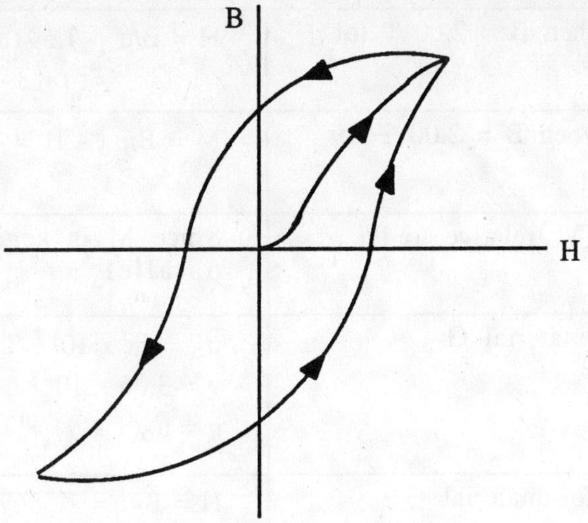

Figure 30.3

The behavior shown in Figure 30.3 is called hysteresis and the curve is called a hysteresis loop.

Practice: Table 30-1 gives the magnetic susceptibility (dimensionless) at 20°C for various materials.

Table 30-1

Material	A	B	C	D	E	F	G	H
χ (x 10^{-5})	-16.6	130.0	-1.00	0.20	-2.20	-9.90	306	2.20

The susceptibility is related to the permeability by

$$\mu = \mu_0(1 + \chi)$$

Determine the following:

1. The materials that are diamagnetic	Those materials with $\mu < \mu_0$ are diamagnetic. This means that any material with a negative susceptibility is diamagnetic. Therefore A, C, E and F are diamagnetic.
2. The materials that are paramagnetic	Those materials with $\mu > \mu_0$ (χ positive) are paramagnetic. Therefore B, D, G and H are paramagnetic
3. The material that is most highly diamagnetic	Material A since it has the most negative value of χ
4. The material that is most highly paramagnetic	Material G since it has the most positive value of χ.
5. The value of μ for material A	$\mu_0 = 4\pi \times 10^{-7}$ T•m•A^{-1} $\chi = -1.66 \times 10^{-4}$ $\mu = \mu_0(1 + \chi) = 1.25643 \times 10^{-6}$ T•m•A^{-1}

6. The value of H when B = 2.00 T for material A	$H = B/\mu = 1.5918 \times 10^6\ A \bullet m^{-1}$
7. The value of M when B = 2.00 T for material A	$M = B/\mu_o - H = -260\ A \bullet m^{-1}$
8. The direction of **M** relative to **B**	Since M is negative, **M** and **B** are anti-parallel
9. The value μ for material G	$\mu_o = 4\pi \times 10^{-7}\ T \bullet m \bullet A^{-1}$ $\chi = 3.06 \times 10^{-3}$ $\mu = \mu_o(1 + \chi) = 1.26048 \times 10^{-6}\ T \bullet m \bullet A^{-1}$
10. The value of H for material G when B = 2.00 T	$H = B/\mu = 1.58670 \times 10^6\ A \bullet m^{-1}$
11. The value of M for material G when B = 2.00 T	$M = B/\mu_o - H = 4850\ A \bullet m^{-1}$
12. The direction of **M** relative to **B**	Since M is positive, **M** and **B** are parallel

Example: Figure 30.4 is the hysteresis curve for a ferromagnetic material. (a) What is the permeability when B = 1.00 T? (b) What is the magnetization when B = 1.00 T? (c) What is B when the material is used as a permanent magnet?

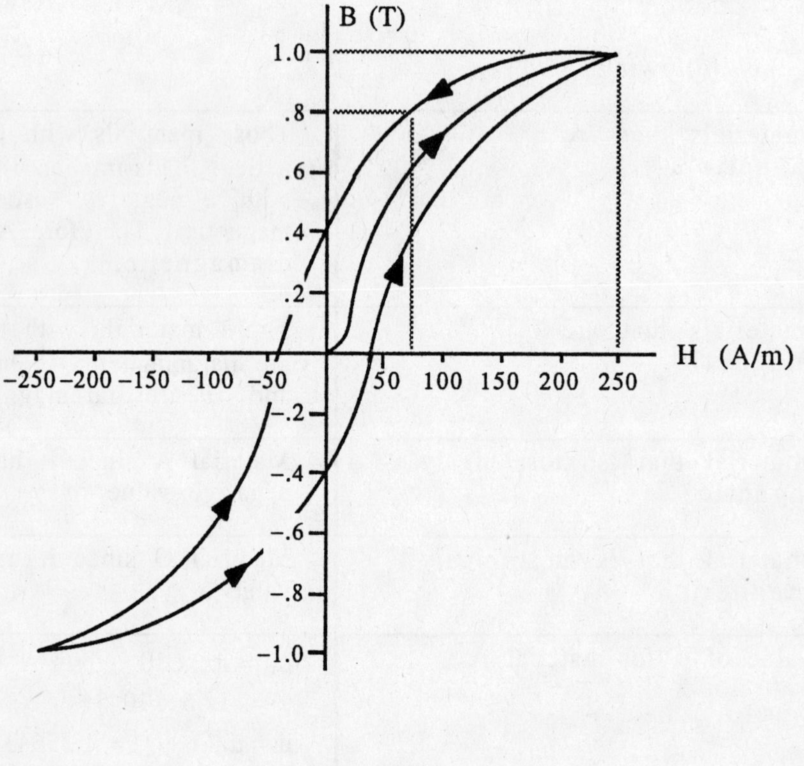

Figure 30.4

Given: The hysteresis curve in Figure 30.4.

Determine: (a) μ when B = +1.00 T (b) M when B = +1.00 T
(c) B when this material is used as a permanent magnet

Strategy: (a) Use $\mu = B/H$ and data obtained from Figure 30.4 to determine μ.
(b) Knowing a value for B, we can determine H from the curve and then M.
(c) Inspection of the curve shows a non zero value of B when H = 0; this is the condition for a permanent magnet.

Solution: (a) When B = 1.00 T, read H = 250 $A{\bullet}m^{-1}$ from the hysteresis curve and

$$\mu = B/H = (1.00\ T)/(250\ A{\bullet}m^{-1}) = 4.00 \times 10^{-3}\ T{\bullet}m{\bullet}A^{-1}$$

(b) The value for M when B = 1.00 T is

$$M = B/\mu_o - H = 7.955 \times 10^5\ A{\bullet}m^{-1}$$

(c) When H = 0 the material is permanently magnetized. From Figure 30.4 we see that this leads to B = 0.400 T.

Related Text Exercises: Ex. 30-16 through 30-20.

PRACTICE TEST

Take and grade this practice test. Doing so will allow you to determine any weak spots in your understanding of the concepts taught in this chapter. The following section prescribes what you should study further to strengthen your understanding.

Refer to Table 30.1 and determine the following:

_________ 1. μ for material B
_________ 2. H for material B when the magnetic field is 2.00 T
_________ 3. M for material B when the magnetic field is 2.00 T
_________ 4. Direction of M relative to the magnetic field for material B
_________ 5. μ for material F
_________ 6. H for material F when the magnetic field is 2.00 T
_________ 7. M for material F when the magnetic field is 2.00 T
_________ 8. Direction of M relative to the magnetic field for material F

Refer to Figure 30.4 and determine the following:

_________ 9. μ when B = 0.800 T
_________ 10. M when B = 0.800 T

(See Appendix I for answers.)

PRINCIPAL CONCEPTS AND EQUATIONS PRESCRIPTION

Your score on the practice test is an excellent measure of your understanding of the chapter. You should now use the following chart to write your own prescription for dealing with any weaknesses the practice test points out. Look down the leftmost column to the number of the question(s) you answered incorrectly, reading across that row you will find the concept and/or equation of concern, the section(s) of the study guide you should return to for further study, and some suggested text exercises which you should work to gain additional experience.

Practice Test Questions	Concepts and Equations	Prescription	
		Principal Concepts	Text Exercises
1	Relation between permeability and susceptibility: $\mu = \mu_0(1 + \chi)$	---	---
2	Magnetic intensity: $\mathbf{H} = \mathbf{B}/\mu$	5	30-6
3	Magnetization: $\mathbf{M} = \mathbf{B}/\mu_0 - \mathbf{H}$	5	30-6
4	Magnetization: $\mathbf{M} = \mathbf{B}/\mu_0 - \mathbf{H}$	5	30-6
5	Relation between permeability and susceptibility: $\mu = \mu_0(1 + \chi)$	---	---
6	Magnetic intensity: $\mathbf{H} = \mathbf{B}/\mu$	5	30-6
7	Magnetization: $\mathbf{M} = \mathbf{B}/\mu_0 - \mathbf{H}$	5	30-6
8	Magnetization: $\mathbf{M} = \mathbf{B}/\mu_0 - \mathbf{H}$	5	30-6
9	Permeability: $\mu = B/H$	5	30-20
10	Magnetization: $\mathbf{M} = \mathbf{B}/\mu_0 - \mathbf{H}$	5	30-20

31 ELECTROMAGNETIC OSCILLATIONS AND AC CIRCUITS

RECALL FROM PREVIOUS CHAPTERS

Previously learned concepts and equations frequently used in this chapter	Text Section	Study Guide Page
Energy stored in a capacitor: $U = QV/2 = CV^2/2 = Q^2/2C$	23-3	23-7
Ohm's Law: $V = IR$	24-2	24-5
Power: $P = I^2R = V^2/R = IV$	25-2	25-4
Energy stored in an inductor: $U = LI^2/2$	29-3	29-9

NEW IDEAS IN THIS CHAPTER

Concepts and equations introduced	Text Section	Study Guide Page
Inductive and capacitive reactance: $X_L = 2\pi\nu L$ and $X_C = 1/2\pi\nu C$	31-4, 5	31-6
Impedance: $Z_{RLC} = [R^2 + (X_C - X_L)^2]^{1/2}$	31-6	31-18
Maximum potential difference: $V_{RLCm} = I_m Z_{RLC}$; $V_{Rm} = I_m R$ $\quad V_{Lm} = I_m X_L$ $\quad V_{Cm} = I_m X_C$	31-3, 4, 5, 6	31-2
Phasors: $\mathbf{V}_{RC} = \mathbf{V}_R + \mathbf{V}_C$ $\quad \mathbf{V}_{RL} = \mathbf{V}_R + \mathbf{V}_L$; $\mathbf{V}_{LC} = \mathbf{V}_L + \mathbf{V}_C$ $\quad \mathbf{V}_{RLC} = \mathbf{V}_R + \mathbf{V}_L + \mathbf{V}_C$	31-3, 4, 5, 6	31-2
Instantaneous voltages and current: $V_s = V_m \sin\omega t$; $i = I_m \sin(\omega t + \phi)$; $V_R = V_{Rm} \sin(\omega t + \phi)$; $V_C = V_{Cm} \sin(\omega t + \phi - \pi/2)$; $V_L = V_{Lm} \sin(\omega t + \phi + \pi/2)$	31-3, 4, 5	31-2
Phase angle: V_R is in phase with i; V_L leads i by $\pi/2$ rad; V_C lags i by $\pi/2$ rad; $\phi = \tan^{-1}(\lvert V_L - V_C\rvert/V_R)$	31-3, 4, 5, 6	31-2
Resonance: $X_L = X_C$ $\quad \nu = 1/2\pi(LC)^{1/2}$	31-6	31-18
Average Power: $\overline{P} = I_{rms}{}^2R = I_{rms}V_{rms}\cos\phi$	31-7	31-22
Power factor: $\cos\phi = R/Z = R/[R^2 + (X_C - X_L)^2]^{1/2}$	31-7	31-22

PRINCIPAL CONCEPTS AND EQUATIONS

1 Phasors (Section 31-3)

Review: A phasor is a rotating vector representing a physical quantity such as electrical potential difference or current. During this general introduction to phasors we will not associate them with a particular physical quantity; instead we will represent them by $\mathbf{P}_1$ and $\mathbf{P}_2$ with magnitude P_{1m} and P_{2m} and arbitrary units u.

Figure 31.1 shows the phasor $\mathbf{P}_1$ of magnitude P_{1m}, initial phase angle ϕ_{1i} and instantaneous value P_1.

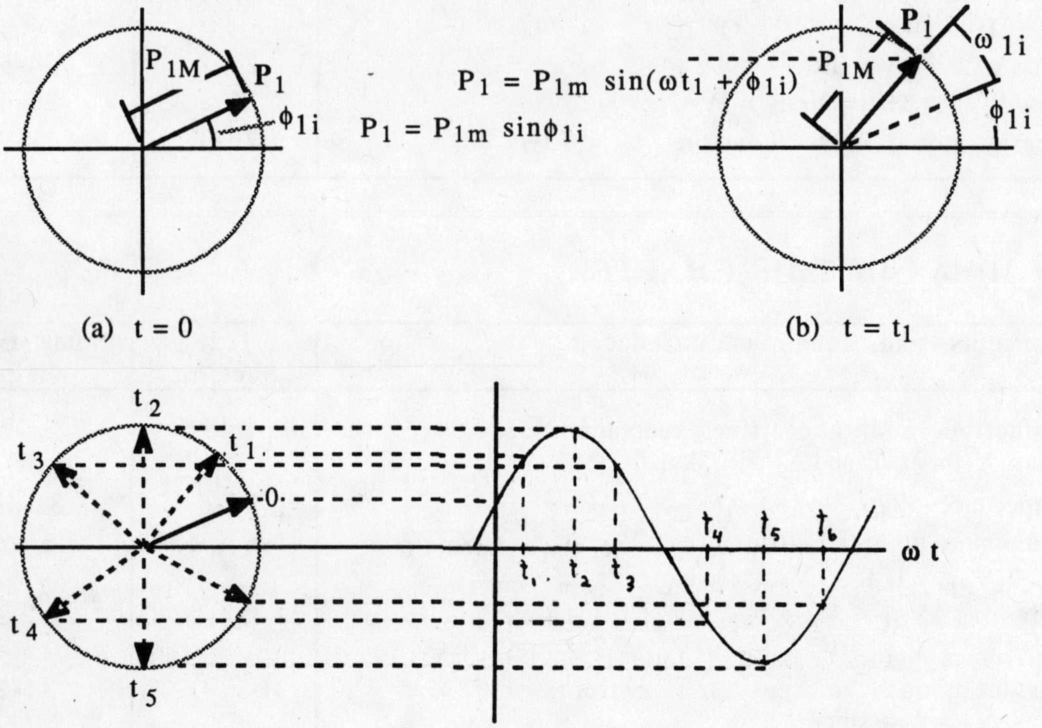

(c) Various t (d) Instantaneous value of $\mathbf{P}_1$

(a) The phasor diagram at $t = 0$
(b) The phasor diagram at $t = t_1$
(c) The phasor diagram at various times
(d) The instantaneous value of $\mathbf{P}_1$ at various times (phasor plot)

Figure 31.1

Figure 31.2 shows phasors $\mathbf{P}_1$ and $\mathbf{P}_2$ with magnitudes P_{1m} and P_{2m}, initial angles ϕ_{1i} and ϕ_{2i}, and a phase angle difference $\phi = \pi/2$.

(a) t = 0 (b) $t = t_1$

Figure 31.2

In general, the analysis will be simpler if one of the phasors has an initial phase angle of zero. The present case will be considerably simpler to deal with if $\phi_{1i} = 0$. Then the initial phase angle of $\mathbf{P_2}$ is just the phase angle difference.

Figure 31.3 shows phasors $\mathbf{P_1}$, $\mathbf{P_2}$, and $\mathbf{P_T}$ with magnitudes P_{1m}, P_{2m} and P_{Tm}

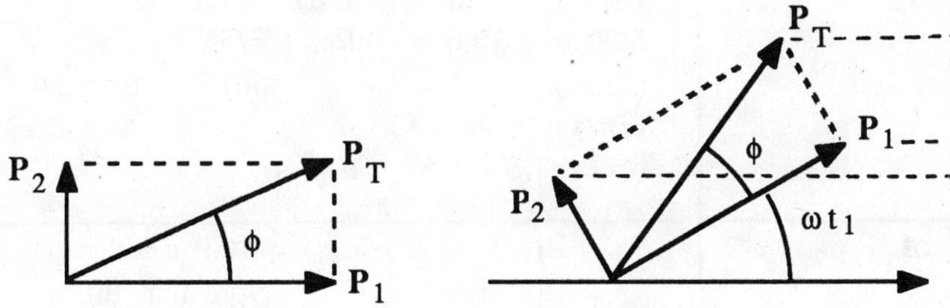

(a) t = 0 (b) $t = t_1$

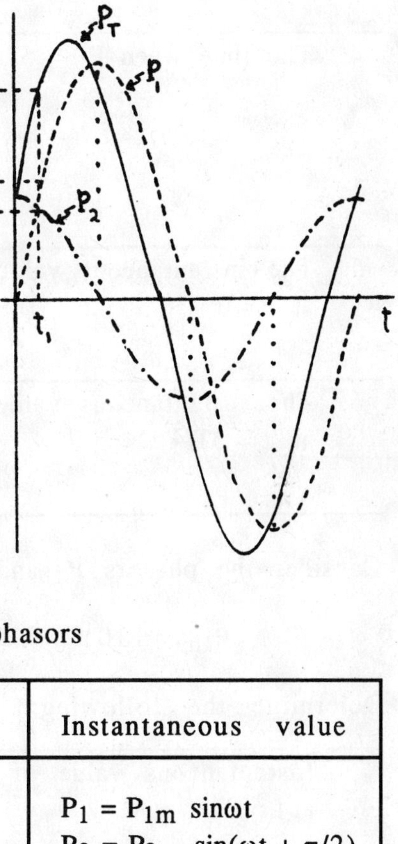

Figure 31.3 (c)

The table below summarizes the information about these phasors

Phasor	Internal phase angle	Magnitude	Instantaneous value
$\mathbf{P_1}$	0	P_{1m}	$P_1 = P_{1m} \sin\omega t$
$\mathbf{P_2}$	$\pi/2$	P_{2m}	$P_2 = P_{2m} \sin(\omega t + \pi/2)$
$\mathbf{P_T} = \mathbf{P_1} + \mathbf{P_2}$	$\phi = \tan^{-1}(P_{2m}/P_{1m})$	$P_{Tm} = (P_{1m}^2 + P_{2m}^2)^{1/2}$	$P_T = P_{Tm} \sin(\omega t + \phi)$

Practice: Consider the $\mathbf{P}_1$ shown in Figure 31.1 and the following data:

$P_{1m} = 10.0$ u $\quad \nu = 60.0$ Hz $\quad P_1 = 8.66$ u at $t = T/4$

Determine the following:

1. Period for the phasor	$T = 1/\nu = 0.0167$ s
2. Angular speed for the phasor	$\omega = 2\pi\nu = 377$ rad/s
3. The initial phase angle ϕ_i	$P_1 = P_{1m}\ \sin(\omega t + \phi_i) = 8.66$ u at $t = T/4$ 8.66 u $= 10.0$ u $\sin[(2\pi/T)(T/4) + \phi_{1i}]$ $0.866 = \sin(\pi/2 + \phi_{1i})$ $\pi/2 + \phi_{1i} = \sin^{-1}(0.866) = \pi/3$ or $2\pi/3$ hence $\phi_{1i} = -\pi/6,\ \pi/6$ Figure 31.1 shows $\mathbf{P}_1$ in the first quadrant at $t = 0$, hence $\phi_{1i} = \pi/6$
4. The time when $P_1 = 5.00$ u	$P_1 = P_{1m}\ \sin(\omega t + \pi/6)$ 5.00 u $= 10.0$ u $\sin(\omega t + \pi/6)$ $(2\pi/T)t + \pi/6 = \sin^{-1}(0.500) = \pi/6,\ 5\pi/6$ $(2\pi/T)t = 0,\ 2\pi/3$ $t = 0,\ T/3 = 5.57 \times 10^{-3}$ s
5. The instantaneous value of $\mathbf{P}_1$ at $t = 0$	$P_1 = P_{1m}\ \sin(\omega t + \phi_{1i}) = 10.0$ u $\sin(\pi/6)$ $= 5.00$ u Note that this is consistent with step 4 of this practice.
6. The instantaneous value of $\mathbf{P}_1$ at $t = 3T/4$	$P_1 = P_{1m}\ \sin(\omega t = \phi_{1i})$ $= 10.0$ u $\sin[(2\pi/T)(3T/4) + \pi/6]$ $= 10.0$ u $\sin(10\pi/6) = -8.66$ u

Consider the phasors $\mathbf{P}_1$ and $\mathbf{P}_2$ in Figure 31.3 and the following data:

$P_{1m} = 10.0$ u $\quad P_{2m} = 5.00$ u $\quad \phi_{1i} = 0 \quad \phi_{2i} = \pi/2 \quad \nu = 60.0$ Hz

Determine the following:

7. Instantaneous value of $\mathbf{P}_1$ and $\mathbf{P}_2$ at $t = 0$	$P_1 = P_{1m}\ \sin\omega t = 0$ at $t = 0$ $P_2 = P_{2m}\ \sin(\omega t + \pi/2) = 5.00$ at $t = 0$
8. Instantaneous value of $\mathbf{P}_1$ and $\mathbf{P}_2$ at $t = T/8$	At the time $t = T/8$, $\omega t = (2\pi/T)(T/8) = \pi/4$ $P_1 = P_{1m}\ \sin\omega t = (10.0\ \text{u})\sin(\pi/4) = 7.07$ u $P_2 = P_{2m}\ \sin(\omega t + \pi/2)$ $= (5.00\ \text{u})\ \sin(3\pi/4) = 3.54$ u

9. The magnitude of $\mathbf{P}_T$	$P_{Tm} = (P_{1m}^2 + P_{2m}^2)^{1/2} = 11.2$ u
10. The phase angle between $\mathbf{P}_T$ and $\mathbf{P}_1$	$\phi = \tan^{-1}(P_{2m}/P_{1m}) = 26.6°$
11. The instantaneous value of P_T at $t = T/8$	At the time $t = T/8$ $\omega t = (2\pi/T)(T/8) = \pi/4 = 45.0°$ $P_T = P_{Tm}\sin(\omega t + \phi)$ $= (11.2\text{ u})\sin(45.0° + 26.6°) = 10.6$ u

Example: Consider the two phasors $\mathbf{P}_1$ and $\mathbf{P}_2$ and the following information.

$P_1 = 10.4$ u at $t = T/3$ $\quad P_{2m} = 5.00$ u $\quad \phi_{2i}$ is positive
$P_2 = 2.50$ u at $t = T/6$ $\quad \phi_{1i} = 0°$

Calculate (a) the magnitude of $\mathbf{P}_1$, (b) the initial phase angle for $\mathbf{P}_2$, and (c) the magnitude and phase angle between $\mathbf{P}_T = \mathbf{P}_1 + \mathbf{P}_2$ and $\mathbf{P}_1$

Given:

$P_1 = 10.4$ u at $t = T/3$ - the instantaneous value of $\mathbf{P}_1$ at the time $t = T/3$
$\phi_{1i} = 0°$ - the initial phase angle of $\mathbf{P}_1$
$P_2 = 2.5$ u at $t = T/6$ - the instantaneous value of $\mathbf{P}_2$ at the time $t = T/6$
$P_{2m} = 5.00$ u - the magnitude of $\mathbf{P}_2$
ϕ_{2i} is positive - some information about the initial phase angle of $\mathbf{P}_2$

Determine:

P_{1m} - The magnitude of $\mathbf{P}_1$
ϕ_{2i} - The initial phase angle of $\mathbf{P}_2$
P_{Tm} - The magnitude of $\mathbf{P}_T = \mathbf{P}_1 + \mathbf{P}_2$
ϕ - The phase angle between $\mathbf{P}_T$ and $\mathbf{P}_1$

Strategy: Knowing the instantaneous value of $\mathbf{P}_1$ at some time and the initial phase angle for $\mathbf{P}_1$, we can determine the magnitude of $\mathbf{P}_1$. Knowing the instantaneous value of $\mathbf{P}_2$ at some time and the magnitude of $\mathbf{P}_2$, we can determine the initial phase angle of $\mathbf{P}_2$. Knowing the magnitude of $\mathbf{P}_1$ and $\mathbf{P}_2$, we can determine the magnitude of $\mathbf{P}_T$ and the phase angle between $\mathbf{P}_T$ and $\mathbf{P}_1$.

Solution: (a) Since $\phi_{1i} = 0$, we can write the general form of P_1 as

$$P_1 = P_{1m}\sin\omega t$$

Next we know that at $t = T/3$

$$\omega t = (2\pi/T)(T/3) = 2\pi/3 \quad \text{and} \quad P_1 = 10.4\text{ u}$$

Inserting values into the general form of P_1, obtain

$$10.4 \text{ u} = P_{1m} \sin 2\pi/3 = 0.866\, P_{1m} \Rightarrow P_{1m} = 12.0 \text{ u}$$

(b) The general form of P_2 is

$$P_2 = P_{2m} \sin(\omega t + \phi_{2i})$$

We know that at $t = T/6$

$$\omega t = (2\pi/T)(T/6) = \pi/3 \quad \text{and} \quad P_2 = 2.50 \text{ u}$$

Inserting values into the general form of P_2, obtain

$$2.50 \text{ u} = 5.00 \text{ u} \sin(\pi/3 + \phi_{2i})$$

or

$$\sin(\pi/3 + \phi_{2i}) = 0.500 \Rightarrow \pi/3 + \phi_{2i} = \sin^{-1}(0.500) = \pi/6,\ 5\pi/6$$

Solving for ϕ_{2i}, obtain

$$\phi_{2i} = -\pi/6,\ \pi/2$$

Since we were given that ϕ_{2i} is positive, we know that

$$\phi_{2i} = \pi/2$$

(c) Knowing P_{1m} (part a) and P_{2m} (given), we can determine the magnitude of $\mathbf{P}_T$ and the phase angle between $\mathbf{P}_T$ and $\mathbf{P}_1$ as follows:

$$|\mathbf{P}_T| = P_{Tm} = (P_{1m}{}^2 + P_{2m}{}^2)^{1/2} = [(12.0 \text{ u})^2 + (5.00 \text{ u})^2]^{1/2} = 13.0 \text{ u}$$

$$\phi = \tan^{-1}(P_{2m}/P_{1m}) = 22.6°$$

Related Text Exercises: Your text contains no exercises which deal just with phasors and their properties. However a thorough understanding of phasors will facilitate your understanding of other sections of this chapter and enhance your ability to work exercises and problems in other sections of this chapter.

2 AC Source Connected to a Resistor, Capacitor, or Inductor

(Sections 31-3, 4 and 5)

Review: Consider an alternating current source which generates a sinusoidal output voltage

$$V_s = V_m \sin\omega t$$

When this voltage source V_s is applied to a resistor R (see Figure 31.4(a)) the potential difference across the resistor is V_R.

$$V_s = V_R$$

but $\quad V_s = V_m \sin\omega t \quad$ and $\quad V_R = iR \quad$ hence

$$V_m \sin\omega t = iR$$

or $\quad i = (V_m/R)\sin\omega t = I_m \sin\omega t$

where $\quad I_m = V_m/R = V_{Rm}/R$

For this case the phasors representing the potential difference across the resistor and the current through the resistor are in phase as shown in Figure 31.4.

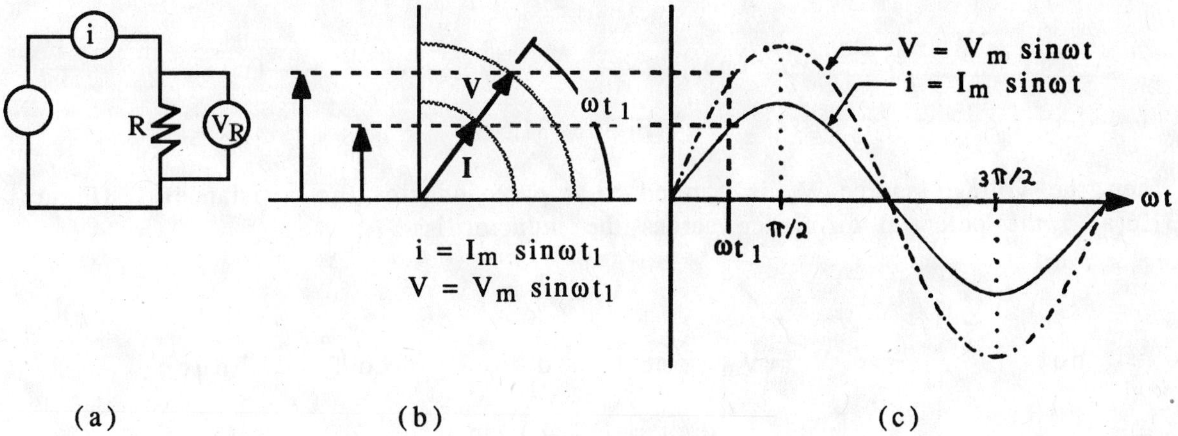

Figure 31.4

When the voltage source V_s is applied to a capacitor C (Figure 31.5(a)) the potential difference across the capacitor is V_C.

$$V_s = V_C$$

but $\quad V_s = V_m \sin\omega t \quad$ and $\quad V_C = q/C \quad$ hence

$$V_m \sin\omega t = q/C$$

or $\quad q = CV_m \sin\omega t$

and $\quad i = dq/dt = \omega C V_m \cos\omega t = I_m \sin(\omega t + \pi/2)$

where $\quad I_m = \omega C V_m = V_m/(1/\omega C) = V_m/X_C = V_{Cm}/X_C$

and $\quad X_C = 1/\omega C \quad$ is called the capacitive reactance

For this case the phasors representing the potential difference across the capacitor and the current to the capacitor are $\pi/2$ radians or 90.0° out of phase. The current leads the voltage by 90.0° as shown in Figure 31.5.

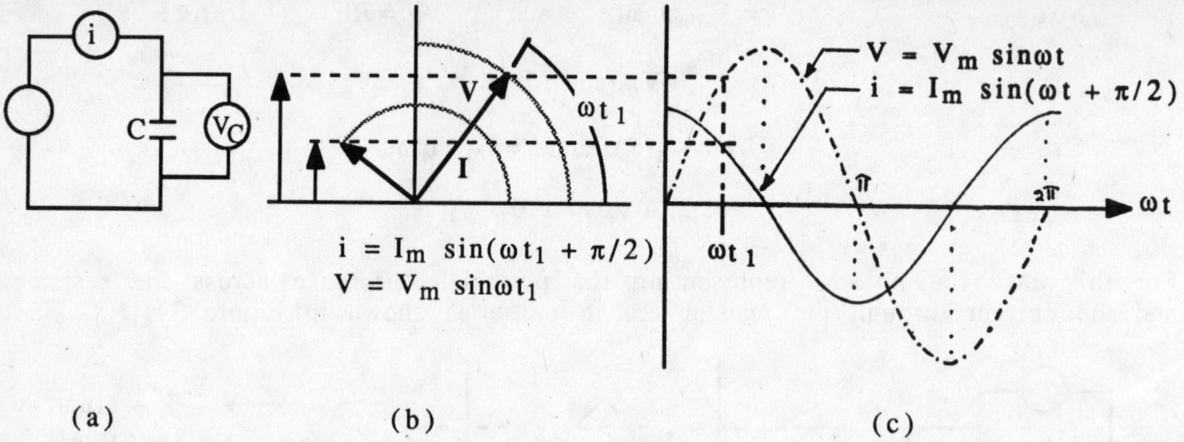

Figure 31.5

When the voltage source V_s is applied to a pure inductor (no resistance) L (Figure 31.6(a)) the potential difference across the inductor is V_L.

$$V_s = V_L$$

but $\quad V_s = V_m \sin\omega t \quad$ and $\quad V_L = L\,di/dt \quad$ hence

$$di/dt = (V_m/L)\sin\omega t$$

or $\quad i = \int di = (V_m/L)\int \sin\omega t\,dt = -(V_m/\omega L)\cos\omega t = I_m \sin(\omega t - \pi/2)$

where $\quad I_m = V_m/\omega L = V_m/X_L = V_{Lm}/X_L$

and $\quad X_L = \omega L \quad$ is called the inductive reactance

For this case the phasors representing the potential difference across the inductor and the current through the inductor are π/2 radians or 90.0° out of phase. The current lags the voltage by 90.0° as shown in Figure 31.6.

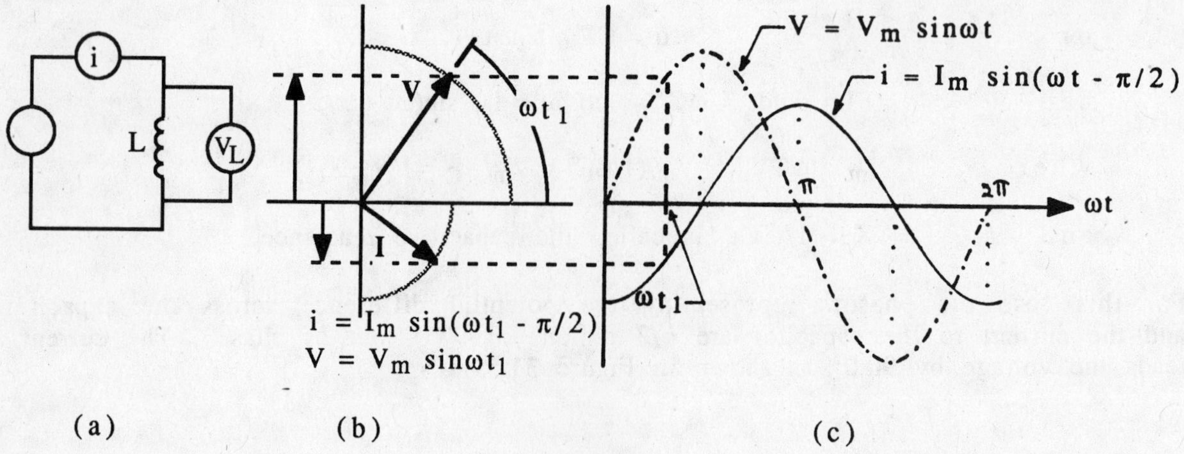

Figure 31.6

We can summarize the above work with the following table.

Circuit element	Instantaneous potential difference across element	Instantaneous current through element	Phase angle between V and i
R	$V_R = V_s = V_m \sin\omega t$	$i = I_m \sin\omega t$	0° V_R and i are in phase
C	$V_C = V_s = V_m \sin\omega t$	$i = I_m \sin(\omega t + \pi/2)$	i leads V_C by $\pi/2$ rad
L	$V_L = V_s = V_m \sin\omega t$	$i = I_m \sin(\omega t - \pi/2)$	i lags V_L by $\pi/2$ rad

If you have trouble remembering when the current leads or lags, the following may help your recall.

ELI the ICE man

That is in an inductive (L) circuit the current lags the voltage and in a capacitive (C) circuit the current leads the voltage.

Practice: Consider the circuit shown in Figure 31.4(a) and the following data.

$V_m = 10.0$ V $R = 20.0\ \Omega$ $\phi_i = 0°$ $\nu = 60.0$ Hz

Determine the following:

1. Phase angle between the potential difference across the resistor and the current through the resistor	The current is in phase with the potential difference, hence $\phi = 0°$
2. General expression for instantaneous values of V_R and i	$V_R = V_s = V_m \sin\omega t$ $i = I_m \sin\omega t$
3. Maximum values for the potential difference across and the current through the resistor	$V_{Rm} = V_{sm} = V_m = 10.0$ V $I_m = V_{Rm}/R = 0.500$ A
4. Instantaneous value for the potential difference across and the current through the resistor at $t = T/4$	$\omega t = (2\pi/T)(T/4) = \pi/2$ $V_R = V_m \sin\omega t = (10.0\text{ V}) \sin\pi/2 = 10.0$ V $i = I_m \sin\omega t = (0.500\text{ A}) \sin\pi/2 = 0.500$ A Notice that this calculation shows that $V_R = V_{Rm}$ and $i = I_m$ at $t = T/4$. This is in agreement with the phasor plot shown in Figure 31.4(c)

Consider the circuit shown in Figure 31.5(a) and the following data

$V_m = 10.0$ V $C = 1.00 \times 10^{-4}$ F $\phi_i = 0°$ $\nu = 60.0$ Hz

Determine the following:

5. Phase angle between the potential difference across the capacitor and the current to the capacitor	The current leads the potential difference by $\pi/2$ rad or 90.0°
6. General expression for instantaneous values of V_C and i	$V_C = V_s = V_m \sin\omega t$ $i = I_m \sin(\omega t + \pi/2)$
7. Capacitive reactance	$X_C = 1/\omega C = 1/2\pi\nu C = 25.5\ \Omega$
8. Maximum value for the potential difference across and the current to the capacitor	$V_{Cm} = V_{sm} = V_m = 10.0$ V $I_m = V_{Cm}/X_C = 0.392$ A
9. Instantaneous value for the potential difference across and the current to the capacitor at t = T/4	$\omega t = (2\pi/T)(T/4) = \pi/2$ $V_C = V_m \sin\omega t = (10.0\text{ V}) \sin\pi/2 = 10.0$ V $i = I_m \sin(\omega t + \pi/2)$ $= (0.392\text{ A}) \sin(\pi/2 + \pi/2) = 0$ A Notice that this calculation shows that $V_C = V_{Cm}$ and $i = 0$ at $t = T/4$. This is in agreement with the phasor plot shown in Figure 31.5(c).

Consider the circuit shown in Figure 31.6(a) and the following data.

$V_m = 10.0$ V $\quad L = 2.00 \times 10^{-1}$ H $\quad \phi_i = 0°$ $\quad \nu = 60.0$ Hz

Determine the following:

10. Phase angle between the potential difference across the inductor and the current through the inductor	The current lags the potential difference by $\pi/2$ rad or 90.0°
11. General expression for instantaneous values of V_L and i	$V_L = V_s = V_m \sin\omega t$ $i = I_m \sin(\omega t - \pi/2)$
12. Inductive reactance	$X_L = \omega L = 2\pi\nu L = 75.3\ \Omega$
13. Maximum value for the potential difference across and the current through the inductor	$V_{Lm} = V_{sm} = V_m = 10.0$ V $I_m = V_{Lm}/X_L = 0.133$ A

14. Instantaneous value for the potential difference across and the current through the inductor at $t = T/4$	$\omega t = (2\pi/T)(T/4) = \pi/2$ $V_L = V_m \sin\omega t = (10.0 \text{ V}) \sin\pi/2 = 10.0 \text{ V}$ $i = I_m \sin(\omega t - \pi/2) = 0.133 \text{ A} \sin 0 = 0 \text{ A}$ Notice that this calculation shows that $V_L = V_{Lm}$ and $i = 0$ at $t = T/4$. This is in agreement with the phasor plot shown in Figure 31.6(c).

Related Text Exercises: Ex. 31-26 through 31-43.

3 AC Source Connected to a RC, RL, or LC Series Combination
(Sections 31-3, 4, 5 and 6)

Review: RC Series Combination - When the voltage source V_s is applied to a series resistor capacitor combination, at any instant the potential difference across the resistor V_R plus the potential difference across the capacitor V_C is equal to the potential difference across the series combination V_{RC} or the potential delivered by the source V_s.

$$V_s = V_{RC} = V_R + V_C$$

Figure 31.7 shows the circuit, phasor diagram and phasor plot for this situation.

$$V_{RC} = V_R + V_C$$

$$V_R = V_{Rm} \sin(\omega t_1 + \phi)$$
$$V_C = V_{Cm} \sin(\omega t_1 + \phi - \pi/2)$$
$$V_{RC} = V_{RCm} \sin\omega t_1$$
$$= V_R + V_C$$

(b)   (c)   (d)

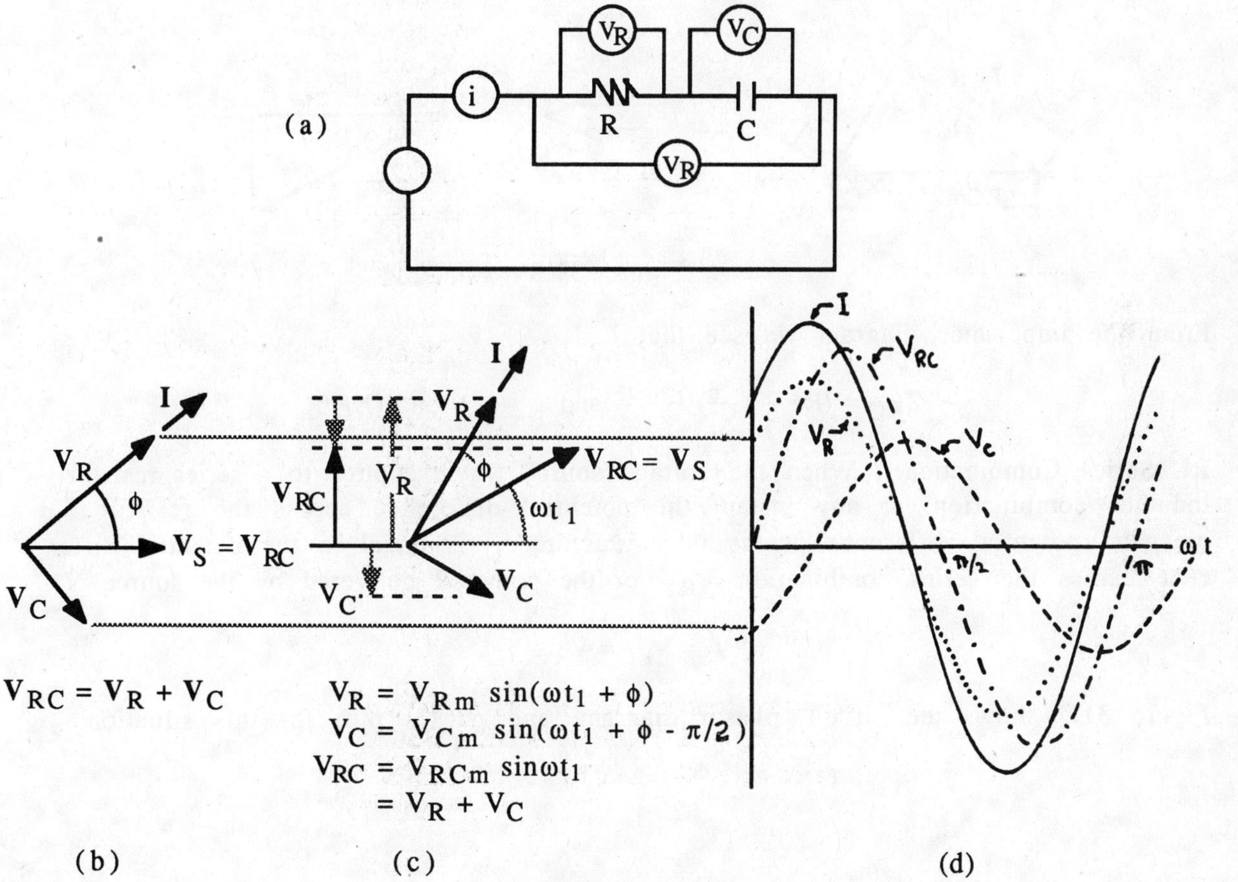

Figure 31.7

The following Table summarizes what we know about this situation.

Phasor	Magnitude	Instantaneous value at t_1
$V_s = V_{RC}$	$V_m = V_{RCm} = I_m Z_{RC}$	$V_{RC} = V_{RCm} \sin\omega t_1$
V_R	$V_{Rm} = I_m R$	$V_R = V_{Rm} \sin(\omega t_1 + \phi)$
V_C	$V_{Cm} = I_m X_C$	$V_C = V_{Cm} \sin(\omega t_1 + \phi - \pi/2)$
I	I_m	$i = I_m \sin(\omega t_1 + \phi)$

Also

$$V_{RCm}^2 = V_{Rm}^2 + V_{Cm}^2$$

Recalling that $V_{RCm} = I_m Z_{RC}$, $V_{Rm} = I_m R$, $V_{Cm} = I_m X_C$

$$I_m^2 Z_{RC}^2 = I_m^2 R^2 + I_m^2 X_C^2$$

or

$$Z_{RC} = (R^2 + X_C^2)^{1/2}$$

From the phasor diagram we can construct an impedance diagram as shown in Figure 31.8

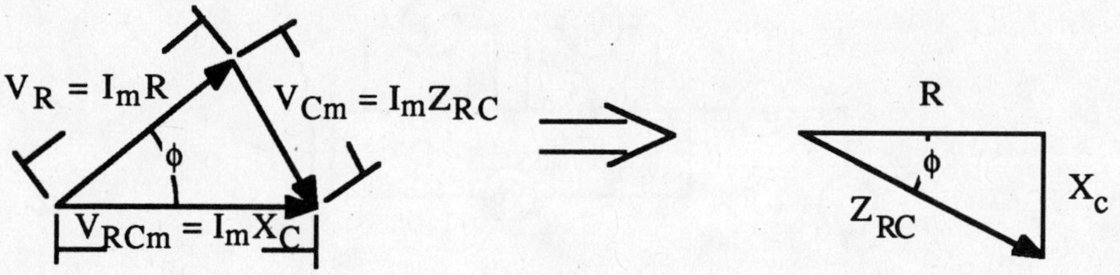

Figure 31.8

From the impedance diagram we see that

$$Z_{RC} = (R^2 + X_C^2)^{1/2} \quad \text{and} \quad \phi = \tan^{-1}(X_C/R)$$

RL Series Combination - When the voltage source V_s is applied to a series resistor inductor combination, at any instant the potential difference across the resistor V_R plus the potential difference across the inductor V_L is equal to the potential difference across the series combination V_{RL} or the potential delivered by the source V_s.

$$V_s = V_{RL} = V_R + V_L$$

Figure 31.9 shows the circuit, phasor diagram, and phasor plot for this situation.

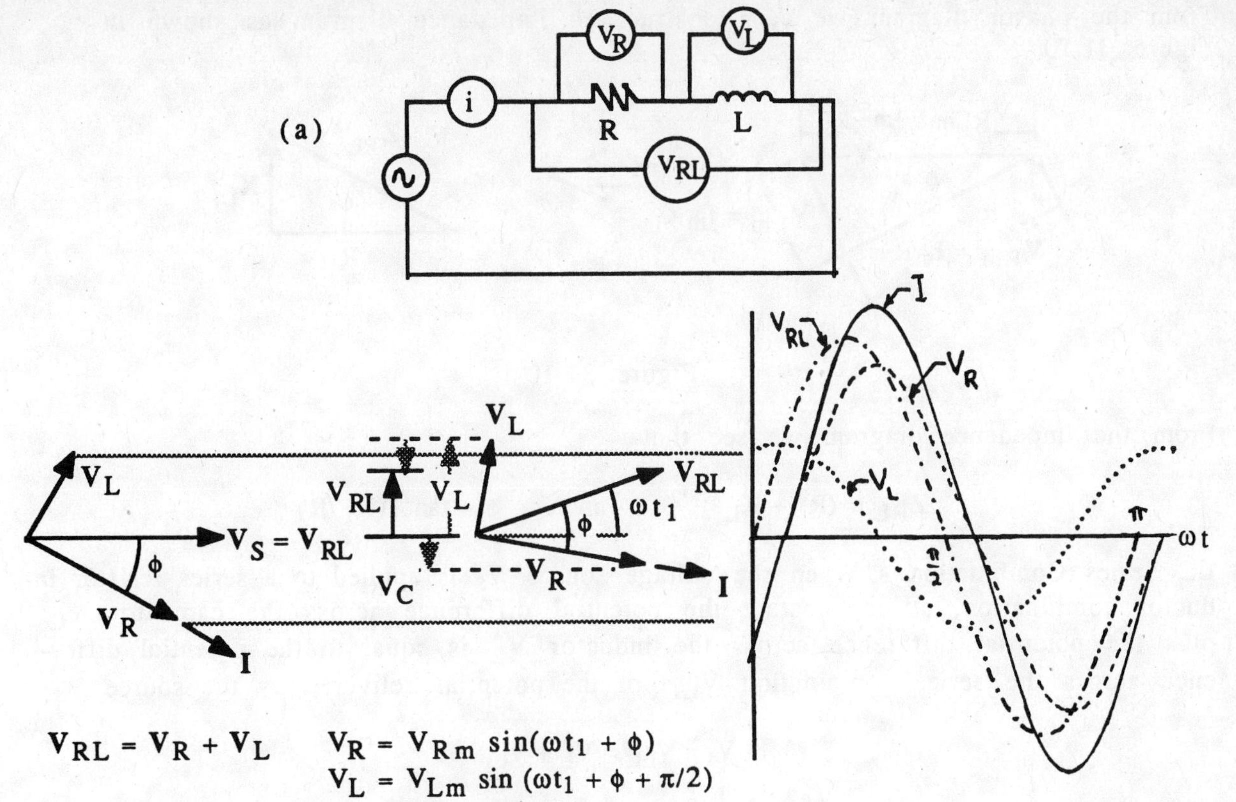

$V_{RL} = V_R + V_L$

$V_R = V_{Rm} \sin(\omega t_1 + \phi)$
$V_L = V_{Lm} \sin(\omega t_1 + \phi + \pi/2)$
$V_{RL} = V_{RLm} \sin\omega t_1$
$= V_L - V_R$

(b) (c) (d)

Figure 31.9

The following Table summarizes what we know about this situation.

Phasar	Magnitude	Instantaneous value at t_1
$V_{RL} = V_s$	$V_{RLm} = V_m = I_m Z_{RL}$	$V_{RL} = V_{RLm} \sin\omega t_1$
V_R	$V_{Rm} = I_m R$	$V_R = V_{Rm} \sin(\omega t_1 + \phi)$
V_L	$V_{Lm} = I_m X_L$	$V_L = V_{Lm} \sin(\omega t_1 + \phi + \pi/2)$
I	I_m	$i = I_m \sin(\omega t_1 + \phi)$

Also

$$V_{RLm}^2 = V_{Rm}^2 + V_{Lm}^2$$

Recalling that $V_{RLm} = I_m Z_{RL}$, $V_{Rm} = I_m R$, $V_{Lm} = I_m X_L$

$$I_m^2 Z_{RL}^2 = I_m^2 R^2 + I_m^2 X_L^2$$

or

$$Z_{RL} = (R^2 + X_L^2)^2$$

From the phasor diagram we can construct an impedance diagram as shown in Figure 31.10.

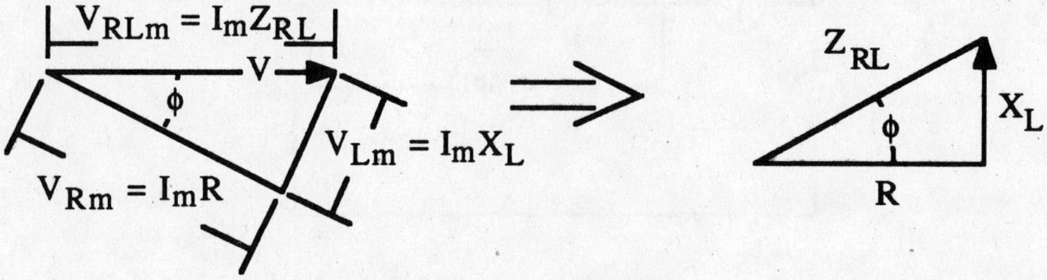

Figure 31.10

From the impedance diagram we see that

$$Z_{RL} = (R^2 + X_L^2)^{1/2} \qquad \text{and} \qquad \phi = \tan^{-1}(X_L/R)$$

LC Series Combination - When the voltage source V_s is applied to a series resistor inductor combination, at any instant the potential difference across the capacitor V_C plus the potential difference across the inductor V_L is equal to the potential difference across the series combination V_{LC} or the potential delivered by the source V_s.

$$V_s = V_{LC} = V_L + V_C$$

Figure 31.11 shows the circuit and phasor diagram for the cases (b) $V_L > V_C$, (c) $V_L = V_C$ and (d) $V_L < V_C$.

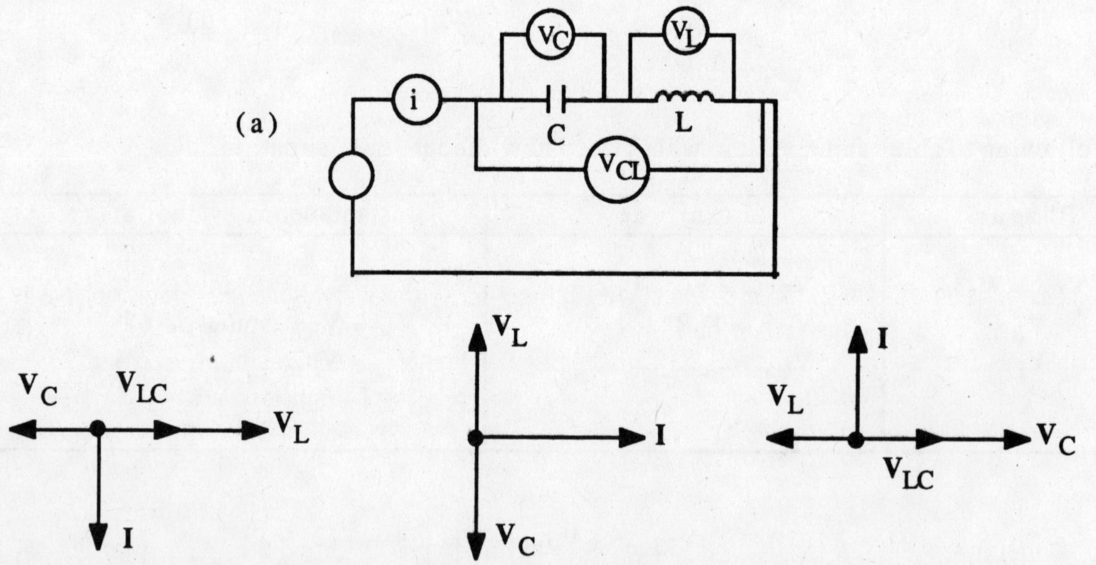

(b) $V_{Lm} > V_{Cm}$ (c) $V_{Lm} = V_{Cm}$ (d) $V_{Lm} < V_{Cm}$

Figure 31.11

The case shown in Figure 31.11(a) is given further consideration in Figure 31.12.

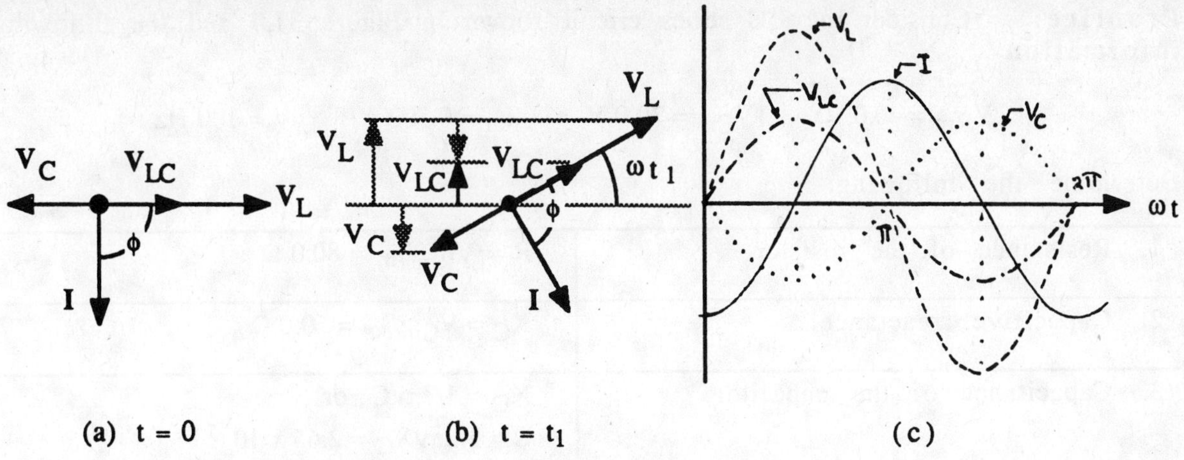

Figure 31.12

The following Table summarizes what we know about this situation (i.e. $V_L > V_C$)

Phasor	Magnitude	Instantaneous Value at t_1
$V_{LC} = V_s$	$V_m = V_{LCm} = I_m Z_{LC}$	$V_{LC} = V_{LCm} \sin\omega t_1$
V_L	$V_{Lm} = I_m X_L$	$V_L = V_{Lm} \sin(\omega t_1 + \phi + \pi/2)$
V_C	$V_{Cm} = I_m X_C$	$V_C = V_{Cm} \sin(\omega t_1 + \phi - \pi/2)$
I	I_m	$i = I_m \sin(\omega t_1 + \phi)$

Also
$$V_{LCm} = V_{Lm} - V_{Cm}$$

Recalling that $V_{LCm} = I_m Z_{LC}$, $V_{Lm} = I_m X_L$, $V_{Cm} = I_m X_C$

$$I_m Z_{LC} = I_m X_L - I_m X_C$$

or

$$Z_{LC} = X_L - X_C$$

From the phasor diagram we can construct an impedance diagram as shown in Figure 31.13.

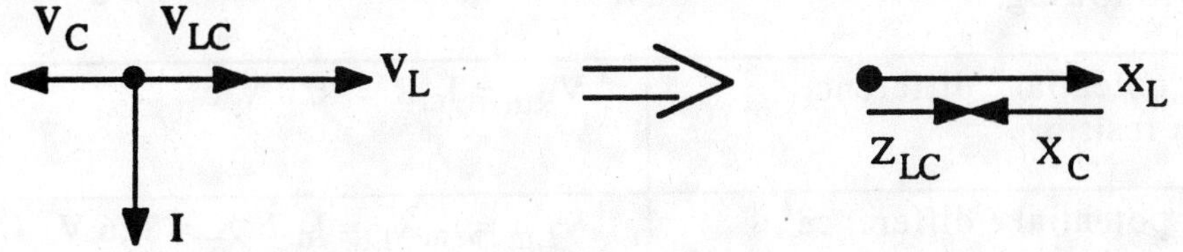

Figure 31.13

From the impedance diagram we see that

$$Z_{LC} = X_L - X_C \quad \text{and} \quad \phi = \pi/2$$

Practice: Consider the RC series circuit shown in Figure 31.7 and the following information

$V_{Rm} = 40.0$ V, $V_{Cm} = 30.0$ V, $I_m = 0.500$ A, $\nu = 100$ Hz

Determine the following:

1. Resistance of the resistor	$R = V_{Rm}/I_m = 80.0\ \Omega$
2. Capacitive reactance	$X_C = V_{Cm}/I_m = 60.0\ \Omega$
3. Capacitance of the capacitor	$X_C = 1/2\pi\nu C$ or $C = 1/2\pi\nu X_C = 2.65 \times 10^{-5}$ F
4. Impedance of the circuit	$Z_{RC} = (R^2 + X_C^2)^{1/2} = 100\ \Omega$
5. Amplitude of the AC voltage source	$V_s = V_{RCm} = I_m Z_{RC} = 50.0$ V or $V_s = V_{RCm} = (V_R^2 + V_C^2)^{1/2} = 50.0$ V
6. Phase angle between the circuit current and the potential difference across the resistor	The current is in phase with the potential difference across the resistor.
7. Phase angle between the circuit current and the potential difference across the capacitor	The current leads the potential difference across the capacitor by $\pi/2$ rad or 90.0°
8. Phase angle between the circuit current and the potential difference across the RC series combination	$\phi = \tan^{-1}(V_{Cm}/V_{Rm}) = 36.9°$ or $\phi = \tan^{-1}(X_C/R) = 36.9°$

Consider the RL series circuit shown in Figure 31.9 and the following information.

$R = 200\ \Omega$, $L = 1.00$ H, $\nu = 100$ Hz, $I_m = 0.200$ A

Determine the following:

9. Maximum potential difference across the resistor	$V_{Rm} = I_m R = 40.0$ V
10. Maximum potential difference across the inductor	$V_{Lm} = I_m X_L = I_m 2\pi\nu L = 126$ V
11. Impedance of the circuit	$R = 200\ \Omega$, $X_L = 2\pi\nu L = 628\ \Omega$ $Z_{RL} = (R^2 + X_L^2)^{1/2} = 659\ \Omega$

12. Amplitude of the AC voltage source	$V_{RLm} = I_m Z_{RL} = 132$ V or $V_{RLm} = (V_{Rm}^2 + V_{Lm}^2)^{1/2} = 132$ V
13. Phase angle between the circuit current and the potential difference across the resistor	The current is in phase with the potential difference across the resistor
14. Phase angle between the circuit current and the potential difference across the inductor	The current lags the potential difference across the inductor by $\pi/2$ rad or 90.0°
15. Phase angle between the circuit current and the potential difference across the RL series combination	$\phi = \tan^{-1}(V_{Lm}/V_{Rm}) = 72.4°$ or $\phi = \tan^{-1}(X_L/R) = 72.4°$

Consider the LC series circuit shown in Figure 31.11 and the following information.

$L = 0.478$ H $\quad C = 1.00 \times 10^{-5}$ F $\quad I_m = 0.500$ A $\quad \nu = 100$ Hz

Determine the following:

16. Capacitive reactance	$X_C = 1/2\pi\nu C = 159\ \Omega$
17. Inductive reactance	$X_L = 2\pi\nu L = 300\ \Omega$
18. Impedance of the circuit	$Z_{LC} = X_L - X_C = 141\ \Omega$
19. Maximum potential difference across the capacitor, inductor, series combination of capacitor and inductor	$V_{Cm} = I_m X_C = 79.5$ V $V_{Lm} = I_m X_L = 150$ V $V_{LCm} = I_m Z_{LC} = 70.5$ V
20. Phase angle between the circuit current and the potential difference across the capacitor inductor series combination	Since $X_L > X_C$, the circuit is inductive. Hence the current lags the potential difference across the series combination by $\pi/2$ rad or 90.0°

Related Text Exercises: Ex. 31-44 through 31-53.

4 AC Source Connected to a RLC Series Circuit (Section 31-6)

Review: When the voltage source V_s is applied to an RLC series combination, at any instant the potential difference across the resistor V_R plus the potential difference across the inductor V_L plus the potential difference across the capacitor V_C is equal to the potential difference across the combination V_{RLC} or the potential delivered by the source.

$$V_s = V_{RLC} = V_R + V_L + V_C$$

Figure 31.14 shows the circuit and phasor diagrams for the cases (b) $V_L > V_C$, (c) $V_L = V_C$, (d) $V_L < V_C$.

(a)

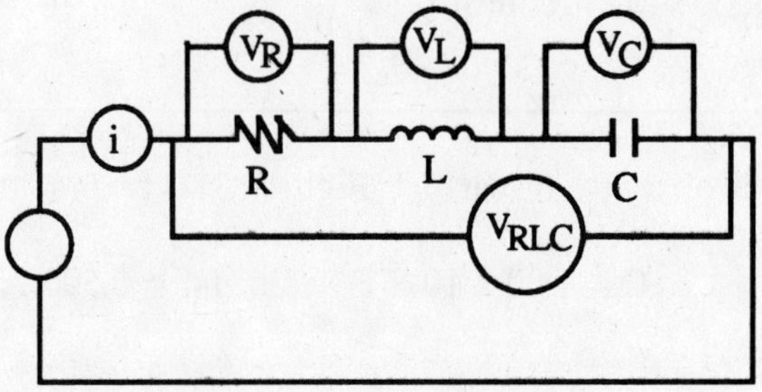

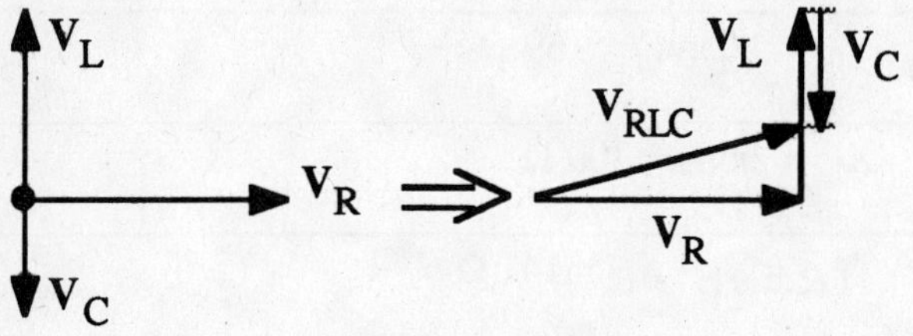

(b) $V_{Lm} > V_{Cm}$

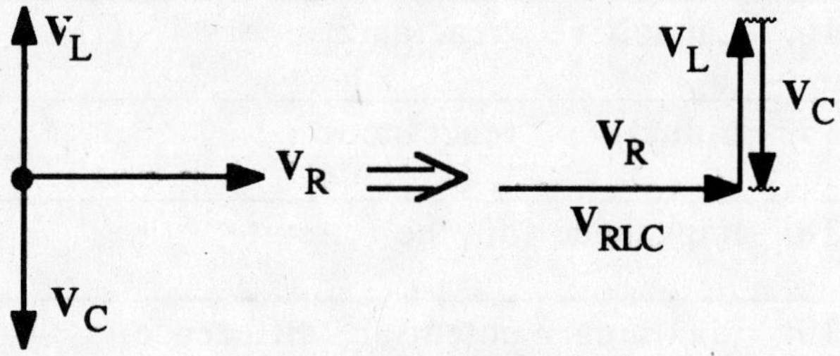

(c) $V_{Lm} = V_{Cm}$

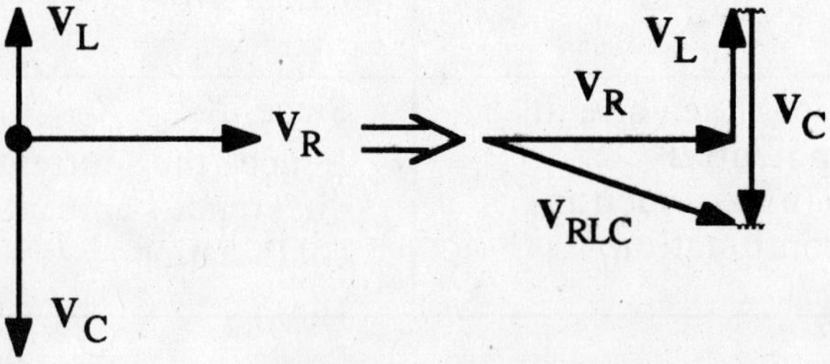

(d) $V_{Lm} < V_{Cm}$

Figure 31.14

The case shown in Figure 31.14(b) is given further consideration in Figure 31.15.

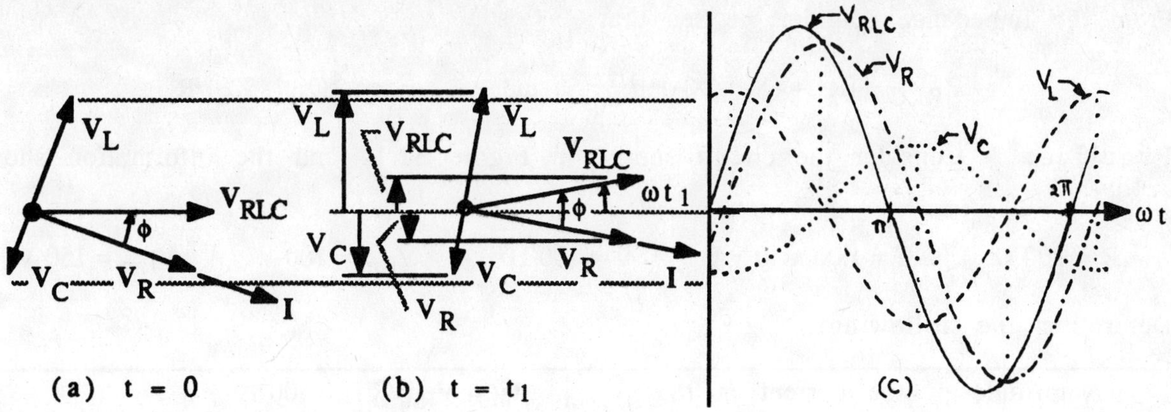

Figure 31.15

The following table summarizes what we know about this situation.

Phasor	Magnitude	Instantaneous value at t_1
$V_s = V_{RLC}$	$V_m = V_{RLCm} = I_m Z_{RLC}$	$V_{RLC} = V_{RLCm} \sin\omega t_1$
V_R	$V_{Rm} = I_m R$	$V_R = V_{Rm} \sin(\omega t_1 + \phi)$
V_L	$V_{Lm} = I_m X_L$	$V_L = V_{Lm} \sin(\omega t_1 + \phi + \pi/2)$
V_C	$V_{Cm} = I_m X_C$	$V_C = V_{Cm} \sin(\omega t_1 + \phi - \pi/2)$
I	I_m	$i = I_m \sin(\omega t_1 + \phi)$

Also $$V_{RLCm}^2 = V_{Rm}^2 + (V_{Lm} - V_{Cm})^2$$

Recalling that

$$V_{RLC} = I_m Z_{RLC} \qquad V_{Rm} = I_m R \qquad V_{Lm} = I_m X_L \qquad V_{Cm} = I_m X_C$$

$$I_m^2 Z_{RLC}^2 = I_m^2 R^2 + I_m^2 (X_L - X_C)^2$$

or

$$Z_{RLC} = [R^2 + (X_L - X_C)^2]^{1/2}$$

From the phasor diagram we can construct an impedance diagram as shown in Figure 31.16

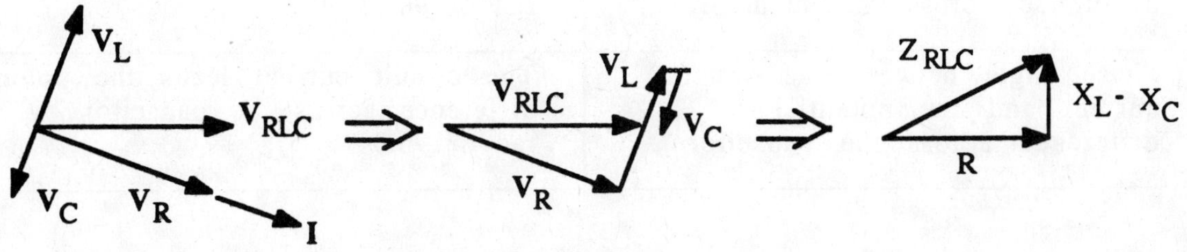

Figure 31.16

From the impedance diagram we see that

$$Z_{RLC} = [R^2 + (X_L - X_C)^2]^{1/2} \quad \text{and} \quad \phi = \tan^{-1}(X_L - X_C)/R$$

Practice: Consider the circuit shown in Figure 31.14 and the information shown below.

$R = 200\ \Omega$ $\quad$ $C = 1.00 \times 10^{-5}$ F $\quad$ $\nu = 100$ Hz $\quad$ $V_{Rm} = 100$ V $\quad$ $V_{Lm} = 150$ V

Determine the following:

1. Amplitude of the current in the circuit	$I_m = V_{Rm}/R = 0.500$ A
2. Capacitive reactance	$X_C = 1/2\pi\nu C = 159\ \Omega$
3. Maximum potential difference across the capacitor	$V_{Cm} = I_m X_C = 79.5$ V
4. Inductive reactance	$X_L = V_{Lm}/I_m = 300\ \Omega$
5. Inductance of the inductor	$L = X_L/2\pi\nu = 0.478$ H
6. Amplitude of the AC voltage source	Notice that the circuit is inductive ($X_L > X_C$), hence looking at Figure 31.16 we write $V_m = V_{RLCm} = [V_{Rm}^2 + (V_{Lm} - V_{Cm})^2]^{1/2} = 122$ V
7. Total impedance	$Z_{RLC} = V_{RLCm}/I_m = 244\ \Omega$ or looking at Figure 31.16 we write $Z_{RLC} = [R^2 + (X_L - X_C)^2]^{1/2} = 244\ \Omega$
8. Phase angle between the circuit current and the potential difference across the resistor	The circuit current is in phase with the potential difference across the resistor
9. Phase angle between the circuit current and the potential difference across the inductor	The circuit current lags the potential difference across the inductor by $\pi/2$ rad or 90.0°
10. Phase angle between the circuit current and the potential difference across the capacitor	The circuit current leads the potential difference across the capacitor by $\pi/2$ rad or 90.0°

11. Phase angle between the circuit current and the potential difference across the RLC series combination	Since we have established that the circuit is inductive, looking at Figure 31.16 we write $\phi = \tan^{-1}[(V_{Lm} - V_{Cm})/V_{Rm}] = 35.2$ or $\phi = \tan^{-1}(X_L - X_C/R) = 35.2°$
12. Frequency at which this circuit would be in resonance	At resonance $X_L = X_C$, $2\pi\nu L = 1/2\pi\nu C$ $\nu = (1/4\pi^2 CL)^{1/2} = 7.28$ Hz

Example: An RLC series circuit has elements with the following values: R = 1.00 x 10^2 Ω, L = 0.200 H, C = 2.50 x 10^{-5} F. The alternating voltage has an amplitude of 50.0 V and a frequency of 100 Hz. Calculate the (a) capacitive and inductive reactance, (b) impedance of the circuit, (c) current amplitude, (d) maximum voltage across the resistor, capacitor, and inductor, (e) phase angle difference between the circuit current and voltage, and (f) amount we must change the inductor to create resonance.

Given: R = 100 Ω L = 0.200 H C = 2.50 x 10^{-5} F V_m = 50.0 V ν = 100 Hz

Determine: (a) X_C and X_L, (b) Z_{RLC}, (c) I_m, (d) V_{Rm}, V_{Cm} and V_{Lm}, (e) ϕ, (f) the value of L to cause resonance.

Strategy: (a) Knowing C, L, and ν, we can determine X_C and X_L. (b) Knowing R, X_C, and X_L, we can determine Z_{RLC}. (c) Knowing Z_{RLC} and V_m, we can determine I_m. (d) Knowing I_m , R, X_C, and X_L, we can determine V_{Rm}, V_{Cm}, and V_{Lm}. (e) Knowing V_{Rm}, V_{Cm} and V_{Lm}, we can determine ϕ. (f) Knowing X_C, we can determine the value of X_L and hence L for resonance.

Solution:

(a) $X_C = 1/2\pi\nu C = 63.7\ \Omega$ and $X_L = 2\pi\nu L = 126\ \Omega$
(b) $Z_{RLC} = [R^2 + (X_L - X_C)^2]^{1/2} = 118\ \Omega$
(c) $I_m = V_m/Z_{RLC} = 0.424$ A
(d) $V_{Rm} = I_m R = 424$ V, $V_{Cm} = I_m X_C = 27.0$ V, $V_{Lm} = I_m X_L = 53.4$ V
(e) $\phi = \tan^{-1}[(V_{Lm} - V_{Cm})/V_{Rm}] = 31.9°$
(f) For resonance we need $X_C = X_L = 2\pi\nu L$ or $L = X_C/2\pi\nu = 0\ 101$ H
The inductance must be reduced by 0.0990 H (0.200 H - 0.101 H)

Related Text Exercises: Ex. 31-44 through 31-53.

5 Power For an RLC Circuit Driven By an AC Source (Section 31-7)

Review: In an RLC circuit driven by an AC source, the source supplies energy and the resistor dissipates it. Energy enters and leaves the capacitor and inductor, but no energy is supplied by or dissipated by these two circuit elements. The rate at which energy is dissipated by the resistor is called the power dissipated.

The instantaneous power dissipated by the resistor is

$$P = i^2R = [I_m \sin(\omega t + \phi)]^2R$$

The average power $\bar{P}$ dissipated by the resistor is

$$\bar{P} = I_m{}^2R\,\overline{\sin^2(\omega t + \phi)} = I_m{}^2R/2 = I_{rms}{}^2R$$

where

$$\overline{\sin^2(\omega t + \phi)} = 1/2 \quad \text{and} \quad I_{rms} = I_m/(2)^{1/2}$$

Another useful way to express $\bar{P}$ is as follows:

$$\bar{P} = I_{rms}{}^2R = I_{rms}(V_{rms}/Z)R = I_{rms}V_{rms}\cos\phi$$

where $\cos\phi = R/Z = R/[R^2 + (X_C - X_L)^2]^{1/2}$ is called the power factor

The power factor ($\cos\phi$) may vary from 0 to 1. For more efficient use of the energy source, the power factor should be as high as possible. A low power factor means the voltage and current are way out of phase. When this happens, a large current is needed for a given voltage in order to supply a large power. This will cause unwanted joule heating losses in transmission lines. In practice the power factor is controlled by the addition of capacitance to the circuit.

Practice: A metal lathe in a machine shop operates on 120 V at 60.0 Hz. While in operation, its power factor is measured to be 0.800 at a power of 480 W. The voltage leads the current.

Determine the following:

1. The rms current in the circuit	$\bar{P} = I_{rms}V_{rms}\cos\phi$ $I_{rms} = \bar{P}/V_{rms}\cos\phi$ $= 480\text{ W}/(120\text{ V})(0.800) = 5.00\text{ A}$
2. The rms voltage across the resistor	Looking at the following diagram 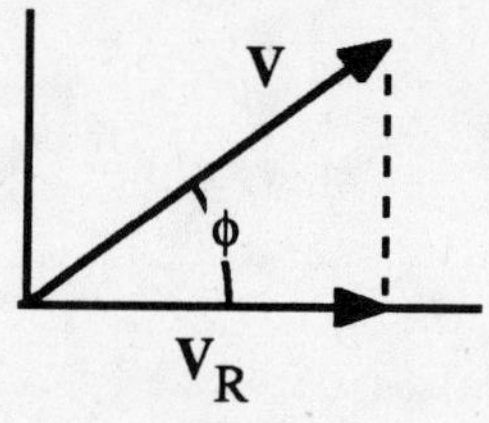

	we see that $(V_R)_{rms} = V_{rms} \cos\phi = (120\ \text{V})(0.800) = 96.0\ \text{V}$
3. Capacitance of the capacitor which when placed in series with the lathe motor will result in a power factor of 1.00	For this case we want $(V_C)_{rms} = (V_L)_{rms} = V_{rms} \sin\phi = (120\ \text{V})(0.600) = 12.0\ \text{V}$ $X_C = (V_L)_{rms}/I_{rms} = 72.0\ \text{V}/5.00\ \text{A} = 14.4\ \Omega$ $C = 1/2\pi\nu X_C = 184\ \mu\text{F}$
4. The new rms current after the capacitor is installed	$R = (V_R)_{rms}/I_{rms} = 96.0\ \text{V}/5.00\ \text{A} = 19.2\ \Omega$ Now that $V_C = V_L$, we have $(V_R)_{rms} = V_{rms} = 120$ V and $I_{rms} = (V_R)_{rms}/R = 6.25$ A
5. The new average power delivered by the source after the capacitor is installed	$\bar{P} = I_{rms}V_{rms} = (6.25\ \text{A})(120\ \text{V}) = 750\ \text{W}$
6. Ratio of the power delivered to the motor with the capacitor in place to the power delivered without the capacitor	$\bar{P}_{old} = 480$ W $\quad \bar{P}_{new} = 750$ W $\bar{P}_{new}/\bar{P}_{old} = 1.56$ The addition of the 184 μF capacitor has allowed the source to deliver 50 percent more power to the lathe motor.

Example: An electric drill with an armature resistance of 10.0 Ω and an inductance of 15.0×10^{-3} H is used in a 120 V 60 Hz circuit. The power developed by the drill can be increased by the use of a capacitor. (a) Calculate the current power factor of the drill. (b) What size capacitor should be installed for maximum power at the drill? (c) What is the ratio of average power developed after inserting the capacitor to average power before insertion?

Given:
$R = 10.0\ \Omega$ resistance of the drill armature
$L = 15.0 \times 10^{-3}$ H inductance of the drill armature
$V_{rms} = 120$ V source potential
$\nu = 60.0$ Hz source frequency

Determine:
(a) Current power factor of the drill
(b) Capacitor size for maximum power at the drill
(c) Ratio of power developed after inserting the capacitor to that before

Strategy: Knowing L and ν, we can determine X_L, then φ, and finally cosφ the power factor. Knowing X_L and that maximum power is developed by the drill when the power factor is one (i.e. when $X_L = X_C$), we can determine the capacitor size. Knowing the definition of average power, we can determine the average power delivered by the drill before ($\bar{P}_b$) and after ($\bar{P}_a$) the capacitor is installed. Finally we may determine the ratio $\bar{P}_a/\bar{P}_b$.

Solution: (a) The inductive reactance is

$$X_L = \omega L = 2\pi\nu L = 2\pi(60.0\ \text{Hz})(15.0 \times 10^{-3}\ \text{H}) = 5.65\ \Omega$$

The phase angle is

$$\phi = \tan^{-1}(X_L/R) = \tan^{-1}(5.65\ \Omega/10.0\ \Omega) = 29.5°$$

The power factor is

$$\cos\phi = \cos 29.5° = 0.871$$

(b) Maximum power will be delivered by the drill when the power factor is one. The power factor will be one when

$$X_C = X_L$$

or

$$1/2\pi\nu C = 2\pi\nu L$$

Hence the size capacitor needed is

$$C = 1/4\pi^2\nu^2 L = 4.69 \times 10^{-4}\ \text{F}$$

(c) The average power delivered by the drill before and after the capacitor is inserted is

$$\bar{P}_b = I_{b\ rms}V_{b\ rms}\cos 29.5° = (V_{b\ rms}/Z_b)V_{b\ rms}\cos 29.5° = (V_{b\ rms}{}^2/Z_b)(0.870)$$

$$\bar{P}_a = I_{a\ rms}V_{a\ rms}\cos 0° = (V_{a\ rms}/R)V_{a\ rms} = V_{a\ rms}{}^2/R$$

The ratio of $\bar{P}_a$ to $\bar{P}_b$ is

$$\bar{P}_a/\bar{P}_b = (V_{a\ rms}{}^2/R)/(V_{b\ rms}{}^2/Z_b)(0.870)$$

But $V_{a\ rms} = V_{b\ rms}$ and $Z_b = (R^2 + X_L{}^2)^{1/2} = 11.5\ \Omega$

$$\bar{P}_a/\bar{P}_b = Z_b/(0.870)R = 1.32$$

Related Text Exercises: Ex. 31-54 through 31-61.

PRACTICE TEST

Take and grade this practice test. Doing so will allow you to determine any weak spots in your understanding of the concepts taught in this chapter. The following section prescribes what you should study further to strengthen your understanding.

The alternating current and voltage in a circuit are $i = 10 \sin(120\pi t + \pi/6)$ and $V = 120 \sin(120\pi t + \pi/2)$.

Determine the following:

_______ 1. Amplitude of the circuit current
_______ 2. Period of the time-varying current and voltage
_______ 3. Initial phase angle that would allow us to write $V = V_m \sin 2\pi \nu t$
_______ 4. Time when V = 60.0 V
_______ 5. Phase angle difference between i and V

The components of an RLC series circuit have the values R = 100 Ω, L = 0.300 H, C = 4.00 x 10^{-5} F, and the circuit is powered by a 120-V, 60.0 Hz source.

Determine the following:

_______ 6. Inductive reactance
_______ 7. Capacitance reactance
_______ 8. Total impedance
_______ 9. Maximum circuit current
_______ 10. Phase angle between circuit current and voltage
_______ 11. Maximum potential difference across the resistor
_______ 12. Maximum potential difference across the capacitor
_______ 13. Maximum potential difference across the inductor
_______ 14. Power factor
_______ 15. Value of the capacitance that brings this circuit into resonance
_______ 16. Maximum circuit current at resonance
_______ 17. Power factor at resonance
_______ 18. Average power delivered to circuit at resonance
_______ 19. Maximum potential drop across the resistor at resonance
_______ 20. Power loss to the capacitor at resonance

(See Appendix I for answers.)

PRINCIPAL CONCEPTS AND EQUATIONS PRESCRIPTION

Your score on the practice test is an excellent measure of your understanding of the chapter. You should now use the following chart to write your own prescription for dealing with any weaknesses the practice test points out. Look down the leftmost column to the number of the question(s) you answered incorrectly, reading across that row you will find the concept and/or equation of concern, the section(s) of the study guide you should return to for further study, and some suggested text exercises which you should work to gain additional experience.

Practice Test Questions	Concepts and Equations	Prescription	
		Principal Concepts	Text Exercises
1	Amplitude of the current: I_m	1	31-44, 45
2	Period: $T = 1/\nu = 2\pi/\omega$	1	---
3	Initial phase angle: ϕ_i	1	31-44, 50
4	Instantaneous voltage: $V = V_m \sin(\omega t + \phi_i)$	1	31-45, 50
5	Phase angle difference: ϕ	1	31-44, 53
6	Inductive reactance: $X_L = 2\pi\nu L$	2	31-37, 38
7	Capacitive reactance: $X_C = 1/2\pi\nu C$	2	31-29, 30
8	Impedance: $Z_{RLC} = [R^2 + (X_L - X_C)^2]^{1/2}$	4	31-44
9	Current: $I_m = V_{RLCm}/Z_{RLC}$	4	31-50, 53
10	Phase angle: $\phi = \tan^{-1}(\lvert X_L - X_C\rvert/R)$	4	31-44, 53
11	$V_{Rm} = I_m R$	2	31-26, 50
12	$V_{Cm} = I_m X_C$	2	31-33, 50
13	$V_{Lm} = I_m X_L$	2	31-41, 50
14	Power factor: $\cos\phi = R/Z_{RLC}$	5	31-58, 60
15	Series resonance: $X_L = X_C$, $\phi = 0$	4	31-46, 52
16	Current at resonance: $I_m = V_m/R$	4	31-46
17	Power factor at resonance: $\cos\phi = 1$	5	31-58, 60
18	Average power at resonance: $\bar{P} = I_{rms}V_{rms}\cos\phi = I_{rms}V_{rms}$	5	31-54, 55
19	$V_{Rm} = I_m R$	4	31-26, 50
20	Power loss to capacitor: $\bar{P}_{cap} = 0$	5	---

32 WAVES

RECALL FROM PREVIOUS CHAPTERS

Previously learned concepts and equations frequently used in this chapter	Text Section	Study Guide Page
Average power: $P = \Delta E/\Delta t$	8-5	8-15
Instantaneous power: $P = dE/dt$	8-5	8-15
The relationship between period, frequency and angular frequency: $\omega = 2\pi\nu = 2\pi/T$	14-1	14-1

NEW IDEAS IN THIS CHAPTER

Concepts and equations introduced	Text Section	Study Guide Page
Characteristics of a wave:	32-1, 2	32-1
amplitude A		
wavelength λ		
wave number $k = 2\pi/\lambda$		
frequency ν		
angular frequency $\omega = 2\pi\nu$		
period $T = 1/\nu = 2\pi/\omega$		
speed $v = \nu\lambda = \omega/k$		
Traveling wave: $y(x,t) = A \sin[(2\pi/\lambda)(x - vt)]$; $y(x,t) = A \sin(kx - \omega t)$	32-3	32-3
Power of a wave: $P = \mu\omega^2A^2v \cos^2(kx - \omega t)$	32-5	32-6
Average power: $\bar{P} = \mu\omega^2A^2v/2$	32-5	32-6
Intensity: $I = (\Delta E/\Delta t)/\Delta S = P_0/\Delta S = P_0/4\pi r^2$	32-5	32-6
Standing waves in a string: $L = n\lambda_n/2$ $\quad v = (F/\mu)^{1/2}$ $\quad \nu_n = v/\lambda_n$	32-6	32-10

PRINCIPAL CONCEPTS AND EQUATIONS

1 Characteristics of Mechanical Waves (Sections 32-1, 2)

Review: A mechanical oscillator moving in simple harmonic motion in a medium creates a sinusoidal disturbance. This disturbance propagates through the medium (away from the source) with a constant velocity and is called a traveling wave. Figure 32.1 shows a section of a transverse wave traveling to the right. It is important to appreciate that the particles of the medium oscillate about their equilibrium position while the disturbance is traveling through the medium.

Figure 32.1

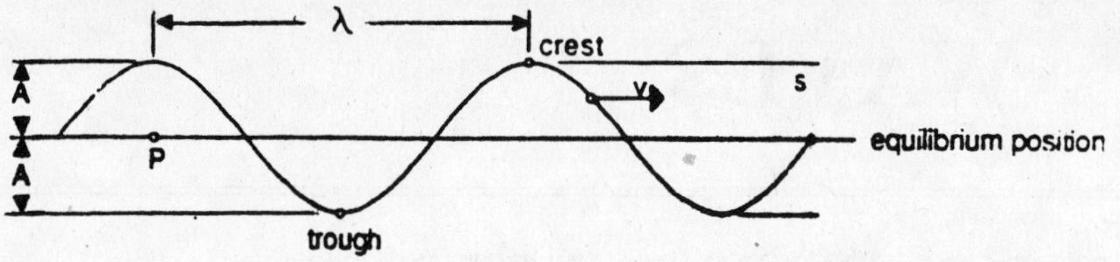

λ = wavelength = minimum distance over which wave repeats itself
A = amplitude = maximum displacement of medium from equilibrium
ν = frequency = number of waves passing some arbitrary point P per second
T = period = time for one wavelength to pass some point P; time for a particle of the medium to undergo a complete cycle of its motion
v = speed of propagation = speed at which wave moves through medium
s = path of a medium particle oscillating about the equilibrium position

Two important relationships for these physical quantities are

$$T = 1/\nu \quad \text{and} \quad v = \nu\lambda$$

Practice: Figure 32.2 shows a series of three rapid succession photos of a traveling wave in front of a grid. The grid is (0.150 m) x (0.150 m).

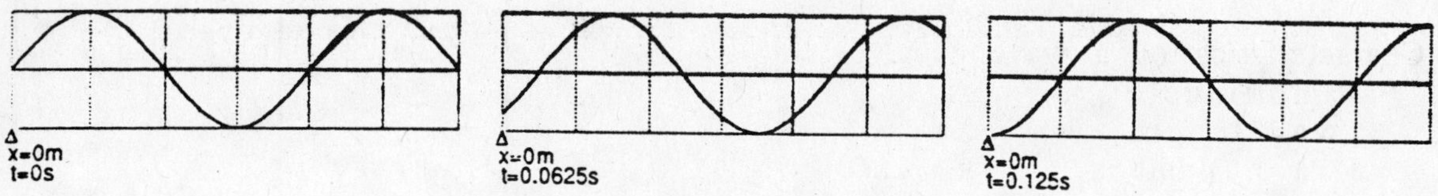

Figure 32.2

Determine the following for the traveling wave:

1. Amplitude	A = 0.150 m
2. Wavelength	λ = 0.600 m
3. Period	Since the wave advances $\lambda/4$ in 0.125 s, T = 4(0.125 s) = 0.500 s
4. Frequency	$\nu = 1/T = 2.00$ Hz
5. Speed of propagation	Since the wave advances $\Delta x = \lambda = 0.600$ m in $\Delta t = T = 0.500$ s, $v = \Delta x/\Delta t$ = (0.600 m)/(0.500 s) = 1.20 m/s; $v = \nu\lambda$ = (2.00 Hz)(0.600 m) = 1.20 m/s

Example: A cello string vibrates with a frequency of 500 Hz. If the speed of sound in air is 340 m/s, calculate (a) the number of times the string vibrates while the sound travels 100 m and (b) the wavelength of the sound wave.

Given: ν_c = 500 Hz = frequency of vibration of the cello string
v = 340 m/s = speed of sound in air
d = 100 m = distance the sound travels

Determine: (a) The number of times the string vibrates while the sound travels 100 m. (b) The wavelength of the sound wave.

Strategy: Knowing d and v, we can determine the time t it takes the sound to travel the distance. Knowing the string's frequency of vibration, we can determine the period T. From t and T, we can determine the number of times the string vibrates while the sound travels the distance d. The frequency of the sound in air is the same as the vibrational frequency of the string. From v and ν for air, we can determine the wavelength of the sound wave.

Solution: The time it takes the sound to travel the 100 m is

$$t = d/v = (100 \text{ m})/(340 \text{ m/s}) = 0.294 \text{ s}$$

The period of oscillation for the vibrating string is

$$T = 1/\nu = 1/500 \text{ Hz} = 2.00 \times 10^{-3} \text{ s}$$

The period T is the time for one vibration of the string. The time t is the total time the string has to vibrate. If during the time t the string vibrates N times, we can write t = NT. Then the number of vibrations of the string while the sound travels the 100 m is

$$N = t/T = (0.294 \text{ s})/(2.00 \times 10^{-3} \text{ s}) = 147$$

The vibrational frequency of the sound in air is the same as the vibrational frequency of the sound source, namely the cello string. The wavelength is

$$\lambda = v/\nu = (340 \text{ m/s})/(500 \text{ Hz}) = 0.680 \text{ m}$$

Related Text Exercises: Ex. 32-1 through 32-5.

2 Traveling Harmonic Waves (Section 32-3)

Review: The following expression describes a harmonic wave traveling to the right in a medium.

$$y(x,t) = A \sin[(\frac{2\pi}{\lambda})(x - vt)]$$

where A = amplitude of the wave
v = speed of propagation of the wave
λ = wavelength of the wave
x = distance from the source
t = time we consider medium at distance x from the source
y(x,t) = transverse displacement of the traveling wave at position x and time t

It is frequently convenient to use the following relationships to write the above expression for a traveling wave in a more concise form.

$$k = 2\pi/\lambda \quad \text{-- wave number}$$
$$\omega = 2\pi\nu = 2\pi v/\lambda \quad \text{-- angular frequency}$$

Inserting k and ω into the above expression for a traveling wave we obtain

$$y(x,t) = A \sin(kx - \omega t)$$

As a result of introducing the parameters k and ω, we have an alternate way of expressing the speed of propagation of the wave though the medium.

$$v = \nu\lambda = (\omega/2\pi)(2\pi/k) = \omega/k$$

The student should be aware of all the things that can be determined with the expression

$$y = A \sin(kx - \omega t)$$

You can determine the following:

1. Wave number k
2. Wave length $\lambda = 2\pi/k$
3. Angular frequency ω
4. Wave frequency $\nu = \omega/2\pi$
5. Wave velocity $v = \nu\lambda = \omega/k$
6. Period of the wave $T = 1/\nu = 2\pi/\omega = \lambda/v$
7. Wave amplitude A
8. Direction of travel $- \Rightarrow$ right and $+ \Rightarrow$ left
9. Transverse velocity of a particle in the medium at any position and any time
$$v_y = \partial y/\partial t = -\omega A \cos(kx - \omega t)$$
10. Transverse acceleration of a particle in the medium at any position and any time
$$a_y = \partial^2 y/\partial t^2 = -\omega^2 A \sin(kx - \omega t)$$
11. A picture of the traveling wave for all values of x at any time t (i.e. a plot of y vs x at a particular time t)
12. A picture of the periodic disturbance of a particle at a particular position x as a function of time t (i.e. a plot of y vs t for a particular value of x)

In order to avoid cluttering the expression for y(x,t), we usually do not show units of the physical quantities involved. It is understood, however that if A, λ and x are in meters, v in meters/second and t in seconds, then y is in meters. If you pick a value for x, the transverse motion of the medium at that location is sinusoidal in time. If you pick a value for t, the transverse displacement of the medium is sinusoidal in x.

Practice: Given the following expression for a traveling transverse wave

$$y = (0.200 \text{ m}) \sin[50\pi(x - 4.00\, t)]$$

Determine the following:

1. Wave number	$k = 50\pi\ m^{-1}$
2. Wave length	$\lambda = 2\pi/k = 2\pi/50\pi = 4.00 \times 10^{-2}$ m
3. Angular frequency	$\omega = 200\pi\ s^{-1}$
4. Wave frequency	$\nu = \omega/2\pi = 100$ Hz
5. Wave velocity	$v = \nu\lambda = \omega/k = 4.00$ m/s
6. Period of the wave	$T = 1/\nu = 2\pi/\omega = \lambda/v = 1.00 \times 10^{-2}$ s
7. Wave amplitude	A = 0.200 m
8. Direction the wave is traveling	- ⇒ to the right
9. Transverse velocity of a particle in the medium which is 1.00 m from the source at the time t = 0.400 s	$v_y = \partial y/\partial t$ $= (-0.800\ m/s)\cos[50\pi(x - 4.00t)]$ $v_y(1.00\ m, 0.400\ s)$ $= (-0.800\ m/s)\cos(-20\pi) = -0.800\ m/s$
10. The appearance of this traveling wave between $x = -4.00 \times 10^{-2}$ m and $x = 12.0 \times 10^{-2}$ m at t = 0 s	$y = (0.200\ m)\sin[50\pi(x - 4.00t)]$, at t = 0, this becomes $y = (0.200\ m)\sin(50\pi x)$. When this is plotted for values of x between $x = -4.00 \times 10^{-2}$ m and $x = 12.0 \times 10^{-2}$ m, we obtain the following figure. 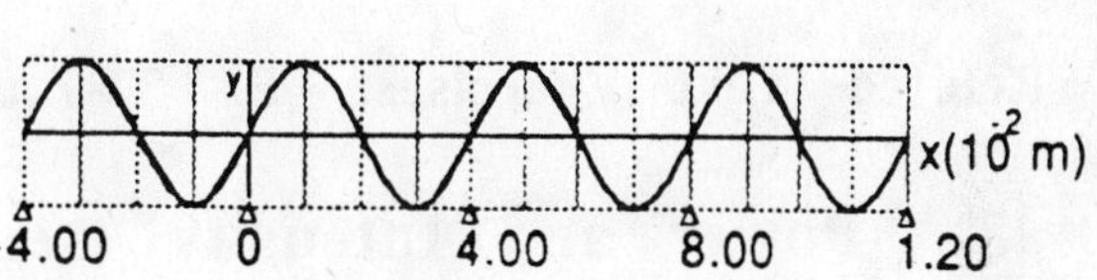 Figure 32.3
11. The time t when the medium at x = 0 has a transverse displacement y = 0.200 m for the first time after the record of time is started	Examining the traveling wave at t = 0 (step 10 Figure 32.3), we see that the medium at x = 0 has a displacement of y = 0.200 m in 3T/4 or 7.50×10^{-3} s. We can also obtain this result analytically by starting with the expression for the traveling transverse wave: $y = (0.200\ m)\sin[50\pi(x - 4.00t)]$. Inserting y = 0.200 m and x = 0 m, we obtain $0.200\ m = (0.200\ m)\sin(-200\pi t)$ which reduces to $-1 = \sin 200\pi t$. This condition is satisfied for the first time when $200\pi t = 3\pi/2$ or $t = 7.50 \times 10^{-3}$ s.

12. The time when the medium at $x = 1.00$ m has a transverse displacement of $y = -0.100$ m for the first time after the record of time is started	$y = (0.200\ \text{m}) \sin[50\pi(x - 4.00\ t)]$ Insert $x = 1.00$ m and $y = -0.100$ m to get $-0.100\ \text{m} = (0.200\ \text{m})\sin[50\pi(1.00 - 4.00t)]$ or $-0.500 = \sin[50\pi(1.00 - 4.00t)]$. This is true for the first time when $50\pi(1.00 - 4.00t) = 7\pi/6$ or $t = 0.244$ s.
13. The transverse displacement of the medium at $x = 0.750$ m and $t = 0.150$ s	$y = (0.200\ \text{m}) \sin[50\pi(x - 4.00t)]$ Insert x and t to obtain $y = (0.200\ \text{m}) \sin[50\pi(0.150)] = -0.200$ m.

Example: A traveling transverse wave has an amplitude of 0.150 m, a frequency of 200 Hz and a speed of propagation of 100 m/s. For this wave, write the equation for the displacement of any point in the medium at any time.

Given: $A = 0.150$ m $\quad \nu = 200$ Hz $\quad v = 100$ m/s

Determine: An expression for the transverse displacement of any point in the medium at any time.

Strategy: Insert the given quantities into the general expression that describes a transverse wave traveling through a medium.

Solution: Combine v and ν to obtain λ: $\lambda = v/\nu = 0.500$ m

Insert A, λ and v into $\quad y = A \sin[(\frac{2\pi}{\lambda})(x - vt)] \quad$ to obtain

$$y = (0.150\ \text{m}) \sin[4\pi(x - 100t)]$$

Related Text Exercises: Ex. 32-7 through 32-20.

3 Power and Intensity (Section 32-5)

Review: Consider a small segment of a string undergoing transverse periodic motion due to a periodic transverse wave.

Figure 32.4

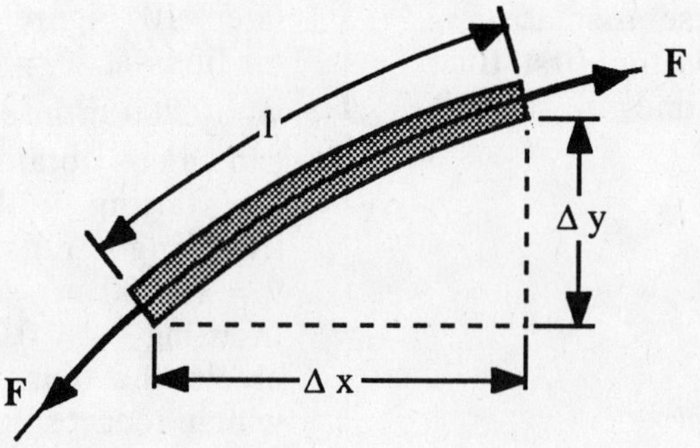

Δx = unstretched length of the string segment
μ = linear mass density of the string
$m = \mu\Delta x$ = mass of the string segment
Δy = displacement of the right end of the string segment relative to the left end
l = stretched length of the string segment
y = displacement of the string segment from its equilibrium position at any time
$v_y = \partial y/\partial t$ = transverse velocity of the string segment at any time

The kinetic energy of the string segment at any time is

$$\Delta K = mv^2/2 = (\mu\Delta x)(\partial y/\partial t)^2/2$$

The kinetic energy density of the wave is

$$\Delta K/\Delta x = (\mu/2)(\partial y/\partial t)^2$$

The stretch of the string segment is

$$\Delta l = [\Delta x^2 + \Delta y^2]^{1/2} - \Delta x$$

for the limiting case of small segments this may be rewritten as

$$\Delta l = \left[[1 + (\partial y/\partial x)^2]^{1/2} - 1]\right]\Delta x$$

Using the binomial expansion $[(1 + z)^n \approx 1 + z/2 + ...]$ and the assumption $(\partial y/\partial x) \ll 1$, we obtain

$$\Delta l = (\partial y/\partial x)^2\Delta x/2$$

If the segment is small we may write the potential energy of the segment due to its stretch as the work done by the constant tension F in stretching the element the amount Δl

$$\Delta U = F\Delta l = F(\partial y/\partial x)^2\Delta x/2$$

The potential energy density of the wave is

$$\Delta U/\Delta x = (F/2)(\partial y/\partial x)^2$$

The total energy density of the wave is

$$\Delta E/\Delta x = \Delta K/\Delta x + \Delta U/\Delta x = (\mu/2)(\partial y/\partial t)^2 + (F/2)(\partial y/\partial x)^2$$

The power of the wave is

$$P = (\Delta E/\Delta x)(\Delta x/\Delta t) = [(\mu/2)(\partial y/\partial t)^2 + (F/2)(\partial y/\partial x)^2]v$$

For a harmonic wave of the form

$$y = A \sin(kx - \omega t)$$

we have

$$\partial y/\partial t = -\omega A \cos(kx - \omega t)$$

and

$$\partial y/\partial x = kA \cos(kx - \omega t)$$

If the medium is a stretched string then

$$v^2 = F/\mu = \omega^2/k^2$$

Using the above we may determine the power associated with the wave described by $y = A \sin(kx - \omega t)$ to be

$$P = [(\mu/2)\omega^2 A^2 \cos^2(kx - \omega t) + (\mu\omega^2/2k^2)k^2 A^2 \cos^2(kx - \omega t)]v$$

or

$$P = \mu\omega^2 A^2 v \cos^2(kx - \omega t)$$

Since the average of $\cos^2\omega t$ over a cycle is 1/2 we may write the average power as

$$\bar{P} = \mu\omega^2 A^2 v/2$$

If the source of waves is a point source and the wave fronts propagate outward in three dimensions as concentric spheres, a quantity which characterizes the energy flow per unit time per unit of area normal to the propagation direction is the intensity

$$I = \frac{\Delta E/\Delta t}{\Delta S} = \frac{P_0}{\Delta S} = \frac{P_0}{4\pi r^2}$$

where P_0 -- is the steady power output or energy delivered per unit time by the source

r -- is the distance from the source

$\Delta S = 4\pi r^2$ -- is the area the power is spread out over at a distance r from the source

$I = P_0/4\pi r^2$ -- is the intensity due to P_0 at a distance r from the source

Practice: Consider a string with a linear mass density $\mu = 2.00 \times 10^{-3}$ kg/m with a harmonic wave propagating down it which may be described by

$$y = 0.100 \text{ m} \sin(10\pi x - 20\pi t)$$

Determine the following at $x = 1.00$ m and $t = 0.250$ s:

1. The wave number, angular frequency and amplitude	$k = 10\pi \text{ m}^{-1}$, $\omega = 20\pi \text{ s}^{-1}$, $A = 0.100$ m
2. The speed of the wave down the string	$v = \omega/k = 2.00$ m/s

3. Tension in the string	$F = v^2\mu = 8.00 \times 10^{-3}$ N
4. Kinetic energy density	$\partial y/\partial t = -2\pi \cos(10\pi x - 20\pi t)$, at $x = 1.00$ m and $t = 0.250$ s obtain $\partial y/\partial t = 2\pi$ $\Delta K/\Delta x = (\mu/2)(\partial y/\partial t)^2 = 4\pi^2 \times 10^{-3}$ J/m
5. Potential energy density	$\partial y/\partial x = \pi \cos(10\pi x - 20\pi t)$, at $x = 1.00$ m and $t = 0.250$ s obtain $\partial y/\partial x = -\pi$ $\Delta U/\Delta x = (F/2)(\partial y/\partial x)^2 = 4\pi^2 \times 10^{-3}$ J/m
6. Total energy density	$\Delta E/\Delta x = \Delta K/\Delta x + \Delta U/\Delta x = 8\pi^2 \times 10^{-3}$ J/m
7. Power carried by the wave	$P = (\Delta E/\Delta x)v = 16\pi^2 \times 10^{-3}$ W, or $P = \mu\omega^2A^2v \cos^2(10\pi x - 20\pi t)$ at $x = 1.00$ m and $t = 0.250$ s obtain $P = \mu\omega^2A^2v = 16\pi^2 \times 10^{-3}$ W
8. Average power carried by the wave	$\bar{P} = \mu\omega^2A^2v/2 = 8\pi^2 \times 10^{-3}$ W

Example: A point source sends energy out into three dimensional space and the intensity 10.0 m from the source is 10.0 W/m². Determine the power output of the source and the intensity 20.0 m from the source.

Given: A point source, $I_1 = 10.0$ W/m² at $r_1 = 10.0$ m from the source

Determine: The power output P_0 of the source and the intensity 20.0 m from the source.

Strategy: Knowing I_1 and r_1 we can determine P_0. Knowing P_0 we can determine I_2 at r_2.

Solution: The power of the source is

$$P_0 = I_1 4\pi r_1^2 = (10.0 \text{ W/m}^2)(4\pi)(10.0 \text{ m})^2 = 4.00\pi \times 10^3 \text{ W}$$

The intensity at any other point may easily be found by recalling that the power is constant (it is just spread out over larger and larger areas, hence the intensity gets smaller), hence

$$P_1 = P_2$$

$$I_1 4\pi r_1^2 = I_2 4\pi r_2^2$$

or

$$I_2 = I_1 r_1^2/r_2^2 = 2.50 \text{ W/m}^2$$

Related Text Exercises: Ex. 32-27 through 32-34.

4 Standing Waves in a String (Section 32-6)

Review: If a harmonic wave of amplitude A, wave number k and angular frequency ω described by the function $y_1 = A \sin(kx - \omega t)$ travels down the string as shown in Figure 32.5, a wave of the same amplitude, wave number, and angular frequency ω described by $y_2 = A \sin(kx + \omega t)$ will be reflected back up the string as shown in Figure 32.5.

Figure 32.5

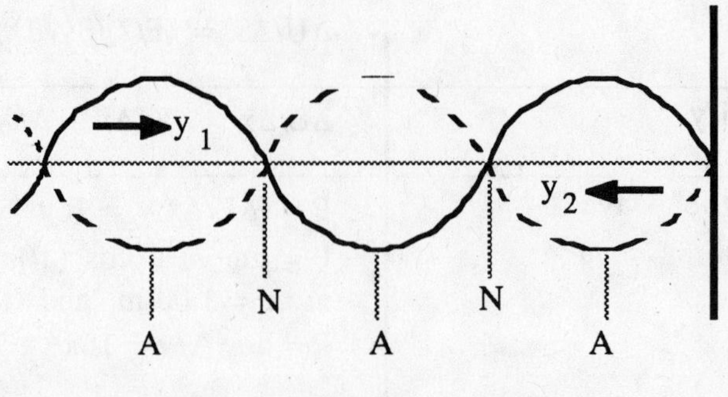

When the wave function y_1 and y_2 are added (principal of superposition) we obtain

$$y(x,t) = y_1(x,t) + y_2(x,t) = (2A \sin kx) \cos\omega t$$

Notice that the function y(x,t) is not in the form of a traveling wave [f(x ± vt)]; it is a standing or stationary wave. The wave pattern does not move, but the elements of the string move transverse to the equilibrium position.

The amplitude of y(x,t) is given by 2A sinkx which has its maxima when sinkx = 1 or

$$kx = \pi/2, \quad 3\pi/2, \quad 5\pi/2, \quad \ldots, \quad (n + 1/2)\pi \qquad n = 0, 1, 2, \ldots$$

Since $k = 2\pi/\lambda$, we have the positions of maximum amplitudes or antinodes located by

$$x_n = (n + 1/2)\pi(\lambda/2\pi) = (n + 1/2)(\lambda/2) \qquad n = 0, 1, 2, \ldots$$

The antinodes are indicated by "A" in Figure 32.5. The amplitude of y(x,t)(2A sinkx) has its minima when sinkx = 0, or

$$kx = 0, \quad \pi, \quad 2\pi, \ldots, \quad n'\pi \qquad n' = 0, 1, 2, \ldots$$

Since $k = 2\pi/\lambda$, we have the positions of minimum amplitudes or nodes located by

$$x_{n'} = (n'\pi)(\lambda/2\pi) = n'\lambda/2 \qquad n' = 0, 1, 2, \ldots$$

The nodes are indicated by "N" in Figure 32.5.

The ends of the string are nodes and an integral number of half wavelengths must fit into the length of the string, hence

$$L = n(\lambda_n/2) \qquad \text{or} \qquad \lambda_n = 2L/n \qquad n = 1, 2, 3, \ldots$$

where

$$\lambda_1 = 2L,\ \lambda_2 = L,\ \lambda_3 = 2L/3,\ \ldots,\ \lambda_n = 2L/n$$

A standing wave cannot have just any wavelength; it can only have one of the specific wavelengths λ_n that fit the condition that a node occur at each end and an integral number of half wavelengths fit into the string.

Since the tension F in the string and the linear mass density μ of the string are constant, the speed at which the periodic disturbance travels down the string is constant and is given by

$$v = (T/\mu)^{1/2}$$

Subsequently, the natural frequencies (i.e. those for which the standing wave condition is satisfied) are given by

$$\nu_n = v/\lambda_n = (T/\mu)^{1/2}/(2L/n) = (n/2L)(T/\mu)^{1/2} \quad n = 1, 2, 3, \ldots$$

The fundamental frequency is given by

$$\nu_1 = (T/\mu)^{1/2}/2L$$

In terms of the fundamental frequency, the natural frequencies are

$$\nu_n = n\nu_1 \qquad n = 1, 2, 3, \ldots$$

The natural frequencies are multiples of the fundamental. The natural frequencies are called harmonics: the fundamental is the first harmonic, $\nu_2 = 2\nu_1$ is the second harmonic and so on.

The above information about standing waves is summarized in Figure 32.6.

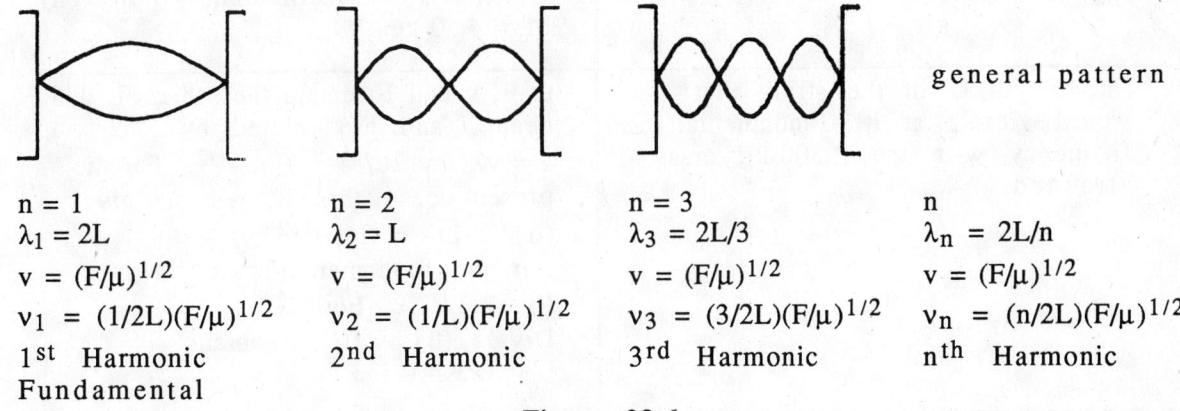

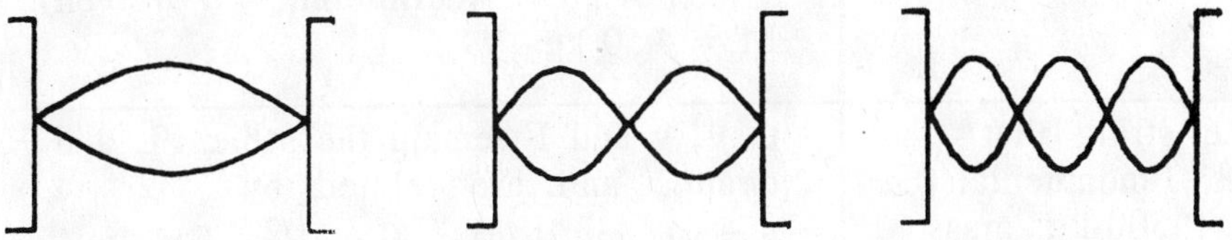

general pattern

$n = 1$	$n = 2$	$n = 3$	n
$\lambda_1 = 2L$	$\lambda_2 = L$	$\lambda_3 = 2L/3$	$\lambda_n = 2L/n$
$v = (F/\mu)^{1/2}$	$v = (F/\mu)^{1/2}$	$v = (F/\mu)^{1/2}$	$v = (F/\mu)^{1/2}$
$\nu_1 = (1/2L)(F/\mu)^{1/2}$	$\nu_2 = (1/L)(F/\mu)^{1/2}$	$\nu_3 = (3/2L)(F/\mu)^{1/2}$	$\nu_n = (n/2L)(F/\mu)^{1/2}$
1st Harmonic Fundamental	2nd Harmonic	3rd Harmonic	n^{th} Harmonic

Figure 32.6

Practice: A lightweight string is attached to a vibrating tuning fork at one end. The other end passes over a pulley and standing waves are set up in the string when a block of mass M is attached as shown in Figure 32.7.

$L = 1.00$ m = length of string
$m = 2.00 \times 10^{-3}$ kg = mass of string
$M = 0.500$ kg = mass of block

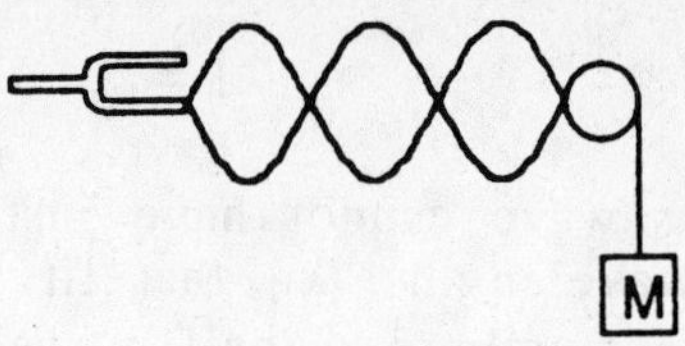

Figure 32.7

Determine the following:

1. Tension in the string	$F = Mg = 4.90$ N
2. Linear mass density of the string	$\mu = m/L = 2.00 \times 10^{-3}$ kg/m
3. Speed of the transverse wave (not to be confused with the transverse motion of the medium) as it travels down the string	$v = (F/\mu)^{1/2} = 49.5$ m/s
4. Wavelength of the transverse wave in the string	$\lambda = 2L/3 = 0.667$ m
5. Frequency of the transverse wave in the string	$\nu = v/\lambda = 74.2$ Hz. This is also the frequency of the tuning fork.
6. The mass M' that will cause the string to vibrate at its fundamental frequency	μ, ν, v and L remain the same. M and λ change and are related by (α) $v = \nu\lambda = (F/\mu)^{1/2} = (Mg/\mu)^{1/2}$ For the new (primed) situation (β) $v' = \nu\lambda' = (F'/\mu)^{1/2} = (M'g/\mu)^{1/2}$ Divide (β) by (α) to obtain $M' = M(\lambda'/\lambda)^2$ $\lambda = 0.667$ m, $\lambda' = 2L = 2.00$ m, $M = 0.500$ kg (step 4), (definition), (given) $M' = 4.50$ kg
7. The length L' of the string that would vibrate at its fundamental frequency with the 0.500-kg mass attached	μ, ν, v and F remain the same. L and λ change and are related by $v = \nu\lambda = \nu(2L/n) = (F/\mu)^{1/2}$. For the present case this expression gives (α) $2\nu L/3 = (F/\mu)^{1/2}$ For the fundamental case it gives (β) $2\nu L'/1 = (F/\mu)^{1/2}$ Divide (β) by (α) to obtain $L' = L/3 = 0.333$ m

Example: The strings of a guitar are 0.600 m long. A particular string has a linear mass density of 6.00×10^{-2} kg/m and is stretched to a tension of 50.0 N. (a) What is the fundamental frequency? (b) How far from the end should the string be pressed in order to cause it to vibrate in its third harmonic?

Given: $L = 0.600$ m $\mu = 6.00 \times 10^{-3}$ kg/m $F = 50.0$ N

Determine: (a) The fundamental frequency of the string. (b) Where to press in order to cause it to vibrate in the third harmonic (i.e., with a frequency three times that of the fundamental).

Strategy: Knowing F and μ, we can determine v. Knowing L, we can determine λ for the fundamental. The frequency for the fundamental can be determined from v and λ. Once the fundamental frequency is known, the frequency of the third harmonic can be determined. Knowing the frequency of the third harmonic and that v is still the same, we can determine the wavelength and hence where to press the string.

Solution: (a) The speed of propagation of the transverse waves down the string is

$$v = (F/\mu)^{1/2} = [50.0 \text{ N}/(6.00 \times 10^{-3} \text{ kg/m}]^{1/2} = 91.3 \text{ m/s}$$

The wavelength for the fundamental is $\lambda_1 = 2L = 1.20$ m. The fundamental frequency is

$$\nu_1 = v/\lambda_1 = (91.3 \text{ m/s})/(1.20 \text{ m}) = 76.1 \text{ Hz}$$

(b) The frequency of the third harmonic is $\nu_3 = 3\nu_1 = 228$ Hz.

The associated wavelength is $\lambda_3 = v/\nu_3 = (91.3 \text{ m/s})/(228 \text{ Hz}) = 0.400$ m.

If the string is 0.600 m long, it will vibrate in the third harmonic (have a wavelength of 0.400 m) if touched 0.400 m from one end.

Related Text Exercises: Ex. 32-39 through 32-46.

PRACTICE TEST

Take and grade this practice test. Doing so will allow you to determine any weak spots in your understanding of the concepts taught in this chapter. The following section prescribes what you should study further to strengthen your understanding.

Water waves approach a buoy in the ocean. There is a distance of 10.0 m between adjacent crests, and a crest reaches the buoy every 5.00 s. The buoy bobs up and down in such a manner that its vertical position varies by 2.00 m.

Determine the following:

________ 1. Amplitude of the waves
________ 2. Frequency of the waves
________ 3. Wavelength of the waves
________ 4. Speed of propagation of the waves
________ 5. Number of waves which pass the buoy in 1.00 h

A traveling transverse wave is described by

$$y = (0.100\ \text{m}) \sin[2\pi(x - 2.00t)]$$

Determine the following:

________ 6. Wave number
________ 7. Wavelength
________ 8. Angular frequency
________ 9. Wave frequency
________ 10. Wave velocity
________ 11. Period of the wave
________ 12. Amplitude of the wave
________ 13. Direction the wave is traveling
________ 14. Transverse displacement of the medium at $x = 20.0$ m and $t = 10.0$ s
________ 15. Transverse speed of a particle in the medium which is 20.0 m from the source at the time $t = 10.0$ s

Consider a string with a linear mass density $\mu = 4.00 \times 10^{-3}$ kg/m with a harmonic wave propagating down it described by

$$y = 0.100\ \text{m} \sin(10\pi x - 20\pi t)$$

Determine the following at $x = 1.00$ m and $t = 0.250$ s:

________ 16. Tension in the string
________ 17. Kinetic energy density of the string
________ 18. Potential energy density of the string
________ 19. Power carried by the wave

A string on an instrument is 1.00 m long, has a mass of 2.00×10^{-2} kg, and is vibrating in the second harmonic with a frequency of 300 Hz.

Determine the following:

________ 20. Linear mass density of the string
________ 21. Wavelength in the string
________ 22. Speed of propagation of transverse waves in the string
________ 23. Tension in the string
________ 24. Tension in the string that would cause it to vibrate in the fundamental mode with a frequency of 300 Hz

(See Appendix I for answers.)

PRINCIPAL CONCEPTS AND EQUATIONS PRESCRIPTION

Your score on the practice test is an excellent measure of your understanding of the chapter. You should now use the following chart to write your own prescription for dealing with any weaknesses the practice test points out. Look down the leftmost column to the number of the question(s) you answered incorrectly, reading across that row you will find the concept and/or equation of concern, the section(s) of the study guide you should return to for further study, and some suggested text exercises which you should work to gain additional experience.

Practice Test Questions	Concepts and Equations	Prescription	
		Principal Concepts	Text Exercises
1	Waves -- amplitude	1	---
2	Waves -- frequency: $\nu = 1/T$	1	32-13
3	Waves -- wavelength	1	---
4	Waves -- speed: $v = f\lambda$	1	32-14, 15
5	Waves -- number: $N = \nu t$	1	---
6	Traveling wave -- wave number	2	32-7, 8
7	Traveling wave -- wavelength: $\lambda = 2\pi/k$	2	32-7, 8
8	Traveling wave -- angular frequency	2	32-7, 8
9	Traveling wave -- frequency: $\nu = \omega/2\pi$	2	32-7, 8
10	Traveling wave -- velocity: $v = \nu\lambda = \omega/k$	2	32-18
11	Traveling wave -- period: $T = 1/\nu = 2\pi/\omega = \lambda/v$	2	32-7, 8
12	Traveling wave -- amplitude	2	32-7, 8
13	Traveling wave -- direction of motion	2	32-17, 18
14	Traveling wave -- displacement	2	32-17, 18
15	Traveling wave -- transverse speed of medium: $v = \partial y/\partial t$	2	32-9, 10
16	Tension in string: $F = v^2\mu$	3	32-42
17	Kinetic energy density: $\Delta K/\Delta x = (\mu/2)(\partial y/\partial t)^2$	3	32-31
18	Potential energy density: $\Delta U/\Delta x = (F/2)(\partial y/\partial x)^2$	3	32-31
19	Power carried by wave: $P = (\Delta E/\Delta x)v$	3	32-27, 31
20	Linear mass density: $\mu = m/L$	3	32-42
21	Standing waves -- wavelength	4	32-40
22	Standing waves -- speed	4	32-42
23	Standing waves -- tension	4	32-44, 45
24	Standing waves -- tension	4	32-44, 45

33 SOUND

RECALL FROM PREVIOUS CHAPTERS

Previously learned concepts and equations frequently used in this chapter	Text Section	Study Guide Page
Power: $P = E/t$	8-5	8-15
Intensity: $I = (E/t)/A = P/A$	32-5	32-6

NEW IDEAS IN THIS CHAPTER

Concepts and equations introduced	Text Section	Study Guide Page
Sound intensity level: $\beta = 10 \log(I/I_o)$	33-2	33-1
Standing waves in pipes --	33-4	33-3
closed-closed: $\lambda_n = 2L/n$ $\quad \nu_n = nv/2L$		
open-open: $\lambda_n = 2L/n$ $\quad \nu_n = nv/2L$		
open-closed: $\lambda_n = 4L/(2n - 1)$ $\quad \nu_n = (2n - 1)v/4L$		
Doppler effect: $\nu_o = \nu_s(v - v_o)/(v - v_s)$	33-6	33-6

PRINCIPAL CONCEPTS AND EQUATIONS

1 Intensity and Sound Intensity Level (Section 33-2)

Review: Intensity of sound is the energy per unit time that the sound wave carries across the imaginary unit area placed perpendicular to the sound. Intensity is measured in $(J/s)/m^2$ or watts per square meter (W/m^2). The lowest intensity that the ear can detect is $I_0 = 1.00 \times 10^{-12}$ W/m^2, and the sound becomes painful at an intensity of 1.00 W/m^2. The sound intensity level for sound of intensity I is defined as:

$$\beta = 10 \log(I/I_0)$$

where log is understood to be to the base 10. Sound intensity level has no units, but we give the name decibel (dB) to sound levels reported in this manner. The reference intensity I_0 is the threshold of hearing, or 10^{-12} W/m^2.

Let us now calculate β for several values of I. In order to do this we need to recall two properties of logarithms.

$$\log 10^n = n \log 10 \quad \text{and} \quad \log 10 = 1$$

$I(W/m^2)$	$\beta(dB)$
10^{-12}	$10 \log(10^{-12}/10^{-12}) = 10 \log 10^0 = (10)(0) \log 10 = (10)(0)(1) = 0$
10^{-11}	$10 \log(10^{-11}/10^{-12}) = 10 \log 10^1 = (10)(1) \log 10 = (10)(1)(1) = 10$
10^{-10}	$10 \log(10^{-10}/10^{-12}) = 10 \log 10^2 = (10)(2) \log 10 = (10)(2)(1) = 20$
⋮	⋮
10^{-6}	$10 \log(10^{-6}/10^{-12}) = 10 \log 10^6 = (10)(6) \log 10 = (10)(6)(1) = 60$
⋮	⋮
10^{-1}	$10 \log(10^{-1}/10^{-12}) = 10 \log 10^{11} = (10)(11) \log 10 = (10)(11)(1) = 110$
10^0	$10 \log(10^0/10^{-12}) = 10 \log 10^{12} = (10)(12) \log 10 = (10)(12)(1) = 120$

The results of the preceding calculations may be summarized in the following table of values.

I	10^{-12}	10^{-11}	10^{-10}	10^{-9}	10^{-8}	10^{-7}	10^{-6}	10^{-5}	10^{-4}	10^{-3}	10^{-2}	10^{-1}	10^0
β	0	10	20	30	40	50	60	70	80	90	100	110	120

You should notice that when the intensity changes by a factor of 10^n, the sound intensity level changes by 10n. For example, when the intensity goes from 10^{-8} to 10^{-4}, it has increased by a factor of 10^4; the corresponding sound intensity level goes from 40 to 80, it has increased by (10)(4) = 40.

Practice: A rock concert is held outside. The slightly elevated stage holds a number of speakers that have a total power output of 400π W. Assume the sound travels outward in a spherical shell and any sound that hits the crowd sitting on the ground is totally absorbed.

Determine the following:

1. The sound intensity heard by observer 1 sitting 100 m away	$I_1 = (E/t)/A_1 = P/4\pi r_1^2 = 10^{-2}$ W/m^2 where $A_1 = 4\pi r_1^2$ is the area of a sphere of radius r_1. The sound is distributed uniformly over a spherical shell and the bottom half is absorbed.
2. The sound intensity level (β) heard by observer 1	$\beta_1 = 10 \log(I_1/I_o) = 10 \log(10^{-2}/10^{-12})$ $= 100$ dB
3. The distance of observer 2 from the speakers in order for her to hear sound 1/16 the intensity of that heard by observer 1	$I = (E/t)/A = P/A$ or $P = IA$, P = constant; hence $P_1 = P_2$ or $I_1A_1 = I_2A_2$, Insert $A_1 = 4\pi r_1^2$, $A_2 = 4\pi r_2^2$ and $I_2 = I_1/16$ to obtain $r_2 = r_1[I_1/(I_1/16)]^{1/2} = 400$ m

4. The sound intensity level (β) heard by observer 2	$\beta_2 = 10 \log(I/I_0) = 10 \log[(I_1/16)/I_0]$ $= 10 \log(6.25 \times 10^{-4}/10^{-12})$ $= 10 \log(6.25 \times 10^{8})$ dB $= 10 (\log 6.25 + 8 \log 10) = 88.0$ dB
5. The distance of observer 3 from the speakers in order for him to hear a sound intensity level 0.80 of that heard by observer 1	$\beta_3 = 0.80\beta_1 = 80.0 \text{ dB} = 10 \log(I_3/I_0)$ or $I_3 = I_0 \log^{-1}(\beta_3/10) = 10^{-4}$ W/m^2 $I_3 = P/A_3 = P/4\pi r_3{}^2$ or $r_3 = (P/4\pi I_3)^{1/2} = 10^3$ m

Example: How far from a 100π-W speaker should an usher seat a person with a 20.0-dB hearing impairment if that person wants to hear sound at the 60.0-dB level?

Given: $P = 100\pi$ W = power of speaker
$\beta = 60.0$ dB = desired sound intensity level
20.0 dB = hearing impairment

Determine: The distance from a 100π-W speaker that a person with a 20.0 dB hearing impairment must be located in order to get a sound intensity level of 60.0 dB.

Strategy: From the desired sound intensity level and the hearing impairment, we can determine the sound intensity level at the person's seat. Knowing the sound intensity level, we can determine the sound intensity at the seat. Finally, we can obtain the distance of the seat from the sound intensity and speaker power.

Solution: Sound intensity level at the seat:

$$\beta = 60.0 \text{ dB} + 20.0 \text{ dB} = 80.0 \text{ dB}$$

Sound intensity at the seat

$$\beta = 10 \log(I/I_0) \quad \text{or} \quad I = I_0 \log^{-1}(\beta/10) = 10^{-4} \text{ W/m}^2$$

Distance from seat to speaker is obtained from

$$I = P/A = P/4\pi r^2 \quad \text{or} \quad r = (P/4\pi I)^{1/2} = 500 \text{ m}$$

Related Text Exercises: Ex. 33-15 through 33-19.

2 Standing Longitudinal Waves in a Pipe (Section 33-4)

Review: Standing longitudinal waves can be set up in pipes closed at both ends (closed-closed), open at both ends (open-open), and closed at one end and open at the other (closed-open). This is reviewed in the following figures.

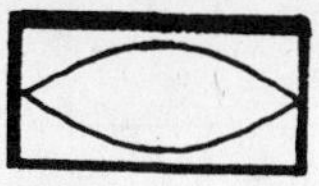		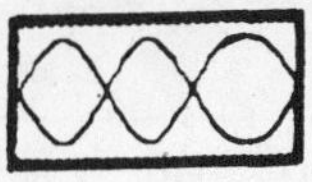	general pattern
$n = 1$ $L = \lambda_1/2$ $\lambda_1 = 2L$ $\nu_1 = v/2L$ 1st Harmonic Fundamental	$n = 2$ $L = \lambda_2$ $\lambda_2 = L$ $\nu_2 = v/L = 2\nu_1$ 2nd Harmonic	$n = 3$ $L = 3\lambda_3/2$ $\lambda_3 = 2L/3$ $\nu_3 = 3v/2L = 3\nu_1$ 3rd Harmonic	$n = 1, 2, 3, \ldots$ $L = n\lambda_n/2$ $\lambda_n = 2L/n$ $\nu_n = nv/2L = n\nu_1$ n^{th} Harmonic

Figure 33.1 -- Closed-Closed Pipes

			general pattern
$n = 1$ $L = \lambda_1/2$ $\lambda_1 = 2L$ $\nu_1 = v/2L$ 1st Harmonic Fundamental	$n = 2$ $L = \lambda_2$ $\lambda_2 = L$ $\nu_2 = v/L = 2\nu_1$ 2nd Harmonic	$n = 3$ $L = 3\lambda_3/2$ $\lambda_3 = 2L/3$ $\nu_3 = 3v/2L = 3\nu_1$ 3rd Harmonic	$n = 1, 2, 3, \ldots$ $L = n\lambda_n/2$ $\lambda_n = 2L/n$ $\nu_n = nv/2L = n\nu_1$ n^{th} Harmonic

Figure 33.2 -- Open-Open Pipes

			general pattern
$n' = 1$ $L = \lambda_1/4$ $\lambda_1 = 4L$ $\nu_1 = v/4L$ 1st Harmonic Fundamental	$n' = 3$ $L = 3\lambda_2/4$ $\lambda_2 = 4L/3$ $\nu_2 = 3(v/4L) = 3\nu_1$ 3rd Harmonic	$n' = 5$ $L = 5\lambda_3/4$ $\lambda_3 = 4L/5$ $\nu_3 = 5(v/4L) = 5\nu_1$ 5th Harmonic	$n' = (2n-1)$ $= 1, 3, 5, \ldots$ $n = 1, 2, 3, \ldots$ $L = (2n - 1)\lambda_n/4$ $\lambda_n = 4L/(2n - 1)$ $\nu_n = (2n - 1)v/4L$ $= (2n - 1)\nu_1$ $(2n-1)^{th}$ Harmonic

Figure 33.3 -- Open-Closed Pipes

Notice from Figures 33.1 and 33.2 the similarities between Closed-Closed Pipes and Open-Open Pipes. Notice from Figure 33.3 that the Open-Closed Pipes have only the odd harmonics.

Practice: Figure 33.4 shows an open-closed pipe of length $L_{oc} = 1.00$ m and an open-open pipe of adjustable length L_{oo}. The speed of sound in air is 340 m/s.

Figure 33.4

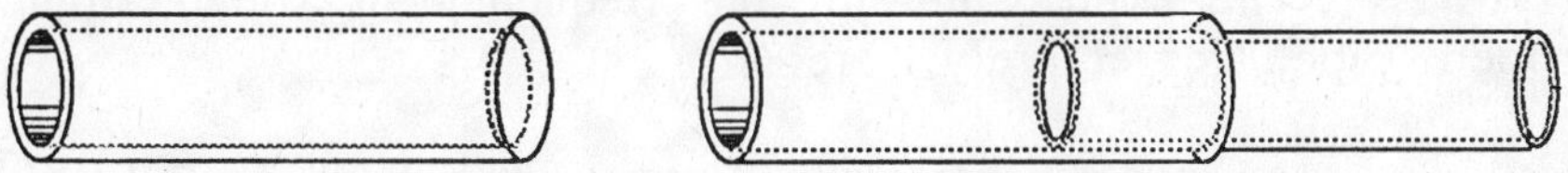

$L_{oc} = 1.00$ m $\qquad$ L_{oo} = adjustable

Determine the following:

1. Fundamental frequency of the open-closed pipe	$\nu_{noc} = (2n - 1)v/4L_{oc}$, $n = 1$ $\nu_{1oc} = v/4L_{oc} = 85.0$ Hz
2. Wavelength associated with the fundamental of the open-closed pipe	$\lambda_n = 4L_{oc}/(2n - 1)$, $n = 1$ $\lambda_1 = 4L_{oc} = 4.00$ m
3. Frequency of the fifth harmonic of the open-closed pipe	Harmonics are multiples of fundamental frequency. For the 5th harmonic, $2n - 1 = 5$, $n = 3$, and $\nu_3 = 5\nu_1 = 425$ Hz
4. Wavelength associated with the 7th harmonic of the open-closed pipe	$\lambda_n = 4L_{oc}/(2n - 1)$, $n = 4$ $\lambda_4 = 4L_{oc}/7 = 0.571$ m
5. Length of an open-open pipe in order for the open-closed pipe fundamental to cause an open-open pipe second harmonic	$\nu_{2oo} = \nu_{1oc} = 85.0$ Hz, $\nu_{noo} = n_{oo}v/2L_{oo}$, $n_{oo} = 2$ for the 2nd harmonic. Hence $L_{oo} = n_{oo}v/2\nu_{2oo} = 4.00$ m
6. Length of an open-open pipe in order for the open-closed pipe third harmonic to cause an open-open pipe third harmonic	For the third harmonic in the open-open pipe, $n_{oo} = 3$. For the third harmonic in the open-closed pipe, $n_{oc} = 2$. $\nu_{2oc} = (2n_{oc} - 1)v/4L_{oc} = 255$ Hz $\nu_{3oo} = \nu_{2oc} = 255 \text{ Hz} = n_{oo}v/2L_{oo}$ $L_{oo} = n_{oo}v/2\nu_{3oo} = 2.00$ m

Example: If you have an open-closed pipe that is twice as long as an open-open pipe, can the two ever create sound of the same frequency?

Given: $L_{oc} = 2L_{oo}$, the open-closed pipe is twice as long as the open-open pipe; $v_{oc} = v_{oo}$ because the speed of sound in air is the same for both the open-closed and open-open pipes.

Determine: Whether or not these two pipes can create sound of the same frequency.

Strategy: Equate expressions for the frequency of sound from the open-closed and open-open pipe; insert the fact that $L_{oc} = 2L_{oo}$; then see if it is possible to satisfy the resulting relationship.

Solution: The expressions for the frequency of sound from a open-closed and an open-open pipe, respectively, are:

$$\nu_{noc} = (2n_{oc} - 1)v/4L_{oc} \quad \text{and} \quad \nu_{noo} = n_{oo}v/2L_{oo}$$

Equating these expressions, we obtain

$$(2n_{oc} - 1)v/4L_{oc} = n_{oo}v/2L_{oo}$$

Cancelling v and inserting $L_{oc} = 2L_{oo}$, we find that the condition that must be satisfied in order to obtain the same frequency sound from both pipes is

$$2n_{oc} - 1 = 4n_{oo}$$

Since n_{oc} and n_{oo} can be any positive integer value, the left side of this expression is always an odd integer and the right side is an even integer. Since the equality cannot be satisfied, it is evident that we cannot obtain sound of the same frequency from the two pipes when $L_{oc} = 2L_{oo}$.

Related Text Exercises: Ex. 33-22 through 33-29.

3 The Doppler Effect (Section 33-6)

Review: If the frequency of sound emitted by a source is ν_s, the frequency of sound heard by an observer is

$$\nu_o = (v - v_o)\nu_s/(v - v_s)$$

where v = speed of sound in air
v_o = speed of the observer (positive if the direction of movement is away from the source and negative if in the opposite direction)
v_s = speed of the source (positive if the direction of movement is towards the observer and negative if in the opposite direction)

Practice: Shown below are six possible situations which could occur for a sound source (a train whistle) and an observer. The frequency of the source is 100 Hz. The source speed is 30.0 m/s, and that of the observer is 20.0 m/s when moving. The speed of sound in air is 340 m/s.

Determine ν_o for each of the following situations:

1.	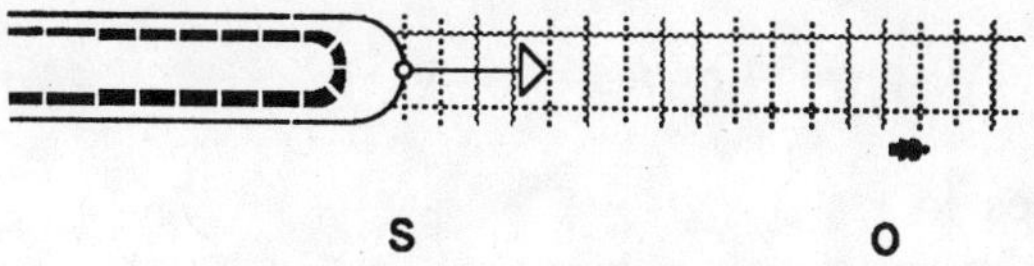	v_s = +30.0 m/s, $v_o = 0$ ν_o = (340 - 0)ν_s/(340 - 30.0) = 110 Hz
2.	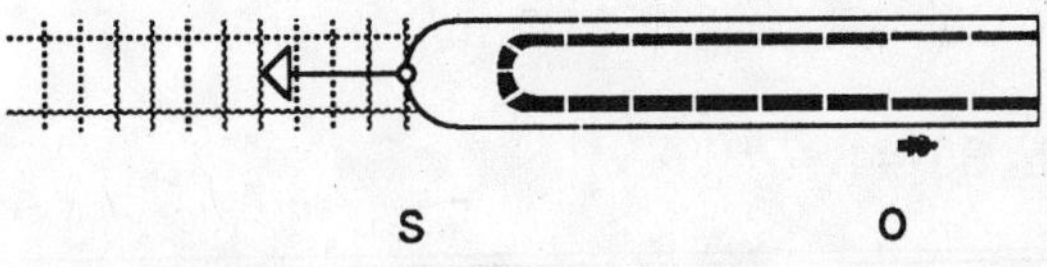	v_s = -30.0 m/s, $v_o = 0$ ν_o = (340 - 0)ν_s/(340 + 30.0) = 91.9 Hz
3.	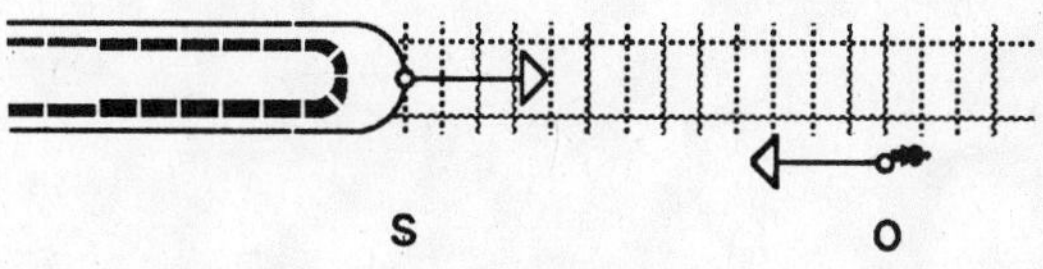	v_s = +30.0 m/s, v_o = -20.0 m/s ν_o = (340 + 20.0)ν_s/(340 - 30.0) = 116 Hz

4.	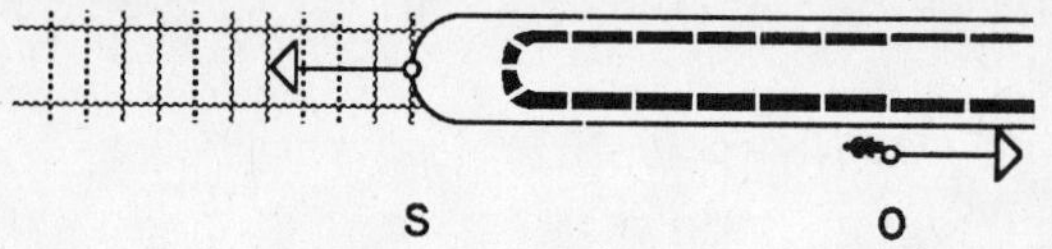	v_s = -30.0 m/s, v_0 = +20.0 m/s $\nu_o = (340 - 20.0)\nu_s/(340 + 30.0) = 86.5$ Hz
5.	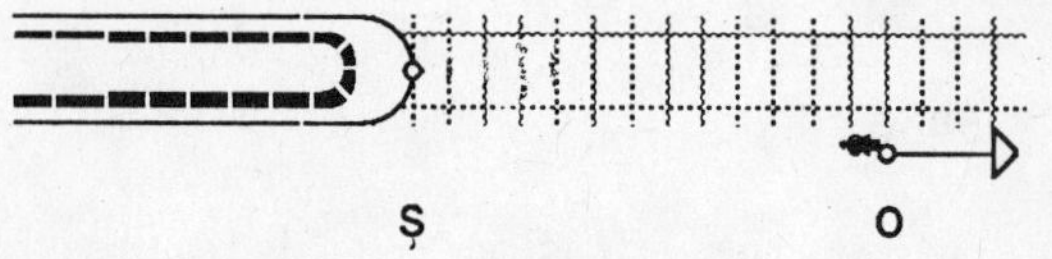	v_s = 0, the source is not moving v_0 = +20.0 m/s $\nu_o = (340 - 20.0)\nu_s/(340 - 0) = 94.1$ Hz
6.	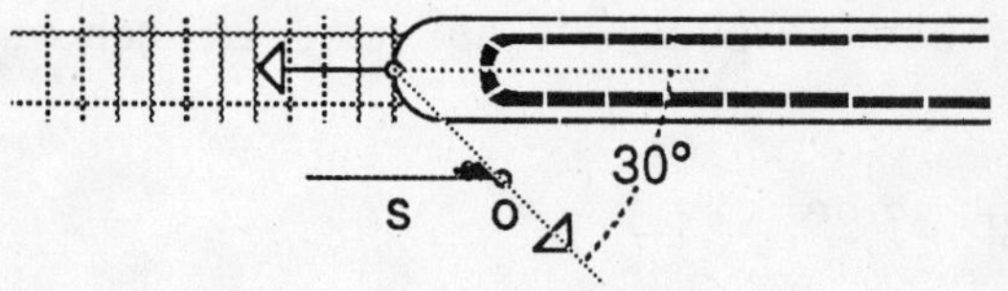	The component of the velocity of the source away from the observer is $v_s = -(30.0 \text{ m/s}) \cos 30° = -26.0$ m/s v_0 = +20.0 m/s $\nu_o = (340 - 20.0)\nu_s/(340 + 26.0) = 87.4$ Hz

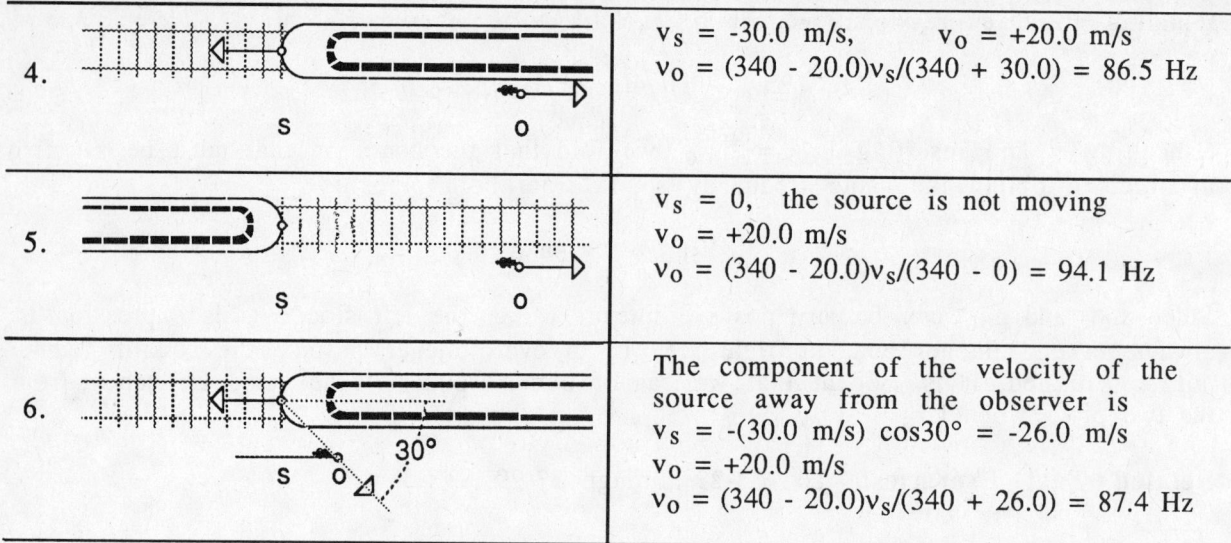

Example: A woman standing beside a railroad track measures a drop of 15 Hz in the pitch of the whistle of a train as it passes her. If the actual frequency of the whistle is 120 Hz, how fast was the train moving?

Given: $\Delta\nu = 15$ Hz $\qquad \nu_s = 120$ Hz $\qquad v = 340$ m/s

Determine: The speed of the train, v_s

Strategy: Write expressions for the frequency of sound heard by the woman as the train approaches and recedes. Set the difference between these two expressions equal to $\Delta\nu$ and solve for v_s.

Solution: Frequency heard by the observer as the train approaches (ν_{oa})

$$\nu_{oa} = (340 - 0)\nu_s/(340 - v_s)$$

Frequency heard by the observer as the train recedes (ν_{or})

$$\nu_{or} = (340 - 0)\nu_s/(340 + v_s)$$

We know that

$$\Delta\nu = \nu_{oa} - \nu_{or} = [340\nu_s/(340 - v_s)] - [340\nu_s/(340 + v_s)]$$

Since $\nu_s = 120$ Hz and $\Delta\nu = 15$ Hz, this reduces to

$$v_s^2 + 5440v_s - 340^2 = 0 \qquad \text{or} \qquad v_s = 21.2 \text{ m/s}$$

Related Text Exercises: Ex. 33-36 through 33-44.

PRACTICE TEST

Take and grade this practice test. Doing so will allow you to determine any weak spots in your understanding of the concepts taught in this chapter. The following section prescribes what you should study further to strengthen your understanding.

Listener A is a distance of 10.0 m from a speaker and she measures the sound intensity level to be 92.0 dB.

Determine the following:

_______ 1. Intensity of the sound heard by listener A

A second speaker (identical to the first) is placed by the first speaker.

Determine the following:

_______ 2. Intensity of the sound heard by listener A after the second speaker is in place

_______ 3. Sound intensity level heard by listener A after the second speaker is in place

_______ 4. Distance from speakers of a second person (listener B) in order to experience one half the intensity of listener A

_______ 5. Sound intensity level heard by listener B at the distance determined in 4. above

An organ pipe open at both ends is 1.00 m long. The frequency of its third harmonic is 510 Hz.

Determine the following:

_______ 6. Wavelength of the standing longitudinal wave set up in the pipe when the air in it is vibrating in its third harmonic

_______ 7. Fundamental frequency for this pipe

_______ 8. Wavelength for the fifth harmonic

_______ 9. Length of a pipe closed at one end that has a fundamental frequency equal to the frequency of the third harmonic of the pipe open at both ends

_______ 10. Length of a pipe closed at both ends that has the frequency of its forth harmonic equal to the frequency of the third harmonic of the pipe open at both ends

A tuning fork of frequency 500 Hz is moved away from an observer and toward a flat wall with a speed of 2.00 m/s. Assume the speed of sound in air is 340 m/s.

Determine the following:

_______ 11. Apparent frequency of the sound waves coming directly to the observer from the tuning fork

_______ 12. Apparent frequency of the sound waves coming to the observer after being reflected off of the wall

_______ 13. The number of beats heard by the observer

The observer now walks toward the wall at a speed of 1.00 m/s.

Determine the following:

_________ 14. Apparent frequency of the sound waves coming directly to the observer from the tuning fork

_________ 15. Apparent frequency of the sound waves coming to the observer after being reflected off of the wall

_________ 16. The number of beats heard by the observer

(See Appendix I for answers.)

PRINCIPAL CONCEPTS AND EQUATIONS PRESCRIPTION

Your score on the practice test is an excellent measure of your understanding of the chapter. You should now use the following chart to write your own prescription for dealing with any weaknesses the practice test points out. Look down the leftmost column to the number of the question(s) you answered incorrectly, reading across that row you will find the concept and/or equation of concern, the section(s) of the study guide you should return to for further study, and some suggested text exercises which you should work to gain additional experience.

Practice Test Questions	Concepts and Equations	Prescription	
		Principal Concepts	Text Exercises
1	Sound intensity level: $\beta = 10 \log(I/I_0)$	1	33-17, 18
2	Intensity: $I_T = I_1 + I_2$	1	33-18, 19
3	Sound intensity level: $\beta = 10 \log(I/I_0)$	1	33-17, 19
4	Power: $P = I_1A_1 = I_2A_2$	3 of Ch 32	32-33, 34
5	Sound intensity level: $\beta = 10 \log(I/I_0)$	1	33-17, 18
6	Standing waves: $L = n\lambda_n/2$	2	33-22, 23
7	Standing waves: $v_n = nv/2L$	2	33-23, 24
8	Standing waves: $L = n\lambda_n/2$	2	33-24, 25
9	Standing waves: $v_{noo} = nv/2L_{oo}$ $v_{noc} = (2n - 1)v/2L_{oc}$	2	33-25, 26
10	Standing waves: $v_{noo} = nv/2L_{oo}$ $v_{noc} = nv/2L_{oc}$	2	33-26, 27
11-16	Doppler shift: $v_o = v_s(v - v_o)/(v - v_s)$	3	33-36 to 33-42

34 MAXWELL'S EQUATIONS AND ELECTROMAGNETIC WAVES

RECALL FROM PREVIOUS CHAPTERS

Previously learned concepts and equations frequently used in this chapter	Text Section	Study Guide Page
Power: $P = W/t$	8-5	8-15
Pressure: $p = F/A$	15-1	15-1
Force on a charged particle due to an electric field: $\mathbf{F} = q\mathbf{E}$	20-4	20-5
Force on a charged particle due to a magnetic field: $\mathbf{F} = q\mathbf{v} \times \mathbf{B}$	26-1	26-1
Energy density of a electric field: $\mu_E = \varepsilon_0 E^2/2$	23-3	23-7
Energy density of a magnetic field: $\mu_B = B^2/2\mu_0$	29-3	29-9
Characteristics of a wave: k, λ, ν, ω, v	32-1, 2	32-1

NEW IDEAS IN THIS CHAPTER

Concepts and equations introduced	Text Section	Study Guide Page
Electromagnetic waves	34-1, 2, 3	34-1
Electromagnetic wave intensity: $S = \mu c = \varepsilon_0 E^2 c = B^2 c/\mu_0$	34-4	34-4
Average wave intensity: $\bar{S} = E_0 B_0/2\mu_0 = \varepsilon_0 E^2 c/2 = cB_0{}^2/2\mu_0$	34-4	34-4
Poynting vector: $\mathbf{S} = \mathbf{E} \times \mathbf{B}/\mu_0$	34-4	34-4
Radiation pressure: $P_{abs} = S/c$ $\quad P_{refl} = 2S/c$	34-5	34-7

PRINCIPAL CONCEPTS AND EQUATIONS

1 Electromagnetic Waves (Sections 34-1, 2 and 4)

Review: Figure 34.1 shows an electromagnetic wave.

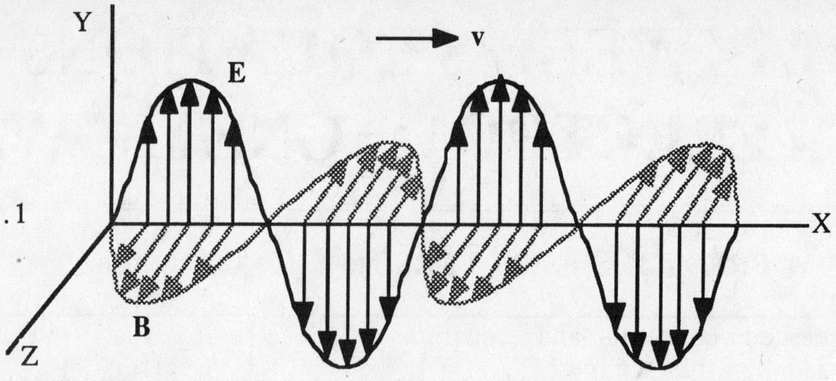

Figure 34.1

Where $\mathbf{E} = E_y(x,t)\mathbf{j}$ and $E_y = E_0 \sin(kx - \omega t)$
$\mathbf{B} = B_z(x,t)\mathbf{k}$ and $B_z = B_0 \sin(kx - \omega t)$
$v = c\mathbf{i}$ and $c = 1/(\mu_0\varepsilon_0)^{1/2}$

Notice that **E** and **B** are mutually perpendicular and they have the same value for the following quantities:

wave number -- k
wavelength -- $\lambda = 2\pi k$
angular frequency -- ω
linear frequency -- $\nu = \omega/2\pi$
phase -- $\phi = 0$
speed -- $c = 1/(\mu_0\varepsilon_0)^{1/2} = \omega/k$

In addition, from Faraday's law we have

$$\partial E_y/\partial x = -\partial B_z/\partial t \quad \text{(see Eq. 34-6 of Text)}$$

For this case

$$\partial E_y/\partial x = kE_0 \cos(kx - \omega t) \quad \text{and} \quad \partial B_z/\partial t = -\omega B_0 \cos(kx - \omega t)$$

or

$$kE_0 = \omega B_0$$

This may be rearranged to obtain

$$E_0 = (\omega/k)B_0 = cB_0$$

We may also arrive at this result using Ampere's law

$$\partial B_z/\partial x = -\mu_0\varepsilon_0\partial E_y/\partial t \quad \text{(see Eq. 34-7 of Text)}$$

Again for this case

$$\partial E_y/\partial t = -\omega E_0 \cos(kx - \omega t) \quad \text{and} \quad \partial B_z/\partial x = kB_0 \cos(kx - \omega t)$$

or

$$kB_0 = \mu_0\varepsilon_0\omega E_0$$

This may be rearranged to obtain

$$E_0 = (k/\omega)(1/\mu_0\varepsilon_0)B_0 = cB_0$$

We also see that since

$$E_y = cB_z$$

reduces to $E_0 = cB_0$, then it must be true.

Practice: Suppose that the electric field amplitude of the wave shown in Figure 34.1 is E_0 = 150 N/c, and that its frequency is $\nu = 5.00 \times 10^7$ Hz.

Determine the following:

1. The magnetic field amplitude B_0	$B_0 = E_0/c = 5.00 \times 10^{-7}$ T
2. The angular frequency ω	$\omega = 2\pi\nu = 3.14 \times 10^8$ rad/s
3. The wave number k	$k = \omega/c$ = 1.05/m
4. The wavelength λ	$\lambda = 2\pi/k = 5.98$ m or $\lambda = c/\nu = 6.00$ m The difference is due to rounding off.
5. An expression for **E**	$\mathbf{E} = E_0 \sin(kx - \omega t)\mathbf{j}$ $= (150\ \text{N/C}) \sin\left[\begin{matrix}(1.05/\text{m})x - \\ (3.14 \times 10^8\ \text{rad/s})t\end{matrix}\right]\mathbf{j}$
6. An expression for **B**	$\mathbf{B} = B_0 \sin(kx - \omega t)\mathbf{k}$ $= \left(\frac{5.00\ \text{T}}{10^7}\right) \sin\left[\begin{matrix}(1.05/\text{m})x - \\ (3.14 \times 10^8\ \text{rad/s})t\end{matrix}\right]\mathbf{k}$

Example: Suppose the electric field part of an electromagnetic wave in a vacuum is

$$\mathbf{E} = (200\ \text{N/c}) \sin[(2.00\ \text{rad/m})y + (6.00 \times 10^8\ \text{rad/s})t]\,\mathbf{i}$$

(a) What is the direction of propagation of the electromagnetic wave?
(b) What is the wavelength of the wave?
(c) What is the frequency of the wave?
(d) What is the amplitude of the magnetic field part of the wave?
(e) Write an expression for the magnetic field part of the wave?

Given: $\mathbf{E} = (200\ \text{N/c}) \sin[(2.00\ \text{rad/m})y + (6.00 \times 10^8\ \text{rad/s})t]\,\mathbf{i}$

Determine: The direction of propagation, λ, ν, B_0 and $\mathbf{B}_0$.

Strategy: We can determine E_0, k and ω from the expression for **E** and then determine λ, ν, B_0 and an expression for **B**. Also knowing that **E** x **B** must be in the direction of propagation of the wave, we can determine the direction in which the wave is traveling.

Solution:
(a) The wave is traveling in the -**j** direction
(b) k = 2.00 rad/m and $\lambda = 2\pi/k = \pi$ m
(c) $\omega = 6.00 \times 10^8$ rad/s and $\nu = \omega/2\pi = 9.55 \times 10^7$ Hz
(d) E_0 = 200 N/c and $B_0 = E_0/c = 6.67 \times 10^{-7}$ T
(e) $\mathbf{B} = (6.67 \times 10^{-7}\ \text{N/c}) \sin[(2.00\ \text{rad/m})y + (6.00 \times 10^8\ \text{rad/s})]\,\mathbf{k}$

Related Text Exercises: Ex. 34-1 through 34-17.

2 Electromagnetic Wave Intensity (Section 34-4)

Review: Figure 34.2 shows an electromagnetic plane wave traveling through space and a volume element ($A\Delta x$) in the path of the wave.

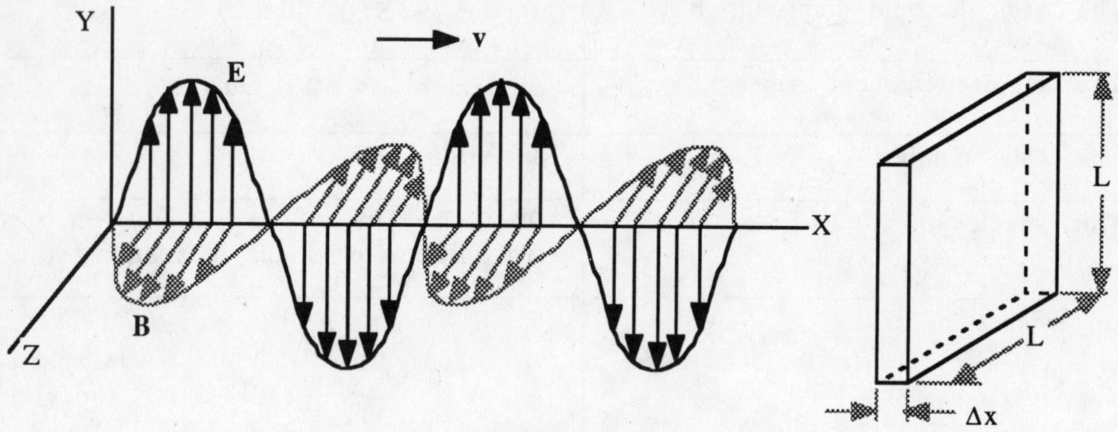

Figure 34.2

The energy density u_E associated with the electric field is

$$u_E = \varepsilon_0 E^2/2$$

The energy density u_B associated with the magnetic field is

$$u_B = B^2/2\mu_0$$

For plane electromagnetic waves, these energy densities are equal, hence

$$u = u_E + \mu_B = 2u_E = 2u_B$$

By using $E_y = cB_z$ and $c^2 = 1/\mu_0\varepsilon_0$, we can express u in several forms. A convenient form is

$$u = \varepsilon_0 E^2$$

The electromagnetic energy within this volume is

$$\Delta U = u(A\Delta x)$$

The rate at which energy passes through this volume is

$$\Delta U/\Delta t = uA\Delta x/\Delta t = uAc$$

The wave intensity S is the rate at which energy passes through this area per unit of area. This may be expressed as

$$S = (\Delta U/\Delta t)/A = (uAc)/A = uc = \varepsilon_0 E^2 c$$

At this time it is convenient to introduce the Poynting vector **S**.

$$\mathbf{S} = \mathbf{E} \times \mathbf{B}/\mu_0$$

We need to know the following two very important things about the Poynting vector:

I. The magnitude of the Poynting vector is the wave intensity, that is the energy per unit time crossing a unit of area perpendicular to the direction of propagation of the electromagnetic wave. This may be proven as follows:

$$\mathbf{E} = E_y\mathbf{j} \quad \text{and} \quad \mathbf{B} = B_z\mathbf{k}$$

$$\mathbf{S} = \mathbf{E} \times \mathbf{B}/\mu_0 = E_yB_z(\mathbf{j} \times \mathbf{k})/\mu_0 = E_yB_z\mathbf{i}/\mu_0$$

Recall

$$c^2 = 1/\mu_0\varepsilon_0 \quad \text{and} \quad E_y = cB_z$$

to get

$$\mathbf{S} = (1/\mu_0)E_yB_z\mathbf{i} = \varepsilon_0 c^2 E_y(E_y/c)\mathbf{i} = \varepsilon_0 E_y^2 c\mathbf{i}$$

Notice that

$$|\mathbf{S}| = S = \varepsilon_0 E_y^2 c$$

which is identical to the previously obtained expression for the intensity of the electromagnetic wave.

Note: We have shown that the magnitude of the Poynting vector is the same as the intensity of the electromagnetic wave.

II. The direction of the Poynting vector is the direction of propagation of the electromagnetic wave. This may be shown as follows:

$$\mathbf{S} = \mathbf{E} \times \mathbf{B}/\mu_0 = E_y\mathbf{j} \times B_z\mathbf{k}/\mu_0 = (E_yB_z/\mu_0)(\mathbf{j} \times \mathbf{k}) = (E_yB_z/\mu_0)\mathbf{i}$$

Notice that the direction of propagation of the wave and the direction of **S** are the same.

Note: We have shown that the direction of **S** is the same as the direction of propagation of the electromagnetic wave.

For the case of a harmonic plane wave

$$\mathbf{E} = E_y\mathbf{j} = E_0 \sin(kx - \omega t)\mathbf{j} \quad \text{and} \quad \mathbf{B} = B_z\mathbf{k} = B_0 \sin(kx - \omega t)\mathbf{k}$$

we get

$$\mathbf{S} = \mathbf{E} \times \mathbf{B}/\mu_0 = (1/\mu_0)E_0B_0 \sin^2(kx - \omega t)\mathbf{i}$$

The average wave intensity is (recall $\overline{\sin^2 A} = 1/2$)

$$\bar{S} = E_0B_0/2\mu_0 = \varepsilon_0 E_0^2 c/2 = cB_0^2/2\mu_0$$

Practice: The electromagnetic wave intensity at a particular instant at a point in a vacuum is 2.00 KW/m^2.

Determine the following at that point and time:

1. Magnitude of the electric field	$S = \varepsilon_0 E^2 c$ $E = (S/\varepsilon_0 c)^{1/2} = 868$ N/c
2. Magnitude of the magnetic field	$S = B^2 c/\mu_0$ $B = (\mu_0 S/c)^{1/2} = 2.90 \times 10^{-6}$ N/A•m
3. Energy density due to the electric field and the magnetic field	$u_E = \varepsilon_0 E^2/2 = S/2c = 3.33 \times 10^{-6}$ J/m^3 $u_B = B^2/2\mu_0 = S/2c = 3.33 \times 10^{-6}$ J/m^3
4. Total energy density	$u = u_E + u_B = 2u_E = 2u_B$ $= S/c = \varepsilon_0 E^2 = B^2/\mu_0 = 6.66 \times 10^{-6}$ J/m^3

Example: An electromagnetic wave in a vacuum is traveling in the +y direction, and its plane of polarization is parallel to the yz plane. The frequency of the wave is 50.0 MHz and its average intensity is 480 W/m^2. Write expressions for **E**, **B** and **S** as functions of y and t.

Given: $\nu = 5.00 \times 10^7$ Hz $\qquad \bar{S} = 480$ W/m^2

Determine: Expressions for the vectors **E**, **B** and **S** as functions of y and t.

Strategy: Knowing ν, we can determine ω and k. Knowing $\bar{S}$, we can determine E_0 and B_0. Knowing E_0, B_0, k and ω, we can write expressions for the magnitude of **E** and **B** (i.e. E and B) as functions of y and t. Knowing that **S** and the velocity of propagation of the wave are in the same direction and knowing that **S** is obtained from the cross product of **E** and **B**, we can determine the directions of **S**, **E** and **B**. We can then write expressions for the vectors **E**, **B** and **S** as functions of y and t.

Solution: Knowing ν, we can determine ω and k.

$$\omega = 2\pi\nu = 3.14 \times 10^8 \text{ rad/s} \qquad \text{and} \qquad k = \omega/c = 1.05/\text{m}$$

Knowing $\bar{S}$, we can determine E_0 and B_0.

$$\bar{S} = \varepsilon_0 E_0^2 c/2 \qquad \text{or} \qquad E_0 = (2\bar{S}/\varepsilon_0 c)^{1/2} = 6.01 \times 10^2 \text{ N/c}$$

$$\bar{S} = cB_0^2/2\mu_0 \qquad \text{or} \qquad B_0 = (2\mu_0\bar{S}/c)^{1/2} = 2.01 \times 10^{-6} \text{ N/A•m}$$

Expressions for the magnitude of **E** and **B** are:

$$E = E_0 \sin(ky - \omega t) = (6.01 \times 10^2 \text{ N/c}) \sin[(1.05/\text{m})y - (3.14 \times 10^8 \text{ rad/s})t]$$

$$B = B_0 \sin(ky - \omega t) = (2.01 \times 10^{-6} \text{ N/A•m}) \sin[(1.05/\text{m})y - (3.14 \times 10^8 \text{ rad/s})t]$$

Since the wave is traveling in the +y direction, then

$$\mathbf{S} = S\mathbf{j} \qquad \text{(S is determined below)}$$

Since the yz plane is the plane of polarization, then

$$\mathbf{E} = E\mathbf{k} \qquad \text{(E is given above)}$$

Since $\mathbf{S} = \mathbf{E} \times \mathbf{B}/\mu_0$, then

$$\mathbf{B} = B\mathbf{i} \qquad \text{(B is given above)}$$

The magnitude of **S** is given by

$$S = |\mathbf{S}| = |\mathbf{E} \times \mathbf{B}/\mu_0| = E_0B_0 \sin^2(ky - \omega t)$$

Related Text Exercises: Ex. 34-18 through 34-29.

3 Radiation Pressure (Section 34-5)

Review: Figure 34.3(a) shows an electromagnetic wave incident normally on a slab of material. Figures 34.3(b and c) show what happens as this wave encounters an electron in the material.

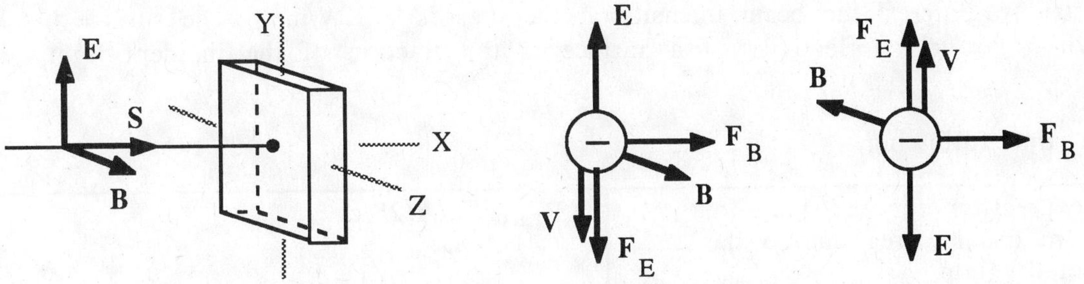

Figure 34.3

As the electromagnetic wave interacts with the material, the oscillating **E** field causes the bound electrons to execute an oscillatory motion. In this manner the electric field does work on the charges but it does not tend to displace the entire object. In contrast (as Figures 34.3(b and c) show), the magnetic field of the wave does no work on the particles ($\mathbf{F}_m$ is perpendicular to **v**) but exerts a force on the object in the direction of propagation of the wave.

Looking at Figure 34.3(b), we can write expressions for the electric and magnetic force on the electron.

$$\mathbf{F}_E = F_E(-\mathbf{j}) = qE_y(-\mathbf{j})$$

$$\mathbf{F}_m = -q(\mathbf{v} \times \mathbf{B}) = -q(-v\mathbf{j} \times B_z\mathbf{k}) = qvB_z(\mathbf{j} \times \mathbf{k}) = qv(E_y/c)\mathbf{i} = F_E(v/c)\mathbf{i}$$

The power delivered to an electron by the electric field is

$$P = W/t = F_Ev$$

The total power delivered to all the electrons in the slab is

$$P_{total} = \Sigma F_E v$$

The magnetic force on an electron is

$$F_m = qvB_z = F_E(v/c) \qquad \text{(See above)}$$

The total magnetic force on the slab is the sum of all these magnetic forces.

$$F_{total} = \Sigma F_m$$

The pressure on a slab of cross sectional area A when the electromagnetic wave is absorbed is

$$P_{abs} = F_{total}/A = \Sigma F_m/A = \Sigma F_E(v/c)/A = (W/tA)/c = S/c$$

If the wave is totally reflected, the pressure is

$$P_{refl} = 2S/c$$

Practice: A beam of light is incident normally on a surface of area $1.00 \times 10^{-2}\,m^2$. At the instant of interest the beam intensity at the area is 5 KW/m^2. The surface has a reflectivity of r (the reflectivity of a surface is the fraction of the incident light reflected).

Determine the following:

1. An expression for the radiation pressure on the area due to the reflected light	$P_{rad\ reflected} = r(2S/c)$
2. An expression for the radiation pressure on the area due to the absorbed light	The fraction of the absorbed light plus the reflected light equal one. $a + r = 1$ or $a = 1 - r$ Hence the radiation pressure due to the absorbed light is $P_{rad\ absorbed} = a(S/c) = (1 - r)(S/c)$
3. An expression for the total radiation pressure	$P_{rad\ total} = P_{rad\ reflected} + P_{rad\ absorbed} = (r + 1)(S/c)$
4. The total radiation pressure for the case of r = 1/5	$P_{rad\ total} = (r + 1)(S/c) = (6/5)(S/c)$ $= 2.00 \times 10^{-5}\ N/m^2$
5. Force on the surface	$F = P_{rad\ total}A = 2.00 \times 10^{-7}\ N$

Example: The average power of a laser beam is 5.00 mW, and the beam has a uniform intensity over an area of 1.00×10^{-6} m^2. If this beam is normally incident on a completely reflecting surface, determine the pressure exerted by the beam on the part of the surface it strikes and the force exerted on the surface by the beam.

Given: $\bar{P} = 5.00$ mW $\quad A = 1.00 \times 10^{-6}$ m^2 $\quad r = 1.00$

Determine: The pressure and force on the surface due to the beam.

Strategy: Knowing the average power $\bar{P}$ and the area A, we can determine the average wave intensity $\bar{S}$. Knowing the average wave intensity $\bar{S}$ and the reflectivity r, we can determine the radiation pressure p. Knowing the radiation pressure p and the area A, we can determine the force on the surface.

Solution: The average wave intensity is

$$\bar{S} = \bar{P}/A = 5.00 \times 10^3 \text{ W/m}^2$$

The radiation pressure is

$$p = \bar{S}/c = \bar{P}/cA = 1.67 \times 10^{-5} \text{ N/m}^2$$

The force on the surface is

$$F = pA = \bar{P}/c = 1.67 \times 10^{-11} \text{ N}$$

Related Text Exercises: Ex. 34-30 through 34-34.

PRACTICE TEST

Take and grade this practice test. Doing so will allow you to determine any weak spots in your understanding of the concepts taught in this chapter. The following section prescribes what you should study further to strengthen your understanding.

Suppose the electric field part of an electromagnetic wave in a vacuum is

$$\mathbf{E} = (3.00 \times 10^2 \text{ N/c}) \sin[(3.00 \text{ rad/m})x - (9.00 \times 10^8 \text{ rad/s})t]\mathbf{j}$$

Determine the following:

_________ 1. Direction of propagation of the wave
_________ 2. Wavelength of the wave
_________ 3. Frequency of the wave
_________ 4. Amplitude of the electric field part of the wave
_________ 5. Amplitude of the magnetic field part of the wave
_________ 6. Maximum intensity of the wave
_________ 7. Average wave intensity of the wave
_________ 8. Maximum energy density of the wave
_________ 9. Plane of polarization of the wave

If this electromagnetic wave is incident normally on a $1.00 \times 10^{-2}\ m^2$ surface which has a reflectivity of $r = 0.200$, determine the following:

_________ 10. Radiation pressure due to absorption
_________ 11. Radiation pressure due to reflection
_________ 12. Force on the surface

(See Appendix I for answers.)

PRINCIPAL CONCEPTS AND EQUATIONS PRESCRIPTION

Your score on the practice test is an excellent measure of your understanding of the chapter. You should now use the following chart to write your own prescription for dealing with any weaknesses the practice test points out. Look down the leftmost column to the number of the question(s) you answered incorrectly, reading across that row you will find the concept and/or equation of concern, the section(s) of the study guide you should return to for further study, and some suggested text exercises which you should work to gain additional experience.

Practice Test Questions	Concepts and Equations	Prescription	
		Principal Concepts	Text Exercises
1	Electromagnetic wave -- direction	1	34-15
2	Wave characteristics -- wavelength: $\lambda = 2\pi/k$	1 of Ch 32	34-11, 15
3	Wave characteristics -- frequency: $\nu = \omega/2\pi$	1 of Ch 32	34-12, 15
4	Wave characteristics -- amplitude	1 of Ch 32	34-13, 16
5	Electromagnetic wave: $B_0 = E_0/c$	1	34-13, 16
6	Wave intensity: $S_{max} = \varepsilon_0 E_0{}^2 c = B_0{}^2 c/\mu_0$	2	34-20, 21
7	Average wave intensity: $\bar{S} = \varepsilon_0 E^2 c/2 = cB_0{}^2/2\mu_0 = S_{max}/2$	2	34-23, 24
8	Energy density: $u = S/c$	2	34-18, 19
9	Electromagnetic wave -- plane of polarization	2	34-26, 27
10	Radiation pressure due to absorption: $P_{abs} = (1 - r)\bar{S}/c$	3	34-31, 32
11	Radiation pressure due to reflection: $P_{refl} = r(2\bar{S}/c)$	3	34-32, 33
12	Relation between pressure and force: $P = F/A$	3 of Ch 15	34-33, 34

35 GEOMETRICAL OPTICS

RECALL FROM PREVIOUS CHAPTERS

You should be able to proceed with this chapter without reviewing any previous concepts and equations.

NEW IDEAS IN THIS CHAPTER

Concepts and equations introduced	Text Section	Study Guide Page
Index of refraction: $n = c/v$	35-1	35-1
Snell's law: $\sin\theta_1/\sin\theta_2 = v_1/v_2 = n_2/n_1$	35-1	35-1
Critical angle: $\sin\theta_c = n_2/n_1$	35-1	35-1
Mirror equation: $1/s + 1/s' = 1/f$	35-2	35-7
Mirror magnification: $m = h'/h = -s'/s$	35-2	35-7
Mirror sign conventions for s, s', f, h, h' and m	35-2	35-7
Images by refraction: $n_1/s + n_2/s' = (n_2 - n_1)/r$	35-3	35-11
Lens makers eq: $n_1/f = (n_2 - n_1)(1/r_a - 1/r_b)$	35-4	35-13
Refractive power: $P = 1/f$	35-4	35-13
Thin lens equation: $1/s + 1/s' = 1/f$	35-4	35-13
Lens magnification: $m = h'/h = -s'/s$	35-4	35-13
Lens sign conventions for s, s', f, h, h' and m	35-4	35-13

PRINCIPAL CONCEPTS AND EQUATIONS

1 Refraction (Section 35-1)

Review: Figure 35.1 shows a ray of light refracted first as it travels from air to glass and again as it travels from glass to air.

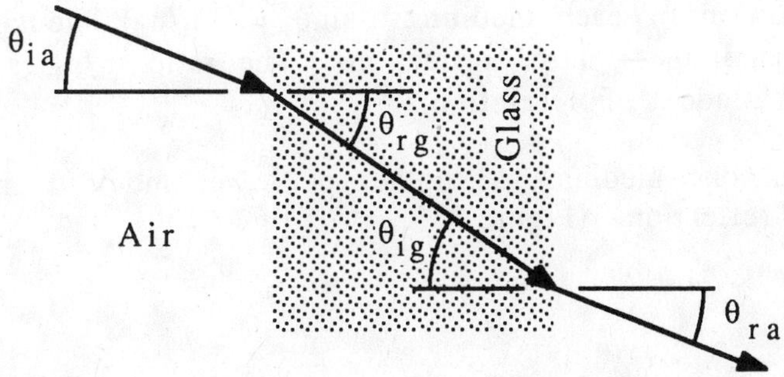

Figure 35.1

The angles of incidence in air and glass are θ_{ia} and θ_{ig}. The angles of refraction in glass and air are θ_{rg} and θ_{ra}.

Note: The authors of your text represent the index of refraction and the angle of incidence in medium 1 by n_1 and θ_1. In like manner they represent the index of refraction and the angle of refraction in medium 2 by n_2 and θ_2. This is fine for a generic discussion of two media, however when the media are for example air and water you will find it convenient to use subscripts a and w rather than 1 and 2. Also in order to avoid any chance of confusion, you might find it convenient to add another subscript to explicitly indicate incidence and refraction. For example, we use the following notation:

θ_{ia} = angle of incidence in air
θ_{rg} = angle of refraction in glass
θ_{ig} = angle of incidence in glass
θ_{ra} = angle of refraction in air

The subscripts make it clear which medium we are considering and whether we are interested in an angle of incidence or refraction.

Refraction is the bending of light as it passes from one transparent medium to another. It is caused by the fact that light travels at different speeds in the two media. The index of refraction of a transparent medium is equal to the ratio of the speed of light in a vacuum to the speed of light in that medium:

$$n = c/v$$

The greater the index of refraction of the second medium, the slower the light travels and hence the more its path is bent. Snell's law,

$$\sin\theta_i/\sin\theta_r = v_i/v_r = (c/n_i)/(c/n_r) = n_r/n_i$$

shows how the angle of refraction is related to:

1. the angle of incidence: $\sin\theta_r \propto \sin\theta_i$, hence large $\theta_i \rightarrow$ large θ_r and small $\theta_i \rightarrow$ small θ_r

2. the speed of light in each medium: $\sin\theta_r \propto (v_r/v_i)$, hence small $(v_r/v_i) \rightarrow$ small $\theta_r \rightarrow$ large refraction

3. the index of refraction of each medium: $\sin\theta_r \propto (n_i/n_r)$, hence small $(n_i/n_r) \rightarrow$ small $\theta_r \rightarrow$ large refraction. The ratio n_i/n_r is called the relative index of refraction.

When light travels from one medium to another of lower index of refraction, we may observe total internal reflection, (Figure 35.2).

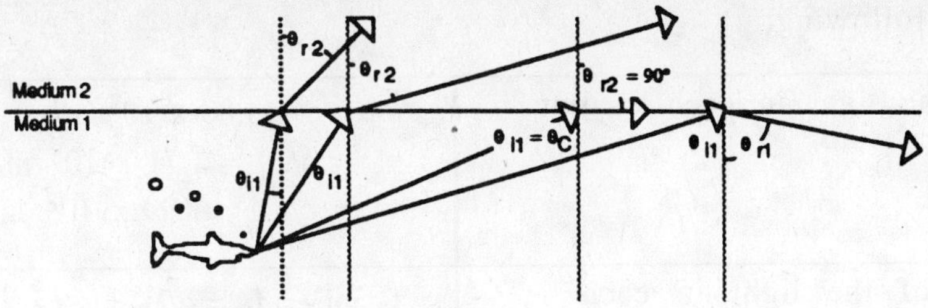

θ_{i1} = angle of incidence in medium 1
θ_{r2} = angle of refraction in medium 2
θ_{r1} = angle of reflection in medium 1
θ_c = critical angle

Figure 35.2

As the angle of incidence in medium 1 increases, the angle of refraction in medium 2 increases. The angle of incidence in medium 1 is equal to the critical angle when the angle of refraction in medium 2 is 90°. We may determine θ_c by applying Snell's Law to this situation.

$$n_1 \sin\theta_c = n_2 \sin 90°$$
$$\sin\theta_c = n_2/n_1$$

Notice that, in order for this to make sense, we must have $\theta_c < 90°$, which means $\sin\theta_c < 1$. This is possible only if $n_2 < n_1$. Subsequently, we expect a critical angle and total internal reflection only when light goes from one medium to another of lower index of refraction.

The index of refraction of a substance depends somewhat on the wavelength of the light being used. If a ray composed of many wavelengths of light is refracted, it will be dispersed into rays whose directions depend on the index of refraction for the various wavelength. The property of having the index of refraction vary with wavelength is called dispersion.

Practice: Figure 35.3 shows a ray of light traveling through air, flint glass and acrylic plastic.

$\theta_{ia} = 60.0°$

$n_a = 1.00$

$n_g = 1.75$

$n_p = 1.51$

$n_a = 1.00$

$\lambda_a = 6.50 \times 10^{-7}$ m

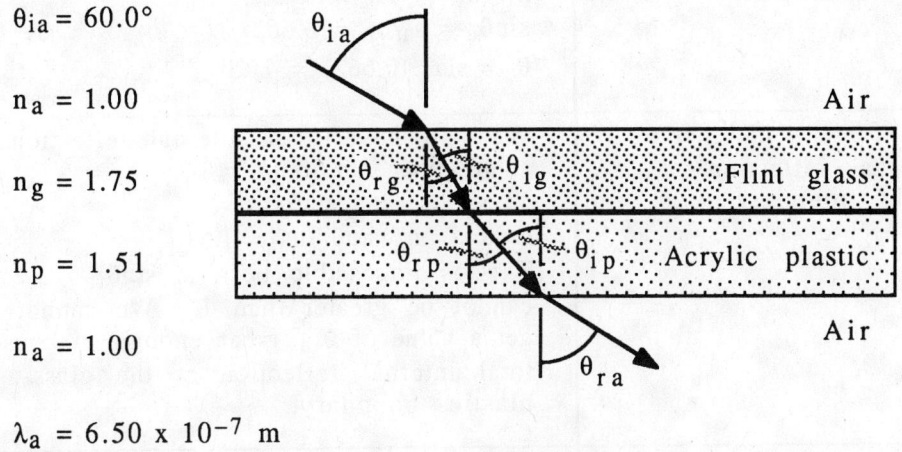

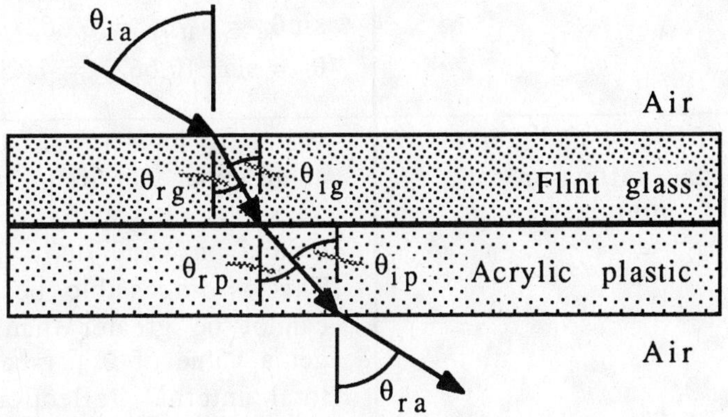

Figure 35.3

Determine the following:

1. Speed of the light in each medium	$v_a = c = 3.00 \times 10^8$ m/s $v_g = c/n_g = 1.71 \times 10^8$ m/s $v_p = c/n_p = 1.99 \times 10^8$ m/s
2. Frequency of the light in each medium	$f_a = f_g = f_p = c/\lambda_a = 4.62 \times 10^{14}$ Hz
3. Wavelength of the light in each medium	$\lambda_a = 6.50 \times 10^{-7}$ m $\lambda_g = v_g/f_g = c/n_g f_g = \lambda_a/n_g$ $= 3.71 \times 10^{-7}$ m $\lambda_p = v_p/f_p = c/n_p f_p = \lambda_a/n_p$ $= 4.30 \times 10^{-7}$ m
4. Angle of refraction in glass	$n_a \sin\theta_{ia} = n_g \sin\theta_{rg}$ $\sin\theta_{rg} = n_a \sin\theta_{ia}/n_g = 0.495$ $\theta_{rg} = \sin^{-1}(0.495) = 29.7°$
5. Angle of refraction in plastic	$\theta_{ig} = \theta_{rg}$ (geometry) $n_g \sin\theta_{ig} = n_p \sin\theta_{rp}$ $\sin\theta_{rp} = n_g \sin\theta_{ig}/n_p = 0.574$ $\theta_{rp} = \sin^{-1}(0.574) = 35.0°$
6. Angle of refraction in air	$\theta_{ip} = \theta_{rp}$ (geometry) $n_p \sin\theta_{ip} = n_a \sin\theta_{ra}$ $\sin\theta_{ra} = n_p \sin\theta_{ip}/n_a = 0.866$ $\theta_{ra} = \sin^{-1}(0.866) = 60.0°$
7. Critical angle for the glass relative to the plastic	$n_g \sin\theta_c = n_p \sin 90°$ $\sin\theta_c = n_p/n_g = 0.863$ $\theta_c = \sin^{-1}(0.863) = 59.6°$
8. Critical angle for the plastic relative to air	$n_p \sin\theta_c = n_a \sin 90°$ $\sin\theta_c = n_a/n_p = 0.662$ $\theta_c = \sin^{-1}(0.662) = 41.5°$
9. Minimum value for θ_{ia} in order to have total internal reflection at the glass-plastic boundary	We will have total internal reflection at this boundary if $\theta_{rg} = \theta_{ig} = \theta_c = 59.6°$. $n_a \sin\theta_{ia} = n_g \sin\theta_{rg}$ $\sin\theta_{ia} = n_g \sin\theta_c/n_a = 1.51$ This is impossible because $\sin\theta_{ia}$ cannot be greater than 1. We cannot get a value of θ_{ia} great enough to cause total internal reflection at the glass-plastic boundary.

Example: A beam of light composed of red and blue is incident on a 30°-60°-90° prism in such a manner that the red light goes through the prism parallel to the base (Figure 35.4).

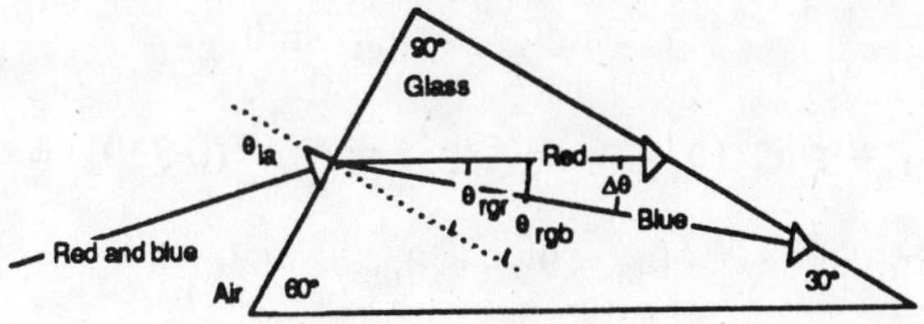

Figure 35.4

The index of refraction in the glass is 1.46 for red light and 1.47 for blue light. Determine the angle of incidence for the incident beam and the angular separation ($\Delta\theta$) for the red and blue light in the prism and establish whether refraction or internal reflection takes place at the second glass-air boundary.

The notation used is

θ_{iar}, θ_{iab}, θ_{ia} = angle of incidence in air for red light, blue light, and the combined beam

θ_{rgr}, θ_{rgb} = angle of refraction in glass for red and blue light

$\Delta\theta$ = angular separation between the refracted red and blue light

n_{gr}, n_{gb} = index of refraction in glass for red and blue light

θ_{cr}, θ_{cb} = critical angle at the glass-air interface for red and blue light

Given: The red light goes through the prism parallel to the base.
The prism is a 30°-60°-90°
n_{gr} = 1.46 = index of refraction in glass for red
n_{gb} = 1.47 = index of refraction in glass for blue

Determine: The angle of incidence for the incident beam, the angular separation for the red and blue light in the prism, and whether refraction or internal reflection takes place at the second glass-air boundary.

Strategy: From the geometry, we can determine the angle of refraction in glass for red light (θ_{rgr}). Knowing this angle and the indices of refraction, we can determine the angle of incidence in air for the red light (θ_{iar}) and hence for the beam (θ_{ia}). Since the blue light is part of the beam, it has the same angle of incidence as the beam. Knowing the angle of incidence in air for the blue light and the indices of refraction, we can determine the angle of refraction in glass for blue light (θ_{rgb}). We can determine the angular separation of the red and blue light from their respective angles of refraction in glass. We can determine the angles of incidence at the second boundary by geometry. Finally, we can determine the critical angles for the red and blue light and compare them to the respective angles of incidence to see if refraction or internal reflection takes place at the second glass-air boundary.

Solution: Looking at Figure 35.4, we see that the angle of refraction in glass for red light (θ_{rgr}) is 30.0°. Knowing this, we can determine the angle of incidence in air for red light (θ_{iar}) and hence the angle of incidence in air (θ_{ia}) for the beam.

$$n_a \sin\theta_{iar} = n_{gr} \sin\theta_{rgr}$$

$$\theta_{iar} = \sin^{-1}(n_{gr} \sin\theta_{rgr}/n_a) = \sin^{-1}(0.730) = 46.9°$$

$$\theta_{ia} = \theta_{iab} = \theta_{iar} = 46.9°$$

The angle of refraction in glass for blue light may now be determined.

$$n_a \sin\theta_{iab} = n_{gb} \sin\theta_{rgb}$$

$$\theta_{rgb} = \sin^{-1}(n_a \sin\theta_{iab}/n_{gb}) = \sin^{-1}(0.497) = 29.8°$$

The angular separation of the red and blue light in the glass is

$$\Delta\theta = \theta_{rgr} - \theta_{rgb} = 0.200°$$

From Figure 35.5 we can determine the angles of incidence in glass at the second glass-air boundary.

$$\theta_{igr} = 90.0° - 30.0° = 60.0° \qquad \theta_{igb} = 90.0° - 29.8° = 60.2°$$

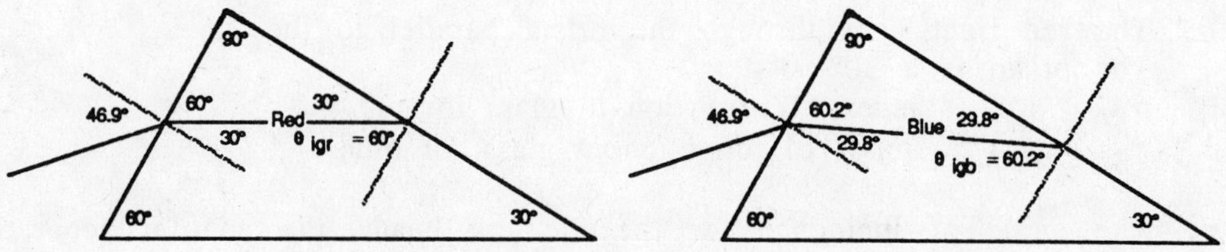

Figure 35.5

The critical angles can now be determined.

$$n_{gr} \sin\theta_{cr} = n_a \sin 90° \quad \text{and} \quad n_{gb} \sin\theta_{cb} = n_a \sin 90°$$

$$\theta_{cr} = \sin^{-1}(1/n_{gr}) \qquad \theta_{cb} = \sin^{-1}(1/n_{gb})$$

$$\theta_{cr} = 43.2° \qquad \theta_{cb} = 42.9°$$

Since $\theta_{igr} > \theta_{cr}$ and $\theta_{igb} > \theta_{cb}$, the red and blue light rays experience total internal reflection at the second glass-air boundary.

Related Text Exercises: Ex. 35-7 through 35-15.

2 Mirrors and Images Formed by Reflection (Section 35-2)

Review: Figure 35.6 reviews reflection and image formation for a planar mirror.

$$\theta_r = \theta_i$$
$$h' = h$$
$$s' = -s$$
$$m = \frac{h'}{h} = \frac{-s'}{s} = 1$$

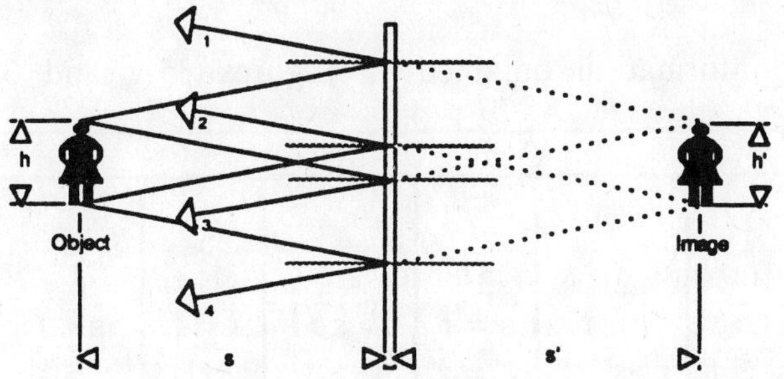

Figure 35.6

We see that, for all rays (only four are shown), the angle of reflection is equal to the angle of incidence. The image size is equal to the object size. The image distance is equal to the object distance. The image is erect (has the same orientation as the object). The image is virtual (the reflected rays do not go through the image).

Figure 35.7 reviews reflection and image formation for a concave mirror.

$$f = \frac{r}{2}$$
$$\frac{1}{s} + \frac{1}{s'} = \frac{1}{f}$$
$$m = \frac{h'}{h} = \frac{-s'}{s}$$

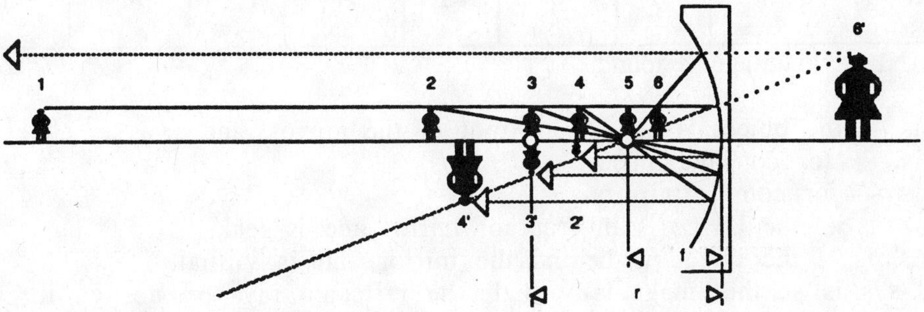

Figure 35.7

Figure 35.8 reviews reflection and image formation for a convex mirror.

$$f = \frac{r}{2}$$
$$\frac{1}{s} + \frac{1}{s'} = \frac{1}{f}$$
$$m = \frac{h'}{h} = \frac{-s'}{s}$$

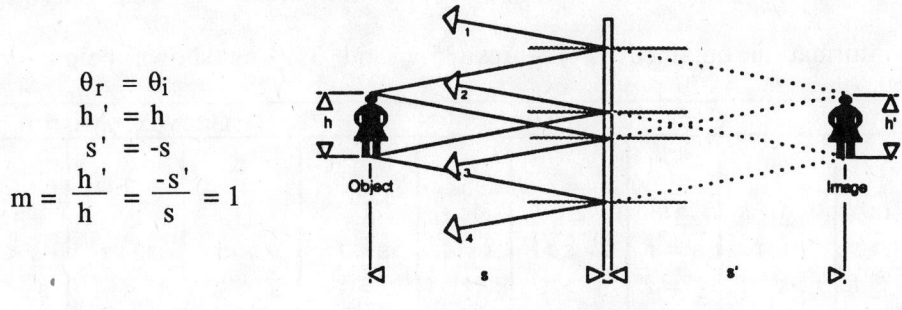

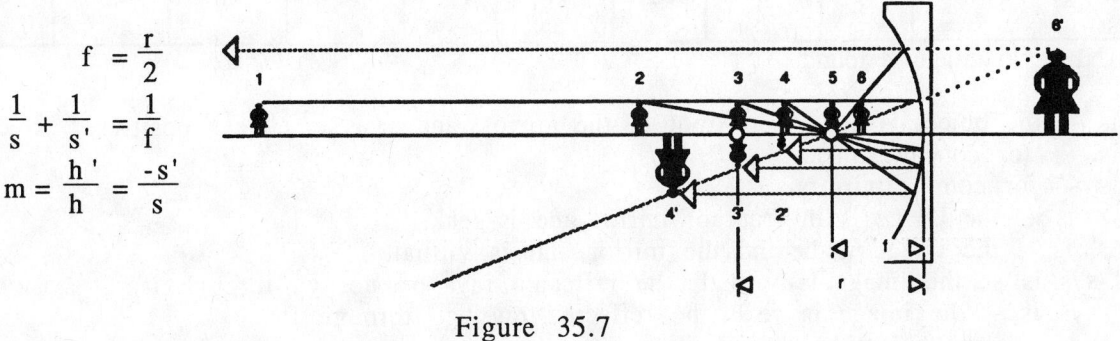

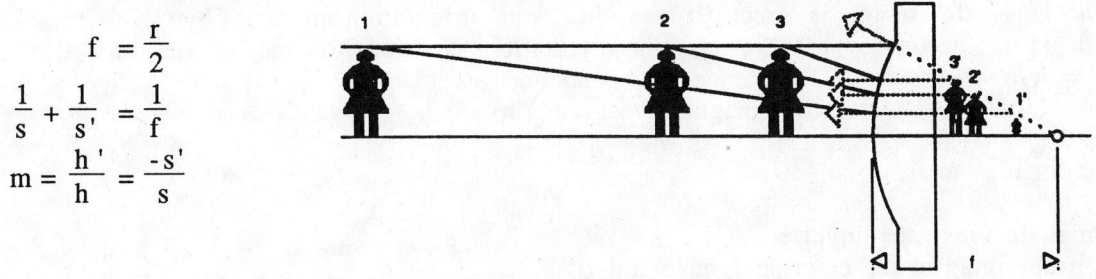

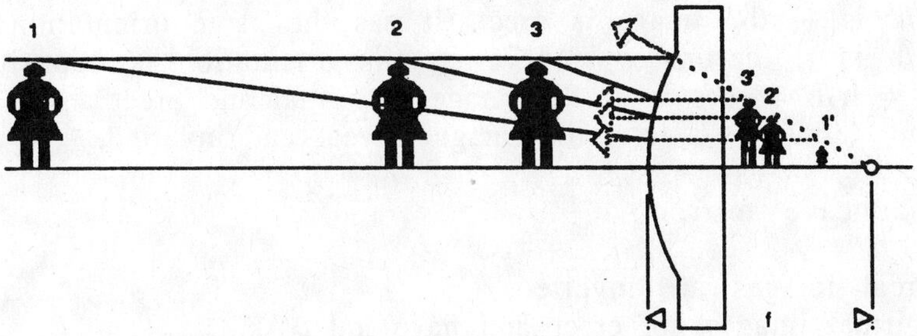

Figure 35.8

Note: The ray diagrams in Figures 35.7 and 35.8 are easy to construct and contain a large quantity of information. In fact, the table, the sign conventions, and the summary statements that follow are all obtained from these two figures. Obviously, you can't hope to memorize all of this information. You have no choice but to learn how to do ray diagrams like those shown in Figures 35.7 and 35.8 and to study them until you can write down all of this information from the diagrams.

A summary of the information obtained in Figures 35.7 and 35.8 is shown below.

	Concave Mirror						Convex Mirror		
Object number	1	2	3	4	5	6	1	2	3
Image number	1'	2'	3'	4'	NI*	6'	1'	2'	3'
Object location	$s \gg r$	$s > r$	$s = r$	$r > s > f$	$s = f$	$s < f$	$s \gg r$	$s \approx r$	$s < r$
Image location	$s' \gtrsim f$	$r > s' > f$	$s' = r$	$s' > r$	NI	$s' > -s$	$s' < -f$	$s' < -f$	$s' < -f$
Sign of s	+	+	+	+	+	+	+	+	+
Sign of s'	+	+	+	+	NI	-	-	-	-
Sign of -s'/s	-	-	-	-	NI	+	+	+	+
$\lvert s'/s \rvert$	$\ll 1$	< 1	$= 1$	> 1	NI	> 1	< 1	< 1	< 1
Sign of h	+	+	+	+	+	+	+	+	+
Sign of h'	-	-	-	-	NI	+	+	+	+
Sign of h'/h	-	-	-	-	NI	+	+	+	+
$\lvert h'/h \rvert$	< 1	< 1	$= 1$	> 1	NI	> 1	< 1	< 1	< 1
Sign of m	-	-	-	-	NI	+	+	+	+
$\lvert m \rvert$	$\ll 1$	< 1	$= 1$	> 1	NI	> 1	< 1	< 1	< 1

* NI = No image is found.

s is +, the object is always in front of the mirror, and so s is always positive.
f is + for concave mirrors.
f is - for convex mirrors.
If s' is +, the image is in front of mirror and is real.
If s' is -, the image is behind the mirror and is virtual.
If -s'/s is +, the image is virtual (the reflected rays do not go through it).
If -s'/s is -, the image is real (the reflected rays go through it).
If h is + , the object height is above the principal axis.
if h is -, the object height is below the principal axis.
If h' is +, the image height is above the principal axis.
If h' is -, the image height is below the principal axis.
If h'/h is +, the image is erect (it has the same orientation as the object).
If h'/h is -, the image is inverted (its orientation is opposite that of the object).
If $m = h'/h = -s'/s$ is +, the image is virtual and erect.
If $m = h'/h = -s'/s$ is -, the image is real and inverted.

For concave mirrors:

All real images are inverted.
All virtual images are erect and have $|m| > 1$.
If $s > r$, then $r > s' > f$, $|m| < 1$, and the image is real and inverted.
If $s = r$, then $s' = r$, $|m| = 1$, and the image is real and inverted.
If $r > s > f$, then $s' > r$, $|m| > 1$, and the image is real and inverted.
If $s = f$, no image is found.
If $f > s > 0$, then $s' > -s$, $|m| > 1$, and the image is virtual and erect.

For convex mirrors:

All images are virtual.
All images are erect.
All images are formed inside the focal length.
All images have $|m| < 1$.

Practice: Figure 35.9 shows objects in front of planar, concave and convex mirrors. Information is also given about focal lengths, object size and object distance.

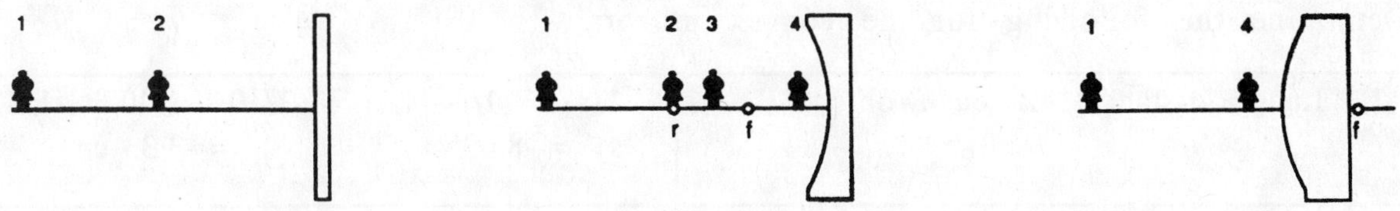

(a) Planar mirror (b) Concave mirror (c) Convex mirror

$|f| = 10.0$ cm
$h_1 = h_2 = h_3 = h_4 = 2.00$ cm
$s_1 = 40.0$ cm, $s_2 = 20.0$ cm, $s_3 = 15.0$ cm, $s_4 = 5.0$ cm

Figure 35.9

Determine the following for the planar mirror:

1. Image distance for each object	$s_1' = -s_1 = -40.0$ cm $s_2' = -s_2 = -20.0$ cm
2. Magnification of each object	$m_1 = -s_1'/s_1 = +1.00$ $m_2 = -s_2'/s_2 = +1.00$
3. Size and orientation of each image	$h_1' = m_1h_1 = +2.00$ cm $h_2' = m_2h_2 = +2.00$ cm + means erect
4. Whether the image is real or virtual for each object	Both images are virtual (the reflected rays do not pass through them).

Determine the following for the concave mirror:

1. Image distance for each object	$1/s_1' = 1/f - 1/s_1 = 1/10 - 1/40 = 3/40$ $s_1' = 40/3 = 13.3$ cm $s_2' = 20.0$ cm, $s_3' = 30.0$ cm, $s_4' = -10.0$ cm
2. Magnification of each object	$m_1 = -s_1'/s_1 = \frac{(-13.3 \text{ cm})}{(40.0 \text{ cm})} = -0.333$ $m_2 = -1.00$, $m_3 = -2.00$, $m_4 = +2.00$

3. Size and orientation of each image	$h_1' = m_1 h_1 = -0.333 \times 2.00$ cm $= -0.666$ cm $h_2' = -2.00$ cm, $h_3' = -4.00$ cm, $h_4' = +4.00$ cm + means erect and - means inverted
4. Whether the image is real or virtual for each object	The image is real for objects 1, 2 and 3 and virtual for object 4.

Determine the following for the convex mirror:

1. Image distance for each object	$1/s_1' = 1/f - 1/s_1 = -1/10 - 1/40 = -1/8$ $s_1' = -8.00$ cm, $s_4' = -3.33$ cm
2. Magnification of each object	$m_1 = -s_1'/s_1 = +0.200$ $m_4 = -s_4'/s_4 = +0.667$
3. Size and orientation of each image	$h_1' = m_1 h_1 = +0.400$ cm $h_2' = m_2 h_2 = +1.33$ cm, + means erect
4. Whether the image is real or virtual for each object	The image is virtual for both objects.

Example: Two students have identical concave mirrors with 30.0-cm focal lengths. Each student places an object in front of her mirror and reports that the image is three times the size of the object. However, when they compare data, they find that their object distances are not the same. What are the object distances?

Given: $f = 30.0$ cm = focal length for each mirror
Image size is three times object size for each case

Determine: Object distance for each case

Strategy: Figure 35.10 shows the circumstances that allow this situation.

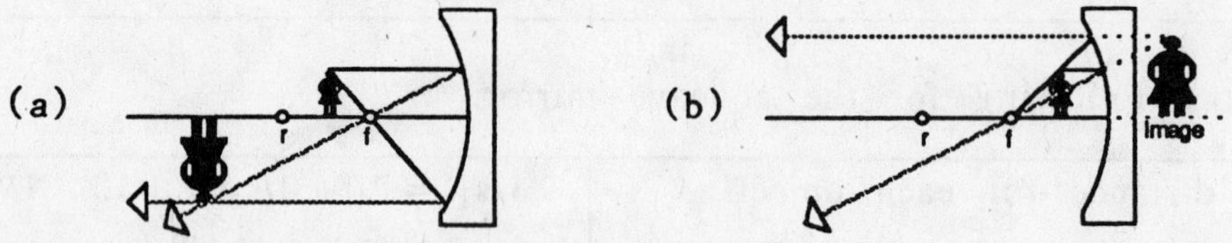

Figure 35.10

We can use the image size and orientation to obtain the magnification. We can use our knowledge of magnification to eliminate the image distance in the mirror equation. Finally, we can solve for the object distance in terms of the focal length.

Solution:

Case a

$h' = -3h$ (image is inverted)

$m = h'/h = -3$

$s' = -ms = +3s$

$$\frac{1}{s} + \frac{1}{s'} = \frac{1}{s} + \frac{1}{3s} = \frac{1}{f}$$

$s = 4f/3 = 40.0$ cm

Case b

$h' = +3h$ (image is erect)

$m = h'/h = +3$

$s' = -ms = -3s$

$$\frac{1}{s} + \frac{1}{s'} = \frac{1}{s} - \frac{1}{3s} = \frac{1}{f}$$

$s = 2f/3 = 20.0$ cm

Related Text Exercises:

Plane mirrors: Ex. 35-16, 17 and 18
Concave mirrors: Ex. 35-20, 22 and 23
Convex mirrors: Ex. 35-19, 21, 22 and 23.

3 Images Formed by Refraction (Section 35-3)

Review: If an object is in front of a refracting medium interface as shown in Figure 35.11, its image may be determined by

$$\frac{n_1}{s} + \frac{n_2}{s'} = \frac{n_2 - n_1}{r}$$

Where: n_1 = the index of refraction of the medium the object is in
n_2 = the index of refraction of the second medium
s = the object distance - the distance of the object from the medium interface. s has a positive value and is in medium 1.
s' = the image distance - the distance of the image from the medium interface
= + if the image (real) is in medium 2
= - if the image (virtual) is in medium 1
r = radius of curvature of the medium interface
= + if the surface curves away from the object
= - if the surface curves toward the object

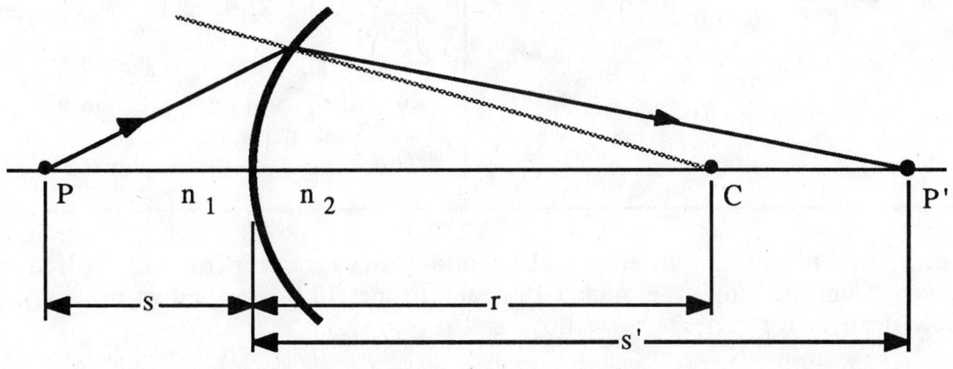

Figure 35.11

Practice: Consider the three situations (A, B and C) shown in Figure 35.12. The radius associated with the curved surface in case A and B is 1.00 m.

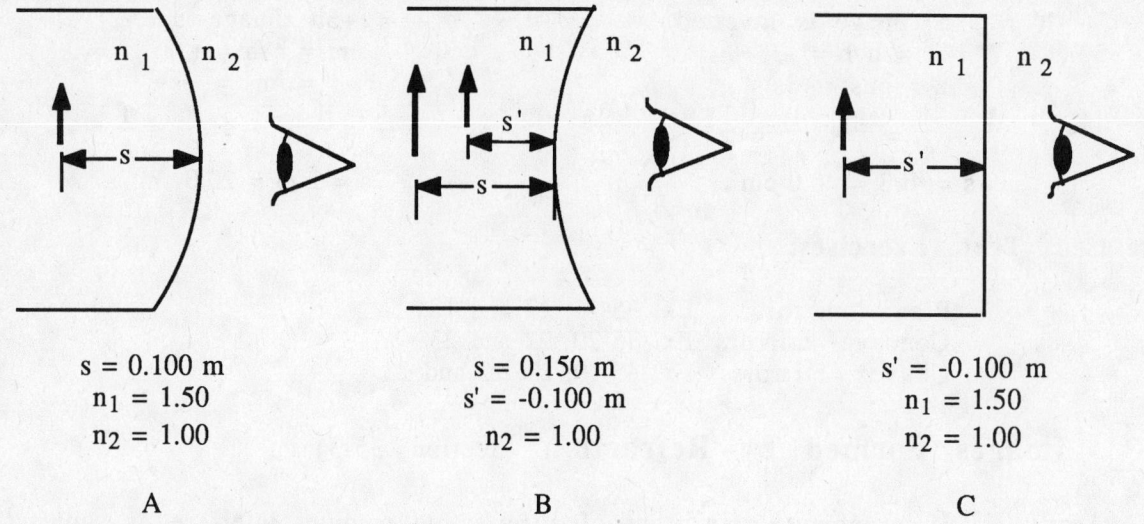

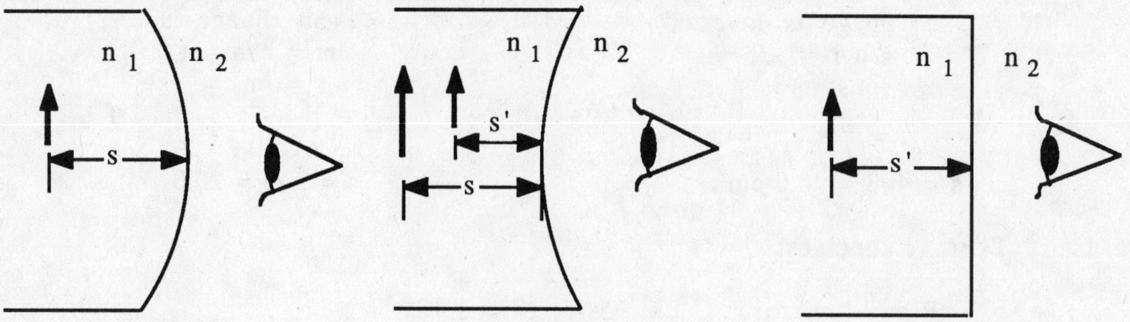

A	B	C
$s = 0.100$ m $n_1 = 1.50$ $n_2 = 1.00$	$s = 0.150$ m $s' = -0.100$ m $n_2 = 1.00$	$s' = -0.100$ m $n_1 = 1.50$ $n_2 = 1.00$

Figure 35.12

Determine the following:

1. Location of the image for case A	$\frac{n_1}{s} + \frac{n_2}{s'} = \frac{n_2 - n_1}{r}$, $n_1 = 1.50$, $n_2 = 1.00$, $r = -1.00$ m, $s = 0.100$ m Inserting values, we obtain $s' = -6.90 \times 10^{-2}$ m The image is formed in medium 1.
2. The index of refraction for medium 1 of case B	$\frac{n_1}{s} + \frac{n_2}{s'} = \frac{n_2 - n_1}{r}$, $n_2 = 1.00$, $s = 0.150$ m, $s' = -0.100$ m, $r = +1.00$ m Inserting values, we obtain $n_1 = 1.43$
3. The object distance for case c	$\frac{n_1}{s} + \frac{n_2}{s'} = \frac{n_2 - n_1}{r}$, $n_1 = 1.50$, $n_2 = 1.00$, $r = \infty$, $s' = -0.100$ m Inserting values, we obtain $s = 0.150$ m The image is in medium 1.

Example: A nickel is placed on the bottom of a beaker partially filled with water. When viewed from the top, the nickel appears to be 10.0 cm below the surface of the water. How deep is the water? $n_a = 1.00$ and $n_w = 1.33$

Given: $n_a = 1.00$ $n_w = 1.33$ the apparent depth of the nickel = 10.0 cm

Determine: The depth of the water.

Strategy: First we need to recognize that $r = \infty$ and $s' = -10.0$ cm for this situation. Knowing n_1, n_2, r and s' we can determine s.

Solution:
$$\frac{n_1}{s} + \frac{n_2}{s'} = \frac{n_2 - n_1}{r}$$

Insert values to obtain
$$\frac{1.33}{s} + \frac{1.00}{-10.0 \text{ cm}} = \frac{1.00 - 1.33}{\infty} = 0$$

$$s = \frac{(1.33)(10.0 \text{ cm})}{1.00} = 13.3 \text{ cm}$$

The water is 13.3 cm deep.

Related Text Exercises: Ex. 35-24, 25 and 26.

4 Thin Lenses and Image Formation (Section 35-4)

Review: A lens causes light to be refracted twice, once as it enters the lens (air to glass interface) and a second time as it exits the lens (glass to air interface). Figure 35.13 shows light entering, traveling through and exiting a lens. Let's agree to call the first interface (air to glass) surface "a" and the second interface (glass to air) surface "b".

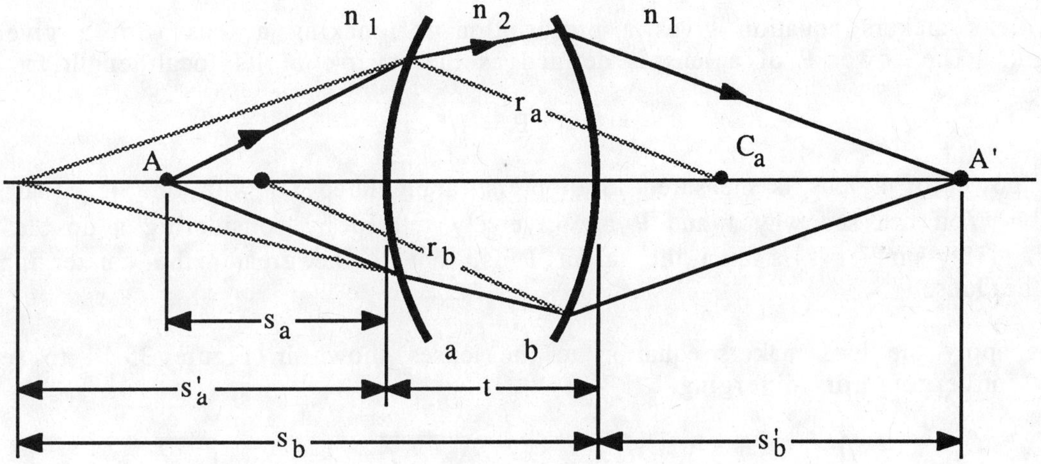

Figure 35.13

For surface a, as light enter the lens we may write

(1)
$$\frac{n_1}{s_a} + \frac{n_2}{s_a'} = \frac{n_2 - n_1}{r_a}$$

r_a = + if the first surface curves away from the object
r_a = - if the first surface curves toward the object

For surface b, as light exits the lens we may write

(2) $$\frac{n_2}{s_b} + \frac{n_1}{s_b'} = \frac{n_1 - n_2}{r_b}$$

r_b = + if the second surface curves away from the object
r_b = - if the second surface curves toward the object

Notice that $s_b = -s_a' + t \approx -s_a'$ if the lens is thin. Equations (1) and (2) may be combined to obtain

(3) $$\frac{n_1}{s_a} + \frac{n_1}{s_b'} = (n_2 - n_1)(\frac{1}{r_a} - \frac{1}{r_b})$$

Now let's agree to let the object distance for the lens-as-whole be $s = s_a$ and the image distance for the lens-as-whole be $s' = s_b'$. This allows us to rewrite equation (3) as

(4) $$\frac{n_1}{s} + \frac{n_1}{s'} = (n_2 - n_1)(\frac{1}{r_a} - \frac{1}{r_b})$$

If we also agree to let the focal length of the lens-as-whole be f and recall that $s' = f$ when $s = \infty$, we get the lens makers equation:

(5) $$\frac{n_1}{f} = (n_2 - n_1)(\frac{1}{r_a} - \frac{1}{r_b})$$

The lens makers equation gives a prescription for making a lens with a given focal length. The power P of a lens is defined as the inverse of its focal length f:

$$P = 1/f$$

The power of a lens is measured in diopters (abbreviated D) with one D equal to one m^{-1}. You can see why f and P are inversely related by considering a double convex lens. The smaller f is then the larger P is, that is the greater the converging power of the lens.

Let's apply the lens makers equation to the lenses shown in Figure 35.14 to see if they are converging or diverging.

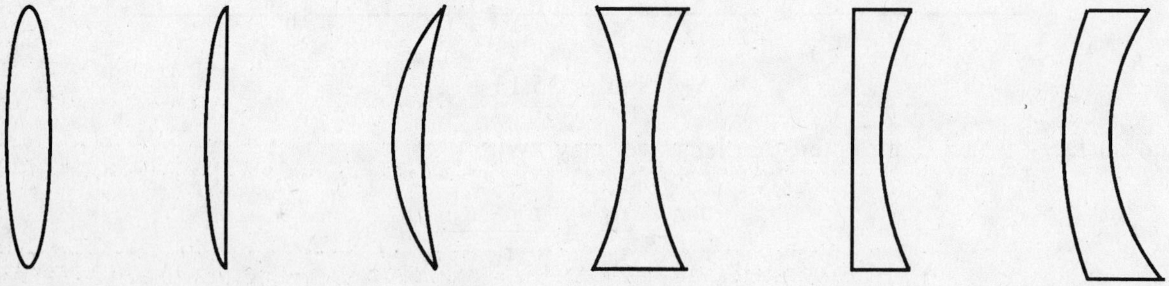

(A)	(B)	(C)	(D)	(E)	(F)
$\lvert r_a\rvert = \lvert r_b\rvert$	$r_a < r_b = \infty$	$r_a < r_b$	$\lvert r_a\rvert = \lvert r_b\rvert$	$\infty = r_a > r_b$	$r_a > r_b$

Figure 35.14

For all cases the object is considered to be to the left of the lens, hence the left side of the lens is surface a and the right side is surface b. Also for all cases the sign of $n_2 - n_1$ is positive. The following table summarizes the sign convention.

Case	Sign of r_a	Sign of r_b	Sign of $\left(\frac{1}{r_a} - \frac{1}{r_b}\right)$	Sign of f	Type lens
A	+	-	+	+	converging
B	+	+	+	+	converging
C	+	+	+	+	converging
D	-	+	-	-	diverging
E	+	+	-	-	diverging
F	+	+	-	-	diverging

When equations (4) and (5) are combined we get the thin lens equation

$$\frac{1}{s} + \frac{1}{s'} = \frac{1}{f}$$

Figure 35.15 reviews image formation for a convex (converging) lens. Note that the image is located by tracing two rays from the tip of the object. One ray is incident parallel to the optical axis and exits the lens toward the focal point. The other ray is incident through the focal point and exits the lens parallel to the optical axis.

$$f = r/2$$

$$\frac{1}{s} + \frac{1}{s'} = \frac{1}{f}$$

$$m = \frac{h'}{h} = \frac{-s'}{s}$$

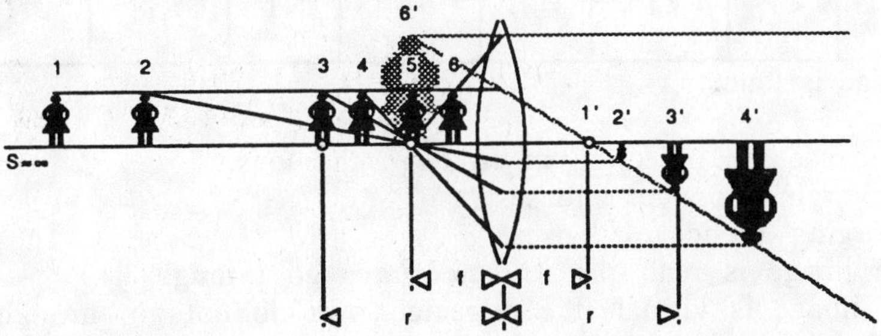

s = object distance
s' = image distance

f = focal length
r = lens radius of curvature

Figure 35.15

Figure 35.16 reviews image formation for a concave (diverging) lens. One ray is incident parallel to the optical axis and exits as though it came from the focal point on the incident side. The other ray is incident toward the focal point on the exiting side but exits parallel to the optical axis.

$$f = r/2$$

$$\frac{1}{s} + \frac{1}{s'} = \frac{1}{f}$$

$$m = \frac{h'}{h} = \frac{-s'}{s}$$

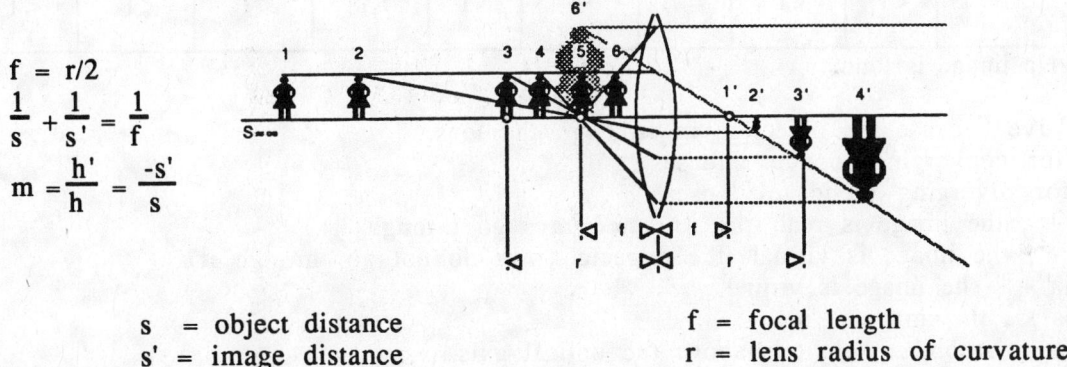

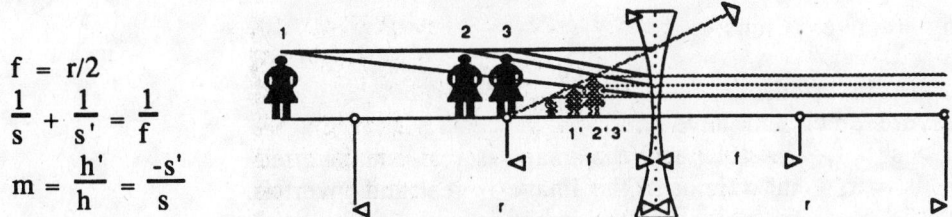

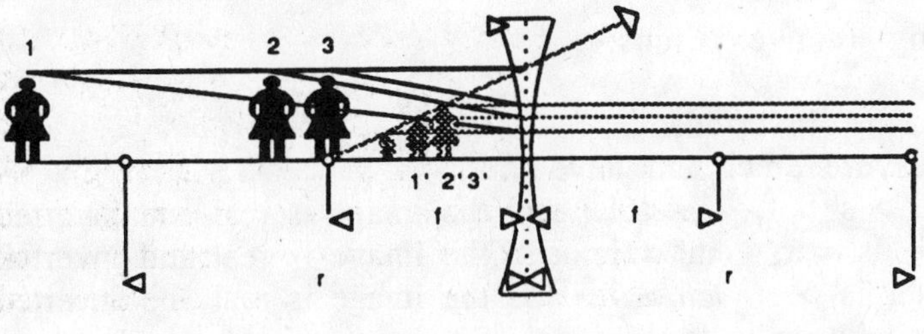

Figure 35.16

Note: The ray diagrams in Figures 35.15 and 35.16 are easy to construct and contain much information. In fact, the table, sign convention, and summary statements that follow are all obtained from these two figures. Obviously, you can't hope to memorize all this information. You have no choice but to learn how to do ray diagrams like those shown in Figures 35.15 and 35.16 and to study them until you can write down all of this information from the diagrams.

A summary of the information obtained in Figures 35.15 and 35.16 is shown below.

	Convex Lens						Concave Lens		
Object number	1	2	3	4	5	6	1	2	3
Image number	1'	2'	3'	4'	NI*	6'	1'	2'	3'
Object location	$s >> r$	$s > r$	$s = r$	$r > s > f$	$s = f$	$s < f$	$s > r$	$r > s > f$	$s = f$
Image location	$s' \gtrsim f$	$r > s' > f$	$s' = r$	$s' > r$	NI	$-s' > s$	$-s' < f$	$-s' < f$	$-s' < f$
Sign of s	+	+	+	+	+	+	+	+	+
Sign of s'	+	+	+	+	NI	-	-	-	-
Sign of -s'/s	-	-	-	-	NI	+	+	+	+
$\lvert s'/s \rvert$	<1	<1	=1	>1	NI	>1	<1	<1	<1
Sign of h	+	+	+	+	+	+	+	+	+
Sign of h'	-	-	-	-	NI	+	+	+	+
Sign of h'/h	-	-	-	-	NI	+	+	+	+
$\lvert h'/h \rvert$	<1	<1	=1	>1	NI	>1	<1	<1	<1
Sign of m	-	-	-	-	NI	+	+	+	+
$\lvert m \rvert$	<1	<1	=1	>1	NI	>1	<1	<1	<1

* NI = No image is found.

s is positive because the object is in front of the lens.
f is + for converging (convex) lenses.
f is - for diverging (concave) lenses.
If s' is + , the image is real (the refracted rays go through it).
If s' is - , the image is virtual (the refracted rays do not go through it).
If -s'/s is + , the image is virtual.
If -s'/s is - , the image is real.
If h is + , the object height is above the optical axis.
If h is - , the object height is below the optical axis.
If h' is + , the image height is above the optical axis.
If h' is - , the image height is below the optical axis.
If h'/h is + , the image is erect.
If h'/h is - , the image is inverted.
If $m = h'/h = -s'/s$ is + , the image is virtual and erect.
If $m = h'/h = -s'/s$ is - , the image is real and inverted.

For a converging (convex) lens:

All real images are inverted.
All virtual images are erect and have $|m| > 1$.
If $s > r$, then $r > s' > f$, $|m| < 1$, and the image is real and inverted.
If $s = r$, then $s' = r$, $|m| = 1$, and the image is real and inverted.
If $r > s > f$, then $s' > r$, $|m| > 1$, and the image is real and inverted.
If $s = f$, then no image is found.
If $f > s > 0$, then $s' > -s$, $|m| > 1$, and the image is virtual and erect.

For a diverging (concave) lens:

All images are virtual.
All images are erect.
All images are formed inside the focal length.
All images have $|m| < 1$.

Practice: Figure 35.17 shows objects in front of converging and diverging lenses. Information is also given about focal lengths, object size and object distance.

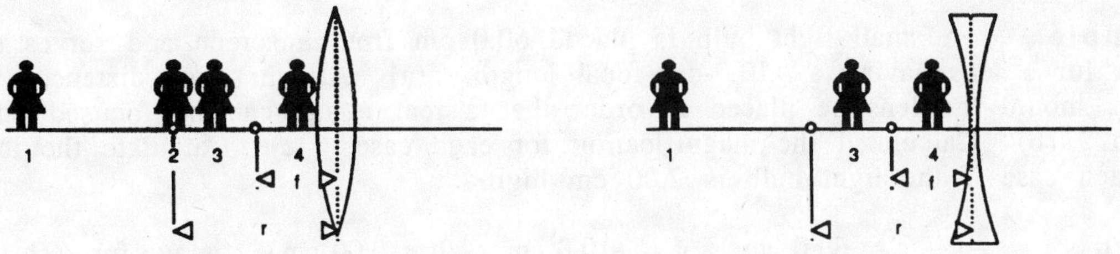

Converging lens

$|f| = 10.0$ cm

$s_1 = 40.0$ cm $\quad s_2 = 20.0$ cm

Diverging lens

$h_1 = h_2 = h_3 = h_4 = 2.00$ cm

$s_3 = 15.0$ cm $\quad s_4 = 5.0$ cm

Figure 35.17

Determine the following for the converging lens:

1. Image distance for each object	$1/s_1' = 1/f - 1/s = 1/10 - 1/40 = 3/40$ $s_1' = 40/3 = 13.3$ cm, $s_2' = 20.0$ cm, $s_3' = 30.0$ cm, $s_4' = -10.0$ cm
2. Magnification for each object	$m_1 = -s_1'/s_1 = -13.3$ cm/40.0 cm $= -0.333$ $m_2 = -1.00$, $m_3 = -2.00$, $m_4 = +2.00$
3. Size and orientation of each image	$h_1' = m_1 h_1 = -0.666$ cm, $h_2' = -2.00$ cm $h_3' = -4.00$ cm, $h_4' = +4.00$ cm + means erect and - means inverted
4. Whether the image is real or virtual for each object	The image is real for objects 1, 2 and 3 and virtual for object 4.

Determine the following for the diverging lens:

5. Image distance for each object	$1/s_1' = 1/f - 1/s_1 = -1/10 - 1/40 = -1/8$ $s_1' = -8.00$ cm, $s_3' = -6.00$ cm, $s_4' = -3.33$ cm

6. Magnification for each object	$m_1 = -s_1'/s_1 = +0.200$ $m_3 = +0.400, \quad m_4 = +0.666$
7. Size and orientation of each image	$h_1' = m_1 h_1 = +0.400, \quad h_3' = +0.800$ cm, and $h_4' = +\ 1.33$ cm, + means erect.
8. Whether the image is real or virtual	All images for diverging (concave) lenses are virtual.

Example: A small light bulb is placed 60.0 cm from a screen and serves as an object for a lens having a +10.0-cm focal length. (a) At what two distances from the screen should the lens be placed in order that a real image can be focused on the screen? (b) Calculated the magnification for each case. (c) Calculate the image size for each case if the light bulb is 2.00 cm high.

Given: $s + s' = 60.0$ cm, $f = +10.0$ cm, $h = 2.00$ cm, image for each case is real

Determine: (a) The two locations of the lens relative to the screen that allow a real image. (b) The magnification for each case (m_1 and m_2). (c) The image size for each case (h_1' and h_2').

Strategy: (a) Since the images are formed on the screen, the two distances of interest (i.e. the distance from the lens to the screen) are the image distances s_1' and s_2'. For this problem, the lens equation amounts to one equation and two unknowns (s and s'). The relationship $s + s' = 60.0$ cm also amounts to one equation and the same two unknowns. We can solve these two equations simultaneously to obtain s and s'. The solution involves a quadratic, the two roots of which are the two possible distances of the lens from the screen. (b) Knowing the object and image distances, we can determine the magnification. (c) Knowing the magnification and the object size, we can determine the image size.

Solution:

(a) We can write the following two equations involving the unknowns s and s'.

$$s + s' = 60.0 \text{ cm} \qquad\qquad 1/s + 1/s' = 1/f$$

When these equations are combined, we obtain

$$\frac{1}{60 - s'} + \frac{1}{s'} = \frac{1}{10} \qquad\qquad s'^2 - 60s' + 600 = 0$$

Solving this quadratic in s', we obtain

$$s' = \frac{-(-60) \pm [(-60)^2 - 4(600)]^{1/2}}{2} = \frac{60 \pm (1200)^{1/2}}{2} = 47.3 \text{ cm} \quad \text{and} \quad 12.7 \text{ cm}$$

The two possible cases are as follows:

Case I. $s_1' = 47.3$ cm, $s_1 = 12.7$ cm   Case II. $s_2' = 12.7$ cm, $s_2 = 47.3$ cm

(b) The magnification for each case is

Case I. $m_1 = -s_1'/s_1 = -3.72$ Case II. $m_2 = -s_2'/s_2 = -0.268$

(c) The size of the image for each case is

Case I. $h_1' = m_1h_1 = -7.44$ cm   Case II. $h_2' = m_2h_2 = -0.536$

Related Text Exercises:

Lens makers equation: Ex. 35-29 through 35-33,
Converging lenses: Ex. 35-27, 34, 35, 36 and 39,
Diverging lenses: Ex. 35-27, 36 and 38.

PRACTICE TEST

Take and grade this practice test. Doing so will allow you to determine any weak spots in your understanding of the concepts taught in this chapter. The following section prescribes what you should study further to strengthen your understanding.

A block of clear glass with opposite faces parallel is placed successively in various transparent, colorless liquids referred to as liquids 1, 2 and 3. The refractive indices are: air = 1.00, glass = 1.50, liquid 1 = 1.30, liquid 2 = 1.50, liquid 3 = 1.70.

Figure 35.18 shows five situations involving the glass and liquids.

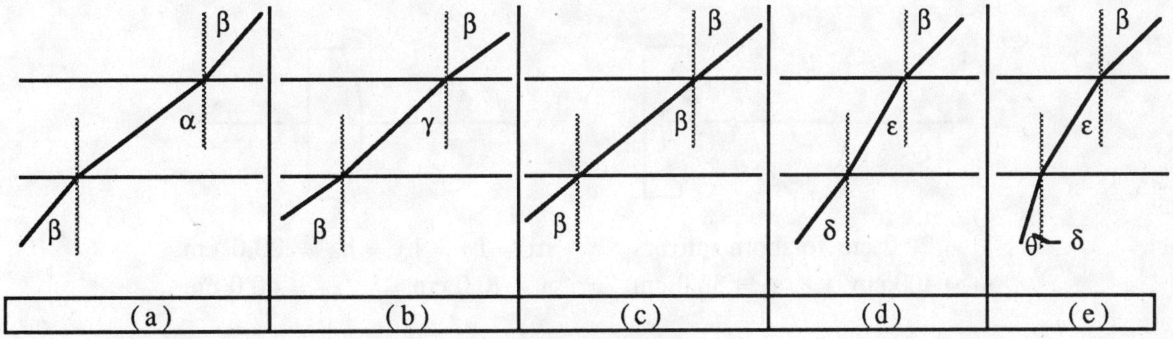

$\alpha > \beta > \gamma > \delta > \varepsilon > \theta$

Figure 35.18

For each of the situations given below, select the diagram that best represents the path of a ray passing through the glass under the conditions described.

_________ 1. The block is submerged in liquid 1
_________ 2. The block is submerged in liquid 2
_________ 3. The block is submerged in liquid 3
_________ 4. The block half submerged in liquid 1 (the top half is in air)
_________ 5. The block half submerged in liquid 3 (the top half is in air)

Figure 35.19 shows a ball bearing embedded in a block of ice. The ray incident on the block undergoes total internal reflection at the second ice-air boundary.

Figure 35.19

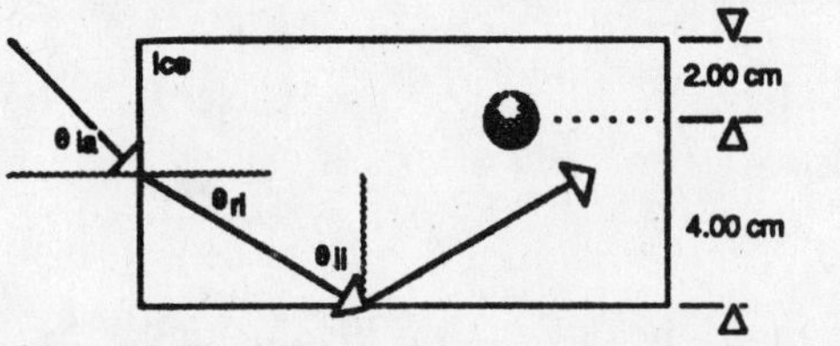

n_i = 1.31 = index of refraction of ice
θ_{ri} = angle of refraction in ice
θ_{ia} = angle of incidence in air
θ_c = critical angle for ice relative to air
$\theta_{ii} = \theta_c + 2.20°$ = angle of incidence in ice

Determine the following:

_________ 6. θ_c
_________ 7. θ_{ri}
_________ 8. θ_{ia}
_________ 9. Apparent depth of the ball bearing when viewed from the top
_________ 10. Speed of light in the ice

Figure 35.20 shows objects in front of concave and convex mirrors.

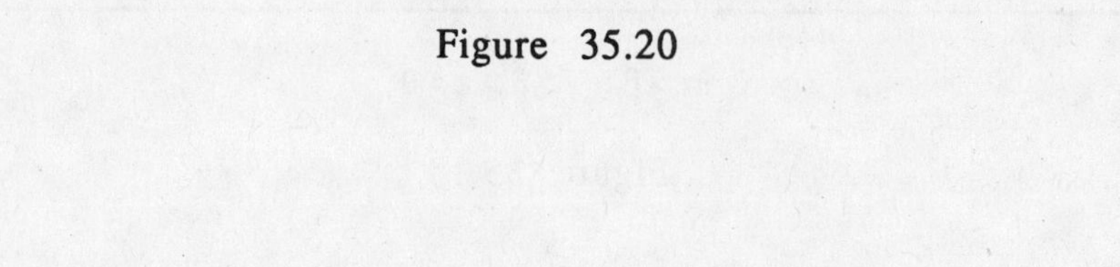

$|f|$ = 20.0 cm for both mirrors $h_1 = h_2 = h_3 = h_4$ = 20.0 cm
s_1 = 100 cm s_2 = 30.0 cm s_3 = 10.0 cm s_4 = 40.0 cm

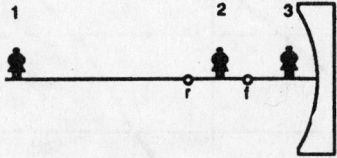

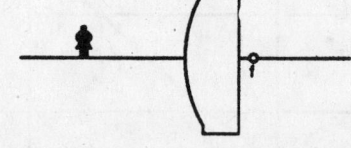

Figure 35.20

Determine the following:

______ 11. Location of the image of object 1
______ 12. Location of the image of object 2
______ 13. Location of the image of object 3
______ 14. Location of the image of object 4
______ 15. Magnification of object 1
______ 16. Magnification of object 3
______ 17. Size and orientation of the image of object 2
______ 18. Size and orientation of the image of object 4
______ 19. Whether the image of object 3 is real or virtual
______ 20. Whether the image of object 4 is real or virtual

Figure 35.21 shows objects in front of converging and diverging lenses.

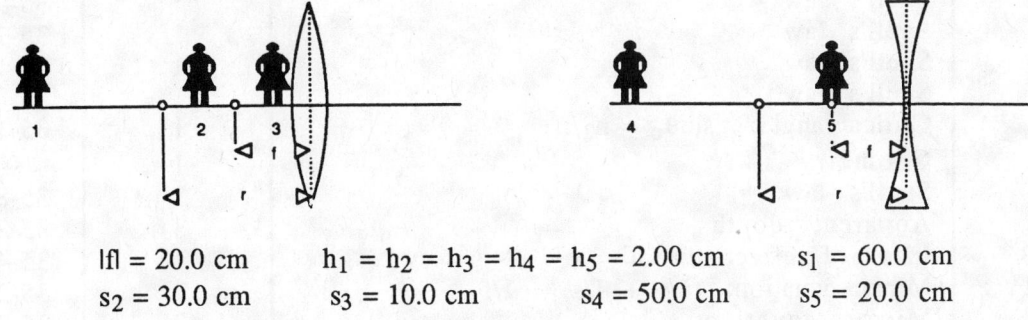

$|f| = 20.0$ cm $h_1 = h_2 = h_3 = h_4 = h_5 = 2.00$ cm $s_1 = 60.0$ cm
$s_2 = 30.0$ cm $s_3 = 10.0$ cm $s_4 = 50.0$ cm $s_5 = 20.0$ cm

Figure 35.21

Determine the following:

______ 21. Image distance for object 1
______ 22. Image distance for object 4
______ 23. Magnification of the image of object 2
______ 24. Magnification of the image of object 5
______ 25. Size and orientation of the image of object 3
______ 26. Size and orientation of the image of object 5
______ 27. Objects with virtual images

(See Appendix I for answers.)

PRINCIPAL CONCEPTS AND EQUATIONS PRESCRIPTION

Your score on the practice test is an excellent measure of your understanding of the chapter. You should now use the following chart to write your own prescription for dealing with any weaknesses the practice test points out. Look down the leftmost column to the number of the question(s) you answered incorrectly, reading across that row you will find the concept and/or equation of concern, the section(s) of the study guide you should return to for further study, and some suggested text exercises which you should work to gain additional experience.

Practice Test Questions	Concepts and Equations	Prescription	
		Principal Concepts	Text Exercises
1	Snell's law: $\sin\theta_i/\sin\theta_r = n_r/n_i$	1	35-8, 9
2	Snell's law	1	35-9, 10
3	Snell's law	1	35-10, 11
4	Snell's law	1	35-11, 12
5	Snell's law	1	35-8, 14
6	Critical angle: $\sin\theta_c = n_2/n_1$	1	35-13, 15
7	Geometry	1	---
8	Snell's law	1	35-8, 9
9	Apparent depth	3	35-24, 25
10	Index of refraction: $n = c/v$	1	35-7
11	Mirror equation: $1/s + 1/s' = 1/f$	2	35-20, 22
12	Mirror equation	2	35-20, 22
13	Mirror equation	2	35-20, 22
14	Mirror equation	2	35-19, 21
15	Magnification: $m = -s'/s$	2	35-20, 22
16	Magnification	2	35-20, 22
17	Image size: $m = h'/h$	2	35-20, 22
18	Image size	2	35-20, 22
19	Real vs virtual	2	35-20, 22
20	Real vs virtual	2	35-20, 22
21	Thin lens equation: $1/s + 1/s' = 1/f$	4	35-27, 34
22	Thin lens equation	4	35-27, 31
23	Magnification: $m = -s'/s$	4	35-34, 35
24	Magnification	4	35-36, 38
25	Image size: $m = h'/h$	4	35-36, 39
26	Image size	4	35-36, 38
27	Virtual images	4	35-27

36 INTERFERENCE AND DIFFRACTION

RECALL FROM PREVIOUS CHAPTERS

Previously learned concepts and equations frequently used in this chapter	Text Section	Study Guide Page
Index of refraction: $n = c/v$	35-1	35-1

NEW IDEAS IN THIS CHAPTER

Concepts and equations introduced	Text Section	Study Guide Page
Double-slit constructive interference when $\Delta r = m\lambda = d \sin\theta_m$ at $x_m = L \tan\theta_m = Lm\lambda/d$ destructive interf. when $\Delta r = (m' + 1/2)\lambda = d \sin\theta_m'$ at $x_m' = L \tan\theta_m = L(m' + 1/2)\lambda/d$	36-1	36-1
Diffraction grating: constructive interference when $\Delta r = m\lambda = d \sin\theta_m$ at $x_m = L \tan\theta_m = Lm\lambda/d$ destructive interf. when $\Delta r = (m' + 1/2)\lambda = d \sin\theta_m'$ at $x_m = L \tan\theta_m' = L(m' + 1/2)\lambda/d$ angular half-width of an interference maxima: $\Delta\theta_{1/2} = \lambda/Nd \cos\theta = 1/N[(d/\lambda)^2 - m^2]^{1/2}$ dispersion: $D = \Delta\theta_m/\Delta\lambda = m/d \cos\theta_m$ angular separation of two maxima: $\Delta\theta_m = (m/d \cos\theta_m)\Delta\lambda$	36-3	36-4
Thin film interference: Case I: neither or both rays change phase constructive $2\tau = m\lambda/n$ destructive $2\tau = (m + 1/2)\lambda/n$ Case II: one or the other ray changes phase constructive $2\tau = (m + 1/2)\lambda/n$ destructive $2\tau = m\lambda/n$	36-5	36-9

PRINCIPAL CONCEPTS AND EQUATIONS

1 Double-Slit (Section 36-1)

Review: Figure 36.1 reviews what happens to a double-slit interference pattern as the width of the slit goes from $w << \lambda$ to $w > \lambda$. We are most interested in the case $w > \lambda$ since this is the area where your laboratory work will be done.

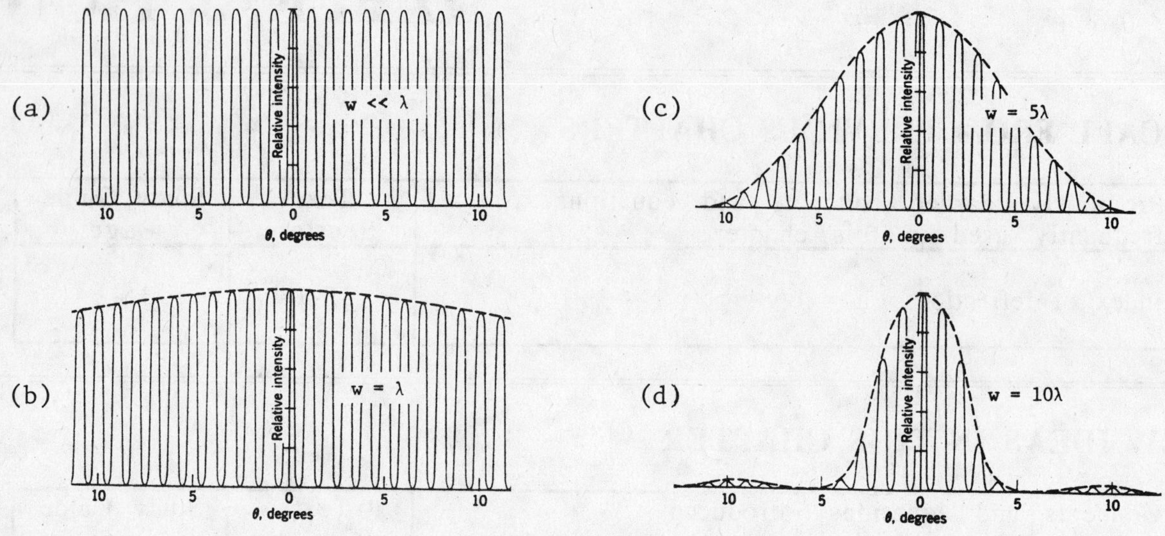

Figure 36.1

Figure 36.2 reviews constructive and destructive interference from a double slit for the case $w > \lambda$.

θ_m = bright fringe angular deviation
x_m = bright fringe linear deviation
m = order for bright fringes
L = distance from slits to screen
λ = wavelength of light
d = slit separation

θ_m' = dark fringe angular deviation
x_m' = dark fringe linear deviation
m' = order for dark fringes
Δr = path difference
w = slit width

Figure 36.2

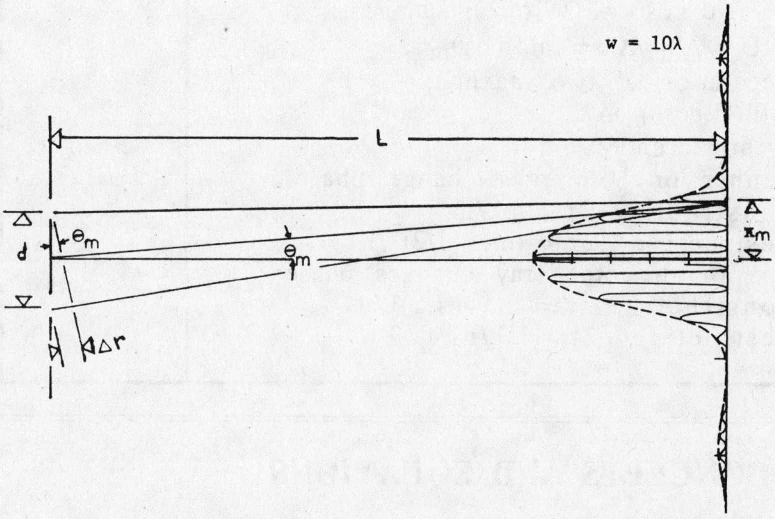

Constructive interference occurs when

$$\Delta r = m\lambda = d \sin\theta_m \qquad m = 0,1,2,3$$

Since the angular deviation is small, constructive interference occurs at

$$x_m = L \tan\theta_m \approx L \sin\theta_m = Lm\lambda/d \qquad m = 0,1,2,3 \ldots$$

Destructive interference occurs when

$$\Delta r = (m' + 1/2)\lambda = d \sin\theta_{m'} \qquad m' = 0,1,2,3 \ldots$$

Since the angular deviation is small, destructive interference occurs at

$$x_m = L \tan\theta_{m'} \approx L \sin\theta_{m'} = L(m' + 1/2)\lambda d \qquad m' = 0,1,2,3 \ldots$$

It should be noted that when m' = 0 we have the first minima ... and so on.

Practice: The following information is given for the situation shown in Figure 36.2.

$$d = 2.00 \times 10^{-5}\ \text{m} \qquad \lambda = 6.00 \times 10^{-7}\ \text{m} \qquad L = 2.00\ \text{m}$$

Determine the following:

1. Angular deviation θ_2 for second order constructive interference	$m\lambda = d \sin\theta_m$, $m = 2$ $\theta_2 = \sin^{-1}(m\lambda/d) = 3.44°$
2. Linear distance between central and second-order maxima	$\tan\theta_m = x_m/L$, $m = 2$ $x_2 = L \tan\theta_2 = 0.120$ m or $x_2 = Lm\lambda/d = 0.120$ m
3. Linear distance between any two adjacent constructive maxima	$x_m = Lm\lambda/d$ $x_{m+1} = L(m + 1)\lambda/d$ $\Delta x = x_{m+1} - x_m = L\lambda/d = 6.00 \times 10^{-2}$ m
4. Where we should reposition the screen so that the linear separation of the two second-order bright bands is 10.0 cm	$x_2 = Lm\lambda/d$, $2x_2 = 0.100$ m $L = x_2 d/m\lambda = 0.833$ m
5. Wavelength of light that would place the third-order maxima 12.5 cm from the central maximum	$\lambda = x_m d/Lm$, $m = 3$, $x_3 = 12.5$ cm $\lambda = 4.17 \times 10^{-7}$ m
6. The wavelength of light that would have fifth-order minima at the site of the present third-order maxima	$x_m = Lm\lambda/d$ and $x_{m'} = L(m' + 1/2)\lambda d$ The third order maxima has $m = 3$ The fifth order minima has $m' = 4$ x_3 (maxima) = x_4 (minima) $3L\lambda/d = (4 + 1/2)L\lambda'/d$ $\lambda' = 2\lambda/3 = 4.00 \times 10^{-7}$ m

Example: Light is incident on a double slit, and the interference pattern is formed on a distant screen. The separation of the slits changes, and the second-order bright fringe occurs where the third-order bright fringe originally occurred. Determine the new slit separation if the old separation was 0.150 mm.

Given: $d_i = 1.50 \times 10^{-4}$ m $\quad x_{3\,initial} = x_{2\,final}$

Determine: The final slit width.

Strategy: We can write an expression for the location of the third-order bright fringe for the initial slit separation d_i. We can write an expression for the location of the second-order bright fringe for the final slit separation d_f. We can equate these two expressions and determine the new slit separation.

Solution: The initial location of the third-order bright fringe is

$$x_{3\,initial} = 3L\lambda/d_i$$

The final location of the second order bright fringe is

$$x_{2\,final} = 2L\lambda/d_f$$

Equating these two expressions and solving for d_f, we have

$$3L\lambda/d_i = 2L\lambda/d_f$$

$$d_f = (2/3)d_i = (2/3)(1.50 \times 10^{-4}\text{ m}) = 1.00 \times 10^{-4}\text{ m}$$

Related Text Exercises: Ex. 36-1 through 36-7.

2 Diffraction Grating (Section 36-3)

Review: Figure 36.3 reviews the interference pattern for two slits and six slits for the case where the slit width is very small relative to the wavelength. The six slits have the same separation as the two slits.

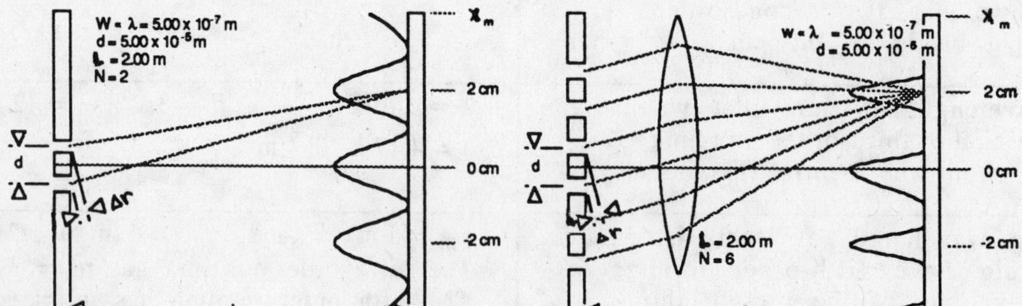

Figure 36.3

Constructive interference occurs when

$$\Delta r = m\lambda = d \sin\theta_m \qquad m = 0,1,2,3, \ldots$$

Since the angular deviation is small, constructive interference occurs at

$$x_m = L \tan\theta_m \approx L \sin\theta_m = Lm\lambda/d \qquad m = 0,1,2,3, \ldots$$

Destructive interference occurs when

$$\Delta r = (m' + 1/2)\lambda = d \sin\theta_{m'} \qquad m' = 0,1,2,3, \ldots$$

Since the angular deviation is small, destructive interference occurs at

$$x_m = L \tan\theta_{m'} \approx L \sin\theta_{m'} = L(m' + 1/2)\lambda/d \qquad m' = 0,1,2,3, \ldots$$

Notice that this is identical to the double slit information. In general, gratings do not have a spacing much less than the wavelength of light and they have one thousand or more slits per cm. Let's see what happens to the interference pattern when we use more typical values.

Typical values for a diffraction grating are:

$n = 5000/\text{cm}$ = lines or scratches per cm on the grating
$d = 1/n = 2.00 \times 10^{-6}$ m = distance between scratches
$w = d = 2.00 \times 10^{-6}$ m = slit width ($w = 4\lambda$ for this case)
$L = 2.00$ m = distance from grating to screen
let's agree to use light of wavelength $\lambda = 5.00 \times 10^{-7}$ m

The angular deviation for successive maxima is

$$\theta_m = \sin^{-1}(m\lambda/d) = 14.5° \text{ for } m = 1$$
$$= 30.0° \text{ for } m = 2$$
$$= 48.6° \text{ for } m = 3$$

The diffraction pattern will look like that shown in Figure 36.4.

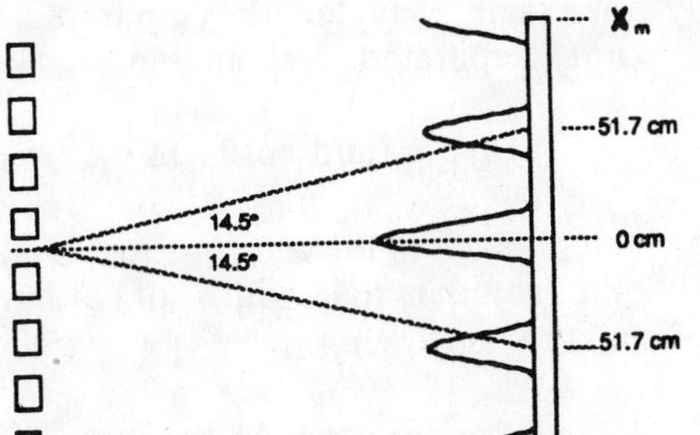

Figure 36.4

For this case, where the slit width is not much smaller than the wavelength (here $w = d = 4\lambda$), the angular deviation is large and the maxima are far apart. Consequently, the location of the site of destructive interference is of little interest to us (actually, it locates the center of the large dark space between the maxima) and it is inappropriate to use the approximation $\tan\theta_m = \sin\theta_m$.

For such a diffraction grating, constructive interference occurs when

$$m\lambda = d \sin\theta_m \qquad m = 0,1,2,3, \ldots$$

and the sites of constructive interference are located at

$$x_m = L \tan\theta_m \qquad m = 0,1,2,3, \ldots$$

From Figure 36.4 we see that as N (the number of slits illuminated) gets large the waves combine to give nearly complete cancellation at all angles except those which correspond to the interference maxima. While the maxima in Figure 36.4 are narrow, they do have a finite width. We define the angular half-width $\Delta\theta_{1/2}$ of an interference maximum as the angle between the center of the maximum and its adjacent minimum as shown in Figure 36.5. The angular half-width of an interference maximum may be written as

$$\Delta\theta_{1/2} = \lambda/Nd \cos\theta \qquad \text{or} \qquad \Delta\theta_{1/2} = 1/N[(d/\lambda)^2 - m^2]^{1/2}$$

Where
- N = number of slits illuminated
- d = spacing between the slits
- λ = wavelength of the light
- m = order of interest
- θ = angular location of line of interest

Notice that as N increases, $\Delta\theta_{1/2}$ decreases, that is the interference maxima get narrower.

A quantity which characterizes the capability of a grating to spatially spread a light beam according to wavelength is called the grating's dispersion D. The dispersion of a grating is defined as

$$D = \Delta\theta_m/\Delta\lambda = m/d \cos\theta_m$$

Thus, two waves of nearly the same wavelength (separated by an amount $\Delta\lambda$) will have their interference maxima separated by an amount

$$\Delta\theta_m = (m/d \cos\theta_m)\Delta\lambda$$

Two waves of nearly the same wavelength can be resolved if their angular separation $\Delta\theta_m$ is greater than the angular half-width of either line. The criterion (Rayleigh) for this results in the following:

$$\Delta\theta_m > \Delta\theta_{1/2} \Rightarrow \Delta\lambda > \lambda/Nm$$

According to this, two unresolved lines may be resolved either by illuminating more slits of the grating or by observing a higher order spectrum.

$\Delta\theta_{1/2}$ = angular half-width of a line

$\Delta\theta_m$ = angular separation of two lines

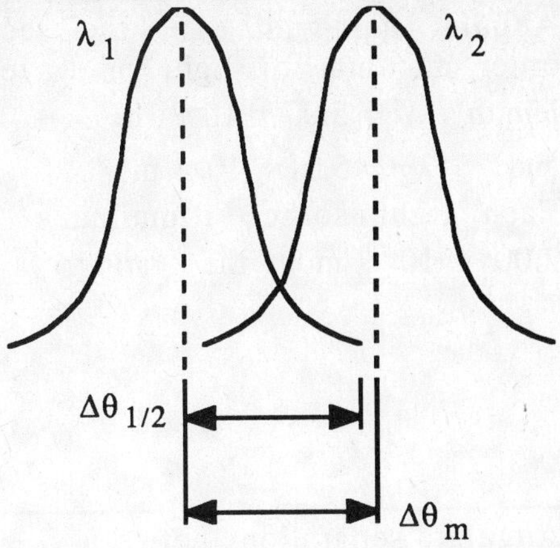

Figure 36.5

Practice: A diffraction grating with 5000 lines/cm is located 2.00 m from a viewing screen.

Determine the following:

1. Grating spacing (i.e., distance between the scratches)	$d = 1/n = 2.00 \times 10^{-4}$ cm
2. Number of orders you may view using light of wavelength 5×10^{-7} m	$m = d \sin\theta m/\lambda$ If we look to the left or right as far as possible ($\theta_m = 90°$), we see all orders. $m = d \sin 90°/\lambda = 4.00$
3. Longest wavelength for which the third order can be observed	$m\lambda = d \sin\theta_m$ $\lambda = d \sin 90°/m = 6.67 \times 10^{-7}$ m
4. Distance on the screen between second-order images for light of wavelength 4.00×10^{-7} m and 6.00×10^{-7} m	$m\lambda = d \sin\theta_m$ $\theta_m = \sin^{-1}(m\lambda/d) = \sin^{-1}(m\lambda n)$ $\theta_1 = \sin^{-1}[(2)(4.00 \times 10^{-7}\text{ m})(5 \times 10^5/\text{m})]$ $= 23.6°$ $x_1 = L \tan\theta_1 = 0.874$ m $\theta_2 = \sin^{-1}[(2)(6.00 \times 10^{-7}\text{ m})(5 \times 10^5/\text{m})$ $= 36.9°$ $x_2 = L \tan\theta_2 = 1.50$ m $\Delta x = x_2 - x_1 = 0.63$ m
5. Spacing for a grating whose third-order image has the same angular deviation as the second-order image of this grating under similar conditions (i.e., same λ and L)	$m\lambda = d \sin\theta_m$ $\sin\theta_2 = 2\lambda/d$ and $\sin\theta_3' = 3\lambda/d'$ $\theta_3' = \theta_2$; hence $\sin\theta_3' = \sin\theta_2 = 2\lambda/d$ substituting and rearranging $d' = 3\lambda/(2\lambda/d) = 3d/2 = 3.00 \times 10^{-4}$ cm

6. Angular half-width for the second order maxima for light of wavelength $\lambda_1 = 5.00 \times 10^{-7}$ m and $\lambda_2 = 5.01 \times 10^{-7}$ m. Each light source illuminates 2.00×10^{-2} m of the grating	$\Delta\theta_{1/2} = 1/N[(d/\lambda)^2 - m^2]^{1/2}$ $N = \left[\begin{array}{l}(5.00 \times 10^3 \text{ lines/cm}) \times \\ (10^2 \text{ cm/m})(2 \times 10^{-2} \text{ m})\end{array}\right]$ $= 1.00 \times 10^4$ lines $d = 2.00 \times 10^{-6}$ m, $m = 2$ $\lambda_1 = 5.00 \times 10^{-7}$ m $\Rightarrow (\Delta\theta_{1/2})_1 = 2.89 \times 10^{-5}$ rad $\lambda_2 = 5.01 \times 10^{-7}$ m $\Rightarrow (\Delta\theta_{1/2})_2 = 2.89 \times 10^{-5}$ rad
7. Angular separation between λ_1 and λ_2	$\Delta\theta_m = (m/d \cos\theta_m)\Delta\lambda$ $m = 2$, $\Delta\lambda = 1.00 \times 10^{-9}$ m $d = 2.00 \times 10^{-6}$ m $\Delta r = m\lambda = d \sin\theta_m$, hence $\theta_m = \sin^{-1}(m\lambda/d) = \sin^{-1}(0.500)$ $\cos\theta_m = 0.866$ $\Delta\theta_m = 1.15 \times 10^{-3}$ rad
8. If these two λs are resolved	Since $\Delta\theta_m > \Delta\theta_{1/2}$, the two lines are resolved. Also note that $\Delta\lambda = 1.00 \times 10^{-9}$ m. $\lambda/Nm = 2.50 \times 10^{-11}$ m Since $\Delta\lambda > \lambda/Nm$, the lines may be resolved.

Example: White light containing wavelengths from 4.00×10^{-7} m to 7.00×10^{-7} m is directed at a planar diffraction grating of 6000 lines/cm. (a) Do the first and second-order spectra overlap? (b) If they do, what is the angular overlap; if they don't, what is the angular separation?

Given:

$\lambda_{min} = 4.00 \times 10^{-7}$ m = minimum wavelength
$\lambda_{max} = 7.00 \times 10^{-7}$ m = maximum wavelength
$n = 6000$ lines/cm = lines/cm on the grating

Determine: (a) Whether or not the first-order ($m = 1$) and second-order ($m = 2$) spectra overlap. (b) The angular overlap or separation.

Strategy: We can determine the angular deviation for the longest wavelength in the first-order and then for the shortest wavelength in the second-order. All desired information may be determined from these angles.

Solution: The angular deviation for any wavelength and order may be determined by

$$m\lambda = d \sin\theta_m \quad \text{or} \quad \theta_m = \sin^{-1}(m\lambda/d) = \sin^{-1}(m\lambda n)$$

The angular deviation for the longest wavelength in the first order is:

$$\theta_{1max} = \sin^{-1}[(1)(7.00 \times 10^{-7}\ m)(6 \times 10^5/m)] = \sin^{-1}(0.420) = 24.8°$$

The angular deviation for the shortest wavelength in the second order is:

$$\theta_{2min} = \sin^{-1}[(2)(4.00 \times 10^{-7}\ m)(6 \times 10^5/m)] = \sin^{-1}(0.480) = 28.7°$$

The first-order and second-order spectra do not overlap, and the angular separation is:

$$\Delta\theta = \theta_{2min} - \theta_{1max} = 3.90°$$

Related Text Exercises: Ex. 36-16 through 36-30.

3 Thin Film Interference (Section 36-5)

Review: Figure 36.6 shows light incident on a thin film and partially reflected from each surface.

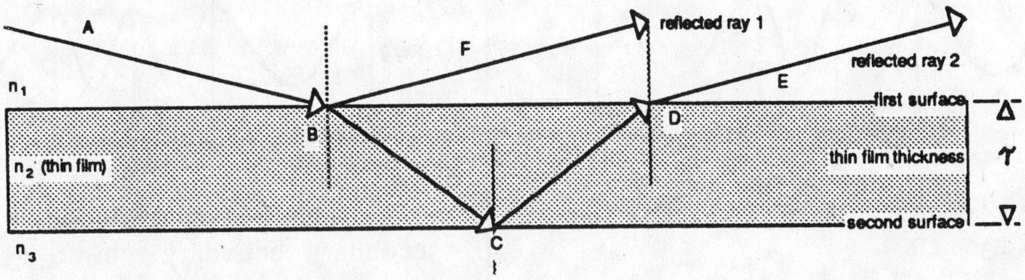

Figure 36.6

Light that reaches the eye is composed of light reflected from the top and bottom surfaces. These two parts, which we consider as separate rays or waves, are superimposed at the eye and produce an interference phenomenon. The interference may be constructive, destructive, or something in between, depending on such factors as

(a) the length of the path difference of the two rays
(b) the wavelength of the light in the thin film
(this depends on the index of refraction of the thin film material)
(c) phase changes at the reflecting surface.

(a) If the length of the path difference is an integral number of wavelengths, constructive interference will occur if neither or both rays change phase on reflection, destructive interference will occur if one or the other of the rays changes phase on reflection. If the length of the path difference is an odd integral number of half wavelengths, constructive interference will occur if one or the other of the rays changes phase on reflection, destructive interference will occur if neither or both rays change phase on reflection.

(b) The wavelength of the light in the film is not equal to its wavelength in a vacuum (or air). As light goes from one medium to another, the frequency remains constant but the speed and wavelength change. Light of wavelength λ in a vacuum has a wavelength of λ/n in a medium with an index of refraction n. This may be written as

$$\lambda_n = \lambda/n$$

(c) If the medium from which the light reflects has a larger index of refraction than the medium in which it is traveling, there will be a 180° phase change (remember: low to high, phase change by π). If the light is reflected from a medium that has a smaller index of refraction than that medium in which the light is traveling, there will be no phase change (remember: high to low, phase change by 0). The preceding is reviewed for two cases.

Case I - neither or both rays change phase on reflection

Neither ray changes phase on reflection: $n_1 > n_2 > n_3$

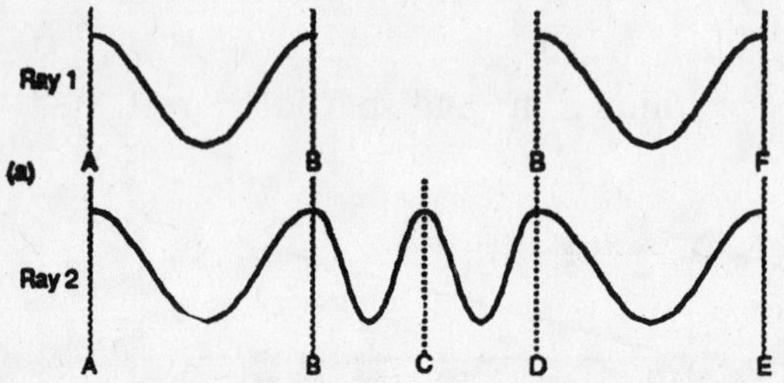

Both rays change phase on reflection: $n_1 < n_2 < n_3$

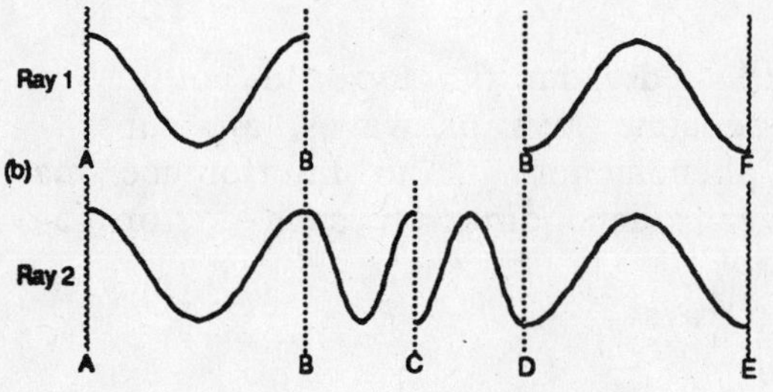

Case I Constructive Interference

$2\tau = m\lambda/n \qquad m = 0,1,2,3, \ldots$

Case I Destructive Interference

$2\tau = (m + 1/2)\lambda/n \qquad m = 0,1,2,3, \ldots$

Case II - one or the other of the rays changes phase on reflection

Phase change on reflection from the first boundary $n_1 < n_2 > n_3$

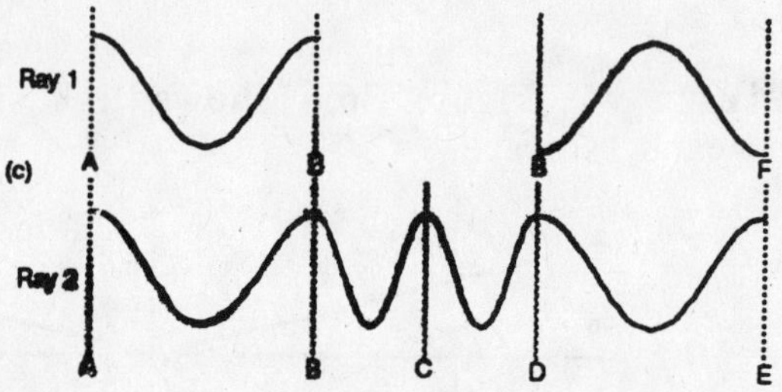

Phase change on reflection from the secondary boundary: $n_1 > n_2 < n_3$

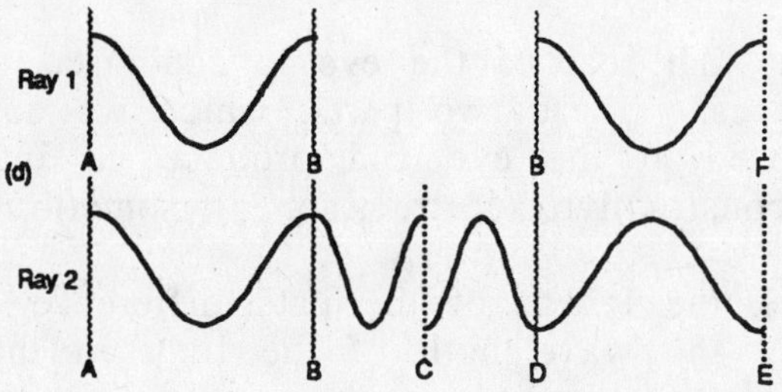

Case II Constructive Interference

$2\tau = (m + 1/2)\lambda/n \qquad m = 0,1,2,3, \ldots$

Case II Destructive Interference

$2\tau = m\lambda/n \qquad m = 0,1,2,3, \ldots$

Figure 36.7

Practice: The following information is given for Figure 36.7.

$n_1 = 1.00 \qquad n_2 = 1.36 \qquad n_3 = 1.65 \qquad t = 4.00 \times 10^{-7}$ m

Determine the following:

1. Phase change for ray 1 and ray 2 at first surface	The reflected ray (1) undergoes a 180° phase change. The transmitted ray (2) does not undergo a phase change.
2. Phase change for ray 2 at the second surface	It experiences a 180° phase change.
3. Wavelengths in the visible range that experience destructive interference	This is a case-I situation (both rays change phase on reflection). Destructive interference occurs for $2\tau = (m + 1/2)\lambda/n_2$, or $\lambda = 2n_2\tau/(m + 1/2)$ $\lambda_0 = 21.8 \times 10^{-7}$ m $m = 0$ $\lambda_1 = 7.25 \times 10^{-7}$ m $m = 1$ $\lambda_2 = 4.35 \times 10^{-7}$ m $m = 2$ $\lambda_3 = 3.11 \times 10^{-7}$ m $m = 3$ Only λ_1 and λ_2 are in the visible range.
4. Wavelengths in the visible range that experience constructive interference	For case I, constructive interference occurs for $2\tau = m\lambda/n_2$, or $\lambda = 2\tau n_2/m$ $\lambda_0 = \infty$ $m = 0$ $\lambda_1 = 10.9 \times 10^{-7}$ $m = 1$ $\lambda_2 = 5.44 \times 10^{-7}$ $m = 2$ $\lambda_3 = 3.63 \times 10^{-7}$ $m = 3$ Only λ_2 is in the visible range.
5. The thinnest the film can be and destructive interference occur for reflected light of 5.00×10^{-7} m wavelength	For case I, destructive interference occurs for $2\tau = (m + 1/2)\lambda/n_2$; hence $\tau = (m + 1/2)\lambda/2n_2$ The thinnest film occurs for $m = 0$, or $\tau = \lambda/4n_2 = 9.19 \times 10^{-8}$ m

Example: A spacer separates one end of two 12.0-cm-long plates of flat glass, and the other ends of the two pieces are touching each other. When light of wavelength 5.46×10^{-7} m is incident on the plates, the distance between consecutive dark fringes is 0.500 cm. What is the thickness of the spacer?

Given: $L = 12.0$ cm = length of the two plates of flat glass
$\lambda = 5.46 \times 10^{-7}$ = wavelength of the incident light
$s = 0.500$ cm = separation of consecutive dark fringes

Determine: T = thickness of the spacer

Strategy: Figure 36.8 shows the air wedge and reflected rays 1 and 2.

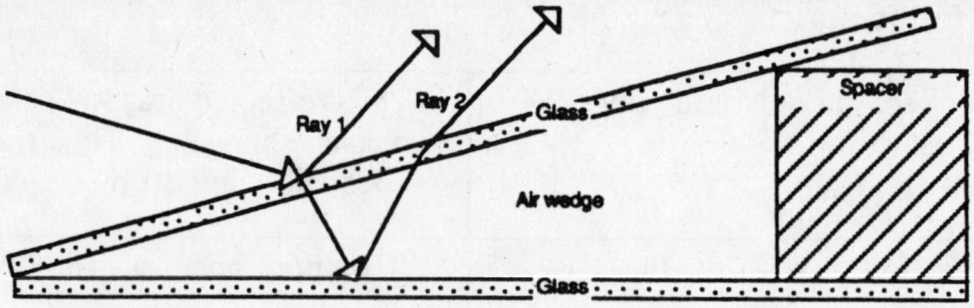

Figure 36.8

First we need to establish that a dark fringe exists as the contact end of the glass plates and that this is a case-II situation. Then we can determine the separation of the plates at the next dark fringe and hence the angular separation of the plates. Knowing the angular separation and the length of the plates, we can determine the thickness of the spacer.

Solution: At the contact end of the glass plates, ray 1 (reflects off the top layer of the air wedge) experiences no phase change and ray 2 (reflects off the top of the second glass plate) experience a 180° phase change. Because of this phase change, a dark fringe occurs at the contact end. Also, since only one of the rays (2) experiences a phase change, this is a case-II situation, for which the thin film thickness (i.e. separation of the glass plates due to the spacer) at which destructive interference occurs is obtained as follows:

$$2\tau = m\lambda/n \qquad m = 0,1,2,3 \ldots$$
$$\tau = m\lambda/2n$$
$$\tau_0 = 0 \quad \text{for} \quad m = 0 \quad \text{(contact end)}$$
$$\tau_1 = \lambda/2n = 2.73 \times 10^{-7}\ \text{m} \quad \text{for} \quad m = 1 \quad \text{(second dark fringe)}$$

The angular separation of the plates is

$$\alpha = \tan^{-1}(\tau_1/s) = (3.13 \times 10^{-3})°$$

The thickness of the spacer is

$$T = L \tan\alpha = 6.56 \times 10^{-6}\ \text{m}$$

Related Text Exercises: Ex. 36-37 through 36-40.

PRACTICE TEST

Take and grade this practice test. Doing so will allow you to determine any weak spots in your understanding of the concepts taught in this chapter. The following section prescribes what you should study further to strengthen your understanding.

Light of wavelength 6.00×10^{-7} m is incident on two slits separated by 0.100 mm, and the resulting interference pattern is viewed on a screen 1.00 m away.

Determine the following:

_________ 1. Angular deviation for the second-order maximum
_________ 2. Distance of the second-order maximum from the central maximum
_________ 3. Angular deviation for the second dark space from the central maximum
_________ 4. Distance of the second dark space from the central maximum
_________ 5. Interference fringe spacing on the screen
_________ 6. Where the screen should be located relative to the slits in order to have the third-order maxima separated by 5.00 cm
_________ 7. Wavelength of light that would have second-order minima at the site of the present second order maxima
_________ 8. Width of the central maxima (at the base) if the slit separation is doubled

Light of wavelength 6.66×10^{-7} m is incident on a planar diffraction grating that has 5000 lines/cm and 2500 lines are illuminated by the beam of light. The resulting diffraction pattern is viewed on a screen 0.500 m away.

Determine the following:

_________ 9. Grating spacing
_________ 10. Number of diffraction images that can be observed
_________ 11. Angular deviation of first-order diffraction image
_________ 12. Distance on screen between first and second-order images
_________ 13. Longest wavelength for which the second order can be observed
_________ 14. Angular half-width for the first order maxima
_________ 15. Width of the first order maxima
_________ 16. Angular separation between the first order maxima of this situation and that for light that has a wavelength 20.0 A° greater
_________ 17. Minimum difference in wavelength that will still allow us to resolve the first order maxima

A thin film of oil on water is shown in Figure 36.9

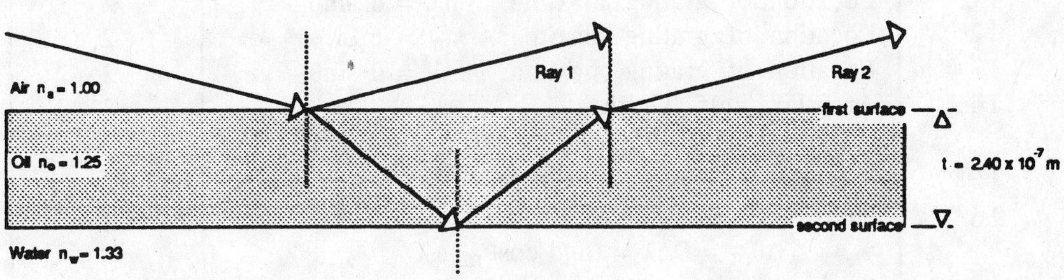

Figure 36.9

Determine the following:

_________ 18. Phase change of ray 1 at the first surface
_________ 19. Phase change of ray 2 at the second surface
_________ 20. Wavelengths in the visible range that experience constructive interference

_________ 21. Wavelengths in the visible range that experience destructive interference

_________ 22. Thinnest film of this oil that can cause destructive interference for light of wavelength $\lambda = 5.00 \times 10^{-7}$ m

(See Appendix I for answers.)

PRINCIPAL CONCEPTS AND EQUATIONS PRESCRIPTION

Your score on the practice test is an excellent measure of your understanding of the chapter. You should now use the following chart to write your own prescription for dealing with any weaknesses the practice test points out. Look down the leftmost column to the number of the question(s) you answered incorrectly, reading across that row you will find the concept and/or equation of concern, the section(s) of the study guide you should return to for further study, and some suggested text exercises which you should work to gain additional experience.

Practice Test Questions	Concepts and Equations	Prescription	
		Principal Concepts	Text Exercises
1	Angular deviation for maxima: $m\lambda = d \sin\theta_m$	1	36-1, 3
2	Linear deviation for maxima: $x_m = mL\lambda/d$	1	36-2, 3
3	Angular deviation for minima: $(m' + 1/2)\lambda = d \sin\theta_{m'}$	1	36-3, 4
4	Linear deviation for minima: $x_{m'} = (m' + 1/2)L\lambda/d$	1	36-4, 6
5	Fringe spacing: $x_{m+1} - x_m = L\lambda/d$	1	36-1, 2
6	Location of maxima: $x_m = mL\lambda/d$	1	36-5
7	Location of maxima and minima	1	36-3, 6
8	Width of central maxima: $\Delta x_{cm} = 2x_0 = \lambda L/d$	1	36-3, 6
9	Grating spacing: $d = 1/n$	2	36-17, 18
10	Location of maxima: $m\lambda = d \sin\theta_m$, $\theta_m = 90°$	2	36-23, 27
11	Location of grating maxima: $m\lambda = d \sin\theta_m$	2	36-16, 19
12	Location of grating maxima: $x_m = mL\lambda/d$	2	36-22, 23
13	Location of grating maxima: $m\lambda = d \sin\theta_m$	2	36-22, 23
14	Angular half width: $\Delta\theta_{1/2} = \lambda/Nd \cos\theta = 1/N[d/\lambda)^2 - m^2]^{1/2}$	2	36-23, 25
15	Width of maxima: $\Delta x = 2L\Delta\theta_{1/2}$	2	---
16	Grating dispersion: $\Delta\theta_m = D\Delta\lambda = (m/d \cos\theta_m)\Delta\lambda$	2	36-28, 29
17	Rayleigh criterion: $\Delta\lambda > \lambda/Nm$,	2	36-26, 30
18	Thin films - phase change	3	36-37, 38
19	Thin films - phase change	3	36-38, 39
20	Constructive interference: $\lambda = 2\tau n_2/m$	3	36-39, 40
21	Destructive interference: $\lambda = 2\tau n_2/(m + 1/2)$	3	36-37, 39
22	Destructive interference: $\tau = \lambda/4n_2$	3	36-38, 40

37 DIFFRACTION AND POLARIZATION

RECALL FROM PREVIOUS CHAPTERS

Previously learned concepts and equations frequently used in this chapter	Text Section	Study Guide Page
Resolving vectors into components	2-3, 4	2-8
Index of refraction	35-1	35-1
Snell's law: $\sin\theta_1/\sin\theta_2 = v_1/v_2 = n_2/n_1$	35-1	35-1

NEW IDEAS IN THIS CHAPTER

Concepts and equations introduced	Text Section	Study Guide Page
Single-slit diffraction:	37-1, 2, 3	37-1
Intensity minima: $a \sin\theta_{m'} = \pm m'\lambda$		
$x_{m'} = L \sin\theta_{m'} = \pm Lm'\lambda / a$		
Intensity maxima: $a \sin\theta_m = \pm(m + 1/2)\lambda$		
$x_m = L \sin\theta_m = \pm L(m + 1/2)\lambda/a$		
Intensity ratio: $I = I_c (\sin^2\beta)/\beta^2$		
$\beta = (\pi a/\lambda) \sin\theta$		
Malus' law: $I_p = I_o \cos^2\theta$	37-6	37-7
Degree of polarization: $P = (I_{\parallel} - I_{\perp})/(I_{\parallel} + I_{\perp})$	37-6	37-7
Brewster's law: $\tan\theta_p = n_2/n_1$	37-7	37-6

PRINCIPAL CONCEPTS AND EQUATIONS

1 Single-Slit Diffraction (Section 37-2)

Review: Figure 37.1 reviews the diffraction pattern of a single-slit.

The pattern consists of a bright central maximum which is flanked by secondary maxima, and the intensity of each succeeding maximum decreases with distance from the center.

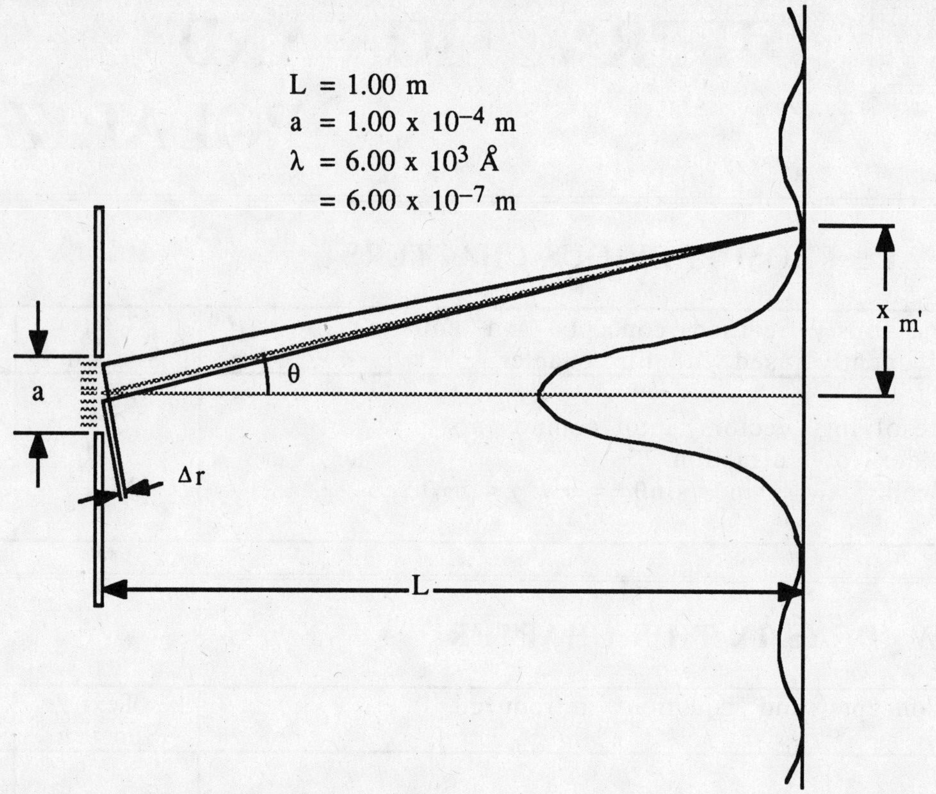

Figure 37.1

Intensity minima occur at the angle $\theta_{m'}$ given by

$$a \sin\theta_{m'} = \pm m'\lambda \qquad (m' = 1,2,3, \ldots)$$

since $L >> x_{m'}$, we may write

$$\sin\theta_{m'} \approx \tan\theta_{m'} = x_{m'}/L$$

Intensity minima occur at $x_{m'}$ on the screen.

$$x_{m'} = L \sin\theta_{m'} = \pm Lm'\lambda/a$$

The secondary intensity maxima occur approximately midway between their adjacent minima. Thus the angles θ_m which locate the secondary maxima are given by

$$a \sin\theta_m = \pm(m + 1/2)\lambda \qquad (m = 1,2,3, \ldots)$$

since $L >> x_m$, we may write

$$\sin\theta_m \approx \tan\theta_m = x_m/L$$

Secondary intensity maxima occur at x_m on the screen.

$$x_m = L \sin\theta_m = \pm L(m + 1/2)\lambda/a \qquad (m \approx 1,2,3, \ldots)$$

Note:

1. For a given slit width a, light of longer wavelength results in a greater amount of diffraction (larger $x_{m'}$ and x_m).

2. For a given wavelength, a narrower slit causes a greater amount of diffraction (larger $x_{m'}$ and x_m).

The light intensity at various positions on the screen is given by

$$I = I_c(\sin^2\beta)/\beta^2$$

Where

$$I_c = \text{light intensity at the center of the screen}$$
$$\beta = (\pi a/\lambda)\sin\theta$$

Combining these, the intensity on the screen for any angle θ is

$$I = I_c \sin^2(\pi a \sin\theta/\lambda)/(\pi a/\lambda)^2 \sin^2\theta$$

since L >> x, we may write

$$\sin\theta \approx \tan\theta = x/L$$

The intensity at any point x on the screen then is

$$I = I_c \sin^2(\pi a x/L\lambda)/(\pi a x/L\lambda)^2$$

Intensity minima occur when

$$a \sin\theta_{m'} = \pm m'\lambda$$
$$\beta = \pi a \sin\theta/\lambda = \pi(\pm m'\lambda)/\lambda = \pm m'\pi$$
$$\sin\beta = \sin(\pm m'\pi) = 0 \qquad (m' = 1,2,3, \ldots)$$
$$I = I_c(\sin^2\beta)/\beta^2 = 0$$

Intensity maxima occur when

$$a \sin\theta_m = \pm(m + 1/2)\lambda$$
$$\beta = \pi a \sin\theta/\lambda = \pm(m + 1/2)\lambda\pi/\lambda = \pm(m + 1/2)\pi$$
$$\sin^2\beta = \sin^2[\pm(m + 1/2)\pi] = 1 \qquad (m = 1,2,3, \ldots)$$
$$I = I_c/[\pm(m + 1/2)\pi]^2 \qquad (m = 1,2,3, \ldots)$$

When m = 1 (first secondary maxima)

$$I = I_c/(9\pi^2/4) = 0.0450\ I_c$$

m = 2 (second secondary maxima)

$$I = I_c/(25\pi^2/4) = 0.0162\ I_c$$

Practice: Consider the single slit information shown in Figure 37.1.

Determine the following:

1. Angular location of the first dark region	Intensity minima occur at $\theta_{m'}$ $a \sin\theta_{m'} = \pm m'\lambda$ (m' = 1,2,3, ...) The first dark region has m' = 1 $\theta_{m'} = \pm\sin^{-1}(\lambda/a) = \pm 0.343°$
2. Position on the screen (relative to the center of the central maxima) of the first dark region	since x << L, we may write $\sin\theta_{m'} \approx \tan\theta_{m'} = x_{m'}/L$ $x_{m'} \approx L \sin\theta_{m'} = L(\pm m'\lambda/a)$ $\approx \pm m'L\lambda/a = \pm 6.00 \times 10^{-3}$ m
3. Width of the central maxima on the screen	$\Delta x_{cm} = 2x_{1'} = 1.20 \times 10^{-2}$ m
4. Angular location of the first secondary maxima	Intensity maxima occur at θ_m $a \sin\theta_m = \pm(m + 1/2)\lambda$ (m = 1,2,3, ...) The first secondary maxima occurs at m = 1 $\theta_m = \pm\sin^{-1}[(m + 1/2)\lambda/a]$ $= \pm\sin^{-1}(3\lambda/2a) = \pm 0.516°$
5. Position on the screen of the first secondary maxima	since x << L, we may write $\sin\theta_m \approx \tan\theta_m = x_m/L$ $x_m \approx L \sin\theta_m = L[\pm(m + 1/2)\lambda/a]$ $\approx \pm(m + 1/2)L\lambda/a = \pm 9.00 \times 10^{-3}$ m
6. Width of the first secondary maxima on the screen	From step 3 we see that the minima are located by $x_{m'} = \pm m'L\lambda/a$. The first secondary maxima is bounded by the $x_{1'}$ and $x_{2'}$ minima. Hence $\Delta x_{\text{first secondary maxima}} = x_{2'} - x_{1'} = L\lambda/a = 6.00 \times 10^{-3}$ m
7. Relative intensity of the first and second secondary maxima compared to the intensity of the central maxima	$I = I_c/[\pm(m + 1/2)\pi]^2$ For the first secondary maxima m = 1 $I = I_c/(9\pi^2/4) = 0.0450\ I_c$ For the second secondary maxima m = 2 $I = I_c/(25\pi^2/4) = 0.0162\ I_c$
8. Relative intensity of the diffraction pattern compared to the intensity of the central maximum at a distance of 1.50 cm from the center of the central maximum	$x = 1.50 \times 10^{-2}$ m From the review we have $I/I_c = \sin^2(\pi a x/L\lambda)/(\pi a x/L\lambda)^2$ $\pi a x/L\lambda = 2.50\ \pi$ $\sin(\pi a x/L\lambda) = 1.00$ $I/I_c = 1/(2.50\pi)^2 = 0.0162$

9. Wavelength of light that would have its third secondary maximum occur at the site of the present second minimum	Notice that this will amount to a smaller amount of diffraction, hence we expect a smaller λ. From the review, maxima and minima are located on the screen by $x_{m'} = \pm Lm'\lambda/a$ minima $x_m = \pm L(m + 1/2)\lambda/a$ maxima We want $(x_3)_{new} = (x_2)_{present}$ or $\pm L(3 + 1/2)\lambda_{new}/a = \pm L(2)\lambda_p/a$ $\lambda_{new} = (4/7)\lambda_{present} = 3.43 \times 10^{-7}$ m
10. Amount we would have to narrow the slit in order to double the width of the central maximum on the screen	The central maximum is bounded by the first minimas. We can double the width of the central maxima by doubling $x_{1'}$, that is we want $(x_{1'})_{new} = 2(x_{1'})_{present}$. According to the review $x_{m'} = \pm Lm'\lambda/a$. Hence, we have $L\lambda/a_{new} = 2(L\lambda/a_{present})$ or $a_{new} = a_{present}/2$.
11. New location of the screen in order to move the third secondary maxima into the present site of the second minimum	The maxima and minima are located respectively by $x_m = \pm L(m + 1/2)\lambda/a$ max $x_{m'} = \pm Lm'\lambda/a$ min We want $(x_3)_{new} = (x_{2'})_{present}$ or $L_{new}(7/2)\lambda/a = L_p(2)\lambda/a$ or $L_{new} = (4/7)L_p = 5.71 \times 10^{-1}$ m
12. Maximum number of fringes expected on either side of the central maximum if the slit width is 10λ	Minima occur at $a \sin\theta_m = \pm m'\lambda$ the maximum value for $\theta_{m'}$ is 90.0° $(10\lambda) \sin 90.0° = \pm m'\lambda$ or $m' = \pm 10$ We expect to see a maximum of 10 minima and hence 9 secondary maxima on either side of the central maximum.

Example: Light of wavelength 600 nm is incident on a slit 0.100 mm wide and the resulting diffraction pattern observed on a screen 2.00 m away. (a) What is the distance between consecutive dark fringes? (b) If the entire apparatus is immersed in water, what is the distance between consecutive dark fringes?

Given: $\lambda = 6.00 \times 10^{-7}$ m $a = 1.00 \times 10^{-4}$ m $L = 2.00$ m

Determine: (a) The dark-fringe spacing when the apparatus is in air. (b) The dark fringe spacing when the apparatus is immersed in water.

Strategy: Using our knowledge of single-slit theory, we can develop an expression for the location of two successive minima and hence the fringe spacing. We can use this expression and the given information to determine the fringe spacing in air. Next we can calculate the wavelength in water and then use the previously developed expression to determine the fringe spacing in water.

Solution: For a single slit, the mth dark space occurs at

$$x_{m'} = \pm m'L\lambda/a$$

and the next dark space occurs at

$$x_{m'+1} = \pm(m' + 1)L\lambda/a$$

Hence the fringe spacing is given by

$$\Delta x = x_{m'+1} - x_{m'} = L\lambda/a = 1.20 \times 10^{-2} \text{ m}$$

If the apparatus is immersed in water, L and a do not change but the wavelength becomes

$$\lambda_n = \lambda/n = 4.51 \times 10^{-7} \text{ m}$$

The fringe spacing in water is

$$\Delta x_n = L\lambda_n/a = L\lambda/na = \Delta x/n = 9.02 \times 10^{-3} \text{ m}$$

Related Text Exercises: Ex. 37-1 through 37-14.

2 Polarization (Section 37-5, 6 and 7)

Review: Figure 37.2 shows an electromagnetic wave and its plane of polarization.

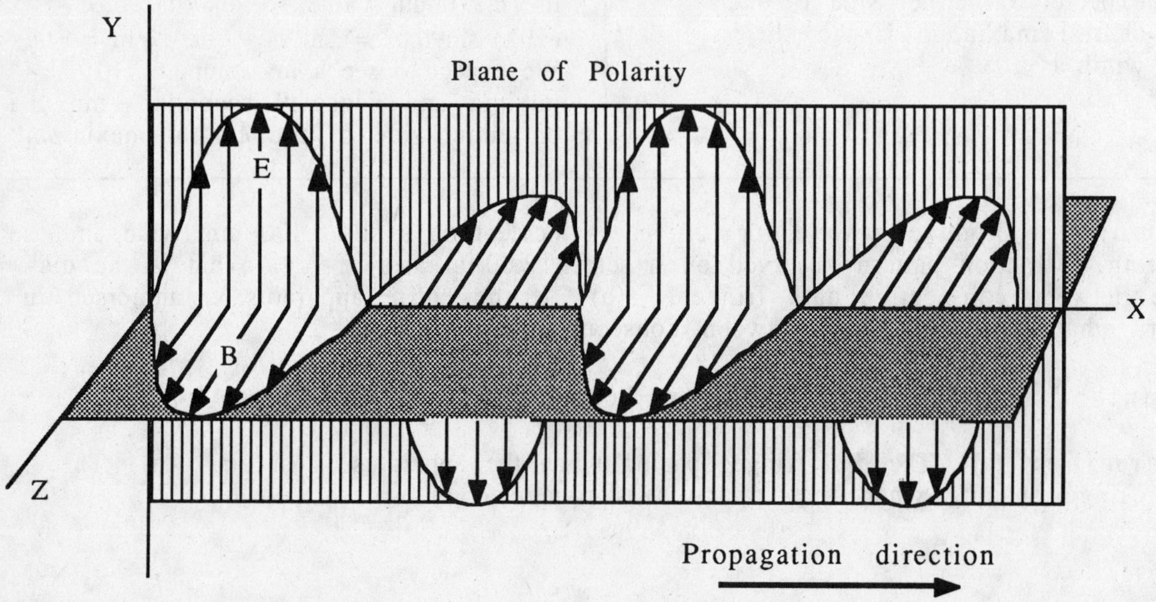

Figure 37.2

Unpolarized light amounts to many such waves with the **E** vectors (we are concerned here with only the **E** vector) oriented in all directions.

Figure 37.3 shows an unpolarized light incident upon a polarizer. A polarizer is an optical device that selectively transmits light that has its plane of polarization parallel to the polarizer's transmission axis. Light that has its plane of polarization perpendicular to the transmission axis is blocked out by absorption or reflection.

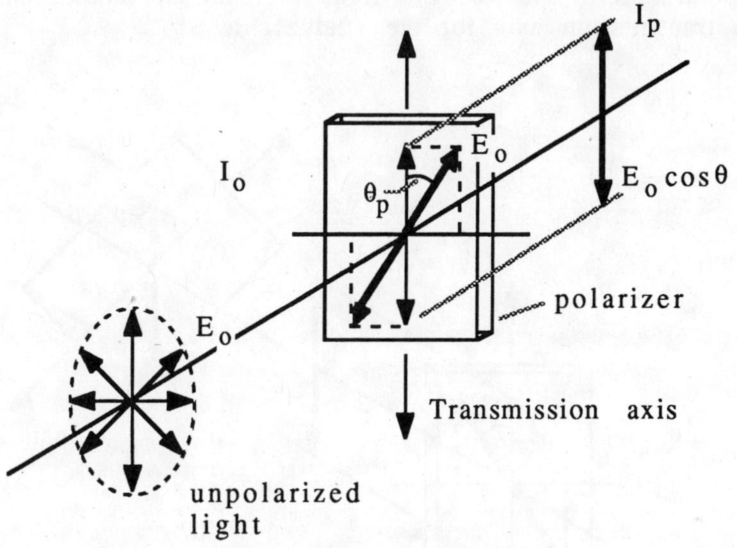

Figure 37.3

Of all the incident $\mathbf{E}_0$ vectors originating from the light source, let's consider just the one labeled in Figure 37.3. This particular electromagnetic wave is incident upon the polarizer with its plane of polarization oriented at an angle θ_p off of the direction of polarization (direction of the transmission axis of the polarizing material). $\mathbf{E}_0$ may be divided into components parallel ($E_0 \cos\theta_p$) and perpendicular ($E_0 \sin\theta_p$) to the direction of polarization. Only the parallel component will be transmitted, the perpendicular part will be reflected or absorbed.

The intensity of light is proportional to the square of the electric field vector, hence

$$I_0 \propto E_0{}^2, \text{ intensity of light from the light source}$$

$$I_p \propto E_0{}^2 \cos^2\theta_p, \text{ intensity of light after the polarizer}$$

This gives us the following expression for the contribution to the intensity from this one vector $\mathbf{E}_0$.

$$I_p = I_0 \cos^2\theta_p \qquad \text{(Malus' Law)}$$

We get the total intensity of the light after the polarizer by adding the contribution from all waves. However, because there is no positive sense associated with the transmission axis, all possible orientations of the vector $\mathbf{E}_0$ are in the range $\theta_p = 0$ to $\theta_p = \pi/2$ rad.

Since $\cos^2\theta$ averaged over $\theta = 0$ to $\theta = \pi/2$ radians is 1/2, we have

$$I_p = I_0/2$$

Next, let the polarized light (I_p, the light after the polarizer) be incident upon an analyzer (just another polarizer, but called an analyzer to differentiate it from the original polarizer). If the analyzer is oriented as shown in Figure 37.4, the angle between the polarization plane of the light incident upon the analyzer and the direction of the transmission axis for the analyzer is θ_a.

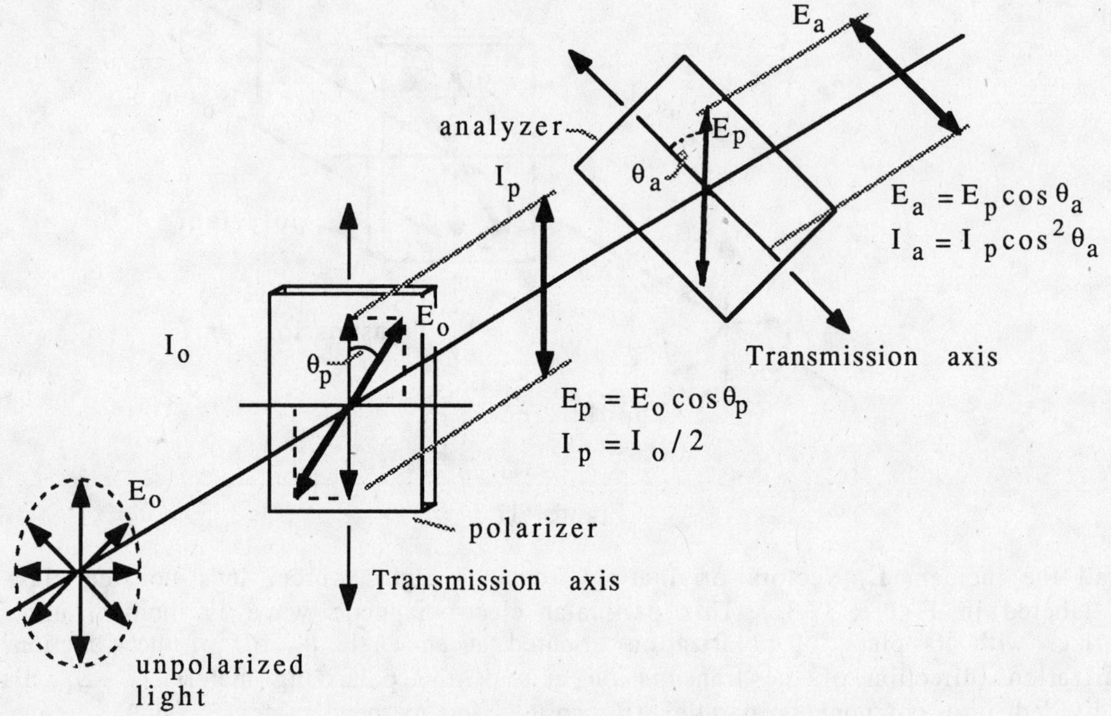

Figure 37.4

Then

$$I_a = I_p \cos^2\theta_a = (I_0/2) \cos^2\theta_a$$

Consider unpolarized light incident on a polarizer. Next think of all the **E** vectors resolved into components parallel and perpendicular to the transmission axis (i.e. axis of polarization) of the polarizing material. Since the polarizing material is real (i.e. not ideal), some light will get through perpendicular as well as parallel to the transmission axis. We wish to define the degree of polarization P such that

$P = 0$, if no polarization occurs, i.e. $I_{||} = I_{\perp} = I_T/2$
$P = 1$, if the polarization is total, i.e. $I_{||} = I_T$ and $I_{\perp} = 0$
$P = 0.5$, if half the light is polarized, i.e. $I_{||} = I_T/2$

Notice that if P is defined as shown below, it accomplishes the above

$$P = (I_{||} - I_{\perp})/(I_{||} + I_{\perp})$$

Notice that if no polarization occurs ($I_{||} = I_{\perp} = I_T/2$) then $P = 0$

if polarization is total ($I_{||} = I_T$ and $I_{\perp} = 0$) then $P = 1$

if half the light is polarized then the polarized part contributes $I_{||} = 0.5I_T$ and the unpolarized part contributes $I_{||} = 0.25I_T$ and $I_{\perp} = 0.25I_T$. Hence we have $I_{||} = 0.75I_T$ and $I_{\perp} = 0.25I_T$ which gives us the result $P = 0.5$.

When unpolarized light is incident upon a nonmetallic material, the beam becomes partially polarized by reflection. For one particular angle of incidence, called the polarizing angle θ_p (or Brewsters angle), the reflected beam is at right angles to the refracted beam and complete polarization by reflection is achieved. The direction of the polarization of the reflected beam is parallel to the reflecting surface. Figure 37.5 shows this situation. The polarizing angle is given by

$$\tan\theta_p = n_2/n_1$$

This expression is known as Brewster's law.

Figure 37.5

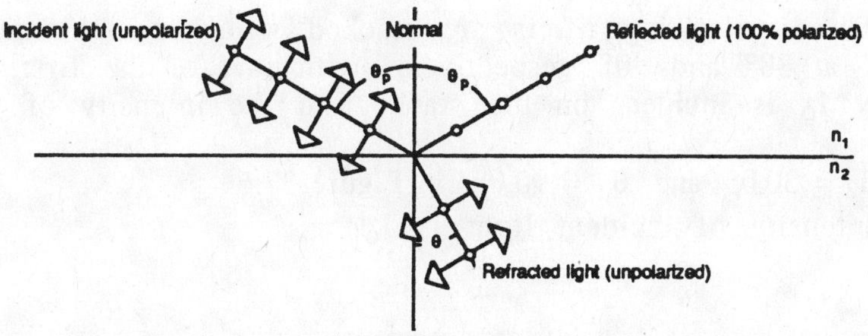

Practice: Refer to Figure 37.4 and determine the following:

1. Intensity of light after the polarizer	$I_p = I_0/2$
2. Angle of orientation of analyzer in order for I_a to equal $I_p/8$	$I_a = I_p \cos^2\theta_a$, $I_a = I_p/8$ Combine these to obtain $\theta_a = \cos^{-1}[(1/8)^{1/2}] = 69.3°$
3. Final intensity (I_a) if the polarizer is vertical as shown and the angle θ_a for the analyzer is 30°	I_0 = intensity of source $I_p = I_0/2$ = intensity after polarizer (step 6) $I_a = I_p \cos^2\theta_a = (I_0/2) \cos^2 30° = 0.375I_0$

Refer to Figure 37.5 and determine the following:

4. Polarizing angle for light reflected off crown glass	$\tan\theta_p = n_{glass}/n_{air} = (1.58/1.00) = 1.58$ $\theta_p = \tan^{-1}(1.58) = 57.7°$

5. For light traveling in water, angle of incidence that causes complete polarization when reflected off flint glass	$\tan\theta_p = n_{glass}/n_{water}$ $\theta_{incidence} = \theta_p = \tan^{-1}(1.75/1.33) = 52.8°$
6. Index of refraction of a medium for which the Brewster angle for light incident in air is 60°	$\tan\theta_p = n_{medium}/n_{air}$ $n_{medium} = n_{air} \tan\theta_p = 1.73$
7. Polarizing angle of a medium for which the reflected beam is completely polarized when the refraction angle is 35°	If the refraction angle is 35°, the reflection angle is 55° for complete polarization $\theta_p = \theta_{incidence} = \theta_{reflection}$ $= 90 - \theta_{refraction} = 55°$
8. Index of refraction of a medium for which the reflected beam is completely polarized when the refraction angel is 35°	For this case, $\theta_p = 55°$ (step 4) $n_{medium} = n_{air} \tan\theta_p = 1.43$

Example: Three polarizing filters are stacked with the polarizing axes of the second and third at 30° and 90°, respectively, with that of the first. If unpolarized light of intensity I_0 is incident on the stack, find the intensity of the emerging light.

Given: $\theta_2 = 30.0°$ and $\theta_3 = 90.0°$ in Figure 37.6
Intensity of incident light is I_0

Figure 37.6

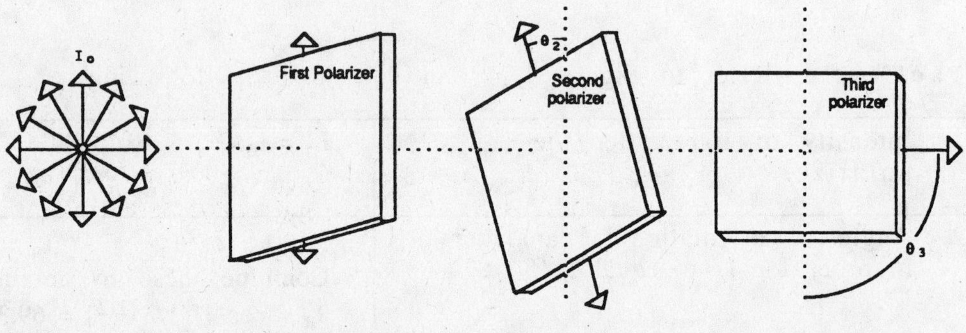

Determine: Intensity of the emerging light.

Strategy: We can establish the intensity of light (I_1) after the first polarizer. Knowing this and θ_2, we can establish the intensity of light (I_2) after the second polarizer. Knowing this and θ_3, we can establish the intensity of light (I_3) after the third polarizer.

Solution: The light incident on the first polarizer can be resolved into components perpendicular and parallel to the direction of polarization. Because the incident light is a random mixture of all directions of polarization, these two components are equal and hence half the light gets through. We can write the following.

$$I_1 = I_0/2$$

Applying the law of Malus, we obtain

$$I_2 = I_1 \cos^2 30.0° = I_1(0.750)$$

$$I_3 = I_2 \cos^2 60.0° = [I_1(0.750)](0.250) = 0.188 I_1 = 0.188(I_0/2) = 0.094\ I_0$$

Example: A beam of light is incident at an angle of 53.0° on a reflecting non metallic surface. The reflected beam is completely linearly polarized. Determine the angle of refraction of the transmitted beam and the refractive index of the reflecting material.

Given: $\theta_p = 53.0°$ and the incident medium is air.

Determine: The angle of refraction (θ_r) and the index of refraction (n) of the reflecting material.

Strategy: Knowing θ_p and that the angle between the reflected and refracted beams is 90.0°, we can determine θ_r. Knowing the polarization angle and the index of refraction of the medium for the incident beam, we can determine the refractive index of the reflecting medium.

Solution:

$$\theta_p = \theta_{reflection} = 53.0° \qquad \theta_{refraction} = 90.0° - 53.0° = 37.0°$$

$$\tan\theta_p = n/n_{air} \qquad n = n_{air} \tan\theta_p = \tan 53.0° = 1.33$$

Alternatively, using Snell's law, $n_{air} \sin\theta_p = n \sin\theta_r$,

$$n = n_{air} \sin\theta_p/\sin\theta_r = \sin 53.0°/\sin 37.0° = 1.33$$

Related Text Exercises: Ex. 37-24 through 37-36.

PRACTICE TEST

Take and grade this practice test. Doing so will allow you to determine any weak spots in your understanding of the concepts taught in this chapter. The following section prescribes what you should study further to strengthen your understanding.

Light of wavelength 6.00×10^{-7} m is incident on a single slit 0.100 mm wide, and the resulting diffraction pattern is viewed on a screen 2.00 m away.

Determine the following:

_________ 1. Angular position of the first dark space
_________ 2. Distance of the first dark space from the location of the central maximum
_________ 3. Width of the central maximum on the screen
_________ 4. Angular deviation between the first and second dark spaces
_________ 5. Approximate distance of the first secondary maximum from the central maximum
_________ 6. Width of the first secondary maximum on the screen
_________ 7. Where the screen should be relocated relative to the slit in order to double the width of the central maximum
_________ 8. Wavelength of light that would have its first dark space at the site of the present second secondary maximum
_________ 9. The amount we would have to change the slit width in order to double the width of the central maximum on the screen
_________ 10. The ratio of the intensity of the first secondary maximum to the central maximum
_________ 11. The ratio of the intensity of the second secondary maximum to the first secondary maximum

Unpolarized light of intensity I_0 is incident on a stack of three polarizers that have an angle of 45.0° between the directions of polarization of each successive polarizer. Determine the following:

_________ 12. Intensity of light after the first polarizer
_________ 13. Intensity of light after the second polarizer
_________ 14. Intensity of light after the third polarizer

Unpolarized light traveling in air is incident upon a liquid with a refractive index of 1.60 at Brewster's angle. It is established that 10.0% of the incident beam is reflected and 90.0% is refracted. Determine the following:

_________ 15. Polarizing angle for this liquid
_________ 16. Angle of reflection for the reflected beam
_________ 17. Angle between the reflected and refracted beam
_________ 18. Angle of refraction for the refracted beam
_________ 19. Degree of polarization for the reflected beam
_________ 20. Degree of polarization for the refracted beam

(See Appendix I for answers.)

PRINCIPAL CONCEPTS AND EQUATIONS PRESCRIPTION

Your score on the practice test is an excellent measure of your understanding of the chapter. You should now use the following chart to write your own prescription for dealing with any weaknesses the practice test points out. Look down the leftmost column to the number of the question(s) you answered incorrectly, reading across that row you will find the concept and/or equation of concern, the section(s) of the study guide you should return to for further study, and some suggested text exercises which you should work to gain additional experience.

Practice Test Questions	Concepts and Equations	Prescription	
		Principal Concepts	Text Exercises
1	Angular position of dark space: $a \sin\theta_{m'} = \pm m'\lambda$	1	37-2
2	Location of dark space: $x_{m'} = L \sin\theta_{m'} = \pm Lm'\lambda/a$	1	37-4
3	Width of central maximum: $\Delta x_{cm} = 2x_{1'}$	1	37-6
4	Angular deviation: $\theta_{m'} = \sin^{-1}(m'\lambda/a)$	1	37-2
5	Location of secondary maxima: $x_m = \pm L(m + 1/2)\lambda/a$	1	37-2, 7
6	Width of secondary maximum: $\Delta x = x_{2'} - x_{1'} = L\lambda/a$	1	37-2, 7
7	Width of central maximum: $\Delta x_{cm} = 2x_{1'} = 2L\lambda/a$	1	37-1, 6
8	Location of dark spaces and secondary maxima $(m + 1/2)\lambda_{present} = m'\lambda_{new}$, $m = 2$ and $m' = 1$	1	37-4, 7
9	Width of central maximum: $\Delta x_{cm} = 2L\lambda/a$	1	37-3
10	Intensity ratio: $I = I_c (\sin^{-2}\beta)/\beta^2$	1	37-10, 12
11	Intensity ratio	1	37-10, 12
12	Law of Malus: $I = I_o \cos^2\theta$	2	37-24, 25
13	Law of Malus	2	37-25, 26
14	Law of Malus	2	37-26, 27
15	Polarizing angle: $\tan\theta_p = n_2/n_1$	2	37-29, 31
16	$\theta_{reflection} = \theta_{incidence} = \theta_p$	2	37-30, 31
17	Polarizing angle: 90.0°	2	37-31, 32
18	Angle of refraction: $\theta_{refl} + \theta_{refrac} = 90°$	2	37-32, 34
19	Degree of polarization: $P = (I_{\parallel} - I_{\perp})/(I_{\parallel} + I_{\perp})$	2	37-27, 28
20	Degree of polarization	2	37-28, 36

38 RELATIVITY

RECALL FROM PREVIOUS CHAPTERS

Previously learned concepts and equations frequently used in this chapter	Text Section	Study Guide Page
Relationship between velocity, displacement and time: $v = \Delta s/\Delta t$	3-2	3-5
Kinetic energy: $K = mv^2/2$	8-4	8-13
Linear momentum: $p = mv$	10-3	10-10

NEW IDEAS IN THIS CHAPTER

Concepts and equations introduced	Text Section	Study Guide Page
Time dilation: $T = \gamma T_0$	38-3	38-1
Length contraction: $L = L_0/\gamma$	38-3	38-5
Relativistic velocities $U_x' = (U_x - v)/(1 - U_x v/c^2)$	38-5	38-8
Relativistic mass: $m = \gamma m_0$	38-6	38-10
Rest energy: $E_0 = m_0 c^2$	38-6	38-11
Relativistic energy: $E = mc^2 = \gamma m_0 c^2 = \gamma E_0$	38-6	38-11
Kinetic energy: $K = E - E_0 = (\gamma - 1)E_0$	38-6	38-11
Relationship between energy and momentum: $E^2 = p^2c^2 + E_0^2$	38-6	38-11
Relationship between kinetic energy and momentum: $p^2c^2 = K^2 - 2kE_0$	38-6	38-11

PRINCIPAL CONCEPTS AND EQUATIONS

1 Time Dilation (Section 38-3)

Review: Consider observer O in a frame of reference that is earthbound. Another frame of reference, attached to a spaceship, moves past the earth frame at a speed of 0.900c. Observer O' is in the spaceship. Two events (A and B) take place on the spaceship, and both O and O' record when and where these events takes place. For event A, a passenger on the spaceship raises her hand. For event B, she lowers her hand.

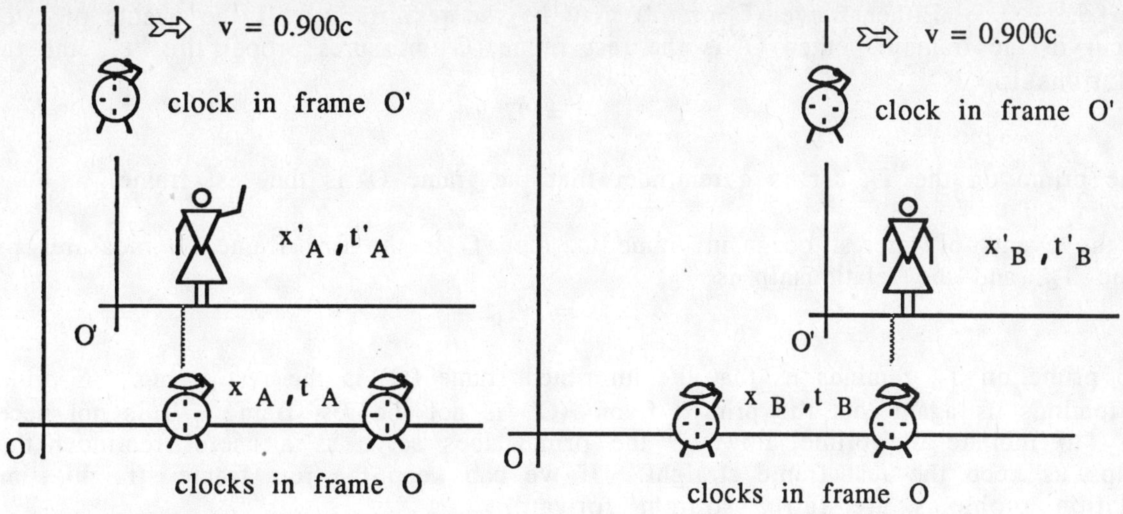

(a) When the girl raises her hand, O says she is at x_A and the time is t_A, but O' says she is at x_A' and the time is t_A'.

(b) When the girl lowers her hand, O says she is at x_B and the time is t_B, but O' says she is at x_B' (note that $x_B' = x_A'$) and the time is t_B'.

Figure 38.1

Since the two events occur at the same location in the frame O' is in, the time interval can be determined with a single clock in this frame. The frame of observer O' is the rest frame for the time interval between events A and B. Your instructor may say that the time interval measured by the clock in the frame of observer O' is proper time. The time between events A and B according to the clock in the frame of O' is

$$T_o' = t_B' - t_A'$$

where t_A' is the specific time the girl raises her hand according to the clock in the frame of observer O'.

t_B' is the specific time the girl lowers her hand according to the clock in the frame of observe O'.

T' represents a time interval according to the clock in the frame of observer O'.

"o" the subscript "o" tells us that the time interval is in the rest frame of the clock, i.e. it tells us that the time interval T' was determined on a clock at rest with respect to events A and B.

The time interval between events A and B as measured by O (who is in motion relative to the ship) is

$$T = t_B - t_A$$

where t_A and t_B are specific times according to the clock in the frame of observer O.

T represents a time interval according to the clock in the frame of observer O. The lack of a subscript "o" tells us that the clocks used to determine this time interval are moving relative to the two events A and B. Notice that two clocks are needed, one for each event.

Note: Either frame (O or O') may be the rest frame. If the events of interest occur in the frame O', then O' is the rest frame, O' measures proper time t'_0, and the relationship is

$$T = \gamma T'_0$$

The prime on the T_0 serves a reminder that the frame O' is the rest frame.

If the events of interest occur in frame O, then O is the rest frame, O measures proper time T_0, and the relationship is

$$T' = \gamma T_0$$

No prime on T_0 reminds us that the unprimed frame (O) is the rest frame. A prime on T reminds us again that the primed frame (O') is not the rest frame. It is not necessary to include the prime, however the prime does serve as a useful reminder that helps us keep the rest frame straight. If we can keep the rest frame straight, time dilation problems are fairly straight forward.

Since in the case under consideration, the primed frame is the rest frame for the time interval of interest, we write

$$T = \gamma T'_0$$

If $T = t_B - t_A = 5.00$ min, then the time between the two events according to O' is

$$T'_0 = T/\gamma = T[1 - (v/c)^2]^{1/2} = (5.00 \text{ min})(0.436) = 2.18 \text{ min}$$

If O' claims that $T'_0 = t'_B - t'_A = 5.00$ min, the time between the two events according to O is

$$T = \gamma T'_0 = T'_0/[1 - (v/c)^2]^{1/2} = (5.00 \text{ min})/0.436 = 11.5 \text{ min}$$

Note: The time dilation formula ($T = \gamma T_0$) is simple to use if you are able to establish the values of T and T_0. Recall that T_0 is rest frame or proper time, it is a time interval measured by a clock at rest with respect to the reference frame in which the events occur. An observer measuring an interval of time between events occurring in his or her own reference frame measures the rest frame or proper time T_0 and this interval is shorter than that measured by an observer in motion with respect to her or him. That is, the moving clock runs slow. This effect, the "stretching out" of time for a moving clock, is called time dilation.

Practice: A student and professor are moving at 0.800c with respect to each other. The student is given an examination to be completed in 1.00 h by the professor's clock.

Determine the following:

1. The value of γ	$\gamma = 1/[1 - (v/c)^2]^{1/2} = 1.67$
2. The two events of interest	Event A - the professor's clock reads hour zero Event B - the professor's clock reads hour one

3. The frame in which events A and B occur at the same location	Events A and B occur at the same location in the professor's frame.
4. The rest or proper frame for the time interval	The professor's frame is the rest or proper frame for the time interval
5. The proper time	$(T_p)_o = 1.00$ h
6. The time the student has to take the exam according to the student's clock	$T_s = \gamma(T_p)_o = (1.67)(1.00 \text{ h}) = 1.67$ h

Let's redo the above steps, only this time the examination is to be completed in 1.00 h according to the student's clock.

7. The two events of interest	Event C - the student's clock reads hour zero Event D - the student's clock reads hour one
8. The frame in which events C and D occur at the same location	Events C and D occur at the same location in the student's frame.
9. The rest or proper frame for the time interval	The student's frame is the rest or proper frame for the time interval
10. The proper time	$(T_s)_o = 1.00$ h
11. The time the student has to take the exam according to the professor's clock	$T_p = \gamma(T_s)_o = (1.67)(1.00 \text{ h}) = 1.67$ h

Example: A light on your spaceship flashes every 12.0 h according to a clock on the spaceship. The light flashes every 24.0 h according to a clock in the earth frame. How fast is the ship moving relative to the earth?

Given: $T_s = 12.0$ h and $T_e = 24.0$ h

Determine: How fast the ship is moving relative to the earth

Strategy: We must first establish which observer measures proper time. Once this is established, we can obtain γ and then v.

Solution: The flashing light is on the spaceship with you. You see both events occurring in your frame at the same place. You measure proper time, and hence we may write

$$(T_s)_o = 12.0 \text{ h}$$

$$T_e = 24.0 \text{ h}$$

The value for γ can be determined by

$$T_e = \gamma(T_s)_0$$

$$\gamma = T_e/(T_s)_0 = 2.00$$

Finally, we can obtain v as follows:

$$\gamma = 1/[1 - (v/c)^2]^{1/2}$$

$$v = c[1 - 1/\gamma^2]^{1/2} = 0.866c = 2.60 \times 10^8 \text{ m/s}$$

Example: A woman has a pulse rate of 70.0 beats/min. What would her pulse rate be according to a doctor moving with a velocity of 0.800c relative to her?

Given: Pulse rate = 70.0 beats/min $\quad v = 0.800c$

Determine: The woman's pulse rate according to a doctor moving with a velocity of 0.800c relative to her

Strategy: We must first establish which observer measures proper time. Knowing that her pulse rate is 70.0 beats/min, we can establish the time between two heart beats in the woman's frame and in the frame moving relative to her. Knowing the time between two heart beats in the frame moving relative to her, we can establish her pulse rate in that frame.

Solution: The two events of interest are two successive heart beats. These two events occur at the same place in the woman's frame, and hence she and any observer at rest relative to her measure proper time.

The time between two heart beats in the woman's frame (the proper frame) is

$$(T_w)_0 = 1/(70.0 \text{ min}) = 1.43 \times 10^{-2} \text{ min}$$

The time between two heart beats in the frame moving relative to the woman (the doctors frame) is

$$\gamma = 1/[1 - (v/c)^2]^{1/2} = 1.67$$

$$T_d = \gamma(T_w)_0 = (1.67)(1.43 \times 10^{-2} \text{ min}) = 2.39 \times 10^{-2} \text{ min}$$

The woman's pulse rate according to the doctor is

$$\text{Pulse rate} = (1 \text{ beat})/T_d = (1 \text{ beat})/(2.39 \times 10^{-2} \text{ min}) = 41.8 \text{ beats/min}$$

Related Text Exercises: Ex. 38 - 21, 23, 24, 26, 27 and 28.

2 Length Contraction (Section 38-3)

Review: The length of an object as measured in a frame of reference in which it is at rest is called its rest frame length or proper length (L_0). If an object is at rest with respect to you and you measure its length to be L_0, then L_0 is its rest frame or proper length. If that object travels past you at a speed v, the apparent length (as seen by you) in the direction of the motion is decreased (length contraction) and you claim that it is

$$L = L_0/\gamma \quad \text{where} \quad \gamma = 1/[1 - (v/c)^2]^{1/2}$$

Note: Either frame (O or O') may be the rest frame. If the object is at rest in frame O', O' is the rest frame, O' measures the rest frame or proper length, and the relationship is

$$L = L_0'/\gamma$$

The prime on L_0 serves as a reminder that O' is the rest frame.

If the object is at rest in frame O, O is the rest frame, O measures the proper or rest frame length, and the relationship is

$$L' = L_0/\gamma$$

No prime on L_0 reminds us that the unprimed frame (O) is the rest frame. A prime on L reminds us again that the primed frame (O') is not the rest frame.

It is not necessary to include the prime, however it does serve as a useful reminder and helps us keep the rest frame straight. If we keep the rest frame straight, length contraction problems are fairly straight forward.

Note: The length contraction formula ($L = L_0/\gamma$) is simple to use if you can establish the value of L_0 and L. Recall that L_0 is proper length. An observer at rest with respect to an object measures its proper length L_0, and this length is greater than that measured by any other observer in motion with respect to the object.

Practice: Observers O and O' have metersticks mounted in their reference frames, as shown in Figure 38.2, and O' moves past O with a speed of v = 0.800c.

Figure 38.2

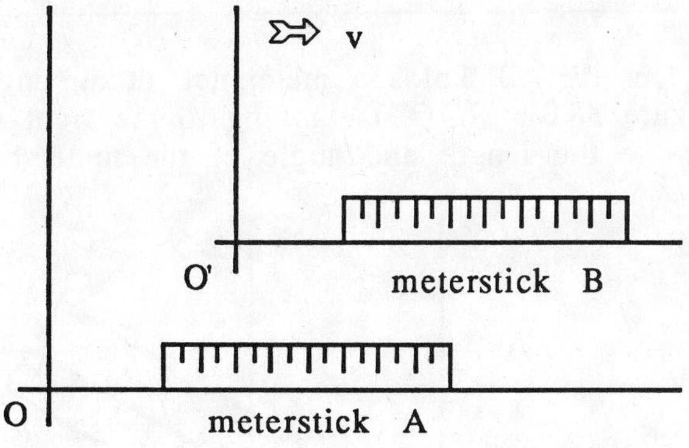

Determine the following:

1. $1/\gamma$	$1/\gamma = [1 - (v/c)^2]^{1/2} = 0.600$
2. The rest or proper frame for meterstick A	A is at rest with respect to O, and so O is its rest or proper frame.

3. Length of A as measured by O	O measures A to be 1.00 m, $(L_A)_0 = 1.00$ m
4. Length of A as measured by O'	$L_A' = (L_A)_0/\gamma = 0.600$ m
5. Proper length of B	B is at rest with respect to O', and so O' measures the proper length of B to be 1.00 m. $(L_B)_0' = 1.00$ m
6. Length of B as measured by O	$L_B = (L_B)_0'/\gamma = 0.600$ m

A spaceship has a length of 100 m in its rest frame and appears to be 80.0 m to an observer in an earth frame.

Determine the following:

7. Proper length of the ship	$(L_s)_0 = 100$ m
8. $1/\gamma$	$1/\gamma = L_e/(L_s)_0 = 0.800$
9. Relative velocity of the reference frames	$1/\gamma = [1 - (v/c)^2]^{1/2} = 0.800$, hence $v = c[1 - 1/\gamma^2)]^{1/2} = 0.600c$
10. Speed spaceship must have in order for its length to appear to be 0.5 $(L_s)_0$	$(L_s)_0 = 100$ m $L_e = (L_s)_0/2$ $L_e = (L_s)_0/\gamma$ $1/\gamma = [1 - (v/c)^2]^{1/2}$ $L_e/(L_s)_0 = [1 - (v/c)^2]^{1/2}$ or $v = (1 - [L_e/(L_s)_0]^2)^{1/2}c = 0.866c$

Example: An observer O' holds a meterstick at an angle of 30° with respect to the positive x' axis (Figure 38.3). If O' is moving to the right with a velocity of 0.800c with respect to O, what are the length and angle of the meterstick as measured by O?

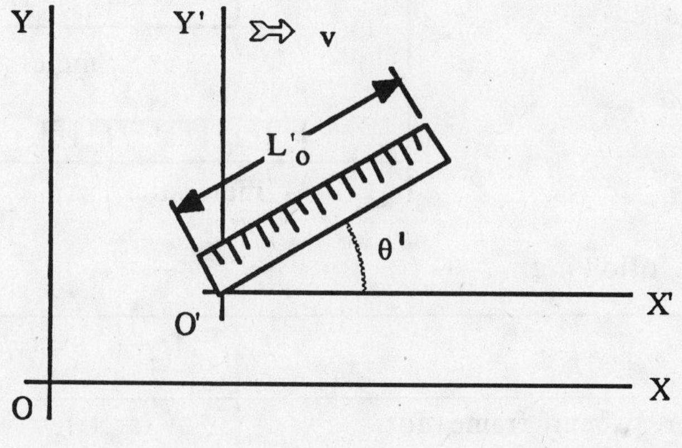

Figure 38.3

Given: $L_0' = 1.00$ m $\quad \theta' = 30.0°$ $\quad v = 0.800c$

Determine: L = length of meterstick according to O
θ = angle according to O

Strategy: Knowing L_0' and θ', we can determine the components of L_0' (x_0' and y_0') according to O'. Knowing x_0' and y_0', we can determine the components of the meterstick (x and y) according to O. Knowing x and y, we can determine the length of the stick and its orientation according to O.

Solution: The x_0' and y_0' components of the meterstick according to O' are

$$x_0' = L_0' \cos\theta' = (1.00 \text{ m}) \cos 30° = 0.866 \text{ m}$$
$$y_0' = L_0' \sin\theta' = (1.00 \text{ m}) \sin 30° = 0.500 \text{ m}$$

The x and y components of the meterstick according to O are

$$x = x_0'/\gamma = x_0'[1 - (v/c)^2]^{1/2} = 0.600x_0' = 0.520 \text{ m}$$

$$y = y' = 0.500 \text{ m} \qquad \left(\begin{array}{c}\text{contraction occurs only}\\ \text{along the direction of motion}\end{array}\right)$$

The length of the stick and its orientation according to O are

$$L = [x^2 + y^2]^{1/2} = 0.721 \text{ m}$$

$$\theta = \tan^{-1}(y/x) = \tan^{-1}(0.962) = 43.9°$$

Related Text Exercises: Ex. 38-21 through 38-25, and 38-28.

3 Relativistic Velocities (Section 38-5)

Review: Figure 38.4 shows two different reference frames and a moving object whose velocity can be measured by observers in either frame.

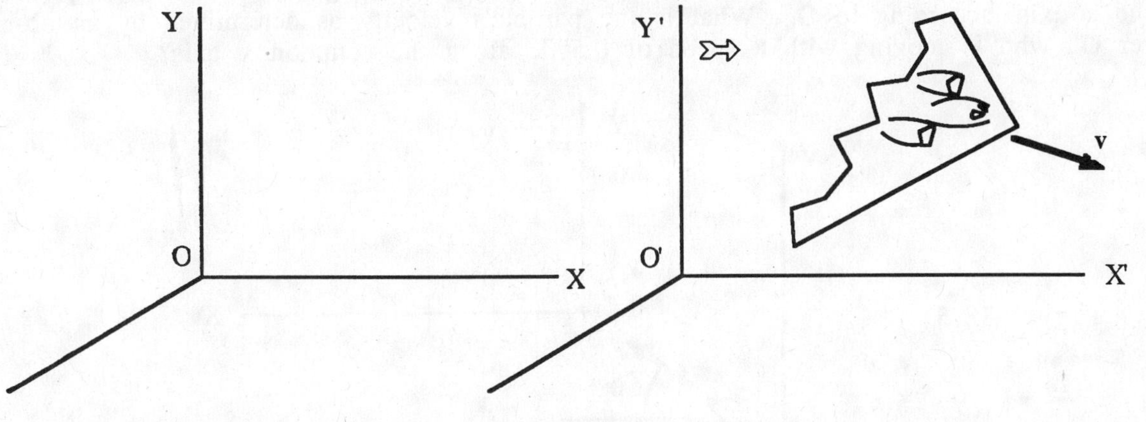

Figure 38.4

The components of the velocity vector according to observer O are U_x, U_y, U_z. The components of the velocity vector according to observer O' are U_x', U_y', U_z'. The transformation equations relating U_x, U_y, and U_z to U_x', U_y', and U_z' are

$$U_x' = (U_x - v)/(1 - U_x v/c^2)$$

$$U_y' = U_y[1 - (v/c)^2]^{1/2}/(1 - U_x v/c^2)$$

$$U_z' = U_z[1 - (v/c)^2]^{1/2}/(1 - U_x v/c^2)$$

Practice: Observer O' is moving relative to observer O with a speed of v = 0.900c. Both observers see spaceships A, B, and D. The speed of A according to O is U_A = 0.700c, the speed of B according to O' is U_B' = -0.800c, and the speed of D according to O is U_D = c.

Determine the following:

1. Speed of A according to O' (U_A')	$U_A' = (U_A - v)/(1 - U_A v/c^2)$ = -0.540c O' sees A approaching at a speed of 0.540c
2. Speed of B according to O (U_B)	$U_B' = (U_B - v)/(1 - U_B v/c^2)$ This may be solved for U_B to obtain $U_B = (U_B' + v)/(1 + U_B' v/c^2)$ = 0.357c
3. Speed of O according to O' (U_O')	$U_O' = (U_O - v)/(1 - U_O v/c^2)$ U_O = 0 (speed of O according to O) U_O' = (0 - 0.900c)/(1 - 0) = -0.900c
4. Speed of D according to O' (U_D')	$U_D' = (U_D - v)/(1 - U_D v/c^2)$ = c

Example: A particle moves with a speed 0.900c at an angle of 30° with respect to the x axis, according to O. What is the particle's velocity as determined by an observer O', who is moving with a speed of 0.500c along the common x axis?

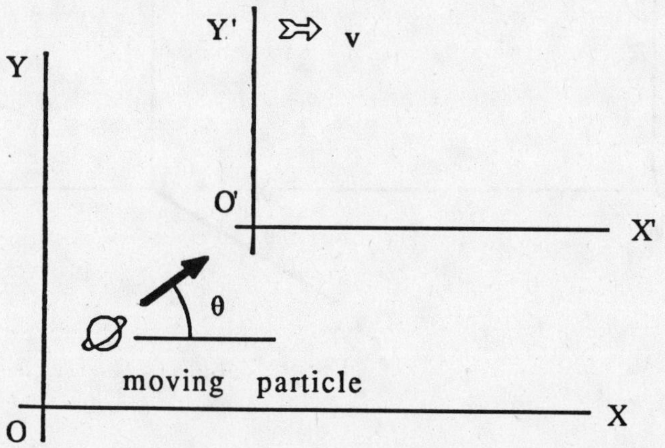

Figure 38.5

Given: $U = 0.900c$ = speed of particle according to O
$\theta = 30°$ = angle according to O
$v = 0.500c$ = speed of O' relative to O

Determine: The velocity (magnitude and direction) of the particle according to O'.

Strategy: Knowing U and θ, we can determine U_x and U_y. Knowing U_x and U_y, we can use the x and y velocity transformation equations to obtain U_x' and U_y'. Knowing U_x' and U_y', we can determine the magnitude and direction of the particle's velocity according to O'.

Solution:

$$U_x = U \cos\theta = 0.779c, \qquad U_y = U \sin\theta = 0.450c$$

$$U_x' = \frac{U_x - v}{1 - U_x v/c^2} = \frac{0.799c - 0.500c}{1 - (0.799c)(0.500c)/c^2} = 0.457c$$

$$U_y' = \frac{U_y[1 - (v/c)^2]^{1/2}}{1 - U_x v/c^2} = \frac{(0.450c)(0.866)}{1 - (0.799c)(0.500c)/c^2} = 0.638c$$

$$U' = [(U_x')^2 + (U_y')^2]^{1/2} = 0.785c \qquad \theta_2 = \tan^{-1}(U_y'/U_x') = 54.4°$$

Related Text Exercises: Ex. 38-29 through 38-32.

4 Relativistic Mass (Section 38-6)

Review: The mass-transformation equation is

$$m = \gamma m_0 = m_0/[1 - (v/c)^2]^{1/2}$$

where m_0 is the rest mass of an object (i.e., its mass as measured in its own frame) and m is the mass of the object as measured by an observer moving at a speed v with respect to the object.

Practice:

Determine the following:

1. Rest mass of an electron	$m_0 = 9.11 \times 10^{-31}$ kg
2. Mass of an electron traveling at a speed of 0.800c past us	$m = \gamma m_0 = m_0/[1 - (v/c)^2]^{1/2}$ $m = m_0/0.600 = 15.2 \times 10^{-31}$ kg
3. Speed of an electron that has an apparent mass increase of $0.100m_0$	$m = 1.10m_0$ $m = m_0/[1 - (v/c)^2]^{1/2}$ $v = c[1 - (m_0/m)^2]^{1/2} = 0.417c$

Example: Determine the percentage increase in the mass of a rocket moving at a speed of 0.800c.

Given: $v = 0.800c$

Determine: The percentage increase in the mass of a rocket traveling at a speed of 0.800c

Strategy: Knowing v and c, we can determine the ratio m/m_0 and then the percentage increase in the mass of the rocket.

Solution: $m = \gamma m_0 = m_0/[1 - (v/c)^2]^{1/2}$

$m/m_0 = 1/[1 - (v/c)^2]^{1/2} = 1.67$

$m = (m_0 + \Delta m) = 1.67m_o \qquad \Delta m = 0.67m_o$

$\Delta m/m_0 = 0.67 = 67\%$

Related Text Exercises: Ex. 38-33 and 38-38.

5 Momentum and Energy (Section 38-6)

Review: An object with a rest mass m_0 has a rest energy

$$E_0 = m_0 c^2$$

If work is done on that object, causing it to move past an observer at a speed v, the observer says the mass, energy, kinetic energy, and momentum of the object are:

$$m = \gamma m_0$$
$$E = mc^2 = \gamma m_0 c^2 = \gamma E_0$$
$$K = E - E_0 = (\gamma - 1)E_0 = (\gamma - 1)m_0 c^2$$
$$p = mv = \gamma m_0 v$$

These quantities are related as follows (see text Exercise 38-36 and Problem 38-6):

$$E^2 = p^2c^2 + E_0^2$$
$$p^2c^2 = K^2 - 2KE_0$$

If the mass of an object decreases by an amount Δm, the energy decreases by an amount

$$\Delta E = \Delta mc^2$$

This energy may be lost by radiation or by a decrease in speed. The above expression is called the law of conservation of mass-energy.

Practice: An electron is accelerated from rest through an electric potential difference of 5.00×10^6 V.

Determine the following:

1. Rest mass of the electron	$m_0 = 9.11 \times 10^{-31}$ kg
2. Rest energy of the electron	$E_0 = m_0 c^2 = 8.20 \times 10^{-14}$ J = 0.513 MeV
3. Kinetic energy of the electron	$K = eV = 5.00$ MeV $= 8.00 \times 10^{-13}$ J
4. Total energy of the electron	$E = K + E_0 = 5.51$ MeV $= 8.82 \times 10^{-13}$ J
5. Relativistic mass of the electron	$M = E/c^2 = 9.80 \times 10^{-30}$ kg
6. Speed of the electron	$E = \gamma E_0, \quad \gamma = 1/[1 - (v/c)^2]^{1/2}$ Combining these equations, obtain $v = c[1 - (E_0/E)^2]^{1/2} = 0.996c$
7. Momentum of the electron	$p = mv = 2.93 \times 10^{-21}$ kg•m/s or $E^2 = p^2c^2 + E_0^2$, which gives $p = (E^2 - E_0^2)^{1/2}/c = 2.93 \times 10^{-21}$ kg•m/s
8. Increase in mass of the electron if the kinetic energy increases by 10.0 keV	$\Delta E = \Delta K = 10.0$ keV $\Delta m = \Delta E/c^2 = 1.78 \times 10^{-33}$ kg
9. Magnitude of the magnetic field required to keep the electron in a circle of radius 1.00 m	$F_C = mv^2/r \quad F_B = qvB$ $F_B = F_C \quad qvB = mv^2/r$ $B = mv/qr = p/qr = 1.83 \times 10^{-2}$ T

Each second, the sun gives off about 4.00×10^{26} J of radiant energy through the conversion of mass to energy by nuclear reactions.

Determine the following:

10. Mass lost by the sun each second	$\Delta E = \Delta mc^2 \quad \Delta E/\Delta t = (\Delta m/\Delta t)c^2$ $\Delta m/\Delta t = (\Delta E/\Delta t)/c^2 = (4.00 \times 10^{26}\ \text{J/s})/c^2$ $= 4.44 \times 10^9$ kg/s
11. Mass lost by the sun in the last 1.00×10^{12} years, assuming the same rate of radiation	$\Delta M = (\Delta m/\Delta t)\Delta t$ $= (4.44 \times 10^9\ \text{kg/s})(1.00 \times 10^{12}\ \text{y})$ $= (4.44 \times 10^{21}\ \text{kg•y/s})(3.15 \times 10^7\ \text{s/y})$ $= 1.40 \times 10^{29}$ kg

12. Percentage of the sun's mass lost due to radiation in the last 1.00×10^{10} years (assume the present mass to be 2.00×10^{30} kg)	$\Delta M = 1.40 \times 10^{29}$ kg for the last 1.00×10^{10} years. The mass 1.00×10^{10} years ago was $M_{then} = M_{now} + \Delta M = 2.14 \times 10^{30}$ The percentage of the sun's mass lost in the last 1.00×10^{10} years is $\frac{\Delta M}{M_{then}} \times 100 = \frac{1.40 \times 10^{29}}{2.14 \times 10^{30}} \times 100 = 6.54\%$

Example: An electron moves in the laboratory with a speed 0.800c. An observer moves with a speed 0.600c in a direction opposite the direction of electron's motion. What is the kinetic energy of the electron according to the observer?

Given:

$m_0 = 9.11 \times 10^{-31}$ kg = rest mass of electron
$U_{EL} = 0.800c$ = speed of electron in laboratory frame
$v = -0.600c$ = speed of observer in laboratory frame

Strategy: We can use the velocity transformation to determine the speed of the electron in the observer's frame. Knowing the speed with respect to the observer, we can establish the total energy and then the kinetic energy according to the observer.

Solution: The velocity of the electron with respect to the observer is

$$U_{EO} = (U_{EL} - v)/(1 - U_{EL}v/c^2) = 0.946c$$

We may now determine γ

$$\gamma = 1/[1 - (U_{EO}/c)^2]^{1/2} = 3.08$$

Finally, the kinetic energy is

$$K = E - E_0 = (\gamma - 1)E_0 = (2.08)(0.511 \text{ MeV}) = 1.06 \text{ MeV}$$

Related Text Exercises: Ex. 38-33 through 38-42.

PRACTICE TEST

Take and grade this practice test. Doing so will allow you to determine any weak spots in your understanding of the concepts taught in this chapter. The following section prescribes what you should study further to strengthen your understanding.

Figure 38.6 shows observer O' moving past observer O with a relative speed 0.800c. A man in frame O' throws a ball of rest mas 5.00×10^{-1} kg against a wall which O claims is 4.00 m away, and it returns to him in 5.00 s according to O'.

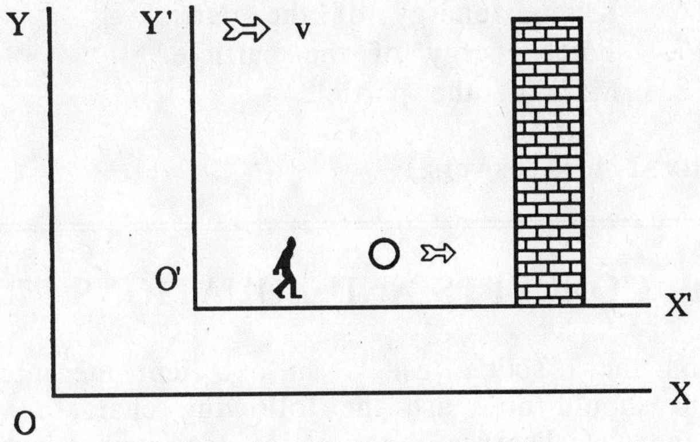

Figure 38.6

Determine the following:

_________ 1. The time it takes the ball to make a round trip according to O
_________ 2. The distance between the man and the wall according to O'
_________ 3. The mass of the ball according to the man
_________ 4. The mass of the ball according to O
_________ 5. The speed of O as measured by O'

Observer O' detects a spaceship traveling 0.900 x in the x' direction

_________ 6. The speed of the spaceship according to O

An electron is accelerated from rest through an electrical potential difference of 1.00×10^6 V.

Determine the following:

_________ 7. Final kinetic energy of the electron
_________ 8. Total relativistic energy of the electron
_________ 9. Relativistic mass of the electron
_________ 10. Speed of the electron
_________ 11. Momentum of the electron

A 2.00×10^3 MW nuclear power plant operating at full power delivers 2.00×10^9 J of electrical energy and 4.00×10^9 J of waste heat every second.

Determine the following:

_________ 12. The rate at which mass is converted to energy
_________ 13. The amount of mass "burned" in one year

A particle of 4.00×10^{-19}-C charge and 5.00×10^{-28}-kg rest mass moving perpendicular to a 0.500-T magnetic field traces out a 1.00-m radius circle.

Determine the following:

________ 14. Magnitude of the velocity of the particle
________ 15. Magnitude of the linear momentum of the particle
________ 16. Kinetic energy of the particle
________ 17. Total energy of the particle
________ 18. Mass of the particle

(See Appendix I for answers.)

PRINCIPAL CONCEPTS AND EQUATIONS PRESCRIPTION

Your score on the practice test is an excellent measure of your understanding of the chapter. You should now use the following chart to write your own prescription for dealing with any weaknesses the practice test points out. Look down the leftmost column to the number of the question(s) you answered incorrectly, reading across that row you will find the concept and/or equation of concern, the section(s) of the study guide you should return to for further study, and some suggested text exercises which you should work to gain additional experience.

Practice Test Questions	Concepts and Equations	Prescription: Principal Concepts	Prescription: Text Exercises
1	Time dilation: $t = \gamma t_o$	1	38-13, 18
2	Length contraction: $L = L_o/\gamma$	2	38-23, 26
3	Rest mass: m_o	4	38-40, 41
4	Relativistic mass: $m = \gamma m_o$	4	38-40, 41
5	Relativistic velocities: $U_x' = (U_x - v)/(1 - U_x v/c^2)$	3	38-6, 8
6	Relativistic velocities	3	38-10, 30
7	Kinetic energy: $K = eV$	2 of Ch 22	22-9
8	Total energy: $E = K + E_o$	5	38-38, 40
9	Total energy: $E = mc^2$	5	38-38, 40
10	Total energy: $E = \gamma E_o$	5	38-38, 40
11	Momentum: $p = mv$, $E^2 = p^2c^2 + E_o^2$, or $p^2c^2 = K^2 + 2KE_o$	5	38-36, 39
12	Mass-energy conversion: $P = \Delta E/\Delta t = (\Delta m/t)c^2$	5	38-41, 42
13	$M = (\Delta m/t)t$	5	38-41, 42
14	Magnetic force: $mv^2/r = qvB$	1 of Ch 26	26-2, 3
15	Momentum: $p = mv = qrB$	5	---
16	Kinetic energy: $K = E - E_o = (\gamma - 1)E_o$	5	38-33, 34
17	Total energy: $E = K + E_o$, $E = \gamma E_o$	5	38-38
18	Relativistic mass: $m = \gamma m_o$	4	38-40, 41

39 QUANTIZATION OF ELECTROMAGNETIC RADIATION

RECALL FROM PREVIOUS CHAPTERS

Previously learned concepts and equations frequently used in this chapter	Text Section	Study Guide Page
Centripetal force: $F_c = mv^2/r$	6-2	6-7
Kinetic energy: $K = mv^2/2$	8-4	8-13
Power: $P = E/\Delta t$	8-5	8-15
Elastic collisions	10-7	10-18
Angular momentum: $L = mvr$	13-1	13-2
Stefan-Boltzmann law: $P = e\sigma AT^4$	16-6	16-8
Coulomb's law: $F = q_1q_2/4\pi\varepsilon_0 r^2$	20-3	20-2
Electric potential energy: $U = q_1q_2/4\pi\varepsilon_0 r$	22-1	22-3
Electrical potential difference: $V = W/q$	22-3	22-3
Intensity: $I = P/A$	32-5	32-9

NEW IDEAS IN THIS CHAPTER

Concepts and equations introduced	Text Section	Study Guide Page
Planck's radiation law: $R_\nu = 2\pi h\nu^3/c^2(e^{h\nu/kt} - 1)$	39-2	39-2
Quantization of energy: $E = nh\nu$	39-2	39-2
Photon energy: $E = h\nu = hc/\lambda$	39-2	39-2
Threshold frequency and work function: $\phi = h\nu_0$	39-3	39-5
Photoelectric equation: $eV_0 = h\nu - \phi$	39-3	39-5
Compton effect: $\Delta\lambda = \lambda' - \lambda = h(1 - \cos\theta)/mc$	39-4	39-8
Line spectra: $1/\lambda = R_H[(1/n_f^2) - (1/n_i^2)]$	39-5	39-12
The Bohr model of Hydrogen	39-6	39-13
Radius of allowed orbits: $r_n = n^2d_0$	39-6	39-13
Speed of electrons in allowed orbits: $v_n = v_1/n$	39-6	39-13
Energy levels of the Hydrogen atom: $E_n = E_1/n^2$	39-6	39-13
Energy and wavelength of emitted photons $E_{photon} = E_f - E_i = -E_1[(1/n_f^2) - (1/n_i^2)]$ $1/\lambda = R_H[(1/n_f^2) - (1/n_i^2)]$	39-6	39-13
Energy level diagrams	39-6	39-18

PRINCIPAL CONCEPTS AND EQUATIONS

1 Cavity Radiation (Section 39-2)

Review: A blackbody is any object that can absorb and emit radiation of all wavelengths. The light emitted by a blackbody can be closely approximated by the radiation emerging from a small opening in a cavity.

The power radiated (rate of emission of electromagnetic energy) from an area A of surface at temperature T is given by the Stefan-Boltzmann law

$$P = e\sigma AT^4$$

If the radiating object is a blackbody then $e = 1$ and

$$P = \sigma AT^4$$

Where $\sigma = 5.67 \times 10^{-8}\ W{\cdot}m^{-2}k^{-4}$ is the Stefan-Boltzmann constant
A is the area of the emitter
T is the absolute temperature

The power radiated per unit area from the cavity is called the radiancy R

$$R = P/A$$

Note that the energy associated with R is at various frequencies. The power radiated per unit area per unit frequency range is the spectral radiancy

$$R_\nu = (P/A)/\Delta\nu$$

The spectral radiancy R_ν integrated over all frequencies gives the radiancy. That is the radiancy is the area under the R_ν vs ν plot, hence

$$R = \int_0^\infty R_\nu d\nu$$

The spectral radiancy for a blackbody appears as shown in Figure 39.1.

Figure 39.1

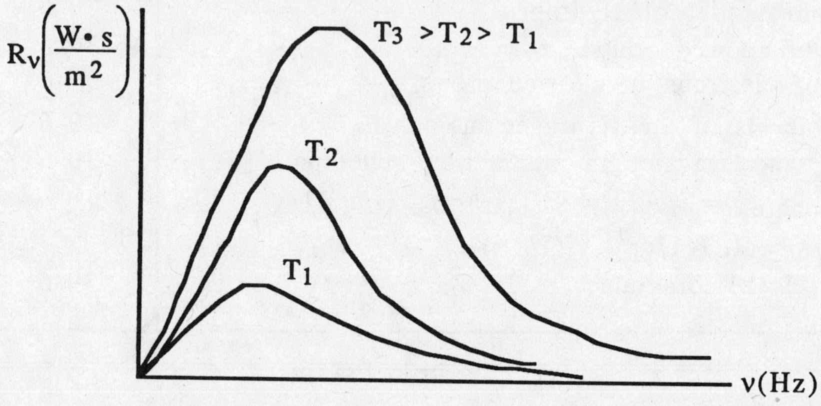

Planck developed a formula that matched this measured frequency distribution of blackbody radiation. To do this he assumed the atoms in the walls of the cavity were small harmonic oscillators with electric charges.

The random thermal motions of the oscillators results in the emission of electromagnetic radiation of all frequencies which fills the cavity and interacts with the oscillators. When thermal equilibrium is reached, the average rate of emission of radiant energy by the oscillators matches the average rate of absorption by the oscillators. Thus, the oscillators share their energy with the radiation in the cavity.

Planck further postulated that the energy of the oscillators is quantized and the only permitted values of the energy are

$$E = nh\nu \qquad n = 1, 2, 3, \ldots$$

The oscillators do not radiate continuously, but rather by jumping from one stationary state to another. That is the energy radiated in a transition from an initial energy state ($E_i = n_i h\nu$) to a lower final energy state ($E_f = n_f h\nu$) is

$$\Delta E = E_i - E_f = (n_i - n_f)h\nu = \Delta n h\nu$$

The energy is radiated in discrete packets and the energy of one of these packets (photons) is

$$E_{photon} = h\nu = hc/\lambda$$

Using this quantization condition, Planck calculated the average energy of the oscillators

$$\langle E \rangle = h\nu/(e^{h\nu/kt} - 1)$$

and then used this to derive his spectral radiancy formula

$$R_\nu = 2\pi h\nu^3/c^2(e^{h\nu/kt} - 1)$$

This expression accounts for all features of the blackbody radiation shown in Figure 39.1.

Practice: Suppose that the spectral radiancy from a cavity radiator is as shown in Figure 39.2. The cavity radiator has an area 1.00×10^{-5} m^2.

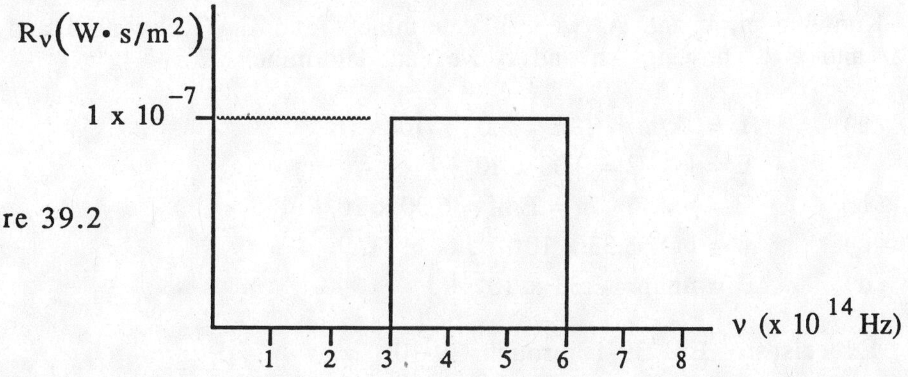

Figure 39.2

Determine the following:

1. Radiancy of the cavity radiator	$R = \int_0^\infty R_\nu d\nu$ Since R_ν is a constant, obtain $R = R_\nu \Delta\nu$ $= (1.00 \times 10^{-7}\ W{\bullet}s/m^2)(3.00 \times 10^{14}/s)$ $= 3.00 \times 10^7\ W/m^2$
2. Power emitted by the cavity radiator	$P = RA$ $= (3.00 \times 10^7\ W/m^2)(1.00 \times 10^{-5}\ m^2)$ $= 3.00 \times 10^2\ W$
3. Temperature of the cavity radiator	$P = e\sigma AT^4$ But e = 1 for a blackbody, hence $T = [P/\sigma A]^{1/4} = 4.80 \times 10^3\ K$

Example: A 1.00 kg mass is attached to a spring of force constant k = 10.0 N/m. The spring is stretched 0.200 m from its equilibrium position and released. (a) Find the frequency of oscillation and the total energy of oscillation according to classical mechanics. (b) Assuming that the energy is quantized, find the quantum number n for the system. (c) Find the amount of energy carried away if the oscillator undergoes a one-quantum change in energy (i.e., it jumps to the next lower quantum state). (d) The change in energy of the oscillator if n increases by 1000.

Given:
$m = 1.00$ kg = mass attached to spring
$k = 10.0$ N/m = spring constant
$A = 0.200$ m = amplitude of oscillation
$\Delta n = 1000$ = change in quantum number

Determine:
f = frequency of oscillation of spring
E = total energy of oscillator
n = quantum number for oscillator
ε = energy carried away by one quantum
ΔE = energy change of oscillator where $\Delta n = 1000$

Strategy: Knowing m, k and A, we can determine f and E. Knowing f and E, we can determine n and ε. Knowing Δn and f, we can determine ΔE.

Solution:

(a) $f = (k/m)^{1/2}/2\pi = 5.03 \times 10^{-1}$ Hz

$E = kA^2/2 = 2.00 \times 10^{-1}$ J

(b) $E = nhf, \quad n = E/hf = 6.00 \times 10^{32}$

(c) $\varepsilon = hf = 3.33 \times 10^{-34}$ J

(d) $\Delta E = \Delta nhf = 3.33 \times 10^{-31}$ J

Related Text Exercises: Ex. 39-1 through 39-10.

2 Photoelectric Effect (Section 39-3)

Review: Figure 39.3 shows photons of frequency ν incident on a photo cell connected to a variable retarding potential.

Figure 39.3

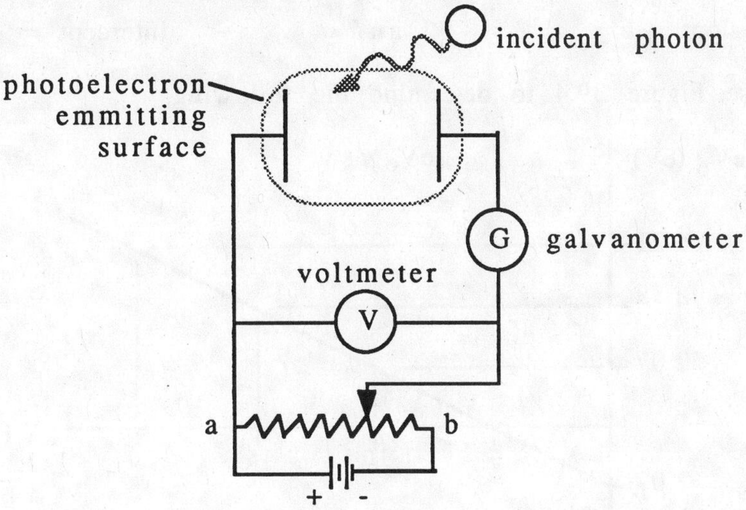

Given either the frequency or wavelength of the incident photons, we can determine their energy.

$$E = h\nu = hc/\lambda$$

Electrons will be emitted by the photoelectron emitting surface if the incident photons can supply them with a minimum amount of energy called the work function ϕ of the surface.

$$E_{min} = h\nu_0 = hc/\lambda_{max} = \phi$$

In this expression, ν_0 is the threshold (or minimum) frequency and λ_{max} is the longest (or maximum) wavelength of the photons which have enough energy to cause an electron to be emitted.

If the incident photon has more energy than the work function (i.e. if $E > \phi$), the excess energy shows up as kinetic energy of the emitted electron. This may be stated mathematically as

$$E = \phi + K$$

As the sliding contact in Figure 39.3 is moved from a to b, the retarding potential is increased (the collector becomes more negative with respect to the emitter) and the electron flow (monitored by the galvanometer) decreases. When the electron flow ceases (i.e. the current drops to zero) the retarding potential is called the stopping potential (V_0). At this point, the kinetic energy of the most energetic electrons (K_{max}) is changed to electrical potential energy (eV_0) and we may write

$$K_{max} = eV_0$$

Combining the above, we may write

$$h\nu = \phi + eV_0$$

or

$$eV_0 = h\nu - \phi$$

Comparing this with the equation for a straight line

$$y = mx + b$$

We see that a plot of eV_o vs ν will be a straight line with

slope = h   and   intercept = $-\phi$

Practice: Use Figure 39.4 to determine the following.

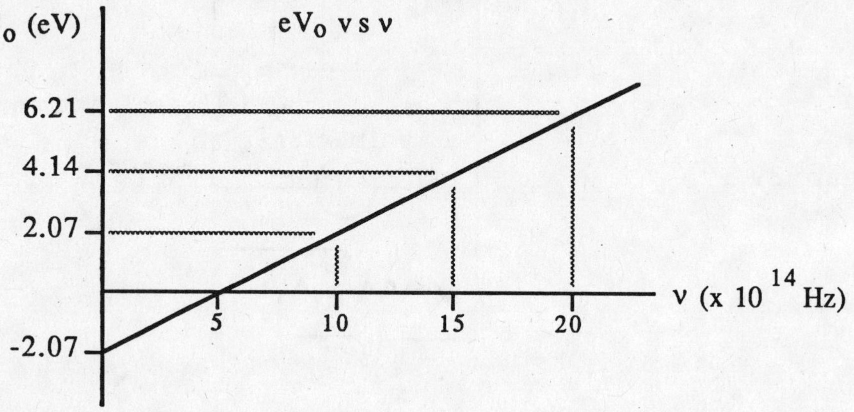

Figure 39.4

Determine the following:

1. Work function of the photo-electron emitting surface	$eV_o = h\nu - \phi$ and $y = mx + b$ Comparing these equations, we see that the intercept of the eV_o vs ν plot is the negative of the work function $-\phi = -2.07$ eV or $\phi = 2.07$ eV
2. Threshold frequency for the photoelectric effect to take place with this surface	The threshold frequency is the minimum frequency of photons which will emit electrons, however the electron will have no kinetic energy. Hence we want the frequency when $K_{max} = eV_o = 0$. From the plot this frequency is $\nu_0 = 5.00 \times 10^{14}$ Hz
3. Maximum of wavelength of light which cause electrons to be emitted	$\lambda_{max} = c/\nu_0 = 6.00 \times 10^{-7}$ m
4. Minimum energy of photons which will cause electrons to be emitted	$E_{min} = h\nu_0 = hc/\lambda_{max} = \phi$ $E_{min} = 2.07$ eV
5. Slope of the eV_o vs ν plot	Slope = 6.21 eV/(15 x 10^{14} Hz) $= 4.14 \times 10^{-15}$ eV•s $= 6.62 \times 10^{-34}$ J•s This value differs slightly from the accepted value for Planck's constant due to the fact that the graph cannot be read with great accuracy.

6. Maximum kinetic energy of electrons emitted by photons of frequency $\nu = 10.0 \times 10^{14}$ Hz	From the plot we see that when $\nu = 10.0 \times 10^{14}$ Hz, then $K_{max} = eV_0 = 2.07$ eV
7. Stopping potential of the electrons emitted by photons of frequency $\nu = 10.0 \times 10^{14}$ Hz	From step 6 we have $eV_0 = 2.07$ eV or $V_0 = 2.07$ V
8. Energy of photon of frequency 10.0×10^{14} Hz	$E = h\nu = 6.63 \times 10^{-19}$ J = 4.14 eV This may also be obtained from the eV_0 vs ν plot by noting that photons of frequency $\nu = 10.0 \times 10^{14}$ Hz cause electrons with $K_{max} = 2.07$ eV to be emitted. Since these photoelectrons had to overcome the work function, the energy of the photons must be $E = \phi + K_{max} = 2.07$ eV + 2.07 eV = 4.14 eV $= 6.63 \times 10^{-19}$ J
9. If photoelectrons will be emitted when the incident radiation has a wavelength of 10.0×10^{-7} m	$\nu = c/\lambda = 3.00 \times 10^{14}$ Hz Since this value is less than the threshold frequency, electrons will not be emitted.
10. Additional energy needed by the photons in step 7 in order to emit electron	The energy of these photons is $E = h\nu = 1.24$ eV, The work function is $\phi = 2.07$ eV The additional energy needed is $E_{add} = \phi - E = 0.83$ eV We can also determine this value from the plot. According to the plot when $\nu = 3.00 \times 10^{14}$ Hz, $eV_0 = -0.83$ eV, that is the energy needed by the photons in order to emit electrons is 0.83 eV

Example: The maximum wavelength for electron emission from a certain metallic surface is 5.50×10^{-7} m. Calculate the binding energy, or work function, of an electron on this surface. Find the stopping potential for electrons emitted from this surface when light of wavelength 2.00×10^{-7} m strikes it.

Given: $\lambda_{max} = 5.50 \times 10^{-7}$ m, $\lambda = 2.00 \times 10^{-7}$ m

Determine: ϕ = work function
V_0 = stopping potential for electrons emitted when the incident light has $\lambda = 2.00 \times 10^{-7}$ m

Strategy: Knowing the maximum wavelength, we can determine the work function for the surface. Knowing the wavelength of the incident light and the work function for the surface, we can determine the stopping potential of the emitted electrons.

Solution: The work function is

$$\phi = h\nu_0 = hc/\lambda_{max} = 3.62 \times 10^{-19} \text{ J}$$

The stopping potential for electrons emitted when light of wavelength $\lambda = 2.00 \times 10^{-7}$ m is incident on the surface may be obtained as follows. The energy of the incident ($\lambda = 2.00 \times 10^{-7}$ m) photons is

$$E = h\nu = hc/\lambda = 9.95 \times 10^{-19} \text{ J}$$

The maximum kinetic energy of the emitted electrons is

$$K_{max} = E - \phi = 6.63 \times 10^{-19} \text{ J}$$

The stopping potential is

$$V_0 = K_{max}/e = 3.96 \text{ V}$$

Related Text Exercises: Ex. 39-11 through 39-20.

3 Compton Effect (Section 39-4)

Review: Figure 39.5(a) shows a photon incident on a stationary electron. This incident photon may be classified by its energy (E), frequency (ν), or wavelength (λ):

$$E = h\nu = hc/\lambda$$

After the photon-electron interaction (Figure 39.5(b)), the electron recoils with a velocity v and a longer wavelength photon is scattered. The scattered photon may be classified by its energy (E'), frequency (ν') or wavelength (λ').

$$E' = h\nu' = hc/\lambda'$$

The electron is initially at rest, hence has only its rest mass energy. After the interaction the recoil electron may be classified by its kinetic energy (K).

$$K = E - E'$$

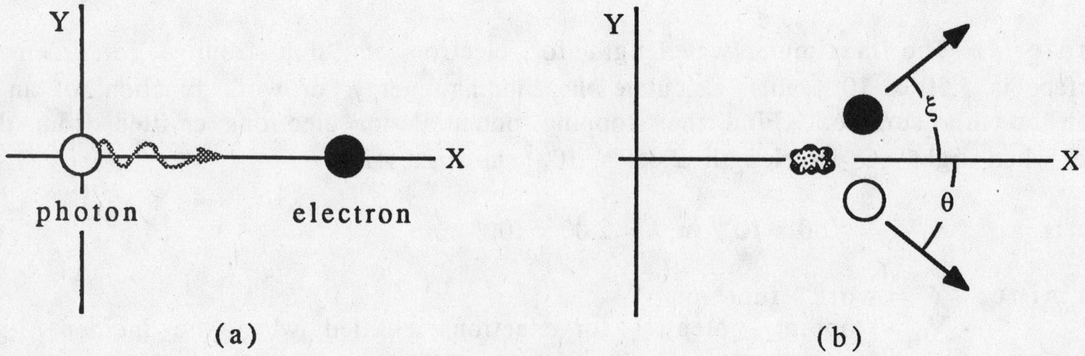

Figure 39.5

Compton observed that the wavelength of the scattered photons was greater than the wavelength of the incident photons and that the wavelength difference depended on the scattering angle ξ. This phenomenon is called the Compton Effect.

Compton theoretically explained this phenomenon by treating both the photons and electrons as particles. Using "billiard ball physics" (conservation of energy and momentum), he derived an expression for the change in wavelength of the two photons in terms of easily measured quantities:

The equation for conservation of energy is

(a) $$hc/\lambda + mc^2 = hc/\lambda' + \gamma mc^2$$

Conservation of momentum in the x direction gives

(b) $$h/\lambda = \gamma mv \cos\xi + (h/\lambda')\cos\theta$$

Conservation of momentum in the y direction gives

(c) $$0 = \gamma mv \sin\xi - (h/\lambda')\sin\theta$$

In the above equations (a, b, and c)

hc/λ = energy of the incident photon
h/λ = momentum of the incident photon
hc/λ' = energy of the scattered photon
h/λ' = momentum of the scattered photon
mc^2 = rest energy of the electron
γmc^2 = total energy of the scattered photon
θ = scatter angle of the photon
ξ = departure angle of the recoil electron

Note that equations a, b, and c contain the four quantities that describe the final state of the system: v and ξ for the electron and λ' and θ for the photon. Compton eliminated v and ξ from these equations and solved for λ' in terms of θ. The final result is

$$\Delta\lambda = (h/mc)(1 - \cos\theta)$$

Notice:

If $\theta = 0°$, $\Delta\lambda = 0$, there is no change in wavelength
$\theta = 90°$, $\Delta\lambda = h/mc$, called the Compton wavelength
$\theta = 180°$, $\Delta\lambda = \Delta\lambda_{max} = 2h/mc$

Practice: The energy of the incident photons in Figure 39.5 is 10.0 keV.

Determine the following:

1. Wavelength of incident photons	$E = hc/\lambda$ $\lambda = hc/E = 1.24 \times 10^{-10}$ m
2. Maximum wavelength shift $\Delta\lambda$	Maximum $\Delta\lambda$ occurs when $\theta = 180°$ $\Delta\lambda = \lambda' - \lambda = h(1 - \cos\theta)/mc$ $\Delta\lambda = 2h/mc = 4.85 \times 10^{-12}$ m
3. Wavelength of photons scattered at $\theta = 60.0°$	$\lambda' = \lambda + h(1 - \cos\theta)/mc = 1.25 \times 10^{-10}$ m

4. Kinetic energy of recoil electrons when photons are scattered at 60.0°	The energy of the incident x-rays is ε = 10.0 keV. The energy of the scattered x-rays is $\varepsilon' = hc/\lambda'$ = 9.94 keV. The kinetic energy of the recoil electrons is $K = \varepsilon - \varepsilon'$ = 60.0 eV.
5. Energy of incident photons that causes scattered photons to exit the target at 135° with an energy of 20.0 keV	ε' = 20.0 keV $\lambda' = hc/\varepsilon' = 6.22 \times 10^{-11}$ m $\lambda = \lambda' - h(1 - \cos\theta)/mc = 5.81 \times 10^{-11}$ m $\varepsilon = hc/\lambda = 3.42 \times 10^{-15}$ J
6. Wavelength of scattered photons when electrons recoils with 5.00 keV of kinetic energy when 10.0 keV x-rays are incident on the target	$\varepsilon' = \varepsilon - K = 5.00$ keV $= 8.00 \times 10^{-16}$ J $\lambda' = hc/\varepsilon' = 2.49 \times 10^{-10}$ m

Example: X-ray photons of wavelength 5.00×10^{-11} m are incident on a copper target, and Compton scattered photons are observed at an angle of 135° with respect to the direction of the incident beam. Find the (a) wavelength of the scattered photons, (b) magnitude of the linear momentum of the incident and scattered photons, (c) kinetic energy of the recoil electrons, and (d) magnitude and direction of the momentum of the recoil electrons.

Given: $\lambda = 5.00 \times 10^{-11}$ m = wavelength of incident photons
θ = 135° = scattering angle for photons

Determine: λ' = wavelength of scattered photons
p and p' = momentum of incident and scattered photons
K = kinetic energy of recoil electrons
p_e = linear momentum of recoil electrons

Strategy: We can use the Compton equation and the known information to determine the wavelength of the scattered photons. We can determine the momentum of the incident and scattered photons from their wavelength. We can conserve energy to determine the kinetic energy of the recoil electrons. We can conserve momentum to determine the magnitude and direction of the linear momentum of the recoil electrons.

Solution: The wavelength of the scattered photons is obtained by using the Compton equation:

$$\lambda' = \lambda + h(1 - \cos\theta)/mc = 5.41 \times 10^{-11} \text{ m}$$

The momentum of the incident and scattered photons is

$$p = h/\lambda = 1.33 \times 10^{-23} \text{ kg·m/s} \qquad p' = h/\lambda' = 1.22 \times 10^{-23} \text{ kg·m/s}$$

The kinetic energy of the recoil electron is determined by conserving energy:

$$E + E_0 = E' + K + E_0$$

$$K = E - E' = hc/\lambda - hc/\lambda' = hc\left(\frac{1}{\lambda} - \frac{1}{\lambda'}\right) = 3.01 \times 10^{-16}\ \text{J}$$

The magnitude and direction of the linear momentum of the recoil electron are determined by conserving momentum, as shown in Figure 39.6.

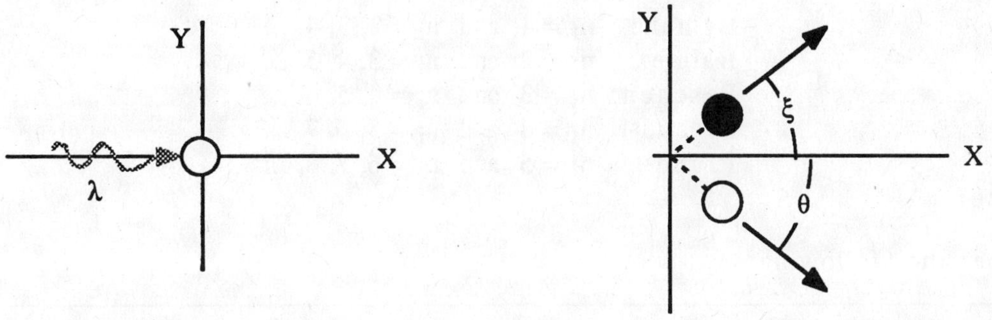

Figure 39.6

Conserving momentum in the x direction, we obtain

(i) $\frac{h}{\lambda} = \left(\frac{h}{\lambda'}\right)\cos\theta + p\cos\xi$ $\qquad p\cos\xi = \frac{h}{\lambda} - \frac{h\cos\theta}{\lambda'} = 2.19 \times 10^{-23}\ \text{kg}\cdot\text{m/s}$

Conserving momentum in the y direction, we obtain

(ii) $0 = -\left(\frac{h}{\lambda'}\right)\sin\theta + p\sin\xi$ $\qquad p\sin\xi = \left(\frac{h}{\lambda'}\right)\sin\theta = 0.866 \times 10^{-23}\ \text{kg}\cdot\text{m/s}$

Dividing (ii) by (i) we obtain

$\frac{p\sin\xi}{p\cos\xi} = \tan\xi = 0.395$ $\qquad \xi = \tan^{-1}(0.395) = 21.6°$

This value for ξ can be inserted into either (i) or (ii) to obtain p.

From (i), $p = \frac{2.19 \times 10^{-23}\ \text{kg}\cdot\text{m/s}}{\cos\xi} = 2.35 \times 10^{-23}\ \text{kg}\cdot\text{m/s}$

From (ii), $p = \frac{0.866 \times 10^{-23}\ \text{kg}\cdot\text{m/s}}{\sin\xi} = 2.35 \times 10^{-23}\ \text{kg}\cdot\text{m/s}$

The final momentum vector for the recoil electron has a magnitude of 8.93×10^{-23} kg•m/s and is directed at an angle $\xi = 75.8°$ from the direction of the incident photon.

Related Text Exercises: Ex. 39-21 through 39-27.

4 Line Spectra (Section 39-5)

Review: As a result of the work of J. J. Balmer and J. R. Rydberg, it was discovered that the wavelengths of the spectral lines of hydrogen can be determined by

$$\frac{1}{\lambda} = R_H\left(\frac{1}{n_f^2} - \frac{1}{n_i^2}\right) \qquad n_i > n_f$$

The various series observed and reported are

Lyman:	$n_f = 1$ and $n_i = 2, 3, 4, \ldots$
Balmer:	$n_f = 2$ and $n_i = 3, 4, 5, \ldots$
Paschen:	$n_f = 3$ and $n_i = 4, 5, 6, \ldots$
Brackett:	$n_f = 4$ and $n_i = 5, 6, 7, \ldots$
Pfund:	$n_f = 5$ and $n_i = 6, 7, 8, \ldots$

Practice:

Determine the following:

1. Longest wavelength in Lyman series	For the Lyman series, $n_f = 1$ and $n_i = 2, 3, 4, \ldots$ The longest wavelength occurs for $n_i = 2$. $R_H = 1.10 \times 10^7/\text{m}$ $\frac{1}{\lambda} = R_H\left(\frac{1}{1^2} - \frac{1}{2^2}\right) = \frac{3R_H}{4} = 8.25 \times 10^6/\text{m}$ $\lambda = 1.21 \times 10^{-7}$ m
2. Shortest wavelength in Lyman series	The shortest wavelength occurs for $n_i = \infty$. $\frac{1}{\lambda} = R_H\left(\frac{1}{1^2} - \frac{1}{(\infty)^2}\right) = R_H = 1.10 \times 10^7/\text{m}$ $\lambda = 9.09 \times 10^{-8}$ m
3. Frequency of light when $n_f = 2$ and $n_i = 4$	$\frac{1}{\lambda} = R_H\left(\frac{1}{2^2} - \frac{1}{4^2}\right) = \frac{3R_H}{16}$ $f = \frac{c}{\lambda} = \frac{3R_Hc}{16} = 6.19 \times 10^{14}$ Hz
4. Energy of photons given off when $n_f = 3$ and $n_i = 4$	$\frac{1}{\lambda} = R_H\left(\frac{1}{3^2} - \frac{1}{4^2}\right) = \frac{7R_H}{144}$ $E = \frac{hc}{\lambda} = \frac{7R_Hhc}{144} = 1.06 \times 10^{-19}$ J

Example: Calculate the wavelength, frequency , and energy of the longest wavelength photon for the Lyman series of hydrogen.

Given: Lyman series: $n_f = 1$ and $n_i = 2, 3, 4, \ldots$

Determine: The wavelength, frequency, and energy of the longest wavelength photon for the Lyman series of hydrogen.

Strategy: We must first determine the value of n_i that gives the longest wavelength photon. Knowing n_f and n_i, we can calculate the wavelength. We can then use the wavelength to determine the frequency and the energy of the photon.

Solution: The wavelength of the various photons of the Lyman series is obtained by

$$\frac{1}{\lambda} = R_H\left(\frac{1}{1^2} - \frac{1}{n_i^2}\right) = R_H[1 - (\frac{1}{n_i^2})]$$

The largest value of λ occurs for the smallest value of n_i. For the Lyman series, the smallest value of n_i is 2. Using $n_i = 2$, we obtain λ, f, ε:

$$\frac{1}{\lambda} = R_H[1 - (\frac{1}{n_i^2})] = R_H[1 - (\frac{1}{2^2})] = \frac{3R_H}{4} \qquad \lambda = \frac{4}{3R_H} = 1.21 \times 10^7 \text{ m}$$

$$f = \frac{c}{\lambda} = 2.48 \times 10^{15} \text{ Hz} \qquad \varepsilon = hf = 1.64 \times 10^{-18} \text{ J}$$

Related Text Exercises: Ex. 39-28 through 39-32.

5 Bohr Model of Hydrogen (Section 39-6)

Review: Shown below are the principal assumptions of Bohr and their consequences.

Assumption 1. The electron revolves about the nucleus in certain non-radiating circular orbits.

In order for an electron to travel in a circular orbit with a speed v and radius r, it must be experiencing a centripetal force given by

(a) $$F_{cent} = mv^2/r$$

For the case of an electron in a Hydrogen atom, this central force is supplied by the Coulomb attractive force between the positively charged proton nucleus and the negatively charged orbital electron.

(b) $$F_{cent} = F_{coulb} = e^2/4\pi\varepsilon_0 r^2$$

Combining (a) and (b), we have

(c) $$mv^2/r = e^2/4\pi\varepsilon_0 r^2$$

When (c) is multiplied by r/2, we recognize the result as the kinetic energy of the orbital electron

(d) $$K = mv^2/2 = e^2/8\pi\varepsilon_0 r$$

If the orbital electron is a distance r from the nucleus, the electrical potential energy of the two charge system is

(e) $$U = -e^2/4\pi\varepsilon_0 r$$

Adding (d) and (e), we obtain the total mechanical energy of the electron-proton system.

(f) $$E = K + U = -e^2/8\pi\varepsilon_0 r$$

Assumption 2. The only allowed orbits are those in which the orbital angular momentum is a positive integer multiple of $h/2\pi$

(g) $$L = \frac{nh}{2\pi} \qquad n = 1, 2, 3, \ldots$$

Recall that the angular momentum of a particle traveling in a circle is

(h) $$L = mvr$$

Combining (g) and (h), we have

(i) $$mvr = \frac{nh}{2\pi}$$

If we combine (c) and (i) in such a manner as to eliminate v, we obtain

(j) $$r_n = \left(\frac{\varepsilon_0 h^2}{\pi m e^2}\right) n^2 \qquad n = 1, 2, 3, \ldots$$

Let us agree to denote the radius of the smallest orbit (the value for r_n when $n = 1$) by a_0

(k) $$a_0 = r_1 = \frac{\varepsilon_0 h^2}{\pi m e^2} = 5.30 \times 10^{-11} \text{ m}$$

Then the radii of the other orbits are given by

(l) $$r_n = n^2 a_0 \qquad n = 2, 3, 4, \ldots$$

If (j) is inserted into (f), we obtain

(m) $$E_n = \frac{-me^4}{8\varepsilon_0^2 h^2 n^2} \qquad n = 1, 2, 3, \ldots$$

Equation (m) gives us the quantized energy of the stationary states of the Hydrogen atom.

If (c) and (i) are combined in such a manner as to eliminate r, we obtain

(n) $$v_n = \left(\frac{e^2}{2\varepsilon_0 h}\right)/n \qquad n = 1, 2, 3, \ldots$$

If we denote the speed of the electron in the smallest orbit by v_1, then

(o) $$v_1 = \frac{e^2}{2\varepsilon_0 h} = 2.18 \times 10^6 \text{ m/s}$$

and

(p) $$v_n = \frac{v_1}{n}$$

Assumption 3. Radiation (a photon) is emitted by the atom when an electron undergoes a transition from one stationary state to another. The energy of the photon is $E_i - E_f$

(q) $$E_{photon} = E_i - E_f$$

Where $E_i = \dfrac{-me^4}{8\varepsilon_0^2h^2n_i^2}$ $\quad E_f = \dfrac{-me^4}{8\varepsilon_0^2h^2n_f^2}$ $\quad E_{photon} = h\nu = hc/\lambda$ or

(r) $$\frac{1}{\lambda} = \left(\frac{me^4}{8\varepsilon_0^2h^3c}\right)\left[\frac{1}{n_f^2} - \frac{1}{n_i^2}\right]$$

Evaluating the constants, we realize that

(s) $$\frac{me^4}{8\varepsilon_0^2h^3c} = R_H$$

Hence, we have

(t) $$\frac{1}{\lambda} = R_H\left[\frac{1}{n_f^2} - \frac{1}{n_i^2}\right] \qquad n_f < n_i$$

Notice that this expression for $1/\lambda$ is developed from theory using first principles and Bohrs assumptions and it agrees exactly with the empirical results of Balmer. As a result of the preceding theory, we have expressions that allow us to determine the following for the one-electron hydrogen atom:

(l) radius of the orbits available to the electron
(p) speed of the electron in the available orbits
(b) centripetal force on an electron in an available orbit
(d) kinetic energy of an electron in an available orbit
(e) electric potential energy of the atom when the electron is in one of the available orbits
(m) total energy of the atom when the electron is in one of the available orbits
(g or h) angular momentum of the electron in any orbit
(q or t) energy, frequency, and wavelength of the photon emitted when an electron jumps to a lower orbit

Practice: Given the following information for a hydrogen atom:

$m = 9.11 \times 10^{-31}$ kg = mass of an electron
$e = 1.60 \times 10^{-19}$ C = charge on an electron
$h = 6.63 \times 10^{-34}$ J•s = Planck's constant
$\dfrac{1}{4\pi\varepsilon_0} = 9.00 \times 10^9 \dfrac{N\cdot m^2}{C^2}$ = Coulomb's law constant
$c = 3.00 \times 10^8$ m/s = speed of light
$a_0 = 5.30 \times 10^{-11}$ m = radius of the first Bohr orbit
$v_1 = 2.18 \times 10^6$ m/s = speed of an electron in the first Bohr orbit
$E_1 = -2.18 \times 10^{-18}$ J = energy of a hydrogen atom when the electron is in the first Bohr orbit
$R_H = 1.10 \times 10^7$/m = Rydberg's constant

Determine the following:

1. Radius of n = 2 orbit	$r_n = n^2 a_0 = 4a_0 = 2.12 \times 10^{-10}$ m
2. Speed of electron in n = 2 orbit	$v_n = v_1/n = v_1/2 = 1.09 \times 10^6$ m/s
3. Kinetic energy of electron in the n = 2 orbit	$K_2 = mv_2^2/2 = 5.41 \times 10^{-19}$ J $K_2 = ke^2/2r_2 = 5.43 \times 10^{-19}$ J The difference is due to rounding off.
4. Electric potential energy of an electron in n = 2 orbit	$U_2 = -ke^2/r_2 = -1.09 \times 10^{-18}$ J
5. Total energy of an electron in n = 2 orbit	$E_2 = K_2 + U_2 = -5.47 \times 10^{-19}$ J $E_2 = E_1/n^2 = E_1/4 = -5.45 \times 10^{-19}$ J The difference is due to rounding off.
6. Centripetal force on electron in n = 2 orbit	$F_c = mv_2^2/r_2 = 5.11 \times 10^{-9}$ N $F_c = ke^2/r_2^2 = 5.13 \times 10^{-9}$ N The difference is due to rounding off.
7. Energy of photon in n = 2 to n = 1 transition	$\varepsilon_{2\to1} = E_2 - E_1 = 1.63 \times 10^{-18}$ J $\varepsilon_{2\to1} = -E_1\left[\frac{1}{n_f^2} - \frac{1}{n_i^2}\right] = 1.63 \times 10^{-18}$ J
8. Frequency of photon emitted in n = 2 to n = 1 transition	$f_{2\to1} = \varepsilon_{2\to1}/h = 2.46 \times 10^{15}$ Hz $f_{2\to1} = (\frac{-E_1}{h})\left[\frac{1}{n_f^2} - \frac{1}{n_i^2}\right] = 2.46 \times 10^{15}$ Hz
9. Wavelength of photon emitted in n = 2 to n = 1 transition	$\lambda_{2\to1} = c/f_{2\to1} = 1.22 \times 10^{-7}$ m $\frac{1}{\lambda_{2\to1}} = R_H\left[\frac{1}{n_f^2} - \frac{1}{n_i^2}\right] = \frac{3R_H}{4}$ $= 8.25 \times 10^{-6}$ /m $\lambda_{2\to1} = 1.21 \times 10^{-7}$ m The difference is due to rounding off.
10. Photon energy that causes atom to undergo n = 2 to n = 3 transition	$E_3 = -E_1/n_f^2 = -E_1/9 = -2.42 \times 10^{-19}$ J $E_2 = -E_1/n_i^2 = -E_1/4 = -5.45 \times 10^{-19}$ J $\varepsilon_{2\to3} = E_3 - E_2 = 3.03 \times 10^{-19}$ J $\varepsilon_{2\to3} = -E_1\left[\frac{1}{n_i^2} - \frac{1}{n_f^2}\right] = 3.03 \times 10^{-19}$ J

11. Photon frequency and wavelength that cause atoms to undergo n = 2 to n = 3 transition	$f_{2\to3} = \varepsilon_{2\to3}/h = 4.57 \times 10^{14}$ Hz $\lambda_{2\to3} = c/f_{2\to3} = 6.56 \times 10^{-7}$ m $\frac{1}{\lambda_{2\to3}} = R_H\left[\frac{1}{n_i^2} - \frac{1}{n_f^2}\right] = 1.53 \times 10^6$/m $\lambda_{2\to3} = 6.54 \times 10^{-7}$ m The difference is due to rounding off.

Example: A hydrogen atom in the n = 4 state decays to the n = 2 state. (a) Determine what happens to the radius (r_n) of the electron's orbit, the linear speed (v_n) of the electron in its orbit, its angular speed (ω_n), the centripetal force (F_{cn}) on it, its kinetic energy (K_n), the electric potential energy (U_n) and the total energy (E_n) of the atom, and the angular momentum (L_n) of the electron. (b) Determine the energy (ε), frequency (f), wavelength (λ), and the angular momentum (L) of the emitted photon.

Given: The atom is a hydrogen atom.
$n_i = 4$ = initial state
$n_f = 2$ = final state

Determine: (a) The change in r_n, v_n, ω_n, F_{cn}, K_n, U_n, E_n, and L_n. (b) ε, f, λ, and L of the emitted photon.

Strategy: (a) As a result of our development of the Bohr theory, we know expressions for r_n, v_n, F_{cn}, K_n, U_n, E_n, and L_n. We can develop an expression for ω_n. We can use these expressions to determine the desired quantities for the initial and final states. We can then take the ratio of the desired quantities for each state in order to determine what happens as the atom decays from n = 4 to n = 2. (b) We can use the Bohr theory to determine ε. Knowing ε, we can use the information about photons to obtain f and λ. Finally, we can use conservation of angular momentum to obtain L.

Solution:

(a) The radius of an acceptable orbit is $r_n = n^2 a_0$. Hence

$$r_4 = 16a_0 \qquad r_2 = 4a_0 = r_4/4.$$

The speed of an electron in an acceptable orbit is $v_n = v_1/n$. Hence

$$v_4 = v_1/4 \qquad v_2 = v_1/2 = 2v_4$$

From v_n and r_n, we can obtain $\omega_n = v_n/r_n = (1/n^3)(v_1/a_0)$. Hence

$$\omega_4 = (1/64)(v_1/a_0) \qquad \omega_2 = (1/8)(v_1/a_0) = 8\omega_4$$

The centripetal force is $F_{cent} = mv_n^2/r_n = (1/n^4)(mv_1^2/a_0)$. Hence

$$F_{c4} = (1/256)(mv_1^2/a_0) \qquad F_{c2} = (1/16)(mv_1^2/a_0) = 16F_{c4}$$

Alternately, the centripetal force is $F_{cn} = ke^2/r_n^2 = (1/n^4)(ke^2/a_0^2)$. Hence

$$F_{c4} = (1/256)(ke^2/a_0^2) \qquad F_{c2} = (1/16)(ke^2/a_0^2) = 16F_{c4}$$

The kinetic energy is $K_n = mv_n^2/2 = (1/n^2)(mv_1^2/2)$. Hence

$$K_4 = (1/16)(mv_1^2/2) \qquad K_2 = (1/4)(mv_1^2/2) = 4K_4$$

The electric potential energy is $U_n = -ke^2/r_n = (1/n^2)(-ke^2/a_0)$. Hence

$$U_4 = (1/16)(-ke^2/a_0) \qquad U_2 = (1/4)(-ke^2/a_0) = 4U_4$$

The total energy is $E_n = (E_1/n^2)$. Hence

$$E_4 = (E_1/16) \qquad E_2 = (E_1/4) = 4E_4$$

The angular momentum is $L_n = nh/2\pi$. Hence

$$L_4 = 4h/2\pi \qquad L_2 = 2h/2\pi = L_4/2$$

(b) $$\varepsilon_{4\to 2} = E_4 - E_2 = -E_1\left[\frac{1}{n_2^2} - \frac{1}{n_4^2}\right] = -(3/16)E_1 = 4.09 \times 10^{-19}\ \text{J}$$

The frequency and the wavelength of the photon are

$$f_{4\to 2} = \varepsilon_{4\to 2}/h = 6.17 \times 10^{14}\ \text{Hz} \qquad \lambda_{4\to 2} = c/f_{4\to 2} = 4.86 \times 10^{-7}\ \text{m}$$

The angular momentum of the photon is determined by assuming that the angular momentum lost by the atom is carried off by the photon:

$$L_{photon} = L_4 - L_2 = h/\pi = 2.11 \times 10^{-34}\ \text{J}\cdot\text{s}$$

Related Text Exercises: Ex. 39-33 through 39-42.

6 Energy Level Diagram (Section 39-6)

Review: The energy of the hydrogen atom in its nth state is

$$E_n = (E_1/n^2) \quad \text{where} \quad E_1 = -2.18 \times 10^{-18}\ \text{J} = -13.6\ \text{eV}$$

Therefore $E_1 = -13.6$ eV, $E_2 = -3.40$ eV, $E_3 = -1.51$ eV, $E_4 = -0.850$ eV, $E_5 = -0.544$ eV, $E_6 = -0.378$ eV, and so on. It is useful to arrange these energy values in an energy level diagram as shown in Figure 39.7.

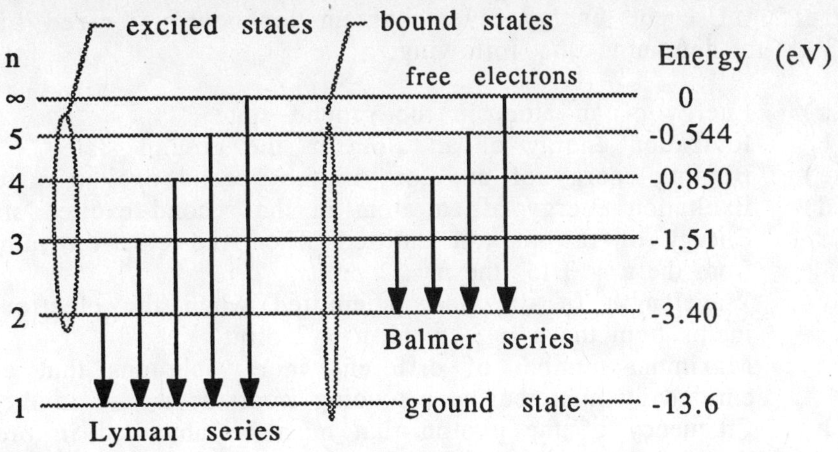

Figure 39.7

Practice: Refer to Figure 39.7.

Determine the following:

1. Energy of an atom in ground state	For the ground state, n = 1, E_1 = -13.6 eV
2. Energy of an atom in second excited state	n = 3 for second excited state E_3 = -1.51 eV
3. Binding and excitation energy for atom when electron is in third excited state	n = 4 for third excited state $E_{binding} = E_4 = -0.850$ eV $E_{excited} = E_4 - E_1 = +12.75$ eV
4. Energy, frequency, and wave-length of lowest-energy photon emitted in Lyman series	$\varepsilon = E_2 - E_1 = +10.2$ eV $f = \varepsilon/h = 2.46 \times 10^{15}$ Hz $\lambda = c/f = 1.22 \times 10^{-7}$ m
5. Maximum number of photons emitted by atom in n = 4 state	The atom can decay to the ground state by any of the following schemes: $n = 4 \rightarrow n = 1$ $n = 4 \rightarrow n = 2$; $n = 2 \rightarrow n = 1$ $n = 4 \rightarrow n = 3$; $n = 3 \rightarrow n = 1$ $n = 4 \rightarrow n = 3$; $n = 3 \rightarrow n = 2$; $n = 2 \rightarrow n = 1$ This amounts to eight possibilities but only six different photons.
6. Wavelength of incident photon that causes atom to go from n = 2 to n = 4 state	$\varepsilon_{2\rightarrow4} = E_4 - E_2 = 2.55$ eV $\lambda_{2\rightarrow4} = hc/\varepsilon_{2\rightarrow4} = 4.87 \times 10^{-7}$ m
7. Whether or not a photon 2.00 eV can excite a hydrogen atom from n = 2 state to the n = 3 state	E_2 = -3.40 eV, E_3 = -1.51 eV, Excitation can occur only with a photon of energy $E_{2\rightarrow3} = E_3 - E_2 = 1.89$ eV. Since this photon has an energy of ε = 2.00 eV, the excitation cannot occur.

Example: Use the energy level diagram for doubly ionized Lithium shown in Figure 39.8 to determine the following:

(a) Energy of an atom in the ground state
(b) Ionization energy of an atom in the ground state
(c) Binding energy of an atom in the second excited state
(d) Excitation energy of an atom in the second excited state
(e) Energy of the photon emitted when the electron jumps from the n = 3 to the n = 2 orbit
(f) Wavelength of the photon emitted when the electron jumps from the n = 3 to the n = 2 orbit
(g) Maximum number of different energy photons that could be emitted by numerous atoms going from the n = 4 to the n = 1 state
(h) Frequency of the photon that must be absorbed in order to cause the electron to jump from the n = 3 to the n = 4 orbit

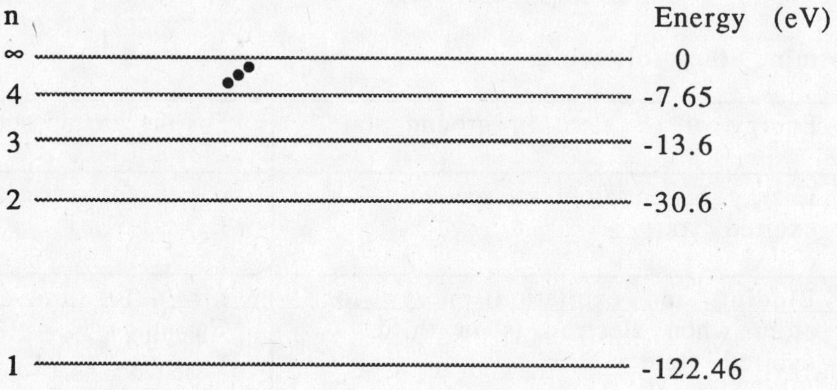

Figure 39.8

Given: Energy level diagram

Determine: Quantities (a) through (h) listed above

Strategy:

(a) The energy of an atom in the ground state is read from the diagram as E_1

(b) The ionization energy of a hydrogen like ionized atom in the ground state is just the energy to free an electron in the n = 1 state. Hence the ionization energy is just the negative of the binding energy of an electron in the n = 1 state

(c) The binding energy of an atom in the second excited state is read from the diagram as E_3. Note that n = 1 is the ground state, n = 2 is the first excited state and n = 3 is the second excited state

(d) An electron in the second excited state has been excited from the n = 1 to the n = 3 state. Knowing this we can determine the excitation energy

(e) Read E_2 and E_3 from the diagram and then calculate the energy of the photon emitted

(f) Knowing the energy of the photon from (e), calculate its wavelength

(g) Look at all possible ways to reach the ground state from the n = 4 state and then determine the number of different energy photons that could be emitted

(h) Read E_3 and E_4 from the diagram, calculate the energy and then the frequency of the photon

Solution:

(a) Energy of an atom in the ground state is $E_1 = -122.4$ eV

(b) Ionization energy is $E_{ion} = 122.4$ eV

(c) Binding energy of an atom in the second excited state is $E_3 = -13.6$ eV

(d) Excitation energy of atom in the second excited state is

$$E_f - E_i = E_3 - E_1 = -13.6 \text{ eV} - (-122.4 \text{ eV}) = 108.8 \text{ eV}$$

(e) Energy of the photon emitted when an electron jumps from n = 3 to n = 2 is

$$E_{photon} = E_i - E_f = -13.6 \text{ eV} - (-30.6 \text{ eV}) = 20.0 \text{ eV}$$

(f) Wavelength of the photon associated with the electron jump in (e) is

$$\lambda = hc/E_{photon} = (1.24 \times 10^{-6} \text{ eVm})/20.0 \text{ eV} = 6.20 \times 10^{-8} \text{ m}$$

(g) The atom can decay to the ground state by any of the following schemes:

$n = 4 \rightarrow n = 1$
$n = 4 \rightarrow n = 2 \rightarrow n = 1$
$n = 4 \rightarrow n = 3 \rightarrow n = 1$
$n = 4 \rightarrow n = 3 \rightarrow n = 2 \rightarrow n = 1$

This amounts to eight possibilities but only six different photon energies

(h) In order to be absorbed, the photon must have an energy

$$E_{photon} = E_4 - E_3 = (-7.65 \text{ eV}) - (-13.6 \text{ eV}) = 5.95 \text{ eV}$$

The frequency of a photon with this energy is

$$\nu = E_{photon}/h = (5.59 \text{ eV})/(4.14 \times 10^{-15} \text{ eV}\cdot\text{s}) = 1.35 \times 10^{15} \text{ Hz}$$

Related Text Exercises: Ex. 39-36 and 39-41.

PRACTICE TEST

Take and grade this practice test. Doing so will allow you to determine any weak spots in your understanding of the concepts taught in this chapter. The following section prescribes what you should study further to strengthen your understanding.

A thin walled metal sphere of radius $R = 3.00 \times 10^{-1}$ m has a circular opening in its surface of radius $r = 3.00 \times 10^{-3}$ m. The sphere is kept at $T = 3.00 \times 10^{3}$ K, and the emissivity of its surface under these conditions is $e = 0.300$.

Determine the following:

_________ 1. Rate at which energy is radiated from the opening
_________ 2. Energy radiated from the opening in an hour
_________ 3. The time it would take the sphere to radiate the same amount of energy as that radiated from the opening in an hour
_________ 4. The rate of emission of photons from the opening assuming that all of the radiation has a wavelength of 10.0 nm

Light of wavelength $\lambda = 3.00 \times 10^{-7}$ m is incident on a photocell that has a work function of 3.00 eV. Determine the following:

_________ 5. Energy of the incident photons
_________ 6. Threshold frequency
_________ 7. Maximum kinetic energy of the emitted electrons
_________ 8. Stopping potential for the emitted electrons

When 10.0 keV x-rays are incident on a thin target, recoil electrons leave the target with a kinetic energy of 60.0 eV. Determine the following:

_________ 9. Wavelength of the scattered x-rays
_________ 10. Angle of scatter for the x-rays
_________ 11. Angle of departure for the recoil electrons
_________ 12. Momentum of the recoil electrons

Determine the following for a hydrogen atom:

_________ 13. Radius of the $n = 3$ orbit
_________ 14. Speed of an electron in the $n = 3$ orbit
_________ 15. Angular speed of an electron in the $n = 3$ orbit
_________ 16. Kinetic energy of an electron in the $n = 3$ orbit
_________ 17. Centripetal force experienced by an electron in the $n = 3$ orbit
_________ 18. Electric potential energy of an atom in the $n = 3$ state
_________ 19. Angular momentum of an electron in the $n = 3$ orbit
_________ 20. Energy of the photon emitted when the atom goes from the $n = 3$ to the $n = 2$ state
_________ 21. Angular momentum of the photon emitted when the atom goes from the $n = 3$ to the $n = 2$ state
_________ 22. Wavelength of the photon emitted when the atom goes from the $n = 3$ to the $n = 2$ state

(See Appendix I for answers.)

PRINCIPAL CONCEPTS AND EQUATIONS PRESCRIPTION

Your score on the practice test is an excellent measure of your understanding of the chapter. You should now use the following chart to write your own prescription for dealing with any weaknesses the practice test points out. Look down the leftmost column to the number of the question(s) you answered incorrectly, reading across that row you will find the concept and/or equation of concern, the section(s) of the study guide you should return to for further study, and some suggested text exercises which you should work to gain additional experience.

Practice Test Questions	Concepts and Equations	Prescription	
		Principal Concepts	Text Exercises
1	Stefan-Boltzmann law and Power: $P = E/t = e\sigma AT^4$	1	39-1, 2
2	Power: $P = E/t$	5 of Ch 8	39-3
3	Stefan-Boltzmann law and Power: $P = E/t = e\sigma AT^4$	1	39-1, 2
4	Energy quantization: $E/t = (n/t)h\nu$	1	---
5	Photon energy: $E = hc/\lambda$	2	39-11, 12
6	Work function: $\phi = h\nu_0$	2	39-13, 14
7	Photoelectric effect: $K = eV_0 = E - \phi$	2	39-17, 19
8	Electrical potential difference: $V_0 = K/e$	2 of Ch 22	39-14, 16
9	Compton effect: $E' = E - K$ and $\lambda' = hc/E'$	3	39-26
10	Compton effect: $\lambda' - \lambda = (h/mc)(1 - \cos\theta)$	3	39-23, 24
11	Compton effect - conservation of momentum	3	39-27
12	Compton effect - conservation of momentum	3	39-27
13	Radius of an orbit: $r_n = n^2 a_0$	5	39-34, 37
14	Speed of electron in orbit: $v_n = v_1/n$	5	39-37
15	Angular speed: $\omega_n = v_n/r_n$	5	---
16	Kinetic energy: $K_n = mv_n^2/2$	5	39-34
17	Centripetal force: $F_n = mv_n^2/r_n$	5	---
18	Electrical potential energy: $U_n = -e^2/4\pi\varepsilon_0 r_n$	5	39-34
19	Angular momentum: $L_n = nh/2\pi$	5	39-40
20	Energy of emitted photon: $E = E_i - E_f$	5	39-35, 36
21	Angular momentum of photon: $L = L_i - L_f$	5	39-40
22	Wavelength of emitted photon: $\lambda = hc/E$	5	39-35

40 QUANTUM MECHANICS

RECALL FROM PREVIOUS CHAPTERS

Previously learned concepts and equations frequently used in this chapter	Text Section	Study Guide Page
Kinetic energy: $K = mv^2/2$	8-4	8-13
Momentum: $p = mv = (2mK)^{1/2}$	10-3	10-10
Electric potential difference: $V = \Delta U/q_o$	22-3	22-3
Energy and momentum of a photon: $E = h\nu$ and $p = E/c$	39-2	39-5
Bragg equation: $m\lambda = 2d \sin\theta_m$	36-4	---
Relativistic mass: $m = \gamma m_o$ $\gamma = [1 - (v/c)^2]^{-1/2}$	38-6	38-10
Single slit diffraction pattern minima: $m\lambda = a \sin\theta_m$, $\sin\theta \approx \tan\theta = y_m/L$	37-2	37-1

NEW IDEAS IN THIS CHAPTER

Concepts and equations introduced	Text Section	Study Guide Page
The de Broglie wavelength: $\lambda = h/p$	40-1	40-1
Heisenberg uncertainty relations: $\Delta x \Delta p_x \geq \hbar/2$ $\Delta E \Delta t \geq \hbar/2$	40-4	40-7
Probability density: $dP/dV = \lvert\psi\rvert^2$	40-5	40-9
Time independent Schrödinger Wave Equation: $(-\hbar^2/2m)(\partial^2\psi/\partial x^2 + \partial^2\psi/\partial y^2 + \partial^2\psi/\partial z^2) + U\psi = E\psi$ $E = \hbar^2k^2/2m$ and $k = 2\pi/\lambda$	40-6	40-11

PRINCIPAL CONCEPTS AND EQUATIONS

1 De Broglie Waves (Section 40-1, 2)

Review: Matter has a particle nature and is characterized by its energy and momentum. Light has a wave nature and is characterized by its wavelength and frequency. This is summarized in Figure 40.1(a).

After Einstein explained the photoelectric effect, physicists realized that light had a particle nature (photons). We can write expressions for the energy and momentum of the photons (Figure 40.1(b)).

Louis de Broglie looked at all this information, considered the symmetry of nature, and hypothesized that, if light has a particle nature, then perhaps matter has a wave nature. We can rewrite the expressions for the energy and momentum of a photon to obtain the expressions for the wavelength and frequency of a particle (Figure 40.1(c)). The de Broglie equation gives the wavelength of a particle in terms of its momentum:

$$\lambda = h/p = h/mv$$

(a)	Particle Nature	Wave Nature	(b)	Particle Nature	Wave Nature	(c)	Particle Nature	Wave Nature
Matter	E P		Matter	E P		Matter	E P	$f = E/h$ $\lambda = h/p$
Light		λ f	Light	$\varepsilon = hf$ $p = \varepsilon/c$ $= h/\lambda$	λ f	Light	$\varepsilon = hf$ $p = \varepsilon/c$ $= h/\lambda$	λ f

(a) Matter has a particle nature. Light has a wave nature.
(b) Matter has a particle nature. Light has a wave and a particle nature.
(c) Matter has a particle and a wave nature. Light has a wave and a particle nature.

Figure 40.1

Experimental evidence for the existence of matter waves came from the Davisson and Germer experiment (Figure 40.2). A monoenergetic (54.0-eV) beam of electrons was incident upon crystalline nickel. Electrons were reflected in such a manner as to have the associated matter waves constructively interfere at an angle of 50.0° with respect to the incident beam.

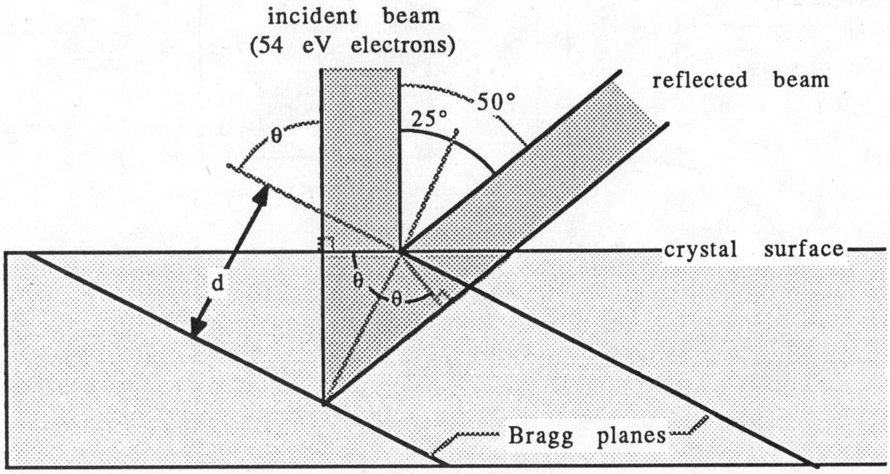

Figure 40.2

The path difference for the two waves is shown in Figure 40.2 to be $2d \sin\theta_m$. Constructive interference of the associated matter waves occurs when the path difference is an integral multiple of the whole wavelengths; hence the Bragg equation gives

$$m\lambda = 2d \sin\theta_m \qquad m = 1, 2, 3, \ldots$$

For crystalline nickel, $d = 9.09 \times 10^{-11}$ m and first-order interference ($m = 1$) occurs for $\theta = 65.0°$, hence

$$\lambda = 2d \sin\theta_m = 2(9.09 \times 10^{-11}\text{ m}) \sin 65.0° = 1.65 \times 10^{-10}\text{ m}$$

A particle with kinetic energy K has a momentum p, and its associated matter wave has a wavelength λ:

$$p = (2mK)^{1/2} \qquad \lambda = h/p = h/(2mK)^{1/2}$$

The matter wave associated with 54.0-eV electrons has a wavelength

$$\lambda = h/p = h/(2mK)^{1/2} = 1.67 \times 10^{-10}\text{ m}$$

Note: The experimental value of λ for 54.0-eV electrons from the Davisson-Germer experiment is in excellent agreement with the theoretical value obtained using the de Broglie hypothesis.

If the electron beam is incident on a polycrystalline foil, some of the crystallites are oriented (with respect to the beam) in such a manner that Bragg's law is satisfied. Each of these will produce a diffracted beam. Together these beams form circular patterns which are characteristic of the atomic spacing in the crystal structure. See Figure 40.3.

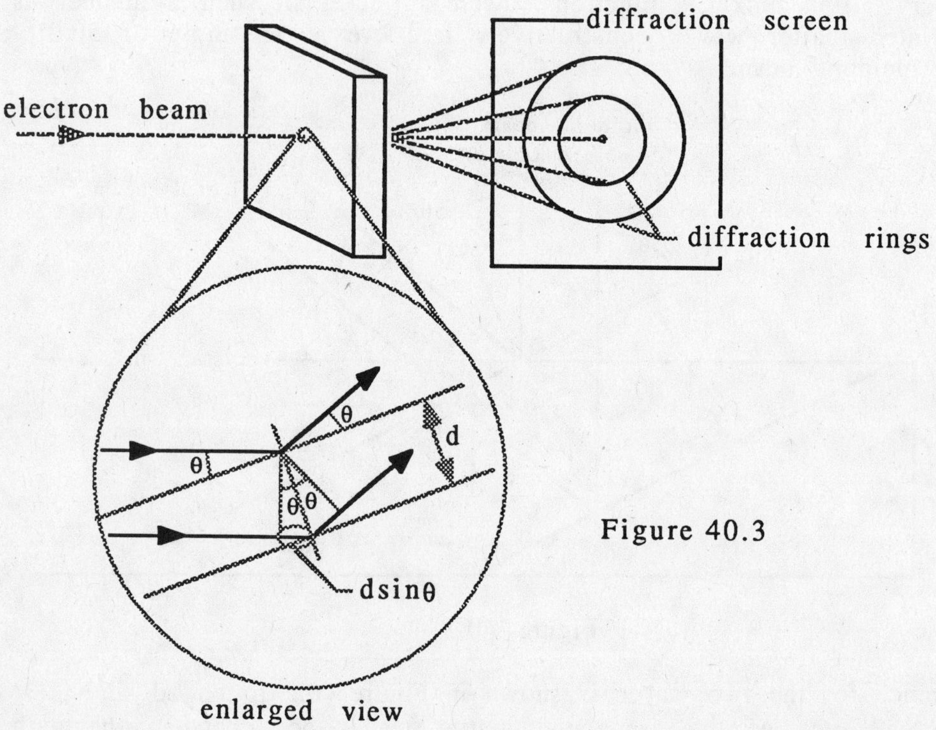

Figure 40.3

As you can see from the telescopic view, constructive interference will occur when the path difference (2d $\sin\theta$) is an integral number of whole wavelengths ($m\lambda$), i.e. when

$$m\lambda = 2d \sin\theta_m \qquad m = 1, 2, 3, \ldots$$

Practice: Determine the de Broglie wavelength of the following:

1. A 4.00×10^3 kg truck moving at 5.00 m/s	$p = mv = 2.00 \times 10^4$ kg•m/s $\lambda = h/p = 3.32 \times 10^{-38}$ m
2. A 0.200 kg ball moving 35.0 m/s	$p = mv = 7.00$ kg•m/s $\lambda = h/p = 9.46 \times 10^{-35}$ m
3. A 10.0 mg dust particle moving 1.00 m/s	$p = mv = 1.00 \times 10^{-5}$ kg•m/s $\lambda = h/p = 6.63 \times 10^{-29}$ m
4. A proton moving 3.00×10^6 m/s	First let's calculate γ $\gamma = [1 - (v/c)^2]^{-1/2} = 1.00005$ Hence $m \approx m_0$ and relativistic effects may be ignored. $m_0 = 1.67 \times 10^{-27}$ kg $p = m_0 v = 5.01 \times 10^{-21}$ kg•m/s $\lambda = h/p = 1.32 \times 10^{-13}$ m
5. A 10.0 KeV neutron	$K = 10.0$ KeV $= 1.00 \times 10^4$ eV $E_0 = m_0 c^2 = 9.40 \times 10^8$ eV Since $K \ll E_0$, then $E = K + E_0 \approx E_0$ and we may ignore relativistic effects, hence $p = (2mK)^{1/2} = 2.32 \times 10^{-21}$ kg•m/s $\lambda = h/p = 2.86 \times 10^{-13}$ m
6. An electron accelerated through a 1.00×10^3 V electrical potential difference	$K = qV = (1e)(1.00 \times 10^3\ V) = 1.00 \times 10^3$ eV $E_0 = m_0 c^2 = 5.11 \times 10^5$ eV, Since $K \ll E_0$, then $E = K + E_0 \approx E_0$ and we may ignore relativistic effects, hence $p = (2mK)^{1/2} = 1.71 \times 10^{-23}$ kg•m/s $\lambda = h/p = 3.88 \times 10^{-11}$ m
7. An electron moving 0.800 c	$v = 0.800\ c = 2.40 \times 10^8$ m/s $\gamma = [1 - (v/c)^2]^{-1/2} = 1.67$, hence relativistic effects are important $m_0 = 9.11 \times 10^{-31}$ kg $m = \gamma m_0 = 1.52 \times 10^{-30}$ kg $p = mv = 3.65 \times 10^{-22}$ kg•m/s, then $\lambda = h/p = 1.82 \times 10^{-12}$ m

	alternately: $E_0 = m_0c^2$ $E = mc^2 = \gamma m_0c^2 = \gamma E_0, \quad E^2 = p^2c^2 + E_0^2$ or $p = (E^2 - E_0^2)^{1/2}/c = (\gamma^2 - 1)m_0c$ $= 3.65 \times 10^{-22}$ kg m/s, then $\lambda = h/p = 1.82 \times 10^{-12}$ m

Note: The de Broglie wavelength associated with the truck, ball and dust particle is so small that we would not expect to find evidence of its existence. The de Broglie wavelength associated with the proton, neutron, and electrons are all greater than 10^{-13} m; hence we should be able to find evidence for its existence by allowing these waves to be incident on a diffraction grating with very fine rulings or a crystal with appropriate Bragg plane spacing.

Electrons that have been accelerated from rest through a 50.0-V potential difference are incident normal to a crystalline surface. First-order constructive interference of the reflected matter waves occurs at an angle of 40.0° with respect to the incident beam.

Determine the following:

8. Final kinetic energy of electrons in eV and joules	$\Delta U = -qV = -50.0$ eV $\Delta K = -\Delta U = 50.0$ eV $= 8.00 \times 10^{-18}$ J
9. Final momentum of electrons	$K = mv^2/2 = m^2v^2/2m = p^2/2m$ $p = (2mK)^{1/2} = 3.82 \times 10^{-24}$ kg•m/s
10. Wavelength of the associated matter waves	$\lambda = h/p = 1.74 \times 10^{-10}$ m
11. Angle between incident beam and Bragg planes	see below

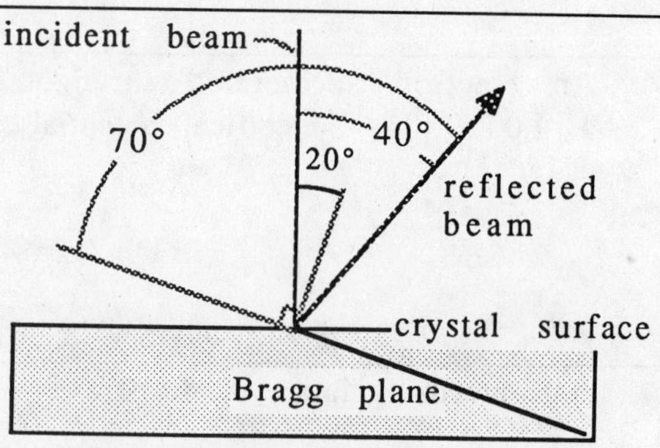

Figure 40.4

40° = angle between incident beam and reflected beam

20° = angle between incident beam and normal to Bragg planes

70° = angle between incident beam and Bragg planes

12. Spacing between Bragg planes	$m = 1$, $\theta = 70.0°$, $m\lambda = 2d \sin\theta$ $d = m\lambda/(2 \sin\theta) = 9.26 \times 10^{-11}$ m
13. Kinetic energy of a neutron with the same wavelength	$\lambda = h/p = h/(2mK)^{1/2}$, $\lambda_n = \lambda_e$ $h/(2m_nK_n)^{1/2} = h/(2m_eK_e)^{1/2}$ $K_n = m_eK_e/m_n = 4.36 \times 10^{-21}$ J
14. Voltage through which you would have to accelerate a proton in order for it to have the same wavelength	$\lambda = h/p = h/(2mqV)^{1/2}$, $\lambda_p = \lambda_e$ $h/(2m_peV_p)^{1/2} = h/(2m_eeV_e)^{1/2}$ $V_p = m_eV_e/m_p = 2.73 \times 10^{-2}$ V

Example: What is the wavelength of the matter wave associated with an electron in the second Bohr orbit of a hydrogen atom? How does this compare with the circumference of the orbit?

Given: An electron in the $n = 2$ state of a hydrogen atom.

Determine: The wavelength associated with the electron and compare it with the circumference of the orbit.

Strategy: We can use the Bohr theory to determine the speed of the electron in its orbit and the circumference of the orbit. Knowing the speed and mass of the electron, we can determine the wavelength of the associated wave.

Solution: From Bohr theory, we know that the speed of an electron in the second Bohr orbit is

$$v_2 = v_1/2 = 1.09 \times 10^6 \text{ m/s} \quad \text{where} \quad v_1 = 2.18 \times 10^6 \text{ m/s}$$

From Bohr theory, we know that the radius of the second Bohr orbit is

$$r_2 = (2)^2 r_1 = 2.12 \times 10^{-10} \text{ m}, \quad \text{where} \quad r_1 = 5.30 \times 10^{-11} \text{ m}$$

The wavelength of the matter wave associated with an electron in the second Bohr orbit is

$$\lambda = h/p = h/mv_2 = 6.68 \times 10^{-10} \text{ m}$$

The circumference of the second Bohr orbit is

$$c_2 = 2\pi r_2 = 1.33 \times 10^{-9} \text{ m}$$

Notice that $c_2 = 2\lambda$, and hence two de Broglie waves can fit into the circumference of the second Bohr orbit.

Related Text Exercises: Ex. 40-1 through 40-17.

2 Heisenberg Uncertainty Relations (Section 40-4)

Review: Heisenberg determined that there are limits on the precision that can be obtained when we simultaneously measure momentum and position (both linear and angular) and energy and time. These statements of uncertainty are stated algebraically as follows:

Linear position and momentum

$$\Delta x \Delta p_x \geq \hbar/2 \qquad \Delta y \Delta p_y \geq \hbar/2 \qquad \Delta z \Delta p_z \geq \hbar/2$$

Angular position and momentum $\Delta\theta \Delta L \geq \hbar/2$

Energy and time $\Delta E \Delta t \geq \hbar/2$

where $\hbar = h/2\pi = 1.05 \times 10^{-34}$ J•s

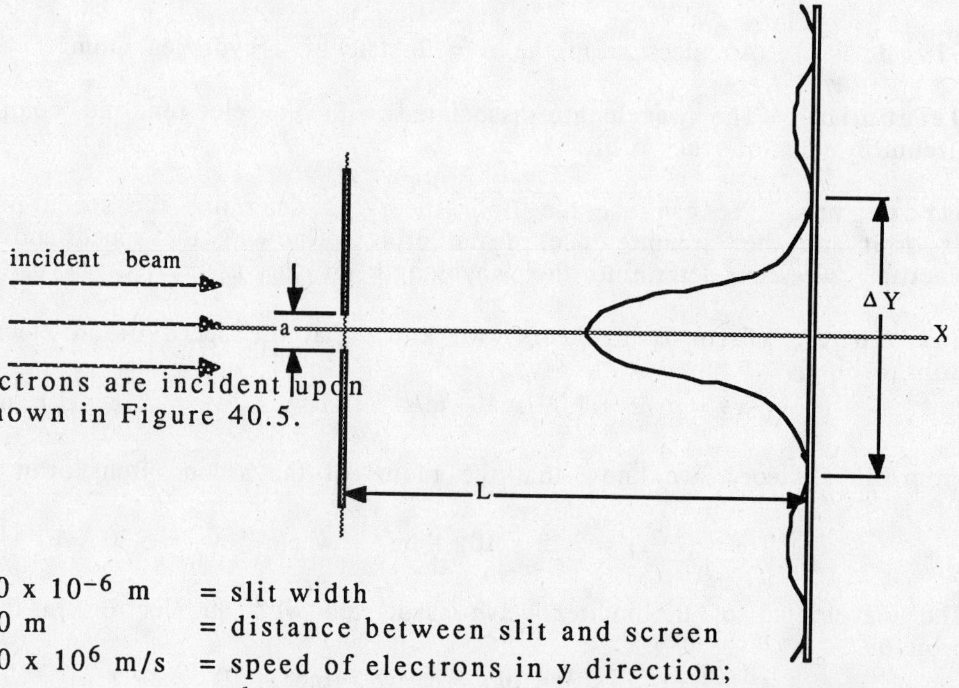

Practice: Electrons are incident upon a single slit as shown in Figure 40.5.

Figure 40.5

$a = 1.00 \times 10^{-6}$ m = slit width
$L = 2.00$ m = distance between slit and screen
$v_x = 1.00 \times 10^{6}$ m/s = speed of electrons in y direction; electrons have no initial speed in x direction

Determine the following:

1. Momentum of electrons in x direction	$p_x = mv_x = 9.11 \times 10^{-25}$ kg•m/s
2. Wavelength of matter waves associated with electrons	$\lambda = h/p_x = 7.28 \times 10^{-10}$ m
3. Location of first minimum in single-slit pattern for this wavelength	$m\lambda = a \sin\theta_m$ $\sin\theta_m \approx \tan\theta_m = y_m/L$, $m = 1$ $y_1 = L \sin\theta_1 = L(\lambda/a) = 1.46 \times 10^{-3}$ m

4. Width of central maximum using single-slit theory	$\Delta Y = 2y_1 = 2.92 \times 10^{-3}$ m
5. Uncertainty in y position for electrons going through slit	$\Delta y \le a = 1.00 \times 10^{-6}$ m
6. Uncertainty in y component of momentum of electrons going through slit	$\Delta p_y \ge \hbar/\Delta y = 5.28 \times 10^{-29}$ kg•m/s
7. Momentum of electrons in y direction after they go through slit	$p_y = \Delta p_y \ge 5.28 \times 10^{-29}$ kg•m/s
8. Speed of electrons in y direction after they go through the slit	$v_y = p_y/m \ge 58.0$ m/s
9. Time it takes electrons to go from slit to screen	$t = L/v_x = 2.00 \times 10^{-6}$ s
10. Maximum displacement of electrons in y direction	$y_{max} = v_y t \ge 1.16 \times 10^{-4}$ m
11. Width of central maximum according to uncertainty principle	$\Delta Y = 2y_{max} \ge 2.32 \times 10^{-4}$ m

Note: As a consequence of the limit on the simultaneous determination of position and momentum of the electrons, when the position is uncertain by an amount less than or equal to 1.00×10^{-6} m (step 5), the momentum is uncertain by an amount greater than or equal to 5.28×10^{-28} kg•m/s (step 7), and the central maximum on a screen 2.00 m away must be at least 2.32×10^{-4} m wide. According to single-slit diffraction theory, the width of the central maximum is 2.92×10^{-3} m (step 11) which is consistent with the uncertainty principle. No ingenious subtlety in experiment design can remove this basic uncertainty.

Example: An electron in an excited state has a lifetime of 1.00×10^{-6} s. What is the minimum uncertainty in measuring the energy of this excited state? What is the minimum uncertainty in the frequency of the observed spectral line?

Given: Maximum value for Δt

Determine: ΔE = minimum uncertainty in energy of this excited state
$\Delta \nu$ = minimum uncertainty in frequency of the observed spectral line

Strategy: Knowing the lifetime, we can determine the maximum value for Δt. Knowing the maximum value for Δt, we can determine the minimum value for ΔE. Knowing the minimum value for ΔE, we can determine the minimum value for $\Delta \nu$.

Solution: Since the lifetime of the excited state is 1.00×10^{-6} s, the uncertainty of this value can be no greater than the value itself. Hence, Δt can be no larger than 1.00×10^{-6} s. This allows us to write

$$\Delta t \leq 1.00 \times 10^{-6} \text{ s}$$

The minimum uncertainty in the energy is then

$$\Delta E \geq (\hbar/2)/\Delta t = 5.28 \times 10^{-29} \text{ J}$$

The minimum value for $\Delta\nu$ is determined from $E = h\nu$ or $\Delta E = h\Delta\nu$; hence

$$\Delta\nu = \Delta E/h = 7.96 \times 10^{4}/\text{s}$$

Related Text Exercises: Ex. 40-21 through 40-29.

3 Interpretation of the Wave Function (Section 40-5)

Review: If we represent a particle by a wave function $\psi(x,y,z)$, then the probability that the particle is inside the volume $dV = dxdydz$ centered at x, y, z is

$$dP = |\psi(x,y,z)|^2 dV$$

The probability density is

$$dP/dV = |\psi(x,y,z)|^2$$

The probability of finding the particle inside the volume V is

$$P_V = \int_V |\psi|^2 dV$$

Since the probability of finding the particle some place in all space is unity, we may write

$$P_{\text{all space}} = \int_{\text{all space}} |\psi|^2 dV = 1$$

The wave function ψ may be the superposition of two or more other wave functions:

$$\psi = \psi_1 + \psi_2$$

then

$$|\psi|^2 = |\psi_1 + \psi_2|^2$$

Practice: The wave function $\psi(x) = (2/L)^{1/2} \sin\pi x/L$ describes a particle moving along the x-axis between $x > 0$ and $x < L$.

Determine the following:

1. An expression for the probability that the particle is inside the length dx centered at x	$dP = \lvert\psi(x)\rvert^2 dx = (2/L)\sin^2 \pi x/L$

2. An expression for the probability density for finding the particle	$dP/dx = \lvert\psi(x)\rvert^2 = (2/L)\sin^2 \pi x/L$
3. The probability density at x = L/2	$dP/dx = (2/L)\sin^2[(\pi/L)(L/2)]$ $= (2/L)\sin^2(\pi/2) = 2/L$
4. The probability of finding the particle somewhere on the x axis	$= (2/\pi)(\pi/2) = 1$ $P_{all\ x} = \int_{-\infty}^{+\infty} \lvert\psi(x)\rvert^2 dx$ $= (2/L)\int_{-\infty}^{+\infty} \sin^2(\pi/L)x\,dx$ $= (2/L)(L/\pi)\int_{-\infty}^{+\infty} \sin^2 u\,du$ This says that the probability of finding the particle somewhere on the x-axis is unity. This certainly makes sense since we have one particle and we know it is on the x-axis between x = 0 and x = L.
5. The probability of finding the particle somewhere between x = 0 and x = L	$= (2/\pi)[u/2 - (\sin 2u)/4]\Big\rvert_0^\pi$ $= (2/\pi)(\pi/2) = 1$ $P_x = \int_0^L \lvert\psi(x)\rvert^2 dx$ $= (2/L)\int_0^L \sin^2(\pi/L)x\,dx$ $= (2/L)(L/\pi)\int_0^\pi \sin^2 u\,du$ This says that the particle must be in the region x > 0 to x < L.
6. The wave function for values of $x \le 0$ and $x \ge L$	From step 5 we see that the particle must be in the region $x > 0$ to $x < L$, hence $\psi(x \le 0) = 0$ and $\psi(x \ge L) = 0$.
7. The probability density for values of $x \le 0$ and $x \ge L$	$\lvert\psi(x \le 0)\rvert^2 = 0$ $\quad\lvert\psi(x \ge L)\rvert^2 = 0$ This says we do not expect to ever find the particle outside the range $0 < x < L$. This is consistent with the fact that the probability of finding the particle in the range $0 < x < L$ is unity -- i.e. it must be in this range hence it can't be in the range $x \le 0$ or $x \ge L$.

Related Text Exercises: Ex. 40-30 through 40-32.

4 The Schrödinger Wave Equation and its Solution

(Sections 40-6, 7)

Review: The time dependent wave equation

(A) $$-\hbar^2/2m(\partial^2\Psi/\partial x^2 + \partial^2\Psi/\partial y^2 + \partial^2\Psi/\partial z^2) + U(x,y,z)\Psi = i\hbar\partial\Psi/\partial t$$

has as its solution the time dependent wave function

$$\Psi(x,y,z,t)$$

If we assume that $\Psi(x,y,z,t)$ is the product of a time independent wave function $\psi(x,y,z)$ and the time dependent function $e^{-i\omega t}$, we have

$$\Psi(x,y,z,t) = \psi(x,y,z)e^{-i\omega t}$$

Inserting this into the time dependent wave equation (A) we obtain

$$(-\hbar^2/2m)(\partial^2\psi/\partial x^2 + \partial^2\psi/\partial y^2 + \partial^2\psi/\partial z^2)e^{-i\omega t} + U(x,y,z)\psi e^{-i\omega t} = i\hbar(-i\omega)e^{-i\omega t}\psi$$

Dividing by $\psi e^{-i\omega t}$ and recognizing that $E = h\nu = (h/2\pi)(2\pi\nu) = \hbar\omega$, we obtain

$$(-\hbar^2/2m)(\partial^2\psi/\partial x^2 + \partial^2\psi/\partial y^2 + \partial^2\psi/\partial z^2)/\psi + U(x,y,z) = E$$

or multiplying through by ψ we obtain the time independent wave equation

(B) $$(-\hbar^2/2m)(\partial^2\psi/\partial x^2 + \partial^2\psi/\partial y^2 + \partial^2\psi/\partial z^2) + U(x,y,z)\psi = E\psi$$

where $E = \hbar^2k^2/2m$ and $k = 2\pi/\lambda$

If we are interested in only one-dimensional problems then the time independent wave equation becomes

(C) $$(-\hbar^2/2m)d^2\psi/dx^2 + U(x)\psi = E\psi$$

Solutions to this one-dimensional, time independent wave equation are of the form

$$\psi(x) = A \sin kx + B \cos kx$$

The constants A and B must be such that

$$\int_{-\infty}^{+\infty} |\psi|^2 dx = 1$$

Practice: Consider a particle of mass m that is confined to a one-dimensional region $0 < x < L$.

Determine the following:

1. Potential energy function for this situation	$U(x \le 0) = \infty$ $U(0 < x < L) = 0$ $U(x \ge L) = \infty$ 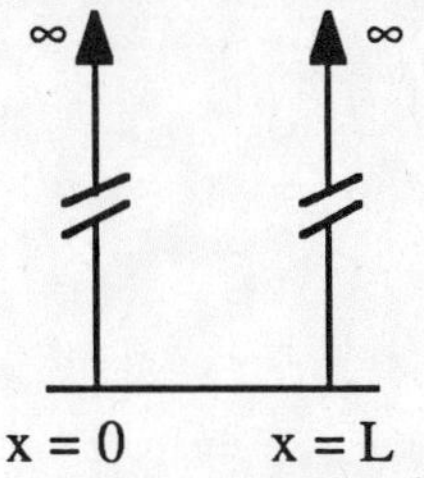 Figure 40.6
2. Value of the wave function for values of $x \le 0$ and $x \ge L$	$\psi(x \le 0) = 0$ and $\psi(x \ge L) = 0$
3. The general form of the wave function for values of x in the range $0 < x < L$	$\psi(x) = A \sin kx + B \cos kx$ But since $\psi(x = 0) = 0$ we have $\psi(x = 0) = A \sin 0 + B \cos 0 = 0$. This happens only if $B = 0$, hence we have $\psi(x) = A \sin kx$, but since $\psi(x = L) = 0$, we have $\psi(x = L) = A \sin kL = 0$. This happens only if $k_n L = n\pi$, hence we have $\psi_n(x) = A \sin k_n x$ where $k_n = n\pi/L$.
4. The normalization constant A	Since the particle must exist between $x = 0$ and $x = L$, the probability of finding the particle in this region is unity. Hence we may write $1 = \int_0^L \lvert\psi_n\rvert^2 dx = \int_0^L A^2 \sin^2 k_n x dx = A^2 L/2$ or $A = (L/2)^{1/2}$ then $\psi_n(x) = (L/2)^{1/2} \sin(n\pi/L)x$, $n = 1,2,3,...$
5. The first three wave functions and the corresponding energies	$\psi_n(x) = (L/2)^{1/2} \sin(n\pi/L)x$ $E_n = \hbar^2 k_n^2/2m = n^2\pi^2\hbar^2/2mL^2$, $k = n\pi/L$ $n = 1$ $\psi_1(x) = (L/2)^{1/2} \sin \pi x/L$ $E_1 = \pi^2\hbar^2/2mL^2$ $n = 2$ $\psi_2(x) = (L/2)^{1/2} \sin 2\pi x/L$ $E_2 = 4\pi^2\hbar^2/2mL^2$ $n = 3$ $\psi_3(x) = (L/2)^{1/2} \sin 3\pi x/L$ $E_3 = 9\pi^2\hbar^2/2mL^2$

6. A plot of the first three wave functions	$\psi_n(x) = (L/2)^{1/2} \sin(n\pi/L)x$ 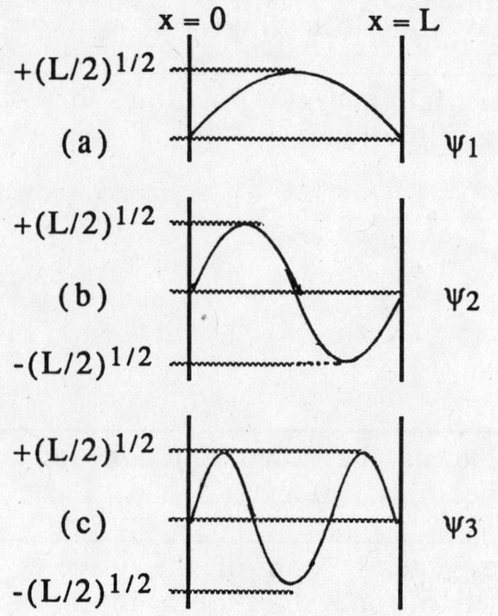 Figure 40.7
7. A plot of the probability density for the first three wave functions	$\|\psi_n(x)\|^2 = (L/2) \sin^2(n\pi/L)x$ 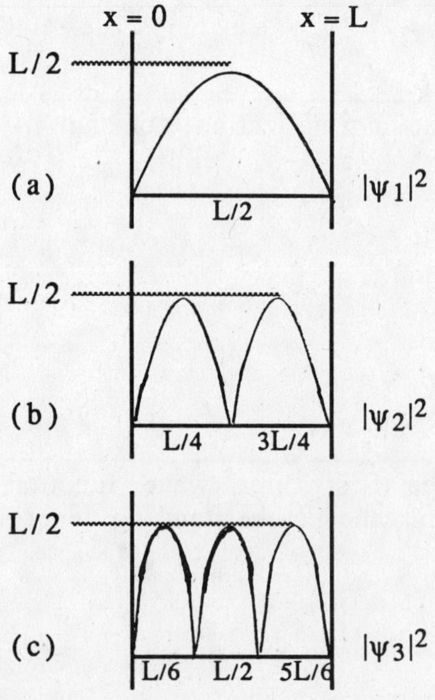 Figure 40.8

8. The site where the particle has the greatest probability of being found when it is in each of the first three states	n = 1: According to Figure 40.8(a) when the particle is in the n = 1 state it has the greatest possibility of being found at $x = L/2$. n = 2: According to Figure 40.8(b) the particle has the greatest probability of being found at $x = L/4$ and $x = 3L/4$. n = 3: According to Figure 40.8(c) the particle has the greatest probability of being found at $x = L/6$, $x = L/2$, and $x = 5L/6$.
9. The probability density at the midpoint of the box (i.e. at x = L/2) for the first three states	$\lvert\psi_1(x = L/2)\rvert^2 = L/2$ $\lvert\psi_2(x = L/2)\rvert^2 = 0$ $\lvert\psi_3(x = L/2)\rvert^2 = L/2$
10. The probability of finding the particle at x = L/20 between x = 0 and x = L/10 when it is in the n = 1 state	$\lvert\psi_1(x)\rvert^2 = (L/2)\sin^2(\pi x/L)$ $P = \int_0^{L/10} \lvert\psi_1(x)\rvert^2 dx$ $= (L/2)\int_0^{L/10} \sin^2(\pi x/L)dx$ Making the change of variable $u = \pi x/L$ $du = \pi dx/L$ when $x = 0$ $u = 0$ $x = L/10$ $u = \pi/10$ $P = (L/2)(L/\pi)\int_0^{\pi/10} \sin^2 u\, du$ $= (L^2/2\pi)[u/2 - (1/4)\sin 2u]\Big\vert_0^{\pi/10}$ $= (L^2/2\pi)[\pi/20 - (\sin 36°)/4]$ $= (1.61 \times 10^{-3})L^2$

Example: The energy of the n = 4 state of a particle in a box is 2.00 MeV. (a) What is the energy of the ground state? (b) If this particle is a proton, then what is the length of the box?

Given: The particle is in the n = 4 state
$E_4 = 2.00$ MeV
The particle is a proton, $m = 1.672 \times 10^{-27}$ kg

Determine: (a) E_1 -- the ground state energy
(b) L -- the length of the box

Strategy: (a) Knowing the energy, E_4, we can determine the energy E_1. (b) Knowing E_1 and the mass of the particle, we can determine the length L of the box.

Solution: (a) An expression for the nth energy state of a particle in a box is

$$E_n = (\hbar^2\pi^2/2mL^2)n^2 = (h^2/8mL^2)n^2$$

Then

$$E_1 = (h^2/8mL^2)(1^2) \quad \text{and} \quad E_4 = (h^2/8mL^2)(4^2)$$

Hence

$$E_1 = E_4/16 = 2.00\text{ MeV}/16 = 0.125\text{ MeV} = 2.00 \times 10^{-14}\text{ J}$$

(b)

$$E_1 = h^2/8mL^2 \quad \text{or} \quad L = h/(8mE_1)^{1/2}$$

$$L = \frac{(6.625 \times 10^{-34}\text{ J}\cdot\text{s})}{[8(1.672 \times 10^{-27}\text{ kg})(2.00 \times 10^{-14}\text{ J})]^{1/2}} = 4.04 \times 10^{-14}\text{ m}$$

Related Text Exercises: Ex. 40-33 through 40-42.

PRACTICE TEST

Take and grade this practice test. Doing so will allow you to determine any weak spots in your understanding of the concepts taught in this chapter. The following section prescribes what you should study further to strengthen your understanding.

Determine the de Broglie wavelength of the following:

_________ 1. A 2.00×10^3-kg vehicle moving at 10 m/s
_________ 2. A 5.00-keV proton
_________ 3. A 5.00-keV photon
_________ 4. An electron accelerated through a 1.00×10^4 V potential difference
_________ 5. A proton moving 0.800 c

Electrons that have been accelerated from rest through a 80.0-V potential difference are incident normal to a crystalline surface. First-order constructive interference of the reflected matter waves occurs at an angle of 70.0° with respect to the incident beam.

Determine the following:

_________ 6. Final kinetic energy of the electrons in joules
_________ 7. Final momentum of the electrons
_________ 8. Wavelength of the associated matter waves
_________ 9. Angle between the incident beam and the reflecting Bragg planes
_________ 10. Spacing between the reflecting Bragg planes
_________ 11. Kinetic energy of a proton with the same wavelength
_________ 12. Kinetic energy of a neutron with the same wavelength
_________ 13. Voltage through which you would have to accelerate a proton in order for it to have the same wavelength

The position of a photon is known to within 1.00×10^{-10} m.

Determine the following:

_________ 14. Minimum uncertainty in its momentum
_________ 15. Minimum uncertainty in its energy
_________ 16. The minimum amount of time that must be taken to make the position measurement

Consider a particle of mass m that is confined to a one-dimensional region $0 < x < 2L$.

Determine the following:

_________ 17. The potential function for all values of x
_________ 18. The wave function for values of $x \leq 0$ and $x \geq 2L$
_________ 19. The general form of the wave function for values of x in the range $0 < x < 2L$
_________ 20. The normalization constant
_________ 21. The wave function and corresponding energy for the $n = 2$ state
_________ 22. The site where the particle has the greatest probability of being found when it is in the $n = 2$ state
_________ 23. Probability density at the point $x = L/2$ when $n = 2$
_________ 24. Probability of finding the particle at $x = 0.5L$ between $x = 0.4L$ and $x = 0.6L$ when $n = 2$

(See Appendix I for answers)

PRINCIPAL CONCEPTS AND EQUATIONS PRESCRIPTION

Your score on the practice test is an excellent measure of your understanding of the chapter. You should now use the following chart to write your own prescription for dealing with any weaknesses the practice test points out. Look down the leftmost column to the number of the question(s) you answered incorrectly, reading across that row you will find the concept and/or equation of concern, the section(s) of the study guide you should return to for further study, and some suggested text exercises which you should work to gain additional experience.

Practice Test Questions	Concepts and Equations	Prescription	
		Principal Concepts	Text Exercises
1	de Broglie wavelength: $\lambda = h/mv$	1	40-2, 7
2	de Broglie wavelength: $\lambda = h/(2mK)^{1/2}$	1	40-3, 4
3	de Broglie wavelength: $\lambda = hc/E$	1	40-4
4	de Broglie wavelength: $\lambda = h/(2meV)^{1/2}$	1	40-6
5	de Broglie wavelength: $\lambda = h/\gamma m_0 v$	1	40-2, 7
6	Electric potential difference: $\Delta K = -\Delta U = qV$	2 of Ch 22	22-3, 20
7	Kinetic energy and momentum: $K = mv^2/2 = p^2/2m$	4 of Ch 8 3 of Ch 10	8-23, 32 10-14, 15
8	de Broglie wavelength: $\lambda = h/p = h/(2mK)^{1/2}$	1	40-4, 5
9	Reflection: $\theta_i = \theta_r$	2 of Ch 35	35-16, 17
10	Bragg equation: $m\lambda = 2d \sin\theta_m$	---	36-35, 36
11	de Broglie wavelength: $\lambda = hc/E$	1	40-4
12	de Broglie wavelength: $\lambda = h/(2mK)^{1/2}$	1	40-5, 6
13	de Broglie wavelength: $\lambda = h(2meV)^{1/2}$	1	40-6
14	Uncertainty principle: $\Delta x \Delta p_x \geq \hbar/2$	2	40-23, 24
15	Energy of a photon: $\Delta E = c\Delta p$	2 of Ch 39	39-12, 20
16	Uncertainty principle: $\Delta E \Delta t \geq \hbar/2$	2	40-27, 28
17	Potential energy function: U	4	---
18	Wave function: $\psi = 0$	4	---
19	Wave function: $\psi = A \sin kx + B \cos kx$	4	40-37
20	Normalization: $\int_o^{2L} \lvert\psi\rvert^2 dx = 1$	4	40-40
21	Wave function: ψ_2	4	40-38
22	Probability density: $\lvert\psi_2\rvert^2$	3	---
23	Probability density: $\lvert\psi_2\rvert^2$	3	---
24	Probability: $\int_{0.4L}^{0.6L} \lvert\psi_2\rvert^2 dx = P$	3	---

41 THE HYDROGEN ATOM AND THE PERIODIC TABLE

RECALL FROM PREVIOUS CHAPTERS

Previously learned concepts and equations frequently used in this chapter	Text Section	Study Guide Page
Orbital angular momentum: $\mathbf{L} = \mathbf{r} \times \mathbf{p}$	13-1	13-2
A conservative force is derivable from a potential energy function: $F_x = -dU(x)/dx$	9-1	9-1
Magnitude of magnetic moment: $m = NIA$	26-2	26-8
Potential energy of a magnetic moment: $U = -\mathbf{m} \cdot \mathbf{B}$	26-2	26-8
Energy, frequency and wavelength of photons: $E = h\nu = hc/\lambda$	39-2	39-5
Wave functions: ψ	40-5	40-11

NEW IDEAS IN THIS CHAPTER

Concepts and equations introduced	Text Section	Study Guide Page
Quantum numbers: n, l, m_l, m_s	41-1, 4	41-1
Orbital angular momentum and its z component: $L = [l(l+1)]^{1/2}\hbar$ $\quad L_z = m_l\hbar$	41-1	41-1
Spin angular momentum and its z component: $S = [s(s+1)]^{1/2}\hbar$ $\quad S_z = m_s\hbar$	41-4	41-1
Zeeman effect: $\mu_{lz} = -\mu_B m_l$, $\quad \Delta E = \mu_B B$, $\quad \Delta\nu = \mu_B B/h$, $\quad \Delta\lambda = \lambda^2\mu_B B/hc$	41-3	41-5
Stern-Gerlach experiment: $F_z = \mu_{sz} dB/dz$ $\quad \mu_{sz} = -g\mu_B m_s$	41-8	41-9

PRINCIPAL CONCEPTS AND EQUATIONS

1 Quantum Numbers (Sections 41-1, 4, 5)

Review: Schrödinger's development of the quantum theory greatly expanded our knowledge of the atom. In the process of solving the wave equation for the electron's wave function, quantum numbers arise. We now review and summarize our knowledge of these quantum numbers.

n = principal quantum number; gives information about the energy (E_n = -13.6 eV/n^2) of the atom and the size of the electron orbit
n = 1, 2, 3, ...

l = angular momentum quantum number; gives information about the shape of the electron orbit
l = 0, 1, 2, ... , n - 1 (n possible values)
The angular momentum is given by $L = [l(l + 1)]^{1/2}\hbar$

m_l = magnetic quantum number; gives information about the orientation of the electron orbit
$m_l = 0, \pm 1, \pm 2, \ldots, \pm l$ (2l + 1 possible values)
The z component of the angular momentum is given by $L_z = m_l\hbar$

s = spin quantum number = 1/2
The spin angular momentum is given by $S = [s(s + 1)]^{1/2}\hbar$

m_s = spin orientation quantum number; gives information about the orientation of the spin of the electron
$m_s = \pm 1/2$
The z component of the spin angular momentum is given by
$S_z = m_s h/2\pi = \pm(1/2)\hbar$

L and L_z are shown in Figure 41.1 for l = 1. S and S_z are shown in Figure 41.2.

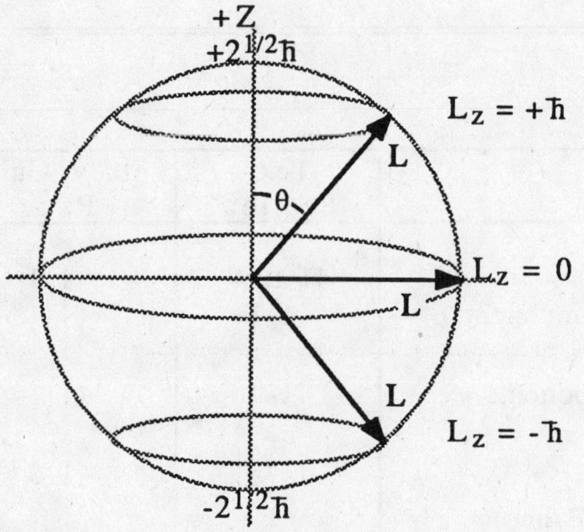

$l = 1$
$L = [l(l + 1)]^{1/2}\hbar = 2^{1/2}\hbar$
$L_z = 0, +\hbar, -\hbar$
$\cos\theta = L_z/L = m_l/[l(l + 1)]^{1/2}$

Figure 41.1

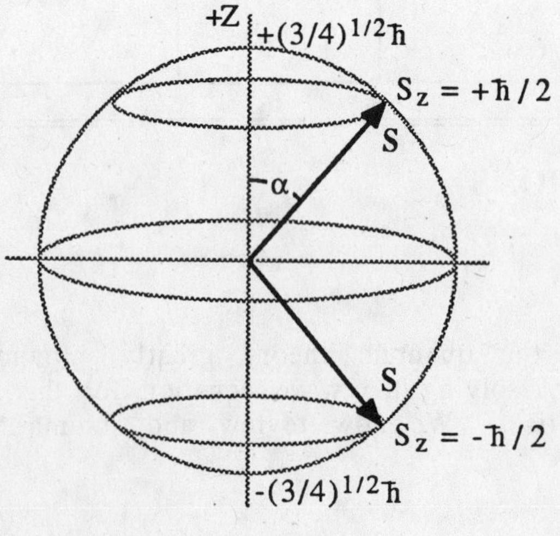

$s = 1/2$
$S = [s(s + 1)]^{1/2}\hbar = (3/4)^{1/2}\hbar$
$S_z = m_s\hbar = \pm\hbar/2$
$\cos\alpha = S_z/S = m_s/[s(s + 1)]^{1/2}$

Figure 41.2

Table 41.1 gives the possible quantum numbers for a 28 electron atom in the ground state.

n	l	m_l	m_s	subshell (No. of e^-'s)	shell (No. of e^{-1}s)	wave functions	
1	0	0	±1/2	l = 0 or 1s (2 e^-'s)	n = 1 or K (2 e^-'s)	$\Psi_{100-1/2}$	$\Psi_{100+1/2}$
2	0	0	±1/2	l = 0 or 2s (2 e^-'s)	n = 2 or L (8 e^-'s)	$\Psi_{200-1/2}$	$\Psi_{200+1/2}$
2	1	-1	±1/2	l = 1 or 2p (6 e^-'s)		$\Psi_{21-1-1/2}$	$\Psi_{21-1+1/2}$
2	1	0	±1/2			$\Psi_{210-1/2}$	$\Psi_{210+1/2}$
2	1	+1	±1/2			$\Psi_{211-1/2}$	$\Psi_{211+1/2}$
3	0	0	±1/2	l = 0 or 3s (2 e^-'s)	n = 3 or M (18 e^-'s)	$\Psi_{300-1/2}$	$\Psi_{300+1/2}$
3	1	-1	±1/2	l = 1 or 3p (6 e^-'s)		$\Psi_{31-1-1/2}$	$\Psi_{31-1+1/2}$
3	1	0	±1/2			$\Psi_{310-1/2}$	$\Psi_{310+1/2}$
3	1	+1	±1/2			$\Psi_{311-1/2}$	$\Psi_{311+1/2}$
3	2	-2	±1/2	l = 2 or 3d (10 e^-'s)		$\Psi_{32-2-1/2}$	$\Psi_{32-2+1/2}$
3	2	-1	±1/2			$\Psi_{32-1-1/2}$	$\Psi_{32-1+1/2}$
3	2	0	±1/2			$\Psi_{320-1/2}$	$\Psi_{320+1/2}$
3	2	1	±1/2			$\Psi_{321-1/2}$	$\Psi_{321+1/2}$
3	2	2	±1/2			$\Psi_{322-1/2}$	$\Psi_{322+1/2}$

Table 41.1

Note: According to the Pauli exclusion principle, no two electrons in the same atom may exist in the same quantum state (i.e. have the same set of quantum numbers).

Practice: Determine the following for a neon atom with 10 electrons:

1. A table showing the ground-state quantum numbers for each of the 10 electrons

n	l	m_l	m_s
1	0	0	±1/2
2	0	0	±1/2
2	1	+1	±1/2
2	1	0	±1/2
2	1	-1	±1/2

2. Number of shells occupied by the electrons	Since n = 2, the electrons occupy two shells.
3. Number of possible spin orientations for each electron	Since $m_s = \pm 1/2$, each electron has two possible spin orientations.
4. Number of possible orientations for an orbit with $l = 1$	If $l = 1$, then $m_l = +1,0,-1$ and the orbit has three possible orientations.
5. Which electrons are, on average, nearer the nucleus	Electrons with n = 1 are, on average, nearer the nucleus
6. Which electrons are in less eccentric orbits	Electrons with $l = 1$ have greater angular momentum and hence less eccentric orbits about the nucleus.
7. Angle between z axis and orbital angular momentum vector if $l = 1$ and $m_l = \pm 1$	$\cos\theta = \frac{L_z}{L} = \frac{m_l \hbar}{[l(l+1)]^{1/2}\hbar} = 0.707$ $\theta = \cos^{-1}(0.707) = 45.0°$
8. Angle between z axis and spin angular momentum vector is $m_s = 1/2$	$\cos\alpha = \frac{S_z}{S} = \frac{m_s \hbar}{[s(s+1)]^{1/2}\hbar} = 0.577$ $\alpha = \cos^{-1}(0.577) = 54.7°$

Example: The wave function for an excited hydrogen atom is represented by $\psi_{32-1+1/2}$. Determine the following:

(a) The principal quantum number
(b) The energy in eV
(c) The orbital angular momentum
(d) The maximum and minimum possible orbital angular momentum
(e) The projection of the orbital angular momentum onto the z axis
(f) The maximum value of the projection of the orbital angular momentum onto the z axis
(g) The spin angular momentum
(h) The projection of the spin angular momentum onto the z axis
(i) The angle between the direction of **L** and the +z-axis
(j) The angle between the direction of **S** and the +z-axis

Given: The wave function $\psi_{32-1+1/2}$

Determine:

(a) n	(e) L_z	(i) θ
(b) E_n	(f) $(L_z)_{max}$	(j) α
(c) L	(g) S	
(d) L_{max}, L_{min}	(h) S_z	

Strategy: Using our knowledge of quantum numbers we can determine the desired quantities.

Solution: The general form of the wave function is $\Psi_{nlm_lm_s}$ and the wave function of interest is $\Psi_{32-1+1/2}$

(a) Comparing the general form and the actual wave function representation, we see that the principal quantum number is n = 3.
(b) $E_n = -13.6\ eV/n^2 = -13.6\ eV/9 = -1.51\ eV$
(c) $L = [l(l + 1)]^{1/2}\hbar$ and for this case $l = 2$, hence $L = (6)^{1/2}\hbar$
(d) For n = 3, $l_{max} = 2$, hence $L_{max} = (6)^{1/2}\hbar$
$l_{min} = 0$, hence $L_{min} = 0$
(e) $L_z = m_l\hbar$, and for this case $m_l = -1$, hence $L_z = -\hbar$
(f) For $l = 2$, $(m_l)_{max} = 2$, hence $(L_z)_{max} = 2\hbar$
(g) $S = [s(s + 1)]^{1/2}\hbar$, and s = 1/2 always, hence $S = (3)^{1/2}\hbar/2$
(h) $S_z = m_s\hbar$, and for this case $m_s = +1/2$, hence $S_z = \hbar/2$
(i) $\theta = \cos^{-1}(L_z/L) = \cos^{-1}[-\hbar/(6)^{1/2}\hbar] = \cos^{-1}(-0.408) = 114.1°$
(j) $\alpha = \cos^{-1}(S_z/S) = \cos^{-1}[(\hbar/2)/(3^{1/2}\hbar/2)] = \cos^{-1}(0.577) = 54.7°$

Related Text Exercises: Ex. 41-1 through 41-6 and 41-22 through 41-29.

2 Zeeman Effect (Section 41-3)

Review: A circulating electron has an orbital angular momentum **L** and an orbital magnetic dipole moment $\boldsymbol{\mu}_l$ which is proportional to **L** (see Figure 41.3).

Figure 41.3

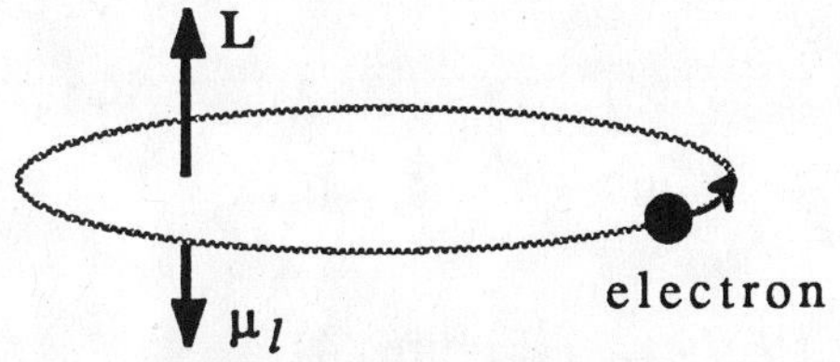

As we studied in chapter 30, the orbital magnetic dipole moment is related to the orbital angular momentum by

$$\boldsymbol{\mu}_l = -(e/2m_e)\mathbf{L}$$

Since **L** is quantized, $\boldsymbol{\mu}_l$ is quantized and its z component is

$$\mu_{lz} = -(e/2m_e)L_z = -(e/2m_e)(m_l\hbar) = -(e\hbar/2m_e)m_l = -\mu_B m_l$$

Since $m_l = 0, \pm 1, \ldots, \pm l$, we see that the z component of the atom's magnetic dipole moment is quantized in increments of $e\hbar/2m_e$. This quantum of magnetic moment

$$\mu_B = e\hbar/2m_e = 5.79 \times 10^{-5}\ eV/T = 9.27 \times 10^{-24}\ J/T$$

is called the Bohr magneton.

If we introduce a magnetic field in the z direction **B** = B**k**, the magnetic dipole moment has a magnetic potential energy

$$U_m = -\boldsymbol{\mu}_l \cdot \mathbf{B} = -\mu_{lz}B = -(-\mu_B m_l)B = (\mu_B B)m_l$$

From the above equation we see that an atom in a magnetic field has a contribution to its energy that depends on the orbital magnetic quantum number m_l. Subsequently when an atom is placed in a magnetic field, the spectral lines emitted by the atoms are split into several lines. This splitting due to the orbital magnetic dipole moment is called the Normal Zeeman effect. This effect is shown in Figure 41.4 for a transition from $l = 1$ to $l = 0$ and for a transition from $l = 2$ to $l = 1$.

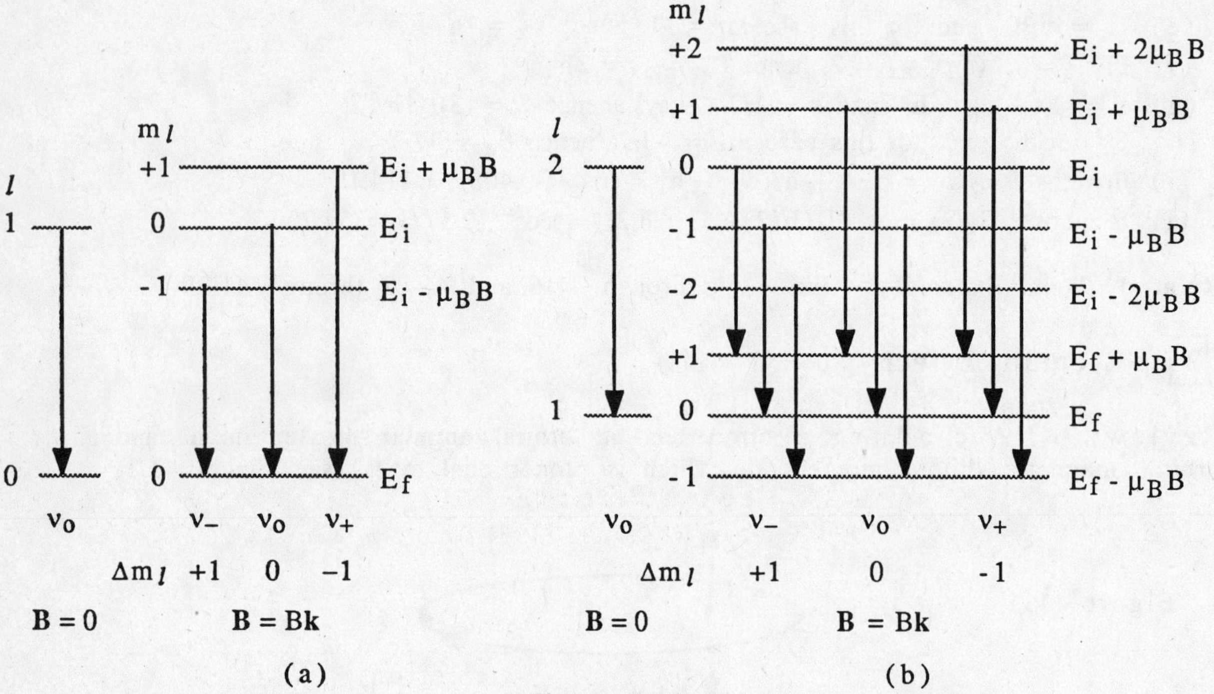

Figure 41.4

Notice that the allowed transitions obey the following selection rules:

$$\Delta l = \pm 1 \qquad \text{and} \qquad \Delta m_l = \pm 1 \text{ or } 0$$

When $\mathbf{B} = 0$ the frequency of the spectral line is

$$\nu_0 = (E_i - E_f)/h = \Delta E_0/h$$

When $\mathbf{B} = \mathrm{B}\mathbf{k}$, the spectral line of frequency ν_0 is split into three lines of frequency ν_+, ν_0, ν_-.

$$\nu_+ = (\Delta E_0 + \mu_B B)/h \qquad \nu_0 = \Delta E_0/h \qquad \nu_- = (\Delta E_0 - \mu_B B)/h$$

that is the spacing $\Delta\nu$ between the spectral lines increases or decreases by the amount

$$\Delta\nu = \mu_B B/h$$

Practice: Consider a one electron atom in the $n = 3$ state.

Determine the following:

1. Possible values of the orbital angular momentum quantum number	For n = 3, the possible values for l are $l = 0, 1, 2$
2. Possible values of the orbital angular momentum	$L = [l(l + 1)]^{1/2}\hbar$ $l = 0 \quad L = 0$ $l = 1 \quad L = (2)^{1/2}\hbar$ $l = 2 \quad L = (6)^{1/2}\hbar$
3. Possible values of the orbital magnetic dipole moment	$\mu_l = (e/2m_e)L = (e\hbar/2m_e)[l(l + 1)]^{1/2}$ $\mu_l = \mu_B[l(l + 1)]^{1/2}$ $l = 0 \quad \mu_l = 0$ $l = 1 \quad \mu_l = (2)^{1/2}\mu_B$ $l = 2 \quad \mu_l = (6)^{1/2}\mu_B$

4. Possible values of the magnetic quantum number	n l m_l
	3 2 -2, -1, 0, 1, 2
	     1 -1, 0, 1
	     0 0

5. Possible values of the z component of the orbital angular momentum	$L_z = m_l\hbar$ l L_z 2 $-2\hbar, -\hbar, 0, \hbar, 2\hbar$ 1 $-\hbar, 0, \hbar$ 0 0
6. Possible values of the z component of the orbital magnetic dipole moment	$\mu_{lz} = \mu_B m_l$ l μ_{lz} 2 $-2\mu_B, -\mu_B, 0, \mu_B, 2\mu_B$ 1 $-\mu_B, 0, \mu_B$ 0 0
7. A sketch of L and μ_l for $l = 0, 1, 2$	$l = 0, L = 0, \mu_l = 0, L_z = 0, \mu_{lz} = 0$ $l = 1, L = (2)^{1/2}\hbar, L_z = -\hbar, 0, \hbar$ $\mu_l = (2)^{1/2}\mu_B, \quad \mu_{lz} = -\mu_B, 0, \mu_B$ 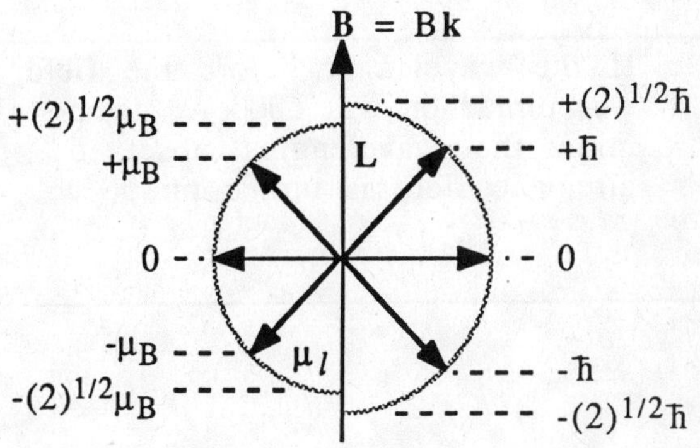

$l = 2$, $L = (6)^{1/2}\hbar$, $L_z = -2\hbar, -\hbar, 0, \hbar, 2\hbar$

$\mu_l = (6)^{1/2}\mu_B$, $\mu_{lz} = -2\mu_B, -\mu_B, 0, \mu_B, 2\mu_B$

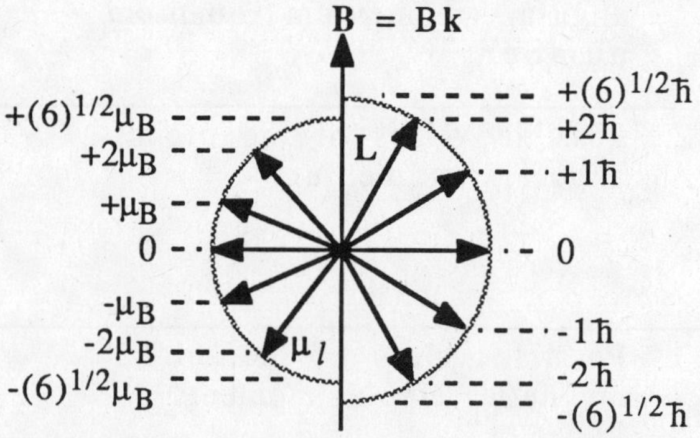

8. If the electron undergoes the transition $l = 2$ to $l = 1$, in the absence of a magnetic field how many spectral lines are emitted

One

9. If the electron undergoes the aforementioned transition, in the presence of an external magnetic field how many spectral lines are emitted and what are the selection rules

Three spectral lines are emitted (refer to Figure 41.4(b)) and the selection rules are

$\Delta l = -1$ $\quad l = 2 \rightarrow l = 1$

$\Delta m_l = +1$ $\quad m_l = 0 \rightarrow +1$; $-1 \rightarrow 0$; $-2 \rightarrow -1$ $\quad \} \ \nu_-$

$\Delta m_l = 0$ $\quad m_l = +1 \rightarrow +1$; $0 \rightarrow 0$; $-1 \rightarrow -1$ $\quad \} \ \nu_o$

$\Delta m_l = -1$ $\quad m_l = +2 \rightarrow +1$; $+1 \rightarrow 0$; $0 \rightarrow -1$ $\quad \} \ \nu_+$

10. If the magnitude of the external field is 0.500 T, what is the spacing between energy levels with the same l but different m_l

$U_m = \mu_B B m_l$

$\Delta U_m = \mu_B B \Delta m_l$

$\Delta m_l = 1$ between two levels

$\Delta U = \mu_B B = (9.27 \times 10^{-24}\ \text{J/T})(0.500\ \text{T})$

$= 4.64 \times 10^{-24}\ \text{J}$

11. If the wavelength before the field was turned on was 6000 Å, determine the wavelengths observed after the field is turned on

$\Delta E = 4.64 \times 10^{-24}\ \text{J}$

$E = h\nu = hc/\lambda$

$\Delta E = -hc\,d\lambda/\lambda^2$

$|d\lambda| = \lambda^2 \Delta E/hc = 0.0841$ Å

$\lambda = 5999.916$ Å, 6000 Å, 6000.084 Å

12. The magnetic field required to observe the normal Zeeman effect if the spectrometer used can resolve spectral lines separated by 0.400 Å at 6000 Å	$E = hc/\lambda$ $dE = -hcd\lambda/\lambda^2$, $dE = (e\hbar/2m_e)B = \mu_B B$ Combining these, obtain $B = hc|d\lambda|/\mu_B\lambda^2 = 23.8$ T

Example: Determine the normal Zeeman splitting (in Å) of the calcium 4226 Å line when the atoms are placed in a 1.00×10^{-2} T magnetic field.

Given: $\lambda_o = 4226$ Å and $B = 1.00 \times 10^{-2}$ T

Determine: $|d\lambda|$, the normal Zeeman splitting

Strategy: Knowing how the energy of a photon depends on the wavelength, we can develop an expression for $|d\lambda|$ in terms of dE. Knowing how dE depends on B for the Zeeman effect, we can determine $|d\lambda|$.

Solution: The relationship between energy and wavelength is

$$E = h\nu = hc/\lambda$$

Differentiating this obtain

$$dE = -hcd\lambda/\lambda^2 \quad \text{or} \quad |d\lambda| = \lambda^2|dE|/hc$$

The energy shift for the Zeeman effect is obtain by

$$|dE| = e\hbar B/2m_e = \mu_B B$$

Combining the last two equations, obtain

$$|d\lambda| = \lambda^2|dE|/hc = \lambda^2\mu_B B/hc = 8.34 \times 10^{-14} \text{ m}$$

Related Text Exercises: Ex. 41-14 through 41-18.

3 Electron Spin And The Stern-Gerlach Experiment (Section 41-4)

Review: Similar to its orbital angular momentum (L) and associated orbital magnetic moment (μ_l) an electron has an intrinsic spin angular momentum (S) and associated spin magnetic moment (μ_s) as shown in Figure 41.5.

Figure 41.5

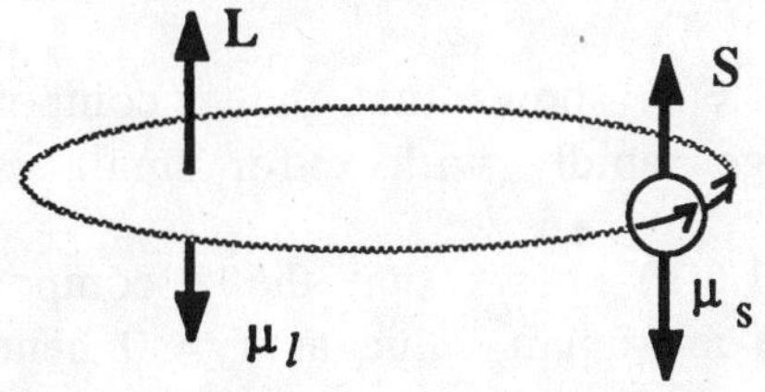

The spin quantum number s has only one value ($s = 1/2$). The spin angular momentum has only one value.

$$S = [s(s + 1)]^{1/2}\hbar = (3)^{1/2}\hbar / 2$$

The spin orientation quantum number (m_s) has two values

$$m_s = \pm 1/2$$

Subsequently the component S_z has two values

$$S_z = m_s\hbar \Rightarrow S_z = -\hbar/2 \quad \text{and} \quad S_z = +\hbar / 2$$

Similar to the orbital case we have

$$\boldsymbol{\mu}_s = -g(e/2m_e)\mathbf{S}$$

and

$$\mu_{sz} = -g(e/2m_e)S_z = -g(e/2m_e)m_s\hbar = -g(e\hbar/2m_e)m_s = -g\mu_B m_s$$

If we send a beam of atoms having zero total orbital angular momentum (e.g. hydrogen atoms in the ground state have $n = 1$ and $l = 0$, subsequently $L = [l(l + 1)]^{1/2}\hbar = 0$) through an inhomogeneous magnetic field as shown in Figure 41.6, the beam will experience a force depending on m_s.

For the magnetic field shown note that $B_y = 0$ and that B_x and B_z depend on x and z. The potential energy of an electron in the magnetic field is

$$U_m = -\boldsymbol{\mu}_s \cdot \mathbf{B} = -(\mathbf{i}\mu_{sx} + \mathbf{j}\mu_{sy} + \mathbf{k}\mu_{sz}) \cdot (\mathbf{i}B_x + \mathbf{k}B_z)$$

or

$$U_m = -\mu_{sx}B_x - \mu_{sz}B_z$$

Now recall the relationship between force and potential energy developed in Text Section 9-2:

$$F_x = -\partial U_m/\partial x = \mu_{sx}\partial B_x/\partial x + \mu_{sz}\partial B_z/\partial x$$

$$F_y = -\partial U_m/\partial y = 0$$

$$F_z = -\partial U_m/\partial z = \mu_{sx}\partial B_x/\partial z + \mu_{sz}\partial B_z/\partial z$$

If the beam is very small, we can write the following

$B_x = 0$	Figure 41.6(c) shows that $B_x \to 0$ as $x \to 0$
$\partial B_x/\partial x \cong 0$	Figure 41.6(c) shows that the x component of B does not change rapidly with x for small values of x
$\partial B_z/\partial x = 0$	Figure 41.6(b) shows that the z component of B goes through a maximum value at $x = 0$ hence $\partial B_z/\partial x = 0$

When the information relevant to a small beam is inserted into the expressions for the components of the force we have

$$F_x \cong 0$$
$$F_y = 0$$
$$F_z = \mu_{sz} dB/dz = -g\mu_B m_s dB/dz$$

Note:

1. If the field is uniform, then $dB/dz = 0$ and there is no force on the beam.
2. If the field is nonuniform, since m_s has two directions, F_z has two directions.

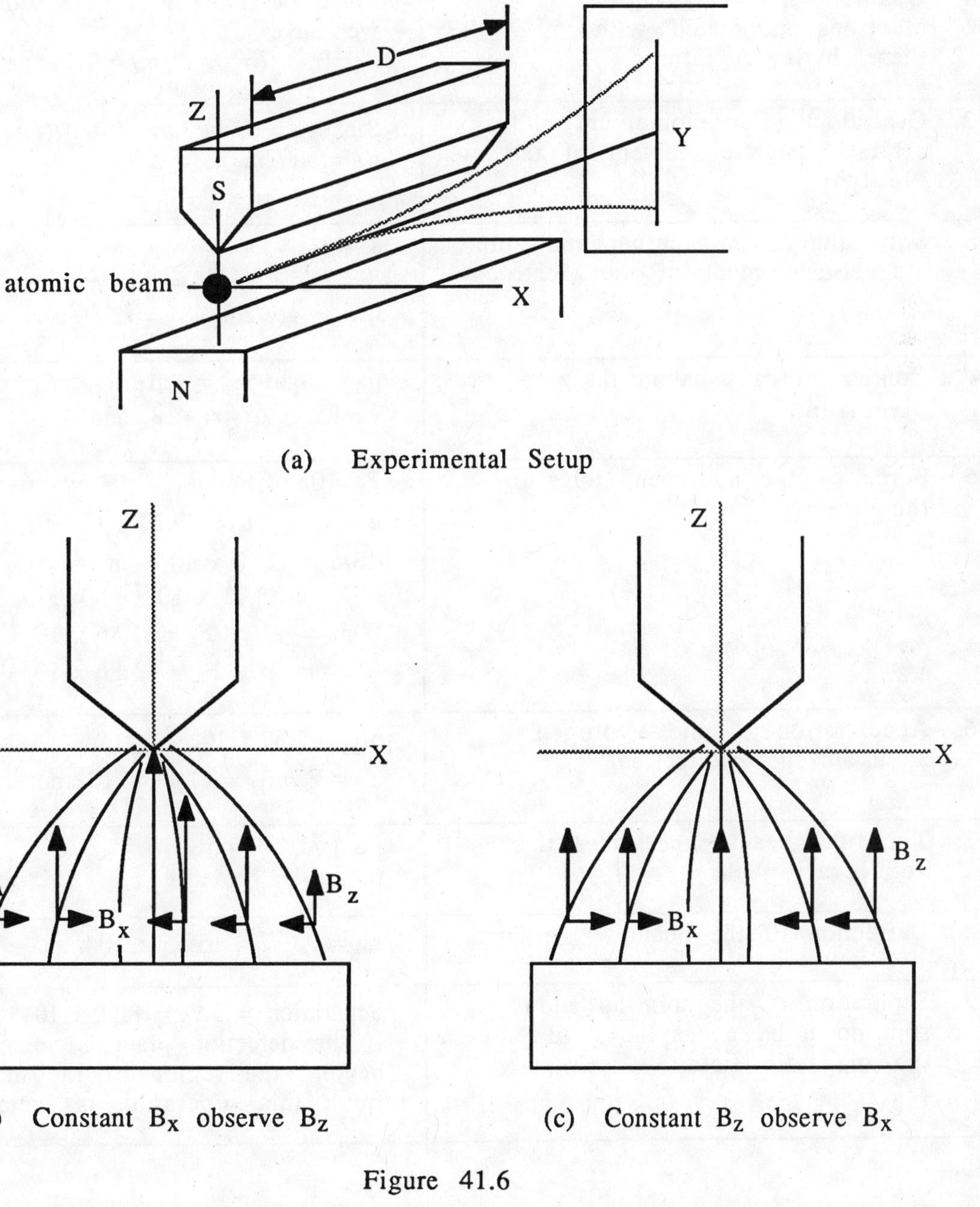

Figure 41.6

Practice: The atomic beam shown in Figure 41.6 is made up of hydrogen atoms in the ground state. The following information is also known about the experimental set up.

$D = 2.00 \times 10^{-1}$ m -- length of the magnetic field
$v_y = 2.00 \times 10^5$ m/s -- speed of the hydrogen atoms through the magnetic field
$dB/dz = 2.00 \times 10^2$ T/m -- magnetic field gradient experienced by the beam

Determine the following:

1. Quantum numbers for the electrons associated with these hydrogen atoms	Since the atom is in the ground state we have $n = 1$, $l = 0$, $m_l = 0$, $m_s = \pm 1/2$
2. Orbital angular momentum and orbital magnetic moment of the electrons	Since $l = 0$, we have $L = [l(l + 1)]^{1/2}\hbar = 0$ $\mu_l = -(e/2m_e)L = 0$
3. Spin angular momentum and spin magnetic moment of the electrons	$s = 1/2$ $S = [s(s + 1)^{1/2}\hbar = (3)^{1/2}\hbar / 2$ $\lvert\mu_s\rvert = g(e/2m_e)\lvert S\rvert = (3)^{1/2}\mu_B$
4. Component of μ_s along the z direction	$\mu_{sz} = -g\mu_B m_s = -2\mu_B m_s$, Since $m_s = \pm 1/2$, we have $\mu_{sz} = +\mu_B$ and $-\mu_B$
5. Force on the hydrogen atoms in the beam	$F_x = 0$, $F_y = 0$, $F_z = -g\mu_B m_s dB/dz$ $g = 2$, $\mu_B = 9.27 \times 10^{-24}$ J/T $dB/dz = 2.00 \times 10^2$ T/m $\lvert F_z\rvert = m_s(3.71 \times 10^{-21}$ N) If $m_s = +1/2$, $F_z = -1.86 \times 10^{-21}$ N If $m_s = -1/2$, $F_z = +1.86 \times 10^{-21}$ N
6. Acceleration of the hydrogen atoms in the z direction	$m_H = 1.67 \times 10^{-27}$ kg $a_z = F_z/m_H = \pm 1.11 \times 10^6$ m/s^2
7. Time the beam experiences this acceleration	$t = D/v_y = 1.00 \times 10^{-6}$ s
8. Deflection of the beam	$\Delta z = a_z t^2/2 = \pm 5.55 \times 10^{-7}$ m
9. Separation of the spin up and spin down beams as they exit the magnetic field	Separation $= 2\Delta z = 1.11 \times 10^{-6}$ m If the detection plate is some distance beyond the region of the magnetic field, this separation is increased.

Example: An electron in the lower energy spin orientation may be caused to "flip" its spin orientation by incident radiation. If electrons are experiencing a magnetic field of 0.500 T, determine the wavelength of radiation that will cause an electron in the lower energy spin orientation to "flip" into the higher energy spin orientation.

Given: B = 0.500 T
An electron in the lower energy spin orientation

Determine: The wavelength of radiation that will cause the electron to "flip" into the higher energy spin orientation.

Strategy: Knowing how the energy of the electron depends on the magnetic field and the spin orientation quantum number, we can determine the energy difference between the two spin orientation states. The energy of the incident photons must be the same as the energy difference between the two spin orientation states. Knowing the energy of the incident photons, we can determine the wavelength of the incident radiation.

Solution: The energy associated with the orientation in a magnetic field is given by

$$U_m = -\boldsymbol{\mu}_s \cdot \mathbf{B} = -\mu_{sz}B = g\mu_B m_s B$$

The energy difference between the two spin orientations is

$$U_m = g\mu_B B\Delta m_s = g\mu_B B[(1/2) - (-1/2)] = g\mu_B B = 9.27 \times 10^{-24}\ \mathrm{J}$$

The energy the incident photons must have in order to cause the desired change in spin orientation or to "flip" the electron is

$$E = \Delta U_m = 9.27 \times 10^{-24}\ \mathrm{J}$$

The wavelength of the incident radiation then is

$$E = h\nu = hc/\lambda$$

or

$$\lambda = hc/E = 2.14 \times 10^{-2}\ \mathrm{m}$$

Related Text Exercises: Ex. 41-19 through 41-21.

PRACTICE TEST

Take and grade this practice test. Doing so will allow you to determine any weak spots in your understanding of the concepts taught in this chapter. The following section prescribes what you should study further to strengthen your understanding.

The wave function for an excited hydrogen atom is represented by $\psi_{21-1+1/2}$.

Determine the following:

_________ 1. The principal quantum number
_________ 2. The energy in eV
_________ 3. The orbital angular momentum quantum number
_________ 4. The orbital angular momentum
_________ 5. The maximum value of the projection of the orbital angular momentum onto the z axis
_________ 6. The spin angular momentum
_________ 7. The projection of the spin angular momentum onto the z axis
_________ 8. The angle between the direction of **L** and the +z-axis
_________ 9. The angle between the direction of **S** and the +z-axis

Consider a one electron atom in the n = 3 state.

Determine the following:

_________ 10. The value of the orbital magnetic dipole moment
_________ 11. The z component of the orbital magnetic dipole moment
_________ 12. If the electron undergoes the transition $l = 1$ to $l = 0$, in the absence of a magnetic field how many spectral lines are emitted?
_________ 13. If the electron undergoes the transition $l = 1$ to $l = 0$, in the presence of an external magnetic field, how many spectral lines are emitted?
_________ 14. If the magnitude of the external magnetic field is a 0.500 T, what is the spacing between the energy levels with the same l but different m_l?
_________ 15. If the wavelength before the field was turned on was 6000 Å, what wavelengths are observed after the field is turned on?

An atomic beam of hydrogen atoms in the ground state passes through a nonuniform magnetic field as shown in Figure 41.6(a). The length of the magnetic field is 1.50×10^{-1} m, the hydrogen atoms enter the magnetic field with a horizontal speed of 1.5×10^5 m/s and the magnetic field gradient experienced by the beam is $dB/dz = 3.00 \times 10^2$ T/m.

Determine the following:

_________ 16. Spin magnetic moment of the electrons
_________ 17. Component of the spin magnetic moment parallel to the z-axis
_________ 18. Force on the atoms in the beam
_________ 19. Acceleration of the atoms in the beam
_________ 20. Separation of the spin up and down beams as they exit the magnetic field

(See Appendix I for answers.)

PRINCIPAL CONCEPTS AND EQUATIONS PRESCRIPTION

Your score on the practice test is an excellent measure of your understanding of the chapter. You should now use the following chart to write your own prescription for dealing with any weaknesses the practice test points out. Look down the leftmost column to the number of the question(s) you answered incorrectly, reading across that row you will find the concept and/or equation of concern, the section(s) of the study guide you should return to for further study, and some suggested text exercises which you should work to gain additional experience.

Practice Test Questions	Concepts and Equations	Prescription	
		Principal Concepts	Text Exercises
1	Principal quantum number: n	1	41-1
2	Energy levels for hydrogen atom: $E_n = -13.6 \text{ eV}/n^2$	1	41-1
3	Orbital angular momentum quantum number: l	1	41-1, 2
4	Orbital angular momentum: $L = [l(l + 1)]^{1/2}\hbar$	1	41-1, 5
5	z component of the orbital angular momentum: $L_z = m_l\hbar$	1	41-1, 2
6	Spin angular momentum: $S = [s(s + 1)]^{1/2}\hbar$	1	---
7	z component of the spin angular momentum: $S_z = m_s\hbar$	1	41-19
8	Angle between **L** and +z-axis: $\theta = \cos^{-1}(L_z/L)$	1	---
9	Angle between **S** and +z axis: $\alpha = \cos^{-1}(S_z/S)$	1	---
10	Orbital magnetic dipole moment: $\mu_l = (e/2m_e)L$	2	41-15
11	z component of orbital magnetic moment: $\mu_{lz} = \mu_B m_l$	2	41-15
12	Zeeman effect	2	41-15, 16
13	Zeeman effect	2	41-15, 16
14	Magnetic potential energy: $\Delta U_m = \mu_B B \Delta m_l$	2	41-15, 16
15	Zeeman effect -- shift in wavelength: $\lvert d\lambda \rvert = \lambda^2(\Delta U_m)/hc$	2	41-17, 18
16	Spin magnetic moment: $\lvert\mu_s\rvert = g(e/m_e)\lvert S\rvert = (3)^{1/2}\mu_B$	3	41-20, 21
17	Component of μ_s: $\mu_{sz} = -g\mu_B m_s$	3	41-21
18	Stern-Gerlach -- force: $F_x = 0$, $F_y = 0$, $F_z = -g\mu_B m_s dB/dz$	3	41-21
19	Newton's second law: $a_z = F_z/m$	2 of Ch 5	41-21
20	Describing motion: $2\Delta z = a_z(D/v_y)^2$	4 of Ch 3	41-21

42 ELECTRONS IN SOLIDS

RECALL FROM PREVIOUS CHAPTERS

Previously learned concepts and equations frequently used in this chapter	Text Section	Study Guide Page
Conductivity and resistivity: $\sigma = 1/\rho$	24-2	24-5
Semiconductors: Intrinsic and Doped	24-4	---
n-type and p-type semiconductors	24-6	---
Energy levels for a particle in a box: $E_n = (\hbar^2\pi^2/2mL^2)n^2$	40-6, 7	40-9

NEW IDEAS IN THIS CHAPTER

Concepts and equations introduced	Text Section	Study Guide Page
Energy of a particle in a 3-D box: $E_{n_1n_2n_3} = (\hbar^2\pi^2/2mL^2)(n_1^2 + n_2^2 + n_3^2)$	42-1	42-1
Quantum number space: $R^2 = n_1^2 + n_2^2 + n_3^2$	42-1	42-1
Number of states for the set of quantum numbers n_1, n_2, n_3: $N = \pi(n_1^2 + n_2^2 + n_3^2)^{3/2}/3$	42-1	42-1
Number of states with energy E: $N = [L^3(2m)^{3/2}/3\pi^2\hbar^3]E^{3/2}$	42-1	42-1
Density of states: $g(E) = dN/dE = [L^3(2m)^{3/2}/2\pi^2\hbar^3]E^{1/2}$	42-1	42-1
Fermi-Dirac distribution: $p(E) = 1/(e^{(E-E_F)/kt} + 1)$	42-2	42-5
Fermi energy	42-2	42-5
Resistivity: $\rho = m/n_e e^2\tau = mv/n_e e^2\Lambda$ $\rho = \rho_v + \rho_i$ and $1/\tau = 1/\tau_v + 1/\tau_i$	42-3	42-8
Electron energy bands	42-4	42-11

PRINCIPAL CONCEPTS AND EQUATIONS

1 Free Electron Model (Section 42-1)

Review: Recall from chapter 40 that the energy levels available for a particle of mass m in a one-dimensional box of length L are

$$E_n = (\hbar^2\pi^2/2mL^2)n^2$$

In like manner the energy levels available for a particle of mass m in a three-dimensional box (a cube of edge length L and volume L^3) are

$$E_{n_1n_2n_3} = (\hbar^2\pi^2/2mL^2)(n_1^2 + n_2^2 + n_3^2)$$

If we construct a "quantum number" space as shown in Figure 42.1, with a set of quantum numbers n_1, n_2, and n_3 representing a point which is a distance R from the origin then

$$R^2 = n_1^2 + n_1^2 + n_3^2$$

and hence

$$E = (\hbar^2\pi^2/2mL^2)R^2 \quad \text{or} \quad R = (2m)^{1/2}(L/\hbar\pi)E^{1/2}$$

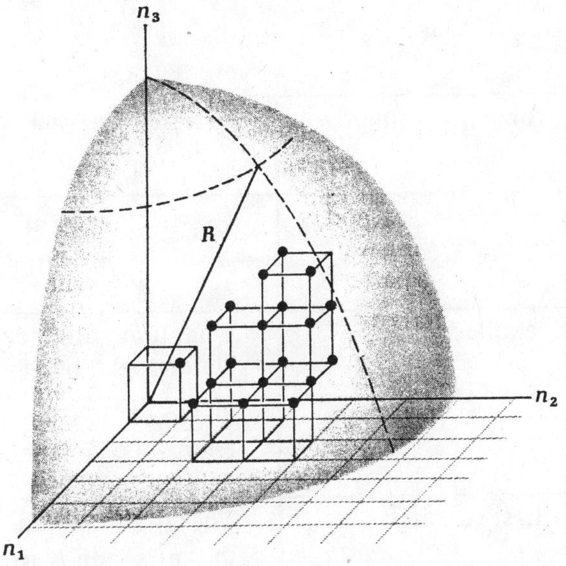

Figure 42.1

Using the concept of "quantum number" space we can make the following statements.

1. Each cell has an edge of length one, hence the volume of a cell is $(1)^3 = 1$.

2. Each cell corresponds to one lattice point, hence there is one lattice point per unit volume in "quantum number" space.

3. The number of points N_{pts} is equal to the volume of that region (this is a consequence of 2. above).

$$N_{pts} = V(R) = (4\pi R^3/3)/8 = \pi R^3/6$$

4. Each point corresponds to a set of quantum numbers (n_1, n_2, n_3). But a state for an electron in a 3-D box is specified by four quantum numbers n_1, n_2, n_3 and m_s. Recall that m_s has two values ($m_s = \pm 1/2$) hence each lattice point corresponds to two electron states.

5. The number of states for the set of quantum numbers n_1, n_2 and n_3 or the radius R is

$$N = 2(\text{Volume of octant of sphere}) = 2(\pi R^3/6) = \pi R^3/3$$

6. The number of states with energy E is

$$N = \pi R^3/3 = (\pi/3)(2m)^{3/2}(L/\hbar\pi)^3E^{3/2} = [L^3(2m)^{3/2}/3\pi^2\hbar^3]E^{3/2}$$

7. If we create the concept of density of states g(E), then g(E) will be the number of states per unit of energy E with energy between E and E + dE, hence

$$g(E) = dN/dE = [L^3(2m)^{3/2}/2\pi^2\hbar^3]E^{1/2}$$

Practice: Consider a free electron in a copper cube of edge 1.00×10^{-3} m.

Determine the following:

1. Quantum numbers for the ground state	$n_1 = n_2 = n_3 = 1$
2. Ground state energy	$E_{111} = (\hbar^2\pi^2/2mL^2)(1^2 + 1^2 + 1^2)$ $= 1.79 \times 10^{-31}$ J
3. Energy of the first excited state	Quantum numbers for the first excited state might be $n_1 = 2$, $n_2 = 1$, and $n_3 = 1$. Hence $(n_1^2 + n_2^2 + n_3^2) = 6$ and the energy is $E_{211} = 3.58 \times 10^{-31}$ J.
4. Degeneracy of the first excited state	The first excited state has the sum of all the n^2's equal to 6. This can occur for the following cases case / n_1 / n_2 / n_3 / Σn^2 1 / 2 / 1 / 1 / 6 2 / 1 / 2 / 1 / 6 3 / 1 / 1 / 2 / 6 But recall that each set of quantum numbers gives us two states (don't forget $m_s \pm 1/2$). We have found three sets of quantum numbers that work, hence we have six states. We say that this state has six fold degeneracy in energy.
5. Spacing (in J) between the ground state level and the first excited state for a free electron	$\Delta E = E_{211} - E_{111} = 1.79 \times 10^{-31}$ J

6. Density of states at 10.0 eV	$g(E) = [L^3(2m)^{3/2}/2\pi^2\hbar^3]E^{1/2}$ $= 1.38 \times 10^{38}/J$
7. Number of states within a 10.0 meV range of 10.0 eV	$\Delta N = g(E)\Delta E$ $= (\frac{1.38 \times 10^{38}}{J})(\frac{10.0\ eV}{10^3})(\frac{1.6\ J}{10^{19}\ eV})$ $= 2.21 \times 10^{16}$
8. Number of states with energy less than or equal to 10.0 eV	$N = [L^3(2m)^{3/2}/3\pi^2\hbar^3]E^{3/2} = 1.45 \times 10^{20}$
9. Quantum number for a state with energy $E = 8.36 \times 10^{-31}$ J	$N = \pi R^3/3 = [L^3(2m)^{3/2}/3\pi^2\hbar^3]E^{3/2}$ $R = [L^3(2m)^{3/2}/\pi^3\hbar^3]^{1/3}E^{1/2} = 3.742$ or $R^2 = 14$, hence the quantum numbers are 3, 2, and 1. It doesn't matter which is n_1, n_2, or n_3.

Example: Determine the number of conduction electron states within a 10.0 meV range of 10.0 eV in a $1.00 \times 10^{-6}\ m^3$ sample of copper.

Given:

$V = 1.00 \times 10^{-6}\ m^3$ -- volume of sample

$E = 10.0\ eV = 1.60 \times 10^{-18}\ J$ -- energy of conduction electrons

$\Delta E = 10.0 \times 10^{-3}\ eV = 1.60 \times 10^{-17}\ J$ -- energy range of conduction electrons

Determine: The number of conduction electron states within a 10.0 meV range of 10.0 eV in the sample.

Strategy: Knowing the volume of the sample and the energy of the conduction electrons, we can determine the density of states. Knowing the density of states and the energy range we can determine the number of conduction electron states within the energy range.

Solution: The density of states at 10.0 eV is

$$g(10.0\ eV) = [V(2m)^{3/2}/2\pi^2\hbar^3]^{1/2}E^{1/2} = 1.36 \times 10^{41}/J$$

The number of states ΔN within a 10.0 meV range of 10.0 eV is

$$\Delta N = g(10.0\ eV)\Delta E = (1.36 \times 10^{41}/J)(1.60 \times 10^{-17}\ J) = 2.18 \times 10^{24}$$

Related Text Exercises: Ex 42-1 through 42-9.

2 Fermi-Dirac Statistics (Section 42-2)

Review: The Fermi-Dirac distribution function is

$$p(E) = 1/(e^{(E - E_F)/kT} + 1)$$

where k is Boltzmann's constant ($k = 1.38 \times 10^{-23}$ J/k)
T is the temperature in K degrees
E_F is the Fermi energy
E is the energy
p(E) is the probability that an electron state of energy E is occupied at temperature T

If we have a system of free electrons at T = 0 K, they will fill the available states (one electron per state) from the lowest energy up until all electrons have been accommodated. Thus all states with energy less than a certain value are occupied and those with a higher energy are not. The Fermi energy E_F is the energy value that divides the occupied states and the unoccupied states. Thus the probability of occupation of a state is given by

$$\left.\begin{array}{lll} p(E) = 1 & \text{for} & E < E_F \\ p(E_F) = 1/2 & \text{for} & E = E_F \\ p(E) = 0 & \text{for} & E > E_F \end{array}\right\} \quad (T = 0\ K)$$

The Fermi-Dirac distribution function for T = 0 K is shown in Figure 42.2(a), notice that it is discontinuous at the Fermi energy E_F. At temperatures greater than T = 0 K, the Fermi-Dirac distribution function is continuous (Figure 42.2(b)) and looks like a "smoothed" version of the discontinuous behavior. As shown in Figure 42.4(b) the value of kT sets a convenient scale for discussing the energy dependence of the Fermi-Dirac distribution function. If $E < E_F$ by several kT then $p(E) \approx 1$ and if $E > E_F$ by several kT then $p(E) \approx 0$.

Figure 42.2

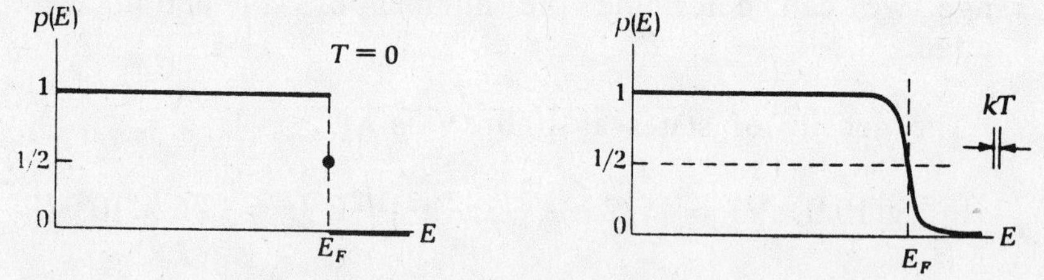

The number of states with energy $E < E_F$ is given by

$$N = [L^3(2m)^{3/2}/3\pi^2\hbar^3]E_F^{3/2}$$

Remember that all of these states are filled at T = 0 K since every state below $E = E_F$ is filled, hence the number of electrons is equal to the number of states. We may also express the number of electrons in terms of an electron density and the volume of the sample. This may be stated mathematically as shown below.

When $\quad E = E_F$ then $N = N_e = n_e L^3$

where $\quad N$ is the number of energy states below $E = E_F$ at $T = 0$ K
N_e is the number of electrons
n_e is the electron density (number of electrons per unit volume)
L^3 is the volume under consideration

Combining $\quad N = [L^3(2m)^{3/2}/3\pi^2\hbar^3]E_F^{3/2}$ and $N = n_e L^3$ obtain

$$E_F(T = 0\ K) = [(3\pi^2)^{2/3}\hbar^2/2m]n_e^{2/3}$$

The above expression shows that E_F depends only on m and n_e. E_F has a slight temperature dependence, however we will ignore it and write:

$$E_F(T) \approx E_F(T = 0\ K)$$

Practice: Consider a sample of copper and assume one free electron per atom. The mass density of copper is $\rho_{Cu} = 8.95 \times 10^3$ kg/m^3. The temperature of the sample is T = 300 K.

Determine the following:

1. Mass of one copper atom	$M = 6.35 \times 10^{-2}$ kg/mole $N_A = 6.02 \times 10^{23}$ atoms/mole $m = M/N_A = 1.05 \times 10^{-25}$ kg/atom
2. Density of copper atoms	$\rho = M/V = Nm/V$ or $n_{atomic} = N/V = \rho/m$ $n_{atomic} = \dfrac{(8.95 \times 10^3\ kg/m^3)}{(1.05 \times 10^{-25}\ kg/atom)}$ $= 8.52 \times 10^{28}$ atoms/m^3
3. Density of electrons	Assuming one free electron per atom, the density of atoms and the density of electrons are essentially the same. $n_e = n_{atoms} = 8.52 \times 10^{28}$ electrons/m^3
4. Fermi energy at T = 0 K	$E_F(T = 0\ K) = [(3\pi^2)^{2/3}\hbar^2/2m]n_e^{2/3}$ $= 1.12 \times 10^{-18}$ J
5. Fermi energy at T = 300 K	$E_F(T = 300\ K) \approx E_F(T = 0\ K) = 1.12 \times 10^{-18}$ J Since the Fermi energy is essentially temperature independent

6. The probability that a state with $E = 0.500E_F$ will be filled	$E = 0.500E_F$ $kT = 4.14 \times 10^{-21}$ J $E_F = 1.12 \times 10^{-18}$ J (step 5) $(E - E_F)/kT = -135$ $e^{(E - E_F)/kt} = 2.34 \times 10^{-59}$ $p(0.500E_F) = 1/(e^{(E - E_F)/kt} + 1) = 1$ Looking at Figure 42.2(b) this is exactly what we would expect since $E = 0.500E_F$ is 135 kT below E_F.
7. The probability that a state with $E = E_F$ will be filled	When $E = E_F$ then $e^{(E - E_F)/kt} = e^0 = 1$ and $p(E_F) = 1/(e^{(E - E_F)/kt} + 1) = 1/2$
8. The probability that a state with $E = 1.50E_F$ will be filled	$E = 1.50E_F$ $E - E_F/kT = 135$ $e^{(E - E_F)/kt} = 4.26 \times 10^{58}$ $p(1.50E_F) = 1/(e^{(E - E_F)/kt} + 1) = 0$ Looking at Figure 42.2(b) this is exactly what we would expect since $E = 1.50E_F$ is 135 kT above E_F.
9. The probability that a state with $E = E_F - 2kT$ will be filled	$E = E_F - 2kT$ or $E - E_F = -2kT$ $e^{(E - E_F)/kt} = e^{-2} = 0.135$ $p(E) = 1/1.135 = 0.881$
10. The energy expressed in kT above the Fermi energy for which the probability of being filled is 0.119	$p(E) = 1/(e^{(E - E_F)/kt} + 1) = 0.119$ $e^{(E - E_F)/kt} + 1 = 8.40$ $e^{(E - E_F)/kt} = 7.40$ $E - E_F = kT \ln(7.40) = 2kT$ At an energy of $E = E_F + 2kT$, that is 2kT above E_F, the probability of a state being filled is 0.119.
11. The Fermi velocity	The Fermi velocity is just the velocity of electrons at the Fermi energy. At T = 300 K $E_F = 1.12 \times 10^{-18}$ J (step 4) $U = 0$ and $K = E_F = mv_F^2/2$ $v_F = (2\ E_F/m)^{1/2} = 1.57 \times 10^6$ m/s
12. The Fermi temperature	The Fermi temperature is the temperature at which thermal motion (kT) is equal to the Fermi energy. $kT_F = E_F$ or $T_F = E_F/k = 8.12 \times 10^4$ K Note that this is greater than the melting point of copper.

Example: For copper, calculate the energy which has an occupancy probability of 0.900 at (a) T = 300 K, (b) T = 600 K, and (c) T = 1200 K.

Given: p(E) = 0.900 and T = 300 K, 600 K and 1200 K

Determine: The energy state that has an occupancy probability of 0.900 at T = 300 K, 600 K, and 1200 K.

Strategy: In the practice section (steps 4 and 5) we determined the Fermi energy for copper ($E_F = 1.12 \times 10^{-18}$ J). Knowing the Fermi-Dirac distribution function, the Fermi energy and the occupancy probability we can determine the energy that has that occupancy probability at a particular temperature.

Solution:

$$p(E) = 1/(e^{(E - E_F)/kt} + 1)$$

or

$$e^{(E - E_F)/kt} = 1/P(E) - 1$$

$$(E - E_F)/kT = \ln[1/P(E) - 1]$$

or

$$E = kT \ln[1/P(E) - 1] + E_F$$

$$E = kT \ln[1/0.900 - 1] + 1.12 \times 10^{-18} \text{ J}$$

$$E = kT(-2.20) + 1.12 \times 10^{-18} \text{ J}$$

$$E = 1.12 \times 10^{-18} \text{ J} - (3.04 \times 10^{-23} \text{ J/K})T$$

at

(a) $T = 3.00 \times 10^2$ K obtain $E = 1.11 \times 10^{-18}$ J
(b) $T = 6.00 \times 10^2$ K obtain $E = 1.10 \times 10^{-18}$ J
(c) $T = 1.20 \times 10^3$ K obtain $E = 1.08 \times 10^{-18}$ J

Related Text Exercises: Ex. 42-10 through 42-17.

3 Conduction In the Free Electron Model (Section 42-3)

Review: Conductivity σ is the reciprocal of the resistivity ρ

$$\sigma = 1/\rho$$

where

$$\rho = m/n_e e^2 \tau = mv/n_e e^2 \Lambda$$

and

- m -- is the electron mass
- e -- is the electron charge
- n_e -- is the number density of free electrons
- τ -- is the relaxation time
- v -- is the speed of the electrons ($v \approx v_F$ The Fermi speed)
- Λ -- is the mean free path between collisions
- ρ -- is the resistivity

Resistance to the flow of electrons (i.e. resistivity) is the result of two different lattice imperfections, in particular

(1) Vibrations -- The vibration of ions about the lattice sites can cause electrons to be scattered. This is characterized by τ_V and contribution to the resistivity is ρ_V. Note that τ_V and ρ_V are temperature dependent $\rho_V \propto 1/\rho_V \propto T$.

(2) Impurities -- The impurities are characterized by τ_i and the contribution to the resistivity is ρ_i. The quantities ρ_i and τ_i are independent of temperature.

Only those electrons within several kT of E_F are important in the scattering process, hence contribute to the resistivity. The total resistivity ρ is the sum of ρ_V and ρ_i, hence

$$\rho = \rho_V + \rho_i$$

Also the total relaxation time is related to τ_V and τ_i by

$$1/\tau = 1/\tau_V + 1/\tau_i \quad \text{or} \quad \tau = \tau_V\tau_i/(\tau_V + \tau_i)$$

Practice: Consider a copper sample and the following information

$T = 300$ K
$n_e = 8.52 \times 10^{28}/m^3$ -- step 3 of practice for concept [2]
$E_F(T = 300 \text{ K}) = 1.12 \times 10^{-18}$ J -- steps 4 and 5 of practice for concept [2]
$\rho = 1.75 \times 10^{-8}\ \Omega \cdot m$ -- CRC Handbook
$m = 9.11 \times 10^{-31}$ kg
$e = 1.60 \times 10^{-19}$ C

Determine the following:

1. Speed of the conduction electrons in copper at T = 300 K	The energy of the conduction electrons is all kinetic and is equal to E_F, hence $E_F = mv_F^2/2$ or $v_F = (2E_F/m)^{1/2} = 2.46 \times 10^6$ m/s.

Note: Do not confuse the Fermi speed with the drift speed of the conduction electrons (typically 10^{-5} m/s). The Fermi speed is the average speed of the electrons between collisions and the drift speed is the average speed at which electrons actually drift through the conductor.

2. Average time between collisions for the conduction electrons in copper	$\rho = m/n_e e^2\tau$ or $\tau = m/n_e e^2\rho = 2.38 \times 10^{-14}$ s
3. Average number of collisions made by each conduction electron per second	$N_{colls} = 1/\tau = 4.20 \times 10^{13}/s$

4. Means free path for the conduction electrons	$\Lambda = v_F\tau = 5.85 \times 10^{-8}$ m
5. Atomic spacing for copper	Let's agree to represent the distance between atoms by d. Then the volume d^3 contains one atom. Knowing that copper has 8.52×10^{28} atoms/m^3 ($n_e = n_{atoms}$) and that the volume d^3 contains one atom, we may set up the ratio $\frac{8.52 \times 10^{28} \text{ atoms}}{m^3} = \frac{1 \text{ atom}}{d^3}$ or $d = (1/8.52 \times 10^{28})^{1/3}$ m $= 2.27 \times 10^{-10}$ m
6. Number of atoms the conduction electrons pass before being scattered	$\Lambda = 5.85 \times 10^{-8}$ m, $d = 2.27 \times 10^{-10}$ m $\Lambda = N_{pass}d$ $N_{pass} = \Lambda/d = 2.58 \times 10^2$

Now suppose it is determine that the relaxation time for impurity scattering (τ_i) is 3.00×10^{-14} s, determine the following:

7. Average separation of impurities	$\Lambda_i = \tau_i v_F = 7.36 \times 10^{-8}$ m = 73.6 nm
8. The relaxation time for vibrations	$1/\tau = 1/\tau_i + 1/\tau_v$ or $\tau_v = \tau\tau_i/(\tau_i - \tau) = 1.15 \times 10^{-13}$ s
9. The contribution to the resistivity due to impurities and vibrations	$\rho_i = m/n_e e^2\tau_i = 1.39 \times 10^{-8}\ \Omega\cdot$m $\rho_v = m/n_e e^2\tau_v = 3.63 \times 10^{-9}\ \Omega\cdot$m, Note: $\rho_i + \rho_v = (1.39 + 0.363) \times 10^{-8}\ \Omega\cdot$m $= 1.75 \times 10^{-8}\ \Omega\cdot$m This is in complete agreement with the given resistivity.

Suppose it is established that the resistivity doubles when the temperature increases to 600 K. Determine the following:

10. The contribution to the resistivity due to impurities and vibrations	ρ_i is temperature independent, hence $\rho_i = 1.39 \times 10^{-8}\ \Omega\cdot$m $\rho_{600} = 2\rho_{300} = 3.50 \times 10^{-8}\ \Omega\cdot$m and $\rho_v = \rho_{600} - \rho_i = 2.19 \times 10^{-8}\ \Omega\cdot$m

Example: In a copper sample the average separation of impurities in 50.0 nm. If the sample is at low temperature, estimate its resistivity. The Fermi energy is $E_F = 1.12 \times 10^{-18}$ J and number density of free electrons for copper is $n_e = 8.52 \times 10^{28}/m^3$.

Given:
$T \approx 0$ K (low temperature)
$\Lambda_i = 5.00 \times 10^{-8}$ m -- average separation of impurities
$E_F = 1.12 \times 10^{-18}$ J -- Fermi energy
$n_e = 8.52 \times 10^{28}/m^3$ -- number density for free electrons in copper

Determine: The resistivity ρ of copper

Strategy: Knowing the Fermi energy E_F, we can determine the Fermi speed v_F. Knowing the Fermi speed v_F and the average separation of impurities Λ_i, we can determine the relaxation time τ_i. Knowing m, e, n_e and τ_i, we can determine ρ_i. Finally, since $T \approx 0$ K we can say $\rho_v \approx 0$ and hence $\rho = \rho_i$.

Solution: The Fermi speed is given by

$$E_F = mv_F^2/2 \quad \text{or} \quad v_F = (2\,E_F/m)^{1/2} = 2.46 \times 10^6 \text{ m/s}$$

The relaxation time τ_i is

$$\tau_i = \Lambda_i/v_F = 5.00 \times 10^{-8} \text{ m}/2.46 \times 10^6 \text{ m/s} = 2.03 \times 10^{-14} \text{ s}$$

Then

$$\rho_i = m/n_e e^2\tau_i = 2.06 \times 10^{-8}\ \Omega\cdot\text{m}$$

Since

$$T \approx 0 \text{ K} \quad \text{we have} \quad \rho_v \approx 0\ \Omega\cdot\text{m} \quad \text{and} \quad \rho \approx \rho_i = 2.06 \times 10^{-8}\ \Omega\cdot\text{m}$$

Related Text Exercises: Ex. 42-18 through 42-21.

4 Electron Energy Bands (Section 42-4)

Review: The electronic properties of a solid depend on the arrangement of the bands and gaps and on how they are populated by the electrons. This approach to describing electrons in solids is called the band theory of solid. Figure 42.3 shows the band structure for a conductor, insulator, intrinsic semiconductor, and n-type and p-type extrinsic semiconductors.

(a) Conductor - a substance that has the Fermi energy within an energy band is a conductor. A conductor has a band with many occupied and many unoccupied states. The band is not full. Since the electrons in such a band are responsible for conduction, the band is often called the conduction band.

(b) Insulator - The valence band is totally occupied at T = 0 K (i.e. the number of electrons occupying the valence band equals the number of states in the band). The energy gap is large (6 eV for the diamond form of carbon) and the conduction band is empty.

(c) Intrinsic semiconductor - the energy gap is small (1.1 eV for Si). The valence band is full and the conductor band empty at T = 0 K. At higher temperatures, a small but significant number of electrons are thermally excited into the conduction band. For each electron excited into the conduction band there is a hole in the valence band. These electrons and holes are the charge carriers, and the current is due to their motion. The concentration of these carriers in a pure or intrinsic semiconductor at a given temperature depends only on the material (hence the name intrinsic).

(d) Extrinsic n-type semiconductor - the energy gap is small and as a result of "doping" with donor impurities it has an unoccupied donor level near the conduction band. Since the impurities contribute electrons (negatively charged) they are called n-type semiconductors.

(e) Extrinsic p-type semiconductor - the energy gap is small and as a result of "doping" with acceptor impurities it has an occupied acceptor level near the valence band. Since the acceptor impurities accept electrons out of the valence band, that is contribute holes (positively charged) they are called p-type semiconductors.

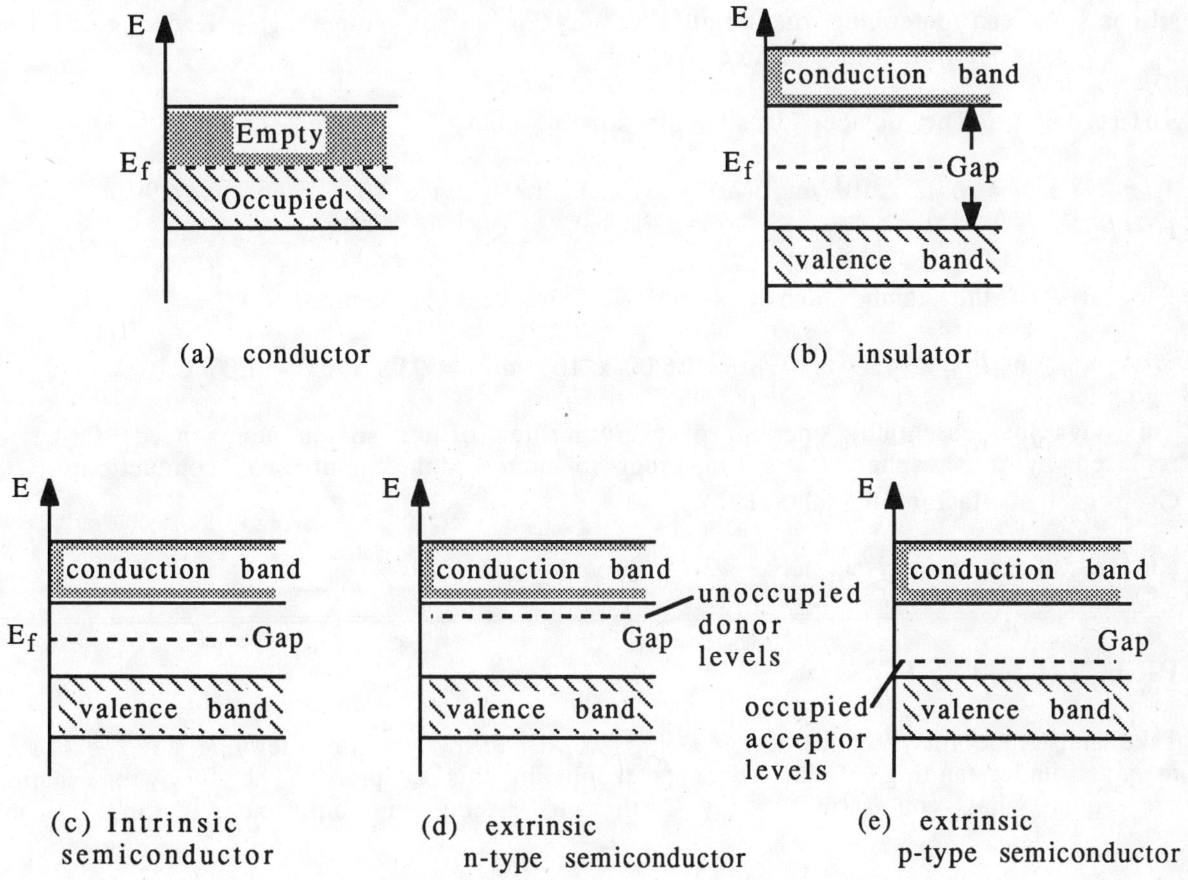

Figure 42.3

Example: The number density of conduction electrons in pure silicon is about 10^{16} m^{-3} at 300 K. Suppose that a silicon semiconductor is doped with phosphorous such that the number of conduction electrons is increased by a factor of 5.00×10^6. What fraction of silicon atoms must be replaced by phosphorus in order to accomplish this increase in conduction electrons? The density and atomic weight of silicon are $\rho = 2.33 \times 10^3$ kg/m^3 and $A = 2.81 \times 10^{-2}$ kg/mol.

Given:

$T = 300$ K

$n_e = 1.00 \times 10^{16}/m^3$ before "doping"

$n_p = (1.00 \times 10^{16}/m^3)(5.00 \times 10^6) = 5.00 \times 10^{22}/m^3$

$\rho = 2.33 \times 10^3$ kg/m^3

$A = 2.81 \times 10^{-2}$ kg/mol

Determine: The fraction of silicon atoms that must be replaced by phosphorus impurity atoms in order to increase the number of conduction electrons by a factor of 5.00×10^6.

Strategy: Knowing Avogadro's number, the mass density and atomic weight of silicon, we can determine the number density of silicon atoms n_{si}. Knowing n_{si} and n_p we can determine the desired fraction.

Solution: The number density for silicon atoms is

$n_{si} = N_A\rho/A = (6.02 \times 10^{23}/\text{mol})(2.33 \times 10^3 \text{ kg/m}^3)/(2.81 \times 10^{-2} \text{ kg/mol}) = 4.99 \times 10^{28}/m^3$

The ratio of the number densities gives us the desired fraction.

$$n_{si}/n_p = (4.99 \times 10^{28}/m^3)/(5.00 \times 10^{22}/m^3) = 9.98 \times 10^6 \approx 1.00 \times 10^6$$

This says that essentially one our of every million of the silicon atoms needs to be replaced by a phosphorus atom in order to increase the number of conduction electrons by a factor of 5.00×10^6.

Related Text Exercises: Ex. 42-18 through 42-21.

PRACTICE TEST

Take and grade this practice test. Doing so will allow you to determine any weak spots in your understanding of the concepts taught in this chapter. The following section prescribes what you should study further to strengthen your understanding.

Consider a free electron in a copper cube of edge 1.00×10^{-3} m.

Determine the following:

______ 1. Ground state energy
______ 2. Energy of the first excited state
______ 3. Spacing (in J) between the ground state and the first excited state
______ 4. Density of states at 10.0 eV
______ 5. Number of states within a 10.0 meV range of 10.0 eV
______ 6. Number of states with an energy less than or equal to 10.0 eV
______ 7. Quantum numbers for a state with energy $E = 8.36 \times 10^{-31}$ J

Consider the following data for a sample of copper:

$$n_e = 8.52 \times 10^{28} \text{ electrons/m}^3 \text{ -- density of electrons}$$
$$T = 3.00 \times 10^2 \text{ K -- temperature of sample}$$

Determine the following:

______ 8. Fermi energy at 300 K
______ 9. The probability that a state with $E = 0.500E_F$ will be filled
______ 10. The probability that a state with $E = 1.50E_F$ will be filled
______ 11. The probability that a state with $E = E_F$ will be filled
______ 12. The probability that a state with energy 2kT below E_F will be filled
______ 13. The Fermi velocity
______ 14. The Fermi temperature
______ 15. The energy which has an occupancy probability of 0.900

Consider the following data for a sample of copper:

$$n_e = 8.52 \times 10^{28} \text{ electrons/m}^3 \text{ -- density of electrons}$$
$$T = 3.00 \times 10^2 \text{ K -- temperature of sample}$$
$$E_F(T = 300 \text{ K}) = 1.12 \times 10^{-18} \text{ J -- Fermi energy at } T = 300 \text{ K}$$
$$\rho = 1.75 \times 10^{-8} \ \Omega\cdot\text{m -- resistivity}$$

Determine the following:

______ 16. Speed of the conduction electrons in the copper
______ 17. Average time between collisions for the conduction electrons in the copper
______ 18. Average number of collisions made by each conduction electron per second
______ 19. Mean free path for the conduction electrons
______ 20. The number of atoms each conduction electron passes before being scattered

(See Appendix I for answers.)

PRINCIPAL CONCEPTS AND EQUATIONS PRESCRIPTION

Your score on the practice test is an excellent measure of your understanding of the chapter. You should now use the following chart to write your own prescription for dealing with any weaknesses the practice test points out. Look down the leftmost column to the number of the question(s) you answered incorrectly, reading across that row you will find the concept and/or equation of concern, the section(s) of the study guide you should return to for further study, and some suggested text exercises which you should work to gain additional experience.

Practice Test Questions	Concepts and Equations	Prescription	
		Principal Concepts	Text Exercises
1	Energy of a particle in a 3-D box: $E_{n_1n_2n_3} = (\hbar^2\pi^2/2mL^2)(n_1 + n_2^2 + n_3^2)$	1	42-1, 2
2	Energy of a particle in a 3-D box	1	42-2, 3
3	Energy between two states: $\Delta E = E_{211} - E_{111}$	1	42-2, 3
4	Density of states at energy E: $g(E) = [L^3(2m)^{3/2}/2\pi^2\hbar^3]E^{1/2}$	1	42-8, 9
5	Number of states within ΔE of E: $\Delta N = g(E)\Delta E$	1	42-8, 9
6	Number of states with energy $\leq$ E: $N = [L^3(2m)^{3/2}/3\pi^2\hbar^3]E^{3/2}$	1	42-8, 9
7	Set of quantum numbers for a particular E: $R^2 = (2mL^2/\hbar\pi^2)E_{n_1n_2n_3}$ and $R \Rightarrow n_1, n_2, n_3$	1	42-1, 2
8	Fermi energy: $E_F = [(3\pi^2)^{2/3}\hbar^2/2m]n_e^{2/3}$	2	42-16, 17
9	Fermi-Dirac distribution function: $p(E) = 1/(e^{(E - E_F)/kt} + 1)$	2	42-11, 12
10	Fermi-Dirac distribution function	2	42-12, 13
11	Fermi-Dirac distribution function	2	42-13, 14
12	Fermi-Dirac distribution function	2	42-14, 15
13	Fermi velocity: $E_F = mv_F^2/2$	2	42-19
14	Fermi temperature: $E_F = kT_F$	2	---
15	Fermi-Dirac distribution function	2	42-11, 13
16	Fermi: velocity: $E_F = mv_F^2/2$	2	42-19
17	Relaxation time: $\tau=m/n_ee^2\rho$	3	42-18, 21
18	Number of collisions: $N = 1/\tau$	3	---
19	Mean free path: $\Lambda = v_F\tau$	3	42-21
20	Atomic spacing	3	---

43 THE ATOMIC NUCLEUS

RECALL FROM PREVIOUS CHAPTERS

Previously learned concepts and equations frequently used in this chapter	Text Section	Study Guide Page
Mass-energy equivalence: $\Delta E = \Delta mc^2$	38-6	38-11

NEW IDEAS IN THIS CHAPTER

Concepts and equations introduced		Text Section	Study Guide Page
Notation: $^{A}_{Z}X$		43-1	43-2
Isotopes: same Z number		43-1	43-2
Isotones: same N number		43-1	43-2
Isobars: same A number		43-1	43-2
Nuclear radius: $R = R_0A^{1/3}$		43-1	43-2
Binding energy: $B = (ZM_H + Nm_n - M_a)c^2$		43-2	43-4
Binding energy per nucleon: B/A		43-2	43-4
Radioactive decay: $N = N_0e^{-\lambda t}$		43-5	43-7
Half-life: $T_{1/2} = (\ln 2)/\lambda$		43-5	43-7
Activity: $R = -dN/dt = \lambda N = R_0e^{-\lambda t}$		43-5	43-7
Alpha decay: $^{A}_{Z}P \rightarrow {}^{A-4}_{Z-2}D + {}^{4}_{2}He$		43-5	43-11
$Q_\alpha = (M_P - M_D - M_{He})c^2$			
Beta decay: $^{A}_{Z}P \rightarrow {}^{A}_{Z+1}D + {}^{0}_{-1}e + \bar{\nu}$	(β^- decay)	43-5	43-11
$Q_{\beta-} = (M_P - M_D)c^2$	(β^- decay)		
$^{A}_{Z}P \rightarrow {}^{A}_{Z-1}D + {}^{0}_{+1}e + \nu$	(β^+ decay)		
$Q_{\beta+} = (M_P - M_D - 2m_e)c^2$	(β^+ decay)		
Gamma decay: $^{A}_{Z}P^* \rightarrow {}^{A}_{Z}P + \gamma$		43-5	43-11
Nuclear reactions: $Q = (M_{reactants} - M_{products})c^2$		43-6	43-15
Nuclear fission		43-6	43-18
Nuclear fusion		43-6	43-21

PRINCIPAL CONCEPTS AND EQUATIONS

1 Nuclear Notation (Section 43-1)

Review: The notation for a particular nucleus or nuclide is

$$^{A}_{Z}X$$

Where X -- represents the symbol for an element. When we talk about a nucleus we are talking about the nucleus of a specific atom, subsequently we use the symbol for the atomic element.

Z -- called the proton number, represents the number of protons in the nucleus and hence tells us the charge on the nucleus.

N -- called the neutron number, represents the number of neutrons in the nucleus.

A -- called the mass number, is the number of mass units in the nucleus. Since each proton and each neutron contributes one mass unit apiece, A is equal to the number of protons plus the number of neutrons in the nucleus. $A = N + Z$

To completely specify a nuclide, we need only two of the three numbers A, Z and N.

Isotopes -- Nuclides with the same Z number, for example $^{12}_{6}C$ and $^{14}_{6}C$ have the same Z number ($Z = 6$). Isotopes are in vertical columns on a N vs Z chart of nuclides.

Isotones -- Nuclides with the same N number, for example $^{31}_{15}P$ and $^{32}_{16}S$ have the same N number ($N = A - Z = 16$). Isotones are in horizontal rows on a N vs Z chart of nuclides.

Isobars -- Nuclides with the same A number, for example $^{14}_{6}C$ and $^{14}_{7}N$ both have $A = 14$. Isobars are diagonals on a N vs Z chart of nuclides.

Figure 43.1 shows a small section of an N vs Z chart of nuclides, isotopes, isotones and isobars.

Nuclei are approximately spherical in shape with the radius given by

$$R = R_0 A^{1/3} \quad \text{where} \quad R_0 = 1.10 \text{ fm} \quad \text{and} \quad 1.00 \text{ fm} = 1.00 \times 10^{-15} \text{ m}$$

The volume of the nucleus is

$$V = 4\pi R^3/3 = 4\pi R_0{}^3 A/3$$

Nuclear forces are very strong attractive forces with a short range (10^{-14} m).

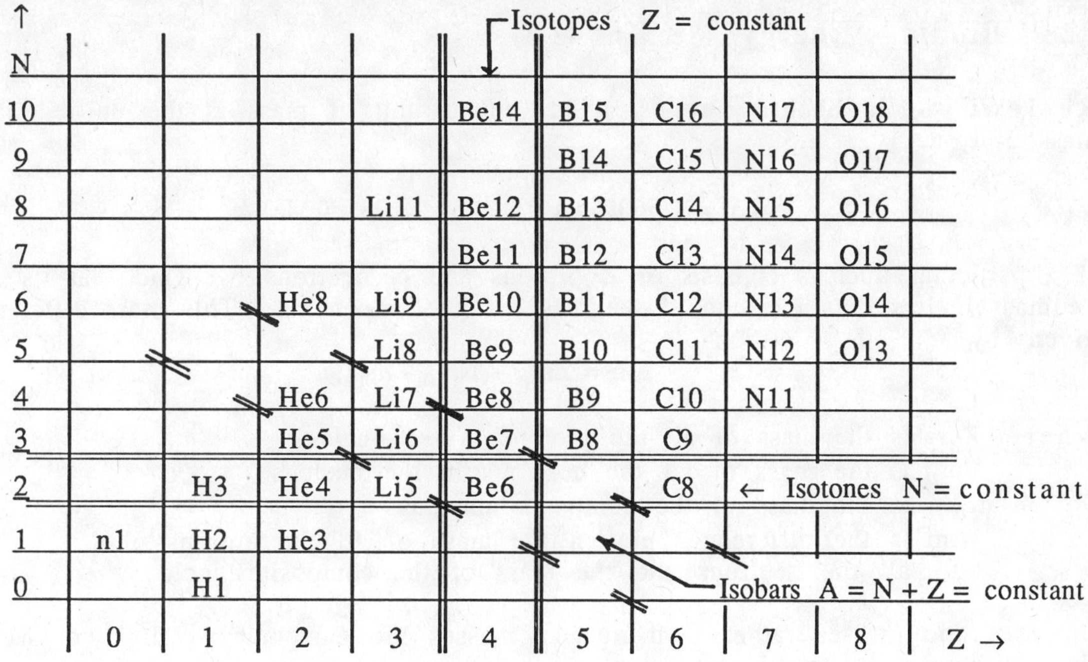

Figure 43.1

Practice: Consider the section of the chart of the nuclides shown in Figure 43.1. Determine the following:

1. Proton number, mass number, and neutron number (Z, A, N) for ${}^{6}_{4}Be$	$A = Z + N$ or $N = A - Z$, ${}^{A}_{Z}X \Rightarrow {}^{6}_{4}Be$ $Z = 4$, $A = 6$, $N = 6 - 4 = 2$
2. The isotopes of Be	${}^{6}_{4}Be$, ${}^{7}_{4}Be$, ${}^{8}_{4}Be$, ${}^{9}_{4}Be$, ${}^{10}_{4}Be$, ${}^{11}_{4}Be$, ${}^{12}_{4}Be$, ${}^{14}_{4}Be$
3. The isotones of ${}^{6}_{4}Be$	${}^{3}_{1}H$, ${}^{4}_{2}He$, ${}^{5}_{3}Li$, ${}^{6}_{4}Be$, ${}^{8}_{6}C$
4. The isobars of ${}^{6}_{4}Be$	${}^{6}_{2}He$, ${}^{6}_{3}Li$, ${}^{6}_{4}Be$
5. The radius of the ${}^{6}_{4}Be$ nucleus	$R_0 = 1.10$ fm $= 1.10 \times 10^{-15}$ m $A^{1/3} = (6)^{1/3} = 1.82$ $R = R_0A^{1/3} = 2.00 \times 10^{-15}$ m
6. Range of the nuclear forces	Range $\cong 10^{-14}$ m

Related Text Exercises: Ex. 43-1 through 43-6.

2 Binding Energy (Section 43-2)

Review: In nuclear physics, a convenient unit of mass is the unified atomic mass unit, u.

$$1\ u = 1.660559 \times 10^{-27}\ kg = 931.5\ MeV/c^2$$

If a particular nucleus consists of Z protons and N neutrons, we find that its composite mass is less than the sum of the mass of its components. This mass difference is given by

$$\Delta m = Zm_p + Nm_n - m_{nuc}$$

where Zm_p is the mass of all the protons in the nucleus
Nm_n is the mass of all the neutrons in the nucleus
m_{nuc} is the mass of the composite nucleus
Δm is the difference between the mass of all the protons plus all the neutrons and the mass of the composite nucleus

Since atomic masses rather than nuclear masses are customarily tabulated, we do the following substitutions

$Zm_p = ZM_H - Zm_e$ $\quad$ M_H is the atomic mass of ^1H

$m_{nuc} = M_a - Zm_e$ $\quad$ M_a is the atomic mass of the nuclide

to obtain

$$\Delta m = (ZM_H - Zm_e) + Nm_n - (M_a - Zm_e)$$

Notice that the mass of the Z electrons cancel to obtain

$$\Delta m = ZM_H + Nm_n - M_a$$

The energy equivalence of this mass difference is called the binding energy of the nucleus.

$$B = \Delta mc^2 = (ZM_H + Nm_n - M_a)c^2$$

The binding energy per nucleon is a useful criterion for comparing the binding energies of different nuclides and is given by

$$\text{Binding energy per nucleon} = B/A$$

Note: When doing nuclear calculations, it is frequently necessary to use more than three significant figures. For example, if we use only three significant figures, a proton and a neutron have the same mass. When available, we will record all information to the sixth decimal place, do the calculations, and then round off to three significant figures, unless the rounding off will obscure some of the nuclear information.

Practice: Given the following masses:

$m_e = 0.000548$ u = mass of electron, $M_H = 1.007825$ u = mass of neutral ^1_1H atom

$m_n = 1.008665$ u = mass of neutron, $M_{He} = 4.002603$ u = mass of neutral ^4_2He atom

$m_p = 1.007277$ u = mass of proton, $M_O = 15.994915$ u = mass of neutral $^{16}_8O$ atom

Determine the following:

1. Expression for mass difference (Δm) between individual nucleons and the composite ^4_2He nucleus	$\Delta m = ZM_H + Nm_n - M_{He}$
2. Mass difference for the above case	$ZM_H = 2(1.007825 \text{ u}) = 2.015650$ u $Nm_n = 2(1.008665 \text{ u}) = 2.017330$ u $M_{He} = 4.002603$ u $\Delta m = ZM_H + Nm_n - M_{He} = 0.030377$ u
3. Energy required to break the ^4_2He nucleus up into individual nucleons. That is, the binding energy for ^4_2He	$B = \Delta mc^2$ $= (0.030377 \text{ u})(931.5 \text{ MeV/c}^2 \text{ u})c^2$ $= 28.30$ MeV
4. Energy given off when two neutrons and two protons are brought together to form a ^4_2He nucleus	This energy is equal to the binding energy of the ^4_2He nucleus (step 3.) $E_{\text{given off}} = B = 28.30$ MeV
5. Binding energy per nucleon for ^4_2He	$B/A = 28.30 \text{ MeV}/4 = 7.075$ MeV
6. Expression for mass difference between the individual nucleons and composite $^{16}_8O$ nucleus	$\Delta m = ZM_H + Nm_n - M_O$ $Z = 8, \quad N = 8$
7. Mass difference for the above case	$ZM_H = 8M_H = 8.062600$ u $Nm_n = 8m_n = 8.069320$ u $M_o = 15.994915$ u $\Delta m = ZM_H + Nm_n - M_O = 0.137005$ u
8. Binding energy for $^{16}_8O$ nucleus	$B = \Delta mc^2$ $= (0.137005 \text{ u})(931.5 \text{ MeV/u c}^2)c^2$ $= 127.6$ MeV

9. Binding energy per nucleon for $^{16}_{8}O$	$B/A = 127.6\ MeV/16 = 7.98\ MeV$

Example: The binding energy per nucleon for for $^{238}_{92}U$ is 7.570 MeV. Determine the total binding energy for this nucleus and the mass of the nucleus in atomic mass units.

Given: $B/A = 7.570\ MeV$ for $^{238}_{92}U$

Determine: The total binding energy for the $^{238}_{92}U$ nucleus and the mass of this nucleus.

Strategy: Knowing that the nucleus of interest is $^{238}_{92}U$, we can determine the number of nucleons involved. Knowing the binding energy per nucleon and the number of nucleons, we can determine the total binding energy. Knowing the total binding energy and the mass-energy equivalence, we can determine Δm for this atom. Knowing Δm, Z, A, M_H, m_n and m_e, we can determine the mass of the nucleus.

Solution: The nucleus has 238 nucleus (i.e., A = 238). Hence the total binding energy is

$$B = (B/A)A = \left(\frac{7.570\ MeV}{nucleon}\right)(238\ nucleons) = 1.802 \times 10^3\ MeV$$

The mass difference is determined by using the mass-energy equivalence:

$$\Delta m = B/c^2 = (1.802 \times 10^3\ MeV)(1\ u\ c^2/931.5\ MeV)/c^2 = 1.935\ u$$

The mass of a neutral $^{238}_{92}U$ atom is determined by

$$\Delta m = ZM_H + Nm_n - M_u \quad \text{or} \quad M_u = ZM_H + Nm_n - \Delta m$$

or

$$M_U = (92)(1.007825\ u) + (146)(1.008665\ u) - 1.935\ u = 238.050\ u$$

The mass of the nucleus may be determined by subtracting the mass of 92 electrons

$$(M_U)_{nucleus} = (M_U)_{atom} - 92\ m_e = 238.050\ u - 92(0.000548\ u) = 238.000\ u$$

Example: Which nucleus is more stable, $^{19}_{7}N$ (19.016747 u) or $^{14}_{7}N$ (14.003074 u)?

Given: $^{19}_{7}N$ (19.016747 u) and $^{14}_{7}N$ (14.003074 u)

Determine: Which of these two nuclei is more stable?

Strategy: We can determine the binding energy per nucleon for each nucleus. The nucleus with the larger binding energy per nucleon is more stable.

Solution: First let's obtain Δm, B and B/A for each nucleus.

$^{19}_{7}N$ $\Delta m = 7M_H + 12m_n - M_{N-19} = 7.054775\ u + 12.103980\ u - 19.016747\ u$
$= 0.142008\ u$

$B = \Delta mc^2 = (0.142008\ u)(931.5\ MeV/1\ u\ c^2)c^2 = 132.3\ MeV$

$B/A = 132.3\ MeV/19 = 6.963\ MeV$

$^{14}_{7}N$ $\Delta m = 7M_H + 7m_n - M_{N-14} = 7.054775\ u + 7.060655\ u - 14.003074\ u$
$= 0.112356\ u$

$B = \Delta mc^2 = (0.112356\ u)(931.5\ MeV/1\ u\ c^2)c^2 = 104.7\ MeV$

$B/A = (104.7\ MeV)/14 = 7.479\ MeV$

Since the binding energy per nucleon is greater for $^{14}_{7}N$, it is the more stable of the two nuclei.

Related Text Exercises: Ex. 43-7 through 43-13.

3 Kinematics of Radioactive Decay (Section 43-5)

Review: The equation that describes radioactive decay is

$$N = N_o e^{-\lambda t}$$

where N = number of radioactive nuclei present at any time t
N_o = number of radioactive nuclei present at time t = 0
λ = the decay constant
= probability per unit time that a particular nucleus will decay
t = time

The half-life $T_{1/2}$ of a radioactive sample is the time it takes for one half of the nuclei to decay.

If at $t = 0$ we have $N = N_o$, then after one half-life we have

$$t = T_{1/2} \qquad N = N_o/2$$

That is, the number of radioactive nuclei left after a time equal to one half-life is just one half of the original number. As shown below, when this information is inserted into the radioactive decay equation, we obtain a relationship between $T_{1/2}$ and λ.

$$N = N_o e^{-\lambda t}$$

Insert $N = N_o/2$ at $t = T_{1/2}$ to obtain

$$N_o/2 = N_o e^{-\lambda T_{1/2}}$$

which may be rearranged to obtain

$$2 = e^{\lambda T_{1/2}}$$

Taking the natural log of this equation, obtain

$$\ln 2 = \lambda T_{1/2}$$

or

$$T_{1/2} = \ln 2/\lambda$$

Using this expression for $T_{1/2}$ in terms of λ, we see that after one half-life (i.e., $t = T_{1/2}$), we have

$$N = N_0 e^{-\lambda T_{1/2}} = N_0 e^{-\lambda \ln 2/\lambda} = N_0 e^{-\ln 2} = N_0/2$$

After two half-lives (i.e., $t = 2T_{1/2}$), we have

$$N = N_0 e^{-\lambda(2T_{1/2})} = N_0 e^{-\lambda 2 \ln 2/\lambda} = N_0 e^{-2 \ln 2} = N_0(e^{-\ln 2})^2 = N_0/2^2 = N_0/4$$

After n half-lives (i.e., $t = nT_{1/2}$), we have

$$N = N_0 e^{-\lambda(nT_{1/2})} = N_0 e^{-\lambda n \ln 2/\lambda} = N_0 e^{-n \ln 2} = N_0(e^{-\ln 2})^n = N_0/2^n$$

Notice that after seven half-lives,

$$N = N_0/2^7 = N_0/128 = 0.78\%\ N_0$$

we have less than 1% of the original number of radioactive nuclei left. Consequently, we say that the exponential decay process is essentially complete after seven half-lives.

The activity R of a radioactive sample is the rate at which the nuclei are decaying (i.e. the rate at which the nuclei are decreasing - hence the minus):

$$R = -dN/dt = -(-\lambda)N_0 e^{-\lambda t} = \lambda N$$

or

$$R = -dN/dt = -(-\lambda)N_0 e^{-\lambda t} = R_0 e^{-\lambda t}$$

where $R_0 = \lambda N_0$ is just the activity at $t = 0$. Activity may be measured in any of the following:

emissions/min		
emissions/s		
becquerel (Bq)	where	1 Bq = 1 emission/s
curie (Ci)	where	$1\ \text{Ci} = 3.7 \times 10^{10}$ Bq

Practice: A sample of radioactive material shows a measured activity of 5.00×10^3 emissions per minute when first measured and 3.00×10^3 emissions per minute when measured 1 h later.

Determine the following:

1. Activity at $t = 0$	$R_0 = 5.00 \times 10^3$ emissions/min
2. Activity after $t = 3600$ s	$R = 3.00 \times 10^3$ emissions/min
3. Decay constant	$A = A_0 e^{-\lambda t}$ $\frac{3.00 \times 10^3}{\text{min}} = (\frac{5.00 \times 10^3}{\text{min}})e^{-\lambda(3600\text{ s})}$ $0.600 = e^{-\lambda(3600\text{ s})}$ $1.67 = e^{\lambda(3600\text{ s})}$ $\ln(1.67) = \lambda(3600\text{ s})$ $\lambda = \ln(1.67)/3600\text{ s}$ $= 1.42 \times 10^{-4}\text{ s}$
4. Half-life	$T_{1/2} = \ln 2/\lambda = 4.88 \times 10^3$ s
5. Number of radioactive nuclei present at $t = 0$	$R_0 = \lambda N_0$ $N_0 = \frac{R_0}{\lambda} = \frac{(5.00 \times 10^3\text{ dis/min})}{(1.42 \times 10^{-4}\text{ s})}$ $= (\frac{3.52 \times 10^7\text{ s}}{\text{min}})(\frac{\text{min}}{60.0\text{ s}}) = 5.87 \times 10^5$
6. Activity after three half-lives	$R = R_0 e^{-\lambda t}$; when $t = 3T_{1/2} = 3\ln 2/\lambda$ $R = R_0 e^{-\lambda(3\ln 2/\lambda)} = R_0 e^{-3\ln 2}$ $= R_0/(2)^3 = 5.00 \times 10^3/8 = 625$ dis/min
7. Activity after 2 h	$R = R_0 e^{-\lambda t}$ $\lambda t = (1.42 \times 10^{-4}/\text{s})(7.20 \times 10^3\text{ s}) = 1.02$ $R = (5.00 \times 10^3\text{ dis/min})e^{-1.02}$ $= (5.00 \times 10^3\text{ dis/min})(0.361)$ $= 1.81 \times 10^3$ dis/min
8. Number of radioactive nuclei left after 2 h	$R = \lambda N$, $N = R/\lambda = \frac{(1.81 \times 10^3/\text{min})}{(1.42 \times 10^{-4}/\text{s})}$ $N = (\frac{1.27 \times 10^7\text{ s}}{\text{min}})(\frac{\text{min}}{60.0\text{ s}}) = 2.12 \times 10^5$ Also $N = N_0 e^{-\lambda t}$ $\lambda t = 1.02$ (step 7), $N_0 = 5.87 \times 10^5$ (step 5) $N = 5.87 \times 10^5\, e^{-1.02} = 2.12 \times 10^5$

Example: A sample of strontium 90 ($T_{1/2} = 28.0$ years) has an activity of 2.50×10^3 Bq. (a) What is the initial mass (m_0) of the sample? (b) How long will it be until only 10% of the sample is left?

Given: $T_{1/2} = 28.0$ years = half-life for strontium 90
$R_0 = 2.50 \times 10^3$ Bq = initial activity of sample
$m_f = 0.100\ m_0$ = final mass of sample

Determine: (a) Initial mass of the sample and time when the mass of the sample is 10% of the initial mass.

Strategy: Knowing the half-life, we can determine the decay constant. Knowing the decay constant and the initial activity, we can determine the initial number of nuclei and hence the initial mass of the sample. Knowing the initial mass and the final mass, we can determine the time.

Solution:

$$\lambda = \frac{\ln 2}{T_{1/2}} = \left(\frac{\ln 2}{28.0\ \text{year}}\right)\left(\frac{1\ \text{year}}{3.65 \times 10^2\ \text{day}}\right)\left(\frac{1\ \text{day}}{8.64 \times 10^4\ \text{s}}\right) = \frac{7.85 \times 10^{-10}}{\text{s}}$$

$$N_0 = \frac{R_0}{\lambda} = \frac{2.50 \times 10^3/\text{s}}{7.85 \times 10^{-10}/\text{s}} = 3.18 \times 10^{12}\ \text{nuclei}$$

If we initially have N_0 nuclei present, we have N_0 atoms present. The initial number of moles of the sample and the initial mass are

$$n_0 = \frac{N_0}{N_A} = \frac{m_0}{M} \qquad \text{or} \qquad m_0 = N_0 M/N_A$$

where n_0 = initial number of moles of sample
N_0 = initial number of nuclei and atoms
N_A = Avogadro's number
m_0 = initial mass of sample
M = molecular mass of sample

Inserting values, we obtain

$$m_0 = \frac{N_0 M}{N_A} = \frac{(3.18 \times 10^{12}\ \text{atoms})(87.6\ \text{g/mole})}{(6.02 \times 10^{23}\ \text{atoms/mol})} = 4.63 \times 10^{-10}\ \text{g}$$

Since the mass of the sample is directly proportional to the number of nuclei present, we can write

$$m = m_0 e^{-\lambda t}$$

We are interested in the value of t when $m = 0.100 m_0$

$$0.100 m_0 = m_0 e^{-\lambda t} \quad \Rightarrow \quad 10.0 = e^{\lambda t} \quad \Rightarrow \quad \ln 10.0 = \lambda t$$

$$t = \ln 10.0/\lambda = 2.30/(7.86 \times 10^{-10}/\text{s}) = 2.93 \times 10^9\ \text{s}$$

Related Text Exercises: Ex. 43-22 through 43-26.

4 Radioactive Decay (Section 43-5)

Review: The nuclei of some nuclides are inherently unstable or radioactive. Such a nucleus may eject a particle spontaneously, without any external stimulus. When this happens, the nucleus changes in some manner. We say that the nucleus changes in some manner. We say that the nucleus is naturally radioactive and that it has undergone the process of radioactive decay. The particles emitted are called radioactive emissions. The radioactive emissions of primary concern are alpha particles, beta particles, and gamma rays. These particles and their characteristics are listed below and the details of the decay process follow.

Alpha particle (α). An alpha particle is just like the nucleus of a He atoms. It consists of two protons and two neutrons, hence Z = 2 and A = 4. The symbol for an α-particle is ${}_{2}^{4}He$.

Beta particle (β^- and β^+). A β^- particle is just like an electron, hence Z = -1 and A = 0. The symbol for a β^--particle is ${}_{-1}^{0}e$. A β^+ particle is just like a positron, hence Z = +1 and A = 0. The symbol for a B^+-particle is ${}_{+1}^{0}e$.

Gamma ray (γ). A gamma ray is just like any other photon or electromagnetic wave. Because it has no charge or mass, the symbol for a gamma ray is γ.

Note: If an alpha particle is just like the nucleus of a helium atom, a beta particle just like an electron or positron, and a gamma ray just like any other photon or electromagnetic wave, why do we give them special names? You cannot, for example, tell a β^--particle from an electron or a β^+-particle from a positron. However the label "beta-minus particle" or "beta-plus particle" tells us that a given electron or positron originated from a nuclear event rather than being an atomic (outside the nucleus) particle. In like manner, the label "alpha particle" allows us to distinguish between a helium atom stripped of its electrons and something that looks and acts just like it but originated from a nuclear event. The same is true of a gamma ray, as opposed to any x-ray, photon, or other electromagnetic wave.

In addition to the above particles we need to be familiar with neutrinos (ν) and antineutrinos ($\bar{\nu}$). These particles have zero electric charge, zero rest mass and intrinsic spin of 1/2. The neutrino is associated with β^+-decay and the antineutrino is associated with β^--decay.

Alpha decay

When a parent nucleus emits an alpha particle, the daughter nucleus has two fewer protons and a mass number that is four less.

$${}_{Z}^{A}P \rightarrow {}_{Z-2}^{A-4}D + {}_{2}^{4}He$$

Conserving mass-energy we write

$$M_P c^2 = M_D c^2 + M_{He} c^2 + Q_\alpha$$

where M_P, M_D and M_{He} are the rest (atomic) masses of the parent, daughter and α-particle and Q_α is the sum of the kinetic energies of the daughter nucleus and the excitation energy of the parent. Then

$$Q_\alpha = (M_P - M_D - M_{He})c^2$$

This means that α-decay is energetically possible only if

$$M_P > M_D + M_{He} \quad \text{or} \quad Q_\alpha > 0$$

Beta decay (β^- and β^+)

When a parent nucleus emits a β^- particle, the daughter nucleus has one more proton and one less neutron hence the same mass number (the parent and daughter are isobars):

$$^{A}_{Z}P \rightarrow {}^{A}_{Z+1}D + {}^{0}_{-1}e + \bar{\nu}$$

During β^- emission, we believe the following takes place in the nucleus

$$^{1}_{0}n \rightarrow {}^{1}_{1}p + {}^{0}_{-1}e + \bar{\nu}$$

Conserving mass-energy we write

$$(M_P - Zm_e)c^2 = [M_D - (Z + 1)m_e]\,c^2 + m_ec^2 + Q_{\beta-}$$

or

$$Q_{\beta-} = (M_P - M_D)c^2$$

This means that β^--decay is energetically possible only if

$$M_P > M_D \quad \text{or} \quad Q_{\beta-} > 0$$

When a parent nucleus emits a β^+ particle, the daughter nucleus has one less proton and one more neutron hence the same mass number (the parent and daughter are isobars).

$$^{A}_{Z}P \rightarrow {}^{A}_{Z-1}D + {}^{0}_{+1}e + \nu$$

During β^+ emission, we believe the following takes place in the nucleus:

$$^{1}_{1}p \rightarrow {}^{1}_{0}n + {}^{0}_{+1}e + \nu$$

Conserving mass-energy we write

$$(M_P - Zm_e)c^2 = [M_D - (Z - 1)m_e]\,c^2 + m_ec^2 + Q_{\beta+}$$

or

$$Q_{\beta+} = (M_P - M_D - 2m_e)c^2$$

This means that β^+ decay is energetically possible only if

$$M_P > M_D + 2m_e \quad \text{or} \quad Q_{\beta+} > 0$$

Gamma decay

If the parent nucleus is in an excited state, it can dissipate its excitation energy by emitting a gamma ray:

$$^{A}_{Z}P^{*} \rightarrow {}^{A}_{Z}P + \gamma$$

where the asterisk indicates the excited state of a nuclide.

Note: In the process of looking at nuclear events and predicting results, we must conserve nuclear charge and mass.

Practice:

Determine the following:

1. $^{14}_{6}C \rightarrow \square + {}^{0}_{-1}e + \bar{\nu}$	Conserving charge and mass, we see that the unknown nucleus must have Z = 7 and A = 14. The nucleus is $^{14}_{7}N$.
2. $^{60}_{27}Co^{*} \rightarrow \square + \gamma$	$^{60}_{27}Co$
3. $^{210}_{83}Bi \rightarrow \square + {}^{4}_{2}He$	$^{206}_{81}Tl$
4. $^{206}_{81}Tl \rightarrow {}^{206}_{82}Pb + \square + \bar{\nu}$	$^{0}_{-1}e$
5. $^{4}_{2}He + {}^{14}_{7}N \rightarrow {}^{17}_{8}O + \square$	$^{1}_{1}H$
6. $^{214}_{84}Po \rightarrow {}^{210}_{82}Pb + \square$	$^{4}_{2}He$
7. $^{226}_{88}Ra \rightarrow {}^{222}_{86}Rm + \square$	$^{4}_{2}He$
8. $^{29}_{12}Mg \rightarrow {}^{0}_{-1}e + \square + \bar{\nu}$	$^{29}_{13}Al$
9. $^{47}_{21}Sc \rightarrow {}^{47}_{21}Sc + \square$	γ
10. $^{236}_{92}U \rightarrow {}^{131}_{53}I + 3{}^{1}_{0}n + \square$	$^{102}_{39}Y$

11. ${}^{4}_{2}He + {}^{9}_{4}Be \rightarrow {}^{12}_{6}C + \square$	${}^{1}_{0}n$
12. ${}^{10}_{5}B + \square \rightarrow {}^{7}_{3}Li + {}^{4}_{2}He$	${}^{1}_{0}n$

Given the following atomic masses:

${}^{4}_{2}He = 4.00260$ u	${}^{12}_{6}C = 12.00000$ u
${}^{7}_{3}Li = 7.01600$ u	${}^{12}_{7}N = 12.01861$ u
${}^{7}_{4}Be = 7.01693$ u	${}^{208}_{81}Tl = 207.98200$ u
${}^{12}_{5}B = 12.01435$ u	${}^{212}_{83}Bi = 211.99128$ u

Determine the following:

13. Is the following reaction possible ${}^{12}_{5}B \rightarrow {}^{12}_{6}C + {}^{0}_{-1}e$	β^--emission can occur if $M_P > M_D$ $M_P = M_B = 12.01435$ u $M_D = M_C = 12.00000$ u Since $M_P > M_D$, β^--emission is possible.
14. Is the following reaction possible ${}^{12}_{7}N \rightarrow {}^{12}_{6}C + {}^{0}_{+1}e$	β^+-emission can occur if $M_P > M_D + 2m_e$ $M_P = M_N = 12.01861$ u $M_D = M_C = 12.00000$ u $2m_e = 2(0.00055\text{ u}) = 0.00110$ u $M_D + 2m_e = 12.00110$u, Since $M_P > M_D + 2m_e$, β^+-emission is possible.
15. Is the following reaction possible ${}^{212}_{83}Bi \rightarrow {}^{208}_{81}Tl + {}^{4}_{2}He$	α-particle emission can occur if $M_P > M_D + M_{He}$ $M_P = M_{Bi} = 211.99128$ u $M_D = M_{Tl} = 207.98200$ u $M_{He} = 4.00260$ u $M_D + M_{He} = 211.98460$ u Since $M_P > M_D + M_{He}$, α-particle emission is possible.

Example: Thorium-229 undergoes alpha decay, and its daughter nucleus undergoes beta decay. What are the daughter and granddaughter nuclei? What is the Q value for each reaction.

Given: ${}^{229}_{90}Th$ - the parent nucleus, The parent decays by alpha emission. The daughter decays by beta emission.

Determine: The daughter and the granddaughter nuclei and the Q values for each reaction.

Strategy: By conserving nuclear charge and mass, we can determine the daughter and granddaughter nuclei. By knowing the atomic masses associated with the nuclei involved, we can determine the Q values.

Solution:

$$^{229}_{90}\text{Th} \rightarrow {}^{4}_{2}\text{He} + {}^{225}_{88}\text{Ra}$$

$$^{225}_{88}\text{Ra} \rightarrow {}^{0}_{-1}\text{e} + {}^{225}_{89}\text{Ac}$$

The daughter nucleus is $^{225}_{88}$Ra, and the granddaughter nucleus is $^{225}_{89}$Ac.

The atomic masses are: $M_{Th} = 229.03176$ u $\quad M_{Ra} = 225.02360$ u

$M_{Ac} = 225.02322$ u $\quad M_{He} = 4.00260$ u

The Q values are

$$Q_\alpha = (M_P - M_D - M_{He})c^2 = (M_{Th} - M_{Ra} - M_{He})c^2$$
$$= (0.00556 \text{ u})(931.5 \text{ MeV/u } c^2)c^2$$
$$= 5.18 \text{ MeV}$$

$$Q_{\beta-} = (M_P - M_D)c^2 = (M_{Ra} - M_{Ac})c^2$$
$$= (0.00038 \text{ u})(931.5 \text{ MeV/u } c^2)c^2$$
$$= 0.358 \text{ MeV}$$

Related Text Exercises: Ex. 43-27 through 43-35.

5 Nuclear Reactions (Section 43-6)

Review: We have previously studied radioactive decay, that is nuclear events which occur spontaneously (Principle Concept 4). We now consider nuclear processes which are caused by an external influence. In a typical nuclear reaction, an incident particle a is projected toward a target nucleus X which is converted to a residual nucleus Y and an emergent particle b. (See Figure 43.2.)

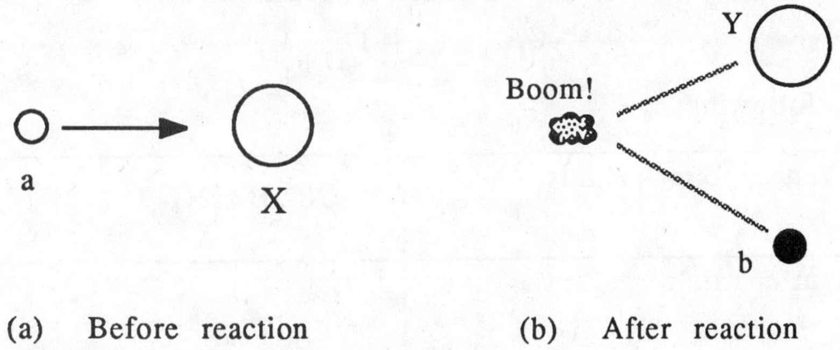

Figure 43.2

The reaction may be written as

$$a + X \rightarrow Y + b$$

or in abbreviated form

$$X(a,b)Y$$

Where a is the incident particle
X is the target nucleus
a and X are the reactants of the reaction
b is the emergent particle
Y is the residual nucleus
b and Y are the products of the reaction

We can determine the reaction energy or Q value of the reaction by using conservation of mass-energy. In the rest frame of the target nucleus, we have

$$(m_a c^2 + K_a) + M_X c^2 = (m_b c^2 + K_b) + (M_Y c^2 + K_Y)$$

or

$$Q = (K_Y + K_b) - K_a = [(m_a + M_X) - (m_b + M_Y)]c^2$$

From this we see that the reaction energy or Q value may be written as the difference in the kinetic energy of the products and reactants

$$Q = (K_Y + K_b) - K_a$$

or the difference in mass of the reactants and products

$$Q = [(m_a + M_X) - (m_b + M_Y)]c^2$$

IF $Q > 0$ -- The process is exothermic, hence energy is released. The kinetic energy of the products is greater than the kinetic energy of the incident particle.

If $Q < 0$ -- The process in endothermic, hence energy is absorbed The process cannot occur unless the incident particle carries kinetic energy greater than the value of Q.

Practice: Consider the following nuclear reaction

$${}^{4}_{2}He + {}^{14}_{7}N \rightarrow {}^{17}_{8}O + {}^{1}_{1}H$$

Determine the following:

1. The abbreviated form of this reaction	${}^{14}_{7}N({}^{4}_{2}He, {}^{1}_{1}H){}^{17}_{8}O$
2. The incident particle	${}^{4}_{2}He$
3. The target nucleus	${}^{14}_{7}N$

4. The reactants	$^{4}_{2}He$ and $^{14}_{7}N$
5. The emergent particle	$^{1}_{1}H$
6. The residual nucleus	$^{17}_{8}O$
7. Mass of the reactants	$^{4}_{2}He = 4.002603$ u $^{14}_{7}N = 14.003074$ u $M_{reactants} = 18.005677$ u
8. Mass of the products	$^{1}_{1}H = 1.0078252$ u $^{17}_{8}O = 16.999131$ u $M_{products} = 18.006956$ u
9. The Q value	$Q = (M_{reactants} - M_{products})c^2$ $= -(1.2790 \times 10^{-3}\ u)(931.5\ MeV/u\ c^2)c^2$ $= -1.19$ MeV This process will not occur unless the α-particle brings more than 1.19 MeV of kinetic energy to the reaction.

Example: Compute the Q of the reaction $^{7}_{3}Li(p,\alpha)^{4}_{2}He$.

Given:

$^{7}_{3}Li = 7.016004$ u	Target nucleus	reactants	
$^{1}_{1}H = 1.007825$ u	incident particle		
$^{4}_{2}He = 4.002603$ u	emergent particle	products	
$^{4}_{2}He = 4.002603$ u	residual nucleus		

Determine: The Q value for this reaction

Strategy: Knowing the mass of the reactants and the mass of the products, we can determine the Q value for this reaction.

Solution: $^{7}_{3}Li$ 7.016004 u $\quad$ $^{4}_{2}He$ 4.002603 u

$^{1}_{1}H$ 1.007825 u $\quad$ $^{4}_{2}He$ 4.002603 u

$M_{reactants} = 8.023829$ u $\quad$ $M_{products} = 8.005206$ u

$$\begin{aligned} Q &= (M_{reactants} - M_{products})c^2 \\ &= (0.018623 \text{ u})(931.5 \text{ MeV/u } c^2)c^2 \\ &= 17.3 \text{ MeV} \end{aligned}$$

This reaction is exothermic, with 17.3 MeV of energy released. The total kinetic energy of the outgoing α-particles will exceed the kinetic energy of the incoming proton by 17.3 MeV.

Related Text Exercises: Ex. 43-36 through 43-41.

6 Nuclear Fission (Section 43-6)

Review: Fission refers to the process of a heavy nucleus splitting into two lighter nuclei and emitting two or more neutrons. It occurs spontaneously in only a few of the very heavy elements, such as U-238. Fission can be induced by a neutron. The isotopes that fission with either a fast or a slow neutron are U-233, U-235, Pu-239 and Pu-241. An example of such a fission process is

$$^{235}_{92}U + ^{1}_{0}n \rightarrow ^{236}_{92}U \rightarrow ^{139}_{56}Ba + ^{95}_{36}Kr + 2\,^{1}_{0}n$$

We can determine the amount of energy released in this fission by finding the difference in mass (Δm) of the reactants and products and then using mass-energy equivalence ($E = \Delta mc^2$):

Mass of reactants	Mass of products
$^{235}_{92}U = 235.043925$ u	$^{139}_{56}Ba = 138.908830$ u
$^{1}_{0}n = 1.008665$ u	$^{95}_{36}Kr = 94.897331$ u
sum = 236.052590 u	$2\,^{1}_{0}n = 2.017330$ u
	sum = 235.823491 u

The mass difference is

$$\Delta m = m_{reactants} - m_{products} = 0.229099 \text{ u}$$

The energy released is

$$\Delta E = \Delta mc^2 = (0.229099 \text{ u})(931.5 \text{ MeV/u}) = 213 \text{ MeV}$$

Practice: Consider the following neutron induced fission of a U-235 nucleus in which two neutrons are ejected.

$$^{235}_{92}U + ^{1}_{0}n \rightarrow ^{96}_{40}Zr + \boxed{} + 2\,^{1}_{0}n + 4\,^{0}_{-1}e + \text{energy}$$

Determine the following:

1. Atomic number of the unknown isotope	Let Z_X represent the atomic number of the unknown isotope. Conserving Z number, we have $92 + 0 = 40 + Z_X + 0 + (-4)$ $Z_X = 92 - 36 = 56$
2. Atomic mass number of the unknown isotope	Let A_X represent the atomic mass number of the unknown isotope. Conserving A number, we have $235 + 1 = 96 + A_X + 2$ $A_X = 236 - 98 = 138$
3. The unknown isotope	For the unknown isotope, we have Z = 56 and A = 138; hence it is barium-138.
4. Mass of the reactants	${}^{235}_{92}U = 235.043925$ u ${}^{1}_{0}n = 1.008665$ u sum = 236.052590 u = $m_{reactants}$
5. Mass of the products	${}^{96}_{40}Zr = 95.908286$ u ${}^{138}_{56}Ba = 137.905000$ u $2\,{}^{1}_{0}n = 2.017330$ u sum = 235.830616 u = $m_{products}$
6. Δm for the reaction	$\Delta m = m_{reactants} - m_{products}$ $= 236.052590$ u $- 235.830616$ u $= 0.221974$ u
7. Energy released	$E = \Delta m(931.5 \text{ MeV/u}) = 207$ MeV
8. Number of moles in 1 kg of U-235	$n = m/AM$ $= \dfrac{(1.00 \text{ kg})(10^3 \text{ g/kg})}{235 \text{ g/mol}} = 4.26$ mol
9. Number of nuclei in 1 kg of U-235	$N = nN_A$ $= (4.26 \text{ mol})(6.023 \times 10^{23} \text{ nuclei/mol})$ $= 2.57 \times 10^{24}$ nuclei
10. Total energy released when 1.00 kg of U-235 fissions	$E_{total} = N(E/\text{fission})$ $= (2.57 \times 10^{24})(2.07 \times 10^2 \text{ MeV})$ $= 5.32 \times 10^{26}$ MeV

11. Mass converted to energy when 1.00 kg of U-235 fissions	$E = 5.32 \times 10^{26}$ MeV $= 8.51 \times 10^{13}$ J The mass converted to energy is $\Delta m/kg = E/c^2$ $= 8.51 \times 10^{13}$ J/(9.00×10^{16} m^2/s^2) $= 9.46 \times 10^{-4}$ kg
12. Mass of U-235 that must fission to meet annual energy consumption on the earth (4.00×10^{20} J)	$\frac{E}{kg} = 5.32 \times 10^{26} \frac{MeV}{kg} = 8.51 \times 10^{13} \frac{J}{kg}$ $m = E_{total}/(E/kg)$ $= 4.00 \times 10^{20}$ J/(8.51×10^{13} J/kg) $= 4.70 \times 10^{6}$ kg
13. Mass converted to energy in meeting the earth's annual energy need by this fission process	$\Delta m/kg = 9.46 \times 10^{-4}$ kg (step 11) $\Delta m_{total} = (\Delta m/kg)$(number of kg) $= (9.46 \times 10^{-4}$ kg$)(4.70 \times 10^{6})$ $= 4.45 \times 10^{3}$ kg, or $\Delta m_{total} = E_{total}/c^2$ $= 4.00 \times 10^{20}$ J/(9×10^{16} m^2/s^2) $= 4.44 \times 10^{3}$ kg

Example: Determine the mass of U-235 fuel needed per year by a 50% efficient 200-MW power plant. Assume an energy release of 200 MeV/fission.

Given: E/fission = 200 MeV/fission

$P = 200 \times 10^{6}$ W = power output of plant

${}^{235}_{92}U$ = fissionable material used as fuel

$\varepsilon = 0.500$ = efficiency of plant

Determine: Mass of U-235 needed per year by a 50% efficient 200-MW power plant.

Strategy: Knowing the power, time, and efficiency of the power plant, we can determine the total amount of energy released. Knowing the total amount of energy released and the energy released per fission, we can determine the number of fissions that must occur. Knowing the number of fissions, we can determine the mass of U-235 needed.

Solution: The total amount of energy to be released in 1 year is

$$E_{released} = \text{(power)(time)}/\varepsilon$$
$$= (2.00 \times 10^{8}\ W)(1\ y)(3.15 \times 10^{7}\ s/y)/0.500 = 1.26 \times 10^{16}\ J$$

The energy released per fission is

$$E/\text{fission} = 200\ MeV = 3.20 \times 10^{-11}\ J$$

The number of fissions that must occur is

$$E_{released} = N(E/\text{fission})$$

$$N = E_{released}/(E/\text{fission}) = 1.26 \times 10^{16}\ \text{J}/3.20 \times 10^{-11}\ \text{J} = 3.94 \times 10^{26}$$

The mass of U-235 needed is

$$m = N(AM)/N_A = (3.94 \times 10^{26})(0.235\ \text{kg/mol})/(6.02 \times 10^{23}/\text{mol}) = 154\ \text{kg}$$

Related Text Exercises: Ex 43-38 and 43-39.

7 Nuclear Fusion (Section 43-6)

Review: Fusion refers to the process of merging (fusing together) several light nuclei to form one larger nucleus and release energy. There are two methods for determining the amount of energy available.

Method I - mass-energy equivalence. When using this method, we find the mass of the reacting nuclei that fuse, the mass of the product nucleus, the mass of any other products, and the mass difference (Δm) between the reactants and the product(s) and then use mass-energy equivalence ($E = \Delta mc^2$) to determine the energy released.

Example: $${}^2_1H + {}^2_1H \rightarrow {}^4_2He + \gamma$$

Note: We will use m_H, m_D and m_T to represent the mass of an atom of ${}^1_1H$, ${}^2_1H$ and ${}^3_1H$ respectively.

Mass of the reactants is

$$2m_H = 2(2.014102\ \text{u}) = 4.028204\ \text{u}$$

Mass of the product nucleus is

$$m_{He} = 4.002603\ \text{u}$$

The mass difference is

$$\Delta m = 2m_H - m_{He} = 0.025601\ \text{u}$$

The binding energy is

$$B = \Delta mc^2 = (0.025601\ \text{u})(931.5\ \text{MeV/u}\ c^2)c^2 = 23.8\ \text{MeV}$$

Method II - binding energy per nucleon. When using this method, we find the binding energy per nucleon before fusion and the binding energy per nucleon after fusion. We then multiply the difference in binding energies by the number of nucleons involved to obtain the energy released.

Example: $${}^2_1H + {}^2_1H \rightarrow {}^4_2He + \gamma$$

The binding energy for ${}^{2}_{1}H$ is

$$B_D = (1M_H + 1m_n - M_D)c^2$$
$$= (1.007825\ u + 1.008665\ u - 2.014102\ u)(931.5\ MeV/u\ c^2)c^2$$
$$= (0.002388\ u)(931.5\ MeV/u\ c^2)c^2$$
$$= 2.22\ MeV$$

The binding energy per nucleon for ${}^{2}_{1}H$ is

$$(B/A)_D = 2.22\ MeV/2\ nucleons = 1.11\ MeV/nucleon$$

The binding energy for ${}^{4}_{2}He$ is

$$B_{He} = (2M_H + 2m_n - M_{He})c^2 = 28.3\ MeV$$

The binding energy per nucleon for ${}^{4}_{2}He$ is

$$(B/A)_{He} = 28.3\ MeV/4\ nucleons = 7.07\ MeV/nucleon$$

The total energy released is

$$Energy = (7.07\ MeV/nucleon - 1.11\ MeV/nucleon)(4\ nucleons) = 23.8\ MeV$$

Note: In general, you will find method I more useful except for very simple cases of fusion (such as this example) where we can easily determine the binding energy per nucleon before and after fusion. If the binding energies are not already known, it is much quicker to use the mass-energy equivalence method.

Practice: Consider the fusion reaction

$${}^{2}_{1}H + {}^{2}_{1}H \rightarrow {}^{4}_{2}He$$

Determine the following:

1. Initial kinetic energy of each deuteron that allows them sufficient energy to get close enough to fuse together (A nuclear radius is of the order of 1.00×10^{-14} m.)	Assume they will fuse if they get within 2.00×10^{-14} m of each other. Also assume the initial potential energy of the deuteron pair is zero and they come to rest just as they fuse. Using conservation of energy, we have $\Delta K + \Delta U = 0$ $(K_f - K_i) + (U_f - U_i) = 0$ $K_i = U_f = ke^2/r = \dfrac{(9.00 \times 10^9\ N\cdot m^2/c^2)(1.60 \times 10^{-19}\ c)^2}{2.00 \times 10^{-14}\ m}$ $K_i = 1.15 \times 10^{-14}\ J = 0.0720\ MeV$ $(K/deuteron)_i = 5.75 \times 10^{-15}\ J$ $= 0.0360\ MeV$

2. Absolute temperature required to give deuterons this amount of energy	$K = 3kT/2; \quad T = 2(K)/3k$ $T = \frac{2(5.75 \times 10^{-15}\ J)}{3(1.38 \times 10^{-23}\ J/K)} = 2.78 \times 10^{8}\ K$
3. Energy released	$M_D = 2.014102\ u; \quad M_{He} = 4.002603\ u$ $\Delta m = 2M_D - M_{He} = 0.025601\ u$ $E_{released} = \Delta mc^2 = 23.8\ MeV$
4. Ratio of energy released to energy invested	$E_{released} = 23.8\ MeV$ (Step 3) $E_{invested} = 0.0720\ MeV$ (Step 1) $\frac{E_{released}}{E_{invested}} = \frac{23.8\ MeV}{0.0720\ MeV} = 331$
5. The number of such fusion reactions required to supply the annual energy consumption of the earth (4.00×10^{20} J)	$E_T = 4.00 \times 10^{20}\ J = 2.50 \times 10^{33}\ MeV$ Let N represent the number of fusion reactions. $E_T = N(E/reaction)$ $N = \frac{E_T}{E/reaction} = \frac{2.50 \times 10^{33}\ MeV}{23.8\ MeV/reaction}$ $N = 1.05 \times 10^{32}$ reactions
6. Number of deuterium nuclei required to cause this number of reactions	Since each reaction requires 2 deuterium nuclei, we need $(1.05 \times 10^{32}\ \text{reactions})(\frac{2\ ^{2}_{1}H\ \text{nuclei}}{\text{reaction}})$ $= 2.10 \times 10^{32}$ nuclei
7. Number of water molecules in 1.00 kg of water	The number of moles of water in 1.00 kg is $n = (\frac{1.00\ kg}{18.0\ g/mol})(\frac{10^3\ g}{kg}) = 55.6\ mol.$ The number of molecules in 55.6 mol of water is $N = nN_A = 3.35 \times 10^{25}$ molecules.
8. Number of hydrogen nuclei in 1.00 kg of water	Since water molecules each contain two hydrogen nuclei, we have $2(3.35 \times 10^{25}) = 6.70 \times 10^{25}$ nuclei.
9. Number of deuterium nuclei in 1.00 kg of water	One out of every 6700 hydrogen nuclei in water is deuterium. Therefore $(6.70 \times 10^{25}\ H)(\frac{1\ \text{deuterium}}{6.70 \times 10^{3}\ H})$ $= 1.00 \times 10^{32}$ deuterium nuclei

10. Mass of water needed to supply the deuterium for a sufficient number of fusion reactions to provide the earth's annual energy consumption (4.00×10^{20} J) by this fusion process	This energy may be provided by $N = 1.05 \times 10^{32}$ reactions of the type being studied. This requires 2.10×10^{32} deuterium nuclei. Since 1.00 kg of water contains 1.00×10^{22} deuterium nuclei, the mass of water needed is $M = 2.10 \times 10^{10}$ kg.

Example: Two deuterium nuclei fuse to produce tritium and a proton. Calculate the rate at which deuterium is consumed to produce 10.0 MW of power. (Assume all energy from the reaction is available.)

Given: $^2_1H + ^2_1H \rightarrow ^3_2H + ^1_1p$ $\qquad P = 10.0$ MW

Determine: The rate of consumption of deuterium to produce 10.0 MW of power.

Strategy: We can look up the mass of the reactants and products and then determine the energy released per fusion. Knowing the energy released per fusion and the rate at which energy must be produced ($P = 10.0$ MW $= 1.00 \times 10^6$ J/s $= E/t$), we can determine the rate at which the fusion reactions must occur and subsequently the rate at which deuterium is consumed.

Solution: The mass of the reactants and products is

$$m_p = 1.007277 \text{ u} \qquad M_D = 2.104102 \text{ u} \qquad M_T = 3.016050 \text{ u}$$

The energy released per fusion is

$$E_{release}/\text{fusion} = [2M_D - (M_T + m_p)]c^2$$

$$= [2(2.014102 \text{ u}) - (3.01650 \text{ u} + 1.007277 \text{ u})](931.5 \text{ MeV/u } c^2)c^2 = 4.54 \text{ MeV/fusion}$$

The rate at which energy must be supplied is

$$E/t = P = 10.0 \text{ MW} \left(\frac{10^6 \text{ W}}{\text{MW}}\right)\left(\frac{\text{J/s}}{\text{W}}\right)\left(\frac{\text{eV}}{1.60 \times 10^{-19} \text{ J}}\right)\left(\frac{\text{MeV}}{10^6 \text{ eV}}\right) = 6.25 \times 10^{19} \text{ MeV/s}$$

The rate at which the fusion reactions must occur is

$$\text{Fusion/s} = (6.25 \times 10^{19} \text{ MeV/s})\left(\frac{\text{Fusion}}{4.54 \text{ MeV}}\right) = 1.38 \times 10^{19} \text{ fusions/s}$$

The rate at which deuterium nuclei are consumed is

$$\text{Deuterium nuclei/s} = \left(\frac{1.38 \times 10^{19} \text{ fusions}}{\text{s}}\right)\left(\frac{2 \text{ nuclei}}{\text{fusion}}\right) = 2.76 \times 10^{19} \text{ nuclei/s}$$

The rate at which deuterium mass is consumed is

$$\left(\frac{\Delta m}{\Delta t}\right)_D = \left(\frac{2.76 \times 10^{19}\ \text{nuclei}}{s}\right)\left(\frac{1\ \text{mol}}{6.02 \times 10^{26}\ \text{nuclei}}\right)\left(\frac{2.01\ g}{\text{mol}}\right)\left(\frac{kg}{10^3\ g}\right)$$

$$\left(\frac{\Delta m}{\Delta t}\right)_D = 9.22 \times 10^{-11}\ kg/s$$

Related Text Exercises: Ex. 43-39.

PRACTICE TEST

Take and grade this practice test. Doing so will allow you to determine any weak spots in your understanding of the concepts taught in this chapter. The following section prescribes what you should study further to strengthen your understanding.

Determine the missing nucleus or particle for each of the following nuclear reactions:

_________ 1. ${}^{40}_{19}K \rightarrow {}^{40}_{20}Ca + \square + \bar{\nu}$

_________ 2. ${}^{214}_{84}Po \rightarrow \square + {}^{4}_{2}He$

_________ 3. ${}^{61}_{28}Ni^* \rightarrow {}^{61}_{28}Ni + \square$

_________ 4. ${}^{12}_{5}B \rightarrow \square + {}^{0}_{-1}e + \bar{\nu}$

_________ 5. ${}^{10}_{5}B + {}^{1}_{0}n \rightarrow {}^{7}_{3}Li + \square$

_________ 6. ${}^{210}_{84}Po^* \rightarrow \square + \gamma$

Given the following masses:

m_e = 0.000548 u = mass of electron, M_H = 1.007825 u = mass of neutral ${}^{1}_{1}H$ atom

m_n = 1.008665 u = mass of neutron, M_{He} = 4.002603 u = mass of neutral ${}^{4}_{2}He$ atom

m_p = 1.007277 u = mass of proton, M_C = 13.003354 u =mass of neutral ${}^{13}_{6}C$ atom

Determine the following:

_________ 7. The mass difference between four nucleons (two protons and two neutrons) and the ${}^{4}_{2}He$ nucleus

_________ 8. The amount of energy given off if you brought two neutrons and two protons together to form a ${}^{4}_{2}He$ nucleus

________ 9. The binding energy per nucleon for ${}^{4}_{2}He$

________ 10. The mass difference for ${}^{13}_{6}C$

________ 11. The binding energy for ${}^{13}_{6}C$

A sample of radioactive material is monitored by a counting system that has an efficiency of 50%. The system records 6.00×10^5 counts/min when the sample is first placed in the counting chamber and 4.00×10^5 counts/min after 30.0 min.

Determine the following:

________ 12. Count rate (in s^{-1}) at $t = 0$
________ 13. Activity (A_o) at $t = 0$
________ 14. Activity (A) after 30.0 min
________ 15. Decay constant for these radioactive nuclei
________ 16. Half-life for these radioactive nuclei
________ 17. Number of radioactive nuclei present at $t = 0$
________ 18. Number of moles of the sample at $t = 0$
________ 19. Count rate after 4 h
________ 20. Time at which 10% of the sample is left

Consider the following fission reaction.

$${}^{235}_{92}U + {}^{1}_{0}n \rightarrow \boxed{} + {}^{101}_{44}Ru + 3\,{}^{1}_{0}n + 6\,{}^{0}_{-1}e + \text{Energy}$$

Mass of ${}^{235}_{92}U = 235.04393$ u $\qquad {}^{1}_{0}n = 1.00867$ u

${}^{101}_{44}Ru = 100.90558$ u $\qquad {}^{0}_{-1}e = 0.00055$ u

________ 21. Atomic number of the unknown isotope
________ 22. Atomic mass of the unknown isotope
________ 23. Mass difference between reactants and products
________ 24. Energy release for this fission reaction
________ 25. Number of moles in 1 kg of U-235
________ 26. Number of nuclei in 1 kg of U-235
________ 27. Total amount of energy (in MeV) released when 1 kg of U-235 fissions in this manner
________ 28. Mass (in kg) converted to energy when 1 kg of U-235 fissions in this manner
________ 29. Rate of consumption of U-235 if this fission reaction is to supply 1000 MW power reactor that is 40% efficient

(See Appendix I for answers.)

PRINCIPAL CONCEPTS AND EQUATIONS PRESCRIPTION

Your score on the practice test is an excellent measure of your understanding of the chapter. You should now use the following chart to write your own prescription for dealing with any weaknesses the practice test points out. Look down the leftmost column to the number of the question(s) you answered incorrectly, reading across that row you will find the concept and/or equation of concern, the section(s) of the study guide you should return to for further study, and some suggested text exercises which you should work to gain additional experience.

Practice Test Questions	Concepts and Equations	Prescription	
		Principal Concepts	Text Exercises
1	Beta decay: ${}^{A}_{Z}P \rightarrow {}^{A}_{Z+1}D + {}^{0}_{-1}e + \bar{\nu}$	4	43-21, 27
2	Alpha decay: ${}^{A}_{Z}P \rightarrow {}^{A-4}_{Z-2}D + {}^{4}_{2}He$	4	43-21, 27
3	Gamma decay: ${}^{A}_{Z}P^{*} \rightarrow {}^{A}_{Z}P + \gamma$	4	43-21
4	Beta decay: ${}^{A}_{Z}P \rightarrow {}^{A}_{Z+1}D + {}^{0}_{-1}e + \bar{\nu}$	4	43-21, 30
5	Alpha decay: ${}^{A}_{Z}P \rightarrow {}^{A-4}_{Z-2}D + {}^{4}_{2}He$	4	43-21, 28
6	Gamma decay: ${}^{A}_{Z}P^{*} \rightarrow {}^{A}_{Z}P + \gamma$	4	43-21
7	Mass difference: $\Delta m = ZM_H + Nm_n - M_a$	2	43-9, 11
8	Binding energy: $B = \Delta mc^2$	2	43-11, 12
9	Binding energy per nucleon: B/A	2	43-11
10	Mass difference: $\Delta m = ZM_H + Nm_n - M_a$	2	43-9, 11
11	Binding energy: $B = \Delta mc^2$	2	43-12, 13
12	Count rate	3	---
13	Activity: $R_0 = CR_0/\varepsilon$	3	43-22, 23
14	Activity: $R = CR/\varepsilon$	3	43-22, 23
15	Decay constant: $R = R_0 e^{-\lambda t}$	3	43-24, 25
16	Half life: $T_{1/2} = \ln 2/\lambda$	3	43-24, 25
17	Activity: $R_0 = N_0\lambda$	3	43-25, 26
18	Moles of sample: $n = N/N_A$	3	43-23
19	Radioactive decay: $CR = CR_0 e^{-\lambda t}$	3	43-22, 23
20	Radioactive decay: $N = N_0 e^{-\lambda t}$	3	43-24, 25
21	Conservation of atomic number (Z)	4	43-27, 28
22	Conservation of atomic mass number (A)	4	43-29, 30
23	Mass difference: $\Delta m = M_{reactants} - M_{products}$	5	43-37
24	Mass energy equivalence: $\Delta m = E/c^2$	#	43-7, 8
25	Number of moles: $n = \mathcal{M}/M$	6	43-38, 41
26	Number of nuclei: $N = nN_A$	6	43-38, 41
27	$E_{total} = (E/\text{fission})(N)$	6	43-38, 41
28	Mass energy equivalence: $\Delta m = E/c^2$	6	43-7, 8
29	$\Delta M/\Delta t = (E_{total}/t)/(E/M)$	6	43-38, 41

44 THE PHYSICS OF STARS

RECALL FROM PREVIOUS CHAPTERS

Previously learned concepts and equations frequently used in this chapter		Text Section	Study Guide Page
Central force:	$F_c = mv^2/r$	6-2	6-7
Gravitational attraction:	$F = GMm/r^2$	7-1	7-1
Work:	$W = \int \mathbf{F} \cdot d\mathbf{r}$	8-3	8-7
Kinetic energy:	$K = mv^2/2$	8-4	8-13
Gravitational potential energy:	$U_g = -W_g$	9-1	9-1
Equation of state of an ideal gas:	$PV = nRT$	17-1	17-1
Kinetic energy and temperature:	$K = 3NkT/2$	18-1, 2	18-1
Nuclear Fusion		43-6	43-21

NEW IDEAS IN THIS CHAPTER

Concepts and equations introduced	Text Section	Study Guide Page
Kinetic, Potential and total energy of a dust particle: $K = GMm/2r$; $U_g = -GMm/r$; $E = -GMm/2r$	44-2	44-2
Virial theorem: $E_{tot} = -K_{tot} = U_{tot}/2$	44-2	44-3
Gravitational potential energy of the sun: $U_s = -0.72GM_{sun}{}^2/R$	44-3	44-7
Average Internal temperature of a main-sequence star: $\langle T \rangle = 0.124 m_H GM_{sun}/kR \approx 3 \times 10^6$ K	44-4	44-14
Central pressure for a main-sequence star Using gas law: $P_c = 3M_{sun}kT_c/2\pi m_H R^3 \approx 3 \times 10^9$ atom Using balance of forces: $P_c = 1.24(45/36\pi)(GM^2/R^4) \approx 6 \times 10^9$ atom	44-4	44-15
Luminosity of the Sun: $L_s = 3.85 \times 10^{26}$ J/s	44-1	44-8
Luminosity and mass of main sequence stars $L \propto M^3$	44-4	44-15
Lifetime and mass of main sequence stars $\mathcal{L} \propto 1/M^2$	44-4	44-15

PRINCIPAL CONCEPTS AND EQUATIONS

1 Relationship between Kinetic Energy, Potential Energy and Total Energy in Stars (Section 44-2)

Review: A star begins its life as a widely dispersed dust cloud with essentially zero gravitational potential energy. As time progresses, gravitational forces begin pulling this stellar mass together and the continual struggle between inward gravitational forces and outward forces due to internal pressure begins. Figure 44.1 shows an artists conception of the tremendously large dust cloud, a more dense central mass made up of dust which has already been pulled together and packed gravitationally and a single dust particle orbiting the massive center.

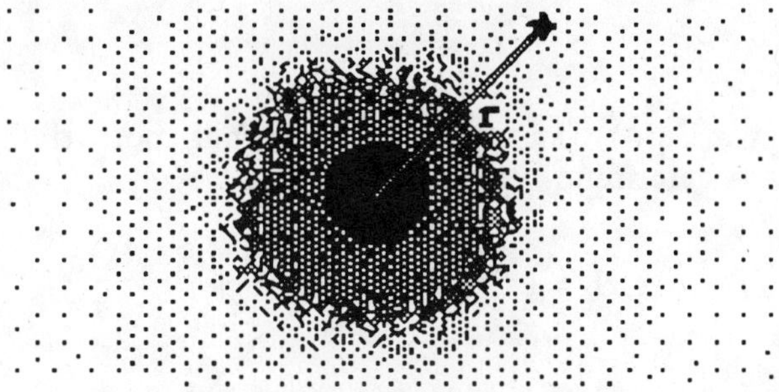

r = distance of dust particle from center of the dense central mass
m = mass of the dust particle
M = all mass inside the orbital radius of the dust particle
G = universal gravitational constant
v = orbital speed of the dust particle

Figure 44.1

Since the dust particle has a mass m and is in an orbit of radius r with speed v, it must be experiencing a central force

(1) $$F_c = mv^2/r.$$

This central force is supplied by the gravitational attraction between the central mass M (all mass inside the orbital radius of the dust particle) and the dust particle of mass m.

(2) $$F_c = F_g = GMm/r^2$$

Inserting the central force expressed in Eq(2) into Eq(1) we have

$$mv^2/r = GMm/r^2$$

or

(3) $$mv^2 = GMm/r.$$

From Eq(3) we may write the kinetic energy of the dust particle as

(4) $$K = mv^2/2 = GMm/2r.$$

Next we know that the gravitational potential energy of the dust particle is just the negative of the work by gravity to bring it from infinitely far away to its present orbit a distance r from the center.

$$U_g = -W_g = -\int_{\infty}^{r} \mathbf{F}_g \cdot d\mathbf{r} = -\int_{\infty}^{r} \left(\frac{GMm}{r^2}\right)(-\hat{\mathbf{r}}) \cdot dr\hat{\mathbf{r}}$$

or

$$U_g = GMm \int_{\infty}^{r} \left(\frac{dr}{r^2}\right) = GMm\left(\frac{-1}{r}\right)\Bigg|_{\infty}^{r}$$

or

(5) $$U_g = -GMm/r$$

Given the kinetic energy (Eq(4)) and the gravitational potential energy (Eq(5)) of the dust particle, we can determine its total energy.

(6) $$E = K + U_g = \left(\frac{GMm}{2r}\right) - \left(\frac{GMm}{r}\right) = \frac{-GMm}{2r}$$

Summarizing we have

(7) $$K = GMm/(2r); \qquad U = -GMm/r; \qquad E = -GMm/(2r)$$

or

(8) $$K = -U/2; \qquad E = U/2; \qquad E = -K$$

Your text shows that you may replace the instantaneous values of E, K and U with their time averages to obtain

$$\langle E\rangle = \langle K\rangle + \langle U\rangle$$

(9) and

$$\langle K\rangle = -\langle U\rangle/2; \qquad \langle E\rangle = \langle U\rangle/2; \qquad \langle E\rangle = -\langle K\rangle$$

Your text also shows that the time average values of the kinetic, potential and total energy for one particle (i.e. $\langle K\rangle$, $\langle U\rangle$ and $\langle E\rangle$) are related to the system values (i.e. K_{tot}, U_{tot} and E_{tot}) by

(10) $$\langle K\rangle = K_t/N; \qquad \langle U\rangle = U_t/N; \qquad \langle E\rangle = E_t/N$$

where N is the number of particles in the cloud.

Combining Eqs. 9 and 10 we have

(11) $$E_{tot} = -K_{tot} = \left(\frac{1}{2}\right)U_{tot}$$ (Virial Theorem)

The Virial Theorem (Eq(11)) tells us that if the total energy of a system bound together by an inverse square force decreases, the kinetic energy must increase (because of the negative sign) by the same amount and the potential energy must decrease by twice as much.

According to the Stefan - Boltzmann law, in order for the star to glow it must radiate or lose energy; so its total energy E_{tot} decreases. According to the Viral theorem (Eq(11)) if the total energy E_{tot} decreases by some amount the total kinetic energy K_{tot} will increase by the same amount and the total gravitational potential energy will decrease by twice this amount.

Recall that the gravitational potential energy varies as -1/r. So in order for U to decrease (that is become more negative) r must decrease.

Also, knowing how the internal temperature of the gravitationally collapsed mass is related to the kinetic energy of the particles that make up the mass (Eq(13)), we know that the star must increase in temperature.

(13) $$K_{tot} = \left(\frac{3}{2}\right)Nk\langle T\rangle$$

where $\langle T\rangle$ is the average temperature of the dust cloud.

As the dust cloud radiates it losses energy, this energy comes at the expense of the gravitational potential energy. According to the Virial Theorem as the gravitational potential energy decreases a certain amount the kinetic energy increases half this amount and this in turn raises the temperature of the star. The other half of the decrease in the potential energy is emitted as radiation.

It is essential that you note that the star cannot cool off by radiating energy away; instead when it radiates a certain amount of energy the same amount of energy goes into an increase in kinetic energy which shows up as an increased temperature. It appears that we have identified a crucial feature of the evolution of stars. That is the more a star radiates, the hotter it becomes, and the hotter it becomes, the more it radiates.

Practice: Consider a star early in its life (that is it is still forming) and a bit of stellar material orbiting the central core of the star.

Let r = distance of the bit of material from the center of the core
m = mass of the bit of material
M = mass of all material inside the orbit of the bit of material
ν = frequency of radiation emitted by the star
G = universal gravitational constant

Determine the following in terms of the above:

1. Kinetic energy of the bit of stellar material	$K = mv^2/2 = GMm/(2r)$
2. Gravitational potential energy of the bit of stellar material	$U_g = -GMm/r$

3. Total energy of the bit of stellar material	$E = K + U_g = -GMm/(2r)$
4. Average temperature of the bit of stellar material	From $K = (3/2)k\langle T\rangle$, we obtain $\langle T\rangle = 2K/(3k)$.

The bit of stellar mass is now gravitationally pulled into the new position $r_{final} = (3/4)r_{initial}$ and the frequency of the radiation emitted is ν.

Determine the following in terms of the above:

5. How much does its gravitational potential energy change	$U_{initial} = -GMm/r_{initial}$ $U_{final} = -GMm/(\frac{3r_{initial}}{4}) = (\frac{4}{3})U_{initial}$ $\Delta U = U_f - U_i$ $= (4/3)U_{initial} - U_i$ $= U_i(1/3)$ $= (1/3)U_i = (-1/3)(GMm/r_i)$ The minus sign tells us the potential energy has decreased.
6. How much does the kinetic energy of the bit of stellar mass change	$\Delta K = -(1/2)\Delta U$ $= -(1/2)(-1/3)(GMm/r_i)$ $= (1/6)(GMm/r_i)$ The positive sign tells us the kinetic energy has increased.
7. How much does the average temperature of the bit of stellar mass change	$K = (3/2)k\langle T\rangle$ or $\Delta\langle T\rangle = \frac{2\Delta K}{3k} = \frac{2}{3k}\left((\frac{1}{6})(\frac{GMm}{r_i})\right)$ $= (GMm)/(9kr_i)$
8. How much energy is radiated	$E_{rad} = -\Delta E_{total} = -(\Delta U_{total}/2)$ $= -(-1/6)(GMm/r_i)$ $= GMm/(6r_i)$
9. Wavelength of this radiated energy	From $\nu = c/\lambda$, we obtain $\lambda = c/\nu$.
10. Number of protons radiated	$E_{rad} = h\nu N_{photons}$ $N_{photons} = E_{rad}/(h\nu) = (GMm/6r_i)/(h\nu)$

Related Text Exercise: Ex. 44-1.

2 Sources of Energy in Stars (Section 44-3)

As a result of our work with the previous section, we are aware that a source of energy for a star is its decrease in gravitational potential energy as it collapses.

In this section we will consider the star nearest us (the Sun) and we will find that the energy from gravitational collapse is not sufficient to account for the total power emitted (Luminosity) over the known lifetime of the planet earth. We know that life has existed on the planet earth for at least 3 billion years and that its survival has been dependent upon the Luminosity being constant and having the same value as today.

To determine the power available from the Sun due to gravitational collapse, we start with a widely dispersed dust cloud of essentially zero gravitational potential energy and calculate its loss in gravitational potential energy as it collapses to our present Sun. We know (by the Virial theorem) that half of this energy loss is radiated and that the other half goes into increasing the kinetic energy and consequently the temperature of the particles in the stellar mass.

To determine the loss in gravitational potential energy of the Sun, let's imagine building up the Sun by bringing in infinitesimal pieces of the dust cloud from an infinitely far away position, one piece at a time. Figure 44.2 shows the idea: a piece of matter of mass dm is brought up to the partially assembled sun of mass m and is spread around it in a thin spherical shell of thickness dr.

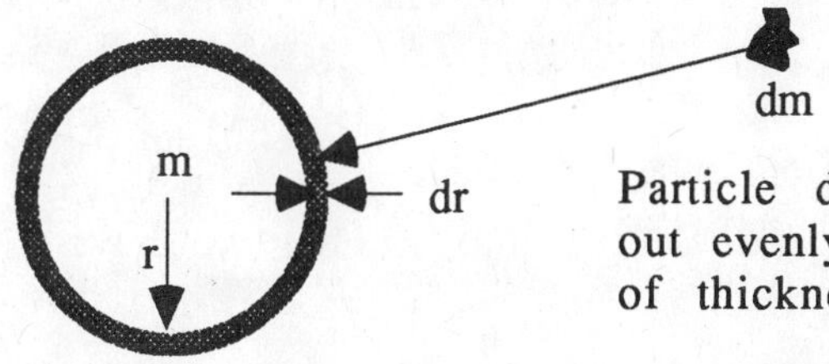

Figure 44.2

The change in the Suns potential energy as a result of bringing the mass dm in from infinitely far away is:

(14) $$dU_s = U_{dm}(r) - U_{dm}(\infty) = -Gmdm/r$$

The total change in the potential energy of the Sun is obtained simply by adding up the change for each layer. This may be done by integrating from 0 to R, the present radius of the Sun.

(15) $$U_s = \int dU_s = -G\int mdm/r = -G\int m\rho dV/r$$

$$dV = 4\pi r^2 dr \quad - \quad \text{Volume of the shell of radius r}$$

At this point it should also be pointed out that the mass and density are functions of r, that is Eq(15) should be written as

(16) $$U_s = -4\pi G\int m(r)\rho(r)rdr$$

The density of the Sun $\rho(r)$ varies from nearly zero at the surface to a maximum value ρ_c at its center. Let's agree to use a linear approximation for $\rho(r)$.

(17) $$\rho(r) = \rho_c(1 - r/R)$$

Notice that this approximation gives use $\rho(R) = 0$ and $\rho(0) = \rho_c$.

Now that we have agreed on the form of $\rho(r)$ we can determine the form for m(r) as follows:

(18) $$m(r) = \int_0^r dm = \int_0^r \rho(r')dV = \int_0^r \rho_c(1 - r'/R)4\pi r'^2 dr'$$

or

(19) $$m(r) = 4\pi\rho_c \int_0^r (1 - r'/R)r'^2 dr' = 4\pi\rho_c \int_0^r [r'^2 - (r'^3/R)]dr'$$

or

(20) $$m(r) = 4\pi\rho_c \left[\left(\frac{r'^3}{3} \right) - \left(\frac{r'^4}{4R} \right) \right]_0^r = \left(\frac{4}{3}\right)\pi\rho_c r^3 \left(1 - \frac{3r}{4R} \right)$$

We may determine the constant ρ_c by noting that

$$m(R) = M_{sun} = \left(\frac{4}{3}\right)\pi\rho_c R^3 \left(1 - \frac{3}{4}\right) = \pi\rho_c R^3/3$$

or

(21) $$\rho_c = 3M_{sun}/(\pi R^3)$$

Inserting Eq(17) and Eq(20) into Eq(16) we have

$$U_s = -4\pi G \int_0^R \left[\left(\frac{4}{3}\right)\pi\rho_c r^3 \left(1 - \frac{3r}{4R} \right)\right][\rho_c(1 - r/R)]r dr$$

After simplifying the integrand we have

$$U_s = \left(\frac{-16\pi^2 G\rho_c^2}{3}\right)\int_0^R \left[r^4 - \left(\frac{7r^5}{4R}\right) + \left(\frac{3r^6}{4R^2}\right)\right]dr$$

Integrating and inserting the limits we have

$$U_s = \left(\frac{-16\pi^2 G\rho_c^2}{3}\right)\left(0.0155R^5 \right)$$

Inserting ρ_c and simplifying we have

(22) $$U_s = \left(\frac{(-16\pi^2 G)(0.0155R^5)}{3}\right)\left(\frac{3M_{sun}}{\pi R^3}\right)^2 = \left(\frac{-0.744GM_{sun}^2}{R}\right) = (-1.24)\left(\frac{3}{5}\right)\left(\frac{GM_{sun}^2}{R}\right)$$

Note: The final form of the coefficient in Eq(22) may seem awkward to you at this point, however in this form it is easy to compare to the example problem where the same calculation is done assuming the Sun's density has the constant value ρ_c.

Eq(22) tells us the total loss of potential energy by the dust cloud which has resulted in our Sun as it collapsed from dust particles to its present size.

By the Virial theorem we have established that half of this energy went into increased kinetic energy of the stellar material (consequently the increase in the temperature) and the other half was radiated with a small fraction of it reaching the planet Earth. The amount of energy radiated is

(23) $$E_{rad} = (1/2)U_s = (1/2)(-0.744GM_{sun}^2/R) = 1.41 \times 10^{41} \text{ J}$$

We have experimentally determined that the rate at which the Sun radiates energy (its Luminosity) is

(24) $$L_s = 3.85 \times 10^{26} \text{ J/s}$$

We have experimental evidence that leads us to believe that the Sun has radiated energy at this rate for at least three billion years.

$$t_{rad} = 3.00 \times 10^9 \text{ y} = 9.46 \times 10^{16} \text{ s}$$

If this is the case, then the total energy radiated so far is

(25) $$\begin{aligned} E_{total\ rad} &= (\text{Rate of radiation})(\text{Time of radiation}) \\ &= (L_s)(t_{rad}) = (3.85 \times 10^{26} \text{ J/s})(9.46 \times 10^{16} \text{ s}) \\ &= 3.64 \times 10^{43} \text{ J} \end{aligned}$$

Notice that the total energy radiated from gravitational collapse (Eq(23)) is less than 1% of the total energy radiated (Eq(25)).

$$\left(\frac{E_{rad}}{E_{total\ rad}}\right) \times 100\% = \left(\frac{1.41 \times 10^{41} \text{ J}}{3.64 \times 10^{43} \text{ J}}\right) \times 100\% = 0.39\%$$

If only 0.39% of the total energy radiated can be accounted for by gravitational collapse, there must be some other very large source of energy.

Recall that half of the collapsing dust clouds loss in gravitational potential energy goes into increasing the kinetic energy of the material and hence its temperature. Once the temperature of the central core becomes sufficiently high, the fusion of protons to form Helium will serve as a nuclear source of energy. This nuclear process is summarized below.

First a proton decays into a neutron by positron emission.

$$ {}^1_1H \rightarrow {}^1_0n + e^+ + \nu$$

This neutron very quickly reacts with a proton to form a deuteron and some energy is released in the process.

$$_{0}^{1}n + _{1}^{1}H \rightarrow _{1}^{2}H \quad \text{with release of energy (0.42 Mev)}$$

The above two reactions are frequently combined to read

$$_{1}^{1}H + _{1}^{1}H \rightarrow _{1}^{2}H + e^{+} + \nu \quad \text{with release of energy (0.42 Mev)}$$

This deuteron reacts with another proton to form Helium, a gamma ray and more energy.

$$_{1}^{2}H + _{1}^{1}H \rightarrow _{2}^{3}He + \gamma \quad \text{with release of energy (5.49 Mev)}$$

This Helium nucleus reacts with another Helium nucleus (formed in the same manner) to produce a different isotope of Helium, two new protons and more energy.

$$_{2}^{3}He + _{2}^{3}He \rightarrow _{2}^{4}He + _{1}^{1}H + _{1}^{1}H \quad \text{with release of energy (12.86 Mev)}$$

Now let's do some accounting to see how much energy is available as the result of one complete reaction.

two protons to a deuteron	0.42 Mev
deuteron plus proton to Helium-3	5.49 Mev
	5.91 Mev
double this number since this	x 2
process takes place twice	11.82 Mev
two Helium-3 nuclei to one Helium-4	
plus two protons	12.86 Mev
	24.70 Mev

The positron (e^{+}) created in the first reaction quickly combines with an electron (e^{-}) and then mass annihilation occurs to give 1.02 Mev. Since this takes place twice we have an additional 2.04 Mev.

    2.04 Mev
    26.74 Mev

This entire process is summarized below.

$$_{1}^{1}H \rightarrow _{0}^{1}n + e^{+} + \nu$$
$$\llcorner + e^{-} \rightarrow (1.02\ \text{Mev})$$
$$\llcorner + _{1}^{1}H \rightarrow _{1}^{2}H + (0.42\ \text{Mev})$$
$$\llcorner + _{1}^{1}H \rightarrow _{2}^{3}He + \gamma + (5.49\ \text{Mev})$$

$$_{1}^{1}H \rightarrow _{0}^{1}n + e^{+} + \nu$$
$$\llcorner + e^{-} \rightarrow (1.02\ \text{Mev})$$
$$\llcorner + _{1}^{1}H \rightarrow _{1}^{2}H + (0.42\ \text{Mev})$$
$$\llcorner + _{1}^{1}H \rightarrow _{2}^{3}He + \gamma + (5.49\ \text{Mev})$$

$$\llcorner + \llcorner \rightarrow _{2}^{4}He + _{1}^{1}H + _{1}^{1}H \ (12.86\ \text{Mev})$$

Adding up all of the energies we obtain the energy available per reaction.

$$E_{reaction} = 26.74 \text{ Mev}$$

If we assume the Sun is primarily composed of Hydrogen, then the number of protons available for this reaction is:

$$N = \frac{\text{Mass of Sun}}{\text{Mass of Hydrogen atom}} = \frac{M_S}{m_H} = \frac{1.99 \times 10^{30} \text{ kg}}{1.67 \times 10^{-27} \text{ kg}} = 1.19 \times 10^{57}.$$

Since there is a net loss of 4 protons each time this reaction occurs, the number of possible reactions is

$$N_{reacts} = N/4 = (1.19 \times 10^{57})/4 = 2.98 \times 10^{56}.$$

Since each reaction gives 26.74 Mev of energy the maximum amount of energy available is:

$$\begin{aligned} E_{max} &= (E_{reaction})(N_{reacts}) \\ &= (26.74 \text{ Mev})(10^6 \text{ ev/Mev})(1.6 \times 10^{-19} \text{ J/ev})(2.98 \times 10^{56}) \\ &= 1.27 \times 10^{45} \text{ J.} \end{aligned}$$

Recall that in the last three billion years we have radiated some of this energy ($E_{total\ rad} = 3.64 \times 10^{43}$ J) so the energy available is

$$E_{avail} = E_{max} - E_{total\ rad} = 127 \times 10^{43} \text{ J} - 3.64 \times 10^{43} \text{ J} = 1.23 \times 10^{45} \text{ J.}$$

At the present rate of radiating this energy, it will last for a time of:

$$\begin{aligned} t_{remaining} &= \frac{E_{avail}}{L_S} = \frac{1.23 \times 10^{45} \text{ J}}{3.85 \times 10^{26} \text{ J/s}} = 3.19 \times 10^{18} \text{ s} \\ &= (3.19 \times 10^{18} \text{ s})\left(\frac{\text{h}}{3.60 \times 10^3 \text{ s}}\right)\left(\frac{\text{d}}{2.40 \times 10^1 \text{ h}}\right)\left(\frac{\text{y}}{3.65 \times 10^2 \text{ d}}\right) \\ &= \left(\frac{31.9 \times 10^{17}}{31.5 \times 10^6}\right) \text{y} \approx 1 \times 10^{11} \text{ y} \\ &\approx 100 \text{ billion years.} \end{aligned}$$

It appears that lack of energy from the Sun will not be an immediate problem.

Practice

Determine the following:

1. Rate at which the Sun is radiating energy	$L_S = 3.85 \times 10^{26}$J/s

2. Rate at which the Earth receives this radiated energy	The energy radiated by the Sun is uniformly spread out over a sphere. By the time this energy gets to the earth it is spread out over a sphere of radius R_{se} (R_{se} stands for the distance between the Sun and the earth, and $R_{se} = 1.5 \times 10^{11}$ m) so the rate at which energy is received per square meter is: $\frac{L_s}{4\pi R_{se}^2} = \frac{3.85 \times 10^{26} \text{ J/s}}{4\pi(1.5 \times 10^{11} \text{ m})^2} = 1.36 \times 10^3 \text{ Js}^{-1}\text{m}^{-2}$ This radiation sees the earth essentially as a disk of radius $R_e = 6.38 \times 10^6$ m, and area $A_{disk} = \pi R_e^2 = 1.28 \times 10^{14} \text{ m}^2$. So the rate at which the earth receives this energy is $(1.36 \times 10^3 \text{ Js}^{-1}\text{m}^{-2})(1.28 \times 10^{14} \text{ m}^2) = 1.74 \times 10^{17}$ J/s.
3. Minimum time t_{min} the earth has been receiving energy at this rate	$3 \times 10^9 \text{ y} = 9.46 \times 10^{16}$ s
4. Total energy radiated by the Sun in this time	$E_{total\ rad} = L_s t_{min} = (3.85 \times 10^{26} \text{ J/s})(9.46 \times 10^{16} \text{ s}) = 3.64 \times 10^{43}$ J
5. Total energy radiated by the Sun due to gravitational collapse during the past 3 billion years	From Eq(23) we have $E_{rad} = 1.41 \times 10^{41}$ J
6. Percent of the total energy radiated which is due to gravitational collapse	$\% = (E_{rad}/E_{total\ rad}) \times 10^2\% = \frac{(1.41 \times 10^{41} \text{ J})(10^2\%)}{3.64 \times 10^{43} \text{ J}} = 0.39\%$
7. Number of complete proton-proton cycle reactions which occur in the Sun each second.	The Sun is radiating energy at the rate: $L_s = 3.85 \times 10^{26}$ J/s. Each reaction produces: $26.74 \text{ Mev} \left(\frac{10^6 \text{ ev}}{\text{Mev}}\right)\left(\frac{1.6 \times 10^{-19} \text{ J}}{\text{ev}}\right) = 4.28 \times 10^{-12}$ J. So the number of reactions per second must be: $(N)(4.28 \times 10^{-12} \text{ J}) = L_s$ $N = \frac{L_s}{4.28 \times 10^{-12} \text{ J}} = \frac{3.85 \times 10^{26} \text{ J/s}}{4.28 \times 10^{-12} \text{ J}} = (0.900 \times 10^{38})\text{/s} = (9.00 \times 10^{37})\text{/s}$.

8. Mass of the Sun consumed each second by proton-proton cycle nuclear reactions	The net effect of each cycle is the loss of 4 protons $m_{4-protons} = 4(1.672 \times 10^{-27}\text{ kg})$ $= 6.69 \times 10^{-27}\text{ kg}$ In step 7 above we calculated that the rate at which the reaction occurs is $(9.00 \times 10^{37})/\text{s}$ The mass consumed each second is $M_{1/s} = (\frac{\text{mass loss}}{\text{reaction}})(\frac{\text{reactions}}{\text{s}})$ $= (6.69 \times 10^{-27}\text{ kg})(9.00 \times 10^{37}\text{ s}^{-1})$ $= 6.02 \times 10^{11}\text{ kg/s}$
9. At this rate what percent of the Sun's mass is consumed in one billion years	$M_{consumed} =$ $=$ (rate of consumption)(time) $= (6.02 \times 10^{11}\text{ kg/s})(10^9\text{ y})(3.15 \times 10^7\text{ s/y})$ $= 1.90 \times 10^{28}\text{ kg}$ $\% = (\frac{M_{consumed}}{M_{sun}}) \times 10^2\%$ $= (\frac{1.90 \times 10^{28}\text{ kg}}{2 \times 10^{30}\text{ kg}}) \times 10^2\%$ $= 0.950$
10. Anticipated lifetime of the Sun	$\frac{M_{sun}}{M_{cons/billion\ years}}$ $= \frac{1.99 \times 10^{30}\text{ kg}}{(1.90 \times 10^{28}\text{ kg})/(\text{billion years})}$ $= 1.05 \times 10^2$ billion years.

Example: Assume that the density of the Sun is constant and that it is ρ_c. Determine the decrease in gravitational potential energy of the dust cloud as it collapses to the size of our present Sun.

Given: The density of the Sun has the constant value ρ_c.

Determine: The decrease in gravitational potential energy of the dust cloud as it collapses to the size of our present Sun.

Strategy: Starting with the expression for the potential energy of the Sun as a function of its size and assuming the density has the constant value ρ_c, we can determine the decrease in gravitational potential energy of a dust cloud that has collapsed to the size of our present Sun.

Solution: Starting with Eq(15)

$$U_s = -G\int m(r)\rho(r)dV/r$$

Inserting $\rho(r) = \rho_c$; $m(r) = (\frac{4}{3})\pi r^3\rho_c$; $dV = 4\pi r^2 dr$:

$$U_s = -G\int_0^R \left((\frac{4}{3})\pi r^3\rho_c\right)\left(\frac{(\rho_c)(4\pi r^2 dr)}{r}\right)$$

Pulling all constants out of the integrand:

$$U_s = \left(\frac{-16\pi^2 G\rho_c^2}{3}\right)\int_0^R r^4 dr$$

Integrating and inserting limits:

$$U_s = \frac{-16\pi^2 G\rho_c^2 R^5}{15}$$

Inserting the value for ρ_c (Eq(21)), obtain

$$U_s = \frac{-3GM^2}{5R}$$

Notice how this result compares (see Eq(22)) to that calculated using the variable density $\rho_c = \rho_c(1 - r/R)$.

Related Text Exercises: Ex. 4-2 through 4-10.

3 Internal Temperature, Pressure, Luminosity and Lifetime of the Sun and Other Main-Sequence Stars (Section 44-4)

Review: Internal Temperature - According to the Virial Theorem the decrease in gravitational potential energy and the increase in kinetic energy are related by

(26) $$2K_{tot} = -U_{tot}$$

For a star with a linear density dependence:

$$\rho(r) = \rho_c(1 - r/R)$$

it has been determined that

(27) $$U_{tot} = -(1.24)(3/5)(GM_{sun}^2/R)$$

and

(28) $$K_{tot} = (3/2)Nk\langle T\rangle$$

In this expression for K_{tot}, N stands for the total number of particles. Having assumed the Sun to be entirely composed of hydrogen atoms, we established earlier that the number of atoms is

$$N_{atoms} = M_{sun}/m_H$$

Since each atom gives us an electron and a proton, the total number of particles is

(29) $$N = 2N_{atoms} = 2M_{sun}/m_H$$

Inserting Eq(27) and Eq(28) into Eq(26) we have

$$2[(3/2)Nk\langle T\rangle] = -[-(1.24)(3/5)(GM_{sun}^2/R)]$$

or

(30) $$\langle T\rangle = 0.248GM_{sun}^2/(NkR)$$

Inserting the number of particles N from Eq(29) into Eq(30) we may obtain the average temperature of the sun.

(31) $$\langle T\rangle = 0.124m_HGM_{sun}/(kR) = 2.90 \times 10^6 \text{ K}$$

As shown in the text, if one assumes the temperature of the sun decreases linearly from the core temperature T_c to zero at the surface, that is

$$T(r) = T_c(1 - r/R)$$

then one may show that the average temperature and the core temperature of the Sun are related by

$$T_c = 4\langle T\rangle = 4(2.90 \times 10^6 \text{ K}) \approx 12.0 \times 10^6 \text{ K}$$

For a T(r) which is more descriptive of the Sun one obtains:

$$T_c \approx 14 \times 10^6 \text{ K}$$

and the accepted value is:

$$T_c = 15 \times 10^6 \text{ K}$$

Internal Pressure - Starting with the equation of state of an ideal gas

$$PV = nRT$$

where n is the number of moles and may be expressed as

$$n = N/N_A.$$

Solving for the pressure, we obtain

$$P = nRT/V = (N/N_A)RT/V = (N/V)(R/N_a)T$$

Recalling that

$$k = R/N_A$$

and agreeing to represent the particle density N/V by n

$$n = \frac{N}{V} = \frac{2M_{sun}/m_H}{(4/3)\pi R^3}$$

we have

$$P = nkt.$$

Combining these we see that if the central temperature of a star is T_c, its central pressure P_c is given by

$$P_c = nkT_c = (\frac{3}{2})(\frac{M_{sun}kT_c}{\pi m_H R^3}) = 3.1 \times 10^{14}\ N/m^2 = 3.1 \times 10^9\ atm.$$

The pressure at the center of the Sun is several billion atmospheres.

As shown in the text, if one does a more rigorous treatment using the balance of inward gravitational forces and net outward force due to pressure {assuming a linear density distribution $\rho(r) = \rho_c(1 - r/R)$} one obtains the central pressure of the star as

(33) $$P(r = 0) = (\frac{45}{36\pi})(\frac{GM^2}{R^4})$$

If one uses a density distribution which more closely approximates the actual composition of the sun one obtains

(34) $$P(r = 0) = 1.24(45/36\pi)(GM^2/R^4) = 5.5 \times 10^{14}\ N/m^2 = 5.5 \times 10^9\ atm$$

Note that this value is essentially the same as the value obtained from the ideal gas law and T_c.

Luminosity and Lifetime - The luminosity and lifetime of a main-sequence star depend strongly on its mass, as shown below.

(35) Luminosity $L \propto M^3$

(36) Lifetime $\mathcal{L} \propto \frac{1}{M^2}$

This means that more massive stars are brighter and have a shorter lifetime.

Practice: Consider four main-sequence stars (our Sun and stars A, B and C) and the following information.

Star	Mass	Luminosity	Lifetime	Average Temperature	Central Pressure
Sun	M_S	L_S	$\mathscr{L}_S$	$<T>_S$	P_S
A	$M_A = 10M_S$				
B	$M_B = M_S/10$				
C		$L_C = 8L_S$			

Determine the following:

1. Luminosity of star A compared to the Sun	$L_S \propto M_S^3$; $L_A \propto M_A^3 = (10M_S)^3 = 10^3\ M_S^3$ These proportionality statements may be made into equality statements by introducing proportionality constants. The proportionality constant is the same for each star (say K). Then we have: $L_S = KM_S^3$; $L_A = KM_A^3 = K10^3\ M_S^3$. When we take the ratio of these two equality statements the proportionality constants cancel to give: $L_A/L_S = 10^3$ or $L_A = 10^3\ L_S$ Star A is one-thousand times brighter than the Sun.

Note: In much of the following work we will use the same procedure as the above, however since the constants will always cancel ... we will go from proportionality statements to the ratio equality statement and not bother with the constants.

2. Luminosity of star B compared to the Luminosity of star A	$L_A \propto M_A^3 = (10M_S)^3 = 10^3\ M_S^3$ $L_B \propto M_B^3 = (10^{-1}M_S)^3 = 10^{-3}\ M_S^3$ $L_B/L_A = (10^{-3}\ M_S^3)/(10^3\ M_S^3) = 10^{-6}$ The Luminosity of star B is one millionth that of star A.
3. Lifetime of B compared to the Sun	$\mathscr{L}_S \propto 1/M_S^2$ $\mathscr{L}_B \propto 1/M_B^2 = 1/(10^{-1}M_S)^2 = 10^2/M_S^2$ $\mathscr{L}_B/\mathscr{L}_S = (10^2/M_S^2)/(1/M_S^2) = 10^2$ $\mathscr{L}_B = 10^2\ \mathscr{L}_S$ The lifetime of B is one-hundred times that of the sun.

4. Mass of star C compared to the mass of the Sun	$L_s \propto M_s^3$; $L_C = 8L_s \propto M_C^3$ $L_s/L_C = L_s/8L_s = (M_s^3)/(M_C^3)$ or $(M_C/M_s)^3 = 8 = 2^3$; or $M_C = 2M_s$ The mass of star C is twice that of the sun.
5. Average temperature of star B compared to star A	According to Eq(31) $\langle T \rangle \propto M$ $\langle T \rangle_B \propto M_B = 10^{-1} M_s$ $\langle T \rangle_A \propto M_A = 10\ M_s$ $\langle T \rangle_B / \langle T \rangle_A = (10^{-1}\ M_s)/(10\ M_s) = 10^{-2}$ $\langle T \rangle_B = 10^{-2} \langle T \rangle_A$ The average temperature of star B is one hundredth that od star A.
6. The central pressure of star C compared to the central pressure of star B	According to Eq(34) the central pressure in the star is proportional to M^2 $(P_c \propto M^2)$ $(P_c)_C \propto M_C^2 = (2M_s)^2 = 4M_s^2$ (step 4) $(P_c)_B \propto M_B^2 = 10^{-2}\ M_s^2$ $(P_c)_C/(P_c)_B = (4M_s^2)/(10^{-2}\ M_s^2) = 400$ $(P_c)_C = 400(P_c)_B$ The pressure at the center of C is 400 times the pressure at the center of B.

Related Text Exercises: Ex. 44-11 through 44-14.

PRACTICE TEST

Take and grade this practice test. Doing so will allow you to determine any weak spots in your understanding of the concepts taught in this chapter. The following section prescribes what you should study further to strengthen your understanding.

The sun is orbited by a 1.00×10^{-3} kg dust particle in an orbit which has a radius $r = 3R_{sun}$. The particle spirals inward to $r = 2R_{sun}$ before it reaches equilibrium. (i.e. the net outward force due to pressure is equal to the inward force due to gravity). During this process radiation of frequency $\nu = 5.10 \times 10^{14}\ s^{-1}$ is emitted. You are also given the following information:

$M_{sun} = 1.99 \times 10^{30}$ kg $\qquad R_{sun} = 6.96 \times 10^{8}$ m $\qquad G = 6.67 \times 10^{-11}\ Nm^2kg^{-2}$

Determine the following.

________ 1. Initial (i.e. when $r = 3R_{sun}$) orbital speed of the particle

________ 2. Initial kinetic energy of the particle

________ 3. Initial gravitational potential energy of the particle

________ 4. Change in gravitational potential energy of the particle as it spirals inward from $r = 3R_{sun}$ to $r = 2R_{sun}$

________ 5. Change in kinetic energy of the particle as it spirals inward from $r = 3R_{sun}$ to $r = 2R_{sun}$

________ 6. Amount of energy radiated as the particle spirals inward from $r = 3R_{sun}$ to $r = 2R_{sun}$

________ 7. Wavelength of the energy radiated as the particle spirals inward from $r = 3R_{sun}$ to $r = 2R_{sun}$

The following information about the Sun is given.

$L_s = 3.85 \times 10^{26}$ J/s	Luminosity of the Sun
$E_{reaction} = 26.7$ Mev	Energy available from each proton-proton cycle fusion reaction
$M_{sun} = 1.99 \times 10^{30}$ kg	Mass of the Sun
$m_H = 1.67 \times 10^{-27}$ kg	Mass of a hydrogen atom

Assume the sun is composed entirely of hydrogen

Determine the following.

________ 8. Number of Hydrogen atoms in the sun

________ 9. Number of protons available for nuclear fusion

________ 10. Number of protons consumed by each proton-proton cycle fusion reaction

________ 11. Maximum possible number of proton-proton cycle fusion reactions

________ 12. Maximum total energy available from each of these reactions

________ 13. Approximate amount of time it would take the Sun to radiate the energy found in question 12 assuming it continued to radiate at its present rate.

Given the following information about two different main-sequence stars:

Star A: $M_A = 100\ M_{sun}$ and Star B: $M_B = 10\ M_{sun}$,

determine the following.

________ 14. Ratio of the mass of star A to star B

________ 15. Ratio of the average temperature on star A to star B

________ 16. Ratio of the central pressure on star A to star B

________ 17. Ratio of the luminosity of star A to star B

________ 18. Ratio of the lifetime of star A to star B

(See Appendix I for answers.)

PRINCIPAL CONCEPTS AND EQUATIONS PRESCRIPTION

Your score on the practice test is an excellent measure of your understanding of the chapter. You should now use the following chart to write your own prescription for dealing with any weaknesses the practice test points out. Look down the leftmost column to the number of the question(s) you answered incorrectly, reading across that row you will find the concept and/or equation of concern, the section(s) of the study guide you should return to for further study, and some suggested text exercises which you should work to gain additional experience.

Practice Test Questions	Concepts and Equations	Prescription	
		Principal Concepts	Text Exercises
1	Central force	1	6-36, 37
2	Kinetic energy: $K = GMm/2r$	1	---
3	Gravitational potential energy: $U_g = -GMm/r$	1	---
4	Gravitational potential energy	1	---
5	Virial theorem: $E_{tot} = -K_{tot} = U_{tot}/2$	1	44-1
6	Virial theorem	1	44-1
7	Wavelength and frequency: $c = \nu\lambda$	2 of Ch 39	39-35
8	Number of atoms: $N_{atoms} = M_{sun}/m_H$	2	44-6
9	One proton per atom	---	44-12
10	Proton-proton cycle	2	---
11	$N_{reaction} = N_{protons}/4$	2	---
12	Energy = $(E_{reaction})(N_{reactions})$	2	44-8, 9
13	Energy = $L_s t$	2	44-8, 9
14	Ratio of masses	---	---
15	Dependence of $\langle T \rangle$ on mass: $\langle T \rangle \propto M$	3	---
16	Dependence of P_c on mass: $P_c \propto M^2$	3	---
17	Dependence of L on mass: $L \propto M^3$	3	---
18	Dependence of $\mathscr{L}$ on mass: $\mathscr{L} \propto 1/M^2$	3	---

45 ENERGY SOURCES FOR THE FUTURE

RECALL FROM PREVIOUS CHAPTERS

Previously learned concepts and equations frequently used in this chapter	Text Section	Study Guide Page
Power: $P = E/t$	8-6	8-15
Efficiency of any heat engine: $\eta = W/Q_H = 1 - \lvert Q_c\rvert/Q_H$	19-1, 2	19-1
Heat pump coefficient of performance: $K_{hp} = \lvert Q_H\rvert/W = \lvert Q_H\rvert/(\lvert Q_H\rvert - Q_c)$	19-1, 2	19-1
Efficiency of a Carnot engine: $\eta_c = 1 - (T_c/T_H)$	19-4	19-6
Intensity: $I = P/A = E/At$	32-5	32-6

NEW IDEAS IN THIS CHAPTER

Concepts and equations introduced	Text Section	Study Guide Page
Exponential growth in the rate of consumption of energy: $P = P_0 e^{(t/\tau)}$	45-4	45-6
Energy consumption-rate model: $P = P_m/(1 + x^2)$ where $x = (t - t_m)/\Delta t$	45-4	45-11
Solar power incident on a collector: $P_{inc} = I\hat{s}\cdot A\hat{n}$	45-5	45-13
Solar Intensity: $I = (1.10 \times 10^2 \text{ W/m}^2)e^{-(0.170/\sin\beta)}$	45-5	45-13
Sun's declination: $\delta = (23.45°)\sin\omega t$	45-5	45-13
Solar altitude at noon: $\beta_N = 90° + \delta - \lambda$	45-5	45-13
Optimum collector angle: $\alpha = 70° - \lambda$	45-5	45-13

PRINCIPAL CONCEPTS AND EQUATIONS

1 Application of the Laws of Thermodynamics to the Use of Energy (Section 45-1)

Review: The first law of thermodynamics is a statement of conservation of energy:

Energy cannot be created or destroyed, it can only be transferred.

Energy transfers are of two types: work W and heat Q.

The second law of thermodynamics may be stated as:

> In any real process that involves the transfer of energy, the total energy of all systems involved in the process is degraded.

Energy is degraded when work is converted to heat or when heat becomes available from a lower temperature difference between reservoirs. In general, work is more valuable than heat.

From chapter 19, let's recall the schematics of a heat engine and a heat pump.

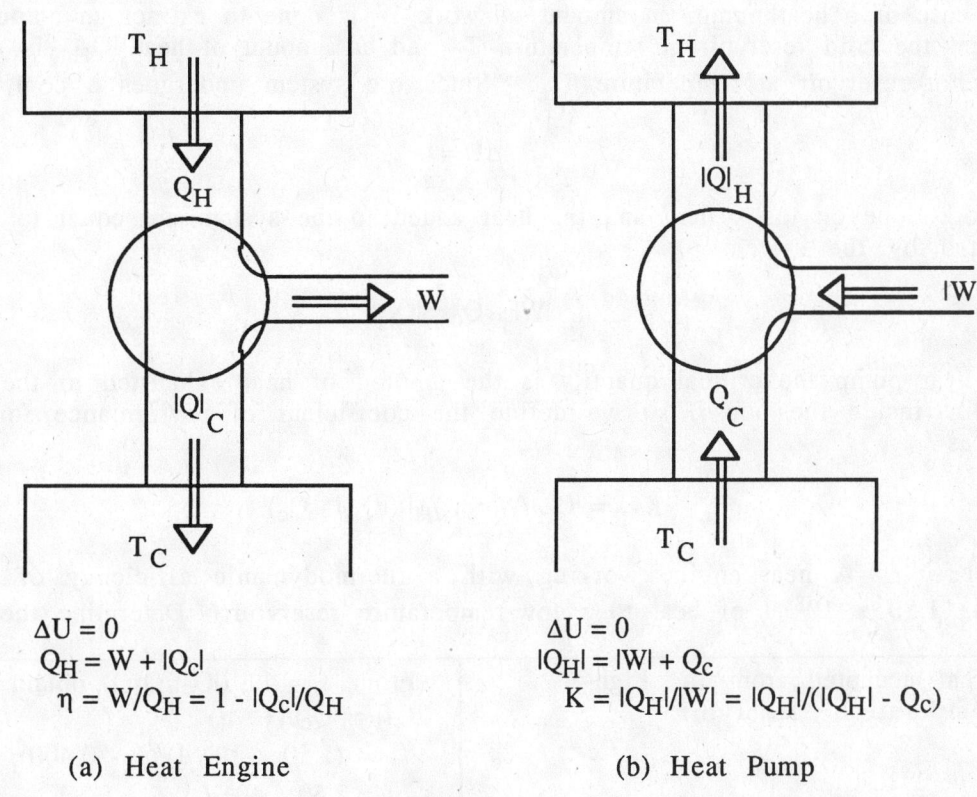

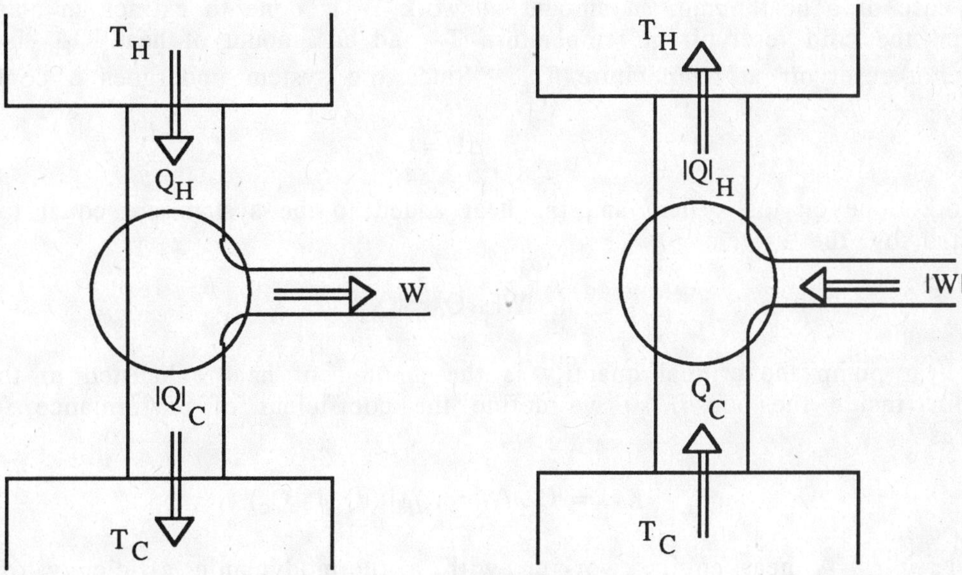

$\Delta U = 0$
$Q_H = W + |Q_c|$
$\eta = W/Q_H = 1 - |Q_c|/Q_H$

(a) Heat Engine

$\Delta U = 0$
$|Q_H| = |W| + Q_c$
$K = |Q_H|/|W| = |Q_H|/(|Q_H| - Q_c)$

(b) Heat Pump

Figure 45.1

In the case of a heat engine, an amount of heat Q_H is added to the system (from the hot reservoir) at temperature T_H, the system does an amount of work W, and an amount of heat $|Q_c|$ is exhausted to the cold reservoir at temperature T_c. Since the system undergoes a complete cycle, we have

$$\Delta U = 0$$

The heat added to the system either does work or is exhausted to the cold reservoir.

(1) $$Q_H = W + |Q_c|$$

The efficiency η is the amount of work we get out of the engine per unit of heat added.

(2) $$\eta = W/Q_H = (Q_H - |Q_c|)/Q_H = 1 - |Q_c|/Q_H$$

A Carnot engine is an idealized heat engine with the highest efficiency of any engine operating between the same two reservoirs. The efficiency of a Carnot engine is

(3) $$\eta_c = 1 - T_c/T_H$$

For most engines T_c = 300 K (ambient temperature of the earth's surface atmosphere). This means that the practical way to increase the efficiency of the engine is to make T_H as large as possible.

In the case of a heat pump, an amount of work W is done to extract an amount of heat Q_c from the cold reservoir at temperature T_c and an amount of heat $|Q_H|$ is exhausted to the hot reservoir at temperature T_H. Since the system undergoes a complete cycle, we have

$$\Delta U = 0$$

The work done on the system and the heat added to the system are equal to the heat exhausted by the system

(4) $$|W| + Q_c = |Q_H|$$

For a heat pump the critical quantity is the amount of heat exhausted to the hot reservoir (inside the house) so we define the coefficient of performance for a heat pump as

(5) $$K_{hp} = |Q_H|/W = |Q_H|/(|Q_H| - Q_c).$$

Practice: A heat engine working with a thermodynamic efficiency of 0.400 exhausts 1.50 x 10^8 J of heat to a low-temperature reservoir. Determine the following:

1. Heat accepted from the high-temperature reservoir.	From $\eta = 1 - (|Q_c|/Q_H)$ obtain $Q_H = |Q_c|/(1 - \eta)$ $= (1.50 \times 10^8\ J)/(1 - 0.400)$ $= 2.50 \times 10^8\ J$
2. Work done by the engine	From $Q_H = W + |Q_c|$ obtain $W = Q_H - |Q_c| = 2.50 \times 10^8\ J - 1.50 \times 10^8\ J$ $= 1.0 \times 10^8\ J$ Also from $\eta = W/Q_H$ obtain $W = \eta Q_H = (0.400)(2.50 \times 10^8\ J)$ $= 1.0 \times 10^8\ J$

A Carnot engine operates first between high temperature reservoir A (T_{HA} = 600 K) and a low temperature reservoir (T_c = 300 K) and then between high temperature reservoir B (T_{HB} = 400 K) and the same low temperature reservoir. In both cases the engine absorbs 4 J of heat (i.e. Q_H = 4.00 J) from the high temperature reservoir.

Determine the following.

3. Efficiency of the Carnot engine when operating with high temperature reservoir A, and then also for reservoir B	$\eta_A = 1 - (T_c/T_{HA}) = 1 - (300\text{ K}/600\text{ K})$ $= 0.500$ $\eta_B = 1 - (T_c/T_{HB}) = 1 - (300\text{ K}/400\text{ K})$ $= 0.250$

Note: The above practice step shows why we can make the statement that the practical way to increase the efficiency of a Carnot engine is to make T_H larger.

4. Heat converted to work when the Carnot engine is operating with high temperature reservoir A, B	From $\eta = W/Q_H$ obtain $W_A = \eta_A Q_H = (0.500)(4.00\text{ J}) = 2.00\text{ J}$ $W_B = \eta_B Q_H = (0.250)(4.00\text{ J}) = 1.00\text{ J}$

Note: The above practice step shows why we can make the statement that heat transferred from a high temperature reservoir is more valuable (of a higher grade) the higher the temperature of the high temperature reservoir.

5. Amount of heat exhausted to the low temperature reservoir A, B	From $Q_H = W + \lvert Q_c\rvert$ obtain $\lvert Q_{cA}\rvert = Q_H - W_A = 4.00\text{ J} - 2.00\text{ J} = 2.00\text{ J}$ $\lvert Q_{cB}\rvert = Q_H - W_B = 4.00\text{ J} - 1.00\text{ J} = 3.00\text{ J}$

Consider a heat pump with a coefficient of performance $K_{hp} = 3.00$. Determine the following.

6. Amount of heat delivered to the house when 10.0 J of work is used to run the heat pump	From $K_{hp} = \lvert Q_H\rvert/\lvert W\rvert$ obtain $\lvert Q_H\rvert = K_{hp}\lvert W\rvert = (3.00)(10.0\text{ J}) = 30.0\text{ J}$

Note: With the heat pump 10 J of work gave us 30 J of heat (the other 20 J was absorbed from the low temperature reservoir). This practice step shows why we can make the statement that work is more valuable than heat.

A heat pump has a coefficient of performance of 2.30 and the electric motor that runs its compressor uses 970 W of electric power. Determine the following.

7. Rate at which the heat pump provides heat to the house.	From $K_{hp} = \lvert Q_H\rvert/\lvert W\rvert$ obtain $\lvert Q_H\rvert = K_{hp}\lvert W\rvert$ and then $d\lvert Q_H\rvert/dt = K_{hp}d\lvert W\rvert/dt$ $= (2.30)(970\text{ W}) = 2.23\text{ kW}$
8. Rate at which the heat pump absorbs heat from the environment outside the house.	From $\lvert W\rvert + Q_c = \lvert Q_H\rvert$ obtain $Q_c = \lvert Q_H\rvert - \lvert W\rvert$ and then $dQ_c/dt = d\lvert Q_H\rvert/dt - d\lvert W\rvert/dt$ $= 2.23\text{ kW} - 0.970\text{ kW} = 1.26\text{ kW}$

Example: Suppose you own a heat engine with an efficiency η that is one-half that of a Carnot Engine:

$$\eta = 0.500\,\eta_c$$

You intend to convert heat to work with your engine, and the low-temperature reservoir that you must use is the engine's environment with a temperature $T_c = 300$ K. You may buy heat of two different grades at the following prices:

(a) heat from a reservoir at 500 K for \$16/GJ or
(b) heat from a reservoir at 750 K for \$18/GJ

You are to determine the cost of the work in \$/GJ for each case.

Given:

$\eta = 0.500\,\eta_c$	Efficiency of the heat engine
$T_c = 300$ K	Temperature of the low-temperature reservoir
$T_{HA} = 500$ K	Temperature of high-temperature reservoir A
c_A = \$16/GJ	Cost of heat from high-temperature reservoir A
$T_{HB} = 750$ K	Temperature of high-temperature reservoir B
c_B = \$18/GJ	Cost of heat from high-temperature reservoir B

Determine: Cost of work in \$/GJ when the heat engine is connected to high temperature reservoir A, B.

Strategy: Knowing T_c and T_H for each case, we can determine the efficiency of a Carnot engine and then your engine when operating between the hot and cold reservoirs for each case. Knowing the efficiencies, we can determine the amount of heat we must buy to produce 1 GJ of work for each case. Finally knowing the cost of the heat, we can determine \$/GJ for each case.

Solution: First let's determine the efficiency of a Carnot engine (η_c) when operating between the high and low-temperature reservoir for each case.

Case A	**Case B**
Temperature of the reservoirs:	
$T_c = 300$ K and $T_{HA} = 500$ K	$T_c = 300$ K and $T_{HB} = 750$ K
Efficiency of a Carnot engine operating between these reservoirs:	
$(\eta_c)_A = 1 - (T_c/T_{HA}) = 0.400$	$(\eta_c)_B = 1 - (T_c/T_{HB}) = 0.600$
Efficiency of your engine operating between these reservoirs:	
$\eta_A = (0.500)(\eta_c)_A = 0.200$	$\eta_B = (0.500)(\eta_c)_B = 0.300$
Heat needed to produce 1 GJ of work:	
$Q_{HA} = W/\eta_A = 1\text{ GJ}/0.200 = 5.00\text{ GJ}$	$Q_{HB} = W/\eta_B = 1\text{ GJ}/0.300 = 3.33\text{ GJ}$
Cost of heat:	
$c_A = \$16/\text{GJ}$	$c_B = \$18/\text{GJ}$
Cost for heat needed to produce 1 GJ of work:	
$(\$/\text{GJ})_A = (Q_{HA})(c_A)$	$(\$/\text{GJ})_B = (Q_{HB})(c_B)$
$= (5.00\text{ GJ})(\$16/\text{GJ})$	$= (3.33\text{ GJ})(\$18/\text{GJ})$
$= \$80.00/(\text{GJ of work})$	$= \$60.00/(\text{GJ of work})$

Heat from high temperature reservoir B is a better energy buy.

Related Text Exercises: Ex. 45-1 through 45-8.

2 Exponential Growth in the Rate of Energy Consumption
(Section 45-4)

Review: First we need to recall that the rate of consumption of energy is just power P.

(6) $$P = E/t$$

If the rate of consumption of energy at some time we arbitrarily choose to be zero is P_0 and if the rate of consumption is increasing exponentially, then the rate of consumption of energy P may be described by

(7) $$P = P_0 e^{t/\tau}$$

In Eq(7)

P_0 = rate of consumption of energy at the time we choose to be zero
t = any time
τ = mean time (time for the power to increase from P_0 to $P_0 e$)
P = rate of consumption of energy at any time t

When dealing with exponential growth, a useful concept is the doubling time (T). The doubling time is the time required for the rate of consumption to double. The doubling time is related to the mean time as shown below.

$P = 2P_0$ (that is it has doubled) when $t = T$ (the doubling time)

Inserting this information into Eq(7) we obtain the relationship between the doubling time T and the mean time τ.

$$P = P_0 e^{t/\tau}$$
$$2P_0 = P_0 e^{T/\tau}$$
(8) $$2 = e^{T/\tau}$$
$$\ln 2 = T/\tau$$

or

(9) $$T = \tau \ln 2 = 0.693\tau$$

If we take the natural logarithm of Eq(7) we obtain

$$\ln P = \ln P_0 + \ln e^{t/\tau}$$

or

$$\ln P = (1/\tau)t + \ln P_0$$

Comparing this with the equation for a straight line

$$y = mx + b$$

we see that a plot of lnP vs t will be a straight line with a slope $1/\tau$ and an intercept $\ln P_0$. Also rather than looking up the natural log of all the P values, we could just plot P vs t on semilogarithmic graph paper.

Figure 45.2 shows a plot of P vs t.

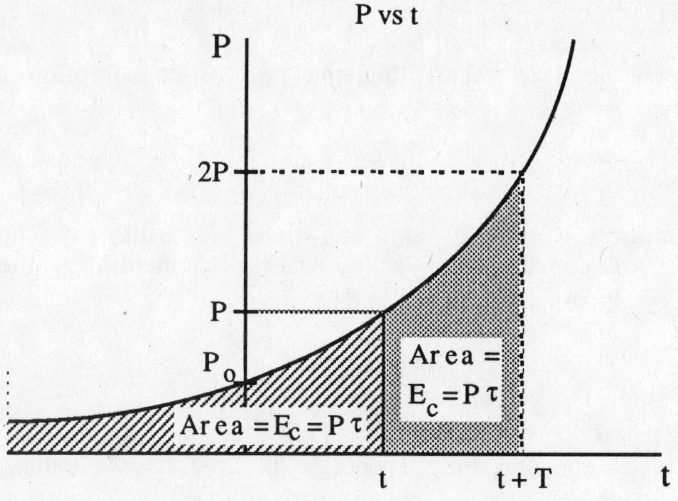

Figure 45-2

Note: $P = P_0$ at $t = 0$ If the power is P at t then it will be 2P at t + T. The area under the curve between any two times is the energy consumed in that time interval. E_c the cumulative consumption of energy up to some time is the area under the curve up to that time. E_c doubles after each doubling time.

The energy consumed between any two times is just the area under the curve during that time interval.

$$E\begin{pmatrix}\text{consumed}\\ \text{between}\\ t_1 \text{ and } t_2\end{pmatrix} = \int_{t_1}^{t_2} P(t)dt = \int_{t_1}^{t_2} P_0 e^{(t/\tau)} dt \tag{11}$$

The cumulative consumption of energy up to some time t may then be found by:

$$E_c(-\infty \to t) = P_0 \int_{-\infty}^{t} e^{(t/\tau)} dt = \tau P_0 e^{(t/\tau)}\Big|_{-\infty}^{t} = \tau P_0 e^{(t/\tau)} = \tau P \tag{12}$$

The energy consumed during all time prior to some time t (i.e. the cumulative consumption) is equal to the energy consumed during the next doubling time. This statement is proven below.

$$E_c = \int_{t}^{t+T} P_0 e^{(t/\tau)} dt = \tau P_0 e^{(t/\tau)}\Big|_{t}^{t+T} = \tau P_0\left(e^{[(t+T)/\tau]} - e^{(t/\tau)}\right) \tag{13}$$

$$E_c = \tau P_0\left(e^{(t/\tau)} e^{(T/\tau)} - e^{(t/\tau)}\right) = \tau P_0 e^{(t/\tau)}\left(e^{(T/\tau)} - 1\right)$$

From Eq(8) $e^{T/\tau} = 2$, hence

$$E_c = \tau P_0 e^{(t/\tau)}(2 - 1) = \tau P$$

Summarizing, we may write

$$E_c = \int_{-\infty}^{t} P(t)dt = \int_{t}^{t+T} P(t)dt = \tau P \tag{14}$$

Practice: Shown below is some petroleum consumption data over the past sixty years and plots of P vs t (Plot I), lnP vs t (Plot II) and P vs t on semilogarithmic graph paper (Plot III). As a matter of convenience lnP is given in the data table and 1930 has been chosen as t = 0.

Petroleum Data

Year	P(EJ/y)	lnP	t(y)
1930	19	2.94	0
1940	27	3.30	10
1950	37	3.61	20
1960	50	3.91	30
1970	70	4.25	40
1980	95	4.55	50
1990	130	4.87	60

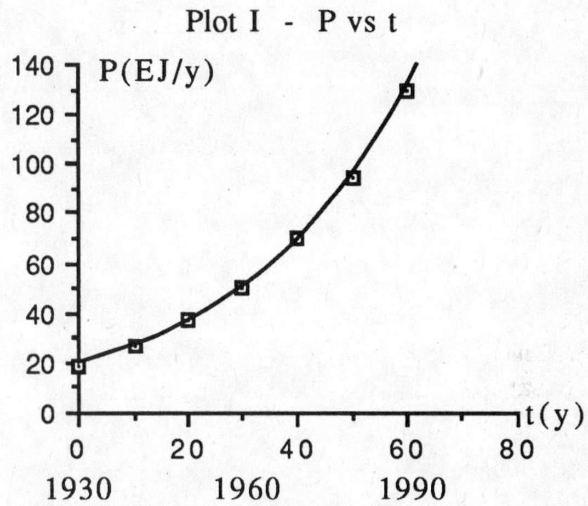

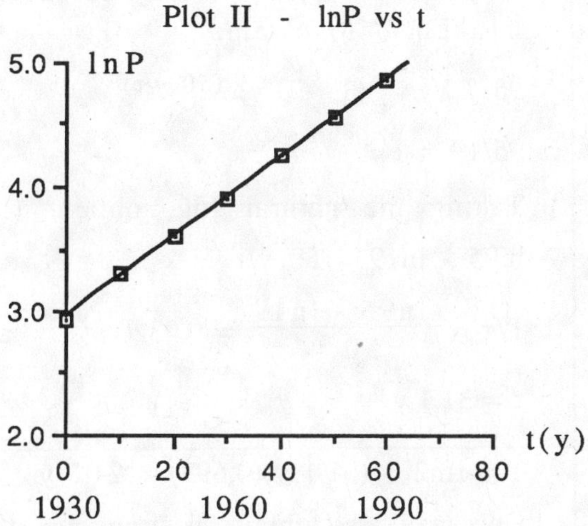

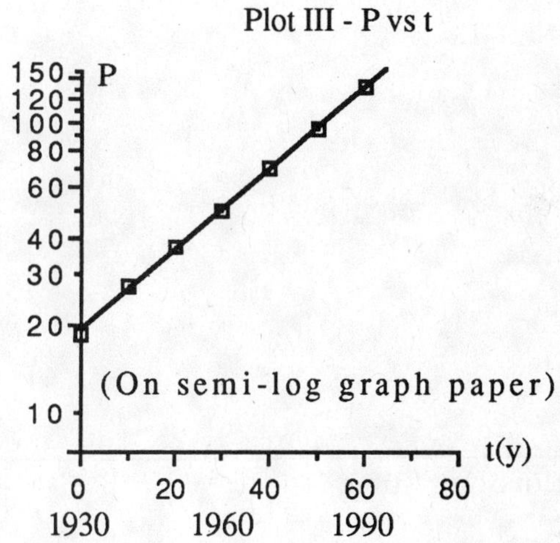

Determine the following:

1. Rate of consumption of energy in 1975	According to the data table t = 45 corresponds to 1975. Looking at plots I or III we see that at t = 45 y the rate of consumption of energy is P = 83 EJ/y.

2. In what year was the rate of consumption of energy equal to 110 EJ/y	Looking at plots I or III we see that P = 110 EJ/y at t = 54 y which corresponds to 1930 + 54 = 1984.
3. How much did the rate of consumption of energy increase between 1955 and 1985	According to the data table at 1955 t = 25 and at 1985 t = 55. According to plots I and II $P(25) = 43$ EJ/y; $P(55) = 110$ EJ/y $\Delta P = P(55) - P(25) = 67$ EJ/y
4. Mean time τ	From Eq(10) we see that $1/\tau$ is the slope of the lnP vs t or the P vs t plot on semi-log graph paper. Working with plot III: $\frac{1}{\tau} = \text{slope} = \frac{\ln 95 - \ln 19}{50 - 0} = 0.0320$ and $\tau = 1/0.0320 = 31.1$ y. Using the same two points from plot II: $\frac{1}{\tau} = \text{slope} = \frac{4.55 - 2.94}{50 - 1} = 0.0320$ and $\tau = 1/0.0320 = 31.1$ y. We may also just use the information given in the data table and Eq(7). Knowing $P_0 = 19$E J/y and that P = 95 EJ/y when t = 50 we may substitute into Eq(7) to obtain: $P = P_0 e^{t/\tau}$ 95 EJ/y = (19 EJ/y)$e^{(50\ y/\tau)}$ or $95/19 = e^{(50\ y/\tau)}$ Taking the natural log, obtain $\ln 95 - \ln 19 = 50\ y/\tau$ or $1/\tau = \frac{\ln 95 - \ln 19}{50\ y} = 0.0320\ y^{-1}$ and $\tau = 31.1$ y
5. Doubling time	$T = \tau \ln 2 = (31.1\ y)(0.693) = 21.6$ y One may also obtain T from any of the plots. Just pick a value for P (lets use plot I) and its corresponding value of t. For example $P_1 = 27$ EJ/y at $t_1 = 10$. Now $P_2 = 2P_1 = 54$ EJ/y at $t_2 = 32$ y. Then $t = \Delta t = t_2 - t_1 = 32\ y - 10\ y = 22$ y This number is close to the value calculated and the accuracy is limited by how accurately we can read the plots.

6. Total petroleum energy consumed prior to 1950	The quantity we wish to determine is called the cumulative energy consumption. According to Eq(14) $E_c(\text{prior to 1950}) = P(1950)\tau$ $= (37\ \text{EJ/y})(31.1\ \text{y}) = 1150\ \text{EJ}.$
7. Total petroleum energy consumed between 1950 and 1972	The time interval of interest is 22 y or essentially a doubling time. According to Eq(14) the cumulative energy consumption doubles with each doubling time. This means that in the time interval 1950 → 1972 we will have used as much energy as was consumed prior to 1950. $E_c(\text{prior to 1950}) = 1150\ \text{EJ}$ $E(1950 \rightarrow 1972) = 1150\ \text{EJ}$
8. Energy consumed between 1955 and 1965	$P_0 = 19\ \text{EJ/y}$ at $t = 0$ $t = 25\ \text{y}$ at 1955; $t = 35\ \text{y}$ at 1965 $E = \int_{25\,y}^{35\,y} P_0 e^{(t/\tau)} dt = P_0 \tau e^{(t/\tau)} \Big|_{25\,y}^{35\,y}$ $= P_0\tau\left(e^{(35\ y/\tau)} - e^{(25\ y/\tau)}\right)$ At this point we can insert numbers to calculate E or we can rewrite the last expression as $E = \left(P_0 e^{(35\ y/\tau)} - P_0 e^{(25\ y/\tau)}\right)\tau$ $= [P(35) - P(25)]\tau.$ Reading these values off the graph, obtain $E = (60\ \text{EJ/y} - 44\ \text{EJ/y})\tau$ $= (16\ \text{EJ/y})(31.1\ \text{y}) = 498\ \text{EJ}.$ The last expression reminds us that we could also obtain the desired quantity by determining the cumulative energy for 1955 and 1965 and subtracting. $E(1955 \rightarrow 1965) = E_c(1965) - E_c(1955)$ $= [P(35) - P(25)]\tau$

Example: Suppose a resource is being consumed at a rate of 25 EJ/y during 1990 and the consumption rate is increasing exponentially with a doubling time of 18.0 y. (a) What is the mean time for this growth? (b) If the consumption rate continues to be exponential, what will it be during the year 2000?

Given: $P = 25$ EJ/y during 1990
$T = 18.0$ y
Consumption rate increases exponentially

Determine: τ - the mean time for growth
P - the rate of consumption of this energy source in the year 2000

Strategy: Knowing the doubling time T, we can determine the mean growth time τ. If we call 1990 t = 0, then P_0 is the consumption rate during 1990. Knowing P_0 and τ, we can determine the consumption rate in the year 2000, that is P(t=10.0).

Solution:
(a) Knowing the doubling time T, we can determine the mean growth time τ.

$$\tau = T/\ln 2 = (18.0\ y)/(0.693) = 26.0\ y$$

(b) Knowing P_0, t and τ we can determine P(t=10.0) which is the consumption rate in the year 2000.

$$P(10.0\ y) = P_0 e^{(t/\tau)} = (25\ EJ/y)e^{(10.0\ y/26.0\ y)} = 36.7\ EJ/y$$

The rate of consumption of this energy source in the year 2000 will be 36.7 EJ/y.

Related Text Exercises: Ex. 45-11 through 45-15.

3 Energy Consumption-Rate Model (Section 45-4)

Review: The rate of consumption of a depletable resource is expected to follow a curve like that shown in Figure 45.3.

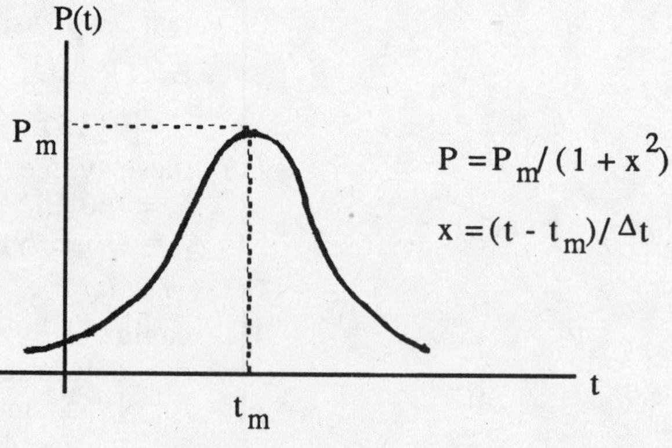

Figure 45.3

The ultimate recoverable energy from this resource E_∞ is total area under the curve.

$$E_\infty = \int_{-\infty}^{+\infty} P(t)dt \qquad (15)$$

The maximum rate of consumption is P_m and this occurs at the time t_m. Knowing E_∞ and E_c and P for some time, we can determine P_m and t_m as follows. Since these expressions are derived in detail in the text, they are just listed here.

(16) $$x = \tan\left(\frac{\pi E_c}{E_\infty} - \frac{\pi}{2}\right)$$

(17) $$P_m = P(1 + x^2)$$

(18) $$t_m = t - \left(\frac{xE_\infty}{\pi P_m}\right)$$

Practice: The 1990 data for coal is:

$P = 100$ EJ/y; $E_c = 6{,}400$ EJ $E_\infty = 23{,}000$ EJ

Determine the following:

1. x	$x = \tan\left(\frac{\pi E_c}{E_\infty} - \frac{\pi}{2}\right) = \tan\left(\frac{\pi 6{,}400 \text{ EJ}}{23{,}000 \text{ EJ}} - \frac{\pi}{2}\right)$ $= -0.837$ Looking at Eq(18) we see that the value of x will be negative for all time $t < t_m$. This means that the rate of consumption of coal had not peaked out by 1990.
2. P_m the maximum rate of consumption of coal	$P_m = P(1 + x^2) = [100 \text{ EJ/y}][1 + (-0.837)^2]$ $= 170$ EJ/y
3. The year that coal consumption will peak, i.e. t_m	$t_m = t - \left(\frac{xE_\infty}{\pi P_m}\right) t_m$ $= 1990 \text{ y} - \left[\frac{(-0.837)(23{,}000 \text{ EJ})}{\pi(170 \text{ EJ/y})}\right]$ $= 1990 \text{ y} + 36 \text{ y} = 2026 \text{ y}$

Example: In the practice section we used a model for the depletion of the world coal supply that included the estimated $E_\infty = 23{,}000$ EJ. We found that $P_m = 170$ EJ/y and $t_m = 2026$. Determine the amount t_m would increase if we discover we have twice the coal reserves we originally thought. That is E_∞ increases from 23,000 EJ to 46,000 EJ.

Given: $E_\infty = 46{,}000$ EJ
$E_c = 6{,}400$ EJ and $P = 100$ EJ/y for 1990

Determine: t_m - the time when coal is being consumed at the maximum rate.

Strategy: Knowing E_c and E_∞ we can determine x. Knowing P and x, we can determine P_m. Knowing t, E_∞ and P_m for 1990 we can determine t_m.

Solution: Knowing E_c and E_∞ we can obtain x.

$$x = \tan\left(\frac{\pi E_c}{E_\infty} - \frac{\pi}{2}\right) = \tan\left[\pi\left(\frac{6{,}400}{46{,}000} - 0.500\right)\right] = -2.14$$

Notice that this value for x is larger than the value for x obtained in step 1 of the practice. This means we are now farther from t_m ... which we should be since we doubled the reserves.

Knowing P and x, we can obtain P_m.

$$P_m = P(1 + x^2) = (100\ \text{EJ/y})(1 + [2.14]^2) = 558\ \text{EJ/y}$$

Since the rate of consumption in 1990 was 100 EJ/y, this says that in that year we were below the maximum rate of consumption.

Knowing t, P_m and E_∞ we can obtain t_m.

$$t_m = t - \left(\frac{xE_\infty}{\pi P_m}\right) = 1990 - \left(\frac{(-2.14)(46{,}000\ \text{EJ})}{\pi(558\ \text{EJ/y})}\right) = 1990 + 56\ \text{y} = 2046$$

It should be noted that by doubling the reserves we added only 20 years until time of maximum rate of consumption.

Related Text Exercises: Ex. 45-16 through 45-19.

4 Solar Energy (Section 45-5)

Review: Figure 45.4 shows a flat plate solar collector facing South.

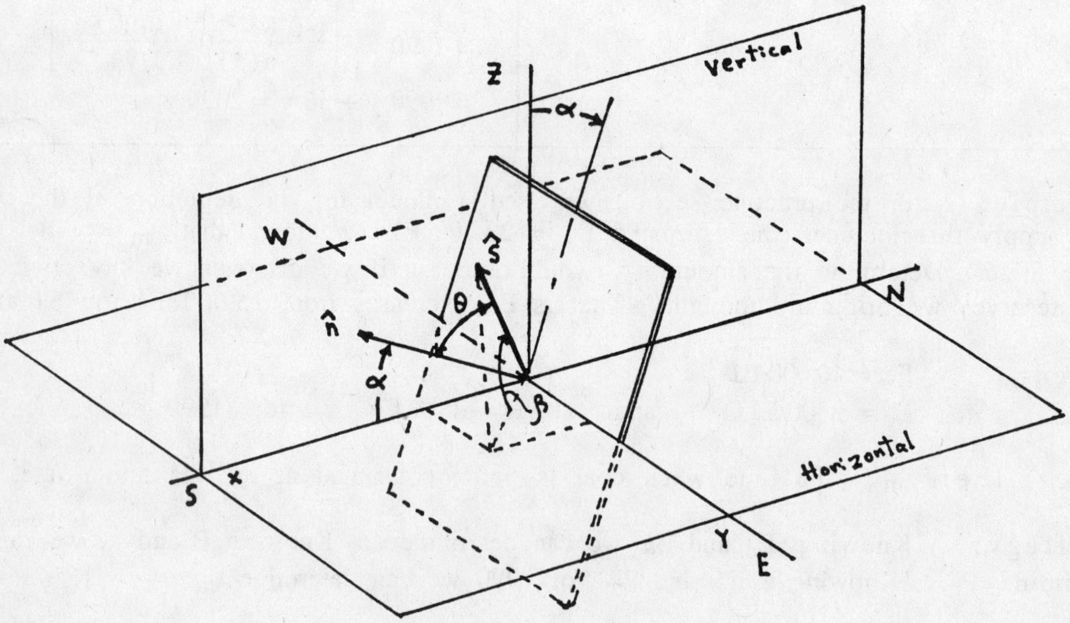

Figure 45.4

Looking at Figure 45.4 we may make the following statements:

- The collector faces South
- The collector has been leaned at an angle of α from the vertical z-y plane.
- $\hat{n}$ is normal to the plane so it points at an angle α above the horizontal x-y plane. Since the collector is fixed, $\hat{n}$ is always in the vertical x-z plane oriented at the angle α.
- $\hat{s}$ is a unit vector pointing directly to the sun. This unit vector changes directions during the day.
- β is the solar altitude angle. That is the angle between the horizontal plane x-y and the direction of $\hat{s}$.
- θ is the angle between $\hat{n}$ and $\hat{s}$.
- At solar noon, both $\hat{n}$ and $\hat{s}$ are in the vertical x-z plane
- At solar noon, $\beta = \alpha + \theta$.

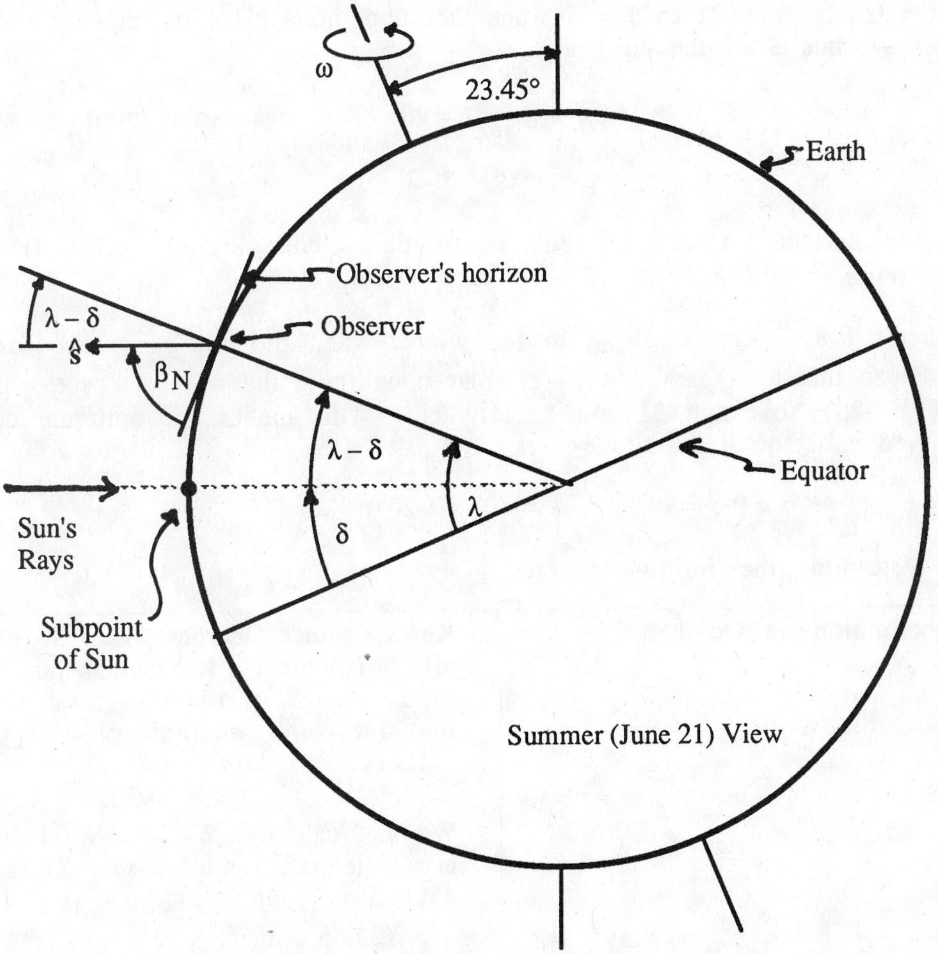

Figure 45.5

The solar power incident on the collector in Figure 45.5 is given by

(19) $$P_{inc} = I\hat{s} \cdot A\hat{n} = IA\hat{s} \cdot \hat{n} = IA\cos\theta$$

where A is the surface area of the collector and I is the average solar intensity incident upon the collector.

The average solar intensity at sea level on a sunny day is given by

(20) $$I = (1.10 \times 10^2\ W/m^3)e^{-(0.170/\sin\beta)}$$

In order to collect solar energy efficiently, we need to be concerned with the orientation of the collector with respect to the horizontal - that is the direction of $\hat{n}$ or the value of the angle α.

Figure 45.5 can help us establish the proper value for α.

$\delta = (23.45°)\sin\omega t$ - sun's declination - latitude of the sun's subpoint

ω = angular speed of the earth in its orbit about the sun

= 360°/365.25 day = 0.98563°/day

t = 0 on March 21 so δ = + when the subpoint is N of the equator and - when S of the equator

$$\beta_N + (\lambda - \delta) = 90° \quad \text{at solar noon}$$

or

(21) $$\beta_N = 90° + \delta - \lambda$$

A sun-following collector tracks the sun continuously changing its orientation so that $\hat{n}$ and $\hat{s}$ remain parallel.

A fixed collector has the largest P_{inc} in the winter when the energy is needed most if it is oriented so that $\hat{n}$ is parallel to $\hat{s}$ at solar noon (note that this makes $\alpha = \beta_N$) on dates when δ = -20 (November 21 and January 21). This means the optimum collector angle for a fixed collector at a latitude λ is

(22) $$\alpha = \beta_N \doteq 90° + \delta - \lambda = 90° + (-20°) - \lambda = 70° - \lambda$$

Practice: Determine the following:

1. Sun's declination on October 21	Knowing that October 21 is 30 days after September 21, we can just look at Figure 45-18 of the text to determine that the sun's declination is approximately $\delta \approx -10°$. We can calculate δ since we know ω = 0.98563°/d and that Oct. 21 is 210 days after March 21. $\delta = (23.45°)\sin\omega t = 23.45° \sin(0.98563°/d)(210\ d) = 23.45° \sin 207° = -10.6°$

2. Sun's altitude at solar noon for an observer located at 40° N latitude	At solar noon the sun's altitude is $\beta_N = 90 + \delta - \lambda = 90° + (-10.6°) - 40°$ $= 39.4°$
3. Sun's intensity for this observer on a clear day at solar noon.	$I = (1.10 \times 10^3 \ W/m^2)e^{(-0.170/\sin\beta)}$ $= (1.10 \times 10^3 \ W/m^2)e^{(-0.170/\sin 39.4°)}$ $= (1.10 \times 10^3 \ W/m^2)(0.765)$ $= 8.42 \times 10^2 \ W/m^2$
4. Optimum collector angle for a fixed collector.	$\alpha = 70° - \lambda = 70° - 40° = 30°$
5. Angle between $\hat{n}$ and $\hat{s}$ at solar noon if the collector is set at the optimum collector angle	At solar noon: $\beta_N = 39.4°$ (step 2) The optimum collector angle is: $\alpha = 30°$ (step 4) At solar noon α and β_N are in the same plane and $\theta = \beta_N - \alpha = 39.4° - 30.0° = 9.4°$
6. Power incident on the collector at solar noon if the area of the collector is 5.00 m^2	$P = I\hat{s} \cdot A\hat{n} = IA\cos\theta$ Obtain I from step 3 and θ from step 5 to obtain $P = (8.42 \times 10^2 \ W/m^2)(5.00 \ m^2)\cos 9.4°$ $P = 4.15 \times 10^3 \ W = 4.15 \ kW$

Example: Consider a coordinate system and a collector like the one shown in Figure 45.4. The normal vector for the collector is given by $\hat{n} = 0.866\,\hat{i} + 0.500\,\hat{k}$. At 9:00 AM the unit vector pointing toward the sun is given by $\hat{s} = 0.433\,\hat{i} + 0.250\,\hat{j} + 0.866\,\hat{k}$. The collector is a flat plate of 6.00 m^2 area and the sun's intensity is 200 W/m^2. At this time (i. e. 9:00 AM) determine the power incident on the collector.

Given: $\hat{n} = 0.866\,\hat{i} + 0.500\,\hat{k}$ - unit vector normal to the solar collector
$\hat{s} = 0.433\,\hat{i} + 0.250\,\hat{j} + 0.866\,\hat{k}$ - unit vector pointing from the solar collector to the sun at 9:00 AM.
$A = 6.00 \ m^2$ - area of the solar collector
$I = 700 \ W/m^2$

Determine: The power incident on the collector.

Strategy: Knowing $\hat{n}$ and $\hat{s}$ we may determine θ. Knowing I, A and θ we can determine P_{inc}.

Solution:

$$\hat{n} \cdot \hat{s} = (0.866\,\hat{i} + 0.500\,\hat{k}) \cdot (0.433\,\hat{i} + 0.250\,\hat{j} + 0.866\,\hat{k}) = \cos\theta$$

$$\cos\theta = (0.866)(0.433)\,\hat{i}\cdot\hat{i} + (0.500)(0.866)\,\hat{k}\cdot\hat{k}$$

$$\cos\theta = 0.375 + 0.433 = 0.808$$

$$P_{inc} = I\hat{s}\cdot A\hat{n} = IA\cos\theta = (700\ W/m^2)(6.00\ m^2)(0.808) = 3.39 \times 10^3\ W = 3.39\ kW$$

Related Text Exercises: Ex. 45-22 through 45-28.

PRACTICE TEST

Take and grade this practice test. Doing so will allow you to determine any weak spots in your understanding of the concepts taught in this chapter. The following section prescribes what you should study further to strengthen your understanding.

You own a. heat engine which has an efficiency which is 30% that of a Carnot engine. You operate this engine between two reservoirs of temperature 700 K and 300 K and the engine exhausts 2.00×10^8 J of heat to the cold reservoir. Determine the following:

_________ 1. Efficiency of a Carnot engine operating between these two reservoirs
_________ 2. Efficiency of your engine operating between these two reservoirs
_________ 3. Heat accepted from the high-temperature reservoir
_________ 4. Work done by your engine

The rate of consumption of an energy source was 50 EJ/y in 1950 and had increased to 120 EJ/y by 1990. Determine the following:

_________ 5. Mean time for this energy source
_________ 6. Doubling time for this source
_________ 7. Rate of consumption of this energy source in the year 1975
_________ 8. Cumulative energy consumption prior to 1950
_________ 9. Energy consumed during the second doubling time after 1950

Consider a solar collector located at a latitude of 40° N. Determine the following.

_________ 10. Sun's declination on October 21
_________ 11. Sun's altitude at the site of the collector at solar noon on October 21
_________ 12. Sun's average intensity at the collector site on a clear day at solar noon
_________ 13. Optimum collector angle at the collector site
_________ 14. Angle between $\hat{n}$ and $\hat{s}$ at solar noon on October 21 if the collector is set at the optimum collector angle
_________ 15. Power incident on the collector at solar noon

(See Appendix I for answers.)

PRINCIPAL CONCEPTS AND EQUATIONS PRESCRIPTION

Your score on the practice test is an excellent measure of your understanding of the chapter. You should now use the following chart to write your own prescription for dealing with any weaknesses the practice test points out. Look down the leftmost column to the number of the question(s) you answered incorrectly, reading across that row you will find the concept and/or equation of concern, the section(s) of the study guide you should return to for further study, and some suggested text exercises which you should work to gain additional experience.

Practice Test Questions	Concepts and Equations	Prescription	
		Principal Concepts	Text Exercises
1	Efficiency of a Carnot engine: $\eta_c = 1 - (T_c/T_H)$	1	45-4
2	Efficiency: $\eta = 0.3\eta_c$	1	---
3	Q_H for heat engine: $Q_H = \lvert Q_c \rvert/(1 - \eta)$	1	45-7
4	W by heat engine: $W = \eta Q_H$	1	45-7
5	Mean time: $P = P_0 e^{(t/\tau)}$	2	45-11, 12
6	Doubling time: $T = \tau \ln 2$	2	45-12, 13
7	Rate of consumption of energy: $P = P_0 e^{(t/\tau)}$	2	45-11, 12
8	Cumulative energy consumption: $E_c = P\tau$	2	45-14, 15
9	Energy consumed in T after E_c: E_c	2	45-14, 15
10	Sun's declination: $\delta = (23.45°)\sin\omega t$	4	45-24
11	Sun's altitude at solar noon: $\beta_N = 90° + \delta - \lambda$	4	45-25
12	Sun's average intensity: $I = (1.10 \times 10^3\ W/m^2)e^{-(0.170/\sin\beta)}$	4	45-23
13	Optimum collector angle: $\alpha = 70° - \lambda$	4	45-27
14	Angle between $\hat{n}$ and $\hat{s}$: $\theta = \beta_N - \alpha$	4	45-26
15	Solar power incident on collector: $P = I\hat{s} \cdot A\hat{n}$	4	45-26

APPENDIX I

ANSWERS TO PRACTICE TESTS

1

1. $kg \cdot m \cdot s^{-1}$
2. $kg \cdot m^2 \cdot s^{-2}$
3. m
4. $kg \cdot m \cdot s^{-2}$
5. Yes
6. Yes
7. No
8. Yes
9. 60.0 km
10. $53.33
11. 8.33×10^{-2} l/km
12. 5.00 l/h
13. 60.0 km/h
14. 3.14 m^2
15. 0.524 m^3
16. 4.02 kg/m^3
17. 17.9 m/s
18. 1.29×10^3 cm^2
19. 22.4 mi/h
20. 3.94×10^6 in
21. 3.0×10^6
22. 2.8
23. 102.60
24. 8.40

2

1. $A_x = 4.33$ m
   $A_y = 2.50$ m
2. $B_x = 4.00$ m
   $B_y = -6.93$ m
3. $C_x = 2.00$ m
   $C_y = -4.00$ m
4. $D_x = -3.00$ m
   $D_y = +2.00$ m
5. $(\mathbf{A} + \mathbf{B})_x = 8.330$ m
   $(\mathbf{A} + \mathbf{B})_y = -4.43$ m
6. 9.43 m
7. 28.0° S of E
8. 90.0°
9. (0.833 m)**i** - (0.470 m)**j**
10. 4.47 m
11. 63.4° S of E
12. 150°
13. $(\mathbf{C} - \mathbf{D})_x = 5.00$ m
    $(\mathbf{C} - \mathbf{D})_y = -6.00$ m
14. 7.81 m
15. 50.2° S of E
16. $(\mathbf{A} + \mathbf{B} + \mathbf{C} - \mathbf{D})_x = 13.3$ m
    $(\mathbf{A} + \mathbf{B} + \mathbf{C} - \mathbf{D})_y = -10.4$ m
17. 16.9 m
18. 38.0° S of E

3

1. 24.0 m
2. 72.0 m
3. -16.0 m
4. 80.0 m
5. -1.33 m/s
6. 0 m/s
7. 6.67 m/s
8. 6.00 m/s
9. -12.0 m/s
10. 12.0 m/s
11. 0 m/s^2
12. 2.00 m/s^2
13. 36.0 m **i**
14. 58.0 m **i**
15. 51.0 m
16. (21.0 m/s)**i**
17. 21.0 m/s
18. (29.0 m/s)**i**
19. 17.0 m/s
20. (8.00 m/s^2)**i**
21. +4.47 m/s
22. 1.02 m
23. 0.456 s
24. -9.80 m/s^2
25. 5.00 s
26. -44.5 m/s

4

1. 5 m **i** + 9 m **j**
2. 10.3 m
3. 60.9° ccw wrt +x axis
4. 27 m **i** + 17 m **j**
5. 22 m **i** + 8 m **j**
6. (11 m/s)**i** + (4 m/s)**j**
7. (7 m/s)**i** + (4 m/s)**j**
8. 8.06 m/s
9. (4 m/s^2)**i**
10. 15.0 m/s, 26.0 m/s
11. 0 m/s^2, -9.80 m/s^2
12. 2.65 s
13. 84.5 m
14. (15.0 m/s)**i**
15. (-9.80 m/s^2)**j**
16. 5.30 s

17. 15.0 m/s, -26.0 m/s
18. 6.81 s
19. 43.4 m/s
20. 69.8° below horizontal
21. 102 m
22. v_0^2/r_0
23. $2v_0^2/r_0$
24. $4v_0^2/r_0$
25. $v_0^2/32r_0$
26. 45.4 s
27. 68.1 m
28. 2.66 m/s 34.3° down stream
29. 43.0° up stream
30. 1.61 m/s across

5

1. 433 N
2. 433 N
3. 250 N
4. 50.0 N
5. 0.980 m/s^2
6. 1.98 m/s
7. 2.02 s
8. 86.6 N
9. 0.289 m/s
10. 28.9 N
11. 57.8 N
12. 686 N
13. 873 N
14. 499 N
15. 406 N
16. 966 N

6

1. 980 N
2. 100 N
3. 0 N
4. 855 N
5. 85.5 N
6. 131 N
7. 1.08×10^3 N
8. 173 N
9. 0 N
10. 849 N
11. 84.9 N
12. 405 N
13. 948 N
14. 94.8 N
15. 222 N
16. 22.9 N
17. 19.4 N
18. 12.5 N
19. 7.68 N
20. 3.10 N

7

1. -1.33×10^{-10} N i
2. 1.33×10^{-10} N i
3. -2.66×10^{-10} N i
4. -3.99×10^{-10} N i
5. -3.33×10^{-11} N i
6. -2.00×10^{-10} N i
7. 2.83 m
8. $(-1.94 \times 10^{-10}$ N/kg)i
9. $(1.94 \times 10^{-10}$ N/kg)i
10. 0
11. 0
12. 15.7 m/s^2
13. 1.10×10^3 N
14. 3.74×10^8 m
15. 1.99×10^{-2} kg

8

1. 162 N
2. 16.2 N
3. -500 J
4. -162 J
5. 0
6. 1.30×10^3 J
7. 638 J
8. 63.8 N
9. 638 J
10. 638 J
11. 11.2 m/s
12. 6.27 m/s^2
13. 1.79 s
14. 726 W
15. (2.00 m)i
16. (4.00 m)i
17. (2.00 m)i
18. (2.00 m/s)i
19. 10.0 J
20. 10.0 W

9

1. 1.00×10^4 J
2. 2.00×10^3 J
3. 0
4. 0
5. -500 J
6. 1.50×10^3 J
7. 1.50×10^3 J
8. 5.42 m/s
9. 1.15×10^4 J

10. 1.00×10^4 J
11. -1.00×10^4 J
12. -1.73×10^3 J
13. 8.27×10^3 J
14. 8.27×10^3 J
15. 9.77×10^3 J
16. 1.00×10^2 N
17. 0.980 m/s^2
18. 97.7 m

10

1. $x_{cm} = 7.83$ m, $y_{cm} = 14.8$ m
2. $(v_{cm})_x = 4.83$ m/s, $(v_{cm})_y = 12.7$ m/s
3. $(a_{cm})_x = 1.67\ m/s^2$, $(a_{cm})_y = 6.00\ m/s^2$
4. $(F_{cm})_x = 10.0$ N, $(F_{cm})_y = 36.0$ N
5. $P_{bi} = 4.00$ kg•m/s
6. $(v_{bi})_y = -20.0$ m/s
7. $(v_{bf})_y = +15.7$ m/s
8. 24.4°
9. 3.80 kg•m/s
10. 3.57 kg•m/s
11. 35.7 N
12. 1.62 m/s
13. 0
14. 1.62 m/s
15. 4.54 m/s
16. 2.18 m/s
17. 5.72 m/s
18. -0.333 m/s
19. -0.333 m/s
20. 9.09 m/s
21. -8.26 m/s^2
22. -1.82 N
23. 0.844

11

1. -866 N
2. -433 N
3. 1299 N
4. -500 N
5. +250 N
6. +250 N
7. 0
8. Yes
9. -866 N•m
10. -346 N•m
11. +1212 N•m
12. 1212 N
13. +606 N
14. -350 N
15. 40.0 N•m
16. Clockwise
17. 24 N•m
18. Counterclockwise
19. -16 **k**
20. Clockwise

12

1. $(5.00 \times 10^{-3}$ rad/s)**k**
2. $(1.50 \times 10^{-2}$ rad/s)**k**
3. $(5.00 \times 10^{-3}\ m/s^2)$**k**
4. 2.00×10^{-2} rad
5. 17.0 rad
6. 14.0 rad/s
7. 6.00 rad/s^2
8. $4.00 \times 10^{-2}\ m/s^2$
9. 4.00 rad/s^2
10. 0.800 m/s^2
11. 0.200 m/s
12. 20.0 rad/s
13. 4.00 m/s
14. 0.500 m
15. 7.96 rev
16. 10.0 m
17. $2h/t$
18. $2h/tr$
19. $2mh^2/t^2$
20. mgh
21. $mgh - 2mh^2/t^2$
22. $mr^2[(gt^2/2h) - 1]$

13

1. $mr^2/2$
2. $mr^2\omega_i/2$
3. $mr^2\omega_i^2/4$
4. $mr^2/4$
5. $3mr^2/4$
6. $mr^2\omega_i/2$
7. $2\omega_i/3$
8. $mr^2\omega_i^2/6$
9. $-mr^2\omega_i/12$
10. Distortion of putty and friction
11. $2h/t$
12. $2h/rt$
13. $2mh^2/t^2$
14. mgh
15. $mgh - 2mh^2/t^2$
16. $mr^2[(gt^2/2h) - 1]$
17. $2h/t^2$
18. $2h/rt^2$
19. $m[g - (2h/t^2)]$
20. $mr[g - (2h/t^2)]$
21. $mr^2[(gt^2/2h) - 1]$
22. mgh/t

14

1. 200 N/m
2. 0.200 m
3. 10 rad/s
4. 1.59 Hz
5. 0.628 s
6. 2 m/s
7. 20 m/s^2
8. 270°
9. 0
10. -2 m/s
11. 0
12. 0
13. 4.00 J
14. 0
15. 4.00 J
16. 0.0287 m
17. -1.98 m/s
18. -2.87 m/s^2
19. 0.0824 J
20. 3.92 J

15

1. 1.94×10^4 N/m^2
2. 1.21×10^{-6}
3. 2.42×10^{-8} m
4. 3.77×10^{-5} m^2
5. 1.88×10^4 N
6. 4.90×10^4 Pa
7. 9.80×10^4 Pa
8. 2.48×10^5 Pa
9. 5.00 m
10. 7.00×10^2 kg/m^3
11. 8.62×10^{-1} N
12. 8.62×10^{-1} N
13. 8.26×10^{-1} N
14. 1.05×10^3 kg/m^3
15. 6.17×10^{-1} N
16. 6.17×10^{-1} N
17. 5.00×10^2 kg/m^3
18. 7.85×10^{-3} m^3/s
19. 1.00 m/s
20. 16.0 m/s
21. 1.53×10^5 Pa
22. 1.15 m

16

1. 50.0°C
2. 60.0°S
3. 343 K
4. 4.40×10^{-4}
5. 8.80×10^7 N/m^2
6. -23.8°C
7. 7.50 cm^3
8. 67.8°C
9. 1.54 cal/s
10. 38.9°C
11. 1.00×10^4 BTU/h
12. 9.60×10^3 BTU/h
13. 9.38×10^3 BTU/h
14. 6.96×10^5 BTU
15. $3.48
16. 16.3 W
17. 11.6 W
18. 4.70 W
19. 2.82×10^3 J
20. 9.16 min

17

1. 49.0 J
2. 11.7 cal
3. 2.34×10^{-2} C°
4. 10.0×10^5 N/m^2
5. 4.01 mol
6. 20.0×10^5 N/m^2
7. 60.0×10^5 N/m^2
8. 24.93×10^2 N/m^2
   1.00 m^3, 300 K
9. 24.93×10^2 N/m^2
   2.00 m^3, 600 K
10. 1.00 mol
11. 2.49×10^3 J
12. 6.23×10^3 J
13. 3.74×10^3 J
14. 12.465×10^2 N/m^2
    4.00 m^3, 600 K
15. 3.46×10^3 J
16. 0
17. 3.46×10^3 J
18. 12.465×10^2 N/m^2
    1.00 m^3, 150 K
19. -3.74×10^3 J
20. -9.35×10^3 J
21. -5.61×10^3 J
22. 0
23. 1.87×10^3 J
24. 1.87×10^3 J
25. 0
26. 2.21×10^3 J
27. 2.21×10^3 J

18

1. 4.00 mol
2. 6.64×10^{-27} kg
3. 2.41×10^{24}
4. 1.60×10^{-2} kg
5. $2.25 \times 10^{6}\ m^2/s^2$
6. 7.47×10^{-21} J
7. 1.80×10^{4} J
8. 1.80×10^{4} J
9. $4.00 \times 10^{-3}\ m^3$
10. 4.00×10^{3} J
11. 9.995×10^{3} J
12. 5.996×10^{3} J
13. $8.00 \times 10^{-3}\ m^3$
14. 11.09×10^{3} J
15. $12.11 \times 10^{-3}\ m^3$
16. 364.8 K
17. -5,806 J
18. -9,100 J
19. -13,684 J
20. -22,810 J
21. 0
22. 11,770
23. 11,770

19

1. 400 J
2. 0.200
3. 2.33
4. 2.00×10^{3} J
5. 0.500
6. 4.00 KW
7. $3.20
8. 4.453×10^{3} J
9. +5.76 J/K
10. -3.301×10^{3} J
11. -5.76 J/K
12. 0
13. 0.259
14. 2.87
15. 22.23°C
16. +6.34 J/K
17. +0.439 J/K
18. -5.986 J/K
19. +0.793 J/K
20. 0
21. 0.793 J/K

20

1. 1.2×10^{-2} N **i**
2. -1.2×10^{-2} N **i**
3. -1.35×10^{2} N **j**
4. 1.35×10^{-2} N **j**
5. (-1.20 **i** + 1.35 **j**) $\times 10^{-2}$ N
6. (-4.00 **i** + 4.50 **j**) $\times 10^{3}$ N/C
7. (-1.15, 1.29) m
8. (6.40 **i** - 7.20 **j**) $\times 10^{-16}$ N
9. (7.03 **i** - 7.91 **j**) $\times 10^{14}\ m/s^2$
10. (17.5 **i** + 0.96 **j**) $\times 10^{2}$ N/C
11. (35.0 **i** + 1.92 **j**) $\times 10^{-4}$ N
12. $k\int_0^L \left(\frac{\lambda(x')(x - x')dx'}{[(x-x')^2 + y^2]^{3/2}}\right)$
13. $k\int_0^L \left(\frac{\lambda(x')y\,dx'}{[(x-x')^2 + y^2]^{3/2}}\right)$
14. $-(kQ/L)\{[x^2+y^2]^{1/2} - [(x-L)^2+y^2]^{1/2}\}$
15. $\left(\frac{kQ}{Ly}\right)\left(\frac{x}{[x^2+y^2]^{1/2}} - \frac{x-L}{[(x-L)^2+y^2]^{1/2}}\right)$
16. $2\pi bkx\ \mathbf{i}[(x^2 + R_o{}^2)^{1/2} + x^2/(x^2 + R_o{}^2)^{1/2} - 2x]$ for $x > 0$

21

1. $3.37 \times 10^{-11}\ N{\bullet}m^2/C$
2. $-5.65 \times 10^{-11}\ N{\bullet}m^2/C$
3. $-6.74 \times 10^{-11}\ N{\bullet}m^2/C$
4. $1.12 \times 10^{-11}\ N{\bullet}m^2/C$
5. $3.37 \times 10^{-11}\ N{\bullet}m^2/C$
6. $3.37 \times 10^{-11}\ N{\bullet}m^2/C$
7. 0
8. 0
9. $4.00\ N{\bullet}m^2/C$
10. $4.00\ N{\bullet}m^2/C$
11. 0
12. 0
13. $8.00\ N{\bullet}m^2/C$
14. 7.12×10^{-11} C
15. $1.80 \times 10^{10}/r^2$
16. 0
17. $6.30 \times 10^{10}/r^2$
18. $-0.039\ C/m^2$
19. $0.035\ C/m^2$
20. $k(4 + r^3)/2r^2$
21. 0
22. $11k/r^2$
23. $-0.119\ C/m^2$
24. $0.0550\ C/m^2$
25. $kQ_or^3/R_o{}^5$
26. kQ_o/r^2

22

1. xy^2z
2. 30 J
3. $-2e^{-3}\mathbf{i} - e^{-3}\mathbf{j} + 2e^{-3}\mathbf{k}$
4. $-4e^{-3}\mathbf{i} - 2e^{-3}\mathbf{j} + 4e^{-3}\mathbf{k}$
5. $(4e^{-3} - 2e^{-1})$ J
6. -9.00×10^5 V
7. 9.55×10^5 V
8. 1.80×10^6 V
9. 1.86×10^6 V
10. 557 J
11. 570 J
12. 17.6 m/s
13. $\{-18/[(x^2 + y^2)^{1/2}] + 27/[(x - 2)^2 + y^2]^{1/2} + 36/[(x-2)^2+(y-2)^2]^{1/2} + 27/[x^2 + (y - 2)^2]^{1/2}\} \times 10^5$
14. $(kQ/L) \ln \{[L + (L^2 + y^2)^{1/2}]/|y|\}$
15. $2k[(L^2 + y^2)^{1/2} - |y|]$
16. $\pi kb\{R(R^2 + x^2)^{1/2} + x^2\ln(|x|/[R + (R^2 + x^2)^{1/2}])\}$
17. kQ/r^2
18. kQr/R^3
19. Graph
20. kQ/r
21. $\left(\frac{3kQ}{2R}\right) - \left(\frac{kQr^2}{2R^3}\right)$
22. Graph
23. kQ/r^2
24. 0
25. Graph
26. kQ/r
27. kQ/R
28. Graph

23

1. $C \approx 2\pi\varepsilon_0 R$
2. (60/47) μF
3. (600/47) μC
4. V_1 = (200/47)V
   V_2 = (150/47)V
   V_3 = (120/47)V
5. 12.0 μF
6. Q_1 = 30.0 μC, Q_2 = 40 μC
   Q_3 = 50 μC
7. $V_1 = V_2 = V_3 = V$ = 10.0 V
8. (35/12) μF
9. Q_1 = (150/12) μC
   Q_2 = (200/12) μC
   Q_3 = (350/12) μC
10. $V_1 = V_2$ = (50/12) V
    V_3 = (70/12) V
11. 4.43 pF
12. 106.2 pC
13. 22.2 pF
14. 531.0 pC
15. 531.0 pC
16. 120 V
17. $Q_1 = Q_2$ = (1200/7) μC
18. Q_{1F} = (7200/49) μC
    Q_{2F} = (9600/49) μC
19. V_{1F} = (2400/49) V
    V_{2F} = (2400/49) V
20. $U_{1I} = 4.90 \times 10^{-3}$ J
    $U_{2F} = 3.70 \times 10^{-3}$ J
21. $U_{1F} = 3.60 \times 10^{-3}$ J
    $U_{2F} = 4.80 \times 10^{-3}$ J

24

1. 1.55 Ω
2. 0.500 A
3. 0.278 A
4. 2.50 V
5. 120 Ω
6. 2.40×10^4 m
7. 121.5 Ω
8. 9.83×10^{-7} Ω•m
9. 133.3 Ω
10. 246.7°C
11. .0300 Ω
12. 4970 Ω
13. 0.2505 A
14. $8.46 \times 10^{28}/m^3$
15. 2.80×10^{-4} m/s
16. 2.60×10^{-14} s
17. 1.33 A
18. 0.670 A
19. 4.40×10^{-4} m/s
20. 2.2×10^{-4} m/s 21. 4.20×10^3 A/m^2
22. 2.10×10^3 A/m^2
23. 1.47×10^{-15} s

25

1. 3.00 V
2. 2.00 V
3. 2.00 V
4. 2.00 Ω
5. 1.00 Ω
6. 10.0 C
7. 3.00 W
8. 2.00 W
9. 1.00 W
10. 2/3
11. 30.0 J
12. 1.71 A
13. 2.57 A
14. 0.86 A
15. 0.86 V
16. 10.26 W
17. 11.70 W
18. 15.42 W
19. 13.20 W
20. 5.16 W

21. 5.92 W
22. 10.0 V
23. 2.00×10^{-3} s
24. 0.018 A
25. 86.5 μC

26

1. Q_1
2. Q_3
3. Q_2 and Q_4
4. 2.00×10^{-13} N
5. 4.00×10^{-19} C
6. 1.33×10^{-4} m
7. 1.50×10^{-17} kg•m
8. 1.50×10^{-10} N
9. 2.00×10^{-13} N
10. 1.50×10^{-23} kg
11. 3.06 A
12. left to right
13. counterclockwise viewed from +x-axis
14. 5.00 A
15. 80.0π N•m
16. 0
17. j
18. -j
19. 160π J

27

1. $\mu_0 I/4R$
2. $\mu_0 I/2(5)^{1/2}\pi R$
3. $[\mu_0 I/4(5)^{1/2}\pi R] \times [(5)^{1/2}\pi + 2]$
4. $[\mu_0 I/2R]\ [1 + (5)^{-3/2}]$ To the right
5. $[\mu_0 I/2R]\ [1 + (5)^{-3/2}]$ To the right
6. $\mu_0 I/(2)^{3/2}R$
7. $\mu_0 Ir/2\pi a^2$
8. $(\frac{\mu_0 I}{2\pi r})(\frac{b^2 - r^2}{b^2 - a^2})$
9. 0
10. $\mu_0 Kr^3/4$
11. $\mu_0 Ka^4/4r$
12. $+2.50 \times 10^{-5}$ T
13. -2.00×10^{-5} T
14. 0.667 cm
15. 2.00×10^{-5} N towards #2
16. Make $I_1 = 4.00$ A to the right
17. -3bc
18. 2ac
19. -ab
20. 0
21. μ_0
22. $-2\mu_0$
23. $-4\mu_0$
24. 0

28

1. 20πt T•m^2
2. 20π V
3. 5π A
4. left to right
5. 7.86×10^{-4} T **k**
6. 3.14×10^{-3} T/s
7. 1.58×10^{-5} T•m^2/s
8. 9.9×10^{-7} T•m^2/s
9. 9.90×10^{-7} V
10. 9.9 μV
11. 0
12. 3mV ccw
13. 0
14. 3 mV cw
15. 0
16. Graph
17. 0
18. 1.5 mA ccw
19. 0
20. 1.5 mA cw
21. 0
22. 0
23. 2.25×10^{-4} N left
24. 0
25. 2.25×10^{-4} N left
26. 0
27. 1.50×10^{4} V

29

1. 2.52×10^{-3} T
2. 2.52×10^{-6} T•m^2
3. 6.30×10^{-6} T•m^2/s
4. 6.30×10^{-6} V
5. 6.30×10^{-4} V
6. 6.30×10^{-5} V
7. 1.26×10^{-5} V•s/A
8. 1.26×10^{-4} V•s/A
9. b → a
10. 2.52×10^{-4} J
11. 0.333 s
12. 0.144 A
13. 4.32 V
14. 0.680 V
15. 6.80×10^{-2} A/s
16. 0.075 A
17. 200
18. 1.65 KW
19. 0.135 W
20. 8.20×10^{-5}

30

1. 1.258274×10^{-6} T•m/A
2. 1.589479×10^{6} A/m
3. 2066.7 A/m
4. **M** and **B** are parallel
5. 1.256516×10^{-6} T•m/A
6. 1.591703×10^{6} A/m
7. -157.3 A/m
8. **M** and **B** are antiparallel
9. 0.0107 T•m/A
10. 6.366×10^{5} A/m

31

1. 10.0 A
2. 1.67×10^{-2} s
3. 0°
4. 2.78×10^{-3} s and 1.39×10^{-2} s
5. 60.0°
6. 113 Ω
7. 66.3 Ω
8. 110 Ω
9. 1.09 A
10. 25.0°
11. 109 V
12. 72.3 V
13. 123 V
14. 0.906
15. 2.35×10^{-5} F
16. 1.20 A
17. 1.00
18. 144 W
19. 120 V
20. 0

32

1. 1.00 m
2. 0.200 Hz
3. 10.0 m
4. 2.00 m/s
5. 720
6. 6.28/m
7. 1.00 m
8. 12.6 rad/s
9. 2.00 Hz
10. 2.00 m/s
11. 0.500 s
12. 0.100 m
13. right
14. 0 m
15. -1.26 m/s
16. 1.6×10^{-2} N
17. $8\pi^2 \times 10^{-3}$ J/m
18. $8\pi^2 \times 10^{-3}$ J/m
19. $3.2\pi^2 \times 10^{-2}$ J/s
20. 2.00×10^{-2} kg/m
21. 1.00 m
22. 3.00×10^{2} m/s
23. 1.80×10^{3} N
24. 7.20×10^{3} N

33

1. 1.58×10^{-3} W/m^2
2. 3.16×10^{-3} W/m^2
3. 95.0 db
4. 14.1 m
5. 92.0 db
6. 66.7 cm
7. 170 Hz
8. 40.0 cm
9. 16.7 cm
10. 1.33 m
11. 497 Hz
12. 503 Hz
13. 6/s
14. 499 Hz
15. 504 Hz
16. 5/s

34

1. +x direction
2. 2.09 m
3. 1.43×10^{8}/s
4. 3.00×10^{2} N/C
5. 1.00×10^{-6} N/A•m
6. 2.39×10^{2} W/m^2
7. 1.20×10^{2} W/m
8. 7.97×10^{-7} J/m^3
9. xy - plane
10. 3.20×10^{-7} N/m^2
11. 1.60×10^{-7} N/m^2
12. 4.80×10^{-9} N

35

1. b
2. c
3. a
4. d
5. e
6. 49.8°
7. 38.0°
8. 53.8°
9. 1.53 cm
10. 2.29×10^8 m/s
11. 25.0 cm
12. 60.0 cm
13. -20.0 cm
14. -14.3 cm
15. -0.250
16. 2.00
17. -40.0 cm
18. 5.72 cm
19. Virtual
20. Virtual
21. 30.0 cm
22. -14.3 cm
23. -2.00
24. 0.500
25. 4.00 cm, erect
26. 1.00 cm, erect
27. 3, 4, 5

36

1. 0.688°
2. 1.20×10^{-2} m
3. 0.516°
4. 9.00×10^{-3} m
5. 6.00×10^{-3} m
6. 1.39 m
7. 8.00×10^{-7} m
8. 6.00×10^{-3} m
9. 2.00×10^{-6} m
10. 3.00
11. 19.4°
12. 1.66×10^{-1} m
13. 1.00×10^{-6} m
14. 1.41×10^{-4} rad
15. 1.41×10^{-4} m
16. 1.06×10^{-3} rad
17. 2.66×10^{-10} m
18. 180°
19. 180°
20. 6.00×10^{-7} m
21. 4.00×10^{-7} m
22. 4.80×10^{-7} m

37

1. 0.344°
2. 1.20×10^{-2} m
3. 2.40×10^{-2} m
4. 0.344°
5. 1.80×10^{-2} m
6. 1.20×10^{-2} m
7. 4.00 m
8. 1.50×10^{-6} m
9. $a_{new} = a_{old}/2$
10. 0.0450
11. 0.360
12. $I_0/2$
13. $I_0/4$
14. $I_0/8$
15. 58.0°
16. 58.0°
17. 90.0°
18. 32.0°
19. 1.00
20. 0.111

38

1. 8.35 s
2. 6.67 m
3. 5.00×10^{-1} kg
4. 8.33×10^{-1} kg
5. -2.40×10^3 m/s
6. 0.988 C
7. 1.00 MeV
8. 1.51 MeV
9. 2.69×10^{-30} kg
10. 2.82×10^8 m/s
11. 7.59×10^{-22} kg•m/s
12. 6.67×10^{-8} kg/s
13. 2.10 kg
14. 2.40×10^8 m/s
15. 2.00×10^{-19} kg•m/s
16. 3.02×10^{-11} J
17. 7.52×10^{-11} J
18. 8.35×10^{-28} kg

39

1. 1.30×10^2 W
2. 4.68×10^5 J
3. 3.00×10^{-1} s
4. 6.53×10^3/s
5. 6.63×10^{-19} J
6. 7.24×10^{14}/s
7. 1.83×10^{-19} J
8. 1.14 V
9. 1.25×10^{-10} m
10. 54.0°
11. 62.5°
12. 4.84×10^{-24} kg•m/s
13. 4.77×10^{-10} m
14. 7.27×10^5 m/s
15. 1.52×10^5 rad/s

16. 2.41×10^{-19} J
17. 1.01×10^{-9} N
18. -4.83×10^{-19} J
19. 3.16×10^{-34} J•s
20. 3.03×10^{-19} J
21. 1.06×10^{-34} J•s
22. 6.57×10^{-7} m

40

1. 3.31×10^{-38} m
2. 4.05×10^{-13} m
3. 2.49×10^{-10} m
4. 1.23×10^{-11} m
5. 9.88×10^{-16} m
6. 1.28×10^{-17} J
7. 4.83×10^{-24} kg•m/s
8. 1.37×10^{-10} m
9. 55.0°
10. 8.36×10^{-11}
11. 6.99×10^{-19} J
12. 6.98×10^{-19} J
13. 4.37 V
14. 5.27×10^{-25} kg•m/s
15. 1.58×10^{-16} J
16. 3.34×10^{-19} s
17. $U(x \le 0) = \infty$
    $U(0 < x < L) = 0$
    $U(x \ge L) = \infty$
18. $\Psi(x) = 0$
19. $\Psi_{nx} = A \sin k_n x$,
    $k_n = n\pi/L$
20. $(L/2)^{1/2}$
21. $\Psi_2(x) = [(L/2)^{1/2} \sin 2\pi x/L]$
    $E_2 = 4\pi^2\hbar^2/2mL^2$
22. L/4 and 3L/4
23. 0
24. 5.00×10^{-2} L^2

41

1. 2
2. -3.4 eV
3. 1
4. $(2)^{1/2}\hbar$
5. $\hbar$
6. $3^{1/2}\hbar/2$
7. $\hbar/2$
8. 45.0°
9. 54.7°
10. $2^{1/2}\mu_B$
11. $-\mu_B$
12. 1
13. 3
14. 4.69×10^{-24} J
15. 5999.916 Å
    6000.000 Å
    6000.084 Å
16. $3^{1/2}\mu_B$
17. $\pm\mu_B$
18. $\pm 2.78 \times 10^{-21}$ N
19. $\pm 1.66 \times 10^{6}$ m/s^2
20. 1.66×10^{-6} m

42

1. 1.79×10^{-31} J
2. 3.58×10^{-31} J
3. 1.79×10^{-31} J
4. 1.38×10^{38}/J
5. 2.21×10^{16}
6. 1.45×10^{20}
7. 3, 2, and 1
8. 1.12×10^{-18} J
9. 1
10. 0
11. 1/2
12. 0.881
13. 1.57×10^{6} m/s
14. 8.12×10^{4} K
15. 1.11×10^{-18} J
16. 2.46×10^{6} m/s
17. 2.38×10^{-14} s
18. 4.20×10^{13}/s
19. 5.85×10^{-8} m
20. 2.58×10^{2}

43

1. ${}^{0}_{-1}e$
2. ${}^{210}_{82}Pb$
3. γ
4. ${}^{12}_{6}C$
5. ${}^{4}_{2}He$
6. ${}^{210}_{84}Po$
7. 0.030377 u
8. 28.3 MeV
9. 7.08 MeV
10. 0.104251 u
11. 97.1 MeV
12. 1.00×10^{4}/s
13. 2.00×10^{4}/s
14. 1.33×10^{4}/s
15. 2.27×10^{-4}/s
16. 3.07×10^{3} s

17. 8.85×10^7
18. 1.47×10^{-16}
19. 386/s
20. 1.02×10^4 s
21. 54.0
22. 132
23. 0.216866 u
24. 202 MeV
25. 4.26 mol
26. 2.56×10^{24} nuclei
27. 5.17×10^{26} MeV
28. 9.19×10^{-4} kg
29. 3.02×10^{-5} kg/s

44

1. 2.52×10^5 m/s
2. 3.18×10^7 J
3. -6.36×10^{10} J
4. -3.18×10^{10} J
5. 1.59×10^{10} J
6. 1.59×10^{10} J
7. 5.88×10^2 nm
8. 1.19×10^{57}
9. 1.19×10^{57}
10. 4
11. 2.98×10^{56}
12. 1.27×10^{45} J
13. 1.00×10^{11} g
14. 10
15. 10
16. 10^2
17. 10^3
18. 10^{-2}

45

1. 0.571
2. 0.171
3. 2.41×10^8 J
4. 4.10×10^8 J
5. 45.7 y
6. 31.7 y
7. 86.4 EJ/y
8. 2285 EJ
9. 4570 EJ
10. -10.6°
11. 39.4°
12. 8.42×10^2 W/m^2
13. 30°
14. 9.4°
15. 4.15 kW

APPENDIX II

CONVERSION FACTORS, CONSTANTS AND UNITS

Factors for Converting to and from SI Units

One Non-SI Unit	Equals in SI
acre	4.047×10^{3} m^2
angstrom	1.000×10^{-10} m
astronomical unit (AU)	1.496×10^{11} m
atmosphere (standard)	1.013×10^{5} N/m^2
atomic mass unit (u)	1.661×10^{-27} kg
British thermal unit	1.054×10^{3} J
calorie	4.184 J
day	8.640×10^{4} s
dyne	1.000×10^{-5} N
electronvolt	1.602×10^{-19} J
erg	1.000×10^{-7} J
foot	3.048×10^{-1} m
gallon	3.785×10^{-3} m^3
gauss	1.000×10^{-4} T
horsepower	7.457×10^{2} W
hour	3.600×10^{3} s
inch	2.540×10^{-2} m
light-year	9.461×10^{15} m
liter	1.000×10^{-3} m^3
mile (statute)	1.609×10^{3} m
mile (nautical)	1.852×10^{3} m
ounce (avoirdupois)	2.780×10^{-1} N
ounce (troy)	3.050×10^{-1} N
ounce (U.S. fluid)	2.957×10^{-5} m^3
pound	4.448 N
quart	9.464×10^{-4} m^3
slug	1.459×10^{1} kg
ton (metric), tonne	1.000×10^{3} kg
torr (mmHg 0°C)	1.333×10^{2} N/m^2
yard	9.144×10^{-1} m

Often-Used Physical Constants

Quantity	Symbol	Magnitude	Unit
Avogadro number	N_A	6.022×10^{23}	mol^{-1}
Bohr radius	r_1	5.292×10^{-11}	m
Boltzmann constant	k	1.381×10^{-23}	J/K
Coulomb constant	k_e	8.988×10^{9}	$N \cdot m^2/C^2$
Electric permittivity (vacuum)	ε_o	8.854×10^{-12}	F/m
Electron rest mass	m_e	9.110×10^{-31}	kg
Elementary charge	e	1.602×10^{-19}	C
Gas constant	R	8.314	J/(mol·K)
Gravitational constant	G	6.672×10^{-11}	$N \cdot m^2/kg^2$
Mass of earth	M_e	5.979×10^{24}	kg
Neutron rest mass	m_n	1.675×10^{-27}	kg
Planck's constant	h	6.626×10^{-34}	J·s
Proton rest mass	m_p	1.673×10^{-27}	kg
Radius of earth (av)	R_e	6.376×10^{6}	m
Rydberg constant	R	1.097×10^{7}	m^{-1}
Speed of light (vacuum)	c	2.998×10^{8}	m/s
Standard gravity	g	9.807	m/s^2
Stefan-Boltzmann constant	σ	5.670×10^{-8}	$W/(m^2 \cdot k^4)$

Derived SI Units (Common)

Quantity	Unit Name	Symbol	Expressed in Fundamental Units	Expressed in Other SI Units
Capacitance	farad	F	$A^2 \cdot S^4/(kg \cdot m^2)$	
Electric charge	coulomb	C	A·s	
Electric potential	volt	V	$kg \cdot m^2/A \cdot s^3$	J/C
Electric resistance	ohm	Ω	$kg \cdot m^2/A^2 \cdot s^3$	V/A
Force	newton	N	$kg \cdot m/s^2$	
Frequency	hertz	Hz	s^{-1}	
Inductance	henry	H	$kg \cdot m^2/(A^2 \cdot s^2)$	V·s/A
Magnetic field intensity	tesla	T	$kg/(A \cdot s^2)$	N·s/(C·m)
Power	watt	W	$kg \cdot m^2/s^3$	J/s
Pressure	pascal	Pa	$kg/(m \cdot s^2)$	N/m^2
Viscosity	poiseuille	Pl	kg/(m·s)	$N \cdot s/m^2$
Work and energy	joule	J	$kg \cdot m^2/s^2$	N·m